Natural Resource Conservation

Emerging Issues & Future Challenges

Natural Resource Conservation

Emerging Issues & Future Challenges

Editors :

M.Madhu
Praveen Jakhar
P.P. Adhikary
H.C.Hombe Gowda
V. N. Sharda
P. K. Mishra
M.K. Khan

SATISH SERIAL PUBLISHING HOUSE

403, Express Tower, Commercial Complex, Azadpur, Delhi-110033 (India)
Phone : 011-27672852, Fax : 91-11-27672046
E-mail : info@satishserial.com, hkjain1975@yahoo.com
Website : www.satishserial.com

Published by :

SATISH SERIAL PUBLISHING HOUSE

403, Express Tower, Commercial Complex, Azadpur, Delhi-110033 (INDIA)
Phone : 011-27672852 Fax : 91-11-27672046
E-mail : info@satishserial.com, hkjain1975@yahoo.com

ISBN 978-93-81226-66-7

Typeset at: Harminder Kharb *for* Laxmi Art Creation # 9971212252
Printed at: Salasar Imaging Systems, Delhi-110035

भारत सरकार
कृषि अनुसंधान और शिक्षा विभाग एवं
भारतीय कृषि अनुसंधान परिषद
कृषि मंत्रालय कृषि भवन, नई दिल्ली – 110 001

Government of India
Department of Agricultural Research & Education
and Indian Council of Agricultural Research,
Ministry of Agriculture, Krishi Bhavan
New Delhi - 110 001, Tel. : 23382629, 23386711
Fax : 91-11-23384773 E-mail : dg-icar@nic.in

डा. एस. अय्यप्पन
सचिव एवं महानिदेशक

Dr. S. Ayyappan
Secretary &
Director General

Foreword

Agricultural development has enabled us to increase food production through enhanced productivity levels since Indian independence, but we cannot afford to become complacent in the area of food, nutritional and environmental security in view of our increasing population. However, change in global economic and trade scenario demands a paradigm shift in agriculture to make farming and farm produce are profitable and cost-effective to meet the quality of global market.

Natural resources like land, water and biodiversity are the key to sustain agricultural productivity and govern our agricultural systems holistically. Land degradation constitutes one of the greatest threats for food and environmental security. Degradation of soil and vegetative cover is caused by a variety of anthropogenic pressures including nutrient mining, soil erosion, increasing water scarcity, adverse impacts of climate change and accumulation of toxic elements in soil and water, deforestation, over-razing, unsustainable agricultural practices and industrial activities. All of these factors combined with increased rate of land degradation are contributing towards decline in agricultural productivity leading to food insecurity.

Management of natural resources are enormous and the task of raising productivity and production of food, fodder, fuel, fiber, fish and meat from our finite resources for meeting the demands of the burgeoning human and livestock populations is gigantic. Conservation of natural resources in an essential prerequisite to achieve sustainable agricultural development for food and environmental security in the country through integrated resource management approaches.

The edited bood "Natural Resource Conservation : Emerging Issues & Future Challenges" has amply covered issues and strategies on natural resource conservation for enhancing agricultural productivity.

I congratulate the authors for their efforts in bringing out this vital publication in the form of a book. I am sure, this book would be immensely useful reference for researchers, administrators, students, planning and developmental agencies in the country.

Dated the 10th January, 2013
New Delhi

(S. Ayyapan)

Foreword

Preface

Degradation of natural resources and ecosystems under the influence of biotic and abiotic factors is perhaps the biggest challenge being faced by the humankind. Anthropogenic pressures and natural processes are the main causal agents triggering degradation. Increasing anthropogenic activities driven by the escalating population growth, the rates and kinds of degradation especially, natural resources (land, water and forests) have increased manifold in the country. The north eastern and eastern region comprising 13 states of the country is bestowed with abundant natural resources supporting a total population of 315.7 M and 30% of tribal population of the country. This region is unique and is characterized with diverse natural resources (physiographic, edaphic, bio-diversity, water resources and climatological parameters like rainfall, bright sunshine hours etc.), human resources (majority of tribal population) and the socio-economic aspects (low level of literacy, low per capita income etc.). Majority of the population, predominantly tribal, is dependent on agriculture and land-based activities. About 14% of the TGA of the region (15.5% in NEH and 12% in Eastern region) has been treated under various programmes against treated area of 17.2% of TGA of the country under different programmes and schemes. This book is the outcome of the conference on Livelihood and Environment Security through Resource Conservation in Eastern Region (LESRC-2012) was organized by the Indian Association of Soil and Water Conservationists (IASWC) Dehradun in collaboration with CSWCRTI, Research Centre, Sunabeda, Koraput, Odisha and OUAT, Bhubaneswar, Odisha held at Bhubaneswar during April 5th to 7th, 2012.

The papers included in this volume are divided into six major sections *viz.*, (i) Natural resources inventorization, (ii) Climate change effects, adaptation and mitigation strategies, (iii) Conservation technologies for livelihood and environmental security, (iv)Livelihood security and watershed management, (v)Regional experiences and future thrust and (vi) Socio economic and policy issues. The papers included in this volume under different sections are multidisciplinary and covering a wide range of information for better understanding of the diverged problems in the field of Natural Resource Management.

The three days long deliberations on up-to-date information by the delegates on the main theme and sub-themes of the conference across the broad spectrum of Natural Resource Conservations to enhance Environmental and Livelihood Security concluded with the following recommendation: (1) Soil and water associated bio-resources of the region need to be properly inventorized and characterised using modern tools and techniques, evaluated and a repository to be identified. State Land Use Boards need to be activated to make them functionally effective, (2) Integrated approaches to identify deficiencies and toxicity of essential plant nutrients and agro-chemicals due to problems of soil acidity, salinity & alkalinity to enhance productivity, (3) Concerted effort is required to assess the effect of climate change on various natural resources for identifying & developing adaptation and mitigation mechanisms for climate resilient agriculture. Strategic crop planning through hydrological drought analysis is necessary at the micro level, (4) Assessment of carbon sequestration potentials of different production systems, rehabilitated degraded areas with resource conservation practices is necessary, (5) Eastern India is endowed with much diversified and potentially rich natural resources but is subjected to many land degradation process and natural disasters which affects the lives of a large population. There is a strong need for conservation of biological hotspots and improve wastelands by adopting an integrated approach, (6) Promoting micro water harvesting and other water storage structures for enhancing productivity by resorting to multiple use of water resources and flood control with emphasis on water quality, (7) Alternate enterprises including livestock & fishery for marginalized sections of society needs to be identified and capacity building for small and marginal farmers should be a priority area, which requires synergy with various Institutions, (8) Greater intervention through agroforestry and horticulture is necessary to discourage unbalanced and unsustainable short cycle of jhuming and develop opportunities to encourage settled hill agriculture with resource conservation measures. Institutional mechanisms for the management of CPR's need to be evaluated and modified to suit weaker sections of society including landless farmers and women, (9) Identification and evaluation of IFS models to all sections of the farming community in the region, (10) To encourage development of bio-industrial watersheds on cluster approach with emphasis on farm mechanization and processing facilities and provide immediate benefits to stakeholders, (11) To document ITKs on resource conservation and modify them to MTKs by addressing researchable issues, (12) Participatory research and extension is essential for watershed plus activities by involving farmers, NGO's, GO's, CBO's in a convergence mode. Effective and continued monitoring and feedback mechanism should be in-built during the implementation phase for impact evaluation and identifying areas for future research and (13) Mechanisms need to be evolved for linking front-end activities of agricultural supply chain with back-end activities of farm production and value addition. Public-private partnership is to be strengthened for generation and dissemination of scientific resource management technologies.

In view of the concerns expressed by the experts on the significance of the various issues and urgency in updated information, this volume is undertaken to be published. This edited volume contains a series of commissioned articles by experts in various fields under the themes of the conference. The comprehensive scientific information on wide range of issue related to Natural Resource Management, Soil and Water Conservation, Climate Change, Bio-diversity, Remote Sensing and GIS, Land Use and Land Cover, Watershed Management, Sustainable Agriculture and Food Security, Socio-Economical aspects etc. The contributions published in this document pertain to leading technologists and scientists, planners, administrator, policy makers, program implementing official etc. The aim of the publication is not only to discuss and to provide up-dated information but also to encourage awareness among the stakeholders to achieve a desired goal of food and environmental security through resource conservation.

We acknowledge the support of IASWC and other sponsors viz., DoLR, Ministry of Rural Development, Indian Council of Agricultural Research (ICAR), National Rainfed Area Authority (NRAA), Plant Protection and Farmers Right Authority (PP&FRA), Planning Commission, New Delhi and NABARD, Mumbai. We are grateful to Dr S. Ayyappan, Secretary, DARE & DG,ICAR for his constant encouragements in organising the conference and publication of this book. We express our deep sense of gratitude to Dr A.K. Singh the then DDG (NRM), ICAR, New Delhi, Dr D.P.Ray, Vice-Chancellor, OUAT, Bhubaneswar, Dr O.P. Chaturvedi, Chief Editor, Indian Journal of Soil Conservation, Dr D. Mandal, Secretary, IASWC for their support and guidance. Our specially thank to staff of CAET, OUAT, Bhubaneswar. We also place on record the contributions of different committee members and staff of the CSWCRTI, Research Centre, Sunabeda, Koraput.

The book will serve as a reference material to the researchers, students, planners, programme implementing officers, Administrator and other interested in judicious management of natural resources for sustainable development ultimately to achieve food and environmental security. We express our sincere thanks to the contributors and publisher for understanding the vital importance of natural resources to meet the future challenges and in bringing out this publication.

Editors

Contents

Section - I

Section - II

Section - III

Section - V

Section - VI

Section - I

Chapter 1

Status of Natural Resources for Livelihood & Environmental Security in North-Eastern and Eastern Region of India

V.N. SHARDA

INTRODUCTION

The north Eastern and Eastern region comprising 13 states of the country, namely Arunachal Pradesh, Assam, Manipur, Meghalaya, Mizoram, Nagaland, Sikkim, Tripura, Bihar, Jharkhand, West Bengal and Odisha, respectively, is bestowed with abundant natural resources supporting a total population of 315.7 M and 30% of tribal population of the country (Fig. 1). This region is unique and is characterized with diverse natural resources (physiographic, edaphic, bio-diversity, water resources and climatological parameters like rainfall, bright sunshine hours etc.), human resources (majority of tribal population) and the socio-economic aspects (low level of literacy, low per capita income etc.). Majority of the population, predominantly tribal, is dependent on agriculture and land-based activities. The agricultural production system in the region is mostly rainfed and mono-cropped at subsistence level. Slash and burn agriculture is still practiced in almost all the states. The region, once richly endowed with genetic diversity of plants, has been denuded due to human interferences and adoption of unscientific land use systems. With rapid increase in human and livestock population and the rising demand of food, feed, fuel, fodder, fiber, timber and the other developmental activities, the farmers have been forced to exploit forestland and water resources at sub-optimal level in complete defiance of the inherent potential. This has resulted in progressive decrease of forest cover, loss of biodiversity, serious soil erosion leading to depletion of plant nutrients, gradual degradation and decline in land productivity and its carrying capacity, silting of major river basins causing frequent floods in the plains, and drying up of perennial steams as well as ecological imbalances. Gradual degradation of

Member, Agricultural Scientists Recruitment Board (ASRB), KAB-I, Pusa, New Delhi these resources is of primary concern and calls for location-specific measure to efficiently conserve, utilize and manage these resources for optimizing production on sustained basis without adversely affecting its quality.

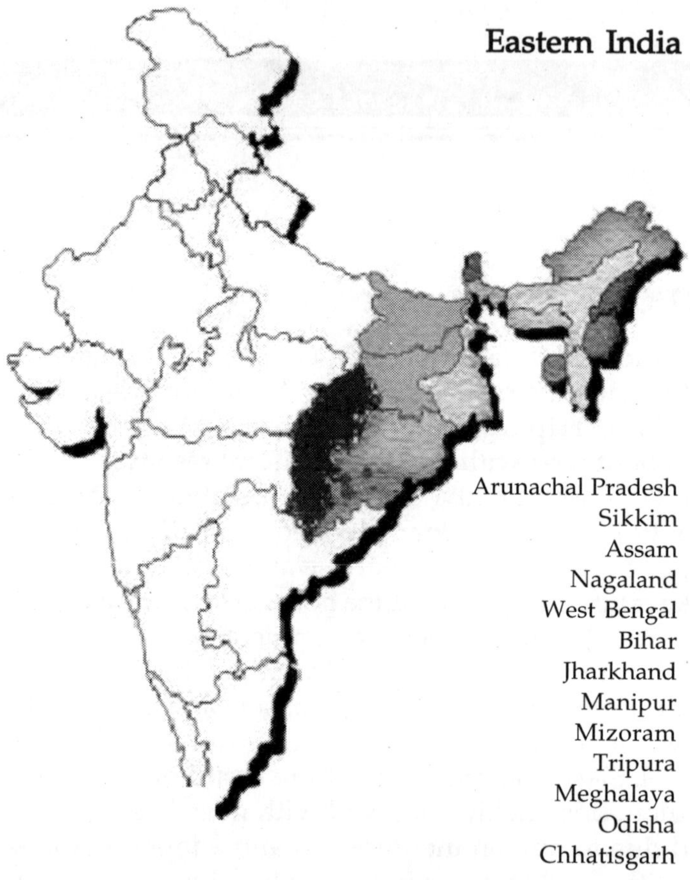

Fig. 1 : Eastern States in India

Agriculture is the main economic activity in the region and despite major impact of green revolution in the irrigated areas of the country; modernization of agriculture has escaped this region as evidenced by poor adoption of modern technologies, low consumption of fertilizers and other indicators of growth. One of the most important reasons for this slow growth is the lack of location-specific and system-based technologies. Apart from this, the dwindling resources of soil, water, flora, fauna and increasing concern for environmental safety has drawn the attention of the planners and policy makers at regional as well national level. Historically, the eastern region was the most prosperous agricultural tract of the country. However, with the advent of the input intensive new agricultural technology based Green Revolution, the region lagged behind the other parts of the country. Public investments in agriculture in this region for this period were grossly inadequate to be compatible with the requirements of input intensive agriculture and as such this region has not benefited by the Green Revolution. Thus, there is an imperative need to improve the food grain production from this resource-rich region through adoption of multi-pronged strategies to efficiently conserve and utilize the natural resources and ensure livelihood security to the rural and tribal community of the region.

Demography Features and Agricultural Economy

The Eastern and north Eastern region, spreading over thirteen states, covers a geographical area of 81.65 M ha, equivalent to around 25% of the total geographic area of the country and it supports about 30% (336 M) population of the country (Table 1).

Table 1 : Geographic and demographic features in NEH and Eastern regions of India

State/UT	TGA (M ha)	Rainfall (mm)	Population (M)	ST Population		Population Density (No. per sq. km)	Decadal Growth (%)	Literacy Rate (%)
				In lakh	In %			
Arunachal Pradesh	8.37	2782	1.4	3.52	64.2	17	25.9	66.9
Assam	7.84	2818	31.2	16.78	12.4	397	16.9	73.1
Manipur	2.23	1881	2.7	3.74	34.2	122	18.7	79.8
Meghalaya	2.24	2818	3.0	9.97	85.9	132	27.8	75.4
Mizoram	2.11	1881	1.1	4.23	94.4	52	22.8	91.5
Nagaland	1.66	1881	2.0	9.13	89.1	119	-0.5	80.1
Sikkim	0.71	2739	0.6	0.57	20.6	86	12.4	82.2
Tripura	1.05	1881	3.7	5.04	31.0	350	14.8	87.7
NEH	**26.21**	**2335**	**45.6**	**52.98**	**54.0**	**159**	**17.3**	**79.6**
Bihar	9.42	1186	103.8	3.93	0.9	1102	25.1	63.8
Chhattisgarh	13.6	1430	20.79	66.17	31.8	154	-	65.0
Jharkhand	7.97	1326	33.0	35.66	26.3	414	22.3	67.6
Odisha	15.57	1489	41.9	40.67	22.1	269	14.0	73.4
West Bengal	8.88	2089	91.3	22.24	5.5	1029	13.9	77.0
ER	**55.44**	**1504**	**290.8**	**168.67**	**17.3**	**593**	**18.8**	**69.4**
Total (NEH + ER)	**81.65**	**2015**	**336.4**	**180.98**	**39.8**	**326**	**17.8**	**75.7**
India	**328.60**	**1190**	**1210**	**843.26**	**8.2**	**325**	**17.6**	**74.0**

The Eastern region is densely populated (593 per sq.km) compared to the NEH region (159 per sq.km). Hoverer, decadal population growth trend is almost similar to that of the country. Average literacy rate of the region is 76 % which is higher than the national average of 74 % with highest rate in Mizoram (91%) in NEH region and West Bengal (77%) in Eastern region while the state of Bihar has the lowest literacy rate of 64% as per census, 2011. The NEH region is blessed with good amount of average annual rainfall against the average of eastern region and the country as a whole.

Land Utilisation Pattern

The land use pattern of NEH and ER is strongly influenced by the altitude, climate and mountainous terrain, especially in the field of agriculture and forestry. Land use, particularly in the NEH region, is the result of techniques and customs adopted by different ethnic groups in migrations as well as their subsequent adaptation to the location specific nature of soils and landscapes of settlements. Out of the total area, ER states account for more than 50% of the reporting area of the region. Forest is the main land use in these states (Fig. 2 and Table 2) and they are under varying forest cover densities followed by scrub forest, alpine barren land, snow and glaciers. Except Sikkim and Tripura, all states have the forest cover above 70% of their respective geographical areas. However, reserve forest area is below 60% of the reported area in most of the states except Arunachal Pradesh and Mizoram (above 70%).

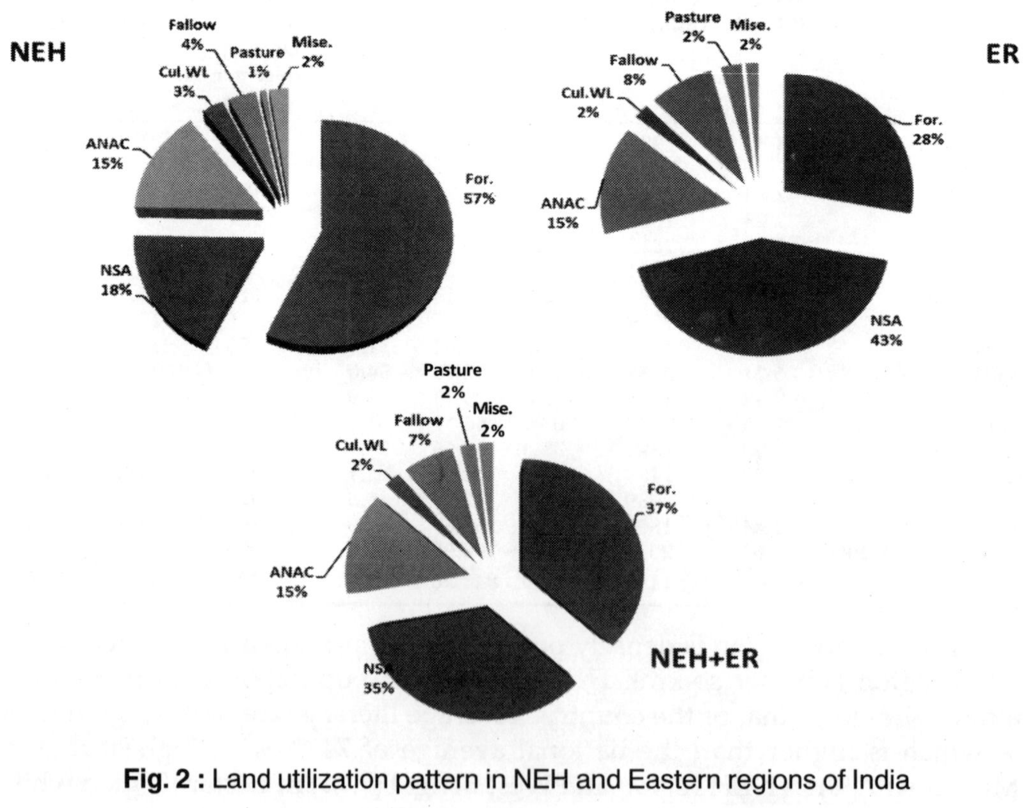

Fig. 2 : Land utilization pattern in NEH and Eastern regions of India

Table 2 : State wise land utilization pattern in NEH and Eastern regions of India (Area in '000 ha)

State	Reporting area for land utilization	Forest	Area not available for cultivation	Permanent pasture	Misc. tress	Culturable wasteland	Fallow other than current fallow	Current fallow	Net sown area
Arunachal Pradesh	5659	5154	65	18	37	67	73	37	209
Assam	7850	1954	2512	160	209	77	60	127	2753
Manipur	1951	1693	27	1	6	1	0	0	224
Meghalaya	2227	942	228	0	158	450	169	68	213
Mizoram	1946	1594	134	5	10	5	82	24	92
Nagaland	1595	863	75	0	120	57	76	82	322
Sikkim	728	319	250	4	5	2	30	5	112
Tripura	1049	606	134	0	27	1	1	1	280
NEH	**23005**	**13125**	**3425**	**188**	**572**	**660**	**491**	**344**	**4205**
Bihar	9360	622	2085	16	241	46	119	569	5662
Chhttisgarh	13787	6300	1020	848	1	344	232	248	4779
Jharkhand	7970	2333	1365	88	113	274	783	1244	1769
Odisha	15571	5813	1842	443	482	392	434	426	5739
West Bengal	8684	174	1754	5	58	34	22	341	5296
ER	**55372**	**15242**	**8066**	**1400**	**895**	**1090**	**1590**	**2828**	**23245**
Total (NEH + ER)	**78377**	**28367**	**11491**	**1588**	**1467**	**1750**	**2081**	**3172**	**27450**

The net cultivated area varies widely from 3.37% (Arunachal Pradesh) to 26.41% (Tripura) and is approximately 6.8% of the total reported area of the states. Among all the states, maximum culturable wasteland (4.5 M ha) is in Meghalaya which is under the age old practice of shifting cultivation. The area under pasture is decreasing which is a concern for the sizable livestock population, particularly in the ER. Under the present system of un-controlled grazing of livestock, broad-leaved trees are grazed or lopped frequently during seedling due to which they die as a result of overgrazing and trampling. The deforestation, which includes the cutting of agricultural terraces on steeper and more marginal mountainous slopes and *jhum* in north eastern hills, has led to a catastrophic soil erosion and loss of productive land through accelerated landslide incidence. The low food production and hunger in Eastern states is directly related to low net sown area, *viz;* Jharkhand and Chhattisgarh.

About 58% of the workforce of the region is engaged in agriculture with 252 lakh operational holdings (Table 3). The average size of land holdings in NEH and Eastern regions is 1.95 and 1.70 ha, respectively which is higher than the national average. It indicates ample scope for agricultural development in this region which is waiting for second green revolution. The percent contribution of agriculture to NDP is relatively higher for Eastern region (39) as compared to NEH (35) and the national average (34).

Table 3 : State-wise important farm indicators in NEH and Eastern regions of India (1999-2000)

State	No. of Districts	No. of Operational Holdings (in Lakh)	Gross Cropped Area (Lakh ha.)	Work Force in Agri-culture (%)	Agri-culture as (%) of NDP	Avg. Size of Holding (ha)
Arunachal Pradesh	13	0.9	2.40	65	29	3.71
Assam	23	25.2	39.84	69	34	1.27
Manipur	9	3.0	1.68	66	32	0.57
Meghalaya	7	1.7	2.39	74	30	1.70
Mizoram	8	1.3	0.82	58	21	0.49
Nagaland	8	0.4	2.10	80	55	5.00
Sikkim	4	0.6	0.62	26	-	1.94
Tripura	4	3.2	4.43	5	43	0.97
NEH	**76**	**36.3**	**54.28**	**55**	**35**	**1.95**
Bihar + Jharkhand	55	118.0	102.10	79	40	0.87
Odisha	30	39.5	86.45	67	50	1.34
West Bengal	17	58.8	87.18	53	27	0.88
ER	**102**	**216.3**	**275.73**	**66**	**39**	**1.03**
Total (NEH + ER)	**178**	**252.6**	**330.01**	**58**	**36**	**1.70**
India	**606**	**1060**	**1849.67**	**54**	**34**	**1.77**

The region as a whole accounts for about 18% of the gross cropped area of the country with maximum share from the Eastern region (27.5 M ha, 15%) against only 5.4 M ha (3%) in the NEH region. Both NEH and Eastern regions account for 24% of the total number of operational holdings of the country. Land holdings of < 2 ha account for 27% of the country land holdings in the region and it clearly shows that small and fragmented land holdings are the major hindrance for agricultural development in the region as they account for 24% of the operational holdings of the country.

Natural Resources in the Region

Soil resources: Traditionally, soils of the region are generally categorized as alluvial soils, black soils and red -laterite soils. Taxonomically, these soils are classified under five orders in north-eastern region (Table 4); Inceptisols (48.06%), Entisols (29.71%), Ultisols (17.22%), Alfisols (4.68%) and Mollisols (0.33). On the other hand, soils of eastern region belong to Inceptisol (45.64%), Alfisol (31.96%), Entisol (20.06%), and Vertisol (2.34%). It is evident from the table that Inceptisols are the dominant soils ranging from 42.69% to 89.81% in all the states of north eastern region excepting Mizoram followed by Entisols, whereas, within eastern states, Inceptisols as dominant soils are followed by Alfisols. However, in Mizoram, Ultisols cover 38.64% of the total geographical area of the state followed by Inceptisols (37.27%), Entisols (21.43%) and Alfisols (2.66%). In general, deep and light textured flood plain alluvial soils have been classified under Entisols. These soils were further classified as Fluvaquents, Haplaquents and Ustifluvents, Ustorthents based on aquic moisture regime, high clay content with decreasing organic carbon with depth and Ustic moisture regime, respectively. The soils of *Diara* land occurring in alluvial fans and flood plains were young and stratified with AC profiles and had very faint and poor pedogenic manifestations, which were classified as Typic Ustifluvents. The red and laterite soils of Machkund catchment in Orissa have been classified as Udic haplustalfs, Ultic paleustalfs and Ultic rhodustlafs. The red, black and yellow coloured soils of Rajmahal traps of Bihar have been classified under red soils as Alfisols, black soils as Inceptisols, Vertisols and Entisols and yellow soils as Inceptisols.

Bihar and eastern U.P. have large areas under *Tal, Diara and Chaurs* lands. The soils of *Diara* land occurring in alluvial and flood plains of U.P. and Bihar are young and stratified with AC profiles and have very faint and poor pedogenic manifestations, which are classified as Typic Ustifluvents. The *Diara* lands in Bihar constitute an area of 11.59 lakh ha. The topography of this land is undulating with confused patterns of upland and low land. The *Chaurs* is a large tract of land in North Bihar which remains water logged for a variable period and represents a fragile and unstable eco-system. About 4 lakh hectares

Table 4 : State-wise area under different soil orders in NEH and Eastern regions of India

State/UT	Area under different soil orders (M ha)							
	Alfisol	Aridisol	Entisol	Inceptisol	Mollisol	Vertisol	Ultisol	Total
Arunachal Pradesh	0.023 (0.31)	0 (0.00)	2.986 (40.78)	3.126 (42.69)	0 (0.00)	0 (0.00)	1.188 (16.22)	7.323
Assam	0.887 (12.30)	0 (0.00)	2.640 (36.60)	3.250 (45.05)	0 (0.00)	0 (0.00)	0.437 (6.06)	7.214
Manipur	0.004 (0.18)	0 (0.00)	0.516 (23.55)	0.860 (45.05)	0 (0.00)	0 (0.00)	0.811 (37.02)	2.191
Meghalaya	0.081 (3.61)	0 (0.00)	0.240 (10.70)	1.024 (45.65)	0 (0.00)	0 (0.00)	0.898 (40.04)	2.243
Mizoram	0.056 (2.66)	0 (0.00)	0.452 (21.43)	0.786 (37.27)	0 (0.00)	0 (0.00)	0.815 (38.64)	2.109
Nagaland	0.048 (2.90)	0 (0.00)	0.121 (7.30)	1.489 (89.81)	0 (0.00)	0 (0.00)	0 (0.00)	1.658
Sikkim	0 (0.00)	0 (0.00)	0.237 (33.38)	0.391 (55.07)	0.082 (11.55)	0 (0.00)	0 (0.00)	0.710
Tripura	0.047 (4.49)	0 (0.00)	0.084 (8.03)	0.846 (80.88)	0 (0.00)	0 (0.00)	0.069 (6.60)	1.046
North East Region	**1.146** (4.68)	**0** (0.00)	**7.276** (29.71)	**11.772** (48.06)	**0.082** (0.33)	**0** (0.00)	**4.218** (17.22)	**2.4494** (100)
Bihar (Jharkhand)	5.992 (35.46)	0 (0.00)	4.741 (28.05)	6.060 (35.86)	0 (0.00)	0.107 (0.63)	0 (0.00)	16.900
Odisha	5.219 (33.52)	0 (0.00)	1.583 (10.17)	7.910 (50.80)	0 (0.00)	0.859 (5.52)	0 (0.00)	15.571
West Bengal	2.003 (22.57)	0 (0.00)	1.972 (22.22)	4.900 (55.21)	0 (0.00)	0 (0.00)	0 (0.00)	8.875
Eastern Region	**13.214** (31.96)	**0** (0.00)	**8.296** (20.06)	**18.870** (45.64)	**0** (0.00)	**0.966** (2.34)	**0** (0.00)	**41.346** (100)

is *Chaur* land with varying depths of water during different periods of the year. The *Tal* land in Bihar extends from Fatuah to Lakhisarai over an area of about 1034 sq. km with a length of about 105 km. The topography of *Tal* land is such that rain water from the area gets accumulated up to a depth of about 5 metres. Since the area is low lying and doesn't have drainage facilities, it remains inundated for a period of 3 to 4 months in a year, covering almost the entire *kharif* season. High rainfall and longer fallow season results in high runoff and soil loss leading to severe land degradation problems as well as groundwater table depletion every year.

Land capability classes of eastern region: Most of alluvial plains present in Bihar, eastern U.P, West Bengal and Assam fall under very good lands. Most of the region in Jharkhand and Chhattisgarh and parts of West Bengal are moderately good lands. Land with severe limitations for crop growth is seen in western parts of Jharkhand and eastern parts of Chhattisgarh and southern parts of Odisha. Arid lands with limited supply of water are spread over southern tip of Odisha and parts of Jharkhand (Fig.3). Most of alluvial plains present in Bihar and eastern U.P., fall under very good lands while most of the areas in Jharkhand and Chhattisgarh have moderately good lands. Besides, intensive agriculture practiced with high use of selective agro-chemicals has become unsustainable due to imbalances in soil fertility, multi-nutrient deficiencies and deterioration of land quality and soil health. Declining trends in productivity, as recently observed in many existing

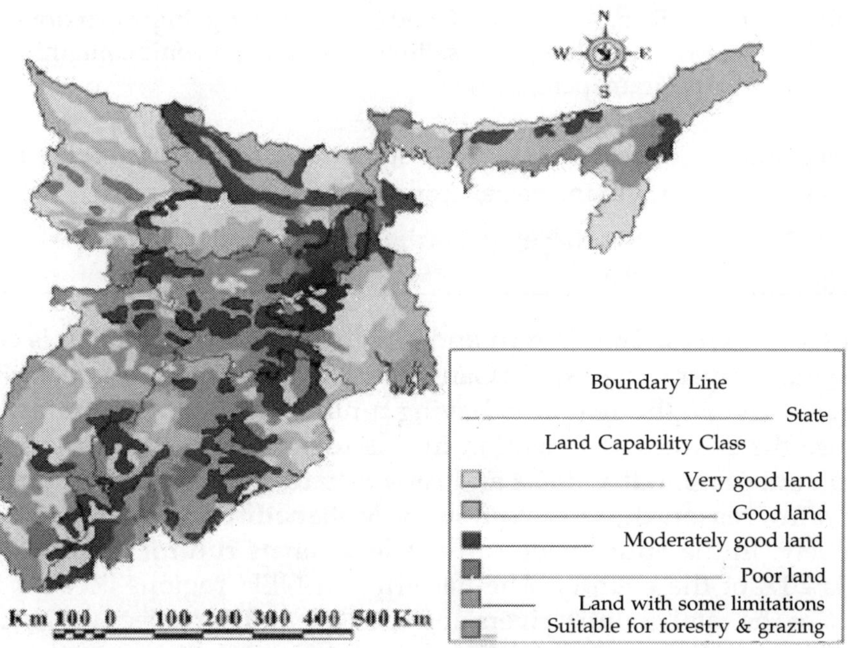

Fig. 3 : Land capability classification of eastern region

cropping systems, have resulted in declining fertilizer use efficiency and organic matter levels in the soils. The most critical region specific constraints which are responsible for deceleration in productivity are presented in Table 5.

Table 5 : Region specific factors for deceleration in productivity

Region/State	Principal Crop	Region Specific Constraint
Eastern Himalayan Region: Assam, N.E. States, Sikkim	Rice, wheat, maize, ragi	Al- toxicity and soil acidity, soil erosion and floods, shifting cultivation, low SRRs, non-availability of electricity, poor roads, poor input delivery system and communication infrastructure.
Lower and Middle Gangetic Plains Region: West Bengal, Bihar, Eastern UP	Rice, wheat, maize, jowar, bajra, ragi	Flood/ water logging, improper drainage, salinity/alkalinity, arsenic contamination, low SRRs, non-availability of electricity, high population growth, poor roads and communication infrastructure.
Eastern Plateau & Hills Region: Odisha, Jharkhand, Chhattisgarh	Maize, sorghum, pulses	Moisture stress, drought and soil acidity, iron toxicity, low SRRs, non-availability of electricity, high population growth, poor roads, poor Input delivery and communication infrastructure.
Bihar and Eastern U.P.	Rice, wheat, maize, jute, gram, peas, sugarcane	Flood/water logging, improper drainage, salinity/alkalinity, arsenic contamination
Jharkhand and Chhattisgarh	Rice, wheat, soybean, maize, ragi, pigeon-pea, potato	Moisture stress, drought and soil acidity, iron toxicity

***Water resources*:** The Eastern and NEH region of the country is covered by three major river systems, *viz;* Ganga, Bramaputra, and Mahanadi. Besides these major rivers, the region is having a number of small tributary rivers which are the sources of irrigation in this region. Both the regions receive good amount of rainfall. Odisha and Assam states experience frequent floods during rainy season due to overflow of Mahanadi and Bramaputra rivers, respectively. In the entire region, rivers and canals run for about 38774 km which is 20% of the country. The Eastern and NEH regions account for 9% and 11% of the total country rivers and canals lengths, respectively (Table 6). The entire region accounts for 34% of the total water bodies of the country out of which NEH and Eastern region contributes about 8% and 26%, respectively to the country water bodies. The NEH region is having only 22% of the total water bodies of the region against 78% in the Eastern region. The major water bodies found in this region are reservoirs followed by tanks, lakes and ponds, floodplain lakes and derelict water. Brackish water is found only in the Eastern region.

Table 6 : State-wise inland water resources in NEH and Eastern regions of India

State	Rivers & Canals (Length in kms.)	Water Bodies (lakh ha)				
		Reservoirs	Tanks, Lakes and Ponds	Floodplain Lakes and Derelict Water (Lakh ha)	Brackish Water	Total
Arunachal Pradesh	2000	-	2.76	0.42	-	3.18
Assam	4820	0.02	0.23	1.10	-	1.35
Manipur	3360	0.01	0.05	0.04	-	0.10
Meghalaya	5600	0.08	0.02	Neg	-	0.10
Mizoram	1395	-	0.02	-	-	0.02
Nagaland	1600	0.17	0.5	Neg	-	0.67
Sikkim	900	-	-	0.03	-	0.03
Tripura	1200	0.05	0.13	-	-	0.18
NEH	**20875**	**0.33**	**3.71**	**1.59**	-	**5.63**
Bihar	3200	0.60	0.95	0.05	-	1.60
Chhattisgarh	3573	0.84	0.63	-	-	1.47
Jharkhand	4200	0.94	0.29	-	-	1.23
Odisha	4500	2.56	1.14	1.80	4.3	9.80
West Bengal	2526	0.17	2.76	0.42	2.1	5.45
ER	**17999**	**5.11**	**5.77**	**2.27**	**6.4**	**19.55**
Total (NEH+ ER)	**38874**	**5.44**	**9.48**	**3.86**	**6.4**	**25.18**
India	**195210**	**29.07**	**24.14**	**7.98**	**12.4**	**73.59**

Abbr. : Neg : Negligible.
Source: Ministry of Agriculture, Govt. of India (2005-06)

Ultimate Irrigation Potential (UIP) of both the regions through major, medium and minor irrigation schemes including ground water accounts for about 25% of the country UIP (Table 7). Contribution from NEH and Eastern region is 3.1% and 21.9% of the country UIP, respectively. A similar trend has been observed for UIP through the Major, Medium and Minor irrigation schemes. Between the two regions, the Eastern region has greater irrigation potential as compared to the NEH region. The data shows that there is a good scope to bring more area under irrigation and improve agriculture production.

Table 7 : State-wise Ultimate Irrigation Potential in NEH and Eastern regions of India (Area in lakh ha)

State	Major & Medium Surface Water	Minor Irrigation			Grand Total
		Surface Water	Ground Water	Total	
Arunachal Pradesh	0.0	1.5	0.2	1.7	1.7
Assam	9.7	10.0	9.0	19.0	28.7
Manipur	1.4	1.0	3.7	4.7	6.0
Meghalaya	0.2	0.9	0.6	1.5	1.7
Mizoram	0.0	0.7	0.1	0.7	0.7
Nagaland	0.1	0.7	0.1	0.8	0.9
Sikkim	0.2	0.5	0.0	0.5	0.7
Tripura	1.0	1.0	0.8	1.8	2.8
NEH	**12.6**	**16.2**	**14.4**	**30.6**	**43.2**
Bihar	52.2	15.4	41.2	56.6	108.9
Chhattisgarh	11.5	0.8	4.9	5.7	17.2
Jharkhand	12.8	3.5	8.3	11.8	24.6
Odisha	36.0	10.0	42.0	52.0	88.0
West Bengal	23.0	13.0	33.2	46.2	69.2
ER	**135.5**	**42.8**	**129.6**	**172.4**	**307.9**
Total (NEH + ER)	**148.0**	**59.0**	**144.0**	**203.0**	**351.0**
India	**584.7**	**173.4**	**640.9**	**814.3**	**1398.9**

Source: Central Water Commission, Govt. of India (2008)

The state wise ground water availability, utilization and stage of development in NEH and Eastern region of India are exhibited in Table 8. The total recharge from rainfall and other sources is about 137.14 BCM Yr^{-1} (31.7%) for the region out of which NEH and Eastern regions share about 7.8 and 23.9%, respectively. Net annual ground water availability after the discharge during monsoon season is 125.8 BCM, out of which only 35.76 BCM is being utilized for irrigation, domestic and industrial purposes. A similar trend was observed for the natural discharge during non-monsoon season and net annual availability of ground water in the entire region.

The projected demand upto 2025 for domestic and industrial use is higher in Eastern region compared to NEH region which accounts for only 4.7% of the country's demand. Ground water availability for future irrigation in the region is more than half (54.3%) of India's ground water availability of 161.06 BCM Yr^{-1}. However, ground water availability is much higher in Eastern region (39%) than the NEH region (15.5%) of the country's total ground water. Among the states, Assam, Bihar, Odisha and West Bengal are having higher quantity of surplus ground water resource and thus these states can easily meet the future demand for domestic and Industrial sectors with the present ground water resources. The abundant ground water resource is mainly due to high rainfall during monsoon season and perennial mega rivers like Ganga and Brahmaputra. The percent ground water development is only 28.42% for the entire region with NEH and Eastern regions accounting for about 18.17% and 31.77%, respectively. Poor Agricultural growth in NEH region is also attributed to poor ground water development as compared to the Eastern region.

Table 8 : State-wise ground water resources availability, utilisation and stage of development in NEH and Eastern regions of India (BCM Yr^{-1})

State	Total Recharge from Rainfall and other Sources	Natural Discharge during Non-monsoon Season	Net Annual Ground Water Availability	Annual Ground Water Draft (Total)	Projected Demand for Domestic and Industrial Uses upto 2025	Ground Water Availability for Future Irrigation	Stage of Ground Water Devel-opment (%)
Arunachal Pradesh	2.56	0.26	2.30	-	0.01	2.29	0.04
Assam	27.23	2.34	24.89	5.44	0.98	19.06	22.00
Manipur	0.38	0.04	0.34	-	0.02	0.31	0.65
Meghalaya	1.15	0.12	1.04	-	0.10	0.94	0.18
Mizoram	0.04	0	0.04	-	-	0.04	0.90
Nagaland	0.36	0.04	0.32	0.01	0.03	0.30	3.00
Sikkim	0.08	-	0.08	0.01	0.02	0.05	16.00
Tripura	2.19	0.22	1.97	0.17	0.20	1.69	9.00
NEH	**33.99**	**3.02**	**30.98**	**5.63**	**1.36**	**24.68**	**18.17**
Bihar	29.19	1.77	27.42	10.77	2.14	16.01	39.00
Chhattisgarh	14.93	1.25	13.68	2.80	0.70	10.67	20.00
Jharkhand	5.58	0.33	5.25	1.06	0.56	3.99	20.00
Odisha	23.09	2.08	21.01	3.85	1.22	16.78	18.00
West Bengal	30.36	2.9	27.46	11.65	1.24	15.32	42.00
ER	**103.15**	**8.33**	**94.82**	**30.13**	**5.86**	**62.77**	**31.77**
Total (NEH+ ER)	**137.14**	**11.35**	**125.80**	**35.76**	**7.22**	**87.45**	**28.42**
India	**432.43**	**33.73**	**398.7**	**230.41**	**29.09**	**161.06**	**58.00**

Abbr. :BCM Yr^{-1} : Billion Cubic Metre per Year.

Percent decline in ground water level upto 10 m was higher for the entire region as compared to the country as a whole (Table 9). Percent of wells showing declining trend was much higher in NEH region compared to Eastern region with respect to 0-2 m and 2-5m water depths but it was higher in Eastern region by 48% for the higher depth range of 5-10m as compared to NEH and country as a whole. But higher depth range per cent of wells showing decline in ground water level was relatively less in all the states of the region compared the national figure. The data reveals that, in both the regions (NEH and ER), less ground water recharge during post monsoon period coupled with higher rate of withdrawal of ground water for different purposes leads to declining ground water tables. To maintain sustainable water yield in the wells, appropriate measures for groundwater recharge through water harvesting structures in the region are very much needed.

Table 9 : State-wise number of wells showing decline in water level and their percentage in NEH and Eastern regions of India (pre monsoon 2001)

State	No of wells analysed	Percentage of wells showing decline					
		0-2 m	2-5 m	5-10 m	10-20 m	20-40	>40 m
Arunachal Pradesh	12	17	42	25	17	0	0
Assam	209	19	58	20	2	0	0
Manipur	-	-	-	-	-	-	-
Meghalaya	27	15	78	7	0	0	0
Mizoram	-	-	-	-	-	-	-
Nagaland	-	-	-	-	-	-	-
Sikkim	-	-	-	-	-	-	-
Tripura	27	11	59	30	0	0	0
NEH	**275**	**18**	**60**	**20**	**3**	**0**	**0**
Bihar	269	3	33	56	8	0	0
Chhattisgarh	360	3	22	56	19	0	0
Jharkhand	180	1	9	59	31	0	0
Odisha	873	9	41	45	4	0	0
West Bengal	520	2	30	42	23	2	0
ER	**2202**	**5**	**32**	**48**	**14**	**1**	**0**
Total (NEH + ER)	**2477**	**7**	**35**	**45**	**13**	**1**	**0**
India	**10274**	**6**	**28**	**39**	**21**	**4**	**2**

Forest Vegetation and Bio-diversity

NEH and Eastern India, is a treasure house of plant genetic diversity and 6-7 thousand plant species constituting approximately 40% of the total flora of the country exist in this region. Forests found in these regions are unique and have high number of flora and fauna. The forests in the region are extremely diverse in structure and composition and combine tropical and temperate forest types, alpine meadows and cold deserts. Northeast India forms one of the major regions of tropical forests in India, especially the species-rich tropical rain forests. The tropical semi-evergreen and moist deciduous forests in the lowlands of this region extend south and west into the subcontinent, and east into Southern China and Southeast Asia. The sub-tropical forests of the region follow the foothills of the Himalayas to the west; also extend into Southeast China in the east. Himalayan temperate and sub-alpine zone forests extend from northern Pakistan and adjacent Afghanistan through Northeast India to Southwest China.

Other natural vegetation comprises of multipurpose tree species, fruit plants, medicinal shrubs and herbs, bamboos and canes. A large number of economically important plants such as food, fodder, spices, oilseeds, fruits, vegetables, medicinal and aromatic, orchids and other flowering plants grow wildly in nature. The region is either primary or secondary centre of origin for many crop plants. Among all states, NEH region is having highest number of Bamboo species of India. Bamboo, wild life and other important flora and fauna is a source of livelihood for a large number of people especially in the tribal areas of NEH region. Besides generating revenue and employment, the forests have a critical bearing on water resources in the region. About 52.85% of the total geographical area of both the regions is under forests constituting about 46.4% of the total forest area of the country (Table 10). The states of Arunachal Pradesh in NEH region and Odisha in Eastern region have maximum forest area of 67,410 and 48,903 sq. km, respectively. The forests in NEH and Eastern regions account for about 66.07% and 25.77% of the TGA of the region, respectively.

Table 10 : Area under forests in NEH and Eastern regions of India

State	TGA (sq. km)	Forest Cover (sq. km) Very dense	Moderate	Open	Total	% to TGA
Arunachal Pradesh	83,743	20,868	31,519	15,023	67,410	80.50
Assam	78,438	1,444	11,404	14,825	27,673	35.28
Manipur	22,327	730	6,151	10,209	17,090	76.54
Meghalaya	22,429	433	9,775	7,067	17,275	77.02
Mizoram	21,081	134	6,086	12,897	19,117	90.68
Nagaland	16,579	1,293	4,931	7,094	13,318	80.33
Sikkim	7,096	500	2,161	698	3,359	47.34
Tripura	10,491	109	4,686	3,182	7,977	76.04
NEH region	**262,184**	**25,511**	**76,713**	**70,995**	**173,219**	**66.07**
Bihar	94,163	231	3,280	3,334	6,845	7.27
Chhattisgarh	136030	4163	34911	16600	55674	41.18
Jharkhand	79,714	2,590	9,917	10,470	22,977	28.82
Odisha	155,707	7,060	21,366	20,477	48,903	31.41
West Bengal	88,752	2,984	4,646	5,365	12,995	14.64
ER	**554366**	**17058**	**74120**	**56246**	**147394**	**25.77**
Total (NEH + ER)	**816550**	**42539**	**190042**	**166887**	**320613**	**52.85**
India	**3287263**	**83510**	**319012**	**288377**	**690899**	**21.02**

Source: India State of Forest Report-2011

During the last decade, forest cover increased by 3700 sq. km in Eastern and NEH region (Fig. 2) which constitutes about 12% of the total forest cover increase in the country. Among NEH states, increase in forest cover (0.77%) was observed during 2001-05 but it decreased (-0.05%) during 2005-11, mainly due to increased conversion of forests for shifting cultivation and other biotic pressures. Among Eastern states, increase of forest cover was negligible (0.27%) during the first five years and higher (1.8%) in the second set of five years in the decade. Among the Eastern and NEH states, higher forest cover increase was observed in Eastern region states than NEH states. Increase in the forest cover is mainly on account of effective protection measures by the village forest protection committees, regeneration of shifting cultivated area and enhanced plantation activities both inside and outside the forest areas. But still majority of the shifting cultivated areas are still under *Jhum* cultivation. This leads to loss of biodiversity and environmental degradation. Hence, vigorous efforts are needed to curb the *Jhum* cultivation and enhance the forest cover of the mega biodiversity hot spot of India.

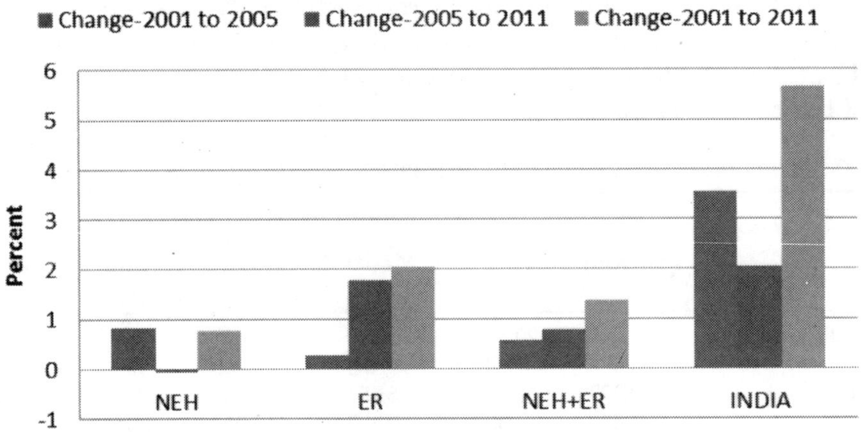

Fig. 4 : Change in forest cover during 2001 to 2011

Biodiversity Conservation

The conservation and sustainable use of natural ecosystems and biodiversity is essential to support sustainable development with biological resources providing raw materials for livelihoods, sustenance, trade, medicine and industrial development. Protecting forests and other natural ecosystems can provide a host of services that reduce human vulnerability to natural hazards along with the benefits of global value such as carbon sequestration, hydrological and geochemical cycling of essential elements etc. Protected areas, which form the corner stone of biodiversity and the natural habitats within them, can protect watersheds and regulate the water flow, prevent soil erosion, influence the rainfall regime and local climate, conserve genetic reservoirs etc. Richness of biodiversity is fundamental to agricultural production and food security besides being a valuable ingredient of environmental conservation. India has the distinction of being as one of the 12 Mega centres of plant diversity in the world with 1, 26, 188 species.

Eastern and north eastern regions of the country are rich in biodiversity because of their diverse physiographic and climatic conditions. Variability in climatic and edaphic conditions brings about significant difference in the forest types in the region. The Northeast is among the 34 hot spots of the world, identified in India, the other being the Western Ghats. Northeast India is one of the richest biomes of the world, high in endemism and rare species which is now under constant threat. Northeast India is thus the geographical 'gateway' for much of India's flora and fauna. It is in this lowland-highland transition zone that the highest diversity of biomes or ecological communities can be found, and species diversities within these communities are also extremely high. Cherrapunjee in the State of Meghalaya holds the record for the highest rainfall in a single month (9,300 mm) as well as in a year (26,461 mm) in India, while the nearby Mawsynram has the world's highest average rainfall (11,873 mm).

Predominant pattern of agricultural growth, ever increasing industrial development and growth of urban centres have caused erosion of biodiversity in the region. The greater demand of water for all these enterprises has led to the destruction of natural aquatic habitat and erosion of aquatic biodiversity. The conflict between agriculture and biodiversity is inevitable which can be overcome by evolving strong institutional mechanisms. India is a signatory to conservation of biological diversity which stipulates safety of genetic resources through a regulatory control mechanism.

Agricultural Scenario

The region is dominated with rain dependent agriculture which accounts for about 72.15% of the total cropped area in the region against only 57% at the national level. Between the regions, NEH region is highly depended on monsoon (78%) compared to the Eastern region (63.4%) (Table 11). Net rainfed area to the net area sown follows the similar trend that of gross rainfed area to the gross cropped area.

Table 11 : State-wise irrigated, rainfed and per capita availability of resources and investment in NEH and Eastern regions of India

State	Cropping Intensity (%)	Net Irrigated area to Net area sown (%)	Net Rainfed area to Net area sown (%)	Total Rainfed area to Total cropped area (%)	Net Sown Area	Net Irrigated Area	Per capita	
							Income	Investment
Arunachal Pradesh	123	41	59	57	0.17	0.04	8945	5958
Assam	143	5	95	96	0.10	0.00	5909	719
Manipur	100	22	78	78	0.10	0.02	6729	2590
Meghalaya	120	25	75	74	0.09	0.02	8305	1981
Mizoram	100	14	86	84	0.10	0.02	10669	5418
Nagaland	120	22	78	75	0.15	0.03	9927	2526
Sikkim	110	8	92	90	0.19	0.01	9056	6442
Tripura	106	21	79	67	0.08	0.02	7407	2185
NEH	**115**	**20**	**80**	**78**	**0.12**	**0.02**	**8046**	**3477**
Bihar	137	59	41	42	0.06	0.03	3330	279
Chhattisgarh	121	28	72	73	0.25	0.06	6904	351
Jharkhand	119	9	91	89	0.06	0.01	6821	312
West Bengal	178	58	42	44	0.15	0.05	8955	395
Odisha	150	31	69	69	0.06	0.04	5457	811
ER	**141**	**37**	**63**	**63**	**0.12**	**0.04**	**6111**	**430**
NEH+ER	**125**	**26**	**73**	**72**	**0.12**	**0.03**	**7508**	**1954**
India	**135**	**42**	**58**	**57**	**0.13**	**0.05**	**7626**	**2468**

The net irrigated area is much lower (26.4%) in the region compared to the national level (42%). However, NEH region is lagging behind in terms of area under irrigation compared to that of Eastern states (37%). The average cropping intensity is only 125% in the region as a whole compared to the national average of 135%. The Eastern states account for higher cropping intensity owing to better irrigation sources as compared to all the NEH states where single crop particularly during the monsoon is mostly prevalent. Even though north eastern states fall under humid zone, net irrigated area over net area sown is very low *i.e.,* 20% and it is lower than the national average, mainly due to non-development of irrigation facility because of steep and undulating topography in these states except in the state of Assam. Per capita availability of net sown area in these regions is comparable with the national average but per capita availability of irrigated area is much lower. However, per capita income and investment is slightly higher in NEH region compared to ER and at national level but per capita investment is significantly lower in all the eastern states. Agriculture development is much lower than its potential in all the states of both the regions, thus limiting employment in the agriculture sector and causing a large section of the population to remain below poverty line.

Agriculture Production and Productivity

The north eastern region receives an annual average rainfall of 2000 mm accounting for around 10% (42.50 M ha-m) of the country's total precipitation of 420.00 M ha-m. The soil of the region is acidic to strongly acidic in reaction. The low pH of the soil is basically due to the leaching of the bases under the influence of high rainfall. The soils are, however, rich in organic matter. The depth of the soil varies from shallow in inceptisols and antisols to very deep in alluvial soils. The region, by and large, is characterized by fragility, marginality, inaccessibility, cultural heterogeneity, ethnicity and rich biodiversity. The region is having more diversified cultivated crops in comparison to the other regions. The more diversified crops are attributed to the variations in climate and rainfall. Due to high rainfall and waterlogged conditions, rice is the predominant *kharif* crop of the entire region occupying nearly 70-90% area under both irrigated and rainfed conditions. At some places, pulses and oilseeds are also grown during *kharif* season. Wheat, potato, sugarcane, pulses and oilseeds are the major *rabi* crops while jute is also grown as pre-monsoon crop. Major crops grown in red lateritic soils of the region are rice, soybean, sorghum, maize, rice, pigeon pea, wheat and chickpea whereas in inceptisol soils sugarcane, maize, wheat, potato, jute and oilseeds are mainly grown. The state wise and region wise area, production and productivity of cereals, pulses and oilseeds are presented in Table 12. The region as a whole contributes 27 %, 14% and 8% of area under cereals, pulses and oilseeds, respectively to the average area under these crops in the country. The production of these crops also follows a similar trend to that of the area. The average productivity of cereals and oilseeds is lesser by 13 and 36%, respectively over the average national productivity. However, the productivity of pulses is higher (+14%) for the region as compared to the average productivity of 593 kg ha^{-1} in the country. Between the two regions, Eastern region showed higher productivity of cereals and oil seeds but productivity of pulses is higher in NEH region (716 kg ha^{-1}).

Table 12 : State-wise area, production and productivity of major crops in NEH and Eastern regions of India (2000-09)

State/UT	Cereals			Oil seeds			Pulses		
	Area ('000ha)	Prod. ('000t)	Yield (kg ha^{-1})	Area ('000ha)	Prod. ('000t)	Yield (kg ha^{-1})	Area ('000 ha)	Prod. ('000 t)	Yield (kg ha^{-1})
Arunachal Pradesh	193	240	1247	27	26	962	8	8	1078
Assam	2405	3407	1417	266	139	523	113	63	558
Manipur	166	391	2363	2	1	429	15	7	497
Meghalaya	127	229	1803	10	7	670	4	3	9
Mizoram	66	107	1609	4	1	229	5	3	529
Nagaland	267	432	1616	76	68	896	35	42	1189

Sikkim	69	100	1460	9	8	872	13	12	928
Tripura	240	629	2616	4	3	675	7	5	691
NEH	**3532**	**5534**	**1566**	**397**	**251**	**633**	**199**	**142**	**716**
Bihar	6421	10367	1615	141	138	979	608	497	818
Chhattisgarh	4213	4714	1119	313	141	449	868	450	518
Jharkhand	1764	2498	1397	124	69	553	410	302	736
Odisha	4630	7760	1676	323	197	608	859	384	446
West Bengal	6281	15071	2419	707	705	997	186	148	793
ER	**23309**	**40410**	**1645**	**1609**	**1249**	**717**	**2931**	**1780**	**662**
Total (NEH + ER)	**26841**	**45955**	**1720**	**2005**	**1500**	**680**	**3130**	**1922**	**676**
India	**99037**	**193337**	**1950**	**24893**	**23187**	**925**	**22282**	**13240**	**593**

Source: www.indiastat.com

Productivity of Crops

The average productivity of selected crops in all the states of both NEH and Eastern regions is presented in Table 13 which reveals that, average productivity of all the crops is significantly lower than the national average productivity except for rice, maize, total food grains in West Bengal and vegetables in Eastern region and rice and Maize in Manipur. West Bengal is having higher productivity of Rice (2458 (kg ha^{-1}), Maize (2785 kg ha^{-1}), Wheat (2355 kg ha^{-1}) and Ragi (1175 kg ha^{-1}). The poor yield of crops in this region is attributed to adoption of traditional crops and varieties under rainfed conditions as well as the age old practice of shifting and other unscientific methods of cultivation. Agriculture development is much lower than its potential in the region, thus threatening the livelihood security of rural masses. For example, NEH states occupy about 2.74% of the total cultivated area and 3.77% population of the country to produce only about 2.6% (5.92 M tonnes) of the total food grain during 2010-11.Integrated farming systems in rice-based rainfed ecosystem in the region through introduction of fish, duck and pig culture have tremendous potential in the region by taking into consideration the socio-economic background of the farming community. There is a scope for bridging the gaps in productivity of all the crops in NEH and eastern region through integrated production-cum-resource conservation technologies.

Table 13 : State-wise average productivity (kg ha^{-1}) of selected crops in NEH and Eastern regions of India (2000-2009)

State	Rice	Maize	Wheat	Ragi	Total food grains	Vegetables
Arunachal Pradesh	1204	1383	1575	-	1212	432
Assam	1482	721	1132	-	1427	11692
Manipur	2383	2548	-	-	2294	7142
Meghalaya	1174	1476	1760	-	1686	8392
Mizoram	1460	1747	-	-	1510	8767

Nagaland	1644	1716	1692	-	1552	8367
Sikkim	1461	1510	1342	896	1357	4625
Tripura	2360	993	2357	-	2294	11633
NEH	**1646**	**1511**	**1643**	**896**	**1666**	**7631**
Bihar	1419	2420	1960	804	1598	15141
Chhattisgarh	1455	1351	1010	264	1008	13188
Jharkhand	1481	1462	1652	642	1347	13633
Odisha	1360	1683	1415	605	1250	11841
West Bengal	2458	2785	2355	1175	2431	15091
ER	**1635**	**1940**	**1679**	**698**	**1527**	**13779**
Total (NEH + ER)	**1642**	**1677**	**1659**	**731**	**1613**	**9996**
India	**2033**	**2016**	**2729**	**1394**	**1726**	**14742**

Self Sufficiency of Food

The self-sufficiency of cereals, pulses and oilseeds were calculated by taking into account the latest population and average area, production and productivity of crops during 2000-09. There is a huge gap between the production and consumption of the food grains in this region (Table 14). Both the regions are deficit in cereals, pulses, oilseeds and food grains thus majority of the population suffers from undernourishment and depends on other states for their requirements. There is a deficit of 7.3%, 82%, 82% and 15 % for cereals, pulses, oil seeds and total food grain production, respectively in NEH region. On the other hand in Eastern region, the deficit was 7.3%, 64.8%, 88.2% and 2.03% for cereals, pulses, oilseeds and food grain production, respectively. States like Arunachal Pradesh, Nagaland, Sikkim and Tripura in NEH region, Chhattisgarh, Odisha and West Bangal in Eastern region are surplus in cereals. Pulses and oil seeds are deficit in all the states of both the regions except Chhattisgarh state. To achieve the self-sufficiency in production of cereals, pulses and oil seeds, an additional area of about 3.04 M ha, 5.38 M ha and 10.88 M ha has to be brought under cereals, pulses and oil seeds cultivation, respectively or by increasing the productivity levels of crops significantly through modern production technologies coupled with increasing area under inter crops, irrigation and by efficient management of inputs and natural resources.

Livestock

The Eastern and NEH region has over 1190 lakh livestock population constituting 20% of total livestock population of the country (Table 15). However in NEH region, the livestock population is very poor except in the state of Assam. In Eastern region, the maximum population of cattle (104.38 lakhs), buffaloes (57.3 lakhs) and goats (94 lakhs) is in Bihar. Whereas, NEH region has wide diversity in the livestock population with rearing of yak, mithun, rabbit etc. as it is part of tribal lives rituals and customs. In spite of a

Table 14 : Requirement, actual production, surplus/ deficit of food grains and oil seeds in NEH and Eastern regions of India (in M tonnes)

State	Cereals & Millets			Pulses			Oilseeds			Food grains		
	Req.	Prodn.	S/D	Req.	Prodn.	S/D	Req.	Prodn.	S/D	Req.	Prodn.	S/D
Arunachal Pradesh	0.17	0.23	0.06	0.02	0.01	-0.01	0.04	0.03	-0.01	0.19	0.24	0.05
Assam	4.21	3.72	-0.49	0.49	0.06	-0.43	0.97	0.14	-0.83	4.70	3.78	-0.91
Manipur	0.38	0.40	0.02	0.04	0.00	-0.04	0.09	0.00	-0.09	0.42	0.40	-0.02
Meghalaya	0.37	0.22	-0.15	0.04	0.00	-0.04	0.09	0.01	-0.08	0.41	0.22	-0.19
Mizoram	0.14	0.10	-0.05	0.02	0.01	-0.01	0.03	0.00	-0.03	0.16	0.10	-0.06
Nagaland	0.31	0.38	0.07	0.04	0.03	0.00	0.07	0.05	-0.03	0.35	0.41	0.06
Sikkim	0.09	0.10	0.01	0.01	0.01	0.00	0.02	0.01	-0.01	0.10	0.10	0.01
Tripura	0.50	0.58	0.08	0.06	0.01	-0.05	0.12	0.01	-0.11	0.56	0.59	0.02
NEH	**6.17**	**5.72**	**-0.45**	**0.71**	**0.13**	**-0.59**	**1.43**	**0.25**	**-1.18**	**6.88**	**5.85**	**-1.04**
Bihar	16.40	11.87	-4.54	1.89	0.50	-1.40	3.79	0.14	-3.65	18.30	12.36	-5.93
Chhattisgarh	3.29	4.71	1.43	0.38	0.45	0.07	0.76	0.14	0.62	3.66	5.16	1.50
Jharkhand	5.21	2.70	-2.51	0.60	0.30	-0.30	1.20	0.07	-1.13	5.81	3.00	-2.81
Odisha	6.63	7.97	1.34	0.77	0.38	-0.38	1.53	0.20	-1.33	7.39	8.35	0.96
West Bengal	14.43	15.34	0.90	1.67	0.15	-1.52	3.33	0.71	-2.63	16.10	15.48	-0.62
ER	**45.96**	**42.58**	**-3.37**	**5.31**	**1.78**	**-3.53**	**10.62**	**1.25**	**-9.37**	**51.26**	**44.36**	**-6.90**
Total (NEH+ER)	**52.13**	**48.30**	**-3.82**	**6.02**	**1.91**	**-4.11**	**12.04**	**1.50**	**-10.55**	**58.15**	**50.21**	**-7.94**

Source: www.indiastat.com

Req. = Requirement (Cereal and millet @ 158 kg $capita^{-1}$ $annum^{-1}$, pulses @ 18.25 kg $capita^{-1}$ $annum^{-1}$, oil seeds @ 36.5 kg $capita^{-1}$ $annum^{-1}$)

desired aptitude towards animal husbandry practices, per capita availability of milk, meat, egg and fish per annum is only 31.53 litres, 9.36 kg, 33.50 numbers and 4.12 kg, respectively. Hence, the region has enormous potential to enrich commercialization through scientific techniques, improved breeds and management of pastures and common property resources for quality fodder production.

Table 15 : Livestock population in NEH and Eastern regions of India (in lakh)

State	Cattle	Sheep	Goats	Buffaloes
Arunachal Pradesh	4.89	0.19	2.95	0.28
Assam	80.97	1.70	29.87	6.78
Manipur	3.97	0.06	0.33	0.77
Meghalaya	6.91	0.18	3.27	0.18
Mizoram	0.34	0.01	0.17	0.06
Nagaland	4.31	0.04	1.75	0.34
Sikkim	1.56	0.06	1.24	0.02
Tripura	7.47	0.03	4.72	0.14
NEH	**110.42**	**2.28**	**44.30**	**8.56**
Bihar	104.38	3.82	94.90	57.03
Chhattisgarh	84.60	1.20	23.40	15.90
Jharkhand	74.68	6.80	50.31	13.43
Odisha	134.21	16.20	58.03	13.94
West Bengal	182.74	15.25	187.74	10.86
ER	**580.61**	**43.27**	**414.38**	**111.16**
Total (NEH + ER)	**691.03**	**45.55**	**458.68**	**119.72**
India	**1804.60**	**614.70**	**1246.30**	**978.70**

Source: Livestock census -2009

Horticulture

The regions is endowed with diverse climate, and thus offers ideal agro-climatic conditions for cultivation of a wide variety of tropical and temperate fruits from low lands to hills and valleys. The dominant horticultural crops of the region include pineapple, citrus, banana, apple, guava, mango, papaya and sapota among fruits, brinjal, cabbage, cauliflower, bhindi, peas, tomato, onion and tapioca among vegetables, cashew nut, aracanut, tea and coffee among important plantation crops grown in the region. The state wise area and production of total horticultural crops is presented in Table 16. Both NEH and ER contribute about 28.5% area and 33% production of the country. The share of NEH region to the country area and production of total horticultural crops is only 5% and 4.5%, respectively as against about 23.5% area and 28% production from the Eastern region. This shows that, NEH region accounts for only 17% area and 14% production in the both the regions of the country. Orange is a unique crop of this region; however, average productivity is poor except in Assam and Meghalaya. High productivity from well-maintained orchards and trees bearing 2500-3000 fruits has been reported. Orange is a highly remunerative crop in *Jampui* hills of Tripura where an organised effort made by the state department has paid rich dividends. Productivity of pineapple in Assam compares well with that of all India average but it is much lower than the neighbouring West Bengal where production technologies, *viz;* lower inducements, balanced nutrition and high density plantations have been adopted in North Bengal situations. These technologies need to be tried at a large scale in NEH region to improve the productivity. Productivity of banana in Assam and Meghalaya (12-13 t ha^{-1}) is, however, at par with that of the neighbouring states like West Bengal and Bihar. Ginger and turmeric are well-established cash crops of Meghalaya. Ginger is grown in Mizoram too with high productivity. Large-scale production of planting material and marketing support has encouraged ginger cultivation.

Table 16 : State-wise area and production of Horticultural crops in NEH and Eastern Regions of India

State	Area ('000 ha)	Production ('000 Metric tons)
Arunachal Pradesh	89.8	190.3
Assam	488.7	6327.6
Manipur	67.3	511.5
Meghalaya	107	799.7
Mizoram	68.9	610.2
Nagaland	49.5	342.3
Sikkim	67.6	207.9
Tripura	83.6	1049.2
NEH	**1022.4**	**10038.7**
Bihar	1141.1	17386.4
Chhattisgarh	500.6	4889.2
Jharkhand	251.5	4068.8
Odisha	1346.1	11306.9
West Bengal	1679.8	25287.3
ER	**4919.1**	**62938.6**
Total (NEH + ER)	**5941.5**	**72977.3**
India	**20875.7**	**223089.0**

Land Degradation

The Central Soil and Water Conservation Research and Training Institute, Dehradun in collaboration with National Bureau of Soil Survey and Land Use Planning, Nagpur computed potential erosion rates for different states of the country considering six erosion categories, viz; very slight, slight, moderate, moderately severe, severe and very severe with potential erosion rates of < 5, 5-10, 10-15, 15-20, 20-40 and > 40 t ha^{-1} yr^{-1}. Table 17 presents the potential erosion rates of more than 10 t ha^{-1} yr^{-1} categories. As evident, in north-eastern region, about 61.7% area suffers from potential erosion rate of more than the permissible limit of 10 t ha^{-1} yr^{-1} against national average of about 39%. Similarly, about 32.43% area in Eastern region has potential erosion rate of more than the permissible limit. Among the north Indian states, Sikkim has the least (36.86%) area under more than the permissible rate category while Nagaland has the highest percentage (87.27%). Within eastern region, Bihar has the least (12.97%) area under more than the permissible rate category while Jharkhand has the highest percentage (64.77%).

Table 17 : Area under different potential soil erosion rate categories in Eastern and North-Eastern states

State/UT	TGA (Sq. km)	Soil loss (>10 t ha^{-1} yr^{-1})	Soil erosion class area (%)				
			Moderate (10-15 t ha^{-1} yr^{-1})	Moderately (15-20 t ha^{-1} yr^{-1})	Severe (20-40 t ha^{-1} yr^{-1})	V. Severe (40-80 t ha^{-1} yr^{-1})	Extra Severe(>80 tha^{-1} yr^{-1})
Arunanchal Pradesh	83743	72.65	5.10	5.42	23.65	27.31	11.17
Assam	78438	65.79	4.58	18.08	14.83	28.30	0.00
Manipur	22327	53.29	15.24	11.43	26.61	0.00	0.00
Meghalaya	22429	77.90	14.77	10.21	26.25	13.86	12.80
Nagaland	16579	87.27	4.09	3.80	15.96	28.48	34.94
Sikkim	7096	36.86	0.90	1.19	7.82	10.93	16.20
Tripura	10486	38.40	7.10	7.00	6.50	8.60	9.20
NEH	**241098**	**67.58**	**6.66**	**10.37**	**19.56**	**22.64**	**8.35**
Bihar	93979	12.97	6.33	3.53	2.73	0.58	0.00
Jharkhand	79898	64.77	15.75	11.64	20.95	12.20	4.73
Odisha	155707	31.77	10.29	6.69	9.58	4.43	1.08
West Bengal	88752	20.19	11.89	4.28	3.67	0.37	0.00
Chhattisgarh	134805	66.30	7.99	6.45	18.32	13.69	19.07
ER	**553141**	**39.90**	**10.10**	**6.42**	**11.24**	**6.50**	**5.63**
NEH + ER	**794239**	**48.30**	**9.06**	**7.62**	**13.76**	**11.40**	**6.46**
India	**3281270**	**38.73**	**11.22**	**6.46**	**9.92**	**7.14**	**3.99**

Land degradation constitutes one of the greatest threats for food and environmental security. Degradation of soil and/or vegetative cover is caused by a variety of anthropogenic pressures including development of big hydropower projects, deforestation, over-grazing, unsustainable agricultural practices and industrial activities. It is estimated that a major part of the total area is degraded due to various factors like soil erosion, water logging, salinity/ alkalinity and acidity *etc.* (Table 18) with erosion by water being the chief contributor followed by chemical degradation and physical degradation. In NEH region of the country, about 54.20 % area suffers from degradation due to deforestation and *jhum* cultivation followed by water erosion (18.08%). It is also observed that around 62.53% of the total geographical area in eastern region is degraded exclusively by water induced soil erosion which in conjunction with salt-affected and acid soils works out to be 73.88%. Among the Eastern states, Jharkhand has maximum (71.65%) TGA under water erosion followed by Bihar (59.81 %), Odisha (58.46 %), and West Bengal (54.53 %). Thus water induced soil erosion poses a serious threat to sustainable production in the region. Extent of acid soils (having pH< 5.5) affected area is much more in NEH region (7.90% of TGA) than that in Eastern region (5.30% of TGA). Among the north-eastern states, Meghalaya has maximum area (33.43% TGA) under acid soils followed by Arunachal Pradesh (13.93% of TGA), Tripura (12.87% of TGA), Assam (8.99% of TGA) Manipur (6.45% of TGA), Sikkim (3.33% of TGA), and Nagaland (1.10% of TGA).

Table 18 : Soil degradation in NEH and Eastern regions of India

State/UT	Area (%)				
	Saline & alkaline	Water eroded	Acid soils	Degraded forest	Others
Arunachal	0.0	7.66	13.93	54.92	23.49
Assam	0.0	42.11	8.99	15.36	33.54
Manipur	0.0	2.04	6.45	85.41	6.10
Meghalaya	0.0	7.33	33.43	79.39	10.28
Nagaland	0.0	0.06	1.10	95.74	3.10
Sikkim	0.0	3.33	3.33	21.67	71.67
Tripura	0.0	3.33	12.87	72.98	10.83
NEH Region	**0.0**	**18.08**	**7.9**	**54.20**	**19.82**
Bihar	9.99	59.81	1.39	16.41	12.40
Jharkhand	0.00	71.65	5.73	11.95	10.68
Odisha	3.52	58.46	2.87	32.16	2.98
WB	19.07	54.53	11.21	5.14	10.05
ER	**6.05**	**62.53**	**5.30**	**17.92**	**8.21**

Erosion Risks-Priority Classes for Soil Erosion Control in Eastern and North-Eastern Regions

The entire region is covered by three agro-ecological regions namely, Agro-Ecological Region 15 (AER 15), Agro-Ecological Region 16 (AER 16) and Agro-

Ecological Region 17 (AER 17). Agro-Ecological Region 15 (AER 15) comprising of the plains of the Brahmaputra and the Ganga Rivers covers parts of the States of Assam and West Bengal with an area of 12.3 M ha, representing 3.7% of the total geographical area of the country. Data pertaining to potential erosion rates indicate that overall soil erosion through water is not a serious problem in majority of the areas in the region as nearly 59.6% area has soil loss of less than 10 t ha^{-1} yr^{-1} (Table 19). Soil erosion has affected only about 21.3% area of the total land mass with severe and very severe erosion rates. Data pertaining to soil loss tolerance limits indicate that the T value varies between 7.5 and 12.5 t ha^{-1} yr^{-1} across the region. Around 68.4 % area of the region falling under alluvial plains has the T- value of 12.5 t ha^{-1} yr^{-1}. This is attributed to good soil depth and better soil resistivity to erosion. Estimates based on comparison between prevailing and permissible rates of erosion indicate that, nearly 61.9% area is very much stable and, therefore, no conservation treatment is required which is at par with area having soil loss of less than 10 t ha^{-1} yr^{-1} (59.6%) Therefore, based on all these assessment, it is proposed that about 62-68% territory of the region is very stable with respect to soil erosion. The spatial distribution of different priority classes in the region is presented in Fig. 5.

Table 19 : Area under different potential erosion rates, tolerance limits and priority classes in the Agro-Ecological Region 15

	Erosion categories (t ha^{-1} yr^{-1})						
Area (M ha) under different erosion categories	< 5 (very low)	5 to 10 (low)	10 to 20 (moderate)	20 to 40 (severe)	> 40 (very severe)	Others	
	3.31 (26.89)*	4.03(32.73)	1.82(14.77)	0.98(7.96)	1.64(13.32)	0.53(4.32)	
	T-value (t ha^{-1} yr^{-1})						
Area (M ha) under different T values	2.5	5.0	7.5	10.0	12.5	Rock/ unreported	
	-	-	0.65(5.29)	2.33(18.98)	8.42(68.4)	0.89(7.31)	
	Priority class (t ha^{-1} yr^{-1})						
	1	2	3	4	5	6	others
Area (M ha) under different priority classes	>35	25-35	15-25	5-15	0-5	<0	-
	0	1.62 (13.21)	0.92(7.51)	0.73(5.97)	1.01 8.2)	7.61(62.0)	0.38 (3.16)

*Values in the parentheses are in percentage

Agro-Ecological Region 16 encompasses northern hilly parts of West Bengal, northern parts of Assam, Arunachal Pradesh and Sikkim States. It occupies an area of 9.6 3 M ha, representing 2.9 % of the total geographical area of the country. Analysis of soil erosion data revealed that soil erosion rates varied enormously across the region, ranging from less than 5 t ha^{-1} yr^{-1} to more than 40 t ha^{-1} yr^{-1} (Table 20). About 41.1% area of the region is affected by very severe erosion with erosion rate of >40 t ha^{-1} yr^{-1} followed by severe erosion (15.96%), moderate erosion (12.00%), low erosion (10.74%), and very low erosion (5.59%). The results further revealed that due to strong resistance capacity, T-values were more or less consistent at around 7.5 and 12.5 t ha^{-1} yr^{-1}, although some parts of the region, especially the Darjeeling hills (0.25%), have lower T-value of 5.0 t ha^{-1} yr^{-1}. By comparing the potential erosion rates with permissible rates, it was observed that about 40.93% land area has high erosion risk. However, practically no area falls under very high erosion risk category under priority class 1. The analysis further revealed that land area of nearly 20.93% is very stable and, therefore, no conservation measure is required in such areas. By considering severe and very severe rates of erosion, it is observed that about 57.06% area is very critical, however, based on the priority classification only 68.49% area was found to be vulnerable. Owing to the preponderance of forests, barren and uncultivated

Fig. 5 : Priority classes for soil erosion control in Agro-Ecological Region 15

lands, per capita availability of land for cultivation in the hill areas of the region is lowest in India (0.12 ha). The spatial distribution of erosion risk areas classified under various priority classes is presented in Fig.6.

Table 20 : Area under different potential erosion rates, tolerance limits and priority classes in the Agro-Ecological Region 16

	Erosion categories (t ha^{-1} yr^{-1})					
	< 5 (very low)	5 to 10 (low)	10 to 20 (moderate)	20 to 40 (severe)	> 40 (very severe)	Others
Area (M ha) under different erosion categories	0.54 (5.59)*	1.03(10.74)	1.16 (12.00)	1.54 (15.96)	3.96 (41.1)	1.41 (14.62)

	T-value (t ha^{-1} yr^{-1})					
	2.5	5.0	7.5	10.0	12.5	Rock/ unreported
Area (M ha) under different T values	-	0.02 (0.25)	2.92 (30.40)	4.81 (50.04)	0.77 (8.06)	1.08 (11.25)

	Priority class (t ha^{-1} yr^{-1})						
	1	2	3	4	5	6	others
Area (M ha) under different priority classes	>35 -	25-35 3.94 (40.93)	15-25 1.51 (5.75)	5-15 0.89 (9.31)	0-5 0.24 (2.50)	<0 2.01 (20.9)	- 1.01 (10.58)

*Values in the parentheses are in percentage

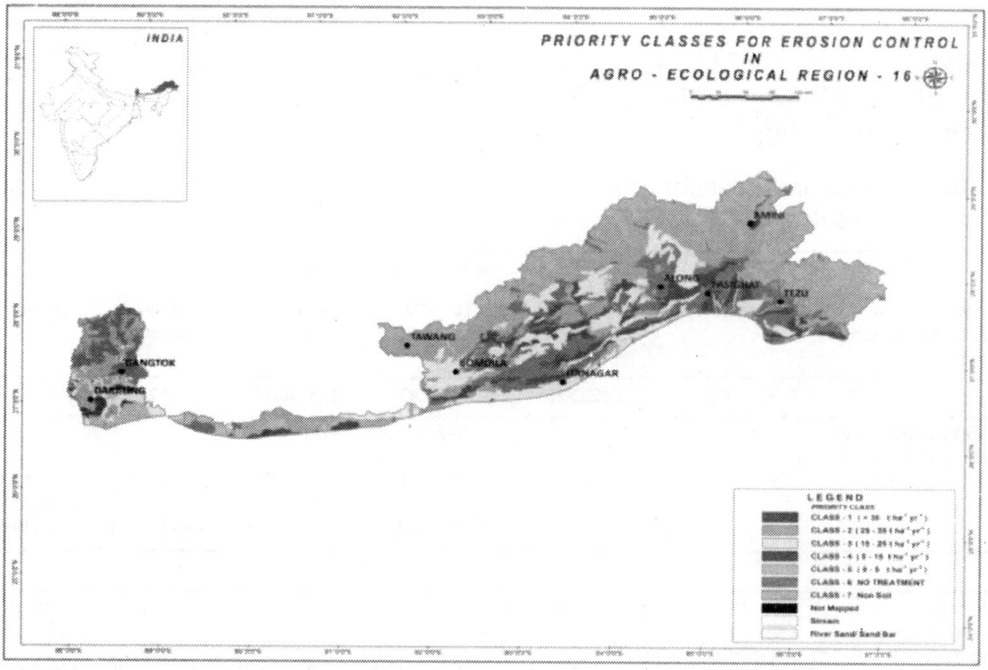

Fig. 6 : Priority classes for soil erosion control in Agro-Ecological Region 16

Agro-Ecological Region 17, known as north-eastern hills (Purvachal) agro-ecoregion, comprises of hilly States of Nagaland, Meghalaya, Manipur, Mizoram and southern Tripura. The region covers an area of 10.66 M ha, representing 3.3% of the total geographical area of the country. Soil erosion data with different categories, corresponding T -values and priority area of AER 17 are given in Table 21. Analysis of soil erosion data revealed that soil erosion rates varied enormously across the region, ranging from less than 5 t ha^{-1} yr^{-1} to more than 40 t ha^{-1} yr^{-1}. About 24.39% area in the region falls under the category of very severe erosion, whereas the percentage of very low, low, moderate and severe erosion classes is 5.29, 22.72, 17.43 and 10.16%, respectively. The results further revealed that due to strong resistance capacity, T-values were more or less consistent at around 7.5 and 12.5 t ha^{-1} yr^{-1} except for about 0.44% area under T value of 5.0 t ha^{-1} yr^{-1}, which indicates substantial contribution of soil depth in deciding the T-values. The areas of major concern are soils on ridges and side slopes which can tolerate only a low soil loss of up to 7.5 Mg ha^{-1} yr^{-1}. Based on the analysis, it is projected that about 52.08% of total geographical area (TGA) of the region requires different degrees of erosion management and 28.39% of TGA falls under no treatment category due to higher values of permissible erosion limits (Table 21). As evident from the table, though 34.55% area of the region falls under severe and very severe erosion categories, 11.05% has high priority for conservation treatment as the soil loss tolerance limit is less than 10.0 t $ha^{-1}yr^{-1}$. However, the assessment based on comparison between prevailing and permissible rates of erosion indicate that nearly 24.39% area of the region is very critical as classified upto priority class 2. The spatial distribution of priority areas under different classes is presented in Fig. 7.

Table 21 : Area under different potential erosion rates, tolerance limits and priority classes in the Agro-Ecological Region 17

	Erosion categories (t ha^{-1} yr^{-1})					
Area (M ha) under different erosion categories	< 5 (very low)	5 to 10 (low)	10 to 20 (moderate)	20 to 40 (severe)	> 40 (very severe)	Others
	0.56 (5.29)*	2.42(22.72)	1.86(17.43)	1.08(10.16)	2.60 24.39)	2.13(20.0)
Area (M ha) under different T values	T-value (t ha^{-1} yr^{-1})					
	2.5	5.0	7.5	10.0	12.5	Rock/ unreported
	-	0.04 (0.44)	1.13 (10.61)	5.98 (56.15)	1.40 (13.17)	2.09 (19.63)
	Priority class (t ha^{-1} yr^{-1})					
Area (M ha) under different priority classes	1	2	3	4	5	6	others
	>35	25-35	15-25	5-15	0-5	<0	-
	0.004 (0.04)	2.59(24.35)	1.07(10.11)	1.46 (13.77)	0.40 (3.81)	3.02(28.39)	2.08 (19.53)

*Values in the parentheses are in percentage

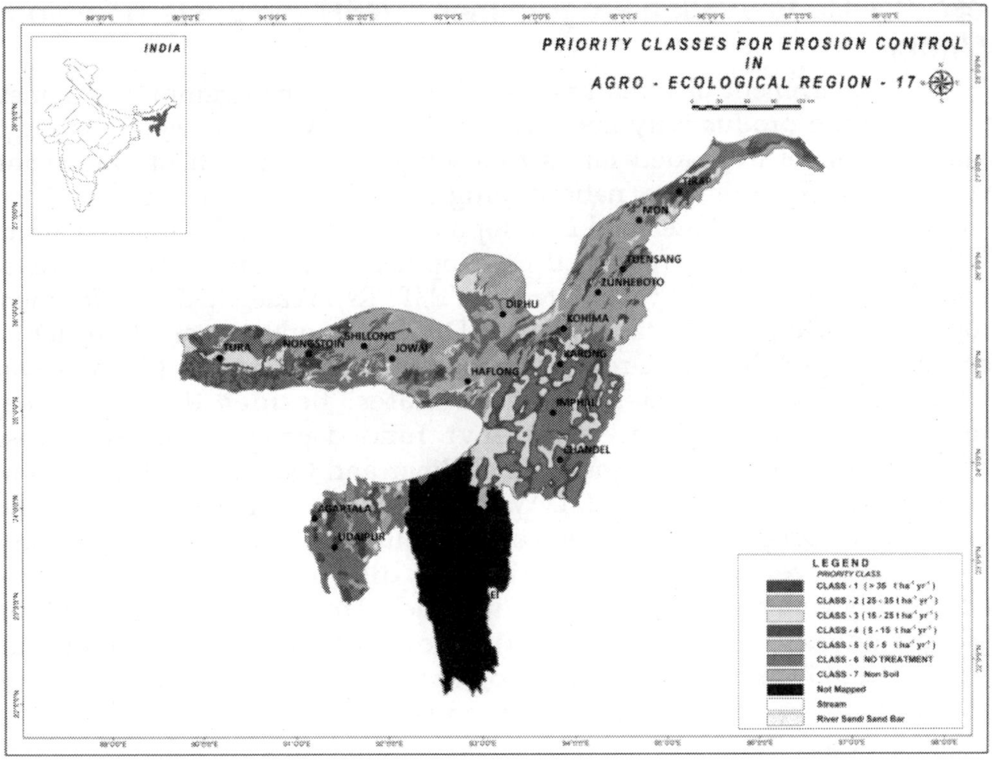

Fig. 7 : Priority classes for soil erosion control in Agro-Ecological Region 17

In general, the areas where the prevailing erosion rate is more than the permissible limits require to be addressed on priority through appropriate conservation measures for bringing the erosion rate within the permissible limits. Cultivation has been extended to marginal lands through extensive deforestation resulting in high soil erosion and low productivity. Indiscriminate deforestation and practice of *Jhum* cultivation lead to accelerated erosion for which proper conservation measures need to be established especially on very steep slopes. Cultivation has been extended to marginal lands through extensive deforestation resulting in high soil erosion (>40 t ha^{-1} yr^{-1}) and low productivity. Valley plain is degraded by the deposition of debris from upper hill slopes carried by high runoff water. Scouring action of torrents is also very common in the region. Therefore, these areas were found to be most sensitive and need special care during soil disturbance. *Jhum* cultivation and indiscriminate deforestation lead to accelerated erosion for which proper vegetation cover need to be established on very steep slopes.

Resource Conservation Programmes in NEH and Eastern Regions of India

The Government of India has launched various programmes/schemes, to increase the productivity and livelihood security in the region through rehabilitation of degraded lands following the approach of watershed management. The prominent national programmes implemented in both NEH and ER are National Watershed Development Programme for Rainfed Areas (NWDPRA), Integrated Watershed Development Programme (IWDP), River Valley Project and Flood Prone Rivers (RVP/FPRs), Watershed Development Project for Shifting Cultivation Areas (WDPSCA) and National Afforestation Programme (NAP). The Drought Prone Area Programme (DPAP) was implemented only in Eastern region states. Besides these national programmes/schemes, many state Govt. funded programmes were also implemented particularly in the state of Bihar and Odisha in ER and some states in NEH region (Table 22). With the launch of massive watershed development programmes in the country during 1990's, all the previous programmes were converged to develop different areas by adopting participatory watershed management approach. The maximum area has been treated under IWDP (43.94 lakh ha) followed by DPAP (25.9 lakh ha), NWDPRA (13.7 lakh ha) and other programmes (₹ 13.55 lakh ha). Similarly, the maximum expenditure has been made under NAP (₹ 708.88 crores) followed by NWDPRA (₹ 547.11 crores). A total area of 107.29 lakhs ha has been treated by incurring an investment of ₹ 3051.83 crores in both the regions.

Table 22 : Physical and financial achievements under different watershed programmes in NEH and Eastern regions of India (Physical in lakh ha and financial in crore ₹)

Programme	Bihar		Jharkhand		Odisha		West Bengal		Chhattisgarh		NEH	
	Phy	Fin	Phy	Fin	Phy	Fin	Phy	Fin	Phy	Fin	Phy	Fin
NWDPRA	0.44	13.84	0.71	39.93	4.49	128.73	1.58	47.78	1.34	56.4	5.14	260.43
DPAP	2.72	32.5	7.98	64.35	6.6	93.96	2.8	24.32	5.8	86.05	-	-
IWDP	3.21	28.85	1.36	12.45	5.17	111.00	1.18	13.76	3.84	84.6	29.18	512.24
RVP/ FPR	1.64	41.36	0	0	2.59	45.76	1.48	39.21	0.31	14.64	0.78	44.74
NAP (NAEB)	0.12	21.93	0.4	81.92	0.57	91.29	0.23	51.64	0.44	103.55	1.64	358.55
Others	5.65	64.63	0	0	3.98	187.27	0	0	0	0	3.92	294.15
Total	**13.78**	**203.11**	**10.45**	**198.65**	**23.4**	**658.01**	**7.27**	**176.71**	**11.73**	**345.24**	**40.66**	**1470.11**

Source: Watershed Development in India: Status and Perspective (2008)

Statewise Area Treated and Expenditure

State wise total area treated, total expenditure incurred and percent of total geographical area treated are exhibited in Table 23. A total area of 9.55

M ha has been treated by spending an amount of ₹ 2706.6 crores in both the regions which accounts for about 16.9% of the total treated area and 14% of the total expenditure of the country under various resource conservation programmes. About 14% of the TGA of the region (15.5% in NEH and 12% in Eastern region) has been treated under various programmes against treated area of 17.2% of TGA of the country. Among the states in the region, maximum TGA has been treated in the state of Chhattisgarh (36.2%) followed by Nagaland (35.2%), Sikkim (32.4%), Mizoram (31.4%) and Manipur (19.7%).

Table 23 : State-wise total area treated and expenditure made under watershed development programmes in NEH and Eastern regions of India (up to March, 2007)

State	Area treated	Expenditure (₹ in crores)	% of TGA treated (M ha)
Arunachal Pradesh	0.50	116.39	6.0
Assam	1.09	273.21	14.0
Manipur	0.44	162.27	19.7
Meghalaya	0.33	115.35	14.8
Mizoram	0.66	273.34	31.4
Nagaland	0.58	316.77	35.2
Sikkim	0.23	101.68	32.4
Tripura	0.23	111.10	21.6
NEH	**4.06**	**1470.11**	**15.5**
Bihar	1.38	203.12	14.6
Chhattisgarh	4.89	1429.28	36.2
Jharkhand	1.04	198.65	13.1
Odisha	2.34	658.01	15.0
West Bengal	0.73	176.71	8.2
ER	**6.66**	**1581.73**	**12.0**
Total (NEH + ER)	**9.55**	**2706.60**	**14.0**
India	**56.54**	**19470.57**	**17.2**

Source: Watershed Development in India: Status and Perspective (2008)

Production and Monetary Losses Due to Water Erosion

State-wise production and monetary losses due to water induced soil erosion in NEH and Eastern regions are exhibited in Table 24. In the entire region, loss in production of cereals, total pulses, total oil seeds and total of all annual crops accounts for 31%,14%,6% and 23% of the production loss in the country, respectively. Monetary loss also follows a similar trend to that of production loss in the entire region. The major share is from the Eastern region where it accounts for a loss of 63% in cereals, 88% in total pulses, 59% in total oilseeds and 65% total of all the crop production losses of the entire region and the monetary loss also follows a similar trend. About 28.2 lakh tons of total production amounting to ₹ 19882 million is being lost annually in the region thus threatening the self sufficiency of food grains. The per capita annual food grains and monetary loss works out to 10 kg and ₹ 51.5, respectively and this monetary loss is equivalent to 1% of the per capita income and 2% of the per capita investment made in the entire region. Therefore, to prevent further land degradation and to increase the agricultural production and income of the community, resource conservation-cum-livelihood programmes need to be implemented on top priority following integrated watershed approach.

Table 24 : Production and monetary losses due to water erosion in North-Eastern and Eastern regions of India

State	Production loss (Lakh tons)				Monetary loss (₹ in millions)			
	Total Cereals	Total Pulses	Total Oilseeds	Total Cereals	Total Pulses	Total Oilseeds	Total	Total
Arunachal Pradesh	0.5	0.0	0.1	0.7	321	30	178	529
Assam	5.6	0.1	0.3	6.1	3398	209	518	4125
Manipur	0.7	0.0	0.0	0.7	402	12	0	414
Meghalaya	0.4	0.0	0.0	0.4	235	25	44	304
Mizoram	0.2	0.0	0.0	0.3	128	33	27	188
Nagaland	1.0	0.1	0.2	1.3	548	138	244	930
Sikkim	0.2	0.0	0.0	0.2	132	26	40	198
Tripura	0.2	0.0	0.0	0.3	40	6	9	55
NEH	**8.9**	**0.3**	**0.7**	**9.9**	**5204**	**478**	**1060**	**6742**
Bihar	1.0	0.2	0.0	1.2	590	253	24	867
Chhattisgarh	9.0	0.8	0.4	10.2	5083	1102	504	6689
Jharkhand	2.8	0.3	0.1	3.2	1592	396	80	2068
Odisha	8.3	1.5	0.8	10.6	4672	2212	1134	8017
West Bengal	3.1	0.1	0.2	3.4	1763	150	276	2188
ER	**15.2**	**2.1**	**1.0**	**18.3**	**8617**	**3011**	**1513**	**13140**
Total (NEH+ER)	**24.1**	**2.4**	**1.7**	**28.2**	**13821**	**3489**	**2573**	**19882**
India	**77.8**	**17.3**	**28.1**	**123.2**	**49613**	**25000**	**62724**	**137337**

General Problems and Constrains Related to Resource Conservation and Livelihood Security

- Practice of shifting cultivation in hilly slopes of the region causing severe land degradation leading to loss of top fertile soil, forest cover, biodiversity and environmental pollution.
- Lack of appropriate data base related to availability of various natural resources including soil and water in the region which is a major hindrance for the planning.
- Poor socio-economic status leading to low adoption of modern production-cum-conservation practices resulting in high poverty. Rural

population of this region accounts for more than 75% and is characterized by fragility, marginality, inaccessibility, cultural heterogeneity and ethnicity.

- Low production and low productivity of crops due to topographical and inaccessibility particularly in NEH region coupled with use of local varieties, un-scientific and faulty methods of cultivation and low fertilizer consumption.
- Soils of this region are acidic to strongly acidic in reaction leading to deficiency and toxicity of essential plant nutrients thereby adversely affecting the crop growth and productivity.
- Poor water resource management and inadequate technological backup for harvesting and utilisation of surface runoff for crop production.
- Large scale deforestation, destruction of native germplasm plants through burning for shifting cultivation leading to loss of biodiversity.
- Deforestation and the resultant loss of soil, especially in the hilly areas, are leading to increased siltation of rivers and streams. The deep pools that are the favoured habitats of many species are rapidly becoming shallow and choked with silt, leading to a decline in habitat. At the same time, swamps, marshes, and other wetlands are increasingly being reclaimed for urban and agricultural expansion.
- Non-availability of adequate and timely agricultural inputs like, quality seeds, fertilizers *etc.*
- Lack of technical skills and high illiteracy rate and poor management of resources.
- Very limited information on advanced farming and cropping systems suited to different locations.
- Poor infrastructural development like connectivity, education, health, transport, market and other basic facilities in the interior areas of the eastern and NEH states leading to poor over all development including agriculture.
- Fragmentation of holdings and right of land ownership issue particularly in NEH region threatening livelihood security due to poor resource conservation.
- Inadequate agricultural mechanization due to the hilly terrain, lack of investment and mechanized transfer of technology. Majority of the farmers are below poverty line.

- Loss of interest in agriculture especially among the youth with resultant increase in rural-urban migration as well as preference for non-farm sector.
- The region shall also remain dependent on other states of the country for food which might also record an increase in insurgency related activities.
- Increasing rural poverty due to the non renumarative agriculture production.
- Lack of policy frame work for channelization of production-processing-marketing components.
- Large gap between the potential and productivity of major crops. This leads to gap between food production and the requirement at the regional level.
- Sedimentation of hydro-electric / irrigation reservoirs and tanks due to unscientific methods of cultivation.
- The region is much vulnerable to recurrent natural disasters like cyclones, flooding, landslides *etc.*
- Lack of co-ordination among different developmental / research agencies working in the region leading to poor implementation of resource conservation programmes with low impacts.
- Extensive mining activities leading to land degradation particularly in some states of eastern region.

REFERENCES

Annual Report. 2010. Central Water Commission, Ministry of Water Resources. Govt. of India, New Delhi.

Anonymous.1993. Annual Report. Indian Council of Forestry Research and Education, Dehradun.

Anonymous. 2010. Statistical Abstracts of India. Central Statistical Organisation, Ministry of Statistics and Programme Implementation, Govt. of India, New Delhi.

Census of India. 2011. Directorate of Census, Ministry of Home Affairs, Government of India, New Delhi.

Annual Report of Hydrology Project 2006-07. Central Ground Water Board, Ministry of Water Resources, Govt. of India, Faridabad.

Ground Water Scenario of India 2010. Central Ground Water Board, Ministry of Water Resources. Government of India, Faridabad.

NBSSLUP. 2009. Soil Resource Management Reports. National Bureau of Soil Survey and Land Use Planning, Nagpur.

Patiram and Rajesh Kumar. 2005. Land husbandry for restoration of degradation in north-eastern hills. *Himalayan Ecology*: 13 (1):1-7.

Patiram, Awasthi, R.P., Pradhan, Y. and Prasad, R.N. 1994. Soil fertility researches in north-eastern hill region: A review. *J. Hill Research*, 7: 1-8.

Policy Paper 3. 2001. Harnessing and Management of Water Resources for Enhancing Agricultural Production in the Eastern Region. National Academy of Agricultural Sciences, New Delhi.

Policy Paper 9. 2001. Strategies for Agricultural Research in the North-East. National Academy of Agricultural Sciences, New Delhi.

Sharda, V.N., Dogra, P. and Prakash, C. 2010. Production and monetary losses due to water erosion in rainfed areas of India. CSWCRTI, Dehradun, 206 p.

Sharda, V.N., Juyal, G.P. and Naik, B.S, 2008. Watershed Development in India: Status and Perspective. Allied Printers, Dehradun, 219 p.

Sharda. V.N and Mandal, D. 2011. Priority classes for erosion risks areas in different states and regions of India. Technical bulletin, CSWCRTI, Dehradun, 172p.

Sharma, K.P., Moore, B. and Vorosmarty, C.J. 2000a. Anthropogenic, climatic and hydrologic trends in the Kosi Basin, Himalaya. Climate Change, 47 (1-2): 141-165.

Sharma, K.P., Vorosmarty, C.J and Moore, B. 2000b. Sensitivity of the Himalayan hydrology to land use and climatic changes. Climate Change, 52: 287-313.

State of Forest Report. 2011. Forest Survey of India, Dehradun.

Vision 2030. 2011. Central Soil and Water Conservation Research and Training Institute, 218, Kaulagarh Road, Dehradun, Uttarakhand.

Vision 2030. 2011. Directorate of Water Management, Bhubaneswar, Odisha.

Vision 2030. 2011. ICAR Research Complex for NEH Region, Barapani, Meghalaya.

Wasteland Atlas of India. 2005. National Remote Sensing Agency. Department of Land Resources. Ministry of Rural Development, Govt. of India.

www.indiaagristat.com

www.indiastat.com

www.nhb.com

□□□

Chapter 2

Bioindustrial Watershed Management for Livelihood and Environmental Security

J.S. BALI

INTRODUCTION

Livelihood and environmental security is a challenging task to achieve in present circumstances. The level of livelihood in India is enormous and mainly concerned about the livelihoods in the rural India where 71% of 1210 M population lives. It means livelihood of 859 M people which is more than the combined population of USA and Europe. Livelihood does not mean continuous availability of a daily wage of ₹ 200 per day, for digging earth with a spade in hand but it is the total security - social and economical.

What is the income which will be called a decent livelihood for a normal Indian? Directorate of Economics and Statistics reveals that the national average annual income per capita is ₹ 46492 only, against the Delhi per capita income of ₹ 116886. Average monthly per capita expenditure is ₹ 1984 for India and 2811 for Delhi. Average monthly expenditure per family in Delhi is ₹ 12055. We may adopt a minimum monthly income of ₹ 10000 per family as the minimum level of livelihood in the rural areas which our planning should aim at. The present position is less than ₹ 1200 per month for a family of five, for about 400 M persons below the poverty line. The situation is even more serious for the rural communities.

Why is it so? The fact is that farming has become profitless. What is the future of livelihoods of educated young men in this scenario of profitless farming? And farming is the predominant profession of rural Indians. But in present scenario farming leads to debt. Farmers are borrows money from the money lenders with exorbitant interest rates with mortgaging properties mainly land. Farmers are forced to sale their cultivable lands in order to come out

---

Ex-Joint Commissioner (Soil Cons.), 157, Madhuban, New Delhi- 110 092 from the debt traps. There are some farmer-friendly legislation which prohibited land, house, bullocks and implements to be mortgaged and forfeited against debts. But the penetration of alternative cooperative or bank credit for farming in India is still less than 10 %. Dependence of the farmers upon local money lenders, unfortunately, continues.

MGNREGA and Livelihood Security

Lack of continuous work at a decent daily wage is less than the demand in the rural areas. Even the MGNREGA works which assure 100 days of casual employment, do not offer decent wages. MGNREGA works should be tied to soil and water conservation in order to mitigate drought and other natural adverse conditions, and create permanent productive assets. Incidentally, MGNREGA is causing extreme shortage of farm labour in India.

Soil & Water Conservation and Livelihood Security

The marginal farmers and land less labours are continuously migrating towards the city for better livelihood. They would be more than satisfied with a monthly assured income of ₹ 10000 in his village. But that was only a dream for the marginal farmer. For the landless the situation is much worse. It is here in the rural areas that poverty is endemic. It is here that soil and water conservation provides the much needed employment, while creating productive assets.

Even during the British days when famines were frequent, soil conservation works came handy to provide work and wages while making bunds and contour bunds and repairing or building new ponds. After independence, plans, programmes of soil and water conservation have been providing regular daily wage employment while making contour bunds, ponds and other conservation structures. Since inception up to 2011, Government programmes, mostly of the Ministries of Agriculture and Rural Development, treated 56.9 M ha with soil and water conservation works and spent Rs. 27496 crores, with average per ha expenditure of Rs.4832. Figuring out on the basis of only about 9000 cft of earthwork per acre for contour bunds of 12.5 f^2 section, and 75 f^3 of earthwork per man day, it works out to 121 man-days $acre^{-1}$ or 300 man-days of employment $hectare^{-1}$. That would come to 17017 M man-days of employment created by soil and water conservation up to 2010-11.

Considering 250 working days in a year and 30 years of productive work for a man, every 7500 man-days of employment created would amount to one livelihood created. On this basis we can say that soil and water conservation works directly created 2.27 M livelihoods, from inception up to 2010-11.

But such calculations are over-simplifications and hide many anomalies. Out of 56.9 M ha of the area treated with conservation measures, 21.3 M ha were treated more intensively, by a trained soil conservation organization of the State Agriculture Departments. The remaining 35.6 M ha, shown against the Ministry of Rural Development, were treated through untrained persons of the Panchayats; and the intensity of treatment also could not be taken as proportional to the problems of the problematic land.

As far as achievements on the environmental front are concerned, we can say that most of the soil and water conservation works promote environmental protection directly and indirectly. We have to keep in mind that the soil and water conservation works are the foundation of agriculture. Environmentally safe agriculture cannot be practiced unless the land is first surveyed and its soil and slope problems are addressed. The sound agriculture and the livelihoods created as the superstructure of agriculture and other land uses could not last without a sound foundation of soil and water conservation. Soil and Water conservation can thus claim credit for all agricultural livelihoods.

Post Green Revolution Era and Livelihood Security

The present day legacy of Indian farming is the green revolution. But the success of green revolution is dependent on many factors. Irrigation, especially groundwater exploitation with privately owned tube wells and pumps, seed multiplication by the newly set up Seed Corporations, fertilizers and pesticides-indigenously produced and imported-provided the technology and administrative backdrop of the Green Revolution.

But nothing is perfect in the human context. The Green Revolution could be successfully harnessed only by the well-to-do farmers. The poor were left out. Food grains were there, but where was the purchasing power with the poor? We have today stocks of food grains out-flowing from our warehouses, but hunger is persisting. Poverty and plenty exist together. The poor do not have the money to buy the food grains. Only those with irrigation facilities could benefit from the high-yielding seed varieties. Dry land farmers-who constitute more than two-third of the farmers-could not participate in the new found prosperity.

Repeated sowing of high yielding varieties took out high amounts of nutrients, and exhausted the soil. Deficiencies of nutrients like zinc and some micro nutrients appeared. Wheat and rice crop sequence and excessive use of water, resulted in inefficient use of the scarce water resources and destroyed soil physical properties. Genetically also, the hybrid vigour decreased and yields declined. But the fact is that agriculture in India has become by and large, profitless. Who will take up a profitless profession? The talk of livelihoods in a profitless profession is unreal. Only those who have no alternative will continue to practice agriculture forcibly. A survey shows that youth are running away from agriculture and majority of farmers have no alternative. The situation is even more serious for marginal farmers. Rainfed farmers are in worse condition. Even when all the irrigation potential has been exploited, half of the land will remain, effectively, under rainfed farming (some prefer to call it, dry farming). And, what about the landless rural people?.

How can we mend this situation of profitless-ness in agriculture? Before answering this question, let us remember that the situation of farmers is the same all over the world. But the degree of this problem in India is more acute. The rich men of the world are not farmers.

Approaches to Solve the Farmer's Drudgery

The West solves this problem of their farmers with heavy subsidies. America gives about 40 % subsidies; Europe gives 40 to 50 % where France gives 80 % to their farmers. India gives less than 10 % and still some shout, it is too much. Either you increase agricultural prices by 100 or more or give heavy subsidies. Increase of food prices will put the landless poor, both rural and the urban, in worse state. No government in the world likes to let food prices go up. Hence, there is no way out, but to subsidize farming and the farmers, heavily.

But our government pleads lack of funds, and is reluctant to increase farm subsidies. Government neither increases food prices, nor farm subsidies. The result is that the farmer is compelled, by his circumstances, to sell below his cost of production. *The poor farmer subsidizes the food of the well-to-do. The poor subsidize the rich in India!* But, what is the way out. India is so full of problems, that even a half-wit can count half a dozen straightaway on his fingers. The difficult question is: what is the way out?

Bio-Industrial Watershed and Livelihood Security

There is a way out. It involves mixing processing industries and its profits with farming. It means integrating water development, land development, bio-production, and bio-industries in upstream areas. It means projectizing agriculture and forestry in the watersheds, and fix targets of production and net income enhancement of the farmers of each watershed. It means watershed management in which processing is an essential component. More and more of processed food will be consumed in the country in the future. I call this system of watersheds as *Bioindustrial Watershed Management*. Only Bio-industrial Watershed Development holds promise for skilled as well as unskilled employment of a continuous nature, which we can call, livelihoods.

The whole country should be divided into 500-1000 watershed projects. Processing industries must go green, use non-conventional energy sources, and must go upstream and keep farmers' share in these processing industries at least 40 %. The system would be on the basis of public-private ventures. Cost of water and land development, and tree planting must be invested by the government including the cost on research and pilot projects. Agri-production and agro-processing within the watershed would, however, be in private hands. Government should come forward to provide training, credit and other facilities to the private entrepreneurs, including rural entrepreneurs.

If the above pattern of *Bioindustrial Watershed Management* is adopted, we shall be able to end India's endemic rural poverty. Soil and water conservation trained personnel can play a vital role in Bio-industrial Watershed Management. Water development in the upstream areas is an essential job. It includes construction of millions of ponds for surface water development and millions of tube wells for groundwater development. Processing industries in the upstream areas will involve food processing as well as processing of wood and other non-timber forestry products.

Bio-processing, within *Bio-industrial Watershed Projects,* is a vast challenge and opportunity for the soil and water conservationists of the country. Scientific Storage is a vital and hitherto less developed part of the chain of watershed development practices. *Bio-industrial Watershed Projects* have the potential of converting presently profitless farming into profitable ventures, and provide stable livelihoods while protecting the environment. It will end rural poverty once for all. Poverty is just another name for lack of livelihoods.

Components of Bio-Industrial Watershed

The following minimum components (they can vary from watershed to watershed) will be there in a Bio-industrial Watershed Project:

(i) Intensive land and Water Development where three practices should predominate, *viz.,* Land Leveling and Tube wells (along with groundwater recharge practices) and planting of farmer-friendly trees ,ii) Bio-production systems, iii) Bio-processing industries, iv) Scientific storage, v) Farmer-friendly marketing, vi) Credit-tie up at low interest rates, vii) Research, training and other supporting measures and viii) Infrastructure of power, roads, communication etc., and others.

Rural Entrepreneur training should be initiated to generate livelihoods in the rural areas, based upon sustainable use of the natural resources.Public-Private-Partnerships should be developed where the components of resource conservation will be on the Government, but production, processing and marketing will be in private hands, backed by farmer-friendly government policies. All development activities should be integrated within Bio-industrial Watershed Projects, under the overall control of Catchment Area Commissioners.

Bioindustrial Watershed Project Planning and Implementation

India is a "soft state" – we plan well, implement poorly.

Product Choice: The first step in planning would be the choice of a bio-product which can grow in that area or is already established there naturally. For example, suitable climate or facilities may exist in an area for growing of vegetables or fruits as in the mountains. It may be medicinal and aromatic herbs as in Uttaranchal. It may be milk and milk products. It may be paper and paper pulp based on the forests. It may be rice husk which can yield many industrial products through processing.It may be simply packaging of wheat, rice, maize, sorghum and other food grains. It might be a wheat area in which processed wheat products could be manufactured, packed and marketed.

Processing of medicinal and aromatic herbs is a profitable business for the rural areas. Uttaranchal has been declared by the Government as an herbal State. Profitable industries based on herbs have already been set up in the 100 % export zone of Noida and many other places.Milk and milk products are a promising line for Bioindustrial Watersheds. There is already a "White revolution" in India. But many cities do not still have scientific milk schemes producing safe milk and milk products. AMUL has already set a good example. Given determination and zeal for social service many more AMUL's can be set up profiting millions of our small and marginal farmers whose livelihood is dependent upon cows and buffaloes.

Scope is tremendous. The technologies developed by the Department of Science and Technology of India which can be purchased at nominal cost. Many more agricultural processing technologies are available. Central Food Technology Research Institute at Mysore has developed many processing technologies for entrepreneurs. Processing and Post-harvest technology have developed into definite branches of study.

Feasibility Reports: Planning for the Bioindustrial Watershed Projects should begin with the preparation of a Preliminary Feasibility Report and then a Detailed Feasibility Report. The two feasibility reports should establish the viability of the venture. Once the basic feasibility of the selected product for processing and marketing is established, action on the Project Components should be started. An outline of major project components is given below:

Project components

I. *Basic Infrastructure:*
- Health facilities
- Education facilities
- Power
- Roads and transport
- Communications

II. *Foundation Practices*
- Forest and wildlife conservation and development
- Water development and management (Irrigation & Drainage)
- Land development and management (Soil & Water Conservation)

III. *Sustainable Production Systems - Based on Plants*
- Agriculture
- Horticulture (Floriculture, Fruit and Vegetable cultivation etc)
- Agro-Forestry
- Other Systems

Based on animals
- Poultry
- Milk and milk products
- Meat
- Other systems

IV. *Processing Industry*
- Preliminary feasibility report
- Detailed feasibility report
- Land, water, power and other basic needs of the industry
- Plant erection and buildings
- Operational components

V. *Storage*
- Grain storage
- Cold storage
- Low cost storage

VI. *Marketing*
- Local marketing
- National marketing
- International marketing

VII. *Finance*

- Bank credit
- Institutional credit
- Government liabilities
- Subsidies

VII. *Building Partnerships with*

- National companies
- Multinationals
- Government local, State and Central
- NGOs

IX. *Implementation Steps*

- Building infrastructure and arranging finance
- Ecology: Apply soil and water conservation in the watershed
- Setting up agreement and contracts for sustainable production systems based on plants
- Animal shelters and facilities for sustainable production systems based on animals
- Farm buildings for agricultural production for consumption and for processing
- Processing industry installation
- Building storage structures
- Setting up marketing facilities
- Building management structures

X. *Policy Package*

Bioindustrial Watershed Management

The economic and social benefits of watershed management could be up-scaled further by bringing industry to the watersheds and transforming them to the industrial watersheds (Bali, 2008). The new system would have the bio-produce processed and value added at the watershed level, thereby, bringing urban profits of industry and commerce to the rural poor. The watershed people through cooperatives or corporates should wholly or substantially own these bio-industries. Even the landless people of the watersheds would be partners of the bio-industry. The fresh or processed watershed produce will be sold directly to the consumers by these cooperatives, doing away with the middlemen. An industrialist or entrepreneur ready to invest for the bio-industry should be acceptable provided the person provides 40 % of share to the cooperative.

The bio-industry would take upon itself the role of running the affairs of bio-industrial watersheds. It would arrange for grants from the Government for soil and water conservation works and build-up of infrastructure and credit from financial institutions for production purposes. It would develop backward linkages to provide production inputs like seeds, planting materials, fertilizers, manures, pesticides, drip / fertigation systems and technical know-how *etc.*, and forward linkages in finding markets and remunerative prices for the produce. The bio-industrial watershed management is different from the contract farming in the sense that it does not only focus on production and buy back of the produce, but takes care of natural resource conservation and protection also to have sustainable production. It would be prudent to have all the Central and State governments' watershed management programmes converted into bio-industrial watersheds.

There are a number of watershed level activities related to agriculture, horticulture, agroforestry, livestock, fisheries, poultry, sericulture, mushroom farming, honey bee keeping and vermi-composting *etc.*, which could form agri-enterprises for bio-industrial watershed development. We need to choose such enterprises suiting to varied socio-economic-environmental conditions of different rural areas for greater livelihood, nutritional and environmental security. The operation of one thousand per annum capacity vermi-compost bio-industry involving one hundred households of a watershed, could generate net income of about Rs. 50 lakhs besides creating employment to rural households, especially women. There is growing demand for vermi-composting for high value fruit, plantation, vegetable and organically grown crops. The country with the existing vegetation wealth is capable of sustaining 150 M bee colonies, providing 1.5 lakh tonnes of honey. The value addition to the produce at the watershed level would fetch more income to the farmers while generating employment for 15 M rural and tribal families. Likewise, the mushroom farming is big livelihood generator in rural areas. The horticulture-based food processing industry under bio-industrial watersheds is going to offer immense economic gains and livelihood opportunities besides food and nutritional security to rural people. Horticulture is emerging as a single largest category in world agricultural trade accounting for over 20 % of share in recent years. The great demand for fresh fruits and vegetables worldwide is stimulating international trade in horticulture in India. The Government has been spearheading horticultural development through National Horticulture Mission in providing suitable infrastructure in terms of handling, transportation, storage, markets, processing and value addition. India loses more than Rs 58,000 crores worth the agricultural produce due to lack of post-harvest and food processing infrastructure.

The bio-industrial watershed management system could be viewed on the parlance of Amul cooperatives in Gujarat, Kuppam vegetable farming and processing cooperative in Andhra Pradesh, Kokam fruit concentrate production cooperatives in Maharashtra and Goa, sugar mills co-operatives in Maharashtra and Gujarat, Leh Berry fruit concentrate production cooperative in Leh (J&K) and many other ventures in different parts of the country. As animals are essential component of farming system in rural India, the bio-industry based on milk, poultry, wool, meat and other animal products could be a great success. The development of cold storage / cold chain to store and transport perishable produce to the markets at appropriate time to avoid distress sales should be part of the bio-industrial development programme. The programme fits very well in the framework of various rural development schemes, including Mahatma Gandhi National Rural Employment Guarantee Act (MGNREGA), launched by the Government of India. The programme also has great promise in fulfilling the United Nations Millennium Development Goals (MDGs) of eradicating extreme poverty and hunger, promoting gender equality and empowerment of women and ensuring environmental sustainability. The greater institutional support to the programme through Government departments, financial institutions and Institutions would allow tapping of a multitude of potential benefits of the bio-industrial watershed management programme. The National Rainfed Area Authority established recently to bring in much needed convergence of resources with different Ministries and Departments could provide fillip to the programme.

We need to strengthen micro-credit system for establishing micro-enterprises in the watershed areas. The bigger financial institutions would seldom come to the help of small enterprises with little capacity to pay back the loans at higher interest rates. The micro finance institutions, to the contrary, would extend small loans at low interest rates and flexible time limits. The setting up of more Banks and Cooperative Societies in the rural sector is desired besides strengthening National Bank for Agriculture and Rural Development (NABARD).

CONCLUSIONS

Soil and Water Conservation and the strategy of integrated watershed management are the essential tools of rural development in India. Daily paid casual employment generated in the construction phase of soil and water conservation programme of the Government, cannot be strictly called livelihoods.Livelihoods should come from the profitable use of land, water and other associated resources. Major land use where 80 % of the rural population is involved is agriculture. Unfortunately, agriculture, especially the dry land agriculture, has become unprofitable. Even where almost full irrigation coverage is available, as in Punjab, farmers are under debt, and the youth want to run away from it. Farmers are committing suicides in almost all the States. It can be said that agriculture, as practiced today, does not and cannot provide livelihoods to meet the aspirations of the educated youth of India.

The way out is to increase agricultural prices to double the present levels or to subsidize agriculture, heavily, as done by Europe and USA. But Governments in India are not in a position to afford it. The way out is through Bioindustrial Watershed Management which would include the bio- processing industries in the upstream where farmers will be provided a stake and a definite share in them.At present the prices which the consumer pays for agricultural production are two to four times of the prices which the farmers get. This situation will be remedied in the Bioindustrial Watersheds. Marketing models successfully tried by many Corporate Bodies will be adopted for the benefit of the farmers.

REFERENCE

Bali, J. S. 2008. Bioindustrial watershed management in rainfed areas for rural employment and prosperity. *In: Anand-Swarup et al. (ed) Conservation Farming: Enhancing Productivity and Profitability of Rainfed Areas.* Soil Conservation Soc India, NASC Complex, NewDelhi, India, pp 379-398.

□□□

Chapter 3

Soil Health and Water Quality Issues for Sustainable Agricultural Production in the Eastern Region

D.K. KUNDU, RAVEENER SINGH[1] AND B.S. MAHAPATRA

INTRODUCTION

Area situated between latitudes of 17° N and 29° N and longitudes of 80° E and 97° E constitutes the eastern region of India. The region includes Assam, Bihar, Chhattisgarh, Jharkhand, Odisha, West Bengal and eastern part of Uttar Pradesh. Total geographical area of the eastern region is 73.60 M ha, which is 22% of the total geographical area of the country. Out of the 73.60 M ha geographical area of the eastern region, only about 45% (33.60 M ha) is the net cultivated area. The region is inhabited by about 34% of total population and contributes to the food grain production of 58 million tonnes (34.6% of the total production of the country). Food grain productivity in the region for year 2004-05 was the highest in West Bengal (2.5 t ha^{-1}) followed by eastern UP (2.1 t ha^{-1}), Bihar (1.5 t ha^{-1}), Assam (1.4 t ha^{-1}), Odisha (1.3 t ha^{-1}), Jharkhand (1.2 t ha^{-1}), and Chhattisgarh (1.0 t ha^{-1}). The cropping intensity in West Bengal, Odisha, eastern UP, Assam, Bihar, Jharkhand and Chhattisgarh is 177, 152, 151, 145, 142, 117 and 115%, respectively (Table 1). The average cropping intensity in eastern India is 142%. The climate of the region is tropical, hot and sub-humid to humid with high rainfall. The average annual rainfall in this region varies from 1091 to 2477 mm with an average of 1525 mm, which is sufficient and substantial for growing any crop. Bulk of the rain (about 80%) occurs during the monsoon season. The rainfall has erratic temporal and spatial distribution with considerable year-to-year variations. Even in years when the total rainfall is adequate, long dry spells and inadequate rainfall at the crucial stages of the crop growth adversely contribute to instability in agricultural production. The plains of this region invariably suffer from the excess of the stagnating water causing water logging

---

Central Research Institute for Jute & Allied Fibres, Barrackpore, Kolkata- 700 120
[1]Division of Agricultural Physics, IARI, New Delhi- 110 012

and crop damage during monsoon. On the contrary, the undulating topography and sloping lands suffer from excessive runoff and nutrient losses. The coastal areas are vulnerable to seawater intrusion and cyclones.

This region is richly endowed with two basic natural resources *viz.*, soil and water. Ultimate irrigation potential from major and medium irrigation projects in the eastern region has been estimated at 19.66 M ha-m as against 58.48 Mha-m for the whole country. Historically, it was the most prosperous agricultural tract of the country, and its agriculture maintained a lead over other regions. However the present day situation indicates the contrary facts. Agricultural development is much below its potential in this region. For improving and sustaining agricultural production in this region, it is necessary to analyze characteristics of the soil and water resources, and protect these two natural resources from degradation.

Table 1: Natural resources endowment of different states of eastern region

State	TGA ('000 ha)	Net cultivated area ('000 ha)	Cropping intensity (%)	Food grain produc-tivity (t ha^{-1})	Average annual rainfall (mm)	Net irrigated area ('000 ha)	Ground water resou-rce ('000 ha)	Ground water develop ment (%)
Assam	7843	2751	145	1.40	2477	114	2471	9
Bihar	9359	5603	142	1.52	1204	4550	2699	46
Chhattisgarh	13780	5480	115	0.98	1430	1263	1607	6
Eastern UP	10402	6328	151	2.10	1091	4197	8112	47
Jharkhand	7970	1807	117	1.24	1277	258	653	33
Odisha	15571	6165	152	1.32	1451	2143	2099	23
West Bengal	8687	5427	177	2.50	1750	1911	2309	38
Eastern Region	73613	33563	142	1.58	1525	14436	19950	29

Source: Singh *et al.* (2009a)

Characteristics of Natural Resources in Eastern India

I. Soil resources

Out of total eight different soil orders reported in India, six soil orders occur in the eastern region. Aridisol and Oxisol are not found in this region (Singh *et al.* 2009a). Area covered under six soil orders in different states of the eastern region is presented in Table 2. Eastern region is dominated by Inceptisol covering about 34 Mha. In general, 22-75% area in different states is under Inceptisol. Soils of the eastern UP, West Bengal and Odisha have more than 50% area under Inceptisol. Alfisols cover about 20 Mha area of the eastern region. About 10-50% areas in different states are under Alfisol. In Chhattisgarh state, 6.9 Mha (50%) area is under Alfisol. A considerable area (14.5 Mha) of this region comes under the soil order Entisol. More than 40% (3.76 Mha) area of Bihar is covered by Entisol. About 3 Mha area of the eastern region, largely in Chhattisgarh, is covered by Vertisol. Hilly areas of Chhattisgarh and eastern UP have 0.09 and 0.07 Mha, respectively under the soil order Mollisol. In Assam, 0.39 Mha area is covered by the soil order Ultisol.

Physical properties of major soil sub-groups occurring in this region were determined by Singh and Kundu (2008a, 2008b, 2010) and Singh *et al.* (2009b, 2010). The soils are largely sandy loam to clay loam in texture which retain good amount of moisture. These soils exhibit medium to high available water capacity which helps in retaining moisture for post-monsoon crops. Saturated hydraulic conductivity of the soils is low indicating poor drainage conditions of the soils. Average erosion index of the soils indicates moderate erodibility. However, Entisols, Inceptisols, Vertisols and Mollisols have high to very high erosion index, indicating an urgent need for adoption of soil conservation measures.

Table 2: Soils of the eastern region and their distribution

State	Soil orders											
	Entisols		Inceptisols		Alfisols		Vertisols		Mollisols		Ultisols	
	Area (Mha)	%	Area (Mha)	%	Area (Mha)	%	Area (Mha)	%	Area (Mha)	%	Area (Mha)	%
West Bengal	1.97	23	4.69	54	1.96	23	-		-		-	
Bihar	3.76	40	4.48	48	1.07	11	-		-		-	
Jharkhand	2.21	28	3.18	40	2.61	33	-		-		-	
Odisha	1.52	10	7.80	50	5.27	34	0.86	6	-		-	
Chhattisgarh	1.64	12	3.10	22	6.89	50	2.08	15	0.09	<1	-	
Assam	1.81	30	2.96	50	0.79	13	-	-	-	-	0.39	7
Eastern UP	1.55	15	7.71	74	1.08	10	-	-	0.07	<1	-	
Eastern Region	14.46		33.92		19.67		2.94		0.16		0.39	

Source: Halder *et al.* (1992, 1996), Sarkar *et al.* (1998), Singh *et al.* (2003), Tamgadge *et al.* (1999)

II. Water Resources

a) *Surface water resources :* Total available surface water in different river basins of the country is 188 M ha-m per year. Water resource potential of the major river basins of eastern region is presented in Table 3. Three major river basins in the eastern region excluding Assam are Mahanadi, Brahmani-Baitarini and Subarnarekha. The annual surface water available in these three basins is 11.2 Mha-m and the total annual monsoon runoff is 11.7 Mha-m. Annual utilizable surface water in the three basins is estimated at 8.8 Mha-m. It has been observed that due to i) non-availability of suitable storage sites in hills and plains, and ii) extreme variability in precipitation, which disallows storage of flash and peak flows, all the stream flows can not be stored in the reservoirs. The live storage capacity under completed projects in the country is 20.65 Mha-m, out of which the share of eastern region is 2.92 Mha-m. The annual per capita water availability in Mahanadi, Brahmani-Baitarini and Subarnarekha basin is 2067, 2388 and 1982 m^3, respectively. Groundwater potential of Ganga, Brahmaputra, Barak and other basins has been estimated at 171725, 27857 and 1795 Mm^3, respectively. In Subarnarekha, Brahmani-Baitarini and Mahanadi basins, groundwater potential is estimated at 2185, 5879 and 21293 Mm^3, respectively.

b) *Ground water resources* : The region also has a large groundwater potential. Total utilizable groundwater resource for irrigation is estimated at 19.5 Mha-m $year^{-1}$. However, average utilization of this resource is only 29%. It varies from 6% in Chhattisgarh and 9% in Assam to 46% in Bihar and 47% in eastern UP. Thus, there is ample scope for further exploitation of groundwater resource. However, utmost caution is needed in coastal belts to avoid seawater intrusion. In several areas endowed with thick but relatively less pervious clay layer at the land surface (*e.g.*, Midnapore district of West Bengal), over-exploitation of groundwater has caused substantial fall in water table. The actual ground water exploitation level is less than 20% in 25 districts of Odisha, 6 districts in Bihar, all the districts of Chhattisgarh and Assam, 6 districts of West Bengal and 12 districts of Jharkhand (Table 3). Under-exploitation of groundwater and rising water table in the canal commands has reduced the underground space for water storage. Consequently, there is reduction in groundwater recharge. The reduced amount of natural recharge adds to the runoff excess, thereby aggravating floods and flood-related damages.

Table 3: Water resources potential of major river basins of eastern India

S. No.	River basins	Catchment area (km^2)	Water resources potential (Mm^3)		Groundwater potential (Mm^3)
			Average	75% dependable	
1	Ganga-Brahmaputra-Meghna				
	(a) Ganga	>861452	525023	436312	171725
	(b) Brahmaputra	>194413	537240	491736	27857
	(c) Barak & other	>41723	48357	-	1795
2	Subarnarekha	29196	12368	9855	2185
3	Brahmani-Baitarani	51822	28477	20051	5879
4	Mahanadi	141589	66879	53786	21293

Source: Anonymous (1993)

Irrigation status : Net irrigated area in eastern India is 14.89 Mha, out of which 4.2 Mha is in eastern UP, 2.51 Mha in Bihar, 2.69 Mha in Odisha, 3.91 Mha in West Bengal, 1.26 Mha in Chhattisgarh, 0.20 Mha in Jharkhand and 0.17 Mha in Assam. As most of the cultivated area in eastern region does not have provision for assured irrigation, even a short spell of drought adversely affects the stability of agricultural production, thereby resulting in low productivity. This has led to agricultural development much below its potential, limited employment in agricultural sector, and a large proportion of the population below poverty line with high incidences of malnutrition. On an average, 60% of the created irrigation potential is utilized for irrigation. In Assam and Jharkhand, utilizable irrigation potential is below 50% level.

Ultimate irrigation potential from major and medium irrigation projects in eastern region has been estimated at 19.66 Mha-m as against 58.48 Mha-m for the whole country. Major irrigation projects of the plain region on Sharda, Gandak, Kosi, Sone and Mahanadi rivers are barrage systems and therefore suffer from the problems of untimely supply of water due to fluctuating flow at the source. Several medium irrigation projects of these plains are pump (lift), canal systems. In eastern region, an ultimate minor irrigation potential of 33.07 Mha has been assessed as compared to 81.48 Mha for the whole country. Thus, the eastern region possesses total irrigation potential of 53 Mha, which constitutes about 38% of the country's total potential of 140 Mha. The region has an extensive network of rivers, *nallas* and lakes from which water could be lifted for irrigation by installing appropriate water lifting systems.

III. Climate resources

The climate of the region is tropical, hot and humid to sub-humid with high rainfall. The average annual rainfall in this region varies from 1091 to 2477 mm with a regional average of 1526 mm, which is sufficient and substantial for growing any crop. Assam gets mean annual rainfall of 2477 mm, which is the highest in the region, followed by West Bengal (1750 mm), Odisha (1451 mm), Chhattisgarh (1430 mm), Jharkhand (1277 mm), Bihar (1204 mm), and eastern UP (1091 mm). Bulk of the rain (about 80%) occurs during the monsoon season. It has erratic temporal and spatial distribution with high year-to-year variation. High intensity and heavy rains cause frequent floods in the eastern region, especially in plains, and coastal deltaic areas. The coastal areas are also vulnerable to seawater intrusion and cyclones.

Average annual potential evapotranspiration of eastern region varies between 1742 and 1986 mm with an average value of 1850 mm, which is higher than the country's average (1755 mm). Maximum and minimum potential evapotranspiration values are observed in Odisha and Assam, respectively. In all the states of eastern region, average potential evapotranspiration exceeds average annual rainfall, except in Assam.

Soil and Water Related Constraints

I. Soil related constraints

Soil degradation : Extent of degraded areas in eastern region under various categories is reported in Table 4. Chhattisgarh and Assam states show severe soil degradation problems in 53.7 and 76.8% of the total land area, respectively. Loss of top-soil is a common result of degradation in all the states of eastern region. Waterlogging and flooding are the other major concerns. Singh *et al.* (2003) and Singh and Kundu (2009) dealt on various prospects of managing the salt affected soils of the eastern region.

Table 4: Percentage of areas under various soil degradation problems in eastern states

States	Water erosion		Physical deterioration		Chemical deterioration (Salinization)	Land affected by more than one problem			Total degraded land (%)
	Wt	Wd	Water logging	Flooding		(Cs + Pw)	(Cs + Pf)	Wt+Cn, Pf, Pw, Cn	
West Bengal	19.7	-	6.9	1.1	-	1.9	1.3	-	30.9
Bihar and Jharkhand	23.3	-	11.5	-	1.3	-	-	-	36.1
Odisha	34.4	-	4.4	-	0.5	-	-	-	39.3
Chhattisgarh	50.2	-	0.5	0.4	-	-	-	2.6	53.7
Assam	17.7	-	2.5	10.5	-	-	-	45.9	76.8
Eastern UP	27.8	2.3	2.5	5.5	4.5	-	-	9.5	52.1

Source: Haldar *et al.* (1992, 1996), Sarkar *et al.* (1998), Singh *et al.* (2003), Tamgadge *et al.* (1999)

Cs= Chemical deterioration (salinization), Wt= Loss of top soil, Pw= Physical deterioration (waterlogging), Wd= Terrain deformation, Pf= Physical deterioration (flooding), Et= Wind erosion

Soil acidity : Soil acidity is an important constraint responsible for low productivity in the eastern region. Data on the extent and degree of soil acidity in different states are presented in Table 5. In eastern region, 27.44, 10.47 and 0.17 Mha areas are classified under slightly, moderately and strongly acidic soils. Strongly acidic soils occur in Chhattisgarh and Assam states only. More than 50% of the total area of eastern region is affected by the problem of soil acidity.

Table 5: Extent and degree of soil acidity in different states of eastern India (Mha)

States	Strongly acid (pH<4.5)	Moderately acidic (pH 4.5 - 5.5)	Slightly acidic (pH 5.5 - 6.5)	Total
West Bengal	-	0.56	4.20	4.76
Bihar	-	0.04	2.32	2.36
Odisha	-	0.26	8.41	8.67
Chhattisgarh	0.15	6.30	4.39	10.84
Assam	0.02	2.31	2.33	4.66
Eastern U.P.	-	-	0.02	0.02
Jharkhand	-	1.00	5.77	6.77
Eastern India	0.17	10.47	27.44	38.08
Total India	6.19	24.81	58.94	89.94

Source: Sharma and Sarkar (2005)

Fertilizer consumption status : Fertilizer consumption in this region is the lowest in the country. The consumption in the states of West Bengal, Bihar, Eastern UP, Jharkhand, Odisha, Chhattisgarh and Assam is 122.4, 130.68, 100.7, 68.2, 41.4, 46.5 and 46.6 kg NPK ha^{-1}, respectively. In four out of seven states, average fertilizer consumption is less than the national average of 90 kg NPK ha^{-1}. Primary cause for low fertilizer consumption in eastern states is waterlogging by surface flooding during *kharif* season and *rabi fallow* due to non-availability of irrigation water. Studied have clearly indicated that while fertilizer use in *kharif* rice has been negligible, farmers often use more than recommended doses of fertilizers in irrigated rice during *rabi* season (Table 6).

Table 6: Fertilizer consumption in different states of eastern India (Mha)

States	Fertilizer consumption in *kharif* + *rabi* (kg ha^{-1})			
	N	P_2O_5	K_2O	Total
West Bengal	63.8	33.4	25.2	122.4
Bihar	110.3	12.5	7.7	130.6
Jharkhand	40.7	24.9	2.5	68.2
Odisha	26.7	8.5	6.3	41.4
Chhattisgarh	30.7	11.9	3.9	46.5
Assam	22.2	12.7	11.7	46.6
Eastern U.P	78.2	18.6	3.9	100.7
Eastern India	53.2	17.5	8.8	79.5
All India	59.2	22.1	8.5	89.8

Source: Fertiliser Statistics (2003-2004), Fertilizer Association of India, New Delhi

Heavy metal pollution of soils : The state of Odisha has large deposits of ores and minerals of heavy metals Fe, Mn, Cr, Pb and Ni (Table 7). The soils around the mining areas contain 100-300 ppm Fe, 60-100 ppm Mn, 60-75 ppm Cu, 30-100 ppm Cr, 50-150 ppm Ni, 10-20 ppm Pb and 200-250 ppm Zn. Combustion of coal through thermal plants adds large amounts of pollutant elements to the soil and water. The thermal plant at Talcher daily burns 9000 T of coal from which 3600 T of fly ash is produced. This fly ash contains total 424 ppm Fe, 34 ppm Zn, 39 ppm Cu, 241 ppm Mn, 33 ppm Pb, 13 ppm Cr and 4.5 ppm Cd. Thus total fly ash deposits daily 1526 kg Fe, 122.9 kg Zn, 140 kg Cu, 867 g Mn, 119 kg Pb, 47 kg Cr and 16 kg Cd (Sahu, 2003). About 15% of these heavy metals from the ash pond seep into the flow of Nandira river to pollute its water. Use of fly ash @ 5 to10 t ha^{-1} to improve physical conditions of light-textured soils thus causes heavy metal pollution of the soils.

Surface soil samples collected from 18 villages situated within 2 to 15 km of the chromite mines at Sukinda of Jajpur district in Odisha were found contaminated with Cr^{6+}. All the soils were moderately acidic and they contained on an average 0.404 ppm Cr^{6+}. Considering 0.05 ppm of Cr^{6+} toxic to plants and animals, 143 out of total 167 soil samples analysed were polluted. Villages lying within 3 km radius from the mines have soils polluted with Cr^{6+} (Sahu and Nayak, 2002). Medium and lowlands adjacent to the lateritic uplands as well as the lands close to the Fe-ore mining areas in Odisha suffer from Fe toxicity. About 30,000 ha of Fe-toxic lands have been identified in Odisha (Sahu, 2003). These soils contain >300 ppm of DTPA extractable Fe. Rice grown in such lands suffer from Fe-toxicity with symptoms of bronze coloration of leaves, stunted growth, shy tillering, emergence of defective panicles & chaffy spikelets. Affected rice plants contain more than 300 ppm Fe. Grain yield of 17 rice genotypes varied from 4.0 to 5.5 t ha^{-1} in normal soil and 1.3 to 3.3 t ha^{-1} in Fe-toxic soils of Odisha.

Table 7: Heavy metal deposits in Odisha

Ores/minerals	Suspected pollutant elements	Total deposits (Million tonnes)	Share of country (%)
Iron	Fe, Mn, Al	3567	27.1
Manganese	Mn	50	28.4
Chromium	Cr, Ni, Cd	183	98.4
Lead	Pb	9	1.8
Nickel	Ni	270	91.8
Coal	Zn, Fe, Cu, Cr, Ni, Pb, Co, U, Th	47889	28.6

Source: Directorate of Mines, Odisha

The large scale mining and allied activities going on in the Jharkhand region has caused severe damage to the land resources of the area. Vast areas of rich forests and agricultural land belonging to the indigenous people have been laid waste because of haphazard mining. Eliminating of existing vegetation and alteration of soil profile due to open cast mining operations, including shifting of overburden and reject dumps, have caused severe soil erosion and silting of adjoining courses and degraded the productive capacity of the lands in the area. Excessive underground mining, especially of coal, is causing subsidence of land in many areas as a result of which such lands have been rendered unsafe for habitation, agriculture and grazing. The Jharkhand government has granted 524 leases for mining various major minerals 206 of which have been granted for coal alone. In addition, 2,717 leases covering an area of 8,426 ha have been granted for extracting minor minerals in the state. Mining ruins the land, water, forests and air. The loss or pollution of natural resources degrades the quality of human life in these areas. Increasingly, mineral-based production units like coal-fired power plants, steel plants and cement factories are located near the mines. A cluster of thermal power stations are planned near the major rivers of Jharkhand.

A precise estimate of the amount of agricultural land lost to mining is not readily available. Every mining enterprise, however, means the conversion of land to such purposes as roads, railways and ropeways for mineral transport, townships for housing miners and manager, infrastructure for administrative purposes, land for stockyard and preliminary processing operations. In effect, the total land affected by mining is many times larger than the simple lease area. Several strategies for the management of over burden and mine spoils have been suggested by Sikka *et al.* (2009).

Impact of fallow-rice cropping system on soil health and groundwater quality : Most of the agricultural farms in the eastern region practice fallow-rice cropping system. The fields remain fallow through the dry season due to inadequate water availability, when 30 to 90 kg nitrate N per ha is formed and accumulated in the soil profiles. Kundu *et al.* (2004) studied influence of three different field management practices of dry season on nitrate buildup in 25 soil profiles in Khurda district of Odisha. Dry-season nitrate N buildup in 0-60 cm soil profile ranged from 52.5 to 86.0 kg ha^{-1} (mean 66.8 kg ha^{-1}) for ploughed clean, 35.2 to 57.0 kg ha^{-1} (mean 46.3 kg ha^{-1}) for non-ploughed weedy, and 20.0 to 41.4 kg ha^{-1} (mean 30.4 kg ha^{-1}) for horsegram grown fields. When these fields are flooded and puddled for planting *kharif* rice, nitrate gets lost from the soil profiles through denitrification and leaching. Such loss of nitrate every year from agricultural farms not only depletes N fertility of soils but also contributes to the pollution of ground water (Table 8). Drought-tolerant, short-duration legumes grown during dry season can assimilate part of the nitrate N accumulated in the soil profile and conserve it against undesirable losses (Kundu *et al.*, 2002).

Table 8: Effect of dry-season field management on N loss from soil profiles* under rice-based cropping system. Khurda district, Odisha. 2003.

Dry-season management of field	NO_3 N accumulated in 0-60 cm soil profile through dry season (kg N ha^{-1})	NO_3 N remaining in 0-60 cm soil profile after field preparation for planting *kharif* rice (kg N ha^{-1})	N loss from the soil profile (kg N ha^{-1})
Ploughed, clean	66.8	14.5	52.3
Non-ploughed, weedy	46.3	12.1	34.2
Horse gram grown	30.4	9.7	20.7

Source: Kundu *et al.* (2002, 2004, 2006). * Data represent means of 25 soil profiles.

II. Water related constraints

A substantial portion of the eastern region comprises of hills and valleys. The intense rains during monsoon season generate rapid overland flows on the unprotected hills resulting into sheet, rill and gully erosion. Deforestation and shifting cultivation in the region further accelerate the problems. Water-related constraints are different for different geographical entities. From water management point of view, the eastern region may be categorized into three distinct geographical entities: (i) Plains of eastern UP, Bihar, West Bengal and Assam; (ii) Hilly tracts of Vindhyas in eastern UP, Kamoor hills of Bihar, Bihar plateau region spread over Bihar, West Bengal and Odisha and Chhattisgarh region; and (iii) Coastal plains of West Bengal and Odisha.

Groundwater quality problems and affected districts in different states of eastern region are presented in Table 9. Groundwater quality is not a problem in eastern UP and Jharkhand states. However, excess concentration of iron is observed in 4, 5, 3 and 11 districts in West Bengal, Odisha, Chhattisgarh and Bihar, respectively. Iron problem is also witnessed along the northern bank of Brahmaputra in Assam. Salinity problem is witnessed in coastal areas of Odisha and West Bengal; Durg, Dantewada and Raipur districts of Chhattisgarh; and Begusarai district of Bihar. Arsenic contamination beyond critical level is reported from Malda, south 24-Parganas, Nadia, Hooghly, Murshidabad, Bardhaman and Howrah districts of West Bengal; Rajnandgaon district of Chhattisgarh; and Bhojpur and Patna districts of Bihar. Problem of excess concentration of fluoride in groundwater is reported from Birbhum district of West Bengal; Bolangir district of Odisha; and Giridih, Jammui and Dhanbad districts of Jharkhand. Adverse effects of different contaminants are visible not only on the crops grown in the region but also on the health of human beings and animals.

Table 9: Groundwater quality problems and affected districts in different states of eastern region

Contaminants	Affected states and districts (in part)
West Bengal	
Salinity (EC > 3000 μs cm^{-1} at 25 ° C)	Howrah, Midnapore, South 24 Parganas,
Fluoride (>1.5 mg l^{-1})	Bankura, Bardhaman, Birbhum, Dakhindinajpur, Malda, Nadia, Purulia, Uttardinajpur
Chloride (> 1000 mg l^{-1})	South 24 Parganas, Howrah
Iron (>1.0 mg l^{-1})	Bankura, Bardhaman, Birbhum, Dakhindinajpur, E. Midnapur, Howrah, Hugli, Jalpaiguri, Kolkatta, Murshidabad, N-24praganna, Nadia, S-24pragannas, Uttardinajpur, West Midnapur
Nitrate (>45 mg l^{-1})	Bankura, Bardhaman
Arsenic (>0.05 mg l^{-1})	Bardhaman, Hooghly, Howrah, Malda, Murshidabad, Nadia, North 24 Praganas, South 24 Pragannas
Odisha	
Salinity (EC > 3000 μs cm^{-1} at 25 ° C)	Cuttack, Balasore, Puri
Iron (> 1.0 mg l^{-1})	Balasore, Bargarh, Bhadrak, Cuttack, Deogarh, J.Singhpur, Jajpur, Jharsuguda, Kalahandi, Kandmahal, Keonjhar, Khurda, Koraput, Mayurbhanj, Nayagarh, Puri, Rayagada, Sambalpur, Sundergarh, Suvarnapur
Fluoride (> 1.5 mg l^{-1})	Bolangir, Angul, Balasore, Bargarh, Bhadrak, Boudh, Cuttack, Deogarh, Dhenkanal, Jajpur, Keonjhar, Suvarnapur
Nitrate (> 45 mg l^{-1})	Angul, Balasore, Bargarh, Bhadrak, Bolangir, Boudh, Cuttack, Deogarh, Dhenkanal, Gajapati, Ganjam, J.Singhpur, Jajpur, Jharsuguda, Kalahandi, Kendrapara, Keonjhar, Khurda, Koraput, Malkangiri, Mayurbhanj, Nawapada, Nayagarh, Phulbani, Puri, Sambalpur, Sundergarh, Suvarnapur
Jharkhand	
Fluoride (>1.5 mg l^{-1})	Bokaro, Giridih, Godda, Gumla, Palamu, Ranchi
Iron (>1.0 mg l^{-1})	Chatra, Deoghar, East Singhbhum, Giridih, Ranchi, West Singhbhum

Nitrate (>45 mg l^{-1})	Chatra, Garhwa, Godda, Gumla, Lohardega, Pakur, Palamu, Paschimi Singhbhum, Purbi Singhbhum, Ranchi, Sahibganj
Chhattisgarh	
Nitrate (>45 mg l^{-1})	Bastar, Bilaspur, Dantewada, Dhamtari, Jashpur, Kanker, Kawardha, Korba, Mahasamund, Raigarh, Raipur, Rajnandgaon
Iron (>1.0 mg l^{-1})	Bastar, Dantewada, Kanker, Koriya,
Fluoride (>1.5 mg l^{-1})	Bastar, Bilaspur, Dantewada, Janjgir-Champa, Jashpur, Kanker, Korba, Koriya, Mahasamund, Raipur, Rajnandgaon, Surguja
Arsenic (>0.05 mg l^{-1})	Rajnandgaon
Bihar	
Fluoride (>1.5 mg l^{-1})	Aurangabad, Banka, Buxar, Bhabua(Kaimur), Jamui, Munger, Nawada, Rohtas, Supaul
Iron (>1.0 mg l^{-1})	Aurangabad, Begusarai, Bhojpur, Buxar, Bhabua(Kaimur), East Champaran, Gopalganj, Katihar, Khagaria, Kishanganj, Lakhiserai, Madhepura, Muzafferpur, Nawada, Rohtas, Saharsa, Samastipur, Siwan, Supaul, West Champaran
Nitrate (>45 mg l^{-1})	Aurangabad, Banka, Bhagalpur, Bhojpur, Bhabua, Patna, Rohtas, Saran, Siwan
Arsenic (>0.05 mg l^{-1})	Begusarai, Bhagalpur, Bhojpur, Buxar, Darbhanga, Katihar, Khagaria, Kishanganj, Lakhiserai,Munger, Patna, Purnea, Samastipur, Saran, Vaishali
Assam	
Fluoride (>1.5 mg l^{-1})	Goalpapra, Kamrup, Karbi Anglong, Nagaon,
Iron (>1.0 mg l^{-1})	Cachar, Darrang, Dhemaji, Dhubri, Goalpapra, Golaghat, Hailakandi, Jorhat, Kamrup, Karbi Anglong, Karimganj, Kokrajhar, Lakhimpur, Morigaon, Nagaon, Nalbari, Sibsagar, Sonitpur
Arsenic (>0.05 mg l^{-1})	Dhemaji

Source: Central Ground Water Authority (2010)

Kundu and Singh (2004, 2006), Kundu *et al.* (2005, 2009a, 2009b) and Panda *et al.* (2009) determined quality of groundwater in inland as well as coastal blocks of Odisha and assessed their suitability for use in irrigation. Analyses of ground water quality in Assam (Sarkar *et al.*, 2006) showed that high iron content is a problem for its large scale use for irrigation. Occurrence of iron toxicity and high-iron induced P and K deficiency in lowland rice limit crop productivity in irrigated areas. Iron precipitated from ground water used for irrigation often clog soil pores and impedes drainage and aeration capacity of the soils, especially heavy-textured soils. To overcome the problem of iron toxicity, following soil and water management practices are suggested: (i) aeration of paddy soil with weeder, (ii) storage and aeration of ground water in a tank for precipitating iron before its use for irrigation, (iii) use of modest dose of lime (Sikka *et al.* 2009).

Water pollution by hexavalent chromium in Odisha: Water samples collected from various sources of 16 villages situated within 2 to 15 km distance from the chromite mines of Sukinda of Jajpur district in Odisha were analyzed for chromium pollution. Data in Table 10 show that most of the surface water samples were contaminated with Cr^{6+} while ground water samples were less polluted due to low mobility of Cr^{6+} in soil profile (Sahu and Nayak, 2002).

Table 10: Hexavalent chromium content of water samples studied in villages around chromite mines of Jajpur district, Odisha.

Water source	No. of samples analyzed	No. of samples containing >0.05 ppm Cr^{6+}	% of Cr^{6+} contaminated samples
Mines drain	3	3	100
Rivulet	5	5	100
Floodwater in paddy fields	5	3	60.0
Pond	7	4	57.1
Open well	9	2	22.2
Tube well	16	1	6.25

Source: Sahu and Nayak (2002)

Impact of mining activities on water quality in Jharkhand: Mining wastes pollute streams and rivers. Ore fines and toxic substances carried by rain water into nearby water courses, alters their chemistry and often makes the water unfit for human use. By locating mineral treatment facilities near the mines, water pollution problems get worse. These units use enormous quantities of water for washing the ores. The untreated effluents, slimes or tailings are often released into neighbouring streams or lakes. In many cases, the latter are the sources of water supply to the population.

The large scale mining operations going on in the region have adversely affected groundwater table in many areas with the result that yield of water from the wells of adjoining villages has drastically reduced. Further, effluents discharged from mine sites have seriously polluted the streams and under groundwater of the area. Acid mine drainage, liquid effluents from coal handling plants, colliery workshops and mine sites and suspended solids from coal washeries have caused serious water pollution in the region, adversely affecting fish and aquatic life.

Damodar and Subernrekha river valleys are the cradle of industrialization in Chhotanagpur plateau region. Damodar is the most polluted amongst Indian rivers and ironically almost all polluting industries are government owned. About 130 million litres of industrial effluents and 65 million litre of untreated domestic water finds way to Damodar drainage system every day. A study of the area showed that one coal washery alone was discharging about 45 tonnes of fine coal into the Damodar every day and there are as many as eleven coal washeries in the region with an installed capacity of 20.52 million tonnes annually.

Today the picture of Damodar or Damuda, considered a sacred river by the local tribals, is quite like a sewage canal shrunken and filled with filth and rubbish, emanating obnoxious odors. Other major rivers of the region are also seriously polluted. The Karo river in the West Singhbhum is polluted with red oxide from the iron ore mines of Noamundi, Gua and Chiria. The Subernrekha shows a different type of pollution which is even more hazardous in nature. Metallic and dissolved toxic wastes from TISCO, Jamshedpur and HCL Ghatsila and radioactive wastes from the uranium mill and tailings ponds of the uranium corporation of India limited at Jaduguda flow into Subernrekha and its tributaries.

CONCLUSIONS

Soil and water resources of eastern India are really under threat although the area is very rich in soil diversity. Out of total eight different soil orders reported in India, six soil orders occur in the eastern region. Major soil-related constraints to agricultural production in the eastern region are: soil degradation (soil erosion), inadequate implementation of soil and water conservation measures in hilly tracts, soil acidity, and use of low level of fertilizer inputs. Major water-related constraints are inadequate and unreliable surface irrigation system, meager utilization of groundwater resource, excess water during *kharif* creating unfavourable edaphic environment for plant growth. Climate-induced natural disasters and extreme climatic aberrations in the form of natural calamities like cyclone, flood, droughts and earthquakes drastically affect the agricultural scenario of the region. In addition to this, lack of mechanization and low energy inputs, non-availability of adequate quantity of good quality seeds of high yielding varieties of crops, lack of proper marketing infrastructure for agricultural products are some of the other constraints. The release of different toxic metals like As, Hg, Cr, Ni, etc. from the coals and mine spoil areas in different rivers have caused severe damage to water quality. Continuous dewatering by underground mines also affects water resources. These mines annually pump out millions of litres to drain mine galleries and release it into nearby water courses. This has caused flooding, silting, water logging and pollution in the mining areas. They have also reduced the surrounding water table, and also reduced the available groundwater.

REFERENCES

Anonymous.1993. Reassessment of Water Resources Potential of India. Central Water Commission, Govt. of India, New Delhi.

Haldar, A.K., Thampi, C.J. and Sehgal, J. 1992. Soils of West Bengal for optimizing land use. NBSS Publ. 27b (Soils of India series), National Bureau of Soil Survey and Land Use Planning Nagpur, India, p 48.

Haldar, A.K., Srivastava, R., Thampi, C.J., Sarkar, D., Singh, D.S., Sehgal, J. and Velayutham, M. 1926. Soils of Bihar for optimizing land use. NBSS Publ. 50b (Soils of India series), National Bureau of Soil Survey and Land Use Planning Nagpur, India, p 70.

Kundu, D.K., Singh, R. and Kannan, K. 2002. Nitrate buildup in soil during post-rice dry season and its conservation. *In* Abs. International Rice Research Congress 2002, Beijing, China, pp 421.

Kundu, D.K. and Singh, R. 2004. Assessment of ground water quality in Khurda district, Orissa. *E-Planet Journal* 2 (1): 76-78.

Kundu, D.K., Singh, R., Mishra, A., Roychowdhury, S. and Kannan, K. 2004. Water regime influence on plant nutrient availability in soils. *In* Annual Report 2003-2004, Water Tech Centre for Eastern Region, Bhubaneswar

Kundu, D.K., Singh, R. and James, B.K. 2005. Excessive chloride found in groundwater of coastal Orissa. *WTCER News* 6(2):11

Kundu, D.K., Singh, R. and Kannan, K. 2006. Sustenance of soil N fertility under fallow-rice cropping system. *WTCER News* 7(1):15.

Kundu, D.K. and Singh, R. 2006. Assessment of groundwater quality in three coastal blocks of Orissa. *J. Indian Soc. Coastal Agric. Res.* 24: 83-86.

Kundu, D.K., Singh, R. and James, B.K. 2009. Assessment of ground water quality in selected coastal and inland blocks of Orissa. Pages 57-66 *in* Proc. of Workshop on Groundwater Scenario and Quality in Orissa, 6-7 March 2009, Bhubaneswar (Eds. GC Pati *et al.*). Central Ground Water Authority & CGWB, SE Region, Ministry of Water Resources, Govt of India.

Kundu, D.K., James, B.K. and Kumar, A. 2009. Assessment of ground water quality in Balipatna block of Khurda district, Orissa. Pages 85-89 *in* Proc. Workshop on Water Quality Management in Eastern Region, India, 17-18 Dec 2009, Bhubaneswar, (Eds. G. C. Pati *et al.*). Central Ground Water Authority & CGWB, SE Region, Ministry of Water Resources, Govt. of India.

Panda, D.K., Kundu, D.K., James, B.K. and Kumar, A. 2009. Groundwater quality scenarios in an intensive groundwater irrigated region of Orissa. Pages 103-106 *in* Proc. of Workshop on Groundwater Scenario & Quality in Orissa, 6-7 Mar 2009, Bhubaneswar (Eds. GC Pati *et al.*). CGWA & CGWB, SE Region, Ministry of Water Resources, Govt of India.

Sahu, S.K. 2003. Influence of heavy metals on environment and health problem. *In* Winter School on Characterization and Sustainabe Management of Acid Soils of Eastern India. OUAT, Bhubaneswar

Sahu, S.K. and Nayak, S. 2002. Report of ICAR adhoc project on delineation of chromium in soil, water and crop plants of Orissa. OUAT, Bhubaneswar

Sarkar, A.K., Kundu, D.K., Das, D.K. and Baruah, T.C. 2006. Soil and water quality vis-à-vis agricultural management practices in eastern India. Pages 200-215 *in* Proc. International Conf. on Soil, Water and Environmental Quality Issues and Strategies organized at New Delhi on Jan 28-Feb 1, 2005. Indian Society of Soil Science, IARI, New Delhi 110012, India.

Sarkar, D., Thampi, C.J., Sehgal, J. and Velayutham, M. 1998. Soils of Orissa for optimizing land use, NBSS Pub. 49b (Soils of India series), NBSS&LUP, Nagpur, India, p 82.

Sharma, P.D. and Sarkar, A.K. 2005. Managing acid soils for enhancing productivity. Tech. Bull. NRM Division, ICAR, New Delhi.

Sikka, A.K., Kumar, A., Upadhyaya, A., Kundu, D.K., Dey, P., Sarkar, A.K. and Islam, A. 2009. Development and policy issues for optimum use of soil and water in eastern region. Pages 23-54 *in* Soil and Water Management for Agricultural Transformation in Eastern India. ISSS Bulletin No. 26, Indian Society of Soil Science, New Delhi

Singh, R., Kundu, D.K., Ghosh, S., Chaudhari, S.K. and Kumar, A. 2009a. Natural resources and social capabilities of eastern India: an analysis of opportunities and constraints for efficient management of water. Pages 1-22 *in* Soil and Water Management for Agricultural Transformation in Eastern India. Bull. 26, Indian Society of Soil Science, New Delhi

Singh, R., Kundu, D.K. and Kumar, A. 2009b. Characterisation of dominant soil subgroups of eastern India for formulating water management strategies. WTCER Research Bulletin No. 44. Water Technology Centre for Eastern Region, Bhubaneswar, India. 55 p.

Singh, R. and Kundu, D.K. 2010. Physico-chemical and hydraulic characteristics of soils of major sub-groups of eastern India. *J. Indian Soc. Soil Sci.* 58: 267-278.

Singh, R., Kundu, D.K., Chakraborty, D., Tomar, R.K. and Garg, R.N. 2010. Prediction of hydraulic characteristics of dominant soil subgroups of eastern India. *Indian J. Soil Cons* 38(1): 1-9.

Singh, R., Kundu, D.K. and Verma, H.N. 2003. Alternate uses of salt-affected lands in the eastern region. Pages 419-434 *in* Alternative Land Use Systems for Sustainable Production (Ed. SP Mittal). Oriental Enterprises, 31-B Rajpur Road 2nd Floor, Dehra Dun 248001, India.

Singh, R. and Kundu, D.K. 2009. Management of salt-affected soils in the eastern region. *Journal of Soil Salinity & Water Quality* 1(1&2): 85-90.

Singh, R. and Kundu, D.K. 2008a. Prediction of water storage capacity of major soil groups of eastern India. *Journal of Water Management* 16: 1-9.

Singh, R. and Kundu, D.K. 2008b. Erodibility of major soil sub-groups of eastern region of India. *Indian J. Soil Cons.* 36(3): 172-178.

Singh, S.P., Ram, J., Walia, C.S., Sachdev, C.B., Dhankar, R.P., Sehgal, J., Velayutham, M. and Gajbhiye, K.S. 2003. Soils of Uttar Pradesh for optimizing land use. NBSS Pub. 68 (Soils of India Series), NBSS&LUP, Nagpur, India, p 91.

Tamgadge, D.B., Gaikwad, S.T., Nagabhushana, S.R., Gajbhiye, K.S., Deshmukh, S.N. and Sehgal, J. 1999. Soils of Madhya Pradesh for optimizing land use. NBSS Pub. 59b (Soils of India Series), NBSS&LUP, Nagpur, India, p 182.

□□□

Chapter 4

Soil Quality under Different Cropping Systems in the Recent Alluvium of South Bihar

K. RAJAN AND SANJEEV KUMAR[1]

INTRODUCTION

Agriculture in Bihar involves both traditional and modern cultivation practices. Utilization of natural resources for agricultural production to match the growing population of the country poses heavy pressure on soil resources. Preparatory and intercultural operations are the important practices that bring major changes in the soil systems. Application of organic residues and agrochemicals are also affects the soil qualities especially the soil biological quality. Use of only chemical fertiliser reduces the soil biological activity such as soil respiration compared to unfertilized control (Kang *et al.*, 2005). Tillage, irrigation, nutrient management, use of biocides etc., influence the soil physical, chemical and biological properties in varying degrees. Such a management practices may bring positive effect on some soil properties and negative effective on some properties. In Inceptisols, soil quality faced 60 % negative change under long term cultivation of sugar beet compared to virgin soil (Rezapour and Samadi, 2012).The soil properties, which are involving in soil functions, are selected as soil quality indicators. Considering few indicators for the soil quality assessment under any agricultural environment will not yield proper inference. A comprehensive quantitative assessment involving physical, chemical and biological indicators is necessary for quality assessment of any agro-eco systems. Soil quality indexing is one way of doing it. With this view, soil quality assessed was done in major crop and cropping systems in the young alluvium of south Bihar under hot sub-humid (Dry) region.

Central Soil and Water Conservation Research and Training Institute, Research Center, Udhagamadalam -643 004, Tamil Nadu, India

[1]ICAR Research Complex for Eastern Region, PO: BV College, Patna – 800 014, Bihar

MATERIALS AND METHODS

Site selection :Agro Ecological Region (AER) Map published by National Bureau of Soil Survey and Land Use Planning, Nagpur was used to identify the AER in Bihar. Major part of the Bihar comes under Agro Ecological Region (AER) 13 followed by region 9 (Sub regions 13.1 & 9.2 resp.). Out of total geographical area of 94163 sq. km 63.5 % area is classified under Agro-Ecological Region 13 and 31.6 % under AER 9.

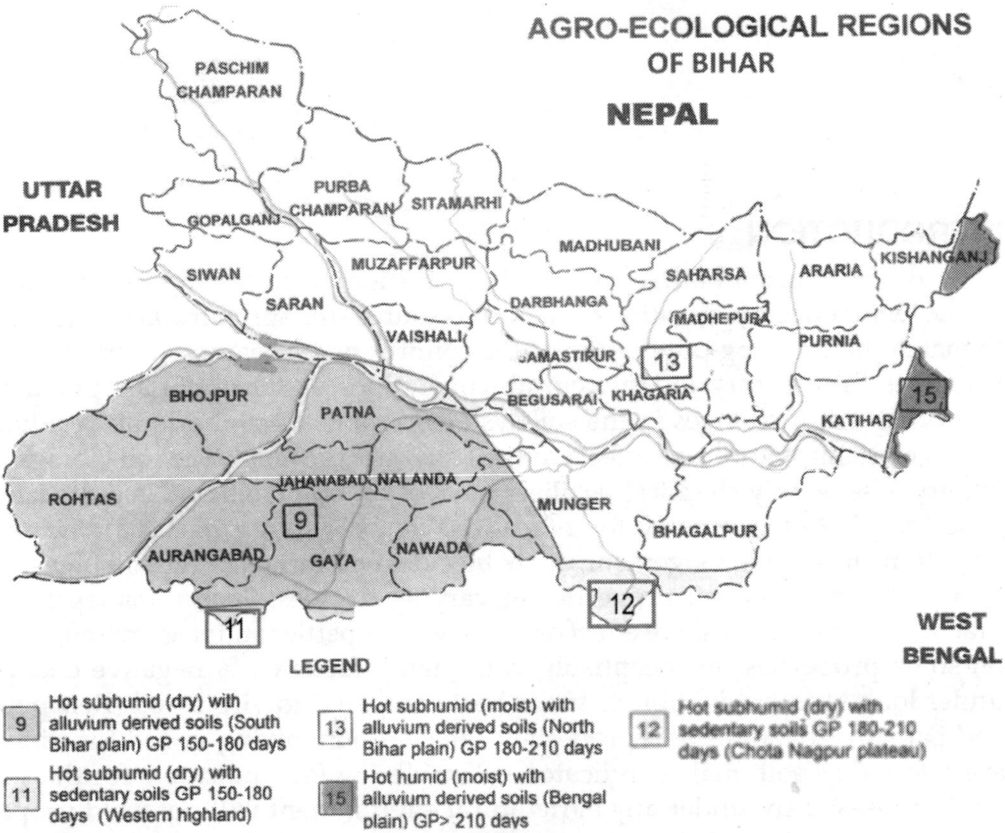

Fig. 1: Agro ecological regions of Bihar

Remaining 4.9 % is under AER – 11, 12 and 15 (Figure 1). In south Bihar, AER -9 occupies 29125 sq. km which is highest among other AERs. Mean annual rainfall is ranging from 700 to 1000 mm and Potential evapo-transpiration ranging from 1300 to 1500 with mean temperature from 24 to 26 °C. Soil map of Bihar (Anonymous, 1986) published by Rajendra Agricultural University, Bihar was used to identify the recent alluvium under

AER 9 in south Bihar (Figure 2). Recent alluvium of Ustifluent has spreaded along the River Ganga from west to east covering three district namely Buxar, Bhojpur and Patna with an area of 1685 sq. km.

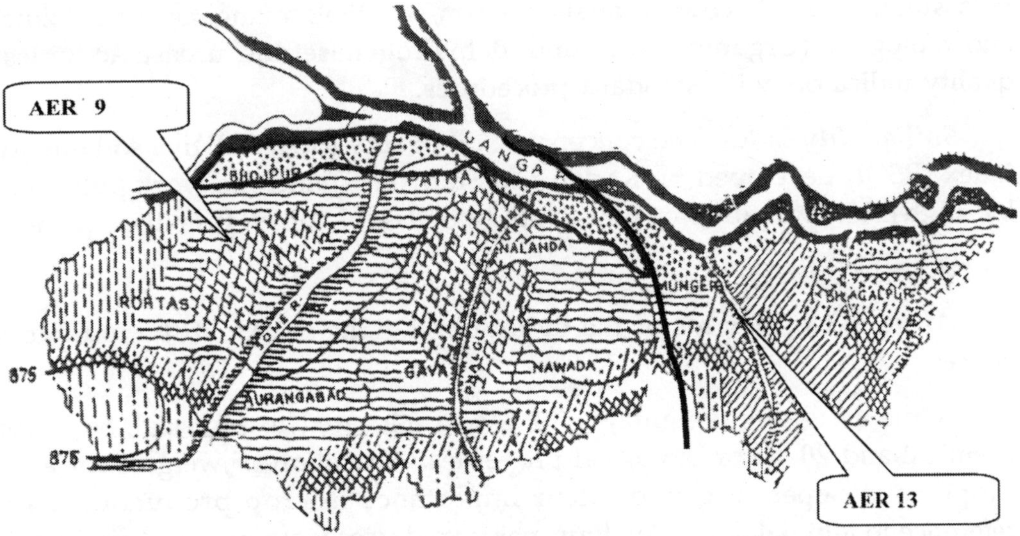

Fig. 2: Recent alluvium of AESR 9.2 in south Bihar

Based on the field survey, it was identified that there were four major cropping systems *viz.,* Rice-Wheat-Fallow(R-W-F), Bajra-Wheat-Fallow (B-W-F), Red gram and Vegetables cultivation in the region. Rice-What-Fallow system is followed by the farmers traditionally in the low lands with heavy soils. Uses of organics are very limited and even the crop residues left in the field after harvest is burnt. Fertilisers are used and mostly nitrogenous fertiliser alone in the form of Urea. Some farmers use phosphatic fertiliser as SSP and very rarely potassic fertiliser. Bajra-Wheat-Fallow system is followed by the farmers in the upland under marginal lands. It is mainly rainfed cultivation. Soil in this area is light in texture. Input application is very poor. Red gram cultivation is profitable one in this region. Part of the agricultural land is grown with this crop. Only one crop is grown and the land is left as fallow for the remaining period. Red gram is grown under medium texture alluvial soil. Commercial cultivation of vegetables is done with heavy investment of inputs such as manures and fertilisers, timely irrigation and optimum tillage

Soil sampling and analysis : Sampling sites were selected based on the long period cultivation of same crops. Cultivation history was collected from the farmers. Ten sites were selected for each major crop/cropping system. Soil samples were collected from surface layer (0-15). Core samples were collected for bulk density where soil samples collected. Soil samples were analysed in the laboratory for 18 soil physical (Soil texture, bulk density, available soil water, micro-aggregates, total porosity), chemical (Soil reaction, electric conductivity, cation exchange capacity, available phosphorus, available potassium, available copper, available iron, available manganese and zinc) and biological (Organic carbon and dehydrogenase and urease activities) quality indicators with standard procedures.

Soil quality index :The concept of Soil quality (SQI) Relative soil quality index (RSQI) developed by Karlen and Stott (1994) and the work published by Singh (2007) was adopted for this study. SQI is calculated from the equation:

$$SQI = \Sigma W_i I_i$$

Where, Wi = Weights of the indicators Ii = the marks of the indicators classes

While assigning weightage, 30 % was allocated to physical, 50 % for chemical and 20 % for biological properties. Allocation of weights for each property is depending upon their importance in crop production with reference to alluvial soils. The four major land uses were assessed for its soil quality and classified them. SQI of every indicator was calculated separately by multiplying weight of indicators and marks allotted to each class. In this way, SQI for every indicator was calculated (Table 1). Summing up of all eighteen indicators produces the SQI value for the soils of crop/cropping systems under study.

There were four classes *viz.*, class I, class II, class III and class IV created from best to poor by fixing ranges of values for each quality indicators. Analytical value for each soil quality indicator for each cropping systems was interpreted with the soil quality index and value was obtained. The equation for calculating RSQI value is given below:

$$RSQI = (SQI/SQIm) \times 100$$

SQI = Soil quality index; SQIm = Maximum value of SQI.

According to the RSQI values, soils were classified into 5 classes from best to worst, Category I rated for the best quality the value from 90-100, Category II rated as very good value ranging from 80-90, category III rated as good (70-80), category IV rated as average (60-70) and the remaining land use falling below category IV kept under category V and rated as poor. These values were fitted in different categories based on the cumulative marks and graded the crop and cropping systems.

Table 1: Soil quality index

Soil quality indicators			Weights	Class I	Class II	Class III	Class IV
Physical (30 %)	1	Bulk density(Mg m^{-3})	8	1.3-1.4	1.4-1.5	1.5-1.6	>1.6
	2	Texture	10	Loamy	Clay loam, Silt loam, Silty clay, Silty clay loam, sandy clay	Clay, Silt, Sandy clay loam,	Sandy loam, Loamy sand, Sand
	3	Available soil water	6	>25	16-25	8-16	<8
	4	Microaggregates(%)	4	>30	15-30	5-15	<5
	5	Porosity (%)	2	45-50	40-45	>50, 35-40	<35
Chemical (50 %)	6	pH	4	6.5-7.5	7.5—8.5 5.0-6.5	8.5-9.5 4.0-5.0	>9.5 <4.0
	7	Electrical conductivity (dSm^{-1})	4	<0.5	0.5-0.2	2.0-4.0	>4.0
	8	CEC (Cmol(p+) kg^{-1})	4	>25	16-25	8-16	<8
	9	Avail. N (kg ha^{-1})	8	>400	300-400	200-300	<200
	10	Avail. P_2O_5 (kg ha^{-1})	8	>22	11-22	5-11	<5
	11	Avail.K_2O (kg ha^{-1})	8	>280	180-280	90-180	<90
	12	Avail.Cu (mg kg^{-1})	2	10-30	0.5 - 5.0	0.2-0.5	<0.2
	13	Avail. Mn (mg kg^{-1})	3	>40	4-40	2-4	<2
	14	Avail. Fe (mg kg^{-1})	4	>40	5.8-50	2.8-5.8	<2.8
	15	Avail. Zn (mg kg^{-1})	5	>10	1-10	0.5-1.0	<0.5
Biological (20 %)	16	OC(g kg^{-1})	12	>10	7.5-10	5-7.5	<5
	17	Dehydrogenase Activity (μg g^{-1} day^{-1})	4	>100	60-100	30-60	<30
	18	Urease activity μg of NH_3 g^{-1}	4	>75	75-30	30-10	<10
Marks			100	4	3	2	1

RESULTS AND DISCUSSION

Soil quality indicators : The soil quality indicators were found better in soils under vegetable cultivation followed by red gram and R-W-F system. Soil quality was poor in B-W-F systems in the recent alluvium. Soil bulk density ranged from 1.30 to 1.63 Mg $m^{-3}$was found highest in R-W-F cropping system (1.55 Mg m^{-3}) and the lowest was under red gram (1.33 Mg m^{-3}). Soil texture was varying from loamy sand to sandy clay loam under different systems. Available soil water was varying from 11.3 to 13 cm. Highest soil microaggregates percentage was found under vegetable cultivation (Figure 3). Frequent ploughing and heavy application of organic manures for vegetable cultivation might have supported high amount of microaggregates in the soil. Mohanty *et al.* (2007) observed a positive correlation between soil organic matter and the structural stability of both macro and microaggregates. Soil porosity was found highest under vegetable cultivation and the lowest percentage was under R-W-F systems where the soil is heavy in nature; wet ploughing for rice cultivation might have reduced the pore spaces severely by the destruction of soil aggregates. Similar result was observed less porosity in paddy soil than non-paddy soil by Raheb and Heidari (2011).

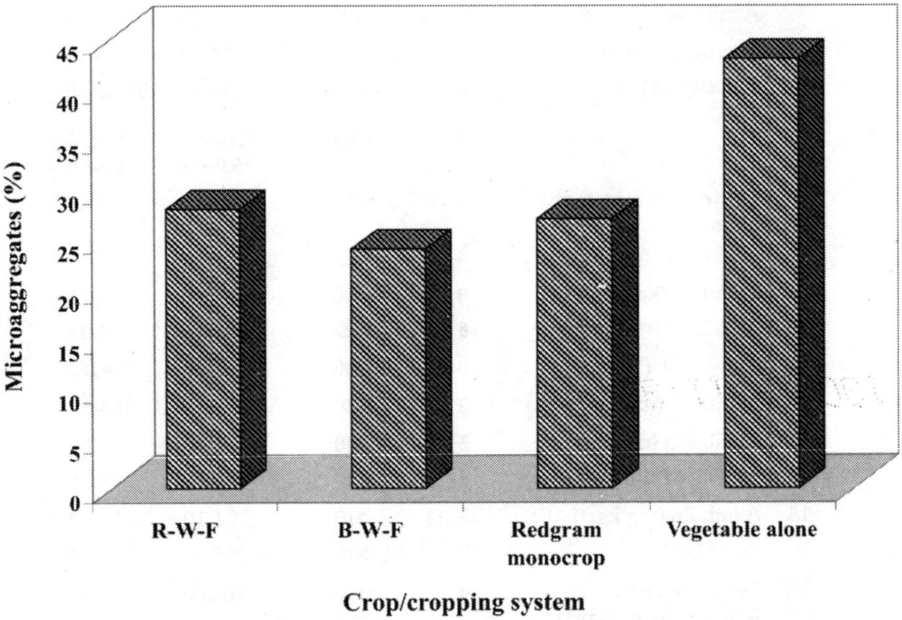

Fig. 3: Soil microaggregates under different crop and cropping system (%)

Soil reaction was in neutral range in all the crops and electrical conductivity is also in safe range (<4 dSm^{-1}). Available soil nitrogen was found highest in vegetable cultivation (232.0 kg ha^{-1}) and lowest in B-W-F system (180.7 kg ha^{-1}). Available phosphorus was also found highest in vegetable cultivation and lowest in red gram cultivated soils (Figure 4)

Highest available potassium was recorded in B-W-F (416.2 kg ha^{-1}) followed by vegetable cultivation (311. 0 kg ha^{-1}) and R-W-F (256.0 kg ha^{-1})the lowest potassium was recorded with red gram cultivation (123.7 kg ha^{-1}). In case of micronutrients, copper was highest in vegetable and red gram cultivation (2.5 mg kg^{-1}) followed by B-W-F (2.1 mg kg^{-1}). The lowest available copper was recorded with R-W-F (1.2 mg kg^{-1}). Available zinc is in near sufficient quantity with some soils (Figure 5). It is about to reach deficient level. Vegetable cultivation recorded the highest zinc content of

2.3 mg $kg^{-1}$and the lowest available zinc was in red gram cultivation of 0.9 mg kg^{-1}. Soil organic carbon status was poor under B-W-F (4.0 g kg^{-1}). Highest organic carbon content was recorded in vegetable cultivation (8.0 g kg^{-1}) followed by red gram (7.0 g kg^{-1}) and R-W-F (5.0 g kg^{-1}). Soil microbial activity was measured through dehydrogenase activity in different systems which was ranging from 24.0 to 76.6 µg g^{-1} day^{-1} (Figure 6). Highest activity in vegetable cultivation attributed to high organics application in the soil. Similar result was observed by Subrahmanyam *et al.* (2011).

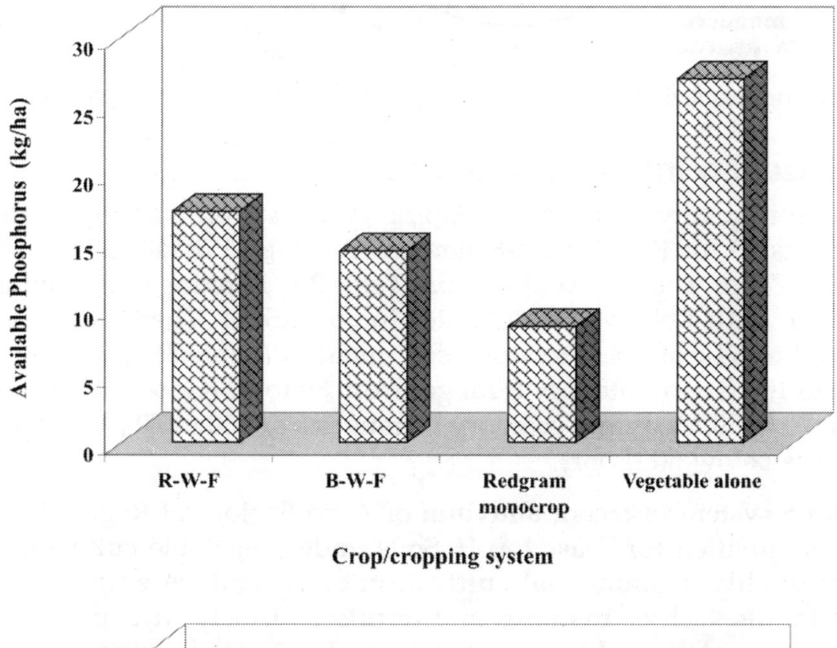

Fig. 4: Soil available phosphorus under different crop and cropping system (kg ha^{-1})

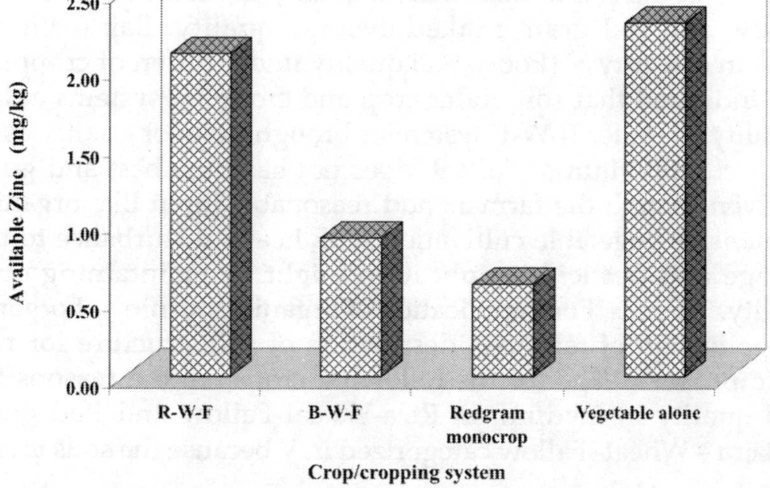

Fig. 5: Soil available zinc under different crop and cropping system (mg kg^{-1})

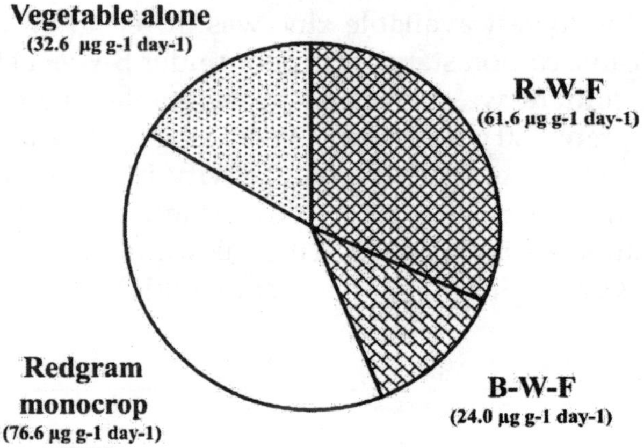

Fig. 6: Dehydrogenase activity under different crop and cropping system (µg g^{-1} day^{-1})

Relative Soil Quality Index and Ranking

As per the soil quality index, the cropping systems were varying in its quality and it was 63 for Rice-Wheat-Fallow, 57 for Bajra-Wheat-Fallow, 66 for red gram and 74 for vegetable cultivation (Table 2). Ranking of cropping system was done with Relative Soil Quality Index (RSQI). There were six categories from I to VI from best to worst soil qualities (Table 3). Category I range from 90 to 100 (Best), category II ranges from 80 to 90 (Good), category III ranges from 70 to 80(Medium), category IV ranges from 60 to 70 (Average) and category V is below 60 (Poor).

In AER 9, no system in recent alluvium of Agro Ecological Region 9 of south Bihar has qualified for Class I & II. Soils under vegetable cultivation rated medium quality. Commercial cultivation of vegetables with heavy investment of inputs such as manures and fertilisers, timely irrigation and optimum tillage helped the soil to maintain its quality in medium condition. Rice-Wheat-Fallow and red gram ranked average quality. Bajra-Wheat-Fallow was found in category V (Poor). Soil quality under different cropping system in AER 9 indicated that soil under crop and cropping systems either inherent poor quality as under B-W-F system or brought to poor quality as in R-W-F system. Recent alluvium of AER 9 does not have the best and good quality of soils. Even though the farmers add reasonable input like organics and fertilisers, intensive vegetable cultivation with heavy disturbance to the soil through tillage and pesticide application might be maintaining only medium soil quality. There is Poor application of organics, burning of organic residues, poor application of fertiliser, destruction of soil structure for rice cultivation and improper tillage for the following crops are the reasons for maintaining soil quality in medium in Rice-Wheat-Fallow and Red gram cultivation and Bajra – Wheat- Fallow categorized in V because the soils under this cropping system is in upland. Very low organic manure and fertiliser application and poor irrigation and tillage practices followed by the farmers might be the reason for the poor soil quality.

Table 2: Relative soil quality index for different crop and cropping systems

Soil quality indicators			R-W-F	B-W-F	RG	Vegetables
Physical	1	Bulk density	16	16	32	32
(30 %)	2	Texture				
	20	20	32	20		
	3	Available soil water	8	8	8	8
	4	Micro aggregates	12	12	12	16
	5	Porosity	8	4	4	8
Chemical	6	pH	16	16	16	16
(50 %)	7	Electrical Conductivity	16	16	16	16
	8	Cation Exchange Capacity	8	8	8	8
	9	Available N	8	8	8	16
	10	Available P	24	24	16	32
	11	Available K	24	32	16	32
	12	Available Cu	8	8	8	8
	13	Available Mn	12	12	12	12
	14	Available Fe	8	8	12	12
	15	Available Zn	16	16	16	16
Biological	16	Organic carbon	24	12	24	24
(20 %)	17	Dehydrogenase activity	12	4	12	12
	18	Urease Activity	12	4	12	8
Cumulative marks			252	228	264	296
RSQI			63	57	66	74

Table 3: Ranking of soil qualities under different crop and cropping systems

Soil quality Classifications	Class ranges (Values of RSQI)	Crop/ cropping systems
I (Best)	90-100	-
II (Good)	80-90	-
III (Medium)	70-80	Vegetable cultivation
IV (Average)	60-70	Rice-Wheat-Fallow & Redgram monocrop
V (Poor)	<60	Bajra-Wheat-Fallow

CONCLUSIONS

Assessment of recent alluvium of AER 9 through Soil Quality Index has categorized the soils under major crops and cropping system. No soil under crops and cropping systems in this region graded under best and very good qualities. Commercial cultivation of vegetables with heavy investment of inputs such as manures and fertilisers, timely irrigation and optimum tillage helped the soil to maintain its quality in medium condition. Soil is poor in quality either by its inherent nature or acquired by human activities, its quality need to be improved for sustainable productivity. Improving organic matter in the soil using in-situ and locally available of organic residues increases the soil productivity. Maintenance of soil nutrient in optimum level through Integrated Plant Nutrient Supply system supports congenial soil ecosystem for higher productivity. Capacity building to the farmers especially for soil health management by integrating organic residues will benefit the soil ecosystem. Soil testing facilities for the farmers of Bihar are the need of the hour to supply balanced nutrient. Timely availability of chemical fertiliser, bio-fertilisers and organic manures encourages the farmers to maintain optimum soil conditions.

REFERENCES

Anonymous. 1986. Soils of Bihar - Their properties and classification. Rajendra Agricultural Univerisity, Pusa, Samastipur, Bihar.

Kang, G. S., Beri, V., Sidhu, B. S and Rupela O. P. 2005. A new index to assess soil quality and sustainability of wheat-based cropping systems. *Biol Fertil Soils.,* 41: 389–398.

Karlen, D.L. and Stott. 1994. A framework for evaluating physical and chemical indicators of soil quality. In: *Defining Soil Quality for a Sustainable Environment* (J.W. Doran, D.C. Coleman, D.F. Bezdicek, and B.A. Stewart, Eds). pp 53-72. SSSA Special Publication. No. 35, *Soil Science Society of America, American Society of Agronomy,* Madison, WI.

Mohanty M, Painuli, D. K., Misra, A.K., Ghosh, P.K. 2007. Soil quality effects of tillage and residue under rice–wheat cropping on a Vertisol in India. *Soil Till Res.,* 92:243–250.

Raheb A. and Heidari A. 2011. Evaluation of porosity in some paddy and non paddy soils by image analysis technique. *International Journal of Science and Advanced Technology.* 1: 47-50 (www.ijsat.com).

Rezapour, S and Samadi, A. 2012. Assessment of Inceptisols soil quality following long-term cropping in a calcareous environment. *Environ. Monit. Assess.* 184: 1311-1323.

Singh, A.K.2007. Evaluation of soil quality under integrated nutrient management. *J. Indian Soc. Soil Sci.,* 55:58-61.

Subrahmanyam, G., Archarna, G and Chamyal, L.S. 2011. Soil microbial activity and its relation to indigenous properties in semi arid alluvial and estuarine soils Mahi river basin, Western India. *International Journal of Soil Science,*6:224-237.

□□□

Chapter 5

Farm Planning for Resource Conservation based on LCC and Soil Health Criteria using GIS : A Case Study

PARTHA PRATIM ADHIKARY, RAMESHA M.N, S.P. TIWARI
H. BISWAS, D.G. DURBUDE AND DEV NARAYAN

INTRODUCTION

The relative scarcity of land for agricultural use because of multiple other uses of land to get physical comfort of human being and insufficient food security compelled us to use the land resources in an optimum way. The irrigated area is limited to only 757 M ha and 91% of the developing nations (excluding China) are mainly dependent on dry farming accounting an area of about 2573 M ha (FAO, 1991). The situation in the arid and semi-arid region is even worse and the problem of optimal land use is highly relevant. In a semi-arid region of India like Bundelkhand, where the resources are scarce and hungry stomachs are more effective planning of individual farm lands based on soil and climatic factors is the need of the day.

The farm planning for resource conservation involves making knowledgeable decisions about the optimum land use for sustainable production and the environment. Soil information is a vital component in the planning process, reflecting directly upon land-use suitability (Coleman and Galbraith, 2000). The land evaluation system was based on land qualities required to grow individual crops productively for a long period based on experiences in tropical areas (Sys *et al.*, 1993). This type of land planning based on land evaluation system was earlier adopted to assess land suitability in Seoni district of Madhya Pradesh (Bobade *et al.*, 2010).

Matching the existing environmental conditions of a place with the preferences of crops as to soil properties and climate gives one a useful tool for determining the suitability of an area to crop production or whether a crop can be grown productively in an area. Soil fertility evaluation, on the

Central Soil and Water Conservation Research and Training Institute, Research Centre, Gwalior Road, Datia (MP) – 475661, India other hand, will be a useful tool in assessing the fertilization requirement of a crop. Albeit, soil fertility does not guarantee crop productivity, when treated as part of crop management, it can certainly augment crop productivity and increases soil productivity.

For the sustainable use of the natural resources, a detailed charter of land resources giving its potential and constraints becomes pre-requisite for planning. The land capability classification is the grouping of a land unit (s) into defined class (es) based on its capability. It is a broad grouping of soils based on their limitations and is designed to emphasize the hazards in different kinds of soils. It serves as a guide to assess suitability of the land for arable crops, grazing and forestry. Dent and Young (1981) stated that the fundamental purpose of land evaluation is to predict the consequences of change. Prediction is needed on the suitability of the land for different forms of production of their benefits, and the consequences of such changes upon the environment. The land capability sub-class of Ustorthents varied from VIe to VIIe due to severe erosion while Ustic Torriorthents had IIe land capability sub-class due to moderate erosion in Punjab (Sidhu *et al.*, 1994). Thangasamy *et al.* (2005) assessed the suitability of soils of Shivagiri microwatershed for arable crops, grazing and forestry.

Recent developments of advanced techniques have the capability to handle enormous amount data which help to plan preciously. Computer program, that includes decision support systems (DSS) and Geographic Information Systems (GIS) have contributed to the speed and efficiency of overall planning process. GIS based mapping of soil resources is an important component of a farm planning strategy. It can reveal potential fertile zones and indicate suitability of soils for growing of various crops (Salem *et al.*, 2008). It can show, quantitatively or qualitatively, certain characteristics of the soil environment that determine vulnerability of soil to erosion (Lakaria *et al.*, 2010).

The present study uses LCC and GIS technique to plan the farming activities for resource conservation in a semi-arid tropic in central India. The spatial variation of soil properties as it currently exists was mapped using GIS, utilizing the data generated from physic-chemical analysis of soil samples collected from the area under study.

MATERIALS AND METHODS

Study area : The study was conducted at the research farm of Central Soil and Water Conservation Research and Training Institute, Research Centre, Datia in the state of Madhya Pradesh. The research farm, covering an area of

188.7 ha is situated between 78°25'47.52" - 78°26'37.43" E longitude and 25°26'16.36" - 25°42'43.89" N latitude. The farm lies on the Sind-Pahuj Doab with rugged terrain. The climate is subtropical semiarid with average annual rainfall and evaporation is about 827 and 2565 mm, respectively. The average annual temperature is 25.6°C with January being the coldest month having mean daily maximum temperature of 21.3°C and minimum of 7.3°C. The temperature of May can go up to 48°C (Adhikary and Biswas, 2011). The average value of the meteorological parameters is presented in Table 1.

Table 1: Average value of the meteorological parameters collected from the meteorological observatory of the Research Farm

S. No.	Meteorological parameter	Value
1	Rainfall	
	Amount (mm)	827
	Intensity (cm h^{-1})	7.6 (Maximum, July)
	Distribution	91% in the months of June to September
2	Temperature (°C)	
	Maximum	36.9 (May)
	Minimum	3.7 (December)
3	Relative Humidity (%)	
	Maximum	86 (September)
	Minimum	26 (April)
4	Wind velocity (km h^{-1})	
	Maximum	6.2 (June)
	Minimum	1.1 (November)
5	Daily Evaporation (mm)	
	Maximum	9.7 (June)
	Minimum	2.0 (December, January)
6	Sunshine hours	
	Maximum	10.7 (April)
	Minimum	5.0 (August)

Geomorphology and soils : The landscape is rugged, featuring undulating terrain with low rocky outcrops, narrow valleys, and plains. Surface rocks are predominantly granite of the Lower Pre Cambrian/Archaen period. The study area abounds in open scrub and thin forest. The major soil groups in the region are red (rakar and parwa), black (kabar and mar) and mixed red and black soils, belonging to the Orders Entisol, Inceptisol and Vertisol (in patches).

Natural vegetation, agriculture, horticulture and agro-forestry : The natural vegetation of the study area consists of open scrub forest with scattered trees. Grasses, climbers and thorny shrubs are found in the lower and middle storey. The principal species in the open scrub area is Anogeissus pendula (kardhai). Thick forest of Kardhai is mainly localized at the hill tops. The forest vegetation in the farm is broadly grouped under tropical deciduous forest, sub grouped as 5B/C2 - Northern Dry Mixed Deciduous Forest, and 6B - Northern Tropical Thorn Forest.

There are two well established and organized orchards in the farm. They are aonla and ber orchards. Scattered and sparsely distributed fruit trees are found on farm bunds. Agro forestry practices are well established and practiced within the farm. Naturally regenerating trees do exist on the bunds of farm lands and on fallow land. The commonly observed species are *Madhuca latifolia, Azadirachta indica, Ziziphus jujube, Zizyphus mauritiana, Emblica officinalis, Mangifera indica, Butea monosperma, Acacia senegal, Acacia nilotica, Prosopis juliflora* and *Prosopis cineraria.*

Soil survey for land capability classification : The farm area has been divided into various plots as demarcated for various agricultural and forestry land uses. Under each plot, the slope of the plot was measured using Dumpy level; the effective soil depth was measured using soil auger; the soil texture was measured by Hydrometer method (Bouyoucos, 1962) and soil erosion was estimated using Universal Soil Loss Equation (USLE) (Weischmeir and Smith, 1978). All these data was used to determine the land capability class of that particular plot. Likewise the LCC of the total farm area was determined. All the land capability classes were digitized using Arc GIS to generate the LCC map of the whole farm.

Soil sampling and laboratory analysis : Seventy five surface (0-15 cm) soil samples were collected randomly from the research farm. During selecting the sampling locations it was kept in mind that the samples should represent all the existing land uses in the farm. The locations of the sampling sites were determined with the help of a global positioning system (GPS). A few locations were also cross-checked with a differential GPS. The locations of the soil sampling sites in the research farm are presented in Figure 1.

Collected soil samples were brought to the laboratory and shade dried. The dried soil samples were grinded and passed through 2 mm sieve. The materials retained above the sieve were used for coarse fragment analysis and the sieved materials were chemically analysed in the laboratory to measure the concentration of the quality parameters using standard procedures

(Table 2). The software GS+ 8 (Gamma Design Software, 1988) was used for geostatistical analysis of data and Arc GIS 9.3 (ESRI, 2004) for generation of maps.

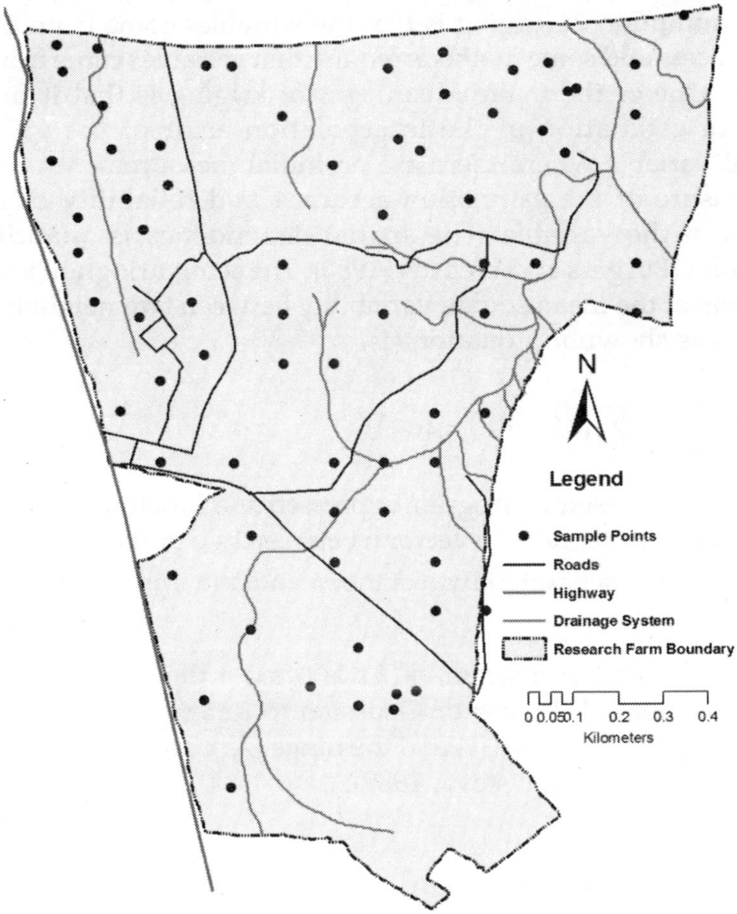

Fig. 1: Map of the study area showing the location of the soil sampling points

Table 2: Specific methods of estimation of different chemical parameters of surface soil in the study area

Parameter	Method
pH	Potentiometric method (Glass electrode pH meter)
EC (Electrical Conductivity)	Conductivity Bridge method (Richards, 1954)
Oxidizable SOC (Soil Organic Carbon)	Walkley and Black method (Walkley and Black, 1934)
Texture	Hydrometer method (Bouyoucos, 1962)
Available P	Olsen method (Olsen *et al.*, 1954)
Available K	Flame Photometric method (Osborn and Johns, 1951)

Geo-statistical approach in development of spatial variability maps : Kriging is a geo-statistical approach for interpolation at unsampled locations. It is flexible and permits investigation of spatial autocorrelation of the variables. The basic assumption in kriging is that the variables come from a stationary stochastic process and some methods require that variables come from a normal distribution. One of the main advantages of kriging is that it presents the possibility of estimation of the interpolation error of the values of the regionalized variable where there are no initial measurements. This feature offers a measure of the estimation accuracy and reliability of the spatial distribution of the variable. The spatial dependence is quantified using semivariogram (Burgess and Webster, 1980). The semivariogram is a graphical representation of the mean square variability between two neighboring points of distance h as shown in Equation (1),

$$\gamma(h)=\frac{1}{2N(h)}\sum_{i=1}^{N(h)}[z(x_i+h)-z(x_i)]^2 \qquad \text{... (1)}$$

Where $\gamma(h)$ is the semivariogram expressed as a function of the magnitude of the lag distance or separation vector h between two pints, $N(h)$ is the number of observation pairs separated by distance h and $z(x_i)$ is the random variable at location x_i.

The experimental variogram, $\gamma(h)$ is fitted a theoretical model such as Spherical, Exponential, Linear or Gaussian to determine three parameters, such as the nugget (c_0), the sill (c) and the range (A_0). These models are defined as follows (Issaks and Srivastava, 1989).

Spherical model:

$$\begin{aligned}\gamma(h) &= c_0 + \left[1.5\left(\frac{h}{A_0}\right) - 0.5\left(\frac{h}{A_0}\right)^3\right] && h \le A_0 \\ \gamma(h) &= c_0 + c, && h > A_0\end{aligned} \qquad \text{... (2)}$$

Exponential model:

$$\gamma(h) = c0 + c\left[1 - \exp\left(-3\frac{h}{A_0}\right)\right] \qquad \text{... (3)}$$

Gaussian model:

$$\gamma(h) = c0 + c\left[1 - \exp\left[-\left(\frac{3h}{A_0}\right)^2\right]\right] \qquad \text{... (4)}$$

Linear model:

$$\gamma(h) = c0 + h\left(\frac{c}{A_0}\right) \quad \text{... (5)}$$

Soil quality data were subjected to Exploratory Data Analysis (EDA) for identifying the outliers and carry out different transformations such as log normal, square root to ensure normal distribution. Then different semivariogram models were generated using geo-statistical software GS+. The best-fit semivariogram model was obtained based on the highest R^2 and lowest RSS values of the fitted model. The corresponding sill, nugget, and range values of the best fitted semivariograms were observed. Subsequently, ordinary kriging was performed to generate the spatial variability map for the soil quality parameters (pH, EC, SOC, P, K) and soil textural parameters (sand, silt, clay) based on the cutoff values of the parameters for different uses.

RESULTS AND DISCUSSIONS

Spatial distribution of land capability classes within the study area : Land capability classification shows, in a general way, the suitability of soils for most kinds of field and forest crops. Crops that require special management are excluded. The soils are grouped according to their limitations for field crops, the risk of damage if they are used for crops, and the way they respond to management. The criteria used in grouping the soils do not include major and generally expensive landform that would change slope, depth, or other characteristics of the soils, nor do they include possible but unlikely major reclamation projects. Capability classification is not a substitute for interpretations designed to show suitability and limitations of groups of soils for rangeland, for forestland, or for engineering purposes. But in spite of all these limitations land capability classification system has widely been used for farm planning as it is very simple and address the general limitations faced by the crops for their successful growth.

LCC map shows the presence of five classes within the study area covering class II, III, IV, VI and VII (Figure 2). Among these five classes, three classes those are II, III and IV are distributed mainly at north-western, north-eastern and southern part of the study area comprising 105.8 ha land of the study area. That means to say more than 50% of the study area is under arable condition. Non arable lands are mainly concentrated at the middle part of the farm. Class II, III and VI lands are dominant in the farm covering 24, 19.6 and 19% of the study area, respectively.

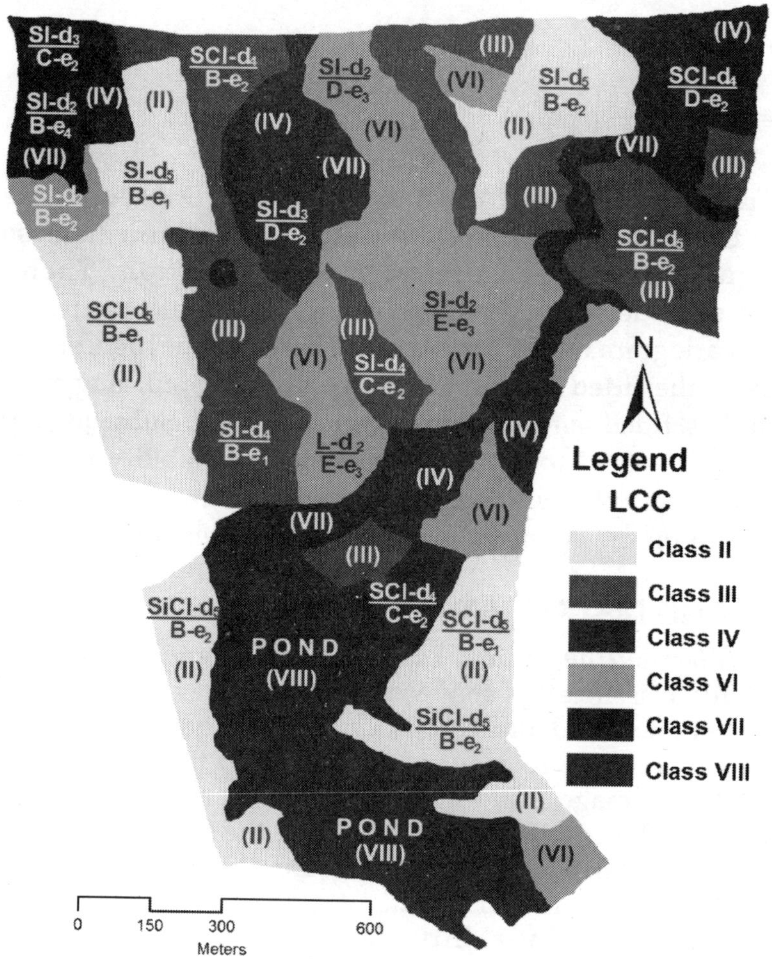

Fig. 2: Map showing the spatial distribution of land capability classes in the study area

Spatial distribution of soil texture within the study area : Soil texture is a qualitative classification tool used in both the field and laboratory to determine classes for agricultural soils based on their physical texture. The classes are distinguished in the field by the 'textural feel' which can be further clarified by separating the relative proportions of sand, silt and clay using the standard laboratory procedure *ie* Particle Size Distribution (PSD). The class is then used to determine crop suitability and to approximate the soils responses to environmental and management conditions such as drought or lime requirements. Though it is a qualitative rather than a quantitative tool but it is a fast, simple and effective means to assess the soils physical characteristics. Although the USDA. system uses 12 classes whilst the U.K.-ADAS uses just 11 but both the systems are mutually compatible.

Sand, silt and clay content varies from 41.9 – 87.1, 4.2 – 37.6 and 3.5 – 44.5 %, respectively. The spatial variation maps of sand, silt and clay indicate three clear cut gradients. Sand content increases from south towards north and north-west of the study area (Figure 3), whereas, clay content shows the opposite trend (Figure 4). The silt content increases from west to east direction in the study area (Figure 5). The sand, silt and clay maps were overlaid based on the criteria set by USDA to get the thematic map of the soil texture. Six textural classes are identified (sand, loamy sand, sandy loam, loam, sandy clay loam and clay). Sandy loam established itself as the most prominent texture covering 45.6 % (86 ha) of the study area. Coarse textured soils are evidenced in the northern side of the study area on and around the hillock, whereas, relatively heavy textured soils are found at the southern part where swampy condition prevails during rainy season.

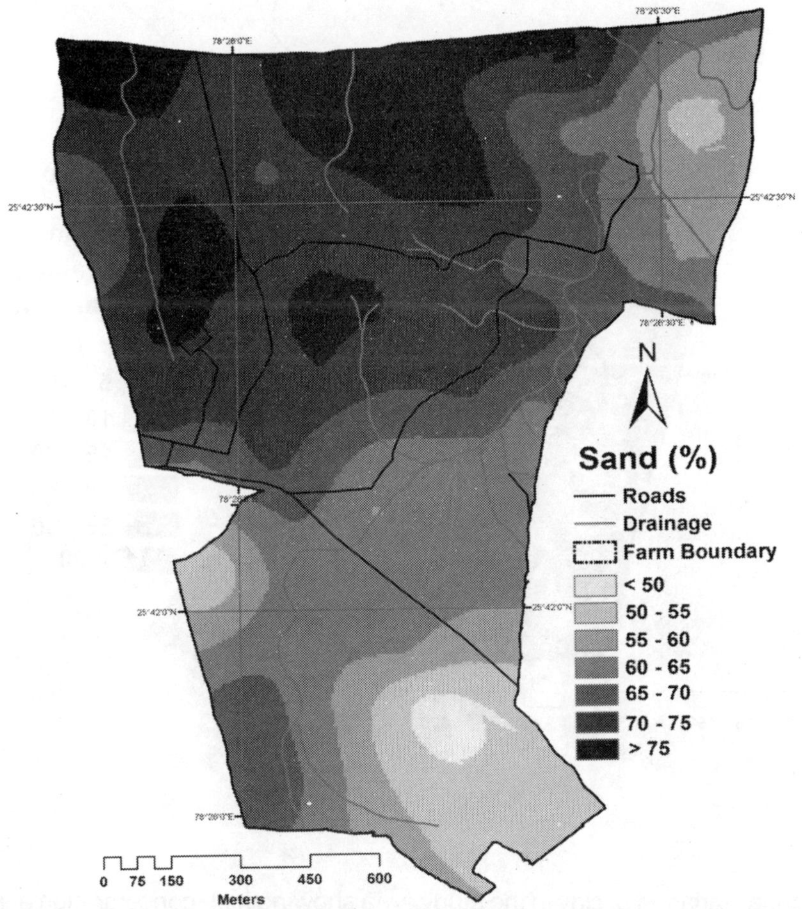

Fig. 3: Spatial variation of sand in the study area showing high concentration at north side

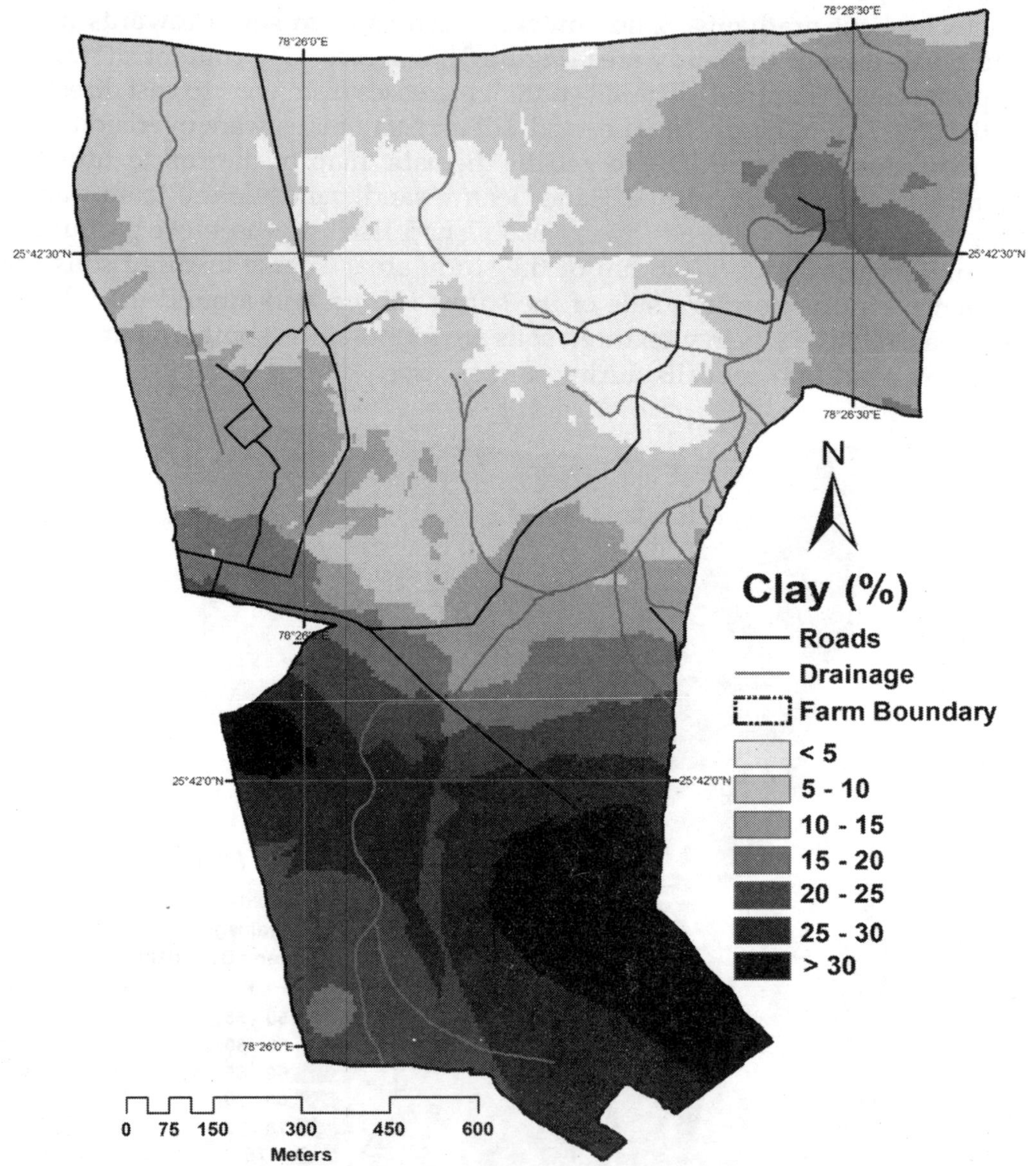

Fig. 4: Spatial variation of clay in the study area showing high concentration at south side

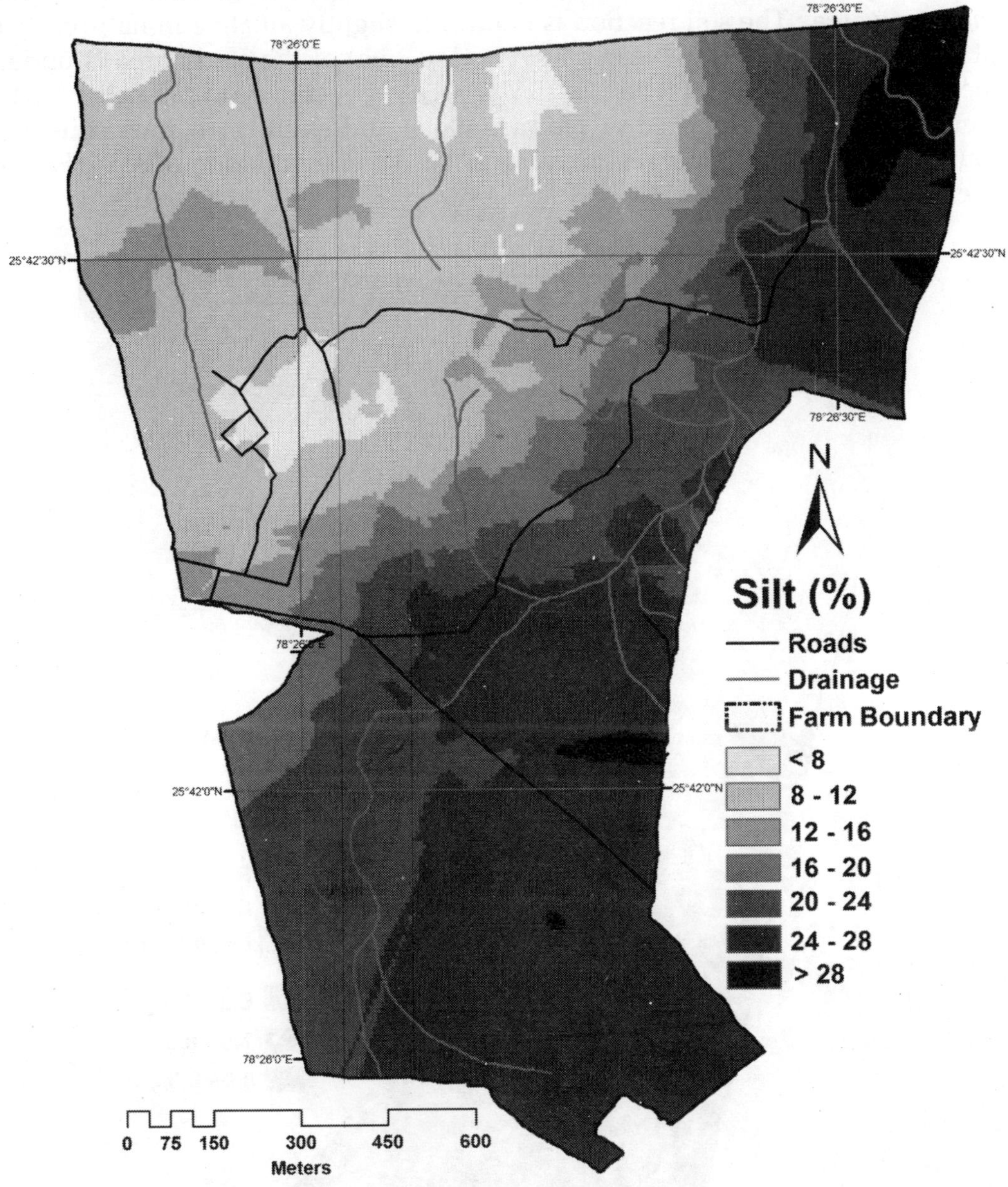

Fig. 5: Spatial variation of silt in the study area showing high concentration at east side

Soil chemical and nutrient status and their spatial distribution within the study area : The soil reaction is neutral to slightly alkaline in nature with pH varies from 6.8 – 8.9. Nearly 74.7 % (140.9 ha) of the study area is under the pH range of 7.5 – 8.0 whereas 15.9 % area is covered with pH value > 8.0. Neutral soils are evidenced at *Hectare* block and slightly alkaline soils are found at *Sirsi* block of the research farm. The spatial variation of soil pH has been shown in Figure 6.

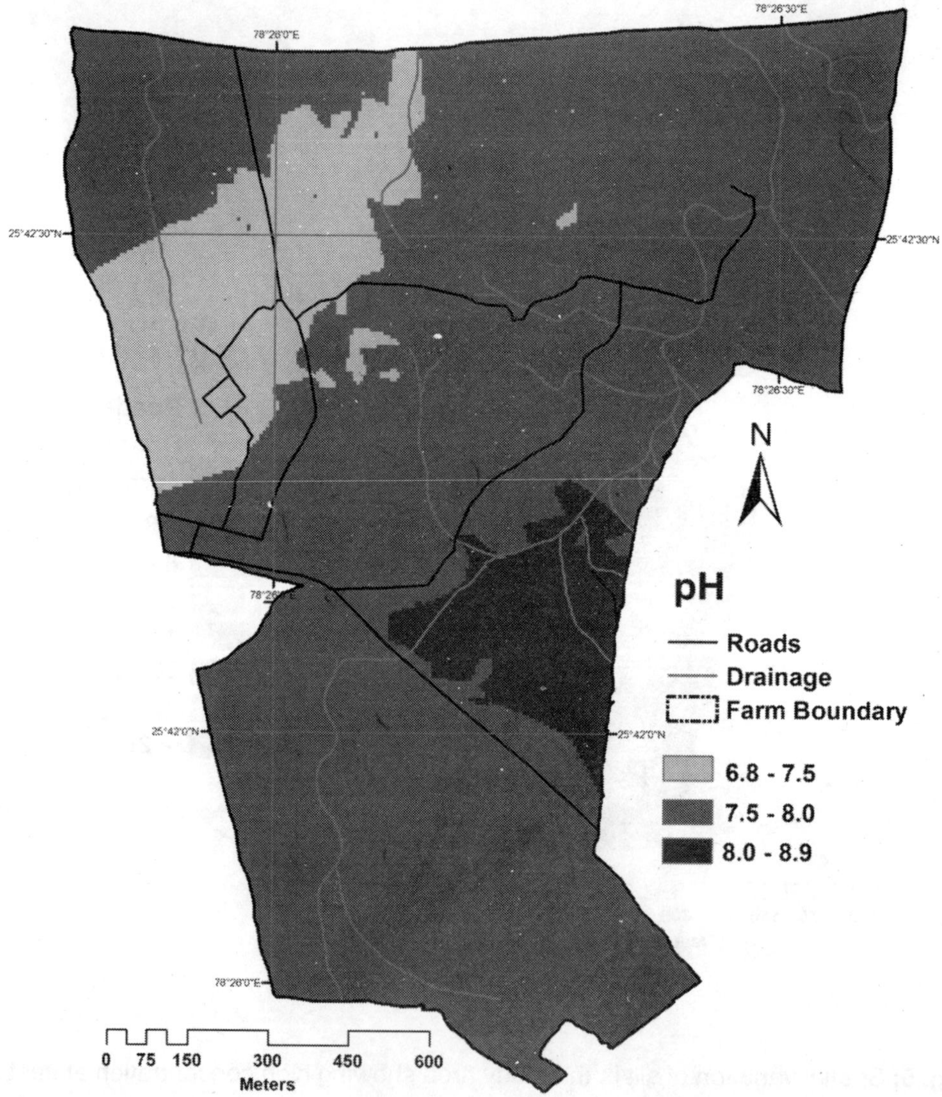

Fig. 6: Spatial distribution of soil pH in the study area showing it's neutral to alkaline nature

The soils of the farm are non saline to slightly saline as evidenced from very low EC. EC varies from 0.14 – 0.89 dSm^{-1}. A few patches at north and north western side of the farm covering 8.7 % (16.5 ha) of the study area showed slight soil salinity development (Figure 7). These areas are highly fertilized areas of *Hectare* block where experimental plots are situated.

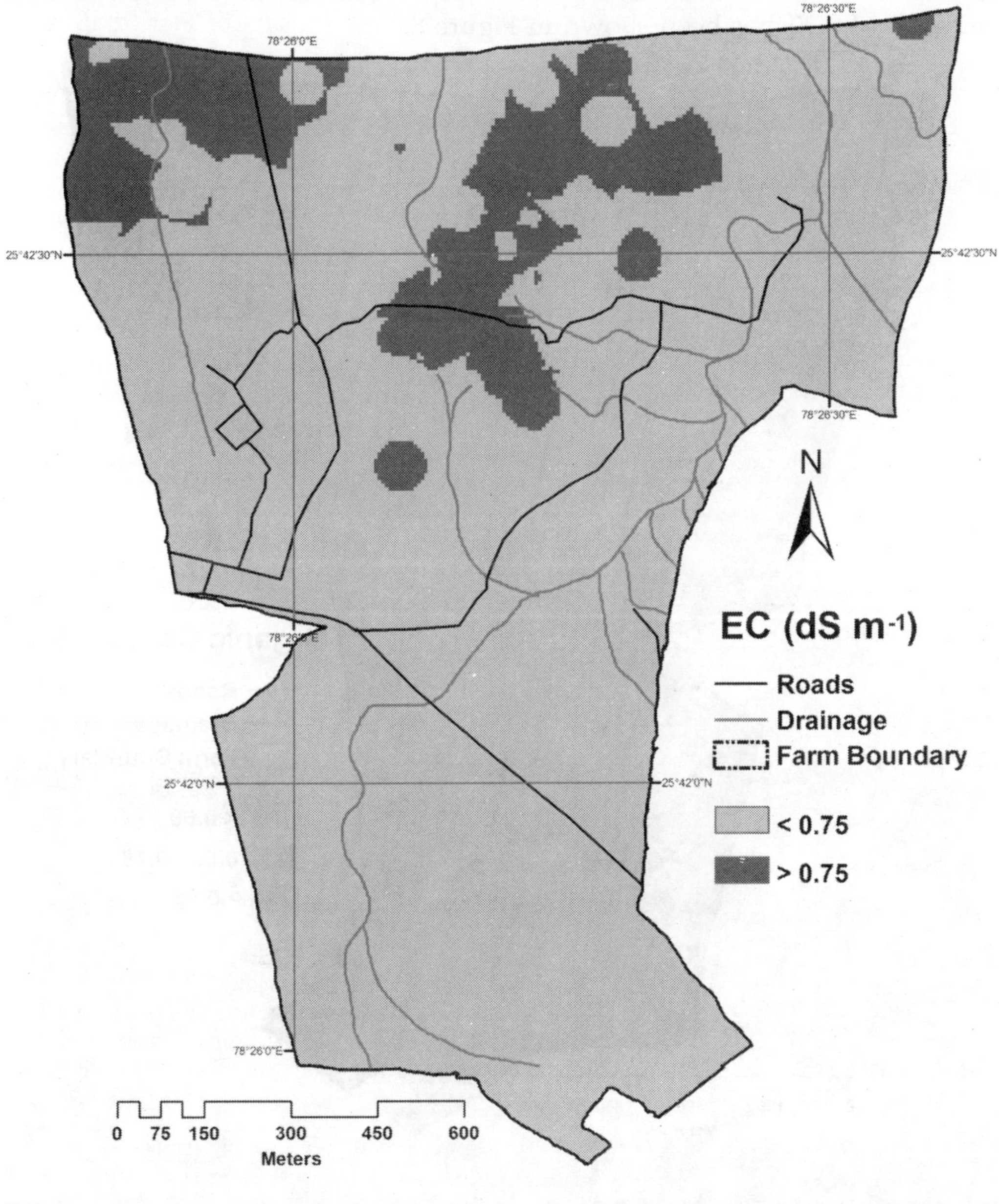

Fig. 7: Spatial distribution of soil EC in the study area showing it's non-saline nature

The soil organic carbon is low to medium with only 16.1 % (30.4 ha) area at south-western border of the farm shows high SOC content. This is due to permanent vegetation cover in depressions with low biotic interference. Low SOC content is found at the northern part of the study area covering 30.6 % (57.7 ha) area of the farm. Poor vegetation cover on the hillocks, high soil erosion and high run off at this area are responsible for low SOC. The spatial variation of SOC has been shown in Figure 8.

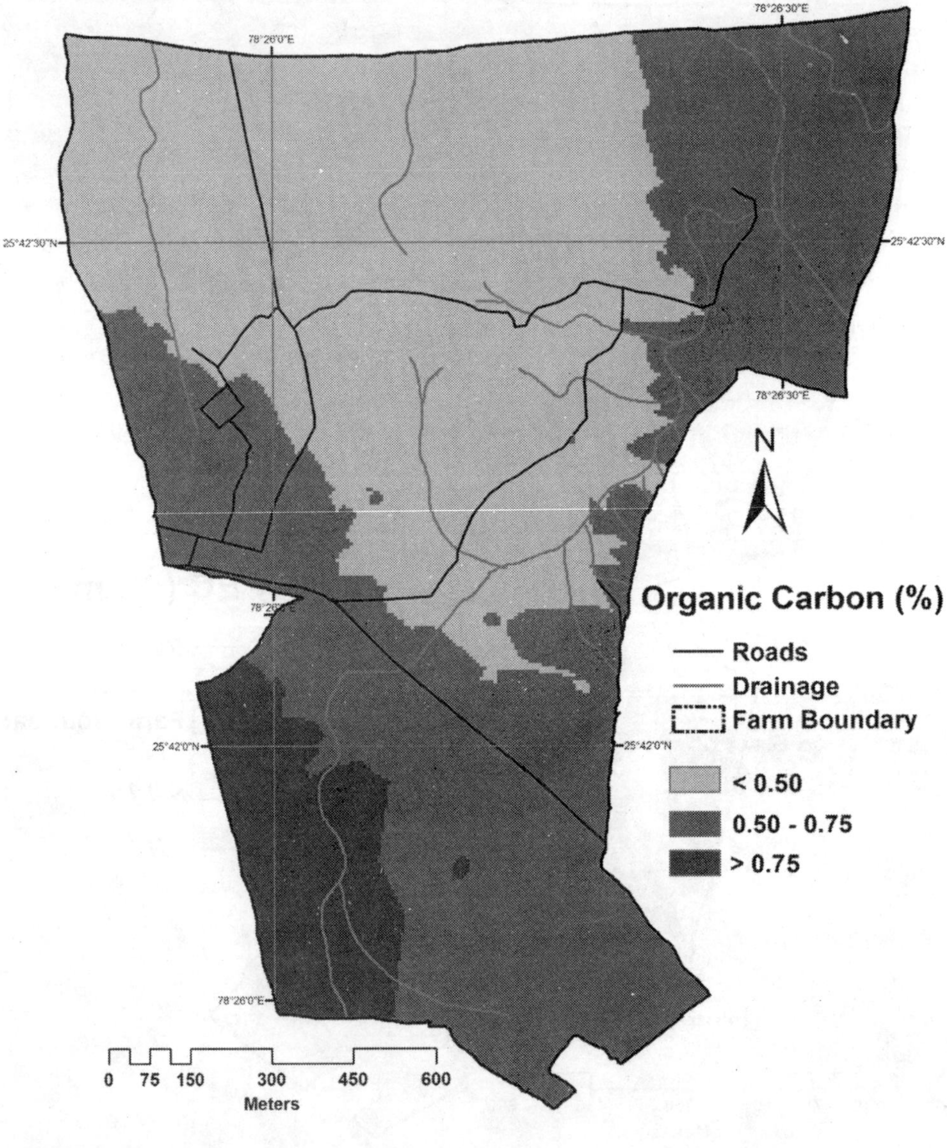

Fig. 8: Spatial distribution of soil OC in the study area indicating that medium to high OC is available at the periphery of the farm

Available P in the soil of the farm is low which varies from 3.3 – 11.5 kg ha^{-1} with average value of 6.6 kg ha^{-1}. Only at southern fringe of the farm covering 11.4 % (21.4 ha) area, medium range of available P is found (Figure 9). P deficiency is one of the main reasons for low crop yield and which can be overcome by application of sufficient amount of manures and fertilizer.

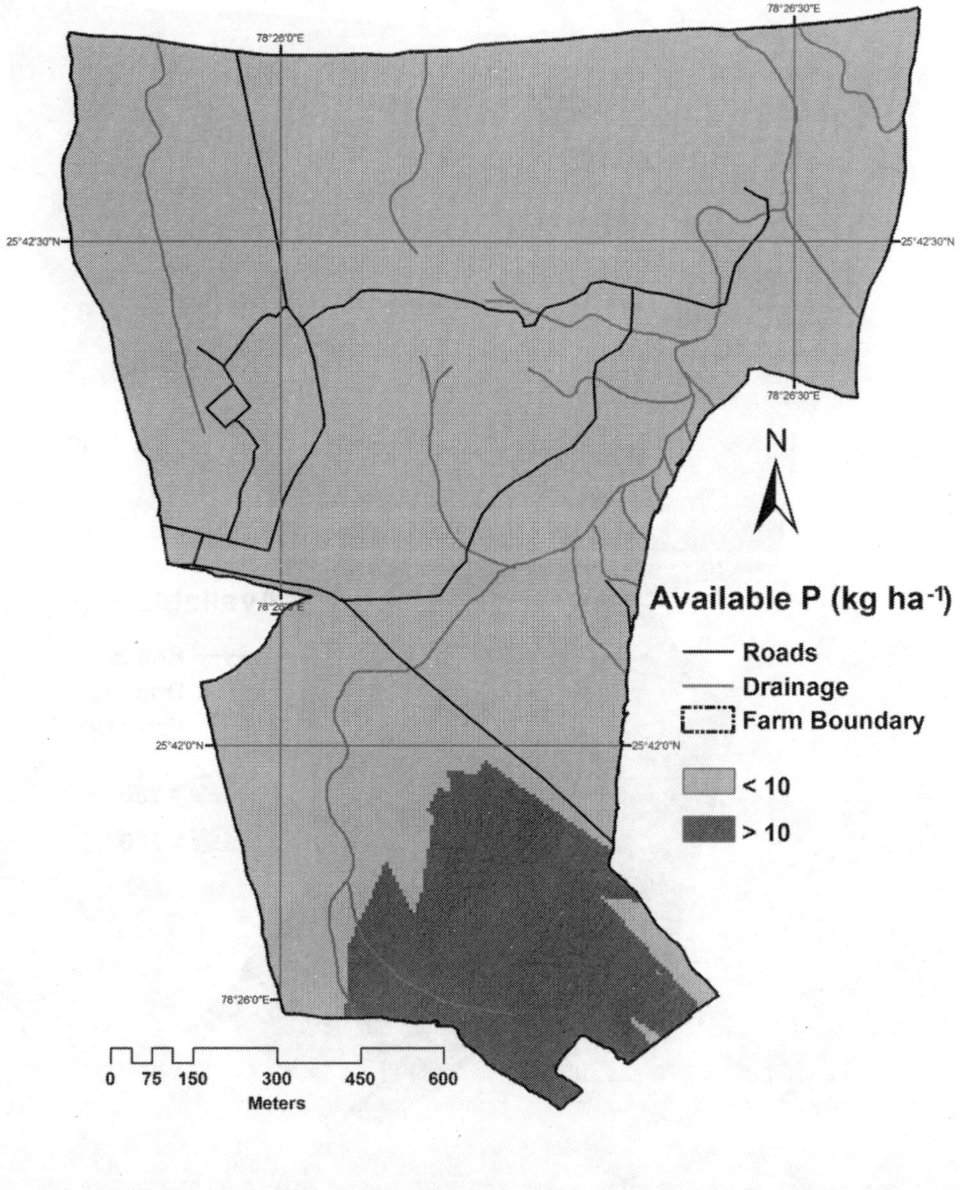

Fig. 9: Spatial distribution of soil available P in the study area indicating that medium level of P is available at the southern part of the farm

Contrary to available P, the available K content in the study area is medium to high, ranges between 131 and 759.4 kg ha^{-1} with average value of 304.5 kg ha^{-1}. High range of available K comprising 132.9 ha area is found at south and north-east side of the farm whereas medium K covering 55.8 ha area is concentrated at north and north-west side (Figure 10).

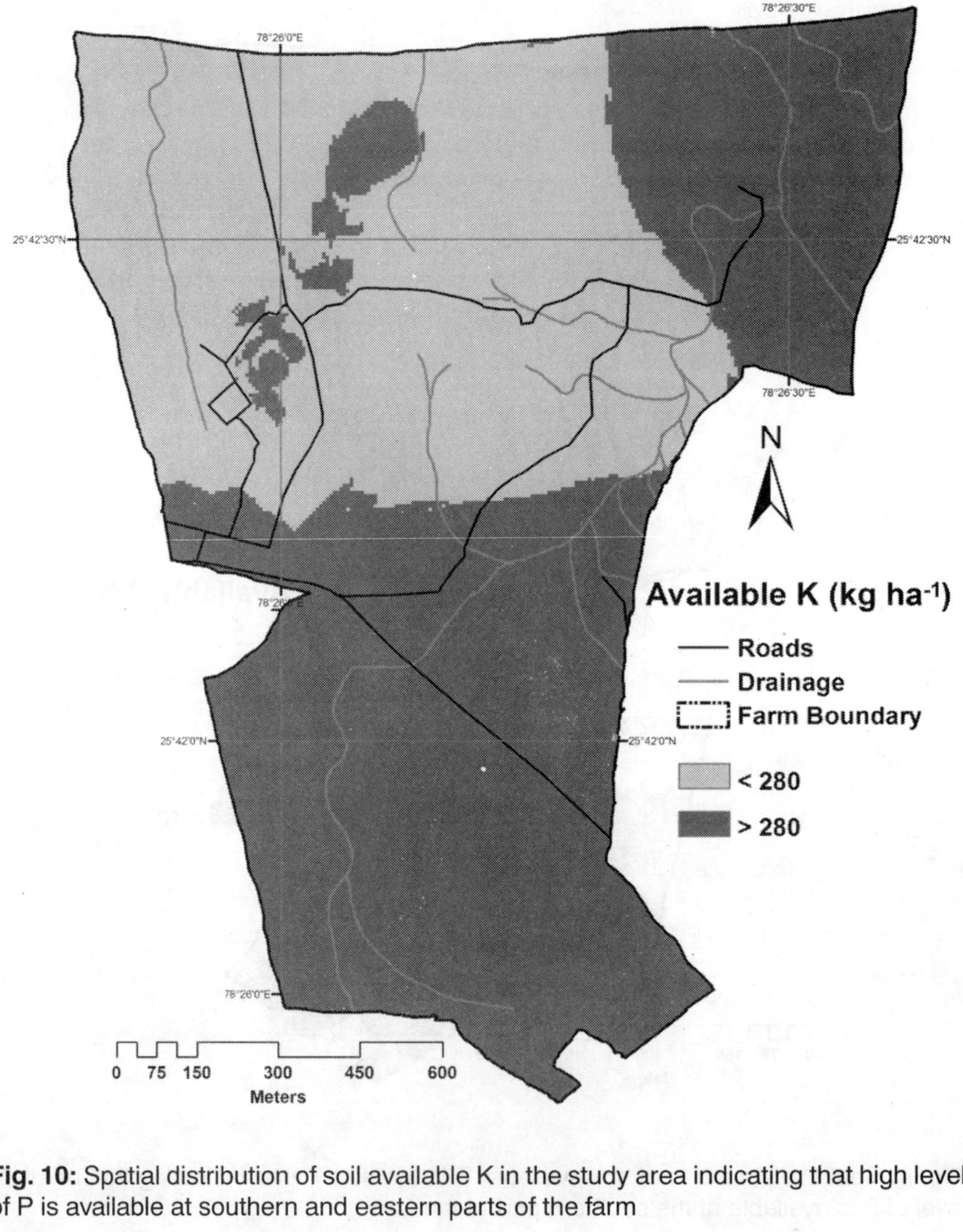

Fig. 10: Spatial distribution of soil available K in the study area indicating that high level of P is available at southern and eastern parts of the farm

Overall, the soil health of the farm is found to be normal; however the *Hectare* block, situated at north-western part of the farm is gradually degrading and need careful conservation. The *Sirsi* and *Shankar* blocks, situated at the southern and north-eastern part of the farm, are found capable for potential exploitation. Agricultural crops of high yielding varieties with integrated nutrient management will be the focus in these two blocks for natural resource conservation and sustainable return.

CONCLUSIONS

The study examined the various soil properties under different land capability classes for resource conservation and sustainable production. In a resource poor region like Bundelkhand region of central India to get maximum profit in a sustainable way planning of farm resources for judicious utilization is a must. The study clearly indicated that the area where intensive cultivation is a common practice, the natural resources is degrading gradually. Therefore to prioritise the area for resource conservation measures, this type of planning will be very helpful.

REFERENCES

Adhikary, P.P. and Biswas, H. 2011. Geospatial assessment of groundwater quality in Datia district of Bundelkhand. *Indian Journal of Soil Conservation*, 39(2): 108-116.

Bobade, S.V., Bhaskar, B.P., Gaikwad, M.S., Raja, P., Gaikwad, S.S., Anantwar, S.G., Patil, S.V., Singh, S.R. and Maji, A.K. 2010. A GIS-based land use suitability assessment in Seoni district, Madhya Pradesh, India. *Tropical Ecology*, 51(1): 41-54.

Bouyoucos, G.J. 1962. Hydrometer method improved for making particle size analysis of soils. *Agronomy Journal*, 54: 464-465.

Burgess, T.M. and Webster, R. 1980. Optimal interpolation and isarithmic mapping of soil properties, I: The semivariogram and punctual kriging. *European Journal of Soil Science*, 31(2): 315-331.

Coleman, A.L. and Galbraith, J.M. 2000. *Using GIS as an Agricultural Land Use Planning Tool*. Bulletin No. 2, Department of Crop and Soil Environmental Science, Virginia Tech. Blacksburg, Virginia.

Dent, D. and Young, A. 1981. *Soil Survey and Land Evaluation*. George Allen and Unwin Publication, London.

E. S. R. I. 2004. *ArcGIS 9*, Environmental Systems Research Institute, Redlands, California, New York, USA.

F.A.O. 1991. *Production Yearbook*. Vol. 45, FAO, Rome.

Gamma Design Software. 1988. *LLC, GS+*, Michigan, USA.

Isaaks, E.H. and Srivastava, R.M. 1989. *An Introduction to Applied Geostatistics*. Oxford University, New York.

Lakaria, B.L., Mandal, D. and Biswas, H. 2010. Permissible soil erosion limits under different landscapes of Chhattishgarh. *Indian Journal of Soil conservation,* 38(3): 148-154.

Olsen, S., Cole, C., Watanabe, F. and Dean, L. 1954. *Estimation of available phosphorus in soils by extraction with sodium bicarbonate.* USDA Circular Nr 939, USDA, Washington, D.C.

Osborn, G.H. and Johns, H. 1951. The rapid determination of sodium and potassium in rocks and minerals by flame photometry. *Analyst,* 76: 410 – 415.

Richards, L.A. 1954. *Diagnosis and improvement of saline-alkali soils.* Agricultural Handbook No. 60, USDA.

Salem, M.Z., Ageeb, G.W. and Rahim, I.S. 2008. Land suitability for agricultural of certain crops in Albostan area, Egypt. *Research Journal of Agriculture and Biological Sciences,* 4: 485-499.

Sidhu, P.S., Raj Kumar and Sharma, B.D. 1994. Characterization and classification of Entisols in different soil moisture regimes of Punjab. *Journal of the Indian Society of Soil Science,* 42: 633-640.

Sys, C., Ranst, V., Debaveye, J. and Beenaert, F. 1993. *Land Evaluation, Part-III Crop Requirements.* Agri. Publication No.7, ITC, Ghent.

Thangasamy, A., Naidu, M. V. S., Ramavatharam, N. and Reddy, R.C. 2005. Characterization, classification and evaluation of soil resources in Sivagiri micro-watershed of Chittoor district in Andhra Pradesh for sustainable land use planning. *Journal of the Indian Society of Soil Science,* 53: 11–21.

Walkley, A. and Black, I.A. 1934. An examination of Degtjareff method for determining soil organic matter and a proposed modification of the chromic acid titration method. *Soil Science,* 27: 29-38.

Wischmeier, W.H. and Smith, D.D. 1978. *Predicting rainfall-erosion losses – A guide to conservation farming.* U.S. Department of Agriculture, Agriculture Handbook No. 537.

□□□

Chapter 6

Effect of Application of Fly Ash on Yield and Physico-Chemical Properties of Soil in Paddy

T. USHA RANI AND K. PRASAD RAO

INTRODUCTION

Fly ash is chemically a ferroalumino silicate and a major solid waste of the industrial areas and it is spherical in size with medium around 4 mμm. Huge amount of fly ash is formed due to low grade of coal and inferior model of electrostatic precipitators used. Fly ash contains many nutrient elements essential for plant growth, which can be used to increase crop productivity (Page *et al.*, 1979 and Adriano *et al.*, 1980). Fly ash is a byproduct of thermal power plant which is degrading the productive land because of inappropriate disposal. It's rich in silica and contain some essential plant nutrients such as phosphorus, calcium, megnesium and micronutrient (Aitken *et al.*, 1984 and Kene *et al* ., 1995) and high concentration of toxic heavy metals such as Cu, Zn, Cd, Pb, Ni, and Cr etc (Rautaray *et al.*,2003). Fly ash production depends on quality of the coal which contains high proportion of ash that leads to 10-30% of fly ash formation (Meij, 1995; Singh and Siddiqui, 2003). Fly-ash consists of minute glass-like particles and its deposition on leaves inhibits the normal transpiration and photosynthesis of plants. It also contains various radionuclides such as uranium and thorium. Fly ash has usually high surface area and light texture due to the presence of large, porous and carbonaceous particles. Fly ash application to the soil changes the physical properties of soil such as bulk density, water holding capacity, hydraulic conductivity and particle size distribution. Fly-ash also affects the physicochemical characteristics of soil because it is generally very basic, rich in various essential and non-essential elements, but poor in both nitrogen and available phosphorus (Dharmendra *et al.*, 2007).

Anurag Singh *et al.* (2008) reported that application of fly ash at higher levels caused severe deficiency of N in soil, which becomes an important factor

Regional Agril. Research Station, ANGRAU, Anakapalle- 531001, Andhra Pradesh, India for suppressed growth and reduction in yield of crops. The decreased yield may be due to accumulation of heavy metals in different plant parts. Fly ash application in small quantities are suitable for better management of crops and its use is uncommon due to non essential elements that adversely affects crops, soil and ground water quality. In view of the large number of coal and steel industries in India, it was considered worthwhile to study the possibilities of agricultural utilization of fly ash. This paper presents the application of fly ash on grain and straw yields of paddy.

MATERIALS AND METHODS

A field experiment was conducted on clay loam soils during *kharif* season from 2004-2006 on Paddy at Sugarcane Research Station, Vuyyuru, Andra Pradesh. The paddy variety 'MTU2077' was used in this study. Soils of the experimental site was clay in texture, well drained, very deep, non-calcareous, neutral in reaction (pH 7.62), normal in EC(0.26 dSm^{-1}), low in available nitrogen 216kg ha^{-1} high in available phosphorus (48kg ha^{-1}) and potassium(513kg ha^{-1}). A total of five treatments *viz.*, T_1=120kg N + 40kg P_2O_5 + 40kgK_2O ha^{-1}, T_2 = T_1 + Fly ash @ 20t ha^{-1} at basal, T_3 = Recommended dose of fertilizer (80kg N + 40kg P_2O_5+ 40kg K_2O ha^{-1}), T_4 = T_3 + Fly ash @ 20t ha^{-1} at basal, T_5 =75% recommended dose of fertilizer + Fly ash @ 20t ha^{-1} at basal were tested in randomized block design with three replications. Twenty one days old rice seedlings of '*MTU2077*' were transplanted by placing two or three seedlings per hill with a crop geometry of 20cm x15cm. Fly ash was applied as per the treatments before transplanting of rice seedlings. Nitrogen was applied in three equal doses *viz.*, after puddling, tillering stage and panicle initiation stage. Phosphorus was applied as basal dose in the form of super phosphate. Potash was applied in the form of muriate of potash in two equal doses as basal at planting and at panicle initiation stages. Soil physical and chemical properties of soils were analyzed by following standard procedures Tandon (1995). The crop was harvested at maturity and the yield components were computed using the standard procedures. The data was statistically analysed by following standard statistical method (Panse and Sukhatme, 1978).

RESULTS AND DISCUSSION

Initial chemical properties of soil and fly ash: An initial chemical property of soil and fly ash is presented in Table 1. Soils of the experimental site was clay in texture, well drained, very deep, non-calcareous, neutral in reaction (pH 7.62), normal in EC(0.26 dsm^{-1}), low in available nitrogen 216kg ha^{-1}, high in available phosphorus (48kg ha^{-1}) and potassium(513kg ha^{-1}). The fly ash collected from thermal power station, Vijayawada contains slightly alkaline in pH, normal in conductivity, traces of heavy metal content and also sufficient levels of major nutrients.

Yield parameters of paddy: The number of tillers was maximum with the application of fly ash @ 20t ha^{-1} along with 120kg N + 40kg P_2O_5 + 40kg K_2O ha^{-1}. Significant increase in number of tillers by increasing levels of fertilizers might be due to increased availability of nutrients coupled with favourable soil conditions by integrated use of fly ash. Plank and Martens (1974) Page *et al* .,(1979) reported that application of flyash along with additional dose of nitrogen increased the number of tillers in paddy.

Grain yield: The grain yields of rice was maximum with the application of fly ash @ 20 t ha^{-1} along with 120kg N, 40kg P_2O_5 and 40kg K_2O ha^{-1} (79q ha^{-1}) which was significantly higher over the application of inorganic fertilizers alone (74.4q ha^{-1}). Similar trend was recorded with the application of fly ash @ 20t ha^{-1} along with recommended dose of fertilizer (80kg N + 40kg P_2O_5 and 40kg K_2O ha^{-1}) 75 q ha^{-1} over application of recommended dose of fertilizers alone (71.9q ha^{-1}). But application of 75% recommended dose of fertilizer along with fly ash @ 20 t ha^{-1} recorded significant lower yields (68.8 q ha^{-1}) over all other treatments. According to Lee *et al* (2006) application of fly ash did not show any toxic effects and provided good supplement to other soil amendments to improve the nutrient balance in paddy soils but did not result in an excessive uptake of heavy metals in the submerged paddy soils and thus increased the yield of the paddy crop. Significant higher grain and straw yields were also observed with the application of fly ash by Sudhir K Sharma *et al* (2006). On the other hand, moderate rates of fly-ash application (2–4% w/w) had a beneficial effect on the dry-matter yield of paddy, but a higher level (8% w/w) had a significant depressing effect (Sikka and Kansal, 1995). Fly ash application @ 0-80 t ha^{-1} increased the yield of paddy from 61.82 q ha^{-1} to 63.58 q ha^{-1} (Md. Wasim Aktar, 2011). In rice- groundnut cropping system application of fly ash @ 10t ha^{-1} along with inorganic fertilizers increased the grain yield over the application of chemical fertilizers alone. The increase in yield under fly ash at 25 t ha^{-1} might be due to increase in nutrient availability to the crops (Mishra *et al.*, 2007 and Yeledhalli *et al.*, 2008)

Straw Yield: The straw yield of rice was maximum with the application of fly ash @ 20t ha^{-1} along with 120kg N + 40kg P_2O_5 + 40kg K_2O ha^{-1} (92q ha^{-1}) which was significantly higher over the application of inorganic fertilizers alone (87.5 q ha^{-1}). Application of recommended dose of fertilizer along with fly ash @ 20t ha^{-1} recorded higher straw yield (86 q ha^{-1}) over the application of recommended dose of fertilizers alone (83.9 q ha^{-1}). But application of 75% recommended dose of fertilizer along with fly ash @ 20t ha^{-1} recorded the significant lower yields over all others. Significantly higher grain and straw yields might be due to the supply of silica resulting in favorable physical environment leading to better aeration, root activity and nutrient adsorption would have resulted in higher grain and rice yields of rice (Mattee et al.,1990).

Effect of fly ash on soil properties: Physical and chemical properties of soil as influenced by application of fly ash to paddy field is presented in Table 3.

Physical properties: Fly ash amended soils have lower bulk density, higher water holding capacity and lower hydraulic conductivity. Fly ash addition generally decreased the bulk density of the soil which intern improves soil porosity and better workability and enhances water retention capacity. The cementing effects of fly ash could possibly impede root development by creating hard areas near the soil fly ash interface in a closed ash disposal system. Non judicial use may lead to deterioration of soil texture and structure mainly in the upper soil layer, surface crust formation impeding the water intake capacity of the soil (Sudhir K Sharma *et al.*, 2006).

Chemical properties: The concentration of all elements was higher in fly ash than in soil except nitrogen. It increased the organic carbon content but not increases the water availability to plants. It obstructs the development of root system by forming the hard layer. The pH of soil was influenced significantly by ash application and there was a marked reduction in pH with increase in dose of fly ash. The pH alone does not determine the role of fly ash as amendment but under certain situations composition of fly ash play a useful role. EC decreases after harvesting of Paddy and this may be due to leaching of salts along with water as reported by Kalra *et al* (2000). Fly ash has the tendency to accumulate B, Se. Though fly ash contained appreciable amount of P but its application to soil decreased the available P content. In rice based cropping system application of fly ash @ 10t ha^{-1} along with inorganic sources of fertilisers improved physical properties of soil through decrease in bulk density, increasing water holding capacity and porosity.

Uptake of nutrients : Availability of S, Fe, B, and Mo also increased whereas K and Mn remained almost unchanged. The application of fly ash increased Si, P and K uptake by the rice plants but did not result in an excessive uptake of heavy metals in the submerged paddy soil. Uptake of N, P, K by rice increased with fly ash addition (Table 4 and 5). Sikka and Kansal (1995) reported addition of fly ash resulted in a significant increase in the contents of N, S, Ca, Na and Fe and a significant decrease in P and Zn contents of rice. The amount of available B increased to a maximum of 2.57 mg kg^{-1}. The B content of the rice plants increased to a maximum of 52–53 mg kg^{-1} (Lee *et al.*,

2006). In rice based cropping system application of fly ash @ 10t ha^{-1} along with inorganic sources improved the physic- chemical properties of the soil and also the uptake of the soil.

CONCLUSIONS

Fly ash could be a good supplement to other inorganic soil amendments to improve the nutrient balance in paddy soils. The high concentration of elements like K, Na, Zn, Ca, Mg and Fe in fly ash increases yield of paddy. Due to fine nature of fly ash it improves the water holding capacity of sandy soil removing the compaction of clay soils. Fly ash improves the physical and chemical properties of soil as well as biological properties of soils which improve the productivity of the crops. Application of fly ash particularly un-weathered soils shows a tendency of accumulating elements like B, Mo, Se and Al whose toxic levels are responsible for reduction in crop yields and consequently influence on animal and human health. However, all these to be carefully assessed while recommending application of fly ash.

Table 1: Characteristics of initial soil at experimental site and fly ash

Soil			Fly ash		
Parameter	Unit	Value	Parameter	Unit	Value
pH		7.62	pH		8.2
EC	dsm^{-1}	0.26	EC	dSm^{-1}	0.12
OC	%	0.82	P	%	0.006
Ava. N	kg ha^{-1}	216	K	%	0.37
Ava. P	kg ha^{-1}	48	Zinc	ppm	20
Ava. K	kg ha^{-1}	513	Cupper	ppm	12
Zinc	mg kg^{-1}	1.75	Iron	ppm	2425
Cupper	mg kg^{-1}	7	Manganese	ppm	67
Iron	mg kg^{-1}	12	Cadmium	ppm	Trace
Manganese	mg kg^{-1}	50	Nickel	ppm	8.8
			Cobalt	ppm	10.4

Table 2: Effect of fly ash on yield and yield parameters of paddy

Treatment	No.of tillers m^{-2}	Plant height (cm)	Grain yield (q ha^{-1})	Increase over T_5 (%)	Straw yield (q ha^{-1})	Increase over T_5 (%)
T_1	515	112.4	74.43	8.3	87.5	9.4
T_2	528	120.8	79.00	14.9	92.0	15.0
T_3	456	113.8	71.93	4.62	83.9	4.9
T_4	470	113.0	75.00	9.1	86.0	7.5
T_5	425	112.3	68.75	--	80.0	--
Mean	479	114.5	73.82		85.9	
CD	46.0	10.7	7.80		8.2	
S.E.m	20.1	3.6	2.80		3.0	
CV	9.2	10.2	9.70		9.6	

Table 3: Soil physical and chemical properties of soil as influenced by application of fly ash after harvesting of paddy

Treatment	BD ($Mg\ m^{-3}$)	HC ($cm\ d^{-1}$)	FC (%)	AW ($mm\ cm^{-1}$)	Na (ppm)	Ca	pH	EC (dsm^{-1})	OC (%)	N	P_2O_5	K_2O ($kg\ ha^{-1}$)
T_1	1.5	6.7	15.7	1.8	925.0	1810.0	7.4	0.3	0.8	215	46	511
T_2	1.4	6.4	17.2	1.6	917.0	1780.0	7.3	0.2	0.7	211	41	512
T_3	1.5	6.6	15.1	1.7	915.0	1710.0	7.5	0.2	0.8	211	44	505
T_4	1.4	6.5	16.5	1.6	908.0	1679.0	7.5	0.2	0.8	209	41	509
T_5	1.4	6.6	16.9	1.7	906.0	1688.0	7.4	0.2	0.8	204	39	511
Mean	1.4	6.6	16.3	1.7	914.0	1733.0	7.4	0.2	0.8	212	42	510
S.E.m	0.1	0.3	0.6	0.1	30.0	55.7	0.2	---	0.0	7	1.34	16.3
CD	0.2	0.7	1.6	0.2	90.4	167.0	0.7	NS	0.1	20.9	4.1	48.8
CV	9.1	9.0	9.4	9.8	9.0	9.0	9.6	9.8	9.2	10.2	9.8	11

Table 4: Effect of fly ash on micronutrients ($mg\ kg^{-1}$ soil) of soil after harvesting of Paddy

Treatment	Zn	Cu	Fe	Mn	B
T_1	1.86	7.2	12.2	48.3	1.89
T_2	1.98	8.3	13.4	50.0	2.57
T_3	1.77	7.1	12.1	47.9	1.56
T_4	1.91	7.9	12.8	49.9	2.10
T_5	1.89	7.6	12.2	49.0	2.20
Mean	1.88	7.6	12.5	49.0	2.06
S.E.m	0.07	0.23	0.43	1.60	0.07
CD	0.17	0.69	1.3	4.8	0.20
CV	9.1	9.4	9.3	10.1	10.7

Table 5: Effect of fly ash on uptake of nutrients by Paddy

Treatment	Nitrogen	Phosphous	Potassium
		($kg\ ha^{-1}$)	
T_1	49.80	16.23	18.20
T_2	57.90	19.10	20.95
T_3	53.78	17.65	19.80
T_4	55.25	18.72	20.18
T_5	50.17	17.25	18.79
Mean	54.98	17.79	19.52
S.E.m	1.9	0.53	0.67
CD	5.7	1.6	2.0
CV	10.1	10.2	10.6

REFERENCES

Anurag Singh, Rajesh Kumar Sharma, Shashi Bhushan Agarwal. 2008. Effects of fly ash incorporation on heavy metal accumulation growth and yield of beta vulgaris plants. *Science Direct Bioresource Technology*, 99 : 7200-7207.

Aitken R.L., Bell L.C. 1984. Plant uptake and phytotoxicity of boron in Australian flyashes. *Plant & Soils*. 84: 245-257.

Adriano D.C, Page A.L, Elseewi A.A, Chang A.C and Satraughan I A 1980 Utilisation and disposal of flyash and other coal residues in terrestrial ecosystem, *A Rev J.Envir Qual.*, 9 333-344.

Dharmendra, K., Gupta., Upendra, N., Rai, Rudra, D., Tripathi. and Masahiro. 2007. Impacts of fly-ash on soil and plant responses. *Journal of Plant Research*, 401-409.

Kene, D, R., Lanjewar, S.A. and Ingole, B.M. 1995. Effect of application of flyash on physic-chemical properties of soils. J. *Soils and Crops*, 1: 11-18.

Lee, H, Ha, H.S., Lee, C.H., Lee, Y.B and Kim, P.J. 2006. Fly ash effect on improving soil properties and rice productivity in Korean paddy soils. *Bioresource Technology*. 97(13): 1490-1497.

Lee., Ho Sung Ha., Chang Hoon Lee., Yong Bok Lee. and Pil Joo Kim. 2006. Fly ash effect on improving soil properties and rice productivity in Korean paddy soils *Bioresource Technology*, 97(13): 1490-1497.

Matte, S.S, Mukhopadhyay, T.M., Gupta, S.K. and Banerjee, S.K .1990. Evaluation of fly ash as a useful material in agriculture. *J. Ind. Soc. Soil Sci.*, 38 : 342-344

Md. Wasim Aktar. 2011. Fly ash use in Agriculture Environment articles Amazines. *Environment Research*, 3: 123-127

Meij, R., Swaine D.J. and Goodarzi, F . 1995. The distribution of trace elements during the combustion of coal. *In*: (Eds.), Environmental Aspects of Trace Elements in Coal. Kluwer Academic Publication, Dordrecht, The Netherlands, pp111–127.

Mishra,M., Sahu, R.K. and Padhy, R.N.2007. Growth, yield and elemental status of rice (*Oryza sativa*) grown in fly ash amended soils. *Ecotoxicology*, 16(2): 271-278.

Naveen Kalra, R., Harit, C. and Sharma, S. K. 2000. Effect of fly ash incorporation on soil properties of texturally variant soils. *Bioresource Technology*, 75(1): 91-93.

Page, A.L., Eiseewi, A.A. and Straughan, I.R. 1979. Physical and chemical properties of fly ash from coal fired power plants with reference to environmental impacts. *Residue review*, 71: 83-120

Panse, V.G. and Sukhatma.1978. Statistical methods for Agricultural works, II Edn. ICAR, New Delhi India.pp145.

Plank, C.O. and Martens, D.C. 1974. Boron availability as influenced by application of flyash to soil. *Proc. Soil Sci. Am.*, 38: 974-977.

Rautaray, S.K., Ghosh, B.C. and Mitra, B.N. 2003. Effect of fly ash, organic waste and chemical fertilizers on yield, nutrient uptake and heavy metal content and residual fertility in a rice–mustard cropping sequence under acid lateritic soils. *Bioresour Technol.*, 90:275–283.

Siddarth Singh, Dadhibal Prasad Gond, Amit Pal, Tewary, B.K. and Sinha. 2011 Performance of several crops grown in fly ash amended soil. *World of Coal Ash Conference*, May 9-12 2011.

Sikka,R. and Kansal, B.D. 1995. Effect of fly-ash application on yield and nutrient composition of rice, wheat and on pH and available nutrient status of soils. *Bioresource Technology*, 51(2-3): 199-203.

Singh, L.P.and Siddiqui, Z.A. 2003. Effects of fly ash and Helminthosporium oryzae on growth and yield of three cultivars of rice. *Bioresour Technol.*, 86: 73–78.

Sudhir K Sharma and Naveen Karla. 2006. Effect of fly ash incorporation on soil properties and productivity of crops A review. *Journal of Scientific and Industrial research*, 65: 383-390.

Tandon Ed. L.S. 1995. Methods of Analysis of soils, plants, waters and fertilizers. Fertiliser development and consultation organization. New Delhi.

Tiwari, S., Kumari, B., Singh, S.N., 2007. Evaluation of metal mobility/immobility in fly ash induced by bacterial strains isolated from the rhizospheric zone of Typha latifolia growing on fly ash dumps. *Bioresour Technol.*, doi:10.1016/j.biortech.2007.02.01.

Yeledhalli, N.A., Prakash, S. S., Ravi, M. V. and Narayana Rao, K. 2008. Long-term effect of fly ash on crop yield and soil properties. *Karnataka J. Agric. Sci.*, 21(4):507-512.

❑❑❑

Chapter 7

Application of SWAT Model in Lower Himalayan Region of India

V. K. BHATT, A. K. TIWARI AND D. R. SENA*

INTRODUCTION

The Soil and Water Assessment Tool is a river basin model that was developed for the USDA Agricultural Research Service, by Blackland Research Center in Texas (http://www. brc. tamus. edu/blackland/). Soil and Water Assessment Tool (SWAT) has been extensively used since 1993 mainly by hydrologists for watershed hydrology related issues (Srinivasan *et al.*, 1998; Santhi *et al.*, 2001; Cao *et al.*, 2006; Schuol and Ambaspour, 2007; Keshta *et al.*, 2009). SWAT a river basin model that is currently one of the worlds leading spatially distributed hydrological models. SWAT as a distributed parameter continuous time model developed originally by the USDA Agricultural Research Service (ARS) and Texas A&M

University (Arnold *et al.*, 1996 & 1998). It divides a catchment into smaller discrete calculation units for which the spatial variation of the major physical properties are limited, and hydrological processes can be treated as being homogeneous. The total catchment behaviour is a net result of several small sub-basins. The soil map and land use map within sub-basin boundaries are used to generate a homogeneous physical property, *i.e.*, Hydrological Response Unit (HRU). The water balance for HRUs is computed on a daily time step. Hence, SWAT subdivides the river basin into units that have similar characteristics in soil and land cover and that are located in the same sub-basin. SWAT model has been tested for predicting runoff (Srinivasan & Arnold, 1994). However, application of SWAT for prediction of runoff from micro-catchment is limited. Data of several years is required for development of a

Central Soil and water Conservation Research and Training Institute, Research Centre Chandigarh, India

* Central Soil and water Conservation Research and Training Institute,Dehradun Uttarakhand, India

long term plan for homogeneous watersheds. The hydrologic component of SWAT is based on the following water balance equation:

$$SW_t = SW + \sum_{i=1}^{t}(R_i - Q_i - ET_i - P_i - QR_i) \quad \text{... (1)}$$

where SWt is the final soil water content (mm),SW is the water content available for plant uptake, defined as the initial soil water content minus the permanent wilting point water content (mm), t is time in days, R is rainfall (mm), Qi is surface runoff (mm), ETi is evapotranspiration (mm), Pi is percolation (mm) and QRi is return flow. SWAT incorporates some of the most common hydrological equations for the simulation of flow. For the accurate implementation of these equations, detailed input data are needed. Of significant value to the simulation are the digital elevation model (DEM) of the watershed, the soil and land use data and the climatic data of the area. The importance of land uses in the operation of the model lies mainly in the computation of surface runoff with the help of the SCS curve. The model includes in its database 102 different land use types, with each one assigned to a CN-II value (Curve Number for hydrological condition II). The user is required to link each of the land uses that appear in the watershed, to the ones that the model can identify. The success of the simulation depends highly on the accuracy of soils and land uses.

In India, like most of the developing countries in the world, the majority of the basins are either sparsely gauged or not gauged at all. This necessitates the application of a robust model for estimation of runoff and sediment. In the present study the Arc View - SWAT interface (AVSWAT-X version 2005) was used to delineate the catchment boundary and the burning option to derive drainage network of choe gauging watershed. The main objective was to evaluate the applicability and performance of the model in predicting yearly water yield. In order to achieve the objectives sensivity analysis, calibration and validation of the model were essemtial steps for model testing as well as extending the application area.

Study area: The Soil and Water Assessment Tool (SWAT) was applied to mixed forest micro-watershed of lower Hamalayan region of India. This watershed is named as Choe gauging watershed and having an area of 21 ha. The watershed is located at an altitude of 34. 74^0N and longitude of 76. 86^0E. It is located at an elevation of 350 m. Land use of watershed comprises deciduous forest, range forerst and water. Nine years (1973-81) monthly data was taken up to model the hydrological output. The model was calibrated for the period 1973-1978 using the parameters based on sensitivity analysis and validated for the period 1979-1981. Loamy sand is the dominant soil types in the area. Meteorological station is located at near the watershed.

MATERIALS AND METHODS

SWAT is a physically based hydrologic model which requires physically based data. Obtaining physical based data for hydrological modeling is often difficult, even in developed countries where data of high quality are generally collected and analyzed. In this study input data was collected from various sources. Meteorological input data include daily precipitation, maximum and minimum air temperature, wind speed and relative humidity. Spatial input data used were Digital Elevation Model (DEM), Land use/land cover and soil. DEM and subwatersheds are shown in Figure 1. Land use ans soil data were incorporated into SWAT model and used for reclassifying the land use and soil data. Various GIS data preprocessor modules which involve watershed delineation, input map characterization and processing ,stream and outlet definition, the computation of the geomorphic parameters, and characterization of the lans use/land cover and soil were developed in the course of modeling the catrchment. The simulation option of the rainfall runoff modelling was performed on the basis of previous modeling techniques. These include using a curve number method for calculating surface runoff (SCS,1972), a first order Marcov Chain Skewed Normal to determine rainfall distribution,computing potential evaporation by using Penman Monteith method and Muskingum routing method for routing water through the channel networks.

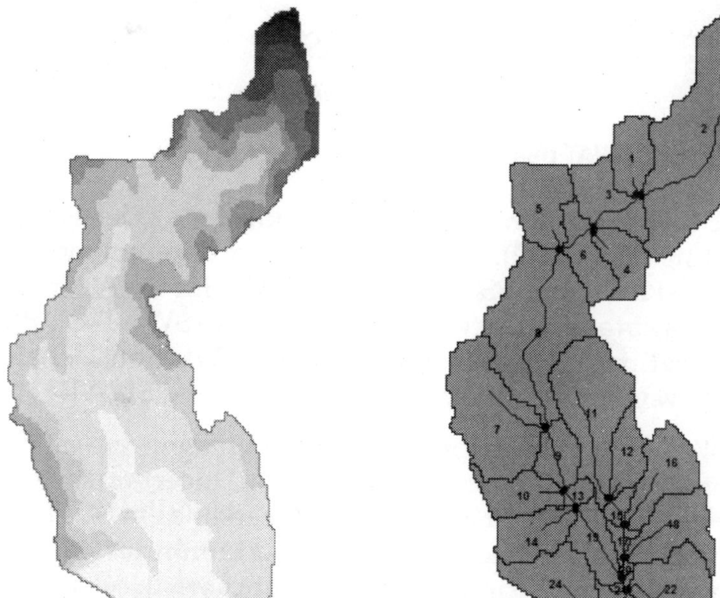

Fig. 1: DEM and sub-basins delineated for Choe watershed

After simulation process sensitivity analysis was performed involving the parameters, deep aquifer percolation (rchrg_dp), minimum depth of water in soil for base flow to occur, (GWQ_MIN), initial SCS -CN for AMC-II (CN2),available water capacity of soil layer (SOL_AWC) etc (Figure 1) manually and by auto-calibration using SWAT CUP software version 3.1.3. Curve number and base flow parameter (Alpha_BF and Alpha_BNK) were found most sensitive parameter for the study watershed. Ranges of parameters used for modeling are shown in Table 1 Calibration and validation were performed for the period 1973-78 and 1979 -81, respectively.

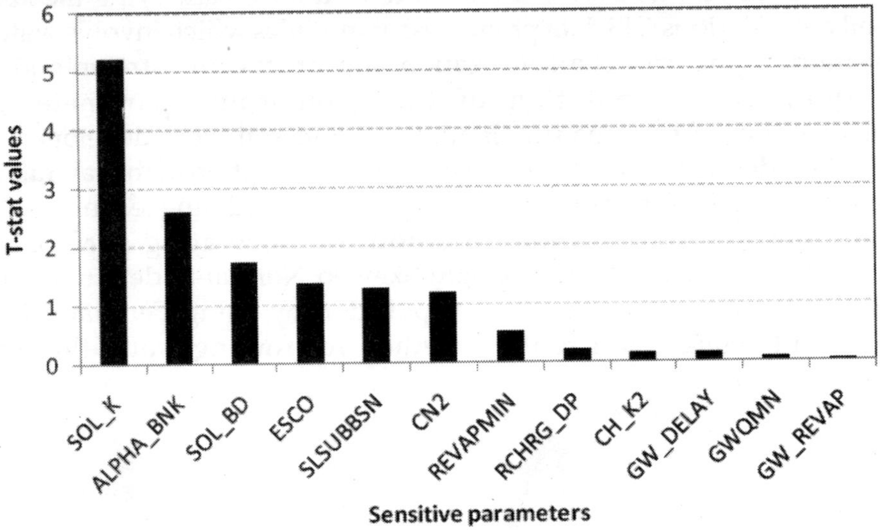

Figure 2 : SWAT parameters considered for sensitivity analysis

SWAT-CUP 2009_version was used to calibrate the model using Sequential uncertainty fitting (SUFI ver2). SUFI-2 is one of five different modules (SUFI2, ParaSol, GLUE, MCMC and PSO) that are linked with SWAT in the package called SWAT Calibration Uncertainty Programs (SWAT-CUP). Its main function is to calibrate SWAT and perform validation, sensitivity and uncertainty analysis for a watershed model created by SWAT.

Various SWAT parameters for estimation discharge were estimated using the SUFI-2 program (Abbaspour *et al.*, 2007). Uncertainty is defined as discrepancy between observed and simulated variables in SUFI-2 where it is counted by variation between them. SUFI-2 combines calibration and uncertainty analysis to find parameter uncertainties while calculating smallest possible prediction uncertainty band. Hence, these parameters uncertainty reflect all sources of uncertainty, *i.e.*, conceptual model, forcing inputs (*e.g.*, temperature) and the parameters themselves. In SUFI-2, uncertainty of input parameters is depicted as a uniform distribution, while model output uncertainty is quantified at the 95% prediction of uncertainty (95PPU). The cumulative distribution of an output variable is obtained through Latin hypercube sampling.

Table 1: Range of parameter values used for modeling of Choe gauging watershed

S. No.	Parameter code	Description	Fitted value	Min. value	Max. value	Location
1	SOL_K	Saturated Hydraulic Conductivity (mm hr^{-1})	0.8957	0.64	0.95	*.sol
2	ALPHA_BNK	Baseflow alpha factor for bank storage	0.2707	0.17	0.30	*.rte
3	SOL_BD	Soil Bulk density	0.4187	0.34	0.43	*.sol
4	ESCO	Soil Evaporation Compesation Factor	0.0412	0.03	0.06	*.bsn
5	SLSUBBSN	Average slope length of basin	0.2550	0.18	0.30	*.hru
6	CN2	SCS Runoff Curve Number	0.4025	0.37	0.47	*.mgt
7	REVAPMIN	Threshold depth of water in the shallow aquifer for"revap" to occur	1.6575	1.50	2.20	*.gw
8	RCHRG_DP	Deep Aquifer Percolation Factor	0.6087	0.54	0.65	*.gw
9	CH_K2	Effective hydraulic conductivity of main channel	193.0725	150.00	203.00	*.rte
10	GW_DELAY	Ground Water Delay (days)	63.0025	61.00	70.00	*.gw
11	GWQMN	Theshold depth of water in the shallow aquifer for flow to occur (mm)	2128.0075	1980.00	2150.00	*.gw
12	GW_REVAP	Ground Water "Revap" Coefficient	2.0545	2.50	3.10	*.gw

SUFI-2 starts by assuming a large parameter uncertainty within a physically meaningful range, so that the measured data initially fall within 95PPU, then narrows this uncertainty in steps while monitoring P_factor and R_factor. The P_factor is the percentage of data bracketed by 95% prediction uncertainty (95PPU) and R_factor is the ratio of average thickness of 95PPU band to the standard deviation of the corresponding measured variable. A p-factor of 1 and R-factor of zero is a simulation that exactly corresponded to measured data. In the each iteration, previous parameter ranges are updated by calculating the sensitivity matrix and the equivalent of a Hessian matrix, followed by the calculation matrix. Parameters are then updated in such a way that new ranges are always smaller than previous ranges and are centered on the best simulation (Abbaspor *et al.* , 2007). These two measured factors can be used as statistical analysis instead of the usual equations such as coefficient of determination (R2), Nash-Sutcliffe (Nash and Sutcliffe, 1970) which only compares two signals. Other statistical analyses in this study are coefficient of determination R^2 multiplied by the coefficient of the regression line and Root Mean Square Error (RMSE).

The objective function used to test the model performance were the Nash and Sutcliffe model efficiency (η) and Root Mean Square Error (RMSE). These functions are given as follows:

$$\eta = ((FIV-FRV)/FIV)*100 \quad \text{... (2)}$$

$$\text{Where, FIV = Initial variance} = \sum_{i=1}^{n}(Q_i - \bar{Q})^2 \quad \text{... (3)}$$

$$\text{FRV = Remaining variance} = \sum_{i=1}^{n}(Q_i - \hat{Q}_i)^2 \quad \text{... (4)}$$

$$\text{RMSE} = \left(\sum_{i=1}^{n}(Q_i - \hat{Q}_i)^2 \Big/ n\right)^{1/2} \quad \text{... (5)}$$

Where Q_i and $\hat{Q}_i$ are i^{th} observed and computed values of the rainfall series, $\bar{Q}$ is mean of observed rainfall series and is length of data.

RESULTS AND DISCUSSION

Month wise simulation was carried out for both watersheds for different periods. Calibration and validation of SWAT model output with observed values are shown for the watershed in Figure 3a to Figure 4b. For Choe gauging watershed monthly observed data of monsoon season from 1973 to 1977 was taken for calibration and three years data from 1978 to 1981 was taken for validation of model. Calibrated and validated SWAT model outputs found to be reasonably simulating the observed runoff values. Regression slope and coefficient of determination R^2 (Figure 3a and 3b) being close to 1 indicate that there is close relationship between observed and simulated runoff. Nash and Sutcliffe Efficiency were found to be 80.2% for calibration and 73.3% (Table 2) for validation as such these values can be considered reasonably well for any model. Similarly RMSE was calculated as 5.8 and 12.3, respectively for calibration and validation.

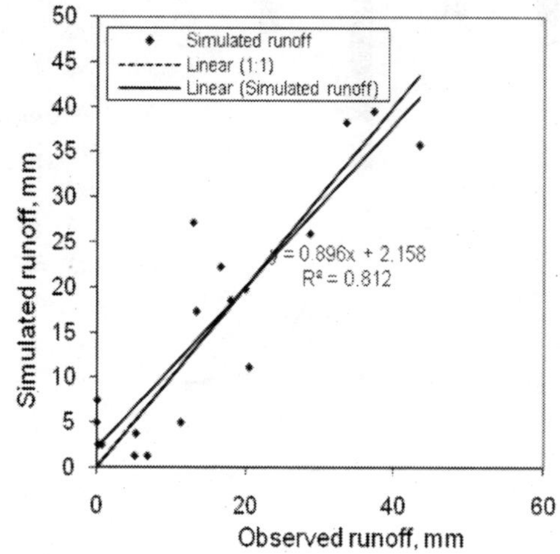

Fig. 3a: Calibration of model output with observed values (1973-78) for choe gauging watershed values.

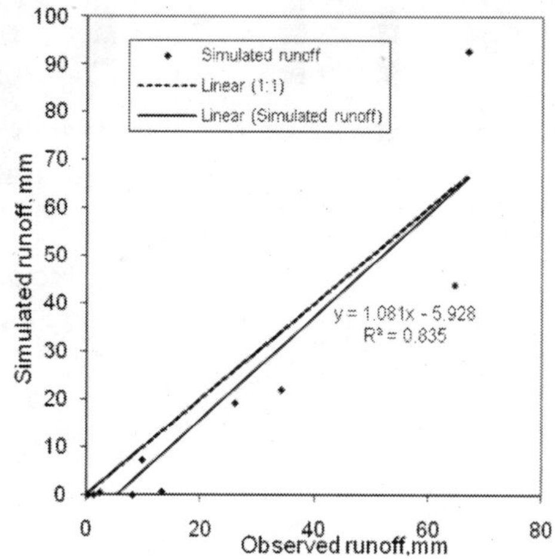

Fig. 3b: Validation of model output with observed (1979-81) for choe gauging watershed.

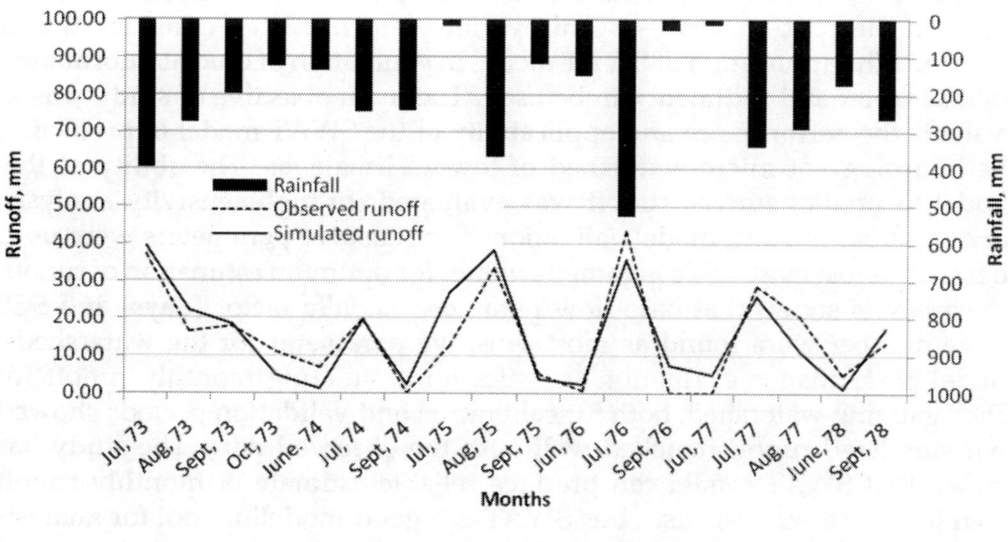

Fig. 4a: Simulated and observed runoff for Calibration period

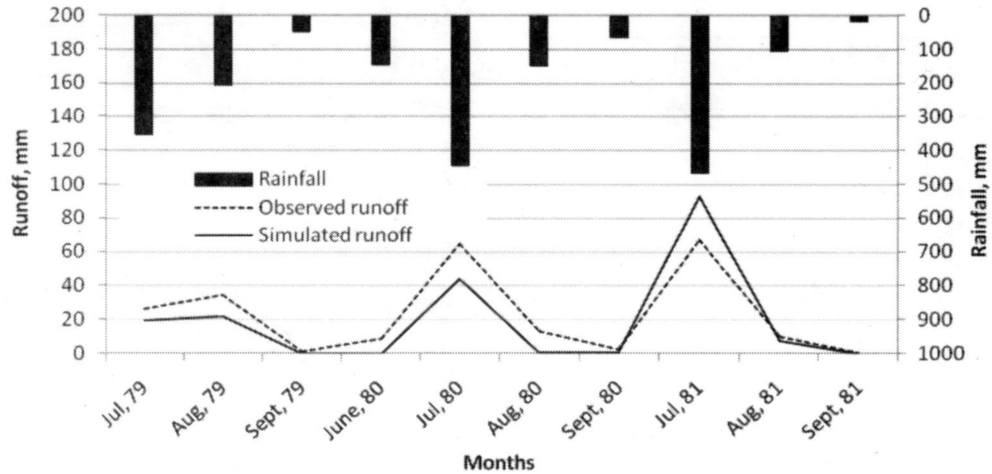

Fig. 4b: Simulated and observed runoff for Validation period (1979-81)

Table 2 : Performance evaluation of model for calibration and validation

Performance measure	Calibration	Validation
Nash and Sutcliffe Efficiency (%)	80.2	73.2
RMSE	5.8	12.3
R^2	0.81	0.83

CONCLUSIONS

Although the model generates detailed outputs at the spatial and temporal scale, in the present analyses only water yield has been considered and reported. The application of SWAT model in generation of crucial information such as water and sediment can be used. The main objective of study was to evaluate the performance and applicability of the SWAT model in predicting the hydrology of micro-watershed of lower Himalayas. The ability of this model to predict surface runoff was evaluated through sensivity analysis, model calibration and model validation. The sensitive parameters were used to find the most reasonable parameter values for optimum estimation of runoff. The analysis shows that base flow parameter *i.e.* alfa factor (days) and SCS curve number were found as most sensitive parameter for the watersheds. Model performance evaluation statistics for simulating monthly runoff for Choe gauging watershed, both for calibration and validation periods showed that simulated runoff matched well with the observed data. The study has shown that SWAT model can produce reliable estimate of monthly runoff even for micro watersheds. Thus SWAT is a good modeling tool for analysis of hydrological processes and water resource planning.

REFERENCES

Abbaspour, K. C., Yang, J., Maximov, L., Siber, R. and Bogner, K., *et al.*, 2007. Modelling hydrology and water quality in the pre-alpine/alpine Thur watershed using SWAT. *J. Hydrol.*, 333: 413-430.

Arnold, J. G., Srinivasan R., Muttiah, R. S. and William, J. R. 1998. Large area hydrological modeling and assessment: Part I: Model Development, J. *Am. Water Resour. As.*, 34(1) : 73-89.

Arnold, J. G., William, J. R., Srinivasan R. and King, K. W. 1996. In Soil and Water Assessment Tool, User's manual. USDA, Agricultural Rsearch Service, Grassland, Soil and Water Research Laboratory, 808 East Blackland Road Temple TX76502.

Cao, W., Bowden, B. W. and Davie, T. 2006. Multi-variable and Multisite calibration and validation of SWAT in a large mountainous catchment with high spatial variability, *Hydrol. Process*, 20: 1057-1073.

Keshta, N., Elshorbagy,A. and Carey,S. 2009. Ageneric system dynamics model for simulating and evaluating the hydrological performance of reconstructed watwersheds, *Hydrol. Earth Syst. Sci.*, 13: 865-881.

Nash, J. E. and Sutcliffe, J. V. 1970. River flow forecasting through conceptual models. *J. Hydrol.*, 10: 282-290.

Santhi, C., Arnold, J. G., Williams, J. R., Dugas, W. A. and Hauck, L. 2001. Validation of the SWAT model on a large River basin with point and nonpoint sources, J. *Am. Water Resour. As.*, 37 (5): 1169-1188.

Schuol, J. and Abbaspour, K. C. 2007. Using monthly weather statistics to generate daily data in a SWAT model application to West Africa, *Ecol. Model.*, 201:301-311.

Srinivasan, R., Ramanarayanan, T. S., Arnold J. G. and Bednarz, S. T. 1998. Large area hydrologic modeling and assessment part II: model application, *J. Am. Water Resour. As.*, 34(1): 91-101.

Srinivasan, R. and Arnold, J. G. 1994. Integration of basin scale water quality model with GIS. *Water Resour. Bull. AWRA*, 30(3):453-462.

SCS. 1972. *National Engineering Handbook* Section 4, Hydrology. USDA-SCS, Washington, DC, USA.

□□□

Chapter 8

Decision Support System for Contour Trenching

R.S. KUROTHE, A.K.VISHWAKARMA, D.R. SENA
GOPAL KUMAR, B.K. RAO AND V.C. PANDE

INTRODUCTION

Depleting natural resources and environmental pollution has become the biggest problem of mankind. Land degradation together with pressure of ever growing demands for food, fuel and fiber has further aggravated the problem by increasing the need for intensive cultivation and use of inorganic, non-degradable inputs for production, particularly from degraded lands. It has been estimated by various agencies that out of 329 M ha total geographical area of the country; the area under degraded lands has been estimated between 63.9 M ha- 187.0 M ha, however a general consensus agrees at about 120.72 M ha (Maji *et al.*, 2010). Degraded lands are characterized by their low productivity potential due to various kinds of constraints related to rainfall management, soil, topography and biotic interference. Trenching is one of the most efficient technologies for restoration of degraded lands which brings desirable changes though in situ conservation of moisture, soil, nutrients and energy fluxes. A contour trench is a micro depression or tiny reservoir constructed across the slope without a spillway with the objective to trap runoff, sediment, nutrient, organic matter etc. and provides opportunity to the collected runoff either to get infiltrated or evaporated before the occurrence of the next rainfall. It is an efficient technology for rehabilitation of degraded lands, particularly in semi-arid or arid regions where rainfall intensities are high, sandy soils with relatively high permeability, in denuded watershed slopes where re-vegetation is proposed and for development of orchards on sloping lands by reducing runoff and sediment loss and helps in obtaining higher biomass productivity from watersheds due to improvement in physical and chemical properties of the soil (Samra *et al.*, 2004). Contour trenching in non-arable lands constitutes

Central Soil and Water Conservation Research and Training Institute, Research Centre, Vasad District Anand, Gujarat 388 306, India the single most popular measure and has been given the highest priority for the treatment of sloping lands in the entire national and international watershed rehabilitation programs.

Although, visually it appears to be a very simple technique of soil and water conservation but, practically it is a very complex process and involves lot of computation based on watershed information for design and layout under different rainfall conditions to derive maximum benefit of the technology. A number of agencies, such as government organizations (forest departments, agriculture departments, soil and water conservation department), agriculture universities and NGOs are engaged in implementing watershed programmes on large scale. Its importance can be realized by allocation of ₹ 19251.22 crores in 11th plan (Planning Commission, 2007) by Govt. of India. A major share has been allocated for trenching and other measures for in situ moisture conservation and to improve the establishment and growth of vegetation to meet rural requirements. However, in the absence of clear cut guidelines as well as tools for optimum designing of trenches (spacing, cross section and density), many such programmes could not come up to expectations. Significant variations in trench specifications (horizontal interval, density of trenching, cross section) have been observed under different agro climatic regions of the country, which are bound to occur due to varied land forms, diversification in vegetation and these factors needs to be considered for optimum designing.

Accurate estimations of runoff generated from watersheds, estimations of vertical and horizontal spacing, volume of earth work, cost estimations and layout of trenches in the field is very tedious and time taking process and often suffers due to insufficient technical knowhow at end users. Even today under field conditions the estimates are based upon thumb rules and hence they are either over designed, resulting in higher costs and in other cases they have failed because of lack of understanding of the unknowns involved in the design.

DSS is an interactive user friendly system which incorporates knowledge and expertise within the framework of the decision support mechanism. The DSS is an integrated assembly of models, data, interpretive routines and other relevant information which efficiently processes input data, runs the model and displays the result in an easy to interpret format (George *et al.*, 2002, Rao and Rajput, 2009). DSS help in analyzing the facts and situations, decision making process, to understand the problem and explore various alternative courses of actions and identification of the most appropriate decision. (Rao and Rajput, 2009). Keeping these considerations in view the present decision support system is developed as a readymade tool using standard methods and techniques used for runoff estimation from fields, trench design and provide opportunity to the user to get rid of tedious and time consuming calculations, avoid chances of mistakes and helps in making accurate decisions for various options that can be applicable for implementing the technology. It is expected to bridge the knowledge gap between experts and actual field level users of the technology.

MATERIALS AND METHODS

Design Procedure

To begin with the decision support system and making estimations on runoff one needs some basic data as input for the area for which estimations have to be made this includes:

Design storm: *It* is the maximum rainfall (mm) recorded during the desired return period in the given area for which estimations for trench design has to be prepared. Considering the life span, objective of making a trench and availability of data for daily maximum rainfall for a return period of five years is appropriate for designing a trench. Maps for this value are available from India Meteorological Department.

Slope :It is the general slope (%) of the area for which trenches are to be designed.

*Estimation of direct runoff from watershed area:*Direct runoff (Q) estimation from watersheds based on input data for design of trenches is estimated by the software based on the runoff equation developed by Soil Conservation Service (USDA, 1972).

$$Q = (P - Ia)^2 / (P - Ia + S)$$

Where, Q = direct runoff (mm)

P = rainfall (mm)

S = potential maximum retention (mm)

Ia = initial abstraction during the period between beginning of rainfall and runoff = 0.2S

Under Indian condition initial abstractions has been estimated to be 0.3S for various soils except in black soil for which it is estimated as 0.1S (Samra *et al.,* 2004). The above equation has been used for estimation of direct runoff in the software as:

$Q = (P - 0.3S)^2 / (P+ 0.7S)$ for all other soils

$Q = (P - 0.1S)^2 / (P+ 0.9S)$ for black soil

The runoff curve number (CN) is an empirical parameter used in hydrology for predicting direct runoff or infiltration from rainfall excess. It is widely used and is an efficient method for determining the approximate amount of direct runoff from a rainfall event in a particular area. The runoff curve number is based on the area's hydrologic soil group, land use, treatment and hydrologic condition.

The potential maximum retention S in mm and curve number (CN) are related by:

$$CN = 25400 / (254+S)$$

Where, CN is standard curve number for given soil. Runoff is greatly influenced by antecedent moisture condition. Moisture conditions are expressed in terms of AMC-I, AMC-II and AMC-III based on total rainfall of preceding five days. In the software the values of CN for AMC II are computed based on physical features and hydrologic conditions of the watersheds as provided by the user based on procedure described by (Chow, 1964).

Since trenches are designed to store runoff expected under wet conditions (AMC III) the values of CN given for AMC II are converted to AMC III by using the formula (Samra *et al.,* 2004):

$$CN_3 = CN_2 \times e^{0.00673\,(100-CN2)}$$

Effect of slope (Sl) is considered using following equations.

$$CN_{2Sl} = (1/3)\,(CN_3 - CN_2)\,[1 - 2e^{(-13.86Sl)}] + CN_2$$

$$CN3sl = CN2sl \times e^{0.00673\,(100-CN2Sl)}$$

Where, Sl is the average land slope in percent

Trench cross section: The trench cross section is used either as default input or as an input variable in the software and user has to submit the desired values for trench dimensions *viz.,* trench width and depth depending upon soil type, infiltration rate, rainfall amount and intensity. In general it is recommended to have the trench depth ranging between 40-50 cm in soils with reasonable depth, and for shallow soils, trench depth may be reduced up to 25-30 cm. The cross section of the trench usually varies from 1000 cm^2 to 2500 cm^2 (Samra *et al.,* 2004). In areas of low rainfall 30 cm x 30 cm cross section is adequate while in areas of high rainfall, cross section of 50 cm x 50 cm is appropriate. After fixing the depth (D) and width (W) of trench, the horizontal spacing is determined.

Trench spacing: Spacing is calculated in terms of horizontal and vertical interval. The definition sketch of a contour trench is presented in Figure 1.

Horizontal spacing: Trench cross-sections of the trapezoidal, rectangular and triangular shapes are usually constructed for runoff impoundment.

Relationship between horizontal spacing of contour trenches, runoff depth and dimensions of trenches is expressed as:

Spacing = Cross sectional area of the trench / Expected runoff = A / Q

For rectangular trenches horizontal interval has been estimated using equation,

HI = W x D / (100 x Q)

Where, W = width of trench, (cm)

Q = depth of expected runoff from the watershed area (cm)
D = depth of trench (cm)
HI = horizontal interval (meters)

The above equation is used to find spacing for continuous contour trench. In case of staggered trenches, HI is expressed as:

HI = W x D / [100 Q x (1 + X/L)], (Samra *et al.*, 2004)
Where, X = gap in between the trenches (m)
L = length of the trench (m)

Vertical spacing: After determining the horizontal spacing the vertical spacing is determined by using the equation:

VI = S / 100 x HI

Where, VI = Vertical interval (m), S = Land slope, (%) and HI = horizontal interval (m)

Volume of earth work :If 100% runoff of the design storm is to be collected then

Volume of earth work (m^3) = 100Q

In case a fraction of direct runoff is to be trapped, then

Volume of earth work (m^3) = QV

Where, Q = Direct runoff (cm) and

V = % of direct runoff to be trapped

Cost of earth work :The cost of earth work involved in execution of trenching depends upon two factors *viz.* cost of earth work per cubic meter and the quantity of earth work required to be done so as to store desired quantity of runoff.

Cost of trenching per ha = Volume of earth work (m^3) x cost of earth work per cubic meter

Design options for various constraints : Based on the results obtained by running the software the user will get detailed information about horizontal interval, vertical interval, number of trenches, length of trench, volume of earth work, cost of earth work and the surface area that will be covered by trenches. The user may wish to redesign the trench specifications for a specific condition for following situations.

100 % runoff : The software initially computes all the specifications with a view to trap 100% runoff that is expected to be generated from the area. Based on the results obtained the user may have to decide on what percent of runoff he may target to harvest so that acceptable specifications may be obtained.

Soil Depth : A major concern while designing is trench depth which may become a limiting factor under certain circumstances such as soil depth is less than the suggested trench depth which may not be acceptable or practically possible to implement. Similarly the calculated trench depth may exceed certain critical depth which is difficult to execute under field conditions and hence the user may desire to opt for a different dimensions for trench.

Area under trenches : By excavating trenches on the field a definite portion of surface area will be occupied by the trenches. The software will calculate the per cent of surface area that is occupied by trenches and will display it for user's consideration. If the user thinks that the proposed surface area occupied by trenches is exceeding the area that is acceptable to him, he may opt for change in surface area. It is required to submit the desired area that user intends to devote as against the area calculated by the software and choosing the dimension that needs to be modified i.e. either depth alone or depth as well as width of the trenches and by submitting the input the software will recalculate the trench dimensions to harvest the desired runoff as well as for obtaining the desired surface area and the results will be displayed to the user.

Plantation spacing : The software will provide results based on inputs provided by the user. If the user wants to raise a plantation so that the conserved soil moisture in the trenches will be effectively utilized for establishment of vegetation, then he can provide the spacing of plantation and the DSS will make the changes and the layout and size of trenches will be adjusted such that each plant will be planted near the trench as the prevailing practice in case of CST.

Trench type :The user may opt for the type of trench he is willing to construct to fulfill the specific purpose of trenching *viz* continuous contour trenches with rectangular or trapezoidal shape or staggered trenches with rectangular or trapezoidal shape. The user may select the type of trenches based on his choice and purpose of constructing trenches. If more runoff is to be stored CCT may be preferred to CST.

Cost :Another important aspect in trench designing is the amount of funds available for trenching. It may be possible that the cost of trenching as per the inputs provided by the user may exceed the amount of fund available with the implementing agency and therefore it is required to redesign the trenches to adjust the cost constraint. The user can fill the amount available with him for execution of the trenching and submit for recalculation. The software will compute the amount of runoff that can be trapped by spending the available amount and present the revised results as per the user's choice.

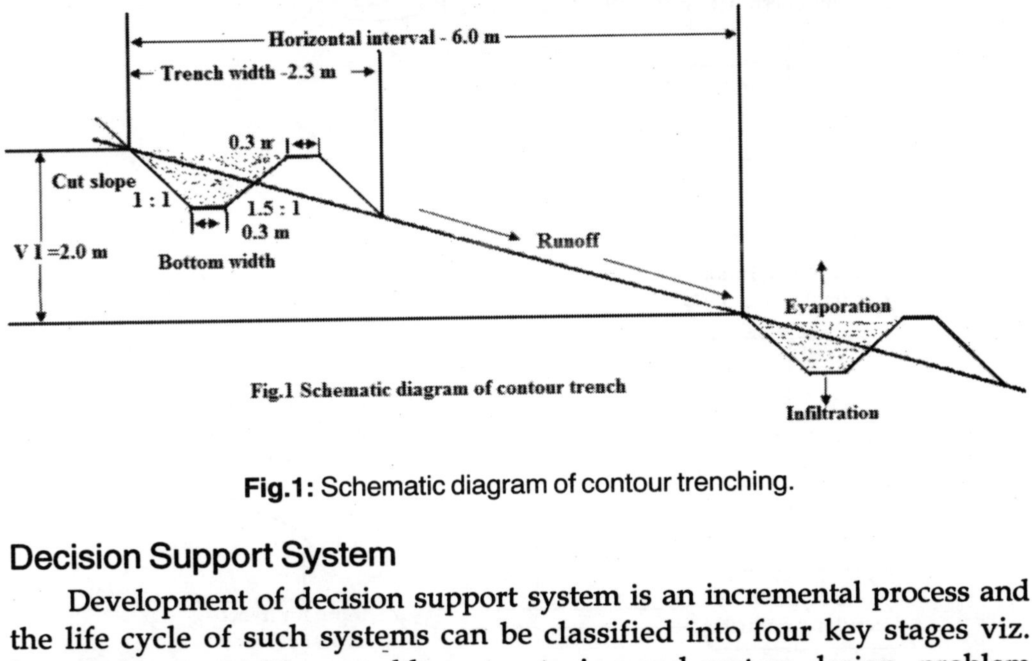

Fig.1: Schematic diagram of contour trenching.

Decision Support System

Development of decision support system is an incremental process and the life cycle of such systems can be classified into four key stages viz. knowledge acquisition, problem structuring and system design, problem encoding and system testing (Buchanan and Shorfliffe, 1984). A number of sources of information and knowledge has been referred and employed in the development of the decision support system. These sources include textbooks, manuals, research articles and personal experience. Using the above information a user friendly decision support system has been developed at CSWCRTI, Research Centre Vasad. The DSS is developed in the form of a computer programme using the input interactive controls and algorithms of

Visual Basic 6.0 programming. The nested 'If…Then…Else' statements were extensively used to construct interactive algorithm for the generation of alternative decisions using the input information. The software is written in very simple and user friendly manner so that any person with little computer knowledge having required data intending to use the technology can easily run the programme and obtain desired results without wasting time in search for literature and tedious estimations.

Basics and steps used in formulation of decision support system for design, layout and alignment of contour trenches are described below in Figure 2.

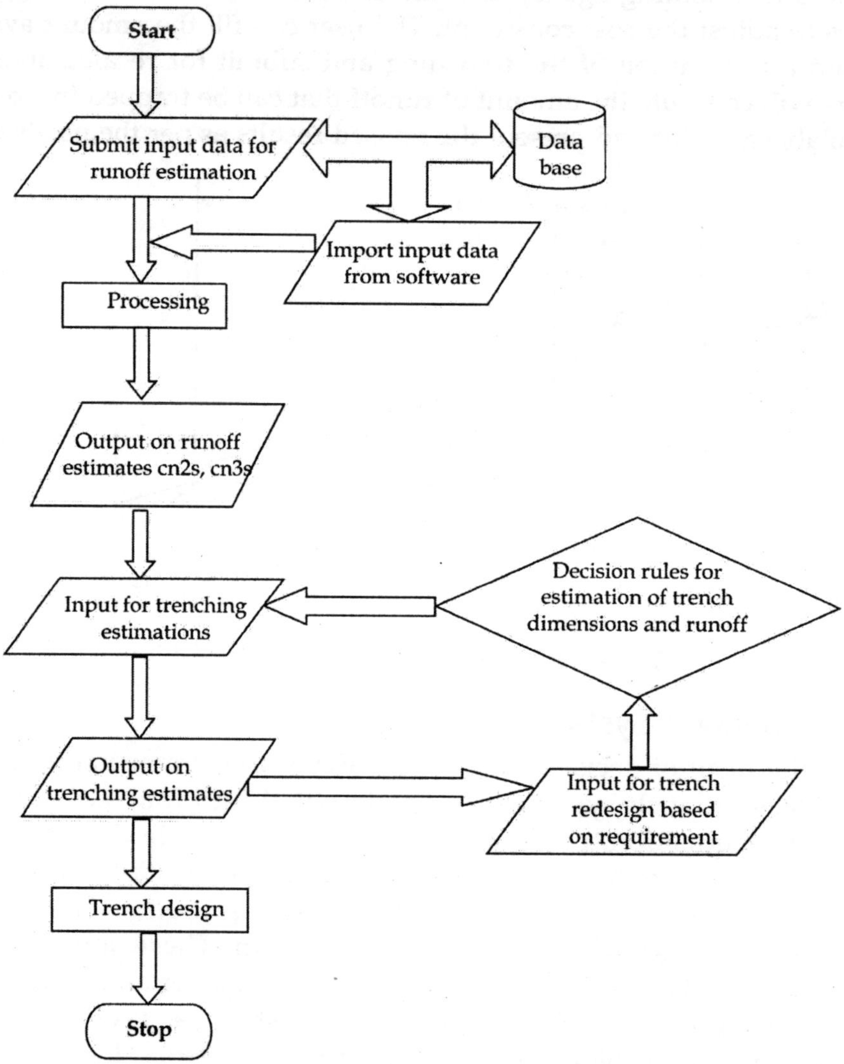

Fig. 2: Flow chart of decision support system for trenching estimations

DSS operation and results: The software provides opportunity to the user to compute direct runoff (cm) under two different kind of soil i.e. coarse textured soils (other soils) and fine textured soil (black soils) where trenching has to be carried out. The DSS provides trench specifications and layout for various options. The operation and results obtained by running the software have been discussed below.

Direct runoff estimations: The operation of the DSS is explained below through an example. In the present example the input data on rainfall is 125 mm for the desired return period, slope percent, 5 and curve number, 58 and average depth of soil (100 cm) of the site. This information is entered into the input screen as shown in Figure 3. If these parameters are not available user has the option to use the inbuilt data base (28 locations) to determine rainfall for four years return period for which the trench has to be designed as suggested by (Samra *et al.*, 2004). If the user does not have the value of CN_2 from other sources readily available he can determine it by providing information of land use/ cover treatment/ practice, hydrologic condition and hydrologic soil group. First three steps are self explanatory. Further if soil hydrologic group can be determined by providing information on runoff potential, soil type and infiltration rate. Curve number will be determined on clicking the button and filled in the appropriate place.

In the above example plot size of one hectare (width = 100 m and length = 100 m) is adopted. The user may fill the actual dimensions of his plot. The DSS can make the design considering the spacing of plantation under continuous contour trenches and staggered contour trenches. This is explained in detail in section 4.4.

After furnishing these information user has to click on the soil type here in this case (other soil) and direct runoff value of 5.86 cm has been generated by the software along with choice of trenches (CCT/SCT with rectangular and trapezoidal cross section. Similarly this information for fine textured soil (black soil) can be determined by adopting the same procedure and pressing the black soil button provided in the form for the same set of data, gives direct runoff of 7.12 cm (Figure 3).

Trench design: Clicking on the desired type of trench *viz.* continuous contour trenches either with rectangular or trapezoidal sections or staggered contour trenches either with rectangular or trapezoidal sections; estimations can be made based on plot size the amount of runoff to be trapped, dimensions of the trenches, and cost of earth work. On clicking the calculate button the information related to horizontal interval, vertical interval, number of trenches, trench length, volume of earth work and cost estimations can be generated for the specific information along with the amount of surface area captured by the trenches.

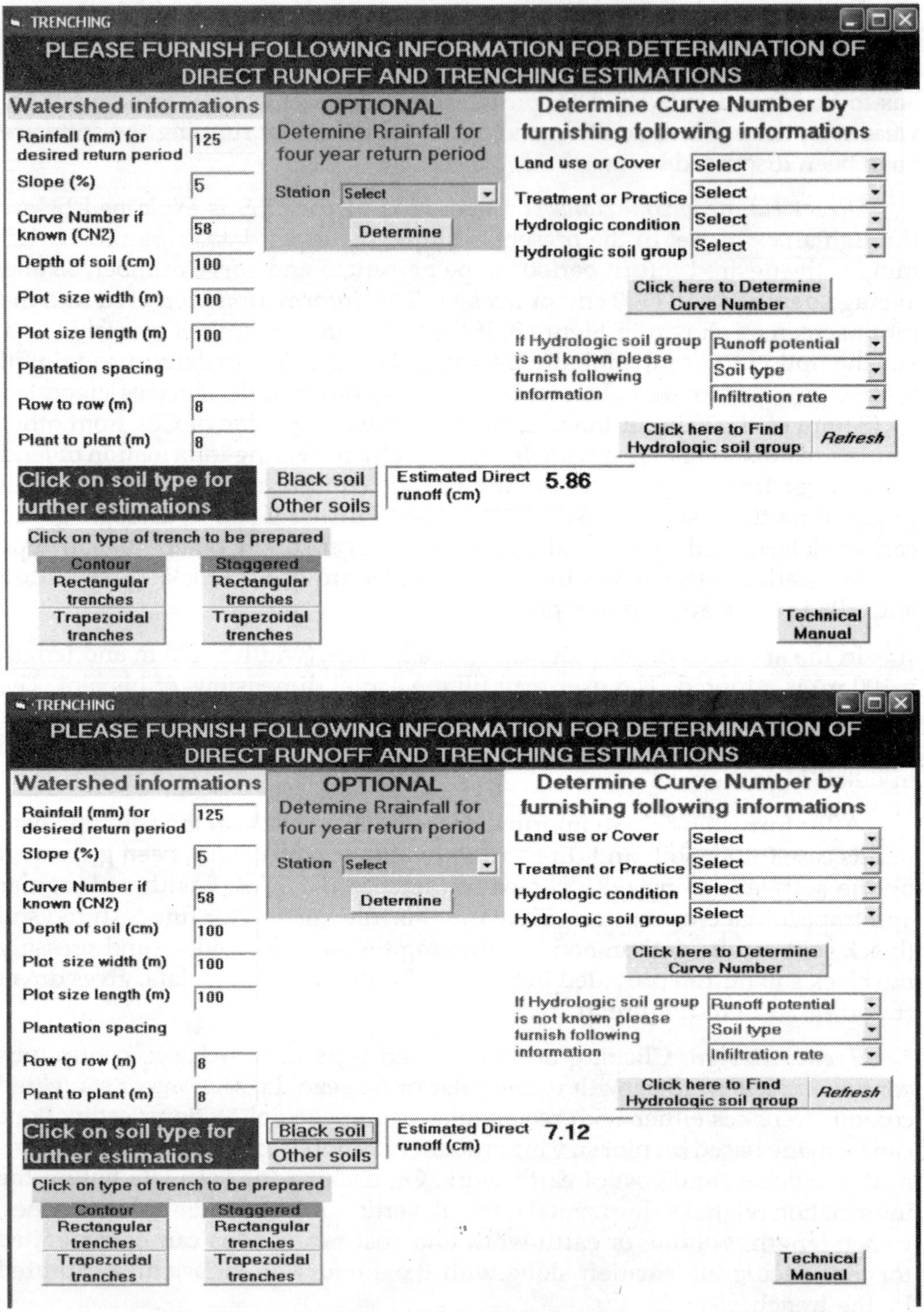

Fig. 3: Input values and runoff expected under different soils

In the above example if the user wants to build continuous contour trenches the screen will show default value of per cent runoff to be stored as 100%, trench width (45 cm), trench depth (45 cm). If the direct runoff depth exceeds 10 cm, the software will ask the user to limit the per cent runoff to be trapped by restricting it to a maximum of 10 cm and proceed for further calculations for trench dimensions based on this value. The user may change these values for his site conditions. He then fills the cost of earth work be (Rs 50 cubic meter^{-1}) and clicks on calculate button. The results for various parameters for CCT are 3.46 m for horizontal interval, vertical interval 0.17 m, number of trenches 29, trench length 2890.17 m, volume of earth work 585.26, cost Rs 29263 and percent of surface area covered by trenches 13.01% for the given dimensions and the amount of runoff to be stored (Figure 4).

Similarly information's for staggered trenches can also be generated. The results for various parameters for staggered trenches having values of per cent runoff (100), width of trench (45 cm), depth of trench (45 cm), gap between trenches (2 m) and length of trench (2 m), are 1.73m for horizontal interval, vertical interval 0.09 m, number of trenches 1445, trench length 2890 m, volume of earth work 585.22, cost Rs 29261 and percent of surface area covered by trenches 13.00% for the given dimensions and the amount of runoff to be stored (Figure 5).

Trench sketch and layout : After making all the estimations, the user can generate a representative sketch of the layout showing all the vital information related to dimensions of trenches to be constructed in the field for different kind of trenches can be generated by pressing the trench design button provided in the form (Figure 6 and 7).

Options for Various Requirements

The procedure described above gives the trench design for general purpose. The DSS also provides options for other requirements like surface area covered by the trenches, plantation spacing for a species and availability of funds (cost constraint).

Changing Surface Area

By modifying depth only: If the user wants to decrease or increase the surface area covered by the trenches, the software is provided with a window where the user can furnish parameters which he wants to modify and can redesign the trenches to trap the desired quantity of runoff. In the above example the surface area coverage with trenches is 13.01% of the total surface area and if the user wishes to reduce the surface area coverage to 10% then on submitting this information in the window along with the dimension to be modified i.e. depth to be modified which user has to select and has to submit

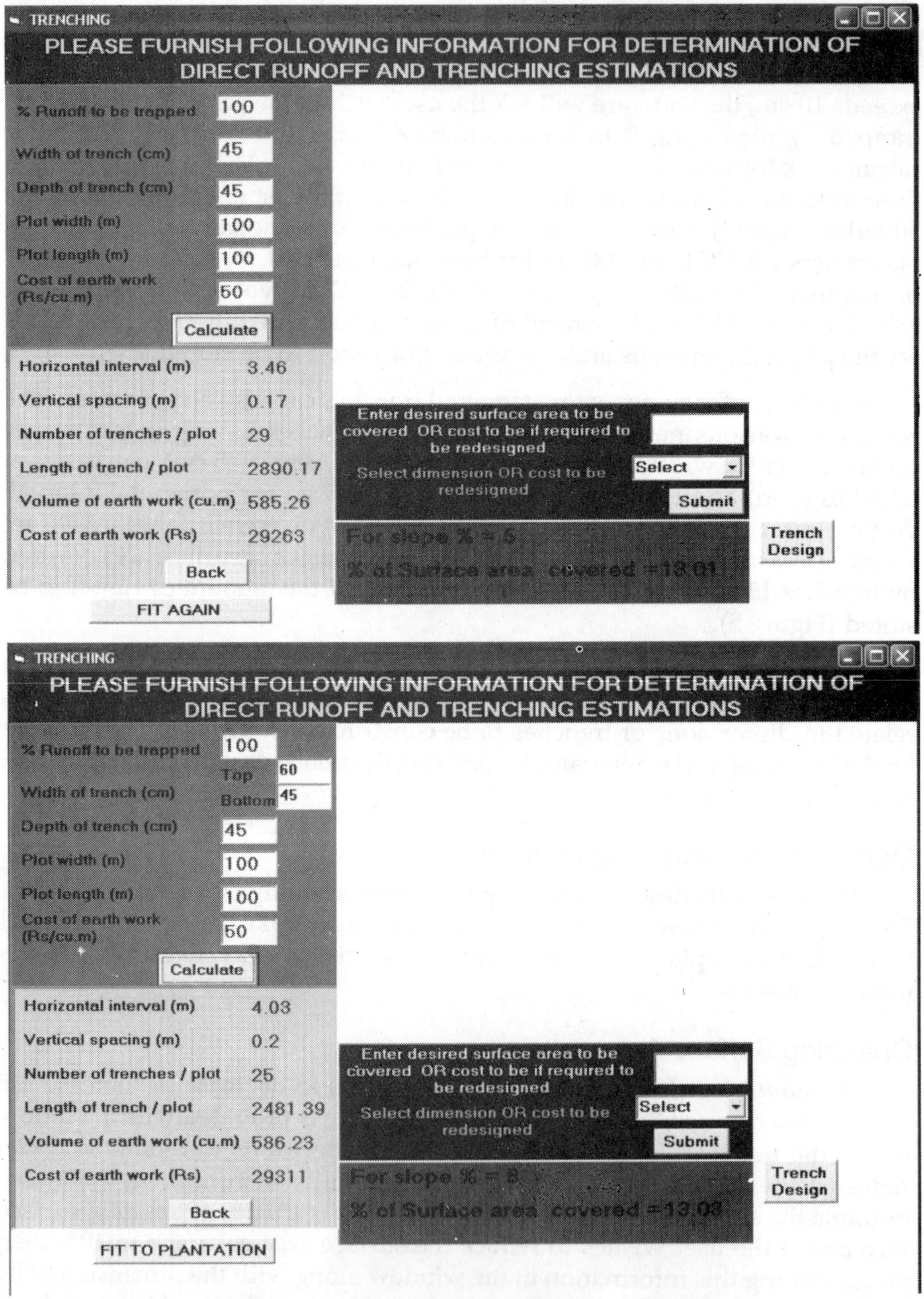

Fig. 4: Results obtained for CCT with rectangular and trapezoidal cross section

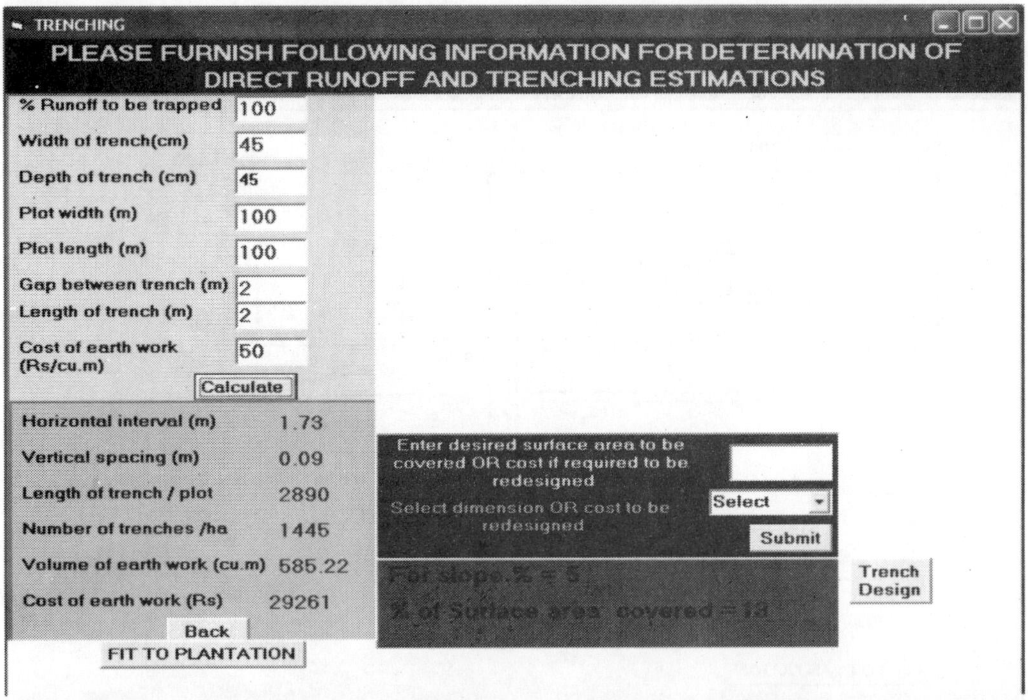

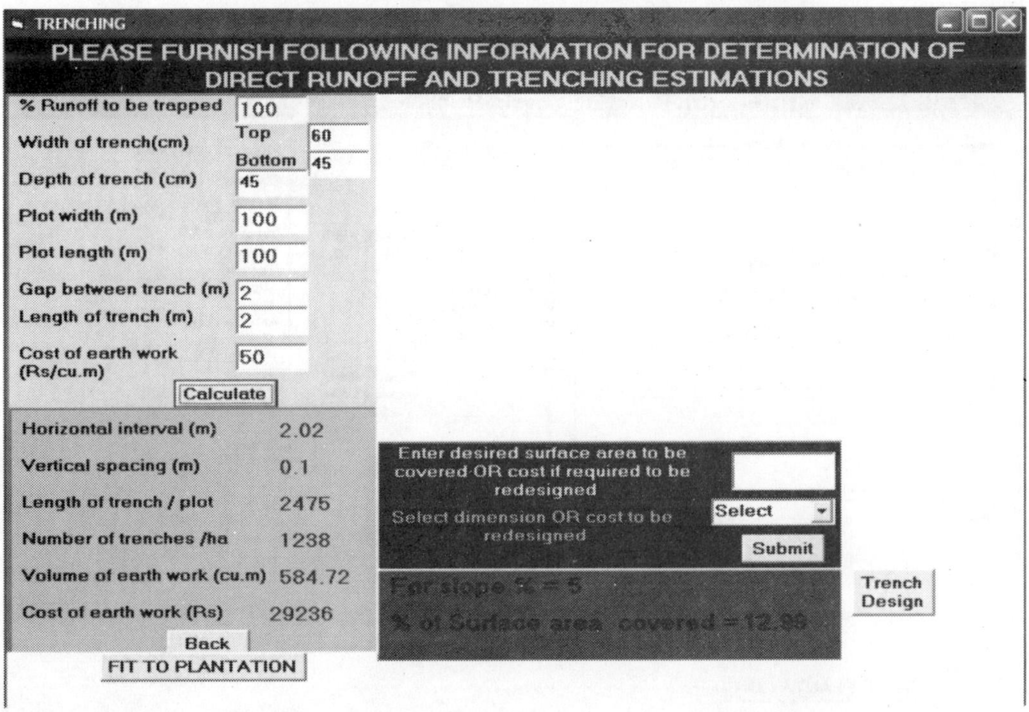

Fig. 5: Results obtained for staggered trenches with rectangular and trapezoidal cross section

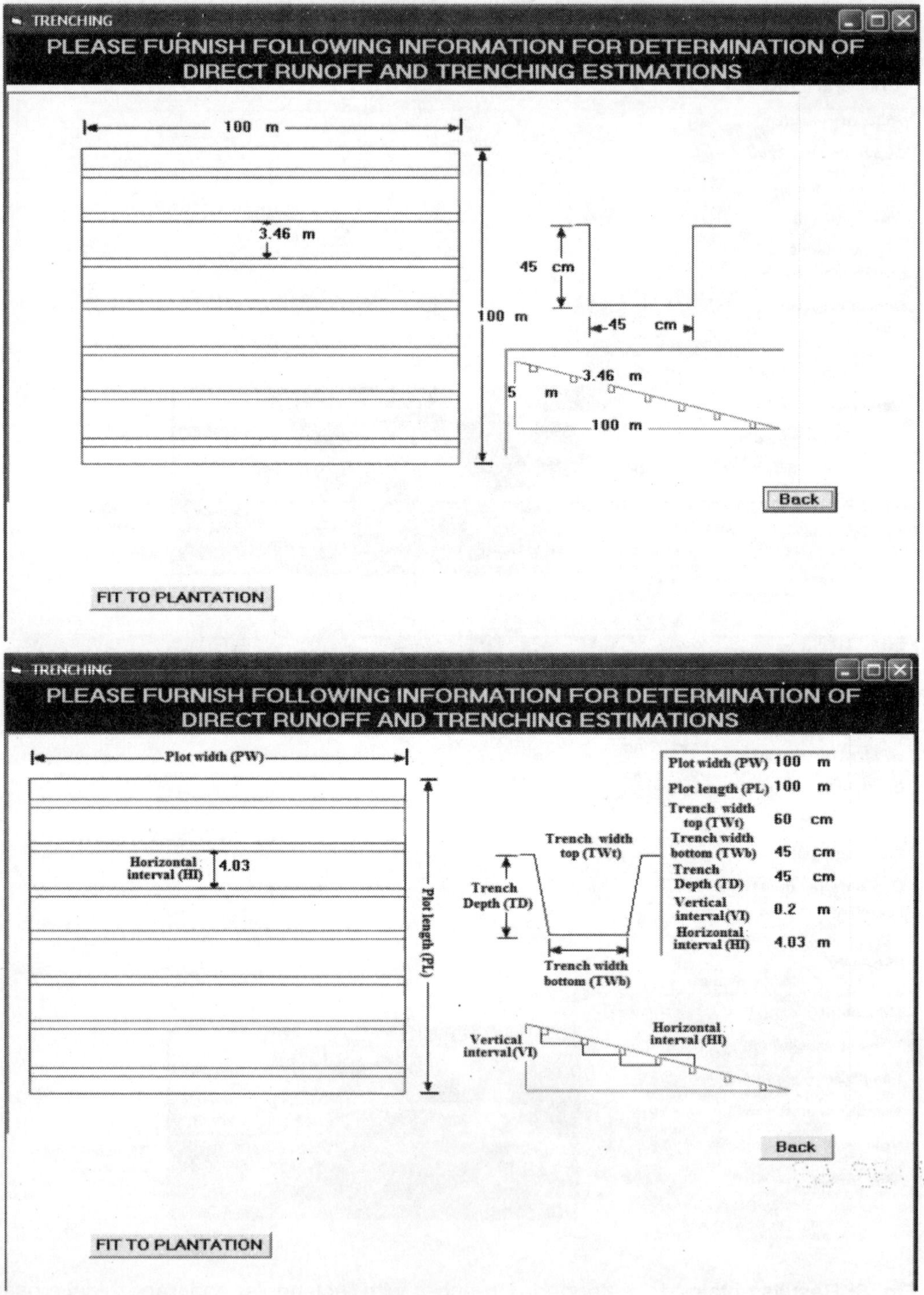

Fig. 6: Showing layouts for CCT with rectangular and trapezoidal cross section

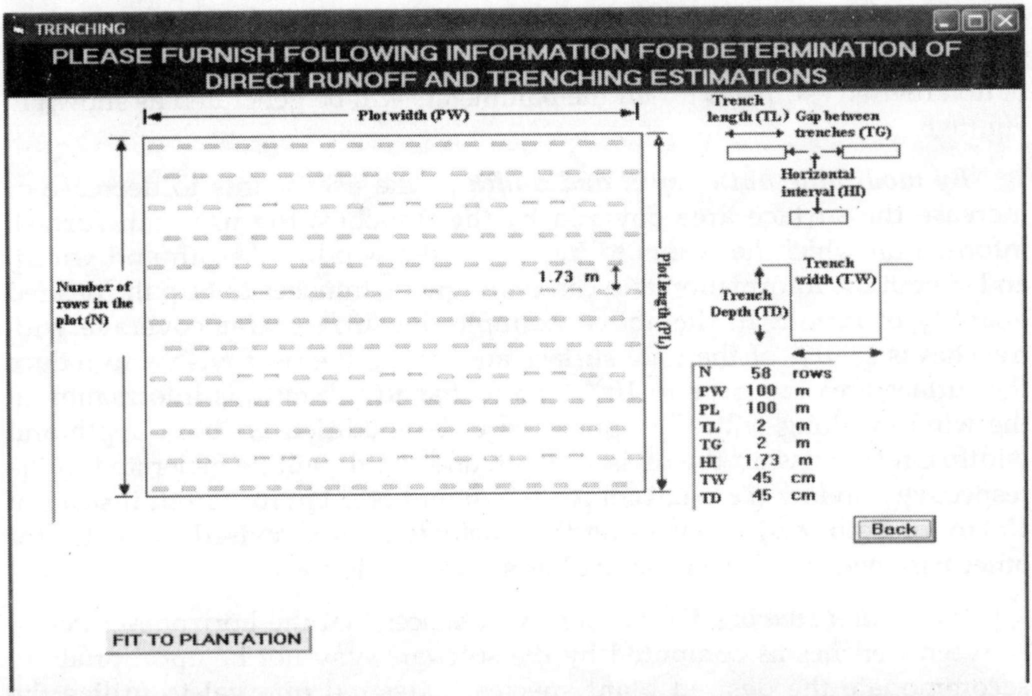

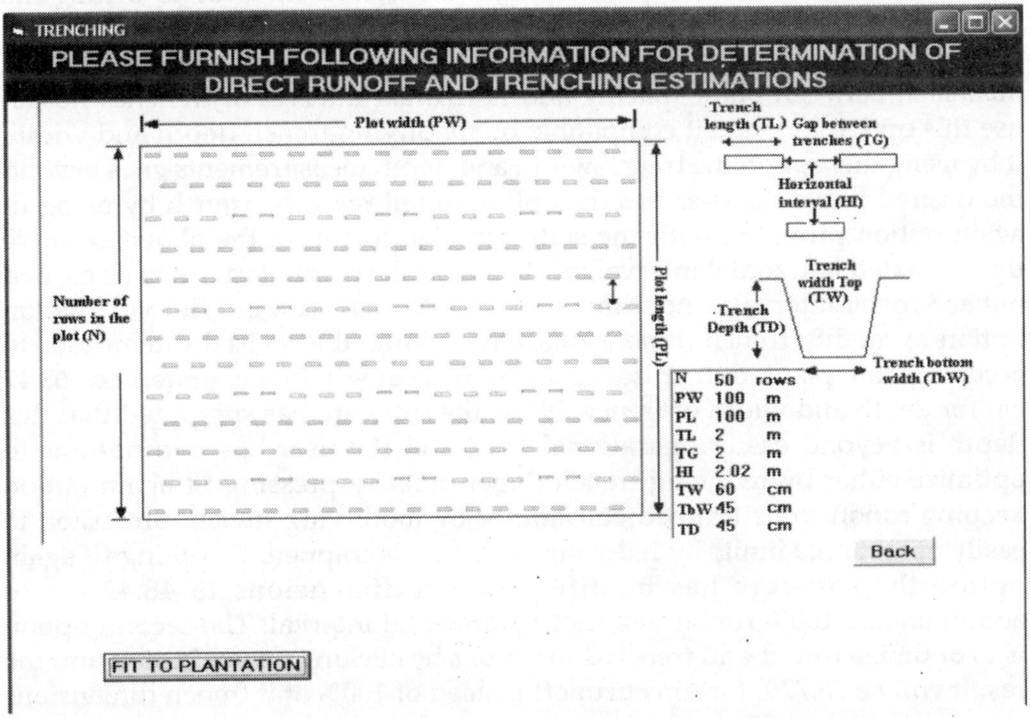

Fig.7: Showing layouts for staggered trenches with rectangular and trapezoidal cross section

the information. On submission the revised depth will be generated in the respective window (*i.e.* 58.50 cm instead of 45 cm) and on pressing the calculate button revised estimates for all the parameters will be generated as shown in Figure 8.

By modifying both depth and width :If the user wants to decrease or increase the surface area covered by the trenches, the user can furnish information which he wants to modify in the window (depth-and width) and submit the information and can redesign the trenches to trap the desired quantity of runoff. In the above example the surface area coverage with trenches is 13.01% of the total surface area and if the user wishes to reduce the surface area coverage to 10% then he has to submit this information in the window along with the dimension to be modified i.e. both depth and width. On submission the revised depth and width will be generated in the respective window (*i.e.* 51.31cm for width and 58.5 cm for depth instead of 45 cm for both) and on pressing the calculate button revised estimates for other parameters will be generated as shown in Figure 9.

Plantation spacing: Often there are chances that the horizontal interval between trenches as computed by the software may not be appropriate to accommodate the desired plant species at desired interval to utilize the moisture conserved through trenches. To solve this problem the software is provided with a fit to plantation option. In case the user finds that there is a mismatch between plant spacing and horizontal interval of trenches he can use this option to make it compatible by modifying trench depth and width. If by using this option the trench width and depth measurements goes beyond the desired limits the user has the option to redesign the trench by using fit again option provided with the software (Figure 10). In the above example the estimated horizontal interval is 3.46 m and the plantation has to be carried out at 8 meter interval. Under this condition the user can use fit to plantation option to modify trench dimensions. On clicking, the revised dimensions to accommodate plantation at desired 8 m interval will be generated *i.e.,* 68.47 cm for depth and width of trench. Now, the software has suggested that this depth is beyond easily operational limit and the user has two options to optimize either by reducing trench dimensions by pressing fit again option keeping runoff to be trapped constant or by modifying trench dimension to easily operational limits by reducing runoff to be trapped. By opting fit again option the software has modified trench dimensions to 48.42 cm to accommodate 100% runoff at 4 meter horizontal interval. The second option is to optimize runoff and trench dimensions by clicking on yes button and the result will be 76.77% for direct runoff instead of 100% and trench dimensions were modified to 60 cm for both depth and width instead of 68.46 cm to accommodate plantation at 8 m interval along with revised estimates for other

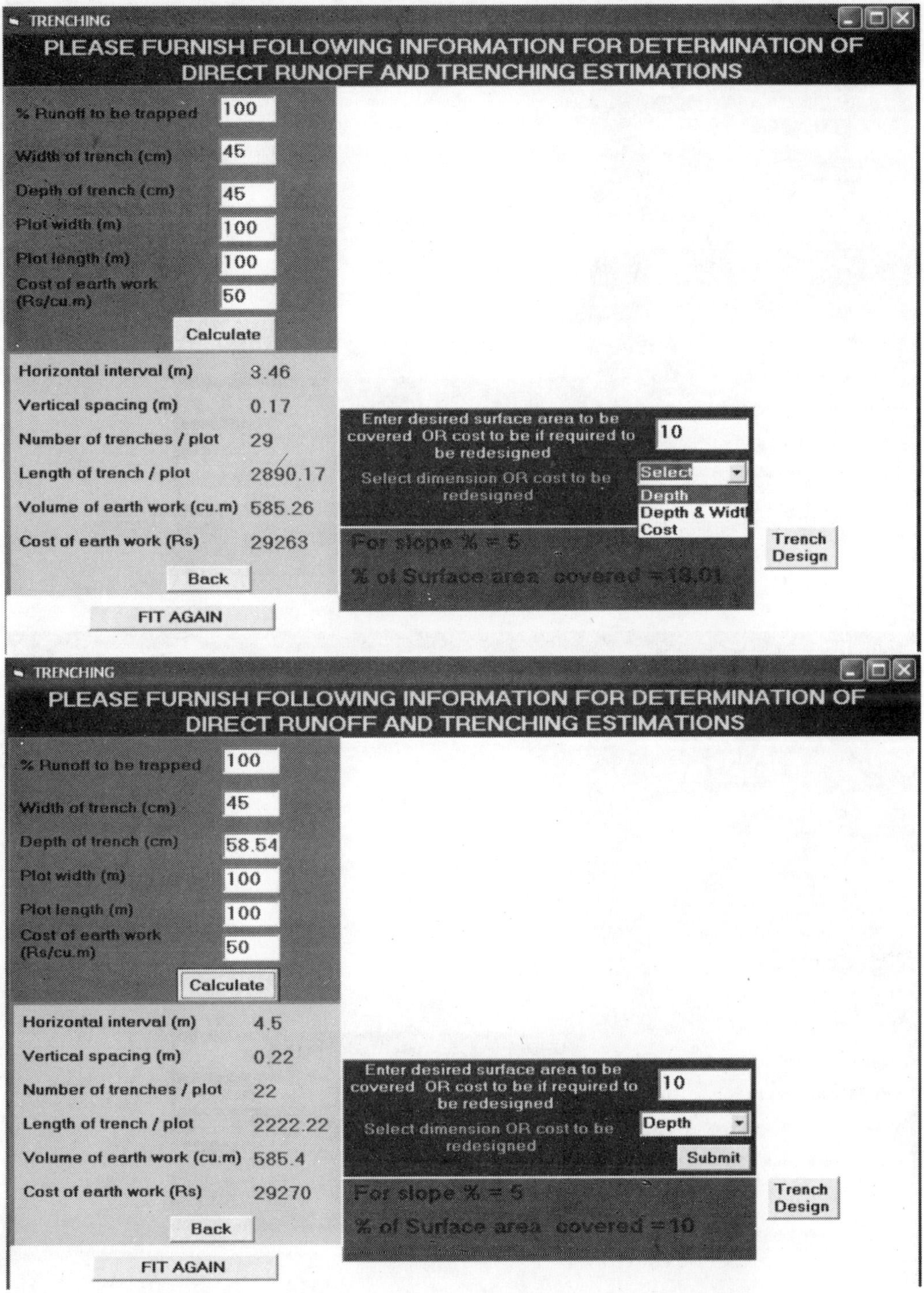

Fig. 8: Procedures for changing surface area by modifying depth

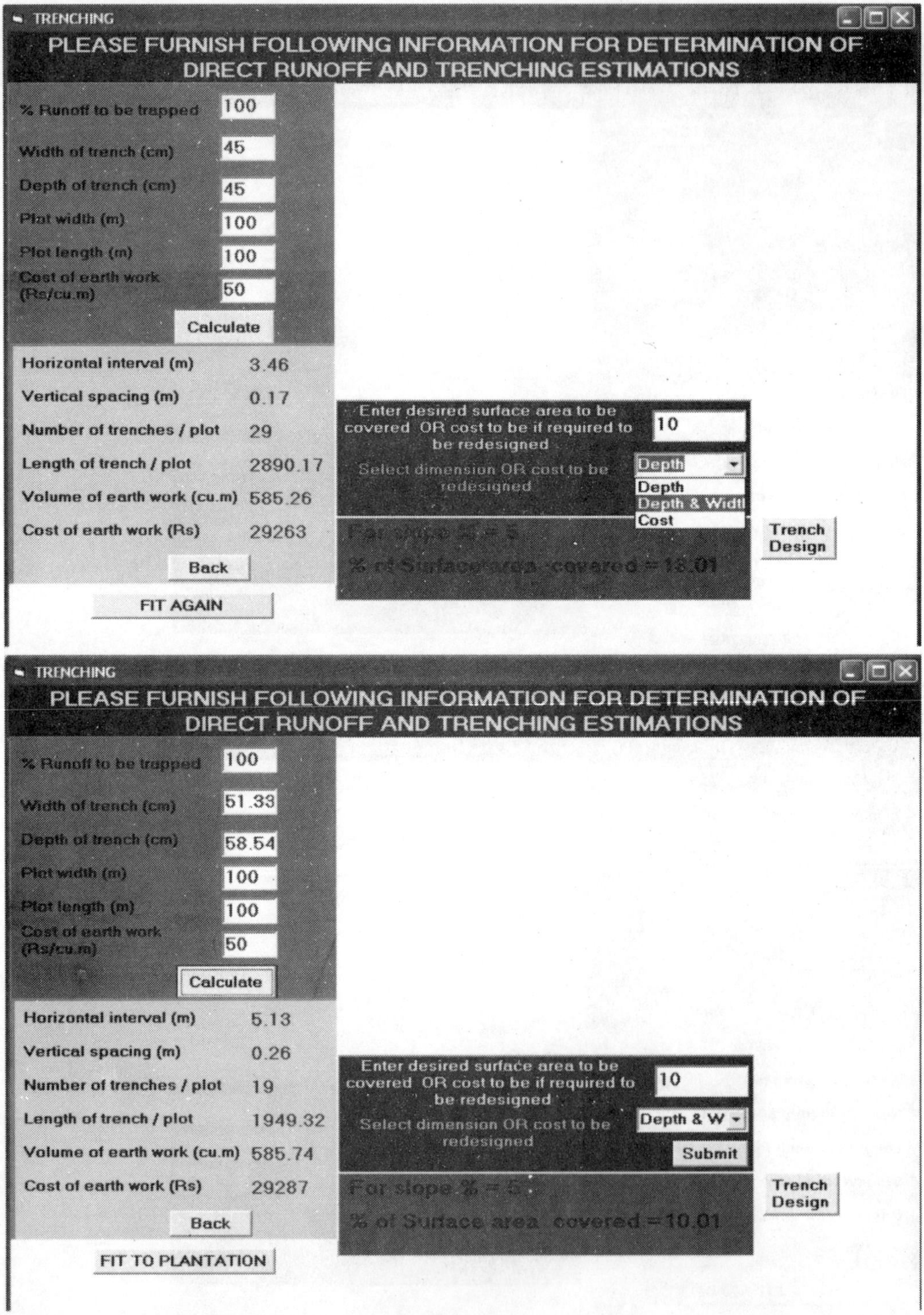

Fig. 9: Procedure for changing surface area by modifying both depth and width dimension

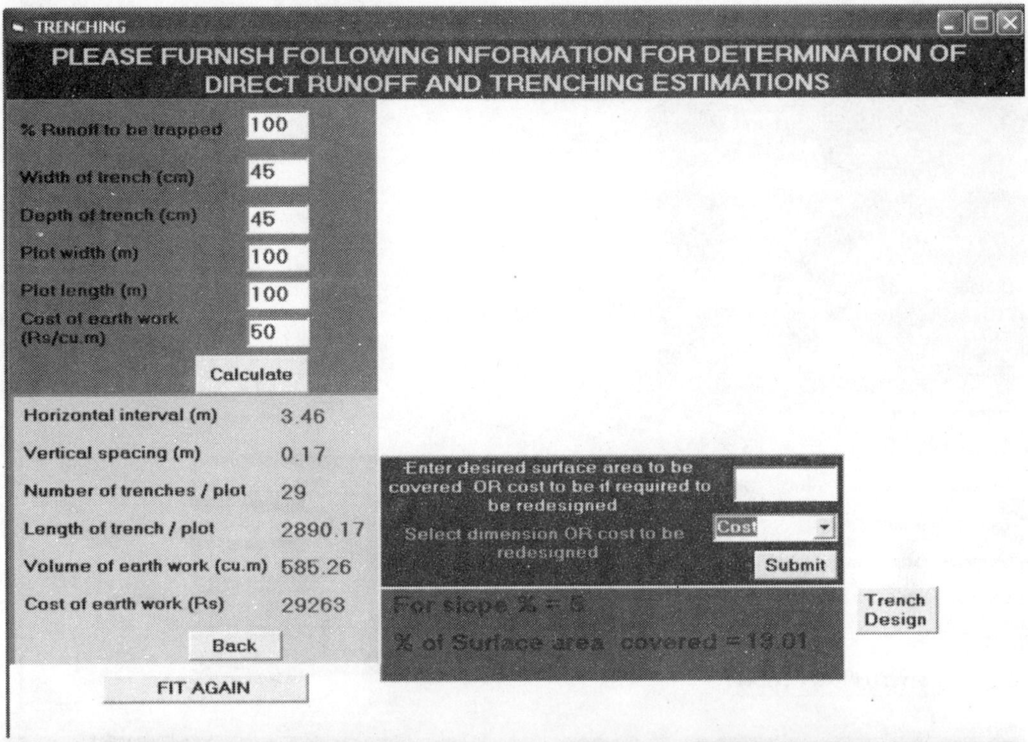
TRENCHING
PLEASE FURNISH FOLLOWING INFORMATION FOR DETERMINATION OF DIRECT RUNOFF AND TRENCHING ESTIMATIONS
% Runoff to be trapped 100
Width of trench (cm) 45
Depth of trench (cm) 45
Plot width (m) 100
Plot length (m) 100
Cost of earth work (Rs/cu.m) 50
Calculate
Horizontal interval (m) 3.46
Vertical spacing (m) 0.17
Number of trenches / plot 29
Length of trench / plot 2890.17
Volume of earth work (cu.m) 585.26
Cost of earth work (Rs) 29263
Back
FIT AGAIN
Enter desired surface area to be covered OR cost to be if required to be redesigned
Select dimension OR cost to be redesigned
Cost
Submit
For slope % = 5
% of Surface area covered = 13.01
Trench Design

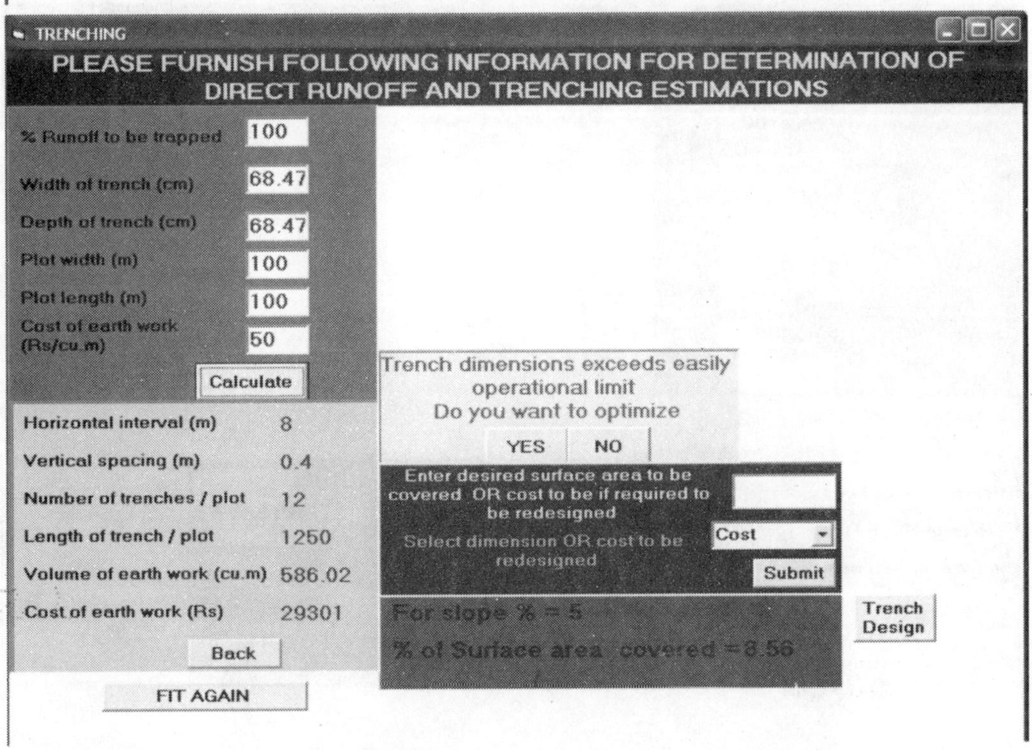
TRENCHING
PLEASE FURNISH FOLLOWING INFORMATION FOR DETERMINATION OF DIRECT RUNOFF AND TRENCHING ESTIMATIONS
% Runoff to be trapped 100
Width of trench (cm) 68.47
Depth of trench (cm) 68.47
Plot width (m) 100
Plot length (m) 100
Cost of earth work (Rs/cu.m) 50
Calculate
Horizontal interval (m) 8
Vertical spacing (m) 0.4
Number of trenches / plot 12
Length of trench / plot 1250
Volume of earth work (cu.m) 586.02
Cost of earth work (Rs) 29301
Back
FIT AGAIN
Trench dimensions exceeds easily operational limit
Do you want to optimize
YES
NO
Enter desired surface area to be covered OR cost to be if required to be redesigned
Select dimension OR cost to be redesigned
Cost
Submit
For slope % = 5
% of Surface area covered = 8.56
Trench Design

contd...

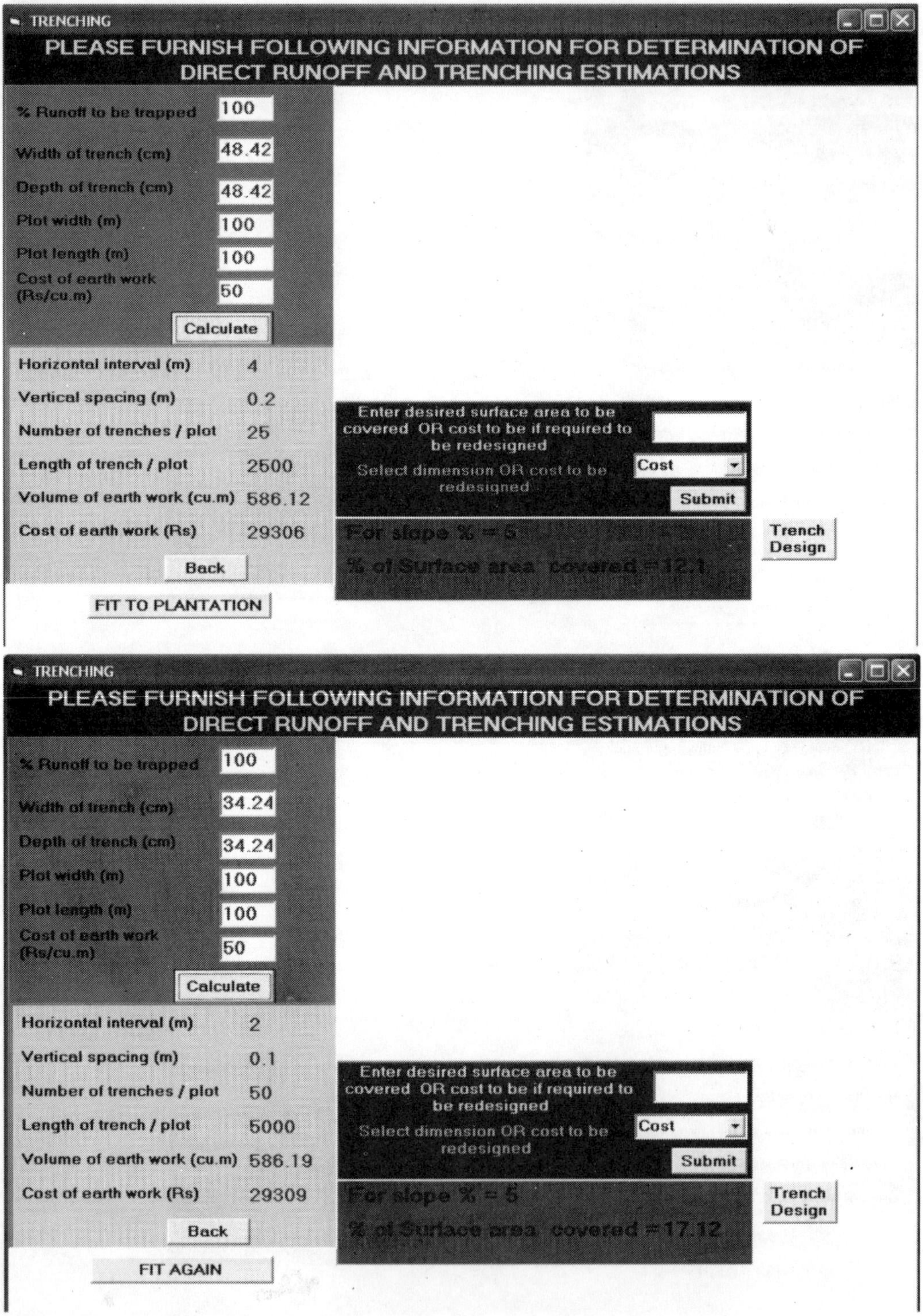

Fig. 10: Procedure for redesigning trenches for plantation spacing

parameters for trench dimensions on clicking calculate button. However, the first option is better as it collects 100% runoff and allows plantation at 8m x 8m spacing by skipping alternate row. Second option may be better if user wishes to reduce the surface area covered and also has the limitations of fund availability.

Availability of funds (cost constraint) : Similarly in cases where there is constraint of funds the user can decrease the runoff to be harvested. The user can furnish information and submit to reduce the quantity of runoff to be trapped by construction of trenches within available funds. In the above example the cost of trenching is ₹ 29263/- for harvesting 100% of the direct runoff and if the available fund is ₹ 25000/- then he has to submit this information in the window along with the parameter to be modified i.e. cost. On submission the revised quantity of runoff that can be stored by construction of trenches with available funds will be generated in the respective window (*i.e.,* 85.43% of runoff instead of 100%) and on pressing the calculate button revised estimates for other parameters will also be generated as shown in Figure 11.

Alert system for critical input data:The decision support system is equipped with a warning system for crucial input data that will become automatically active every time the user furnishes unsuitable trench dimensions. Similarly DSS will guide the user to redesign trenches, if calculated dimensions exceed the critical values as shown in Figure 12.

CONCLUSIONS

Trenching is the most efficient practice for in situ conservation of moisture, soil, nutrients and energy fluxes for restoration of degraded non-arable lands. A user friendly software has been developed using visual basic programming, written in very simple and user friendly manner so that any person with little computer knowledge, having required data can easily run the programme and obtain desired results without wasting time in search for literature and tedious calculations involved in estimations. It is possible to make accurate estimations for trench design parameters (horizontal interval, vertical interval, density of trenching, cross section, earth work, cost *etc.*) for construction of contour and staggered trenches under different conditions to optimize cost and derive maximum benefit from scarce resources. It is expected to be of immense importance and practical use for persons working in watersheds programmes, State government agencies, NGOs and forest departments implementing forestry activities on watershed basis.

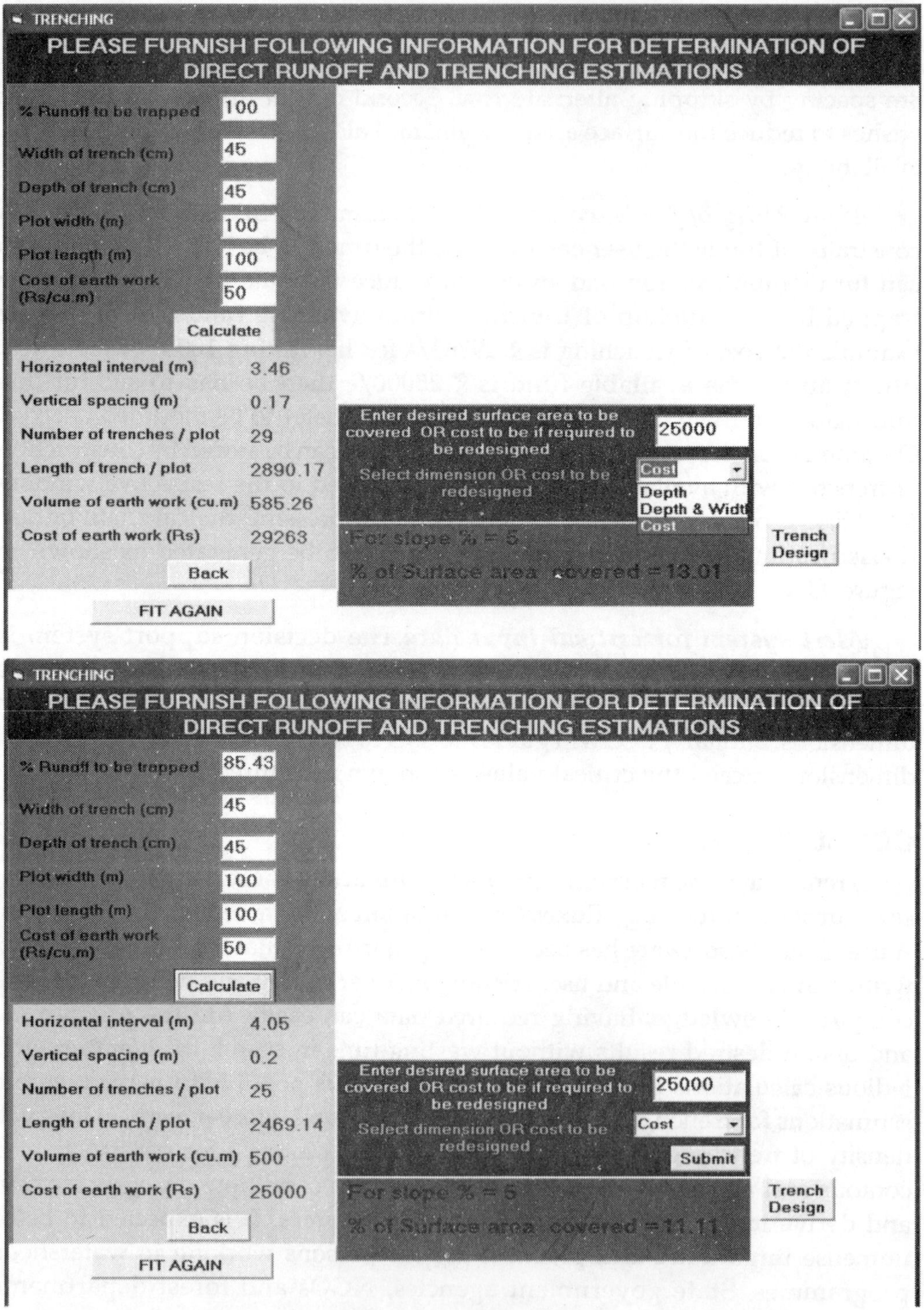

Fig. 11: Procedure for redesigning under cost constraint

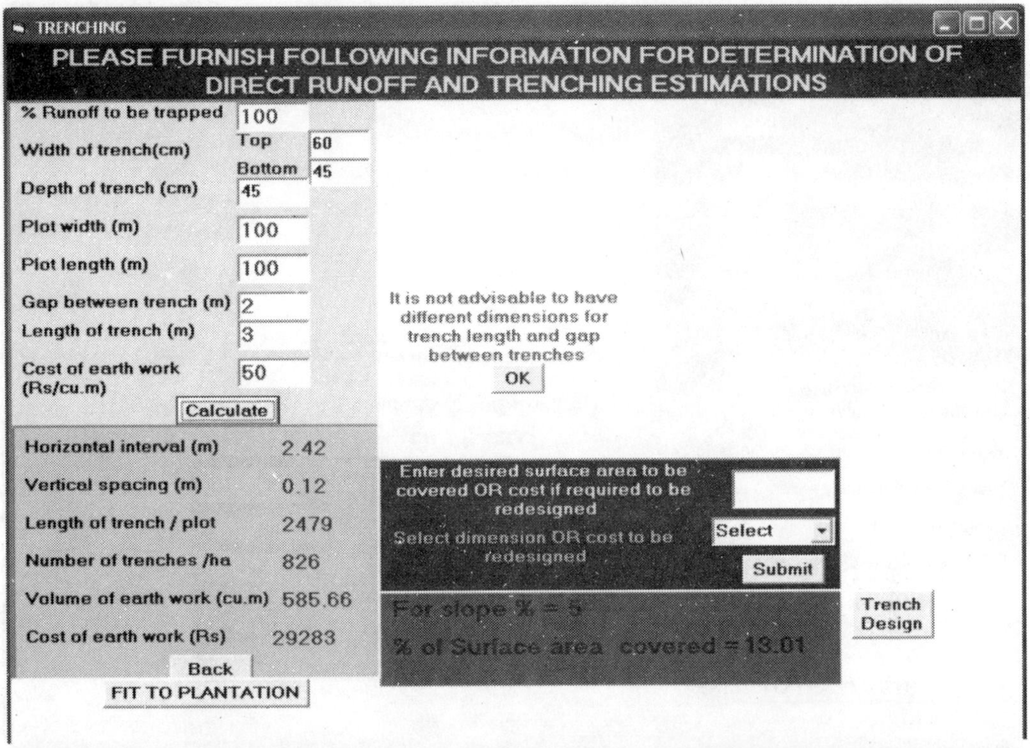
TRENCHING
PLEASE FURNISH FOLLOWING INFORMATION FOR DETERMINATION OF DIRECT RUNOFF AND TRENCHING ESTIMATIONS
% Runoff to be trapped 100
Width of trench(cm) Top 60 Bottom 45
Depth of trench (cm) 45
Plot width (m) 100
Plot length (m) 100
Gap between trench (m) 2
Length of trench (m) 3
Cost of earth work (Rs/cu.m) 50
Calculate
It is not advisable to have different dimensions for trench length and gap between trenches
OK
Horizontal interval (m) 2.42
Vertical spacing (m) 0.12
Length of trench / plot 2479
Number of trenches /ha 826
Volume of earth work (cu.m) 585.66
Cost of earth work (Rs) 29283
Back
FIT TO PLANTATION
Enter desired surface area to be covered OR cost if required to be redesigned
Select dimension OR cost to be redesigned
Select
Submit
For slope % = 5
% of Surface area covered = 13.01
Trench Design

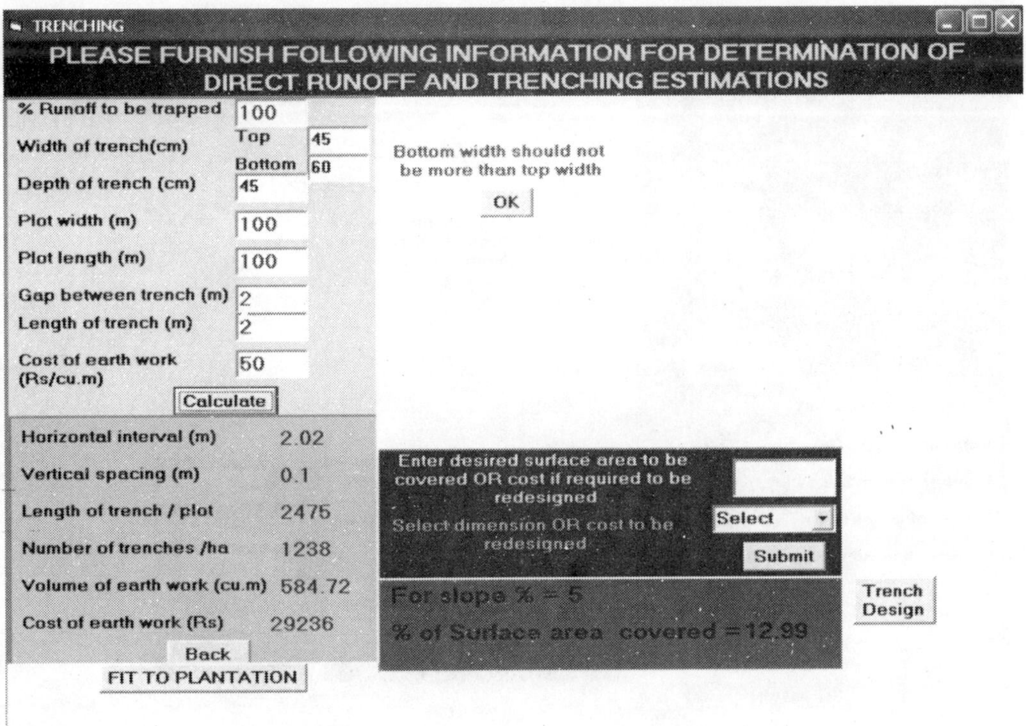
TRENCHING
PLEASE FURNISH FOLLOWING INFORMATION FOR DETERMINATION OF DIRECT RUNOFF AND TRENCHING ESTIMATIONS
% Runoff to be trapped 100
Width of trench(cm) Top 45 Bottom 60
Depth of trench (cm) 45
Plot width (m) 100
Plot length (m) 100
Gap between trench (m) 2
Length of trench (m) 2
Cost of earth work (Rs/cu.m) 50
Calculate
Bottom width should not be more than top width
OK
Horizontal interval (m) 2.02
Vertical spacing (m) 0.1
Length of trench / plot 2475
Number of trenches /ha 1238
Volume of earth work (cu.m) 584.72
Cost of earth work (Rs) 29236
Back
FIT TO PLANTATION
Enter desired surface area to be covered OR cost if required to be redesigned
Select dimension OR cost to be redesigned
Select
Submit
For slope % = 5
% of Surface area covered = 12.99
Trench Design

contd...

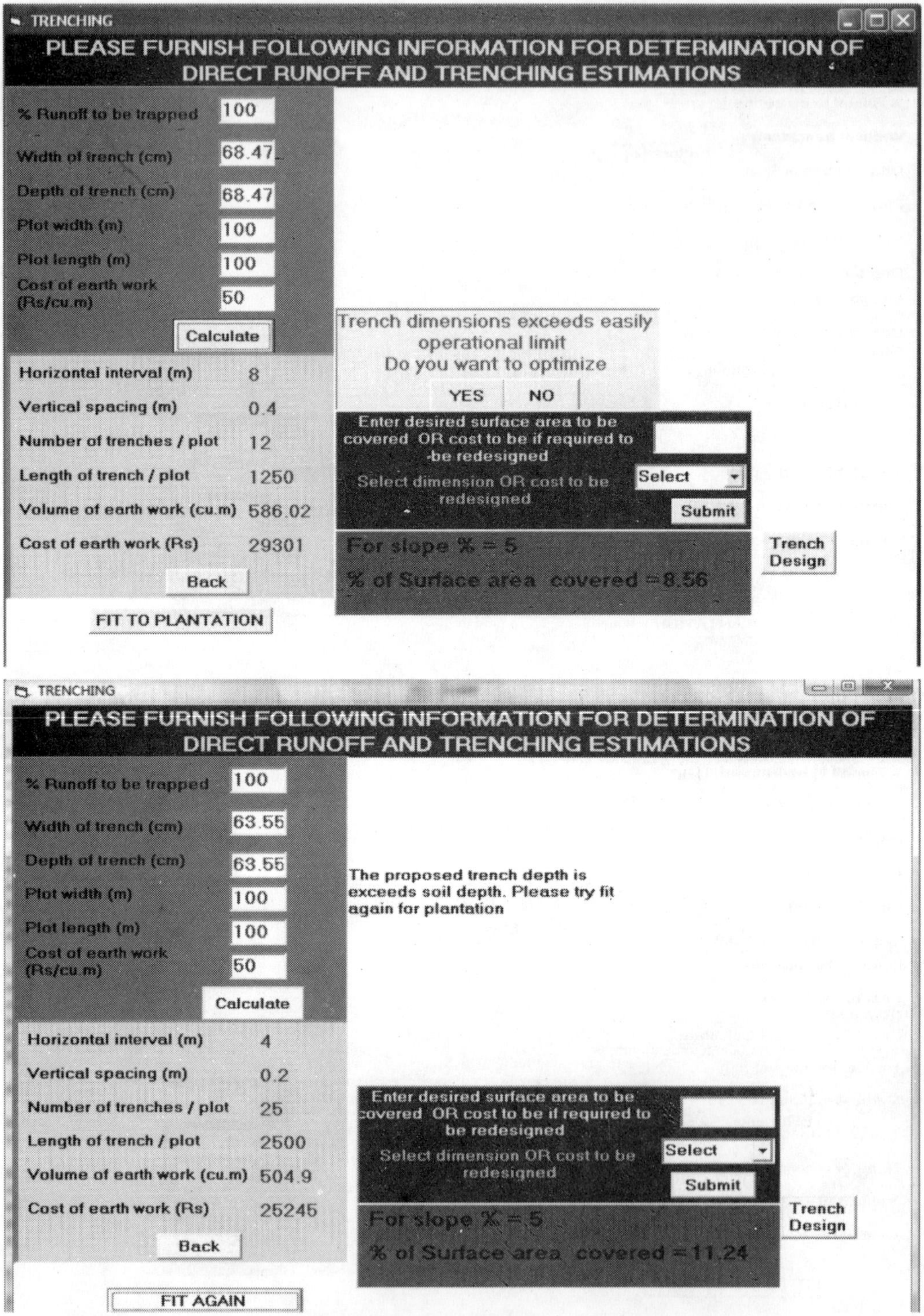

Fig. 12: Showing alerts for various critical levels

REFERENCES

Buchanan, B.G. and Shorfliffe, E. H. 1984. Rule based expect systems. The MYCIN Experiments of the Stanford Heuristic Programming Project, Addison Wesley, Massachusetts, 1984.

Chow, V.T. 1964. Handbook of Applied Hydrology, McGraw Hill Book Co., New York.

George, B.A., Reddy, B.R.S. Raghuvanshi, N.S. and Wallender, W.W. 2002. Decision support system for estimating reference evapotranspiration. *J. Irrig. Drain. Eng. Div.*, 2002, 128-10.

Maji, A.K., Reddy, G.P.O. and Sarkar, D. 2010. Degraded and wasteland of India- Status and spatial distribution, published by ICAR pages1-167 ,25p.

Rao, B.K. and Rajput, T.B.S. 2009. Decision support system for efficient water management in canal command areas. *Current Science*, Vol.97, No.1.

Samra, J.S., Chandra Prakash, Sastry, G. and Raizada, A. 2004. Technical manual on contour trenching for rehabilitation for degraded lands, Published by the Director, CSWCRTI Dehradun.

Soil conservation service, Hydrology, sec. 4 of National Engineering Handbook 1972, Soil conservation service, U.S. Department of Agriculture, Washington, D.C.

Working Group Report of Natural Resource Management for Eleventh Five Year Plan, constituted by Planning Commission, New Delhi, Feb., 2007. http://www.agricoop.nic.in/Nrm/STATNRM.pdf

□□□

Chapter 9

Comparative Performance of Various Rainfall-Runoff Mathematical Models in Nethravathi Catchment in Karnataka State

H.V. GANESH AND KIRTI MALHOTRA[1]

INTRODUCTION

Among several factors, rainfall is an important input component deciding the amount and duration of runoff from a watershed. The other basic parameters such as basin area, slope, its vegetation, land use pattern, climatic parameters such as temperature, relative humidity, wind speed, etc. play their due roles in influencing the runoff from a basin. The rainfall-runoff process is highly nonlinear, dynamic and complex process, during which, the rainfall of the catchment translates in time due to variable source area of the watershed contributing runoff at the outlet and also attenuates due to storage characteristics of the watershed (Ashu Jain and Prasad Indurthy, 2003). Application of mathematical models to relate rainfall to accrued river discharge in respective catchment is a vital tool in water resources management (Yawson *et al.*, 2005). The present study deals with an attempt to estimate daily runoff from a catchment by utilizing its daily mean areal rainfall. Several workers have met with various degrees of success at such attempts. The advantage of utilizing minimum data like rainfall, catchment area, etc, is their ready availability for almost all catchments. (O'Connor, 2001).

MATERIALS AND METHODS

Nethravathi River catchment is located between 12° 30′ to 13° 10′ north latitude and between 75° 02′ and 75° 50′ east longitude. It is drained by river Nethravathi which is a tributary of River Tapi, located in western part of Karnataka. The shape of the catchment is oblong, its topography is hilly, and

Department of Agricultural Engineering, College of Agriculture, GKVK, Bangalore- 560 065, Karnataka, India

[1]Department of Civil Engineering, Acharya Institute of Technology, Soladevanahalli, Bangalore-560 090, Karnataka, India

has plenty of vegetation in the form of thick forests. The major soils of the catchment are lateritic and red loam (Venkata Rao, 1968). The total length of the longest stream from the farthest point to stream gauging station is 1121 km and the elevation difference is 1.333 km between these two extreme points. Hence the overall catchment slope is 0.12 %. The Nethravathi catchment has a circulatory ratio of 0.264 and its stream order is 7. The catchment has twelve influencing rain gauge stations in and around its watershed area. The total area of Nethravathi catchment up to Bantwal gauging station is 3184 km^2. The average annual rainfall of the catchment is 4510 mm (based on 11 years data from 1996 to 2006) and its average annual runoff is 3438 mm (based on ten years data from 1996-2005). The stream gauging station is located at Bantwal in Mangalore taluk of Dakshina Kannada District of Karnataka State (Figure 1). The rainfall is distributed from the June to September, with peaks during July to August months.

For calibration of the mathematical rainfall-runoff models, 8 years rainfall and runoff data on daily basis from 1996 to 2003 was used and for validation of the models, next 2 years data from 2004 to 2005 was used. The rainfall data for the various rain gauges stations were obtained from the Department of Economics and Statistics, Govt. of Karnataka, Bangalore. The river discharge data of the catchment measured at Bantwal stream gauging station was procured Central Water Commission, Bangalore. The catchment map with scale 1:330,000 was drawn using ARC MAP, ARC GIS (ESRI) software at Karnataka State Remote Sensing Application Centre, Govt. of Karnataka, Bangalore. Using Thiessen polygon method, the mean areal rainfall was computed. With the two sets of data namely, the daily mean area rainfall in mm and the daily values of stream discharges in mm of the catchment the rainfall-runoff mathematical models were developed and evaluated.

Regression technique was employed in developing simpler models to complex models involving one day's rainfall to several previous days' rainfall values influencing that particular day's runoff. Since the previous days' catchment rainfall also influence the present day's discharge, the rainfall values of the consecutive previous days were added one by one to evaluate their effect on the model performance in terms of R^2 and correlation values. According to Kohler and Linsley (1951), the antecedent soil moisture index, which is influenced by the previous days' rainfall, has profound effect on future days' runoff. When the influence of the previous days' rainfall stopped or decreased to an insignificant level, the process of addition of previous days' rainfall values to the model building was stopped. Thus, if up to 'n' days previous rainfall values showed influence on current day's stream discharge, and n+1th previous day rainfall did not improve the model performance in terms of R^2 values / correlation values, the process of addition of previous days rainfall values would be stopped. This would result in a multi-linear model of the form : $Qi = C + m_1 P_i + m_2 P_{i-1} + m_3 P_{i-2} \ldots\ldots.. + m_{n+1} P_{i-n}$, where, P_i = i[th] day mean areal rainfall of the catchment, P_{i-1} = mean areal rainfall of the catchment one day prior to i[th] day, P_{i-2} = mean areal rainfall of the catchment two days prior to i[th] day, P_{i-n} = mean areal rainfall of the catchment 'n' days prior to i[th] day.

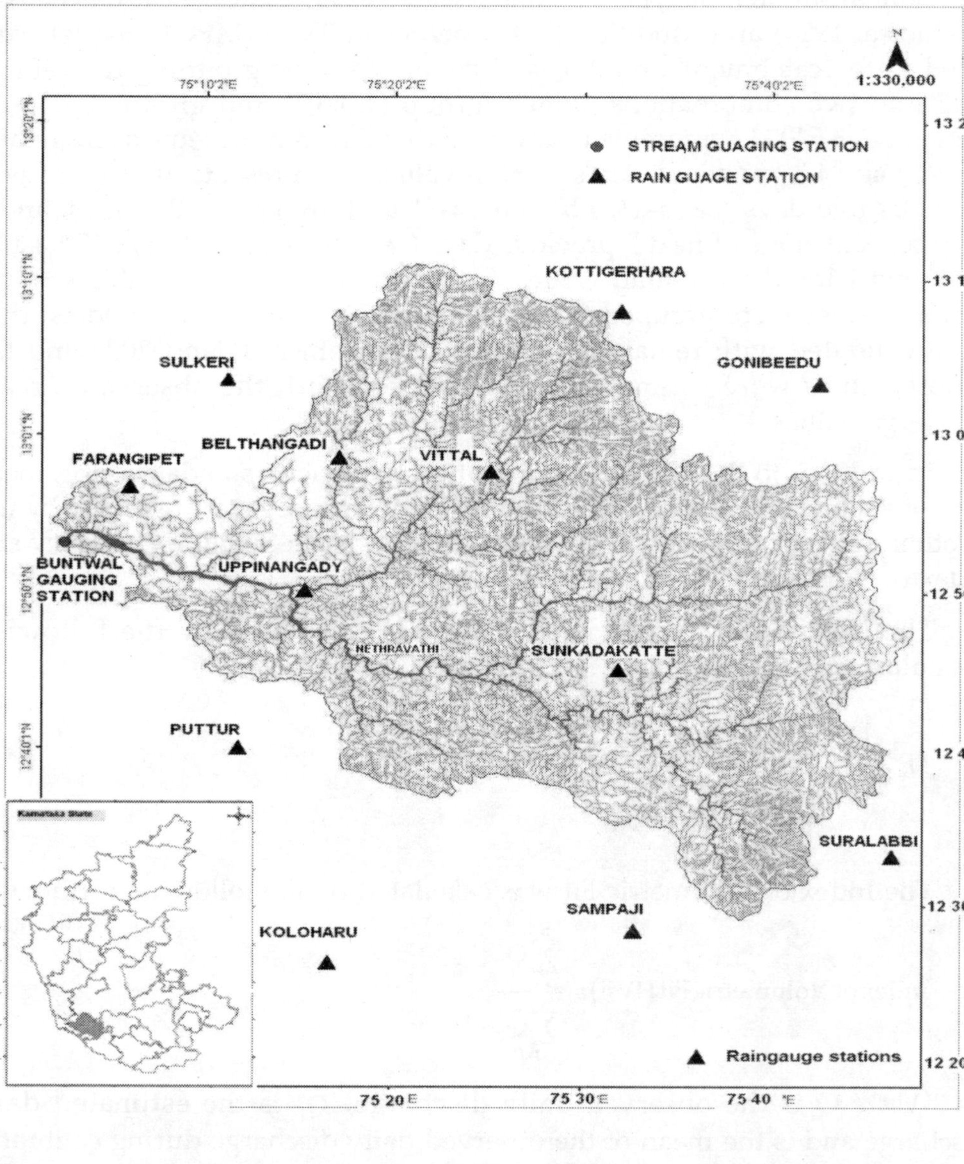

Fig. 1: Nethravathi catchment up to Bantwal gauging station and location of rain gauge stations

Similar exercise was done by raising the various past days' rainfall values to different exponent values ranging from 0 to 2, with increments of 0.1 considering, the non-linear behaviour of rainfall and runoff relationship of a catchment and this resulted in polynomial model of the form : $Q_i = C + m_1 P_i^{n1} + m_2 P_{i-1}^{n2} + ... + m_n P_{i-n-1}^{nn}$ where, $^{n1,\, n2,\, ...nn}$ are the exponents to be determined, m_1, m_2, ...m_n, and C are the regression coefficients and constant respectively to be developed. Further, additional three models, namely Posynomial Summation type (PST) model, Path Coefficient Type (PCT) models (Malhotra, 1994) and Modified Path Coefficient Type (MPCT) model were fitted to the catchment's rainfall and runoff data, by grouping the rainfall data into two combinations namely group (1,3,5,7) and group (1,2,3,4,6). Here, in (1,3,5,7), 1 represents i^{th} day rainfall value, 3 represents average of i-1st, i-2nd and i-3rd previous day's rainfall values, 5 represents average of next previous five days (*i.e.*, i-4th, i-5th, i-6th, i-7th and i-8th) rainfall values, and 7 represents average of next 7 previous days (*i.e.*, i-9th, i-10th, i-11th , i-12th, i-13th, i-14th and i-15th day) rainfall values. Similarly, for the group (1,2,3,4,6), the rainfall values were grouped. After calibration of the various models, they were validated with remaining 2 years data from 2004 and 2005 and the runoff values were estimated and compared with the observed stream discharge values.

In addition to the R^2 values, correlation coefficient, standard error, mean sum of squares, F-values *etc.* the models' performances were assessed by use of other indices of the model efficiency such as Nash-Sutcliffe efficiency and Index of Volumetric Fit (Yawson *et al.*, 2005).

The Nash-Sutcliffe efficiency (N_e) was calculated by the following formula:

$$N_e = \left(1 - \frac{\sum_{i=1}^{n}\left(Q_i - \hat{Q}_i\right)^2}{\sum_{i=1}^{n}\left(Q_i - \bar{Q}_i\right)^2}\right) \times 100$$

The Index of Volumetric Fit was calculated by the following formula:

$$\text{Index of Volumetric Fit}(IVF) = \frac{\sum_{i=1}^{n} Q_i}{\sum_{i=1}^{n} \hat{Q}_i}$$

Where Q_i is the observed daily discharge, $\hat{Q}_i$ is the estimated daily discharge and is the mean of the observed daily discharge during calibrated period and n is the number of data points.

RESULTS AND DISCUSSION

The various models developed by regression techniques for Nethravathi catchment are shown in Table 1. The final coefficient and regression constants obtained for various models at the final step are shown in the comparative statement in Table 2. Table 3 gives the summary of the statistical parameters for regression and correlation analysis performed for various rainfall-runoff models for Nethravathi catchment at Bantwal. In all these statistics, the same trend is noticed with respect to correlation coefficient, R^2 values, standard error, mean sum of squares, F value *etc.* for the eight models applied on Nethravathi catchment.

Table 1: Final discharge governing equations developed for all models for Nethravathi catchment up to Bantwal gauging station

Name of the model	Final discharge governing equation
Multilinear	$Q_i = C + C_0 P_i^1 + C_1 P_{i-1}^1 + C_2 P_{i-2}^1 + C_3 P_{i-3}^1 + C_4 P_{i-4}^1 + C_5 P_{i-5}^1 + C_6 P_{i-6}^1 + C_7 P_{i-7}^1 + C_8 P_{i-8}^1 + C_9 P_{i-9}^1 + C_{10} P_{i-10}^1 + C_{11} P_{i-12}^1 + C_{12} P_{i-12}^1 + C_{13} P_{i-13}^1 + C_{14} P_{i-14}^1 + C_{15} P_{i-15}^1$
Polynomial	$Qi = C + C_0 P_i^1 + C_1 P_{i-1}^{0.5} + C_2 P_{i-2}^1 + C_3 P_{i-3}^{0.5} + C_4 P_{i-4}^{0.5} + C_4 P_{i-5}^{0.5} + C_6 P_{i-6}^1 + C_7 P_{i-7}^{0.5} + C_8 P_{i-8}^1 + C_9 P_{i-9}^1 + C_{10} P_{i-10}^{0.5} + C_{11} P_{i-11}^1 + C_{12} P_{i-12}^{0.5} + C_{13} P_{i-13}^{0.5} + C_{14} P_{i-14}^{0.5} + C_{15} P_{i-15}^{0.5}$
PST 1357	$Q_i = C + C_0 A_0^{0.8} A_1^{0.3} A_2^{0.3} A_4^{0.3} + C_1 A_1^{0.4} A_2^{0.1} A_3^{0.2} + C_2 A_2^2 A_3^{0.3} + C_3 A_3^{0.4}$
PST 12346	$Q_i = C + C_0 A_0^{0.8} A_1^{0.2} A_2^{0.2} A_3^{0.2} A_4^{0.2} + C_1 A_1^{0.4} A_2^{0.1} A_3^{0.1} A_4^{0.2} + C_2 A_2^2 A_3^2 A_4^{0.2} + C_3 A_3^{0.4} A_4^{0.2} + C_4 A_4^{0.4}$
PCT 1357	$Q_i = C + C_0 A_0^{0.8} + C_1 A_0^{0.8} A_1^{0.2} + C_2 A_0^{0.8} A_2^{0.3} + C_3 A_0^{0.8} A_3^{0.3} + C_4 A_1^{0.6} + C_5 A_1^{0.6} A_2^{0.2} + C_6 A_1^{0.6} A_3^{0.2} + C_7 A_2^{0.4} + C_8 A_2^{0.4} A_3^2 + C_9 A_3^{0.6}$
PCT 12346	$Q_i = C + C_0 A_0^{0.8} + C_1 A_0^{0.8} A_1^{0.2} + C_2 A_0^{0.8} A_2^{0.2} + C_3 A_0^{0.8} A_3^{0.4} + C_4 A_0^{0.8} A_4^{0.3} + C_5 A_1^{0.6} + C_6 A_1^{0.6} A_2^{0.2} + C_7 A_1^{0.6} A_3^{0.2} + C_8 A_1^{0.6} A_4^{0.2} + C_9 A_2^{0.4} + C_{10} A_2^{0.4} A_3^2 + C_{11} A_2^{0.4} A_4^{0.3} + C_{12} A_3^{0.} + C_{13} A_3^{0.8} A_4^2 + C_{14} A_4^{0.4}$
MPCT 1357	$Q_i = C + C_0 A_0^{0.8} A_1^{0.3} + C_1 A_0^{0.8} A_2^{0.4} + C_2 A_0^{0.8} A_3^{0.6} + C_3 A_1^{0.8} A_2^{0.1} + C_4 A_1^{0.8} A_3^{0.2} + C_5 A_2^{0.4} A_3^{0.2}$
MPCT 12346	$Q_i = C + C_0 A_0^{0.8} A_1^{0.2} + C_1 A_0^{0.8} A_2^{0.3} + C_2 A_0^{0.8} A_3^{0.5} + C_3 A_0^{0.8} A_4^{0.5} + C_4 A_1^{0.8} A_2^{0.1} + C_5 A_1^{0.8} A_3^{0.2} + C_6 A_1^{0.8} A_4^{0.2} + C_7 A_2^{0.4} A_3^{0.1} + C_8 A_2^{0.4} A_4^{0.2} + C_9 A_3^{0.3} A_4^2$

Q_i = Discharge for the ith day,

$A_0 = P_i/1$, $A_1 = (P_{i-1} + P_{i-2} + P_{i-3})/3$, $A_2 = (P_{i-4} + P_{i-5} + P_{i-6} + P_{i-7} + P_{i-8})/5$ and $A_3 = (P_{i-9} + P_{i-10} + P_{i-11} + P_{i-12} + P_{i-13} + P_{i-14} + P_{i-15})/7$ (for 1357 rainfall grouping)

$A_0 = P_i/1$, $A_1 = (P_{i-1} + P_{i-2})/2$, $A_2 = (P_{i-3} + P_{i-4} + P_{i-5})/3$, $A_3 = (P_{i-6} + P_{i-7} + P_{i-8} + P_{i-9})/4$ and $A_4 = (P_{i-10} + P_{i-11} + P_{i-12} + P_{i-13} + P_{i-14} + P_{i-15})/6$ (for 12346 rainfall grouping)

P_i, P_{i-1}, P_{i-2}, ...P_{i-15} refers to rainfall for the ith day, i-1thday, i-2th day to i -15th day rainfall values.

C = Constant

C_0......C_9 = Coefficients

Table 2: Final regression constants and coefficients developed for all models in case of Nethravathi catchment

	Name of the Model →	Multi-linear	Poly-nomial	PST 1357	PST 12346	PCT 1357	PCT 12346	MPCT 1357	MPCT 12346
	Constant (C)	-0.275	-1.122	-0.1148	0.10418	0.46362	0.56182	0.3181	0.5338
Coefficients	C_0	0.2651	0.3014	0.03885	0.05603	-1.0979	-0.9962	0.0537	-0.0646
	C_1	0.1094	0.2107	0.87783	0.53604	0.33396	0.27624	0.0990	0.1383
	C_2	0.0613	0.1202	-0.0003	0	0.16404	0.1301	0.0446	0.0386
	C_3	0.0294	-0.033	1.08883	0.23127	0.32578	0.10872	-0.2554	0.0759
	C_4	0.0177	0.0930		0.88269	-0.6078	0.24184	0.4104	-0.3869
	C_5	0.0388	0.2457			0.16966	-0.6143	0.6247	0.2287
	C_6	0.0206	0.0486			0.60014	0.19034		0.2920
	C_7	0.0391	0.0747			0.5229	0.09528		-0.6211
	C_8	0.0091	0.0242			-0.0002	0.41777		1.1641
	C_9	0.0270	0.0322			0.41792	-0.0193		-0.0002
	C_{10}	0.0119	-0.062				-0.0001		
	C_{11}	0.037	0.0454				0.33795		
	C_{12}	0.021	0.1667				0.13919		
	C_{13}	0.0260	0.1925				0.0000		
	C_{14}	0.0285	0.2048				0.4344		
	C_{15}	0.0415	0.4336						

The variables and the coefficients to be handled are maximum (1+1+15) in case of both Multilinear model and Polynomial model but the R^2 values and correlation coefficient values are comparatively lower compared to the rest of the models. The number of coefficients to be handled are least (1+4) in case of PST 1357 model followed by (1+5) in case of PST 12346 model and MPCT 1357 model (1+6) (Table 2). PST 12346 model showed lowest Mean Square Deviation (MSD) of 31.53 compared to other models and highest MSD is found in case of Polynomial model (42.14). Similarly, standard error of estimate is lowest in case of PST 12346 model (5.61) and highest in Polynomial model (6.49). However, the correlation coefficient during calibration is highest (0.9209) in PST 12346 and these values are comparatively lower (0.8975 / 0.8931) in Multilinear and Polynomial models. The remaining six models have recorded comparative values of more than 0.91.

Table 3: Summary of statistical parameters for regression and correlation analysis performed for various rainfall-runoff models for Nethravathi catchment up to Bantwal

Name of the Model	Sum of Squares attributed to Regression	Mean Squares attributed to Regression	F - Value	Sum of Squares of Deviation about Regression	Mean Square of Deviation	Std Error of Estimate	Correlation Coefficient (Calibrated)	Coefficient of Determination (R^2)	Correlation Coefficient (Estimated)
Multilinear	461721.06	28857.57	712.16	111554.13	40.52	6.37	0.89750	0.8054	0.8967
Polynomial	457262.96	28578.93	678.19	116012.23	42.14	6.49	0.89310	0.7976	0.8882
PST-1357	483585.15	120896.29	3727.04	89690.04	32.44	5.70	0.91850	0.8436	0.9345
PST-12346	486137.02	97227.40	3084.03	87138.18	31.53	5.61	0.92090	0.8480	0.9379
PCT-1357	481987.94	48198.79	1456.73	91287.25	33.09	5.75	0.91690	0.8408	0.9290
PCT-12346	483812.83	32254.19	992.91	89462.37	32.48	5.70	0.91870	0.8440	0.9300
MPCT-1357	479159.99	79860.00	2344.50	94115.20	34.06	5.84	0.91424	0.8358	0.9271
MPCT-12346	481199.27	48119.93	1441.88	92075.93	33.37	5.78	0.91618	0.8393	0.9302

The daily runoff estimated by the various models developed by regression techniques are compared with the observed daily runoff from the catchment and observed daily mean areal rainfall of the catchment during 2004 and 2005 and it was seen that there is good performance by each of the individual models in better prediction of the catchment's daily discharges at the outlet. Estimation of runoff by all the models is comparatively good and exhibit conformity even in somewhat under prediction during high peaks in later part of the rainy season and better prediction during the rest of the time in the year. The graphical comparison of the estimated and observed runoff for year 2004 only is shown in Figure 2 for limitation of space.

It can be seen that, the Multilinear and Polynomial models gave Nash-Sutcliffe efficiency of 79.71 and 78.50 % while other models gave more than 85% Nash-Sutcliffe efficiency. Similarly the Index of Volumetric Fit was around 1.00 for all the models except PCT 12346 which gave 0.87. Such variations are common due to the fact that the nature and quality of the rainfall and runoff data used in model calibration and other unknown factors interfering with the hydrological process in the watershed may affect the model performance. On the whole, the prediction ability of all the models developed is good for Nethravathi River catchment. The calculated runoff by the different models is presented for a sample period in Table 5, for better appreciation of the model performance.

CONCLUSIONS

Thus, it can be summarized that for Nethravathi River catchment the mathematical rainfall-runoff models fitted quit well and it is possible that these regression models can be used as working models in estimation of runoff for most of the practical purposes in planning and actual operation of water resources projects in watershed developmental programmes. Further, it is concluded that using minimum data such as catchment rainfall, stream discharge, catchment area, *etc*, one can develop models and use the same for prediction of the runoff from those catchments or similar catchments. Earlier workers such as Singh and Frevert (2002) have successfully made similar attempts and came out with models for watershed runoff prediction, for water resources management.

ACKNOWLEDGMENTS

The authors acknowledge with thanks the Central Water Commission, Bangalore and Department of Economics and Statistics, Govt. of Karnataka, Bangalore for providing the required data for the study.

Table 4: Model performance parameters of various rainfall-runoff models developed for Nethravathi catchment.

Name of the model	Nash-Sutcliffe Efficiency (%)	Index of Volumetric Fit (IVF)
Multilinear	79.71	1.03
Polynomial	78.55	1.03
PST-1357	86.60	1.01
PST-12346	87.62	0.98
PCT-1357	85.73	0.99
PCT-12346	85.82	0.87
MPCT-1357	85.29	1.00
MPCT-12346	85.98	0.99

Table 5: Estimated runoff by various models developed by regression method along with the observed runoff and rainfall for Nethravathi catchment up to Bantwal in mm/day

Dates in July 2004	Observed		Estimated Runoff by the model							
	Rainfall	Runoff	Multi-linear	Poly-nomial	PST 1357	PST 12346	PCT 1357	PCT 12346	MPCT 1357	MPCT 12346
1	21.9	15.9	17.7	20.2	18.3	17.2	18.7	21.2	18.4	19.2
2	6.7	10.9	11.6	13.3	12.9	12.5	12.4	15.4	12.1	12.4
3	54.0	21.6	22.0	25.1	19.5	20.3	21.0	23.6	22.6	23.2
4	66.5	47.7	29.2	29.1	23.7	25.5	25.3	28.4	26.9	26.2
5	20.2	32.8	20.8	19.9	16.9	16.5	17.1	18.8	16.9	17.4
6	3.7	23.1	13.6	14.8	12.8	12.9	12.5	15.6	12.2	13.3
7	0.1	15.0	9.1	9.0	10.3	8.6	10.0	12.4	9.2	6.8
8	3.9	12.3	9.1	8.1	9.0	7.9	9.7	13.3	8.8	7.9
9	29.3	13.4	16.4	18.7	11.8	14.1	15.7	20.0	17.5	19.3
10	32.9	17.1	21.0	22.3	18.9	20.6	21.3	23.8	21.3	21.7
11	13.6	15.2	16.9	17.6	16.8	16.1	17.1	17.3	16.4	18.3
12	10.4	12.9	13.6	16.2	16.4	15.8	16.2	17.2	15.6	15.6
13	9.2	11.4	12.0	12.8	15.1	14.3	14.8	17.3	14.1	13.9
14	18.8	13.0	14.8	16.4	15.6	14.9	16.6	19.5	15.8	17.5
15	42.4	16.8	21.8	24.9	20.7	20.9	23.5	25.5	23.0	24.8
16	58.1	29.6	28.2	29.4	27.5	27.7	30.2	32.1	29.2	30.1
17	22.5	26.2	21.7	20.3	21.6	21.8	21.9	24.4	21.5	23.9
18	5.5	17.5	15.1	17.9	16.3	16.1	15.8	19.0	16.0	17.0
19	1.3	13.2	9.9	10.7	12.6	10.6	11.7	15.0	11.3	9.6
20	7.8	11.2	9.1	9.8	10.1	8.9	10.5	13.6	9.9	9.9
21	30.6	17.3	15.8	17.5	12.0	14.2	14.4	18.3	16.2	17.6
22	15.1	15.2	14.0	13.9	12.8	14.5	13.5	16.1	13.2	14.4
23	17.0	12.9	14.4	16.3	15.0	15.0	15.5	16.5	15.1	15.6
24	29.0	19.5	16.8	19.0	19.5	18.7	20.6	21.6	20.0	19.4
25	109.0	50.3	39.1	43.4	35.0	38.0	41.3	43.6	40.0	41.3
26	57.7	41.4	35.1	29.9	32.1	32.2	34.7	37.3	33.3	34.5
27	50.8	37.7	33.4	38.2	33.2	33.1	34.9	38.0	34.0	35.8
28	32.7	33.4	27.5	25.3	29.9	28.9	30.4	35.0	30.4	29.8
29	21.2	32.6	22.2	22.4	24.2	24.4	24.8	31.3	24.2	24.7
30	37.6	33.5	26.7	26.8	28.8	28.8	30.0	35.5	29.2	29.4
31	42.0	40.8	27.4	29.2	27.9	31.6	29.9	35.8	29.5	30.9

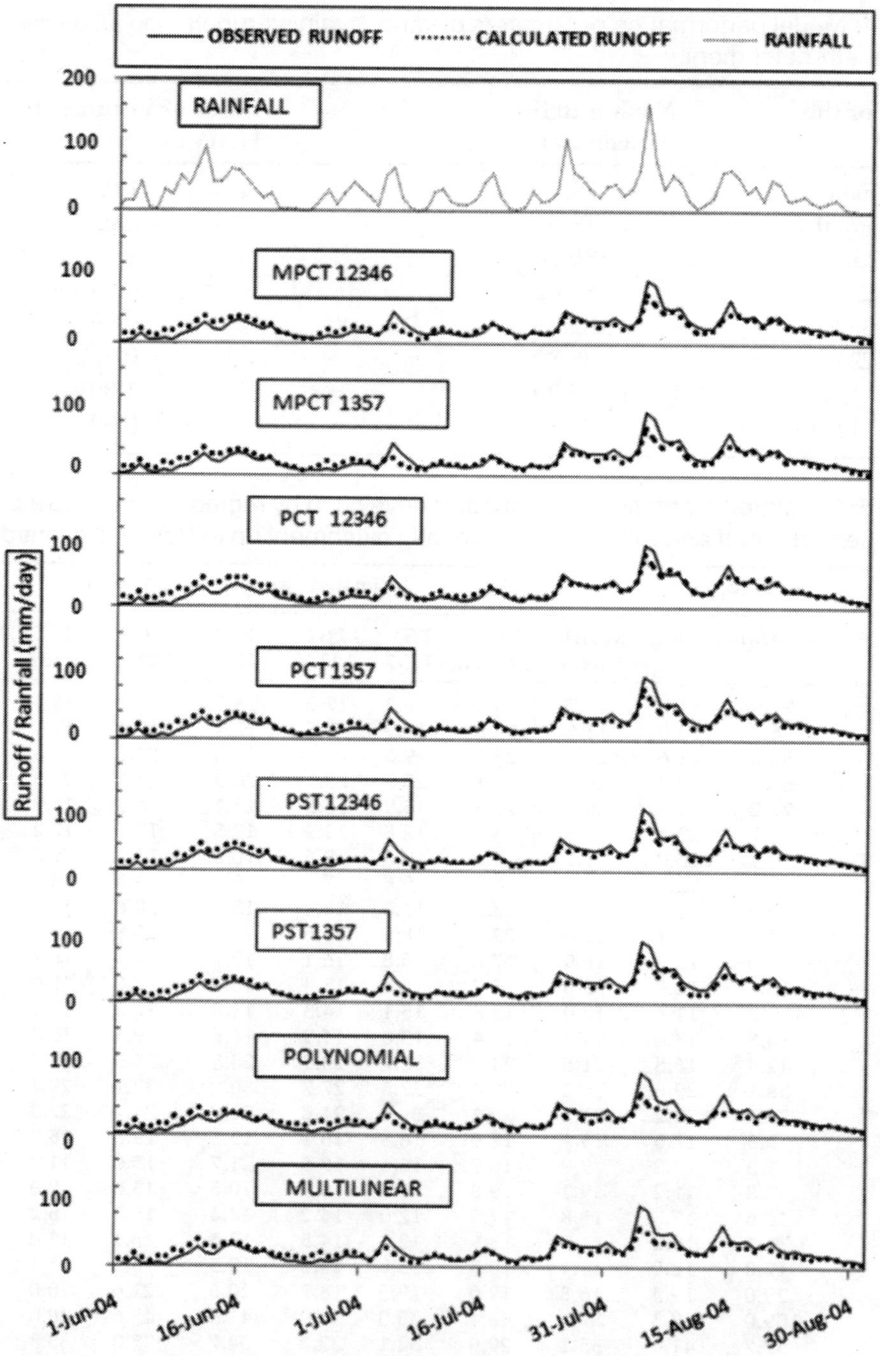

Fig. 2: Comparison of observed runoff with estimated runoff by different models developed through regression technique for Nethravathi catchment during 2004

REFERENCES

Ashu Jain and Prasad Indurthy, S.K.V. 2003. Comparative analysis of event based rainfall-runoff analysis – Deterministic, Statistical and Artificial Neural Networks. Technical Note, *Journal of Hydrological Engineering*, ASCE, March/April, 2003, pp93-98.

Kirti Malhotra 1994. A study of the Rainfall-Runoff models using regression and optimization techniques. *PhD thesis* submitted to Indian Institute of Science, Bangalore.

Kohler, M. A. and Linsley, R. K. 1951. Predicting the Runoff from Storm Rainfall, U. S. Weather Bureau, Washington , D . C., Research Paper No. 34

O'connor, K.M., Goswami, M., Liang G.C., Kachroo, R.K., and Shamseldin, A.Y. 2001. The development of Galway Real-Time River Flow Forecasting System (GFFS). In *Proc. 19th European Regional Conference on Irrigation and Drainage (of ICID)* (4-8 June 2001, Brno & Prague, Czech Republic).

Singh V.P. and Frevert O.K. 2002. *Mathematical Models of Large Watershed Hydrology and Application*, Water Resources Publication, LLC Highlands Ranch, Colorado, USA.

Venkata Rao, B.V. 1968. *Soil Resources of Mysore.*, UAS Extension Series No.2. University of Agricultural Sciencies, Hebbal, Bangalore : 13p.

Yawson, D.K., Kongo, V.M. and Kachroo R.K. 2005. Application of linear and Non linear techniques in river forecasting in Kilombero River basin, Tanzania. *Hydrological Sciences*, 50(5): 783-796.

□□□

Chapter 10

Optimum Dyke Height for Rainwater Conservation and Rice Yield Improvement in Rainfed Medium Land Ecosystem of Odisha

S. K. SAHU, S. C. SENAPATI[1] AND A. P. SAHU[1]

INTRODUCTION

Land and water are the two basic and essential resources which require proper planning for successful agriculture. Odisha is blessed with plenty of rainfall which occurs during June to September. During this period about 50 % of annual rainfall comes from few intense storms resulting into high runoff losses. In addition to this, the erratic nature of rainfall creates water stress at various growth stages of rice. Rice is the predominant crop grown by the people of Odisha during rainy season. Application of recommended doses of fertilizers and other cultural practices, the productivity of rice in the State is still less than the national average. Further, productivity of the rainfed rice eco-system, can be increased by adoption of improved water management techniques. Thus the rain water plays an important role in rice production under rainfed condition during rainy season. The area cropped with rice in the states of Odisha, West Bengal, Bihar and Assam is about 17.87M ha accounting 41.12 % of rice lands of the country (Anon, 1996). Therefore to increase the productivity of rice in the eastern parts of India and particularly in Odisha, effective utilization of rain water is necessary.

In a study at Bangladesh Rice Research Institute (BRRI, 1984) it was observed that with a supplemental irrigation of about 6cm can increase the rice yield up to 59 % and a pond size of about 5 % cultivable land is enough to store water for a supplemental irrigation. A well managed 15cm levee height field can effectively store 91 %of the rainfall and thereby reduce the yield loss

Ex-Sr. Scientist and Directorate of Research, Orissa University of Agriculture and Technology, Bhubaneswar, Odisha, India

[1]Department of Soil and Water Conservation Engineering, College of Agricultural Engineering and Technology, Orissa University of Agriculture and Technology, Bhubaneswar-751 003, Odisha, India

by 21 % (Islam and Mondal, 1992). Similar study at CRRI, Cuttack with different levee heights indicates that the effectiveness of rainfall was 70 to 96 % higher with a levee height of 15cm when compared to 7.5cm and without weir treatments (Mahapatra *et al.*, 1996). The above studies also indicated that in rainfed rice lands, dyke height around the rice fields is a deciding parameter in conserving the rainwater in-situ. Determination of optimum dyke height for conservation of rainwater, sediment, control of declining ground water table and improvement in rice yield has drawn attention of researchers in the recent past (Islam and Mondal, 1992; Mishra *et al.*, 1998; Khepar *et al.*, 2000). A study on *in-situ* conservation of rain water in rainfed rice fields by Mishra *et al.* (1998) revealed that weir height of 6cm and 30cm could store 57 and 99 % of rain water, respectively. An exponential relationship was observed between the runoff and weir height. They observed that conservation of rain water minimized the supplemental irrigation water requirement during dry spells as well as drainage needs of the catchments. They found that 20 to 25 cm weir height proved to be optimum weir height for *in-situ* conservation of rain water, nutrients and sediments under sandy loam soil for paddy fields in Bhubaneswar region.

Rainwater conservation and management is highly essential to improve the rice yield in rainfed conditions. Adoption of such technologies by the farmers is possible only if it is seen by them practically in the field. The present paper focusing on determination of optimum dyke height for rainwater conservation and enhancing rice yield in rainfed medium land ecosystem of Odisha.

MATERIAL AND METHODS

The work was carried out for four consecutive years (2001 to 2004) in the farmers' field near village Gourdiha of Shamakhunta block of Mayurbhanj district of Odisha state. The geographical situation of the site is 21° 45′ N latitude and 86° 48′ E longitude and comes under the Ojhrakhal watershed in the Shamakhunta block.

Soil texture of the experimental site varies from sandy loam to sandy clay loam. Hydraulic conductivity of the surface soil varies from 2.8 to 3.1cm hr^{-1} in the medium land conditions. Bulk density of 1.6 to 1.75gm cc^{-1} and field capacity varies from 0.26 to 0.41%. The location is characterized by warm and dry climate having hot and semi-humid summer and moderate winter. Average annual rainfall is 1441.5mm. The average maximum temperature varies from 33 to 42°C during hottest months and the minimum from 10 to 13°C during colder months.

Five experimental plots in the farmers' field were selected and leveled with a longitudinal slope of 0.1 %. The width of each plot was kept 20m and the length varied from 40m to 55.2m. The dyke height maintained in each plot was 30cm top width, 75cm bottom width and height 45cm. Five weir heights of 10, 15, 20, 25, and 30cm were provided at the down stream side of the experimental plots (Figure 1). Small refuges were constructed at the end

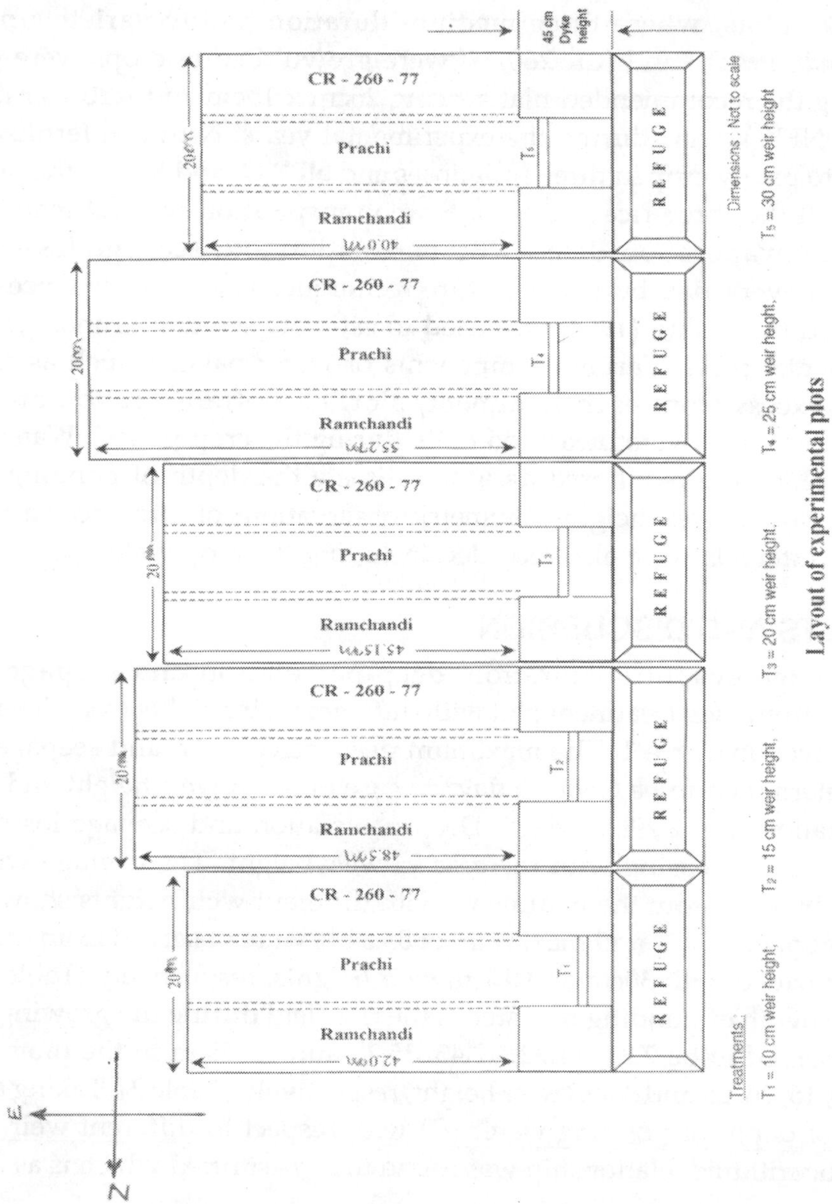

Fig.1: Layout of the experimental plots

of each plot but below the weirs. The length of the refuge was kept equal to the width of the paddy plot *i.e.,* 20m. Depth of the refuge was kept as 1.8m with bottom width of 1m and top width 4.6m. The refuge was constructed excluding net rice plot size of 25m x 20m located at the up stream side of the refuge. Five such plots along with refuges at the down stream side were laid with five weir heights (Figure 1). Each treatment plot was divided into three equal sub plots, where three medium duration paddy varieties namely *'Ramchandi, Prachi'* and *'CR 260-77'* were grown. Paddy crops were grown following the recommended plat spacing 20cm x 15cm and fertilizer dose of 60:30:30 NPK kg ha^{-1} during the experimental years. Nitrogen fertilizer was applied to paddy crop in three split doses and all P_2O_5 and K_2O at the planting time to all the three rice varieties. Evapotranspiration was estimated from daily pan evaporation data. Deep percolation and seepage losses were measured every day by drum culture technique. The water balance model developed by Mishra (1998) was used to estimate the daily runoff from the experimental plots. Various components of water balance such as rainfall, rainfall excess from each treatment/plot, evapotranspiration and deep percolation losses were measured daily during the crop period. Water levels in each plot were monitored daily to estimate the depth of ponding in rice fields at various weir heights. Biometric observations of three rice varieties at periodic intervals were also recorded including the crop yield.

RESULTS AND DISCUSSION

Average evapotranspiration, deep percolation and seepage losses occurred from each treatment plot with different weir heights were determined and is given in Table 1. The maximum deep percolation and seepage losses were determined to be 6.68mm day^{-1} in case of 30cm weir height and evapotranspiration was 2.29mm day^{-1}. Deep percolation and seepage losses were found to be increasing with increase in weir height. The average weighted runoff obtained from the treatments with different weir heights showed that a minimum of 9.82 % and maximum of 53.34 % were observed from treatment plots provided with 30cm and 10cm weir heights, respectively (Table 2). The average depth of ponding of water in the rice field during the growing period was observed to be 7.81, 11.72, 12.43, 15.75 and 19.75cm in the plots treated with 10, 15, 20, 25 and 30cm weir height, respectively (Table 3). Taking average values of depth of ponding water (D) with respect to different weir heights (H), a logarithmic relationship was found to be best fitted which is as follows,

$$D = 9.95 \, Ln(H) - 15.619 \text{ and } R^2 = 0.92$$

Table 1: Average evapotranspiration, deep percolation and seepage losses for different weir heights

Weir height (cm)	ET (mm day^{-1})	Deep percolation and seepage losses (mm day^{-1})	Total losses (mm day^{-1})
10	2.29	2.23	4.53
15	2.29	3.26	5.55
20	2.29	6.02	8.31
25	2.29	6.39	8.93
30	2.29	6.88	9.18

Table 2: Average weighted runoff for different weir heights

Weir height (cm)	Average weighted runoff (%)	Percentage retained (%)
10	53.34	46.66
15	43.20	56.80
20	18.29	81.71
25	12.52	87.48
30	9.82	90.18

Table 3: Average depth of water stored in rice field for different weir heights in different years during cropping season

Weir height	Depth of water stored (cm)				Average
(cm)	2001	2002	2003	2004	
10	8.35	7.41	7.66	7.83	7.81
15	12.83	10.79	11.65	11.62	11.72
20	12.54	11.02	13.46	12.71	12.43
25	16.43	13.26	17.48	15.82	15.75
30	20.77	16.67	21.61	19.95	19.75

Yield and yield attributes of rice varieties with different weir height treatments have been observed and found to be maximum in case of 20cm weir height (Table 4). All the three verities *i.e., 'Ramchandi, Prachi'* and *'CR 260-77'* gave highest yield of 5250 kg ha^{-1}, 5133.28 kg ha^{-1} and 4909.29 kg ha^{-1}, respectively for the same weir height. Based on yield and weir height, mathematical models were developed for all the rice varieties as given below.

For rice variety *'Ramachandi'* $Y = 3024.23 + 191.23H - 4.40H^2$ and $R^2 = 0.81$

For rice variety *'Prachi'* $Y = 3390.84 + 121.85H - 2.44H^2$ and $R^2 = 0.98$

For variety *'CR-260-77'* $Y = 3278.93 + 140.01H - 3.33H^2$ and $R^2 = 0.67$

Table 4: Yield and yield attributing characters for different rice varieties and weir heights

Rice variety	Weir height (cm)	Plant height (cm)	Panicle length (cm)	Panicles per m^2 (nos.)	Number of filled grains per panicle (nos.)	1000 grain weight (gm)	Yield (kg ha^{-1})
'Ramchandi'	10	78	23	315	127	16	4540.31
	15	79	26	328	138	20	4760.49
	20	82	30	342	145	25	5250.34
	25	77	25	308	136	22	4979.92
	30	80	22	297	130	18	4810.04
'Prachi'	10	81	21	298	245	17	4389.65
	15	78	24	307	272	19	4622.12
	20	83	28	318	287	23	5133.78
	25	79	22	302	267	21	4914.93
	30	77	20	289	258	18	4841.84
'CR-260-77'	10	79	19	245	119	16	4345.35
	15	81	23	272	128	18	4562.35
	20	84	27	287	138	22	4909.26
	25	77	21	267	130	20	4557.01
	30	78	28	258	126	19	4532.79

The optimum weir heights to be provided in the plots grown with different rice varieties were found out by differentiating the yield (Y) with respect to weir height (H) and equating to zero *i.e.*, $dy/dH=0$. Thus the optimum weir heights were determined to be 21.73cm, 24.97cm and 21.02cm for the plots grown with rice varieties *'Ramchandi'*, *'Prachi'* and *'CR 260-77'*, respectively. Irrespective of the rice varieties, the mathematical equation related to yield vs. weir height was developed as

$Y = 3231.33 + 151.03H - 3.39H^2$ and $R^2 = 0.85$ and the optimum dyke height was determined to be 22.28cm.

Considering the relationship model, the optimum weir height for efficient management of rain water in medium lands was found to be 22.28cm. Allowing 5cm free board over the weir height the optimum dyke height is determined to be 27.28cm which can be rounded off to 28cm.

CONCLUSIONS

Based on the results on seepage and percolation loss, depth of ponding, biometric characteristics and yield of rice crops in general, the optimum dyke height in medium lands was found to be 28cm for proper rain water management in medium land eco-system of Odisha.

REFERENCES

Anonymous. 1996. Agricultural Statistics at a Glance. Directorate of Economics and Statistics. Department of Agriculture and Cooperation, Ministry of Agriculture, Govt. of India.

BRRI.1984. Annual internal review report for 1983. Agricultural engineering Division, BRRI, Gazipur. Bangladesh:74p.

Islam, M.D. and Mandal, M.K.1992. Water management strategies for increasing monsoon rice production in Bangladesh. *Agric. Water Management*, 22:335-343.

Khepar, S. D., Yadav, A. K., Sondhi, S. K. and Siag, M.2000.Water balance model for paddy fields under intermittent irrigation practices. *Irrig. Sci.*, 19: 199-208.

Mahapatra, I.C., Sahoo,N., Bhuiyan, S. I. and Dinesh Chandra.1996. Management and effective use of water in low land rice ecology. *In*:Int. Symposium on rainfed rice for sustainable food security. 23-25 Sept., CRRI, Cuttack, India.

Mishra, A., Ghorai, A.K. and Singh, S. R.1998. Soil and nutrient conservation in rainfed rice lands in Eastern India. *Agric. Water Management*, 38: 45-57.

□□□

Chapter 11

A Simple Design Software for Drip Irrigation System

ANURADHA BEHERA, ASHWANI KUMAR[1] AND RANU RANI SETHI[1]

INTRODUCTION

Water is a main source for living. Water contributes a vital part in food production in form of irrigation. To achieve required food production with increasing population, India has to enhance the current irrigation potential otherwise the country will be water stressed in the coming years. Therefore hand in hand with technologies for water harvesting and storage, technologies for precision water application methods need to be adopted. Micro-irrigation technology is the most efficient irrigation method available as at present. Micro-irrigation mainly deals with drip and sprinkler irrigation. Drip irrigation is becoming popular day by day due to the use of simple techniques. Drip irrigation is one of the latest methods of irrigation which is increasingly popular in water scarcity areas. Drip irrigation is the slow application of water directly to the plant's root zone. This irrigation method can maintain optimum moisture level in the soil at all times, resulting in less water loss due to solar radiation and wind. Design of drip irrigation system depends on several factors like land, soil, crop, climate *etc*. Drip irrigation used for all types of crops. But the design of drip irrigation system is a measure problem in case of a farmer. To overcome this problem the project work was carried out for the development of simple software for drip irrigation system.

MATERIALS AND METHODS

Drip irrigation software has been developed for the easy design and installation. Material required can be calculated by using the software which is useful for design purpose. This software has been developed during the "In-plant training" conducted by College of Agricultural Engineering and

College of Agricultural Engineering and Technology, Bhubaneswar, Odisha, India
[1] Directorate of Water Management, Bhubaneswar, Odisha, India

Technology, OUAT, Odisha at Directorate of Water Management, Bhubaneswar, Odisha. It was a 4 month duration work from 1st august -31 November 2012. In partial fulfilment of Bachelor of Technology degree the work has been conducted. This paper gives a precise idea about the drip irrigation system design. This includes the layout, different design formulae and theoretical consideration which have been used for the software development. It also related with the configuration of the system and information about the language used to develop the software.

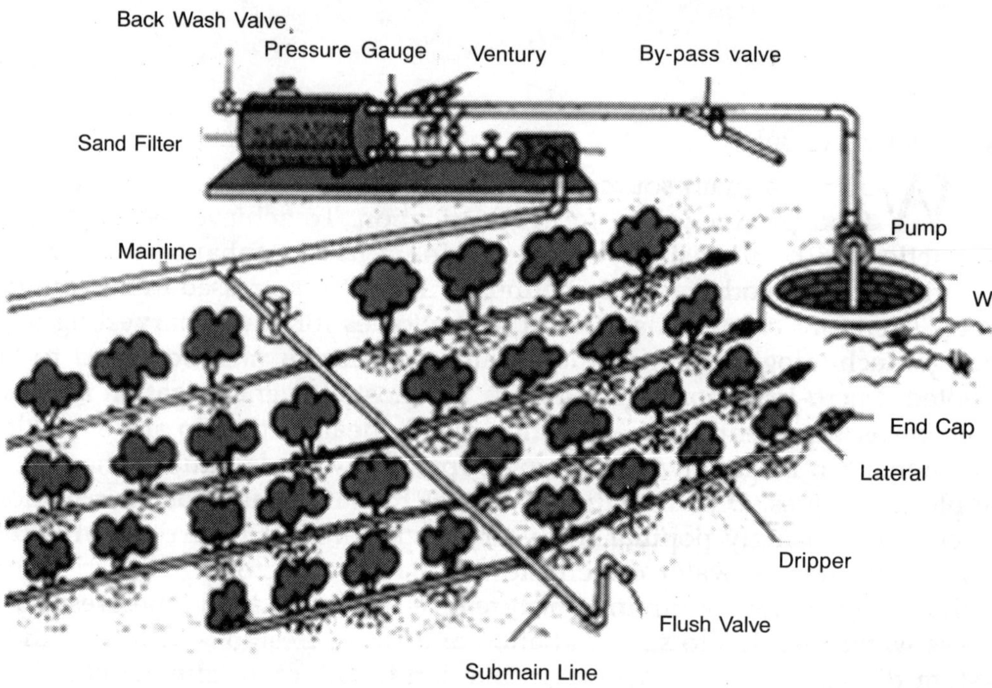

Fig. 1: Layout for Drip Irrigation System Design

Design of Drip Irrigation System :The data are to be collected before the design are survey of land, type of crop, soil type, land slope, climatic record, source of water & water quality. As per the layout shown in Figure 1, which indicate that the mainline passes through the half of the width of the land and act as the same sub-main line again which is equal to the length of the land, which is directly connected to the lateral lines which prohibits the water loss in the design as well as the frictional head loss.

Design of Drip Irrigation System :The data are to be collected before the design are survey of land, type of crop, soil type, land slope, climatic record, source of water & water quality. As per the layout shown in Figure 1, which indicate that the mainline passes through the half of the width of the land and act as the same sub-main line again which is equal to the length of the land, which is directly connected to the lateral lines which prohibits the water loss in the design as well as the frictional head loss.

Configuration of the system : HP Pavilion dv6 Notebook PC, Intel(R) Core(TM) i3 CPU M370 @2.40GHz 2.40 GHz, 3.00 GB (2.86 GB Usable) - RAM, 64-bit Operating System, Microsoft Visual Basic 6.0. The software for the drip irrigation design has developed in Visual basic languages following the steps to operate the software has been given in Figure 2.

The design steps are as followed which has been calculated in this software

Determination of Crop Water Requirement

Peak water requirement,

$$q = \frac{A \times PE \times PC \times KC \times w}{Eu} \quad \text{... (I)}$$

Where,

A = Area of the crop (m^2) = rr (m) × cc (m);

PE = Maximum pan evaporation, mm/day;

PC = Pan Coefficient, approximately taken as 0.7 to 0.8;

KC = Crop coefficient, the value of which for any crop depend upon foliage characteristics, stage of growth of crop, environment & geographical location suggested by FAO.

w = Wetted area of the crop, %. It is the area which is shaded due to its canopy cover when sun is over head, which depends upon stage of growth of plant. Percentage wetted area covered of crop varies in between 1/3rd to 2/3rd of area of the plant.

E_u = Emission uniformity of drip system, decimal.

Selection of Emitter or Drippers

No. of dripper or emitter required per tree,

$$n = \frac{w \times A}{\text{Effected wetted area by one Emitter}} \quad \text{... (II)}$$

$$\text{Total no. of plant, } Tp = \frac{A1}{A} \quad \text{... (III)}$$

Where,

A1 = Area of the farm, m^2

Total no. of Dripper, N = Tp × n ... (IV)

Selection of Design of Lateral

As per the layout design the sub-main passes through the centre of the plot

Hence the length of the lateral is equal to half of the length of the plot.

$$\text{Length of Lateral} = \frac{L1}{2} \quad \text{... (V)}$$

Where,
L1 = Length of the farm, m

$$\text{No. of plant on one side of the Lateral} = \frac{L_{lateral}}{cc} \quad \text{... (VI)}$$

Where,
$L_{lateral}$ = Length of Lateral, m
cc = Crop to crop spacing, m

Total no. of dripper on one lateral =
Number of plant in one lateral × Number of dripper per plant ... (VII)

Discharge rate of one lateral,

Q_l =Number of dripper in one lateral ×
Discharge through each dripper ... (VIII)

Diameter of the lateral is selected from the available sizes of pipe diameter 12mm, 16mm and 20mm.

Considering the thickness of the pipeline the inside diameter of the pipeline is selected.

Selection & Design of Sub-main

According to the design adopted for the drip irrigation submain passes through the middle of the field.

Length of the sub-main (L_{sub}) = Length of the farm (L)

Number of lateral on one side of the sub-main

$$N_{sl} = \frac{(L)}{(rr)} \quad \text{... (IX)}$$

Where,
L = Length of farm, m
rr = Row to row spacing, m

Total number of lateral on sub-main = $N_l \times 2$ ... (X)

N_l = Total number of lateral on sub-main

Discharge rate of sub-main, lit/sec $Q_{sub} = Q_l \times N_l$

Where,

Q_l = Discharge rate of one lateral,

N_l = Total number of lateral on sub-main

Selection & Design of Main line

As per the design layout

$$\text{Length of the mainline, Lm} = \frac{\text{Width of the land}}{2}$$

Discharge rate of mainline, Q_m = Discharge rate of submain Q_{sub}

Friction Loss Equation

The friction loss for pipeline can be computed from the Hazen-Williams or Darcy-Weisbach equation. From Watters & Keller (1978) the Darcy-Weisbach equation for smooth pipes in micro-irrigation system when combined with the Blasius equation for the friction factor,

Head Loss in lateral

$$\Delta H_l = K \times L \times Q^{1.75} \times D^{-4.75} \quad \text{... (XI)}$$

Where,

DH_l = Friction loss in lateral, m;

K = a constant = 7.89×10^5 S I unit for water at 20;

L = Length of the pipeline, m;

Q = Total flow in pipeline, lit sec^{-1};

D = Inside diameter of the pipe, mm.

If the frictional head loss is less than 1.1 m then the diameter for the pipeline selected is safe for the design (Lateral).

If the frictional head loss is less than 0.9 m then the diameter for the pipeline selected is safe for the design (Submain and mainline).

Selection of Pump

Total pressure head drop in meter due to friction,

$$H_f = \Delta H_l + \Delta H_{sub} + \Delta H_m \quad \text{... (XII)}$$

Total head in m,

$$H = Hs + Hd + h + H_f + Hl + D \quad \text{... (XIII)}$$

Where,

$\Delta H1$ = Head loss in lateral, in m

ΔH_{sub} = Head loss in submain, in m

ΔH_m = Head loss in main, in m

Hs = Suction head in m

Hd = delivery head in m

h = Operating pressure head of drip system in m

Hf = Total pressure head drop due to friction in m

Hl = Filter loss + Fitting loss + Ventury loss in m

D = Elevation difference in m

Power requirement of pump,

$$Hp = \frac{Q \times H}{75 \times \eta_{motor} \times \eta_{pump}} \quad \text{... (XIV)}$$

Where,

Hp = Power of the pump in horse power

H = Total Head in meter

η_{motor} = Motor efficiency in %

η_{pump} = Pump efficiency in %

Irrigation Time

It is the ratio of water requirement to the discharge of all the drippers per plant.

The software has been developed following his flow chart given below in Figure 2.

RESULTS & DISCUSSION

Software includes following steps for design of drip irrigation system

- Calculation of water requirement
- Selection of dripper
- Selection and design of lateral
- Selection and design of Submain and mainline
- Selection of pump and Irrigation time
- Final output design

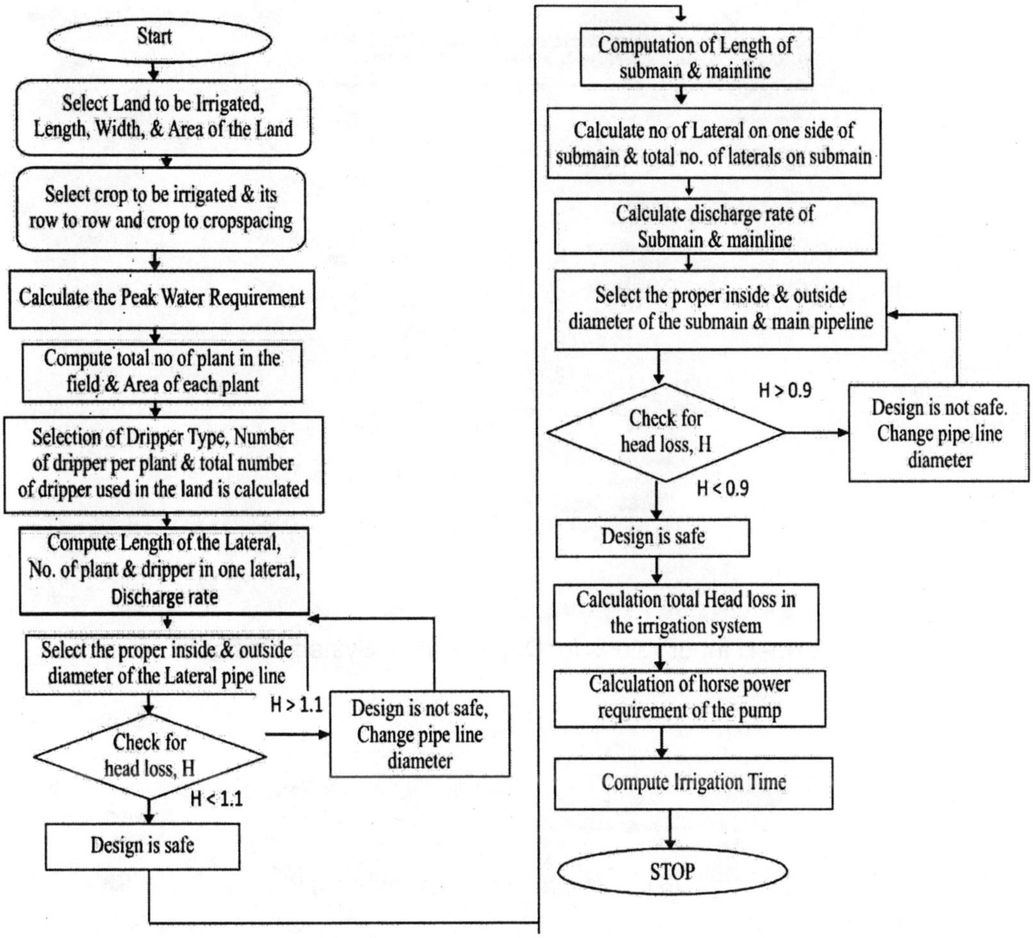

Fig. 2: Flow Chart for Drip irrigation Design (A Simple Design Software for Drip Irrigation)

For these steps five different windows have been shown with a trust model in a square land of 1ha with sapota plant with 6m x 6m spacing.

Instructions

- Here we have to enter the length and width of land, crop type and spacing.
- The maximum pan evaporation, pan coefficient, crop coefficient, wetted area; emission uniformity has been adopted from the prescribed model.
- For this input we can obtain the outputs for area of the tree and peak water requirement (higher maturity stage of crop)
- Click on Next button for further proceedings.

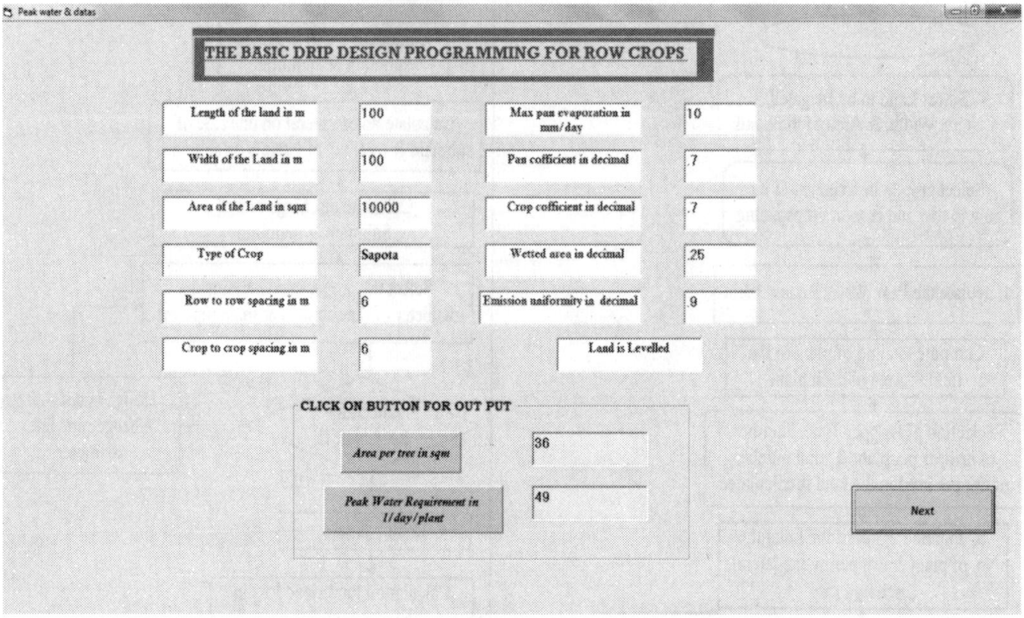

Form-1: Input values for Drip Irrigation System Design

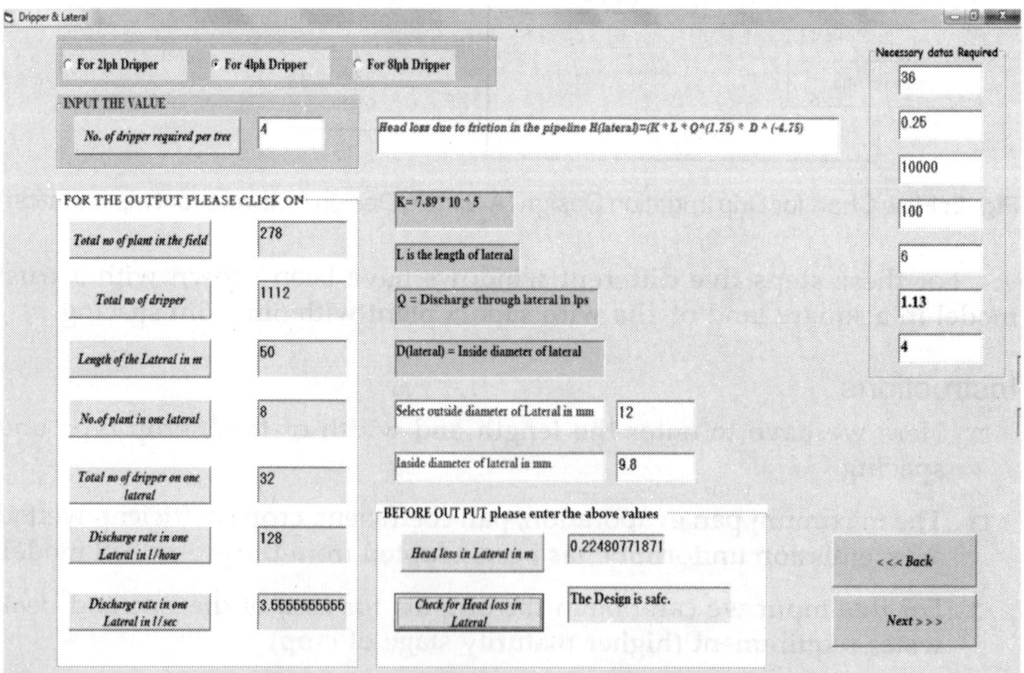

Form-2: Calculated values for dripper and Lateral design

Instructions

- Here we have to select for the dripper type from 2lph, 4lph, 8lph. Then we to select the number of dripper per plant it can be calculated by the formulae or it can be approximately taken from practical experience.
- Then the output obtained for total no. Of plant, total no. Dripper used in the field, length of lateral and discharge rate of lateral.
- Then we have to input for the inside and outside diameter of the lateral line according to the available size in market i.e. 12mm, 16mm, 20mm diameter.
- Then according to the recommended thickness of the lateral input diameter to be given.
- Head loss is calculated to check it is sufficient to carry the flow discharge or not.
- Click on Next >> button to go next form and for previous click on <<Back button.

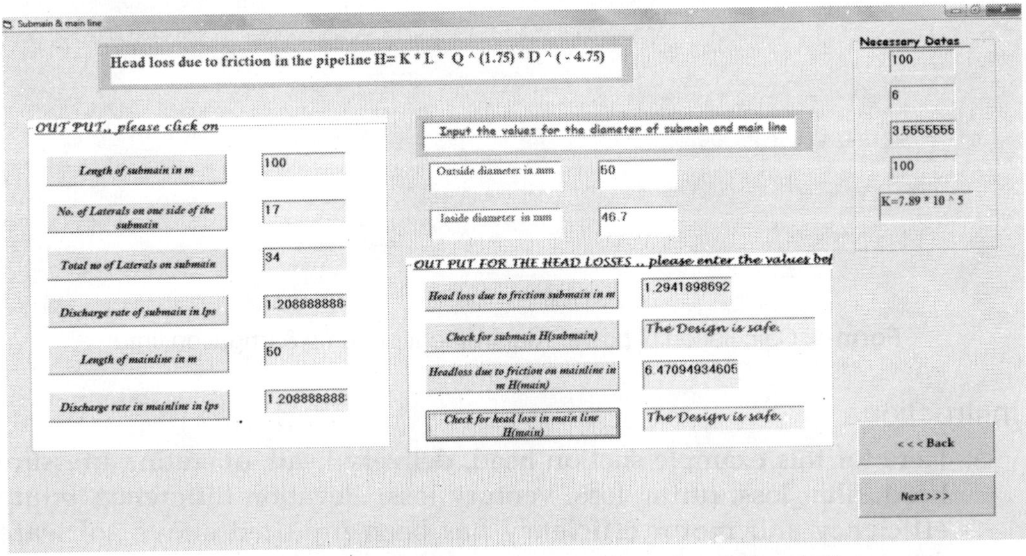

Form-3: Calculated values for submain and mainline design

Instructions

- Here first we can get the output for length of mainline and Submain line and both discharge rate.
- Then according to the calculated discharge rate the diameter of the main and Submain line is to be selected from the available diameter sizes in the market *i.e.* 32mm, 40mm, 50mm, 63mm, 75mm.

- Considering the thickness we can input for the inside diameter of the pipeline(For this design diameter of mainline and Submain line is equal because the mainline act as the Submain line here fig1).
- Head loss is calculated to check the diameter is sufficient to carry the flow discharge or not.
- Click on Next button >> go next form and for previous click on <<Back button.

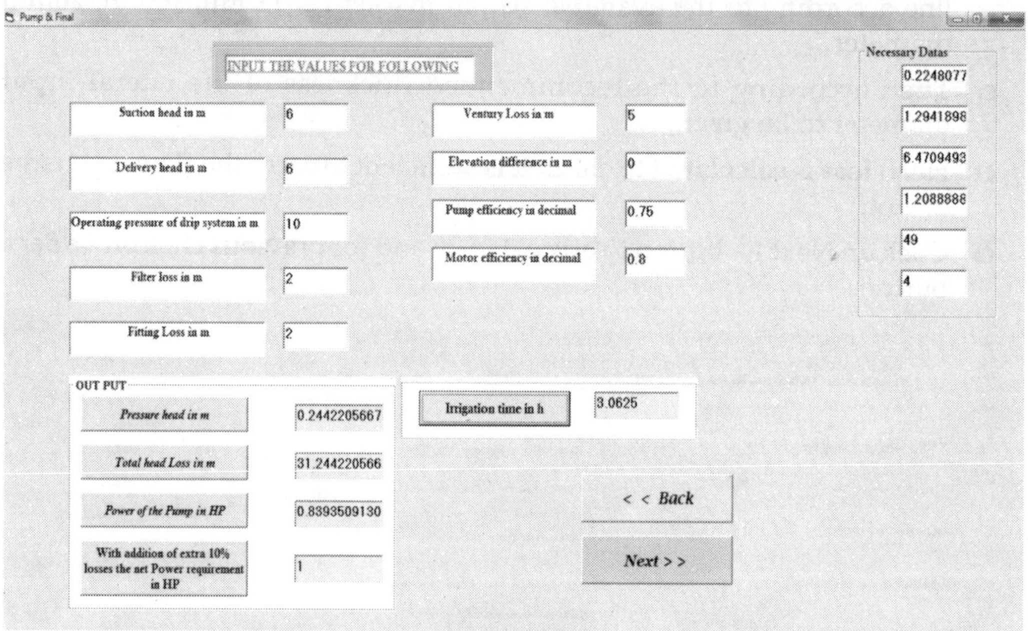

Form-4: Calculation of power requirement of Pump & Irrigation time

Instructions

- Here for this example suction head, delivery head, operating pressure head, filter loss, fitting loss, ventury loss, elevation difference, pump efficiency and motor efficiency has been inputted above software windows.
- The outputs obtained here are for head for the irrigation system, power of the pump and irrigation time.
- Click on Next >> button to go next form and for previous click on <<Back button.

This the final output pages for the drip irrigation system design where from we can obtain overall idea about the design and installation.

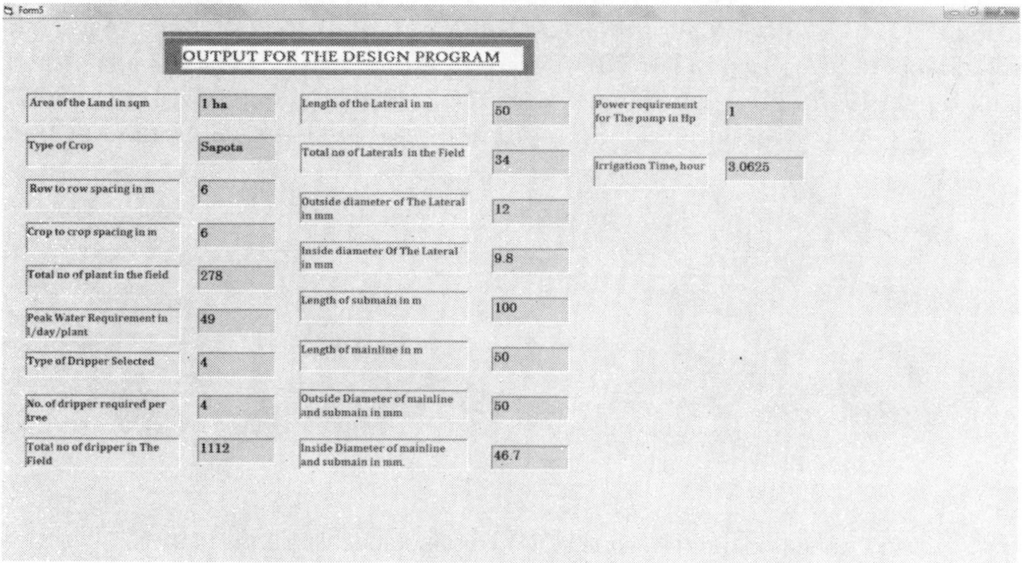

Form-5: Final output For the Drip Irrigation Design

CONCLUSIONS

The software was developed to in a way that it can not to accept any incorrect information. It is easy to design for drip irrigation system with help of this simple software. Estimation of material requirement can be worked out and as well as the cost of installation. The software can be used for all type of crops that can be irrigated by drip irrigation. It is a friendly tool for engineers and even for Progressive farmers.

REFERENCES

Bombale, V.T., Popale, P.G. and Magar, A.P. 2011. Development of software design of drip irrigation system. *International Journal of Agricultural Engineering*,4 (2): 170-175.

Mane, M.S., Ayare B.L. and Magar, A.P. 2006. Principle of Drip Irrigation System. Jain Brothers Limited

Michael, A. 1978. Irrigation Theory and Practices. Vikas Publication.

Petroutsos, E. 1998. Master in Visual Basic 6.0, BPB Publication, New Delhi, pp75-80.

Schwab, G.O., Fangmeir D. D., Elliot, W.J., and Frevert. 1992. Soil Water Conservation Engineering. John Wiley & Sons, Inc. Publication. Fourth Edition

□□□

Chapter 12

Assessment of Groundwater Resource in Eastern Odisha - A Case Study

B.K. SETHY, SHAKIR ALI, R.K.SINGH AND K.N.SHARMA[1]

INTRODUCTION

Groundwater is a replenishable dynamic resource that plays a major role in ensuring livelihood security across the world, especially in economies of developing countries. The socio-economic dependency on groundwater is explained over a range of factors by Burke and Moench (2000). They explained the intricacy in managing groundwater resources. At the same time, groundwater systems have become the "lender of last resort" and depletion of renewable groundwater stocks is taken as the first indicator of water scarcity (Shah and Indu, 2004). Moreover, groundwater is considered to be less vulnerable than surface sources to climate fluctuations and can therefore help to stabilize agricultural production and reduce the need for farmers to migrate when drought threatens agricultural livelihoods (Moench, 2002). In other words, groundwater resources provide a reliable drought buffer in large regions of the world (Calow *et al.*, 1997). The ability to access groundwater plays a major role in reducing risk and increasing incomes (Moench, 2003), especially when other modes of irrigation are not possible or difficult.

India has emerged as one of the largest users of groundwater in the world. Coastal areas of the country have the maximum concentration of population, centers of urbanization, industrialization and intensive agricultural activities. Increasing dependence on groundwater has resulted in indiscriminate extraction in some parts of the coastal region and there are some areas, where groundwater development is sub-optimal. As per estimate, Odisha state

Central Soil and Water Conservation Research and Training Institute, Research Centre Kota-324002 (Rajasthan), India

[1]College of Agricultural Engineering and Technology, Orissa University of Agriculture and Technology, Bhubaneswar, Odisha, India

receives an annual average replenishable groundwater resources 23.09 M ha m of which 2.08 M ha m goes as natural discharge, 3.85 M ha m as groundwater draft and 21.01 M ha m as net annual groundwater availability with stage of development 18.31% only. A huge quantity of available groundwater remains unutilized which necessitates study regarding groundwater potential in different areas of the state. The groundwater balance of the eastern parts (un-divided Cuttack district of Odisha) has been estimated to be 2.461 M ha m for future development (Anonymous, 2010). The future expansion of groundwater uses will continue because of population growth and subsequent growth in agriculture and industry. Hence, periodic monitoring of its quantity and quality are essential to identify the groundwater regime, identification of problems and ascertaining the causes; selection of suitable management scheme to cope up with the prevailing groundwater related problems, etc. Delineation of aquifers is the pre-requisite for the assessment of regional/local groundwater potential. The chemical quality of groundwater in any area depends upon the local lithology, climatic conditions including rainfall, topography and varies from place to place and time to time. The area under study is having abundant utilizable and replenishable groundwater resources and lies in sub-optimal development status. The area falls under coastal region of the state, the problems associated with sea water intrusion and salinity hazards may be reason for under development of groundwater resources. Considering the above facts, Bayalishmouza under Cuttack Sadar Block in Cuttack district was selected for study with the aims to delineate different level of groundwater availability, possible aquifer position and assessing the groundwater quality data in eastern Odisha through a case study.

MATERIALS AND METHODS

Study area : Bayalishmouza area comprising of 42 villages of Cuttack district in eastern Odisha was selected for study in 2011 which spreads 25 sq.Km area and lies between $20^{0}18^{\times}$ to $20^{0}23^{\times}$ N latitude and $86^{0}32^{\times}$ to $86^{0}39^{\times}$ E longitude. The river Kathajori and all distributaries of the river Mahanadi, bound the area. Though, the region is surrounded by the networks of rivers but the people of the area use groundwater for agriculture, domestic and industrial uses having the level of development up to 18.64% (Anonymous, 2010). The climate of the area is tropical having normal mean annual rainfall 1482 mm with highest rainfall 2269.5mm (1991) and lowest 1052.0 mm (1987). About 78% of annual rainfall is received during the month of June to September from southwest monsoon. The mean annual evapotranspiration loss of the area was estimated at 1885 mm. Malianala that passes centrally through the study area helps in reducing inland water congestion. The geomorphology of the area is having flat topography with Entisol and pH range of 6.25 to 7.61.

It is underlain by the coastal alluvium consisting of sand, gravel, clay, pebbles, boulders and Kankar *etc.*, ranging in age from tertiary to recent. The sediments have been deposited under fluvial to marine environment. Groundwater occurs both under unconfined and confined aquifers. The unconfined aquifer is saline in general, but fresh groundwater under unconfined condition occurs over different parts of the area. The major cereal crop is Paddy during Kharif followed by pulses, oil seeds and vegetables in rabi. A strip of *Casurina, Eucalyptus* and Cashew nut spread along the river Kathajori of varying width of 0.5-1.0 km. (Anonymous, 2010).

Data and Data Source : A map of the area to the scale of 1:15840(4½=1 mile) was collected from the department of Odisha Lift Irrigation Corporation Ltd., Cuttack. Out of total 169 numbers of tube wells, 57 tube wells were randomly selected for the study. Locations of tube wells are presented in Figure 1. Lithologs and discharges of these tube wells were collected from records available with Odisha Lift Irrigation Corporation (OLIC) and department of Odisha Groundwater Survey and Investigation (GWSI). In order to assess groundwater qualities, groundwater samples were collected randomly from 14 tube wells. The groundwater samples were analyzed for pH, sodium percentage (Na%), electrical conductivity (EC), sodium adsorption ratio (SAR), total dissolved solids (TDS), chloride content (Cl^-), chloride and bi-carbonate ($Cl: HCO_3$) ratio and magnesium and calcium (Mg:Ca) ratio with the standard established procedures (Todd, 1980; Garg, 1982).

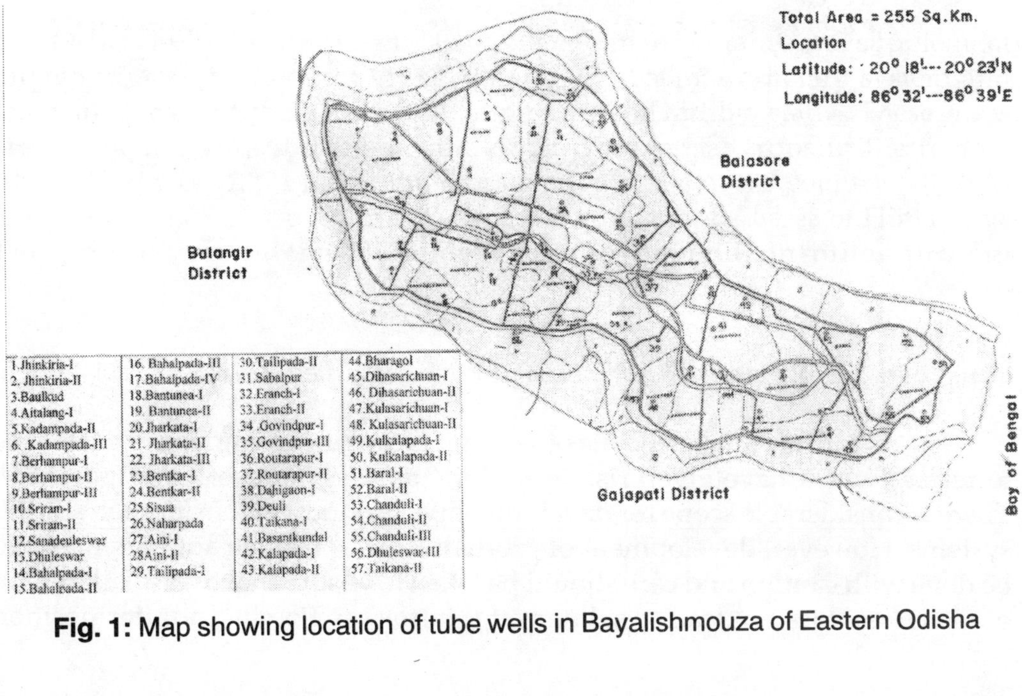

Fig. 1: Map showing location of tube wells in Bayalishmouza of Eastern Odisha

RESULTS AND DISCUSSIONS

Aquifer thickness and discharge of tube wells were arranged in ascending order separately from lithologs and pumping test data respectively for analysis. The groundwater quality data was analyzed based on standard methods and assessed for its suitability for irrigation.

Aquifer thickness: The analysis was carried out by arranging the thickness of aquifers at locations of tube wells in ascending orders. From lithologs of tube wells, aquifer thickness were classified into five zones namely Zone A (thickness <20 m), B (>20-<30m), C (>30-<40 m), D (>40-<50 m) and E (>50 m) Based on thickness of aquifers of tube wells, iso-aquifer thickness lines for 20,30,40, and 50m were drawn on the map (Figure 2)and presented in Table 1. It is observed from the lithologs of aquifer that most of the area of Bayalishmouza may be considered as zone C (43.88%) having a thickness of 30-40 m. Zone C is occupies maximum area followed by zones B (19.32%), A (15.8%), E (12.28%) and D (8.72%). Coarse sand is the major source of the aquifer responsible for groundwater yield.

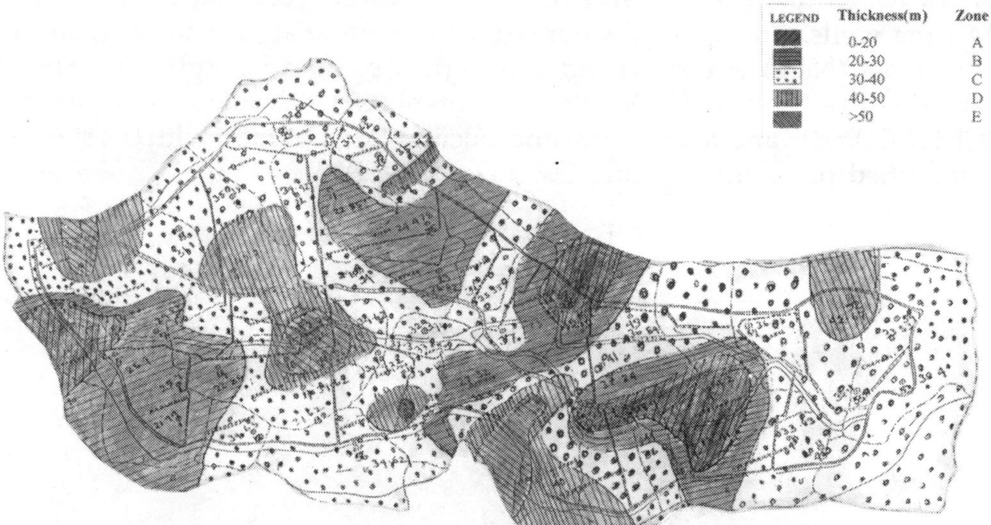

Fig. 2: Delineation of Aquifer position in the study area (Thickness basis)

The region has thick deposits of sediments ranging in age from Pleistocene to recent, which have given rise to multi-aquifer systems of good potential. There is considerable scope for development of groundwater from such aquifer systems. However, development of groundwater from such aquifers needs to be done with caution and care should be taken to ensure that over-exploitation of resources does not lead to saline water intrusion. Possible aquifer position in between two existing tube wells lying consecutive to each other may be predicted by interpreting the aquifer positions surrounding it.

Table 1: Tube well of different aquifer thickness zones, their number and location in Eastern Odisha

Aquifer thickness Zone with Area (km^2)	Tube well Numbers	Location in villages
A (3.95)	2, 3, 9, 36, 42, 43, 45, 46 and 50	Jhinkiria, Baulkud, Berhampur, Routrapur, Kalapada, Kalapada, Dihasarichuan, Dihasarichuan and Kulakalapada
B (4.83)	4, 5, 6, 7, 14, 21, 23, 24, 25, 38 and 48	Aitalang, Kadampada , Kadampada, Berhampur, Bahalpada, Jharkata, Bentkar, Bentkar, Sisua, Dahigaon and Kulasarichuan
C (10.97)	8,12, 13, 17, 18, 19, 22, 26, 27, 28, 30, 31, 32, 33, 35, 37, 41, 44, 47, 49,51, 53, 54, 55 and 56	Berhampur, Deuleswar, Deuleswar, Bahalpada, Bantunea, Bantunea, Jarakata, Naharpada, Aini, Aini, Tailepada,Sabalpur, Eranch, Eranch, Govindpur, Routrapur, Basantakundal, Bhargoal, Kulasarichuan, Kulakalapada , Baral, Chanduli, Chanduli, Chanduli and Dhuleswar
D(2.18)	11, 34, 40 and 52	Sriram, Govindpur, Taikana and Baral
E (3.07)	1, 15, 16, 20, 29, 39 and 57	Jhinkiria, Bahalpada, Bahalpada, Jharakata, Tailepada, Deuli and Taikana

The aquifer position varies from 3.05 m at Jharkata-II to 88.68 m below ground level at Kulakalapada-I with minimum thickness of aquifer 4.57m at Berhampur-I and maximum thickness of aquifer of 56.39m at Jharkata-I. Lower water table depth may lead to water-logging problems in due course of time by gradual rise of groundwater levels in some areas due to over and frequent flood irrigation without due regard to environmental considerations. Such areas offer good scope for further groundwater development as the shallow water table in such areas can be lowered down to six meters or more without any undesirable environmental consequences. The problems related to inferior quality of water in such areas can be solved by mixing them with available canal water. Judicious development through integrated use of surface and groundwater resources can greatly reduce the menace of water-logging and salinity in tube well irrigated areas. There are some areas where water table occurs in deep-seated aquifer levels. Fresh groundwater has been reported to occur beyond a depth of about 70 m below ground level (Sethy, 1999; Anonymous, 2007). Urgent action is required to augment the groundwater in the deep-seated aquifer areas. Large diameter dug wells, filter point wells and shallow tube wells are groundwater abstraction structures best suited for such aquifers. Radial wells and infiltration galleries can also be constructed in areas where the requirement of water is large. Further, sanctuary wells need to be constructed in hydrogeologically suitable areas to meet the unforeseen situations during cyclonic disasters as well as Tsunamis. Free flow of groundwater due to artesian conditions exists in some areas of eastern Odisha. As no energy is required for extraction of groundwater from such aquifers, development of groundwater from these auto-flow zones is both economically viable and eco-friendly.

Discharge : Based on measured discharge by pumping test from tube wells, iso-discharge lines of 20, 30, 40, and 50 litres per second (lps)were drawn on the map (Figure 3)and the area was divided into five discharge zones as A, B, C, D and E. The area was classified into different zones (Table 2) in ascending order namely, Zone A (discharge <20 lps), B (>20-<30lps), C (>30-<40 lps), D (>40-<50 lps) and E (>50 lps). Most of the area (50.48 %) of Bayalishmouza is considered in zone C and yields discharge of 30-40 lps. Zone C is better groundwater potential zone and can be utilized safely for groundwater exploration followed by zones B (21.32%), D (12.8%), E (10.6%) and A (4.8%), respectively.

Table 2: Tube well of different discharge zones, their number and locations in Eastern Odisha.

Discharge Zone with Area (km^2)	Tube well Numbers	Location in villages
A (12)	3, 7, 15, 23 and 24	Baulkud, Berhampur, Bahalpada, Bentkar and Bentkar.
B (5.33)	2, 4, 8, 10, 13, 18, 21, 29, 30, 31, 32, 33, 37, 40, 44, 45, 46, 49 and 53	Jhinkiria, Aitalang, Berhampur, Sriram, Dhuleswar, Bantunea, Taikana, Tailepada, Tailepada, Sabalpur, Eranch , Eranch, Routarapur, Taikana, Bhargoal, Dihasarichuan, Dihasarichuan, Kulakalapada and Chanduli.
C (12.62)	1, 5, 6, 14, 16, 17,19, 22, 25, 26 , 34,35, 43, 47, 48, 50, 51 and 54	Jhinkiria, Kadampada, Kadampada, Bahalpada, Bahalpada , Bahalpada, Bantunea, Jharkata, Sisua,

		Naharpada, Govindpur, Govindpur, Kalapada, Kulasarichuan, Kulasarichuan, Kulasarichuan, Baral and Chanduli.
D(3.2)	9, 20, 38, 52 and 55	Berhampur, Jharakata, Dahigaon, Baral and Chanduli.
E (2.65)	11, 12, 27, 39, 56 and 57	Srirama, Sandeuleswar, Aini, Deuli, Dhuleswar and Taikana.

Results indicated that area is a good repository of groundwater and offers excellent scope for development. Groundwater levels in this tract are mostly shallow, leaving little room for accommodating the monsoon recharge, a major portion of which flows down to the river as surface (flood) and sub-surface runoff (Base flow and recharge). A planned management of water resource in these tracts can capture the surplus monsoon runoff, which otherwise goes waste. The strategy for management in this area involves controlled withdrawal of groundwater from the flood plains during non-monsoon season to create additional space in the unsaturated zone for subsequent recharge/ infiltration during rainy season. There are two distinct conditions. The first condition involves setting up a hydraulic connection between the aquifer and the river as recharge boundary by heavy exploitation of groundwater and expansion of cone of depression. The hydraulic connection between the river and the aquifer ceases as soon as pumping is stopped. The second scenario is more common in case of rivers during intermittent flows; the loose sediments

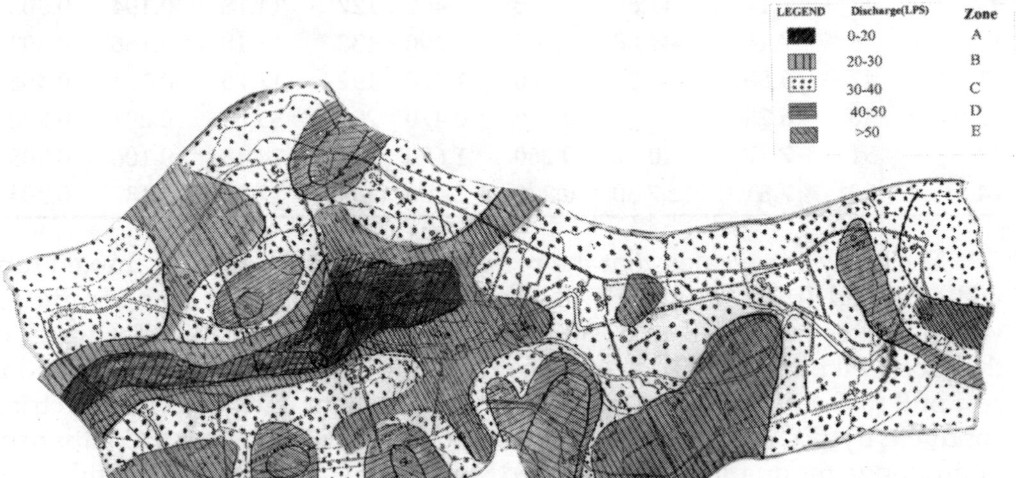

Fig. 3: Delineation of different level of Groundwater availability in the study area

in the flood plains are more or less saturated resulting into shallower groundwater level. The heavy withdrawal of such flood plain aquifers during the non-monsoon creates ample space in the groundwater reservoir which gets recharged by the river during the flood season. In absence of such created space the river water would overflow. This condition is more prevalent in eastern Odisha and provides opportunity for augmentation of groundwater reservoir through induced recharge.

Groundwater quality: The result of groundwater quality is presented in Table 3 and is compared with standard quality parameters for verifying its suitability for irrigation purpose (Todd, 1980; Garg, 1982).

Table 3: Groundwater quality of Eastern Odisha

S.No.	Tube well no.	pH	Na (%)	EC (mMhos cm^{-1})	SAR	TDS	Cl^- (ppm)	Cl: HCO_3	Mg:Ca
1	7	7.20	11.80	0.118	0.312	99	17.72	0.290	0.122
2	9	7.20	24.60	0.350	0.400	182	17.72	0.182	0.684
3	12	6.81	40.16	0.260	1.760	190	14.18	0.116	0.304
4	22	7.30	39.90	1.620	0.400	178	13.20	0.022	0.434
5	31	7.14	41.03	0.260	1.760	195	14.18	0.106	0.075
6	34	7.50	19.71	0.170	0.600	105	17.72	0.290	0.110
7	41	7.20	18.50	0.160	0.600	89	17.72	0.323	0.101
8	43	7.56	53.75	0.350	1.930	252	35.45	0.323	0.293
9	44	7.17	41.80	0.170	1.400	127	14.18	0.194	0.303
10	46	7.18	41.80	0.180	1.290	138	14.18	0.166	0.303
11	47	6.80	43.20	0.360	1.270	192	17.75	0.121	0.456
12	50	6.25	15.12	0.360	0.470	236	35.50	0.291	0.228
13	51	7.27	40.12	0.260	1.670	190	14.18	0.106	0.303
14	52	7.61	57.30	0.270	2.800	204	17.72	0.132	0.304

The analysis revealed that the pH value of groundwater of sampled tube wells ranged from 6.8 to 7.61. Based on pH values, Na% was computed, which ranges from 11.8 to 57.30. The pH and Na% indicates that water of the tube well numbers 43, 44, 46, 51 and 52 fall under permissible limit and in rest of 9 (65%) of tube well number under excellent category. The electrical conductivity (EC) values ranged from 0.16 to 1.62 mMhos cm^{-1}. Comparison of groundwater quality with standard values showed that water in tube well numbers of 7,34,41,44 and 46 is excellent (35%); tube well number 22 (7%) under permissible category and rest (58%) under good category for irrigation purposes. The SAR ranged from 0.312 to 2.80 ppm which indicated that water in all the tube wells is excellent for irrigation. The Total Dissolved Solids (TDS) values varied from 89.00 to 252.00 ppm. On comparison with the standard chart, it is found that water from all the tube wells is safe and suitable for irrigation purpose. Due to low values of TDS, there is no danger for sea water intrusion in the area.

The Chloride (Cl^-) content in the water samples varied between 13.2 and 35.45 ppm. According to the classification of Dalas & Day (1993), chloride ions are non-cumulative toxins; however, excessive amounts of Cl^-, if applied over a long period of time, can constitute a soil health hazard. Water in the tube well no 43 and 50, recorded relatively higher chloride levels compared to the other tube wells (Table 3). This will originate from natural sources such as the intrusion of sea-water, because of the nearness of the area to the coast. It may also be due to anthropogenic activities such as the intrusion of sewage effluent. The high chloride values in the waters of the tube well no. 43 and 50 may increase the corrosive ability of the waters in these areas, particularly in low alkalinity waters. The Chloride and bicarbonate ($Cl:HCO_3$) ratio of water samples ranged from 0.022 to 0.291 and due to the lower $Cl:HCO_3$ ratio, it is inferred that no sea water has intruded into the area and there is no overexploitation of groundwater in the area (Singh and Lawrence, 2007). The values of Magnesium and Calcium (Mg:Ca) ratio ranges from 0.075 to 0.6835 which indicates, the water in all the tube wells is excellent for irrigation purposes.

CONCLUSIONS

Management of groundwater resources in eastern Odisha is an extremely complex proposition as it deals with the interactions between the human society and the physical environment. The highly uneven distribution of groundwater availability indicates that no single management strategy can be adopted for the region as a whole. Any strategy for scientific management of groundwater resources involves a combination of supply side measures and demand side measures. The Supply side measures are aimed at increasing the groundwater availability, taking the environmental, social and economic factors into consideration under 'structural measures', which involves scientific development and augmentation of groundwater resources. Development of additional groundwater resources through suitable means and augmentation of the groundwater resources through artificial recharge and rainwater harvesting fall under this category. Though, the flood plains along the major river courses of the region offer good scope for groundwater development. The groundwater development is mostly demand driven; it can be geared up through proper agricultural, credits, subsidy and energy support policies along with creation of suitable markets. Similarly, there are some areas in the region with artesian condition, which can be mapped and suitable development plans formulated. There is an urgent need for coordinated efforts from various Central and State Government agencies, non-Governmental and social service organizations, academic institutions and the stakeholders for evolving and implementing suitable groundwater management strategies in the region.

REFERENCES

Anonymous. 2007. Preliminary Report on the Development of Groundwater Resources in Odisha, Water Resources Department, Government of Orissa.

Anonymous. 2010. Groundwater resources in Orissa. A Report by Department of Groundwater Survey and Investigation. (Government of Odisha), Bhubaneswar.

Burke, J .J. and Moench, M. 2000. Groundwater and Society. Resources, Tensions and Opportunities.New York, United Nations, 217p.

Calow, R. C., Robins, N. S., Macdonald, A. M. , Macdonald, D.M. Gibbs, J. B. R., Orpen,W. R. G., Mtembezeka, P., Andrews, A. J. and Appiah, S. O. 1997. Groundwater Management in Drought-prone Areas of Africa. *Int. J. Wat Res. Dev.*, 13(2): 241-62.

Dalas, H. F. and Day J. A. 1993. The Effect of Water Quality Variables on River and Ecosystem. Water Research Commission Report No. TT 61/93.

Garg, S.P., 1982. Groundwater and tube wells, Oxford and IBH publication. Second edition, 1982.

Moench, M. 2002. When Management Fails: Evolutionary Perspectives and Adaptive Frameworks for Responding to Water Problems. In: M Moench and A Dixit (ed.), Understanding the Mosaic, p 45.

Moench, M. 2003. Groundwater and Poverty: Exploring the Connections. In: R L lamas and E Custodio (ed.), Intensive Use of Groundwater: Challenges and Opportunities, pp 441-56.

Sethy, B.K. 1999. Groundwater Resources Investigation-A Case study of Bayalishmouza, Cuttack, post graduate thesis of Department of Soil and Water Conservation Engineering, College of Agricultural Engineering & Technology, OUAT, Bhubaneswar, 1999.

Singh, D.A.H. and Lawrence, J.F., 2007. Groundwater quality assessment of shallow aquifer using Geographical Information System in part of Chennai City, Tamil Nadu. *J. Geol. Soc. India*, 69, pp1067-1076.

Shah, T. and Indu, R. 2004. Fluorosis in Gujarat: A Disaster Ahead. IWMI-Tata Programme Annual Partner's Meet, Anand.

Todd, D.K., 1980. Groundwater Hydrology, John Willey publication, 1980.

□□□

Chapter 13

Identification and Characterization of Orchids belongs to North Costal Andhra Pradesh

K. RAVINDRA KUMAR AND M.RAJKUMAR[1]

INTRODUCTION

The Eastern Ghats are spread over four states from the Northeast to Southwest along the East Coast, covering an area of about 75,000 km, with an average width of 200 km in the north and 100 km in the south (Bhairavamurthy 1982). The average rainfall ranges from 1200 mm to 1500 mm. and mean annual temperature between 21^{0}C and 26^{0}C (Dani, 1982). The forest types that predominantly persist in the Eastern Ghats are the tropical semi-evergreen forests, moist deciduous and dry deciduous types inhabited by a number of terrestrial and epiphytic orchids (Jonathan *et al.,* 2005). Agency tracts of Visakhapatnam, Vizianagaram, Srikakulam and East Godavari are considered as one of the Bio Diversity hot spot for many terrestrial and epiphytic orchids.

Orchids are considered as the most fascinating plants because of their diversity and specialization in floral characteristics and morphological features. These are herbaceous plants that are classified in their own family, the orchidaceae, and are considered as the largest family among monocotyledons, containing 600 – 800 genera. In peninsular India there are about 200 species are present (Joseph, 1987) of these 67 are found in the state of Andhra Pradesh (Pullaiah, 1997). Padma *et al.* (1997) reported 19 epiphytic and 14 terrestrial orchids from the Eastern Ghats of Andhra Pradesh. Aside from the many wild species of this family the cultivated orchids constitute an order of royalty in the world of ornamental plants. Economically, they are of immense importance in horticulture. The wild species, on the other hand have been used as indicators of regions that have a healthy ecosystem (Mishra, 2004).

Horticultural Research Station, Chintapalli-531111, Andhra Pradesh, India.
[1]Fruit Research Station, Dr.Y.S.R.Horticultural University Sangareddy, Medak Andhra Pradesh, India.

However, one often over looked value of the plants of this family lies in the role that they have played in medicine. Ethnic groups of *Chenchus, Erukalas, Konda reddis, Koyas, Lamdas, Nukadoras, Valmikis* and *Yanadis* of East Godavari Visakhapatnam, Vizianagaram and Srikakulam districts are using Orchids as traditional drugs for a very long time (Reddy *et al.*, 2005). It has been reported that orchid molecules are important in reducing fevers, serving as anti-impotence aids, Increasing the white blood cell count, curing eye diseases, treating fatigue and head aches and most importantly, functioning as anti-cancer agents (Bulpitt, 2005). In this article, some terrestrial and epiphytic orchids have been addressed with reference to their occurrence, status, botany, medicinal and other horticultural uses.

MATERIALS AND METHODS

An ethnobotonical survey of the orchids of North Costal Andhra Pradesh agency tract was under taken during the years 2009 - 2011. This hill range lies between $17^0.13'$ North latitude and $84^0.33'$ East latitude. The altitude of explored area ranges from 700 - 1200 mts and the average annual rain fall is 1200 mm. Frequent exploration trips conducted to know the flowering time, place of Distribution etc.,. In this study, data on Vernacular name of the Plant, the Parts and Proportions that were used, the methods of preparation were collected from local tribes.

Enumeration

1. *Acampe ochracea*

Synonym: Saccolabium ochraceum

Place of Collection: Lothugedda, Chintapalli

Description: Stem 20 - 60 cm tall, producing slender, cord like roots all along the stem. Leaves fleshy, pale green, 10 - 15 cm long, 1.5 - 2.0 cm wide. Inflorescence axillary branched panicles 10 - 15 cm long, growing parallel to the leaves. Flowers numerous. Sepals and petals pale yellow barred transversely with red lip, white with a greenish yellow, spur 0.5 cm long and straight. Blade of the tip tongue shaped and margin crinkled. Column short. Flowering observed in November and having mild fragrance.

2. *Acampe praemorsa*

Synonyms: Epidendrum Praemorsum, Acampe wightiana

Place of Collection: Vangasara, Chintapalli

Description: Robust plant with Stout stem, Ca 30 cm long, covered by sheathing bases of leaves, with persisting old inflorescence axis and long stout aerial roots from among the leaves. Leaves alternate distichous, large, coriaceous, 12 - 28 X 2.3 cm, oblong, channeled, unequally deeply cleft at apex. Inflorescence short, erect, corymbose, panicles 6 - 10 cm long, leaf opposed. Flowers yellow, mildly sweet scented. Flowering observed in May.

Uses

1) Whole plant paste used for setting of fractures.
2) Root paste used for curing arthritis.

3. *Aerides odoratum*

Synonym: Aerides cornatum

Place of Collection: Chintapalli, Rajupakala, R.V.Nagar and Vangasara

Description: One of the most beautiful among the *Aerides spp*. It is an epiphytic herb. Leaves oblong, lanceolate. Inflorescence a raceme, axillary, slightly longer than leaves, 15-25 cm long, pendulous, laxly many flowered. 1 - 3 in each plant. Flowers sweet scented, white with rose purple at the tips of the perianth segments, about 2.5 cm across. Lip 3 lobed, spur large, curved forward. Capsule oblong, clavate. Flowering observed in 3rd week of May, June.

Uses

1) Can be used as excellent cut flower.
2) Leaf decoction is used against vermin fuse.

4. *Aerides ringense*

Synonym: Aerides radicosum, A.lineare

Place of Collection: Vangasara

Description: Robust epiphyte. Stem thick with persisting sheathing bases of the leaves. Leaves alternate, distichous, articulate and sheathing at base, 6.5 - 24.5 X 1.0 - 2.5 cm, linear, unequally cleft at apex, conduplicate, coriaceous. Inflorescence racemes, 7.35 cm long, simple or branched in the axils of the upper leaves, erect, rigid, lax flowered. Flowers pink or mauve, long pedicellate. Bracts small triangular, acute. Lateral sepals broader than the dorsal sepal. Lip trilobed, spurred, dark pink, side lobes small, erect, obtuse. Fruits 3 cm long, obovoid, oblong (ribbed). Flowering observed in the month of June.

5. *Bulbopyllum neligherrence*

Place of Collection: R.V.Nagar, G.K.Veedhi

Description: Small herbs, Pseudo bulbs conical - ovoid. Leaves tapering into short petiole, oblong - elliptic, obtuse, emarginate. Inflorescence racemes, shorter than leaves, arising from the base of the pseudo bulb. Flowers dense, dull yellow. Lip tongue shaped. Flowering observed during the month of September.

Use

1) Tuber extracts used for good health.

6. *Coelogyne breviscapa*

Place of Collection: Lankapakala, G.K.Veedhi, Sapparala and Darakonda

Description: Pseudo bulbs erect, narrowly oblong, serially arranged on horizontal creeping stem of short intervals, covered by sheating scales. Leaves 2 - 3 on mature bulbs, linear - elliptic, acute at tip, narrowed into a short, keeled petiole. Inflorescence racemes, arising from the base of the pseudo bulb at the apex of the creeping stem. Flowers white slightly scented. Fruits ellipsoid, with marcescent perianth. Flowering observed during the month of April-May.

7. *Cymbidium aloifolium*

Synonym: *Epidendrum aloifolium*

Place of Collection: G.K.Veedhi, Sileru and Vangasara

Description: Locally called as Manu Vajinika or Pedda Vajinika. An epiphytic herb, pseudo bulb 4 – 5 cm long, ovoid. Leaves linear – oblong, 30 – 60 X 2 – 4 cm. Scapes and racemes 20 – 50 cm long, erect below and decurved above. Flowers yellowish red, lateral lobes mottled with purple. Inflorescence long pendulous racemes. Flowering observed during the months of April, May, June and July.

Uses

1) Aerial root paste used for setting of fractures and applying to foot cracks.

8. *Dendrodium aphyllum*

Synonym:*Limodorum aphyllum, Dendrobium Pierardi*

Place of Collection: Chintapalli, Lothugedda, Rajupakala and Vangasara.

Description: Epiphytic herbs, slender stem, pendulous. Leaves 3.5 – 7 X 1 – 2.2 cm, Lanceolate, acuminate, flowers pale rose on short peduncles. Inflorescences drooping with three flowers from each node, flowers have long stalks, petals are pale rose/ light violet coloured, lip golden yellow at top with dark violet blotch at centre. Observed flowering during the months of April and May. Popularly called as "Thousand bloom orchid".

Uses

1) This orchid can be used as a garden plant.

9. *Dendrobium herbaceum*

Synonym:*Dendrobium graminifolium*

Place of collection: Darakonda

Description: Epiphytic orchid, erect, bunchy herb. Stems caespitose, yellowish, shining. Leafless during flowering. Inflorescence a condensed raceme. Flowers creamy white to yellow. Lip attached to the foot of the column, refluxed at the base, five nerved. Observed flowering during the period of January to March. Fruiting observed during the month of June to July.

Uses

1) The tender tip juice is used for ear ache.
2) Fresh leaf paste is used for treating syphilis.

10. *Dendrobium heterocarpum*

Synonym:*Dendrobium aureum*

Place of collection: Chintapalli

Description: Stem occurring in tufts, clavate, thick, 15-25 cm long, 1-2 cm across, very slender below, green, brown or yellow in colour, covered with ribbed sheaths. Leaves membranous, 6-7 cm long, 3cm wide. Flowers in pairs from the top most nodes, appearing after the leaves have fallen off. Yellowish brown very fragrant. Flowering observed during the months of December to February.

Uses

1) This orchid can be used as a garden plant.

11. *Dendrobium microbulbon*

Synonym:D. humile

Place of collection: Chintapalli, G.K.Veedhi and Darakonda.

Description: Herbs in clusters. Pseudo bulbs globose, oblong, 2 or 3 noded, 3-4 leaved. Leaves 2.5 - 6.5 x 0.5-1.4cm, narrowly elliptic – lanceolate, acute, sheathing at base. Inflorescence terminal racemes, lax flowered, 12 cm long, filiform with few sterile bracts. Flowers white with purplish green lips. Flowering observed during the months of June to August.

12. *Dendrobium pierardi*

Synonym:D. cucullatum

Place of Collection: Chintapalli

Description: Stems pendulous, glabrous, leaves ovate - lanceolate, acute, flowers in pairs, forming a spurious raceme. Sepals acuminate, membraneous, petals larger than the upper sepal, acuminate, lip dilated, cucullate, some what truncated, citiated, flowers pale yellow. Flowering observed during the month of August and fruiting observed in September.

13. *Eulophia epidendraea*

Synonym: *Serapias epidendraeae*

Place of Collection: Dharakonda, Sileru ghat roads.

Description: Popularly called as Sega domma gadda. A terrestrial herb, pseudo bulb green. Leaves linear, 30-60 cm long. Flowers yellowish green in spikes. Lip white with pink veins, nerves and crests red. It is found in rocks.

Uses

1) Tuber extracts are used for curing anorexia.

14. *Geodorum densiflorum*

Synonym: *Limodorum densiflorum, Geodorum purpureum*

Place of Collection: Chintapalli

Description: Popularly called as 'Donthula gadda'. A terrestrial herb, stout, with swollen root stock. Leaves elliptic - lanceolate. Inflorescence crowded raceme. Flowers purplish white. Lip purple streaked within. Flowering observed during the month of August.

Uses

1) Tuber extract can be used for controlling cattle fever.

2) Root paste can be used for regularizing menstrual cycle in women.

15. *Hebenaria elliptica*

Place of Collection: Chintapalli

Description: Herbs, 20 – 38 cm. Tubers 1 – 2, ovoid oblong. Stem with 1 or 2 sheaths at the base. Leaves 3 - 4, elliptic - linear, lanceolate acute at tip, sheathing at base. Inflorescence upto 25 cm with the scape. Many flowered, dense. Flowers bracteate, short pedicelled. Observed flowering during the month of September.

16. *Hebenaria furcifera*

Place of Collection: Chintapalli

Description: Popularly called as 'Nela Jimmi gadda'. A terrestrial herb, up to 50 cm tall. Leaves oblong – lanceolate to broadly elliptic, sheating at base. Raceme lax, many flowered. Flowers small, green. Lip trifid from the base. Spur slightly longer than the ovary.

Uses

1) The tuber paste is used as an ointment for cuts, wounds and poisonous bites.

17. *Hebenaria plantaginea*

Place of Collection: Chintapalli

Description: Popularly called as 'Kusuma gedda'. A slender, terrestrial herb. Leaves radical. Stem up to 30 cm long. Flowers pure white. Lip much longer than the sepals, spur green, 15 – 25 mm, longer than ovary.

Uses

1) The tuber paste can be used for controlling chest pain and stomach pain.
2) The tuber paste used for applying scorpion sting and also used for cuts and wounds.

18. *Luisia zeylanica*

Synonym:*L.teretifolia*

Place of Collection: Chintapalli

Description: Popularly called as 'Gundu Bandanica'. Herbs with terete stems, sheathed. Leaves alternate, sheathing at base. Inflorescence condensed racemes, extra axillary. Flowers 2 – 4, greenish Pink, Lip concave towards the base, trilobed with dark maroon horizontal bands. Fruits oblong, winged. Flowering observed during the month of April to July.

Uses

1) Whole plant extract is used for setting of fractures.
2) Tender shoots paste is used for curing pus forming wounds.

19. *Malaxis rheedii*

Synonym:M.versicolor

Place of Collection: Chintapalli, R.V.Nagar.

Description: Popularly called as 'Jeeva Gadda'. Terrestrial, tuberous herb to 30 cm high. Stem errect, stoloniferous, leaves ovate - oblong to elliptic. Flowers yellow. Lip reniform, reflexed. Flowering observed during the months of June - July.

Uses

1) The tuber paste is applied externally for insect bite.
2) The tuber paste also used for curing rheumatism in cattles.

20. *Nervilia aragoana*

Synonym:Pogonia flabelliformis, P.carinata

Place of Collection: Chintapalli

Description: Popularly called as 'Vantaelu dumpa'. Terrestrial, Small tuberous herbs. Leaf appearing after the flowers, orbicular, cuspidate, base cordate. Sepal and petals pale green, veined with purple. Lip 3 - lobed, white. Flowering observed during the month of May - June.

Uses

The leaf paste / tuber paste is used as an ointment for wounds.

21. *Phaius grandifolius*

Place of Collection: Chintapalli

Description: Leaves often 40 - 60 cm long. Spikes stout, many flowered, 90 - 120 cm long, with 10 cm wide flowers. Sepals and petals yellow - brown, with silvery white back. Lip tubular, yellow brown with rosy purple mouth. Flowering observed during the month of November to March.

22. *Pholidota pallida*

Synonym:P. imbricta

Place of Collection: Lankapakalu, G.K.Veedhi, Sapparla and Darakonda.

Description: Pseudo bulbs in clusters, large, swollen, green, oblong - conical, covered by large scaly bracts. Leaves elliptic - oblanceolate, acute, narrowed into short petiole. Inflorescence racemes, pendulous. Flowers white, arranged in two rows. Lip deeply saccate, sac with 3 longitudinal ridged within, 3 - lobed. Flowering observed during the month of July. Fruiting observed during the month of October.

Uses

1) Pseudo bulb Paste is used as an external application for rheumatic swellings.

23. *Rhynchostylis retusa*

Synonym: *Epidendrum retusum, saccolabium guttatum*

Place of Collection: Vangasara and Araku.

Description: An epiphytic, stout stemmed herb. Leaves coriaceous, linear, 14 - 35 X 1.5 - 2 cm, strap - shaped, recurved, margin entire, open unequally lobed. Flowers showy, dark pinkish - violet drooping raceme, about 20 cm long. Flowers about 1.2 cm across, sepals broad, petals smaller, lip adnate to the column, spur saccate. Capsules obovate, 6 - ridged, to 3 X 1.5 cm. Observed flowering in the month of April to July. Fruiting observed in August - September.

Uses

1) One of the most elegant of the epiphytic orchids with long pendulous dense inflorescence. This orchid can be used as cut flower.
2) Fresh plant extract used as an external application for inflammation of wounds and centipede bite.

24. *Vanda tessellata*

Synonym: *Epidendrum tessellatum, Vanda roxburghii*

Place of Collection: Chinnagedda, Ebulam and Chintapalli.

Description: Popularity called as 'Badanika', 'Kodikalla Chettu', 'Ippa Vajinika'. A stout epiphyte. Leaves linear oblong, thick, coriaceous. Raceme 4 - 10 flowered. Flowers large, yellowish green, with bluish tinge, scented. Lip shorter than perianth, yellowish, with pink mid lobe, trilobed. Obśerved flowering during the month of June.

Uses

1) Root paste is applied externally to control rheumatism.
2) Extract of velamen tissue is used for control dysentery.

25. *Vanda testacea*

Synonym: *Aerides testaceum, V.parviflora.*

Place of Collection: Chintapalli, G.K.Veedhi and Darakonda.

Description: An epiphytic, long, stout, stemmed herb on huge trunks. Leaves fleshy, distichous, linear, keeled, at tip. Inflorescence racemes, 2 or more per plant, axillary, erect, fruiting axis elongates, scape thick, terete with few sterile sheating bracts. Flowers yellow, long pedicellate. Lip trilobed, reflexed. Observed flowering during the month of February to May.

Uses

Leaf paste used for setting of bone fractures in cattle's.

CONCLUSIONS

The exploration of Eastern Ghats of Andhra Pradesh revealed rich orchid germplasm wealth which can be utilized in various fields like pharmacology, floriculture and other, as yet unexplored fields of science. About twenty five orchid plants differing in their plant characters, therapeutic usage and economic importance were collected and described as per information provided by tribal farmers. This study is very much useful for creating awareness among people about orchid plants and also to conserve them, as some of the species have become endangered due to continuous loss of forestland; uncontrolled grazing, forest fires and unscientific collection by local tribes and other persons. The habitats of the orchids of the hill ranges are presently under threat of upcoming mining operations. Bearing in mind the rich biodiversity of this region, long term conservation measures will have to be taken to preserve this rich orchidarium for the state.

REFERENCES

Bhairavamurthy, P.V.1982. Environmental aspects of Eastern Ghats. Proc. National Seminar on Resources, Development and Environment of the Eastern Ghats, pp135-138, Andhra University, Visakhapatnam.

Bulpitt,C.J.2005. The uses and misuses of orchids in medicine, *Q J Med*, 98:625-631.

Dani,C.S. 1982. Vegetation of Eastern Ghats of Orissa. Proc. National Seminar on Resources, Development and Environment of the Eastern Ghats, pp43-46, Andhra University, Visakhapatnam.

Dash,P.K., Santilata Sahoo and Subhasisa Bal. 2008. Etnobotanical Studies on orchids of Niyamagiri hill ranges, Orissa, India, Ethnobotanical Leaflets, Vol 2008, Issue I, Article 9.

Jonathan. K.H and Solomon Rau.A.J. 2005. Terrestrial and Epiphytic Orchids of Eastern Ghats. ENVIS Newsletter, Vol II, No.3:2-4.

Joseph,J. 1987. Orchids of Nilagiris, BSI, Calcutta.

Mishra, S. 2004. Orchids of Orissa. Bishen Singh Mahendra Pal Singh, 23-A, New Connaught Place, Dehra Dun, India.

Padma, M and Rajkumar, M. 1997. Orchids of high altitude tribal area of Andhra Pradesh. *Journal of Research, ANGRAU*, 25(3):22-24.

Padma, M and Rajkumar, M. 1999. Terrestrial orchids in the eastern ghats of Andhra Pradesh. *Journal of Research, ANGRAU*, 27(3):12-14.

Pullaiah, T. 1997. Flora of Andhra Pradesh, Scientific Publishers, Jodhpur.

Reddy, K.N., Subba Raju, G.V., Reddy, C.S and Raju, V.S. 2005. Etnobotany of certain orchids of Eastern Ghats, Andhra Pradesh, *ENVIS Newsletter*, Vol II (3,5,9).

□□□

Chapter 14

Impact of Lantana Removal on Soil Erosion and Vegetation Cover in a Forest Watershed in Lower Shivalik Region

PAWAN SHARMA, A.K.TIWARI AND V.K.BHATT

INTRODUCTION

Lantana is a major weed in many tropical and sub-tropical regions where it invades natural and agricultural ecosystems and forest watersheds and has been nominated as among 100 of the "World's Worst" invaders (Day *et al.*, 2003). In disturbed native forests, it can become the dominant under-storey species, disrupting succession and decreasing biodiversity due to its allellopathic qualities (Luna *et al.*, 2007). In a watershed, the dense stands of *Lantana camara* are reported to reduce the ground grass cover thereby lowering its capacity to absorb rain, increasing the runoff and soil erosion (Fensham *et al.*, 1994; Day *et al.*, 2003). In an earlier study conducted in degraded forest watershed at Research farm, CSWCRTI, Chandigarh, bioengineering measures resulted in gradual decrease in soil loss from 37 Mg ha $^{-1}$ yr^{-1} to just 1.0 Mg ha^{-1} yr^{-1} during the period 1964 to 1985. However, this reducing trend was reversed after the invasion of *Lantana camara* during the period 1990-2000, along with suppression of native vegetation and a drastic reduction in everness index and Shannon biodiversity index (Sharma *et al.*, 2009). The available chronological data on vegetation and soil loss from this forest watershed suggests that the reduction in ground cover due to dominance of Lantana has resulted in increase in soil erosion, thereby increasing the threat to forest ecosystem stability and water resource management (Sharma *et al.*, 2009). The paper presents the impact of Lantana removal on vegetation cover, changes in runoff and soil loss in four lantana infested micro-watersheds compare with that of grassed watershed.

Central Soil and water Conservation Research and Training Institute, Research centre. Sector 27 A, Madhya Marg, Chandigarh-160 019, India.

MATERIALS AND METHODS

Site description : The study was conducted in five micro-watershed (WS_1, WS_2, WS_3, WS_4 and WS_5) located in Shivalik Himalayan region at CSWCRTI, Research Farm, Mansa Devi, Panchkula, Haryana (Latitude 30°-45′ N and Longitude 70°-45′ E, 370m above msl) with characteristics as given in Table 1. The soil was sandy-to-sandy loam in texture with full of boulders, low in nutrients and water holding capacity, and was classified as class VII e and light textured hyperthermic, Udic Ustocrept (Grewal *et al.*, 1996). The area was highly degraded due to frequent soil erosion and it receives an annual rainfall of more than 1100 mm. The watersheds had slope between 30 to 50% and area ranged from 0.81 ha to 4.75 ha. Bioengineering measures resulted in stabilization of these watersheds with improvement in vegetation cover and reduction in soil erosion. The invasion of ingress and proliferation of *Lantana camara* L (*verbenaceae*) during late eighties, resulted in poor penetration of light through heavy lantana canopy causing drastic reduction of ground grass cover and increase in runoff and soil loss (Sharma *et al.*,2009). The initial vegetation survey in forest watersheds showed that *Acacia catechu* was the most dominant tree in top tier with the highest Importance Vegetation Index (IVI), while *Lantana camara* and *Murraya koenigii* became the most dominant shrubs in middle tier and the ground cover was very poor with *Adhatoda vasica* and little or no grasses (Table 1). The grassed watershed (WS_4) was taken as control and had a luxurious ground cover of *Eulaliopsis binata.*

Table 1: General characteristics of five micro-watersheds

Water-shed	Slope (%)	Area (ha)	Vegetation type	% Cover area of three tier vegetation in year 2005			IVI of dominant vegetation in year 2005				
				Top	Middle	Ground	*Acacia catetchu*	*Lant-ana camara*	*Adhat oda vas ica*	*Murr-aya koen-ingii*	*E.binata*
WS_1	44.9	4.5	natural mixed forest	30.4	53.6	18.8	70.9	71.5	39.2	37.7	0.0
WS_2	32.1	2.0	-do-	15.4	74.0	12.4	71.7	114.3	30.7	27.7	0.0
WS_3	48.9	4.7	-do-	43.9	68.8	28.8	44.3	60.7	64.5	24.1	0.0
WS_4	50.4	0.8	Grasses	1.9	13.3	75.5	0	16.0	0	0	151.2
WS_5	50.6	1.7	natural mixed forest	38.2	29.8	27.5	56.6	56.7	72.0	26.6	0.0

Vegetative manipulation : To reduce and maintain vegetation cover for obtaining an optimum runoff and reduction in soil loss, vegetation manipulation was started from 2005-06 onwards with the removal of *Lantana camara* from all the micro-watersheds. Fifty per cent crown of each tree was also removed to allow the light to penetrate to the ground floor to encourage the natural vegetation and to improve the ground cover. The canopy cover in the top and the middle tier was maintained by continuous removal of regenerated canopy every year.

Measurement of run-off and soil loss : Rainfall, run-off and soil loss data were monitored every year in the watershed during the period 2005-10. All the micro-watersheds are being gauged by 0.6m deep 2:1 broad crested triangular weirs. The runoff was measured by automatic water level recorders. The rainfall was recorded by recording type of rain gauge and the run-off was gauged by 0.6 m deep 2:1 broad crested triangular weirs. The soil loss was determined by analyzing the run-off samples collected from time to time. All the watersheds were calibrated for initial two years (2005-06).

Vegetation survey : The vegetation survey was done following quadrant method (Mishra, 1968) with quadrants equally and randomly distributed at the three physiographic positions of top, middle and lower reaches of the hill slope. The ecological succession of shrubs and trees was measured through the IVI obtained through summing up of the relative dominance, relative density and relative frequency according to standard procedures given by Mishra (1968). Since soil erosion is mainly a function of total canopy cover area, the potential threat posed by huge canopy cover of Lantana could be underestimated by IVI alone. Therefore, the total vegetation cover area in top middle and ground tier of the three tier vegetation at different locations was measured by planimeter method (Mishra, 1968). The relative cover area occupied by each plant species in a 10m x10 m quadrant and was accordingly plotted on a graph paper. The area on graph paper was computed by planimeter and the percentage of canopy area occupied by different plant species in three tier vegetation was calculated separately.

Soil and microbial analysis : Forest floor was analyzed for litter fall and top soil properties. Composite samples of fresh and decomposed leaf litter were collected from one square feet area from different locations in the watersheds and litter fall per hectare was computed based on mean dry weight of the litter. Composite soil samples were collected at 30cm soil depth and analyzed for different soil parameters *i.e.,* soil organic carbon (Walkley and Black, 1934), and water stable aggregates (Elliot and Gambardella, 1991).

Soil respiration was measured by incubating soil in sealed flasks containing a vial of NaOH to at 28°C for 24 hrs, followed by estimating the evolved CO_2-C trapped in NaOH by addition of an excess of 1.5 mol. L^{-1} $BaCl_2$, and titration with standardized HCl (Zibilske, 1994). Spores of Vesicular Arbuscular Mycorrhiza (VAM) were obtained by wet sieving and decanting of the fresh soil samples and the spores were counted in a graduated petri plate under a dissecting microscope (Daniels and Skipper, 1982).

RESULTS AND DISCUSSION

Litter fall and soil properties : Lantana has a high litter fall resulting in improvement in soil organic carbon and high respiration rate due to litter decomposition (Table 2). There is more improvement in soil properties in Lantana infested watershed. The lower and middle reaches had higher litter fall and soil organic carbon as compared to top reaches (Table 2), possibly due to washing away of litter from top reaches. The grassed watershed had

Table 2: Initial soil properties under lantana infested forest watershed and the grassed watershed in the top, middle and lower reaches.

Watershed	Location on the slope	Initial properties of forest floor					
		Litter fall (q ha^{-1})		Soil properties (0-30 cm soil depth)			
		Fresh litter	Decompo sed litter	OC(%)	VAM spores No. 10g $soil^{-1}$	Soil respire-ation (mg CO_2 C g^{-1} soil 24 hr^{-1})	Water stable aggre-gates (%)
Lantana infested water-sheds (WS^1, WS_2, WS_3 & WS_5)	Upper reaches	26.9	63.6	1.02	5	1.45	52.3
	Middle reaches	38.4	136.5	1.32	7	1.08	57.1
	Lower reaches	31.8	116.8	1.38	10	0.86	51.7
Grassed watershed (WS4)	Upper reaches	--	--	0.48	19	0.81	67.6
	Middle reaches	--	--	0.60	23	0.90	54.9
	Lower reaches	--	--	0.66	27	1.41	56.8

more water stable aggregates and higher number of spores of VAM as compared to Lantana infested watershed (Table 2). The dependency of plants on VAM to survive under degraded environment and the improved soil aggregation under VAM colonized roots has been widely reported (Azcon and Barea, 1997; Franzluebbers *et al.*, 2000). High proliferation of grass roots combined with network of VAM hyphae may play an important role in soil stabilization and reducing the soil loss.

Changes vegetation cover during 2005 to 2009 : The dominant plant species found were *Khair* (*Acacia catechu*), *Basuta* (*Adhatoda vasica*) and *sweet neem* (*Murraya koenigii*) in all the watersheds. On comparing plant canopy in upper storey between 2005 and 2009, it was found that the tree canopy has improved in all the forest watersheds and highest tree canopy was observed in WS_1 possibly due to ingress and fast growth of *Prosopis* in this watershed (Figure 1). In the middle tier, the continuous removal of Lantana during 2006 onwards from all the watersheds, there was a drastic reduction in canopy of lantana from 80-90% to 5-10% in 2009 (Figure 2). As a result, the light could penetrate to reach the ground vegetation which has significantly improved growth of ground cover. As a result, *Basuta* (*Adhatoda vasica*) and *sweet neem* (*Murraya koenigii*) which formed ground cover in 2005 have become as the principal species in middle storey in 2009. The area occupied by *bhabhar* (*Eulaliopsis binata*), *dholu* (*Chrysopogon fulvus*) and local grasses have also improved in 2009 as compared to 2005 (Fig.3). The local native grasses have formed major ground cover in WS_1, WS_2, WS_3 and WS_5. The top reaches in WS_1 have the poorest ground cover, possible due to high density of tree cover in this watershed. But there is good grass cover in the middle and lower reaches of this watershed.

Runoff and soil loss : Runoff from all the watersheds for last four years (Table 3) indicates that watershed WS_2 gave lowest runoff among all the watersheds. On comparing soil loss from all watersheds it was found that in general soil loss has decreased from 2007 to 2010.Vegetation manipulation has shown clear impact on soil loss as there is reduction in soil loss from 2007 to 2010 even after higher rainfall in 2010 (Table 3). Calibration curve and validation curves indicate that runoff have slightly increased from all watersheds during year 2010 in comparison to two years of calibration period from 2005-2006. Total soil loss was also found minimum in WS_2 watershed, which may be due to highest improvement in middle as well ground grass cover in top, middle and lower reaches.

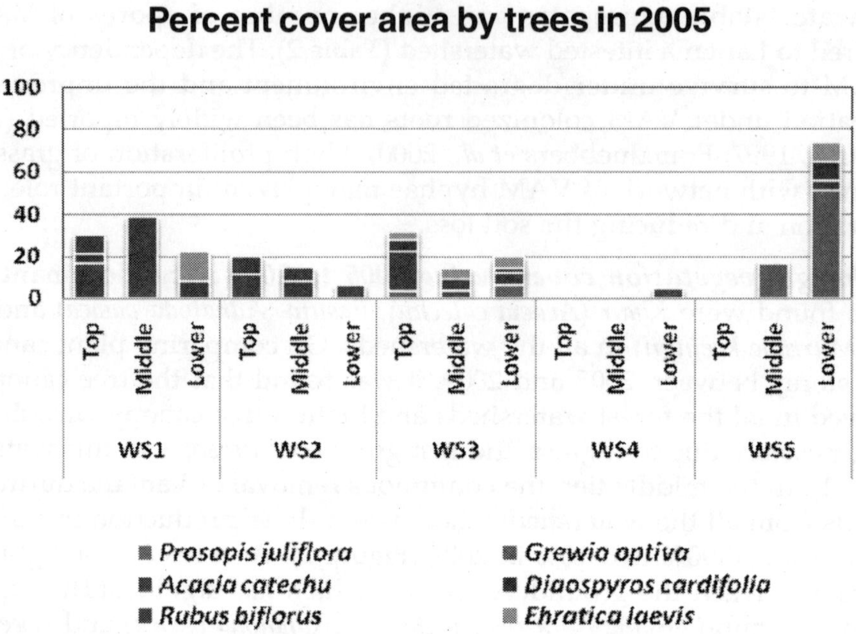

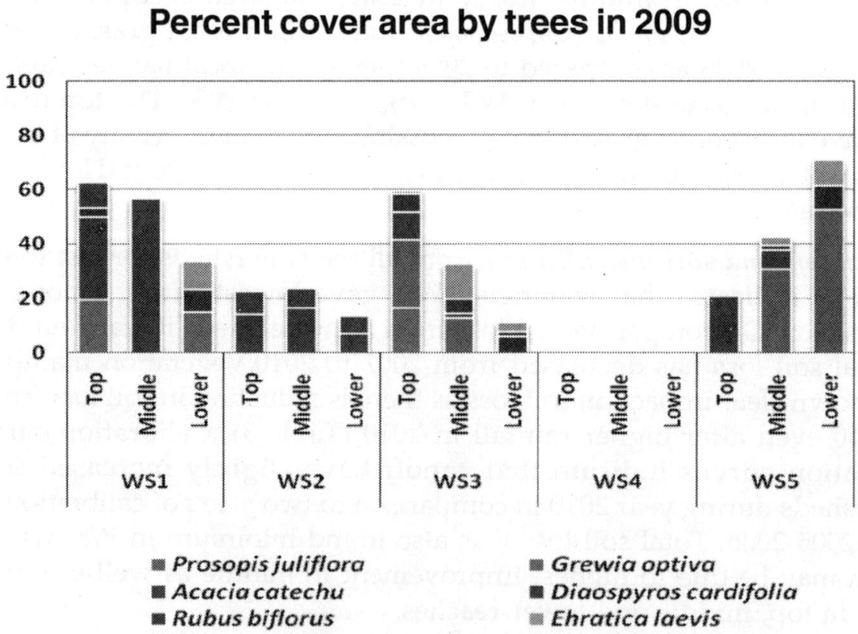

Fig.1: Vegetation cover area in top tier by trees in year 2005 and 2009.

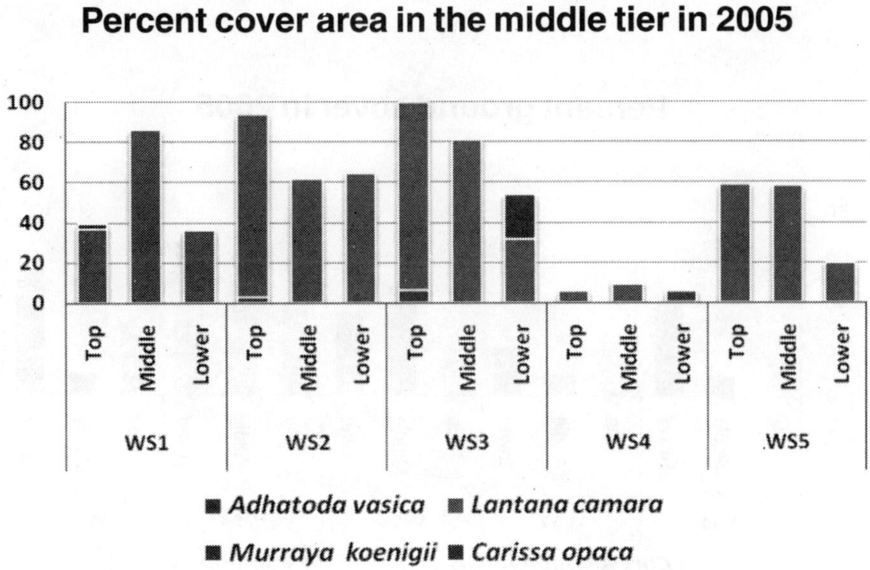

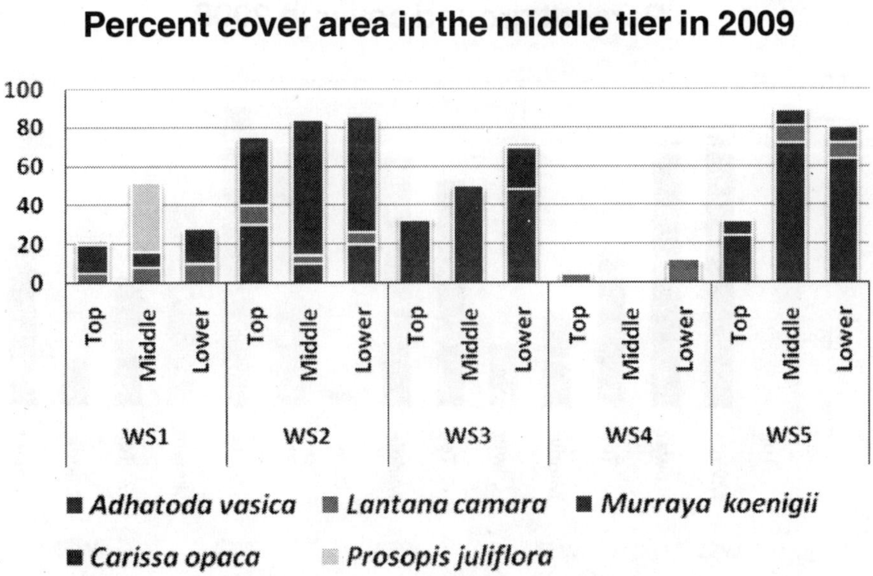

Fig. 2: Vegetation cover area in middle tier in year 2005 and 2009.

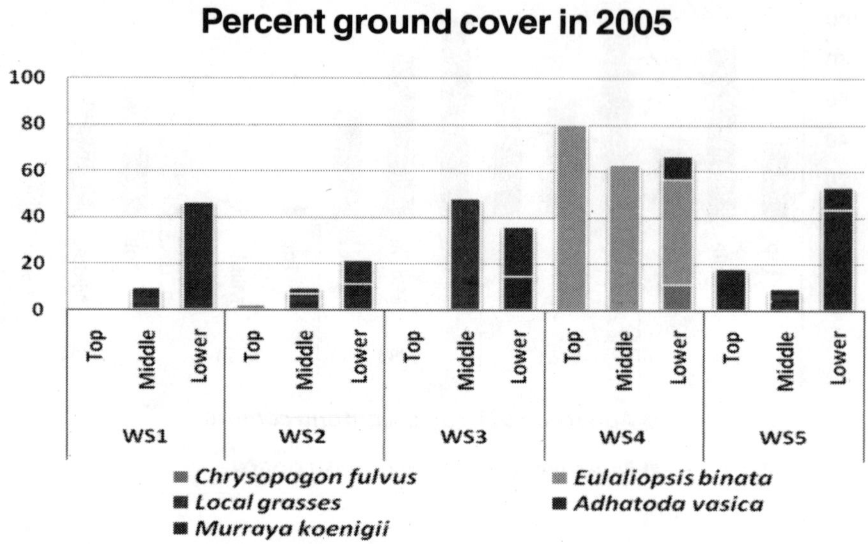

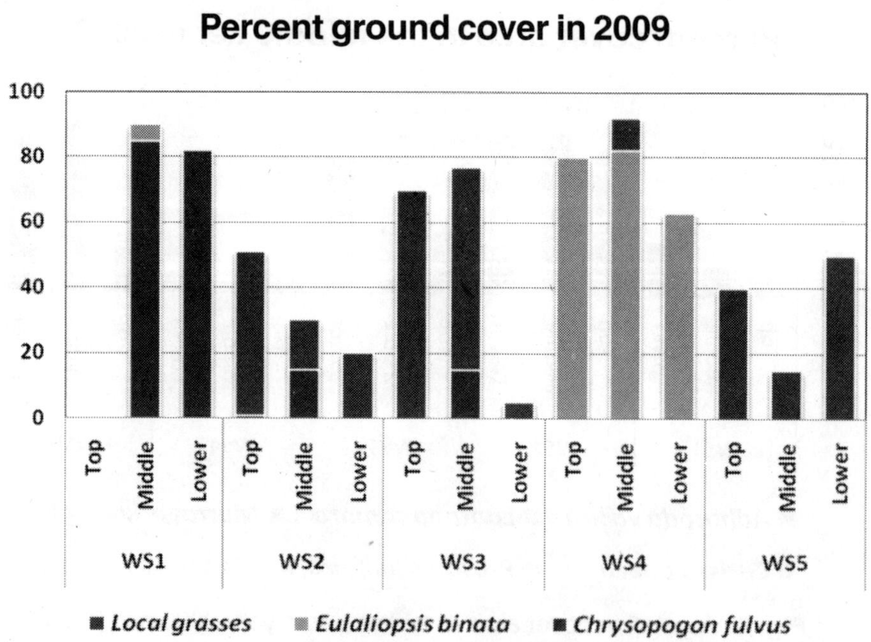

Fig. 3: Ground cover area in 2005 and 2009

Table 3: Soil loss from watersheds during 2007 and 2010

Watershed	Runoff (%)			Soil loss (kg ha^{-1})		
	2007 (Initial)	2010	Change (%)	2007 (Initial)	2010	Change (%)
WS_1	8.5	14.8	+74.1	357	239.1	-33.0
WS_2	2.1	1.9	-9.5	52	1.1	-97.9
WS_3	21.7	8.0	-63.1	2137	42.4	-98.0
WS_4 (Control)	8.7	8.8	+1.2	442	123.0	-72.2
WS_5	11.5	8.3	-27.8	5521	42.6	-99.2

Runoff values were found maximum in WS_1 in 2010 with 74% improvement in runoff as compared to that in 2005-06, while the other watersheds (WS_2, WS_3 and WS_5) showed a reduction from 9.5 to 63%. The reduction in soil loss was also minimum in WS_1 (33%), while it was maximum (97-99%) in WS_2, WS_3 and WS_5, reaching the levels of control grass watershed WS_4. The highest runoff and soil loss in WS_1 may be attributed to poor ground cover in WS_1 in top reaches of the watershed. The luxurious ground cover obtained in WS_2, WS_3 and WS_5 in all the reaches may have resulted in reduction of soil loss and runoff.

CONCLUSIONS

It is concluded that the ground cover vegetation plays major role in hydrological behavior of the watershed. The reduction in ground cover due to poor light availability under lantana canopy may be the major cause of increased soil erosion after lantana invasion. Since the top reaches are the main sources of run off and soil loss, the maintenance of ground cover in top reaches has a more important role to play. Further in-depth studies are required on interactions between three tier vegetation and the maintenance of optimal vegetation cover for improving water quality from a watershed.

Acknowledgement: The authors are grateful to Director, Central Soil and Water Conservation Research & Training Institute, Dehradun for providing all necessary support and valuable guidance. Technical support given by Mr. Harish Sharma and Mr. A.N. Gupta is gratefully acknowledged.

REFERENCES

Azcon, R. and Barea, J.M. 1997. Mycorrhizal dependency of a representative plant species in Meditranean Shrublands (*Levandula spica* L.) as a key factor to its use for revegetation strategies in desertification threatened areas. *Applied Soil Ecology*, 7(1): 83-92.

Daniels, B.A. and Skipper H.D. 1982. Methods of recovery and the quantitative estimation of propagules from the soil. In: *Methods and principals of mycorrhizal research.* (Ed. Schenck N.C.), American Phytopathological Society, St Paul, Minnesota, pp29-35

Day, M., Wiley, C.J., Playford, J. and Zalucki, M. P. 2003. *Lantana: current management status and future prospects* – (Australian Centre for International Agricultural Research: Canberra), ACIAR Monograph 102.

Elliot, J. E.T and C.A. Cambardella. 1991. Physical separation of organic matter. Agric. Ecosyst. Environ. 34:407-419. n, M.L. 1967. *Soil Chemical Analysis*. Prentice Hall India, New Delhi. 86-96.

Fensham, R.J., Fairfax, R.J. and Cannell, R.J.1994. The invasion of *Lantana camara* L. in Forty Mile Scrub National Park, North Queensland. *Australian Journal of Ecology*, 19: 297-305.

Grewal, S.S., Singh, K., Juneja, M.L., and Singh, S.C. 1996. Relative growth, fuel wood yield, litter accumulation and conservation potential of seven Acacia species and an under-storey forestry grass on a slopping bouldry soil. *Indian Forestry*, 19 (2): 174-182.

Franzleuebbers, A.J, Wright, S.F and Steudemann.2000. Soil aggregation and glomalin under pastures in the Southern Piedmont,U.S.A. *Soil Sci.Soc.Am.J.*,64:1018-1026.

Luna, R. K., Minhas, R.K., and Kamboj, S.K.2007. Effect of *Lantana camara* on herbaceous diversity. *Indian Forester*, 133: 109-122.

Mishra, M.1968. *Ecology Workbook*. Oxford & IBH Publishing co. New Delhi, India.

Sharma, P., Singh, P and Tiwari, A.K. 2009. Effects of *Lantana camara* invasion on plant biodiversity and soil erosion in a forest watershed in lower Himalayas, India. *Indian J of Forestry, vol.32, pp369-374.*

Walkley, I.A. and Black 1934. An examination of Degtjareff method for determining soil organic matter and a proposed modification of the chromic acid titration method. *Soil Sci.*, 37: 29-37.

Zibilske, L.M. 1994. Carbon mineralization, *in: Methods of soil analysis. Part 2. Microbiological and biochemical properties*. (Eds. R.W. Weaver, J.S.Angle, and P.S.Bottomley) SSSA Book Ser. 5. SSSA, Madison, WI, pp835–863.

□□□

Chapter 15

Hydro-Ecological Zonation of Eastern Ghats for Strengthening Research Activities on Resource Conservation

D.V. SINGH, A.K. SIKKA[1], M. MADHU[2] AND H. GADEKAR[2]

INTRODUCTION

Eastern Ghats are spread over four States *viz.*, Andhra Pradesh, Tamil Nadu, Orissa and Karnataka with an average width of 125 km (ranging from 40 to 240 km) over a length of 1600 km between the rivers Mahanadi and Vaigai along the east coast. The northern boundary is marked by the Mahanadi basin, the southern boundary by the Nilgiri hills, the western boundary touches the tips of Bastar, Telangana and Karnataka plateaus, and Tamil Nadu uplands and the east coastal plains mark its eastern limit (Chauhan, 1998). Eastern Ghats supports the alluvial coastal plains and thus plays a major ecological role which needs to be conserved while providing livelihood security. Though both the Western and Eastern Ghats form important Southern hilly ecosystem of India, development of Eastern Ghats appears to have received less attention as compared to Western Ghats (Sikka *et al.*, 2001).

Eastern Ghats, located between 77°22′ and 85°20′ E longitude and 11°30′ and 20°00′ N latitudes and characterized by scattered, broken and of much lower altitude than Western Ghats, though geologically they seem to be distinctly older than the Western Ghats. Beginning in north Orissa, they pass through the coastal region of Andhra Pradesh to Tamil Nadu cutting across Karnataka. Their average elevation is about 610 m. This hilly region receives lesser rainfall (600 to over 1400 mm) than that of Western Ghats. Because of their lower elevation and their broken character, the Eastern Ghats have not been as great an obstacle to intercourse between the plateau and the coastal

Central Soil and Water Conservation Research and Training Institute, Research Centre, Udhagamandalam – 643 004, Tamilnadu, India
[1]National Rainfed Area Authority, New Delhi, India.
[2] Central Soil and Water Conservation Research and Training Institute, Research Centre, Sunabeda, Koraput, Odisha, India

plain as the Western Ghats. Almost all the major streams like Mahanadi, Godavari, Krishna, Pennar and Cauvery taking their rise from the Western Ghats have cut extensively through the Eastern Ghats all the way down to the Bay of Bengal. Encompassed between these ranges are numerous gorges, waterfalls and wide alluvial valleys, structural and erosional basins, making the Ghats rich in minerals and hydrologically, vegetatively and agriculturally a prospective region (Sikka *et al.*, 2000).

Farming in such areas is generally of traditional and subsistence level with a scope for better land use systems and crop diversification (Sikka *et al.*, 2004). In view of this, it was felt necessary to study the characteristic features of Eastern Ghats in order to extend and/or strengthen the area based research activities on soil and water conservation and watershed management covering different zones of Eastern Ghats. The paper presents characteristic features of Eastern Ghats, their zonation into different hydroecological zones/subzones and problems and suggests the need of infrastructure/research linkages required to extend and/or strengthen research and technology development on soil and water conservation and watershed management in the Eastern Ghats.

MATERIALS AND METHODS

The Eastern Ghats boundary was delineated from the Atlas of National Atlas Organization (NATMO, 1974) and was transferred on to map of South India at a scale of 1:2 M (*i.e.*, 1 cm = 20 km). A base map of Eastern Ghats region showing districts and major *talukas* was thus prepared to find out area and identify districts and *talukas*/blocks covered under Eastern Ghats region. Respective district officials were contacted for statistical, agricultural, hydrogeological and forestry situations for analyzing geographic details, demographic features, climatic parameters, land use pattern/cropping pattern, soil type and other miscellaneous information in these districts initially by writing them and later personally contacting most of them.

Besides the respective District Officers and State line department Officers, other institutions like Directors of Research of different State Agriculture Universities and Directors of different I.C.A.R. and other Institutes in the region were also contacted. Personal discussions were held with retired Scientist and Officials and notes were made for constraint identification and potential of the area.

The information/maps obtained from NATMO were largely used in association with soils, forest and rainfall information from National Bureau of Soil Survey & Land Use Planning (NBSSLUP), Forest Survey of India and India Meteorological Department (IMD) reports/maps, respectively. Thematic maps/atlases of different regions for soils, slope groups, elevation, rainfall, temperature, land use and vegetation, soil fertility status, etc., were used as input for preparing thematic maps of Eastern Ghats in the scale of 1:2 M. The maps and tables were analyzed and on the basis of elevation, rainfall, land use and vegetation, different zones were delineated on maps by manually superimposing different data layers. The following criteria was used to classify rainfall and elevation for different categories:

Rainfall (mm) Category

< 600 Very low
600 - 800 Low
800 - 1200 Medium
1200 - 1400 High
> 1400 Very High

Elevation (m) Category

< 600 Lower
600 - 900 Mid
> 900 High

On the basis of elevation, rainfall, land use and vegetation, different homogenous units were delineated on a map by superimposing different data layers. Finally the entire stretch of Eastern Ghats was classified into eight hydro-ecological zones.

RESULTS AND DISCUSSION

Geographic and demographic details

Out of the total mountainous area of our country, 19.36 per cent (18.02 M ha) area is occupied by the Eastern Ghats as per Das (1982). However, the area planimetered from the Eastern Ghats boundary map generated in this study comes to 19.76 M ha which is the next major mountain mass to Himalayas. This shows its ecological importance in controlling physical processes. The Eastern Ghats region covers 42 districts and 214 *talukas*/blocks in four States of Andhra Pradesh, Tamil Nadu, Orissa and Karnataka and supports a population of about 38 M. Overall literacy level is low at the 39 % in the Eastern Ghats region with Tamil Nadu having the maximum literacy percentage of 54.38.

Physiography

Elevation/Topography: Topographically the Eastern Ghats region can be characterized into hills, plateau, pediment and plains. The elevation in general ranges from 150 m to over 1600 m above the MSL with a general increasing trend from south towards northern region. In the northern Eastern Ghats region, hill ranges run parallel to each other in the form of chain while in central and southern region these occur as isolated hillocks and eroded buttes, with their elevation ranging from 300 to 1680 m above MSL. On the basis of elevation, the plateau surfaces are divided into upper plateau (600 to 1200 m) and lower plateau (300 to 600 m). Steep to moderately sloping surfaces associated at the foot of hills and plateaus at an elevation of 150 to 300 m slowly merging into the plains are the pediment surfaces. The region with elevation less than 150 m is delineated as plains. It stretches from Chilkalake in the remote north to Pulicat lake in the south. Down below the pediment surfaces and hills starts a boundary of moderately sloping alluvial plains extending to the boundary of the coastal plains towards the lower ridge. The average elevation ranges from 50 to 150 m and it is widened mostly in the east-west by perennial rivers Godavari, Krishna, Mahanadi and Pennar by depositing the heavy load of sediments.

Slope: It generally varies from a low of 1-2 % in plains to a high of 50-60 % in the hilly areas of Eastern Ghats. In southern parts, majority of area falls in slope range of 1-2 % (flat land) and the central part is dominated with flat land (1-2 %) to gentle sloping land (2-8 %). Gentle slopes (2-8 %) to steep slopes (15-60 %) are dominated in northern parts of Eastern Ghats.

Soils: The Eastern Ghats region is characterized by a wide variety of soil types (12 soil types). The region predominantly consists of red sandy soils followed by red loamy soils, red gravelly soils, mixed red and black soils and deep black soils, red earths, medium black soils, alluvial soils, lateritic soils, coastal alluvial soils, deltaic alluvial soils and laterite soils. Generally in medium to high rainfall areas, soils are predominantly red soils whereas in high to very high rainfall areas soils are laterite and lateritic with deep to medium black soils mostly in low to very low rainfall areas. Generally the red soils are found to have low to medium level of NPK except in the forest land use where N is medium to high. Soil erosion is generally moderate to severe in the hilly and/or red soil dominated area occurring over a large portion of Eastern Ghats. Slight to medium erosion could be noticed in the areas with black soils and mixed red and black soils.

Climate

Rainfall: The climate in the region is of tropical monsoon type. It receives rainfall from both the south-west (sharing about 87% of the total) and the north-east monsoons. The rainfall varies in the range of 1200-1600 mm in the northern part, 600-1000 mm in the central and southern parts of Eastern Ghats indicating sub-humid and semi-arid climates, respectively. The eastern portion, especially the coastal plains are characterized by heavy rains with cyclonic storms while rain shadow area of Tamil Nadu and Andhra Pradesh receive very low rainfall (<600 mm).

Temperature: The mean annual temperature of Eastern Ghats region generally varies in a range from 25 to 27.5°C and above with very small portion of the area having mean annual temperature of 20 to 25°C in Satyamangalam, Theni and Koraput hill ranges. The mean temperature in January ranges between 20 to 25°C and shoots up to 41°C during hot months and goes down to even 2°C at few places in high elevation ranges during winter. Due to high elevation and maritime locations, temperature does not go to extremes in the interior parts.

Land use and vegetation : Eastern Ghats is a reservoir of floristic diversity consisting of about 2000 species of plants belonging to angiosperms. The well-known gymnosperms such as Cycas*beddomei, C. circinalis* and *Gnetumscandens* and about 30 species of ferns including *Cythiagigantea* – a tree fern. A rare pteredophyte *Psilotum-nudum* is also reported to occur. *Atylostacajanitolia* a wild relative for *Cajanuscajan* is also existing in a wild state. Likewise Eastern Ghats are endowed with rich flora and fauna and diversified land use. Distribution of land use under different categories indicates that major area is under agricultural land use (57.19 %) followed by forest (30.71 %) and grazing land. The majority of Eastern Ghats is dominated by tribal population. In upper ridges of the hills these tribal families resort to shifting cultivation in reserved and protected forests. The crops cultivated on shifting cultivated lands are upland paddy, *ragi,* niger, *dongarrani, kulthi,* sorgham, etc. The Integrated Tribal Development Agency (ITDA) in Andhra Pradesh has been planting such shifting cultivation ravaged areas with *Anacardiumoccidentale* (cashew) extensively for the benefit of tribal population. Under the social forestry, extensive planting of *Acacia auriculiforimes,* Eucalyptus, etc., are being taken up. In high rainfall, high elevation zone of Tamil Nadu, Forest Department has extensively planted coffee in *podu*ravaged areas. Numerous crops are being cultivated in relation to topo sequence and availability of irrigation potential.

Northern Eastern Ghats region is dominated by rice, millets, pulses and oil seeds with rice as predominant crop in the coastal belt of central and northern Eastern Ghats. Crops like millets, oil seed, pulses and rice are common in central and northern Eastern Ghats region with millets and pulses being predominant in the districts of Dharmapuri and Thiruvannamalai in Tamil Nadu plains.

Among the fruits, the common crops are mango, banana, guava, citrus, grapes, pineapple, pears, jackfruit, *etc.* The plantation crops such as cashew (in coastal/hilly areas), coconut (coastal and lower plains), tea (Tamil Nadu hills), coffee (hilly area) are the major crops along with some spices like black pepper, cardamom, cinnamon, *etc.*

Hydrology and Water Resources : The hydrology of Eastern Ghats consists of varied hydrological phenomena of mountain hydrology, deltaic hydrology, coastal hydrology and ground water hydrology. Typically, the Ghat areas and uplands are characterized by mountain hydrology with high runoff potential and lack of ground water development. Eastern Ghats ranges are traversed by a number of rivers and streams that provide water resources to plains and coastal region of these Ghats. Major rivers like Mahanadi, Godavari, Krishna, Pennar and Cauvery, which have their origin in Western Ghats drain through the Eastern Ghats. Surface irrigation through major and medium irrigation is rather very less and there is increased dependence on ground water, especially in drought affected/prone areas. Minor irrigation through tanks is there to a limited extent. Exploitation of ground water is quite large in Tamil Nadu, with over 50 % of area being irrigated from ground water in this State. The water balance analysis of number of stations in the region indicates moderate to severe water deficit for about 6 to 8 months in the low as well as high rainfall areas of the Eastern Ghats which clearly reveals a good scope of harvesting rain water surplus and storing it as surface or sub-surface storage and/or augment ground water recharge to tide over moisture stress conditions.

Hydrogeologically, Eastern Ghats region is composed of khondalite suit of rocks, granites, granite gneisses and charnockites besides valley fills. The ground water occurrence is moderate, and these resources are developed mostly by means of dug wells (10 to 15 m and 10 to 40 m depth in Andhra Pradesh and Tamil Nadu, respectively) with low to moderate yield. Depth of bore wells ranges from 40 to 60 m in Andhra Pradesh and over 100 m in Tamil Nadu region of the Eastern Ghats. The valley fills in Eastern Ghats generally comprise clay, fine to coarse sands grading to gravel, pebbles at places and admixtures of sandy loams in Andhra Pradesh. Ground water development in these formations is mainly by shallow tube wells and filter points (Rao, 1998). In the State of Andhra Pradesh, ground water development in Eastern Ghats districts is 18 per cent and in the tribal hilly tracts it is only 10 % (Rao, 1998). This suggests development of local level water resources through water harvesting, ground water recharge and water conservation in rainfed/dry zones of Eastern Ghats in the States of Andhra Pradesh, Tamil Nadu and Orissa.

Zonation

The characteristic features as described in the previous section present a wide heterogeneity in physiographic, climatic, edaphic, land use, vegetation and socio-economic conditions in the Eastern Ghats. In view of these variations Eastern Ghats region has been classified into similar zones and sub zones for better representations and application of research problems and their solutions in relation to natural resource conservation and management.

The Eastern Ghats were initially classified into 11 hydro-ecological zones. Some of the zones were further sub-divided into sub zones based on soils, land use and vegetation and slope steepness. The names of these zones and their area are shown in Table 1. The hydro-ecological zones at S.No.6, 8 and 9 being very small in area, were merged with zone numbers 5, 7 and 11, respectively as their sub-zones. Thus, finally Eastern Ghats was divided into a total of eight zones. These zones and sub-zones are depicted in Map 1.

Predominant hydro-ecological zones are south eastern plains with medium rainfall and south eastern plains of Tamil Nadu and Andhra Pradesh with low rainfall followed by northern plains of Andhra Pradesh and Orissa and northern high hills with very high rainfall, rain shadow area of Tamil Nadu with low rainfall and southern plateau with low rainfall with least area under north eastern lower Ghats and southern hills of Tamil Nadu uplands.

Table 1: Area under different zones and sub-zones of Eastern *Ghats*

S.No	Name of the Zone	Area	
		km²	%
1	Rain shadow plains and lower *Ghats* of Tamil Nadu uplands and parts of AP with very low rainfall	22120	11.21
	1a. Red soil	5080	
	1b. Back soil	15240	
2	Southern plains of Tamil Nadu and AP with low rainfall	33140	16.77
	Red Soil	12440	
	Mixed red & black soil	13040	
3	Southern plateau and some *Ghats* lands with low rainfall	17240	8.73
4	South-Eastern plains with medium rainfall	57280	28.99
	4a. Red gravelly soil + forest	7120	
	4b. Thick forest	14040	
5	Southern hills of Tamil Nadu uplands with medium rainfall	6000	3.03
	5a. High rainfall	440	

6	North Eastern lower *Ghats* with high rainfall.	9280	4.69
	6a. Forest	2880	
7	Northern plains and *Ghats* of AP and Orissa with very high rainfall	28980	14.67
	7a. Black soil	1180	
	7b. Lateritic soil	2480	
	7c. Forest	11320	
8	Northern high hills with very high rainfall	23560	11.92
	8a. High rainfall	13120	
	8b. Forest + Grass land	7320	
	Total	197600	100

Problems and Constraints of Resource Management

- Soil Erosion due to shifting cultivation the area of which is difficult to determine because of inaccessible nature of terrain.
- Large scale deforestation, destruction of native germplasma plants and erosion of biodiversity.
- Poor fertility status of soils and low fertilizer application gives low yield per ha.
- The village common lands and grazing lands are highly degraded and gullied due to large scale exploitation.
- There is lack of information on farming and cropping systems suited to different locations.
- Lack of improved water management systems and practices in the area.
- The streams bed cultivation for paddy leading to downstream siltation.
- Incidence of fire due to phases of burning during process of shifting cultivation.
- Low economic status and low adoption pattern of the rural population.
- Fragmentation of holdings.

From the Resource Management point of view, constraints are identified on the basis of three types of categories of lands.

Category-I Hill Region

- High rainfall, steep slopes and heavy runoff leading to severe soil erosion.
- Soil degradation due to heavy erosion on account of shifting cultivation.

- Soils are acidic in reaction and need amendments.
- Low soil fertility status and productivity.
- Want of suitable agro-forestry systems for degraded lands.
- Lack of information about grasses and fodder trees for silvi-pastoral systems.
- Lack of moisture conservation methods.
- Lack of irrigation water and water harvesting systems.
- Depleting mountain streams.
- Landslides and toe cutting/removal.
- Uncontrolled mining activities leading to land degradation.
- Sedimentation of hydro-electric/irrigation reservoirs.
- High degree of backwardness among tribals.
- Lack of technical skill and high illiteracy rate, poor management of natural resources.
- Low level of motivation by concerned agencies.
- Non-availability of improved seeds, fertilizers, etc., to the tribals in time.
- Lack of research support for managing various cultivation related activities.

Category-II Plateau & plains region with low to medium rainfall

This area includes lower elevations of Tamil Nadu, Andhra Pradesh and Karnataka having low rainfall. Following major problems are identified in the area.

- Land degradation due to soil erosion and faulty land use management practices.
- Periodic occurrence of droughts
- Heavy percolation and seepage losses
- Deforestation and lack of protective vegetation
- Poor water holding capacity of soil and shallow soil depth
- Inefficient water management in spite of scarcity of water for irrigation
- Low fertility status
- Declining ground water table due to excessive use/withdrawal
- Late onset and early withdrawal of monsoon and intermittent long dry spells hampering crop production

- Lack of water harvesting/water conservation measures
- Lack of appropriate and viable cropping systems including alternate land use systems.

Category-III Lower ghats (Coastal) region with high rainfall

The region occurs near the coastal zones and receives high rainfall. The major problems can be summarized as below:

- The high intensity rains coupled with faulty land use and management practices cause erosion.
- There are periodic droughts and floods due to heavy runoff and siltation of river beds
- Poor water management due to improper use
- Excessive ground water withdrawal leading to declining water table and salinity problem due to ingress of sea water
- Excessive leaching of nutrients from the soil under heavy rainfall regime
- Clear felling of forests and illicit cutting of trees
- Acidic soils needing amendments for better growth of crops.
- Inappropriate land use system

Apart from these area specific problems, some other problems are common in most parts of the Eastern Ghats. These include increasing population pressure on land, encroachment of forest area, unauthorized settlements, loss of productivity and rapid loss of habitat and genetic diversity, widespread poverty and loss of indigenous knowledge. As a result, these ecosystems are experiencing environmental degradation. Continued changes in the hydrology and water balance as a result of altered land use and management practices are matter of serious concern and these effects are being reflected in low flows, low levels in reservoirs, depleting ground water resources sedimentation and water quality problems.

On the other hand, some of the major constraints for sustainable development in Eastern Ghats are undulating and steep topography, shallow soil depth, poor soil fertility, lack of information pool, poor economic conditions, low literacy level and social structure.

Recommended Infrastructure/Research Linkages

Area specific research being carried out by a number of ICAR Institutes/ Research Centers such as Central Soil & Water Conservation Research & Training Institute/ Research Centers, Koraput and Udhagamandalam; NRC

Oil Palm, Eluru. Central Tobacco Research Institute, Rajamundry; NRC Banana, Trichy, *etc.*, is either specific to a particular region/area or specific to a crop. The SAUs and their NARP/other Research Stations are located in different agro-climatic regions of their respective States, spread throughout the Eastern Ghats region. Research, education and extension activities are being managed by these SAUs and their Stations. Their research activities are generally crop/cropping system specific to the respective areas.

However, the emphasis on integrated resource management has received less attention in different regions of Eastern Ghats. The technology developed at Central Soil & Water Conservation Research & Training Institute, Research Centers, Koraput and Udhagamandalam is mostly specific to high hill of Northern and Southern Eastern Ghats in Orissa and Tamil Nadu, respectively. The high hills area constitutes only about 15 % of the total Eastern Ghats area in the States of Andhra Pradesh, Orissa and Tamil Nadu. In the high hills of Andhra Pradesh, Research Station of APAU at Chitapalle is engaged in some land and water management activities.

In view of the variations in characteristic features, problems, constraints and issues of different hydro-ecological zones in the Eastern Ghats, research activities with regard to integrated natural resource management for sustainable production system may need to be extended/strengthened through a network of projects in this region. Based on the extent of area under each hydro-ecological zone, the strengthening/establishing of Research infrastructure and linkages/networking of the following stations is suggested for integrated natural resource management research and technology development.

Zone 1	Kurnool - Regional Agriculture Research Station, Nandiyal, Dist. Kurnool (A.P.)
Zone 2	Utukur – Regional Research Station, Dist. Cuddapah (A.P.) Erode/Coimbatore – Tamil Nadu Agricultural University
Zone 3 & 5	Yerkad – Horticultural Research Station, Dist. Salem (T.N.)
Zone 4	Horticultural Research Station, Thadiyankudisai, Dist. Dindigul (T.N.) Perumallapalli – Agricultural Research Station,Dist. Chittur (A.P.) Madhira – Agricultural Research Station, Dist. Kammam (A.P.)
Zone 6	Agricultural Research Station, Sethempet (A.P.)
Zone 7	Agricultural Research Station, G. Udhagari, Dist. Phulbani (Orissa)
Zone 8	Central Soil & Water Conservation Research & Training Institute, Research Centre, Semiliguda, Dist. Koraput (Orissa)

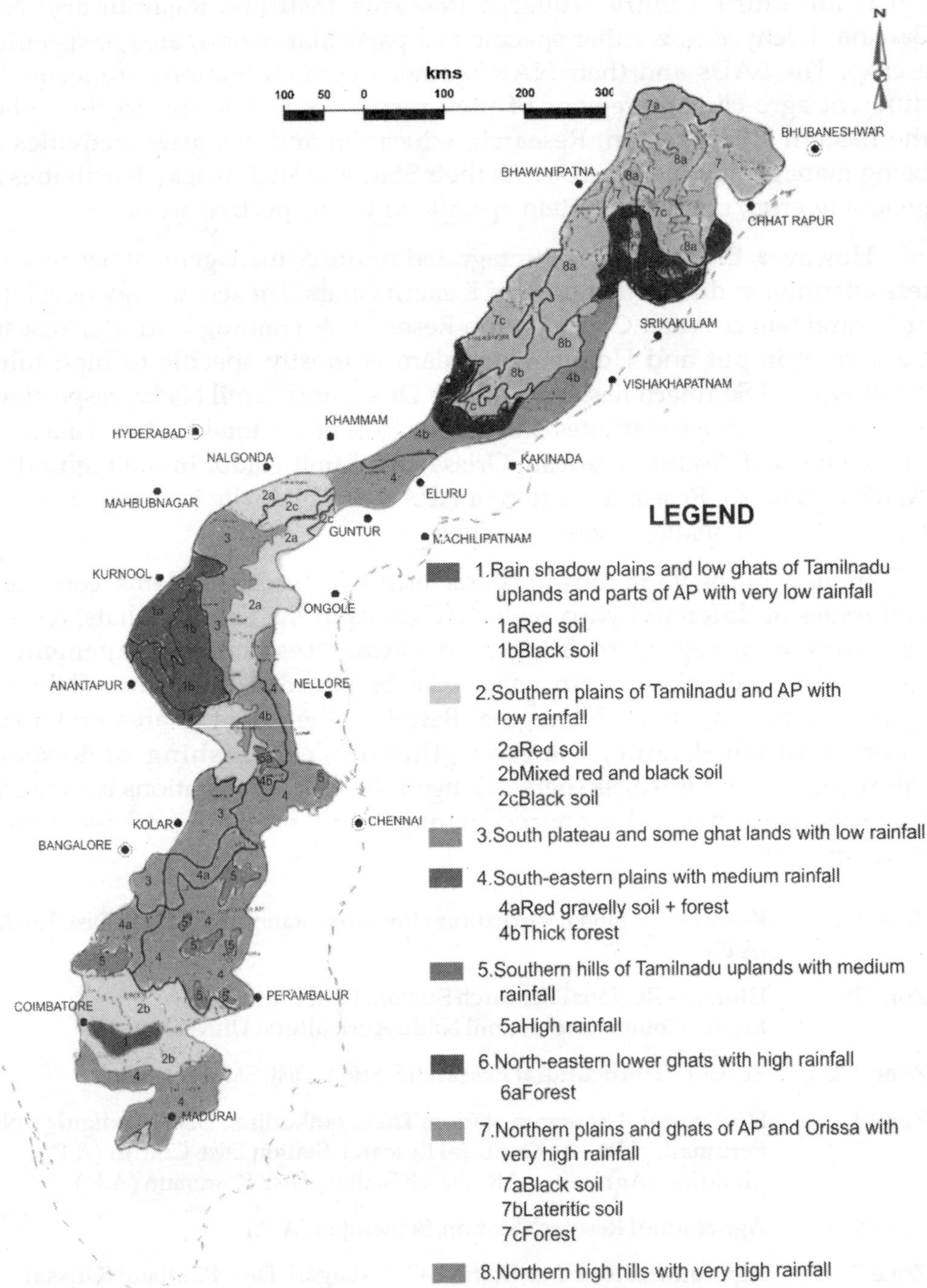

Map.1. Hydro-ecological zones of Eastern Ghats

Central Soil and Water Conservation Research and Training Institute, Research Centre, Udhagamandalam is located at the junction of Eastern and Western Ghats at the Southern end of the Eastern Ghats. The services and advise of this Research Centre may be utilized for strengthening research and technology development programmes in the lower Ghats area of Coimbatore, Erode and southern high hills of Tamil Nadu.

Research and technology development with regard to natural resource conservation and management for sustainable production system may be strengthened at the above Research Stations through inter-institutional linkages. The possibility of establishing Operational Research Projects on watershed basis may be considered to develop and demonstrate integrated technological package through multi-locational and multi-institutional linkages.

REFERENCES

Chauhan, K.P.S. 1998. Framework for conservation and sustainable use of biological diversity.Action plan for the Eastern Ghats Region. The Eastern Ghats, Proceedings of National Seminar on Conservation of Eastern Ghats, Andhra University, Visakapatnam, India.

Das, D.C. 1982. Development Needs of the Southern Hilly Regions. Proceedings of the Working Group Meeting of Soil & Water Conservationists held at the CSWCRTI, Research Centre, Udhagamandalam (Tamil Madu). 198p.

NATMO 1974. National Atlas Organisation, Tourist Atlas of India. Issued by Dept. of Science & Technology, Govt. of India.

Rao, P. Babu.1998. Ground Water Scenario of Eastern Ghats and its future trends in Andhra Pradesh.In proceedings of National Seminar on Conservation of Eastern Ghats, Andhra University, Visakapatnam, India.

Sikka, A. K., Singh, D. V. and Samra, J. S. 2001. Characteristic features, zonation and problems of Western Ghats for natural resource management. *Indian J. Soil Cons.*, 29(2): 107-117.

Sikka, A. K., Singh, D. V., Gadekar, H. and Madhu, M. 2004. Characteristic features, zonation and problems of Eastern *Ghats* (Hills) for natural resource management in India. *Indian J. Soil Cons.*, 32(2): 129-138.

Sikka, A.K., Singh D.V. Gadekar, H. and Madhu M. 2000. Research Report on Diagnostic Survey and Zonation of Eastern Ghats for Natural Resource Conservation and Management. NATP-CSWCRTI, Research Centre, Ooty. 82p.

□□□

Section - II

Chapter 16

Climate Change : Adaption Strategies in Rainfed Agriculture

P.K. MISHRA

INTRODUCTION

Climate Change is the much talked subject of the twenty first century which is drawing attention of every one in one or the other context, be it human and livestock health; or, damage / devastation of infrastructure and life of human being and livestock due to flash flood/ extreme low or high temperature; extinction of some fauna and flora; inundation of coastal low lying areas due to increase in sea level; melting of glaciers and snow bond regions; frequent occurrence of drought and floods and most importantly decrease in food production due to increase in carbon dioxide and temperature; and change in precipitation. These projections are based on the predictions generated through various General Circulation Models or Global Climate Model (GCM) and other models developed by different researcher world wide.

Among all the adverse predictions resulting from climate change, agriculture related predictions draws attentions of one and all. As we all understand, agriculture is most affected by precipitation and more precisely with rainfall. Its predicted that uncertainties will be more with rainfall characteristics *viz.*, increase in one day rainfall and rainfall intensity; change in the onset and offset of monsoon; change in the pre-monsoon, post monsoon, winter and monsoon rainfall; increase in the dry spell between two rain storm period during monsoon and so on.

These uncertainties associated with rainfall need to be looked into for developing strategies for not only maintaining the productivity but enhancing it too as simultaneously the population is also expected to increase. There are various measures/practices to combat the adversaries related with climate

Director, Central Soil & Water Conservation Research and Training Institute
Dehradun - 248 195, Uttarakhand.

change and especially with rainfall uncertainties, however soil and water conservation measures(engineering, biological and agronomical) executed on field scale and on watershed basis have proved to be equally useful under both drought and flood situations arising due to less or high rainfall.

Climate Change in India

Rainfall and temperature :The major weather related risks in agriculture (Rao *et al*, 2010) could be as follows: Monsoons in India exhibit substantial inter-seasonal variations associated with a variety of phenomena such as passage of monsoon disturbances associated with active phase and break monsoon periods whose periodicities vary from 3-5 and 10-15 days, respectively. It is well noticed that summer monsoon rainfall in India varied from 604 to 1020 mm. The inter-seasonal variations in rainfall cause floods and droughts, which are the major climate risk factors in Indian agriculture.

During the period 1871 to 2009, there were 24 major drought years, defined as years with less than one standard deviation below the mean. Another important adverse effect of climate change could be unprecedented heat waves. Heat waves generally occur during summer season where the cropped land is mostly fallow and, therefore, their impact on agricultural crops is limited. However, these heat waves adversely affect orchards, livestock, poultry and rice nursery beds. The heat wave conditions during 2003 May in Andhra Pradesh and 2006 in Odisha are recent examples that have affected the economy to a greater extent. Occurrence of heat waves in the northern parts during summer is common every year resulting in quite a good number of human deaths. Further, the water requirements of summer crops grown under irrigated conditions increases to a greater extent. Another adverse effect of climate change is cold waves which mostly occur in northern states. The Northern states of Punjab, Haryana, U.P., Bihar and Rajasthan experience cold wave and ground frost like conditions during winter months of December and January almost every year. The occurrence of these waves has significantly increased in the recent past due to reported climatic changes at local, regional and global scales. Site-specific short-term fluctuations in lower temperatures and the associated phenomena of chilling, frost, fogginess and impaired sunshine may sometimes play havoc in an otherwise fairly stable cropping/ farming system of a region.

Rainfall is the key variable influencing crop productivity in rainfed farming. Intermittent and prolonged droughts are a major cause of yield reduction in most crops. Long term data for India indicate that rainfed areas witness 3-4 drought years in every 10-year period. However, no definite trend is seen on the frequency of droughts as a result of climate change so far. For any R&D and policy initiatives, it's important to know the spatial distribution of drought events in the country. A long term analysis of rainfall trends in India (1901 to 2004) using Mann Kendall test of significance by CRIDA indicates significant increase in rainfall trends in West Bengal, Central India, coastal regions, south western Andhra Pradesh and central Tamil Nadu. Significant decreasing trend was observed in central part of Jammu & Kashmir, Northern M.P., Central and Western parts of U.P. and Northern and Central Parts of Chattisgarh. Analysis of number of rainy days based on the IMD grid data from 1957 to 2007 showed declining trends in Chattisgarh, Madhya Pradesh and Jammu & Kashmir. In Chattisgarh and eastern Madhya Pradesh, both rainfall and number of rainy days are declining which is a cause of concern as this is a rainfed rice production system supporting large tribal population who have poor coping capabilities.

Studies related to changes in rainfall over India have shown that there is no clear trend of increase or decrease in average annual rainfall over the country. The examination of trend of annual rainfall over India has indicated that 5 year running mean has fluctuated from normal rainfall within ± one standard deviation. Though the monsoon rainfall in India is found to be trendless over a long period of time, particularly on the all India scale, but there are pockets of significant long-term rainfall changes.

A comprehensive study using the monthly rainfall data for 306 stations distributed over India showed that areas of north-east peninsula, north-east India and north-west peninsula indicate widespread decreasing trend in the Indian summer monsoon rainfall. On the other hand, a widespread increasing trend in monsoon rainfall over the west coast, central peninsula and northwest India was found. The decreasing trend ranges between -6 to - 8 % of the normal per 100 years while the increasing trend is about 10 to 12 % of the normal per 100 years. Though these trends are statistically significant, but they account for a relatively small part of the total variance in the rainfall. Trend analysis carried out on sub-divisional rainfall data of 135 years (1871-2005) indicated no significant trend for annual, seasonal and for any monthly rainfall on all-India basis. Annual and monsoon rainfall decreased and pre-monsoon, post-monsoon and winter rainfall increased over the years, with maximum increase in pre-monsoon season. Further analysis on sub-divisional basis (30 sub-divisions) indicated that half of the sub-divisions have shown increasing trend in annual rainfall whereas remaining half sub-divisions showed decreasing trend in annual rainfall.

Trend and variability of seasonal and annual rainfall on the basin scale for the northwest and central parts of India showed that all the nine studied basins except Mahanadi have increasing trend in mean annual rainfall ranging from 2-19% of mean per 100 years. The Indus (lower) basin experienced maximum increase (19%). Ganga, Sabarmati and Brahamani and

Subarnarekha river basins have experienced stable annual rainfall. For the whole northwest and central India, the annual rainfall has increased by 5.2 % of mean per 100 years. Seasonal rainfall analysis showed that maximum increase in rainfall was observed in the post-monsoon season followed by the pre-monsoon season. India has skewed pattern of rainfall distribution, receiving 50 % of its annual rainfall in just 15 days. According to Biswas (2004), Cherranpunji, with highest rainfall in India receives its annual rainfall of 10820 mm between June and August in about 120 hours, but faces a water shortage problem during the dry months.

Due to increase in CO_2 levels the earth is warming up. In the last 15-20 years, there has been a sharp rise in the global temperature. While there are varying projections of temporal variations in temperature, there is a near unanimity in its direction and trend. In India too, the overall mean temperature is showing an increasing trend in most parts of the country. The central and western parts showed decreasing trend in maximum temperature while increasing trend of minimum temperature was observed in east, north and southern parts. Until last year, 2009 was the warmest year on record since 1901 (+0.913°C) above the normal of 24.64°C and now 2010 have surpassed it (+0.93°C). The other warmer years on record in order are 2002 (0.708), 2006 (0.6), 2003 (0.560), 2007 (0.553), 2004 (0.515), 1998 (0.514), 1941 (0.448), 1999 (0.445), 1958 (0.435), 2001 (0.429), 1987 (0.413) and 2005 (0.410).

The Himalayas have a total glaciated area of around 33,000 sq km (Eriksson *et al.*, 2009) which provides important short and long term water storage facilities. "There is about 12,000 cubic km of fresh water stored in the glaciers throughout the Himalayas – more fresh water than in Lake Superior" (Thompson, 2007). Compared to glaciers in other mountain ranges, the Himalayan glaciers are retreating at higher rates, and these rates are accelerating. Projections of glaciers retreat in the region (IPCC, 2007) suggest that the projected increase in the mean annual temperature for high Asia of 1.0C to 6.0°C by 2100 is likely to result in an extensive diminishing of glacial coverage.

The mean surface temperature in Tropical Asia has increased 0.3 - 0.8°C over the past 100 years, although there has been no trend in mean rainfall for the past thirty years in the area. There is also no discernible change in the number, frequency, or intensity of the tropical cyclones over the past 100 years, although some decadal patterns do exist (McLean *et al.*, 1998). Temperature predictions for India due to anthropogenically-caused climate change for the year 2040 versus 1980 are for a 0.7°C and 1.0°C increase for maximum and minimum surface air temperatures, respectively. However, this warming will be less prominent during the monsoon season. Winter diurnal temperature range is predicted to decrease (McLean *et al.*, 1998). With a doubling of pre-industrial levels of CO_2 in the atmosphere, the United Kingdom Meteorological Office (UKMO) GCM predicts a temperature increase for India of 16.2%, the Goddard Institute for Space Studies (GISS) GCM predicts an increase of 10%, and the Geophysical Fluid Dynamics Laboratory (GFDL) GCM predicts an increase of 23.5%. Some recent studies, however, have indicated that the GCMs temperature predictions are too high (Dinar *et al.*, 1998). Soil moisture is also supposed to be altered due to changes in precipitation, runoff, percolation, evaporation and rainfall distribution. However, soil moisture is very hard to predict due to the many factors involved and the uncertainty associated in making any climate change prediction (Dinar *et al.*, 1998).

Water Resources and Agriculture

Sensitivity analysis using hypothetical scenarios indicate that Brahmani basis, which is located in sub-humid tropical climate, stream flow is more sensitive to changes in rainfall than temperature. The findings revealed that the 2°C rise in temperature alone and with 10 % decrease in rainfall could decrease the annual water yield by about 5 to 30 %. This coupled with increased evaporative demands would further widen the gap between water supply and water demand during the periods with such a climatic variability. There is need of preparedness in crop, agronomic practices, and land & water management strategies for minimizing the adverse impact of climatic variability on agriculture.

Simulation modeling using HadCM3 and PRECIS generated scenario indicated likely increase in seasonal as well as annual stream flow. An increase of 26, 28 and 53 % in annual stream flow was projected during 2080 AD under HadCM3, A2a; HadCM3, B2a and PRECIS generated scenario, respectively. Even though all the scenario indicated likely increase in annual stream flow in the Brahmani basin, a decrease in stream flow is projected during the winter (December-February) and June in most of the cases. This has implications for *rabi* crops and as well as meeting the peak water demands during summer. Greater increase in magnitude of stream flow during the monsoon could also lead to flooding and water logging. This shows that there is need for development of different adaptation strategies for winter and summer crops, including improved water management interventions (Sikka *et al.*, 2009).

As far as Indian agriculture is concerned, temperature rise during *rabi* is of more significance. For minimum temperatures, most of the locations in India are showing an increasing trend. This is a cause of concern for agriculture as increased night temperatures accelerate respiration, hasten crop maturity and reduce yields. The increasing trend is more evident in central and eastern zones where rainfall is also showing a declining trend which makes this area more vulnerable and requiring high attention for adaptation research.

Climate change impacts on agriculture are being witnessed all over the world, but countries like India are more vulnerable in view of the high population depending on agriculture, excessive pressure on natural resources and poor coping mechanisms. The warming trend in India over the past 100 years is estimated to be 0.60°C. The projected impacts are likely to further aggravate yield fluctuations of many crops with impact on food security. There are evidences of negative impacts on yield of wheat and paddy in parts of India due to increased temperature, increasing water stress and reduction in number of rainy days. Modelling studies project a significant decrease in cereal production by the end of this century. Parts of western Rajasthan, Southern Gujarat, Madhya Pradesh, Maharashtra, Northern Karnataka, Northern Andhra Pradesh and Southern Bihar are likely to be more vulnerable in terms of extreme events. Irrigation requirement in arid and semi-arid regions is estimated to increase by 10 % by every 1°C rise in temperature.

The climate change (CC) is leading to global warming resulting in changing precipitation patterns. Water that flows over the land surface finally reaching the sea or that is stored in ponds, lakes, wetlands and other water bodies constitute the surface water. Over centuries, human beings adapted to changes in the climate (Pandey *et al.*, 2002). Soil water is the least understood and difficult to manage. The groundwater (GW) is the primary source of drinking water to ½ the world population. It provides irrigation to 70-80 % food production (Ramachandra and Nagarathna, 2010). With climate change leading to variability in rains, the GW uses are increasing. In fact, GRAPHIC (2009) points out that GW is being depleted @ 4.0 ± 2.0 cm year^{-1}. The hot spots identified in our country are western and peninsular regions. Water was taken for granted (as was the case with the soils).

There is a relationship between the intra-annual rainfall variability in a country and its level of prosperity. Countries with low rainfall variability typically have high GDPs (Gross Domestic Products), while countries struggling with large seasonal variability in water availability typically have low GDPs (Brown and Lall, 2006). Increased storage capacity and reduced seasonal differences in availability may help to reduce this gap.

Effect of Climate Change over Runoff and Soil loss on Watershed basis

Tripathi and Sharda (2007) and Tripathi *et al.* (2007) studied the effect of changed rainfall during 21st century on runoff and soil loss on 7 watersheds of the country representing different agro-ecological regions .They concluded that the annual and monsoon rainfall is projected to change by 3.4 to 46.9 % and -8.9 to 42.2 %,respectively. The annual and monsoon rainy days are also projected to change by -2.8 to 86.7 % and -16.0 to 91.4 %, respectively. The annual and monsoon runoff is projected to change between 55.9 % to 1211.1 % and 48.0 to 350.00 %, respectively. The changed (mostly increased in the study watersheds) pre-monsoon , post monsoon and winter monsoon rainfall and runoff will be beneficial for crop production, soil moisture storage and surface and sub-surface water harvesting with adoption of soil and water conservation measures on watershed basis. However, the monsoon rainfall and runoff will be much damaging leading to flood and loss of not only the soil fertility but movement of soil from field to unwanted locations namely drainage network and water storage structures (reducing their carrying capacity and storage capacity, respectively).

The frequency of land slide and slips will also be on higher side. The annual runoff of the magnitude of over 300 mm is projected to increase between 26.1 to 200 % thus increasing the frequency of flood. The annual runoff of less than 50 mm is expected to reduce by 10 to 100 % thus increasing the frequency of drought. It's very likely that high intense rainstorm may be very common leading to frequent drought and flood .The soil loss is expected to increase by 3.7 to 757 %. The most feasible and relevant mitigation measure will be to adopt the watershed management programme more earnestly in combination of planting the permanent vegetation *i.e.*, tree, shrubs & grasses in 10-70 % of the total watershed area of about 500 ha, maintaining at least 10% of cultivated land of individual farmer under agro/horti/ forestry , 10-15 % of common land under social forestry and another 15-70 % of undisturbed forest in every watershed of about 500 ha depending on its location in plain/hilly/ mountainous region.

Efforts under Soil and water Conservation at the National Level

That soil conservation is necessary for sustenance has been realized by the Government. Programs like River Valley Project (RVP), Flood Prone Area (FPR) program were initiated way back in 1962-63 and 1981-82, respectively (Venkateswarlu, 2010). Soil conservation in arable lands was mainly executed as a governmental program. Initially it was largely formation of contour bunds followed by check dam subsequently. Both the programs have not yielded the desired results and the land owners were only mute spectators. In most cases, they dismantled the bunds. Even the Government could rarely follow scientific principles in forming contour bunds. It culminated in strengthening the existing field bunds across the slope which was accepted by the land owners. The logic was that the existing bunds already occupied about 5 % of their land and the additional contour bunds would cover another 8 % area. Further, these bunds were cutting across the fields making their agricultural operations difficult. In any case, conservation of soil that belongs to farmers needs their active participation. And the technologies need be simple, doable and socially acceptable.

India was among the first few countries to have taken timely cognizance of the enormity of the soil erosion problems. Many policy interventions and legislations were enacted from time to time in different states to combat land degradation. The main objective of many sections of the Forest Act, 1878 was primarily aimed to preserve land by regulating the use of common and private land resources. The Punjab Land Preservation Act (Choe Act) was passed in 1900, Ravine Reclamation Project was undertaken in the 1920s in the United Province (now U.P. State) and a Dry Farming Research Project was launched in 1930 at Sholapur, Bijapur, Hageri, Raichur (Karnataka) and Rohtak (Haryana) to stabilize rainfed yields through soil and water conservation strategies in semi-arid regions. In 1948-49, Damodar Valley Corporation Act was passed to moderate floods by integrated watershed management in the Damodar river catchment.

Watershed management, though less focused earlier, has a history of about 60 years in independent India. Multiple agencies have been involved in the growth and development of watershed programmes. Biophysical aspects of watershed management were first started in the Damodar-Barakar basin under Damodar Valley Corporation, Hazaribagh (Bihar) by establishing a Soil Conservation Department during 1949-50. The focus on watershed research was further sharpened with the establishment of a network of Soil Conservation Research, Demonstration and Training Centres (which was latter renamed as Central Soil & Water Conservation Research and Training Institute) in various part of the country by the Central Soil Conservation Board under Union Ministry of Food and Agriculture (Agriculture Department) during 1st and 2nd Five year Plan (1952-1957) to provide necessary research back up and trained manpower.

Paradigm Shift from Soil Conservation to Land Husbandry

Retrieving degraded land which is largely soil erosion, calls for a paradigm shift from the traditional soil conservation through mechanical structures to land husbandry. Here, we like to quote the doyen in soil conservation, Hudson (1992). He said "Current changes include even the term soil conservation will probably fade away to be replaced by land husbandry because that better describes the fundamentals of the new approach. The idea of the care of crops and their management and improvement has for years been called crop husbandry. Animal husbandry has described the care and management of livestock. Soil conservation was appropriate when we were mainly concerned with increasing knowledge and awareness of soil degradation and learning how to decrease the process. But that was mainly a defensive strategy and what we now seek is a positive approach where care and improvement of the land resource comes first and control of erosion as a result of good land husbandry".

Thus, the primary objective of land management should be improved and sustainable production through good land husbandry. Control of soil erosion follows as a consequence. This is a reversal of the previous idea that it is necessary to conserve the soil in order to get better crops. Instead, improve the soil condition for root growth and crop production and in so doing achieve better conservation of water and soil. And to achieve this goal, use of biological measures is more relevant. Among others, they include agroforestry, soil cover with good vegetation (plant cover) and improving soil organic matter (SOM). All these measures sequester more carbon, an ecological protection service.

Adaptation Strategy in Agriculture for Climate Change Scenario

Some of the common tradition crop growing practices which the farmers have been adapting to combat the aberration in the climate are listed below. These practices were evolved by the farmers over period of time. Under climate change scenario we are very likely to witness similar anomalies. What we need is to have systematic logical reasoning and a crop calendar for each situation falling under different agro ecological region.

- Conservation Tillage and Land preparation (Preparatory tillage operation)
- Mulching and Residue Management
- In-Situ Mulching
- Cover management
- Optimization of Sowing Time
- Plant Population
- Crop Geometry and Orientation
- Strip-cropping
- Mixed cropping
- Intercropping
- Contour farming
- Vegetative barriers
- Agroforestry for Soil and Water Conservation

Charles Darwin, long ago, said "It's not the strongest of the species that survive, or the most intelligent person that survives", it is the one that is most adaptable to change. We are aware that, depending on the ecology 1 kg of wheat needs 500-4000 litres of water while meat needs 5000-20,000 litres (Lindqvist *et al.*, 2008). Options available are (i) improve water use efficiency, (ii) go towards more to less water consuming products, and (iii) adopt best practices in biomass production like direct seeding mulch based cropping systems and SRI cultivation of rice.

While dealing with land and water management issues, Kerylenstierna (2009) pointed out "There are certain things that are so simple that 10 years of intensive university studies are required to misunderstand them". Scarcity of water, food and energy besides economic security are all linked to land and water issues and thus to climate change adaptation. Many water issues are "local" while most of the discussion is at "global scale".

For instance, water is a state subject and groundwater is managed on individual basis. The Centre for World Solidarity (CWS) has shown that under limited groundwater (due to over-exploitation), the community can come together to share water from the functional wells and this participatory groundwater management has been accepted by Government of Andhra Pradesh (GoAP) and being put on ground in the state on a pilot scale. In any case, water harvesting for upgrading rainfed agriculture is a large window of opportunity (Falkenmark *et al.*, 2001). Similarly, we preach proper land use systems without considering that the poor largely own only class IV and lower lands. The Deccan Development Society (DDS) and the National Agriculture Technology Project (NATP) study undertaken by Central Research Institute for Dry land Agriculture (CRIDA) have clearly shown that in such marginal lands, farmers practice mixed cropping. They attempt to improve their soils overtime with the leaf litter and the sloughed off burrows of deep rooted crops like niger, castor, *etc.* besides adding fine soil like silt. Thus, "the best way to have a good idea is to have lots of ideas".

While discussing the options to face the dangers of climate change, Holdren (2007) pointed out three options - mitigation, adaptation and suffering the adverse impacts. Neither of these alone can be a solution. 'Mitigation only' strategy will not work because it is too late to avoid substantial climate change. 'Adaptation only' strategy won't work because these measures are too expensive and less effective as the magnitude of climate change is becoming larger. So win-win approach is the need.

Measures to improve water conservation and water management have great value even in the absence of climate-driven increases in stress on water systems. Similar is the case with enhancing soil organic carbon, which makes the soil drought-resistant and also productive besides reducing erosion levels. No-till is a case in point. While discussing the Indian climate policy, Mahanta

(2009) reminded us that the failing monsoon of 2009 followed by the unprecedented rains in early October in the southern states, established the need for addressing such anomalies more frequently than ever. We must realize that annual rainfall renewability is finite and subject to uncertainty. Similarly, products like petroleum and phosphate rock are also finite. And we all know our land is also finite.

With climate change, rainfall and its intensity as well as distribution stand highly skewed. Only 15-20 % of rainfall is generally used by crops. With degradation of the soils due to climate change, this would be only 5%. With loss of top soil, the productivity will be further accentuated. We have already burnt ½ of known global resources of oil in 125 years or so. The rock phosphate situation is no better. It price has shot by 7 times over the last 14 months. Restrictions by way of enhanced tariffs in countries like China make the problems more serious. So we have to choose a new path or one will be forced upon us. Organic Agriculture (OA) that saves about 27 % energy is one route. Organic Agriculture sequesters more carbon (55 to 80% of global Green House Gases due to agriculture). Zero- is another example.

However, as rainfall intensity and frequency of storms increases, the above practices may not provide adequate protection to the soil. The researchers, producers and soil conservationists have to aggressively pursue soil management practices to protect the soil under these changing rainfall conditions, as pointed by Hatfield and Prueger (2004). Land husbandry preached by Hudson (1992) could be one possible solution. All such systems are in some form or other under practice by smallholders. In fact, Society for Elimination of Rural Poverty (SERP) of GoAP adopted a paradigm shift from 'external input driven agriculture' to 'knowledge, skill-based and local natural resources' model in their community managed sustainable agriculture (CMSA-Eco-agriculture). This is unlike the findings of Shackleton *et al.* (2002) of Overseas Development Institute (ODI) who felt it more as rhetoric. SERP strongly believed that this CMSA will lead smallholders from agricultural distress to viable farming. All such systems do cool the climate and protect top soil which is the most strategic resource. Business as usual (making it dirt) is NOT an option when it comes to soil, food and people. It's time for a greener revolution (Montgomery, 2010). And those who practice non-chemical farming need be paid for the ecological protection service rendered by them. It's the time that is now a scarce resource. Think globally but act locally.

Farmers' Opinion about Climate Change and Strategy Options

Though the farmers also observe the change in the climate but at times he mixes it with the normal change. He is not able to isolate the change which is otherwise due to aberration in climate change with climate change. In fact the normal historical climatological data analysis also at times is not able to correctly interpretate it. Farmers also have contingency plan to nullify the erratic rainfall. When an assessment was made on the accuracy of their understanding about the trends in the rainfall amounts, and the strategy for mitigating the adverse impact of climate change, the following conclusions were drawn (Rao *et al.,* 2011):

- Farmers from Akola (Maharashtra) could sense the decrease in rainfall (both annual and seasonal) correctly as the long term (1871-2008) rainfall data analysis showed a slight decreasing trend. The decadal averages also showed a declining trend as the annual rainfall of 817 mm in 1971-1980 period decreased to 678 mm by 2001-2009.
- Farmers at Anantapur (Andhra Pradesh) failed to sense the changes in rainfall pattern as the decadal analysis of rainfall for the period 1969-2008 did not indicate any increasing or decreasing trend.
- Farmers of Faizabad (Uttar Pradesh) sensed the declining trend correctly as the long term analysis (1970-2009) of rainfall data showed a declining trend though it is not statistically significant.
- Farmers of Parbhani (Maharashtra) of the low input group sensed an increase in the annual rainfall however the long term analysis (1961-2009) of the rainfall data did not support their opinion.
- Farmers of Jammu region (Rakh Dhiansar) expressed declining summer temperatures whereas the analysis of the long term (1983-2009) indicated otherwise.
- Farmers from the semi arid region like Solapur (Maharashtra) expressed an increasing trend in annual rainfall whereas statistical analysis for the period 1971-2009 proved that they were wrong in their opinion. However, they could successfully notice an increasing trend in the summer temperature but failed to observe an increasing trend in the minimum temperatures as the minimum temperatures are on rise at the rate 0.02°C per year.
- An early withdrawal of monsoon was stated to have been observed by farmers of Rakh Dhiansar (Jammu) and Anantapur (Andhra Pradesh).
- The commencement and withdrawal of south west monsoon is very crucial for Indian agriculture across all the regions. This is evident from the priority given by the responding farmers through out the country.
- Farmers try to adapt to the variability in weather parameters and market forces by changing their cropping / farming systems. This adaption is vital in mitigating the effects of weather abnormalities as well as

fluctuations in the farm prices. The farmers of the Akola (Maharashtra) were united in expressing the change in their cropping system in recent times to cope up the changes in the weather as well as to overcome the pest incidence in cotton. They shifted from mono-cropping of cotton to a diversified cropping system involving soybean, pigeon pea and chickpea. Similar opinion was expressed by farmers from Bijapur (Karnataka), Faizabad (Uttar Pradesh), Bangalore (Karnataka) and Solapur (Maharashtra). Market driven changes in the cropping system were reported by farmers of Bijapur rather than the weather abnormalities alone. Non-profitability in the existing cropping system was the prime cause to opt for a change in the cropping system as expressed by farmers of Solapur and Faizabad.

- The farmers of several regions are divided on the identification of the prime factor responsible for change in the cropping system. The rice-wheat system of some parts of the Indo-Gangatic region is stated to have been replaced by rice/maize followed by mustard/gram /lentil/ ginger/vegetables. Non adaption of this technique of change in the cropping system was reported by the farmers of Rakh Dhiansar.

Programme under Investigation to Address the Problem of Climate Change in Indian Agriculture

The country has already taken the following programme to combat the adverse effect of climate change:

(i) Massive watershed development programme for developing water resources at micro level (500-3000 ha) in rural areas. This has been observed to be equally effective under both deficit and excessive rain fall conditions. It has reduced soil loss considerably thus maintaining crop productivity.

(ii) Trees in association with agriculture as additional sink for carbon sequestration

(iii) Minimum and zero tillage for maintaining soil carbon

(iv) Development of drought resistant crop verities

(v) Integrated fertilizer management programme

(vi) Integrated pest and disease management programme

Future Thrust Areas

Based on the experiences gained, the future thrust of study is likely to be:

(i) Popularization of efficient micro irrigation system

(ii) Development of heat, salt, drought and flood resistant crop verities

(iii) Development of heat resistant live stock and aquaculture or selection of suitable breed

(iv) Feeding requirement and behavior of live stock and aquaculture

(v) Development of improved efficient agronomical practices

(vi) Improvement of fertilizer intake efficiency

(vii) Development of bio inset pest management

Mitigation Strategies

The mitigation strategies for climate change in agriculture sector needs to be tackled by developing (i) Short & medium term and, (ii) Long term (IPCC, AR4, WG III).

(i) Short & medium term (2030) mitigation strategies

Soil carbon sequestration has vast mitigation potential in agriculture sector which if managed appropriately can reduce vulnerability to climate change. This soil carbon is also likely to be lost with faulty land management practices. Some of the agriculture systems can also be improved to reduce emission of methane and nitrous oxide. However, there is no single mitigation practice to achieve this and practices need to be evaluated for individual agricultural systems and settings.

Biomass from agricultural residues and dedicated energy crops can be an important bio-energy feedstock, but its contribution to mitigation depends on demand for bio-energy from transport and energy supply, on water availability, and on requirements of land for food and fibre production. Widespread use of agricultural land for biomass production for energy may compete with other land uses and can have positive and negative environmental impacts and implications for food security. India, being such a country where about two third populations are still depending on agriculture and live stock (for energy, milk and meat) can not afford to divert the biomass from agriculture residues towards bioenergy feedstock.

Agricultural practices collectively can make a significant contribution at low cost to increasing soil carbon sinks, to GHG emission reductions, and by contributing biomass feed stocks for energy use. The mitigation in the short and medium term (2030) towards improvements of crops yields is as given below.

1. Improved crop and grazing land management to increase soil carbon storage needs to be developed.

2. Restoration of cultivated peaty soils and degraded lands

3. Improved rice cultivation techniques and livestock and manure management to reduce CH_4 emissions.
4. Improved nitrogen fertilizer application techniques to reduce N_2O emissions.
5. Dedicated energy crops to replace fossil fuel use.
6. Improved energy efficiency

Long term (after 2030) mitigation strategies

Investments in and world-wide deployment of low-GHG emission technologies as well as technology improvements through public and private Research, Development & Demonstration (RD&D) would be required for achieving stabilization targets as well as cost reduction. It's said that the lower the stabilization levels, especially those of 550 ppm CO_2-eq or lower, the greater the need for more efficient RD&D efforts and investment in new technologies during the next few decades will be needed. This requires that barriers to development, acquisition, deployment and diffusion of technologies are effectively addressed.

However, the long term mitigation strategy is to be tackled by developing technological innovations and by framing appropriate policies towards implementation of the available technologies and providing suitable budget for this. Financial incentives and regulations for improved land management, maintaining soil carbon content, efficient use of fertilizers and irrigation may encourage synergy with sustainable development and with reducing vulnerability to climate change, thereby overcoming barriers to implementation. The governments frame the policy based not on technological requirement but on their respective national circumstances and an understanding of their interactions. The policies framed in a country may not work in another country because of the varied sociological, economical, and geographical variations. It is also observed that some corporations, local and regional authorities, NGOs and civil groups are adopting a wide variety of voluntary actions. These voluntary actions may limit GHG emissions, stimulate innovative policies, and encourage the deployment of new technologies. On their own, they generally have limited impact on the national or regional level emissions.

Mitigation Options and Sustainable Development

Making development more sustainable by changing development paths can make a major contribution to climate change mitigation, but implementation may require resources to overcome multiple barriers. There is a growing understanding of the possibilities to choose and implement mitigation options in several sectors to realize synergies and avoid conflicts with other dimensions of sustainable development. Agriculture sector is no different. Adaptation measures are necessary irrespective of the scale of mitigation measures. Addressing climate change can be considered an integral element of sustainable development policies.

Reducing both loss of natural habitat and deforestation can have significant biodiversity, soil and water conservation benefits, and can be implemented in a socially and economically sustainable manner. Forestation and bioenergy plantations can lead to restoration of degraded land, manage water runoff, retain soil carbon and benefit rural economies, but could compete with land for food production and may be negative for biodiversity, if not properly designed.

We would like to share the quote of Eleventh Commandment of Lowdermilk (1994). "Thou shalt inherit the Holy Earth as a faithful steward, conserving its resources and productivity from generation to generation". Thou shall safeguard the fields from soil erosion, thy living waters from drying up, thy forests from desolation, and protect thy hills from overgrazing by the herds, that they descendants may have abundance forever. If any shall fail in the stewardship of the land, thy fruitful fields shall become sterile stony ground and washing gullies, and they descendants shall decrease and live in poverty or perish from off the face of the earth".

CONCLUSIONS

The climate change is very likely to adversely affect the agricultural production system especially under rainfed system. The traditional agricultural system had managed the variability in rainfall with various practices and measures. We have to strengthen our knowledge base of traditional system coupled with innovative research to mitigate the adverse effect of climate change under rainfed system.

REFERENCES

Agarwal, P.K. (Editor). 2009. GLOBAL CLIMATE CHANGE AND INDIAN AGRICULTURE, Case Studies from the ICAR Network Project, Publisher: Indian council of agricultural Research, New Delhi.

Anonymous. 2006. Greenhouse Gas Mitigation in India: Scenarios and opportunities through 2031. The Energy and Resources Institute (TERI), New Delhi, India.

Biswas, A.K. 2004. Dams: Cornucopia or disaster?. *International Journal of Water Resources development,* 20(1):3-14.

Brown, C. and Lall,U. 2006.Water and economic development: The role of variability and a framework for resilience. *Natural resources Forum,* 30:306-317.

Dinar A, Mendelsohn, Evenson, R., Parikh, J., Sanghi, A., Kumar, K., McKinsey, J. and Lonergan, S. 1998. Measuring the Impact of Climate Change on Indian Agriculture. Washington, DC: World Bank.

Eriksson, M., Xu, J., Shrestha, A.B., Vaidya, R.A., Nepal, S. and Sandsrom, K. 2009.The Changing Himalayas: Impact of Climate Change on Water Resources and Livelihood in the Greater Himalayas. Kathmandu; ICIMOD.

Falkenmark, MFox, P., Person, G. and Rockstorm, J. 2001. Water harvesting for upgrading rainfed agriculture, SIWI, 94p.

GRAPHIC. 2009. Groundwater Resource Assessment under the pressure of humanity and climate change, GRAPHIC project of UNESCO, IPH Programme.

Hatfield, J.L. and Prueger, J.H. 2004. Impacts of changing precipitation patterns on water quality. *J. Soil Water Cons.*, 59: 51-58.

Holdren, J.P. 2007. Meeting the climate challenge: Avoiding the unmanageable and managing the avoidable. UN Commission on sustainable development, Harward Univ., New York.

Hudson, N. 1992. Land Husbandry, Belfast, London.

IPCC website; Assessment Reports (Working Group I, II and III of Firs, Second, Third and Fourth Assessment reports.

IPCC. FAR, SAR, TSR and FAR.

IPCC.2007. Climate Change 2007. The physical science basis. Contribution of Working Group I to the Fourth Assessment Report of the Intergovernmental Panel on Climate Chang. Cambridge University Press.

Jehangir M., Zokaib S., Merz J., Dongol B. S. and Verma P. K. 2005. Assessment of runoff and soil loss from the selected watersheds in the Hindu-Kush Himalayan Region. *Sarhad Journal of Agriculture*, 21(3):437-444.

Kerylenstierna, J. 2009. Can climate change adaptation support the integration of land and water management issues? Land day, UN Water, FAO, Bonn.

Lowdermilk, W.C. 1994. Conquest of the land through seven thousand years. USDA Information Bulletin, 99.

Lundqvist, D., De Fraiture, C. and Molden, D. 2008. Saving water from field to fork – cutting losses and wastage in the food policy. SIWI Policy brief.

Mahanta, C. 2009. Climate change threats to India's water resources and emerging policy responses. In: Indian Climate Policy: Choices and challenges. Eds. D. Michel and A. Pandya STIMSON, 19-23.

McLean RF, Sinha SK, Mizra MQ and Lal M. 1998. Chapter 11: Tropical South Asia. In: The Regional Impacts of Climate Change: An Assessment of Vulnerability: A Special Report of IPCC Working Group II. [Eds: Watson RT, Zinyowera MC, and Moss RH] Cambridge: Cambridge University Press. pp381-407.

Montgomery, D.R. 2010. Soil-Opinion-2020 Vision, *Nature*, 463/7, 31.

Nearing, M., Prusk, F.F. and O'Neal, M.R. 2004. Expected climate change impacts on soil erosion rates. *J. Soil Water Cons.*, 59: 43-50.

Ouyang, Hua. 2009. The Himalayas-water storage under threat. Sustainable Mountain Development No. 56, ICIMOD, Winter 2009:3-5.

Pandey, D.N., Gupta, A.K. and Anderson, D.M. 2002. Rainwater harvesting as an adaptation to climate change. *Curr. Sci.*, 85, 46-59.

Pruski, F.F. and Nearing, M.A. 2002. Runoff and soil-loss responses to changes in precipitation: A computer simulation study. *Journal of Soil and Water Conservation*, Jan., 2002.

Ramachandra, T., and Nagarathna, A.V. 2010. Review on climate change and ground water. *Curr. Sci.*, 92: 103-105.

Rao, V.U.M., Bapuji Rao, B., Khandgonda, I.R., Rao, A.V.M.S., Vijaya Kumar, P., Dagar, J.C. and Venkateswarlu, B. 2011. Perception of Indian Farmers on Climate Change – An Assessment and Awareness Programme. Central Research Institute for Dryland Agriculture, Santoshnagar, Hyderabad, Andhra Pradesh, India. 33 p.

Rupa Kumar, K., Sahai, A.K. and others. 2006. High Resolution Climate Change scenario for India for the 21st Century. *Current Science*, 90:334-345.

Schackleton, S., B. Campbell, E. Wollenberg and D. Edmunds. 2002. Devolution and Community based NRM: Creating space for local people to participate and benefit. ODI Natural Resources Perspectives, Nov. 76.

Sikka, Alok, K., Islam Adlul, Saha, B. and Anamika. Impact of climate change on stream flow in the Brahmani Basin in Eastern India; In Agarwal, P.K. (Editor). 2009. GLOBAL CLIMATE CHANGE AND INDIAN AGRICULTURE, Case Studies from the ICAR Network Project, Publisher: Indian council of agricultural Research, New Delhi.

Thompson, Lonnie. 2007. New Tibetan ice cores missing A-bomb blast makers; suggest Himalayan ice fields haven't grown in the last 50 years. *Science Daily*, Dec., 13, 2007.

Tripathi, K.P. and V.N. Sharda 2007. Rainfall, Runoff and Soil Loss in selected watersheds of India during 21st century". Paper presented at the Workshop on "Climate change and its impact on farming systems and natural resources" held at Y.S.Parmar University of Horticulture and forestry, Solan, H.P. held during 02.03.2007 to 03.03.2007.

Tripathi, K.P., Sharda, V.N., Shahid Ali, Md, and Mudgal Ashish. 2007. Impact of Climate change on Rainfall, Rainy days, Runoff and Soil Loss in Selected Watersheds of India during 21st Century. Paper presented at the National conference on "Impacts of Climate Change with Particular Reference to Agriculture "Organized at ACRC, CSCMS, TNAU, Coimbatore during Aug. 22-24, 2007.

Vaidya, Ramesh Ananda. 2009. The role of Water Storage in Adaptation to Climate Change in the HKH Region. Sustainable Mountain Development No. 56, ICIMOD, Winter 2009:10-13.

Venkateswarlu, J. 2010. Climate change: Soils on Global Agenda. In: Lead Papers, National Symposium on Climate Change and Rainfed Agriculture, (Eds. Venkateswarlu *et. al.*), CRIDA, Hyderabad, pp106-123.

Williams, J.R., Sharpley, A.N. (Eds.), 1989. EPIC – Erosion/Productivity Impact Calculator: 1. Model documentation, USDA Technical Bulletin No. 1768.

Zhang, X.C., Nearing, M.A.,Garbrecht, J.D. and Steiner, J.L. 2004. Downscaling Monthly Forecasts to Simulate Impacts of Climate Change on Soil Erosion and Wheat Production. *Soil Sci. Soc. Am. J.*, 68:1376–1385.

□□□

Chapter 17

Climate Change Scenario and Implications on Agriculture in Eastern India

LALU DAS

INTRODUCTION

The physically based climate models can be used to generate estimate of future climate. The most comprehensive models available are the General Circulation Models (GCMs) which simulate the circulation and physical properties of the atmosphere and surface meteorological variables over a three dimensional spatial grid boxes that are typically 250 km in width and 600 km in length (Feenstra *et al.*, 1998), representing the earth-climate system. Most of the GCMs used for climate change experiments to date perform rather poorly when judged by their ability to simulate the regional details of present day precipitations (Legates and Willmott, 1992). In spite of extensive use of GCMs to provide potential climatic change scenarios, they have significant inherent limitations in projecting regional climatic patterns (*e.g.*, Houghton *et al.*, 1992). The most significant limitations includes: (i) poor spatial resolution (ii) inadequate coupling of atmospheric and oceanic processes (iii) poor simulation of cloud processes and (iv) inadequate representation of the biosphere and its feedback. Several studies have been conducted in different regions all over the globe, *e.g.*, over Mediterranean basin (Palutikof *et al.*, 1992), East Asia (Hulme *et al.*, 1994), Norway (Benestad, 2002) and over Scandinavia and New Zealand (Kidson and Thompson, 1998) using GCMs. It is generally recognized that the GCM offers best potential for the development of regional climate change scenarios though they are unable to give a good reproduction of local climatic conditions (Grotch and MacCracken, 1991). Further, increasing demand of the impact analysts to have finer resolution climate change scenarios, several methodologies have been developed (*e.g.*, Wilby *et al.*, 1998; Wilby and Wigley, 1997; Murphy, 1998) as GCMs output is

Department of Agricultural Meteorology and Physics, Bidhan Chandra Krishi Viswavidyalaya Mohanpur, Nadia, West Bengal - 741 252 constrains by its course resolution. Among the various methodologies, statistical downscaling methods (Wigley *et al.*, 1990), seeking relationships between the GCM simulated variables and those required by the impact researchers, offer acceptable scenarios (*e.g.*, Von Storch *et al.*, 1993; Crane and Hewitson, 1998; Huth and Kysely, 2000).

The Indian climate is dominated by the great wind system called "monsoon". So monsoon studies have become one of the most important aspects of atmospheric research in India. Now in the context of climate change, a variety of studies are being reported on the intensity of monsoon. Instrumental records indicate that the monsoon rainfall is trendless on all India scale (Mooley and Parthasarathy, 1984; Kripalani *et al.*, 2003) though pockets of significant long-term trend exist in various parts of India (Kolli *et al.*, 1992). It has been shown that the coastal Gangetic West Bengal (GWB) in the east has an increasing trend of 10-12 % in 100 yrs (Kolli *et al.* 2002). Instrumental records also indicate a warming of 0.4°C in 100 yrs in mean annual surface air temperature for the country as a whole. Seasonal variation shows less warming (0.1°C 100 yrs^{-1}) in the monsoon and more warming in the winter (0.4°C 100 yrs^{-1}) and postmonsoon (0.5°C 100 yrs^{-1}) seasons. On the other hand, greenhouse gas warming simulations indicate an increased intensity of the Asian summer monsoon (Chakraborty and Lal, 1994; Hirakuchi and Giorgi, 1995) with some exceptions (Lal *et al.*, 1994). There exist a high degree of uncertainty with all the impact predictions due to limited understating of critical processes in the climate system, existence of multiple climatic and non-climatic systems, regional scale variations and non-linearity. In another study, future climatic scenarios for India for the period 1980-2039 relative to 1920-1979 reveal that ECHAM4 indicates an increasing monsoon rainfall by 13 % while HadCM2 shows a reduction by 6 % (Kolli *et al.*, 2002) though the annual temperature increases by about 1.0°C for both the models. Seasonal variation shows more warming in the winter and pre-monsoon seasons. All these studies, based on either instrumental records or GCM output for all-India scale, do not have any seasonal composite scenarios or high resolution down scaled composite scenarios to explore the spatial details of climate change in a local scale.

The present chapter has the main aims of constructing seasonal high-resolution composite scenarios for a part of the Indian subcontinent, *i.e.*, GWB and its neighborhood, using statistical downscaling technique to infer more information for the impact study, as there is no evidence of availability of finer resolution scenarios using some kind of downscaling methods in the literature.

Data and Study Area

Five GCMs, namely, ECHAM4, HadCM2, CSIRO-Mk2b, CGCM1 and GFDL-R15 are selected for the present purpose of the study as those models are constantly providing outputs for impact research starting from IPCCs second assessment report (SAR, 1995), third assessment report (TAR, 2001) the most recent forth assessment report (AR4, 2007). For developing high resolution downscaled scenarios, a part of the Indian subcontinent, *i.e.,* the GWB and its neighborhood, extending from 20° to 26°N and 83° to 89°E in the eastern India have been considered (Figure 1). How the resolution of each models are increasing over the study area is shown in Figure 2. For the purpose The monthly mean temperature and rainfall data sets for 'IS92a' emission scenarios for GHG and GHG+S forcing are collected from the IPCC data distribution centre for the periods 1961-90, 2010-2039 and 2040-2069. Observational records of monthly mean temperature and rainfall for all available stations within GWB and its neighborhood are collected for the period 1961-2000 from India Meteorological Department (IMD) and Agriculture Department of West Bengal. After scrutinizing the data, stations with the required length of record have been retained.

Model Validation

Before we can precede for the construction of higher resolution downscaled climate change scenarios for the study region, it is important to investigate the acceptability of the GCMs outputs over the study region. This is done through the validation of the GCMs outputs by comparing their simulation results of current climate ($1XCO_2$) with the observations. The better the simulation results of current climate, the more confidence we may put in its ability to generate realistic future regional climate change scenarios. Clearly, if the models fail to simulate current climate satisfactorily, one can place reduced confidence in its prediction for future.

The distribution of rainfall is highly variable in both time and space. Most of the GCMs used in this study, have grid-size considerably less than the scale of many rain producing systems though the regional patterns of rainfall are strongly influenced by local topographic details which is not resolved by these GCMs. Therefore, the control run GCMs results may not agree well with the observed rainfall patterns of such a small area. Validation exercise is being performed by calculating (i) spatial correlation coefficient between GCM output interpolated to station level and observed station values for the period 1961-90 and (ii) comparing seasonal cycles of the GCM control run result with the observed station values over the study area.

Spatial Correlations

One convenient measure of similarity between two climates (*e.g.*, a modeled and an observed climate) is the spatial pattern correlation coefficient (Freenstra *et al.*, 1998). The outputs of the control climate of five GCMs are interpolated first to all stations available in study area. Then spatial correlation coefficients are calculated for the four seasons for rainfall and temperature (Table 1).

Table 1: Spatial correlation coefficients between GCMs run and observed precipitation and temperature

Season	HadCM2		CSIRO-MK2b		CGCM1		GFDL-R15		ECHAM4	
	Rain	Temp	Rain	Temp	Rain	Temp	Rain	Temp	Rain	Temp
Winter	0.22	0.58	0.22	0.77	0.54	0.74	0.12	0.71	0.26	0.54
Pre-monsoon	0.28	0.46	0.30	0.60	0.07	0.62	0.002	0.62	0.17	0.08
Monsoon	0.54	0.46	0.30	0.44	0.65	0.42	0.41	0.43	0.37	0.12
Post-monsoon	0.77	0.17	0.39	0.38	0.28	0.35	0.16	0.20	0.34	0.33

Rain - Rainfall, Temp - Temperature

Each model under investigation shows a better correlation coefficient for temperature with a maximum (0.77) for CSIRO - Mk2b in the winter season. However a low value of 0.08 for ECHAM4 in monsoon season has also been observed. Although there is large variation of spatial correlation from season to season and from model to model, most of the correlation coefficients are greater than 0.3. It is noticed that the correlation coefficient for HadCM2 is more than 0.45 for all the seasons except in the post-monsoon season for temperature. As per expectation, correlation coefficient for rainfall shows a random variation with a maximum for HadCM2 in post-monsoon season (0.77) and a minimum value for GFDL-R15 in pre-monsoon season (0.002). The suspicious high and low values of spatial correlation coefficients indicate the problems associated with the data sets. These are (i) additive bias (ii) susceptible to the outliers *etc.* The later has been checked by the rank correlation coefficient for the same data sets and the results are given in the Table 2.

Table 2: Rank correlation coefficients between GCMs run and observed precipitation and temperature

Season	HadCM2		CSIRO-MK2b		CGCM1		GFDL-R15		ECHAM4	
	Rain	Temp	Rain	Temp	Rain	Temp	Rain	Temp	Rain	Temp
Winter	0.08	0.60	0.35	0.75	0.57	0.70	0.35	0.72	0.28	0.56
Pre-monsoon	0.16	0.43	0.36	0.49	0.08	0.61	0.04	0.59	0.11	0.09
Monsoon	0.37	0.48	0.25	0.52	0.54	0.39	0.41	0.40	0.27	0.16
Post-monsoon	0.20	0.31	0.50	0.57	0.25	0.46	0.24	0.32	0.27	0.47

No significant difference has been found between the two Tables (1 & 2), which have been generated by different methodologies. However, no model shows higher correlation for rainfall in all the seasons, rather correlation coefficients are low irrespective of the seasons. This indicates the limitations of the models to represent the spatial variation of rainfall.

Seasonal Cycle of Rainfall and Temperature

The study region has pronounced and distinctive seasonal cycles of rainfall and temperature with two peaks, one in the monsoon season while the other in the pre-monsoon season. A successful model must be able to reproduce this seasonal cycle that play a vital role in the ecology and economic strategies of the area. A comparison study between the observed values of regional monthly mean precipitation and temperature for the current climate (1961-90) with the simulated GCMs values has been performed for all the selected models. It is interesting to note that in our validation study, monthly regional-averaged rainfall and temperature have been calculated as follows. The number of observation stations have been identified first which lie within 5° x 5° grid box centered on the respective model grid location. Then the simple average is calculated to get the required regional averaged values of rainfall and temperature for that particular GCM. The details of GCMs outputs considered in our validation study are 9 points for every GCM shown in Fig. 1 along with the number of observational stations. This figure (Figure 1) also sketched to count the number of observations stations which fall within the 5°x5° grid boxed centered on respective model outputs. So our regional-averaged values for rainfall and temperature are changing from model to models as the number of stations fall within the grid box centered on the respective GCM output are varying. The fact should be clearly understand first that the number of GCM output which fall in our study areas are more than one normally five/six which again varying depending upon horizontal resolution of the GCMs. Some output fall within study very well and other are outside the region and few cases on the boundary (Figure 1). We considered such an output of GCM for our comparison study, which is well inside the study region, and number of observational stations fall in it is maximum. The observed and model simulated rainfall and temperature for five GCMs are displayed in Figure 2 and Figure 3. All models more or less able to simulate the flow pattern of observed seasonal cycles, which have a maximum amount of rainfall in monsoon season and the second peak in the pre-monsoon seasons. However, the models are inadequate to simulate the unusual occasion of rainfall properly during the winter season (western disturbances) and pre-monsoon season (meso-scale activities). ECHAM4 seems to simulate the observed seasonal cycle better while others fail particularly for the winter and pre-monsoon seasons. On the other hand, regional-averaged values of temperature have also been calculated using the same method for rainfall and results. Compared to rainfall, the seasonal cycle is more adequately simulated in case of temperature in all the seasons for CGCM1, ECHAM4 and HadCM2 but completely fail to simulate the temperature in case of GFDL-R15 model, which may be due to very poor horizontal resolution of the model.

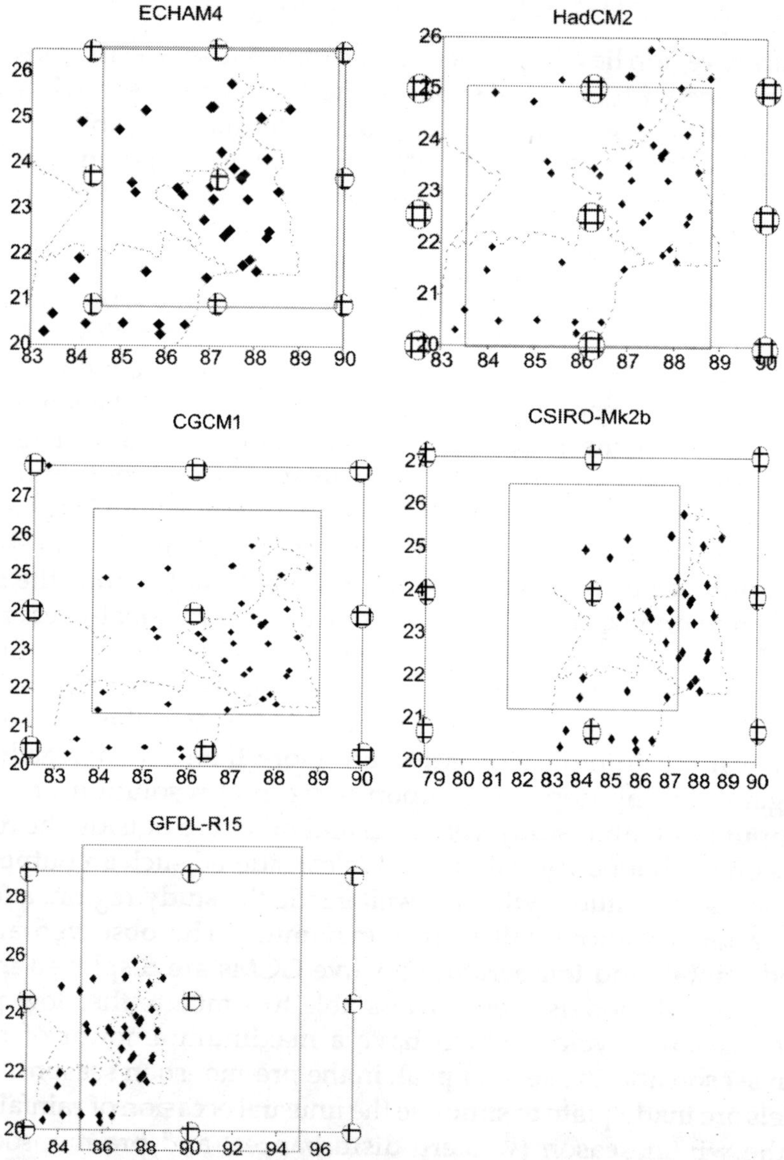

Fig. 1: Showing the distribution of stations (small squares) and the GCMs grid points (big circles) in the study area. 5° x 5° square box centered on a model grid location along with the number of stations, which are inside the square box for respective GCMs.

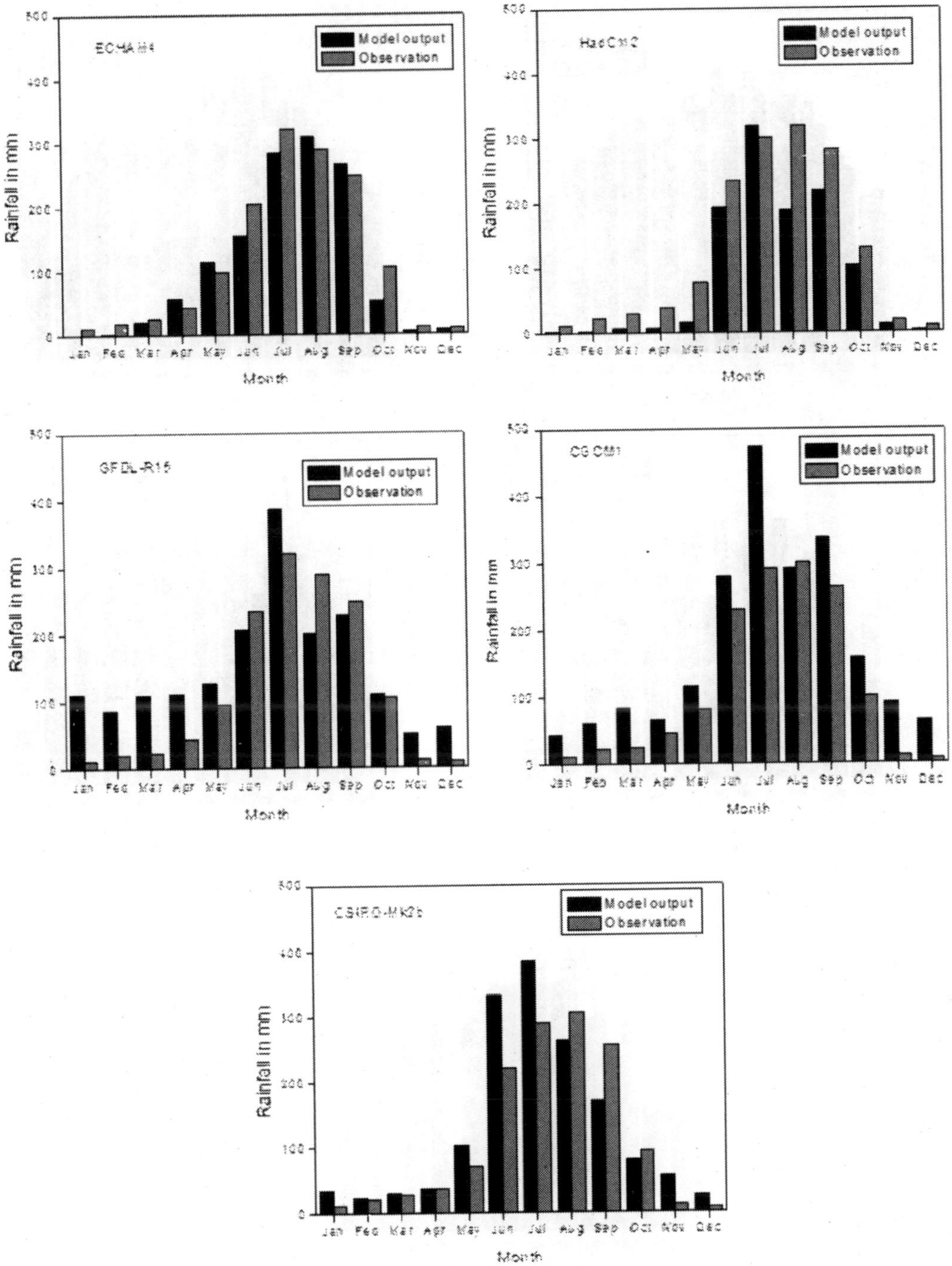

Fig. 2: Comparison of model output and observed rainfall for different models. (Observed values are based on the stations which are within a grid box of size 5°x5° centered around well inside model grid point)

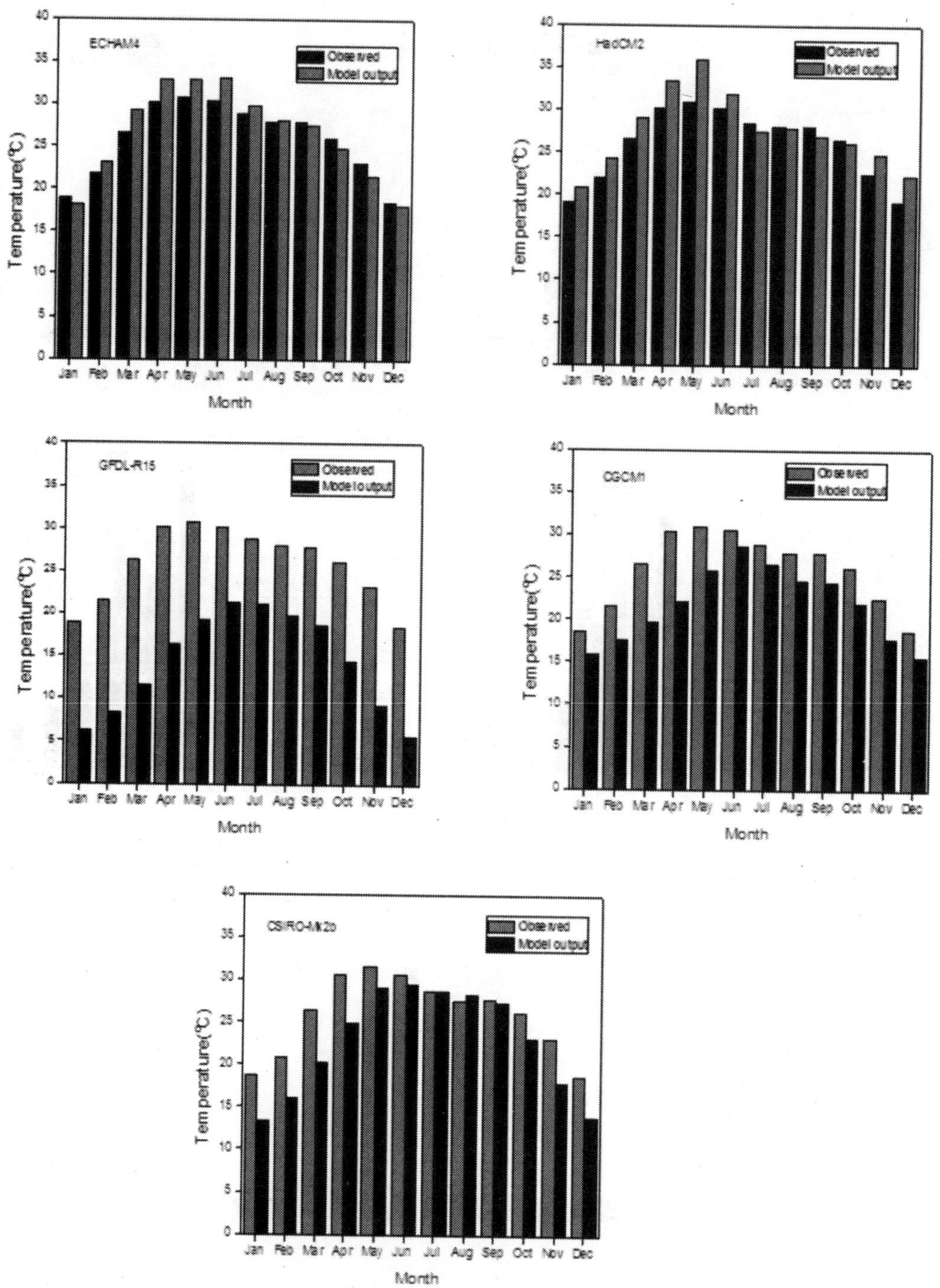

Fig. 3: Comparison of model output and observed mean temperature for different models. (Observed values are based on the station, which are within a grid box of size 5°x5° centered on well inside grid point).

Downscaling GCMs Output Using Regression analysis

The usefulness of the scenarios developed directly from the GCMs output and interpolated to a common grid for all India region in chapter 3, constrains by the coarse resolution of the models, currently limited to a horizontal resolution of about 300 km (*e.g.*, Johns *et al.*, 1997). The GCMs used in our studies having the resolution of high values of 2.8° x 2.8° in ECHAM4 and the poorest resolution 4.5° x 7.5°. The level of this resolution is inadequate for many regional climate change studies where information requires at network of point locations (Robinson and Finkelstein, 1991), implying the need of using a suitable 'downscaling' procedure. This is particularly true when the model results are required for the use of agricultural crop models (*e.g.*, Mearns *et al.*, 1997). To develop higher resolution scenarios, suitable for use in impact studies, the problem is to find out a suitable downscaling procedure. Here, an attempt has been taken to develop sub-grid scale scenarios over GWB and its neighborhood using statistical downscaling method as explained in the following sub-section.

METHODOLOGY

The two most important works by Kim *et al.* (1984) and Wigley *et al.* (1990) provide a clear workable methodology to extract local-to regional scale climatic information from GCM grid point output. These two methods have been modified by Palutikof *et al.* (1992) and Palutikof and Wigley (1996) for developing climate change scenarios for Mediterranean region. In this present study, the same methodology is applied in the GWB and its neighborhood region to generate the high-resolution climate change scenarios. For the development of regression models on seasonal basis, the following procedure is adopted.

First, time series data sets of observed monthly mean temperature and monthly rainfall within the study area are formed for the period 1961-2000. The data series spanning the period 1961-90 have been used for the construction of statistical models while the rest, *i.e.*, data for the period 1991-2000, is used for validation of the constructed models. Then the temperature anomalies for each station, have been calculated using, where is the mean monthly temperature for the ith year and j^{th} month while is the long-term mean monthly temperature for the jth month. Similarly, the rainfall anomalies for each station, Ap_{ij}, have also been calculated using, where p_{ij} is the mean monthly rainfall for the ith year and j^{th} month while P_j represents the long-term mean monthly rainfall for the jth month. The individual station anomalies are then used to calculate regional anomalies (Palutikof *et al.*, 1992; Palutikof and Wigley, 1996). For temperature, the regional anomaly for each station has been calculated by taking the average of all individual station anomalies within a 5°x5° grid box with the corresponding station as the center of the grid box. The concerned station has been left out while taking the average of individual station anomalies in order to make the regional station anomaly independent of the station anomaly itself. While for rainfall, the same procedure as in the case of temperature has been used. However, an area-weighted value of rainfall has been calculated using Thiessen polygon method instead of ordinary average as the rainfall data has substantial spatial variability.

Finally, data sets are made for regional anomaly and station anomaly of mean monthly temperature and rainfall corresponding to each station. Regression analyses are then built, using station temperature anomalies and rainfall anomalies as predictands, and their corresponding regionally averaged anomalies as predictors. The analyses have been performed on a seasonal basis. Thus for each station in the analysis, eight regression equations are formed as there are four seasons. With the assumption that GCM grid-point temperature and rainfall values are equivalent to regionally averaged values, the GCM results have been interpolated from the model grid to the station locations for each model and substituted in the regression equations as predictors. The predicted change in temperature is then divided by the corresponding global mean equilibrium temperature change.

Climate Change Scenarios for 2010-2039 with GHG and GHG+S Forcing

Rainfall change scenarios : The projected seasonal rainfall scenarios based on the above regression equations for two different forcing are shown in Fig.4 and Fig 5 for ECHAM4 and HadCM2 ,respectively. Maximum change in rainfall is seen in the winter season for both the models. The values are ranging from 10% to 25% in HadCM2 and from 5% to 35 % in ECHAM4 for GHG forcing alone where as 5-15% in HadCM2 and 0-25% in ECHAM4 for the GHG+S forcing which indicate that the sulphate aerosol forcing reduces the rainfall change. The pattern indicates a maximum over the coast of the Bay of Bengal and gradually decreases towards the inland in HadCM2 while opposite is the case in ECHAM4. In the pre-monsoon season, negative changes of rainfall are seen almost all through the study region in HadCM2 (–13% to +1%) with the maximum change over the coastal region while slightly positive changes are seen in ECHAM4 (1% to 6%) due to GHG forcing. But with GHG+S forcing, slightly higher vales (0-20%) in ECHAM4 and (+5 to –25%) in HadCM2 with minimum change in southern part and maximum change in northern side. It is interesting to note that the changes are minimum in the monsoon season (–2% to +0% for HadCM2 and 1% to 5 % for ECHAM4 for GHG forcing and -6 % to 0% for HadCM2 and –1% to +3% for ECHAM4 for the combined effect of GHG plus sulphate aerosol forcing). A variation of –

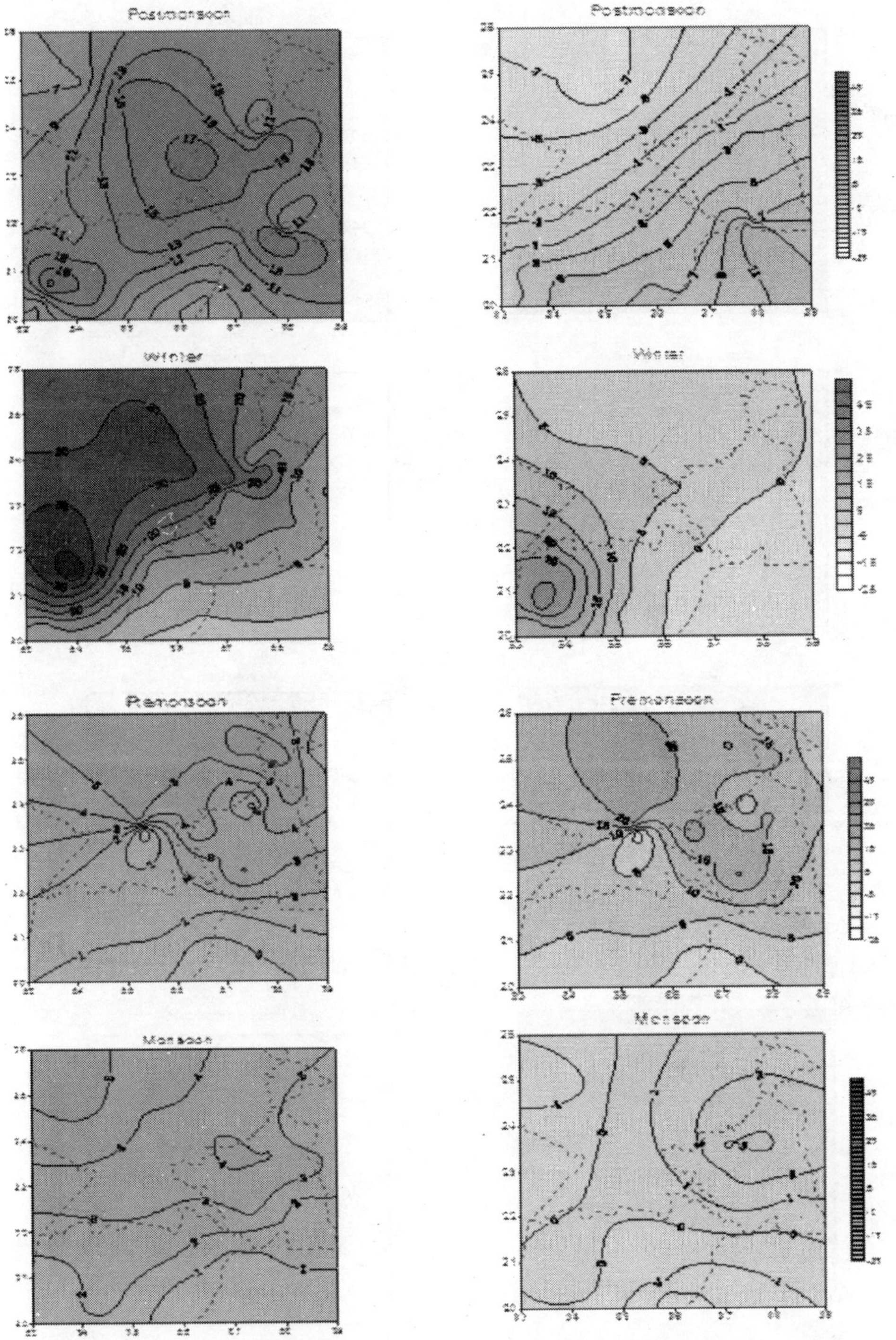

Fig. 4: Seasonal rainfall change (in %) scenarios for 2010-2039 *w.r.t.*, baseline for GHG (left column) and GHG+S (right column) forcing conditions (ECHAM4).

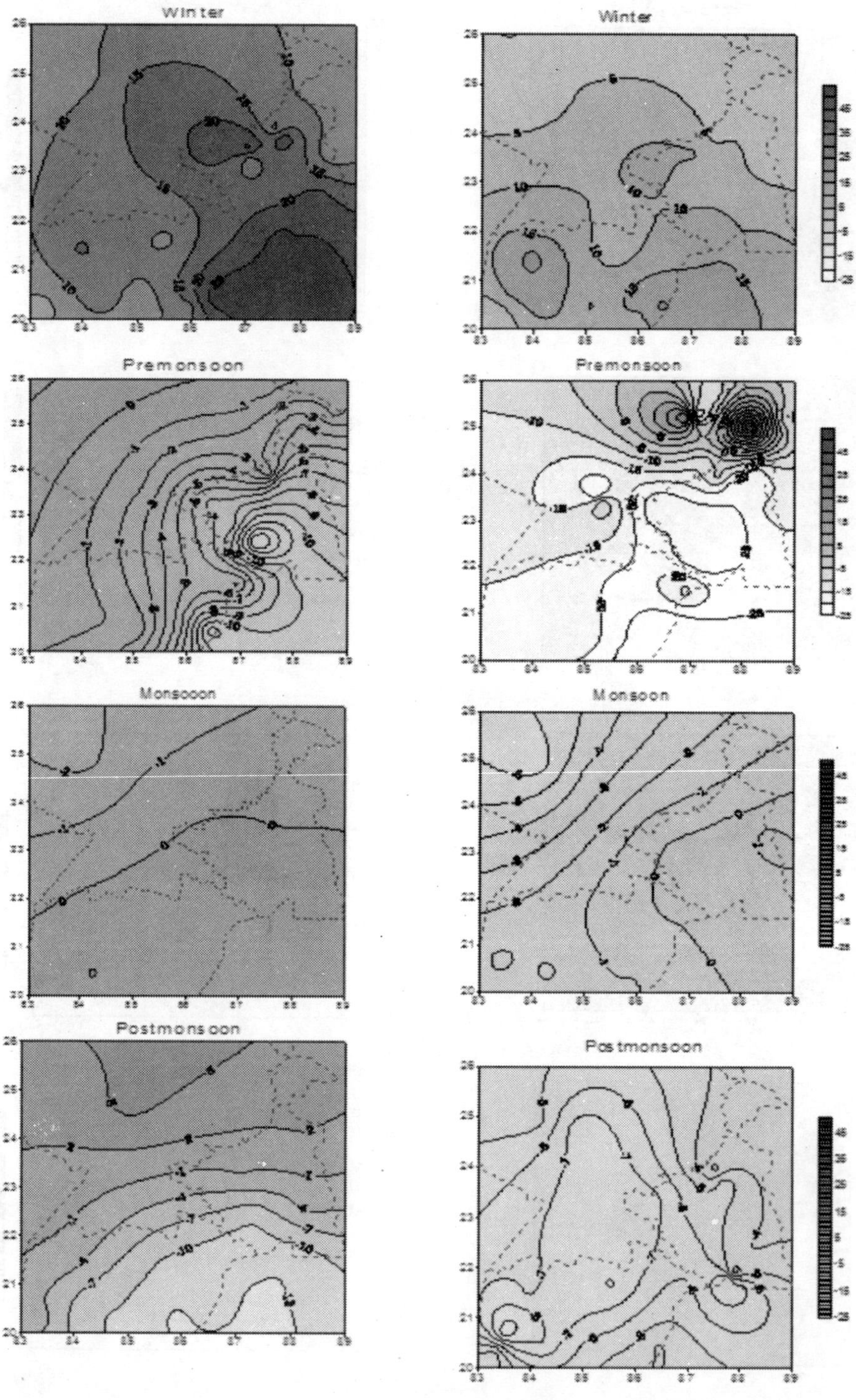

Fig. 5: Seasonal rainfall change (in %) scenarios for 2010-2039 *w.r.t.,* baseline for GHG (left column) and GHG+S (right column) forcing conditions (HadCM2).

13% to +5% in case of HadCM2 is noticed in the post-monsoon season while it is more (7% to 17%) in ECHAM4 for GHG forcing. When the forcing includes sulphate with GHG, the variation is -7% to +11% for ECHAM4 with a strong increasing gradient from NW part (inland) to SE part of the study area while a variation of -4% to -8% for HadCM2 is noticed with out any prominent gradient for the said post-monsoon season.

As stated elsewhere that, no single GCM can be identified is being consistently better at simulating current and future climate change scenarios, composite scenarios have been developed using HadCM2 and ECHAM4 for the best estimation to the impact analysts. The combined rainfall change for the winter season is noticed to be more (10-25 %) in the study region due to GHG forcing and similar changes (2% to 23%) for GHG+S forcing with the maximum change in the SW corner of targeted area. Scenarios in the pre-monsoon season indicate a dipole structure with a positive value 2% in the inland and negative changes (-5%) over the coast for the left column and same dipole (+10%) to (-15%).

Temperature Change Scenarios

The projected seasonal temperature change scenarios have been developed like rainfall scenarios and are shown in Fig. 6 and Fig. 7 for ECHAM4 and HadCM2, respectively considering the GHG (left column) and GHG+S (right column) forcing. A warming of 0.3-0.7 °C is noticed in all the seasons all over the area except monsoon season in HadCM2 due to only GHG effect. The similar change (0.2-0.7°C) is also noticed incase of GHG +S but everywhere the changes is slightly less compared to other one. The maximum change in temperature (more than 1°C is seen in the NW part of the study area for both cases) is observed in the monsoon season that indicates that change crossed the limit of global warming during the period 2010-2039 for HadCM2 model but in case of ECHAM4, no such abnormal worming is projected in any seasons. The spatial distribution shows more change in the eastern and NW part of study region for both the models. In pre-monsoon season, the changes are between 0.3 to 0.6°C with less value over the coastal belt for ECHAM4. HadCM2 indicates a variation of about 0.4 - 0.7°C with a maximum over the central part of the study area for GHG forcing. After the addition of sulphate forcing with green house gas, the result shows changes in temperature (0.15-0.3°C) in ECHAM4 and (0.2 to 0.4°C) in HadCM2. In the monsoon season, HadCM2 shows a variation of 0.3-1.1°C with a positive gradient from the sea towards inland. ECHAM4 also shows a similar gradient though the changes are less with GHG induced forcing as well as GHG+S forcing. Both he models indicate a maximum change of 0.7°C in the post monsoon season with a minimum in the SW and a maximum in the NW parts of the study area in

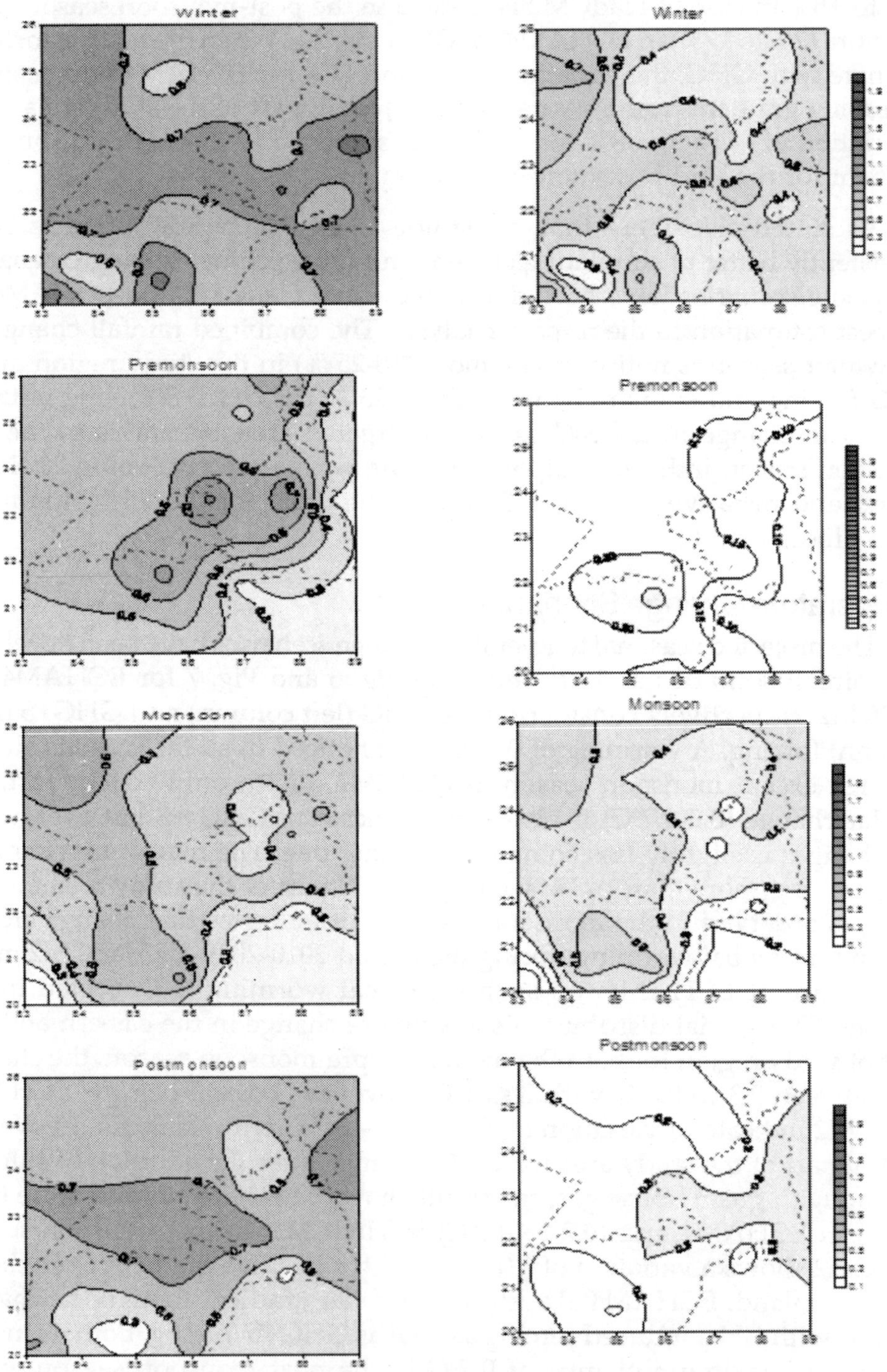

Fig. 6: Seasonal temperature change (°C) scenarios for 2010-2039 *w.r.t.,* baseline for GHG (left column) and GHG+S (right column) forcing conditions (ECHAM4).

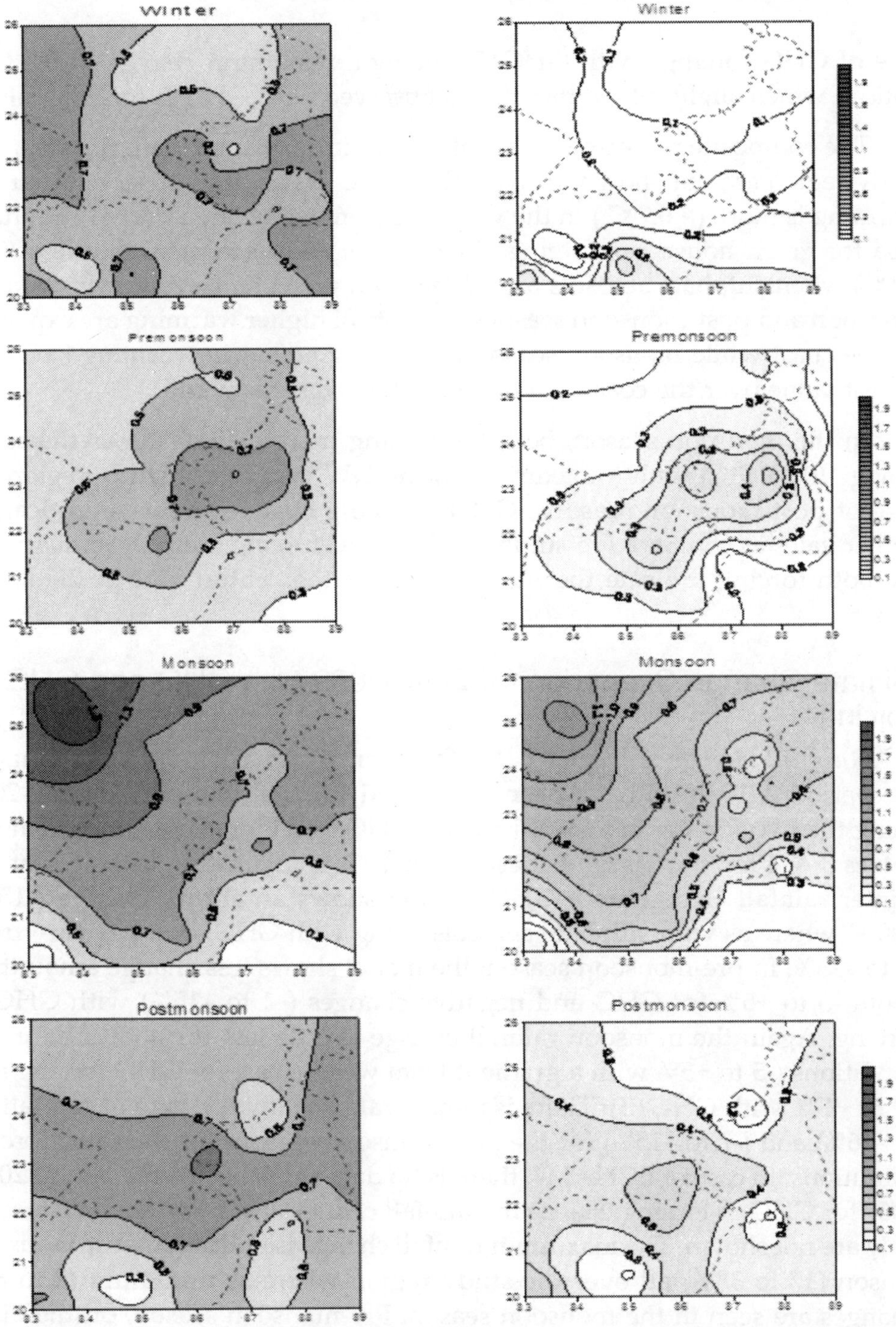

Fig. 7: Seasonal temperature change (°C) scenarios for 2010-2039 w.r.t. baseline for GHG (left column) and GHG+S (right column) forcing conditions (HadCM2).

case of GHG forcing. With GHG+S forcing a maximum change of 0.5°C is noticed which slightly less than that is observed with GHG is forcing only.

The composite projected seasonal temperature change scenarios over the study region are not shown for both the cases. It has been observed that the warming is more (> 0.7°C) in the winter season over a large part of the study area for green house gas forcing alone where as it is comparably less (0.2-0.4°C) when sulphate aerosols are added with green house gases. In the pre-monsoon and post-monsoon seasons, pockets of higher warming are expected off the coast while monsoon season reveals a gradient in warming from the lowest value over the coast to a higher value over the inland.

In the monsoon season, both the forcing more or less shows the same change (0.2-0.9°C) with a maximum in the NW part of the study region. In case of post-monsoon seasons, GHG inducing change shows significantly higher vales as compared to sulphate plus GHG forcing but only similarity is the both forcing indicate the more change in the central part of the study area.

Climate Change Scenarios for 2040-2069 with GHG and GHG+S Forcing

Rainfall change scenarios : Rainfall change scenarios, similar to that in the period 2010-2039, have been developed for the time period 2040-2069 using the two forcing namely GHG and GHG+S(Figures not shown but see Tables 3 & 4 for summary of those results). It is interesting to note that the winter rainfall change with GHG forcing shows an abrupt change (-15 to +80%) where as with sulphate aerosols along with GHG, changes vary from -5 to -13%. In pre-monsoon season, the model shows less change amounting about -3 to +5% for GHG and negative changes (–1 to –15%) with GHG+S forcing. Again the monsoon rainfall change is little less for both the forcing conditions (-5 to +5% with a gradient from west to east for GHG forcing and -9 to +1 % with GHG+S). There is a moderate negative change of rainfall (-4 to –16%) and (-7 to –15%) for the post-monsoon seasons for the same forcing conditions. In case of ECHAM4, there is no data available for the period 2040-2069 for GHG+S forcing. Hence the rainfall changes (in %) with GHG forcing only are not shown. The maximum rainfall change is noticed in post-monsoon season (17 to 38%) all over the study region where as minimum (2 to 6%) changes are seen in the monsoon season. Pre-monsoon season, on the other hand, the rainfall change shows a wider range of variation (-10 to +17%) with maximum in the northeastern part and minimum over the extreme south. As there is no data available for 2040-2069 for GHG plus sulphate forcing, the composite rainfall change scenarios are presented here for GHG forcing only. It is found that the composite rainfall change in winter season is maximum (-7 to +35%) and the maximum lies over the southwest part of the study region. In pre-monsoon season, the maximum change (11%) in noticed over the northern part while the minimum change (-7%) is over the southern region of the study area. In case of monsoon season, the change is less and varies from +2 to +6% with increasing gradient from west to east. Positive changes are noticed in post-monsoon season that cover a values ranging from +1 to +17% with maximum value in the sea.

Temperature Change Scenarios

The high resolution downscaled temperature change scenarios for 2040-2069 using GHG and GHG+S have been developed and results are not ploted. First of all, it is observed that the values of changes with sulphate forcing for temperature during 2040-69 is comparable with the values change during 2010-39 without sulphate aerosol. The projected change in temperature during this time slice, is definitely crossed the global mean temparature change for GHG forcing because everywhere in every seasons, the values of change is rainging from 0.5 to 2.5°C per °C global change for HadCM2 model. In all season,the maximum change is shown in the northwest part of the study area and minimum in the sea areas indicating that the SST is not change as faster as land surface temperature. If we consider, the GHG+S forcing which is to some extend the real situation of change in the environment, then the all season except monsoon, show a less change(0.3 to 0.8°C) in temperature for the whole area, indicating less vulnerable to global worming. In monsoon season, it crossed the global limit in some part of the area.

In case of ECHAM4, the projected change in temperature is shown only for GHG foring because GHG+S foring is not available. It is obserbed that the the expected change in temperature for the peroid 2040-2069 is more than 1°C in every season except for monsoon season where the change is from 1.0°C to 0.5°C for the ECHAM4 model.

The composite temperature change scenarios per °C change in global temperature for green house gas forcing is generated but not shown. The projected change is more than 1°C for all seasons and for the most of study area, indicating the vulnerable situation for the period 2040-2069. The winter season shows the most adverse situation by projecting a change of 1.2 - 2.2°C. The monsoon season change varies from 0.5°C to 2.0°C while 0.5°C to 1.4°C changes are noticed in premonsoon season. In post0monsoon season the changes are comparable with the premonsoon season.

Table 3: Rainfall change (%) scenarios for 2010-2039 and 2040-69

Season	HadCM2		ECHAM4		Composite	
	GHG	GHG+S	GHG	GHG+S	GHG	GHG+S
Winter						
2010-39	+10 to +25	5 to 15	5 to 35	0 to 25	10 to 20	2 to 23
2040-69	-15 to +5	-13 to -5	-21 to -1	--	-7 to +40	--
Pre monsoon						
2010-39	-10 to 0	-25 to 10	0 to 6	0 to 20	-5 to +2	-15 to +10
2040-69	-3 to +5	-13 to -1	-10 to +14	--	-7 to +11	--
Monsoon						
2010-39	-2 to 0		1 to 5	-1 to +3	01 to 02	-3 to +1
2040-69	-5 to +5	-8 to +1	+2 to +6	--	+2 to +6	--
Post monsoon						
2010-39	-13 to +5	-4 to -8	7 to 17	-7 to +11	-3 to +9	-7 to +1
2040-69	-16 to -4	-15 to -5	17 to +38	--	+1 to +17	--

Table 4: Temperature change (°C) scenarios for 2010-2039 and 2040-69

Season	HadCM2		ECHAM4		Composite	
	GHG	GHG+S	GHG	GHG+S	GHG	GHG+S
Winter						
2010-39	0.5 to 0.7	0.1 to 0.3	0.5 to 0.7	0.3 to 0.7	0.4 to 0.8	0.2 to 0.4
2040-69	1.2 to 2.0	0.2 to 0.4	1.2 to 2.2	--	1.2 to 2.2	--
Pre-monsoon						
2010-39	0.3 to 0.7	0.2 to 0.4	0.3 to 0.7	0.1 to 0.2	0.3 to 0.7	0.2 to 0.3
2040-69	0.8 to 1.6	0.3 to 0.5	0.6 to 1.4	--	0.6 to 1.6	--
Monsoon						
2010-39	0.3 to 1.3	0.3 to 1.1	0.3 to 0.6	0.2 to 0.5	0.3 to 0.9	0.2 to 0.8
2040-69	0.8 to 2.8	0.5 to 1.7	0.6 to 1.0	--	0.6 to 2.0	--
Post-monsoon						
2010-39	0.3 to 0.7	0.3 to 0.5	0.5 ton0.2	0.2 to 0.4	0.3 to 0.7	0.2 to 0.4
2040-69	0.8 to 2.4	0.4 to 0.7	0.6 to 1.2	--	0.6 to 1.6	--

CONCLUSIONS

The ability of these five GCMs to represent the spatio-temporal variation of rainfall and temperature over GWB and its neighborhood has been verified through a comparison of model values with the observations. It is noticed that all the models are unable to represent seasonal cycle of rainfall properly.

ECHAM4 appears to be better compared to the other models for the region of interest. Again in case of temperature, considerable differences are noticed in the winter and pre-monsoon seasons though the models except GFDL better represent the monsoon season. The influences of the passage of western disturbances over the northern India in winter and the mesoscale convective activities in the pre-monsoon season (Lohar and Pal, 1995) are unlikely to be represented in GCMs. Except for winter season, CSIRO-Mk2b model simulate the observed rainfall pattern satisfactorily for the other three seasons. In the monsoon season, the ratios (model to observed rainfall) show less than 1 value indicates slightly underestimation Student's t-test using the model output and observed seasonal rainfall also indicates that the HadCM2 and ECHAM4 are more reasonable over GWB and its neighborhood. On the other hand, the ability of the models to represent the spatial patterns is checked through the spatial correlation coefficients. In general, correlation coefficients are higher in case of temperature than rainfall. Except in the monsoon season, low values of correlation coefficients indicate that the models fail to simulate the spatial patterns properly which agrees well with the findings of Hulme *et al.* (1994). The poorest horizontal resolutions are seen in the GFDL-R15. model, which simulated the observed seasonal cycles of temperature inadequately. All seasons except the monsoon season, the model estimate high temperature more than 1°C to a maximum 10°C compared to observed normal in the most parts of the study region. The rainfall is reproduced well by calculated the model to observed ratios very closed to one in the monsoon and post-monsoon season, other seasons it slightly overestimate.

Construction of sub-grid scale scenarios for the mean temperature and rainfall through statistical downscaling reveals that the projected mean temperature is far more than the case when the GCM output is directly interpolated. HadCM2 indicates a warming of 0.3-1.3°C while ECHAM4 shows a warming of 0.3-0.7°C. The former one also indicates a reduction in rainfall in some pockets in different seasons while the later shows an excess of rainfall in all the seasons. Composite seasonal scenarios based on these two models reveal a warming (0.3-0.9°C), which is higher than the projection derived directly from the GCM output over the GWB and its neighborhood. In general, an increase in rainfall is estimated in all the seasons except for the pre-monsoon season. Rainfall scenarios indicate a variation of –5% to 9% except in the winter where it exceeds 20%. This estimation is also higher than that of the projected changes directly derived from the GCMs. The projected rainfall for 2010-2039 with the two forcing is summarized in Tables 3 & 4 which reveals that an increasing trend of rainfall for both the forcing in the winter seasons whereas other three season shows positive as well as negative changes in rainfall for two forcing. Temperature change for both forcing in the same time period is presented in Table 3, indicates a warming is bellow the global scenarios (less than 1°C) all over the GWB and its neighborhood except in monsoon season where it crossed the global limit (more than 1°C).

A brief results of rainfall change scenarios for 2040-2069 period using two forcing namely GHG and GHG+S (Table 4) indicates that all seasons shows both pocket of positive and negative patterns with a maximum (+38%) and a minimum (-21%) in ECHAM4 for GHG forcing. Everywhere GHG+S forcing showed slightly less change of rainfall. Compared to the period 2010-2039, the temperature change for 2040-2069, shows an alarming indication with more than 1°C in all season for both forcing, which is presented in Table 4 HadCM2 shows more warming (upto 2.8°C) and ECHAM4 shows slightly less warming (upto 2.2°C) irrespective to forcing and seasons.

REFERENCES

Benestad, R.E. 2002. Empirically downscaled multimodal ensemble temperature and precipitation scenarios for Norway. *J. Clim.*, 15: 3008-3027.

Chakraborty, B. and Lal, M. 1994. Monsoon climate and its change in a doubled CO_2 atmosphere simulated by CSIRO9 model. *TAO*, 5: 515-536.

Crane, R.G. and Hewitson, B.C. 1998. Double CO_2 precipitation changes for the Susquhanna Basin: Downscaling from the GENESIS General Circulation Model. *Int. J. Clim.*, 18: 65-76.

Feenstra, J.F., Burton, I., Smith, J.B. and Tol, R.S.J. (Eds.). 1998. Handbook on Methods for Climate Change Impact Assessment and Adaptation Strategies. Nairobi and Amsterdam: UNEP and Institute for Environmental Studies/Vrije Universiteit.

Goodess, C.M. and Palutikof, J.P. 1996. "Daily rainfall scenarios for Southeast Spain." Abstract Book, 28th International Geographical Congress, The Hague, August 4-10, p157

Grotch, S. and MacCracken, M. 1991. The use of General Circulation Models to predict regional climate change. *J. Clim.*, 4:286-303.

Hirakuchi, H. and Giorgi, F. 1995. Multiyear present day and 2x CO_2 simulations of monsoon climate over eastern Asia and Japan with a regional climate model nested in a general circulation model. *J Geo Res*, 100: 21105-21125.

Houghton, J.T., Callander, B.A., Varney, S.K. 1992. Climate change 1992 -the supplementary report to the IPCC scientific assessment. WMO/UNEP intergovernmental panel on climate change. Cambridge University Press, Cambridge

Hulme, M., Zhao, Z. and Jiang, Y. 1994. Recent and future climate change in East Asia. *Int. J. Clim.*, 14: 637-658.

Huth, R. and Kysely, J. 2000. Constructing site specific climate change scenarios on a monthly scale using statistical downscaling. *Theor. Appl. Cli.*, 66:13-27.

IPCC 2001. Climate change 2001: Science of climate change. Inter-governmental Panel on Climate Change (IPCC). Cambridge University Press, Cambridge.

Johns, T. C., Carnell, C. E., Crossley, J. F., Gregory, J. M., Mitchell, J. F. B., Senior, C. A., Tett, S. F. B., and Wood, R. A.. 1997. The second Hadley Centre coupled ocean – atmosphere GCM: Model Descriptions, spinup and validation. *Clim. Dyn.*, 13: 103-134.

Kidson, J. W. and Thompson, C.S.1998. A comparison of statistical and model based downscaling techniques for estimating local climate variation. *J. Clim.*, 11: 735-753.

Kolli, R., Pant, G.B., Parthasarathy, B. and Sontakke, A. 1992. Spatial ånd sub-seasonal patterns of long- term trends of India summer monsoon rainfall. *Int. J. Climatol .*, 12:257-268.

Kolli, R.K., Kumar, K.K., Ashrit, R.G., Patwardhan, S.K. and Pant, G.B. 2002. Climate change in India – observations and Model Projection. *In*: Shukla, P.R., Sharma, S.K., and P.V. Ramana , (eds): Climate change and India- Issues, Concerns and Opportunities, Tata McGraw- Hill Publishing Company Limited, New Delhi, 24-75.

Kripalani, R.H., Kulkarni, A., Sabade, S.S. and Khandekar, M.L. 2003. Indian Monsoon variability in a global warming scenario. *Nat. Haz.* , 29: 189-206.

Lal, M., Cubasch, U. and Santer, B.D. 1994. Effect of global warming on Indian monsoon simulated with a coupled ocean-atmosphere general circulation model. *Current Science*, 66: 430-438.

Legates, D. R. and Willmott, C. J. 1992. 'A comparison of GCM-simulated and observed mean January and July precipitation'. *Global Planet. Change*, 97, 345-363.

Lohar, D. and Pal, B. 1995. The effect of irrigation on premonsoon season precipitation over south West Bengal, India. *J. Clim.*, 8: 2567-2570

Mearns, L.O., Rosenzweig, C. and Goglberg, R. 1997. Mean and variance changes in climate secnarios: Methods, Agriculture applications, and measure of uncertainty. *Clim. Change*, 35: 367-396.

Mooley, D.A. and Parthasarthy, B. 1984. Fluctuation of all India summer monsoon rainfall during 1871-1978. *Clim Change*, 6:287-301.

Murphy, J.1998. An Evaluation of Statistical and Dynamical Techniques for Downscaling Local Climate. *J. Clim.*, 12: 2256-2284.

Palutikof, J.P., Guo, X., Wigley, T.M.L. and Gregory, J.M.1992. Regional changes in climate in the Mediterranean Basin due to global greenhouse warming. MAP, Technical Report Series, 66, Mediterranean Action Plan, UNEP, Athens.

Robinson, P. J. and Finkelstein, P. L.1991. 'The Development of Impact-Oriented Climate Scenarios', Bull. *Amer. Meteorol. Soc.* 72, 481–490.

SAR. 1995. Climate Change- 1995: The Science of Climate Change: Contribution of Working Group I to the Second Assessment Report of the Intergovernmental Panel on Climate Change (Eds:John T. Houghton, L. G. Meiro Filho, B. A. Callander, N. Harris, A. Kattenburg, K. Maskell). Cambridge University Press, 557p

Von Storch, H., Zorita, E. and Cubash, U. 1993. Downscaling of global climate change estimating to regional scale: An application to Iberain rainfall in wintertime. *J. Climatol.*, 6:1161-1171.

Wigley, T.M.L., Jones, P.D., Briffa, K.R. and Smith, G. 1990. Obtaining Sub-Grid-Scale Information From Coarse-Resolution General Circulation Model Output. *J. Geophys. Res.* 95, D2, 1943-1953.

Wilby, R. L., Wigley, T.M.L., Conway, D., Jones, P.D., Hewitson, B.C., Main, J. and Wilks, D.S. 1998. Statistical downscaling of general circulation model output: A comparison of methods. *Water Resour. Res.*, 4: 2995-3008.

Wilby, R.L. and Wigley, T.M.L. 1997. Downscale general circulation model output: A review of methods and limitations. *Prog. Phys. Geogrphy.*, 21: 530-548.

Wilby, R.L. and Wigley, T.M.L. 2000. Precipitation predictors for downscaling: observed and general circulation Model relationships. *Int. J. Climatol.*, 20:641-661.

Wilby, R.L., Thomlinson, O.J. and Dawson, C.W. 2003. Multi-site simulation of precipitation by conditional resampling. *Clim. Res.*, 23:183-194.

□□□

Chapter 18

A Comparative Study of Drought Assessment Methods for Koraput Region : A Case Study

SHAKIR ALI, B.K. SETHY, R.K. SINGH AND ASHOK KUMAR

INTRODUCTION

Drought is a recurrent feature with unpredictable frequencies in different parts of word and India, especially in arid and semi-arid regions. It occurs due to delayed onset or early withdrawl and prolongs dry spells during the cropping seasons, and highly rainfall pattern. The drought has more than 150 published definitions (Wilhite and Glantz, 1985). The most common are meteorological, hydrological and agricultural drought. Meteorological droughts are associated with climatological variables as rainfall, temperature and evapotranspiration. Hydrological drought associated with water supply variable such as stream flow, reservoir storage, soil moisture and groundwater levels. Agricultural drought related to soil moisture in root zone to support healthy crop growth during crop growing season. Agricultural droughts do not necessarily coincide with period of meteorological and hydrological droughts. Rainfall is a widely used and useful drought indicator, but evapotranspiration is also an important variable. Difficulties in quantifying evapotranspiration rates suggest that a general classification scheme is best limited to a simple measure of rainfall. Rainfall can provide a direct measurement of water supplies, it influences agricultural and hydrological droughts, and it can reflect drought impacts over different time periods and sectors. The drought indices based solely on precipitation data perform well when compared with more complex hydrological indices (Oladipio, 1985). National commission on Agriculture (NCA, 1976) defines "the meteorological drought is a situation when actual rainfall is significantly lower than the climatologically expected rainfall over a wide area". The long meteorological drought has adverse impact not only to agricultural crops but also other water

---

Central Soil and Water Conservation Research and Training Institute,
Research Centre, Kota-324 002, Rajasthan, India

based industries. The farmers are worst effected community especially tribal districts of where starvation takes place due to drought causes deaths and lead to distress migration of the poor tribal during non-agricultural season.

There are various methods and indices for assessing the degree of severity, frequency and duration of droughts in abroad and India (Morid *et al.*, 2006; Ali, 2004; Lohani *et al.*, 1998; Nkemdrim, and Weber, 1999; Oladipio, 1985.). The meteorological drought method include; percent of normal rainfall(PNR) (Walker, 1912 quoted by Ali, 2004), percent deviation of rainfall deficiency(PDR) (IMD, 1971; IC, 1981), cumulative rainfall anomalies (Ravikumar and Karmagam, 1992), rainfall deciles (Gibbs and Maher, 1967), standardized precipitation index(SPI) (McKee *et al.*, 1993, 1995); effective drought index (EDI) (Byun and Wilhite, 1999); standardized clustered precipitation index (SCPI) and cluster precipitation distributions (CPDs) . The agro-meteorological drought indices are; Palmer drought severity index (PDSI) (Palmer, 1965), Palmer hydrologic drought index (PHDI), crop moisture index (CMI), (Palmer, 1968), Palmer moisture anomaly index (Z index) (Karl, 1986), crop specific drought index (CSDI), (Meyer, 1993). The remote sensing based indices are; normalized difference vegetation index (NDVI), (Kogan, 1995) and crop model DSSAT (Tsuji *et al.*, 1998). Each method has its own merits and demerits and application of these methods varied from country to country and within the country according to the purpose and area of interest of the investigators. For example, percent of normal rainfall (PNR) method can be useful for analyzing a single region or a single season, yet it is easily misunderstood and gives different values depending on the location and time period. Using the PNR can make it difficult to link a value of a departure with a specific impact occurring as a result of the departure, and thus to design appropriate drought mitigation and responses (Willeke *et al.*, 1994). Deciles method can address limitations with the percent of normal approach. The method of Deciles was selected for the many countries because of simplicity, consistency, and understandability over the PDSI (Smith *et al.*, 1993). One challenge, though, is that a long climatological record with consistent observation stations is needed to calculate the deciles accurately. Also, Deciles can be difficult to apply if officials and the public are not familiar with the Decile system. The PDSI has been one of the most widely used in the U.S., even though the SPI has many advantages of statistical consistency and ability to reflect both short-term and long-term drought impacts (Guttman, 1998; Hayes *et al.*, 1999). The PDSI does not properly reflect the drought conditions in a tropical country like India (Bhalme and Mooley, 1991). Bhalme and Mooley(1991) modified PDSI suitable to Indian condition by modifying in the weighing factor equation to describe the moisture anomaly index. The SPI method was developed by McKee *et al.*(1993) quantifies the precipitation deficit over desired region at a specific time scales. It gives the better representation of dryness than Palmer's indices and proposed an alternative to Palmer's indices (Heim Jr, 2002). The SPI can be used to assess drought condition on single and/or multiple time scales, such as for 1, 3, 6, 9, and 12 month periods. These different time scales are designed to reflect the impacts of precipitation deficits on different water resources. For instance, soil moisture conditions respond to precipitation anomalies on a relatively short scale, whereas groundwater, stream flow, and reservoir storage reflect longer-term precipitation anomalies. This temporal flexibility allows the SPI to be useful in both short drought as agricultural drought and long term drought as hydrological drought applications. The agro-meteorological drought methods and indices not only spatial, but also the temporal resolution plays an important role. These models performed better when they were adapted to the most drought sensitive phenological periods of crops. Very recently, an evaluation of common indicators, according to six criteria of performance, indicates strengths of the SPI and Deciles over the PDSI (Keyantash and Dracup, 2002). Percent deviation of rainfall (PDR) proposed by IMD (Indian Meteorological Department) (IMD, 1971, NCA, 1976) method is adopted by several investigators and Government Departments to express the meteorological drought (Kanakasabai *et al.*, 2004). The PDR is based solely on rainfall and requires only the computation of its deviation from long term mean (normal). In this method, the rainfall of a specific time period is considered for assessing the drought severity.

Reconnaissance of various drought assessment methods suggest that a method that holds good for a specific meteorological region, may or may not perform satisfactorily for other regions. It often becomes very difficult to select the most appropriate single drought method. Recognizing the above concerns and needs, the present study is envisaged to identify and derive reliable and promising method for assessing the drought in Koraput by comparing performances of most commonly used methods. Most commonly used drought indices, whose degree of reliability has been reported within a desired limit, are; (i) standardized precipitation index (SPI) and (ii) PDR method. In the present analysis, these two methods are considered for performance evaluation from the long term historical time series rainfall data.

MATERIAL AND METHODS

Study area : Koraput is a representative district of the Eastern Ghats Region of Orissa and located between 18^0 20′ 10″-18^0 41′ 35″ north latitude and 82^0 34′ 46″-83^0 03′ 04″ east longitude at an elevation of 883m above mean sea level. Population of area is mostly tribal and dependent on agriculture as means of livelihood. Highly erratic distribution of rainfall, undulating terrain with scattered naked hillocks and light soil texture and low organic matter made area mono cropped where *kharif* is main crop. Even a small break in monsoon causes substantial reduction in rice crop or in worst case complete failure of crop during drought year.

Data and data source : Historical daily rainfall data for a period of 41 years (1970-2010) were obtained from Agricultural Meteorological observatory of mixed farm, Koraput (1970-1994) and Central Soil and Water Conservation Research and Training Institute, Research Centre (1995-2010), Semiliguda, Koraput (Orissa). A year is divided into 4 seasons *viz.*, pre-monsoon (April-May), monsoon (June-Sept), post monsoon (Oct-Nov) and winter (Dec-March) seasons. The monthly, seasons and annual rainfall amount was computed from daily rainfall data.

METHODOLOGY

Standardized precipitation index (SPI) : The estimation of the SPI is based on a long-term precipitation record, typically at least 30 years. This long-term record was then choosing to a probability distribution (*e.g.*, Gamma distribution, incomplete beta distribution (McKee *et al.*, 1993, 1995) and Pearson III distribution (Guttman, 1998) that reliably fit the long-term rainfall time series data. The gamma distribution has been widely used, reliable and the best fitted distribution for rainfall data (Ali and Sethy, 2007). In this study, gamma distribution has been assumed as the reliable and the best fitted distribution for Koraput rainfall data. The percentile value from the best fitted PDF is then transferred corresponding to the normal (Gaussian) distribution so that a percentile on the fitted distribution corresponds to the same percentile on a normal (Gaussian) distribution and mean and standard deviation of the SPI for a certain station and time period equal to 0 and 1, respectively. That percentile is then associated with a Z-score for the standard Gaussian distribution and the Z-score is the value of the SPI. The positive values indicate wet condition greater than median rainfall, whereas negative values the dry condition less than the median rainfall. The categories of the SPI and drought/ wet condition according to McKee *et al.* (1993) are as follows:

SPI Values	Wet/Drought categories	Cumulative frequencies
2.00 or more	Extremely wet	<2.3%
1.50 to 1.99	Severely wet	2.3-6.7%
1.00 to 1.49	Moderately wet	6.8-15.9%
0 to 0.99	Mild wet	16-50%
0 to -0.99	Mild drought	16-50%
-1.00 to -1.49	Moderate drought	6.8-15.9%
-1.50 to -1.99	Severe drought	2.3-6.7%
-2.00 or less	Extreme drought	<2.3%

Percent deviation of rainfall (PDR) method : This method is proposed by the Indian Meteorological Department (IMD,1971) for assessing the meteorological drought on the basis of percent deviation of rainfall from long term mean rainfall and defined as:

$$D_i = \frac{P_i - \bar{P}}{\bar{P}} \times 100 \qquad \text{... (1)}$$

Where D_i is the percent deviation of rainfall for a specific time period (%), P_i is the actual rainfall for a that time period (mm), 'P is the long term mean(mm) for that time period(mm). The percent deviation of rainfall and drought categories proposed by the IMD is given below:

Percent deviation of rainfall (D_i)	Drought class
>0	No drought
0 to -25%	Mild drought
-25 to -50	Moderate drought
<-50	Severe drought

RESULTS AND DISCUSSION

Rainfall distribution : Annual, seasonal and monthly rainfall characteristic of Koraput for the period of 1970-2010 was analyzed statistically and presented in Table 1. The results showed that the average annual rainfall of district was 1379.7 mm with standard deviation of 266.3mm, whereas average monsoon seasonal rainfall (June-September) was 1076.1mm with standard deviation of 234.6 mm (Table 1). Results exhibited that about 78% of total annual rainfall amount concentrated during monsoon season(June-September) followed by pre-monsoon(March-May) (10.8%) and post-monsoon(Oct-Nov) (9.5%). The contribution of winter rainfall is nearly 1.7% of average annual rainfall. During the monsoon season, the highest rainfall amount was received during July (25.2% of annual) followed by August (23.9%), September (15.6%) and June (13.4%). Order of rainfall received in pre-monsoon period was May (6.1%) >April (3.3%) >March(1.3%) and in post monsoon season was October (2.1%) and November(0.4%). The rainfall variation during monsoon season was lowest in July (35.5%) followed by August (37.37%), September (49.08%) and June(70.11%), while highest in post and pre-monsoon as October (79.39%) followed by May (78.71%) and April (72.70%). For this study, annual, monsoon season (June-Sept) and monthly scale from June to September were considered. Since annual scale is important from hydrological point of view while monsoon season and monthly scales during monsoon period is main *kharif* cropping season specially for paddy and upland *Ragi* in the region.

Table 1: Rainfall characteristics over Koraput district

Month	Mean (mm)	SD (mm)	CV (%)	% of annual
Jan.	9.2	20.1	219.2	0.7
Feb.	8.9	18.3	206.1	0.6
Mar	18.3	21.6	117.8	1.3
April	46.1	33.5	72.7	3.3
May	84.5	66.5	78.7	6.1
June	184.3	129.2	70.1	13.4
July	347.3	123.3	35.5	25.2
Aug	329.9	123.3	37.4	23.9
Sept	214.6	105.3	49.1	15.6
Oct	101.4	80.5	79.4	7.3
Nov	29.3	41.7	142.5	2.1
Dec	5.9	16.7	282.6	0.4
Annual	1379.7	266.3	19.3	100.0
Pre-monsoon (Mar-May)	148.9	79.4	53.3	10.8
Monsoon (June-Sept)	1076.1	234.6	21.8	78.0
Post-monsoon (Oct-Nov)	130.7	103.2	79.0	9.5
Winter (Dec-Feb)	24.0	31.9	133.1	1.7

Drought pattern : The distribution of the standardized precipitation index (SPI) and percent deviation of rainfall (PDR) on annual scale are shown in Figure 1a. Figure depict that both the SPI and PDR methods had almost similar pattern of drought situations with different magnitudes during past 41years (1970-2010) in Koraput region. The SPI method predicted one extreme (1989), one severe (2009), three moderate (1982, 1997 2008) and 36 mild/nearly normal droughts during the 41 years (1970-2010), while the PDR predicted three moderate (1982, 1989, 2009) and 21 mild droughts. It was noticed that on annual scale, both the methods had capability of capturing mild and moderate droughts, while PDR had unable to detect any severe drought in last 41 years. It also failed to identify the year 1997 and 2008 as a moderate/ serious drought as the SPI did.

The results of the monsoon season (June-September) revealed that there were two severe (1989, 2005) and five moderate (1982, 1987, 1990, 1997, and 2009) and 33 mild droughts occurred as predicted by the SPI method (Figure 1a). No extreme drought was recorded by the SPI method. A similar trend drought situation was also predicted by the PDR method with five moderate (1982, 1989, 1990, 1997, 2005) and 17 mild droughts. Once again, the PDR method was unable to detect any severe drought based on seasonal scale as the SPI did. Figure 1b depicts the comparison of drought situations for annual and monsoon season detected by both the methods. It showed that on seasonal scale, both the method detected three additional severe and/or moderate droughts years (1987, 1997 and 2005) in comparison with annual scale. Results indicated that from agricultural point of view, years 1987, 1997 and 2005 were poor years, since, most of the agricultural activities concentrate in monsoon season (June to Sept). On annual scale, these years (1987, 1997 and 2005) were good years and the shortage of water during monsoon season was compensated by the other seasons like pre and post monsoon and winter rain for filling of reservoirs, river flow *etc.* Analysis also revealed that on annual scale, there was no two consecutive years moderate and/or severe drought except the mild droughts, whereas on seasonal scale, two consecutive years 1989 followed by 1990 was detected as a seasonal drought by the both the method (Figure 1b). The SPI detected the year 1989 as severe drought and 1990 as moderate drought year, while and the PDR method detected both the years as moderate droughts.

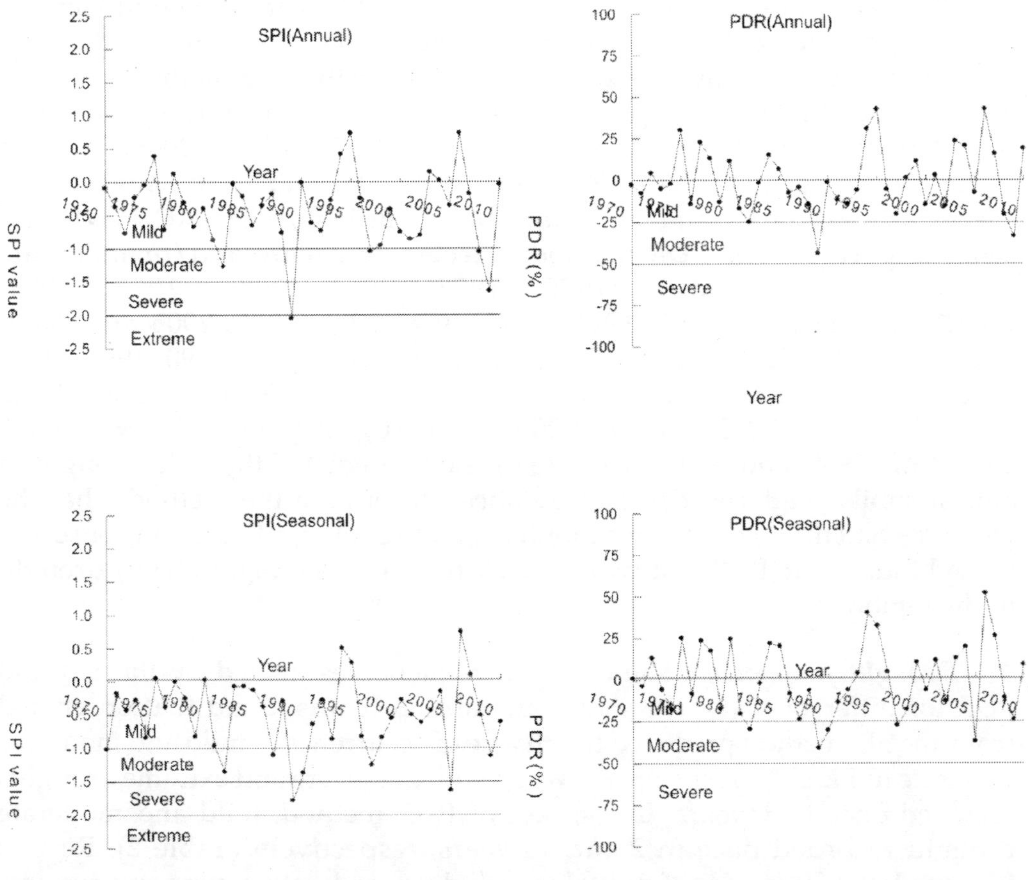

Fig.1a: Distribution of various droughts situations predicted by the SPI and PDR on annual and monsoon season scales

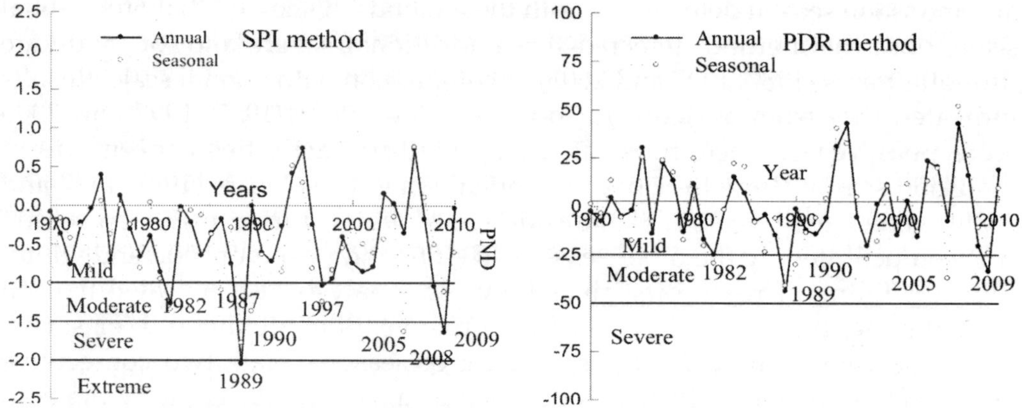

Fig.1b: A comparison of annual and monsoon season drought situations detected by both the methods

The results of monthly scale by both the methods are shown in Figure 2 (a, b). Figures depict that the pattern of drought situations predicted by the SPI was more statistically consistence as compare to the PDR method proposed by IMD. The SPI detected highest droughts including moderate and severe in July (8 in the years 1974, 1979, 1981, 1983, 1989, 1997, 2007 and 2008) followed by the September (6 in 1970, 1989, 1991, 1998, 2002 and 2009), August (5 in 1972, 1975, 1988, 1989 and 1998) and June (5 in 1975, 1976, 2002, 2004 and 2005) (Figure 4). The PDR method detected 18 severe and/or moderate drought in September (1970, 1973, 1974, 1976, 1982, 1984, 1986,1987,1988,1989, 1990, 1991, 1997, 1998, 2001, 2002, 2005 and 2009) followed by the June (14 in 1976, 1977, 1978, 1982, 1989, 1990, 1991, 1995, 19962005, 2006, 2009 and 2010), July (11 in 1974, 1979, 1980, 1981, 1982, 1989, 1990, 1997, 2005, 2007 and 2008) and August (7 in 1972, 1979, 1988, 1989, 1998, 1999, and 2010). Both the methods predicted the mild drought in almost similar pattern. It was also noticed from both the methods that the year 1989 had three consecutive months (July to Sept) moderate and/or severe drought and year 1998 had two consecutive months (Aug. to Sept) drought in the region.

Drought frequency : The drought frequency estimated by the SPI and PDR for different scales are shown in Table 2. It was found that on annual scale, the SPI method predicted the chance of occurrence of mild and moderate was once in 1 and 14, respectively, while both the severe and extreme droughts occurred once in 41years. In the case of PDR method, mild and moderate drought recorded once in 2 and 14 years, respectively (Table 2). Results indicated that frequency of droughts including moderate, severe and extreme droughts was found to be once in 8 and 5years by the SPI and PDR method, respectively. For the monsoon season, drought frequencies predicted by both the method was less compare to the annual scale. The chance of occurrence of moderate, severe and extreme drought was found to be once in 8 and 21 years, respectively by the SPI method, whereas 3 and 8 years, respectively by the PDR method for monsoon season. It was also recorded that the SPI predicted droughts including moderate, severe and extreme more frequent (once in 5 years) than that of the PDR method (8 years) for monsoon season, and *vis-versa* on annual scale (Table 2). It was observed that on monthly scale, drought frequencies predicted by both the methods were almost same order of the annual and monsoon season that varied from month to months. Results revealed that July month had more frequent including moderate, severe and extreme once in 5 years followed by September once in 7 years and both the August and June once in 8 years as predicted by the SPI, while the PDR predicted once in 2, 4, 5, and 6 for the September, July, June and August, respectively (Table 2). The chance of occurrence of mild drought by the SPI and PDR method was 1 and 2 years, respectively.

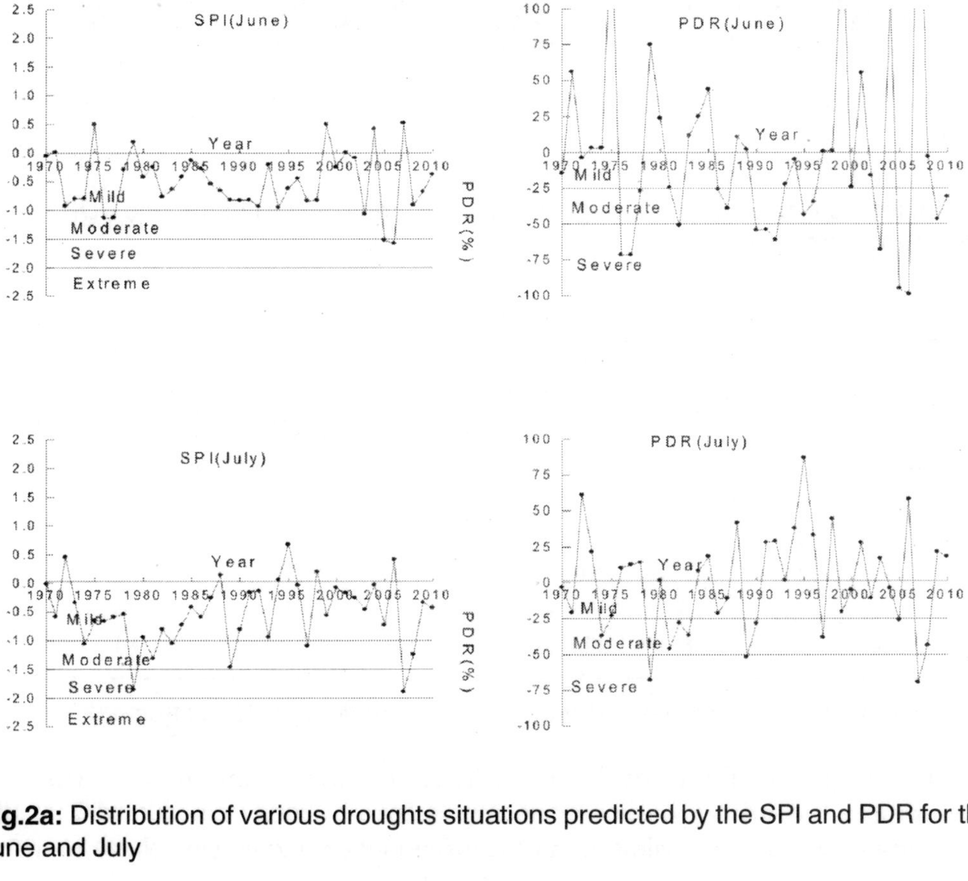

Fig.2a: Distribution of various droughts situations predicted by the SPI and PDR for the June and July

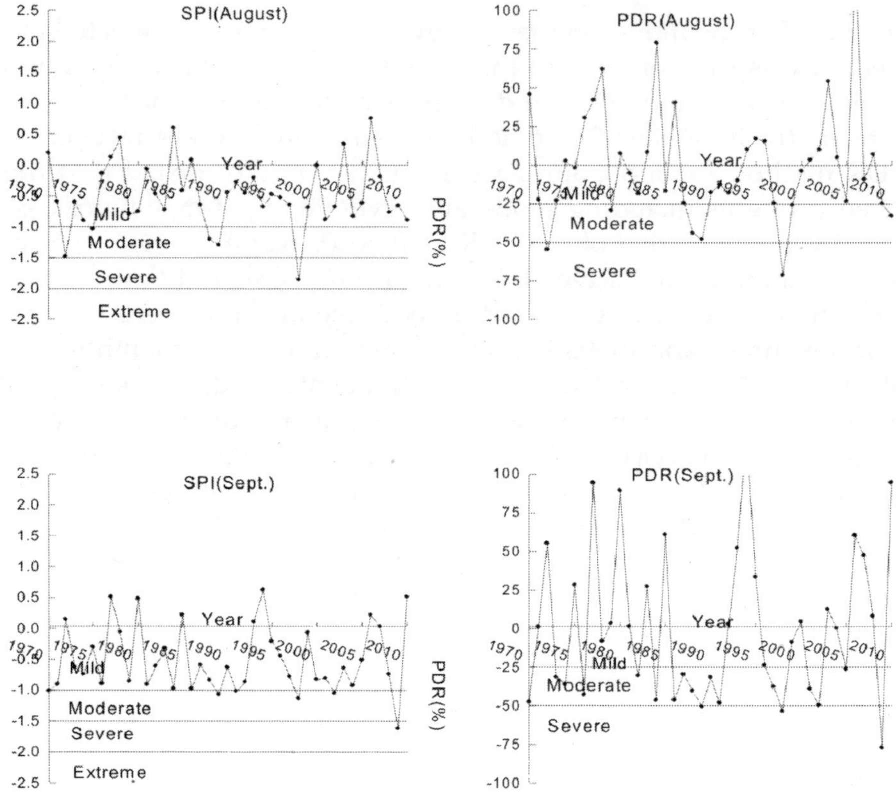

Fig.2b: Distribution of various droughts situations predicted by the SPI and PDR for the August and September

Table 2: Frequency of drought in Koraput district

Time	SPI					DR			
Scales	M	MO	S	E	MO+S+E	M	MO	S	MO+S+E
Annual	1	14	41	41	8	2	14	-	5
Monsoon	1	8	21	-	5	3	8	-	8
June	1	14	20	-	8	5	6	5	5
July	1	7	20	-	5	5	5	14	4
Aug.	1	10	41	-	8	3	8	21	6
Sept.	1	8	41	-	7	14	3	14	2

M= mild drought; MO = moderate drought; S = severe drought; E = extreme drought

This high probability of droughts in agricultural viable months (June to September) has adverse impact on agricultural crops. Since, most of the agricultural activities in tribal area of rainfed performed during these months.

It was noticed that occurrence of frequent droughts in month of June (once in 1(SPI)-3(PDR) years), creates a lot of difficulties for the farmers in field preparation and planting of crop. Moreover, the frequent drought situations in the July (once in 4-5 years) to August (once in 6 -8 years) which is major growth and development of main crop of the region as paddy in low land and *Ragi* on up land could not be mitigated by high rainfall in June or September, where as greater incident of moderate or severe drought in September (once in 2 years) could be affect the production of crops.

CONCLUSIONS

Forty one year rainfall data of Koraput district was to detect the drought by two commonly used drought estimate methods. These methods are; standardized precipitation index (SPI) and percent deviation of rainfall (PDR). Evaluating performances of these methods, the most promising and superior model suitable for Koraput region has been proposed. The following specific conclusions are drawn from the analyses:

- The performances of the SPI and PDR have been found to be nearly complementary to each other for annual and seasonal scales while there was significant deviation for the PDR method on monthly scale.
- Performance evaluation of the two methods suggested the superiority and the effectiveness and statistical consistency of the SPI method over the PDR method.
- The chance of occurrence of mild, moderate and severe droughts by the SPI on monsoon season scale were once in 1, 8 and 21, respectively and no extreme drought was detected by the SPI method.
- The years 1989 and 1997 were detected moderate and/or severe drought by both the methods on all the time scales (*i.e.*, annual, monsoon season and month from June to September).
- The year 1989 had three consecutive months (July to Sept) moderate and/or severe drought and year 1998 had two consecutive months (Aug to Sept) drought in the region.

REFERENCES

Ali, Shakir. 2004. Drought indices in India-A review. *Hydrology Journal*, IAH, **26**(3): 31-40.

Ali, Shakir and Sethy, B.K. 2007. Evaluation of probability density functions for describing rainfall data in south-eastern Rajasthan. *Indian Journal of Soil Conservation, 35(2): 116-121*

Bhalme, H.N. and Mooley, D.A. 1981. On the performance of modified Palmer index. Proc. Int. Symp. *Hydrological aspects of Droughts*, New Delhi. RR-33.

Byun, H.R. and Wilhite, D.A. 1999. Objective quantification of drought severity and duration . *Journal of Climate*, 12:2747-2756.

Gibbs, W.J. and Maher, J.V. 1967. Rainfall deciles as drought indicators. *Bureau of Meteorological Bulletin NO.48*, Commonwealth of Australia, Melbourne.

Guttman,N.B. 1998. Comparing the Parmer drought severity index and standardized precipitation index. *Journal of the American Water Resources Association*, 34(1): 113-121.

Hayes,M. J., Svoboda, M., Wilhite, D.A. and Vanyarkho, O. 1999. Monitoring the 1996 drought using the SPI. *Bulletin of the American Meteorological Society*, 80: 429-438.

Heim Jr., R.R. 2002. A review of twentieth-century drought indices used in the United States. *Bulletin of the American Meteorological Society*, 88(8): 1149-1165.

IMD.1971. Rainfall and droughts in India. Indian Meteorological Department, Pune, India.

IC(Irrigation Commission). 1981. Report of the irrigation commission, Vol.I. Ministry of Irrigation and Power, New Delhi, India.

Kanakasabai, V., Rajendran, M. and Vekatesan, V.2004. A comparative study of meteorological drought assessment methods. *Proceeding of international Conference on the Hydrological Perspectives for Sustainable Development Vol I. Roorkee, India.*

Karl, T.R. 1986. The sensitivity of the Palmer Drought severity index and Palmer's Z-index to their calibration coefficients including potential evapotranspiration. *Journal of Climatic and Applied Meteorology*, 25:77-86.

Keyantash, J. and Dracup, J.A. 2002. The quantification of drought: an evaluation of drought indices. *Bulletin of the American Meteorological Society*, 76(5): 655-668.

Kogan, F.N. 1995. Drought of the late 1980s in the United States as derived from NOAA polar –orbiting satellite data. *Bull. Amer. Meteor. Soc.*, 76:655-668.

Lohani, V.K. Loganathan, G.V. and Mostaghimi, S. 1998. Long-term analysis and short-term forecasting of dry spells by the Parmer drought severity index. *Nordic Hydrology*, 29(1):21-40.

McKee,T.B., Doesken, N.J. and Kliest, J. 1993. The relationship of drought frequency and duration of time scale. Eight conference of applied climatology. *American Meteorology Society*, Jan. 17-23, 1993, Anaheim, CA, pp179-186.

McKee,T.B., Doesken, N.J. and Kliest, J. 1995. Drought monitoring with multiple time scales. Ninth Conference of applied climatology. *American Meteorology Society*, Jan. 15-20, 1993, Dallas TX, pp 233-236.

Meyer, S.J. 1993. A crop specific drought index for corn II. Application in drought monitoring and assessment. *Agronomy Journal*, 85(2): 388-399.

Morid,.S., Smakhtin, V., Moghaddashi, M. 2006. Comparison of seven meteorological indices for monitoring drought in Iran. *Int. J. Climatol.*, 26:971-985.

NCA(National Commission on Agriculture). 1976. Agriculture commission report. Ministry of Agriculture, Government of India, New Delhi.

Nkemdrim, L. and Weber, L.1999. Comparison between the droughts of the 1930s and the 19980s in the Southern prairies of Canada. *Journal of Climate*, 12: 2434-2450.

Oladipio, E.O. 1985. A comparative performance analysis of three meteorological drought indices. *International Journal of Climatology*, 5:655-644.

Palmer, W.C.1965. Meteorological drought. U.S. Weather Bureau Research Paper No. 45.

Palmer, W.C.1968. Keeping track of crop moisture conditions, nationwide: The new crop moisture index. *Weatherwise* 21, 156-161.

Ravikumar, G. and Karmagam, M.1992. Implication of drought definitions. *Proceeding of VIII Congress of Asian and Pacific Division JAHR*, 91-100.

Smith, D.I., Hutchinson, M.F. and McArther, R.J.1993. Australian climatic and agricultural drought: payments and policy. *Drought Network News*, 5(3):11-12.

Tsuji, G., Hoogenboom, G., and Thornton, P.1998. Understanding options for agricultural production. Kluwer Acad. Publ. 339p.

Wilhite, D. A. and Glantz, M.H. 1985. Understanding the drought phenomenon: the role of definition. *Water International*, 10: 111-120.

Willeke, G., Hosking, J.R.M., Wallis, J.R. and Guttman, N.B.1994. The national drought atlas. Institute for Water Resources Report 94-NDS-4, US Army Corps of Engineers.

□□□

Chapter 19

Spatial and Temporal Analysis of Rainfall for Sustainable Agricultural Planning in Barak Valley of North East India

° LAXMI NARAYAN SETHI, HILLOLMOY CHAKRABORTY AND JESMINE LYNGDOH

INTRODUCTION

In India, agricultural production is controlled by monsoon rainfall, which is the primary source for soil moisture for rainfed crops. The pattern and amount of rainfall are among the most important factors that affect agricultural systems. It governs the crop yields and determines the choice of the crops that can be grown. Therefore, a detailed knowledge of rainfall regime is an important prerequisite for agricultural planning. Rainfall is a crucial agro-climatological factor in the seasonally arid parts of the world and its analysis an important perquisite for agricultural planning in India (Gadgil, 1986). India is a tropical country its agricultural planning and utilization water is depends on monsoon rainfall, more than 75% of rainfall accruing during the monsoon season. Monsoon rainfall is uneven both in time and space, so it is important factors to evolving the rainfall analysis.

In agricultural planning, rainfall variability analysis aids to take farm decisions on times of sowing, inter culture operations, fertilizer application *etc.* Many studies have reported the advantages of working out weekly/monthly/seasonally rainfall probabilities for a station or for an agro-climatic region (Zeng, *et al.*, 2000; Cabus, 2008; Deka *et al.*, 2009; Vennila *et al.*, 2008; Rathod and Aruchamy, 2010). However, very few studies have been carried out for the North Eastern States of India. So, there is requirement for rainfall analysis of rainfall data is required for sustainable crop planning for the regions.

Considering the problems associated with rainfall distribution, spatial and temporal analysis of rainfall were carried out to understand the nature

Department of Agricultural Engineering, Assam University (A Central University), Silchar-788 011, Assam, India of rainfall on monthly and seasonal scale as well as scenarios for crop planning and sustainable crop production at different risk levels in the Barak valley.

MATERIALS AND METHODS

Study area: The study area (Barak Valley) is situated in the southern part of the Indian state of Assam. The geographical position of Barak valley is latitude 24°8' and 25°8' N and longitude 92°15' and 93°15' E and with altitude of 31.40 m from the mean sea level. The zone has a geographical area of 6,922 km² (8.84% of state) with three districts, *viz.,* Cachar, Hailakandi, and Karimganj. Figure 1 shows the location of study area and the raingauge stations in the Barak Valley. The study area is covered by nine rain gauges

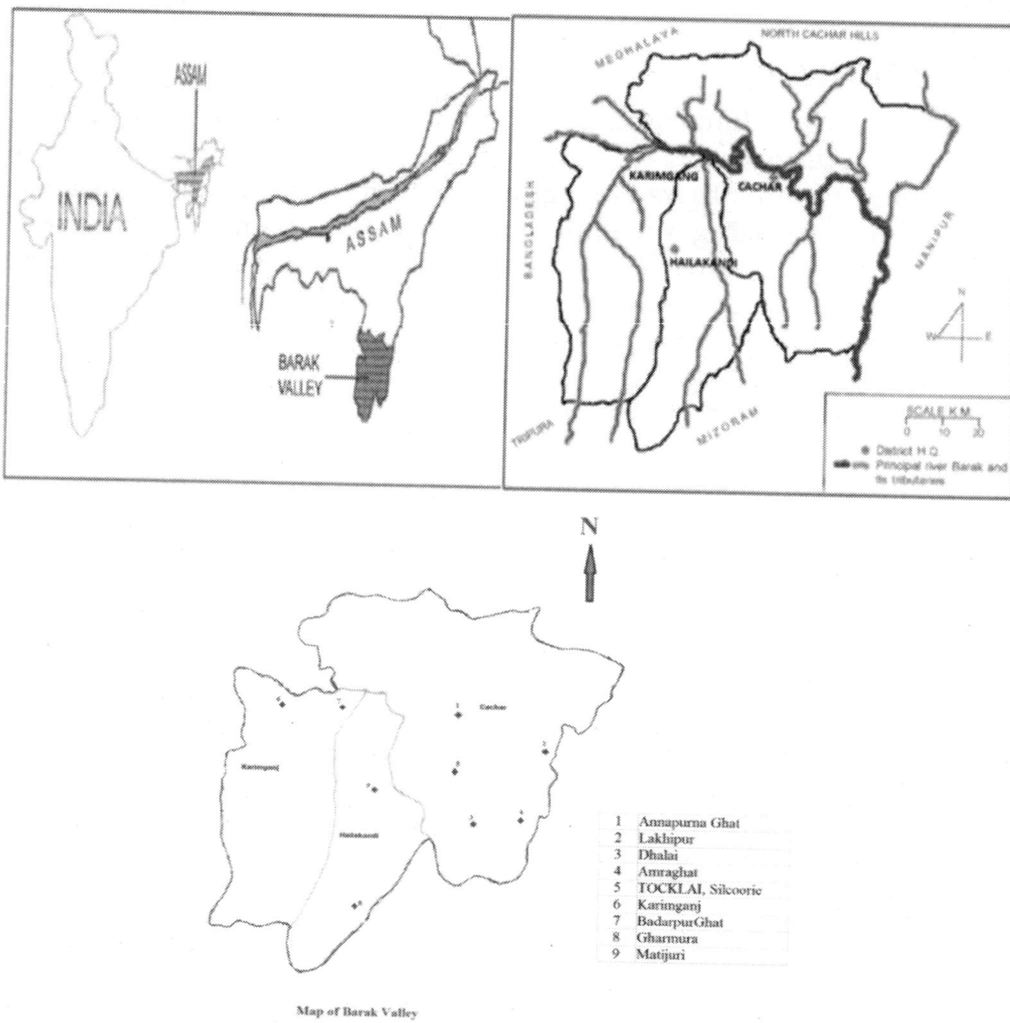

Fig. 1: Location of study area and rain gauge stations

installed by the Central Water Commission, India, District Agriculture Office, Cachar, TOCKLAI, Silcoorie. These raingauge stations in the Barak valley are found at Annapurna Ghat, Lakhipur, Dhalai, Amraghat, Silcoorie located in Cachar district; Karimganj and Badarpur Ghat in Karimganj district; and Gharmura and Matijuri located in Hailakandi district.

Barak Valley zone has 38.58 % of total geographical area under forest, 33.08 % as net sown area and 8.48 % as area sown more than once. About 5.47 % of land is used for plantation crops, miscellaneous trees and other. The climate of the North Eastern region is subtropical, warm and humid. The average rainfall of the region is 3180 mm with average rainy days of 146 days per annum. During the summer months the temperature generally varies from 25 to 40°C, while during the winter season the temperature ranges between 10 to 25°C. The relative humidity varies from 58 to 91% and the sunshine hour varies from 10 to 12 hours.

The Barak valley is characterized by tropical monsoon climate having three distinct seasons *viz.*, Summer (March-May), Rainy /*Kharif* (June-September) and Winter/*Rabi* (October-February). In the last three decades, Silchar and the Barak Valley have been ravaged by three major floods in 1986, followed by the one in 1991, and more recently in 2004. The valley having the rivers for drain water is Barak, Katakhal, and Kushiara.

Agriculture is the main occupation of the people. Rice is the staple diet of the people and cultivation of rice is the main occupation of those engaged in agriculture. The farmers usually grow crops in two seasons' *viz.*, monsoon/*Kharif* (June-September) and winter/*Rabi* (October-January) in both rainfed and irrigated areas. In summer season, there is no cultivation in the study area due to unavailability of water.

Data Monitoring and Rainfall Analysis

The rainfall data are collected on daily basis for nine numbers of raingauge stations in the Barak valley. Based on the availability, the rainfall data for the ten years 2001 to 2010 are considered for spatial and temporal rainfall variability and statistical analysis for sustainable crop planning in Barak Valley. The spatial variability of rainfall considers the pattern of rainfall in different stations, and districts of the valley. However, the temporal variability of rainfall considers the long term variation in monthly, seasonally (Summer, *Kharif* and *Rabi*) and yearly basis.

The statistical analysis such as mean monthly, seasonal and annual rainfall analysis as well as the standard deviation, coefficient variation were also carried out for sustainable crop planning in Barak Valley. Probability analysis was carried out using the SMADA (Stormwater Management and Design Aid, version 6.0) software for predicting the probability of amount of rainfall at different probability levels. The expected values of rainfall on month, season and annual basis at different probability levels were estimated using database SMADA package.

RESULTS AND DISCUSSION

Spatial and Temporal Rainfall Analysis

Temporal analysis for the different stations and the valley was carried out for determining the trends of variation of total and average monthly seasonal and annual rainfall for sustainable crop planning.

Station-wise monthly rainfall analysis: The average monthly rainfall of 10 years (2001 to 2010) for nine raingauge station of three districts of Barak Valley inferred that the variation of rainfall was found in every month, and the amount of rainfall gradually increases from the January to June, and then decreases from July to December (Figure 2). The high intensity trend was noticed in the month of June and July and lowest of rainfall in the month of December-January. However, Karimganj records 735.52 mm in the month of June, 606.27 mm in July, and 444.1 mm in May and it experiences lowest rainfall in the month of January (3.54 mm). The lowest rainfall recorded at Silcoorie which is 5.34 mm in the month of January and its maximum rainfall is 437.97 mm in the month of June.

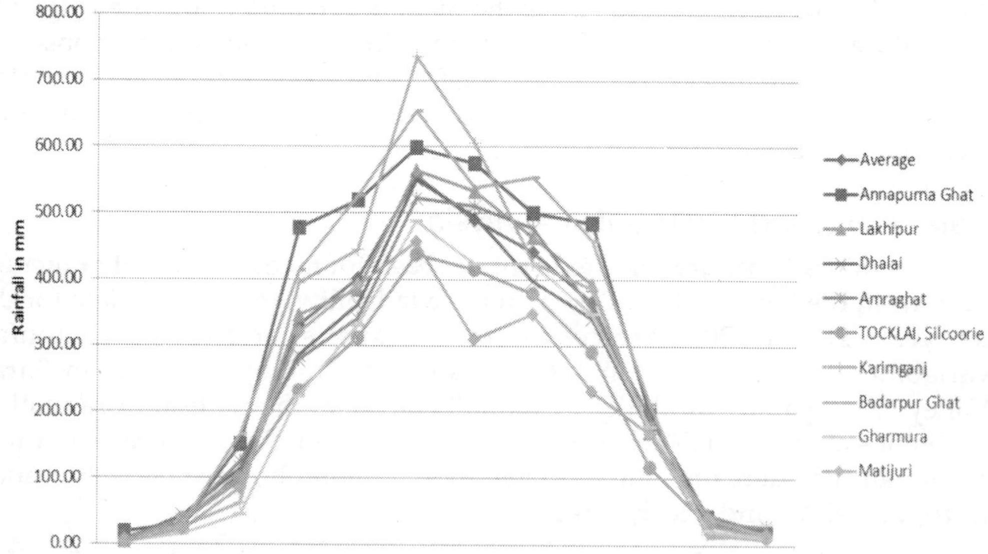

Fig. 2: Variation of monthly rainfall at different raingauge station in Barak Valley

Station-wise seasonal and annual rainfall analysis: The station-wise seasonal and annual average rainfall recorded for the ten years are presented in Figure 3 and Table 1. The maximum and minimum on long term basis for the nine raingauge stations in the Barak Valley are presented in Table 2. The maximum rainfall recorded at Annapurna Ghat, Silchar which is 3614.90 mm and minimum 2373.34mm at TOCKLAI, Silcoorie.

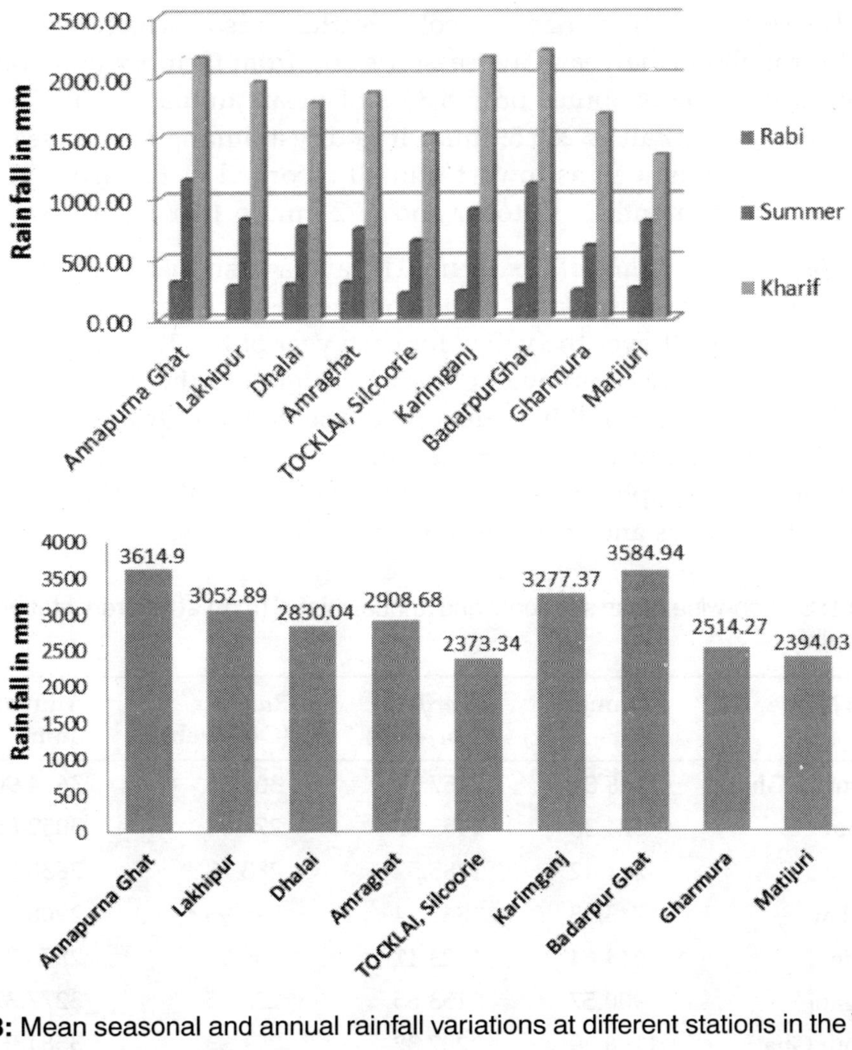

Fig. 3: Mean seasonal and annual rainfall variations at different stations in the Valley

During the summer the amount of rainfall gradually increases and however, the amount of rainfall occurrence in this season is largely due to convection effect, this season contributes 28.35% of mean annual rainfall and this season average rainfall is 836.29 mm. The highest amount of rainfall is recorded at the Badarpur Ghats 525.87mm in the month of May and seasonal average is 1106.29 mm and where the lowest rainfall is recorded in Gharmura 44.57 mm in March and its seasonal average 69.8 mm. During the *kharif* season the valley experiences highest rainfall among the all season. Karimganj receives highest rainfall (735.52 mm) in the month of June and Matijuri receives lowest rainfall (231.9 mm) in the month of September. The average seasonal rainfall was found 1853.10 mm that contributes 62.82% to mean annual rainfall.

Rabi season is characterized by cold weather season which receives least rainfall throughout the year. The season starts from October and continues till February and contributes only 8.83% of mean annual rainfall and this season average rainfall is 260.59 mm. It is driest among the all seasons in which maximum as well as lowest rainfall recorded at Badarpur Ghats of 212.53 mm in the month of October and 1.72 mm in January, respectively.

Valley temporal rainfall analysis: The average annual rainfall has been worked out for the entire Barak Valley on the basis of 9 rain gauge stations spread uniformly all over the valley for each year of the 10-year period from 2001 to 2010. The long term average annual rainfall of the valley was found 2950.05 mm. Average monthly variations and seasonal contribution of rainfall in Barak Valley is presented in Figure 4. The valley is characterized with unique aerial topography in quantum of rainfall due to the influence of three hills from three sides and its western plane of Bangladesh.

Tables 1: Station-wise mean seasonal and annual rainfall (mm) at different stations in the Barak Valley

Station Name	Summer (Mar – May)	*Kharif* (Jun –Sep)	*Rabi* (Oct – Feb)	Annual (mm)
Annapurna Ghat	1148.81	2157.74	308.35	3614.90
Lakhipur	824.38	1953.87	274.00	3052.89
Dhalai	761.42	1785.52	283.10	2830.04
Amraghat	743.38	1865.34	299.96	2908.68
Silcoorie	644.64	1523.17	205.53	2373.34
Karimganj	900.57	2153.65	223.15	3277.37
Badarpur Ghat	1106.29	2207.32	271.33	3584.94
Gharmura	599.52	1684.73	230.02	2514.27
Matijuri	797.63	1346.56	249.84	2394.03
Average	**836.29**	**1853.10**	**260.59**	**2950.05**

Table 2: Station-wise maximum, minimum, and annual average rainfall in the Barak Valley

Raingauge Station	Occurrence of Maximum Rainfall		Occurrence of Minimum Rainfall		Average Annual Rainfall (mm)
	Year	Amount (mm)	Year	Amount (mm)	
Annapurna Ghat	2005	5155.3	2006	2235.9	3614.90
Lakhipur	2010	4674.9	2002	2029.9	3052.89
Dhalai	2010	3644.8	2006	1881.2	2830.04
Amraghat	2010	3809.0	2006	2017.8	2908.68
Silcoorie	2010	3082.7	2006	1875.3	2373.34
Karimganj	2004	4259.4	2008	1596.1	3277.37
Badarpur Ghat	2010	4854.2	2006	2598.5	3584.94
Gharmura	2002	2914.1	2009	1697.8	2514.27
Matijuri	2007	3053.3	2009	1254.5	2394.03

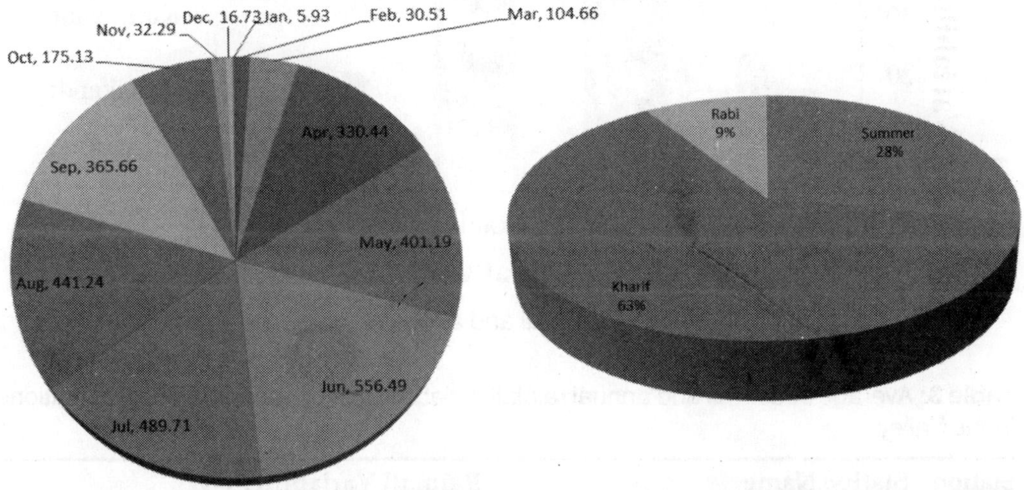

Fig. 4: Average monthly variation and seasonal contribution of rainfall in Barak Valley

Rainfall Variability and Statistics

"Variability defined as the deviation from mean" or "ratio of the standard deviation to the mean rainfall". In the present study the variability of rainfall for both spatial (station, district and valley) and temporal (seasonal and annual) variability of rainfall has been analyzed. Station-wise seasonal and annual rainfall variability are presented in Table 3. However, district-wise and valley average seasonal and annual rainfall variability (%) of rainfall are shown in Figure 5. During summer season, at Lakhipur maximum rainfall variability of 55.40% was observed whereas the lowest variability experience adjacent area at Matijuri (29.36%). However, the variability of rainfall was lower in *kharif* season while comparatively to other season, because high rainfall areas experiences lower fluctuation of variability. During *kharif*, the lowest variability recorded at the stations of region which is Gharmura (14.60%) and maximum at Matijuri (36.13%), these fluctuations indicates the Hailakandi district of the region experiences heavy rainfall. Annapurna Ghat (22.81%) and Silcoorie (15.48%) experiences higher and lower rainfall variability in the Cachar district, respectively. The variability recorded during this winter season is maximum than other seasons. During winter season, maximum variability found at Lakhipur (59.19%), and the lowest variability experience adjacent area at Badarpur Ghat (31.05%).

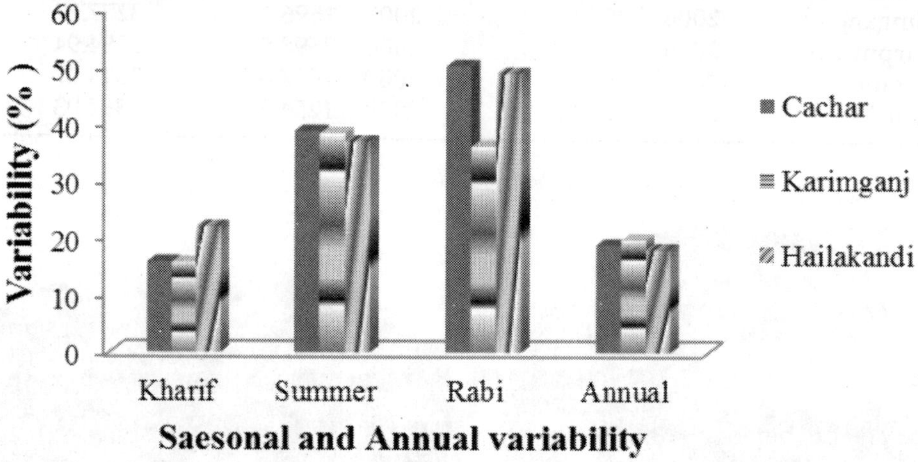

Fig. 5: District-wise seasonal and annual variability of rainfall

Table 3: Average seasonal and annual rainfall variability (%) of rainfall at different stations in the Valley

Station No.	Station Name	Rainfall Variability (%)			
		Summer	*Kharif*	*Rabi*	Annual
1	Annapurna Ghat	44.98	22.81	47.34	26.23
2	Lakhipur	55.40	18.95	59.19	25.24
3	Dhalai	34.52	18.40	52.20	18.85
4	Amraghat	40.38	16.93	55.50	18.65
5	Silcoorie	34.19	15.48	50.27	15.46
6	Karimganj	43.22	21.31	51.87	24.37
7	Badarpur Ghat	35.83	18.21	31.05	19.97
8	Gharmura	51.28	14.60	46.89	16.45
9	Matijuri	29.36	36.13	55.99	24.95

Table 4: Seasonal statistical analysis of three districts of Barak Valley

Parameter	Kharif (Jun-Sep)	Summer (Mar-May)	*Rabi* (Oct-Feb)	Annual
Cachar				
Mean Rainfall (mm)	1857.13	824.53	274.19	2955.97
% of Contribution	62.83	27.89	9.28	100
Standard Deviation	232.62	192.41	40.70	447.53
CV (%)	12.53	23.34	14.84	15.14
Karimganj				
Mean Rainfall (mm)	2180.49	1003.43	247.24	3431.16
% of Contribution	63.55	29.24	7.21	100
Standard Deviation	37.95	145.47	34.07	217.48
CV (%)	1.74	14.50	13.78	6.34
Hailakandi				
Mean Rainfall (mm)	1515.65	698.58	239.93	2454.16
% of Contribution	61.76	28.47	7.78	100
Standard Deviation	239.12	140.08	14.01	85.02
CV (%)	15.78	20.05	5.84	3.46
Barak valley				
Mean Rainfall (mm)	1853.10	836.29	260.59	2950.05
% of Contribution	62.82	28.35	8.83	100
Standard Deviation	299.43	188.20	35.56	475.74
CV (%)	16.16	22.50	13.65	16.13

The mean annual rainfall of Cachar, Karimganj and Hailakandi are 2955.97 mm, 3431.15 mm and 2454.15mm with the annual co-efficient of variation (CV) of 15.14, 6.34 and 3.46 %, respectively. The result indicated that the annual rainfall variability is relatively greater at Cachar than Karimganj and Hailakandi. Annual rainfall variability of region stretches between 15.46% to 26.23% in Cachar district of the region which method that the mean annual rainfall of the Barak Valley is 2950.05mm with a standard deviation of 475.74 mm and a coefficient of variation of 16.13%. It is also observed that out of all these experimental stations Annapurna Ghat, Silchar receives maximum average annual rainfall of 3614.9 mm followed by Badarpur Ghat of 3584.94 mm and also TOCKLAI, Silcoorie receives least average annual rainfall of 2373.34 mm.

It is also observed that during summer and monsoon seasons *i.e.,* from April to September the whole Barak Valley receives maximum rainfall and in post monsoon and winter months *i.e.,* from October to February the valley receives least rainfall. The average annual rainfall values of the 10-year period were examined in order to obtain an idea of the extreme values of annual rainfalls experienced in the valley.

Maximum mean monthly rainfall of Barak Valley varies from 4.74 mm in the month of January to 567.40mm in the month of June. Mean values was found to be 245.59mm. There is large variability among the monthly values of rainfall of different years. The coefficient of variation varies from 25% in the month of July to 205.14% in the month of December. Variance is maximum during *rabi* or winter season and minimum in *kharif* or monsoon season. It indicates atmospheric instability during *rabi* or winter season. Minimum value of standard deviation is 7.94mm during January and maximum is 240.85mm during April. It indicates better weather stability during month of January. Annual rainfall of Barak Valley indicates minimum 2268.67mm during 2006 and maximum 3734.17 mm in the year of 2010 with mean rainfall 2950.05 mm.

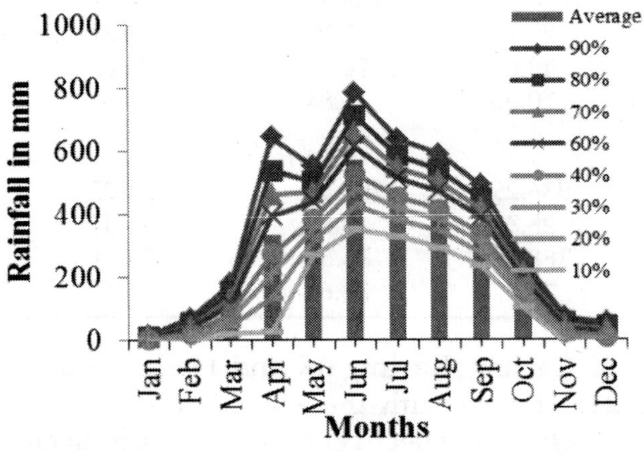

Fig.6: Expected monthly rainfall in Barak Valley

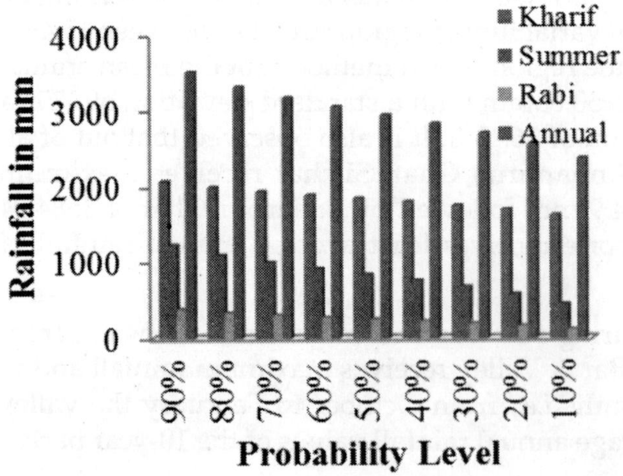

Fig. 7: Expected seasonal and annual rainfall in Barak Valley

Rainfall Probability Analysis

The expected values of rainfall on month, season and annual basis at different probability levels were calculated using database SMADA 6.0 package. The values are entered in the database of package considering Weibull's distribution as reference, fitted to different probability distribution functions such as Normal, Log-Normal, Pearson, Log-Pearson, and Gumbel distribution. From the distribution functions best-fit function was found out to be the normal distribution and the expected rainfall values at different probability of non exceedance (PE) like 10% to 90% were determined. Therefore in general normal distribution is found best suitable distribution for rainy season and non-rainy season of monthly rainfall for three districts of Barak Valley. The expected monthly and seasonal rainfall at different probability levels of the valley are presented in Figure 6 and 7, respectively which indicates that the considerable variation of monthly, seasonal and annual rainfall exists from year to year. It is always being found that at 50% probability level the value arises is the mean value of the month.

CONCLUSIONS

Barak valley receives average rainfall of 2950 mm during 2001-2010 which is above normal rainfall of Assam state (2818 mm). The heavy rainfall experiences at Karimganj district 3431.16 mm followed by Cachar 2955.97 mm and Hailakandi 2454.15 mm. The *kharif* season contributes the highest percentage of rainfall, which is 62.82% (1853.10 mm), summer season 28.35% (836.29 mm) and *rabi* season contributes lowest of which is 8.83% (260.59 mm). The monthly and seasonal rainfall of the three districts are found so high that at all the probability levels starting from 10 %, there is occurrence of rainfall and at 50 % and above probability levels there are chance of occurrence of rainfall more than the average rainfall value. Hence long duration flood resistance crops for the monsoon and short duration cropping pattern for the non-monsoon season could be followed for increasing the agricultural productivity in the Barak Valley.

REFERENCES

Cabus, P. 2008. River flow prediction through rainfall–runoff modelling with a probability-distributed model (PDM) in Flanders, Belgium. *Agricultural Water Management*, 95(7):859-868

Deka, S., Borah, M., Kakaty, S.C, 2009. Distribution of Annual maximum rainfall series of North East India, *European Water*, 27/28: 3-14.

Gadagil, A. 1986. Annual and weekly analysis of rainfall and temperature data analysis for Pune city: a multiple time series approach. *Transactions of the Institute Indian Geographers*, 8(1):16-23.

Rathod, I. M. and Aruchamy, S. 2010. Spatial analysis of rainfall variation in Coimbatore District Tamilnadu using GIS, *International Journal of Geomatics and Geosciences*, 1(2):106-118.

Vennila, G., Subramani, T. and Elango, L. 2007. Rainfall variation analysis of vattamalaikarai sub basin, Tamil Nadu. *Journal of applied hydrology*, 20(3): 51-60.

Zeng, N., Shuttleworth, J.W., Gash, J. H. C. 2000. Influence of temporal variability of rainfall on interception loss: Part-I. Point Analysis. *Journal of Hydrology*, 228: 228-241.

□□□

Chapter 20

Soil and Water Engineering Measures for Mitigating Climate Change Impact and Enhancing Agricultural Productivity

B. P. BEHERA AND B. PANIGRAHI

INTRODUCTION

The soil and water are the two critically important components of life support system. The conservation and efficient management of these resources are therefore, vital for economic growth and development of the country. The indiscriminate use of soil and water by human beings to meet out ever increasing and ever changing needs has deteriorated the capacity and quality to act as life supporting systems in the true sense. The soils are getting impoverished due to intensive use under agriculture without due consideration to soil and water conservation measures and nutrient replenishment. The water is being overused and polluted causing its scarcity. These are the major reasons for low productivity. The crop yields are becoming lower and the costs of various inputs are increasing making agriculture less profitable, resulting not only in their loss, but also in their low input use efficiency. All regions of the world show an overall net negative impact of climate change on water resources and fresh water ecosystem. Areas in which runoff is projected to decline are likely to face a reduction in the value of the services provided by water resources. The beneficial impacts of increased annual runoff in other areas are likely to be tempered in some areas by negative effects of increased precipitation, variability and seasonal shifts on water supply, water quality and flood risks. A changing climate will affect soil and water resources on agricultural land in many ways.

The relationship between Indian agriculture and climate is well known. Droughts, floods, natural calamities, have been frequent in different parts of the country all through the history; have been responsible for many famines, rural poverty and migration despite development of impressive irrigation

Department of Soil & Water Conservation Engineering, College of Agricultural Engineering and Technology, Orissa University of Agriculture and Technology, Bhubanewar, Odisha, India potentials. Similarly temperatures, wind velocity, and humidity during critical stages are known to significantly affect food production due to their effects on various crop growth and yield processes, pest incidences and demand on irrigation resources. Although the contribution of agriculture in the national GDP has now gradually reduced, the climatic variability still plays a very important role in India's food security and economy. With 2.4% of the world area and 4% of fresh water resources, India is supporting about 17% of the world's human population; the per capita availability of land has declined from 0.90 ha in 1951 to 0.36 ha in 2001. The degradation of our natural resources, soil and water in particular, has become a matter of concern for the policy makers and scientists. The estimates of the extent of area suffering from problems of land degradation as reported by the National Rainfed Area Authority are 120.72 M ha and rainfed area is 82.75 M ha out of net sown area. About 5334 M tones of topsoil along with huge quantities of major plant nutrients are washed away every year. About 40% of the cultivated land is under irrigation, which accounts for 55 % of total food grain production, whereas 60 % of the cropped area under rainfed comprising primarily of the arid, semi arid and the drought prone, contribute to only 45% of the production.

The problem : Low levels of productivity and low input consumption characterize rainfed agriculture. Crop production is subjected to considerable instability from year to year due to its dependence on rainfall, which is highly erratic and variant in space and time. More than 200 million of the rural poor live in these regions. These risk prone areas exhibit a wide variation and instability in yields. Gap between yield potential and actual yields is becoming more as compared to the irrigated areas. In rainfed areas of central plain, hill and semi arid or coastal lands call for more crop production is the need of the hour. Hence, in all the conditions, whether irrigated, or rainfed, the need for sustained agriculture is the need of the hour under varied climatic conditions.

Major changes in the temperature, solar radiation and precipitation will affect crop productivity. The magnitude and geographical distribution of such climate induced changes might affect our ability to increase the food production required to feed the more than 10 billion global population projected for the middle of the century. Agriculture sector alone represents 23% of India's gross national product (GNP), plays an important place in the national economy. It sustains the livelihood of nearly 70% of the population. India with the largest area (44.5 M ha) under rice cultivation is largely dependent upon monsoon rains. Some findings tell that the mean temperature in India is projected to increase by 0.1 to 0.3^{0} C in *kharif* and 0.3 to 0.7^{0} C during *rabi* seasons by 2010 and by 0.4 to 2^{0} C during *kharif* and 1.1 to 4.5^{0} C during the *rabi* season by 2070. Similarly mean rainfall is projected not to change significantly till 2010, but to increase by up to 10% during 2070. At the same time, there is an increased possibility of climate extremes, such as the timing of the onset of monsoon, intensities and frequencies of drought and floods. Predictions for the Indian sub continent indicate

a) Increase in ambient temperature
b) Reduction in solar radiation
c) Marginal increase in total precipitation
d) Increase in ambient CO_2 concentration
e) Frequent tropical disturbances in the coastal areas

As per the report of the IPCC, the net increase in temperature is likely to be 5.8^0 C by 2100 AD. This is likely to impact a series of inter related environmental systems like global hydro-eco-system, sea level, crop production and related agricultural activities (Panigrahi *et al*, 2009). As per IPCC (International Panel on Climate Change), warming of the climate system is evident from observations of increase in global average air and ocean temperatures, widespread melting of snow and ice and rising global average sea level. It has also observed that heat waves have become more frequent over most land areas, the frequency of heavy precipitation events has increased over most areas and since 1975, the incidence of extreme high sea level has increased world wide. Global warming refers to increase in the average temperature of the Earth's near surface air and ocean in recent decades and its continuation caused by a steadily thickening blanket of human induced GHGs such as CO_2, CH_4, N_2O, and the fluorinated gases used in refrigeration and semi conductor manufacturing. As per the report, the measured average atmospheric temperature near the Earth's surface rose by 0.74^0 C during the last 100 years (1906-2005). There is a strong evidence to suggest that most of the observed increase in the globally averaged temperature since mid 20th century is attributed to the observed increase in the globally averaged temperature since mid 20th century is attributed to the observed increase in emission of anthropogenic green house gases. Pre-industrial concentration of CO_2 was 280 ppm and this has now risen up to 380 ppm in the beginning of this millennium. Historically, main contributions to global warming have been on account of human induced activities such as fossil fuel consumption, industrial activities, agriculture systems, changing land use pattern, waste decomposition carried out in the developed countries and the economics in transition, which have been responsible for 75% of total GHG emissions. Till now, USA was the largest emitter of green house gases with emission level of 7.07 billion tones. As per the report of the Netherlands Environmental Assessment Agency, China has over taken the US in emission of GHGs. Its CO_2 emissions in 2006 were 8% higher than the US. Though India is among top ten GHG emitters in the world (Table 1) the present GHG emissions from India as a whole in terms of CO_2 and its equivalents are less with about 3% of the Global GHG emissions. Another major cause of global warming is deforestation. Unabated depletion of forest cover the world over is responsible for about a fifth of GHG emissions entering the atmosphere.

Table 1: World's top ten GHG emitters (2005)

S. No	Countries	CO_2 emissions in billion tonnes
1	USA	5.96
2	China	5.32
3	USSR	1.70
4	Japan	1.23
5	India	1.17
6	Germany	0.84
7	Canada	0.63
8	Britain	0.58
9	South Korea	0.50
10	Italy	0.47

Source: Netherlands Environment Assessment Agency

Effect of climate change on soil erosion : Soil erosion due to poor soil conservation practices, aggravated by strong winds and heavy rainfalls in short spells during monsoons, is a serious problem under rainfed agriculture resulting in removal of fertile surface soil, thereby leading to low water holding capacity and depleted fertility status of the lands which in turn would result in poor crop production. Soil erosion and runoff from cropland affects water resources directly by delivering sediment, pollutants attached to sediment, and pollutants in solution to surface water; indirect effects occur through changes in stream channel dynamics and watershed functions. Soil erosion and runoff from agricultural land are major causes and, in many watersheds, the most important causes of non point-source pollution. In addition, the partitioning of precipitation between runoff and infiltration has profound effects on water budgets in watersheds. Runoff volumes, peak flow rates, and timing can dramatically affect stream channels, water supplies, and downstream flooding. Soil erosion and runoff from cropland will respond to changes in climate for a number of reasons, including climate-induced changes in plant biomass, plant residue decomposition rates, soil microbial activity, evapo-transpiration rates, soil surface crusting and sealing, and shifts in land use that occur as adjustments to a changed climate (William *et al.*, 1996; as cited by Nearing, 2001). The most important effect of climate change on soil erosion and surface runoff, however, will come from climate-induced changes in the volume and erosive power of rainfall (Nearing, 2001).

The amount and intensity of precipitation affect the amount of soil that can be eroded and the volume and flow rate of surface runoff. As the accumulated quantity of rainfall increases during a precipitation event, the capacity of the soil to absorb the precipitation (infiltration rate) decreases and surface runoff ensues. Initially, runoff water flows uniformly across the surface; soon, however, runoff forms roughly parallel rills that coalesce into small channels of concentrated flow, which, in turn, coalesce into larger channels, eventually entering a stream. Runoff volume, depth, and flow velocity all increase with increasing volume and intensity of precipitation. Erosion is the process of soil particle detachment and movement. Water flowing across the soil surface,—surface runoff is the most important detachment and transport mechanism. The erosive power of surface runoff is determined by depth of flow, flow velocity, and the number and energy of particles flowing with the water. The capacity for surface runoff to carry soil particles increases as flow depth and velocity increases; this capacity, therefore, increases as the volume and intensity of precipitation increase. Indeed, rainfall erosivity is strongly correlated to the product of total rainstorm energy and the maximum 30-minute rainfall intensity during a storm. (Wischmeier and Smith, 1978).

More than 5 billion tones of soil is washed away every year in our country taking away with it more than 6 M tones of nutrients. Rainfed agriculture has not been given due attention, what it deserves either by the farming communities or by the developmental agencies due to lack of interest to take risks. The immediate necessity is to concentrate on efforts like adoption of proper soil and moisture conservation practices, which would restore and stabilize the productivity of rainfed agriculture that would go a long way to augment crop production in future. If we fail to restore the top surface layer, we certainly have to face disastrous consequences, as the arable lands will be deprived of their fertile surface layers besides silting the nearby water bodies with the eroded soils.

Water erosion and land degradation : Fast degradation and over exploitation of land resources is resulting in loss of its biological and economic productivity, affecting livelihood and environmental securities of millions of people. Of the world's total land area of 13.5 billion ha, about 2 billion ha is degraded due to various forms of land degradation, out of which water erosion alone contributes about 55 % (1.1 billion hq). In India 120.72 M ha area is degraded due to various forms of land degradation, with water erosion being the chief contributor (69%) followed by chemical degradation (20%), wind erosion (10%) and physical degradation (about 1%), (Sharda *et al.*, 2010).

Table 2: Harmonized area statistics of degraded lands/Wastelands of India

S.No	Type of degradation	Arable land (in million ha)	Open forest <40% Canopy (M ha)
1	Water erosion (>10tonnes/ha/year)	73.27	9.30
2	Wind erosion(Aeolian)	12.40	-
	Sub total	85.67	9.30
3	Chemical degradation		
	(a) Exclusively salt affected soils	5.44	-
	(b) Salt affected and water eroded soils	1.20	0.10
	(c) Exclusively acidic soils (pH<5.5)	5.09	-
	(d) Acidic (pH<5.5) and water eroded soils	5.72	7.13
	Sub total	17.45	7.23
4	Physical degradation		
	(a) Mining and industrial waste	0.19	-
	(b) Water logging (permanent) (water table within 2mts depth)	0.88	
	Sub total	1.07	
	Total	104.19	16.53
	Grand total (Arable land and Open forest)	120.72	

(Source: Soil & Water Conservation Today, News Letter, vol.5, No.2. ISSN 0975-4059, April-June 2010)

The major negative effects of land degradation due to water erosion include reduction in crop productivity, nutrient losses, floods/droughts, reduced reservoir capacity, and loss of bio diversity (Singh, 2010). Among these loss in crop productivity is of great significance as it also directly affects food security. Agriculture plays a critical role in food and livelihood security for a large majority of people of India. Frequent climatic extremes in the region have been historically responsible for widespread droughts, floods and poverty. The degradation of soil must not only be checked from proceeding further, but also the lands already at different stages of degradation must be reclaimed and the productivity potentials restored. Water has become the main limiting factor of crop production in India in spite of adequate rainfalls received during most of the years. Normally annual rainfall of around 600 mm and about 1000mm would the sufficient for dry land agriculture and garden land agriculture respectively, provided the rainfall is distributed well throughout the year.

Change in Precipitation Pattern : A change in precipitation regime also produces a change in the level of risk to which agricultural land is exposed. In general, a regime with greater annual precipitation, particularly if increased storm intensity changes more than storm frequency, heightens the risk of soil erosion, runoff, and related environmental and ecological damages. In general, the risk increases at a greater rate than precipitation amount or intensity increases. Whether that new, more risky baseline condition translates into greater soil degradation or a combination of all three outcomes is highly dependent on other factors. Failure of even the dry land farming in many parts of the country is a regular feature due to irregular rainfalls and very few rainy days in a year. Most of the rainfall downpour in short spells during the two seasons of South-West and North-East monsoons. But major proportion of the rainfalls received during the monsoons is allowed to go back to the sea through surface flow. Hence planning for check dams and water harvesting structures should be planned to harvest and conserve the monsoon rains satisfactorily (Thomas, 2011). The irrigation Commission of India has placed the total annual surface water flows in the country at 180 M ha m, including 20 M ha m brought in by streams and rivers from catchments laying outside the country, 45 M ha m regenerated flow from ground water and 115 M ha-m direct contribution by precipitation of which about 10 M ha m is received as snowfall.

Without going for costly major irrigation projects, minor irrigation systems like small and medium sized check dams and restoration of the storage capacities of thousands of ponds, tanks and lakes by desilting and strengthening of bunds must get the top priority attention of all to enable conservation of most of the rainfalls, received during the monsoon and use the conserved water during the lean and drought periods. This will stabilize the farming system, besides recharging the underground aquifers and minimizing water induced erosion and consequent silting of water bodies. Hence in order to save our natural resources like land and water from climate change scenarios, several mitigation measures must be adopted. Rising biotic pressure, expansion of agriculture into fragile eco-systems, over exploitation of ground water, over use of chemicals and several man made activities are declining soil health. The causes of soil degradation are due to soil erosion due to wind and water, water logging, salinity, alkalinity, mining and improper farming practices. Combating climate change with various soil and water conservation measures is discussed in the following paragraphs.

Mitigation of Climate Change : The climate change during 21st century is projected to alter rainfall and consequently the runoff and soil loss. The researchers worldwide are assessing the options for mitigating the impacts of harmful gases (CO_2, CH_4 and others) and their technical feasibility (Kumarswamy and Khan, 2010). Options for efficient sink for gases responsible for global warming and consequently the climate change are being analyzed. The reduction in emission of harmful gases from agriculture sector through land and crop management practices such as minimum and zero tillage and incorporation of crop residue in the soil is being experimented. Crop varieties resistant to higher temperature are also being developed. However, the management of rainfall attributes is a big challenge, which the humanity is going to face in near future. Proper land, soil and crop management practices hold the key for efficient rain water management and reduction in soil loss. The experiences gained by different institutions in India and world wide on watershed management works like strengthening of field bunding, contour bunding and trenching, vegetative barriers, agro forestry, water harvesting and afforestation of denuded lands along with other engineering and biological measures are found to be suitable options to mitigate the adverse impacts of climate change. For combating climate change the following practices may be taken up.

Soil management

- Sustainable soil management can contribute to mitigation (Soil carbon sequestration) adaptation (Increased water and nutrient retention) and agricultural productivity
- Conservation agriculture (minimum tillage, organic matter retention and crop rotation)
- Integrated soil fertility management, including judicious use of fertilizers
- Soil and water conservation measures(for controlling erosion/runoff)

Water management

- Innovative variability of rainfall calls for innovative solutions
- Water harvesting in times of excess rainfall (through percolation ponds or small dams)
- Supplementary (drip) irrigation
- Integrated watershed management

Other technology

- Weather forecasts (disseminated through mobile telephones, internet, TV and radio)
- Disaster management (early warning systems and crop insurance schemes)

Watershed approach to rainfed agriculture : The watershed approach represents the principal vehicle for transfer of rainfed agricultural technology. A watershed is a geographic area that drains to a common point, which makes it an ideal planning unit for conservation of soil and water. A watershed may comprise of one or several villages contain both arable and non arable lands, various categories of land holdings. The watershed approach enables a holistic development of agriculture and allied activities in the area taking into account various kinds of land use based on crops, horticulture, agro forestry, silvi-pasture and forests. The farming systems approach adopted in integrated watershed projects addresses multifarious objectives such as increased drinking water supply, improvement in food, fodder, fuel and fiber (Bhan and Bharati, 2010).

The key attributes of watershed management are in-situ conservation of rainwater and optimization of soil and water resources in a sustainable and cost -effective mode. The watershed approach aims to optimize moisture retention and reduce soil erosion, thus maximizing productivity and minimizing land degradation. Cultivable areas are placed under crops, while erosion prone, less favourable lands are placed under perennial vegetation. Improved moisture management increases the productivity of improved seeds and fertilizer, so conservation and productivity enhancing measures become complementary. The Government of India has accorded the highest priority to the holistic and sustainable development of rainfed areas through this approach. Watershed development addresses the problem of both droughts and floods. In fact there are special programmes such as Drought prone Area Programme (DPAP) for almost one-sixth of the land area in the arid and semi-arid regions of the country. Similarly the special programme of watershed treatment in the catchment of River Valley projects significantly reduces the silting of reservoirs, just as the treatment in the catchments of Flood Prone Rivers. Other schemes are DDP, NWDPRA and NREGA, initiated by the Government from time to time. So far integrated watershed development approach has been used successfully for the holistic development of rainfed areas aimed at improving livelihoods through special focus on increasing agricultural productivity through rain water harvesting and in-situ moisture conservation. Integrated watershed development, specifically as a strategy for adaptation and as a mitigation option, has come to the fore only recently, as scientific evidence predicts increasing in water stress as a result of climate change, especially in arid and semi arid tropics (Tripathy and Sharda, 2011).

Conservation agriculture : Conservation agriculture promotes minimal disturbance of the soil by tillage (Zero tillage), balanced application of chemical inputs (only as required for improved soil quality and healthy crop and animal production), and careful management of residues and wastes. This reduces land and water pollution and soil erosion, reduces long term dependency on external inputs, enhances environmental management, improves water quality and water use efficiency, and reduces emissions of green house gases. It provides direct benefits to environmental issues like land degradation, air quality, climate change, biodiversity and water quality. At present our country is self sufficient in agricultural produce having food grain production more than 230 M t year^{-1}. The food grain requirement in the country would reach about 300 M tonnes by 2020. Considering the impact of population growth, conventional agriculture with chemical inputs, depletion of soil nutrients, scarcity of water and climate change on agricultural productivity, it is a challenge to the scientists and planners to sustain the food security and at the same time maintaining the natural resources. Conservation agriculture has a great role to play in sustaining the agriculture growth and restoring the eco systems (Rath, 2009).

CONCLUSIONS

"The threat from climate change is serious, urgent and it is growing. If we fail to meet it boldly, swiftly and together we risk consigning future generations to an irreversible catastrophe," Barack Obama. Climate change is gathering momentum in present world scenario and is associated with agriculture as a global issue. Agriculture is a significant contributor of green house gases about 10-12%. The UNIPCC estimated that GDP in developing countries would decline by 1.4-3.0% due to climate change. Climate change is a reality and an important part of global warming, which is due to human activity and climate change related disasters have already seriously constrained the development of agriculture, which in turn has reduced the food security. Broadly speaking, conservation agriculture with stress on environmentally curative and preservative practices needs to be given utmost priority at this point of time. Conservative irrigation using micro irrigation systems need to be introduced in various agro-climatic regions of our country with increase in temperature due to climate change. Protected agriculture using shaded climate controlled houses are useful for nursery production through the reduced effect of global warming. Rational use and conservation of forests and land resources, prevention of desertification etc. would go further in reducing the climatic effect.

REFERENCES

Bhan, S. and Bharati, V.K. 2010. Watershed Development-Adaptation Strategy for Climate Change, *Journal of Soil and Water Conservation*,9 (1):65-69.

Kumarswamy, K. and Khan, H.H. 2010. Management for sustainable hi-productive farming, The Hindu, *Survey of Indian Agriculture*, pp66-68.

Nearing, M.A.2001. Potential changes in rainfall erosivity in the U.S. with climate change during the 21st century, *J of Soil and Water Conservation*, 56(3): 229-232.

Panigrahi, B. and Behera, B.P.2009. Impact of climate change on crop production and food security, E-Journal, Earth Sciences in India, vol.2 (III), July, 2009, ISSN 0974-8350.

Rath, B. 2009. Conservation Agriculture, Soil Water Conservation Today, News Letter of the Soil Conservation Society of India, Vol.4, No. 2, April-June, 2009.

Sharda, V.N. 2010. Sustainability and ensuring environmental security, *The Hindu, Survey of Indian Agriculture*, pp 75-77.

Singh, A.K.2010. Current status of degraded lands in India, Soil & Water Conservation Today, News Letter, vol.5, No.2.,ISSN 0975-4059, April-June 2010.

Soil Water Conservation Today, News Letter of the Soil Conservation Society of India, Jan-Mar 2009, Vol. 4, No.1

Thomas, K.V.2011. Climate change and Indian Agriculture, Yojana, Jan 2011, pp16-17.

Tripathy, K.P. and Sharda, V.N.2011. Mitigation of climate change through watershed management, Proceedings of the Souvenir, 45th Annual Convention of ISAE and International symposium in water for agriculture, Jan, 17-19, held at Dr. PDKV, Akola, CA, Nagpur.

William, J, Nearing, M.A, Nicks, A., Skidmore, E., Valentine, C., King, K and Savabi, R.1996. Using Soil erosion models for global change studies, *Journal of Soil and Water conservation*, 51(5): 381-385.

Wischmeir, W.H. and Smith, D.D.1978. Predicting rainfall erosion losses, A guide to conservation planning, U.S. Dept. of Agriculture, Agricultural Hand Book No. 2882.

□□□

Chapter 21

Carbon Sequestration, Crop Productivity and Land Amelioration Potential of Alley Cropping with *Leucaena leucocephala* in sub-humid sub-tropical India

R. K. SINGH, S. S. SHRIMALI, A. K. PARANDIYAL AND ASHOK KUMAR

INTRODUCTION

Promoting Agroforestry is one option many perceive as a major opportunity to deal with problems related to land-use and CO_2-induced global warming. The significance of agroforestry with regards to C sequestration and other CO^2 mitigating effects is being widely recognized. Alley cropping systems with fast growing woody perennials offer an opportunity to improve sustainability of tropical low-input farming systems. In addition to their potential role in meeting fuel and fodder requirements of local inhabitants, perennial hedges can be effectively used to control water erosion and soil fertility maintenance through recycling of nitrogen-rich foliage. Management of perennial hedges in alley cropping systems has an important role to play in optimizing both tree and crop productivity. The ameliorative effect of such systems would be specific to climatic and edaphic factors and there is need to identify and standardize impact assessment method for evaluating management options.

Soil organic matter (SOM) is known to be a key factor in maintaining soil productivity. It regulates plant nutrients and promotes soil aggregation, leading to increased water holding capacity, reduced erosion, and greater infiltration. Rapid loss in SOM in cultivated lands is an almost universal event. Incorporation of fast growing perennials in these low-input farming systems has emerged as a viable option for reversing the trend of land degradation. Because SOM management is viewed as central to the finest scale approaches used to assess the sustainability of soil systems, there is a need to link management options with their long-term influence on SOM. This paper

Central Soil and Water Conservation Research and Training Institute, Research Centre Kota, Rajasthan attempts to evaluate management induces changes in carbon budget of alley cropping system with direct measurement of C-budget components and using prediction model SCUAF (Soil Changes Under Agro-Forestry, Young and Muarya, 1990).

METHODS AND MATERIALS

The study was conducted in the sub-humid subtropical climate of northwestern hill region of India having annual rainfall of 1700 mm. The soil of experimental site is udic haplustalf having pH 5.7 and silty clay loam surface texture with silty loam subsoil.The experiment was laid out with five levels (0, 3.0, 6.0, 9.0 and 12.0 Mg ha^{-1} yr^{-1} dry weight basis) of mulching with *Leucaena* prunings with six replications in a complete randomized block design. Each of these 5x5 m size plots was cropped with a maize-wheat rotation.

Leucaena leucocephala seeds of variety K-8 were planted in five parallel rows in blocks running southeast to northwest. Each block was divided into four 5 m x10 m plots. Each plot had 100 trees. *Leucaena* trees were cut twice a year at sowing of crops in June and October. The harvested biomass was separated into twigs, green foliage and pods. Twigs and green foliage weights were taken row wise for each of the five rows of a plot separately. Pods were weighed plot-wise only. Samples were preserved for moisture determination and nutrient analysis. Bole biomass accumulation was estimated from bole volume increment. At each pruning, 10-15 randomly selected plants in each plot were measured for bole growth and leaf: twig ratio. Three litter traps of 0.25 sq m size were randomly placed in each plot. Litter collected over 1 mm mesh size was weighed bi-weekly during the growing seasons. Biomass production was determined on net area under tree basis (with 1 m x 0.5m spacing=20000 trees ha^{-1}).

On both sides of the hedge rows in open alleys maize - wheat cropping sequence was cultivated. The pruned biomass was spread on the soil surface at the rate of 0, 1.5, 3.0, 4.5 and 6.0 Mg ha^{-1} dry weight basis. Mulch treatments were applied twice a year at sowing of both the crops. During both cropping seasons all the plots were uniformly fertilized with recommended rates of P and K while N application was reduced to half of recommended dose. Soil samples from the plow layer (0-20 cm) of fallow and cropped plots were sampled with a surface core sampler (diameter 7 mm) prior to each mulching in the month of May and October. Within each replicated treatment plot ten sub-samples collected from randomly selected spots were mixed to prepare one composite sample for each plot. Samples were screened a 2-mm sieve without further air drying and refrigerated at 4°C in polyethylene bags until processed. The wet oxidation method without external heating (Anderson and Ingram, 1993) was used for OC determination.

Soil carbon balance in the surface 20 cm depth under a maize-wheat cropping system was determined using the SCUAF carbon model. Carbon inputs and losses, for which direct measurements were not feasible, used either the SCUAF default value or were estimated on the basis of available information. Above ground crop residue were determined by sampling the residual phytomass after crop harvest in a 2m x 2m quadrant for each of five mulch levels and six replications. C addition through weeds was determined at each manual weeding for each plot. Root biomass was estimated on the basis of soil core samples obtained at crop harvest.

RESULTS AND DISCUSSION

Carbon sequestration by *Leucaena:* Growth observed for all components was much higher during 4 month wet season than the 8 month dry season period (Table 1). The foliage and twigs components of the two pruning in the third year after planting constituted 60% of the total annual biomass growth of *Leucaena* hedge rows. During wet season substantial amount of photosynthesis were diverted towards pod formation. The root biomass estimated was about 29% of total above ground biomass. Additionally substantial amount of C sequestration occurred below ground through fine root turns over and root exudates. Fine root biomass of *Leucaena* hedges ranged from 1 to 2 Mg ha^{-1}. The pulses of live and dead root biomass following shoot pruning indicated possibility of pruning Induced shedding of fine roots. A substantial contribution to soil productivity through root litter is expected under alley farming which may be regulated through pruning management. Since prunings are being removed from the system about 17.77 Mg ha^{-1} biomass or 8.8 Mg ha^{-1} C retained annually were estimated to be sequestered by the laucaena hedge rows. Earlier work on biomass production from hedge row intercropping experiments showed strong variations (1-37 Mg ha^{-1}) depending on climate soil type and system management and the estimated values correspond well with the earlier estimates (Motagnini and Nair, 2004).

Table 1: Biomass partitioning during dry and wet season growth of 3 years old Leucaene hedge rows (Mg ha^{-1})

NPP Component	June harvest	October harvest	Total Annual biomass production	% of total biomass
Bole growth	0.87	1.17	2.04	8.20
Triages	2.86	4.33	7.19	29.90
Foliage	2.97	4.83	7.80	31.30
Pads	0.03	1.53	1.56	6.30
Letter	0.32	0.49	0.81	3.20
Root biomass	2.10	3.46	5.56	22.30
Total biomass	**9.15**	**15.81**	**24.96**	**100.00**

Crop yields in alleys: Mulch levels had a strong effect on grain and straw yields of both crops (Table 2 & 3). Maize showed greater response to mulching than wheat. Averaged over two years, maize grain yield increased by 24, 45, 66 and 88% over unmulched control in response to respective mulch rates of 1.5, 3, 4.5 and 6 Mg ha^{-1}. Corresponding increase for com stalk yields were 28, 56, 100 and 155%. With the same mulch rates, wheat grain yields averaged over 3 years increased by 20, 34, 41, and 48%; straw yield increased by 11, 21, 34 and 41%. With increasing rate of mulching, maize grain yield increased linearly whereas wheat grain yield had a quadratic response. Regression relationships were:

Maize grain = 2225+0.3M [R^2 = 0.31; PO.001; n=60].. Eq. 2.1
Maize straw = 3658+1.02M [R^2 - 0.69; PO.001; n-60].. Eq. 2.2
Wheat grain - 2480+0.344M-0.00003M^2 [R^2 = 0.65; PO.001; n=60].. Eq. 2.3
Wheat straw = 3604+0.232M [R^2 = 0.31; PO.001; n=60].. Eq. 2.4

Where crop yields are kg ha^{-1} and M is the rate of *Leucaena* mulch in kg ha^{-1}.

Table 2: Effect of mulch levels on maize yields (kg ha^{-1})

Mulch (Mgha^{-1})	Year 1		Year 2		Pooled over 2 years		
	Grain	Straw	Grain	Straw	Grain	Straw	Total
0	2482	4497	1805	3505	2144	4001	6145
1.5	3291	6100	2328	4169	2809	5135	7944
3.0	3634	7354	2565	5110	3099	6232	9331
4.5	3957	9156	3144	6841	3550	7998	11549
6.0	4285	11152	3767	9279	4026	10216	14241
$LSD_{0.05}$	480	1193	604	1309	368	847	1863

Table 3: Effect of mulch levels on wheat yields (kg ha^{-1})

Mulch (Mgha^{-1})	Year 1		Year 2		Year 3		Pooled over 3 years		
	Grain	Straw	Grain	Straw	Grain	Straw	Grain	Straw	Total
0	2350	3463	2632	3976	2440	3667	2474	3702	6176
1.5	2988	3875	2942	4446	2942	3350	2977	3890	6837
3.0	3267	4275	3208	4841	3437	3450	3304	4189	7493
4.5	3413	4854	3429	5096	3655	4033	3499	4662	8260
6.0	3554	5438	3571	5239	3857	4483	3661	5053	8714
$LSD_{0.05}$	457	1004	269	420	343	NS	200	414	992

When crop yields were adjusted for loss in cropped area for growing different levels of mulch material, total grain yield declined with increasing mulch rates. The leaf fodder from crops increased with 1.5 Mg ha^{-1} mulching then declined. Planting the *Leucaena* hedge rows in 20% of land area would provide about 1.5 Mg ha^{-1} mulch and yield about 4200 kg ha^{-1} of grain yield, 8100 of leaf fodder from crops and 3200 kg ha^{-1} of fuel from trees. There was no significant effect of hedge rows on maize at tree crop interface. However, wheat grains and store yields progressively declined with decreasing distance from hedge row. Overall wheat grain yield reduced by 35% in the vicinity of hedge rows. This reduction was primarily due to competition for soil moisture.

Moisture was not expected to be a limiting factor during the rainy season when maize was cultivated. The hedges were pruned at time of sowing which further minimized the competition for light and moisture. *Leucaena* hedges recovered quickly after pruning in the rainy season and height of hedges reached to 2.5-3m height when the maize was harvested. Under the favorable growing environment (warm temperature and sufficient water supply) during the rainy season, a sharp increase in maize yield in response to increasing rate of mulching with *Leucaena* primings is attributable to increased nutrient availability, and to some extent weed suppression through mulches. A significant improvement in the N concentration of straw and grain of maize and wheat and also in the soil indicated the nutritional advantage under mulched plots. In this study it was not possible to segregate nutritional effects from the weed suppressions effect of mulching.

Wheat was less responsive to *Leucaena* mulching at higher rates because moisture became a more limiting factor during the wheat crop season. Although greater residual water content at wheat sowing was found in sub-surface layers, differences were small and of little advantage to wheat crop. Marginally higher water content observed at subsurface layers (30-75 cm depth) was possibly due to better receptivity of rain water and effective weed control during maize season under the plots mulched at 4.5 Mg ha^{-1} or higher. Estimates based on tree-crop interactions observed in this study show that an alley cropping system with 20% area under *Leucaena* hedge-rows would provide about 12% lower grain yield and 15% higher fodder yield from crops as compared to sole cropping system (Figure 1).

Carbon budget of maize wheat system: The carbon budget of maize-wheat rotation with variable mulch rates is presented in Table 4. In the absence of organic additions annual carbon losses exceed the input through crop residues resulting in a net depletion of soil carbon reserves. Mulching at a rate of 3 Mg ha^{-1} yr^{-1} was able to reverse this trend. There were three sets of data available to relate annual C addition with annual C balance in this study. Observed OC changes in cropped and fallow plots and SCUAF estimated annual C balance, show a linear relationship between C input and soil C balance (Figure 2). According to this relationship a minimum of 3 Mg ha^{-1} yr^{-1} of C incorporation is required to maintain present level of SOM.

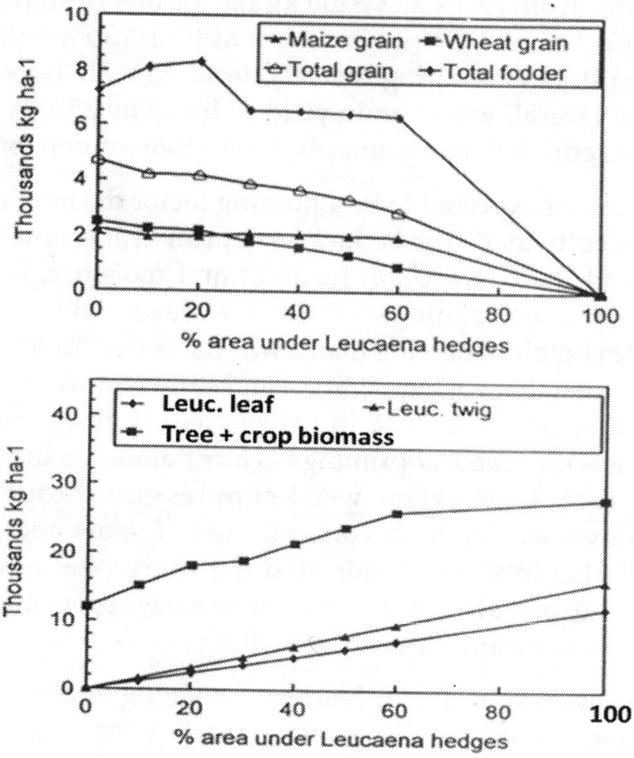

Fig. 1: Grain, fodder and fuel production under different tree: crop ratios

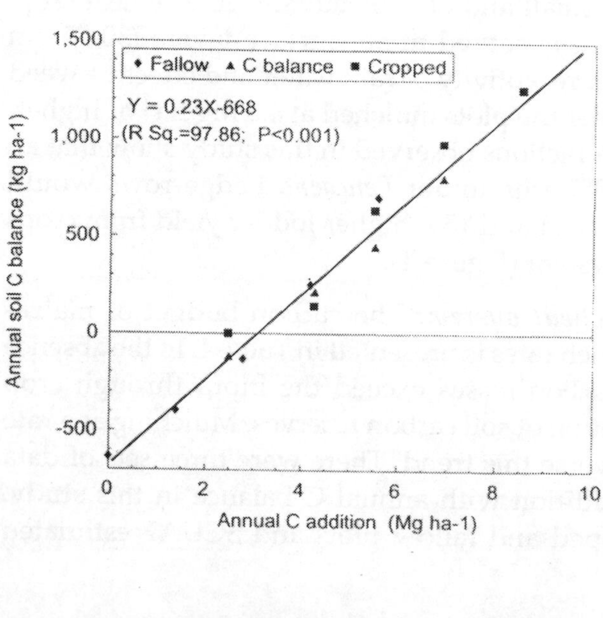

Fig. 2: Relationship between rate of Carbon addition and annual soil C balance as observed in fallow and cropped plots and estimated by C budget of maize-wheat system

Soil organic matter under forest cover in the Doon Valley ranges from about 1% under severe biotic interference, and erosion hazards to 4% under level conditions and protected sites (Singhal *et al.*, 1975 and 1986; Singh *et al.*, 1990). Cultivation on these soils leads to a rapid decline in OC levels. The negative carbon balance of the annual budget of low input farming with maize-wheat rotation indicates that this traditional land use is a degrading system. The annual carbon input to soil in this system is about 2.5 Mg ha^{-1}. Because maize and wheat straw are consumed as fodder, only about 13% of annual C input of the crop residue is recycled to soil. Crop roots (50%) and weeds (36%) are major contributors to soil organic matter. The soil has not reached a new equilibria where present C input is able to sustain the OC level. Soil OC has declined from 1.07% (under adjoining natural Sal forest) to 0.515% during 40 years of known cropping history of this experimental site, which confirmed the trend observed under the traditional cropping practice. Under the favorable growing conditions of this region, recycling of nitrogen-rich foliage of *Leucaena* substantially increased crops yields. Because of increased carbon input through mulching and crop residues, mulching at the rate of only 3 Mg ha^{-1} was able to reverse the trend of SOM depletion.

SCUAF based predictions: SCUAF provided realistic estimates for future changes in SOM in response to recycling of *Leucaena* foliage with different rates. Over a period of 25 years the model predicts decline in OC by 12 % under the no mulch treatment and mulching at the rates of 3, 6, 9, 12 Mg ha^{-1}, as projected by SCUAF, would result in an increase of OC by 16, 50, 94 and 160 % of present OC level, respectively (Figure 3). Rate of loss in soil

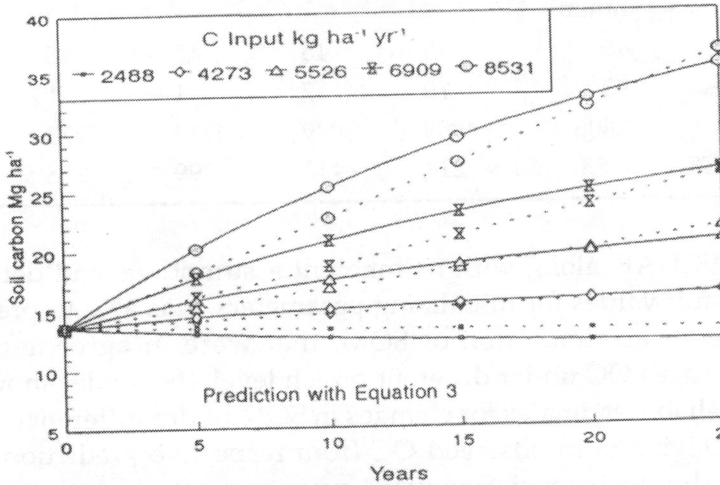

Fig. 3: Changes in soil carbon under different rates of *Leucaena* mulching: SCUAF predicted (solid lines) v/s prediction based on C-balance method (broken lines)

OC projected by SCUAF and annual C budgeting compare well with each other due to procedural similarities in estimating annual losses. The estimate for humus oxidation was comparable with estimates obtained with isotope based studies conducted in tropical locations (Sauerbec and Gonzalez, 1977; Dalai, 1982) and are about ten times higher than the reported estimates for temperate climate (Jenkinson and Rayner, 1977).

Table 4: Annual carbon budget (C kg ha^{-1} yr^{-1}) of maize-wheat systems with different levels of *Leucaena* mulching

Source/Sink		**Mulch levels**				
		No Mulch	**Mulch 3Mg ha^{-1}**	**Mulch 6Mg ha^{-1}**	**Mulch 9Mg ha^{-1}**	**Mulch 12Mgha^{-1}**
Inputs	Straw	314	386	458	578	721
Weed	890	997	685	358	224	
Root	1284	1495	1593	1788	2006	
Leucaena Mulch	0	1395	2790	4185	5580	
Total Input	2488	4273	5526	6909	8531	
Litter to humus conversion losses	Above ground residues	1023	1176	972	796	803
Crop roots	860	1002	1067	1198	1344	
Leucaena mulch	0	1158	2288	3306	4241	
Humus oxidation	Labile humus	615	659	706	764	839
Stable humus	45	45	45	45	45	
Erosion Losses	63	19	7	1	1	
Total Losses	2606	4059	5079	6110	7273	
NET BALANCE	-188	214	447	799	1258	

As the SCUAF, along with its inherent assumptions and determined or selected default values for initializing parameters, was able to predict trends of depletion or accumulation of SOM, that were in agreement with the observed changes OC under different mulch level, the model shows promise to provide reliable estimates for changes in SOM under different carbon input levels. The deviation in observed OC from respective prediction values are most likely due to inconsistent weed management. Also the amount and distribution of rainfall and fertility buildup or depletion influence crop productivity and annual C input through crop residues. These effects are not simulated by SCUAF. However, the model considers feedback factors for calculating annual C input in response to the rise or fall in SOM or nitrogen. As shown in Figure 2, the annual C balance estimated using SCUAF correspond well with observed annual change in soil organic carbon in cropped and fallow plots. This lends support to the idea that the assumptions used for the SCUAF were reasonable. Also, SCUAF predictions for the highest mulch level of 12 Mg ha^{-1} yr^{-1} for SOM accumulation at 100 years was about 2%, which was similar to SOM in native forests of this valley where annual above-ground litter input is about 12 Mg ha^{-1} yr^{-1}.

CONCLUSIONS

This study indicated a declining trend in soil organic matter under a maize-wheat cropping system without recycling of straw or stalks in the sub-tropical climate of northern India. An alley cropping system with paired row of *Leucaena* hedges (1m x 0.5 m spacing) at every 8 m interval in maize-wheat rotation (2:8 tree:crop ratio) would produce about 12% lower grain yield and 15% higher fodder (straw of maize and wheat) yield as compared to sole cropping system. Additionally, trees would produce about 3 and 3.2 Mg ha"1 of leaf mulch and twig yields. With this amount of mulch only marginal increase in soil organic matter (from 0.52% of present level to 0.6% after 25 years) can be expected. *Leucaena* foliage mulching at rates greater than 6 Mg ha^{-1} yr^{-1} resulted in significant increases in total organic carbon and biomass carbon within three years. SCUAF provided reasonable projections for changes in SOM in response to various levels of *Leucaena* foliage recycling. Soils with low organic matter content responded quickly to increased rates of carbon addition through mulches. Over a period of 25 years the model predicts decline in OC by 12 % under the no mulch treatment and mulching at the rates of 3, 6, 9, 12 Mg ha^{-1}, as projected by SCUAF, would result in an increase of OC by 16, 50, 94 and 160 % of present OC level, respectively. It is concluded that alley cropping of maize-wheat system with *Leuceana* hedge-rows has potential arrest land degradation process; however there is little scope of promoting these systems for meeting C-sequestration objectives.

REFERENCES

Anderson, J. M. and Ingram, J. S. I. (eds.). 1993. *Tropical Soil Biology and Fertility: A Hand Book of Methods.* C. A. B. International, Wallingford, UK, pp: 62-69.

Dalai, R. C. 1982. Organic matter content in relation to the period of cultivation and crop yields in some subtropical soils. *Transactions of the Twelfth International Congress of Soil Science,* 6: 59.

Jenkinson, D. S. and Rayner, J. H. 1977. The turnover of soil organic matter in some of the Rothamsted classical experiments. *Soil Sci.,* 123: 298-305.

Motagnini, F. and. Nair, P. K. R. 2004. Carbon sequestration: An unexpected environmental benefit of agroforestry systems. *Agroforestry Systems,* 61: 281-295.

Sauerbeck, D. R. and Gonzalez, M. A. 1977. Field decomposition of carbon-14-labelled plant residues in various soils of the Federal Republic of Germany and Costa Rica. In *Soil organic matter studies.* Vienna : IAEA, pp159-170.

Singh, R., Singh, R. K. and Singh. K. 1990. Effect of different plant covers on soil characterises. *Indian Forester.* 116 (10): 795-802.

Singhal, R. M., Banerjee. S.P. and Pathak, T.C. 1975. Effect of Eucalyptus monoculture on the status of soil organic matter in natural Sal *(Shorea robustd)* zone in Doon Valley. *Indian Forester,* 101 (12): 730-737.

Singhal, R. M. and Sharma, S. D. 1986. Study of soil of Dehradun forest in relation to vegetaion type. *J. of Indian Soc. Soil Sci.,* 34 (2): 350-356.

Young, A. and P. Muraya. 1990. *Soil Changes Under Agroforestry (SCUAF): A predictive model-version* 2. ICRAF, Nairobi, Kenya, 124 p.

□□□

Chapter 22

Nutrient and Water Constraints Affecting Biogeochemical Feedbacks of the CO_2 Fertilization Effect

NARENDRA K. LENKA, RATTAN LAL[1] AND SANGEETA LENKA

INTRODUCTION

The global C balance indicates a net uptake of CO_2 by terrestrial ecosystems and out of the several mechanisms, increment in the biotic or vegetation pool from the CO_2 fertilization effect is prominent. Its significance also stems from the rising atmospheric CO_2 concentration at a rate of about 2 part per million by volume (ppmv) per year with the current concentration of 389 ppm and the likelihood of its increase to 900 to 1100 ppm by the end of 21st century based on the business-as-usual path of fossil fuel consumption (Kiehl, 2011). The net photosynthetic C assimilation by plants through metabolic processes increases with increase in atmospheric CO_2 concentration (eCO_2), resulting in an enhanced net primary production (NPP) which is referred as CO_2- fertilization effect. This is pronounced in C_3 species and the increase is from 14% to 31% enhancement in light-saturated photosynthesis. Growth of C_3 plants (whose photosynthesis is CO_2-limited) generally responds more strongly to elevated CO_2 than that of C_4 plants, particularly when there are strong sinks for C (*e.g.*, reproductive organs or belowground storage organs). Response of C_4 plants is more as a consequence of increased water use efficiency (WUE) under elevated CO_2 scenario.

The CO_2 fertilization though can offset the rise in atmospheric CO_2 concentration substantially, is met with two uncertainties (1) what are the ecological feedbacks of CO_2 fertilization effect with respect to availability of soil nutrients and water (2) whether CO_2 fertilization needs additional requirement of nutrients and water.

Indian Institute of Soil Science, Nabibagh, Bhopal, Madhya Pradesh, India

[1]Carbon Management and Sequestration Centre, Ohio State University, Columbus, USA

Some studies show the increase in NPP to be possible only with optimum soil resources, (*e.g.*, plant nutrients) which determine growth rate and active carbon (C) sinks in plants (Thompson *et al.*, 2004; Reich *et al.*, 2006a). Even response of C_4 plants has been observed only with nutrient supplementation particularly under nutrient limiting environments (Markelz *et al.*, 2011). Such a response is expected as harnessing benefits of the readily available resources needs others at optimum level in accordance with the Liebig's "law of minimum".

To-date, most of the research on effect of CO_2 on plants has focused on physiological aspects and processes and mechanisms at leaf and plant level while maintaining other resources in optimum availability. However, physiological effects must be linked to soil processes moderating nutrient and C dynamics. Studies have shown alteration in tissue chemistry and also in the physiological behavior of plants. The suppressed transpiration of plants results in reduced latent heat flux and increased sensible heat flux from plant canopies, having potential to alter the hydrological processes. However, the extent to which the canopy feedbacks of CO_2 fertilization affect the ecological processes is not fully understood. Thus, this article attempts to compile the research on the effect of CO_2 fertilization on soil nutrients (particularly N and P) and water availability and vice-versa and highlight the possible pedologic constraints to CO_2 fertilization.

CO_2 Fertilization vis-à-vis Soil and Plant Nitrogen

Alteration in plant tissue chemistry : The basic changes in the plant tissue chemistry as a result of eCO_2 are :

1. Decrease in tissue N concentration
2. Increase in accumulation of starch, sugars and non-structural carbohydrates in leaf tissues
3. Increase in lignin content
4. Widening of C : N ratio in plant tissues
5. Reduction in critical leaf N concentration

The decrease in tissue N concentration is likely due to several factors such as dilution effect from increased carbohydrate content, reduction in the Rubisco content, decreased uptake of minerals from soil because of reduction in stomatal conductance, less water uptake by plants (Taub and Wang, 2008), and decrease in the rate of assimilation of NO_3^- into organic compounds (Bloom *et al.*, 2010). The reduction in leaf N concentration can be substantial varying from 10-31% (Ainsworth and Long, 2005). The reduced leaf N concentration also causes lowering of N in grains and plant roots and reduced protein content in major crops. There is a reduction in critical leaf N concentration, *e.g.*, by about 10% in spring wheat (Fangmeire *et al.*, 1999). Even though photosynthesis is down regulated, leaves tend to accumulate more sugar and starch levels resulting in non-structural carbohydrate levels that are 30 - 40% higher on unit leaf area basis (Ainsworth and Long, 2005). Overall, the C: N ratio of leaves increases under eCO_2 conditions in both C_3 and C_4 plants. The other important change in plant tissue is the increase in lignin content in both above and below ground biomass. On an average, there may be up to 6.5% increase in lignin content in leaf tissue under eCO_2 (Norby *et al.*, 2001), and such a trend has also been confirmed from a FACE experiment where eCO_2 produced poor quality litter (high C: N, lignin: N and condensed tannins) (Parsons *et al.*, 2004).

Rate of Tissue Decomposition

Alteration in the plant tissue chemistry has an important role in modifying the soil nutrient availability due to its effect on the mineralization process. Higher non-structural carbohydrate and lignin content along with wider tissue C: N ratio increases the rate of immobilization of soil N and thus reduced availability to plants despite presence of N in sufficient quantity in soil. Thus, the decomposition rates and ecosystem functioning are altered under eCO_2 scenario. The effect on decomposition may be little or none in the short-term. However, in the long-term, the effect of eCO_2 on litter quality and subsequent decomposition may be significant, as has been observed in some FACE study (Parsons *et al.*, 2004). Even, roots of N-fixing legumes decompose more slowly when grown under eCO_2 (de Graff *et al.*, 2011).

Hypotheses of PNL and PDNA

The gradual increase in N limitation has been suggested in form of a hypothesis called "progressive N limitation" (PNL) by Luo *et al.* (2004). The hypothesis assumes that plant growth under eCO_2 sequesters N in wood or in SOM, reducing N availability and creating negative feedback on NPP. Further, any increased C storage in soil may sequester N in SOM progressively decreasing the mineral N available for plant uptake in the long-term. The PNL hypothesis states that without new N input or decreases in N losses, the availability of mineral N declines over time at eCO_2 *vis-a-vis* the ambient CO_2. The PNL concept is built on the biological principle that the formation of organic matter requires a certain amount of N and other nutrients in a relatively fixed ratio with C. The PNL may not occur if the N demand is met through additional N supply under eCO_2 conditions stimulating the NPP (Figure 1).

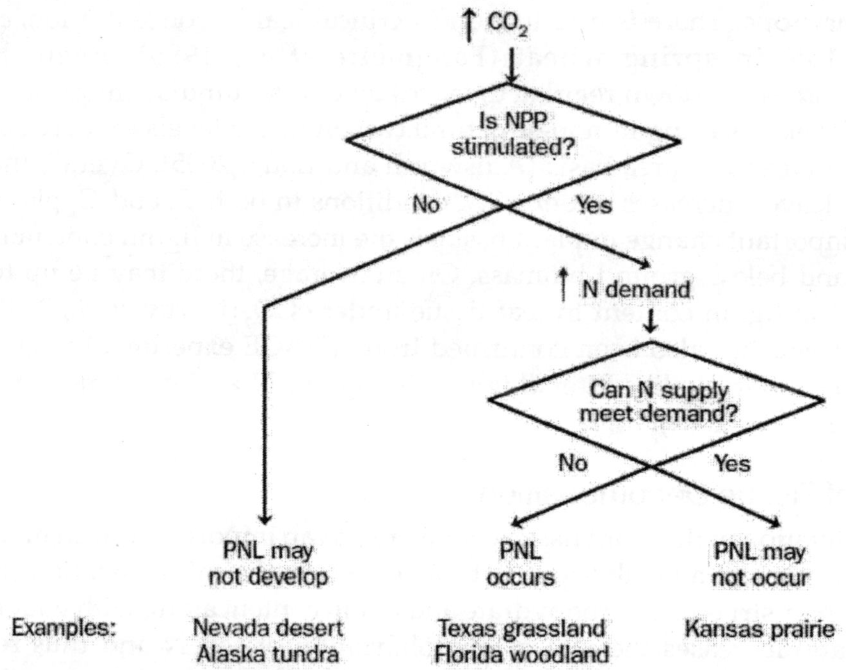

Fig. 1: Schematic diagram of PNL – centred framework showing interaction between C and N under elevated CO_2 (Luo *et al.*, 2004).

There are some experimental evidences to show reduced N availability under eCO_2, attributed to the PNL hypothesis. For instance, a decline in N availability observed in the ORNL-FACE experiment was ascribed to the onset of PNL (Norby and Zak, 2011). Similarly, Hovenden *et al.* (2008) reported a decline in N availability after 4 years of eCO_2 exposure in a grassland FACE study. The PNL effect was also reported to set in after 4 years of CO_2 enrichment at the Duke Forest FACE experiment; but the effect was mitigated with additional input of N (Oren *et al.*, 2001).

Contradicting the above interpretations, Lenka and Lal (2012) argued that "Progressive Decline in N Availability (PDNA)" may be a better explanation for the observed trends of results. The decline in N availability over a period of 4-6 years of eCO_2 exposure may not be necessarily because of the PNL effect through sequestration of C and N in long-term biotic and soil organic carbon (SOC) pools. A long-term response may not be initiated within four years nor can it be alleviated by supplemental application of N as was observed by Oren *et al.* (2001). Short-term responses, over a period of 4-5 years, may show a visible change through an additive effect in N-dynamics and can explain the decline in N availability better than attributing it to a long-term response. It is, therefore, a gradual decline in N availability that seems to be the prime factor reflected in short-term responses, and may be more appropriated PDNA.

Also, formation of a recalcitrant biomass and passive SOC pool may need a longer time frame assuming the turnover periods of active, slow and passive pools. Even an intermediate SOC pool has a turnover period of at least 25 years. Thus, a period of 4-5 years may not be sufficient for formation of a long-term SOC pool and exhibiting long-term response in soil N dynamics. Nonetheless, comparatively less time may be required for the onset of PDNA effect in N-limited ecosystems (*e.g.*, croplands and grasslands) than in forest ecosystems which are often not N-limited (Lenka and Lal, 2012). It is also well known that addition of fresh crop residues or even fresh manures to growing crops can temporarily block the N availability, producing a visible short-term response. It does not imply, however, that long-term responses (Luo *et al.*, 2004) would not affect the N - dynamics under eCO_2.

The temporary withholding of N can affect crop N- nutrition in diverse agro-ecosystems. In general, the crops in tropical regions are sown at the onset of rains when soil temperature also remains high leading to a high rate of microbial activity for decomposition in soil. Thus, any addition of N through basal application may first meet the requirement of microbes rather than for establishment of plants and thus making added N to be less available at least during the initial growth stages. With a poor establishment and slow initial growth, both NPP and crop yield may be adversely affected. Thus in N-limited ecosystems, more N input as fertilizers may be required to produce the same quantity of crop yield. Such a scenario may exacerbate the food insecurity challenges. Apart from the immobilization concern, a reduced rate of decomposition may retard the mineralization process resulting in low addition of N to soil and to a gradual soil impoverishment.

Effect on Mineralization and Immobilization

Experimental data show a decrease in net N mineralization in grassland ecosystems (Gill *et al.*, 2002) and also under in-vitro incubation studies using plant tissues grown under eCO_2 conditions (Torbert *et al.*, 2001). As reviewed by Lenka and Lal (2012), there is a significant decline in soil N in forest ecosystem, 0.34% yr^{-1} under eCO_2 as compared to 0.17% yr^{-1}under ambient conditions (Garten *et al.*, 2011). The decline in inorganic soil N pool may be up to 25-38% as observed in Prairie grassland (Dijkstra *et al.*, 2010). Although gross and net N mineralization rates are not affected in some experiments, the microbial N immobilization is significantly enhanced (de Graaff *et al.*, 2006, Lagomarsino *et al.*, 2008). A study based on synthesis of data from 117 experiments show that gross N immobilization may increase by 22% (de Graaff *et al.*, 2006). In many cases, the effects on N-dynamics are visible only after a certain period of time. For instance, at the Aspen - FACE experiment, the gross N transformation rates were not altered during the first 3 years of the experiment. But, afterwards gross N mineralization doubled and N immobilization increased 4-fold, without noticeable change either in microbial biomass N or soil total N. However in some cases, even after 5 years of eCO_2, there was no impact on rates of N mineralization, immobilization and plant N uptake (Dijkstra *et al.*, 2008).

Additional N Supply May Offset Negative Feedbacks of CO_2 Fertilization Effect

The possible linkages (Figure 2) explaining the C: N dynamics under the eCO_2 scenario indicate the following feedbacks:

Negative feedbacks

i) Continuous exposure to eCO_2 inducing photosynthetic acclimation,
ii) Reduced tissue N concentration,
iii) Increase in C:N ratio,
iv) Increase in non-structural carbohydrates such as lignin and resistant fractions,
v) Reduced decomposition rate,
vi) Higher rate of N immobilization,
vii) Development of a recalcitrant soil carbon pool,
viii) N-mining from deep soil profile due to a low stock replenishment
ix) Possible depletion of soil N profile and
x) Progressive decline in N availability

Positive feedbacks

i) Higher root nodulation,
ii) Increased biological N fixation (BNF),
iii) Increase in N use efficiency

From the experimental evidences as well as from the analysis of possible scenarios, Lenka and Lal (2012) observed a definite increase in N requirement under eCO_2 situation. They enumerated all possible scenarios such as increase, decrease or no change in rate of decomposition and concluded that in any case, additional N supply is required or N losses are to be minimized in order to alleviate the negative feedbacks of CO_2 fertilization effect. Under the 1st scenario, the quick decomposition of soft tissue fractions may help microbes to complete more number of decomposition cycles per unit time, leaving a

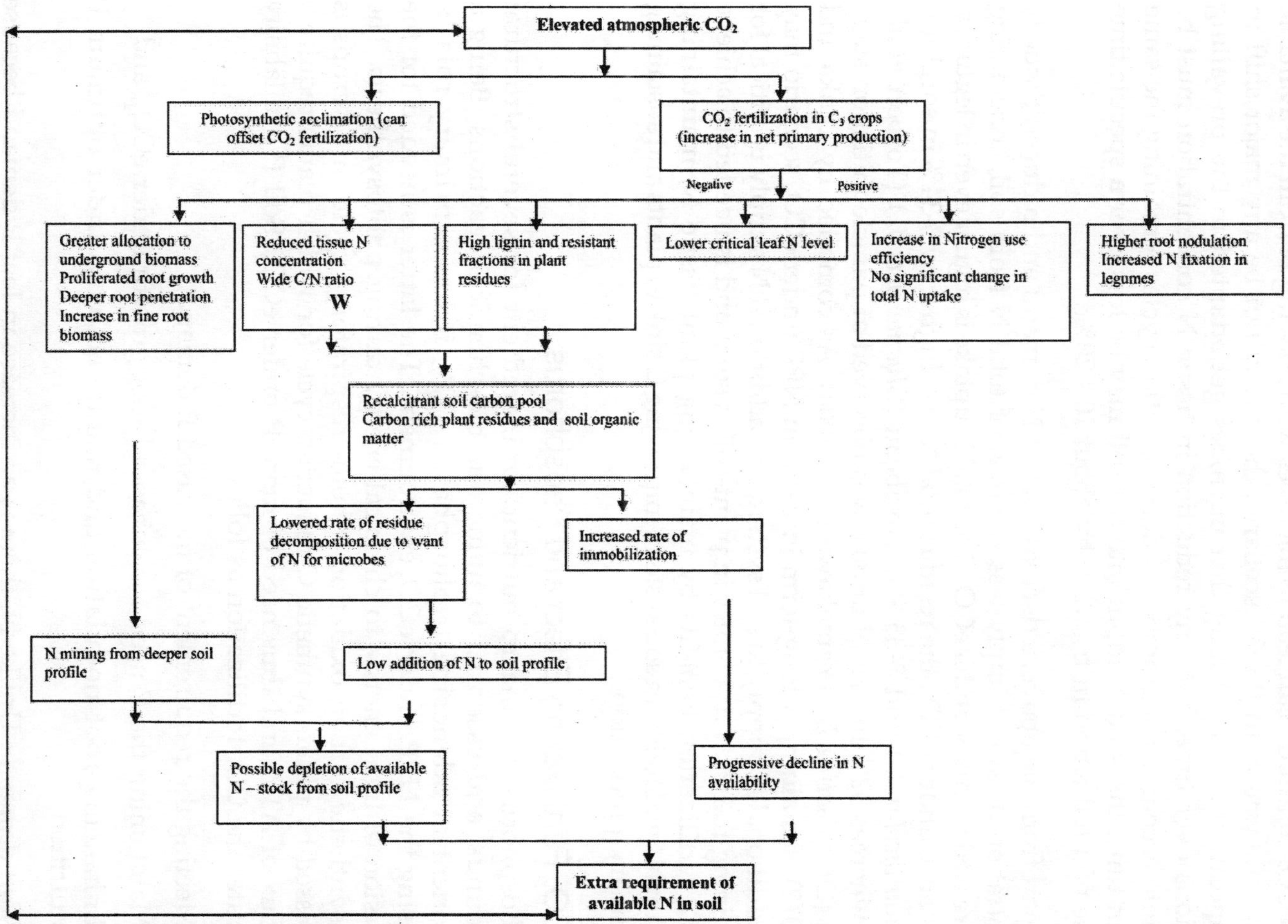

Fig. 2: Possible mechanisms showing positive and negative feedbacks of CO_2 fertilization on soil N availability. This implies that CO_2 benefits can be realized through supplementation of additional N to maintain plant available soil N

higher volume of resistant biomass fractions in the soil. Under the 2nd scenario, the increased availability of C in soil through addition of C-rich plant residues under eCO_2 condition can lead to microbes withdrawing N from the available pool in soil system. In the 3rd scenario, there may not be any major shift in decomposition rate assuming that microbes get adapted to the prevailing conditions over time. Yet, any reduction in tissue N concentration must be taken into account. The number of decomposition cycles remaining the same per unit time, the total N input into the soil may be less over a specific time because of a reduction in tissue N by about 10 - 20%.

Apart from the above, a decrease in leaf N demand and reduced tissue N concentration does not imply less removal of total N from soil, considering the increased biomass under eCO_2. The total N uptake is higher even in legumes. For instance under eCO_2, the residue total N was higher by 3-34% in sorghum (*Sorghum bicolor L.*) and 3-33% in soyabean *(Glycine max L.)* (Torbert *et al.*, 2004), whereas 32% higher N uptake was observed at panicle initiation stage in rice *(Oryza sativa L.)* (Yamakawa *et al.*, 2004). As compiled by Lenka and Lal (2012), sustaining a long-term increase in NPP under eCO_2 scenario may need additional N supply to soil system and additional N supply is critical for maintaining the C: N dynamics in plant-soil system and is needed to achieve the CO_2 fertilization benefits by maintaining plant tissue N, maintaining optimum N availability in soil system, improving C sink in plants and sustaining net primary production.

The CO_2 Fertilization Effect and Phosphorus

Phosphorus is a macro nutrient, an ingredient for several structural compounds, and essential to numerous bio-chemical reactions. Being a component of adenosine triphosphate (ATP), it has a primary role in increasing the NPP under eCO_2 environment. The latter is needed for the conversion of light energy to chemical energy during photosynthesis. The photosynthetic carbon oxidation (photo- respiratory) cycle in C_3 crops is suppressed by photo synthetic C reduction cycle for which plants require a high flux of ATP, and thus need of more P under eCO_2. Soil P availability moderates the CO_2 fertilization as follows:

1. Meeting the requirement of increased P demand,
2. Maintaining the P need for enhanced root growth under eCO_2, and
3. Enhancing root nodulation and nodule activity under optimum P nutrition.

Plant demand : Unlike N, there is an increase in leaf P demand because of a high requirement of ATP which increases the level of critical leaf P concentration in plants. There is a consensus with respect to an increase in total P uptake. Total shoot P uptake in rice at panicle initiation stage is higher by 37 to 58% under eCO_2 than under ambient conditions. Higher uptake by 58% has been observed even with 6% reduction in tissue P concentration (Yamakawa *et al.*, 2004).

Further, the P demand is not influenced by the level of N application or the soil N concentration. Thus, P demand of plants remains high under eCO_2 and may be suppressed under conditions of low P availability. Insufficient P supply also inhibits the photosynthetic activity. In P-limited soils, there is a source-sink imbalance and starch accumulation in leaves (Almeida *et al.*, 2000). Even legumes do not respond to eCO_2 unless P is added (Stocklin *et al.*, 1998).

There also occurs a decrease in the saturation CO_2 concentration under P-limited conditions. A study with lupins indicated no effect of eCO_2 on photosynthesis rate at low P, but exhibited 2 to 2.5 times higher rates at high P with increase in CO_2 concentration from 400-750 μmol mol^{-1} (Campbell and Sage, 2006).

In addition to photosynthesis, P is also essential to enhanced root growth and proliferation. In accord with the principle of resource allocation, a higher proportion of C is transferred to the below-ground than to the above-ground biomass under eCO_2 conditions.

In accord with the effect on the below-ground biomass, eCO_2 increases root nodulation and nodule activity. Such activity is constrained in P-limited environments, which limits the growth of legumes (Stocklin *et al.*, 1998; Almeida *et al.*, 2000). Apart from reducing nodulation, severe P deficiency induces a greater uptake of N from soil leading to a high N accumulation in leaves and other tissues (Almeida *et al.*, 2000), which may also affect N nutrition of associated species. A 6-year study on an N-poor calcareous grassland indicated that P deficiency limited the response of legumes to eCO_2, which in turn limited N transfer to non-leguminous species in the ecosystem (Niklaus and Korner, 2004). Thus, any long-term positive effect of eCO_2 on N fixation can be realized only with optimum P nutrition.

Soil P availability: Deficiency of P is a limiting factor in 40% of world's cropland soils (Vance, 2001), and it is a severe constraint in the acid soils of tropics and calcareous soils of sub-tropical and temperate regions. The P retention potential is low or medium in principal soils of USA, Europe, India and China (USDA, 2003). Deficiency of P is observed in 98% of districts in India (Tiwari, 2001). Based on 9.6 M soil tests, Indian soils are low in available P in 49.3% of districts, medium in 48.8%, and high only in 1.9% (Hasan, 1996). Thus, a high soil P availability is needed to harness CO_2 fertilization effect.

The Co_2 Fertilization Effect and Soil Water

Apart from enhancing photosynthesis, enrichment of CO_2 also affects plant and soil water relations by regulating the stomatal opening. Plants can maintain high photosynthetic rates under eCO_2 with relatively low stomatal conductance. The data from FACE experiments show 22% reduction in the stomatal conductance in case of crops and grassland species (Ainsworth and Rogers, 2007).

The decrease in stomatal conductance may reduce plant water use. However, a gain in water economy under eCO_2 is possible only if the transpiration efficiency observed at individual leaf level translates to field and ecosystem levels. At a higher hierarchy level, the processes and factors which modulate the eCO_2 response also increase. The factors which can confound the direct effect of reduced stomatal conductance are:

Physiological forcing: At the leaf level, a reduced transpiration can increase the leaf temperature by 1°C due to physiological forcing (Kim *et al.*, 2006). The additive effect of increase in leaf temperature by stomatal closure at an ecosystem scale increases air temperature negates any net water saving (Jarvis *et al.*, 1999). The increase in air temperature over land due to eCO_2-physiological forcing may increase by 0.42 ± 0.02 K vis-à-vis 2.86 ± 0.02 K due to CO_2 radiative forcing (Cao *et al.*, 2010).

Higher surface runoff: Due to physiological forcing, the surface hydrology is affected, as the global mean runoff may increase by 6-8.4% by (Betts *et al.*, 2007; Cao *et al.*, 2010). Further, more runoff is possible due to higher antecedent soil moisture due to lowered plant uptake. Suppressed plant transpiration due to stomatal closure is strongly correlated with the increased continental river runoff records observed over the twentieth century (Gedney *et al.*, 2006).

Soil evaporation may be substantial: Significance of soil evaporation under the eCO_2 scenario may be relatively higher than that under ambient conditions. A higher wetness in soil profile due to decreased transpiration may lengthen both stages of evaporation, particularly the 2nd stage. A low amount of rainfall may result in sufficient wetness for the profile to shift to a higher rate of evaporation and thus higher water loss.

The data from FACE experiments substantiate the possibilities described above. There is a decrease in the whole plant water use by 5-20% under eCO_2 with increases in both soil moisture storage and runoff (Leakey *et al.*, 2009). Nevertheless, eCO_2 compensates under mild drought conditions only in both C_3 and C_4 species. The plant water potential remains high under eCO_2 conditions than that under ambient environments. However, the impact is of little significance when water limitation is severe (Reich *et al.*, 2006b). Also drought stress, irrespective of intensity, is not the ideal condition to realize the benefit of CO_2 fertilization. Under well-watered conditions, the feedback is not always significant, and many reports are from controlled or pot cultures (Kim *et al.*, 2006).

CONCLUSIONS

There is a well proven interaction between soil water and plant nutrients. The nutrient limited ecosystems can't harness the benefits of better soil moisture profile, if at all realized, and vice-versa. The elevated CO2 scenario make the things more complex and in all likelihood, the three-way interactions among CO2, plant nutrients and soil water, will decide the future production scenario in major agro-ecosystems in the world.

ACKNOWLEDGEMENT

The paper was prepared when the senior author was undergoing the USDA Norman E Borlaug fellowship program at Ohio State University, Columbus, USA. The financial support of USDA is gratefully acknowledged.

REFERENCES

Ainsworth, E. A. and Rogers, A. 2007. The response of photosynthesis and stomatal conductance to rising CO_2: mechanisms and environmental interactions. *Plant, Cell and Environ.*, 30: 258 -270.

Ainsworth, E. A. and Long, S. P. 2005. What have we learned from 15 years of free-air CO_2 enrichment (FACE)?. A meta-analytic review of the responses of photosynthesis, canopy properties and plant production to rising CO_2. *New Phytol.*, 165: 351-372.

Almeida, J.P.F., Hartwig, U.A., Frehner, M., Nosberger, J. andLuscher, A. 2000. Evidence that P deficiency induces N feedback regulation of symbiotic N fixation in white clover. *J. Expt. Bot.*, 51: 1289-1297.

Betts, R.A., Boucher, O., Collins, M., Cox, P.M., Falloon, P., Gedney, N., Hemming, D.L., Huntigford, C., Jones, C.D., Sexton, D.M.H. Webb, M.J. 2007. Future runoff changes due to climate and plant responses to increasing carbon dioxide. *Nature,* 448: 1037-1042.

Bloom, A. J., Burger, M., Asensio, J.S.R. and Cousins, A.B. 2010. Carbon dioxide inhibits nitrate assimilation in wheat and *Arabidopsis*. *Science,* 328: 899-903.

Campbell, C.D. and Sage, R.F. 2006.Interaction between atmospheric CO_2 content and P nutrition on photosynthesis in white lupin. *Plant Cell Environ.*, 29: 844-853.

Cao, L., Bala, G., Caldeira, K.N., Nemani, R. and Ban-Weiss, G. 2010. Importance of carbon dioxide physiological forcing to future climate change. *Proc. National Acad. Sci. USA.*, 107: 9513-9518.

de Graaff, M. A., Schadt, C. W., Rula, K., Six, J., Schweitzer, J. A. and Classen, A. T. 2011. Elevated CO_2 and plant species diversity interact to slow root decomposition. *Soil Biol. Biochem.*, 43: 2347-2354.

de Graaff, M. A., van Groenigen, K. J., Six, J., Hungate, B. and van Kessel, C. 2006. Interactions between plant growth and soil nutrient cycling under elevated CO_2: a meta-analysis. *Glob. Change Biol.*, 12: 2077-2091.

Dijkstra, F.A., Pendall, E., Mosier, A.R., King, J.Y., Milchunas, D.G. and Morgan, J.A. 2008. Long-term enhancement of N availability and plant growth under elevated CO_2 in a semi-arid grassland. *Funct. Ecol.*, 22: 975–982.

Dijkstra, F.A., Blumenthal, D., Morgan J.A., Pendall, E., Carrillo, Y. and Follett, R.F. 2010. Contrasting effects of elevated $CO_2$and warming on nitrogen cycling in a semi-arid grassland. *New Phytol.*, 187: 426-437.

Fangmeier, A., De Temmerman, L., Mortensen, L., Kemp, K., Burke, J., Mitchell, R., van Oijen, M. and Weigel, H. J. 1999. Effects on nutrients and on grain quality in spring wheat crops grown under elevated CO_2 concentrations and stress conditions in the European, multiple-site experiment 'ESPACE-wheat'. *European J. Agron.*, 10: 215–229.

Garten, C.T., Iversen, C. M. and Norby, R.J. 2011. Litterfall^{15}N abundance indicates declining soil nitrogen availability in a free-air CO2 enrichment experiment. *Ecol.*, 92: 133-139.

Gedney, N., Cox, P.M., Betts, R.A., Boucher, O., Huntingford, C. and Stott, P.A. 2006. Detection of a direct carbon dioxide effect in continental river runoff records. *Nature*, 439: 835-838.

Gill, R.A., Polley, H.W., Johnson, H.B., Anderson, L.J., Maherali, H. and Jackson, R.B. 2002. Nonlinear grassland responses to past and future atmospheric CO_2. *Nature*, 417: 279- 82.

Hasan, R. 1996. Phosphorus status of soils in India. *Better Crops International*, 10: 4-5.

Hovenden, M.J., Newton, P.C.D., Carran, R.A., Theobald, P., Wills, K.E., Vander Schoor, J.K., Williams, A.L. and Osanai, Y. 2008. Warming prevents the elevated CO_2-induced reduction in available soil nitrogen in a temperate, perennial grassland. *Glob. Change Biol.*,14: 1018-1024.

Jarvis, A.J., Mansfield, T.A. and Davies, W.J. 1999. Stomatal behavior, photosynthesis and transpiration under rising CO_2. *Plant Cell Environ.*, 22: 639-648.

Kiehl, J. 2011.Lessons from Earth's Past. *Science*, 331: 158-159.

Kim, S.H., Sicher, R.C., Bae, H., Gitz, D.C., Baker, J.T., Timlin, D. and Reddy, V.R. 2006. Canopy photosynthesis, evapotranspiration, leaf N and transcription profiles of maize in response to CO_2 enrichment. *Glob. Change Biol.*, 12: 588-600.

Lagomarsino, A., Moscatelli, M.C., Hoosbeek, M.R., Angelis, P.D. and Grego, S. 2008. Assessment of soil nitrogen and phosphorous availability under elevated CO_2 and N-fertilization in a short rotation poplar plantation. *Plant Soil*, 308:131-147.

Leakey, A. D. B., Ainsworth, E. A., Bernacchi, C.J., Rogers, A., Long, S.P. and Ort, D.R.2009. Elevated CO_2 effects on plant carbon, nitrogen, and water relations; six important lessons from FACE. *J. Exp. Bot.*, 60: 2859-2876.

Lenka, N.K. and Lal, R. 2012. Soil related constraints to the CO_2 fertilization effect, *Critical Reviews in Plant Sciences*, 31: 342-357.

Luo, Y., Su, B., Currie, W.S., Dukes, J. S. Finzi, A., Hartwig, U., Hungate, B., McMurtrie, R.E., Oren, R., Parton, W.J., Pataki, D.E., Shaw, M.R., Zak, D.R. and Field, C.B. 2004. Progressive nitrogen limitation of ecosystem responses to rising atmospheric carbon dioxide. *Bioscience*, 54:731-739.

Markelz, R.J.C., Strellner, R. S. and Leakey, A.D.B. 2011. Impairment of C_4 photosynthesis is exacerbated by limiting N and ameliorated by elevated CO_2 in maize. *J. Exp. Bot.*, 62: 3235-3246.

Niklaus, P.A. and Korner, C. 2004. Synthesis of a six-year study of calcareous grassland responses to in situ CO_2 enrichment. *Ecological Monographs*, 74: 491-511.

Norby, R.J., Cotrufo, M.F., Ineson, P., O Neill, E.G. and Canadell, J.G. 2001. Elevated CO_2, litter chemistry, and decomposition: a synthesis. *Oecologia*, 127:153-165.

Norby, R.J. and Zak, D.R. 2011. Ecological lessons from free-air CO_2enrichment (FACE) Experiments, *Ann. Rev. Ecol. Evol. Syst.*, 42:181-203.

Oren, R., Ellsworth, D.S., Johnsen, K.H., Phillips, N., Ewers, B.E., Maier, C., Schafer, K.V.R., McCarthy, H., Hendrey, G., McNulty, S.G. and Katul, G.G. 2001. Soil fertility limits carbon sequestration by forest ecosystems in a CO_2-enriched atmosphere. *Nature*, 411:469-472.

Parsons, W. F. J., Lindroth, R. L. and Bockheim, J. G. 2004. Decomposition of *Betula papyrifera* leaf litter under the independent and interactive effects of elevated CO_2 and O_3. *Glob. Change Biol.*, 10: 1666-1677.

Reich, P.B., Hobbie, S.E., Lee, T.Ellsworth, D.S., West, J.B., Tilman, D., Knops, J.M.H., Naeem, S. and Trost, J. 2006a. Nitrogen limitation constrains sustainability of ecosystem response to CO_2. *Nature*, 440: 922-925.

Reich, P.B., Hungate, B.A. and Luo, Y. 2006b. Carbon-nitrogen interactions in terrestrial ecosystems in response to rising atmospheric carbon dioxide. *Ann. Rev. Ecol. Evol. Syst.*, 37: 611-36.

Stocklin, J., Schweizer, K. and Korner, C. 1998. Effects of elevated CO_2 and P addition on productivity and community composition of intact monoliths from calcareous grasslands. *Oecologia*, 116: 50-56.

Taub, D. R. and Wang, X. Z. 2008. Why are nitrogen concentrations in plant tissues lower under elevated CO_2? A critical examination of the hypotheses. *J. Integrative Plant Biol.*, 50:1365-1374.

Thompson, S., Govindasamy, B., Mirin, A., Caldeira, K., Delire, C., Milovich, J., Wickett, M. and Erickson, D.J. 2004. Quantifying the effects of CO_2fertilized vegetation on future global climate and carbon dynamics. *Geophys. Res. Lett.*, 31: L23211.

Tiwari, K. N. 2001. Phosphorous needs of Indian soils and crops. Better Crops International, 15: 6-10.

Torbert, H.A., Prior, S.A. and Rogers, H.H. 2001. Effect of elevated CO_2 and temperature on soil C and N cycling. **In:** *Sustainable Management of Soil Organic Matter.* pp. 309-315. Rees, R.M., Ball, B.C., Campbell, C.D., Watson, C.A., Eds., CAB International, Wallingford, UK.

Torbert, H.A., Prior, S.A., Rogers, H.H. and Runion, G.B. 2004. Elevated atmospheric CO_2 effects on N fertilization in grain sorghum and soybean. *Field Crops Res.*, 88: 47-57.

United States Department of Agriculture. 2003. Phosphorus retention potential map, Natural Resources Conservation Service, USDA, Washington.

Vance, C.P. 2001.Symbiotic N fixation and P acquisition - plant nutrition in a world of declining renewable resources. *Plant Physiol.*, 127: 390-397.

Yamakawa, Y., Saigusa, M., Okada, M. and Kobayashi, K. 2004.Nutrient uptake by rice and soil solution composition under atmospheric CO_2 enrichment. *Plant Soil,* 259: 367-372.

□□□

Chapter 23

Development of Drought Tolerant Single Cross Hybrids of Maize for Enhancing Productivity under Rainfed Areas

R. SAI KUMAR, JYOTI KAUL, RAMESH KUMAR, OM PARKASH BILAL AHMAD AND ANIL YADAV

INTRODUCTION

Significant yield losses in maize (*Zea mays* L.) from drought are expected to increase with global climatic change as temperature rise and rainfall distributions changes in key production area. Drought is one of the most important abiotic stress factor (Bruce *et al.*, 2002) which affects almost every aspect of plant growth (Sadras and Milroy, 1996; Aslam *et al.*, 2006). Drought or more generally, limited water availability is the main factor limiting crop production (Golbashy *et al.*, 2010). Drought is a permanent constraint to agricultural production in many developing countries, and an occasional cause of losses of agricultural products in developed ones (Ceccarelli and Grando, 1996).The best option for crop production, yield improvement and yield stability under drought stress conditions is to develop drought tolerant crop varieties. One of the main goals in breeding programme is selection of the best genotypes under drought stress conditions (Richards *et al.*, 2002). In maize, grain yield reduction caused by drought ranges from 10 to 76%, depending on the severity and stage of occurrence. Keeping in view the importance of genotypes for drought conditions the present experiment was planned to gather information on susceptible / tolerant Single Cross Hybrids of Maize so that tolerant hybrids could be recommended directly for cultivation in the target environment and further drought tolerant inbred lines can be derived by using these hybrids for continuous and gradual process of improvement for drought stress.

MATERIAL AND METHODS

The experimental material consists of ten diverse hybrids of maize *viz.*, three QPM hybrids (HQPM-1, HQPM-5 and HQPM-7), three extra and extra

Directorate of Maize Research, Pusa Campus, New Delhi -110 012, India early hybrids (Prakash, Vivek 17 and Vivek 21) and four medium to late maturity hybrids (HM-4, HM-9, Seed Tech 2324 and Bio-9681), grown in two sets (Control and drought) in randomized block design in three replications by following 60 x 25 cm spacing , at Research Farm Area of Directorate of Maize Research, New Delhi during *Kharif*, 2009 and 2010. Normal package of practice were followed for control whereas, in case of drought, the drought was imposed by withholding the irrigation at most critical stage i.e. flowering stage. The irrigation was withheld 15 days before flowering of the early hybrids and up to 15 days after flowering of the late hybrids. The rainout shelters (Figure 1) were used to avoid any precipitation in drought experiment. The data was recorded for 50% an thesis, 50% silking, anthesis-silking interval and grain yield under normal as well as in drought conditions .The data on yield was recorded at harvesting stage and grain yield was calculated at 15% moisture level and mean of both location was calculated for explanation.

Fig. 1: Rainout shelter to avoid precipitation in drought trial during *kharif*, 2009 & 2010.

RESULTS AND DISCUSSIONS

The considerable gain in performance under drought stress at flowering point to the reduction in anthesis-silking (or increased year growth rate under stress) which is the mechanism for imparting tolerance and in the present study ASI (Table 1) is more or less similar in both years of testing except one (HM-9 during 2010 ASI is 5 days) which is not desirable for stress environments. The shorter ASI is an indication of stress tolerance. Dow *et al.*, 1984 also reported similar results. This further supported by common relationship observed between crop growth rate and yield under both drought and normal environments (Andrade *et al.*, 2002), suggesting that a shorter ASI is indicative of general tolerance to reduced photosynthesis per plant at flowering from many causes (Edmeades *et al.*, 2000).

Table 1: Mean performance of Hybrids under Drought condition (*Kharif* 2009 & 2010) at Delhi

Hybrids	*Kharif* (2009)			*Kharif* (2010)			Yield (*Kharif*, 2009)		Yield (*Kharif*, 2010)	
	Days to 50% Anthesis	Days to 50% Silking	ASI	Days to 50% Anthesis	Days to 50% Silking	ASI	Drought	Control	Drought	Control
HQPM-1	55.33	58.33	3.00	53	55	2.00	4124	4605	3731	4247
HQPM-5	54.00	57.00	3.00	52	54	2.00	3439	4468	1196	2673
HQPM-7	51.00	54.00	3.00	51	53	2.00	3982	4882	2776	2852
HM-4	53.33	56.00	2.67	50	52	2.00	3714	4315	2368	2391
HM-9	53.33	56.33	3.00	51	56	5.00	4300	4420	2397	2837
Vivek 17	45.67	49.33	3.66	44	45	1.00	4260	4394	1777	1957
Vivek 21	46.00	49.33	3.33	46	49	3.00	3782	4273	2748	3641
Prakash	52.00	55.00	3.00	51	52	1.00	4094	4644	2730	2826
Seed Tech 2324	57.33	60.00	2.67	48	50	2.00	4056	4616	4279	5521
Bio 9681	57.00	59.00	2.00	53	56	3.00	4046	4480	3427	5242

Based on mean data of both seasons it was observed that early and extra early hybrids had managed to escape the drought period and performance of hybrids, *viz.*, '*Vivek 17*' (4260 kg ha^{-1})(drought) and 4394 kg ha^{-1} (normal) in 2009 and 1777 kg ha^{-1} (drought) and 1957 kg ha^{-1} (Normal) in 2010 and Prakash (4094 kg ha^{-1} (drought) and 4644 kg ha^{-1} (normal conditions) in 2009 and 2730 kg ha^{-1} (drought) and 2826 kg ha^{-1} (normal) in 2010 was almost at par with that under normal conditions in both years, however, quality protein maize hybrids performed fairly good in drought conditions in both years HQPM-1 (4124 kg ha^{-1} in 2009 and 3731 kg ha^{-1} in 2010); HQPM-7 (3982 kg ha^{-1} in 2009 and 2776 kg ha^{-1} in 2010). Medium late hybrid HM-9 has given exceptional yield (4300 kg ha^{-1}) under drought condition during 2009, whereas, Seed tech 2324 given highest yield (4279 kg ha^{-1}) under drought condition in 2010. Drought stress given at flowering stages has causes maximum yields reduction in some of the hybrids *i.e.*, HQPM-5 2776 kg ha^{-1} (normal) and 1196 kg ha^{-1} (drought) during 2010. Similar results were also observed by Shiri *et al.*, 2010; Golbashy *et al.*, 2010 and Khodarahmpou and Hamidi, 2011. Based on this study we can conclude that quality protein maize hybrid HQPM-1 and late maturing hybrids Seed Tech 2324 and Bio 9681 could be recommended for drought affected areas.

CONCLUSIONS

Considerable gain in performance under drought stress at flowering point to the reduction in Anthesis-silking which is the mechanism for imparting tolerance and in the present study ASI is more or less similar in both years of testing except one (HM-9 during 2010 ASI is 5 days) which is not desirable for stress environments. The shorter ASI is an indication of stress tolerance. Early and extra early hybrids had managed to escape the drought period and performance of hybrids was almost at par with that under normal conditions. Drought stress given at flowering stages has causes maximum yields reduction in some of the hybrids. Based on this study we can conclude that quality protein maize hybrid HQPM-1 and late maturing hybrids Seed Tech 2324 and Bio 9681 could be recommended for drought affected areas.

REFERENCES

Andrade, F.H., Echarte, L., Rizalli, R., Della Maggiora, A. and Casanovas, M. 2002. Kernel number prediction in maize under nitrogen or water stress. *Crop Sci.*, 42:1173-1179.

Aslam, M., Khan, I.A. Saleem, M. and Ali. Z. 2006. Assessment of water stress tolerance in different maize accessions at germination and early growth stage. *Pak. J. Bot.*, 38(5): 1571-1579.

Bruce, B.W., Gregory, O.E. and Barker, T. C. 2002. Molecular and physiological approaches to maize improvement for drought tolerance. *J. Exp. Bot.*, 53:13-25.

Ceccarelli , S. and Grando, S. 1996. Drought as a challenge for plant breeder. *Plant Growth Regul.*, 20: 149-155.

Dow, E.W., Daynard, T.B., Muldoon, J.F., Major,D.J. and Thurtell, G.W. 1984. Resistance to drought and density stress in Canadian and European maize (*Zea mays* L.) hybrids. *Can. J. Plant Sci.*, 64: 575-585.

Edmeades , G.O., Bolanos,J., Elings,A., Ribaut, J.M., Banziger, M. and Westgate, M.E. 2000. The role and regulation of the anthesis-silking interval in maize.In: Westgate,M.E. Boote. K.J.(Eds.),Physiology and modeling kernel set in maize. CSSA, Madison, WI, CSSA Special publication No. 29,pp43-73

Golbashy, M., Khavari, S.K., Ebrahimi, M. and Choucan,R. 2010. Response of corn hybrids to limited irrigation.11th Iranian Crop Science Congress, Tehran, 24-26 July, 2010.University of Shahid Beheshti Persian. 218p.

Khodarahmpou, Z. and Hamidi, J. 2011. Evaluation of drought tolerance in different growth stages of maize (Zea mays L.) inbred lines using tolerance indices. *African J. Biotechnology.*, 10 (62): 13482-13490.

Richards, R.A., Rebetzke, G.J., Condon, A.G. and Van Herwaarden, A.F. 2002. Breeding opportunities for increasing the efficiency of water use and crop yield in temperate cereals. *Crop Sci.*, 42: 111-121.

Sadras, V.O. and Milroy, S. P. 1996. Soil- water thresholds for the responses of leaf expansion and gas exchange: A review. *Field Crops Res.*, 47: 253-266.

Shiri, M., Choukan , R. and Aliyev, R.T. 2010. Drought tolerant evaluation of maize hybrids using biplot method. Trends Appl. *Sci.Res.*, 5(2):129-137.

□□□

Chapter 24

Drought Management by Small Holder Farmer in *Angul* Distric of Odisha

B. SATPATHY AND M. PRUSTY

INTRODUCTION

Droughts and famines have received attention in the recent years. About one third of the country is either drought prone or frequently under the vulnerability of occurrence. The strategies *viz.*, relief to drought hit areas, DPAP and popularization of dry farming worked well. Drought is a recurrent phenomenon and an important constraint to rainfed rice production in Angul district of Odisha. At least 46 thousand ha of rice area (43 % of the total rice area) in the district are subject to drought of different intensities. Rao *et al.* (1988) in their pioneering study on unstable agriculture and droughts found that agriculture in drought prone areas has been highly unstable. They also observed that production instability varied with crops. There is some evidence to show that unstable agricultural incomes tend to raise the incidence of poverty and also vulnerability of poor (Desai *et al.*, 1979). Improved rice technologies that help reduce losses to drought can play an important role in long-term drought mitigation strategy. Important scientific progress is being made in understanding the physiological mechanisms that impart tolerance of drought (Blum, 2005; Lafitte *et al.*, 2006).

Watershed-based approaches that are implemented in drought-prone areas of India provide opportunities for achieving long-term drought proofing by improving overall moisture retention within the watersheds (Rao, 2000). Keeping in mind the above a study was undertaken with the general objective that to understand the impact of drought on household vulnerability in Angul district. The specific objectives are to analyze impact of drought on rural households, to analyze households' adaptation strategy for drought, to

Krishi Vigyan Kendra, Angul, Orissa University of Agriculture and Technology, Bhubaneswar, Odisha, India examine strategy for drought management and to suggest suitable contingent measures for reducing vulnerability.

MATERIALS AND METHODS

Study area : Angul lies between 20° 31' N & 21° 40' N latitude and 84° 15' E & 85° 23' E longitude. The total geographical area of Angul is 6232 sq.kms, thus it is the 11th largest district among 30 districts of Orissa. The relevant secondary data has been collected from 3 blocks of the district selected purposively (Figure 1).

Fig. 1: Map of Angul District where the study has been conducted

***Selection criteria of village*:** Three villages from each block have been selected purposively as per diversification and frequency of drought.

***Selection of sample household* :** Stratified sample selection for household was used to determine the number of household selected from each village. Primary data collection was done through well designed structured questionnaire. Sample information regarding the household is given in Table 1.

Table 1: Sample information regarding selected household

District	Blocks	Villages	Total household	Sample household
Angul	Angul	Shyamsunderpur	125	50
	Chendipada	Handiguda	120	45
	Banarpal	Kulei	85	40

RESULTS AND DISCUSSION

Concept of drought: Droughts are mainly of three types based on their mode of action and target:

Meterological drought: Based on degree of dryness and duration of dry period (when precipitation is less than 25% to its normal.

Agricultural drought: When inadequate soil moisture produces acute crop stress and affects productivity.

Socio-economic drought: The aggregation of the entire above drought when precipitation is not adequate to meet the need of human activities.

In the study area the drought has a direct impact on the area, human and livestock population. In the drought years all these three components decline drastically in comparison to the normal years (Figure 2).

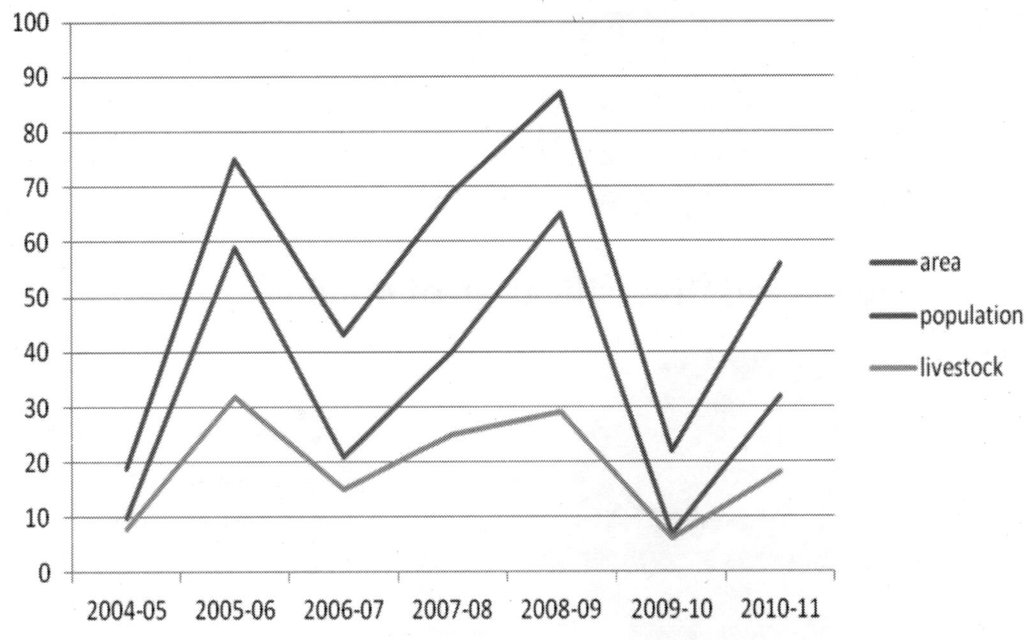

Fig. 2: Area, population and livestock affected by drought in Angul

Livelihood diversification of the villages : During the drought periods the livelihood diversification used to change drastically as compare to the normal years. These changes are the result of the coping strategy adopted by the peoples. Different villages adopt different coping mechanism based on their resources like land, soil, livestock, conservation measures, economy etc. the livelihood diversification of three different villages in the study area are presented in Figure 3.

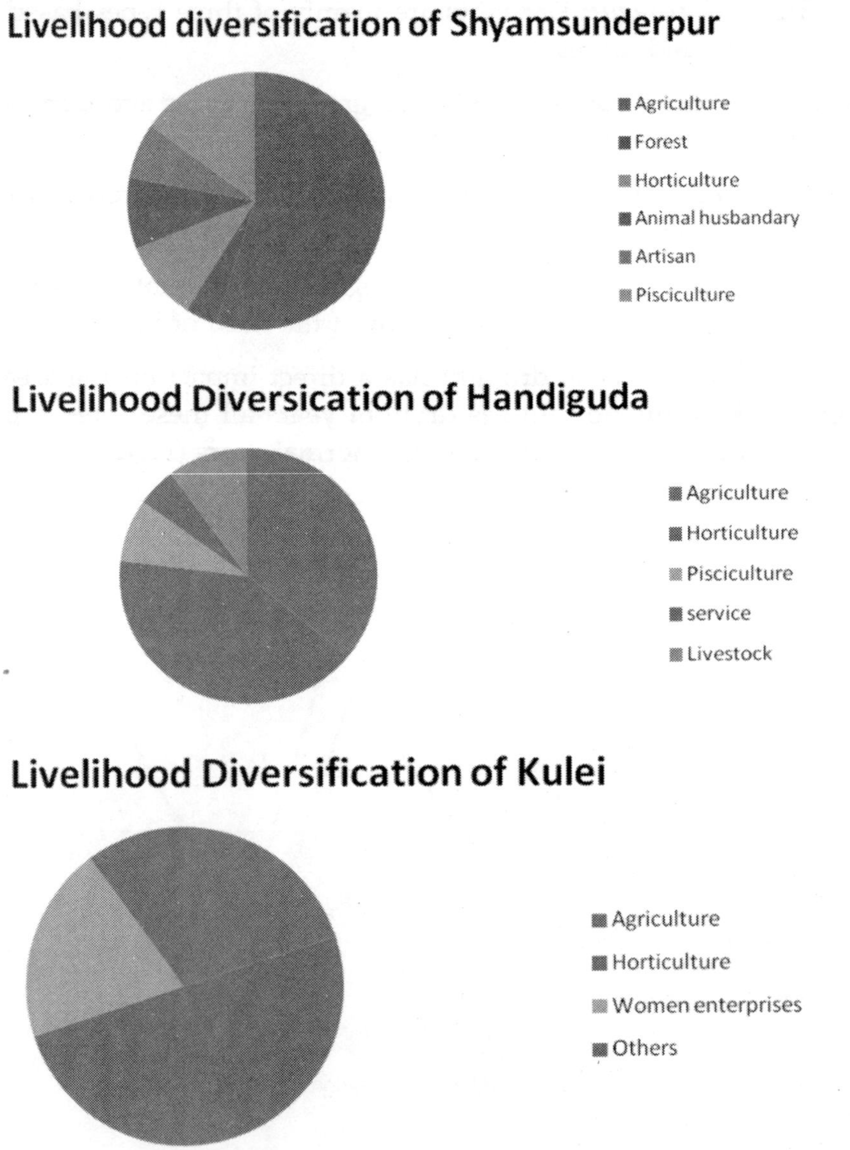

Fig. 3: Livelihood diversification of the villages selected in the Angul district

Figures 3 (a,b,c) indicate occupational diversification of the village. Horticulture takes the centre-stage in the villages. Women based enterprises *viz.,* preservation of fruits and vegetables and mushrooms are into focus. Rural artisans and service household families show diversification. This represents typically arid and semiarid villages.

Household level impact of drought : Table 2 indicate that consumption behaviour of farming households show great variation ranging from reducing quantity (74%) to changing food pattern (46%). During drought years the frequency of taking three meals per day increases from 51% to 77% because of taking more non-conventional feeding stuff and also due to insecurity in life.

Table 2: Household level impact of drought in terms of consumption behaviour of different households in the study area

Consumption beahaviour	Normal year	Drought year
Av. No. of meals per day	2.5	1.8
% of household reducing meals per day during drought year	-	75
Frequency of meals per day (%)	2 meals-49%	
3 meals-51%	2 meals -33%	
3 meals -77%		
% of household reducing quantity of food per meal during drought	-	74
% of household consuming unusual food in drought years not normally eaten	-	46

Outmigration in search of job : The drought prone areas of Angul have traditionally been an area of push factor for outmigration in search of employment and livelihood. More than 21% of total household surveyed reported at least one family member outmigrating in search of livelihood to sustain life.

A significant variation has been observed in the percentage of outmigrating households in drought region. Both the villages V1 and V3 have reported exceptionally high per cent. The per cent migration in Handiguda is only 7%. It is noteworthy to mention that V2 have shown economic diversification and in a way strength to fight natural calamities.

Drought and its impact on production : Drought has tremendous impact on agricultural production. In the study area it influenced on harvest loss of agricultural products (Table 3), total and partial crop failure (Table 4) etc. these impacts vary with different villages based on their coping strategies and strength and resource base.

Table 3: Harvest loss (%) of agricultural produce

Block	Village	0-25		25-50		>50		NR		Total	
		f	%	f	%	f	%	f	%	f	%
Angul	Shyamsunderpur	8	16	11	22	22	44	9	18	50	100
Chendipada	Handiguda	14	31.11	13	28.88	8	17	10	22.22	45	100
Banarpal	Kulei	24	60	7	17.5	5	12.5	4	10	40	100

Table 4 reveals that village one report the maximum loss followed by V2 and then V3.Correspondingly t was also found that V3 has adopted technological interventions which has subsequently resulted in higher yield and coping strength.

Table 4: Household reported total failure of crops in drought area

Villages	Yes	%	No	%	NR	%	Total	%
V1	31	62	16	32	3	6	50	100
V2	9	20	20	44.44	16	35.55	45	100
V3	14	35	26	65	0	0	40	100

Agricultural coping strategy of farmers during drought season : During drought seasons farmers take various measures like growing of drought resistant crops, mix crop, partial mechanization for raising crops and mulching to mitigate the drought effect (Table 5). There are various demonstrations for drought management in the villages (Table 6) to nullify the drought effect.

Table 5: Response on coping strategy of farmers during drought season (Agriculture)

Village	Grow DR crops		Mix crops		Partial Mechanization		Mulching		Total	
	f	%	f	%	f	%	f	%	f	%
V1	6	12	28	56	2	4	14	28	50	100
V2	9	20	21	46.67	5	11.11	10	22.22	45	100
V3	1	2.5	19	47.5	8	20	12	30	40	100

Table 6: Significant Demonstration for drought management

S.No.	Interventions	Farmers	Area (ha)	Av. yield (qha^{-1})	Av. net return	B:C Ratio
1.	Crop Diversification with Arhar + groundnut intercropping	10	2	13.5 17.5	20, 200	2.01
2.	INM in Maize	10	1	29.2	41000	2.5
3.	INM in Sugarcane	10	1	113.8	49,8000	1.78
4.	ICM in groundnut	12	5	19.6	52,400	3.0
5.	ICM in Mung	11	5	6.8	26,600	2.23
6.	ICM in Arhar	15	5	16.6	18,500	2.20
7.	Water management through Drip irrigation in banana	5	2	480	100,000	2.08
8.	IPM in Paddy	10	1	22.5	11,200	1.73
9.	IPM in Sugarcane	10	1	113.8	49,800	1.78
10.	Varietal substitution (short duration Udaygiri)	10	1	17.0	9,000	1.53

***Socioeconomic adaptation to cope up with drought*:** Farmers take various socioeconomic adaptation measures like food reduction, migration to cities, crop diversification and loan borrowing to get rid off the ill effect of drought (Table 7). Still the farmers' income show vary irregularity in agricultural income during drought years (Table 8).

Table 7: Socioeconomic adaptation

Village	Reduction of food		Migration		Sale of assets		Diversification		Borrowing loan	
V1	21	42	23	46	0	0	2	4	4	8
V2	16	35.55	6	13.33	20	44.44	2	4.44	1	2.2
V3	7	17.5	0	0	12	30	4	10	17	42.5

Table 8: Household reported irregularity in agricultural income in drought & normal years

Land holding size (acre)	Reported irregularity in agril. income		Grand total	Average income (₹ yr^{-1})	
	Yes	No	Household No.	Normal yr.	Drought yr.
0-2	36	14	50	15,000	8500
2-4	29	16	45	28,700	12,450
> 4	30	10	40	40,000	21,497
Grand total	95	40	135	83,700	42,447

Contingent crop planning for angul district

- Growing moisture tolerant upland rice varieties Vandana (90 days) and Mandakini (100 days)
- Following partial mechanization to utilize natural resources effectively
- Adopting intercropping system preferably arhar, maize, groundnut and rice based.
- Adopting Integrated farming system
- Adopting mulching to reduce moisture stress
- Vegetables like tomato, cauliflower, radish, brinjal, cowpea can be grown
- The ideal pre rabi crops with residual moisture conditions are black gram, castor, and sesamum
- Provide protective irrigation
- Maize varieties like Ganga-11(100 days) and Navjot (90 days) can be grown.

CONCLUSIONS

Drought has manifold effect to control the livelihood of farmers and the farmers adapt various coping mechanisms to get rid off the ill effect of drought. Drought periods have drastically increased the proportion of food insecure households and lengthened the duration of food insecurity in the area. Since, with climate change, drought periods are predicted to become more frequent in this region in the future, the problem of food insecurity is likely to become even more severe. Ex-ante adaptation strategies are widely practised in the area for drought management.

REFERENCES

Blum, A. 2005. Drought resistance, water-use efficiency, and yield potential: are they compatible, dissonant, or mutually exclusive? *Aust. J. Agric. Res.*, 56(11):1159-1168.

Desai, G.M., Singh, G. and Saha, D.C. 1979. *Impact of scarcity on farm economy and significance of relief operations*, CMA Monograph No. 84. Ahmedabad,IIM.

Rao, H. Ray and Subbarao. 1988. '*Sources of change in crop output*', a project report for NCS unit, Institute of Economic Growth, Delhi.

Lafitte, H.R., Yongsheng, G., Yan, S. and Li, Z.K. 2006. Whole plant responses, key processes, and adaptation to drought stress: the case of rice. *J. Exp. Bot.*, 58(2): 169-175.

Rao, C.H.H. 2000. Watershed development in India: recent experiences and emerging issues. *Econ. Polit. Weekly*, 4 November 2004.

□□□

Section - III

Chapter 25

Water Resources and Multiple Use of Water for Sustainable Production in the Eastern and NEH Region of India

ASHWANI KUMAR

INTRODUCTION

Water is the most crucial input for agricultural production. Globally, agriculture accounts for more than 80% of all freshwater used by humans, most of that is for crop production (Morison *et al.*, 2008). Currently most of the water used to grow crops is derived from rainfed soil moisture, with non-irrigated agriculture accounting for about 60% of production in developing countries. Though irrigation provides only 10% of agricultural water use and covers just around 20% of the crop land, it can vastly increase crop yields, improve food security and contribute about 40% of total food production since productivity of irrigated land is almost three times higher than that of rainfed land. The Food and Agriculture Organization has predicted a net expansion of irrigated land of about 45 M ha in 93 developing countries (for a total of 242 M ha in 2030) and projected that water withdrawals by the agriculture sector will increase by about 14% during 2000-2030 to meet food demand (FAO, 2006). Agriculture sector in India has been and is likely to remain the major consumer of water but the share of water allocated to irrigation is likely to decrease by 10 - 15 per cent in the next two decades. Current use efficiency or productivity of irrigation water is so low that most, if not all, of our future water needs could be met by increased productivity or efficiency alone, without development of additional water resources. Improving water use efficiency by 40% on rainfed and irrigated lands would be required to counter balance the need for additional withdrawals for irrigation over the next 25 years to meet the additional demand for food (Singh *et al.*, 2010). Growing more crops per drop of water use is the key to mitigating the water crisis, and this is a big challenge to many countries. Vagaries of monsoon and declining water able due to over

Director, Directorate of Water Management (ICAR), Bhubaneswar, Odisha-751 023 exploitation have resulted in shortage of fresh water supplies for agricultural use, which calls for an efficient use of this resource. Strategies for efficient management of water for agricultural use involves conservation of water, integrated water use, optimal allocation of water and enhancing water use efficiency by crops.

Historically, water has played a vital role for the growth of predominantly agrarian economy of India and thus, occupies a special place among other natural resources. It is fundamental for sustaining a high quality of life and economic and social development. However, its growing scarcity with wide spatial and temporal variability has raised several sustainability issues and attracted national and international concerns on the efficient management of this vital natural resource. Although, average annual rainfall (120 cm) in India, the main source of water, is higher than the World average (99 cm), its distribution is temporarily and spatially uneven. Entire rainfall in India occurs within the narrow span of few days and varies from 250 mm year^{-1} in western regions to 11,690 mm year^{-1} in Mousinram, Cherapunji of north-eastern (NEH) region of India. Declining per capita availability of water, increasing inter-sectorial competition, changing food habits towards high value agricultural crops, adverse climatic variability and existing negative externalities such as water logging and salinity due to excess surface irrigation necessitate reorientation of water management through institutional restructuring, technological interventions and policy changes.

Irrigation Scenario in Eastern and NEH Region of India

Eastern region constitutes about 22 % of the total geographical area and about 34 % of the total population of the country. The climate of the region is tropical hot and humid/sub-humid with high rainfall. The average rainfall in this region varies from 1091 to 2477 mm with a regional average of 1526 mm, which is sufficient and substantial for growing any crop. Assam gets mean annual rainfall of 2477 mm which is the highest in the region, followed by West Bengal (1750 mm), Odisha (1451 mm), Chhattisgarh (1430 mm), Jharkhand (1277 mm), Bihar (1204 mm) and Eastern UP (1091 mm). Bulk of the rain (about 80%) occurs during the monsoon period. It has erratic temporal and spatial distribution with considerable year-to-year variation. High intensity and heavy rains cause frequent floods in the eastern region, especially in plains, and coastal deltaic area. The coastal areas are also vulnerable to seawater intrusion and cyclones.

Table 1: Irrigation scenario of Eastern and NEH region of India (Up to 10th FYP)

Particular	Eastern Region#	NE region*	India
Ultimate Irrigation Potential (M ha)	33.66	1.45	139.89
Irrigation Potential Created (M ha)	21.83	0.64	123.34
Major & Medium Projects	9.45	0.12	42.28
Minor Surface	3.82	0.50	14.12
Minor Ground	8.56	0.02	66.87
Minor (surface+ground)	12.38	0.52	80.99
Irrigation Potential Utilized (M ha)	14.30	0.41	91.12
Major & Medium Projects	7.32	0.10	34.38
Minor Surface	2.00	0.31	8.34
Minor Ground	4.98	0.01	48.36
Minor (surface+ground)	6.98	0.32	56.70
Utilization (%)	65.52	64.53	73.88
Major & Medium Projects	77.51	78.55	81.32
Minor Surface	52.46	60.90	59.09
Minor Ground	58.13	70.39	72.32
Minor (surface+ground)	56.38	61.18	70.02
Net Annual groundwater Availability (BCM)	120.00	6.09	399.00
Annual Groundwater Draft (BCM)	36.00	0.19	231.00
Irrigation (%)	86.11	42.63	92.20
Domestic and industrial uses (%)	13.88	57.62	7.79
Groundwater Availability for future irrigation (BCM)	82.00	5.62	161.00
Stage of Groundwater Development (%)	27.00	3.19	58.00

comprises of West Bengal, Odisha, Bihar, Chattisgarh, Jharkhand, Assam
*comprises of Arunachal Pradesh, Manipur, Meghalaya, Mizorum, Nagaland, Sikkim, Tripura

Ultimate Irrigation Potential (UIP) in the Eastern region has been assessed as 33.66 ha as compared to 139 M ha in the country (Table 1). Up to the end of tenth Five Year Plan (FYP), 21.8 M ha of irrigation potential has been created in the region. However, utilization of the created irrigation potential (65.52 %) is less than the country's average of 73.88 %. Groundwater development in the Eastern region is poor (27 %) as compared to more than 100 % in many parts of the country. About 86 % of the total groundwater is extracted for irrigation purpose in the region. In NEH region, the wettest region of the country, rainfall is the only source of irrigation. The region constitutes only about one per cent of the country's irrigation potential due to hilly and undulating terrain constraining the development of irrigation infrastructure.

Similarly, groundwater development is only 3.19% and primarily (57.62 %) extracted for domestic and industrial uses. Thus, tapping, conserving and utilization rainwater in efficient way through multiple uses is suitable water management strategy in NEH region of the country.

Emerging Issues and Water Management Strategies in Eastern and NEH Region

Issues	Strategy
• Erratic distribution of rainfall in monsoon period creates drought like situation though total rainfall is very high.	• Bringing more areas under irrigation through the planned development of surface and groundwater resources.
• Non uniforms distribution of water during *kharif* also creates water logging conditions, results unfavourable edaphic environment for plant growth.	• In the hills and plateau regions, adoption of appropriate watershed management technology along with the storage of water in small reservoirs
• Inadequate and unreliable surface irrigation system.	• Need to significantly reduce the use of irrigation water per unit area by radically altering the present on-farm water management practices.
• Meager utilization of groundwater resource.	• Scientific management of water resources to provide reasonably assured good quality irrigation water during dry season and drainage of excess rainwater during monsoon season.
• Inadequate implementation of soil and water conservation measures in hilly tracts.	
• Use of low level of fertilizer inputs.	
• Lack of mechanization and low energy inputs.	
• Non-availability of adequate quantity of good quality seeds of high-yielding varieties of crops.	
• Lack of proper marketing infrastructure for agricultural produce	

Concept of Multiple Use of Water

In the light of low water productivity and large scale inefficiency at different levels of water delivery system, multiple-use of water is being discussed at large scale to improve water productivity and promote the welfare of the society. The concept of multiple-use services (MUS) can be understood as a participatory, integrated and poverty-reduction focused approach, which takes people's multiple water needs as a starting point for providing integrated services, moving beyond the conventional sectorial barriers of the domestic and productive sectors (Kar and Kumar, 2009). The main objective of MUS approach is to meet the full range of people's water needs. These include domestic uses (drinking, cooking, washing, cleaning, sanitation and hygiene), irrigation, livelihood and small enterprises. Multiple use approach involves (1) assessing the range of water needs in collaboration with end users, (2) examining the water resources available-from rainwater to waste water to piped systems, and (3) matching water supplies to needs based on quantity, quality and reliability required for various purposes.

Benefits of Multiple Use of Water Approach

- Generates more income benefits (health, nutrition, time savings, food security and social empowerment),
- Decrease vulnerability by allowing more diversified livelihood strategies,
- More effectively reduce poverty by simultaneously addressing multiple dimensions of poverty,
- Increase sustainability of services: multiple-use services generate enough income to cover on-going operation, maintenance and replacement costs, and, because they better meet the water needs of communities, conflict over water and damage to infrastructure caused by 'illegal' or unplanned uses is decreased and community investment is increased, and
- Improves gender equity: improved water availability also promotes gender equity, as poor women are primarily responsible for fetching water.

Ways to Adopt Multiple Use of Water

Irrigation systems can be adopted by

- Releasing water for household uses and bathing
- Building or reviving community domestic-supply reservoirs
- Building steps in canal banks for laundry and bathing
- Adding pipes, canals and taps to bring water into villages/houses
- Promoting low-cost, point-of-use treatment for drinking water
- Sinking shallow wells to tap cleaner 'seepage' water
- Adding access and crossing points to canals for cattle
- Maintaining flows to preserve fish populations
- Building fish-friendly structures in sluices and canals

Domestic systems can be adopted by

- Increasing pipe diameters/water discharge to allow productive activities
- Providing water without interruptions
- Adding cattle troughs to supply points
- Adding storage tanks
- Adding micro-irrigation systems
- Using different water sources depending on quality needs
- Promoting reuse of household "grey" (waste) water

Use of other water sources can be maximized by

- Working with the private sector to promote the use of affordable pumps and drip irrigation kits (also applies to domestic systems)
- Promoting community/rooftop rainwater harvesting
- Enlarging wells
- Developing gravity-flow systems to pipe in stream-water
- Promoting credit and access to other inputs to enable people to make use of productive water supplies (applies to all)

Technological Interventions by Directorate of Water Management (ICAR) for Better Water Management

Directorate of Water Management (ICAR) has developed many promising technologies for (i) management and conservation of on-farm water resources for sustainable agricultural production, (ii) significant reduction in the use of irrigation water per unit irrigated area, (iii) excess water management in agricultural lands, (iv) development of sustainable cropping systems in relation to the availability of water, (v) devising multiple uses of water to enhance water productivity, (vi) reuse of poor quality groundwater and industrial and municipal waste waters, and (vii) avoiding the contaminations and further degradation of soil and water resources. These technologies have been tested in the farmers' field and being demonstrated through the programmes such as Farmers' Participatory Action Research Programme (FPARP) and Scaling-up of Water Productivity in Agriculture for Livelihoods (SWPA) in different locations. Some of the most promising and bankable technologies to enhance the water productivity in the eastern region are given below:

1. Crop Diversification in Drought Prone Areas

Application: Crop diversification by adopting maize, groundnut-pigeonpea, sole groundnut and sole pigeon pea can be practiced in rainfed upland conditions.

Input needed: Seeds of rice substituted crops (maize, groundnut, black gram, cowpea, green gram), fertilizers and necessary agro-inputs

Output capacity: Higher rice equivalent yield per annum was obtained through maize cob (8125 kg ha^{-1}), groundnut + pigeon pea (5550 kg ha^{-1}), sole groundnut (5640 kg ha^{-1}), sole pigeon pea (5550 kg ha^{-1}).

Specific benefits and impact: Rice equivalent yield of 7.5 t ha^{-1} with average net return of Rs. 15,000 per ha per annum.

Unit cost: Cost of cultivation per unit area is Rs 10,000-14,000 ha^{-1}

2. Raised and Sunken Bed

Application: Crop diversification in medium and low land of eastern India.

Description: The land is converted into alternate sunken and raised beds (1:1) each of 30 m length and 5 m width. Different vegetable crops of local importance may be grown on the raised beds. Sunken beds are used for growing low land rice or other aquatic crops like Colocasia. Fish spawn can also be raised up to fingerling stage in the sunken beds together with rice. Singh *et al.* (2005) reported a 600% enhancement in the use efficiency of water by practicing this technology at small and marginal farms of Khurda district, Orissa (0.36 kgm^{-3} for rice cropping, and 2.62 kgm^{-3} for diversified cropping).

Input needed: Land Area (1 ha), crop input parameters and fish input parameters and 700 man days.

Output capacity: The adoption of the technology increased *kharif* paddy and pointed guard yield from 4.2 t ha^{-1} to 5.2 t ha^{-1} and 4.24 t ha^{-1} to 4.74 t ha^{-1} , respectively. In addition, fish yield of 1.0 t ha^{-1}.

Specific benefits and impact: Additional income per ha per year: Rs. 60,500/-.

Unit cost: Rs. 45,000ha^{-1}.

3. Sub Surface Water Harvesting Structure (SSWHS)

Application: Coastal waterlogged areas

Description: Fresh water floats above the saline water below ground in coastal water logged areas and could be tapped through subsurface water harvesting structure to meet the rabi crop irrigation demand as well as pisciculture. To extract water from these structures, pump up to 2 HP is suitable to avoid saline water ingression in to fresh water layer. The depth of structure should be restricted with in sandy zone below ground up to 5 meter.

Input needed: For 0.1 ha with 4 m depth will create a structure of 4000m^3 involving 550 man days.

Output capacity: The average water productivity of SSWHS involving pisciulture and *rabi* vegetables is Rs. 36 per m^3

Specific benefits and impact: It results in higher area of irrigation and higher cropping intensity and crop productivity. The average benefit: cost ratio of SSWHS was 1.55 in the first year of construction itself. The participatory approach of implementing SSWHS improves the financial status of the several poor farmers living below poverty line in coastal waterlogged areas and also gives better employment opportunity.

Unit cost: The average unit cost for SSWHS construction is Rs. 14 m^{-3}.

4. Micro Level Water Resource Development through Tank Cum Well System

Application: For rainwater harvesting and utilization

Description: The tank cum well system technology along the drainage line in a watershed is recommended for plateau areas having slope of 2 to 5%. The site for the technology should be selected in such a way that the area should have a well defined valley where the runoff flows either as overland flow or channel flow. The well is constructed about 100 to 300 m downstream of the tank to tap the water that is lost by seepage from the tank. The schematic diagram of tank cum well system and multi tank and multi well system in a watershed are shown in Figure 1 and 2, respectively.

Input needed: A set of 15 tanks and wells for a catchment area of 500 ha to irrigate 60 ha area.

Output Capacity: Rs. 30,000/- extra gross income $year^{-1}$ with additional employment generation of 115 man days ha^{-1}.

Specific benefits and Impact: Increase in cropping intensity to 166%.

Unit cost: Rs. 55,000 ha $^{-1}$of net command area

5. Pond Based Farming System for Deep Waterlogged Areas

Application: Deep waterlogged areas (>1m surface water logging)

Description: Due to poor drainage, saucer shaped topography and high monsoon rainfall, some parts of east coast of India remain waterlogged (> 1m surface water logging) and unproductive. To stabilize and enhance net income from such waterlogged ecosystem, pond based farming technology (deep water rice in *kharif* + salt tolerant vegetables like watermelon, ladies finger, spinach, chili in winter + on-dyke vegetables-fruits + fish inside pond) was conceptualized and implemented in representative deep waterlogged areas (1-2.5 m water depth) of coastal Orissa, India.

Input needed: Land (1 ha), required amount of crop and fish inputs

Output capacity: Additional income Rs. 25,000ha^{-1}annum^{-1}, Water Productivity: Rs. 7.2 m^{-3}

Specific benefits and impact: Increased cropping intensity by 200%.

Unit cost: Rs 60,000 ha $^{-1}$.

6. In-situ Rainwater Conservation

Application/ Use: *In-situ* rainwater conservation for multiple use.

Description/ Features: In 8% of the total rice field, small dugout ponds of 2.5 m depth and 1:1 side slope beneficial at downstream. This pond is used for short-duration aquaculture during monsoon and its embankment is used for growing horticultural crops. The conserved rainwater in the pond is used for giving supplemental irrigations to *Kharif* paddy and irrigating *rabi* crops.

Input needed: Size of the pond (800 m^2 (8% of the 1 ha area)), *total cost of the work* (Rs. 66,670/-), *total man days input* (900).

Output Capacity: The *kharif* paddy yield increased from 1.8 tha^{-1} to 4.9 t ha^{-1} with fish yield of 1.4 t ha^{-1}.

Specific Benefit and impact: The cropping intensity increased up to 200%.

Unit Cost: The construction cost of the system is about Rs. 67,000 ha^{-1}.

7. Farm Pond Based Agricultural Diversification Model for Rainfed Areas

Application: In rainfed medium and lowland

Description: Rainwater harvesting system was designed and agricultural diversification model (on-dyke horticulture, fisheries, cultivation of diversifies field crops, short term fruits like papaya, banana, floriculture like marigold, tube rose *etc.*) with harvested rainwater was developed for small and marginal farmers through multiple use of water.

Input needed: Land, Seeds of field crops, vegetables and short duration fruits (papaya, banana, drumstick), fertilizers, necessary agro-inputs, fish seed.

Output capacity: Additional Income: Rs. 25,000-30,000 ha^{-1}

Specific benefits and impact: The same model was implemented in Bahasuni watershed of Dhenkanal, Orissa and cropping intensity increased upto 200%.

Due to harvesting of spring and rainwater water, irrigated areas of two villages of the watershed increased from 3.2 ha to 26.5 ha, where 55 tribal families were benefited. The detailed cost estimates, investment, man day's generation. Production potential, potential gross income generation per year, potential man days generation due to asset created have been computed and technology has been included under 'National Rural Employment Guarantee Act (NREGA) for implementing in watersheds of eastern Indian states.

Unit cost: Rs. 40 m^{-3}

8. Low Cost Management of Acid Soils for Higher Productivity and Water Use Efficiency

Application: Rainfed acid soils

Description: Soil acidity is one of the limiting factors which affects crop production adversely in many parts of the India (nearly 40% soils in India are acidic). It is necessary to ameliorate the acid soils with liming materials along with suitable cropping system and water management to enhance the productivity and water use efficiency of crops in acid soil. Liming materials may be like limestone, dolomite *etc.*, which is costlier and may not be affordable by farmers. But to adopt by farmers, it should be less expensive and available within easy reach of farmers' beside suitability. Hence, low cost liming material like paper mill sludge was used to ameliorate acid soils. This industrial byproduct of paper mill will not only raise agricultural production with low cost but also reduce environmental pollution in the surrounding areas of the factory.

Input needed: Low cost liming materials like paper mill sludge, seeds of rice substituted crops (maize, groundnut, black gram, cowpea, green gram), fertilizers and necessary agro-inputs

Output capacity: Additional income: Rs. 20,000ha^{-1}, Water use efficiency increased by 36-60%.

Specific benefits and impact: Paper mill sludge could ameliorate acidic soils and increased the yield of different crops by 34-68%. The Orissa Government adopted the technology and planned to ameliorate 2.4 lakh hectares acidic land in 2008 by supplying Paper Mill Sludge. From the year 2008-09, the Orissa Government started supply of paper mill sludge at subsidized rate (Rs 10/- for 50 kg bag). During *rabi* 2008-09, the IFFCO, Bhubaneswar adopted the technology of paper mill sludge application in 120 acres areas in Balasore, Orissa for growing groundnut crops by involving 223 farmers.

Unit cost: Rs 12,000 to 15,000 ha^{-1} depending on the soil pH.

9. Deep Water Rice Production Technology for Seasonal Deep Waterlogged Areas

Application/use: Deep waterlogged areas (>1m surface water logging)

Description / features: In some parts of the coastal areas, prolonged water logging (> 1m) during rainy season reduces tillering and growth of normal rice varieties. Sometimes flash flood inundates the standing crop at any stage of growth for 8-10 days at a stretch, resulting in heavy mortality. The crop is damaged completely if this situation occurs at early vegetative stage. So proper establishment of crop before onset of flooding and adoption of deep water logging tolerant rice varieties are of paramount importance to realize net return from the crop. Hence, deep water rice production technology for seasonal deep waterlogged areas was developed.

Input needed: Seeds of deep water rice varieties (*'Hangseswari'*, *'Saraswati'*, *'Ambika'*, *'Sabita'*), fertilizers and necessary agro-inputs

Output capacity: Improved deep water rice cultivar *'Hangseswari'* produced 2.4-2.5 t ha^{-1} yield in *kharif* season in deep waterlogged situation

Specific benefits and impact: Higher and stable income from deep waterlogged areas which represent 3.9 M ha in eastern India. Earlier the seed of improved deep water rice cultivar *'Hangseswari'* was not in State Seed Chain. Now based on recommendation of the institute, the Orissa Government released the variety to put under State Seed Chain and adopted for seed production in State Farm to supply to farmers on large scale. Based on survey in *kharif* 2009 it was revealed that about 180 farmers in 105 ha of land adopted deep water rice.

Unit cost: Cost of cultivation per unit area is Rs 10,000 ha^{-1}

10. Water Chestnut Production Technology for Waterlogged Areas

Application: Waterlogged and marshy areas.

Input needed: Planting materials, fertilizers and necessary agro-inputs

Output capacity: Net income : Rs. 25,344 ha^{-1}.

B:C ratio : 1.72.

Specific benefits and impact: Water chestnut cultivation already been adopted in 900 ha area in Orissa covering six districts. Has a further potential for adoption in 1.10 M ha of non-arable waterlogged areas in the country.

Unit cost: Cos of cultivation: Rs. 18,000ha^{-1}

CONCLUSIONS

Eastern India, though a resource rich region, is considered to be agriculturally backward because of widespread poverty, low crop productivity and poor irrigation development. Due to poor storage capacity, large area is prone to flood in *kharif* season and drought in *rabi* season. Groundwater development is less than 30 % as compared to more than 100 % in many pockets of the country. Poor groundwater development coupled with inefficient surface water irrigation compels farmers to depend on rainfall for irrigation and makes agriculture uncertain and risky. Thus, there is urgent need to discuss water management issues and develop location specific suitable technologies to ensure food security and raise standard of living of the habitants.

REFERENCES

FAO. 2006. Water Use Efficiency in Agriculture: The role Of Nuclear and Isotopic Techniques. Proceedings FAO/IAEA Workshop on Use of nuclear Techniques In Addressing Soil-Water-Nutrient Issues For Sustainable Agricultural Production at 18th World Congress Of Soil Science, Philadelphia, Pennsylvania, USA, 9-15 July 2006.

Kar, G. and Kumar, A. 2009. Water Harvesting In Coastal Waterlogged Area for Its Multiple Use. 60th International Executive Council Meeting & 5th Asian Regional Conference, 6-11 December 2009, New Delhi, India. pp1-11.

Morison, J.I.L., Baker, N.R., Mullineaux, and Davies,W.J. 2008. Improving Water Use In Crop Production. Philosophical Transactions of The Royal society of London B. Biological Sciences, 363(1491): 639-658.

Singh, R., Kundu, D.K., Mohanty, R.K., Ghosh, S., Kumar, A. and Kannan, K. 2005. Raised and sunken Bed Technique For Improving Water Productivity in Lowlands. Water Technology Centrefor Eastern Region, Bhubaneswar, India, Research bulletin No. 28, pp1-42.

Singh. R., Kundu, D.K. and Bandyopadhyay, K.K. 2010. Enhancing Agricultural Productivity through Enhanced Water Use Efficiency. *Journal of Agricultural Physics*,**10**: 1-15.

❑❑❑

Chapter 26

Integrated Farming Systems for Improving Water Productivity in Eastern Region

B P BHATT

INTRODUCTION

Indian agriculture is heavily dependent on the monsoon as a source of water and the failure of the monsoons cause water shortage and thereby results into below-average crop yields. This is particularly true for major drought-prone regions such as southern and eastern Maharashtra, northern Karnataka, Andhra Pradesh, Odisha, Gujarat, South Bihar, Chhattisgarh and Rajasthan. Droughts have periodically led to major Indian famines in the past. Similarly, floods are also the most common natural disasters in India. The heavy southwest monsoon rains cause all the rivers to distend their banks, often flooding surrounding areas. Excess, erratic, or untimely monsoon rainfall may also wash away the fertile soils including other resources or otherwise ruin crops. Almost all of India is flood-prone and extreme precipitation events, such as flash floods, torrential rains have become increasingly common in eastern and central India over the past several decades, coinciding with rising temperatures.

Eastern states of India occupies about 22.5% geographical area and supports to 35% human and 28% livestock population, respectively. Of the total geographical area of 71.84 M ha (Million ha), the net sown area is 31.40 M ha (Million ha) with a cropping intensity of 141%. Alluvial, followed by red and yellow soils are the major soil types in Eastern states. The region has about 2.73 M ha total water area constituting reservoirs, ponds, tanks and beels, oxbow lakes, brackish water *etc.*, besides 15046 km length of rivers and canals. Agriculture is the mainstay of economy in Eastern states since 83% population is rural. However, very high population density (604 persons sq km^{-1} compared to 382 persons at national level), about 116 M below poverty

Director, ICAR Research Complex for Eastern Region, PO: Bihar Veterinary College
Patna – 800 014

line population, lowest per capita income (Rs. 7,875/- to 25,000/- per annum than the national average of Rs. 26,000/-), maximum number of economically most backward districts (69 nos. out of 150 at national level) has put tremendous pressure on natural resources in eastern region. Further, the production levels of agriculture, livestock and fisheries has remained low due mainly to lack of location specific production technologies, dissemination of scientific methods of cultivation, natural calamities like floods, water logging, droughts and social constraints. Mostly farmers depend on unpredictable monsoon for crop production and owing to poor utilization of water resource, the cropping intensity is also low, particularly in rainfed upland ecosystems. They commonly produced per unit area only one-fifth to one-tenth of the food as farms in the industrialized and Green Revolution lands (Pretty, 1995).

In Eastern states of India, immense possibilities exist for integration of aquaculture into crop- animal systems. A vast population of livestock (164.95 M) and birds (176.23 M) thrives in the region. However, most of the excreta of the livestock are not utilized properly. The livestock wastes could be recycled in fishponds through integrated fish farming so as to enhance fish production substantially without using any other fertilizer or supplementary fish feed. A number of studies have shown that average productivity of water can be increased through integrated fish farming. Fish-pig, fish-duck, fish-chicken, fish-cattle and fish-goat are some of the integrations which enhance fish production by 3 to 4-folds compared to fish farming without integrations. Other possible integrations could be fish-buffalo since the region has high population of buffalo, particularly in Eastern U.P. and parts of Bihar. The costs of fish production in these trials were much lower than that of conventional fish culture due mainly to low inputs or efficient utilization of farm resources. Need of the hour is therefore, standardization of area specific technologies keeping in view the farmer's need and socio-economic conditions.

Status of Water Resources in Eastern India

Although, average annual rainfall in India is slightly higher (~1200 mm) than the World average (~1000 mm), its distribution is temporally and spatially uneven (Table 1). Entire rainfall in India occurs with fewer rainy days within a time span of few days, in fact within about 300 hours. The average rainfall in eastern region varies from 1091 to 2477 mm with a regional average of 1526 mm, which is sufficient and substantial for growing variety of crops. On average, Assam gets mean annual rainfall of 2477 mm which is the highest in the region, followed by West Bengal (1750 mm), Odisha (1451 mm), Chhattisgarh (1430 mm), Jharkhand (1277 mm), Bihar (1204 mm) and Eastern UP (1091 mm). However, it has erratic annual and spatial distribution with considerable year-to-year variation.

The region constitutes about 18% of country's utilizable water resources (10% of surface water and 30% of groundwater). The Ultimate irrigation Potential (UIP) for the region has been assessed to be 33.65 M ha compared to 139.89 M ha for the country. However, utilization of the created irrigation potential is only 65.5% in Eastern region compared to 74% at national level. So far in ground water utilization in concerned, it ranges from 18 to 66% in Eastern states compared to 58% at the national level. Ground water utilization is very meagre in the states like Assam, Chhattisgarh, Jharkhand and Odisha.

Table 1: Groundwater (BCM) utilization in eastern states

State	Annual Replenishable GW	Annual GW availability	Annual GW draft	Stage of development (%)	Remarks
Assam	27.23	24.89	5.44	22	All districts – safe
Bihar	29.19	27.42	10.77	39	All assessment units (blocks) – safe
Chhattisgarh	14.93	13.68	2.80	20	Out of 146 assessment units, 8 are semicrtical
Jharkhand	5.58	5.25	1.09	21	All assessment units (blocks) – safe
Orissa	23.09	21.01	3.85	18	6 assessment units –saline; entire state – safe
Eastern UP	76.35	70.18	48.78	66	13 blocks critical; 88 semi-critical
West Bengal	30.36	27.46	11.65	42	1 block critical; 37 block semi-critical
India	433.02	399.25	230.62	58	

Source: Central Ground Water Board, 2006

Water productivity is also very low in Eastern states (0.37 kg m^{-3} of water) compared to 1.01 kg m^{-3} of water in Punjab and other parts of India. Water productivity ranged from 0.21 to 0.29 kg m^{-3} in most of the eastern states except Bihar, Eastern U.P. and West Bengal.

Different Integrated Systems for Improving Water Productivity

***Rice based cropping system*:** Rice is the predominant crop of the region in *kharif* season. The eastern Indo Gangetic plains, has over 20 M ha of rainfed rice area. Drought losses are most severe in the rice bowl states of Chhattisgarh, Madhya Pradesh, Bihar, Jharkhand, Odisha and Uttar Pradesh. Appropriate drought tolerant rice varieties of medium duration like Shahbagi, Vandana, Anjali, CR-Dhan or short duration varieties like Prabhat, Dhanalakshmi and Richaria can be grown in upland areas, where drought is a frequent phenomenon. However, adequate moisture availability needs to be ensured in rice fields for proper panicle exertion and effective filling of grains. In case there is no rainfall, ground water need to be used for the purpose. The other steps required is raising bund height to the tune of 20-25 cm around rice fields so as to arrest rain water and utilize in crop production. Adoption of deficit irrigation strategies such as irrigation at most critical crop growth stages, low depth at frequent interval, use of need based and cost effective water saving measures can be followed for efficient use of available water.

Pre-*rabi* crops such as toria (*Brassica rapa*) cv *'PT 303'* and *'T 9'*, pigeon pea varieties *'Basant'*, *'Bahar'* and *'Sharada'*, early potato and cauliflower can be sown during the first week of September in the areas where there is no chance of excess moisture. Toria and early potato could be harvested by the end of November to mid-December which enable for cultivation of late sown wheat after these crops.

In *rabi* crop, efficient use of soil moisture is possible through adoption of Zero Tillage in wheat, lentil and gram. It will save the input cost and result into higher yield. Ensuring water availability for irrigating crops at critical stages of crop growth is essential for better water use efficiency. Three major crop sequences of Bihar namely rice- wheat – green gram, rice- potato- onion, rice- cabbage- cowpea were compared for water productivity under secondary reservoir multiple water use system and tube well irrigation. Water used in each sequence was worked out taking into consideration water applied in irrigation and rainfall received. Crop water productivity of different crop sequences showed that rice-cabbage-cowpea sequence recorded maximum water productivity values of 1.58-1.79 kg m^{-3} which was significantly superior to Rice- Potato- Onion with 1.37-1.73kg m^{-3} and Rice-wheat- green gram with 0.81-1.21kg m^{-3}.

Crop diversification:Crop diversification by adopting cultivation of maize, groundnut- pigeon pea, sole pigeon pea, black gram, horse gram, green gram, cow pea, and oil seed crops like niger, sesame and safflower can be cultivated in rainfed upland conditions. Higher rice equivalent yield (7.5 t ha^{-1} yr^{-1}) could be obtained through maize cob (8.1 t ha^{-1}), groundnut + pigeon pea (5.6 t ha^{-1}), sole groundnut (5.6 t ha^{-1}), sole pigeon pea (5.6 t ha^{-1}). The cost of cultivation per unit area has been worked out to be ₹ 10,000-14,000 ha^{-1} with net return of ₹ 15,000 ha^{-1} $annum^{-1}$. Government of India has also focused on diversification in rainfed upland ecosystem and, therefore, out of 181 districts of Eastern states, 158 have been identified for pulse production under food security mission. Average pulse productivity is also higher in eastern states compared to national average.

Integrated Farming Systems: Integrated farming system is based on the concept that "there is no waste", and "waste is only a misplaced resource which can become a valuable material for another product" (Edwards *et al.*, 1986). It is generally considered particularly relevant to benefit the rural poor. All the components of agriculture like crop-livestock-fish-forestry-horticulture are integrated in a complimentary way. Soil and water conservation is also given due importance in this type of farming system. Thus, the integrated farming system may provide the environmentally sustainable and economically viable technology that encompasses rational utilization of available resources of the region. The emphasis in such system is on optimizing resource utilization rather than maximization of individual element in the system. The main purposes for such systems are:

- To provide viable alternative to traditional land use practices including shifting cultivation,
- Biodiversity enhancement as the "alternative" feeding system, using mainly local plant resources, which provide comparative advantages to indigenous animal ecotypes and opportunities for greater use of indigenous knowledge,
- Complimentarily of crop-animal-fish-birds-multipurpose trees-horticulture,
- Matching the production system with the available resources, and selecting and promoting crops and farming systems which optimize use of natural resources without depleting them,
- Recognizing that "poverty", and not "food", is the major constraint to equitable development,
- Maximum return per unit area per unit time in terms of produce and money,
- Benefit to the community at large through supply of essential commodities throughout the year, providing year round employment to the farming family and reducing cost of production,
- Insurance against the failure of individual crop/enterprise *etc.*

Various location specific integrated farming system models have been developed by various research organizations including ICAR. Rice-fish-cattle; rice-fish-goat; rice-fish-poultry; rice-fish-duck; rice-fish-pig and rice-fish are some of the potential farming systems practiced by the farmers. These farming systems have been found economically most viable as evident from the data shown in Table 2.

Table 2: Economic status of different integrated farming systems

S.No	Farming system	Cost-benefit ratio	Source	Remarks
1	Agrihortisilvi-pisciculture	1:2.80	Tanjang *et al.*, 2006	Practiced by *Kalita group* in Arunachal Pradesh.
2	Agrihortisilvicultural system	1:1.40	Arunachalam *et al.* 2002	Practiced by *Nyishi* tribes of Arunachal Pradesh.
3	Agripisciculture	1:3.40	Tanjang *et al.*, 2006	Practiced by *Apatani* tribes of Arunachal Pradesh
4	Rice-fish-cattle	1:1.69	Bhatt & Bujarbaruah, 2005	Developed over marshy lands in Meghalaya
5	Rice-fish-goat	1:1.44	Bhatt & Bujarbaruah, 2005	—do—
6	Rice-fish-poultry	1:1.41	Bhatt & Bujarbaruah, 2005	—do—
7	Rice-fish-duck	1:1.31	Bhatt & Bujarbaruah, 2005	—do—
8	Rice-pig-fish	1:1.31	Bhatt & Bujarbaruah, 2005	—do—
9	Agrihorticulture system	1:1.57	Bhatt and Mishra 2003	Rice was intercropped with fruit trees like Assam lemon, guava and peach over sloppy lands in Meghalaya
10	Agropastoral system	1:1.45	Bujarbaruah, 2004	The study was conducted in one of the watersheds under farming system research project at Meghalaya
11	Sole cropping	1:1.04	Bhatt and Mishra, 2003	Sole rice was grown over sloppy lands in Meghalaya

Integrated farming system (IFS) is the viable option in order to achieve food and nutritional security at household level and even at individual level in flood prone areas. Such land use system could provide round the year employment to the farming family. IFS models have been developed for marginal farmers (area of 0.80 ha), inhabiting in lowland and midland irrigated ecosystems. Crop- livestock-fishery was integrated with allied enterprises like duck/ vermicomposting/bee keeping. Cereal crops (*kharif* rice, followed by *rabi* wheat/maize/ lentil/mustard), fruits and vegetables were cultivated besides fish, duck and milch animal. The net monetary gain from such system was estimated to be ₹ 1,80,500 yr^{-1}.

Likewise, the model has been developed for small land holders (size- 0.40 ha) of midland situation. The major components were agri-horti crops, goat, poultry with allied enterprises like mushroom, vermicomposting and poultry. The monetary gain from such model was accounted for ₹ 99,591/-

Area (%) Allocated for Different Enterprises

Enterprises	1 Acre Model	2 Acre Model
Cereals	50	50
Vegetables	22.5	12.5
Goatry	4	-
Livestock	-	1.9
Fodder	12.5	12
Fisheries +Duckry	-	18
Poultry	2.02	-
Veg.Bund	4	4
Mushroom	2.02	-
Manure Pit	2.33	1.5

Annual Return from IFS Models

Model	Gross Income (₹)	Cost of Cultivation (₹)	Net Income (₹)
1 Acre	1,68,412	69,653	98,759
2 Acre	3,41,748	1,39,112	1,61,249

Other successful IFS model is cultivation of makhana (*Euryale ferox*) with fish and water chestnut. It include cleaning of pond before emergence of makhana seedlings, removal of carnivorous fishes by application of mahua oil cake @ 2.5 t ha^{-1}, transplanting and gap filling of Makhana, making refuge area of 10 % of total water body area and integration of different carp fingerlings. Water chestnut is cultivated as a tertiary crop. The technology has a potential for adoption in 1.10 M ha of non-arable waterlogged areas in the country. Traditionally, Makhana (*Euryale ferox Salisb.*) was grown as a sole crop and water bodies were utilized for only seven months - February to August. Fish and water chestnut was grown successfully in combination of makhana in order to ensure higher profitability. With the intervention of makhana-cum-fish and water chestnut farming, the water bodies are utilized throughout the year. A refuge area covering 10% of net water bodies is used as vacant space for integration of Indian major carps. Water Chestnut (*Trapa bispinosa Natans.*) is taken as tertiary crop. Net income from integrated makhana based farming system ranges from ₹ 44, 686 to 51,216 ha^{-1} yr^{-1} whereas income from cultivation of makhana alone is worked out to be ₹ 24,367.0 to ₹ 29,153.0 ha^{-1} yr^{-1}.

Makhana could be cultivated in a cropping system mode in water logged areas (> 0.60 m water depth). While 4-5 months are sufficient for makhana cultivation, other crops could be cultivated successfully from the same piece of land. Makhana could be transplanted in the second week of April and harvested by the second week of August. Short duration varieties of rice could be cultivated thereafter. Wheat can also be sown by mid of December and harvested by the second week of April. Hence, cultivation of three crops per year is possible in such areas, particularly in Bihar.

Multiple Water Use system with Integrated Agriculture and Fishery for Improving Water Productivity

Medium deep waterlogged lands (0.5-1.0 m) can be modified in the form of a reservoir that can be used as a tool for multiple uses with high water and land productivity and profit. About 20 % of farm area may be converted into a secondary reservoir upto 2 meter depth. The excavated soil is spread around the periphery to form a bund of 3-4 m width with crest level at least 50 cm above the peak water level to ensure water do not overtop the bunds. High value horticultural/vegetable production on bunds utilizing seepage water with little supplemented water can produce good profit for the land, which was poorly utilized. In conjunction, good fish production can be achieved with water quality management through water routing for irrigation purposes. The routed water containing good amount of nutrients as well as provide opportunity for applying water in the fields in correct amount and at appropriate time, which enhances yields and quality of agricultural produce. Ducks, poultry, piggery *etc.*, are other components that can be added to have complementary benefits and increase water productivity.This system generates a net income of ₹ 1,32,590 ha^{-1} of land from horticulture, fishery and duckery in addition to irrigation to cereal crops.

Waterlogged areas with more than 1m stagnant water can be used for this technology. Trenches in such areas can be excavated in such a way that excavated soil is filled in alternate strips to make bunds and raising a portion above highest flood level using excavated soil to cultivate vegetable or horticultural crops. Two types of models like meandering type simulating river condition and Island type simulating pond were developed. First model simulates the river flow conditions, in which trenches are made in meandering style, and water may be allowed to enter from one side with enough provision for protecting culturing fishes. The second simulates pond type conditions in which continuous trenches are excavated with island type of raised bed, which may provide enough security to any high value production on the beds, and fishes have access to continuous movement around the island. Banana with vegetables has been grown on the raised beds and composite fish culture in fish trenches. The fish yield to the tune of 1.97 t ha^{-1} and 1.92 t ha^{-1} from meandering and island types, respectively can be obtained by stocking fries@ 15000 ha^{-1}.

Net income from Trenches–cum raised bed system is about ₹ 81000 ha^{-1} yr^{-1}. Fruit crops contributed 54 % to the net income followed by fish (24%) and vegetables (22 %). Other integrations include crop-cattle-fish, crop-fish-chicken, crop-fish-duck, crop-fish-pig, crop-fish-goat, which well work in Eastern and north eastern regions. However, integrated farming systems for rainfed agricultural systems are yet to be evolved.

Restoration of Marshy Lands through Agroforestry Interventions

Agroforestry has the potential to improve livelihood as it offers multiple alternatives and opportunities to farmers to improve farm production and incomes and also provides productive and protective (biological diversity, healthy ecosystems, protection of soil and water resources, terrestrial carbon storage) functions to the ecosystems while protecting the natural environment (Nautiyal *et al.*, 1998). It is promoted widely as a sustainability-enhancing practice that combines the best attributes of forestry and agriculture. Agroforestry has also been recognized by IPCC as having potential for sequestering carbon as part of climate change mitigation strategies (Watson *et al.*, 2000). Agroforestry land use is also vital to rehabilitate the marshy and degraded areas in the eastern states. Raised bed method could be adopted in marshy areas for planting of saplings. Species like *Terminalia arjuna, T. bellirica, T. tomentosa, Populus deltoides, Casuarina equisitifolia, Syzygium cumini, Eucalyptus* spp., *Vitex nogundo, Salix* spp., *Dalbergiasissoo, Moringa* sp., *Morus* sp., *Melia azadirachta, Ficus* spp. *etc.*, have been found suitable for cultivation in such areas besides various bamboo species. The marshy lands could be rehabilitated for cultivation of various agricultural crops besides augmentation of biomass for fuel, fodder and timber from agroforestry interventions.

Fisheries in Flood Plains

The floodplain lakes (Chaurs), saucer shaped water filled depressions 50 to 400 cm in depth is predominantly found in north Bihar, West Bengal, Eastern U.P. and Assam. Fishes inhabiting these water bodies are *magur, singhi, kabai or koi, garai* and other economically unimportant weed fishes. These floodplain lakes are almost unutilized. The fishermen around the chaurs are mostly engaged in capture fishery activity. The fish productivity is approximately 15 kg ha^{-1} yr^{-1}. However, flood plains have the potential to yield 2.0 to 2.5 t ha^{-1} yr^{-1} of fishes in semi intensive culture system with suitable technological interventions. Chaurs can be stocked with naturally collected or hatchery reared fingerlings of Indian major carps. Air breathing fishes can also be reared in chaurs since they do not require much water for their survival and growth. Fish culture in cages (Floating) provides an opportunity for a low cost and confined environment in Chaur (Water bodies) to enhance quality fish seed production from flood plain wetlands.

The ox-bow lake, popularly known as maun (defunct loops of rivers cut off from the main rivers), are also predominant in Bihar, West Bengal and Assam. During floods in river, they receive water. Fish production in mauns ranges from 60 -150 kg $ha^{-1}yr^{-1}$. However, mauns could be used as a capture fishery resource and cage & pen culture practices could be adopted in such water bodies.

Pen Culture of Fish and Prawn in Derelict Water Bodies/Lagoons/ Lakes/ Floodplain Wetlands

Shallow areas of derelict water bodies/lagoons/lakes/floodplain wetlands could be utilized for raising table size fishes and prawns in enclosure (pens). After assessing water depth, duration of water availability and seed availability, pens of suitable size (0.1 to 0.2 ha) and shape could be erected depending on the capacity of water bodies and topography of the area. The selected water bodies should have at least 1 m water depth for a period of 3-4 months. Indian major carps, *viz.,* Catla (*Catla catla*), Rohu (Labeo *rohita*), Mrigal (Cirrhinus *mrigala*) and giant freshwater prawn (*Macrobrachium rosenbergii*) are ideal for rearing in pens.

Paddy-Cum-Fish Culture

Paddy cum fish culture is age old tradition in rainfed as well as irrigated ecosystem of Northeast and Eastern region. Fishes are cultivated in paddy fields either simultaneously or in rotation. Paddy fields of any size can be utilized for fish culture, subject to availability of adequate water in time and space. Both freshwater and saline fields can be utilized for the purpose.

Common carp (*Cyprinus carpio*) is the most preferred species for paddy-fish culture. However, Catla (*Catla catla*), Rohu (*Labeo rohita*), Mrigal (*Cirrhinus mrigala*), Bata (*Labeo bata*), Java punti (*Puntius javanicus*), Silvercarp (*Hypophthalmichthys molitrix*) and giant freshwater prawn (*Macrobrachium rosenbergii*) are some other suitable species for paddy-fish culture.

Major Constraints in Adoption of Farming Systems

Over exploitation of natural resources : In general, 2/3rd of the total geographical area of the hills should be under forests so as to maintain the ecological balance. The rate of soil erosion is also very high in the region compared to other parts of the country. According to one estimate, soil loss is upto 40 t ha^{-1} yr^{-1} in current *jhum* fields. Moreover, in second and third year jhum, soil erosion has been reported to be 146 and 174 t ha^{-1} yr^{-1}, respectively. Hence, restoration of the degraded land is essentially required not only for ecological purposes but also for biomass augmentation (Rao *et al.*, 1999). Hence, first and foremost priority in the region should be rehabilitation of degraded lands through agroforestry interventions.

High biomass requirement : Biomass is the main source of energy in the villages of India. About 87% population use biomass as a source of energy. Further, there is very high requirement of green fodder in the region. Firewood extraction is also one of the significant factors responsible for over exploitation of forests. Recent estimates have shown a very high demand for firewood (Table 3). Likewise, fodder scarcity has been estimated up to 200% in Himalayan states (Pratap, 1999; Sundriyan and Sharma, 1996).

Table 3: Firewood consumption pattern in Himalayan region

Region	Firewood consumption (kg $capita^{-1}$ day^{-1})	References
NEH Region	3.10-10.40	Maikhuri, 1991; Bhatt and Sachan, 2004
Western Himalaya	1.49	Bhatt *et al.*, 1994
Southern India	1.90-2.20	Hegde, 1984
Himalayan range of Nepal	1.23	Mahat *et al.*, 1987

Shifting Cultivation

The farming practices are complex, diverse and risk prone in Eastern and NEH region. Even today, shifting cultivation is mainstay of economy in the Region. According to one estimate, about 4.8 lakh farming families practice *jhum* cultivation over 1.63 M ha of land (Bujarbaruah, 2004). It leads to land degradation particularly during second and third year of *jhum*. Since *jhum* cycle has drastically reduced to 3 to 5 years in most of the north eastern states, there is sharp decline in crop growth and yield. Prevailing land tenure system also restricts the adoption of farming system approach of land use to a great extent.

Potentials

Traditional mountain societies are characterized by the close interconnection that they maintain with nature and natural resources. Hence, conservation linked developmental initiatives should be based upon a value system that the community can understand, appreciate and participate (Ramakrishnan, 2007).

Livestock in hills cannot be seen as a primary producer for milk, meat or dung production. Studies conducted in Eastern Himalayan region have shown that without integration of livestock in farming system, it is very difficult to make the system remunerative and ecologically viable (Bhatt and Bujarbaruah, 2005; Bhatt *et al.*, 2006). Integrated livestock based model on watershed basis; crop-chicken-fish-duck-horticulture-hedgerow spp.; crop-fish-poultry-MPTs; crop-fish-goat-MPTs-hedgerow spp.; crop-fish-pig-bamboo-MPTs-hedgerow spp.; crop-fish-dairy cattle-MPTs-fruit trees-hedgerow-vermicompost-mushroom, and rice-fish etc. have been found promising not only for livelihood support of rural folk but also to maintain ecological balance in the region.

Wide variation in altitude also offers opportunities for cultivation of varieties of agri-horti crops in the region. The traditional varieties/land races have a potential to sustain the farming communities to a great extent. So far in crop composition is concerned, cultivation of as many as 56 crops has been reported from *jhum* fields. Likewise, in western Himalayan region, 10-12 crops are grown simultaneously or sequentially. Such broad genetic base of farming in must be conserved in order to make the long term planning for food security of the people.

Home garden is yet another potential farming system prevalent mainly in Eastern region wherein various crops, MPTs, bamboos, medicinal herbs and vegetables are cultivated besides integrated fish farming, either with duck or poultry or pig. Though very small in area (ranging from 0.2-0.4 ha per family), home garden is most promising for sustainability of the indigenous people.

In totality, integrated farming system models are highly location specific and need the incentives for its horizontal expansion. Consistent monitoring, frequent assessment and refinement are also need of the hour so as to make the farming system models cost effective. Since assessment and refinement is a dynamic process, some of the issues are mentioned below which need attention of the researchers, policy makers and extension functionaries.

Researchable and Other Issues

- IFS as a vehicle particularly in hill and mountain ecosystem to achieve the means for organic agriculture, permaculture and other alternative methods of sustainable food production.
- Standardization of hecterage under IFS to support a family of 5-8 as sole means of livelihood security.
- Role of IFS in addressing the emerging issues like global warming (through studies on carbon sequestration, controlled production of methane etc. from the animal component), quality control (SPS), post harvest technology *etc.*
- Methodology and incentive *etc.*, required for horizontal expansion of IFS technology.
- Information dissemination through electronic media and IT based vehicle for developing rural knowledge/technology centers.
- Ecological and economic efficiency of IFS.
- Ecosystem services provided by IFS.

REFERENCES

Arunachalam, A., Khan, M.L. and Arunachalam, K. 2002. Balancing traditional *juhm* cultivation with modern agroforestry in Eastern Himalaya biodiversity hot spot. *Current Science*, 83: 117-118.

Bhatt, B.P. and Bujarbaruah, K.M. 2005. *Intensive Integrated Farming System: A Sustainable Approach of Land use in Eastern Himalayas.* Technical Bulletin No. 46, ICAR Research Complex for NEH Region, Umiam, Meghalaya, 43p.

Bhatt, B.P., Bujarbaruah, K.M. and Vinod, K. 2006. *Integrated Fish Farming in Eastern Himalayas.* Technical Bulletin No. 47, ICAR Research Complex for NEH Region, Umiam, Meghalaya, 34p.

Bhatt, B.P. and Sachan, M.S. 2004. Firewood consumption pattern of different tribal communities of Meghalaya in North East India. *Energy Policy*, 32: 1-26.

Bhatt, B.P. and Mishra, L.K. 2003. Production potential and cost-benefit analysis of agrihorticulture agroforestry system in North-East India. *J. Sustainable Agriculture*, 22: 99-108.

Bhatt, B.P., Negi, A.K. and Todaria, N.P. 1994. Fuel wood consumption pattern at different altitudes in Garhwal Himalaya. *Energy*, 19: 465-468.

Bujarbaruah, K.M. 2004. Organic farming: Opportunities and challenges. In: Souvenir International Conference on *Organic Food* (Eds. Sarma, B.K.; Bhatt, B.P. and Sahay, C.S.). ICAR Research Complex for NEH Region, Umiam, Meghalaya, 14-17 February, pp13-23.

Edwards, P., Kaewpaitoon, K., McCoy, E.W. and Chantachaeng, C. 1986. Pilot small scale crop/livestock/fish integrated farm. AIT Research report, 184, Bangkok, Thailand, 131 p.

Hegde, M.S. 1984. Fuel problem in villages: Challenges and opportunities. Bulletin of Science, 1987; July-August: 8-13.

Maikhuri, R.K. 1991. Fuel wood consumption pattern of different tribal communities living in Arunachal Pradesh in North East India. *Bioresource Technology*, 35: 291-296.

Mahat, T.B.S., Griffin, D.M. and Shepherd, K.P. 1987. Human impact on some forest of the middle hills of Nepal. Part IV. A detailed study in South-East Sindhu Palanchock and North-East Kabhere Palanchock. *Mountain Research & Development*, 7: 114-134.

Nautiyal, S., Maikhuri, R.K., Semwal, R.L., Rao, K.S. and Saxena, K.G. 1998. Agroforestry systems in the rural landscape – a case study in Garhwal Himalaya, India. *Agroforestry Systems*, 41: 151-165.

Pratap, T. 1999. Sustainable land management in marginal mountain areas of the Himalayan region. *Mountain Research & Development*, 19: 251-260.

Pretty, J.N. 1995. Regenerative agriculture: Policies and Practices for Sustainability and Self Reliance. Earthscam Publication, London, 32 p.

Ramakrishnan, P.S. 2007. Sustainable mountain development: The Himalayan tragedy. *Current Science*, 29:308-316.

Rao, K.S., Maikhuri, R.K. and Saxena, K.G. 1999. Participatory approach to rehabilitation of degraded forest lands: a case study in high altitude village of Indian Himalaya. *International Tree Crops Journal*, 10: 1-17.

Tanjang, S., Deb, S., Brahma, S., Arunachalam, S., Melkania, U., Arunachalam, K. and Srivastava, K. 2006. Traditional agroforestry systems of *Nyishis, Apatanis* and *Kalitas*. *In*: Bhatt, B.P. and Bujarbaruah, K.M. (eds.) *Agroforestry in North East India: Opportunities and Challenges*, ICAR Research Complex for NEH Region, Umiam, Meghalaya, pp313-321.

Watson, R., Noble, I., Bolin, B., Ravindran, N., Verardo, D. and Doken, D. 2000. *Land Use, Land Use Change and Forestry*. Intergovernmental Panel on Climate Change, Cambridge University Press, Cambridge, UK.

□□□

Chapter 27

Impact of Resource Conservation Treatments on Runoff, Soil Loss and Growth Performance of Bamboo for Reclamation of Yamuna Ravine Systems of India

A.K. SINGH, S. KALA, S.K. DUBEY, K.P. MAHOPATRA[1]
B.K. RAO[2] AND M.L. GAUR[3]

INTRODUCTION

Ravine lands are highly degraded dry lands and about 3.97 M ha of ravine lands present in India. In Uttar Pradesh alone, 1.23 M ha (14.5%) of land is under ravines which are found along the bank of river Yamuna and its tributaries. The severity of water erosion is found at the peak along the banks of yamuna and Chambal Rivers in the districts of Agra, Etawah, Kanpur and Fetehpur *etc.*, where terrain has completely deformed into ravines. The problem of subsoil hardness is affecting physical and hydrological properties of the soil. High bulk density and low infiltration rate of these soils lead to low yields and water use efficiency of crops. Vegetation in this region suffers from a variety of unfavorable condition such as nutrient deficiency, moisture stress and biotic interference. The inclement weather conditions coupled with very high summer temperature further aggravate the problem and makes farming an uneconomical position. In such situations, less water and nutrient demanding technologies hold a good promise to sustain the productivity and provide alternative source of income to the farmers. Using bamboo to rehabilitate degraded land after brick mining had devasted land in the villages of Kotwa and Rahimabad at the head of the catchment area of the Ganges and the Yamuna rivers. Impacts of bamboo planting include raised groundwater level, increased land productivity, improved micro-climate and improved socio-economic conditions. About 16 species were tried to identify species that could grow better on marginal soil. After a series of trials,

Central Soil and Water Conservation Research and Training Institute, Research Centre, Chhalesar, Agra, U.P – 282006.
[1]Senior Scientist (Forestry), ICAR Research Complex, Bharapani, Sikkim.
[2]CSWCRTI, Research Centre, Vasad, Gujarat,[3]Dean, CTAE, Godhra, Gujarat.

it was found that the hardy *Dendracalamus strictus* and *Bambusa bambos* worked best and these were the two species that were initially given to the farmers to rehabilitate their land (Singh and Tewari, 2005). *Dendraclamus strictus* is a deciduous densely tufted bamboo. This species occupies 53 % of total bamboo area in India and predominant species of bamboo in Uttar Pradesh, Madhya Pradesh, Orissa, Eastern and Western Ghats regions. In India, it is widely distributed in semi arid and dry zone along plains and hilly tracts usually up to an altitude of 1000 m. It is widely adaptable to temperatures as low as 5° C and as high as 45° C. It is found suitable for reclamation of ravine land. It is used for construction, furniture, as well as raw material in paper mills and cottage industries. Since rain is a major source of water in semiarid region, every drop of it must be conserved and utilized judiciously. Improving vegetative cover in degraded land is prime importance in control of erosion, reduce run-off, improve *in-situ* soil-moisture conservation, increase water-table but also improve productivity as well as profitability. Situation specific cost-effective viable technologies and dissemination of suitable technologies for reclamation and productive utilization of ravine lands in

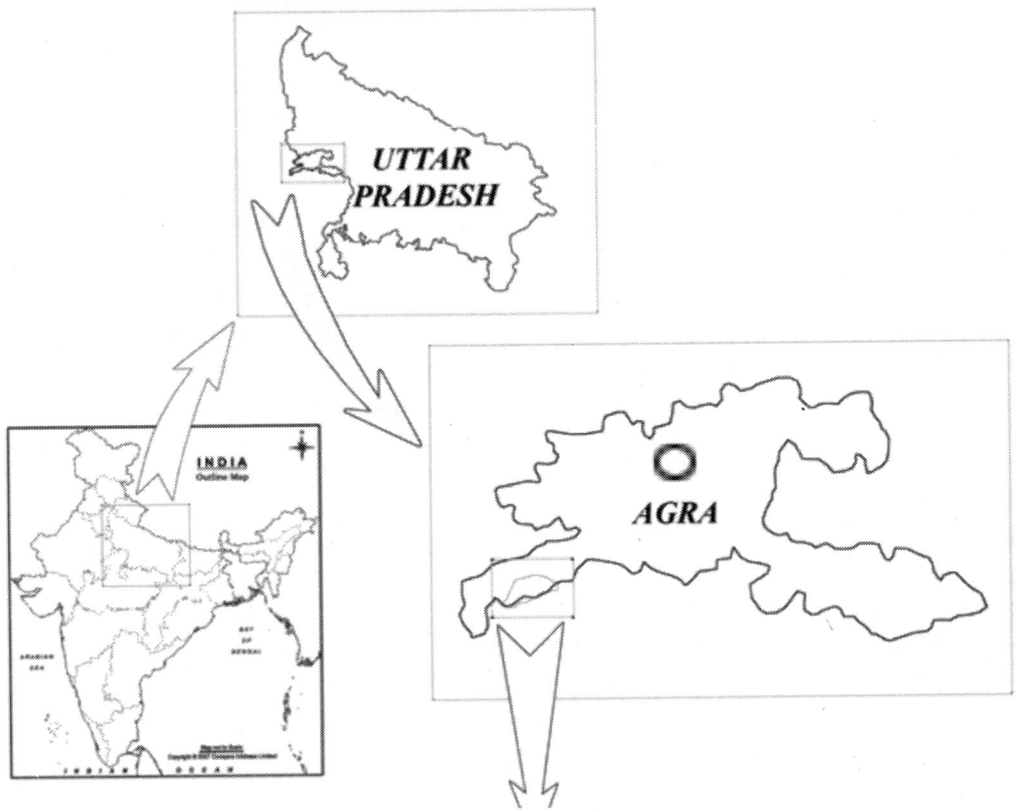

Fig.1: Map Showing the Location of Experiment

India are highly essential for arresting ravine extension, reclamation and improving production ravinous regions. Hence an investigation was carried out at Central Soil and Water Conservation Research and Training Institute, Research Centre, Agra, on Hydrologic and economic evaluation of bamboo plantations in gullied lands under major ravine systems. This paper presents some salient findings of the experimental studies on bamboo plants with moisture conservation techniques in yamuna ravines.

MATERIALS AND METHODS

The experiments were conducted at CSWCRTI, Research Centre, Agra, located at 23^0 52′ to 31^0 28′ N latitudes and 77^0 06′ to 84^0 37′ E longitudes and 168 m above mean sea level.

The climate of the ravinous part of the state is tropical, continental and monsoonic with mild and dry winter and hot uncomfortable summers. The rainfall in the state is orographic in nature. June to September is the rainy season when it receives about 88 % of its total annual rainfall. South-western districts adjacent to Rajasthan receives minimum rainfall of <600mm, the annual rainfall. April, May and June are summer month when hot winds (Loo) blow from the west. May is the hottest month with mean maximum temperature of 41^0 c in the ravine region. January is the coldest month with mean minimum temperature ranges from 6.5^0 to 10^0 c. Relative humidity is least (21-27 %) during summer. Thus, it making the summer dry and hot. It is high during June to September (70-85%). while it remains between 50-70 % during the remaining parts of the year. Ravine lands typically falls under "Northern tropical thorn forest" according to Champion and Seth classification of forest type of India. Natural vegetation of the ravine land is very poor. The predominant vegetation with Vilayati Babul (*Prosopis juliflora*) followed by Babul (*Acacia nilotica*) thorny legumes, Euphorbias and dwarf grass species. There are occasional occurrence of Neem plants (*Azadirachta indica*), Papdi (*Holopteila integrifolia*), Shisham (*Dalbergia sissoo*), Karanj (*Pongamia glabra*) and Chonkra (*Prosopis cineraria*).

This vast tract of existing ravine lands is possessing potential threat to nearby productive lands because of over exploitation and poor management. The ravines are extending at the rate of 8 to 9 m per annum with average soil loss of more than17 tons per ha per year. Therefore, there is an urgent need to detain these problems and protect both the arable and non-arable land from further degradation. Since rain is a major source of water in the region, every drop of it must be conserved in the soil through insitu rainwater conservation. Four small catchments of 3 to 6 ha size were selected and four treatments viz., plantation of rows of bamboo in the upstream and downstream side of small earthen gully plugs (T_1), plantation of bamboo in staggered manner in the gully bed (T_2), two rows of bamboo planted in staggered manner as vegetative barrier (T_3) and control catchment having no intervention (T_4). The selected gully beds were cleared of all vegetation except grasses and each treatment plots was separately gauged using automatic stage level recorders to measure runoff and soil loss. Basic soil parameters of soil pH, EC and organic carbon (%) and plant growth parameters of average culm height (cm), average culm diameter (mm), crown size (cm) and no. of culms clump^{-1} were also recorded in all treatments.

RESULTS AND DISCUSSION

On the basis of three years of experimentation on representative experimental basins of ravine lands. Through this study, it has revealed that changes in land use have a considerable effect on plant growth, runoff and soil loss. The basic soil physio-chemical data have been collected to generate base line information for treatment imposed catchments (Table 1).

Table 1: Basic soil physio-chemical parameters under different treatments

Treatment	pH	EC (millimhos cm^{-1})	Organic carbon (%)	Available N (kg ha^{-1})
T_1	8.55	0.14	0.75	469
T_2	8.63	0.14	0.70	456
T_3	8.63	0.15	0.62	467
T_4	8.69	0.17	0.31	327

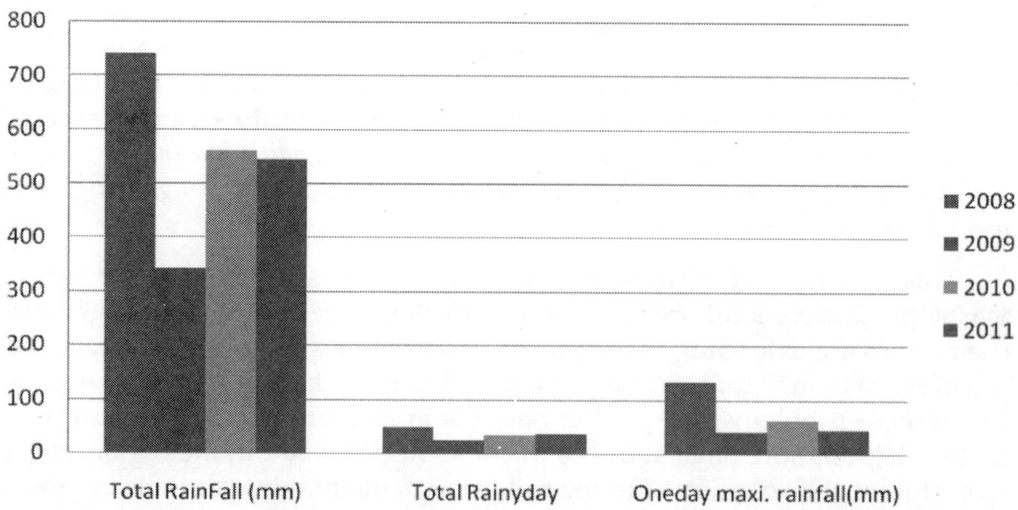

Fig. 2: Distribution of total rainfall, rainy day and oneday maximum rainfall from 2008 to 2011

The results shows that substantially good growth performance of *Bamboo* observed under ravine lands and significantly reduced the runoff and soil loss and improved the soil health in comparison to control catchments. The Table 2 depicted that significant growth performance of bamboo plants under different treatments.

Table 2: Growth performances of bamboo plants under different treatments

Treatment	Av. culm height (cm)	Av. culm diameter (mm)	Av. crown size (cm)	No. of culms $clump^{-1}$
T_1	108.75	7.96	66.82	9.67
T_2	126.20	7.56	72.86	12.00
T_3	99.20	6.20	49.11	6.33
T_4	0	0	0	0

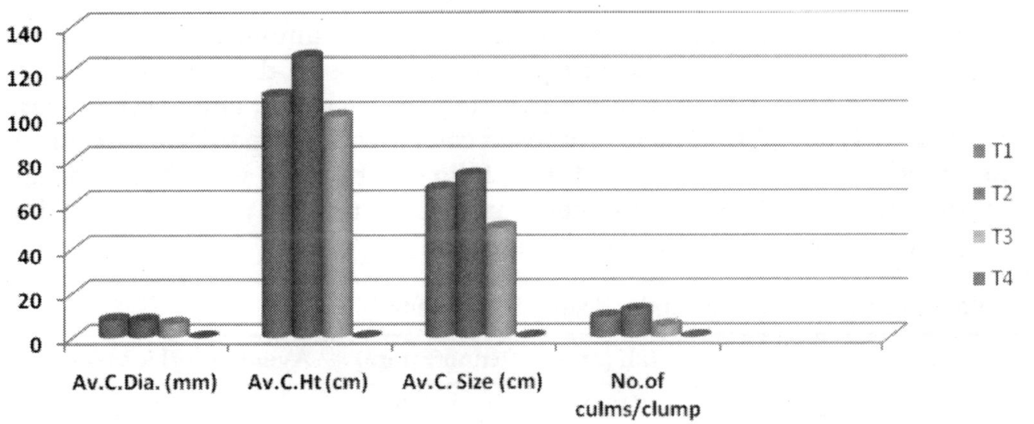

Fig. 3: Growth performances of bamboo plants under different treatments

It is inferred that T_1 (plantation of rows of bamboo in the upstream and downstream side of small earthen gully plugs) has a significantly higher culm height (126.20 cm), culm diameter (7.56 mm), crown size (72.86 cm) and maximum of 12 culms $clump^{-1}$ compare to others. The superior performance of bamboo plants in T_1 due to high insitu moisture availability in subsoil. Severe moisture stress was observed in T_4 (control catchment) due to low moisture status and poor quality soil which may be attributed for 100 % drying of bamboo plants. There was no silt deposition behind the structure was recorded in T_1 and T_2 catchments. But silt deposition behind structure was observed in the following treatments *viz.*, T_3 and T_4 and the values were recorded as 0.348 and 0.410 t ha^{-1} year, respectively.

This moisture conservation technique is substantially quite effective in recharging subsoil moisture through effect of vegetation which helps increase infiltration, reducing runoff volume and soil loss. The better insitu rain water harvesting and moisture conservation techniques helps bamboo plants in sustaining its growth during hot and drought period. The result of the present study were in agreeing with the findings of Tejwani et al. (1975) reported that soil working has an important component on plant establishment, moisture conservation and consequently on plant growth. At Agra staggered contour trenches of 3 m x 45 cm x 45 cm spaced at 3 m apart with 5 m spacing between the rows were found to be the best for raising *Aacia arabica* by direct sowings and it is recommended for afforestation of Yamuna ravines. Yadav (1963) reported that higher level soil moisture content and soil accretion remained under bamboo (*D.strictus*) plantation than other species such as Sal, Teak and Chirpine. Similar findings was also reported by Mathur (1976), that the studies on the effect of land use such as natural fallow (grasses and shrubs) and grasses alone gave the least and negligible runoff from alluvial soils of semi arid regions.

The present study also revealed that significant amount runoff and soil loss were obtained and annual runoff and soil loss in different treatments was tabulated (Table 3). As evident from table 3, no runoff and soil loss were recorded in T_1 and T_2 treatments imposed catchments compare to T_4 (control). The higher soil loss was observed in T_4 followed by T_3. It was observed that the soil loss in 2010 is slightly higher than that of previous year in control due to a greater number of high intensity storms.

Table 3: Runoff and soil loss under different treatments

Treatment	Year	Rainfall (mm)	Runoff (mm)	Average soil loss (t ha^{-1})
T_1	2008	342.2	0	0
2009	265.6	0	0	
2010	456.0	0	0	
2011	226.2		0	0
T_2	2008	342.2	0	0
2009	265.6	0	0	
2010	456.0	12.27	0.77	
2011	226.2	0	0	
T_3	2008	342.2	0	0
2009	265.6	0	0	
2010	456.0	44.0	4.27	
2011	226.2	6.04	0.66	
T_4	2008	342.2	3.23	2.89
2009	265.6	2.18	2.56	
2010	456.0	85.02	4.93	
2011	226.2	35.83	4.40	

Findings from other authors also concomitant with this present study. Due to influence vegetation on soil which increased permeability of the soil reduces surface run-off, evaporation loss, allows better water penetration into the soil and increases drainage capacity of the soil. Sharma *et al.* (1992) have also reported that bamboo conserves soil moisture and mitigates the adverse drought effects on flora and fauna. It is also seen that bamboo based agroforestry models improve ecological parameters of a highly degraded basaltic tract of Jabalpur, Madhya Pradesh (Behari *et al*, 2000).

Rainfall during runoff producing events has only been considered : Soil loss decreased considerably and it has brought change in moisture and silt retention owing to the growth of a dense vegetative cover in upstream and downstream of the gully beds. Bamboo (*D. strictus*) plants roots play effective role to protect soil conservation structures like bunds, side slopes of gully beds. Similar study was reported by Sharda *et al.* (1982) was the effect of contour trenching on runoff and soil loss in 0.2 ha watershed at Agra was investigated. With a view to conserve 100% runoff likely to be generated, 346 trenches (average length 2.35 m) were dug in the watershed and planted with *Cenchrus ciliaris*. The runoff was brought down by contour trenching from 25 % to almost nil. In the same way soil loss was also reduced. Further, the staggered contour trenches of variable length were effective in intercepting the incident rainfall, increasing phytomass production and in preventing gully head extension.

Similarly, Bahadur *et al.* (1980) reported that improved farm practices and land treatment by bamboo, grasses and legumes on bunds, sloppy lands and terraced lands have been found to be effective means of soil and water conservation. Rambabu *et al.* (1978) observed that 40 % reduction in peak rate runoff on account of field bunding alone in agriculture lands of doon valley watersheds. Plantation of *D.strictus* was quite successful in increasing the economic productivity of gullied beds. About 4000 -5000 mature culms ha^{-1} was harvested every year from the bamboo plantations on the gully beds (Anon., 2007).

The results of this treatment have been found to be quite promising in biomass productivity and enhancing the insitu moisture use efficiency. Through this finding we could reclaim major ravine land ecosystem using *Dendrocalamus strictus* along with recommended moisture conservation measures. *D.strictus* have been identified for protective and productive use in ravine lands and this is a potential species for efficient resource utilization in non arable lands, livelihood support systems etc., This technology is applicable to light textured well drained soils including reclaimed ravine land where moisture and nutrient resources are scarce. The technology can suitably be adopted on the cultivable wastelands distributed in the north-western of India and in other areas having similar agro-climatic and soil conditions.

The treatments can be implemented in areas where problem of ravine lands exists. The technology reduces soil erosion, runoff, soil pH and nutrient losses considerably by imposition of trenches, basins and vegetative barriers due to improvement in infiltration rates results in higher water use efficiency.

CONCLUSIONS

It has been concluded through this study that planting of *Dendrocalamus strictus* with suitable moisture conservation practice proved as a viable alternative on ravine lands for gully beds stabilization, control sloping land erosion through good soil binding effect and fast growing vegetative cover. The rainwater optimally harvested through moisture conservation techniques like bunding, trenching, basin lifting and vegetative barrier. It helps in recharging sub-soil moisture status and judiciously reused during critical stages of plant growth in dry spells. The reduction of runoff volume, soil loss and established vegetative cover is effective in mitigating the ravine land extension, restoration of degraded lands and provide multiple benefits to local people through employment and income generation. In ravine lands, *D.strictus* acts as good soil binders owing to their dry hardy nature, peculiar dense clump formation and extensive interlocking fibrous root system, natural capacity to regenerate through its rhizomes which play important role in preventing erosion, water holding capacity and nutrient cycling under gulley beds. This result proved that *D.strictus* is an ideal tropical hardy bamboo species for massive rehabilitation of degraded ravine lands. Therefore, the use of bamboo in watershed management, soil and water conservation, rehabilitation of degraded land could be possible and bamboo farming and material processing are well suited to this region due to twin concerns of livelihood enhancement and environmental protection - the key components for developing these resource poor lands. Hence, an effective organized massive cultivation of this species through community participation could helpful to increase the green cover and establish industries that provide sustainable livelihood development to the local community.

REFERENCES

Anonymous, 2007. Central Soil and Water Conservation Research and Training Institute, Research Centre, bulletin, Agra-6.

Behari Bipin, Rashmi Agarwal, Singh, A.K. & Banerjee, S.K. 2000. Vegetation development in a degraded area under bamboo based agro-forestry system. *Indian Forester*, 126 (7): 710-720.

Mathur, H.N.1976. Effect of clear felling and reforestation on runoff and peak rates in small watersheds. *Indian Forester,* 102(4).

Bahadur Prakash, Satish Chandra and Deep Kumar Gupta. 1980. Hydrological studies on experimental basins in the himalayan region, The influence of man on the hydrological regime with special reference to representative and experimental basins, Symposium-IAHS-AISH Publ.No.130.

Rambabu, Agarwal,M.C., and Gupta,R.K. 1978. Effect of bunding on runoff and peak discharge in agricultural watersheds in Doon valley. Indian *Journal of soil conservation.* 6(2).

Sharma, B. D., Hore, D. K., Pandey, G. & Wadhwa, B. M. 1992. Genetic resources of bamboos in the NE region of India. *Indian J. Forestry,* 15(1):44-51.

Sharda, V.N., Bhushan,L.S and Raghuvir. 1982. Hydrological behaviour of ravineous watersheds under different land uses. Proc. International symposium on Hydrological aspects of mountainous watershed, VI, 14-18.

Singh and Tewari 2005. Uthan Newsletter, centre for sustainable development. Lucknow, Uttar Pradesh

Yadav, J.S.P. 1963. Soil moisture studies under bamboo (*D.Strictus*) plantation. *Indian Forester,* vol.89 (6).

□□□

Chapter 28

Rubber Dam : A New Construction Material for Rain Water Management Structure in Watershed Areas

N. SAHOO, S.K. JENA[1], P.S.B. ANAND[1], A. KUMAR[1]
A. MISHRA[1], D.U. PATIL[1] AND R. BEJ[2]

INTRODUCTION

Indian agriculture is dependent on monsoon. Farming being the principal means of livelihood for around two third of India's population of 110 crore, our agricultural strategy should be pro-nature and pro-small farmer oriented. Water use efficiency in Indian agriculture is one of the lowest in the world despite India having the largest irrigation infrastructure in the world. As per the mid term appraisal of 10th Plan, irrigation efficiencies are 25-30% & the area under irrigation is only about 40%. India's per capita water storage is the lowest at about 207 cubic meter well below in comparison with Russia (6103), Australia (4733), USA (1964), China (1111). To achieve higher and useful storage capacity, more numbers of smaller (rubber) dams are appropriate instead of larger dams. Floods caused by eight major river valleys over 40 M ha of area in the entire country affect a population of 260 M. Similarly drought conditions are there in about 68% of total sown area (16% of total area of the country) spread in 14 states. About 60% of the rainfall is lost through runoff. To arrest the runoff several check dams are required which should be cost effective in comparison to cement concrete structure. Only rubber dam fulfills that purpose. However, this small rubber check dam will again work as a boon for the drought prone area as it is cheap and easy to install (Sahoo *et al.*, 2010). It has also capacity for safe deflation under high flood condition and ability to control water level. India is having no. of small rivers which carry most of the flood water to sea through bigger rivers. Small rubber dams in multiple locations (a novel approach) will help in preventing floods to a great extent, reduce excessive dependence on monsoon, increase

TITE, Bhubaneswear, Odisha, India
[1]Directorrate of Water Management, Bhubaneswar, Odisha, India
[2]Command Area Development Authority, Odisha, India

water storage for release/delivery during drought, even distribution of water, increase water use efficiency, conservation of water, distribution of water to dry wells in drought prone areas and arrest desiltation. The total investment cost of rubber dams will be about 40% lower than the conventional gated regulating structure and the utility of the same is much greater in as much as it addresses the age old problems of flood, droughts, low water use efficiency, excessive dependence on monsoon etc. A series of water harvesting structures and check dam is contemplated in the watershed management plans of different watersheds in various schemes like DPAP, EAS,SGRY & lastly in Prime Minister Gramin Rojagar Yojana. Since concrete check dams are costly and installation is time taking rubber dam will be of much help particularly in poverty stricken area of drought prone tribal dominated watersheds. It has been observed that the impounding type of water harvesting structures fail due to sudden heavy downpour in the catchment. This failure of WHS can be avoided if rubber dams are used in place of earth or concrete because of its capability to deflate.

A rubber dam is an automatically operated gate for water control. It is a nylon-reinforced rubber body secured to a concrete with a unique anchor mechanism. A rubber dam is similar to a sausage-shaped balloon anchored to the stream bed. Flexi dams are fabric reinforced rubber membranes that are securely anchored to concrete foundations and inflated with either water or air to form uniquely versatile dams. When it is inflated, it rises to block water. In deflated condition, the dam body flattens on its foundation to permit free flow of water so that there is no damage to the structure.

Rubber dams are installed to function as weirs or barrages which are relatively low-level dams constructed across a river for the raising of river level for the diversion of flow in full, or a part, into a supply canal or conduit for irrigation, domestic and industrial use.

These types of dams are long tubular-shaped fabrics placed across channels, streams and weir crest to raise the upstream water level when inflated. In open channels, they are commonly used to raise water levels, to increase water storage and to prevent chemical dispersion. The interest in inflatable dams is increasing because of the ease of placement.

A rubber dam is inflatable and deflatable; when it is inflated, it serves as an agricultural weir (low-level dam) and when it is deflated it functions as a flood mitigation device (Sahoo *et al.*, 2010). These dams are mainly used as the dam body for reservoirs for the provision of irrigation water for rice fields.

The ease in which the Flexi dam can be lowered or raised to its design height means it can be used in numerous applications, *i.e.*,

- Raising the crest height of existing dams and weirs to increase their capacity and provide a system of flood control.
- Tidal Barriers to protect rivers from salt contamination.
- Creating recreational lakes without the environmental restrictions of other methods.
- Complex and controllable systems of diversionary and flow control weirs for flood mitigation and waste water treatment works.
- Replace existing steel gate systems.

These dams have a number of advantages, including the fact that they could be provided over a much longer span than those provided by steel gates. Also, rubber dams can be laid flat on river bottom without causing any obstruction to river flow. The simplicity and flexibility of the rubber dam structure and its proven reliability are key consideration in its wide scope of applications. A unique characteristic of the rubber dam is its ability to function as a reliable crest adjustable water gate. When inflated by a medium (air, water or combination of both), it rises to retain water. When deflated by releasing the medium, it collapses flatly down to the foundation; completely opening of channel for free flow of water .the rubber dam can also be adjusted to operate at intermediate heights to meet the needs for different upstream and downstream water levels at different upstream and downstream water levels at different time.

Selection of Site

Baseline/ Bench line surveys were conducted in Kalahandi, Keonjhar and Khurda districts of Orissa. In Keonjhar district, Janheinala of Kheri Tangri and in Khurda district, Kusuminalla,Sankusuminala,Badapokharia impounding structure,Sagadianala,Haladinala and Bajapur Loose Boulder Structure were surveyed. Data related to streams, watersheds and their hydrological behaviour were collected from these sites and analysed. After analysis, five sites *i.e.*, Badapokharia, Baghamari, Chandeswar-1, Chandeswar-2 and Mendhasal (DWM Farm) were selected as best sites for installation of rubber dams. All sites are under Khurda district of Orissa.

Design of Ideal Check dam

Before putting of rubber sheet into the concrete base structure in the field, a check dam was designed considering the optimum values of the parameters.

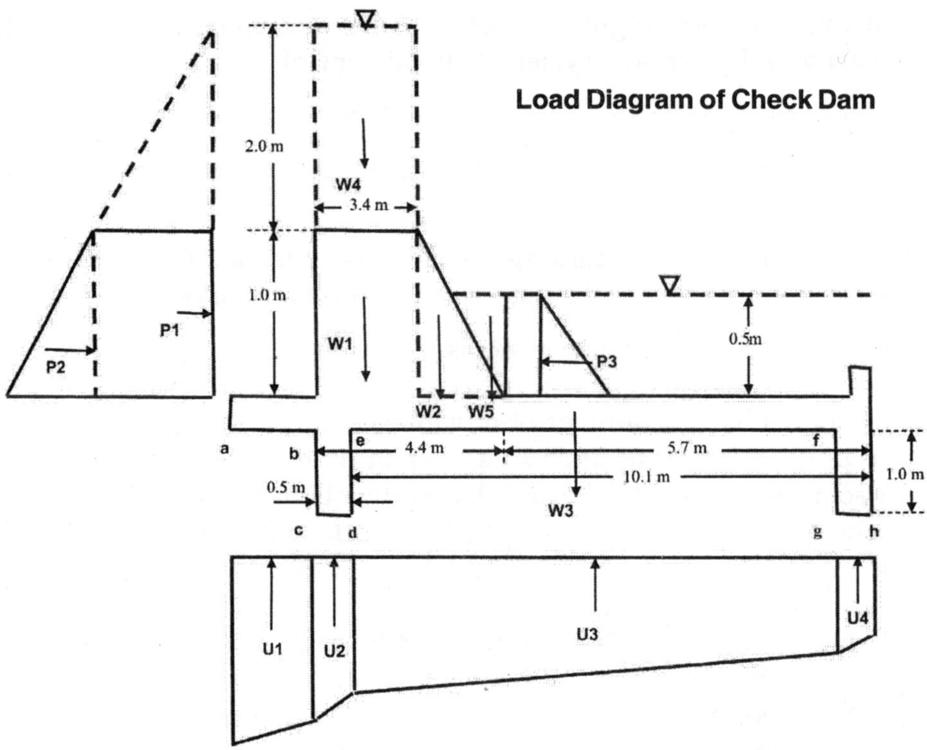

In the above diagram, to determine the different parameters of the structure, flow velocity and peak discharge of the stream were considered as 2.0 m/sec and 32.0 $m^3 s^{-1}$, respectively. The height of head wall is 1.0 m and top width of dam is 3.4m. To maintain stability in dam, the bottom width is to be kept as 3.4 m.Width of cut off wall and toe wall is considered as 0.5 m to avoid sliding of the structure. The forces acting on the dam are due to weight (W), lateral pressure (P) and uplift (U). After calculating these forces, it was found that the structure is safe against overturning and sliding.

The dimensions of the rubber dam installed in five different sites are different. The Chandeswar given in the following table:

Table 1: Dimensions of rubber dam at different places

S. No.	Site	Dimensions (m)		
		Length	Breadth	Height
1.	Baghamari	5.0	1.20	1.0
2.	Badapokharia	2.0	1.50	1.0
3.	Mendhasal(DWM Farm)	2.0	0.56	0.5
4.	Chandeswar I	5.5	1.50	1.5
5.	Chandeswar II	5.6	1.50	1.6

Hydrological Survey

For designing of rubber dam, hydrologic data of the sites are most important. Hence hydrological data like rainfall, runoff, stream flow etc. were collected and analyzed. The rainfall data of ten years and runoff in mm at 80% probability in monsoon months of Khurda district is given in the Table 2 and Table 3.

Table 2: Normal weekly rainfall of khurda district (10 years) and of 2009 in parenthesis in mm

Week/Month	June	July	August	September
1st	100.3(90.8)	283.8(312.5)	102.8(120.5)	72 (73.5)
2nd	5.8 (37.8)	1.4 (65.3)	138.7(143.5)	10.7 (15.3)
3rd	17.4 (16.2)	70.8 (134.3)	152.1(142.7)	104.4(82.1)
4th	65.3 (79.4)	76.3 (95.7)	189.3(178.5)	92.8 (112.7)
Total	188.8(224.2)	432.3(607.8)	582.9(585.2)	279.9(283.6)

Table 3: Runoff (mm) at 80% probability in the site in monsoon months

S. No	Month	Runoff at 80 % probability
1	June	38.1
2	July	129.6
3	August	233.2
4	September	84.1

For sustainability of dam, stream flow data of the rubber dam sites were analyzed. Water level data were taken using levelloger instrument in monsoon months of the year 2009 and 2010.

Fabrication of Rubber Dam

Rubber dam as a hydraulic structure mainly consists of four parts, *i.e.*, (i) a rubberized fabric dam body (ii) a concrete foundation (iii) a control room housing for mechanical and electrical equipments, such as air blower/water pump, automatic inflation and deflation mechanism (iv) an inlet/outlet piping system

For fixing of rubber sheet with the concrete structure, following materials are necessary.

- MS Flat
- Nut and Bolt
- Washers
- Adhesives like araldite, holdtite, MSeal, Sillica gel

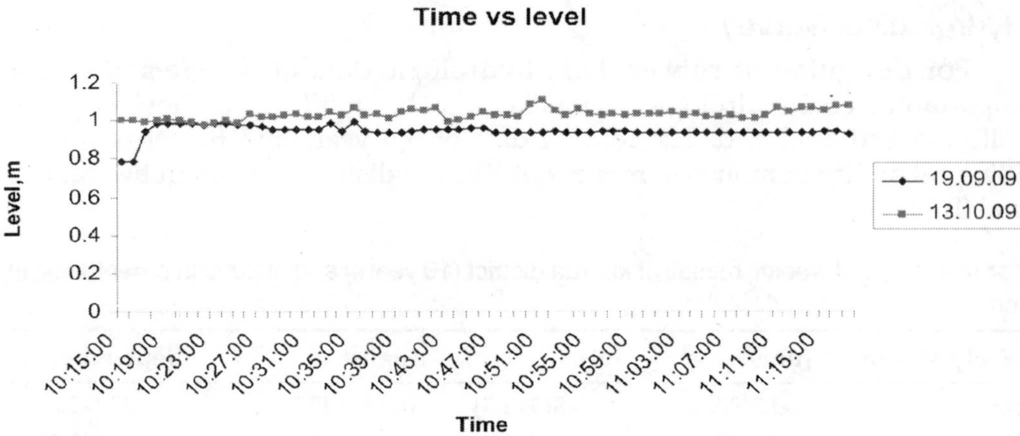

Fig. 1: Water level in the month of September and October of 2009 in Chandeswar-1 of Khurda district

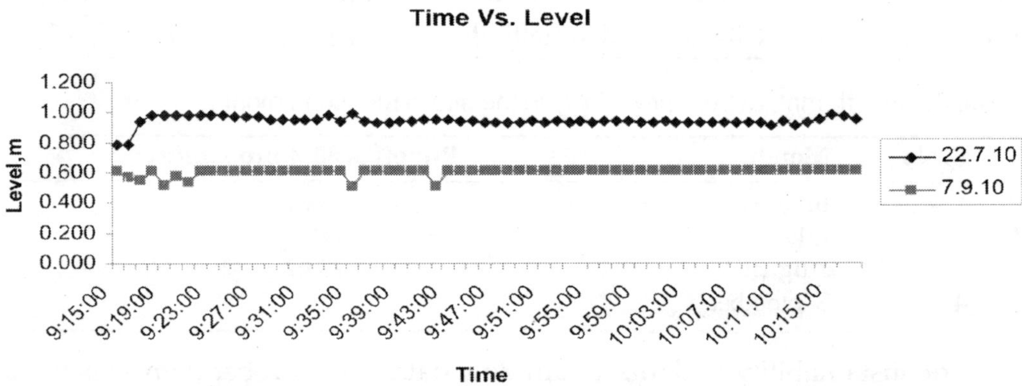

Fig. 2: Variation of water level with time measured in the month of July and September of 2010 at Baghamari

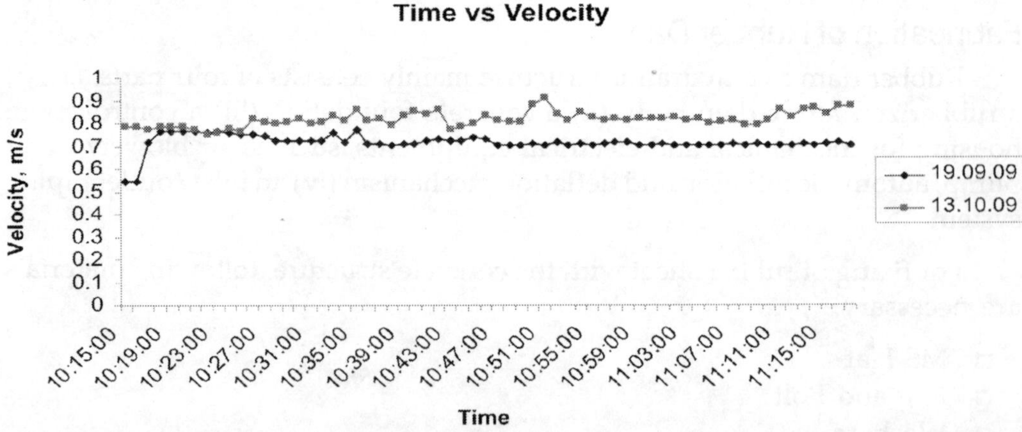

Fig. 3: Velocity of stream in the month of September and October of 2009 in Chandeswar-1 of Khurda district

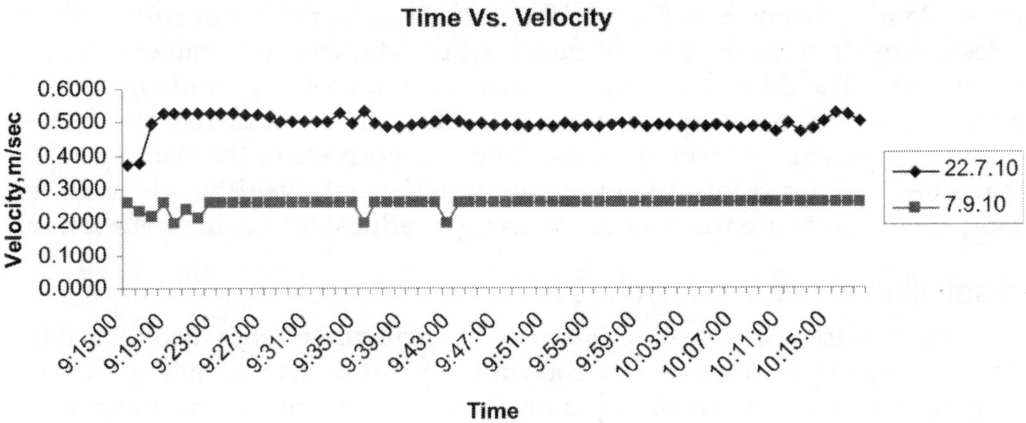

Fig .4: Velocity of stream in the month of September and October of 2010 in Chandeswar-1 of Khurda district

To fabricate different components of rubber dam, MS flats are cut into sixteen different parts of desired dimensions. Four pieces are equal in dimensions which are fitted on the base length. Another four pieces are used to fix the rubber sheet in two side walls of the structure at an angle of 23° to 31° to the base of the structure. Rest smallest four pieces are used to connect the flats of base and side wall. These flats are welded at an angle of 107 ° (Sahoo *et al.*, 2010) with the base flats. The flats cut in a desired dimension are drilled by drill bit of nut and bolt dimension in a drilling machine. The holes

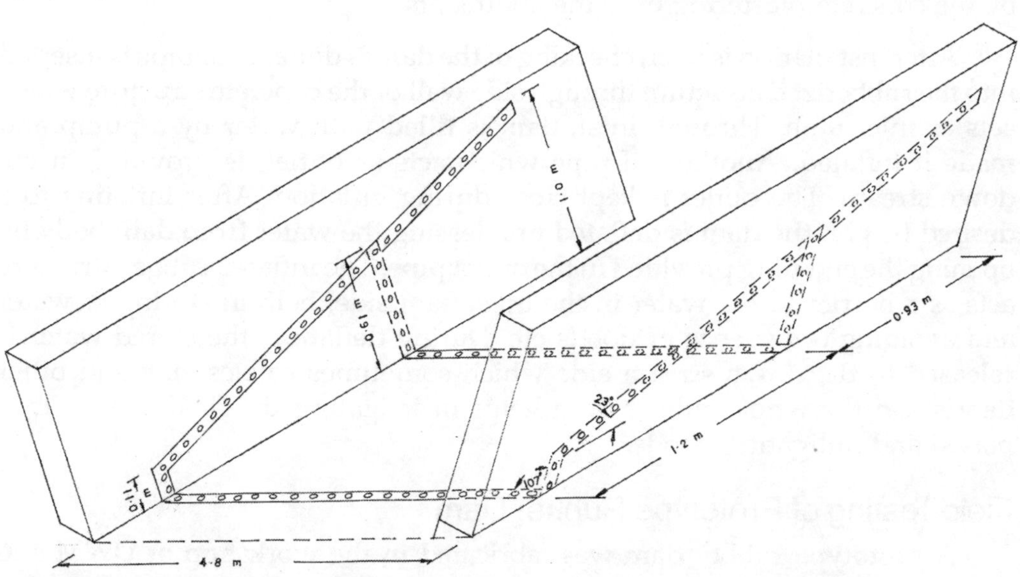

Fig .5: Isometric view of Rubber dam fabrication at Baghamari D/W No.4

are made at a distance of 8 cm in flat as well as in rubber.In rubber sheet, holes are made with the help of punch whose dimension is same as that of nut and bolt. The thickness of rubber sheet used in prototype and big dam are 6mm and 8 mm, respectively .During installation holes in flat and rubber sheet are matched for fixing purpose. The end portions of the flats are made sharpened by files prior to joining the same through welding. Holes in flat and rubber sheets are made clear for using of adhesives during installation.

Installation of Rubber Dam

In this phase, concrete foundation for the structure is renovated by making slots for placing of the iron flat both in the sides as well as along the base length. The base length of the dam varies according to the purpose of construction. The double line anchoring system is fixed in the concrete foundation by concrete itself.

The fabricated rubberized structure which comprises of hloes already being made on it according to the dimensions of the base length, nuts and bolts size are fitted in the double iron flats tightly with the assistance of adhesives. The main hurdle is faced during this phase as because the upper flats are to be detached from the lower flat (fixed in the concrete foundation) by opening the bolts. There requires a little bit of adjustment of the nuts and bolts to put the upper flat over the lower one having the rubber structure in between the two flats. Once the rubber is anchored then mseal, araldite and silica gels are applied to prevent any leakage of water from the rubber structure. Aftermath, the abutments are done properly and with great care, followed by the concrete plastering over the abutments.

After installation is over, checking of the dam is done. A GI pipe is inserted into the rubberized structure through side wall of the concrete structure which acts as inlet pipe. Through inlet, dam is filled with water by a pump and made it inflated. Another GI pipe which acts as outlet, is provided in the down stream. The outlet is kept close during inflation. After inflating to a desired height, the dam is deflated in releasing the water from dam body by opening the end plug provided in the outlet pipe. The inflated rubber structure acts as a barrier to the water in the up stream side, help in storing of water and avoiding occurrence of floods etc. During deflation, the stored water is released to the down stream side which sometimes carries silts and other debris. On the whole rubber dam helps in irrigating the field during lean period and mitigating floods.

Field Testing of Prototype Rubber Dam

A prototype rubber dam was fabricated in the workshop of DWM and installed in DWM Farm at Mendhasal. The length, breadth and height of this type of dam was 2m, 0.56m and 0.5m respectively. The hole to hole distance was maintained at 5 cm. An iron plate was provided over the base of the structure above which the prototype dam was anchored with base. The plate was provided to prevent leakage of water from the dam body. A prototype rubber dam of the above dimension built by IRMRA, Thane was tested in the hydraulics laboratory of IIT,Bombay.This model was run for various possible conditions and the results observed are reported here. Following tests were carried out for the model.

i) Inflation and deflation process.

ii) Pressure variation in the dam after inflation time for inflation and deflation.

iii) Static water level testing to ascertain deformation of the rubber dams at various dam pressures and upstream flow depths.

iv) Dynamic flow test to ascertain deformation of the rubber dams. The experiments conducted were for

- At different dam pressures of 0.3 kg cm^{-2};0.6 kg cm^{-2} and 0.9 kg cm^{-2}
- At different overflowing depths over the dam and discharge rates.

From the above experiments it was shown that, the inflation and deflation time was ranging between 30 to 40 minutes. The inflation/deflation time may be reduced by using more effective pumping, inlet and exit mechanisms. As per available literature, the ideal inflation/deflation time is 20 to 30 minutes. If the inflatable pressure in the rubber dam is higher, there is less deformations when water is stored in static conditions or dynamic conditions of overflow. Hence at least the pressure in the rubber dam may be kept in the range of 0.3 to 0.6 kg cm^{-2}. It was observed from the experiment that the rubber dam performance during the static and dynamic testing was satisfactory. The rubber dam deformations in the static conditions were slightly more than dynamic conditions. In the overflowing dynamic conditions, rubber dam is affected by various hydrodynamic loads of upstream water pressure, downstream water pressure, overflowing water velocity and effects of any debris in the flowing water. Hence care should be taken that the dam body is not affected during the overflowing time.

Field Testing of Large Dam at Baghamari

During filling of water in rubber dam, time was noted at different height. From above figure it was observed that, total time for inflating a height of 1 meter is 17 minutes. (Sahoo *et al.*, 2010).

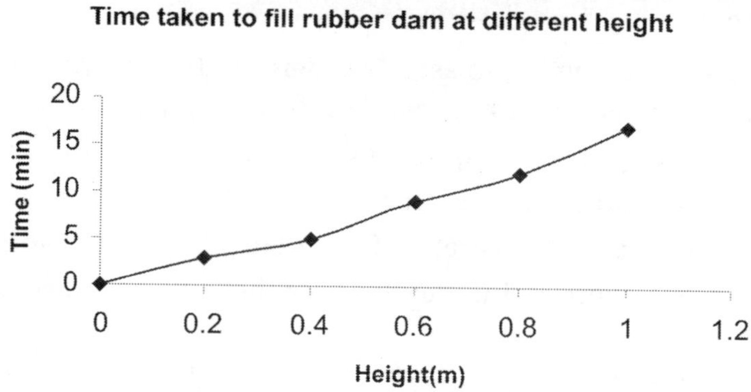

Fig. 6: Inflation height with time

Pressure variation in inflation: For inflation, small low lift Krishak Bandhu Pump was tried. It was possible to inflate in small proto type dam of 2 m length & 6 mm thickness up to 0.5 m. But in case of 5 m span at Baghamari, it could not inflate. So 1.5 H.P kerosene operated pump was used.At 0.6 kg cm^{-2} it was able to inflate up to 0.5 to 0.6 m. For raising dam height up to 1m, pressure required was more than 1kg cm^{-2} *i.e.,* 1.2 kg cm^{-2}. It was measured by pressure gauge.

Silt removal: During 2008-2009, the silt load above the dam varies from 0.1 m to 0.05 m upwards. But due to inflation and deflation in rubber dam, silt load varies from 0.01m to nil upwards.

Wrinkling: In case of prototype rubber dam, wrinklings are more and there was rupture sign which may decrease the life. But in case of 5m span dam at Baghamari, the wrinkling was less and there is no rupture.

Anchoring angle :The anchoring angle in real field condition was maintained from 105° to 110°. When anchoring angle was 110°, the wrinkling are less. In other two dams the angle will be further increased.

Spacing between bolts :The spacing between bolts varied from 5 cm to 8 cm. In prototype it was 10 cm. There was little leakage on corner. Then again between two bolts, another bolt was given reducing the clearance to 5 cm. It worked well. But in case of Baghamari, the spacing was increased to 8 cm. Different gums in different proportion was given to make it water tight. Absolutely there is no leakage found in rubber dam.

Effect of rubber dam : From 15th July' 2010 to 13th Aug'2010, there were no rainfall in the rubber dam project site *i.e.,* Baghamari in Khurda district of Orissa. The fields became dry. But the farmers in Baghamari village got benefit by irrigating their fields with stored water in the up stream side of rubber dam through inflation. Farmers also utilize water in the flowering stage of rice which is very very crucial period for crop, from this rubber dam.

Sample rice yield data of 19 farmers in the Baghamari project covering about 26.75 ha was taken. The area of farming varied from 0.5 ha to 6 ha. Average yield of rice before installation of rubber dam was 2.87t ha^{-1} and it increased up to 4.67t ha^{-1} after installation of the dam. Hence the percentage increase in yield was 62%. During Rabi season the pulse yield was very poor (0.4t ha^{-1}) before installation whereas after installation it increased up to 82%. An area of very little (0.1ha) was covered during summer for vegetables. Hence rubber dam is feasible in this area.

CONCLUSIONS

From the research work on rubber dam, it has been found to be an effective Inflatable Flexible Membrane System for various purposes. Due to its structural simplicity, being inflatable and deflatable, the rubber dam can be put into wide scope of applications such as a dam for storage of water for irrigation, water supply, power generation, fish culture, flood control, environmental improvement and recreation. The rubber dam can be used for various watershed management schemes for rainwater harvesting, soil erosion control and water management practices.

REFERENCES

Sahoo, N., Jena, S.K., Kumar, A., Mishra, A. and Anand, P.S.B. 2010. "Rubber Dam - A new water storage device for flood mitigation and irrigation". Technical Annual, Fifty One Session, The Institution of Engineers (India), Orissa State Centre, Bhubaneswar. pp158-163.

Sahoo, N., Jena, S.K., Kumar, A., Mishra, A., Patil, D.U., Anand, P.S.B. and Bej, R. 2010. "Rubber Dams- A New Concept for Flood Control in Coastal Watershed". *In*:Proc. Of 9th National Symposium of ISCAR, Goa, October 27-30, 2010. pp177-178.

Sahoo, N., Jena, S.K., Kumar, A., Mishra, A., Patil, D.U. and Anand, P.S.B. 2010. "Design, Development and Installation of indigenous Rubber Check Dam in watershed". Proceedings of All India Seminar on "Engineering Interventions to enhance income of small and marginal farmers". Institution of Engineers, New Delhi, July 9-10, 2010. pp183-191.

Sahoo, N., Jena, S.K., Kumar, A., Mishra, A., Patil, D.U., Anand, P.S.B. and Bej, R. 2010. "Rubber Dam: An eco friendly structure for watershed in context to climate change". Proceeding of National Symposium on Sustainable Rice Production System Under Changed Climate, CRRI, Cuttack, November 27-29, pp25-26.

□□□

Chapter 29

Pond Based Integrated Farming System for Yield Stability in Rainfed Areas under Aberant Weather Conditions

S. K. RAUTARAY, A. MISHRA, R.K. MOHANTY, O. P. VERMA
M.S. BEHERA AND A. KUMAR

INTRODUCTION

There is increased demand for production of food, fuel, fibre and fodder with the available land and water resources. Population explosion has led to decreased per capita land and water availability. It's projected that per capita land availability may reach 0.09 ha by 2020 AD and it will be difficult to produce enough food for a family at that time. Further, the income from cropping alone in small and marginal farm is hardly sufficient to sustain the farmer's family. Also, Farmers need regular income for decent living. Hence, the unitary concept of crop, fish or animal husbandry is gradually changing to the integrated system with the view of producing fish, meat, egg, milk, vegetables and other allied products within farm itself. The basic necessity of such integration is not only to make the farm an independent unit, but also to fulfill the demands as input to other structural units. Regarding area under irrigation, there is limited scope for substantial increase. In spite of large-scale expansion of irrigation system, about 60% of the cropped area is under rainfed system and it is likely to continue. Climate change impact has resulted in aberration of weather parameters like rainfall and temperature. Wide variations in rainfall amount and its distribution will intensify the inherent problem of rainfed areas. Integrated Farming System is one of the foreseeable management options for the farmers of rainfed ecology to mitigate the stress under aberrant weather conditions. It emphasizes judicious combination of any one or more of such enterprises and effective utilization of available resources with the small and marginal farmers (Behera and Mahapatra, 1999). It also ensures the recycling of the available resources, by-products and farm wastes within the farm itself (Behera and Mahapatra, 1998). With the

Directorate of Water Management, Chandrasekharpur, Bhubaneswar-23, Odisha, India decreased availability of fresh water, its multiple uses under integrated farming systems are important.

MATERIALS AND METHODS

On-station field experiments were conducted under rainfed lowland situation during during 2002 to 2007 at Regional Rainfed Lowland Rice Research Station, Gerua, Asom (Rautaray *et al.*, 2005) and also, under rainfed upland situation during 2009 to 2011 at the Institute Research Farm of Directorate of Water Management, Bhubaneswar, Odisha to evaluate pond based integrated farming systems for yield stability in rainfed areas.

Farming system under rainfed lowland situation: A farming system model unit was established using one low lying rice field (125m x 40m). The components of rice-fish farming system were *Sali* rice (winter rice grown during June-July to November-December) in wet season, and *Boro* rice (spring-summer rice grown during November-December to May-June) or winter vegetables in the dry season in the main field. Main field occupied 60% of the total area for growing *Sali* rice in wet season while *Boro* rice or winter vegetables in dry season. Dykes occupied 23% area for growing horticulture and agro-forestry plants. The pond refuge (30m x 12 m x 2m) was constructed at the lower end (down slope) of the field occupying 7% area. The two trenches of 2.5 m width occupying 10% of the total area were constructed adjacent to the dykes along the longitudinal side of the field and those connected to the pond refuge at one end. The dug-out soil from the pond refuge and side trenches was used for construction of wide dykes all around with bottom and top width of 4m and 2m, respectively.

Fishes were grown in the refuge pond, trenches, and also in the rice field (at appropriate water level). Fingerlings of catla (*Catla catla*), rohu (*Labeo rohita*), mrigal (*Cirrhinus mrigala*), bata (*Labeo bata*), common carp (*Cyprinus carpio communis*), silver carp (*Hypophthalmichthys molitrix*) and China punti (*Puntius gonionatus*) were released in the pond during May to first week of June. A total of 6,250 fingerlings consisting of catla (24%), silver carp (8%), rohu (12%), bata (12%), mriga1 (24%), common carp (16%), and China-punti (4%) were released. Six platforms of 4mx 3.5m size were constructed from the inner side of the dyke hanging over the trenches for planting the creeper vegetables at the inner side of the dyke and allowing the plants to creep over it. Climbing and creeper vegetables like country bean, bottle gourd, squash gourd and ridge gourd were grown over the platforms for crop intensification over space. Banana, papaya, coconut, arecanut, guava and lemon were grown on dykes. Teak (15 plants) as agro-forestry component was planted only on northern dyke. Vegetables on dyke included snap bean, radish, brinjal, okra, pumpkin, tomato, poi, spinach, coriander and lai sag.

Farming system under rainfed upland situation: The dimension of the pond was 30 m x 30 m. The depth of the pond was 1.8 m. Composite pisciculture was practised in fish pond. Fish seed of *Catla catla, Labeo rohita* and *C. mrigala* were stocked after proper acclimatization @ 15,000 fingerlings ha^{-1} (Mean body weight 3.6 $\pm$ 0.6g). Stocking composition was 20:40:40. Supplemental feeding was provided with a ratio of 65:35 (rice bran: mustard oil cake) @ 5%, 4%, 3% and 2.5% of MBW, twice a day, during 1st, 2nd, 3rd and 4th month before harvesting, respectively. Periodic manuring with raw cow dung @ 500kg ha^{-1} and liming @ 50kg ha^{-1} were carried out at every 15 days interval to maintain plankton population in the eco-system. Fish pond was surrounded by a strong dyke for growing papaya and banana. An apiary unit was established on the dyke. In the main field, rice based cropping system was practiced involving okra, pumpkin, water melon and bottle gourd using the harvested water. Okra was grown under different planting geometry to evaluate differential water requirement under them. For this purpose, uniform row spacing of 50 cm was considered as standard practice and it was compared with paired row planting by reducing the row spacing to 40 cm and increasing the furrow width to 60 cm. Thus, similar plant population was maintained but the number of furrows for irrigation is reduced by about 50%. The third planting pattern was imposed by keeping three rows at 40 cm spacing as one bed alternated with one furrow of 60 cm. In this method, there was increase in plant population. The creeper crops (pumpkin, water melon and bottle gourd) were grown without mulching and mulching with paddy straw).

RESULTS AND DISCUSSION

Farming system under rainfed lowland situation:Wet season (*kharif*) for the year 2006 was a deficit rainfall year with mean precipitation for the growing period (July to November) was 466 mm and it was 51% of the normal. Farmers could not puddle and transplant paddy. As a result, a large chunk of farmers' field remained fallow. In this situation, rice-fish system with a strong dyke (90 cm) all around could act as water harvesting structure. Harvested rainwater was utilized for puddling and the crop was transplanted timely in July. During the period from 2 July to 12 August (standard meteorological week 27 to 32), low rainfall (246.9 mm) coupled with high temperature (weekly mean maximum temperature up to 35.7 ^{0}C), bright sunshine hours (up to 9.3 h) and evaporation (up to 4.4 mm d^{-1}) accentuated the problem in sole rice system. Highest maximum temperature, sunshine hours and evaporation during this period were 37.5 ^{0}C, 11.8 h and 5.4 mm, respectively. However, in rice-fish system, intensity of problem was less due to presence of harvested water. Water with high specific heat allowed slow rise in temperature of the system and presence of horticultural and agro-forestry plants are further useful. A normal rice crop could be harvested (3.71 t ha^{-1}) under rice-fish system as compared to large-scale fallow under sole rice system under farmers' field. Thus, promotion of large number of rice-fish systems may act as a series of miniature water harvesting structures for harvesting the costly plant nutrients, organic matter, soil, and water. With these benefits, the system had potential in stabilizing crop productivity in fragile rainfed lowlands with uncertain weather conditions.

Gross income generated from different components of rice-fish integrated farming system during the period from 2002-03 to 2006-07 is presented in Table 1. The system as a whole generated ₹ 47,810 ha^{-1} in deficit rainfall year 2006-07 to ₹ 92828 ha^{-1} in flash flood affected year 2004 with a mean of Rs 68962 ha^{-1}. This is high compared to generation of ₹ 12496 ha^{-1} from traditional practice of growing high yielding rice variety in *Sali* (wet) season at (Rautaray *et al.*, 2005). Rice accounted 41% of the gross income followed by fish (34%), and horticulture (25%). The fishery component had higher income generation potential as the area allocated for this component was only 17% (10% for pond and 7% for trench). Higher income from fish component suggests the benefit of pond based farming system besides its role as micro-water source. Rice accounted 60% of the area but generated only 41% of the gross income. The mean rice equivalent yield for the period from 2002-03 to 2006-07 was 11.5 t ha^{-1} , which is higher than sole rice cropping 3 to 4 t ha^{-1}and double rice cropping 7 to 8 t ha^{-1} . Income is again expected to increase in the subsequent years from the produce of coconut, areca nut, and teak crops.

Farming system under rainfed upland situation : During *kharif* season, rice crop did not require irrigation. Rice grain yield was 4628 kg ha^{-1}. After harvest of *kharif* rice, different rice based cropping systems were tested. The yield of cabbage was highest (33.82 t ha^{-1}) followed by tomato (15.72 t ha^{-1}), sweet corn (5.72 t ha^{-1}), baby corn (5.33 t ha^{-1}). The two seed crops (sunflower 1.22 t ha^{-1} and horse gram 0.98 t ha^{-1}) yielded low. Net returns were highest under baby corn ₹ 46981 ha^{-1}, followed by cabbage ₹ 41574, sweet corn ₹ 29108, tomato ₹ 8237, sunflower ₹ 3636 and lowest with horse gram ₹ 2248. Fish yield was 191 kg with the productivity of 2.12 t ha^{-1} per 226 days. A strong dyke surrounding the fish was utilized for growing horticultural crops (banana, papaya and coconut). Mean yield of papaya was 53 kg plant^{-1} (₹ 530 plant^{-1}). Apiary unit yielded 4.1 kg honey box^{-1}. Poultry dropping were directly falling on the pond water for direct use as fish feed. Vanraja chicks were grown in the system as broiler bird yielding 2.06 kg bird^{-1}. Paddy straw mushroom was grown during May to September. Each bed required 10 sq.ft area. One bed required expenditure of Rs 40 (spawn Rs 10+10 kg straw ₹ 10 + Gram powder ₹ 6 + Polythene ₹ 9 + Labour ₹ 5) with average yield of 1 kg with market price ₹ 80. Thus, net returns per bed of paddy straw mushroom were ₹ 40 in a cropping cycle of 30 days. Oyster mushroom was grown during October to February. Each bed required 4 sq.ft space and expenditure of ₹ 20 (2 kg straw ₹ 2 + 1 bottle spawn ₹ 9 + Polybag ₹ 2 + 250 g boiled wheat ₹ 4 + Labour charge ₹ 3). Yield of oyster mushroom per bed was 1.5 kg with market price of ₹ 45. Thus, net returns per bed were ₹ 25 in a cropping cycle of 45 days.

Table 1: Gross Income (₹ ha^{-1}) and rice equivalent yield (t ha^{-1}) from different components and, economics of integrated rice-fish system

Year	Gross Income (₹ ha^{-1})			
	Rice	Fish	Horticulture	Total
2002-03	48880	24000	1804	74684
2003-04	22752	25200	14996	62948
2004-05	39078	28320	25430	92828
2005-06	15582	31650	19398	66540
2006-07	14112	8040	25658	47810
Mean	28081	23424	17457	68962
% of gross income	41	34	25	100
Rice Equivalent Yield	4.68	3.90	2.92	11.5

For efficient use of harvested pond water in post rainy season, paired row bed planting was found useful as compared to farmers' practice of ridge planting at uniform row spacing of 50 cm (Rautaray *et al.*, 2012). The results of dry season of 2010 revealed that paired row planting on bed produced similar yield (7.43 t ha^{-1}) with the normal planting but with the less irrigation requirement by 18% (Table 2). Triple row bed planting resulted in lower yield (8.7%) and less irrigation requirement (28.5%). The results suggest that farmers should adopt paired row bed planting to save water without sacrificing yield. However, under water scarcity situation farmers may opt for triple row planting to get higher yield per unit water. The effect of mulching with decomposed paddy straw in the dry season of 2010 revealed that it improved the yields of watermelon and bottle gourd by 12.4 and 14.0 %, respectively as compared to no-mulch.

Table 2: Effect of planting techniques on irrigation requirement and yield of okra (dry season, 2010)

Planting techniques	Yield (t ha^{-1})	Irrigation water use (mm)	Irrigation Water Use Efficiency (kg yield ha^{-1}-mm)
Normal Planting (50 x 25 cm)	7.72	625	12.4
Paired Row (40/60 X 25 cm)	7.43	513	14.5
Three row (40-40/60 x 25 cm)	7.05	447	15.8
CD (0.05)	0.49	-	-

CONCLUSIONS

Pond based farming is beneficial in increasing farm income, creating employment opportunity and improving soil health. The system is useful in minimizing risk, encouraging biodiversity, and improving ecosystem as a whole. Farmers can get fresh, nutritious and balanced food for healthy lifestyle. Promotion of large number of such systems may act as a series of miniature water and nutrient harvesting structures for harvesting the costly plant nutrients, organic matter, soil and water, which would otherwise be dumped into the sea. With the global warming and climate change, a more departure in frequency, intensity and amount of rainfall is expected and thus, utility of these miniature structures may assume greater importance in future, especially in tropical climatic regions. For efficient use of pond water, paired row bed planting of okra, and mulching of water melon and bottle gourd with decomposed paddy straw were effective.

REFERENCES

Behera, U.K. and Mahapatra, I.C. 1998. Income & employment generation for small and marginal farmers through integrated farming system. *Indian Farming,* 48:16-28.

Behera, U.K. and Mahapatra, I.C. 1999. Income & employment generation for small and marginal farmers through integrated farming system. *Indian J. Agronomy,* 44(3): 446-51.

Rautaray, S.K., Das, P.C. and Sinhababu, D.P. 2005. Increasing farm income through rice (*Oryza sativa* L.) - fish based integrated farming system in rainfed lowlands of Assam. *Indian Journal of Agricultural Sciences,* 75 (2): 79-82.

Rautaray, S.K., Mishra, A., Mohanty, R.K., Kumar, A. and Behera, M.S. 2012. Effect of planting techniques and row arrangement on irrigation requirement and yield of okra (*Abelmoschus esculenta*). *In :Abstracts* Global Conference on Horticulture for Food, Nutrition and Livelihood Options held during 28-31May, 2012 at OUAT, Bhubaneswar, 224p.

□□□

Chapter 30

Effect of Integrated Nutrient Management on Soil Properties and Performance of *Aonla (Emblica officinals* Gaertn) based Agri-Horti System in Bundelkhand Region

H. BISWAS, DEV NARAYAN, BRIJ LAL LAKARIA* AND SANT RAJ

INTRODUCTION

Rainfall distribution in Bundelkhand (23° 10′ - 26° 30′ N and 78° 20′ - 81° 40′ E) is erratic and long dry spells during *kharif* season are common. The shallow-depth, coarse-textured red soils (Lithic Ustorthents, loally called *rakar*) frequently undergo moisture stress for most part of the year, forcing the farmers of the region to take up only one successful crop. Some recent studies have shown that it is possible to take up two and even three crops in a year through careful selection and management of limited irrigation water (Narayan *et al.*, 2006). Crops such as green gram and mustard have been found to perform well on red soils with irrigation applied at critical stages (Narayan *et al.*, 2001).

Alternate land use systems, particularly agri-horti systems with the inclusion of *aonla* (*Emblica officinalis* Gaertn.) on marginal, less productive soils, have a tremendous potential in Bundelkhand region (Pathak and Bhatt, 2001). Besides having a nutritional and medicinal value, cultivation of *aonla* is also highly remunerative for small and marginal farmers (Singh and Mishra, 2007).

Farmyard (cattle) manure is the most frequently applied organic fertilizer in Bundelkhand region. In recent years however, focus has shifted to the use of composted manures, especially vermicompost and NADEP compost. No information is available on effect of these manures on the soil properties in the region. The present study was conducted to evaluate the influence of organics applied in conjunction with inorganics on the physico-chemical properties of a less fertile red soil.

Central Soil and Water Conservation Research and Training Institute, Research Centre, Datia (M.P.) – 475 661

* Indian Institute of Soil Science, Nabibagh, Berasia Road, Bhopal

MATERIALS AND METHODS

A field experiment was started at the CSWCRTI Research Farm at Datia (25° 40′N, 78° 28′ E) in *kharif*, 2005 on an *aonla*-based cropping system, with green gram (var. PDM-54) and mustard (var. JM 1) as the field crops grown in rotation on a shallow Lithic Ustorthent (locally known as red soil or *rakar* in Bundelkhand) under rainfed conditions. The soil is a sandy loam, with pH (1:2.5)-7.23, EC-0.13, organic carbon-0.34%, low in available nitrogen, and medium with respect to available phosphorus and potassium. The moisture content at field capacity and permanent wilting point were 14 and 5%, respectively. The crops were grown on 21 m X 14 m plots in a Randomized Block Design following recommended package of practices. Five fertilizer treatments were applied as follows: T_1: Control (No nutrient treatment); T_2: NPK (Recommended dose through chemical fertilizers); T_3: T_2 + FYM (@10 t ha^{-1}); T_4: T_2 + NADEP Manure (@ 5 t ha^{-1}), and T_5: T_2 + Vermicompost (@2 t ha^{-1}). The treatments were replicated four times to thus form 20 plots.

The NADEP compost was prepared in the Research Farm from FYM-soil slurry and farm/forest wastes such as twigs, plant litter and stovers. Vermicompost was also prepared in the Farm with the help of earthworm species *Eisenia foetida*. Soil samples (0-60 cm) were collected every year after the harvest of *rabi* crop and analysed for physico-chemical and chemical properties. Soil moisture samples (0-30 cm) were collected every fortnight after the sowing of the crops till harvest. Soil pH, EC, organic carbon, plant-available nutrients, water stable aggregates, infiltration rate, bulk density and water holding capacity were determined using standard procedures (Jackson, 1973). The results were analysed using standard statistical procedure (Gomez and Gomez, 1984).

RESULTS AND DISCUSSION

The yield pattern of green gram and mustard grown under rainfed conditions over the five-year period has been shown in Figures 1 and 2. The higher yield of green gram in 2005 as compared to the subsequent years is due to better rainfall distribution during the season. During the other years, drought and/or long dry spells during the cropping season brought about the overall yield decline. While the years 2006 and 2007 were drought years, prolonged dry spells ranging from 6 to 14 days, and early cessation of monsoon (7th to 15th September during the years under study) were probably responsible for the decrease in green gram yield. The lowest overall green gram yields during 2007 and 2008 could be attributed to the prolonged dry spells during the last week of August to the first week of September, which coincided with the pre-flowering stage of green gram.

During 2005, the yield of green gram was 1014 kg ha^{-1} under the treatment T_4 (NADEP) followed by T_5 (vermicompost), which recorded a yield of 935 kg ha^{-1}. The former treatment fared 39% better as compared to control (T_1). The lowest overall yield was obtained during 2008, which was the fourth consecutive drought year in the region. Even during that year, T_5 yielded 205 kg ha^{-1}, which was 23% higher as compared to T_2 (100% RDF), and 43% higher as compared to control (T_1). Table 1 shows the status of soil moisture under different treatments over the years of experimentation. The comparison was necessary because of the fact that under rainfed conditions, the residual moisture plays a vital role in the growth of the crops and ultimately their yield. Out of the total number of soil samples collected (per year during the growth stages of both the crops excluding the rainy days) for moisture analysis, the treatments where organic sources of manures were applied, particularly T_4 and T_5, retained available moisture for a longer duration. This is one of the attributing reason for the higher crop yields under these two treatments even during prolonged dry spells. The organic composts, through their higher binding capacity improve the soil aggregation and porosity, which in turn increase the moisture retention capacity of the soil. Increase in green gram yield through application of vermicompost was demonstrated by Rajkhowa *et al.* (2000), Abraham and Lal (2003), and Nagavellamma *et al.* (2004).

Organics applied during *kharif* season are known to have a residual effect on the succeeding crops. This effect was observed in mustard taken up after green gram. The yields of mustard ranged from 529 to 721 kg ha^{-1} during 2006 (year of highest overall yield) and from 110 to 283 kg ha^{-1} during 2009 (lowest overall yield). The treatment T_5 consistently performed better as compared to the rest of the treatments during 2006-09 (as can be seen from Figure 2), sometimes at par with T_4, with the yield enhancement ranging from 36 to 157% over T_1 (control), and from 15 to 72% over T_2 (100% RDF). Singh (2007) reported an improved performance of green gram and mustard grown in a sequence by application of FYM and vermicompost. Residual effect of organics applied to black gram on the succeeding mustard crop was also observed by Kumpawat (2010).

The influence of different treatments on soil characteristics is presented in Table 2. While there was an increase in the soil organic carbon content, percent water stable aggregates and water holding capacity, the bulk density and infiltration rate reduced from their initial values under the vermicompost (T_5) and NADEP (T_4) treatments. The integrated approach of applying organics along with inorganics significantly improved the available nitrogen, and phosphorus contents of the soil as compared to inorganics alone. Vermicompost has been found to increase the aggregate stability of soil particles, resulting in better soil aeration and nutrient cycling (Erich *et al.*,

2002; Ansari, 2008). The same can be also said about NADEP compost. Several authors have reported the improvement in soil organic carbon and available nutrient content under different different crops by the application of vermicompost (Hebbarai *et al.*, 2006; Dwivedi and Singh, 2007).

CONCLUSIONS

From the five year study, it was concluded that *aonla*-based agri-horti system can be taken up under rainfed conditions as exists in Bundelkhand on the shallow, erosion-prone red soils through the improvement in soil physical conditions with the periodic application of organic manures such as vermicompost and NADEP.

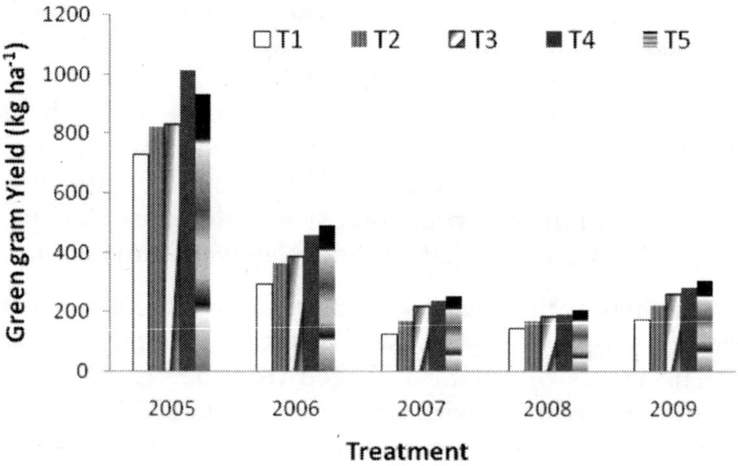

Fig. 1: Yield of green gram from 2005-09

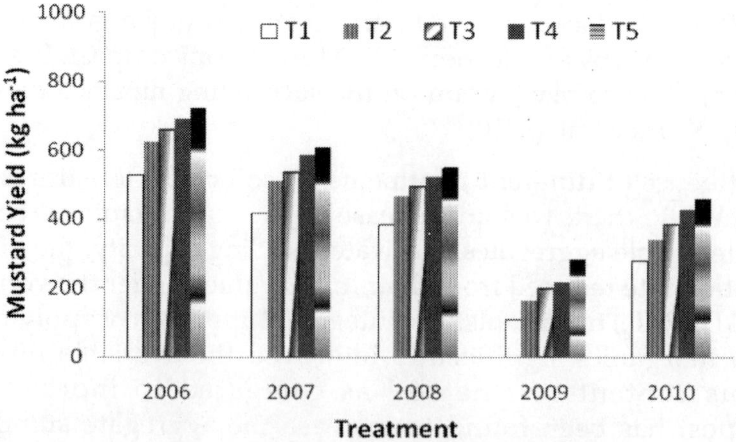

Fig. 2: Yield of mustard from 2006-10

Table 1: Status of surface soil moisture under different treatments during the experimental period

Year	Number of samples per treatment	Number of samples with moisture content in the available range				
		T_1	T_2	T_3	T_4	T_5
2005-06	18	8	12	14	15	15
2006-07	30	14	20	24	23	25
2007-08	28	11	13	18	20	21
2008-09	32	13	19	21	25	25
2009-10	29	15	20	24	24	26

Table 2: Influence of different treatments on the surface (0-15 cm) soil characteristics

Soil property	Initial (2005)	After five years (2010)					
		T_1 (control)	T_2 (NPK)	T_3 (FYM)	T_4 (NADEP)	T_5 (vermi-compost)	CD (*P* = *0.05*)
pH	7.28	7.25	7.29	7.14	7.05	7.08	NS
EC (dS m^{-1})	0.13	0.14	0.25	0.33	0.23	0.28	NS
Bulk density (Mg m^{-3})	1.45	1.47	1.46	1.40	1.40	1.38	0.06
Field capacity (%)	14	13.8	14.1	14.4	14.8	15.1	-
Permanent wilting point (%)	5.0	4.9	5.1	5.3	5.7	5.7	-
Infiltration rate (cm h^{-1})	3.2	3.4	3.1	2.8	2.5	2.5	-
Water stable aggregates (%)	50.7	51.4	50.1	55.9	54.9	58.0	-
Organic C (%)	0.34	0.32	0.35	0.38	0.41	0.44	0.04
Available N (kg ha^{-1})	212	185	236	245	263	268	28
Available P (kg ha^{-1})	14.5	12.9	15.9	18.7	19.8	21.0	3.1
Available K (kg ha^{-1})	234	219	241	251	258	266	NS

REFERENCES

Abraham, T. and Lal, R.B. 2003. Enhancement of productivity potential of greengram (Vigna radiate L.) through integrated nutrient management (INM) in a legume based cropping system. *Madras Agric. J.*, 90: 431-437

Ansari, A.A. 2008. Effect of Vermicompost on the Productivity of Potato (*Solanum tuberosum*), Spinach (*Spinacia oleracea*) and Turnip (*Brassica campestris*). *World J. Agric. Sci.*, 4 (3): 333-336

Dwivedi, D.K. and Singh, S.P. 2007. Effect of nutrient integrated bio-fertilizer, vermicompost and mustard oil cake + inorganic on betelvine (*Piper betle* L.) crop. *The Asian J. Hort.*, 2: 102-105

Erich, M.S., Fitzgerald, C.B. and Porter, G.A. 2002. Biodegradation effect of organic amendments on phosphorus chemistry in a potato cropping system. *Agric. Ecosys. Environ.*, 88: 79-88.

Gomez, K. and Gomez, A. 1984. Statistical Procedures for Agricultural Research. 2nd Edn. John Wiley and Sons, New York

Hebbarai, M, Ganiger, V.M., Reddy B.G.M and Joshi, V.R. 2006. Integrated nutrient management in sapota [*Manilkara zapota*] using vermicompost to increase yield and quality. *Indian J. Agric. Sci.*, 76: 587-590

Jackson, M.L. 1973. Soil Chemical Analysis. Prentice Hall, New Delhi

Kumpawat, B.S. 2010. Integrated nutrient management in blackgram (*Vigna mungo*) and its residual effect on succeeding mustard (*Brassica juncea*) crop. *Indian J. Agric. Sci.*, 80: 76-79

Nagavallemma K.P., Wani, S.P., Stephane Lacroix, Padmaja, V.V., Vineela, C., Babu Rao, M. and Sahrawat, K.L. 2004. Vermicomposting: Recycling wastes into valuable organic fertilizer. Global theme on Agrecosystems Report no. 8. International Crops Research Institute for the Semi-Arid Tropics, Patancheru, Andhra Pradesh, India, 20 p.

Narayan, D., Katiyar, V.S. and Biswas, H. 2006. Water harvesting and recycling for sustainable production in Bundelkhand region. Extended summaries. National Symposium on Conservation Agriculture and Environment, Oct. 26-28, 2006, BHU, Varanasi

Narayan, D., Sharma, A.K., Nitant, H.C. and Lal, B. 2001. Enhancement of productivity of red soils of Bundelkhand by green manuring. International Conference on Nature farming and Ecological balance, March 7-10, 2001, Hisar.

Pathak P.S. and Bhatt R.K. 2001. Revegetation of degraded lands through agroforestry: retrospects and prospects. *(The) Botanica*, 51: 127-142

Rajkhowa, D.J., Gogoi, A.K., Kandal, R. and Rajkhowa, K.M. 2000. Effect of vermicompost on Greengram nutrition. *J. Indian Soc. Soil Sci.*, 48: 207-208.

Singh, G. and Mishra, R. 2007. *Aonla* Rejuvenation. PFDC, CISH, Lucknow, 16p.

Singh, R.A. 2007. Integrated nutrient management in green gram (*Phaseolus radiatus*)-Indian mustard (*Brassica juncea*) - summer groundnut (*Arachis hypogaea*) cropping system. *Int. J. Agric. Sci.*, 3: 263-268.

□□□

Chapter 31

Sizing of Lined and Unlined On-Farm Ponds for Various Crop Substitution Ratios in Rainfed Uplands

B C SAHOO, D. BEHERA AND S N PANDA[1]

INTRODUCTION

Introduction of on-farm reservoir in rice based cropping systems in uplands has definitely enhanced the scope of higher yield from rice with an assured supply of supplemental irrigation (SI). Researchers claimed an increase of more than 50% in rice yield due to SI from the on-farm pond (OFP) during dry spells (Saleh *et al.*, 1996). In Eastern India, the yield has increased up to 58.4 % than purely rainfed conditions under such provisions as reported by Panigrahi *et al.* (2004). However, the provision of SI requirement for winter crops has not been properly addressed in their studies. Only PSI to winter crops seems to be insufficient for achieving higher production from the land. In these circumstances, a higher size of OFP will be required to meet the demand of SI of crops leading to less area left for crop cultivation in the same patch of land. Higher size OFP accompanied with higher investment is not acceptable to resource poor farming community. On the other hand, it may be difficult to justify the investment based on the narrow marginal output from the farm area. Hence, it is proposed to replace the non-remunerative rice crop with remunerative, low water consuming non-rice crops (Kar *et al.*, 2004). The field under changed cropping pattern will generate more surface runoff (SR) for harvesting in the on-farm OFP and in return, it may be able to meet the PSI and SI demand of the winter crops (Rathore *et al.*, 1996). But, the idea of complete replacement of rice has been out-rightly rejected by the upland farmers as reported in some survey reports (Anonymous, 2004). Upland farmers are mostly coming under small and marginal categories and rice is their staple food. Hence, the stand of complete replacement of rice becomes weak in presence of socio-economic aspects. This has led to

College of Agricultural Engineering and Technology, OUAT, Bhubaneswar, Odisha, India
[1]School of Water Resources, Indian Institute of Technology, Kharagpur, West Bengal, India

modification of earlier proposal of complete replacement of rice to partial substitution with a view that the farmer would get some extra return from non-rice crop in addition to meeting his rice requirement. Thus, the crop substitution ratio (CSR) has come up based upon the area of non-rice and rice crop compartments in the same patch of land. Now, proper sizing of OFP for various CSR considered for rainfed uplands comes up prior to introduction of the technology in a risk-prone production system. Hardly, any study related to this emerging area has been taken up in any part of the world. Hence, the present study is undertaken to address the size of OFP, both lined and unlined, for a non-rice based double cropping pattern with assured SI and PSI in both the seasons.

MATERIALS AND METHODS

Theoretical considerations :Based on the proposed CSR, the patch of land is divided into two parts by a dividing dike out of which the upstream (u/s) part is covered with a non-rice crop and the downstream (d/s) part with the existing rice crop in rainy season. In winter, the upstream part again is covered with a pulse crop and the downstream part with an oilseed crop. The OFP is located at the upstream corner of the downstream rice plot as shown in Figure 1. Dynamic root zone water balance model (Hajilal *et al.*, 1998) is used in the present study to simulate the water balance components of the crop compartments. Maximum root zone depth of the crop is assumed as the soil reservoir for all necessary computations.

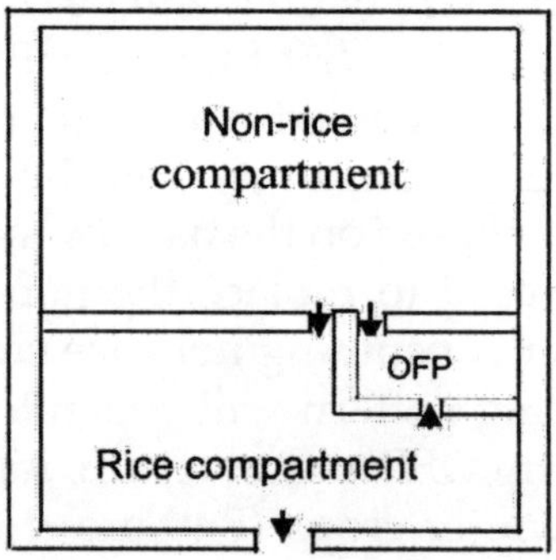

Fig. 1: Plan of farm area with crop substitution ratio and On-farm Pond

The soil reservoir is divided into two imaginary layers basing upon the depth of the root on the day (Figure 2). The soil layer above the root is known as the active layer as both moisture extraction and percolation are assumed to take place from it. The soil layer below actual daily root depth to maximum root zone depth is known as the passive layer as only deep percolation occurs from it. During the period of crop growth these two layers exist distinctly with their relative dimensions determined by the rate of root growth. The day when the roots attain their maximum depth, the passive layer vanishes. The daily water balance in the active and passive layers, ignoring upward flux due to capillary rise, were computed as suggested by Hajilal *et al.* (1998).

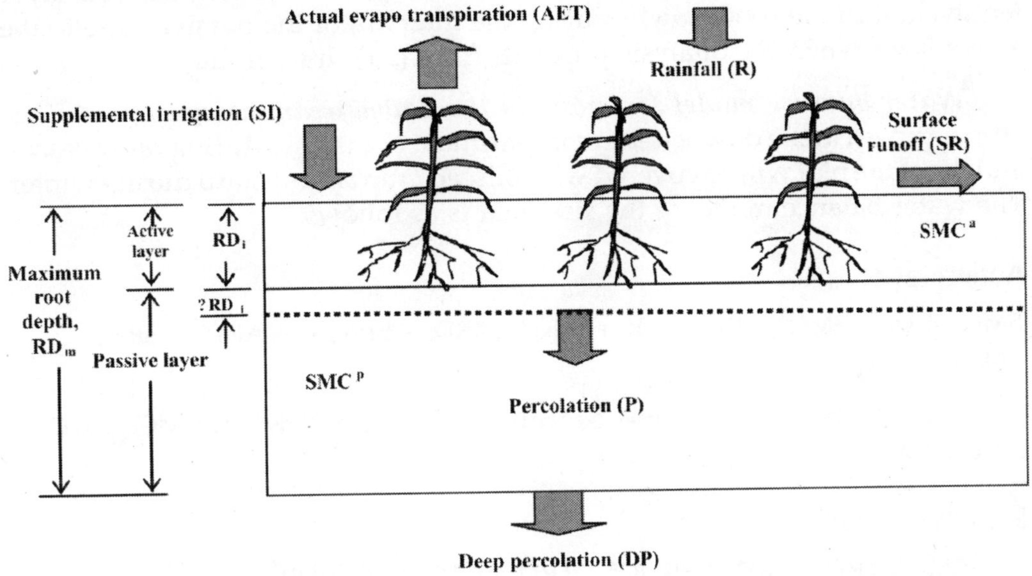

Fig. 2: Schematic diagram of root-zone water balance concept

***Water balance model for crops in the upstream compartment*:**Maize is taken under the study in the upstream compartment of the upland crop field during rainy season and the same compartment is used for a pulse crop *viz.* black gram during winter season.

Active layer

$$SMC^{a}_{ji}.RD_{ji} = SMC^{a}_{ji-1}RD_{ji-1} + R_{i} + SI_{ji} + SMC^{p}_{ji-1}.\delta RD_{ji} - AET^{a}_{ji} - SR_{ji} - P_{ji} \quad ...(1)$$

Passive layer

$$SMC^{p}_{ji}\left(RD_{jm} - RD_{ji}\right) = SMC^{p}_{ji-1}\left(RD_{jm} - RD_{ji}\right) + P_{ji} - DP_{ji} \quad ...(2)$$

where, SMC^{a}_{ji} = soil moisture content in the active layer of jth crop on ith day; SMC^{p}_{ji-1} = soil moisture content in the passive layer of jth crop on i-1 th day; RD_{ji} = Root zone depth of jth crop on ith day; δRD_{ji} =Incremental root growth of jth crop in passive layer on ith day; R_i = Rainfall on i th day; AET^{a}_{ji} = Actual evapotranspiration from active layer of j th crop on ith day; SR_{ji} = Surface runoff from the field of jth crop on i th day; Pji = percolation from the active layer to passive layer of the jth crop on ith day; DPji = Deep percolation from passive layer to down; i = time index taken as 1 day in the study (i=1 means date of sowing and so on) and j = type of crop (j =1 for the crop in rainy season & 2 for the crop in winter season). The term $SMC^{p}{ji-_1}.\delta RD_{ji}$ represents the addition of soil moisture content of a fractional soil layer (equivalent to the root growth during the interval) of the passive zone to the active layer when the analysis proceeds from (i-1) th to ith day.

Water balance model for crops in the downstream compartment :Rice crop is grown in the downstream compartment of the field during rainy season and the same plot will be covered with oilseed crop *viz.* mustard during winter. The water balance model of the rice field is as follows:

Active layer

$$SMC^{a}_{ki.}RD_{ki} = SMC^{a}_{ki-1}RD_{ki-1} + R_{i} + SMC^{p}_{ki-1}\delta RD_{ki} + SRu/s_{ki} - AET^{a}_{ki} - SRP_{ki} - -SRout_{ki} - SP_{ki} - P_{ki} \quad \text{... (3)}$$

Ponding phase starts and continues in the field, when $SMC^{a}_{ki}.\ RD_{ki} > SAT_{ki}.RD_{ki}$

Passive layer

$$SMC^{p}_{ki}(RD_{km} - RD_{ki}) = SMC^{p}_{ki-1}(RD_{km} - RD_{ki}) + P_{ki} - DP_{ki} \quad \text{... (4)}$$

When the oilseed crop will be grown in the compartment, the provision of standing water as has been given for rice crop will be withdrawn from the model.

where, SMC^{a}_{ki} = soil moisture content in the active layer of kth crop on ith day; SMC^{p}_{ki-1} = soil moisture content in the passive layer of kth crop on i-1 th day; RD_{ki} = Root zone depth of kth crop on ith day; RDkm = maximum root depth of kth crop; δRD_{ki} = Incremental root growth of kth crop in passive layer on ith day; R_{i} = Rainfall on i th day; AET^{a}_{ki} = Actual evapotranspiration from active layer of k th crop on ith day; SRu/s_{ki} = surface runoff coming from the u/s jth to d/s kth crop field on ith day; SRP_{ki} = Surface runoff from the field of kth crop to the OFP on i th day; $SRout_{ki}$ = Surface runoff drained out from the field of kth crop on i th day; Pki = percolation from the active layer to passive layer from the kth crop field on ith day; DPki = Deep percolation from passive layer to down; i = time index taken as 1 day in the study (i=1 means date of sowing and so on) and k = type of crop in d/s side of the field (1 for the crop in rainy season & 2 for the crop in winter season).

Water Balance Model for OFP

Two types of OFP are used under the study such as lined and unlined. The shape of the OFP is assumed to be square shaped trapezoidal cross section. Except the area, the other dimensions of the OFP such as depth from field level to bottom (2.4 m), its side slope (1:1), side slope of embankment (1:1), height of embankment (0.3 m), top width of embankment (0.3 m) and berm width (0.3 m) are kept constant in the study. Cross laminated polyethylene tarpaulins of 120 gm^{-2} are used as lining material of OFP.

The water balance model for the OFP is formulated as:

$$FV_i = FV_{i-1} + VR_i + VSRP_{ji} + VSRP_{ki} - VE_i - VSP_i - VSI_i - VSRo_i \quad ...(5)$$

where, FVi = Final volume of storage in the OFP on ith day; FVi-1 = volume of storage in the OFP on i-1th day; VRi= Volume of rainfall on the OFP on ith day; VSRPji = Volume of surface runoff to OFP from jth crop field on ith day; VSRPki = Volume of surface runoff to OFP from kth crop field on ith day; VEi = Volume of evaporation from OFP on ith day; VSPi = Volume of seepage and percolation from OFP on ith day; VSIi = Volume of supplemental irrigation from OFP on ith day and $VSRo_i$= volume of spill over losses from OFP.

Simulation Methods

A computer programme was written in Visual Basic 6.0 for the soil moisture status and SI requirement of the maize crop compartment at the upstream of the farm area and another for the rice compartment at the downstream. Volume balance of OFP was also written in the same language. All the three modules are incorporated in the main model and run simultaneously everyday. Daily rainfall and reference crop evapotranspiration (ETo) values of the study site saved in an excel file were picked up by the programme during the run to compute the variations in soil moisture status, SI requirement of the crops, direct rainfall and surface runoff coming into the OFP. ET_o values were computed by Penman-Monteith equation (Allen *et al.*, 1998). Simulation was carried on for 30 years (1977 to 2006). On any day when SI is required by the crops, the model checks the storage volume of water in the OFP. In case, the availability is less than the demand then the size of OFP is assumed to be undersized and so, provision is made in the model to increase the size by 0.1% and again simulation starts from the 1st day and so on. The simulation in this process continues till a size of OFP is reached which can meet the complete SI and PSI demand of the crops.

RESULTS AND DISCUSSION

Size of lined OFP :The simulation results on OFP size are shown in Table 1. The simulation study revealed that for 70:30 CSR, the size of lined OFP ranged from minimum 6.74 to maximum 10.97% of the farm area (FA) in different years based on the pattern of rainfall. Furthermore, the sizes were found to vary from 6.74 to 15.55% of FA for 60:40 CSR, 6.74 to 20.75% of FA for 50:50 CSR, 6.74 to 27.14% of FA for 40:60 FA and 6.74 to 33.45% of FA for 30:70 CSR. The average size of lined OFP for 70:30, 60:40, 50:50, 40:60 and 30:70 CSR in 1200 m^2 farm area were found to be 7.57, 8.02, 8.71, 9.43 and 10.55% of FA, respectively. It was observed that the size of lined OFP is increasing with decrease in rice substitution. It may be due to more SI requirement of crops leading to requirement of higher storage volume in the OFP. It is also seen from the Table 1 that the SR generation is decreasing with decrease in CSR and the SI requirements of the crops are increasing in rainy season. However, in case of SI demand of winter crops, the requirement was found to be decreasing very slowly as the OFP size increases. As a fixed amount of SI and PSI has been assumed for winter crops so, the decrease in the amount of SI in winter may be due to less area under crop resulting from higher sizes of OFP.

Size of Unlined OFP:Simulation of unlined OFP size was done for 30 years from 1977 to 2006. In none of the years the unlined OFP, irrespective of its size, was found to have adequate storage to meet the SI requirement of the crops taken in winter season. This may be due to rainless days in winter, uncontrolled seepage & percolation loss and evaporation loss from the OFP. However, on cutting off the provision of SI to winter crops after 30 days of germination and subsequent losses due to seepage and percolation, the model started simulating the size of unlined OFP with provision of only PSI to winter crops. Hence, it may be inferred that SI to second crop grown in winter season cannot be ensured with an unlined source of OFP in uplands.

Simulated SR, SI requirement and size of unlined OFP with only PSI to winter crops for various CSR are presented in Table 2. It was found that the size of unlined OFP was varying from 5 to 22.69% of FA for 70:30 CSR in different years of 1977 to 2006. Similarly, the size was varying from 5.15 to 32.14% of FA for 60:40 CSR, 7.15 to 41.63% of FA for 50:50 CSR, 8.73 to 50.3% of FA for 40:60 CSR and 9.84 to 60.16% of FA for 30:70 CSR. The average size of unlined OFP was observed to be 9.78, 11.97, 14.54, 17.13 and 20.60% of FA for 70:30, 60:40, 50:50, 40:60 and 30:70 CSR in 1200 m^2 farm area, respectively, in rainfed uplands. The increase in the size of unlined OFP with decrease in CSR reveals a similar trend as shown in case of lined OFP. However, it may be seen that that the size of unlined OFP required to meet the SI demand of the crops in any CSR is higher than that of the lined OFP. The PSI demand of winter crops met by the unlined OFP was found to be a fraction of that met under the presence of lined OFP. Hence, it implies that a higher size OFP will ultimately reduce the crop area and also may affect the yield rate of the winter crops due to inadequate storage of water.

Table 1: Size of Lined OFP with respect to provision of SI to rainy season crop and PSI & SI to the second crops

CSR		70:30			60:40			50:50			40:60			30:70		
Year	Rainfall (m^3)	Runoff (m^3)	SI Rainy (m^3)	OFP size(%)	Runoff (m^3)	SI Rainy (m^3)	OFP size (%)	Runoff (m^3)	SI Rainy (m^3)	OFP size (%)	Runoff (m^3)	SI Rainy (m^3)	OFP size (%)	Runoff (m^3)	SI Rainy (m^3)	OFP size (%)
1977	1616.05	634.02	126.45	6.74	619.13	141.92	6.74	593.65	159.74	6.81	574.86	181.28	7.08	551.03	198.57	7.44
1978	1918.32	867.65	158.26	6.74	868.05	185.58	7.08	847.85	214.76	8.14	801.63	181.13	6.74	814.80	199.30	7.08
1979	1251.00	510.51	152.21	8.22	498.61	180.33	8.22	493.46	211.51	8.22	498.97	241.82	8.64	491.55	250.70	9.27
1980	1704.20	693.20	127.84	7.75	687.65	173.63	7.75	681.08	197.76	7.75	658.75	223.95	7.75	596.80	182.21	7.75
1981	1343.52	369.58	128.74	6.74	360.59	146.39	6.74	352.59	163.93	6.74	345.91	183.64	7.01	351.81	203.23	6.81
1982	888.72	210.00	169.64	8.22	211.39	202.64	8.31	193.31	182.59	9.17	187.51	204.20	10.65	221.67	297.74	13.80
1983	893.04	149.92	200.59	8.73	153.54	242.22	13.00	153.86	278.13	17.69	127.00	259.25	22.92	124.23	283.30	28.52
1984	1616.05	655.85	125.30	7.08	633.35	143.00	7.08	613.74	157.81	7.08	606.25	178.40	7.08	589.39	196.69	7.08
1985	1203.48	227.75	182.74	7.15	223.15	219.44	8.14	219.93	255.76	9.64	200.28	235.48	13.66	189.77	260.43	20.14
1986	1474.44	441.79	158.13	6.74	414.16	146.75	6.74	411.68	165.60	7.01	410.42	184.06	7.30	406.34	275.98	7.30
1987	1077.30	281.35	178.14	8.64	279.47	211.04	8.73	274.24	245.02	8.73	266.81	279.07	9.74	259.46	313.33	11.19
1988	1264.20	309.88	152.72	6.74	302.22	182.54	6.94	292.87	207.62	8.31	262.14	235.35	6.87	234.73	192.71	7.37
1989	1753.92	795.40	156.99	7.15	784.88	189.71	7.75	769.51	208.11	7.82	765.36	237.59	7.30	754.80	265.42	7.67
1990	2005.56	848.46	109.48	7.23	818.96	122.29	7.15	810.44	135.30	7.08	791.19	148.66	7.01	780.72	160.19	6.94
1991	1128.96	289.21	171.52	6.74	286.06	207.27	6.87	282.00	238.79	8.90	286.34	273.45	8.39	283.63	304.80	11.42
1992	1233.24	389.76	168.27	8.64	390.89	201.56	8.31	378.75	232.38	9.55	355.63	199.35	11.08	341.79	221.69	9.74
1993	2021.28	984.07	179.36	7.08	990.01	213.00	7.37	982.37	250.11	7.52	961.72	277.87	8.47	985.43	310.67	9.45
1994	1450.08	402.43	156.22	7.52	383.90	185.21	7.52	346.72	163.65	8.14	320.82	182.04	7.52	297.69	199.31	7.52
1995	1364.04	454.03	168.92	7.59	453.53	201.30	7.82	451.28	234.14	8.06	439.51	265.67	8.31	433.19	299.09	8.56
1996	1477.44	563.74	144.26	7.37	532.36	124.68	7.37	488.07	138.27	7.37	522.65	211.07	7.37	466.11	163.55	7.44
1997	1383.14	370.24	114.43	7.82	351.95	125.96	7.82	331.89	139.96	7.82	281.13	95.07	7.82	297.30	165.55	7.82
1998	837.90	118.91	192.04	10.97	120.22	230.48	15.55	130.76	271.43	20.75	138.69	310.61	27.14	141.44	349.12	33.45
1999	1380.13	441.79	158.49	7.01	404.20	144.35	7.01	399.91	163.62	7.15	391.93	182.47	7.15	395.75	199.75	7.44
2000	915.62	152.95	162.15	8.22	143.93	190.16	8.90	134.77	222.37	10.24	125.43	252.69	11.53	129.82	281.45	13.93
2001	1489.12	396.91	123.91	7.82	372.46	139.81	7.82	332.20	109.54	7.82	359.67	174.45	7.82	348.27	191.93	7.82
2002	1756.80	714.87	171.41	6.74	696.95	204.07	6.74	678.65	183.81	7.23	658.46	205.49	8.22	644.79	226.67	9.17
2003	1026.96	245.38	173.97	7.90	235.68	205.77	9.27	229.91	241.18	8.64	201.23	212.67	9.55	187.76	234.26	10.55
2004	1285.80	205.61	133.67	6.74	204.90	154.00	6.74	179.85	122.91	6.74	167.27	132.96	6.74	157.70	142.98	6.74
2005	1674.36	610.21	155.59	6.74	592.78	184.49	6.74	565.20	211.34	6.74	562.27	242.34	7.15	548.93	270.71	7.30
2006	1269.60	438.79	184.56	8.31	419.01	226.59	8.31	383.33	205.02	8.31	389.08	237.63	8.90	375.53	268.88	9.74
Mean	**1390.14**	**459.14**	**156.20**	**7.57**	**447.80**	**180.87**	**8.02**	**433.46**	**197.07**	**8.71**	**421.96**	**214.32**	**9.43**	**413.41**	**237.01**	**10.55**

Table 2: Size of Unlined OFP with respect to provision of SI to rainy season crop and only PSI to the second crops

CSR		70:30			60:40			50:50			40:60			30:70		
Year	Rainfall (m^3)	Runoff (m^3)	SI Rainy (m^3)	OFP size (%)	Runoff (m^3)	SI Rainy (m^3)	OFP size (%)	Runoff (m^3)	SI Rainy (m^3)	OFP size (%)	Runoff (m^3)	SI Rainy (m^3)	OFP size (%)	Runoff (m^3)	SI Rainy (m^3)	OFP size (%)
1977	1616.05	634.02	70.61	6.04	619.13	85.90	7.08	593.65	103.57	8.31	574.86	125.23	8.82	551.03	142.56	10.24
1978	1918.32	867.65	101.65	10.55	868.05	128.41	15.39	847.85	157.49	20.75	801.63	121.74	26.87	814.80	139.38	31.51
1979	1251.00	510.51	95.81	16.02	498.61	123.79	16.83	493.46	154.70	18.41	498.97	185.05	20.14	491.55	194.30	20.75
1980	1704 .20	693.20	72.34	8.64	687.65	118.13	8.64	681.08	142.26	8.64	658.75	168.37	9.08	596.80	126.50	9.84
1981	1343.52	369.58	72.66	7.44	360.59	90.08	8.82	352.59	107.38	10.24	345.91	127.02	11.88	351.81	145.91	15.09
1982	888.72	210.00	113.10	16.83	211.39	146.09	17.34	193.31	125.94	21.8	187.51	147.64	27.96	221.67	242.56	34.12
1983	893.04	149.92	144.67	15.55	153.54	188.45	22.24	153.86	226.75	29.39	127.00	210.61	36.95	124.23	237.67	44.63
1984	1616.05	655.85	69.58	6.87	633.35	87.24	7.08	613.74	101.82	8.47	606.25	122.19	9.74	589.39	140.45	9.93
1985	1203.48	227.75	126.01	13.13	223.15	162.48	18.97	219.93	198.75	26.08	200.28	180.30	33.45	189.77	208.93	41.22
1986	1474.44	441.79	102.11	7.08	414.16	90.38	9.17	411.68	108.99	11.77	410.42	127.00	15.7	406.34	217.92	21.59
1987	1077.30	281.35	122.85	11.42	279.47	155.44	13.66	274.24	188.28	20.34	266.81	221.99	26.87	259.46	256.19	33.78
1988	1264.20	309.88	96.30	9.45	302.22	126.00	11.08	292.87	151.78	13.13	262.14	177.86	16.34	234.73	134.60	22.24
1989	1753.92	795.40	100.92	9.27	784.88	133.77	11.19	769.51	151.80	13.66	765.36	181.47	10.24	754.80	209.40	11.31
1990	2005.56	848.46	54.20	5	818.96	66.82	5.75	810.44	79.49	7.37	791.19	92.35	10.03	780.72	103.26	13.39
1991	1128.96	289.21	115.22	8.73	286.06	150.63	11.31	282.00	183.48	12.74	286.34	217.46	14.36	283.63	250.40	18.78
1992	1233.24	389.76	112.84	12.24	390.89	145.70	13.26	378.75	177.37	13.93	355.63	144.90	17.52	341.79	166.93	13.26
1993	2021.28	984.07	123.25	9.17	990.01	156.81	10.97	982.37	193.79	12.37	961.72	222.12	13.39	985.43	255.05	17
1994	1450.08	402.43	100.84	6.87	383.90	129.29	10.03	346.72	108.42	8.82	320.82	126.15	9.84	297.69	143.25	10.87
1995	1364.04	454.03	113.55	7.15	453.53	145.90	8.31	451.28	178.74	9.45	439.51	210.29	10.44	433.19	243.59	12.24
1996	1477.44	563.74	88.72	7.15	532.36	69.13	7.23	488.07	82.59	7.98	522.65	155.23	8.9	466.11	107.59	9.93
1997	1383.14	370.24	58.82	9.55	351.95	70.30	9.84	331.89	84.27	10.03	281.13	39.35	10.24	297.30	109.74	10.76
1998	837.90	118.91	136.62	22.69	120.22	176.98	32.14	130.76	220.33	41.63	138.69	262.95	50.3	141.44	304.64	60.16
1999	1380.13	441.79	102.86	6.04	404.20	88.32	8.39	399.91	107.20	11.31	391.93	125.49	14.64	395.75	142.34	18.41
2000	915.62	152.95	106.93	9.08	143.93	134.69	13.66	134.77	167.00	19.16	125.43	197.30	25.06	129.82	226.76	31.82
2001	1489.12	396.91	68.67	7.37	372.46	84.57	7.37	332.20	54.20	7.98	359.67	118.98	8.73	348.27	136.12	10.76
2002	1756.80	714.87	115.14	8.56	696.95	147.53	10.13	678.65	127.43	11.42	658.46	149.56	13.26	644.79	170.65	18.05
2003	1026.96	245.38	118.13	11.31	235.68	150.23	15.7	229.91	184.20	21.37	201.23	156.83	18.78	187.76	178.16	24.81
2004	1285.80	205.61	78.01	5	204.90	98.31	5.15	179.85	66.88	7.15	167.27	76.46	9.93	157.70	85.89	13.39
2005	1674.36	610.21	99.56	7.15	592.78	128.26	8.31	565.20	154.94	9.36	562.27	186.09	10.34	548.93	214.14	12.87
2006	1269.60	438.79	128.92	12	419.01	170.59	14.07	383.33	149.16	13.26	389.08	182.09	14.07	375.53	213.76	15.39
Mean	**1390.14**	**459.14**	**100.36**	**9.78**	**447.80**	**125.01**	**11.97**	**433.46**	**141.30**	**14.54**	**421.96**	**158.67**	**17.13**	**413.41**	**181.62**	**20.60**

CONCLUSIONS

The size of the OFP required to meet the SI and PSI demand of crops grown in the same patch of land is found to be considerably smaller (22 to 50%) than that in case of unlined OFP in various CSR. The harvested water, both direct rainfall and generated SR from the crop fields, in it can be used effectively in taking two sets of crops successfully in two compartments of the farm area during rainy and winter seasons. However, the unlined OFP, irrespective of its size, is found to be unsuitable to meet the SI demand of winter crops in any of the years. Without carrying out an economic study of both the events, a final conclusion in regards to optimum size of OFP can not be drawn at this stage. However, lined OFP is showing some of the symptoms of an effective source of on-farm irrigation in rainfed uplands of eastern India.

REFERENCES

Allen, R.G., Pereira, L. S., Raes, D. and Smith, M. 1998. Crop evapotranspiration. Guidelines for computing water requirements. FAO Irrigation and Drainage Paper 56, FAO, Rome.

Anonymous. 2004. Survey Report. Krishi Vigyan Kendra (OUAT), Keonjhar, Orissa.

Hajilal, M.S., Rao, N.H. and Sharma, P.B.S. 1998. Planning intraseasonal water requirements in irrigation projects. *Agricultural Water Management*, 37(2): 163-182.

Kar, G., Singh R. and Verma, H. N. 2004. Alternate cropping strategies for assured and efficient in upland rainfed rice areas of eastern India based on rainfall analysis. *Agricultural Water Management*, 67(1): 47-62.

Panigrahi, B., Panda S.N. and Agarawal. A. 2004. Water balance simulation and economic analysis for optimal size of on-farm reservoir. *Water Resources Management*, 30: 1-18.

Rathore, A.L., Pal, A.R., Sahu. R.K. and Choudhary, J.L. 1996. On-farm reservoir and crop management for improving productivity of rainfed areas. *Agricultural Water Management*, 31(3): 253-266.

Saleh, A.F.M. and Bhuiyan. S.I. 1996. Analysis of agricultural drought in a rainfed low land rice system: a case study in North-West Bangladesh. *J. Indian Water Resources Society*, 2 (1): 51-56.

□□□

Chapter

32

Water Budget for Jute Based Cropping System at Barrackpore, West Bengal, India

D. BARMAN, A. R. SAHA, D. K. KUNDU, B. S. MAHAPATRA AND DEBJANI GHOSH

INTRODUCTION

Water management in jute production includes application of irrigation water particularly in early growing period and drains out of excess water from jute field mainly during later stage of its growth. But in India jute is a prevalent rainfed crop, therefore, water budget for jute crop is primarily driven by rainfall amount and intensity and subsequently other important weather parameters like temperature and evaporation. In rainfed agriculture, the total amount of rainfall as well as its distribution affects the plant growth (Sharma *et al.*, 1979; Ram Suresh *et al.*, 1992). Rainfall also determines the potential of any region in terms of crops to be produced, farming system to be adopted, the nature and sequence of farming operations to be followed and to achieve higher agricultural productivity as well (Singh and Dhillon, 1994).

Water budget primarily depends on the type of crop, the type of soil where the crop is grown and the prevailing weather conditions of the particular region. Among both the species of jute, *Deshi* jute (*Corchorus capsularis* L.) is generally grown in the medium high and medium low land and can withstand water logging/excess water for a certain period but, *Tossa* jute (*C. olitorius* L.) is grown in medium high and high land and cannot withstand water logging/excess water (Rahman and Azad, 1991).

Jute is one of the important remunerative crops is widely adaptive cropping systems in West Bengal. But sowing time, length of growing period and retting of jute is highly rainfall and temperature dependant. When evaporative rates are high, frequent irrigations are required to maintain plant available water at levels necessary to maximize yield (Connor *et al.*, 1985;

Central Research Institute for Jute & Allied Fibres, Barrackpore, Kolkata-700 120

Whitefield *et al.*, 1986). Moreover, the question of when to irrigate can be approached in three broad ways: using plant indicators, soil indicators and water balance techniques (Phene *et al.*, 1992). Research result from varying environmental conditions presents recommendations for the schedule of irrigation using the water balance technique and in combination with analysis of historical climate data from a region (Phene *et al.*, 1992). The range of number of irrigation required can be identified using such data and information(Perrier, 1988). In water budget study, prediction of rainfall is necessary for timely agricultural operations. Probability analysis is the most reliable method to predict occurrence of rainfall events based on past behavior of rainfall (Kumar and Kumar, 1989). Rainfall analysis is of great importance for developing and modifying the crop management practices for sustainable production system, particularly in jute based cropping systems.

The amount of rainfall received by hot moist subhumid region of West Bengal is adequate for rainfed farming in *kharif* season. But, timely harvesting of jute and its retting requires precise prediction of onset of monsoon and construction of appropriate water harvesting-cum-jute retting structure. The present study was undertaken to analyze water balance during the jute life cycle for better water management planning in jute based cropping systems.

MATERIALS AND METHODS

Meteorological observatory of Central Research Institute for Jute & Allied Fibres (CRIJAF), Barrackpore is situated at 22°45'16.53" North latitude and 88°25'19.25" East longitude and 9.69 m above mean sea level. Historical rainfall data for the period from 2001-2010 collected from the meteorological observatory at CRIJAF, Barrackpore was converted to weekly, monthly and seasonal rainfall by simple mathematical means. Three agricultural seasons, *viz.*, jute growing season-*zaid*/summer (1 April-15 July), *kharif* (16 July-15 November) and *rabi* (16 November-31 March) were identified according to existing jute based cropping systems in this region. The total rainfall of the corresponding time period of the months of the respective seasons was summed up for its probability analysis. Daily rainfall of the respective year was summed up to calculate yearly total rainfall and average was calculated by dividing the yearly total by the total number of years. Rainy day was considered as rainfall of a particular day if exceeded 2.5 mm.

The date of onset of effective monsoon was defined as the date of commencement of a 7-day spell satisfying the following criteria (Verma and Sharma, 1989; Sharma *et al.*,1979; Sahoo, 1993): (i) first day's rain in 7-day spell is not less than average daily evapo-transpiration (ET), (ii) at least four out of seven days are rainy days with not less than 2.5 mm rain each day and (iii) total rain during the 7-day spell is not less than (5ET+10) mm.

The probability analysis was carried out using Waybill's method (Chow, 1964), which is P = [m/(n+1)]*100, where, P is the plotting position in percent chance; m is the rank number when the data are arranged in descending order and n is total number of years. Jena and Senapati (1981) found that Weibull's equation gave better linear relationship between the rainfall and plotting positions than the other equations. Barman *et al.* (2011) also used the Weibull's equation to predict the rainfall at Sunabeda in Koraput district.

RESULTS AND DISCUSSION

Climatic condition at barrackpore : Decadal average annual rainfall in this region is 1383.2 mm (ranges between 1057.6 mm and 1751.2 mm), 85.7% of which occurs during the months of June to October (monsoon). Mean maximum and minimum temperatures are 31.2°C and 20.5°C, respectively. However, average monthly maximum air temperature is 35.4°C and minimum 24.7°C during May identified as the hottest month followed by April. Similarly January is the coolest month experiencing minimum air temperature 10.5°C and maximum 24.5°C followed by December. The climate is subtropical humid.

Annual and seasonal rainfall and rainy days : Annual average rainfall during the last decade (2001-2010) was 1383.12 mm and 60.8% of which occurred during *kharif* season (July 16 to November 15), 36.2% in *zaid* season (April 1 to July 15) and 2.9% in *rabi* (November 16 to March 31) season (Table 1). Coefficient of variation in seasonal rainfall was 19.1% for *kharif* season, 25.1% for *zaid* and 79.1% for *rabi* season. The number of rainy days (>2.5 mm rainfall in a day) varied from 65 to 97 in a year but average was 80.9 in number with 15.6% coefficient of variation (CV). The occurrence of rainy days in the *kharif* season was 59% followed by *zaid* season 36.1% and then *rabi* season 4.9%. Among the three seasons, the lowest CV for occurrence of rainy day was found in *kharif* season (15.6%), followed by *zaid* (19.0%) but it was found maximum in *rabi* season (58.9%) (Table 1). The lower value of CV depicted more consistent occurrence of rainfall and rainy days annually and for *kharif* and *zaid* season rainfed crops whereas higher value of CV inferred that agriculture in *rabi* season can be practiced by depending on soil residual moisture or assured irrigation due to uncertainty of rainfall. However, trend analysis of annual rainfall during 2001 to 2010 revealed that annual rainfall decreased over the past decade at the rate of 40.9 mm per annum (Figure 1). The amount of rainfall 501.7 mm during the *zaid* season is sufficient for jute cultivation and its retting if proper water harvesting technology is adopted. Marked variation of annual rainfall was observed during the last decade.

Table 1: Annual and seasonal variability of rainfall (mm) and rainy days (Number) at Barrackpore

Year	*Zaid*		*Kharif*		*Rabi*		Annual	
	Rainfall	Rainyday	Rainfall	Rainyday	Rainfall	Rainyday	Rainfall	Rainyday
2001	635.2	37.0	772.6	55.0	38.8	5.0	1446.6	97
2002	684.4	34.0	838.9	47.0	22.2	2.0	1545.5	83
2003	419.4	26.0	1089.2	60.0	29.1	4.0	1537.7	90
2004	578.3	31.0	870.9	56.0	5.9	2.0	1455.1	89
2005	407.1	20.0	729.6	48.0	103.2	8.0	1239.9	76
2006	479.3	33.0	910.3	47.0	0.0	0.0	1389.6	80
2007	586.8	29.0	1099.9	42.0	64.5	4.0	1751.2	75
2008	511.8	34.0	751.8	45.0	70.2	7.0	1333.8	86
2009	254.9	23.0	775.0	41.0	27.7	4.0	1057.6	68
2010	460.2	25.0	579.3	36.0	35.3	4.0	1074.8	65
Mean	501.7	29.2	841.8	47.7	39.7	4.0	1383.2	80.9
SD	126.2	5.5	160.6	7.4	31.4	2.4	215.8	10.1
CV (%)	25.1	19.0	19.1	15.6	79.1	58.9	15.6	12.5

Rainyday = >2.5 mm rainfall in a day

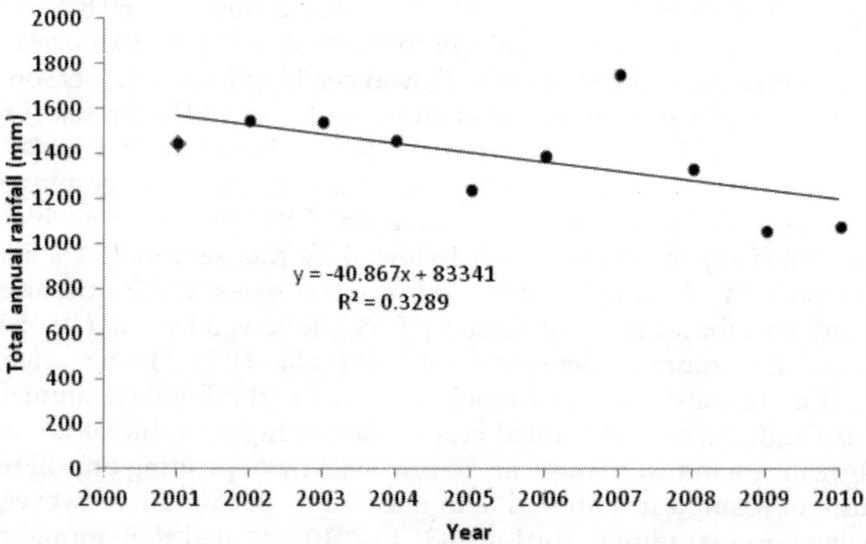

Fig. 1: Trend of annual rainfall during 2001-2010 at Barrackpore

Onset of effective monsoon : The knowledge of mean date of arrival of monsoon is very useful for planning timely *kharif* operations to take maximum advantage of monsoon rain. The probability analysis showed that the onset of monsoon on 23rd standard meteorological week which is extended between

4th to 10th June at more than 95% probability level. India Meteorological Department, Pune predicted that June 10 is the probable date of onset of monsoon which is well matched with this analysis.

Rainfall and rainy day during jute life cycle : Jute life cycle is generally extended from last fortnight of March to first fortnight of July. During this period monthly total rainfall and rainy days increased gradually (Figure 2). The concentration of rainfall was more during the later stage of jute growth as it was coincided with early period of monsoon. Summation of June and July rainfall was 496.9 mm which requires drainage because stagnation of water affects jute growth and fibre quality (Ghorai *et al.*, 2008). High rainfall during the later stage leads to *in-situ* water harvesting for jute retting as well as to use the harvested water in off season cultivation.

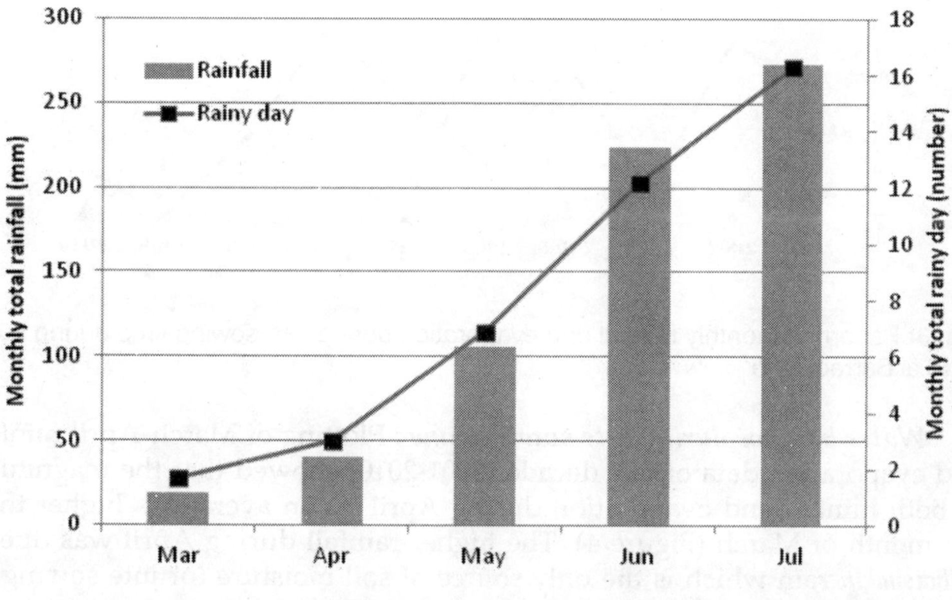

Fig. 2: Monthly distribution of rainfall and rainy day during jute lifecycle at Barrackpore.

Weekly rainfall : The plotting of weekly rainfall and evaporation indicates the surplus and deficit of moisture for cultivation. Weekly rainfall, evaporation and mean temperature were plotted in the Figure 3, to identify water balance scenario for choosing crops to be cultivated and their water management practices. Average weekly rainfall was maximum (58.63 mm) in 39th week followed by 38th week (58.14 mm). From the Figure 3, it was evident that standard week numbers 23rd to 43rd were the monsoon weeks when greater concentration of rainfall was identified. During this period, the total amount of rainfall was 738.2 mm which was more than the evaporative demand, 428.9 mm at Barrackpore region. Therefore, the surplus rain water of 309.3 mm is needed to be managed in soil profile or by harvesting on soil surface so that this surplus water can be used in deficit period. Except the monsoon period, rest of the time in a year water deficit was obvious in the magnitude of 471.8 mm.

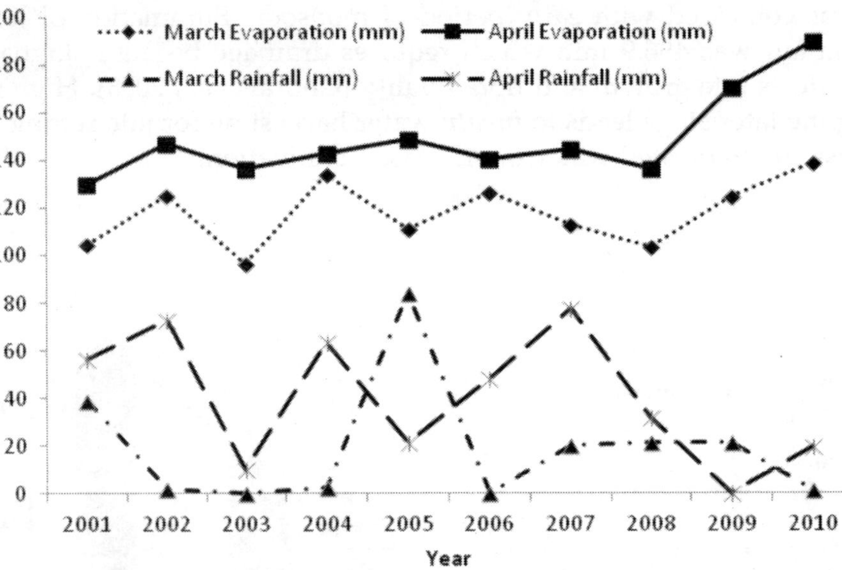

Fig. 3: Pattern of monthly rainfall and evaporation during jute sowing time during 2001 2010 at Barrackpore

Water balance during jute sowing time : Plotting of March-April rainfall and evaporation data of last decade (2001-2010) showed that the magnitude of both rainfall and evaporation during April on an average is higher than the month of March (Figure 4). The higher rainfall during April was due *to kalbaishakhi* rain which is the only source of soil moisture for jute sowing to the small and marginal farmers. But the *kalbaishakhi* rain is erratic and uncertain in nature. Therefore, assured irrigation would be the alternate source of soil moisture for jute sowing.

Jute requires about 418 mm water for its growth and development during its life cycle. In addition, 77 mm water is required for land preparation in respect of better germination of seed. It was calculated that the average rainfall during 10th to 15th week (5th March to 15th April) was 41 mm in the last decade (Figure 3). Therefore, the deficit of water during the land preparation and sowing was 36 mm and the magnitude of which would increase mainly by two phenomena *viz.,* evaporation loss from the soil surface and percolation loss in the soil profile depending upon their various governing factors. This depicted that the requirement of pre-sowing irrigation for jute crop to supplement rain water for better germination and crop growth. Indeed, the requirement of irrigation water would increase as because rainfall in the month of March-April showed a decreasing trend during the last decade (Figure 5).

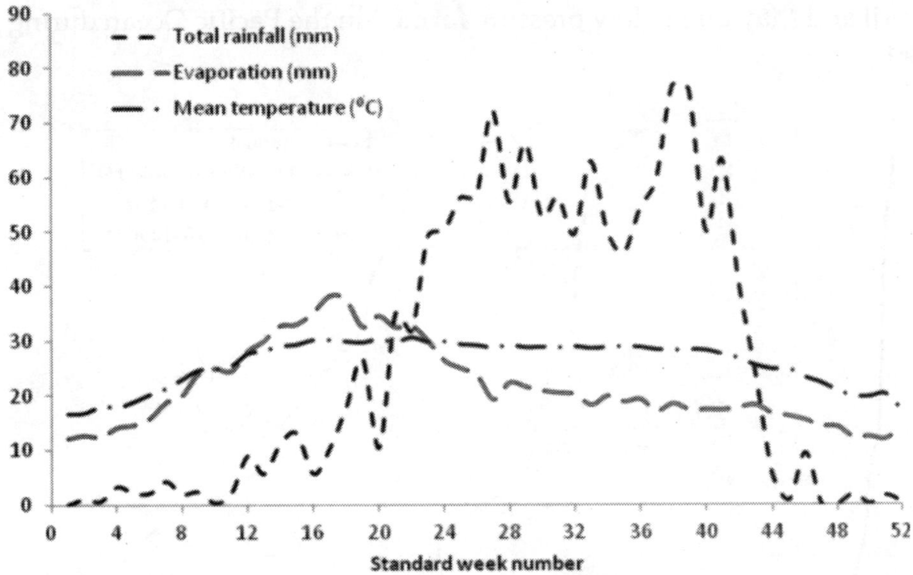

Fig. 4: Weekly distribution of rainfall, evaporation and mean temperature at Barrackpore

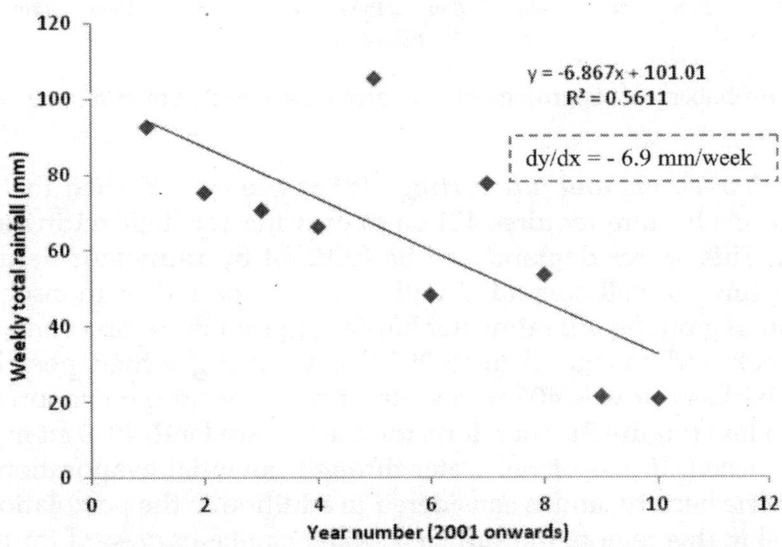

Fig. 5: Weekly rainfall trend in March-April during 2001 - 2010 at Barrackpore

Probability of rainfall : The rainfall data for the past decade (2001-2010) were analyzed and annual and season wise (*kharif, zaid and rabi*) probability of occurrence was predicted (Figure 6). This prediction helps to optimize choice of crops, sowing date and irrigation scheduling for different crops cultivated in this region. The probability of occurrence of rainfall at 80% confidence level was 1193.4 mm per annum, 751.8 mm in *kharif* season, 419.4 mm in *zaid* season and 22.2 mm in *rabi* season. In the *zaid* season, rainfall also occurred during April and May due to low pressure formed in the Pacific Ocean during this period.

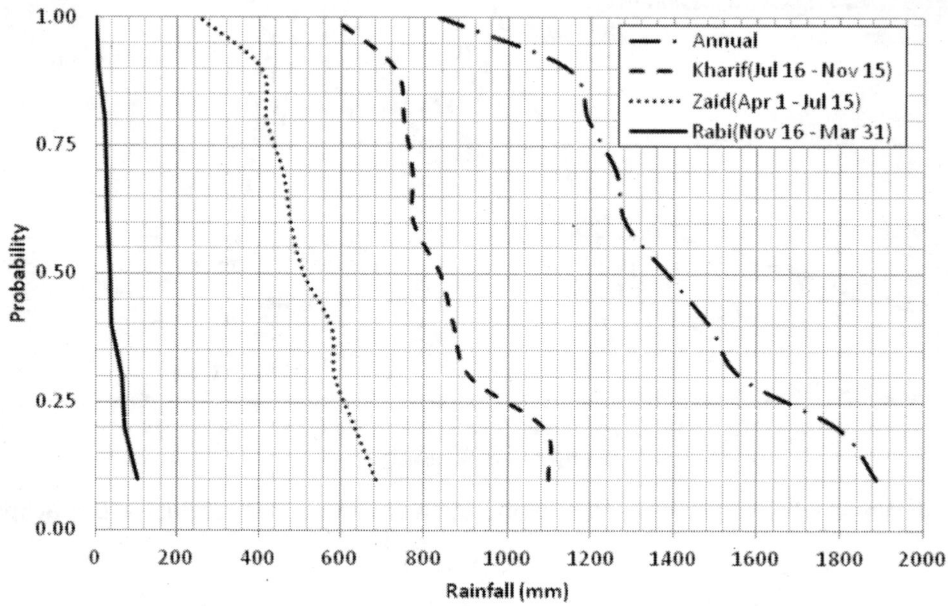

Fig. 6: Probability of occurrence of annual and seasonal rainfall at Barrackpore

Water harvesting and jute retting : It has been established that the jute biomass of one hectare requires 421 cu m of water for their retting and fibre extraction. This water demand can be fulfilled by rainwater because jute harvesting time is well coincided with the early period of monsoon in the Barrackpore region. *In-situ* rainwater harvesting for jute retting can be started on 23rd week and extended unto 29th week during which possibility of rainwater harvesting was 406 mm without considering the evaporation and percolation loss (Figure 3). Therefore, for one hectare land, 4060 cu m of water can be harvested. If loss of rainwater through potential evaporation of 1680 cu m from one hectare land is considered in addition to the percolation loss in silt clay soil in this region, the sufficient water can be harvested for retting of jute *in-situ*. However, the dimension of pond for 421 cu m water harvesting is

30 m in length and 15 m of width for surface with 15 m of length and 7.5 m of width at bottom and 1.5 m depth.

CONCLUSIONS

The knowledge of probable time of arrival of monsoon, which is 23rd week (extended between 4th to 10th June) for Barrackpore region will be helpful for planning *kharif* operations to take maximum advantage of monsoon rain. The probability of 419.4 mm rainfall during *zaid*/summer season at 80% confidence level depicted the requirement of suitable water harvesting structure for jute retting. Similarly, probable (80% confidence level) rainfall of 751.8 mm in *kharif* indicated cultivation of high water demanding crops like paddy in this season. In *kharif* season, rainfall is adequate for rice cultivation; however, irrigation is required for zaid season crops particularly for land preparation and sowing of jute crop. This prediction helps to optimize choice of crops and irrigation scheduling for different crops cultivated in this region.

REFERENCES

Barman, D., Jakhar, P., Hombe Gowda, H.C. and Naik, B.S. 2011. Probability analysis of rainfall characteristics of Semiliguda in Koraput, Orissa. *Indian J. Soil Cons.*, 39(1): 9-13.

Chow, V.T. 1964. Handbook of Applied Hydrology, McGraw Hill Book Co., New York.

Connor D.J., Jones T.R. and Palta J.A. 1985. Response of sunflower to strategies of irrigation in growth, yield and the efficiency of water use. *Field Crops Res.*, 10:15-36.

Ghorai, A.K., Saha, S., Sinha, M.K., Chowdhury, H. and Mahapatra, B.S. 2008. Management of jute crop under waterlogged situations. *Technical Bulletin Series*:**20**. Central Research Institute for Jute & Allied Fibres, Barrackpore, Kolkata-700120.

Jena, D. and Senapati, P.C. 1981. Verification of different probability equation for drainage design for Bhubaneswar. *Indian J. Power and River Valley Dev*. Vol. XXXI, Nov-Dec.

Kumar, D. and Kumar, S. 1989. Rainfall distribution pattern using frequency analysis. *J. Agril. Engg.*, 26(1): 33-38.

Perrier, E. R. 1988. Opportunities for the productive use of rainfall normally lost to cropping for temporal or spatial reasons. *In*: Drought research priorities for the dryland tropics, eds. F. R. Bidinger and C. Johansen. Patancheru, A.P. India: International Crops Institute for the Semi-Arid Tropics

Phene, C.J., Hutmacher R. B., Ayars J.E., Davis K. R., Mead R. M. and Schoneman R. A. 1992. Maximizing water use efficiency with subsurface drip irrigation. Inetrnational Summer Meeting of the Americal Society of Agricultural Engineers, Paper No. 922090. St. Joseph Michigan.

Rahman, M.M. and Azad, A.K., 1991. Drainage in jute fields, Jute and Jute Fabrics, Bangaladesh, 17:7-8.

Ram Suresh, Kumar, D., Prashad, R. and Rai, R.K. 1992. A note on analysis of rainfall for crop planning at Pusa, Bihar. *Indian J. Soil Cons.*, 20(3): 23-27.

Sahoo, M.K. 1993. Analysis of drought phenomenon of pre-divided Kalahandi district of Orissa. *M. Tech. Thesis*, Dept. of Soil and Water Conservation Engineering, Orissa University of Agriculture and Technology, Bhubaneshwar.

Sharda, V.N. and Bhushan, L.S. 1985. Probability analysis of annual maximum daily rainfall for Agra. *Indian J. Soil Cons.*, 13(1): 16-20.

Sharma, H.C., Chauhan, H.S. and Sewa Ram. 1979. Probability analysis of rainfall for crop planning. *J. Agric. Engg.*, XVI (3): 87-94.

Singh, J. and Dhillon, S.S. 1994. Physical determinants of agricultural patterns: In: *Agricultural Geography* (2nd edn.). Tata McGraw Hill Publication Co. New Delhi, pp60-72.

Verma, H.N. and Sharma, P.B.S. 1989. Critical dry spells and supplemental irrigation to rainfed crops. *J. Indian Society of Water Resources*, 9(4):12-16.

Whitefield, D.M., Wright G.C., Gyles, O.A. and Taylor, A.J. 1986. Growth of Lucern (*Medicago sativa* L.) in response to frequency of irrigation and gypsum application on a heavy clay soil. *Irrig. Sci.*, 7: 37-52.

❑❑❑

Chapter 33

Productivity and Water Use of Groundnut (*Arachis hypogaea*) as Influenced by Conjunctive Use of Water Source and Fertilizer Levels

O. P. VERMA, M. DAS, A. KUMAR, R. R. SETHI, R. B. SINGANDHUPE
B. K. JAMES, S. K. RAUTARAY AND M. S. BEHERA

INTRODUCTION

Shortages of surface water supplies necessitate development of groundwater in many canal commands. Ground water can be used optimally in conjunction with surface water in a conjunctive use system (Khare *et. al.*, 2007). The cropping intensity of Orissa is only 163%. This can be improved by increasing ground water use in *rabi* and summer season. To obtain high economic returns per unit use of land and water resources, conjunctive water use planning will be more effective if ground water is utilized efficiently throughout the year in conjunction with canal/surface water source (Sethi *et. al.*, 2006). Groundnut (*Arachis hypogaea*), is a species in the legume family (Fabaceae) native to South America, Mexicoand Central America. Groundnut (*Arachis hypogaea* L.) is an important oilseed crop in India. India is the second largest producer of groundnut accounting for 28% of the total area (6 million ha) and 18.2% production (6.25 M tonnes) of the world. Gujarat ranks first in area and production of groundnut (*Arachis hypogaea*) in India. The groundnut yield in general is higher during *rabi* compared to *kharif*. Hence, farmers prefer cultivation of *rabi* groundnut in vertisols and the crop accounts for nearly 60 % of the total cropped area during *rabi* season (Khare *et. al.*, 2007). However, average productivity is relatively low as groundnut is mostly grown under rainfed condition. Because of high productivity under assured irrigation, groundnut cultivation in *rabi* season is gaining popularity. Optimum scheduling of irrigation led to increase in pod yield and water use efficiency (WUE) (Taha and Gulati, 2001). In eastern India, groundnut (*Arachis hypogaea* L.) is widely grown in winter *(rabi)* under irrigated and unirrigated conditions

Directorate of Water Management, Bhubaneswar-751 023, Odisha, India on residual soil moisture with or without fertilizers. However, information on nutrient-water interaction in *rabi* groundnut under shallow water-table condition is scanty particularly in coastal soils of eastern India. Therefore, the present study was conducted to investigate the effect of conjunctive use of water source and nutrient management in *rabi* groundnut.

MATERIALS AND METHODS

An experiment on groundnut (variety -*Smiriti*) was conducted for two consecutive years (2008-09 and 2009-10) at DWM Research Farm, Deras, Mendhasal, Odisha in factorial randomized block design. The soil is slightly acidic sandy clay loam with low organic carbon content. The treatments consisted of three levels of conjunctive use *viz.*, [C_1-(2 irrigations through canal water only, C_2-two irrigations with canal water + one irrigation with ground water and C_3- 2 irrigations with canal and 2 irrigations with ground water)] and five fertilizer treatments *viz.*, (F_0-no NPK, F_1- 100% NPK, F_2-bio-compost alone, F_3-75% NPK bio-compost and F_4-50% NPK+ bio-compost). The irrigations were scheduled on the basis of critical growth stages of the crop. The fertilizers were applied @ 20, 40 and 40 kg ha^{-1} as N, P_2O_5 and K_2O, respectively. N was supplied through urea, P_2O_5 supplied through DAP and K_2O through Muriate of Potash. "*Smurti*" variety of groundnut was sown on first week of December 2008 and 2009. The crop was harvested on 20 April, 2009 and 24 April, 2010. In *rabi* season, the irrigation schedules were followed at critical crop growth stages in groundnut crop. Conjunctive use planning was done using canal water and ground water source (Table 1).

Table 1: Conjunctive use planning for irrigation in groundnut

Irrigation Treatment	Irrigation depth (cm)	Amount of total water used (m^3)	Percentage of share of water	
			CW	GW
Canal water only (2)	7	2.52	100%	
2 CW+1 GW	7+5	3.42	73.69%	26.31%
2 CW+2GW	7+5	4.32	58.33%	41.67%

RESULT AND DISCUSSION

Pod and haulm yield : Data on groundnut revealed that both pod and haulm yields C_2 and C_3 were significantly higher than C_1. However, C_2 and C_3 were statistically at par (Table 2). The higher pod yield by 33% and 32% in C_2 and C_3 (44.6% and 46%) during 2008-09 and 2009-10, respectively was might be due to better soil moisture availability in soil profile. Regarding fertilizer treatments, highest yield was being in F_4, and the lowest in control. The F_1 and F_4 treatments showed no significant difference in terms of yield

(Table 2). The increase in haulm yield was 41.4% and 39.24 % in C_3 treatment over C_1 in both the years. The 69.3% and 78% increase in haulm yield was observed due to application 75% NPK + Bio-compost compared to control in 2008-09 and 2009-10, respectively. Shinde *et al.* (2000) found that the application of 75 % recommended dose of fertilizer (100 % RDF-25:50:00 kg NPK ha^{-1}) resulted in higher productivity of groundnut at Rahuri in Maharashtra. Hosamani and Janavade (2007) reported that crop receiving four irrigations of 6 cm depth, each scheduled at critical stages (pre-flowering, pegging, pod formation and pod filling stages) and *dhaincha* green manuring along with RDF (25:75:25 kg ha^{-1}, N, P2O5, K2O) recorded significantly higher pod yield. Groundnut under polythene mulch supplemented with irrigation at 0.6:1.0:0.8 IW/CPE ratio resulted in higher yields of dry pod, haulm and kernel (Zagade and Chavan, 2009).

Table 2: Pod and haulm yield of groundnut as influenced by conjunctive use and fertilizer levels

Treatment	Pod yield (kg ha^{-1})		Haulm yield (kg ha^{-1})	
	2008-09	2009-10	2008-09	2009-10
Conjunctive use levels				
C_1	1387	1353	2127	2095
C_2	1845	1786	2717	2637
C_3	2005	1975	3007	2917
CD (P=0.05)	174	191	305	296
Fertilizer levels				
F_0	1308	1268	1962	1834
F_1	2059	2019	3088	3040
F_2	1387	1347	2080	2042
F_3	2214	2174	3321	3263
F_4	1756	1716	2634	2580
CD (P=0.05)	174	191	305	296

Nutrient content and uptake: Nutrient N, P and K content was determined for kernel and above ground biomass of groundnut crop. The nutrient content did not vary significantly due to conjunctive use treatments but significantly influenced due to fertilizer treatments. However, total N, P and K uptake by total plant varied significantly (Figure 1a, 1b and 1c). Chitdeshwari *et al.* (2007) reported that increasing levels of the respective nutrients (34:64:108 kg NPK ha^{-1}) increased the nutrient content in the crop and its uptake. Groundnut under polythene mulch supplemented with irrigation at 0.6:1.0:0.8 IW/CPE ratio resulted in higher yields of dry pod, haulm and kernel over the other interactions, as well as greater total P uptake and K uptake by haulm (Zagade and Chavan, 2009).

Water use by groundnut : The crop water use efficiency was influenced by conjunctive use levels (Table 3). The crop WUE was greater in C_3 than C_2 and C_1 implying a considerable increase in enhancement of WUE. The consumptive use decreased in C_3 compared to C_2. The various fertilizer levels also influenced the consumptive use and WUE of groundnut. The highest CU was in F_1 while WUE was greater in F_4 implying efficient use of irrigation water when 75% of NPK fertilizes were applied in combined with 25% N through biocompost. Higher water use efficiency (WUE) was recorded with four irrigations compared to three irrigations reported by Hosamani and Janavade (2007).

Table 3: Consumptive use and water use efficiency of groundnut as influenced by conjunctive use and fertilizer levels

Treatment	Consumptive use (mm)		Water use efficiency (kg ha-mm^{-1})	
	2009	2010	2009	2010
Conjunctive use level				
C1	769	698	2.27	2.44
C2	773	702	2.22	2.39
C3	766	695	2.36	2.54
Fertilizer levels				
F0	778	709	1.68	1.79
F1	768	698	2.68	2.89
F2	773	704	1.79	1.91
F3	766	695	2.89	3.13
F4	759	689	2.31	2.49

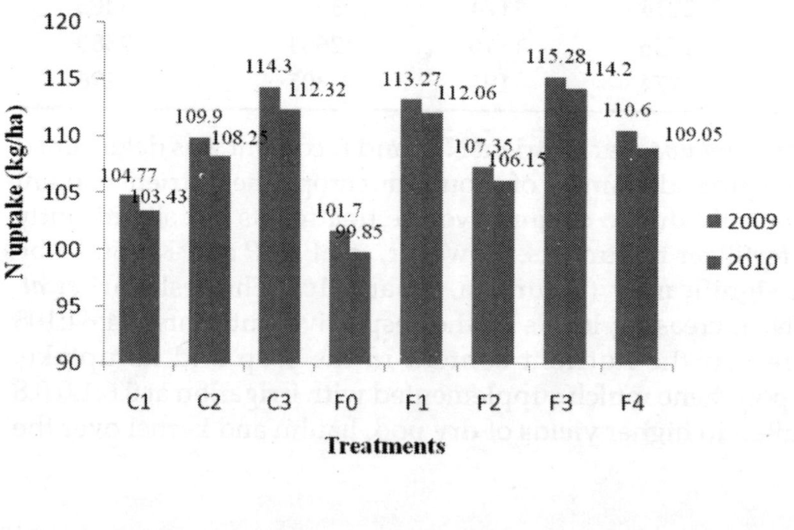

Fig. 1a: Influence of water source and fertilizer levels on N uptake by groundnut crop

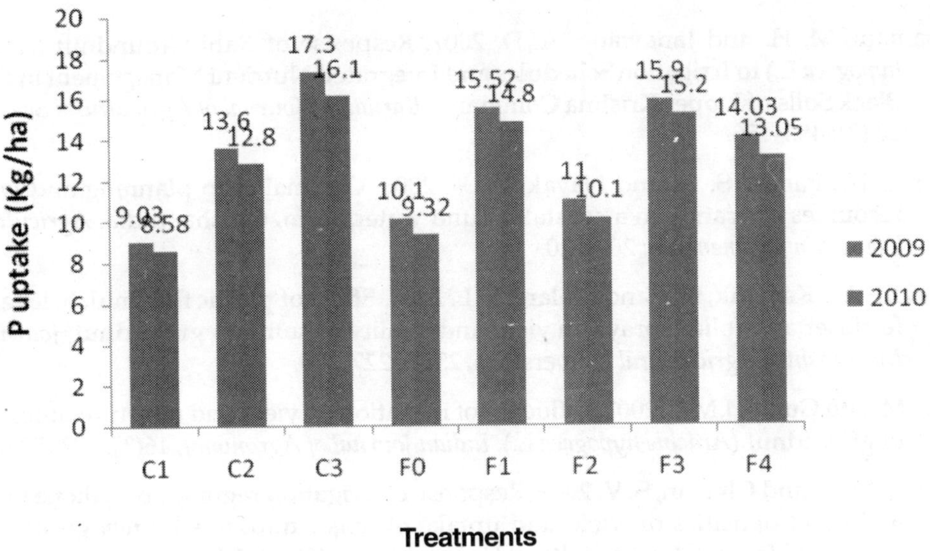

Fig. 1b: Influence of water source and fertilizer levels on P uptake by groundnut crop

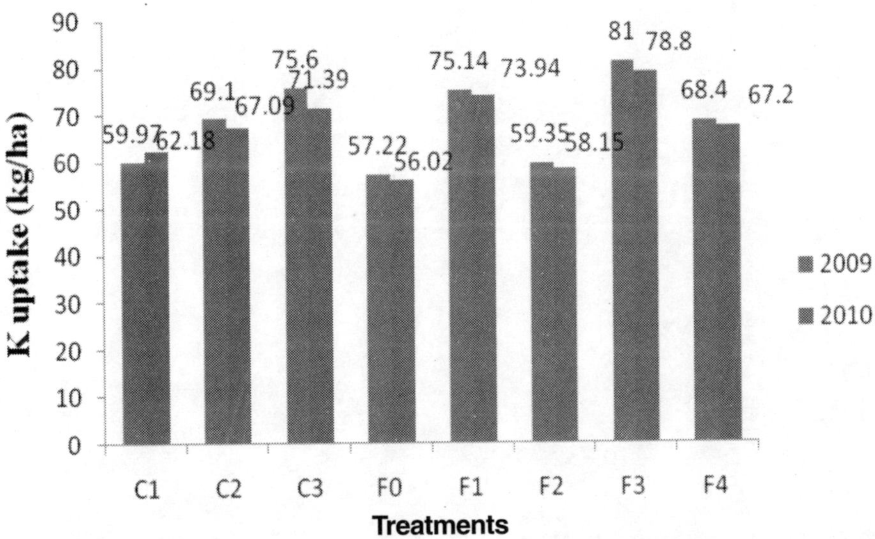

Fig. 1c: Influence of water source and fertilizer levels on K uptake by groundnut crop

REFERENCES

Chitdeshwari, T. P. K. Selvaraj, and Shanmugam, P. M. 2007. Influence of levels and split application of fertilizers on the yield and nutrient uptake by groundnut. *Agricultural Science Digest*, 27(2): 91-94.

Khare, D., Jat, M. K. and Sunder, J. D. 2007. Assessment of water resources allocation options: conjunctive use planning in a link canal command. *Resource Conservation and Recycling*, 5 (2): 487-506.

Hosamani, M. H. and Janawade, A. D. 2007. Response of Rabi Groundnut (*Arachis hypogaea* L.) to Irrigation Schedules and Integrated Nutrient Management in Deep Black Soils of Upper Krishna Command. *Karnataka Journal of Agricultural Sciences,* 20 (3): 453-456.

Sethi, L. N., Panda, S. N. and Nayak, M. K. 2006. Optimal crop planning and water resources allocation in a coastal ground water basin, Odisha, India. *Agricultural Water Management,* 83: 209-220.

Shinde, S. H., Kaushik, S. S. and Bhilare, R. L. 2000. Effect of plastic film mulch, levels of fertiliser and foliar sprays on yield and quality of summer groundnut. *Journal of Maharashtra Agricultural Universities,* 25(2): 227-229.

Taha, M. and Gulati, J.M.L. 2001. Influence of irrigation on yield and moisture utilization of groundnut (*Arachis hypogaea* L.). *Indian Journal of Agronomy,* 46(3): 523–527.

Zagade, M. V. and Chavan, S. V. 2009. Response of irrigation regimes, polythene mulch and plant densities on yield and uptake of major nutrients by *rabi* groundnut. *Journal of Maharashtra Agricultural Universities,* 34 (1): 14-17.

□□□

Chapter 34

Water Management and other Agronomic Practices for Aerobic Rice in Eastern India

K. KANNAN[1], D.K. KUNDU, R. SINGH, A.K. THAKUR, AND S.K.CHAUDHARI

INTRODUCTION

The productivity of Asia's irrigated rice systems is increasingly threatened by water scarcity. Tuong and Bouman (2003) estimated that, by 2025, 2 M ha of Asia's irrigated dry-seasons rice and 13 M ha of its irrigated wet-season rice may experience "physical" water scarcity. Technologies developed for water saving in rice such as alternate wetting and drying, continuous soil saturation and direct seeding may reduce the water requirement of rice to some extent. But a fundamentally different approach to reduce water inputs in rice drastically is necessary for future rice cultivation. Aerobic rice culture is one of the options which not only reduce the water requirement but sustain the rice productivity. Direct-seeded aerobic management eliminates water losses associated with puddling and reduces losses due to evaporation and percolation, reducing total irrigation requirements by 30-50% (Castaðeda *et al.*, 2002). In Asia, upland rice is already grown aerobically with minimal inputs in the upland environment, but mostly as low-yielding subsistence crop to give stable yields under the adverse environmental conditions of the uplands. Upland rice varieties are drought tolerant, but have a low yield potential and tend to lodge under high levels of external inputs such as fertilizer and supplemental irrigation. Alternatively, high yielding lowland rice varieties grown under aerobic soil conditions, but with supplemental irrigation, have been shown to save water but at a severe penalty. Identification of improved upland lowland varieties holds the key for the success of aerobic rice culture.

[1]Central Soil and Water Conservation Research & Training Institute, Research Centre, Fernhill (post), Udhagamandalam-643 004, Tamilnadu, India
Directorate of Water Management, Bhubaneswar

Experiments conducted at Philiphines revealed that a good rice yield (4 to 6 t ha^{-1}) could be harvested from wet as well as dry seasons. Bouman, *et al.* (2005) reported that the highest yields 5.7 t ha^{-1} under aerobic conditions were realized in the dry season with the improved upland variety and the water productivity was 32-88% higher than under flooded condition when the soil moisture was kept between -10 kpa to -30 kpa. In India, researchers currently test the growing of popular lowland rice varieties under aerobic conditions on raised beds. The first results on yields and water use are comparable with those obtained at IRRI (Singh, *et al.*, 2002; Sharma, *et al.*, 2005). The information on aerobic rice in India in general and in the eastern India in particular is meagre. Keeping these points in view, an experiment was laid out in summer to identify the suitable rice varieties and water management practices for aerobic rice culture.

MATERIALS AND METHODS

Field experiments were conducted in the dry season (December to April) 2007 and 2008 at Bhubaneswar ($20^0 30'$ N ,$87^0 48'$ E, 45 msl) in the eastern region of India. The soil of experimental plot consisted of, 18% fine sand, 16% silt and 20% clay and it comes under sandy clay class. Bulk density, hydraulic conductivity and the available soil moisture were estimated as 1.44 gm cc^{-1}, 1.14cm hr^{-1} and 0.18 cm^3 cm^{-3} ,respectively. EC, pH and organic carbon were estimated as 0.25 ds m^{-1}, 5.3 and 0.45 %. An experiment with three replications was set up in a split plot design with the water regime as the main treatment and the variety as the sub plot. Individual subplot size was about 5 m x 5 m. Plots were bunded and separated by 1 m wide strips of bare soil. The same field was used in both the years. The four water regime treatments under aerobic rice (AR) culture were: keeping soil moisture content at 80-90% throughout the crop season (I_1 AR-high moisture), keeping soil moisture content at 60-70% during vegetative and thereafter 80-90% (I_2 AR-Medium moisture) and keeping soil moisture content at 60-70% throughout the crop season (I_3 AR-Low moisture) and conventional cultivation(I_4 Transplanted rice). The amount of water received for each treatment is 700mm, 660mm, 540 mm and 1300 mm, respectively for I_1, I_2 I_3 and I_4 irrigation regimes Variety treatments were consisted of three low land varieties (V_1-*Surendra,* V_2-*Lalat* and V_3- *Kandagiri*) with different duration predominantly grown in this region during dry season under flooded condition. For growing rice varieties under flooded conditions, rice varieties were sown in the wet nursery on the same day when these varieties were sown under aerobic condition and transplanted after 28 days. The flooded plots were puddled and kept continuously flooded from transplanting till 2 weeks before harvest. The water depth was initially 2 cm and gradually increased to 10 cm at full crop development.

The aerobic plots were dry-ploughed and harrowed. The seeds were hand dibbled at 3 cm depth with the spacing of 20 cm x 10 cm during the second week of January. In the AR, phosphorus (40 kg P ha^{-1}as single superphosphate), potassium (40 kg K ha^{-1}as KCl), and zinc (5 kg Zn ha^{-1}as zinc sulfate heptahydrate) were applied and incorporated in all plots 1 day before sowing. Fertilizer N in the form of urea was applied in three splits as top dress only (40 kg ha^{-1}at 10 day after sowing and 20 kg each at maximum tillering and panicle initiation stage). In both years, plots were kept weed-free by an application of pre-emergence herbicide and hand weeding after crop establishment. Pesticides were applied as appropriate for good crop protection. In the FC 21-day-old seedlings from wet-bed nurseries were transplanted into puddle soil at three seedlings per hill at a spacing of 20 cm x 10 cm. The above mentioned quantity of P and K was applied in the field before transplanting. Nitrogen in the form of urea was applied in three splits (40 kg as basal before transplanting, 20 kg each as top dressing during maximum tillering and panicle initiation stages).

The irrigation water input was estimated from the pump operation time and the calibrated pump flow rate. We measured leaf area with LI- 3100C area meter and, after oven drying at 70ºC to constant weight, the dry weight was estimated. At rice maturity all the plants were harvested and dry weights of straw and grain yield were measured. Grain yield is expressed at 14% moisture content. Five samples from every plots were selected randomly before harvest to determine number of panicle, number of filled grain per panicle, and harvest index. Irrigation Expense Efficiency (IEE) (kg grain ha^{-1}cm of water) was calculated as the weight of grains produced per unit of irrigation water input :

$$\text{IEE= Grain yield/ Irrigation water applied} \qquad \text{... (1)}$$

Root samples were collected in between the rows, using a sampling core with 8 cm in diameter and 15 cm long at progressive depths of 15–30 cm depth. The roots were washed and dried in an oven at 65ºC and root weight was record. Total root length (RL) of fresh plant samples was determined by the line intersection method described by Tennant (1975). Root morphological parameters included root diameter (RD), root length density (RLD)(mm cm^{-3}), and root weight density (RWD) (g cm^{-3}) calculated by using the following formula:

$$\text{RLD} = \text{RL} / V \qquad \text{... (2)}$$

$$\text{RWD} = \text{RW} / V \qquad \text{... (3)}$$

Where RW is root weight (g), RL root length (cm), *V* the volume of the soil core.

RESULTS AND DISCUSSION

Plant growth attributes: All the growth attributes *viz.,* plant height; number of tillers per panicle, biomass and root growth was significantly influenced by the irrigation regime and varieties (Table 1 & Figure 1). Plant height and leaf area index (LAI) were significantly higher at 35 days after sowing in aerobically grown rice compared to rice grown under standing water condition (I_4). However, as season advanced, rice under puddle condition produced taller plant (82.4 cm) and higher LAI (3.9) compared to other three irrigation regimes. Higher plant growth characters recorded under aerobic irrigation regime (I_1, I_2 and I_3) during the initial period may be due to the favourable condition created when the seeds are directly sown in the main field with the required spacing in compared to the transplanted rice (I_4) and in the later stages, the condition under transplanted condition was more favour for rice. The highest tiller per hill and biomass) were recorded in conventional system (I_4) and aerobic rice irrigated at 80-90 % of available soil moisture (I_1 AR-high moisture). Rice produced significantly lesser tillers during 90 days after sowing and harvest in the treatments: I_2 AR- Medium moisture and I_3 AR-Low moisture regimes. Among the varieties, the highest tiller number and biomass were recorded under the variety *'surendra'* which was followed by the variety *'lalat'*. However interaction effect of variety and irrigation regimes shows that there was no difference in number of tillers per hill among the variety in medium and lesser water regimes under aerobic cultivation. This may be due to the fact that the long duration rice variety *'surendra'* sustains it tiller production when it is cultivated as transplanted condition or under higher soil moisture regimes when it is cultivated under aerobic condition.

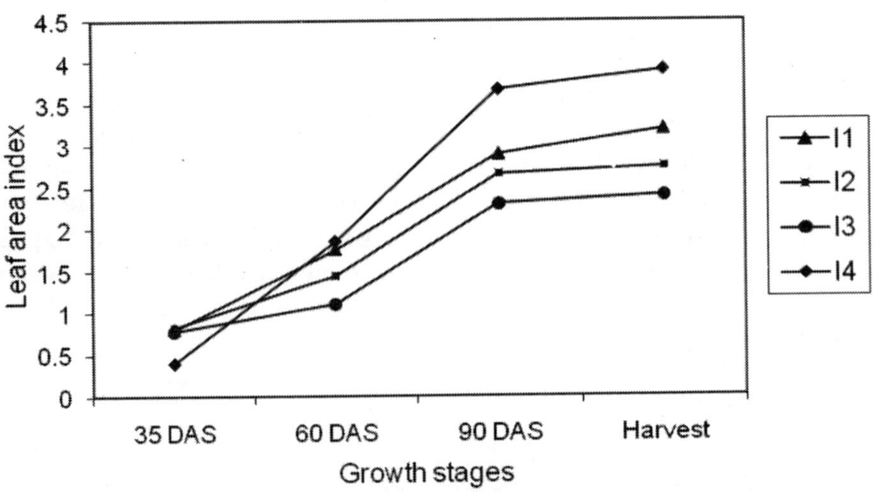

Fig.1: Leaf area Index as influenced by the water regime

Table 1: Plant growth attributes as influenced by irrigation regime and varieties

Treatment	Plant height (cm)		Number of tillers per hill		Biomass (gm per hill) at
	30 DAS	Harvest	90 DAS	Harvest	90 DAS
Irrigation					
I_1 AR-high moisture	28.7	74.9	17.8	20.0	363
I_2 AR- Medium moisture	28.8	72.1	15.4	18.3	282
I_3 AR-Low moisture	28.5	69.8	15.2	17.7	281
I_4-Transplanted rice	19.5	82.4	16.3	21.4	441
CD (p=0.05)	1.1	1.9	0.92	1.7	17
Variety					
V_1 *Surendra*	28.0	74.5	17.3	20.6	375
V_2 *Lalat*	27.0	79.2	16.6	19.3	348
V_3 *Kandagiri*	24.3	70.6	14.6	18.2	302
CD (p=0.05)	0.7	1.3	0.95	0.77	8.0

Method of cultivation caused great changes in root growth characters (Table 2). Root volume was more under aerobic condition. At 90 days after planting, irrespective of the irrigation treatment; aerobic rice produced more root volume. At 90 days the highest root volume was recorded with rice cultivated under aerobic condition compared to conventional method of rice cultivation irrespective of the varieties. Higher root volume observed under aerobic method of rice cultivation may be due to the lesser soil compaction and bulk density compared to the puddle condition. Among the varieties, the highest root volume was observed in *'surendra'* which was followed by *'lalat'*.

Table 2: Root growth of rice as influenced by irrigation regime and varieties

Treatments	Root volume(cc)	Root shoot ratio	RLD (mm cm $^{-3}$)	RWD (g cm^{-3})
Irrigation				
I_1 AR-high moisture	31.0	0.17	3.9	0.39
I_2 AR- Medium moisture	29.3	0.11	3.6	0.35
I_3 AR-Low moisture	22.3	0.13	3.4	0.35
I_4-Transplanted rice	26.7	0.13	4.3	0.21
CD (p=0.05)	2.6	0.01	0.3	0.03
Variety				
V_1 *Surendra*	29.6	0.10	4.0	0.33
V_2 *Lalat*	27.9	0.13	3.9	0.34
V_3 *Kandagiri*	24.7	0.19	3.6	0.30
CD (p=0.05)	2.1	0.02	0.2	0.03

Root to shoot ratio, which is one of the important trait for the suitability of a particular variety to aerobic rice was more in the case of *'lalat'*. Higher root length density and root weight density was also observed under aerobic method of rice cultivation.

Yield and yield attributes and IEE: The highest panicles m^{-2} and number of filled grains per panicle were recorded in conventional method (I_4). When aerobic rice is irrigated with the moisture status of 80-90% of ASM, it equally produced higher panicles and filled grains per panicles compared to other irrigation treatments under aerobic condition (Table 3). However, rice cultivated under transplanted condition produced significantly more filled grains per panicle compared to the rice grown under aerobic condition. There was no significant difference in filled grains per panicle was observed among the varieties. Under aerobic method of rice cultivation, higher moisture regime (I_1 AR-high moisture) produced significantly higher filled grains per panicle compared to I2 AR- Medium moisture and I_3 AR-Low moisture regimes. This may be due to the fact that rice requires higher soil moisture during the flowering to grain filling. Any moisture leads to higher chaff grain and lesser filled grain. Zubaer *et al.* (2007)) reported that when rice was exposed to water stress, there was significant decrease in filled grain per panicles, and drymatter production.

Table 3: Yield attributes of different rice varieties under various irrigation regime

Treatment	Panicles (m^{-2})	Filled grains per panicle	Chaff per panicle	Straw yield (t ha^{-1})	Harvest index
Irrigation					
I_1 AR-high moisture	555	50.3	20.1	3.65	0.48
I_2 AR- Medium moisture	492	44.7	23.5	3.47	0.48
I_3 AR-Low moisture	435	37.7	28.2	2.98	0.44
I_4-Transplanted rice	525	57.6	16.7	4.35	0.48
CD (p=0.05)	22	3.20	2.75	0.44	0.01
Variety					
V_1 *Surendra*	507	51.1	23.9	4.23	0.47
V_2 *Lalat*	496	49.6	21.5	3.85	0.49
V_3 *Kandagiri*	501	42.3	21.1	2.75	0.45
CD (p=0.05)	NS	2.24	1.56	0.27	0.01

Interaction effect of irrigation regimes and varieties on grain yield is given in Table 4. The highest grain yield (4.05 t ha^{-1}) and straw yield (4.35 t ha^{-1}) were observed in the conventional rice cultivation. Among varieties, *'surendra'* produced higher yield and it was on par with the variety *'lalat'*. Reduction in yield due to aerobic cultivation compared to puddle condition was only 19% in case of I_1 treatment (higher moisture regime) and 59 % in case of I_3 treatment (lower moisture regime). Interaction effect of irrigation regimes and variety shows that *'surendra'* and *'lalat'* produced higher grain yield when they were cultivated under transplanted condition. Within the irrigation regimes under aerobic condition, grain yield of all the varieties produced higher grain yield when they are grown under higher soil moisture regime and medium moisture regimes. Higher filled grains per panicles were the most important factor which contributed for higher grain yield obtained from conventional rice cultivation. Even though rice produced higher biomass and panicles per m^2, under higher irrigation regimes under aerobic cultivation (I_1 AR-high moisture), the yield was lesser due to the higher chaffy grains present in the panicle compared to the conventional rice cultivation (I_4). Even though rice cultivated under aerobic condition produced higher root volume initially, it could not sustain the root growth as suberization or roots occurred in the later stages due to unfavourable condition compared to transplanted rice. Less nutrient uptake (data not shown) due to the less feeder root in aerobic condition might contributed in the less biomass and grain in the later stages of rice growth. Suralta and Yamauchi (2008) concluded that when rice is grown under aerobic condition, the inability of roots to acclimate to such changes in soil water regimes may result in reduced growth and function thereby, dry matter production.

Table 4: Interaction effect of irrigation regimes and varieties on yield (t ha^{-1})

Irrigation regimes	Varieties			Mean
	Surendra	Lalat	Kandagiri	
I_1 AR-high moisture	3.95	3.82	2.32	3.36
I_2 AR- Medium moisture	3.62	3.60	2.24	3.15
I_3 AR-Low moisture	2.57	2.63	1.97	2.39
I_4-Transplanted rice	4.88	4.76	2.50	4.05
Mean	3.75	3.70	2.26	

CD (p=0.05) for main plot effect: 0.44; for sub plot effect: 0.21; interaction effect: 0.43

However, higher Irrigation Use Efficiency (IEE) was recorded under the aerobic method of rice cultivation compared to the transplanted rice (Table 5). The highest IEE and optimum yield was recorded in the variety *'surendra'* under aerobic condition when irrigated at 80-90 % available soil moisture (I_1). Bouman *et al.* (2005) observed total water use and water productivity of aerobic rice was 27–51% lower and 32–88% higher than that of flooded rice, respectively.

Table 5: Interaction effect of irrigation regimes and varieties on Irrigation Expense Efficiency (kg grain ha-1 cm of water)

Irrigation regimes	Varieties			Mean
	Surendra	Lalat	Kandagiri	
I_1 AR-high moisture	54.81	53.06	32.27	46.71
I_2 AR- Medium moisture	54.85	54.60	33.94	47.79
I_3 AR-Low moisture	47.59	48.77	36.54	44.30
I_4-Transplanted rice	37.51	36.62	19.23	31.12
Mean	48.69	48.26	30.50	

CD (p=0.05) for main plot effect: 4.98; for sub plot effect: 3.07; interaction effect: 6.15

CONCLUSIONS

The result of the study shows that aerobic rice can be grown in water-short areas that have insufficient water for the production of lowland rice. Since, compared with lowland rice, the water input of aerobic rice is more reduced than the yields, the water productivity of aerobic rice is higher lowland rice (64-88% in our study). Consequently, in areas where water is relatively scarcer than land, total rice production can be maximized by growing aerobic rice. Crop management strategies could be developed to reverse the yield decline of continuous aerobic rice only after the causes of yield decline are identified. New aerobic rice varieties with minimum yield gap compared with flooded rice and crop management strategies that can reverse the yield decline of continuous aerobic rice have to be developed before aerobic rice technology can be adopted in large areas in the tropics.

REFERENCES

Bouman, B.A.M., Peng, S., Castaneda, A.R. and Visperas, R.M. 2005. Yield and water use of irrigated tropical aerobic rice systems. *Agric. Water Manage.*, 74: 87–105.

Castaðeda, A.R., Bouman, B.A.M., Peng, S. and Visperas, R.M.2002. The potential of aerobic rice to reduce water use in water-scarce irrigated lowlands in the tropics. In 'Water-wise rice production. Proceedings of the International Workshop on Water-Wise Rice Production, 8-11 April 2002. Los Baðos, Philippines.' (Eds. B.A.M. Bouman,, H. Hengsdijk, B. Hardy, P.S. Bindraban, T.P. Tuong amd J.K. Ladha). (International Rice Research Institute, Los Banos, Philippines).

Sharma, S. K., Pandey, D.K. Gangawar, K.S. and Tomar, O.K. 2005. Effect of crop establishment methods on performance of rice cultivars and their effect on succeeding wheat (*Triticum aestivum*). *Indian J. Agron.*, 50: 253-255.

Singh, A. K., Choudury, B. U. and Bouman, B. A. M. 2002. The effect of rice establishment techniques and water management on crop-water relations. *In*: *Water-Wise Rice*

Production (Eds. Bouman, Hengsdijk, Hardy, Bindraban, Tuong, Lafitte and Ladha) Los Baños (Philippines): IRRI,

Suralta, R. R. and Yamauchi, A. 2008.Root growth, aerenchyma development, and oxygen transport in rice genotypes subjected to drought and waterlogging. *Environmental and Experimental Botany*, 64 (1): 75-82

Tennant, D. 1975. A test of modified line intersect method of estimating root length. *J. Ecol.* 63, 995–1001.

Tuong, T.P. and Bouman, B.A.M., 2003. Rice production in water-scarce environments. In: Kijne, J.W., Barker, R., Molden, D. (Eds.), Water Productivity in Agriculture: Limits and Opportunities for Improvement. CABI Publishing, UK, pp53–67.

Zubaer, M.A., Chowdhury, A.K.M.M.B., Islam, M.Z., Ahmed, T. and Hasan, A.H. 2007. Efects of water stress on growth and yield attributes of Aman rice genotypes. *Int. J. Sustain. Crop Prod.*, 2(6):25-30.

□□□

Chapter 35

Water and Dissolved Organic Carbon Loss through Bypass Flow as Influenced by Tillage in Vertisols

SANGEETA LENKA, N. K. LENKA, A. K. MISRA, S. K. RAUTARAY[1]
R. S. CHAUDHARY, A. SUBBA RAO

INTRODUCTION

Adverse effect of climate change is evident through high intensity rainfall, dry spells, high temperature, frost etc. have warrant the management of scare water resource. Loss of water through bypass flow is an important phenomenon in black soil that needs to be addressed and estimated correctly in field. Continuity of macropores and/or cracks in structured soils has been documented to determine rates of movement of solutes or contaminants much faster than those that could be explained by considering basic transport and adsorption parameters of the soil material. A flow of water under these conditions has been termed preferential flow or bypass flow (Bouma, 1984) and is defined as "the flow of water through a system of large pores that allows fast flow velocities and bypasses the unsaturated zone" (Beven and Germann, 1982). Sudden high intensity rainfall or irrigation through flooding leads to loss of water through bypass flow in cracking clay soils with more intense swell shrink dynamic.

Black soils in India cover approximately 22.2 % of total geographical area (Sehgal and Abrol, 1994). These soils are characterized by their dark reddish brown to black color and heavy clay texture they are medium deep to deep, the depth at places are often being more than 5 to 6 feet. The soils of this region are clayey with poor structure, low infiltration, highly erodible, poor root zone aeration, a narrow workable water content range and cracks heavily on drying. Cracks are a unique feature in these soils with strong shrink–swell

Indian Institute of Soil Science, Nabibagh, Berasia Road, Bhopal-462 038, Madhya Pradesh, India

[1]Central Institute of Agricultural Engineering, Nabibagh, Berasia Road, Bhopal-462 038 Madhya Pradesh, India

potential. Cracks have a direct bearing on the extent and size distribution of clods formed upon ploughing and thereby influences soil tilth (Ahmad and Mermut, 1997). Srivastava *et al.* (1989) reported that deep and wide cracks arranged in a hexagonal pattern are more likely to produce larger clods, compared to soil with a network of fine cracks. The frequency, size and rate of development of cracks influence the water, air, solute and heat dynamics in soil and hence influence the crop productivity and potential for ground water pollution. Preferential flow has considerable methodological implications for determining solute fluxes and calculating soil element budgets (*e.g.*, Lipsius and Durner, 2005). These soils are irrigated in both summer and winter by flooding. Because of this high application rate mostly 6 cm irrigation, bypass flow that is rapid transport of water and solutes via macropores or cracks to subsoil or to ground water is prevalent during irrigation. Preferential flow increased with irrigation (Gjettermann *et al.*, 1997), rainfall intensity and was stronger for ponded than for sprinkling irrigation (Hamdi *et al.*, 1994).

Because tillage alters the nature and distribution of aggregates and the spaces between them, it can have a major influence on soil hydrology (Armstrong and Harris, 1996) and consequently the pattern and amounts of solutes being leached (Addiscott and Dexter, 1994). Large vertical cracks tend to occur during the summer months as the soil dries and these can provide rapid pathways for the transport of water and solutes to the mole drains (Goss *et al.*, 1993). On the ploughed plots, the connectivity of the macro-pores is interrupted at the base of the plough layer, resulting in lower flow rates and a less peaky hydrograph than for direct drilled plots (Harris *et al.*, 1993). These black soils are difficult to manage and work with. Many studies have documented the effect of management practices on improving nutrient use efficiency, soil physical properties and soil biological properties in vertisols of India. However only few researchers have discussed effect of tillage practices on cracking in black soil in soybean-wheat system and rice wheat system. Further reports on influence of tillage on bypass flow of water and vertical distribution of dissolved organic carbon is negligible. Cracking in vertisols could be managed suitably by adopting appropriate tillage practices, cropping systems, mulching and manure management. Sub-soiling practiced in soybean reduced the magnitude of different crack parameters in comparison to conventional tillage. Similarly Rice grown under puddled condition significantly enhanced different crack Parameters compared to other crops (Bandyopadhya *et al.*, 2003). Also soil conservation measures such as contour furrows and conservation bench terraces and incorporation of amendments help minimize pot-hole formation in black soils (Somasundaram, 2011). No studies appear to have estimated loss of water through bypass flow in vertisols of India and also the probable consequences of tillage and residue management on vertical distribution of DOC. Hence the present investigation was undertaken to estimate the effect of tillage on water and DOC loss through bypass flow in vertisols of India and secondly to study the cracking pattern and behavior with respect to tillage in soybean-wheat systems of India.

MATERIALS AND METHODS

Study area:The investigation was conducted on the research farm of the Central Institute of Agricultural Engineering (CIAE), Bhopal, Madhya Pradesh, India situated at 23°18′ N longitude, 77°24′ E latitude and 485m above mean sea level. The region has a hot sub-humid climate with 1200 mm mean annual rainfall and 1400mm mean annual potential evapo-transpiration. The average maximum monthly temperature is reached in May (40.7°C) and the average minimum monthly temperature is reached in January (10.4°C). The meteorological observations for the study year are presented in Figure 1. Daily weather data including rainfall data were taken from meteorological observatory of CIAE which is the site for study area. Statistical analysis of the data was carried out using the standard statistical techniques (Gomez and Gomez, 1984). The soil is a noncalcareous Vertisol (Isohyperthermic Typic Haplustert) with clay texture (52% clay) bulk density of 1.34Mgm^{-3} at 0.27 g g^{-1} soil water content, available water capacity of 10.16 cm, neutral to alkaline reaction (pH = 7.5), 5.0 g kg^{-1} organic carbon, less than 0.3 dsm^{-1} electrical conductivity and Ca as the dominant exchangeable cation in the Ap horizon.. The initial soil characteristic of the experimental site is presented in Table 1.

Table 1: Initial soil properties of study area

Soil propertie	Soil depth (cm)			
	0-5	5-15	15-30	30-45
Bulk density(g cc^{-1})	1.10	1.26	1.37	1.40
Soil texture	Clayey	Clayey	Clayey	Clayey
Water retention				
FC (0.33 bar, v/v)	39	46	50	53
PWP (15 bar, v/v)	19	19	29	30
TOC (%)	0.96	0.92	0.74	0.59

Experimental details: The experiment started in July 2008 with soybean crop. Soybean is grown in *Kharif* (July to October) and wheat in *rabi* (November to April) in all the years. The Soybean and wheat cultivars sown were JS-335 and HD-1531 respectively. Recommended agronomic practices were carried out for both the crops. The three tillage treatments are no tillage (NT, direct sowing by no till slit drill) and reduced tillage (RT, one pass rotavator and sowing by seed cum fertilizer drill) and conventional tillage (three time ploughing + sowing with seed drill in soybean and 2 time ploughing + sowing with seed drill in wheat) for both soybean and wheat. Fertilizer dose applied was 30:60:30 and 80:40:20 kg ha^{-1} for soybean and wheat respectively. N fertilizer was given in split dose while 100% P and K was applied as basal. The plots were kept weed free by pre-emergence application of Stomp (Pendimethalin), Rule out (Glyphosate) before sowing in Soybean and post emergence application of Pursuit at 20 DAS in No tillage plots. In reduced tillage (RT) and conventional tillage (CT) weeding is done mechanically by tractor drawn weeder at 20 DAS and in wheat crop by application of 2, 4-D at 20-25 DAS in both RT and NT. Also, crops were kept free from insect and pathogen attack.

Bypass flow : By pass flow was measured in field with slight modification in the method reported by Bouma (1984) after harvest of wheat in 2010-11 at two dates of observation 7th April and 7th June, 2011. According to conventional flow theory, water can only leave the columns after saturation has been reached at the base of the column. The top of the column should then have a pressure head of -20 cm. Infiltration rings with 55 cm diameter was inserted in ground with help of hammer. Initial moisture content by gravimetric method was recorded from four places around the ring at 0-15, 15-30, 30-45 and 45-60cm soil depth. Since irrigation by flooding is still commonly practiced by farmers in India so 6 cm water was applied uniformly inside the ring spread with dry wheat straw. It is assumed that black soils will reach field capacity after 24-48 hours after saturation. Therefore in this study after 24 hours of water application gravimetric moisture content were again taken from 4 point inside the ring and averaged out to get the final moisture content in the profile (0-15, 15-30, 30-45 and 45-60cm soil depth). The soil inside the infiltration ring is covered with wheat straw and white polythene sheet to prevent evaporative loss of water. As per conventional flow theory, macropore flow/bypass flow is assessed using the tipping bucket approach. The tipping bucket approach assumes that each layer of soil has a capacity for water called the field capacity or the drained upper limit (Lenka *et al.*, 2009). No water drains from a layer if its volumetric water content (simply called water content henceforth) is less than its field capacity. If the water content exceeds field capacity then, in a given time step, either all of the water in excess of field capacity or some fraction of the water between field capacity and the saturated water content drains into the underlying layer.

Measurement of crack parameters : Observation on cracks was taken on the same day of bypass flow measurements. Each experimental plot was divided into four parts and squares of 1m × 1m were demarcated in each part. Within each of these squares, crack length apparent on the soil surface was measured by a flexible twine run along the crack and measuring the total length of all the cracks in the 1m × 1m (Dasog and Shashidhara, 1993). The average depth and width of cracks were based on measurements made at 0.5m interval along the course of the crack. The crack depth was measured by a 2 mm diameter steel rod inserted until it offered resistance to further penetration, while the width at the same point as that of depth recording was measured with an adjustable divider at 1 cm below the soil surface (Dasog *et al.*, 1988). One centimeter depth was chosen to avoid exaggerated widths caused by surface disturbance. Total volume (V, m^3) and surface area (SA, m^2) of each crack was computed using the following equations assuming triangular shape of the cracks (Sharma *et al.*, 1995):

$$V = \Sigma 0.5wdl \quad \text{... (1)}$$

$$SA = \Sigma 2Cl \quad \text{... (2)}$$

$$C = [(0.5w)^2 + d^2]^{1/2} \quad \text{... (3)}$$

Where w is the mean width of cracks (m), d the mean depth of the crack (m), l the length of the crack (m), and C the parameter based on w and d.

Measurement of dissolved organic carbon (DOC) : Plots were sampled randomly over the whole area from 0-5, 5-15, 15-30, 30-45 cm soil depth using soil core auger after harvest of wheat to determine DOC. In laboratory soil were removed from the cores and mixed thoroughly, quartered to get approximate 500 g soil for each depth. This process was repeated for all the depths. Fresh field moist soil was immediately extracted with distilled water in the ratio 1: 2 (soil: water). The content was shaken in reciprocal shaker for 18 hrs. The extractant was filtered through 0.45μm nylon filter paper and stored in refrigerator at 4°C for analysis of carbon. Carbon in filtrate is determined by oxidation with 0.2 N potassium dichromoate. 10ml of filtrate was oxidized with 0.2 N $K_2Cr_2O_7$ and 10 ml conc. H_2SO_4 and 5 ml of orthophosporic acid and heated at 100 °C for 30 minutes under reflux condition. The content was cooled with 250 ml distilled water immediately. Allow the content to cool to room temperature. The carbon content is determined by titrating the content against 0.005 N ferrous ammonium sulphate using ferroin indicator to get brick red indicator. Moisture correction was carried out as fresh samples were taken for analysis.

RESULTS AND DISCUSSION

Soil cracking: Table 2 summarizes the effect of tillage on cracking pattern of vertisols after third year of soybean-wheat cropping cycle at two sampling times (first at 7th April 2011 *i.e.* immediately after harvest and second at 7th June 2011 two months after first sampling date). At both observation dates tillage significantly modified the surface cracking pattern of vertisols in soybean-wheat system. Crack width, depth and volume were found to be significantly lower in notillage followed by reduced tillage and no tillage. The values of average crack width at first date of observation (7th April, 2011) differed from 0.017 (CT) to 0.027 (NT) m in tillage treatments. Similarly crack depth and volume was significantly higher in no tillage 0.257 m and 167.92 $\times 10^{-4}$ m^3 m^{-2} respectively compared to other tillage treatments. However the total length of crack was significantly higher in conventional tillage (6.78 m) against no tillage (4.84 m). Higher values of crack width, depth and volume in no tilled plots may be attributed to relatively favorable structure in the reduced tillage plots resulting in an increased degree of shrinkage and a decreased degree of swelling in these treatments (Flowers and Lal, 1999). Harris *et al.* (1993) also reported on conventional ploughed plots, the connectivity of the macro-pores is interrupted at the base of the plough layer, resulting in lower flow rates and a less peaky hydrograph than for direct drilled plots. This finding is in agreement with that of Bandyopadhyay *et al.* (2003) and Tomar *et al.* (2009).

Table 2: Effect of tillage on cracking pattern of vertisol after wheat (2010-11) in soybean-wheat system at two observation dates

Treatment	Width (m)	Depth (m)	Length (m)	SA[a] (m^2 m^{-2})	V[b] (m^3 m^{-2} $\times 10^{-4}$)
Date of observation: 7th April 2011					
Conventional tillage	*0.017b (±0.002)	0.164c (±0.019)	6.78a (±0.69)	2.84a	94.50c
Reduced tillage	0.022a (±0.003)	0.192b (±0.021)	5.24b (±0.71)	2.54b	110.67b
No tillage	0.027a (±0.003)	0.257a (±0.024)	4.84c (±0.58)	2.95a	167.92a
Date of observation: 7th June 2011					
Conventional tillage	*0.019b (0.002)	0.241b (±0.028)	7.42a (±0.61)	4.12a	169.88c
Reduced tillage	0.025a (±0.003)	0.294a (±0.031)	5.97b (±0.63)	3.99a	219.40b
No tillage	0.032a (±0.004)	0.401a (±0.039)	5.45c (±0.64)	4.79a	349.74a

[a]Surface area of the cracks

[b]Volume of the cracks

*Means in a column followed by common letter are not significantly different (p=0.05). Values are means with the standard deviation in parenthesis (n = 6)

Irrespective of tillage treatments the cracking parameters *viz.*, width, depth, length and also the volume and surface area of cracks were more at second date of observation was shown in Table 2. The width, depth and length of cracks was more by 12-19%, 47-56% and 9-14% respectively over first date of observation. The increase in cracking intensity *i.e.*, surface area and volume of cracks over the first date of observation could be due to increase in atmospheric air temperature and accompanied evaporation loss. Soil drying and consequently shrinkage is higher if water loss due to evaporation increases by higher air temperature. After harvest of wheat in March last week the field remains fallow for three months (April to June) every year. In India summer starts from March and ends in June (onset of monsoon). The mean monthly maximum temperature and rainfall for the study period is shown in Figure 1. As the hot sunny days advances in summer the air temperatures increases by 7-8°C in April and May from mean maximum temperature of March for *e.g.*, in the year 2010 the mean T_{max} in March, April and May was 36, 41 and 44°C ,respectively. Therefore black clayey soils start showing cracks from the month of March intensity of which increases with advent of time and summer. Which results in formation of deep wide open cracks by end of May and June. With onset of monsoon in June the soil is deep wide open for movement of water through cracks along with other natural processes of soil water movement. The present findings are in conformity with Goss *et al.* (1993) whose work also reported occurrence of large vertical cracks during the summer months as the soil dries and these can provide rapid pathways for the transport of water and solutes to the mole drains.

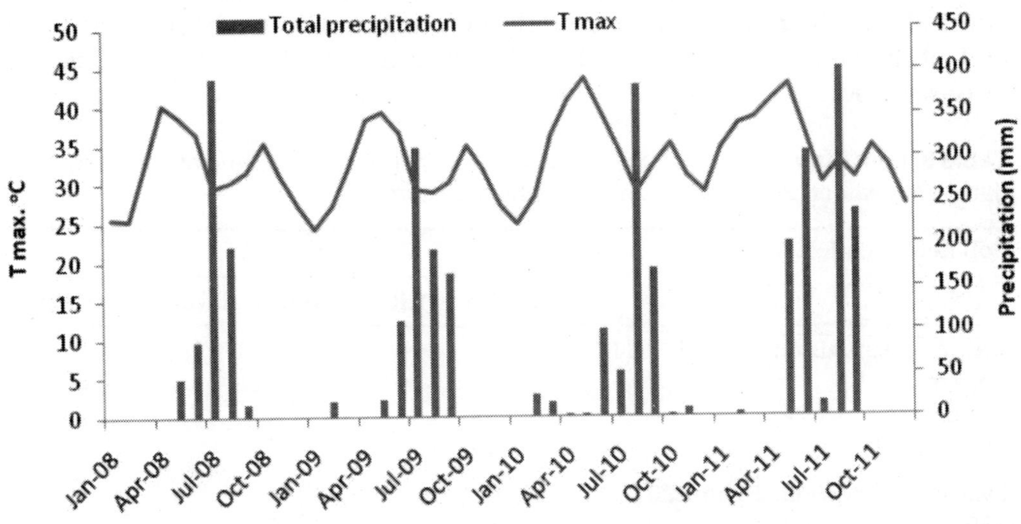

Fig. 1: Mean monthly maximum temperature and total precipitation during crop season 2008-11

Loss of water through bypass flow **:** Table 3 represents loss of water through bypass flow from 60 mm water application in both tillage treatments and at two observation dates. According to conventional flow theory, water can only leave the end of particular soil depth after saturation has been reached at the end. The top of soil column should then have a pressure head of -20 cm. Comparing volumetric water content after water application to water content at pressure head corresponding to -100 and -330 cm. It is seen that volumetric water content after water application still falls short of those corresponding to -100 and -300 cm pressure head. Water gain in respective depths 0-15, 15-30, 30-45 and 45-60 cm was calculated from difference of volumetric water content before and after water application. Thus bypass loss of water beyond 30, 45 and 60 cm was higher in NT *i.e.,* 60, 56 and 43 % of water applied respectively and least in CT *i.e.,* 48, 35 and 20 % of water applied respectively (Table 3). In vertisols tillage breaks the continuity of pore size distribution and macroporosity thus disrupting the crack continuity in top soil as evident from lower crack width, depth and volume in CT as compared to NT in Table 2. Lesser cracking parameters in CT lower loss of water through macropores and cracks than NT. Also tillage alters the nature and distribution of aggregates and the spaces between them, it can have a major influence on soil hydrology and consequently the pattern and amounts of solutes being leached. On the ploughed plots, the connectivity of the macropores is interrupted at the base of the plough layer, resulting in lower flow rates and a less peaky hydrograph than for direct drilled plots. Similar effects of tillage were found at the second date of observation with greater loss of water through bypass flow over the first date. At second date of observation effect of No tillage was more pronounced on loss of water beyond 30 cm, 45 and 60 cm compared to CT *i.e.,*41 to 60 % increase at all depths over first date of observation in NT and 12-15 % increase in CT. This may be attributed to

Table 3: Loss of water (%) through bypass flow as influenced by tillage at two observation dates after wheat (2010-11) in soybean-wheat system

Soil depth (cm)	Treatment		
	Conventional tillage	Reduced tillage	No tillage
Date of observation: 7th April 2011			
>30 cm	48	52	60
>45 cm	35	40	56
>60 cm	20	26	43
Date of observation: 7th June 2011			
>30 cm	54	74	84
>45 cm	40	68	82
>60 cm	23	43	68

higher crack volume and parameters (Table 2) at second date of observation which could have resulted in more loss of water through bypass flow. Large vertical cracks tend to occur during the summer months as the soil dries and these can provide rapid pathways for the transport of water and solutes to the mole drains (Goss *et al.*, 1993).

Vertical distribution of dissolved organic carbon (DOC) : Movement of water through macropores and cracks is also instrumental in transport of solutes, salts, dissolved organic carbon and nitrogen beyond subsurface. Because it is difficult to collect and measure water and dissolved organic carbon (DOC) loss through bypass flow in field. Thus vertical distribution of DOC in soil samples collected from different depths could be used as indicator to measure loss of DOC through bypass flow. Vertical distribution of dissolved organic carbon is shown in Table 4 with respect to different tillage treatments. Tillage has significant effect on vertical distribution of dissolved organic carbon at different soil depth except for 30-45 cm. DOC was highest in NT (43.24 ppm) and was at par with RT (42.07 ppm) and least in CT (31.28 ppm) at surface 0-5 cm soil. However the effect of RT was significantly more at 5-15 cm soil depth followed by NT and CT. Similarly at 15-30 cm soil depth RT was found to have significantly greater DOC and the effect of NT and CT were at par. In the surface 0-5 cm soil depth greater DOC in NT and RT is attributed to retention of crop residue after harvest of wheat crop (12-15 cm standing straw). Residues incorporated by rotavator in reduced tillage (RT) promote the turnover of organic materials. Thus, soil organic carbon (SOC) and labile fractions of SOC such as DOC, water soluble carbon, and microbial biomass carbon etc., were likely more in RT compared to NT and CT. The depth distribution of DOC was impacted by tillage treatments. Placement of residue near the surface in NT resulted in higher DOC in surface 0-5 cm soil depth, but levels significantly decreased with depth. Lesser loss of water through bypass flow and greater retention time for water in soil for RT compared to NT resulted in higher concentration of DOC in subsurface of RT. Dissolved organic C in subsurface soils may be a result of decomposition of crop residues (Lundquist *et al.*, 1999; Wright *et al.*, 2007) or translocation from surface soil (Kalbitz *et al.*, 2000). In a similar study Chen *et al.* (2009) also reported significantly higher Labile C fractions: particulate organic C (POC), permanganate oxidizable C ($KMnO_4$-C), hot-water extractable C (HWC), microbial biomass C (MBC) and dissolved organic C (DOC) in NT (no-tillage with residue cover) and ST (shallow tillage with residue cover) than in CT (Conventional tillage with residue removal) in the upper 15 cm. DOC may directly derive from the decomposition process if it is either unpalatable and therefore remained or washed away before complete incorporation and mineralization by the decomposers. Solid soil organic C can be solubilized by microbial activity. DOC can then be transported down the soil profile with mineral soil serving as a chromatographic column at which DOC is adsorbed replacing older DOC or other anions (Guggenberger and Zech, 1993). Thus, DOC in subsoil may largely derive from the subsoil itself and only in preferential flow paths and under storm flush conditions surface soil DOC directly reaches the subsoil (Kaiser and Guggenberger, 2005; Froberg *et al.*, 2007). Leaching of DOC may account for a significant C flux in terrestrial ecosystems (Siemens, 2003). In the present study RT (reduced tillage) has the same amount of residue as NT however residue incorporation and lesser loss of water through bypass flow in RT would have resulted in greater concentration of DOC in subsoil (5-15 and 15-30 cm) compared to NT and CT. Further CT with residue removal has lesser DOC concentration in spite of least loss of water through bypass flow.

Table 4: Vertical distribution of dissolved organic carbon (ppm) in soil as influenced by tillage after wheat (2010-11) in soybean-wheat system

Soil depth (cm)	Treatment		
	Conventional tillage	Reduced tillage	No tillage
0-5	31.28b (±3.8)	42.07a (±5.4)	43.24a (±5.1)
5-15	22.14c (±2.2)	29.45a (±4.7)	27.32b (±4.2)
15-30	11.34b (±1.8)	13.78a (±2.1)	12.06b (±1.7)
30-45	8.67a (±1.2)	9.04a (±1.1)	8.74a (±1.3)

*Means in a row followed by common letter are not significantly different at p=0.05
Values are means with the standard deviation in parenthesis (n = 6)

CONCLUSIONS

Conservation tillage practices significantly influence the crack parameters *viz.*, width, depth length, surface area and volume. No tillage resulted in greater cracking intensity compared to conventional tillage. This must have contributed to greater loss of water through bypass flow in no tillage followed by reduced tillage and conventional tillage. Our data shows that as the summer heating and drying advances soil shrinkage increases and cracking in vertisols is more irrespective of tillage practice. Further the effect of summer heating and drying is more in notillage compared to other tillage practices. Soil tillage practice also influences the retention of DOC in soil apart from bypass loss of water and soil cracking behavior. No tillage with residue on the surface has maximum DOC concentration in surface 0-5 cm soil layer compared to other tillage practices. However RT with residue incorporation and lower loss of water through bypass flow resulted in higher concentration of dissolved organic carbon in subsoil layers.

REFERENCES

Addiscott, T.M. and Dexter, A.R. 1994. Tillage and crop residue management effects on losses of chemicals from soil. *Soil Till. Res.*, 30: 125-168.

Ahmad, N. and Mermut, A. 1997. Vertisols and technologies for their management. Developments in Soil Sciences. Elsevier, Amsterdam.

Armstrong, A.C. and Harris, G.L. 1996.Movement of water and solutes from agricultural land, the effects of artificial drainage. In: Anderson, M.G., Brooks, S.M. (Eds.), Advances in Hillslope Processes, Vol. 1. Wiley, Chichester, UK, 1996, pp187-211.

Bandyopadhyay, K.K., Mohanty M., Painuli, D.K., Misra, A.K., Hati, K.M., Mandal, K.G., Ghosh, P.K., Chaudhary, R.S. and Acharya, C.L. 2003. Influence of tillage practices and nutrient management on crack parameters in a Vertisol of central India. *Soil Till. Res.*, 71: 133–142.

Beven, K. and Germann, P. 1982. Macropore and water flows in soils.*Water Resour. Res.*, 18: 1311–1325.

Bouma, J. 1984. Using soil morphology to develop measurement methods and simulation techniques for water movement in heavy clay soils. In: Bouma, J., Raats, P.A.C. (Eds.), Water and Solute Movement in Heavy Clay Soils. Int. Soc. Soil Science (ISSS) Symp. Inst. Land Reclamation and Irrigation (ILRI), Wageningen, The Netherland, vol. 37, pp298–316.

Chen, H., Hou, R., Gong, Y., Li, H., Fan, M. and Kuzyakov, Y. 2009. Effects of 11 years of conservation tillage on soil organic matter fractions in wheat monoculture in Loess Plateau of China. *Soil Till. Res.*, 106: 85–94.

Dasog, G.S. and Shashidhara, G.B. 1993. Dimension and volume of cracks under different crop covers. *Soil Sci.*, 155: 424– 428.

Dasog, G.S., Acton, D.I., Mermut, A.R. and de Jong, E. 1988. Shrink– swell potential and cracking in clay soils of Saskatchewan. *Can. J. Soil Sci.*, 68: 251–260.

Flowers, M. and Lal, R. 1999. Axle load and tillage effects on shrinkage. Characteristics of a Mollic Ochraqualf in northwest Ohio. *Soil Till. Res.*, 50: 251–258.

Froberg, M., Jardine, P.M., Hanson, P.J., Swanston, C.W., Todd, D.E., Tarver, J.R. and Garten, C.T., 2007. Low dissolved organic carbon input from fresh litter to deep mineral soils. *Soil Sci. Soc. Am. J.*, 71: 347–354.

Gjettermann, B., Nielsen, K. L., Petersen, C. T., Jensen, H. E. and Hansen, S. 1997. Preferntial flow in sandy loam soils as affected by irrigation intensity. *Soil Technol.*, 11: 139-152.

Gomez, K.A. and Gomez, A.A. 1984. Statistical Procedures for Agricultural Research, 2nd ed. Wiley-Interscience Pub., Singapore.

Goss, M.J., Howse, K.R., Lane, P.W., Christian, D.G. and Harris, G.L. 1993. Losses of nitrate-nitrogen in water draining from under autumn-sown crops established by direct drilling or mouldboard ploughing. *J. Soil Sci.*, 44: 35-48.

Guggenberger, G. and Zech, W. 1993.Dissolved organic-carbon control in acid forest soils of the fichtelgebirge (germany) as revealed by distribution patterns and structural composition analyses. *Geoderma*, 59:109–129.

Hamdi, M., Durnford, D. and Loftis, J., 1994. Bromide transport under sprinkler and ponded irrigation. *J. Drain. Eng.*, 120: 1086-1097.

Harris, G.L., Howse, K.R. and Pepper, T.J. 1993. Effects of moling and cultivation on soil±water and runoff from a drained clay soil. *Agric. Water Mgmt.*, 23: 161-180.

Kaiser, K. and Guggenberger, G., 2005. Storm flow flushing in a structured soil changes the composition of dissolved organic matter leached into the subsoil. *Geoderma*, 127: 177–187.

Kalbitz, K., Solinger, S., Park, J.H., Michalzik, B. and Matzner, E. 2000. Controls on the dynamics of dissolved organic matter in soils. *Soil Sci.*, 165: 277–304.

Lenka, S., Singh, A.K. and Lenka, N.K. 2009. Water and nitrogen interaction on soil profile water extraction and ET in maize–wheat cropping system. *Agric. Water Mgmt.*, 96: 195 - 207.

Lipsius, K. and Durner, W. 2005. Auswirkungen von praferentiellem transport auf die Sickerwasserbeprobung in einem hydrophoben sandboden-Tracerexperimente und Stofftranportsimulationen. Altasten Spektrum/4/2005, 210-216.

Lundquist, E.J., Jackson, L.E. and Scow, K.M. 1999. Wet-dry cycles affect dissolved organic carbon in two California agricultural soils. *Soil Biol. Biochem.*, 31: 1031–1038.

Matthews, A.M., Armstrong, A.C, Leeds-Harrison, P.B., Harris, G.L. and Catt, J.A. 2000. Development and testing of a model for predicting tillage effects on nitrate leaching from cracked clay soils. *Soil Till Res.*, 53: 245-254.

Sehgal, J. and Abrol, I.P. 1994. Soil degradation in India-status and impact. Oxford & IBH, New Delhi Pall.

Sharma, P.K., Verma, T.S. and Bhagat, R.M. 1995. Soil structural improvements with the addition of *Lantana camara* biomass in rice–wheat cropping. *Soil Use Manage.*, 11: 199–203.

Siemens, J. 2003. The European carbon budget: a gap. *Science*, 302: 1681–1681.

Somasundaram, J., Singh, R.K., Prasad, S.N., Sethy, B.K., Kumar, A., Ramesh, K. and Lakaria, B.L., 2011. Management of black Vertisols characterized by pot-holes in the Chambal region, India. *Soil Use and Manage.*, 27: 124–127.

Srivastava, K.L., Smith, G.D. and Jangawad, L.S. 1989. Compaction and shading effect on surface cracking in a Vertisol. *Soil Till. Res.*, 13: 151–161.

Tomar, R.K., Singh, D., Gangwari, K. S., Garg, R.N., Gupta, V.K., Chakraborty, D., Sahoo, R.N., Ranjan, R. and Chakravarty, N.V.K. 2009. Effect of tillage systems and irrigation schedules on soil cracking pattern, water requirement and performance of rice-wheat cropping system in Inceptisols in Semi-arid Regions. *J Soil water Cons.*, 8: 26-33.

Wright, A.L., Dou, F. and Hons, F.M. 2007. Soil organic C and N distribution for wheat cropping systems after 20 years of conservation tillage in central Texas. *Agric. Ecosys. Environ.*, 121: 376–382

Zsolnay, A. 1996. Dissolved humus in soil waters. *In*: Piccolo, A. (Ed.), Humic Substances in Terrestrial Ecosystems. Elsevier Science, Amsterdam, pp171–223.

□□□

Chapter 36

Design and Development of Suitable Technology for Rice-Fish Integration in Medium Land Rainfed Ecosystem of Odisha

S. C. SENAPATI, S. K. SAHU[1] AND A. P. SAHU

INTRODUCTION

Land and water are the two basic and essential resources which require proper planning for successful agriculture. Odisha is blessed with plenty of rainfall which occurs during June to September. During this period about 50 % of annual rainfall comes from few intense storms resulting into high runoff losses. In addition to this, the erratic nature of rainfall creates water stress at various growth stages of rice. Rice is the predominant crop grown by the people of Odisha during rainy season. In spite of application of recommended doses of fertilizers and other cultural practices, the productivity of rice in the State is still less than the national average. In order to enhance the productivity of the rice lands in the State under rainfed eco-system, adoption of improved water management techniques and integrated cultivation system is now becoming essential. Amongst the different integrated systems rice-fish is one aspect and if the rainfall excess obtained from the rice field could be stored for its re-use to the second crop after the harvest of rice crop under this system rice-fish-rabi crop is another suitable component for the rainfed medium and low land.

Rice-fish farming is an age-old practice in India. Still now it has not flourished due to several technical and social constraints. In India out of 43 M ha of rice cultivated land, about 20 M ha is suitable for adoption of integrated rice-fish production (Rao and Singh, 1998). However, only 0.23 M ha is presently under rice-fish production system (Radheshyam, 1998) and the carrying capacity of these suitable lands has not been utilized to the fullest

Department of Soil and Water Conservation Engineering, College of Agricultural Engineering and Technology, Orissa University of Agriculture and Technology, Bhubaneswar-751003, Odisha, India.

[1]Directorate of Research, OUAT, Bhubaneswar

extent. If the area under integrated rice-fish system could be increased, it would help to compensate the economic losses in rice production brought about by natural calamities. This not only enhance the use of land and water resources without bringing about environmental degradation but also fish cultivation in rice fields improves dissolved oxygen, soil quality and fertility (Likangmin, 1988; Mohanty *et al.*, 2002). In addition to enhanced productivity, this system also generates employment, increases farm income and provides nutritional security (Mishra and Mohanty, 2004).

At present the productivity of fish is about 200-800kg per hectare of rice-fish cultivated land has been obtained under rice-fish system (Mukhopadhyaya *et al.*, 1991). A lot of research has been carried out on different aspects of rice-fish integration system (Likangmin, 1988; Lightfoot *et al.*, 1992 and Rao and Singh, 1998), but very little information is available on food preference and intake pattern of cultured species and management strategy to enhance the productivity. Mohanty (2003) reported that increased level of water and concentration of total suspended solids was observed in refuge constructed at the down stream side of the plots with 10cm weir height, followed by 12.5 and 15cm weir height plots. Faster growth rate was recorded in *C. catla* followed by *C. Carpio, C. mrigala, L. rohita* and *M. rosenbergali* and an average productivity of 906.6 to 1282.3 kg ha^{-1} of fish and prawn has been achieved under rice-fish integration system.

In medium lands, Mishra *at al.* (1997) found that rice plots with bund heights of 6cm and 30cm can store 57 % and 99 % of rain water respectively to meet the water demand during *kharif* well as to implement rice fish integration system successfully through fish culture in the refuges during *kharif* and early part of *rabi* season in medium lands. This paper presents the outcome of the design and development of suitable technology for rice-fish integration in medium land rainfed ecosystem of Odisha.

MATERIAL AND METHODS

The work was carried out for four consecutive years (2001 to 2004) in the farmers' field near village Gourdiha of Shamakhunta block of Mayurbhanj district of Odisha state. The geographical situation of the site is 21° 45¢ N latitude and 86° 48¢ E longitude and comes under the Ojhrakhal watershed in the Shamakhunta block (Figure 1). Soil texture of the experimental site varies from sandy loam to sandy clay loam. Hydraulic conductivity of the surface soil varies from 2.8 to 3.1cm hr^{-1} in the medium land conditions. Bulk density is 1.6 to 1.75gm per cc and field capacity varies from 0.26 to 0.41 %. The location is characterized by warm and dry climate having hot and semi-humid summer and moderate winter. Average annual rainfall is 1441.5mm.

The average maximum temperature varies from 33 to 42C during hottest months and the minimum from 10 to 13°C during colder months.

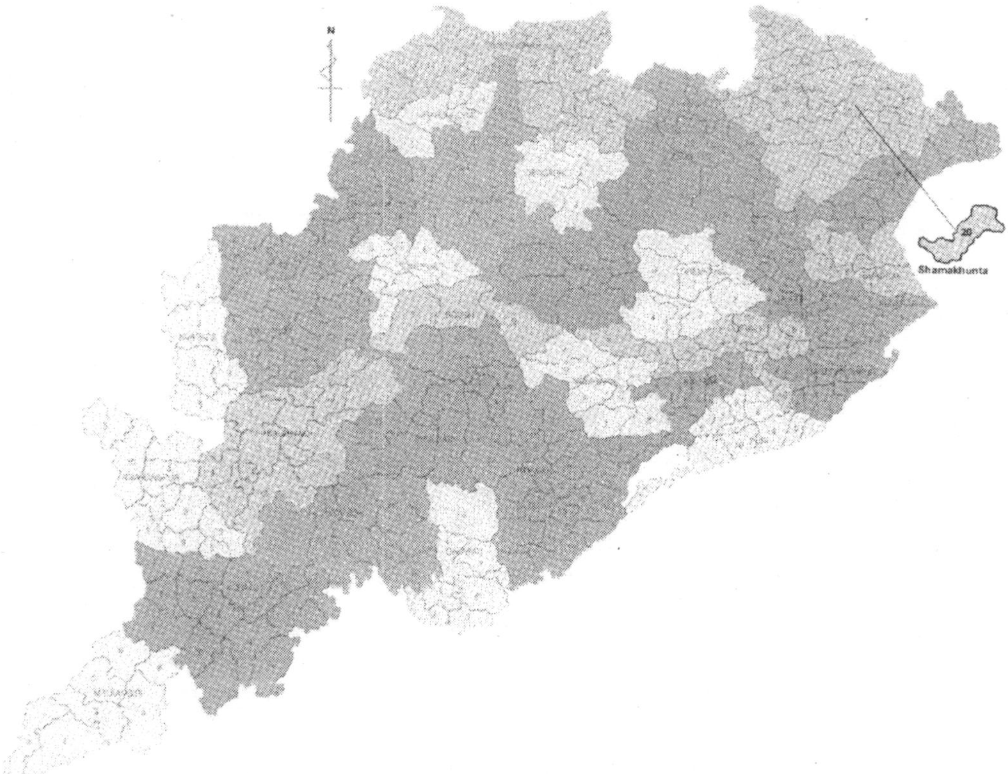

Fig. 1: Location map of experimental site in Shamakhunta block

Five experimental plots in the farmers' field were selected and leveled with a longitudinal slope of 0.1 %. The width of each plot was kept 20m and the length varied from 40m to 55.2m. The dyke height maintained in each plot was 30cm top width, 75cm bottom width and height 45cm. Five weir heights of 10, 15, 20, 25, and 30cm were provided at the down stream side of the experimental plots (Figure 2). Small refuges of length 20m (equal to the width of the plot), bottom width 1.0m, top width 4.6m and depth 1.8m were constructed at the end of each plot but below the weirs heights. Each treatment plot was divided into three equal sub plots, where three medium duration paddy varieties namely *'Ramchandi'*, *'Prachi'* and *'CR 260-77'* were grown. Paddy crops were grown following the recommended plant spacing 20cm x 15cm and fertilizer dose 60:30:30 NPK kg ha^{-1} during the experimental years. Nitrogen fertilizer was applied to paddy crop in three split doses and all P_2O_5 and K_2O at the planting time to all the three rice varieties.

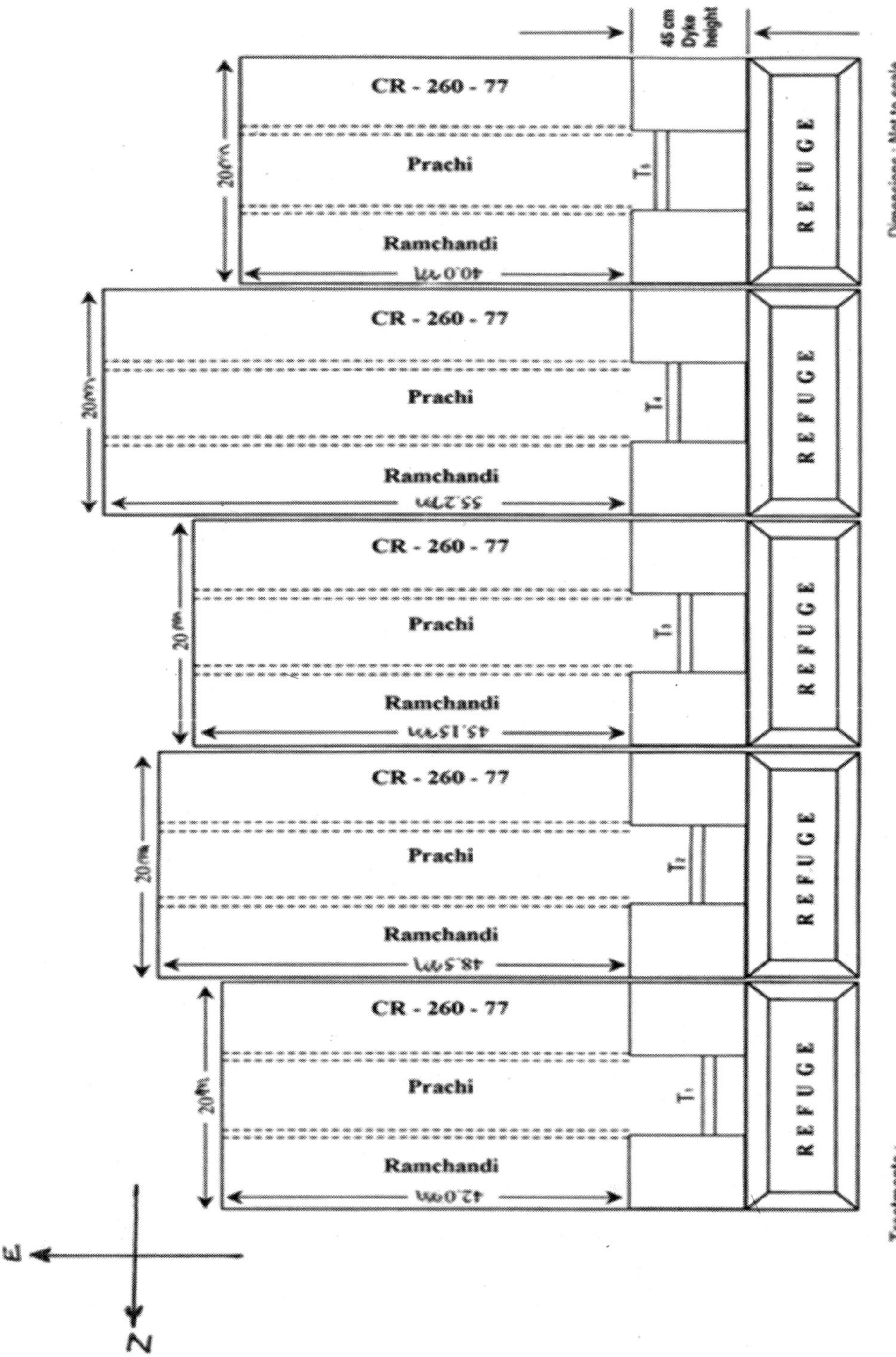

Fig. 2: Layout of the experimental field

Refuge preparation included application of lime at the rate of 2000 kg ha^{-1}, raw cattle dung 5000 kg ha^{-1} and fertilizers (urea and single super phosphate 1:1) at 3 ppm were applied prior to stocking of fingerlings. Fingerlings of *C.catla, L.rohita* and *C.mrigal* at the rate 20000 ha^{-1} of refuge area basis were stocked in the refuges with a species composition of 30:30:40, respectively. Supplement of feed (rice bran:groundnut oil cake, 1:1) was provided twice a day at 4, 6, 8 and 10 % of mean body weight of feeds during 1st, 2nd, 3rd and 4th month till harvesting. Periodic manuring at the rate of 500kg ha^{-1} and liming at the rate of 200kg ha^{-1} were carried out at every fortnight to maintain plankton population in the eco-system. Periodic measurement of body weights and average daily growth were also observed. Water utilised for crop rice was estimated on volumetric basis considering the area under cultivation, seasonal evapotranspiration, deep percolation and seepage losses. Water utilized for fish cultivation in the refuges on volumetric basis was calculated taking into account average losses due to evaporation, deep percolation and seepage and average area of the refuges.

The wheat crop was grown after the harvest of paddy crop in all the five treatment plots with refuges. Moisture available from the soil and amount of water supplied from the refuges were considered for estimation of water utilization to wheat crop.

RESULTS AND DISCUSSION

The average of four years yield of different rice varieties and wheat are given in Table1 and 2,respectively. The result showed that the yield of all the rice varieties which are of same duration were at par. However considering the average yield irrespective of varieties the highest yield of 5097.79 kg ha^{-1} was observed from the plot provided with 20cm weir height followed by 25cm weir height (4817.28kg ha^{-1}) and 30cm weir height (4728.22kg ha^{-1}). Similarly the yield of wheat was observed to be at par in respect to different weir height treatments.

Table 1: Average *kharif* rice yield (2001-2004) of different rice varieties at different weir heights

Treatment:	Rice yield (kg ha^{-1})			
weir heights (cm)	*'Ramchandi'*	*'Prachi'*	*'CR260-77'*	Average
10	4540.31	4389.65	4354.35	4428.10
15	4767.09	4622.12	4562.35	4648.32
20	5250.34	5133.78	4909.26	5097.79
25	4979.92	4914.93	4557.01	4817.28
30	4810.04	4841.84	4532.79	4728.22

Table 2: Average yield of wheat (2001-2004) at different weir height plots

Treatment:	Yield (kg ha^{-1})				
weir heights (cm)	2001	2002	2003	2004	Average
10	2120	2150	2180	2130	2140
15	2040	2210	2220	2150	2150
20	2050	2240	2290	2180	2190
25	2180	2200	2230	2220	2210
30	2240	2230	2200	2250	2230

The yield of fish from each experimental plot was recorded at the time of harvest. Production of fish for different years is given in Table 3. It is revealed that refuge constructed at the down stream side of the plot provided with 10cm weir height gave highest yield *i.e.,* 155.00 kg ha^{-1} of refuge area only and lowest is obtained from the refuge constructed at the lower end of the treatment plot provided with 30cm weir height. This might be due to the increased level of water and presence of high concentration of total suspended solids and nutrients carried by the spilled over high volume of runoff water from the plot of 10cm weir height. This helps a little better growth of fish species.

Rice equivalent yield (REY) of different systems for different weir heights has been presented in Table 4 considering the selling price of rice and fish for the year 2004. The highest REY was recorded in 20cm weir height followed by 25cm and 15cm. The bench mark survey site indicate that the average yield of rice in medium land was 1900 kg ha^{-1} before interventions. After intervention, the yield of rice increased more than 2.5 fold due to rain water conservation measures and adoption of improved rice cultivation practices. When rice-fish integration system was considered the highest REY at 20cm weir height plot is found highest (6472.79kg ha^{-1}), thus increased the productivity more than 3 folds.

Table 3: Average productivity of different fish species in refuges with different weir heights

Treatment: weir heights (cm)	Productivity (kg ha^{-1}180days^{-1})				
	2001	2002	2003	2004	Average
10	153.50	156.75	155.00	154.87	155.00
15	149.54	147.27	148.29	147.16	148.10
20	136.54	138.57	137.02	137.97	137.50
25	136.34	133.26	135.33	134.13	134.75
30	133.8	132.75	134.87	131.25	133.10

Fish spcies: *C. Catla,L. Rohita, C. mrigala*

Table 4: Rice equivalent yield (REY) of different systems for different weir heights

Treatment: weir heights (cm)	Yield (kg ha^{-1})			REY (kg ha^{-1})		
	Rice	Fish	Wheat	Rice	Rice-Fish	Rice-Fish-Wheat
10	4428.10	155.00	2140	4428.10	5978.10	9348.60
15	4648.32	148.10	2150	4648.32	6129.32	9515.57
20	5097.79	137.50	2190	5097.79	6472.79	9922.04
25	4817.28	134.75	2210	4817.28	6164.78	9645.53
30	4728.22	133.10	2230	4728.22	6059.22	9571.47

The economic study of different farming systems under different treatment practices (weir heights) is given in Table 5. The Table showed that net return and benefit-cost ratio observed to be highest in 20cm weir height followed by 25 and 30cm in case of rice system only and rice-fish-wheat system. In the case of rice-fish integration system the net return and B-C ratio was highest at 20cm weir height followed by 15cm and 25cm.

Table 5: Economics of different farming systems for refuges with different weir heights (₹.ha^{-1})

Treatment	Rice System			Rice -Fish System			Rice-Fish-Wheat System		
	Gross return	Net return	B:C ratio	Gross return	Net return	B:C ratio	Gross return	Net return	B:C ratio
T_1	17712	11972	3.08	23912	16996	3.45	37394	23728	2.73
T_2	18593	12853	3.24	24517	17601	3.54	38062	24396	2.78
T_3	20391	14651	3.55	25891	18975	3.74	39688	26022	2.90
T_4	19269	13529	3.35	24659	17743	3.53	38582	24916	2.82
T_5	18913	13173	3.29	24237	17321	3.50	38286	24620	2.80

Cost of Cultivation: ₹ 5740/- ₹ 6916/- ₹ 13666/-

Considering the average depth of ponding water, evaporation, and deep percolation and seepage losses in the respective refuges of different treatments quantity of water utilized for rice, rice-fish and rice-fish-wheat crop were estimated as 9384 m^3, 10349 m^3, and 10349 m^3, respectively. Water available in refuges was utilized both for fish rearing and also irrigation to wheat crop. So water utilized in rice-fish and rice-fish-wheat crop were taken as the same quantity. Based on the water utilized gross water productivity of different integrated farming systems under different weir height treatments were calculated and is given in Table 6. It was revealed from table that the gross water productivity was found to be highest (₹ 2.17 m^{-3}) in 20cm weir height plot followed by 25 and 30cm weir heights for mono crop rice farming system.

Similar trends were noticed in case of rice-fish and rice-fish-wheat farming systems.

Table 6: Gross water productivity for different farming systems with different weir heights

Treatment: weir heights (cm)	Rice			Rice - Fish			Rice - Fish - Wheat		
	Gross return (Rs.)	Water utilized (m^3)	Gross water productivity ($Rs.m^{-3}$)	Gross return (Rs.)	Water utilized (m^3)	Gross water productivity ($Rs.m^{-3}$)	Gross return (Rs.)	Water utilized (m^3)	Gross water productivity ($Rs.m^{-3}$)
10	17712	9384	1.88	23912	10349	2.31	37394	10349	3.60
15	18593	9384	1.98	24517	10349	2.36	38062	10349	3.67
20	20391	9384	2.17	25891	10349	2.50	39688	10349	3.83
25	19269	9384	2.05	24659	10349	2.38	38582	10349	3.72
30	18913	9384	2.01	24237	10349	2.34	38286	10349	3.69

CONCLUSIONS

In medium land, rice-fish integrated farming system 20cm weir height treatment produced highest net return of Rs 18,975 ha^{-1} and B:C ratio (3.74) followed by 25cm weir height with net return of Rs. 17,743 ha^{-1} and B:C ratio (3.53), respectively in comparison to rice crop system alone. Similar tendency was observed giving a net return of Rs. 26022 ha^{-1} (B.C ratio of 2.90) in rice-fish-wheat farming systems.

Higher water productivity of ₹ 3.83 / m^3 was observed in case of rice-fish-wheat crops farming system in comparison to rice alone and rice-fish farming system. This farming system indicated the most efficient way of multiple use of conserved rain water which is nearly double than that of monocropped rice. Thus increase in productivity from unit volume of rain eater and unit area of land, conservation of rainwater, integration of fish culture and adoption of second crop wheat irrigated with the water available from refuges was found to be viable, low input based and easily adaptable technology for small and medium farmers. Rice-fish-wheat integration will also help to generate employment opportunity, increase income for farmers and provide nutritional security in the medium land rainfed ecosystem of Odisha.

REFERENCES

Lightfoot, C., Costa-Pierce, B.A., Bimbao, M.P. and Delacruz, C.R. 1992. Introduction to fish research and development in Asia. ICLARM: 1-10.

Likangmin.1988. Rice-fish culture in China: a review. *Aquaculture*, 71: 173-186.

Mishra, A., Ghorai, A.K. and Singh, S. R. 1997. Effect of dyke height on water, soil and nutrient conservation and rice yield. Research Bulletin, DWM (ICAR), Bhubaneswar: 1-19.

Mishra, A. and Mohanty, R.K. 2004. Productivity enhancement through rice-fish farming using a two-stage rainwater conservation technique. *Agril. Water Management*, 67: 119-131.

Mohanty, R.K., Mishra, A., Verma, H.N. and Bramhanand, P.S. 2002. Rainwater conservation and rice-fish integration for enhancing land and water productivity. Research Bulletin No.11, Water Technology Centre for Eastern Region (ICAR), India: 24p.

Mohanty, R.K. 2003. Feed intake pattern and growth performance of Indian major carps, common carp and fresh water prawn in rice-fish integration system. *Asian Fisheries Science*, 16: 307-316.

Mukhopadhyaya, P.K., Das, D.N. and Roy, B.1991. Deepwater rice-fish farming. Research Bulletin, Deep Water Rice Project, Rice Research Station, Chinsura, West Bengal, India.

Radheshyama.1998. Status of fisheries in India. *In* : Ahmed, S.H (Ed.), Advances in fisheries and fish production. Hindustan Publishing Corporation, New Delhi, India: 309.

Rao, A.P. and Singh, R.1998. Rice-fish farming system. In : Ahmed, S.H (Ed.), Advances in fisheries and fish production. Hindustan Publishing Corporation, New Delhi, India: 309.

□□□

Chapter 37

Sustainability of Integrated Farming Systems in the Eastern Region

SANJEEV KUMAR, S.S. SINGH, UJJWAL KUMAR, A. DEY AND SHIVANI

INTRODUCTION

New agricultural technologies developed and transferred by the conventional research and extension efforts, have not been adopted uniformly by the farmers operating under widely varying agro-climatic and socio-economic environments, when adequate attention is not paid to the agro-climatic and socio-economic milieu in which the farmers operate. Farmers generally seek those technologies which increase their income while keeping risk within reasonable bounds under their own circumstances and management practices. The traditional technology generation and transfer models have been found to be inadequate to meet the specific needs of vast majority farmers. In order to develop and transfer more appropriate technologies to meet the specific client demands, emphasis on farmer's participatory, systemic research and extension effort is now gaining momentum in India and that too in the eastern part of the country. Attempts have been made to integrate the desirable features of farming system research into the mainstream agricultural research so that the technologies developed are relevant, client-oriented and location specific.

In Eastern region about 70% farmers are small and marginal. The average size of the holding is 0.83 ha, with that of small and marginal farmers range from 0.32 to 0.5 ha (Barua, 2001).The poor farmers with small land holdings concentrate only on crop production, mainly cereals with high risks of flood and drought. Due to failure of monsoon and small size of holdings, they hardly get sufficient income to sustain their family. To solve the problems of small resource poor farmers, diverse and risk prone environments has led to the development of a more holistic, resource based, client oriented and interacting

ICAR Research Complex for Eastern Region, Patna- 800 014, Bihar, India approach, popularly known as Integrated Farming System (IFS). These small and marginal farmers can take a suitable crop along with horticulture, animals, fisheries and other components in integrated manner that would minimize risks and provide additional income and employment from the same piece of land per unit time, apart from improving soil fertility over a period of time. The integration is made in such a way that product of one component should be the input for other enterprises with high degree of complimentary effects on each other (Gill *et al.*, 2009). Integrating different components with crop will increase the profitability through recycling of wastes/organic residues of one component into another. IFS assume greater importance for sound management of farm resources to enhance the farm productivity and reduce the environmental degradation, improve the quality of resource poor farmers and maintain sustainability. Therefore, the present investigation was envisaged to generate / to identify an IFS model with suitable combination of components for efficient utilization of natural resources and generate maximum income and employment for the farm family.

MATERIALS AND METHODS

Field experiments were conducted in Integrated Farming Systems (IFS) mode at the main farm of ICAR Research Complex for Eastern Region, Patna during 2007-10 involving poultry, cattle, goat, mushroom, fishery and duckery with crop in all possible combinations, to recycle the residues and byproducts of one component over the others. Seven farming systems were evaluated and each system was allocated an area of 0.8ha. These systems were: (i) crop alone (ii) Crop+fish+poultry (iii) Crop+fish+duck (iv) Crop+fish+goat (v) Crop+fish+ duck+goat (vi) Crop+fish+cattle (vii) Crop+fish+mushroom. The soil of the experimental site was clay loam in texture, having pH 6.6, electrical conductivity 0.44 dsm^{-1}, organic carbon 0.59 (kg m^{-3}), available N 186.0 kg ha^{-1}, P 6.1 kg ha^{-1} and exchangeable K 175.1 kg ha^{-1}. In a 0.8 ha farm, different components of IFS's were allocated different proportions of area which has been shown in Table 1. In another 0.8 ha, only crop activity (three crops/ year) was taken up. Summer maize (*Zea mays* L.)- napier grass (*Pennisetum purpureum* Schum.)- berseem (*Trifolium alexandrinum*) fodder system were followed in 0.1 ha of land.

Under crop activity two cropping systems i) Rice (*Oryza sativa L.*) – wheat (*T. aestivum*) - moong (*Vigna radiata*) and ii) Rice (*Oryza sativa L.*) - maize (*Zea mays L.*)- moong (*Vigna radiata*) were taken. Crops were raised with the combination of organic manures (from different sources) and chemical fertilizers in IFS mode where as in the treatment cropping alone crops were raised with chemical fertilizer only. Poultry and duckery recycled silts were used once in a year *i.e.* in the rice crop whereas other organic sources were used in all the crops in the sequence.

Table 1: Allocation of area under different components of farming systems

Farming system	Area (ha)								
	Crop	Fish	Poultry	Duck	Goat	Cattle	Fodder area	Mush-room	FYM & V. C. pit
Crop alone	0.80	-	-	-	-	-	-	-	-
Crop + fish + poultry	0.66	0.12	Sheltered over fish pond	-	-	-	-	-	0.02
Crop + fish + Duck	0.66	0.12	-	Sheltered over fish pond	-	-	-	-	0.02
Crop + Fish + goat	0.54	0.12	-	-	0.02	-	0.1	-	0.02
Crop + fish + duck + goat	0.54	0.12	-	Sheltered over fish pond	0.02	-	0.1	-	0.02
Crop + fish + cattle	0.54	0.12	-	-	-	0.02	0.1	-	0.02
Crop + fish + mushroom	0.64	0.12	-	-	-	-	-	0.02	0.02

Hundred numbers of poultry birds, 35 ducks sheltered over 2 fish ponds and cattle unit maintained in cattle shed were linked to supplement the feed requirement of polycultured fingerlings (300 numbers) reared in each pond to assess the feasibility of rearing fishes by using different manure as feed. Vermi-pits and FYM pits were also linked with cattle and crops. To sustain the productivity of soil, inorganic fertilizers combined with organic wastes obtained from various components of IFS *viz.*, recycled poultry, duckery and cow dung as FYM, composted residues (veg. residues + cereal residues) and vermicompost each @ 10t ha^{-1} were applied to the crops.

In goatry component, 20 female goats + 1 buck (Black Bengal) were reared for meat purpose and goat droppings were used as goat manure to the crops. In one year, 60 kids were reared and sold @ Rs. 100 kg^{-1} live weight (kids were sold at the age of 9-10 months). Under poultry component, 100 broiler chicks $batch^{-1}$ (9 batches $year^{-1}$) were maintained. Each batch was maintained for 40 days and broilers attained an average weight of 1.5 kg during the period which were sold @ Rs. 70 kg^{-1} live weight. Out of total poultry droppings, 25 % of poultry droppings/litters were used in pond as feed for fishes and 75 % of droppings were used in the crops as manure.

Under fishery component, mixed fish farming was practiced. Fresh water fish, rohu (*Labeo rohita*) as column feeder (30%), catla (*Catla catla*) and silver carp (*Hypophthalmicthys molitrix*) as surface feeder (30%), mrigal (*Cirrhinus mrigala*) and common carp (*Cyprinus carpio var. communis*) as bottom feeder (40%) were raised in both the ponds. At the end of Ist year, the grown up fishes were harvested thrice at 20 days interval. Water in the ponds were drained and dried and selted silts (5 tonnes) were removed and applied as organic source to the first crop in the sequence. In duckery (*Khakhi Campbell*), 30 female and 5 male ducks were integrated with the pond. Droppings of ducks were fed to fishes and no extra feed were provided to the fishes. Number of eggs laid /annum were recorded.

Table 2: Rice grain equivalent yield (RGEY)] of different cropping sequences (mean yield of 2007-10)

Cropping System	RGEY with chemical fertilizer (as farmers practice)	RGEY with Recycled pond silt (poultry) + Poultry manure	RGEY with Recycled pond silt (Duck)	RGEY with Goat manure	RGEY with Cattle manure	RGEY with Vermicompost
Rice- wheat-moong	9,640	11,350	11,130	11292	11,237	11,465
Rice- maize-moong	11,320	13,250	12,857	13,113	12,943	13,070
Average	**10,480**	**12,300**	**11,993**	**12,203**	**12,090**	**12,267**

Year round mushroom production was also included in the system in an area of 0.02 ha by making a small hut with available local materials. From March- September, Paddy straw mushroom (*Volvariella spp.*) and milky mushroom (*Calocybe indica*) whereas, from October to February, Oyster mushroom (*Pleurotus spp.*) were raised by making bamboo racks in the shed. Proper humidity (75- 80 %) was maintained in the hut during the crop season by sprinkling water over the walls of hut and over the bags. Proper agronomic management to all crops, health and hygiene were maintained to animals and birds as per recommendations. Concentrate feed for animals and poultry were purchased from market and expenditures on these items were included in the cost of production. Observations were made on productivity in terms of rice - grain equivalent, economics and employment for different farming systems as well as conventional cropping.

Sustainability of developed IFS models were evaluated based on sustainability index (S.I.) as described by Vittal *et al.* (2002). The S.I. for IFS model were computed as: S.I.= (NR- SD)/(MNR)X 100 where, NR stands for net returns obtained under any model, SD stands for standard deviation of net returns of all models and MNR stands for maximum net returns attained under any model. A suitable and viable IFS model could be identified for their existence based on net return, sustainability index, employment generation and improvement in soil fertility attained over a period of time.

RESULTS AND DISCUSSION

System productivity :The productivity of different components taken in each system was converted as rice grain equivalent yield for making comparison (Table 3). The contribution of crop towards the system productivity ranged from 57.1 to 81.1%, while it ranged from 8.5 to 11.8% for fish; 26.6 to 28.7% for goat; 24.2% for poultry; 7.4 to 10.1% for duck; 34.4% for cattle and 7.2% for mushroom (Figure 1). Results on these combinations for three years revealed that integration of crop +fish +duck + goat resulted in highest system productivity (21200 kg ha^{-1}), average net return (₹ 1,59,485 yr^{-1}) and sustainability index (80%) as compared to other combinations (Table 4). Cropping alone resulted in the lowest system productivity (10,480 kg ha^{-1}) with lowest sustainability index value (27.6 %). Sustainability Index itself clarifies the benefits from different combinations/unit area. Higher the S.I., higher will be the net return Singh *et al.* (2007) also opined that the integration of cropping with other enterprises increased the system productivity per unit area per unit time than the cropping alone due to intensification and diversification of cropping and integration of allied enterprises, and thereby efficient use of resources.

Table 3: Components wise productivity and economics under integrated farming systems (2 acre), (Pooled yield)

Component	RGEY (kg)	Production cost (₹)	Gross returns (₹)	Net return (₹)	B: C ratio
Cropping alone	8384	53,120	1,00,608	47,488	1.9
Crop + poultry manure	9840	52,120	1,18,080	65,960	2.3
Crop + Duck manure	9594	52,004	1,15,128	63,124	2.2
Crop + goat manure	9762	51,886	1,17,144	65,258	2.3
Crop + FYM	9672	51,590	1,16,040	64,450	2.2
Crop + vermicompost	9814	52,100	1,17,768	65,668	2.3
Poultry (100 no. per batch)	4500 (900)	24,600	54,000	29,400	2.2
Duckery (30 + 5)	1556 (6225)	10,990	18,672	7,682	1.7
Goat (20 +1)	5625 (450)	24,814	67,500	42,686	2.7
Cattle (3+3)	7280 (4370 lit)	66,820	87,360	20,540	1.3
Mushroom	1100 (165 kg)	5,620	13,200	7,580	2.3
Fish fed with poultry dropping (0.06 ha)	992 (170 kg)	4,810	11,900	7,090	2.5
Fish fed with poultry dropping (0.06 ha)	817 (140kg)	4,810	9,800	4,990	2.0

Note: Figures in parenthesis denotes actual yield

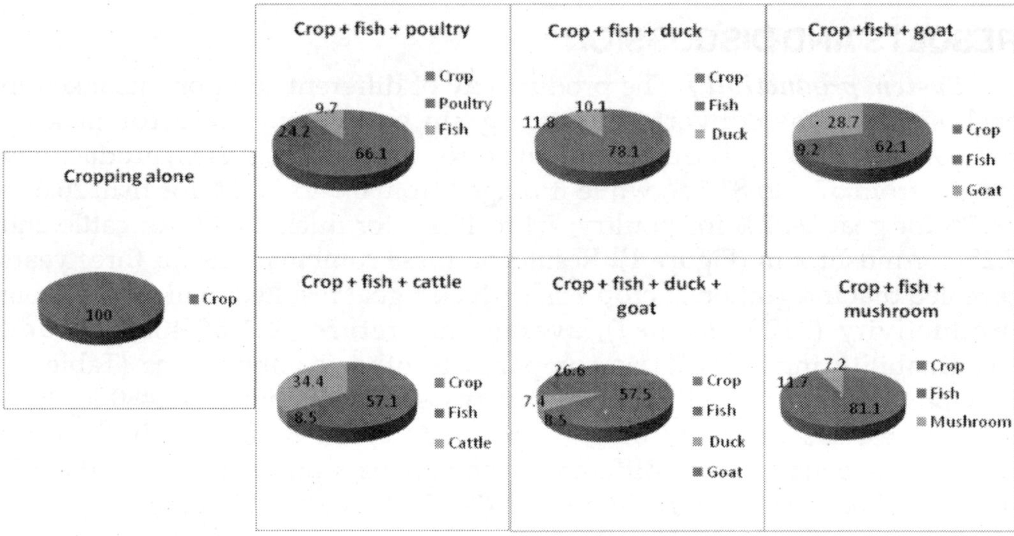

Fig. 1: Percent component productivity of different models

Among different cropping sequences, under IFS compared, rice- maize- moong recorded higher mean yield than that of rice- wheat- moong cropping sequence (Table 2). However, both cropping systems performed better with a combination of both organic and inorganic fertilizer than using inorganic fertilizer alone. 35.2 tonnes of grasses and legume mixture (maize- napier- berseem) was also obtained from 0.1 ha, and was utilized as feed for animals. Crops applied with enriched pond silts having higher nutrients and integration of high value components like fish/poultry/duck/goat/cattle might have contributed for better crop productivity. Similar results of high productivity was also reported by Jayanthi *et al.* (2003) by integrating crop + fish + goat in lowland farming in Tamilnadu.

Among different animal components, goatery fetches highest net return and an average productivity of 5,625 kg was obtained with 20 + 1 goat unit. The goat unit also produced 2,300 kg of goat manure which was recycled within the system., The fishes fed with poultry droppings resulted in higher average fish yield of 170 kg per 0.06 ha over duck fed droppings (140kg per 0.06 ha) (Table 3). A higher level of fish productivity through recycling of poultry manure was reported by Kumar *et al.* (2011a&b) owing to better plankton development as well as direct feed to fishes.

Income and Economics : The highest average net return of Rs. 1,59,485 yr^{-1}. was obtained from an area of 1.0 ha with an average annual expenditure of Rs. 94,915yr^{-1} by integrating crop + fish + duck + goat combination in the system followed by crop + fish + goat (Rs. 151,479 yr^{-1}) combinations (Table 4). Sustainability index was also found higher in these systems (80 and

75.1%, respectively). Poultry (broilers) rearing is economical only when proper care had been taken, otherwise, it is a risky enterprise due to frequent occurrence and breakout of severe pest and diseases leading to 50-100% mortality of the birds which will result in high degree of economic losses. So, proper hygienic conditions should be maintained and birds should be properly vaccinated. The crop + fish + cattle model gave nearly equal rice- grain-equivalent yield as that of crop + fish + duck +goat but due to higher expenditure (purchase of concentrates mixture) this system ranked fourth in terms of net return. Further, if concentrates were prepared at farmer's level with produced materials by the system, expenditure can be lessened by 50% and crop + fish + cattle system can be made more profitable. Kumar *et al.* (2012) also reported increase in net income through integrated farming system due to use of recycled products within the system. Crop + fish + duck + goat model emerged as highest profitable enterprise for irrigated lowlands with an average net return of Rs.437 day^{-1} during the period of experimentation. This was due to the fact that system as a whole provided opportunity to make use of byproduct or waste materials of one component as input on another.

Table 4: Productivity (RGEY) and economics of different farming systems (mean value of 2007-10)

Farming Systems	RGEY (kg ha^{-1})	Production cost (₹ ha^{-1})	Gross return (₹ ha^{-1})	Net return (₹ ha^{-1})	Net return day^{-1} (₹)	Sustainability Index (%)
Cropping alone	10,480	53,120	125760	72640	199	27.6
Crop + fish + poultry	18617	83,945	223405	139460	382	67.4
Crop + fish + Duck	15377	70,219	184520	114301	313	51.5
Crop + Fish + goat	19617	83,925	235404	151479	415	75.1
Crop + fish + duck + goat	21200	94,915	254400	159485	437	80.0
Crop + fish + cattle	21187	125,625	254240	128615	352	60.6
Crop + fish + Mushroom	16556	70,799	198671	127,872	350	60.2
Mean	17576	83221	210914	127693	350	59.2
SD ±	3822	22968	45865	28653	79	
CV (%)	21.75	27.60	21.75	22.44	22.44	

Note: Economics were calculated on prevailing market price of different commodities *viz.*, Rice grain @ Rs. 12 kg$^{-1)}$, poultry @ Rs. 60 kg^{-1}, duck egg@ Rs. 3 egg^{-1} (or say Rs. 30 kg^{-1}), goat meat @ Rs. 150 kg^{-1}, milk @ Rs. 20 lit$^{-1)}$ and fish @ Rs. 70 kg^{-1}.

Employment generation: The data of three year experiments show that 512 mandays ha^{-1} were required for the crop component even if a farmer grows three crops per year. The average employment generation increased to 752 man- days ha^{-1} $year^{-1}$ by integrating crop + fish + duck +goat followed by crop + fish + cattle (722 man- days ha^{-1} yr^{-1}). Keeping in view, the other enterprises like fish, duck and goatry an additional employment of 40, 70 and 130 man-days, respectively had been generated (Figure 2). The integration of different components resulted in efficient utilization of labour, as the labour engaged in crop components managed the dairy, duckery and fishery components, reducing the cost of production with increased profitability. Ravisankar *et al.* (2007) also opined that combining of crops with other enterprises increase labour requirement and thus provide scope to employ more family labours round the year without giving much relaxation during lean season as in traditional agriculture.

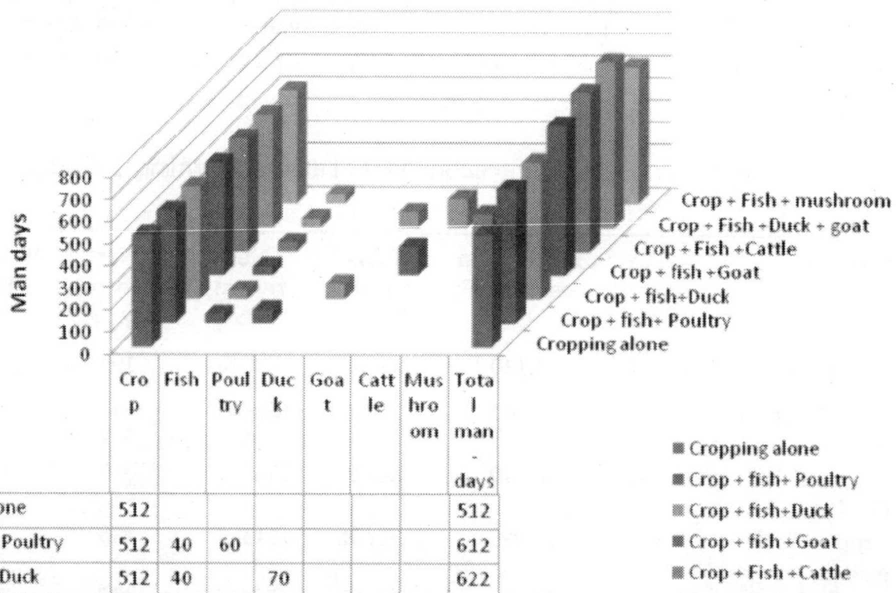

	Crop	Fish	Poultry	Duck	Goat	Cattle	Mushroom	Total man-days
Cropping alone	512							512
Crop + fish+ Poultry	512	40	60					612
Crop + fish+Duck	512	40		70				622
Crop + fish +Goat	512	40			130			682
Crop + Fish +Cattle	512	40				170		722
Crop + Fish +Duck + goat	512	40		70	130			752
Crop + Fish + mushroom	512	40					70	622

Fig. 2: Employment generation (man-days $year^{-1}$) by individual component and different systems

Nutrient recycling: Recycling of by-products and farm wastes has been practiced in this study. Samples of raw droppings of animal and bird, recycled products like FYM, goat manure, vermicompost, poultry manure and silted silt in the ponds were collected and analysed for their nutrient content. Residue recycling revealed that integration of crop + fish + poultry resulted in higher fish productivity over duck dropping fed fishes and resulted in higher net return (₹ 7090 $year^{-1}$) from 0.06 ha of pond (Table 3). Poultry unit had produced 2880 kg of raw droppings, out of which, 25% was fed to fishes and 75% (2160 kg ha^{-1}) was used for preparation of poultry manure and finally applied to the crops while, in case of duck unit, 1500 kg raw dropping were produced in a year and total droppings were allowed to fed to fishes. Recycling of droppings (25% poultry droppings and 100% duck droppings) through fish ponds enhanced the nutrient content by 2-3 folds. Apart from this, poultry unit also provided 80.6, 52.4 and 24.7 kg of N, P_2O_5, K_2O in the form of poultry manure which were used in crops. In case of goat, 1840 kg of goat manure was produced from 2300 kg of raw droppings due to which an additional amount of nutrient *i.e.* 14.2 kg N, 8 kg P_2O_5 and 4.2 kg K_2O was gained. Likewise from 14000 kg of cow dung, which contained 165.2, 100.8 and 156.8 kg of N, P_2O_5 and K_2O when converted into FYM, it produced an additional quantity of 54.3, 80.6 and 53.7 kg of N, P_2O_5 and K_2O. Recycling of plant wastes as vermicompost generated an additional quantity of N: 4.8 kg, P_2O_5 :3.1 kg and K_2O : 3.6 kg which were used within the system as input. Kumar *et al.* (2012) had reported similar benefits due to recycling of different animals' droppings and plant wastes. By analyzing all waste material obtained from animals and plants, it can be interpreted that cattle recycled droppings generated highest P_2O_5 and K_2O while poultry had generated highest N into the system. The additional nutrients gained by recycling of waste/byproducts over raw wastes were also confirmed by Baishya *et al.* (2004) in lowland situation. Recycling of organic manures like FYM, goat manure, vermicompost, poultry and duckery recycled manures obtained from different components added nutrients into the system, thereby minimize the dependency on chemical fertilizers in a sizeable amount and simultaneously provide good soil health for sustainable production of the system.

CONCLUSIONS

Based on the findings, it can be concluded that integrated farming approach with combination of crops and non – crop enterprises : crop+ fish+ goat +duck is sustainable system for Eastern region and far better than cultivating crops alone as it enhances the productivity, economic returns and generate more employment for farm families. Integration of different components with crop depending upon suitability were found encouraging, and it offered enough scope for nutrient recycling within the system, to economize and sustain the system and minimize the dependency on chemical fertilizers for crop production. Addition of organic residues in the form of animal and plant wastes could also help in improving the soil- health and productivity over a longer period of time with lesser environmental hazards.

REFERENCES

Baishya, A., Pathak, A.K., Bhowmick B.C. and Ahmed, S. 2004. Predominant farming system and alternatives in Assam. *Altenate farming systems: Enhanced income and employment generation options for small and marginal farmers* edts. Singh AK., Gangwar B. and Sharma SK. PDFSR, Modipuram (U.P.), 228-237.

Barua, B. 2001. Report of the working group on Agricultural development in eastern and North eastern India for the formulation of the tenth Five year Plan. Govt. of India, planning Commission.

Gill, M.S., Singh, J.P. and Gangwar, K.S. 2009. Integrated farming system and agriculture sustainability. *Indian Journal of Agronomy,* 56(4): 297-304.

Jayanthi C, Balusamy, M., Chinnusamy, C. and Mythili, S. 2003. Integrated nutrient supply system of linked components in lowland integrated farming system. *Indian Journal of Agronomy,* 48 (4): 241-246.

Kumar. S., Singh, S. S., Shivani and Dey, A. 2011a. Integrated farming systems for Eastern India. *Indian Journal of Agronomy,* 56(4): 297-304.

Kumar, S., Singh, S.S., Meena, M.K., Shivani and Dey A. 2011b. Resource recycling and their management under integrated farming system for lowlands of Bihar. *Indian Journal of Agricultural Sciences,* 82(6): 504-510.

Kumar, S., Subhash, N., Singh, S.S., Shivani and Dey, A. 2012. Evaluation of different components under integrated farming system (IFS) for small and marginal farmers under semi- humid climatic environment. *Experimental Agriculture,* 48(3): 399-413.

Ravisankar, N., Pramanik, S.C., Rai, R.B., Shakila, N., Tapan, K.R., Biswas. and Nabisat Bibi. 2007. Study on integrated farming system in hilly upland areas of Bay Islands. *Indian Journal of Agronomy,* 52 (1): 7-10.

Singh, K., Bohra, J.S., Singh, T.K., Singh, J.P., Singh, Y. and Singh, C.S. 2007. Productivity and economics of integrated farming system in irrigated agro-ecosystem of eastern Uttar Pradesh. *Indian Journal of Agronomy,* 52(1): 11-15.

Vittal, K.P.R., Mauthisankar, G.R., Singh, H.P. and Sharma, J.S. 2002. Sustainability index. In: *Sustainability of practices of Dryland Agriculture: Methodology and Assessment,* pp4-9. Central Research Institute for Dryland Agriculture, Hyderabad.

□□□

Chapter 38

Response of Drip Fertigation on Stevia (*Stevia rebaudiana*, Bertoni) in Rice Fallow System

M.S. BEHERA, P.K. MAHAPATRA, O.P. VERMA, R.B. SINGANDHUPE AND A. KUMAR

INTRODUCTION

Stevia is known as *Madhu patra* in Indian market. One hectare of stevia produces sweetener equivalent to 36 ha of sugarcane (Rani *et al.*, 2012). The sweetness in stevia leaves is non-sugar based and harmless. The products of stevia are used in many food items as sweetener, flavor, weight watchers diet, diabetic diets and colour enhancers, antitoxidants and as aging agent and catalyst for alcoholic beverages. It is also used for medicine to treat many ailments including diabetics, high blood pressure, hypertension, depression, digestive disorders, addictions, several skin defects and improving thinking capacity. In medicinal field, stevia has hypoglycemic, oral contraceptive, cardiovascular, antimicrobial activity. Plant sweeteners have attained high significance in the recent years due to their demand for industrial use and higher market price (Lubbe and Verpoorte, 2011). *Stevia* is a small sweet perennial shrub with green leaves that belongs to the family Compositae. It originated in South America (Paraguay and Brazil). The current extract market demand is 1.5 M kg, processed from 12 M kg of *Stevia* leaf (*Stevia* Facts, 2008). Stevioside is one of the active constituents, which is available abundantly in the leaves of *Stevia* (5-10% on dry weight basis) and is 300- 350 times sweeter than sucrose (Zhang *et al.*, 1999).

In view of the above fact, cultivation of stevia is gradually coming into focus in Indian agricultural fields, which is having no optimum agronomic management practices. It naturally grows in low lying areas on poor sandy acidic soils adjacent to swamps and requires constantly wet or shallow water tables. Incorporation of this species in an agricultural production system depends upon a thorough knowledge of the plant and its agronomic potential.

Directorate of Water Management, Chandrasekharpur, Bhubaneswar-751 023 Odisha, India

Studies shown that frequent irrigation is required for stevia to maintain soil moisture above wilting point up to 80% of field capacity and moisture stress reduces leaf production (Lavini *et al.*, 2008). Stevia plant has high requirement of nutrients especially N, P and K and their deficiency is probably the major limitation for the quality biomass production of stevia. Hence, for frequent light irrigation and nutrient application, drip fertigation is considered to be most efficient in improving the yield, quality and saving of water. Hence, the present experiment was conducted to assess the effect of drip fertigation on yield and quality of stevia in the rice fallow where drip irrigation keeps soil in moist condition.

MATERIALS AND METHODS

The experiment was conducted at Research Farm of the Directorate of Water Management, Bhubaneswar, Odisha during *rabi* seasons of 2005-06 and 2006-07 to study the effect of drip irrigation regimes and fertigation levels on stevia (*Stevia rebaudiana*, Bertoni), which was grown in the rice fallow. The experiment was laid out in Factorial Randomized Block Design with three replications. The treatments included three irrigation regimes (I_1- drip irrigation at 100% PE, I_2 at 80% PE and I_3 at 60% PE) and three fertility levels (F_1-100%, F_2 - 75% and F_3 - 50% recommended dose of NPK) with an extra (control) treatment having surface irrigation and soil application of fertilizer. The soil was sandy loam in texture with pH 5.7, low in organic carbon (0.46%), nitrogen (159 kg ha^{-1}); medium in available phosphorus (21 kg ha^{-1}) and potassium (183 kg ha^{-1}). Forty five days old seedlings of *Stevia* variety "*SRB-126*" was planted at a spacing of 60 cm x 30 cm during the evening hours to avoid transplanting shock from higher temperature. The recommended fertilizer dose for stevia was 110-45-45 kg N-P_2O_5-K_2O ha^{-1}. Fertigation was given in equal splits at fortnightly interval starting from 15 days after planting (DAP) up to 30 days before harvest. Full dose of P was applied basally to all the treatments. Required amount of urea and potash was dissolved in water and fed to the drip using ventury system. Fertigation was made by regulating the taps of the laterals by allowing the solution to the specified plots as per the treatments. Different irrigation levels were imposed on the basis of pan evaporation through meteorological approach (Pruitt, 1966).

Water was pumped from a well through a 3 Hp diesel pump and conveyed to the field using PVC pipes of 63 mm dia. after filtering through a screen filter (Figure 1). Ventury was installed for fertigation ahead of the screen filter in the main line. Water was taken to the field through sub mains of 50 mm dia. PVC pipes. In each plot four laterals were placed at a spacing of 0.9 m (Figure 3). Drippers were fixed at 0.6 m. In each plot (4.8 m x 3.3 m) there were 32 drippers. A tap was fixed at the head of each lateral in order to regulate the water supply. Sub mains were closed with end caps and laterals were closed with end plugs. After the installation, trial run was conducted to assess the mean emitter discharge and uniformity coefficient. To estimate approximate time required to operate the system for irrigating each treatment *i.e.,* 100%, 80% and 60% PE, the discharge rate of emitter was measured and then emission uniformity of the system was computed with the following formula:

$$EU(\%) = \frac{q^{-}25\%}{q^{-}} \times 100$$

where,

EU = Emission uniformity (%),

$q^{-}25\%$ = is mean of lowest 25% emitter discharge

q^{-} = mean discharge rate

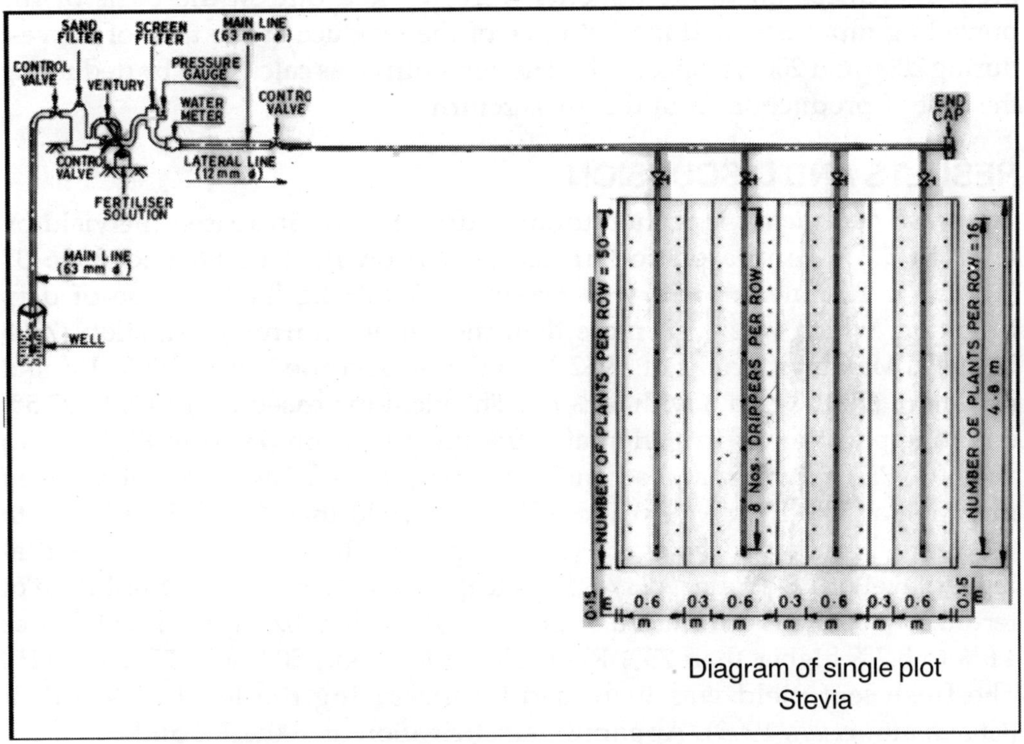

Fig. 1: Layout and design of drip irrigation system for stevia crop

The rating of drip design system was done on the basis of EU value and various statistical parameters like CD and CV%. To irrigate each, treatment, time required to apply a fixed quantity of water was computed based on discharge rate and emission uniformity as per above formulae. In drip irrigation system, discharge rate of drippers was monitored frequently and uniformity coefficients were estimated for operating the system to meet the actual irrigation requirement of a crop. The results revealed that the mean discharge rate were 3.0 stevia with standard deviation of 0.19 lph and coefficient of variation 6.34%. Based on the above discharge rates, emission uniformity was found to be 96.3% for stevia. The above values were quite satisfactory as per the standard design system of drip irrigation. The rating of standard design of drip irrigation system described by Ortega *et al.* (2002) was followed. Extremely high and low discharging emitters were replaced so as to ensure a uniformity coefficient of not less than 90%. The coefficient of variation obtained from the result was quite satisfactory as it was within the desirable limit and our design system was rated as medium in category. Cost of lying out of drip irrigation system and cultivation charges were worked out per hectare. The life of drip system was assumed to be 10 seasons. Cost of production and gross return for all the treatments were worked out on the basis of the prevailing input cost and market price of the produce at the time of harvest during 2006 and 2007, respectively. The net return was calculated by deducting the cost of production from the gross return.

RESULTS AND DISCUSSION

Fresh leaf yield : Irrigation through drip method increased the yield by 4.4% to 5.3% compared to surface irrigation in 2005-06 and 2006-07 (Table 1). The mean yield was maximum (7873 kg ha^{-1}) in case of drip irrigation, which was 4.9% more than the surface (furrow) irrigation (7508 kg ha^{-1}). Maximum yield of 8952 kg ha^{-1} was recorded with 100% PE and minimum (7021 kg ha^{-1}) with 60% PE. The yield increased by 17.1% to 27.5% due to application of irrigation at 100% PE (I_1) as compared to 80% PE (I_2) and 60% PE (I_3) in first and second year, respectively. The crop that received irrigation at 80% PE gave 8.6% to 9.3% more yield than 60% PE during both the years. Fronza and Folegatti (2003) reported that irrigation provided at higher level of PE produced 17% high leaf yields. Application of full dose of fertilizer (100% RD) produced maximum yield of 8207 kg ha^{-1}, which was 4.6% to 8.7% higher than 75% RD (7846 kg ha^{-1}) and 50% RD (7567 kg ha^{-1}). The fresh leaf yield was increased by increasing the level of fertilizer (Chalapathi *et al.*, 1999). Application of irrigation at 100% PE and 100% RD produced maximum fresh leaf yield (9131 kg ha^{-1}) due to availability of adequate moisture and nutrition.

Dry leaf yield: The maximum mean dry leaf yield (2433 kg ha^{-1}) was obtained with drip irrigation, which was 4% higher than surface irrigation (Table 1). Application of irrigation at 100% PE (I_1) produced 15.6% higher yield than 80% PE (I_2) and 24.8% higher than 60% PE (I_3). Irrigation at 80% PE gave 3.6% more dry leaf yield than 60% PE. Maximum dry leaf yield (2525 kg ha^{-1}) was recorded with application of 100% RD (F_1) followed by 75% RD (F_2-2429 kg ha^{-1}) and 50% RD (F_3-2345 kg ha^{-1}). Application of irrigation at 100% PE (I_1) with 100% RD (F_1) produced 2903 kg dry leaf ha^{-1}. Precise availability of moisture and nutrients during early growth helped to have good vegetative growth, which might have enhanced better light utilization resulting in high economic yield. Application of 50: 60: 50 kg NPK ha^{-1} recorded significantly higher dry leaf yield as compared to other treatments (Kumar *et al.*, 2012).

Total biomass and harvest index : Total biomass was affected significantly by the method, level of irrigation and fertility (Table 1). Drip irrigation method produced more biomass (326.7 q ha^{-1}) than the surface irrigation (320.1 q ha^{-1}). Irrigation at 100% PE (I_1) produced maximum quantity of biomass yield (344.5 q ha^{-1}) compared with I_2 (330.2 q ha^{-1}) and I_3 (305.3 q ha^{-1}). Reduction in yield was observed with decrease in the quantity of irrigation water as the crop is very much sensitive to the moisture stress (Aladakatti *et al.*, 2012). Application of 100% recommended dose of fertilizer (F_1) helped to produce maximum biomass yield of 335 q ha^{-1} as compared to 75% (327.5 q ha^{-1}) and 50% RD (317.4 q ha^{-1}). Chalapathi *et al.*, (1999) reported significant increase in biomass yield due to increasing fertilizer levels in ratoon crop of stevia. Application of irrigation at 100% PE with 100% RD produced maximum biomass yield of 35,001 kg ha^{-1}. The biomass yield significantly increased by increasing levels of irrigation and fertilizer due to availability of sufficient moisture and nutrients. Drip irrigation had higher harvest index (0.24) than that of surface irrigation (0.23) (Table 1). Application of less water (80% PE and 60% PE) reduced the harvest index to 0.23 from 0.26 due to lower yield of dry leaf than 100% PE. Fertility also affected the harvest index and it decreased from 0.24 to 0.23 with reduction in fertilizer dose.

Economics : Drip irrigation treatment resulted in highest gross return of ₹ 4, 86,707 ha^{-1} which was 4% higher than the surface irrigation (Table 2). Application of irrigation at 100% PE gave maximum gross return (₹ 5,47,633 ha^{-1}). The crop received full dose of fertilizer (100% RD) gave maximum maximum return of ` 5,05,300 ha^{-1} as compared to 75% RD and 50% RD due to high dry leaf yield. Reduction of 25% to 50% fertilizer from the full dose decreased gross return by 3.9% to 7.2%. Application of 75% RD gave 3.6% more return than 50% RD. Application of irrigation at 100% PE with 100% RD gave maximum gross return of ₹ 5,80,933 ha^{-1}. The net return increased by 5.1% through drip as compared to surface irrigation (Table 2). Irrigating the crop at 100% PE resulted in maximum net return of ₹ 3,18,933 ha^{-1}, which was 29% higher than 80% PE (₹ 2,47,239 ha^{-1}) and 47.6% higher than 60%

Table 1: Effect of irrigation and fertility on fresh and dry leaf yield, biomass yield and harvest index of stevia

Treatment	Fresh leaf yield (kg ha^{-1})			Dry leaf yield (kg ha^{-1})			Biomass yield (kg ha^{-1})			Harvest index (%)		
	2005-06	2006-07	Mean	2005-06	2006-07	Mean	2005-06	2006-07	Mean	2005-06	2006-07	Mean
Method of irrigation												
Control	7585	7431	7508	2363	2315	2339	317.5	322.7	320.1	0.24	0.23	0.23
Drip fertigation	7919	7828	7873	2447	2419	2433	322.9	330.4	326.7	0.25	0.24	0.24
SE (m)±	25.9	4.42	12.96	4.69	2.92	24.40	3.34	2.1	2.2			
CD (0.05)	77.1	13.15	38.5	13.96	8.68	72.51	10.1	6.3	6.7			
Irrigation												
I_1= 100% PE	9045	8858	8952	2766	2709	2738	339.5	349.5	344.5	0.27	0.25	0.26
I_2= 80% PE	7658	7636	7647	2371	2364	2368	328.5	331.8	330.2	0.23	0.23	0.23
I_3= 60% PE	7053	6989	7021	2204	2184	2194	300.8	309.8	305.3	0.23	0.23	0.23
SE (m)±	44.9	7.66	22.45	8.14	5.06	42.27	5.9	3.7	3.9			
CD (0.05)	133.5	22.78	66.7	24.17	15.04	125.6	17.4	10.9	11.6			
Fertility												
F_1= 100% RD	8239	8174	8207	2535	2515	2525	330.8	339.2	335.0	0.25	0.24	0.24
F_2= 75% RD	7888	7804	7846	2442	2416	2429	324.5	330.5	327.5	0.24	0.24	0.24
F_3= 50% RD	7629	7505	7567	2363	2327	2345	313.4	321.5	317.4	0.24	0.23	0.23
SE (m)±	44.9	7.66	22.45	8.14	5.06	42.27	5.9	3.7	3.9			
CD (0.05)	133.5	22.78	66.7	24.17	15.04	125.6	17.4	10.9	11.6			

Table 2: Effect of irrigation and fertility on economics of stevia

Treatment	Gross return (₹ ha^{-1})			Net return (₹ ha^{-1})			Benefit-cost ratio		
	2005-06	2006-07	Mean	2005-06	2006-07	Mean	2005-06	2006-07	Mean
Method of irrigation									
Control	4,72,667	4,63,000	4,67,833	2,51,989	2,43,967	2,47,978	1.14	1.11	1.13
Drip Fertigation	4,89,600	4,83,815	4,86,707	2,61,915	2,59,567	2,60,741	1.16	1.15	1.15
SE (m)±	964	58	500	1072	105	537	0.007	0.001	0.004
CD (0.05)	2865	174	1486	NS	316	1595	0.02	0.003	0.011
Irrigation									
I_1 = 100% PE	5,53,400	5,41,867	5,47,633	3,23,233	3,14,633	3,18,933	1.40	1.38	1.39
I_2 = 80% PE	4,74,622	4,72,822	4,73,722	2,46,622	2,47,856	2,47,239	1.08	1.38	1.39
I_3 = 60% PE	4,40,778	4,36,756	4,38,767	2,15,889	2,16,211	2,16,050	0.99	0.98	0.98
SE (m)±	1670	101	866	1857	182	930	0.01	0.002	0.006
CD (0.05)	4962	301	2573	5518	257	2762	0.04	0.006	0.02
Fertility									
F_1 = 100% RD	5,07,644	5,02,956	5,05,300	2,79,100	2,77,356	2,78,228	1.22	1.23	1.22
F_2 = 75% RD	4,88,467	4,83,178	4,85,822	2,61,078	2,58,767	2,59,922	1.15	1.15	1.15
F_3 = 50% RD	4,72,689	4,65,311	4,69,000	2,45,567	2,42,578	2,44,072	1.11	1.09	1.10
SE (m)±	1670	101	866	1857	182	930	0.01	0.002	0.006
CD (0.05)	4962	301	2573	5518	257	2762	0.04	0.006	0.02

NS = Not significant

*Considered the cost of production and gross return from the produce for the period of 2006 and 2007. During that period the market price of the dry leaf yield was ₹ 200 kg^{-1}. Presently, also the variation in economics is negligible even when compared with recent prevailing wholesale market prices of the produce and the cost of production.

PE (₹ 2,16,050 ha^{-1}). Application of 100% RD gave highest return of ₹ 2,78,228 ha^{-1}, which was 7% to 14% higher than that of 75% and 50% RD. Reduction of 25% (F_2) and 50% RD (F_3) reduced the net return by ₹ 18,306 ha^{-1} and ₹ 34,156 ha^{-1}, respectively as compared to 100% RD (F_1). Application of irrigation and fertilizer at 100% PE and 100% RD gave maximum net return of ₹ 3, 51,150 ha^{-1}. The benefit-cost ratio ranged from 0.98 to 1.4 with different treatments (Table 2). Drip irrigation resulted in higher ratio (1.16) than surface (1.13). Application of irrigation at 100% PE and 80% PE gave equal benefit-cost ratio (1.39), which was more than that of 60% PE (0.985). Application of 100% RD resulted in highest ratio of 1.22, which decreased due to reduction in fertilizer dose. The lowest ratio (1.10) was recorded with 50% RD. Megeji *et al.* (2005) concluded that *Stevia* cultivation was highly remunerative. It could give ₹ 3.75 lakhs net profit ha^{-1} during four years with an average annual income of ₹ 0.93 lakh ha^{-1} with B: C ratio of 1.89.

CONCLUSIONS

It can be concluded that stevia should be irrigated through drip fertigation method at 100% PE with full dose of fertilizers (100% RD) to achieve maximum quantity of dry leaves (2903 kg ha^{-1}), net returns of ₹ 3,51,150 ha^{-1} and benefit-cost ratio of 1.33 was obtained. About 40% water could be saved through drip irrigation in stevia. These findings may be replicated under rice based cropping systems, wherever irrigation facility exists.

REFERENCES

Aladakatti, Y.R., Palled, Y.B., Chetti, M.B., Halikatti, S.I., Alagundagi, S.C., Patil, P.L., Patil, V.C., Janawade, A.D. 2012. Effect of irrigation schedule and planting geometry on growth and yield of stevia (*Stevia rebaudiana* Bertoni.). *Karnataka Journal of Agricultural Sciences*, 25:30-35.

Chalapathi, M.V.S., Thimmegowda, G. Rao, G.E., Devakumar, N. and Chandraprakash, J. 1999. Influence of fertilizer level on growth yield and nutrient uptake of ratoon crop of stevia (*Stevia rebaudiana*). *Indian Journal of Medicinal and Aromatic Plants*, 21: 947-949.

Fronza, D. and Folegatti, M.V. 2003. Water consumption of the *Stevia* (*Stevia rebaudiana* Bert.) crop estimated through microlysimeter. *Scientia Agricola*, 60(3): 595-599.

Kumar, R., Sharma, S., Ramesh, K., Prasad, R., Pathania, V., Singh, B., Singh, R.D. 2012. Effect of agrotechniques on the performance of natural sweetener plant –Stevia (*Stevia rebaudiana* Bertoni) under western Himalayan conditions. *Indian Journal of Agronomy*, 57:74–81.

Lavini, A., Riccardi, M., Pulvento, C., Luca, S.D., Scamosci, M., Riccardo d'Andria. 2008. Yield, Quality and Water Consumption of *Stevia rebaudiana* Bertoni Grown under

Different Irrigation Regimes in Southern Italy. *Italian Journal of Agronomy,* 2: 135-143.

Lubbe, A. and Verpoorte, R. 2011. Cultivation of medicinal and aromatic plants for specialty industrial materials. *Industrial Crops and Products.*

Megeji, N. W., Kumar, J. K., Virendra Singh, Kaul, V. K. and Ahuja, P. S. 2005. Introducing *Stevia rebaudiana-* A natural zero calorie sweetener. *Current Science,* 88 (5):801-804.

Ortega, J. F., Tarjuelo, J. M. and De Juan, J. A. 2002. Evaluation of irrigation performance in localized irrigation system of semi arid regions (Castilla-la Mancha, Spain): Agricultural Engineering International: *CIGR of Scientific Research and Development,* 4:1-17.

Pruitt, W.O. 1966. Empirical methods of estimating evapotranspiration using primarily evaporation pans. Proc. Conf. on evapotranspiration and role in water resources management. *American Society of Agricultural Engineers,* St. Joseph

Rani, B., Maheshwari, R., Chauhan, A.K.,Sharma, P. and Sharma, S. 2012. *Stevia rebaudiana:* its extraction, composition and medicinal applications. *International Journal of Institutional Pharmacy and life sciences,* 2 (2):315-325.

Stevia Facts: Natural *Stevia* Plant Extracts. Web page on the internet. Retrieved on 01-02-08, Cited on 9.2.08. Available from: http://puresweet.com.au/Text/1122341279234-2458

Zhang, S.Q., Kutowy, O., Kumar, A. 1999. *Stevia rebaudiana leaves-* a low calorie source of sweeteners. *L'Actualite Chemique Canadienne,* 5: 22-23.

□□□

Chapter 39

Need and Scope of Rainwater Harvesting in Rainfed Uplands of Eastern India

B. C. SAHOO[1], S. N. PANDA[2] AND B. PANIGRAHI

INTRODUCTION

South-west monsoon aberrations in the eastern region of India during June to September play a crucial role in reducing the production potential of rainfed upland crops. Mean seasonal rainfall of 1200 mm during the season comes in a very erratic manner. Apart from that, unpredictable arrival and withdrawal of monsoon and in-season critical dry spells are some of the common climatic irregularities in the region (Panigrahi *et al.*, 2007). Consequently, the crop season either experiences drought or flood in alternate year leading to uncertainty in crop production. Under these circumstances, rice being a high water requiring crop is never suggested for light textured and highly permeable rainfed uplands. Promotion of rice in such adverse land and climatic situations through intervention of rainwater harvesting reservoirs could not ensure increase in cropping intensity as well as productivity of uplands in eastern India. On the contrary, complete replacement of rice with low duty non-rice crops such as maize (*Zea mays L.*), groundnut (*Arachis hypogaea L.*) and pigeon pea (*Cajanus cajan*) *etc.*, has not been successful in the region as rice is the staple food of people (Kar *et al.*, 2004). On this backdrop, the concept of partial rice substitution in uplands along with rainwater harvesting on-farm ponds (OFP) may be a feasible solution. Combination of a low duty non-rice crop with rice in separate compartments of the same crop field along with rainwater harvesting options through a pond is expected to be able to save more water for the second crop in winter after fulfilling the dry spell need of the rainy season crops. In order to study the feasibility of the concept prior to its application in the field, it is

[1]Associate Professor, Dept. of SWCE, College of Agril. Engg. & Tech., OUAT, Bhubaneswar.
[2]Professor and Head, School of Water Resources, IIT, Kharagpur, (West Bengal)

required to assess the need as well as scope of rainwater harvesting for the proposed cropping pattern in rainfed uplands of eastern India. Possibly, this can be achieved through identification of moisture stress periods and the drainage requirements of the crops during various growth stages and so, emphasized in the present study.

MATERIALS AND METHODS

The study area extends throughout the eastern region of India. The region lies between 18 - 25^0 N and 81 - 89^0 E approximately encompassing four states, namely, Odisha, Chhatisgarh, Jharkhand and West Bengal. Mean annual rainfall received across the region ranges from 1200 - 2000 mm. Rice is the major crop grown in all topo-sequences (upland, medium land and low land) during rainy season. Average soil type of rainfed uplands is light textured, acid lateritic with pH ranging from 4.0 - 6.8. The water holding capacity of the soil is very low and fertility status is very poor due to soil erosion. Field experiments were conducted in the uplands of the experimental farm of Agricultural and Food Engineering Department, Indian Institute of Technology, Kharagpur, India during 2007 and 2008. The site is located at 22^{0}19′ N latitude and 87^{0}19′ E longitude at an altitude of 48 m from mean sea level. The crop fields are bounded by dikes of mean height of 45 cm. The elevation difference between the uplands and low lands in the farm is within 2 m. Basing upon the existing elevation head in the site, an elevation head difference of 5 cm between two crop compartments has been used for developing water balance model in the present study.

Model Development : Diked maize and rice fields of equal dimension placed side by side in a contour in rainfed uplands comprise the initial field setting for the root-zone water balance model. Simulation of soil moisture storage (SMS) in the root-zone of rice and maize has been accomplished through modelling scenario-I and scenario-II, respectively. The term SMS represents the total depth of soil moisture in the layer under consideration. No surface runoff (SR) from- and supplemental irrigation (SI) to- the rice field forms the basis of modelling scenario-I. It implies that the SR likely to be generated from the rice field due to rainfall events is not to be drained out rather arrested in the field itself surrounded by dikes. Spill-over loss from the rice field is ignored assuming that the dike height is sufficient enough to take care of the ponding. The SMS beyond saturation soil moisture content in the root-zone has been considered as the ponding stage in the field. Thus, the simulated daily SMS in the root-zone of the crop is used to assess the drainage need and SI requirement of rice. Modeling Scenario-I depicts the soil moisture dynamics in both the active and passive layers of rice. Two layer water balance model for active and passive layers of rice is expressed as:

Active layer of rice

$$SMC^{a}_{ki}.RD_{ki} = SMC^{a}_{k(i-1)}RD_{k(i-1)} + R_{i} + SMC^{p}_{k(i-1)}.\delta RD_{ki} - AET^{a}_{ki} - P_{ki} - SP_{ki} \quad \dots (1)$$

Passive layer of rice

$$SMC^{p}_{ki}\left(RD^{max}_{k} - RD_{ki}\right) = SMC^{p}_{k(i-1)}\left(RD^{max}_{k} - RD_{ki}\right) + P_{ki} - DP_{ki} \quad \dots (2)$$

where, SMC^{a}_{ki} = soil moisture content in the active layer, cm cm^{-1}; SMC^{p}_{ki} = soil moisture content in the passive layer, cm cm^{-1}; RD_{ki} = root-zone depth of kth crop on ith day, cm; ?RD_{ki} = incremental root growth, cm; R_{i} = rainfall, cm; AET^{a}_{ki} = actual evapotranspiration from active layer, cm; P_{ki} = vertical percolation from the active to passive layer, cm; SP_{ki} = seepage loss through the field dikes, cm; RD^{max}_{k} = maximum root-zone depth of kth crop; DP_{ki} = deep percolation loss from the passive layer, cm; i = time index taken as 1 day in the study (i =1 means day of sowing and so on); and k = index for the crop (rice).

The term $SMC^{p}_{k(i-1)}.\delta RD_{ki}$ in Eq.1 is the SMS of the thin passive layer which is carried over to the active layer with incremental growth of the root system with time. Basing upon the concept of two layer water balance model, when the dynamic root system proceeds from (i-1)th to ith day, an infinitesimally thin passive layer equal to the incremental root growth during the interval becomes a part of the active layer along with the SMS retained in the layer. Thus, in addition to rainfall, the aforementioned term in Eq. 1 becomes another inflow component to the active layer till root growth attains maturity (Hajilal *et al.*, 1998). It is so because after attaining maximum root-zone depth, the passive layer vanishes and the incremental root growth with time becomes zero. The outflow components of the rice field are actual evapotranspiration (AET) from the crop field, vertical percolation that moves below the maximum root-zone depth and seepage (SP) loss through the dikes. The seepage loss from the rice field is assumed to take place through the field dikes under ponding condition only provided required hydraulic head is available. Ponding phase starts in rice field when the SMS of the active layer goes beyond the SMS at saturation. During this phase, the ponding depth is computed as:

$$PD_{ki} = \begin{cases} SMC^{a}_{ki}.RD_{ki} - SAT^{a}_{ki}RD_{ki} & \text{if } SMC^{a}_{ki}.RD_{ki} > SAT^{a}_{ki}.RD_{ki} \\ 0 & \text{if } SMC^{a}_{ki}.RD_{ki} \le SAT^{a}_{ki}.RD_{ki} \end{cases} \quad \dots (3)$$

where, PD_{ki} = ponding depth, cm; and SAT^{a}_{ki} = moisture content on volume basis at saturation in the active layer, cm cm^{-1}.

Unlike rice, the maize crop is very much sensitive to any form of standing water when grown in diked field. So, the concept of Modeling Scenario – I that allows standing water in the crop field is deviated in case of maize field. So, modelling scenario-II is developed for maize field separately based on the concept of instant drainage from- and no SI to- the crop field. Whenever there is generation of SR from the maize field due to excess rainfall, instant drainage is achieved through an outlet in the downstream dike of the field. Thus, the SMS in maize field remains either at saturation for a short while or unsaturated for most of the crop growing season. The water balance model for maize field is expressed as:

Active layer of maize

$$SMC^{a}_{ji}.RD_{ji} = SMC^{a}_{j(i-1)}RD_{j(i-1)} + R_{i} + SMC^{p}_{j(i-1)}.\delta RD_{ji} - SR_{ji} - AET^{a}_{ji} - P_{ji} \quad \text{... (4)}$$

Passive layer of maize

$$SMC^{p}_{ji}\left(RD^{max}_{j} - RD_{ji}\right) = SMC^{p}_{j(i-1)}\left(RD^{max}_{j} - RD_{ji}\right) + P_{ji} - DP_{ji} \quad \text{... (5)}$$

where, SR_{ji} = surface runoff from the crop field, cm; and j = crop index for maize.

The inflow components of modelling scenario – II are almost alike as that of modeling scenario – I. But the outflow components of maize field are different from that of rice in respect of surface runoff (SR) and seepage (SP) loss. While the excess rainfall is drained out instantly in the form of SR, the SP component is ignored in the absence of standing water in the maize field. The SR component of the model shows the drainage need of the maize field during various stages of crop growth. The simulation of SMS in the root-zone of maize on daily basis is likely to explain the irrigation requirement of the crop during rainy season. Linear root growth model (Bhirud *et al.*, 1990) is used to simulate the growth rate of the crops in the model.

Field Experiment : The experiment was designed for growing rice and maize in separate plots under rainfed condition. Since rice requires standing water at different stages of its growth, it is proposed to have a dike of 60 cm height around the rice plots. On the other hand, the surface runoff generated from the maize plot is drained out instantly. The experimental plots each of dimension 60 m x 20 m are replicated twice.

Short duration rice variety *'Anjali'* and hybrid maize of variety *'BIO – 9681'* each of 95 days duration were chosen in the present study. The crops were sown in respective plots on the day of onset of effective monsoon in the region. Seed rate used for broadcast rice was 125 kg ha^{-1} and for line sown maize was 15 kg ha^{-1}. The maize seeds were sown on ridges with plant to plant spacing of 30 cm and row spacing of 45 cm. Recommended doses of chemical fertilizers (N: P: K::60:30:30 for upland rice and 100:40:40 for maize) were applied to both the crops in three splits such as basal, 1st top dressing and final top dressing.

Two non-weighing type cylindrical drum lysimeters, one bottom opened and the other bottom closed, were installed in rice fields to determine the actual evapotranspiration (AET) and deep percolation (DP) under ponding conditions. Soil moisture storage of the active and passive layers was measured in percentage with respect to saturation soil moisture storage by Aqua-pro soil moisture sensor. Assuming the rice field as a basin, two piezometers, one at the upstream (inside the basin) side of the dike and the other at the downstream end, were installed to measure the hydraulic pressure heads during ponding condition. Stage-wise crop coefficients of rice and maize and daily reference crop evapotranspiration (ET_0) in the experimental site were taken into consideration for estimating potential evapotranspiration (PET) of the crops under adequate soil moisture condition. Penman-Monteith method was used to estimate daily ET_0 (Allen *et al.*, 1998).

RESULTS AND DISCUSSION

Ponding during initial and final stages : The initial stage (IS) of both maize and rice extends up to 14 days from the day of sowing including 4 days of germination period. Simulation study carried out for 30 years from 1977-2006 reveals that the ponding stage continues for on an average 5 days during the initial stage of the rice in 90% of the years and obviously, without ponding for only 10% of the years. It needs no mention that any degree of ponding during this succulent stage of the plant is very harmful to the crop from rotting point of view. In the absence of a well managed outlet in the field, the farmer has no other option rather than cutting of the dikes to save the crop which is likely to inundate the low lying areas. This kind of incident is associated with the risk of heavy soil erosion in recently ploughed loose top soil. Thus, the study suggests the need of harvesting of excess rainfall in an on-farm pond in the field itself during the initial stage of rice. The harvested water, in addition to facilitating good germination, saving the succulent seedlings from rotting and reducing erosion of the top soil, can be used to develop standing water in the field for carrying out *beusan* or *biasi* operation in time. Beusan is a secondary tillage operation performed in the rice field under standing water condition at the end of crop establishment or in the beginning of crop development stage with the help of a bullock drawn narrow bottom intercultural plough. The purpose of this operation in rice field is to maintain uniform plant population, suppress weeds and reduce infiltration rate that leads to facilitate standing water in the rice field.

In case of maize field, generation of runoff resembles the condition of ponding in rice field. Unlike rice, ponding at any stage of maize is very much detrimental to the crop as the crop is very much sensitive to even saturation moisture condition. Simulation study of SMS above SAT in maize field during the initial stage reveals that surface runoff is likely to be generated in 90% of the years. The number of runoff producing days is found to range from 2 to 13 days in various years. This excess rainfall needs to be harvested, instead of flowing down as runoff, for supplemental irrigation purpose in other stages of the crop. Diverting excess rainwater to an on-farm pond saves the tender crop in one hand and ensures security against drought at the other. It is also revealed that the runoff producing days during initial stage of maize are more than the ponding days in a rice field. It indicates roughly the chance of harvesting more rainwater from maize field as compared to that from rice field.

Final stage starts 10 days before harvesting of the crop and continues up to the sowing day of the winter crop. Thus, the stage extends for a period of 25 days in the present study. Ponding in rice field during the final stage is found to be for 63% of the years (19 out of 30 years) leaving rest 37% of years unsaturated. On the contrary, out of 30 years of simulation, the runoff generation from the maize field is expected in 53% of the years (16 years out of 30). Standing water in rice field and surface runoff from maize field during their final stages is no way beneficial rather; it facilitates lodging effect in rice and maize crops at harvesting stage. If not drained out quickly, it may also cause delay in carrying out field preparation for the crops to be taken in winter. Hence, it is necessary to drain out the ponding water quickly from both rice and maize fields when the crops are in their final stages.

The ponding days in rice field during the initial and final stages fit well with General Extreme Value (GEV) and Pareto distribution functions, respectively, whereas the runoff producing days of the respective stages of maize fit well with GEV and Normal distribution. Thus, during initial stage the rice field is expected to be under ponding for 4 days whereas 5 runoff producing days are expected from the maize field at 50% PE level (Figure 1). On the contrary, the maize field is expected to generate runoff for 3 days during its final stage, whereas ponding is expected for 4 days in rice field at 50% PE level. Moreover, draining of excess rainfall is essential for both the crop fields during initial and final stages.

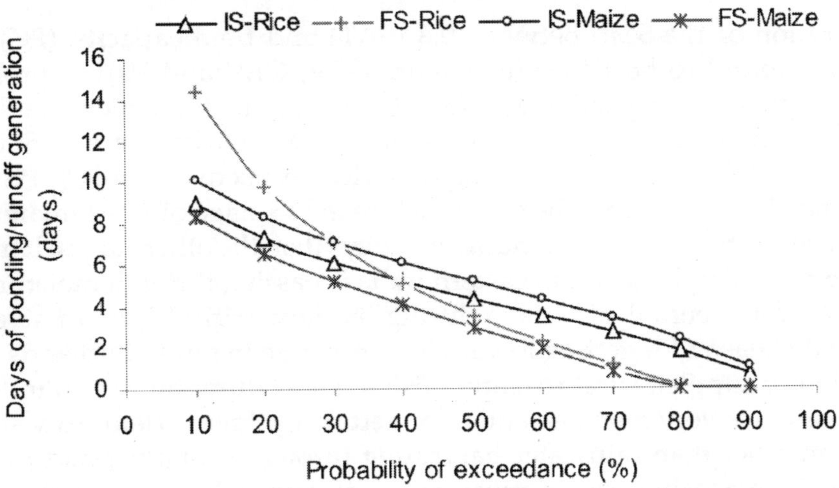

Fig. 1: Ponding days / runoff generation days in rice and maize fields during initial and final stages

Ponding during crop growing season : Number of ponding days in rice field during crop establishment (CE), crop development (CD), mid-season (MS) and late season (LS) stages are found to best fit with Pareto, GEV, GEV and Normal Distribution functions, respectively. At 50% probability of exceedance (PE), the ponding days in CE, CD, MS and LS stages are 6, 11, 10 and 5 days, respectively (Figure 2a). Only 10 days of ponding during the MS stage of rice which extends for a period of 30 days, in rainfed uplands indicates that about 67% of the MS stage will be under unsaturated condition and it may go to inadequate moisture condition on any day. The water harvested in preceding stages can be utilized in case of moisture stress in mid-season stage of rice. Thus, the need for rainwater harvesting in upland rice fields during excess rainfall conditions is verified. On the other hand, the runoff producing days in case of maize during CE, CD, MS and LS stages are found to best fit with Pareto, GEV, Normal and Normal distribution functions, respectively. Thus, at 50% PE level, the expected runoff producing days are 8, 8, 6 and 3 days at CE, CD, MS and LS stages of maize, respectively (Figure 2b). On comparing the number of ponding days in rice field and runoff producing days in maize field, it is understood that the runoff potential of the former is less than the latter.

Unsaturated soil moisture status in rice field : Days of soil moisture storage within 0.8(SAT) to SAT in various crop growth stages of rice are found to best fit with GEV probability density function. The durations for which the SMS lies within 0.8(SAT) to SAT are 4, 5, 12 and 10 days during CE, CD, MS and LS stages of rice, respectively at 50% PE (Figure 3a). Further, the duration of the SMS between 0.8 (SAT) and field capacity (FC) in rice field was found to best fit with Pareto, GEV, GEV and Pareto probability density functions, respectively. Thus, the durations for which the SMS lies below the management allowable depletion (MAD) level are 7, 5, 8 and 4 days during CE, CD, MS and LS stages of rice, respectively, at 50% PE (Figure 3b). It may be noticed that the soil moisture in MS stage of rice remains below MAD level SMS for maximum number of days than the other stages. It indicates that the rainfed upland rice in eastern India is passing through moisture stress condition for a period of 8 days during its most critical growth stage. This may be the reason for getting lower yield from rice in rainfed uplands. Hence, provision of supplemental irrigation (SI) to short duration rice in uplands of eastern India is very much essential for getting optimum yield. Any source of irrigation other than rainwater harvesting through on-farm pond in rainfed uplands is expensive and in most cases infeasible. It indicates the need of rainwater harvesting for rice cultivation in uplands of eastern India.

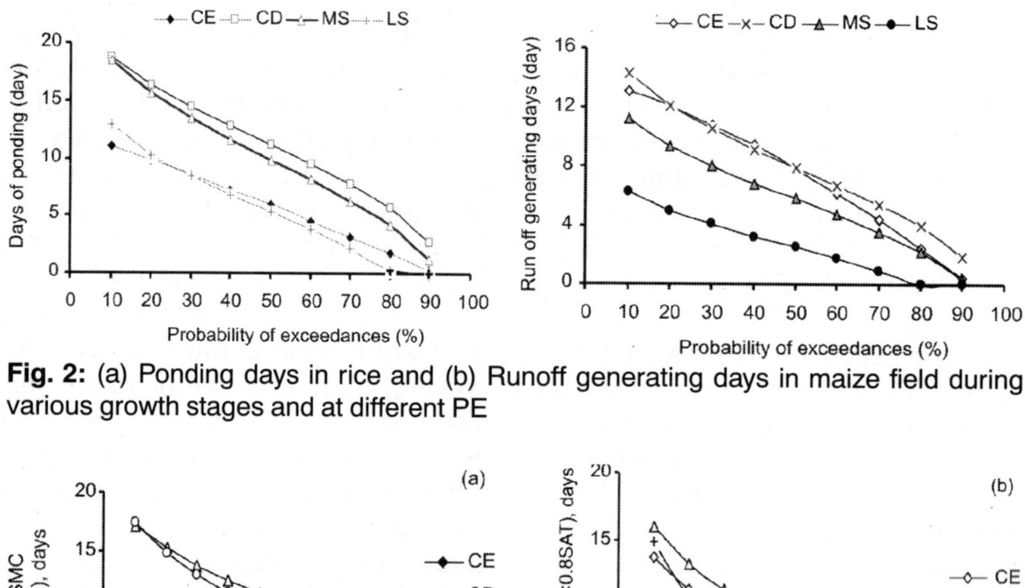

Fig. 2: (a) Ponding days in rice and (b) Runoff generating days in maize field during various growth stages and at different PE

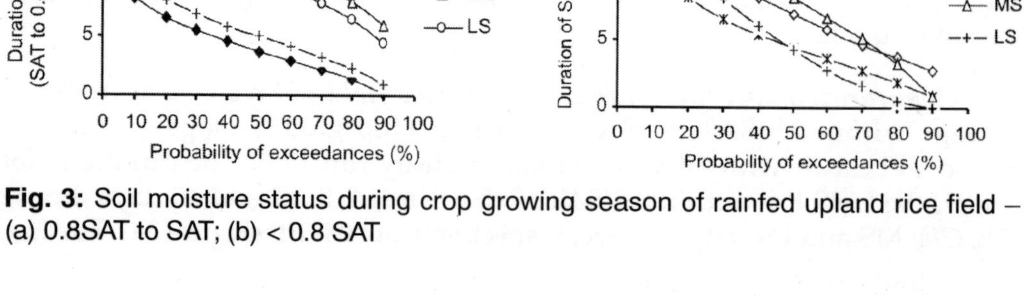

Fig. 3: Soil moisture status during crop growing season of rainfed upland rice field – (a) 0.8SAT to SAT; (b) < 0.8 SAT

Unsaturated soil moisture status in maize field : The durations of SMS within SAT to FC in CE, CD, MS and LS stages of maize continue for 13, 14, 24 and 20 days, respectively (Figure 4a). On the other hand, the SMS in the root-zone depth of maize hardly declines below FC in the entire growing season during monsoon months. Further, at 50% PE, the duration of SMS below FC is less than one in various crop growth stages of maize (Figure 4b). It implies that the soil moisture status hardly declines below the MAD level SMS of maize. It also suggests that there is no need of any source of irrigation for maize in rainy season as far as eastern region of India is concerned. Moreover, it suggests that substitution of rice with maize in uplands of eastern India is a viable alternative to avoid monsoon aberrations. Further, the scope of harvesting rainwater is better from a maize field than that of a rice field and thus, it ensures availability of more water to meet the supplemental irrigation need of the winter crops. It implies that complete or major replacement of rice with maize in rainfed uplands is a better option to increase the cropping intensity.

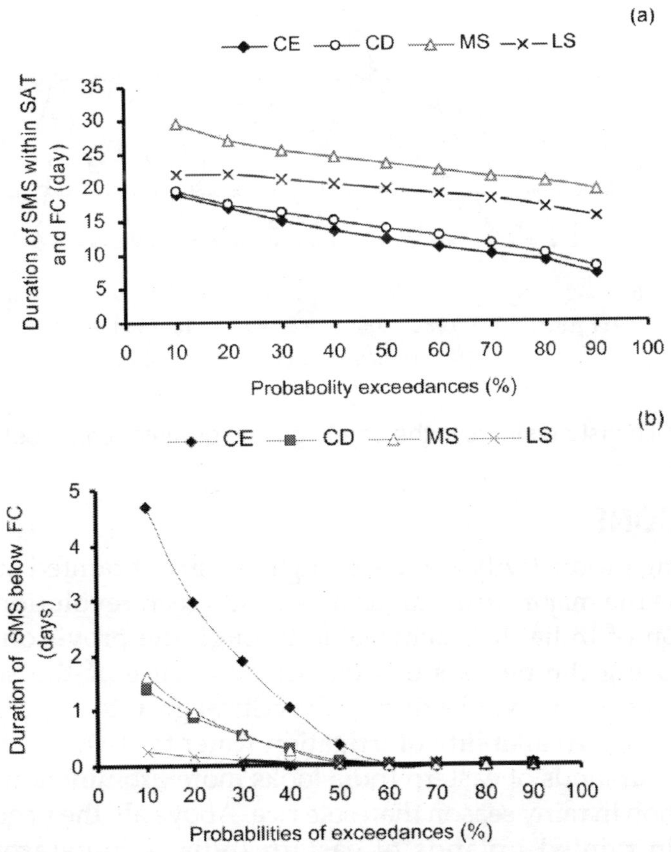

Fig. 4: Unsaturated soil moisture status in maize field – (a) FC < SMS = SAT; (b) SMS < FC

Soil moisture status before sowing of winter crops : It is depicted through Figure 5 that the SMS in the seeding zone lies below germination moisture storage for 18 and 19 years out of 30 years of simulation in rice and maize fields, respectively. It indicates that growing winter crops in rainfed uplands is not feasible for 60 and 63% of years in rice and maize fields, respectively, without the provision of pre-sowing irrigation. This may be the reason why the uplands in rainfed area of the region are lying fallow following rainy season. It is further justifying the need of rainwater harvesting for increasing the cropping intensity in rainfed uplands of eastern region of India. Unnecessary ponding in rice field during the final stage and draining of rainfall excess throughout the crop growing season of maize could be harvested for meeting the pre-sowing as well as supplemental irrigation needs of the winter crops.

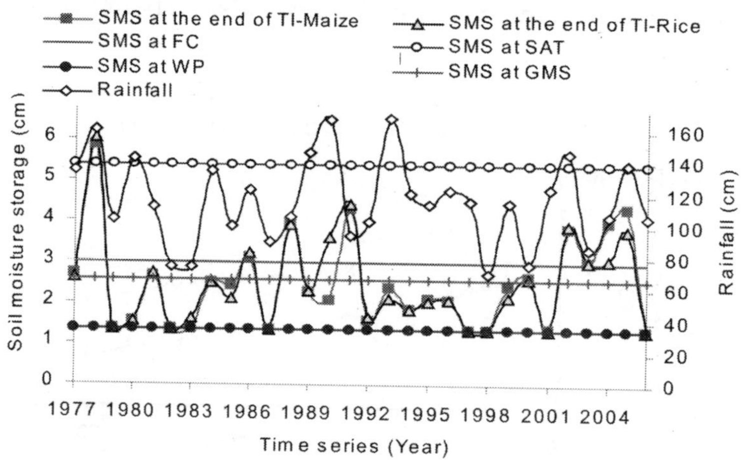

Fig. 5: Soil moisture status in the seeding zone of winter crops before sowing.

CONCLUSIONS

Increasing productivity and cropping intensity of rainfed areas have been considered as the major thrust areas in second green revolution proposed for eastern region of India. It is achievable through the provision of harvesting rainwater during the periods of excesses and reuse of the same in deficit moisture conditions, may it be during the critical growth stage or germination stage of the crops. Availability of irrigation water for two crops in succession exclusively in uplands of eastern India looks more prominent under complete rice substitution in rainy season than sole rice. Above all, the need for rainwater harvesting in rainfed uplands of eastern India is imperative to mitigate droughts, increase cropping intensity and productivity of the land.

REFERENCES

Allen, R.G., Pereira, L.S., Raes, D., Smith, M. 1998. Crop Evapotranspiration. Guidelines for Computing Water Requirements. FAO Irrigation and Drainage Paper 56, FAO, Rome.

Bhirud, S., Tyagi, N.K. and Jaiswal, C.S. 1990. Rational approach for modifying rotational water delivery schedule. *J. of Irrig. and Drain.Engg.* (ASCE), 116 (5): 632-644.

Hajilal, M.S., Rao, N.H. and Sharma, P.B.S. 1998. Planning intraseasonal water requirements in irrigation projects. *Agric. Water Manage.*, 37(2): 163-182.

Kar, G., Singh, R. and Verma, H.N. 2004. Alternate cropping strategies for assured and efficient in upland rainfed rice areas of eastern India based on rainfall analysis. *Agric. Water Manage.*, 67(1): 47-62.

Panigrahi, B., Panda, S.N. and Mal, B.C. 2007. Rainwater conservation and recycling by optimal size on-farm reservoir. *J. of Resour. Conserv. and Recycling*, 50: 459 – 474.

□□□

Chapter 40

Surge Drip Irrigation - An Innovative Approach of Irrigation Methods

D.PARAMJITA, S.C.NAYAK, A.P.SAHU AND N.SAHOO[1]

INTRODUCTION

Water is one of the crucial inputs in farming; its availability is tending to become increasingly scarce and costly. There is scarcity of water in all the rain fed crop growing regions during *rabi*. The introduction of green revolution has not only enhanced the importance of irrigation but also emphasized the need to use limited water resources effectively for maximum output thus paving the way for new technological innovations. Micro irrigation is a suitable method of proper water management and also helps in increasing the crop production to meet the demand of increased population of our country. Among different types of micro irrigation method, the drip irrigation method is now one of the fastest growing technologies in modern agriculture and has proved to be the most efficient. Drip irrigation typically has a decreased wetted volume thus requiring more frequent irrigation. An irrigation regime with an excessively high irrigation frequency can cause the soil surface to remain wet and the evaporation process persists most of the time, resulting in a maximum rate of water loss. This is one of the disadvantages of drip irrigation systems. Advent of sub-surface drip irrigation is in fact an approach to curb this efficiency. The drip irrigation method is now one of the fastest growing technologies in modern agriculture and has proved to be the most efficient one. It has gained worldwide popularity, particularly in water scarce areas, undulating hill areas and saline areas. The present drip irrigation system consists of emitters fitted to laterals and the water ooze out through the emitter drop by drop. Further emitters have been substituted with micro tubes. The

Department of Soil and Water Conservation Engineering, College of Agricultural Engineering and Technology, Orissa University of Agriculture and Technology Bhubaneswar-751 003, Odisha, India,

[1]Directorate of Water Management, Bhubaneswar, Odisha, India

trickle irrigation system consists of emitters, which distribute water drop by drop to the root zone of the crops. Drip irrigation system suffers from clogging of emitter and micro tubes. An alternative approach to the clogging problem is to increase the size of the emitters and micro tubes. However, this may increase the discharge of micro tube emitter and change the pattern of wetting in the soil thus affecting the water availability to the plants. In order to reduce this difficulty Karmelli and Peri (1974) introduced pulse/surge irrigation. Similarly in flow irrigation methods surge irrigation coupled with a long furrow layout can not only minimizes the land losses incurred during layout but would also maximize the irrigation efficiencies. Surge irrigation involves intermittent application of water at specified discharge rates in an ON-OFF irrigation cycling mode into individual furrows and results in accelerated waterfront advance rates and minimum deep percolation losses thus achieving high order of irrigation application, storage and water distribution efficiencies in comparison to continuous flow of water for irrigation. The water use efficiency of drip irrigation is highly dependent on evaporation losses occurring from the constantly saturated soil beneath emitters. Another new irrigation method namely Sand Tube Irrigation (STI) method employs a surface drip system in conjunction with a sand tube column for reducing evaporation significantly. The sand media transmits water into the profile by way of vertical and horizontal flow from the sand tube's base and circumference (Meshkat *et al.*, 1998). Sand tube method is only applicable to permanent tree/vine crops where harvesting and other field operations do not interfere with drip irrigation systems. Keeping the facts stated above in view, the present study has been under taken to study the water front advance by surge drip sand tube irrigation, surge drip gravel tube irrigation and surge drip sand and gravel tube irrigation.

MATERIAL AND METHODS

The study on surge drip irrigation in sand tube, gravel tube and sand & gravel tube was conducted in the Hydraulics Laboratory of the Department of Soil and Water Conservation Engineering, College of Agricultural Engineering and Technology, OUAT,Bhubaneswar during the year 2004-05.

Experimental set up : The experimental set up consists of soil tank model fitted with two fiber glass plates. The dimensions of the soil tank model are Length = 92 cm Breadth = 91.5 cm Height = 61.0 cm Internal angle between flexi glass plates = 1150,Length of flexi glass plates = 58.0 cm (right) and 53.0 cm (left).Bottom of the soil tank and its three vertical sides are made up of metal sheet. Other vertical side has been provided with two flexy glass plates at an angle of 115 degrees. The vertical side provided with flexy glass plates at an angle was used for observing advance of waterfront. A metal mould of

8 cm internal diameter and 15 cm height was used in making sand tube, gravel tube and sand & gravel tube in the soil. The tank was filled with the test soil up to a depth of 46 cm. The metal mould was held vertically in the middle of vertical side provided with flexy glass plates. The space around the metal mould was filled with soil. This was followed by filling the metal mould with sand. The metal mould was lifted gradually holding the handle provided with the mould till a sand tube was made up to the surface of soil in the tank. The procedure was also followed in making gravel tube. In making a sand & gravel tube the lower half of the mould was filled with sand and the upper half of the mould was filled with gravel.

Irrigation system :A plastic bottle of 2 liter capacity was kept in position on a vertical stand. The bottle was provided with a horizontal orifice at its bottom. A piece of plastic micro tube of 50 cm long was fitted to the orifice. At the end of the micro tube an emitter was attached. The emitter was calibrated for its discharge. In order to get higher rate of discharge a capillary tube was fitted to the end of micro tube and discharge of capillary tube was also calibrated. The emitter / capillary tube was placed on top of the sand tube, gravel tube and sand & gravel tube and suitably kept in position. Constant head was maintained in the overhead bottle. Water was allowed to fall on the top of micro tube through the emitter. Supply of water through emitter was considered as the irrigation ON- TIME. The supply was disconnected for certain period of time by removing the micro tube / capillary tube, which is irrigation OFF-TIME.

Experimental procedure :Transparent sheet was fixed on the flexy glass plates of the soil tank model lying in front of the sand tube for demarcating the advance of waterfront. The emitter / capillary tube was placed on the sand tube and other two tubes. Water was supplied from the bottle under constant head through the emitter to the sand tube. To start with, soil in the tank was irrigated at a rate of 5 lph through the sand tube continuously for one hour. The maximum waterfront advances both in horizontal and vertical directions were recorded at the end of one hour. The volume of water supplied during that period was also recorded. The continuous drip irrigation was followed by surge drip irrigation for one hour & forty minutes, keeping the total volume of water applied for irrigation same in both cases. In the surge drip irrigation, the irrigation ON- TIME was kept 20 minutes followed by irrigation OFF TIME of 20 minutes. In similar fashion the waterfront advances were recorded in case of gravel tube and sand & gravel tube under continuous and surge drip irrigation methods.This set of experiment constituted six experiments; two with each tube *i.e.* sand tube, gravel tube and sand & gravel tube.

RESULTS AND DISCUSSIONS

Sand tube :Maximum water front advance in soil was determined with a 5 lph emitter placed on the sand tube installed in the soil tank was taken. In continuous drip irrigation 1.67 liters of water was applied during each 20 minute intervals. Total period of irrigation was 60 minutes and total water applied during irrigation was 5.00 liters. With irrigation of 20 minutes, the horizontal water front advance and the vertical water front advance were observed to be 11.5 cm and 14.3 cm, respectively. With irrigation of 40 minutes, the horizontal water front advance and the vertical water front advance were observed to be 15.7 cm and 22.7 cm, respectively (Figure 3, 5). With irrigation of 60 minutes, the horizontal water front advance and the vertical water front advance were observed to be 20.4 cm and 25.0 cm, respectively (Figure 4, 6).

Fig. 1: Experimental set-up

Fig. 2: Experimental set-up

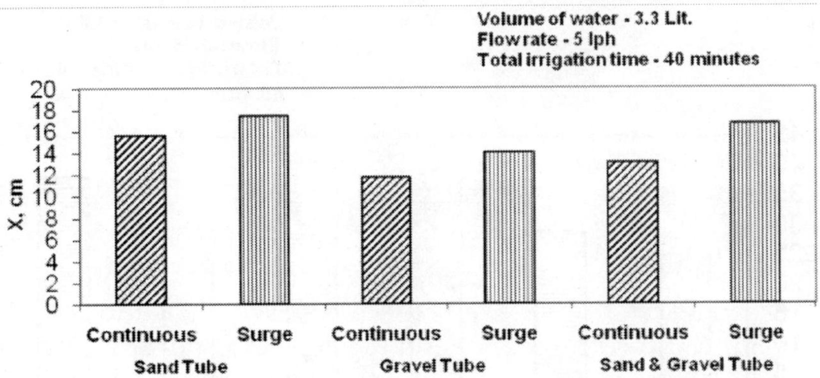

Fig. 3: Maximum horizontal water front advance (X)

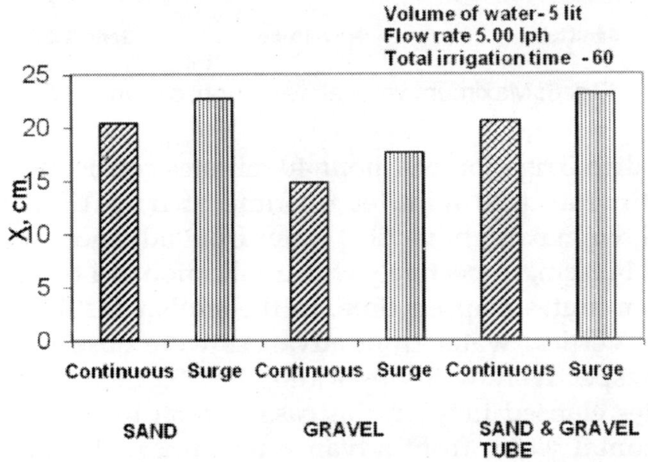

Fig. 4: Maximum horizontal water front advance (X)

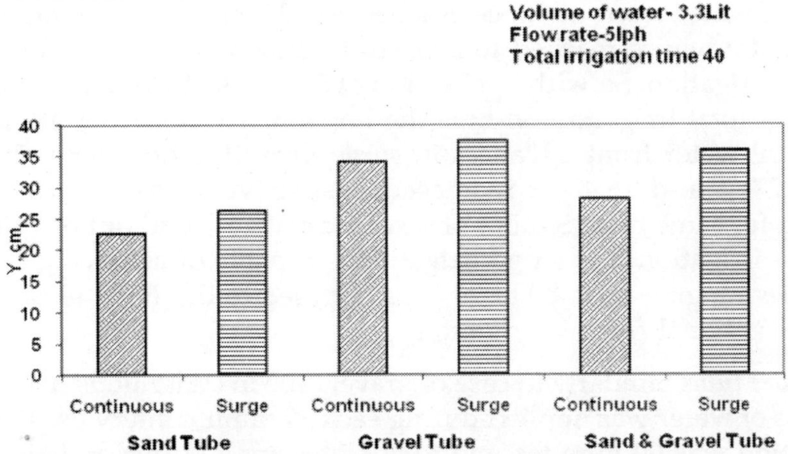

Fig. 5: Maximum vertical water front advance (Y)

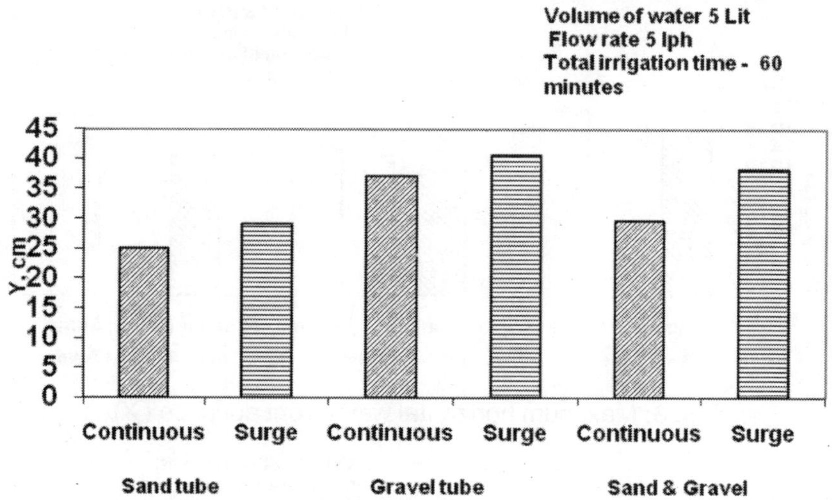

Fig. 6: Maximum vertical water front advance (Y)

For surge drip irrigation of 1 hour 40 minutes consisting of a cycle of 20 minutes on- time and 20 minutes off-time of irrigation, the maximum horizontal and the maximum vertical water front advances were observed to be 11.5 cm and 14.3 cm, respectively with application of 1.67 liters of irrigation water after 20 minutes elapsed time. In the similar fashion, the maximum horizontal and vertical water front advances were observed to be 17.5 cm and 26.3 cm, respectively with application of 3.33 liters of irrigation water after 60 minutes elapsed time. In this case of total irrigation on- time of 40 minutes horizontal water front advance resulted in 11.46 % increase and vertical water front advance resulted in 15.85 % increase over continuous drip irrigation. Then after 100 minutes of elapsed time, the maximum horizontal water front advance reached to 22.7 cm and maximum vertical water front advance reached to 29.1 cm in case of application of 5 liters of water for irrigation. So with application of 5.0 liters of irrigation water for 60 minutes of total irrigation on-time the horizontal water front advances and the vertical water front advances in surge drip irrigation were observed to have 11.27 % and 16.40 % increased, respectively over continuous drip irrigation for same time. Similar observations were found out by (New, 1995) in furrow irrigation that surge irrigation is capable of achieving application efficiencies ranging from 80 to 90 % as opposed to the 10 to 40 % associated with conventional furrow irrigation.

Gravel tube: Similarly in case of gravel tube in continuous drip irrigation 1.67 liters of water was applied during each 20 minute intervals. Total period of irrigation was 60 minutes and total water applied during irrigation was 5 liters. With irrigation of 20 minutes, the horizontal water front advance and the vertical water front advance were observed to be 8.3 cm and 26.4 cm, respectively. With irrigation of 40 minutes, the horizontal water front advance and the vertical water front advance were observed to be 11.8 cm and 34.1 cm, respectively. With irrigation of 60 minutes, the horizontal water front advance and the vertical water front advance were observed to be 14.9 cm and 37.1 cm, respectively.

For surge drip irrigation of 1 hour 40 minutes consisting of a cycle of 20 minutes on- time and 20 minutes off-time of irrigation, the maximum horizontal and the maximum vertical water front advances were observed to be 8.3 cm and 26.4 cm, respectively with application of 1.67 liters of irrigation water after 20 minutes elapsed time. In the similar fashion, the maximum horizontal and vertical water front advances were observed to be 14.0 cm and 37.5 cm, respectively with application of 3.33 liters of irrigation water after 60 minutes. It was also observed that after 100 minutes of elapsed time, the maximum horizontal water front advance reached to 17.6 cm and maximum vertical water front advance reached to 40.6 cm in case of 5 liters of irrigation water application.

In this tube, the horizontal and vertical water front advances in surge drip irrigation were observed to be 18.60 % and 9.97 % increase, respectively over continuous drip irrigation for total irrigation on-time of 40 minutes. Whereas in case of total irrigation on-time of 60 minutes with water application rate 5.0 lph,both the horizontal and vertical water front advances were observed to be 18.12 % and 9.4 % more in surge drip irrigation over continuous drip irrigation, respectively. This type of comparison made by Levin *et al.* (1979) between the results of laboratory and field experiments regarding prediction of soil moisture distribution from a point source on the one hand, and those theoretically calculated by a computer simulation model (Bresler,1975) on the other hand. It was found that when the same total amount of water (12 l) was added to the soil with 2, 4, or 8 lph continuously applied trickle source, a loss of 26 % of the total amount applied was affected below 60cm depth after 12 hours. Under identical conditions, 1 lph continuously applied and 2 lph surge irrigation resulted in 12 % loss.

Sand and gravel tube : Maximum water front advance in soil was determined with a 5 lph emitter placed on the sand & gravel tube installed in the soil tank was taken. In continuous drip irrigation 1.67 l of water was applied during each 20 minute intervals. Total period of irrigation was 60 minutes and total water applied during irrigation was 5.00 l. With irrigation of 20 minutes, the horizontal water front advance and the vertical water front advance were observed to be 9.5 cm and 23.1 cm, respectively. With irrigation of 40 minutes, the horizontal water front advance and the vertical water front advance were observed to be 13.1 cm and 28.1 cm, respectively

(Figure 3, 5). With irrigation of 60 minutes, the horizontal water front advance and the vertical water front advance were observed to be 20.6 cm and 29.7 cm, respectively (Figure 4, 6).

For surge drip irrigation of 1 hour 40 minutes consisting of a cycle of 20 minutes on- time and 20 minutes off-time of irrigation, the maximum horizontal and the maximum vertical water front advances were observed to be 9.5 cm and 23.1 cm, respectively with application of 1.67 l of irrigation water after 20 minutes elapsed time. In the similar fashion, the maximum horizontal and vertical water front advances were observed to be 16.7cm and 35.9 cm, respectively with application of 3.33 liters of irrigation water after 60 minutes. It was also observed that after 100 minutes of elapsed time, the maximum horizontal water front advance reached to 23.2 cm and maximum vertical water front advance reached to 38.4cm in case of 5 liters of irrigation water application.

In sand and gravel tube, in similar condition *i.e.*, for total irrigation on-time of 40 minutes, respective increase of 12.70 % and 27.75 % in horizontal and vertical water front advances in surge drip irrigation were observed over continuous drip irrigation. Again for total irrigation on-time of 60 minutes, the horizontal and the vertical water front advances were observed to increase by 12.62 % and 29.29 %, respectively over continuous drip irrigation. Similar comparison made by Shock *et al.* (1993) in furrow irrigation in view of two parameters *i.e.*, infiltration and runoff. In this trial, both the conventional furrow irrigation and surge irrigation were applied in Treasure Spring Wheat crop on 12 one-half-acre plots. Both systems were operated simultaneously five times during the season. Conventional irrigation applied 24.7 ac-in of water with runoff of 5.6 ac-in and infiltration of 19.1 ac-in whereas surge irrigation applied 12.0 ac-in with 1.7 ac-in of runoff and 10.3 ac-in of infiltration.

CONCLUSIONS

The experiments conducted with surge drip irrigation lead to a distinct conclusion that both the horizontal water front advance and vertical water front advance is higher than those with continuous drip irrigation. Again from surge drip irrigation conducted in different tubes, it was observed that with application of 3.33 l of water at the rate of 5 lph for irrigation on-time of 40 minutes, the horizontal water front advance of sand tube and sand & gravel tube was observed to have 25 % and 19.28 % of increase, respectively over gravel tube. With same volume of water and application rate for 40 minutes the surge drip irrigation through gravel tube and sand & gravel tube was observed to have 42.58 % and 36.50 % of increase in vertical water front advances, respectively compared to sand tube. With application of 5 liters of water at the rate of 5 lph during irrigation on-time of 60 minutes the surge drip irrigation through sand & gravel tube and sand tube resulted in 31.80 % and 17.60 % increase in horizontal water front advances compared to surge drip irrigation with gravel tube. With same volume of water applied at the same application rate for same irrigation on-time of 60 minutes the surge drip irrigation through gravel tube and sand & gravel tube resulted in 39.51 % and 31.95 % increase in vertical water front advance compared to that through sand tube. Therefore in rainfed area where water scarcity is a major problem for agriculture, surge drip irrigation can be employed in farmer's field to meet the crop water requirement with low availability of water.

REFERENCES

Bresler, E.1975. Two dimensional transport of solutes during non steady infiltration from a trickle source. *Soil Sci. Soc.Am.Proc.*,39: 604-613.

Karmelli, D. and Peri, G. 1974. Basic principles of pulse irrigation, ASCE, *Proceedings of the Irrigation and Drainage Division Journal*, 100:309-319

Levin,I.; Van Rooyen, P.C. and Van Rooyen, F.C.1979.The effect of discharge rate and intermittent water application by point-source irrigation on the soil moisture distribution pattern. *Soil Sci.Soc.Am.Proc.*,43:8-16.

Meshkat, M., Warner,R.C. and Workman, S.R.1998. Comparison of water and temperature distribution profiles under sand tube irrigation, *Trans. ASCE*, 41(6):1657-1664

New, Leon.1995. Opportunities to Maximize the Utilization of Waters by Irrigators, Texas Agricultural Extension Service.

Shock, C.C., Saunders, L.D., English, M.J., Mittelstadt, R.W. and Shock, B.M. 1993. Surge irrigation of wheat, to increase irrigation efficiency and reduce sediment loss, OSU, Malheur Experiment Station Special Report, 936: 157-161.

□□□

Chapter 41

Constraints in Adoption of Technologies for Vegetable Production in Jharkhand

K. KUMARI

INTRODUCTION

Jharkhand is predominantly inhabited by tribal people. The plateau state endowed with varying climatic conditions is suitable for growing large number of horticultural crops. Yet potential has remained unexploited. The state has much potential for increasing the vegetable production and in the prevailing agro -climatic condition vegetables mature earlier than the plain areas. Increasing average productivity and profitability of vegetables is urgent need. The technology related to plant protection is critical issues to develop the vegetable sector as it experiences major technology gap. Chitnis and Bhilegaonkar (1985) in their study on technological gap in dry farming system reported that major technological gap was found in plant protection measures 73.32 %. A considerable technological gap was observed in inter-cropping, crop management in aberrant weather, use of fertilizer and sowing technique. Least technological gap was observed in varietal recommendation (8.70 %). Vegetable agro ecosystems comprise a dynamic mix of biotic (living) components, abiotic (non-living) components and the interactions within and between them. Any change in one component or interaction is likely to affect another component or interaction Atteh (1984). The reactions of agro-ecosystems to change can potentially be disastrous to farming communities, and may not be entirely predictable. In order to achieve good vegetable pest management, the vital components of agro ecosystems should be correctly identified, their interrelationships fully understood, and the agro ecosystem treated as a functional unit (Bentley and Andrews 1991).

There arise many problems, when a technology is put into field condition particularly in the new areas. So, the technology should not be introduced at

Krishi Vigyan Kendra, Saraikela-Kharsawan, Birsa Agricultural University, Ranchi once rather it should be assessed and tailored to farmers' need and specific to the situational demand. The constraints faced by farmers in applying technology at field level should be taken into consideration before formulation of the dissemination programme (FAO, 1988) and then only the technology becomes adaptable and appropriate to situational demand. Keeping in view the study was carried out to present how farmer's perceptions on different attributes of technology affects its applicability at farmer's field with the objectives to find out the technological gap and constraints in vegetable production, and to find out correlation between the selected socio-personal factors and technological gap in vegetable production.

MATERIALS AND METHODS

The study was conducted in purposively selected state of Jharkhand. Three agro-climatic zones of the state *i.e.,* North Eastern Central Plateau (zone-IV), Western plateau (zone-V), and South-eastern Plateau (zone-VI) were taken into account. To represent of agroclimatic zone in totality three districts one from each agro-climatic and three blocks namely Angara, Senha and Ghatsila were selected randomly from each selected district *i.e.,* Ranchi, Lohardaga and East Singhbhum, respectively. Two categories of respondents *i.e.,* respondents belonging to progressive village (progressive cultivator) and less progressive village(less progressive cultivator) constituted the sample. Fifty (50) farm families from each of the six villages under study were selected randomly. Thus, the whole sample constituted of 300 respondents (150 progressive cultivators and 150 less progressive cultivators)

Technology Gap was operationalized as the proportion of recommended technology for vegetable cultivation not adopted by the vegetable growers. Six component as recommended by scientists regarding vegetable production technology *e.g.,* (1) Nursery management, (2) Use of variety, (3) Sowing/ planting, (4) Fertilizer management, (5) Plant protection, and (6) Post harvest operations were identified and considered to compute technology gap in improved method of vegetable production technology. To measure the technological constraints respondent's responses were derived on the technical constraints regarding its availability and its attributable constraints. The data were collected by personally interviewing the respondents through a structured scheduled. Apart from the use of schedule detailed information were collected through informal discussion with the respondents and by critically scrutinizing the practices followed for vegetable production. Use of PRA tool, field observation and non-participant observation techniques were thoroughly used. Field observation was done at vegetable field of progressive village and less progressive village."Focused Group Discussion" of PRA was also followed to gather data on the constraints faced in adoption of technologies. After collection, the data were systematically arranged and tabulated for analysis and interpretation. The statistical techniques used for analysis of data under the study included frequency, percentage, mean, chi-square and t-test.

RESULTS AND DISCUSSION

Technology gap Index in adoption of vegetable production technologies : Maximum technological gap was found with respect to post harvest technology of vegetable; followed by plant protection measure, use of proper varieties, nursery management and fertilizer management with the scores 90.15, 72.09, 59.69, 58.48, and 57.70, respectively (Table 1 and Figure 1). In progressive villages and less progressive villages technology gap was maximum with respect to post harvest technology whereas the least gap was noticed in fertilizer management. It was due to unawareness of farmers to value addition of vegetables and also due to poorly defined extension strategy that maximum technological gap index were obtained in post harvest operations. The computed row square value was found to be non significant (Rho value = 0.46) which signifies that there was significant differences between the technological gap index calculated for the progressive village and less progressive village.

Table 1: Technology gap index in adoption of vegetable production

S. No.	Component	Technology gap index (%)					
		Progressive (N=150)	Rank	Less Progressive (N=150)	Rank	Pooled (N = 300)	
1.	Nursery management	42.20	V	74.76	IV	58.48	IV
2.	Use of variety	57.08	II	62.28	V	59.69	III
3.	Sowing/ Planting	42.82	IV	47.03	VI	44.92	VI
4.	Fertilizer management	31.35	VI	84.06	III	57.70	V
5.	Plant protection measures	56.51	III	89.32	II	72.09	II
6.	Post-harvest operations.	82.02	I	98.28	I	90.15	I

Rho value = 0.46 (Non- significant)

Correlation of the socio-personal factors and technological gap in vegetable production technology : It was learned through various references that there is some relations among the technological gap perceived and the socio-economic variables. An effort has been taken to find out the correlation among the technology gap and socio-economic variables of the respondents and is presented in the Table 2. As it revealed in table that there is negative and significant relation between education level of the respondents, their level of participation to the extension activity, family income they incurred and risk taking ability of the respondents. The family income, economic motivation and cosmopolitanisms of the respondents were found to be highly significant and negatively related to the technology gap in the application of improved cultivation practices of vegetable production. R^2 calculated was 0.872 *i.e.,* the variability taken for was to the extent of 87 %.

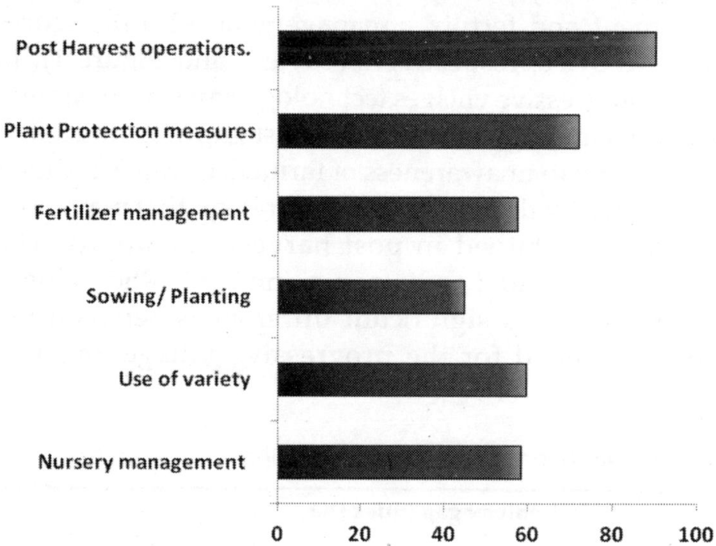

Fig.1: Technology gap index in adoption of vegetable production technologies

Table 2: Correlation of the socio-personal factors and technological gap

Variable	Correlation Coefficient
Caste	0.07310
Education	0.8439*
Size of holding	0.02168*
Occupation	0.0397
Farm power	0.1057
Extension participation	-0.01132*
Material possession	0.1198*
Family type	0.3792
House type	0.0251
Family income	0.03977**
Cosmopolitanisms	0.7385**
Innovation proneness	0.8500
Risk taking ability	0.3406*
Economic motivation	0.1166**

Maximum no (98.67) respondent vegetable growers opined that use of disease resistant variety is simple to practice although it is not available in local market and the soil sterilization fetch very low percentage (14.50). Regarding seed treatment with bavestine 64.67 % of vegetable growers opined that it is flexible to practice although it is not available in local market and application of boron @ 8 kg ha^{-1} along with 1.5 kg molybdenum in soil during land preparation application fetches very low percentage (13.34) regarding its flexibility to use (Table 3). Regarding increasing spacing for late crop 49.34 % of vegetable growers opined that it is compatible to house hold resources although application of boron @ 8 kg ha^{-1} along with 1.5 kg molybdenum in soil during land preparation observed very low percentage (13.34) regarding it. About 94.64 % respondent vegetable growers opined that application of born @ 8 kg ha^{-1} along with 1.5 kg molybdenum in soil during land preparation compatible to existing farming system although application of 6 kg bleaching powder + 12 kg ha^{-1} lindane dust before 15 days of transplanting fetch very low percentage (17.34) regarding it. Almost all respondents opined that use of disease resident variety is economically viable. Seed treatment although had technological gap but still fetch 94 % regarding its visible benefits. Regarding input availability use of disease resident variety had least percent of respondents and regarding risk involvement maximum respondents (60%) opined that application of boron @ 8 kg ha^{-1} along with 1.5 kg molybdenum in soil during land preparation involves risk.

Table 3: Technological constraints in integrated pest management practices of vegetable production

Technology	Application of boron @ 8 kg ha^{-1} along with 1.5 kg. molybdenum in soil	Use of disease resident variety	Use of season specific variety	Soil sterilization for nursery	Use of shade net in nursery	Increasing spacing for late crop	Seed treatment withbavistin @.5mg kg^{-1} of seed	Application of 6 kg. Bleaching powder + 12 kg lindane dust
Simple to practice	49.32	98.6	41.34	14.50	22.71	54.65	94.67	73.34
Flexible to use	13.34	58.64	29.34	16	24	31.28	64.67	43.34
Divisible in application	53.34	94.64	32.00	22.71	21.41	14.70	94.67	37.34
Compatible to Household resource	13.34	21.32	14.5	16	18.64	49.34	42.67	16
Compatible to farming system	94.67	93.34	17.34	22.71	14.70	64.67	54.64	17.34
Viable Economically	72	100	74.67	37.40	36	73.34	41.34	43.34
Benefit visible	41.34	56	37.34	12	12	54.64	94	41.34
Input available	6.67	8	9.4	42.64	29.34	90.67	17.34	14.67
Risk involvement minimum	60	18.67	42.60	12	28	32.67	32.64	28

From the above discussion it can say that the first user of the technology have their own criteria to use the technology. Full use of technology largely depends upon its compatibility to the existing farming systems as well as to the household resources. Availability of required input and visible benefits affects the use of technology. The technologies like application of boron @ 8 kg ha^{-1} along with 1.5 kg molybdenum in soil during land preparation although simple in practice but as per respondents perception it involves risk so it was not favored by the respondents. Similar findings were reported by different authors. The technology and practices under different sectors may be technically quite sound but could not be adopted by the clientele unless the generally observed deficiencies such as lack of material inputs, financial constraints, inadequate supply and service facilities *etc.* were met with. These were especially true in case of economically and socially backward people who were the majority of the targeted population.

CONCLUSIONS

In designing extension strategy regarding transfer of integrated pest management practices to enhance the profitability of vegetable production the findings of this study will helps scientists and extension functionaries finalizing suitable and appropriate technology option according to farming situations and socio-economic conditions of the farmers so that basket of choices could be provided to different categories of farmers. Integrated Pest Management (IPM) concept shows a remarkable change in its technical contents to consideration of agro-ecological concern. In order to give IPM a new and more practical and realistic meaning specially when dealing with farmers of and complex diverse risk prone system as in North East states the focus should be on IPM as a management practice and much less as a technology.

REFERENCES

Atteh, O. D. 1984. "Nigerian farmers' perception of pests and pesticides." *Insect Science and its Application*, 5:213 - 220.

Bentley, J. W. and Andrews, K. L. 1991. Pests, peasants, and publications: anthropological and entomological views of an integrated pest management program for small-scale Honduran farmers. *Human organization*, 50:113 - 124.

Chitnis, D. H. and Bhilegaonkar, M. G. 1985. Technological gap in dry farming system: An analysis. *Maharashtra J. Ext. Edu.*, 4: 73-76.

FAO. 1988. World Agriculture towards 2000. Belhave Press, FAO, Rome.

□□□

Chapter 42

Resource Conservation and Food Security through Soil and Water Conservation Practices

DEV NARAYAN, H. BISWAS, S. P. TIWARI AND P.P. ADHIKARY

INTRODUCTION

Soil erosion is one of the most serious environmental problems in the world today, because it seriously threatens agriculture and the natural environment. Adequate food supplies depend on productive land. At a time when agricultural efforts are focused on increasing food production, soil degradation worldwide is increasing. The dimensions of land degradation are alarming, it affects from 30 to 50% of the earth's land surface. Over one thousand million humans today are malnourished - more than ever before in the history of human society. Today 5.5 thousand million humans exist on earth, and an additional 250000 humans are born each day and must be fed. More than 97% of the world's food comes from the land rather than the oceans and other aquatic systems, with an ever-increasing portion being derived from steep lands in tropical climates. In India water erosion is prevalent in almost all parts of the country with varying degree of severity. It is estimated that out of 120.72 M hectares degraded area in the country, about 68% is exclusively affected by water erosion and the annual production loss in major rainfed crops due to water erosion in the country has been assessed as 15.7% of total production of cereals, oilseeds and pulses which is equivalent to ₹ 162 billion considering minimum support price of 2008-09 (Anonymous, 2010). Therefore, the control of soil erosion for a sustainable agriculture and environment is essential for all programs of food security and environmental conservation. The Bundelkhand region with a geographical area of 7.04 M ha in Central India is characterized with undulating terrain, scare vegetation cover, hostile climate, lack of irrigation facilities and unfavourable edaphic conditions. About 70% of the total area in the region has been affected by

Central Soil and Water Conservation Research and Training Institute,
Research Centre, Datia-475 661, Madhya Pradesh, India

varying degree of erosion hazards (Tiwari and Narayan, 2010). The extensive wastelands in the form of ravine tracts, rocky wastelands in the form of ravine tracts, rocky wastes and scrub land are the results of continued soil degradation and biotic interference. Red soils occupy more than 50% area in the region and are shallow, low water retentive, prone to water erosion and fail to support crops on residual soil moisture. About 53% area comes under rainfed agriculture and crop intensity is about 111% in the region. The yield potential of crops is low and agriculture is not economically viable until scientific management of land and water is taken care of to have a self sustainable agriculture. The surface and ground water availability for drinking, agriculture and industrial purposes is very limited because of less numbers of perennial flow rivers and stony underground strata. This paper highlights the effectiveness of soil and water conservation practices in resource conservation and increasing the food production in red soils in Central India.

MATERIALS AND METHODS

The Bundelkhand region with a geographical area of 7.04 M ha is situated at 23^0 10′ - 26^0 30′ North latitude and 78^0 20′ - 81^0 40′ East longitude in the semi-arid tract of central plateau of India. This region receives about 80% annual rainfall during south-west monsoon. Annual rainfall ranges from 700 to 1150 mm with a mean value of 853 mm. The potential evapo -transpiration is quite high in the range of 1400 to 1700 mm with moisture index value of -40 to -50. The pattern of the rainfall is highly erratic and more than 90 % of the total rainfall occurs within 10 weeks between July to mid September accompanied by intermittent long dry spells. The entire rainfall is received in less than 50 rainy days. Winter showers are rare and uncertain. The frequent drought occurs in entire region which cause crop failure and low crop yield on account of considerable soil moisture deficit. Usually, monsoon commences by the last week of June but sometimes delayed to the first week of July. The active monsoon often withdraws up to the mid of September or end of the August. The mean annual temperature of the region is generally high with high degree of variation between maximum and minimum temperatures. Sometimes maximum temperature in the summer months of May and June touches to 48^0C which is the peak of summer season. Thus, high temperature accompanied with hot wind causes high potential evapo-transpiration. The minimum temperature goes to as low as -1^{0}C during winter months of December - January. Soils exhibit greater heterogeneity with regard to texture, depth and fertility status, which leads to operational as well as management problems. Red soils are shallow, prone to water erosion, low fertile, low water retentive and fail to support crop on residual moisture, on the other hand, black soils are poorly drained and poses workability problems. Deep ravenous terrain is undulating with complex slopes which lead to increase runoff and low moisture storage in soil profile. The results of some of the field experiments pertaining to resource conservation for obtaining sustainable crop production through soil and water conservation practices conducted at the Central Soil and Water Conservation Research and Training Institute, Research Centre, Datia (25^0 40′ N, 78^0 28′ E, and 342.42 m above mean sea level), Madhya Pradesh have been presented and discussed below.

RESUTS AND DISCUSSION

***Contour cultivation and live barrier*:** Method of cultivation plays an important role in resource conservation and increasing crop yield. In contour cultivation, the ridges and rows of the plants placed across the slope form a continual series of miniature barrier to the water moving over the soil and help in reducing runoff, soil loss and conserving higher soil moisture. The live vegetative barriers of different grass species (*Cenchrus ciliaris, Heteropogan contortus* and *Vetiveria zizaniodes*) were found effective in reducing runoff and soil loss from sloppy lands. Vegetative barrier of *Vetiveria zizanioides* planted on sloppy agricultural field in sorghum at 11 m horizontal spacing was found effective in reduction of runoff and soil loss to the extent of 29.6% and 4.5 t ha^{-1}, as against of 45.2% and 10.9 t ha^{-1} runoff and soil loss, respectively, from cultivated fallow plots. Vegetative barriers slow down and disperse surface runoff causing deposition of eroded sediments while allowing some runoff to pass through (Dabney *et al.*, 1999).

***Green manuring for resource conservation and increasing productivity*:** Growing of sunnhemp during rainy season for green manuring not only reduces erosion losses but also improves soil fertility and yield of succeeding *rabi* season crops over traditional practice of keeping land fallow during rainy season. The sunnhemp crop reduced 38% runoff and about 4 times soil loss over traditional practice at 2 % slope in red soils. The results of a field experiment conducted during 1994-95 to 1998-99 in red soils to evaluate the effect of green manuring on soil fertility and yield of wheat indicated that green manuring of sunnhemp added 18.4 t ha^{-1} green bio-mass, 45.4 kg nitrogen, 3.5 kg phosphorus and 51.9 kg potassium ha^{-1} on incorporation into the soil (Table 1).

Table 1: Biomass and nutrients addition by green manuring of sunnhemp in red soils

Parameter	Quantity
Bio-mass addition (t ha^{-1})	
Green bio-mass	18.4
Dry matter yield	4.7
Nutrient addition (kg ha^{-1})	
Nitrogen	45.4
Phosphorus	3.5
Potassium	51.9

The grain and straw yield and water use efficiency of succeeding wheat crop increased with green manuring of sunnhemp (Table 2). The grain and straw yield of wheat increased by 35 and 30 %, respectively, with green manuring over traditional practice (no green manuring). The water use efficiency of wheat also increased by 30 % under green manuring over no green manuring (control).

Table 2: Yield and water use efficiency of wheat as influenced by green manuring of sunnhemp

Parameter	Green manuring	
	Without green manuring	With green manuring
Yield (kg ha^{-1})		
Grain	1477	1991
Straw	2826	3676
Water use efficiency (kg ha^{-1} mm)	5.22	6.76

Further the economic analysis revealed that green manuring of sunnhemp was beneficial over traditional practice of allowing fields fallow during *kharif* season. An additional net return of ₹ 5854 per hectare was recorded with green manuring of sunnhemp over traditional practice (control) besides improving the soil physical (improvement in bulk density, porosity, water stable aggregates and infiltration) and chemical properties (increase in organic carbon, nitrogen, phosphorus and potassium) of the soil. Saha *et al.* (2000) have also reported beneficial effect of green manuring on growth and yield of succeeding wheat.

Tillage and surface mulching: The red soils are found mostly on higher elevation in topo-sequences hence major part of rainfall is lost as surface runoff. Rain water cannot enter in to deeper soil layer because of presence of hard and compact layer in sub-soil or due to crusting and hard setting behavior. As a result crop growth is restricted and runoff, soil loss, nutrient loss and growth of weeds are increased resulting in poor yields under rainfed conditions. Under present situation improved tillage has been viewed as a central management practice of loosening of impeded sub-soil layer coupled with use of soil surface covers such as crop residue, grass mulch and green manure crops for improvement of soil environment and water conservation. A runoff plot based study was conducted during 1998 to 2003 to evolve a suitable tillage and surface mulching practice for higher rain water conservation for augmenting the crop growth and yield of sorghum [*Sorghum bicolor* (L.) Monech.] in red soils under rainfed conditions. The results indicated that improved tillage (20 cm depth) and *in-situ* surface mulching of sunnhemp was helpful in higher rainwater conservation over traditional tillage (7 cm depth) practice. There were 18 % higher rainwater conservation with improved tillage which further increased to 34 % with *in-situ* surface mulching of sunnhemp over traditional tillage practice (Narayan and Tiwari, 2011). The grain yield of sorghum increased by 33 % with improved tillage and it further increased to 47 % with *in-situ* surface mulching of sunnhemp over control. The stover yield also increased by 26 and 36 % under improved tillage and improved tillage + *in-situ* surface mulching of sunnhemp, respectively, over traditional tillage. The highest yield in this treatment was attributed to maximum rainwater conservation, infiltration, pore space, water holding capacity and water stable aggregates and minimum bulk density. Further economic analysis revealed that improved tillage as well as *in-situ* surface mulching of sunnhemp was beneficial over traditional tillage in increasing profit (Table 3). An additional net return of ₹ 4,610 per hectare was recorded with improved tillage which increased to the tune of ₹ 7,437 per hectare with the introduction of *in-situ* surface mulching of sunnhemp over traditional tillage. In red soils (alfisols), improved tillage (20 cm) coupled with *in-situ* surface mulching of sunnhemp is helpful for enhancing rain water conservation as well as its retention and utilization for achieving higher yield of sorghum and profit over traditional tillage practice under rainfed conditions. Patil *et al.* (2004) also reported superiority of deep tillage over traditional tillage in resource conservation and increasing the yield of sorghum under rainfed conditions in semi-arid tropics.

Table 3: Impact of tillage and surface mulching on yield of sorghum and net additional returns

Treatment	Yield (kg ha^{-1})		Net additional returns over control (₹ ha^{-1})
	Grain	Stover	
Traditional tillage (control)	2249	7358	-
Improved tillage	2991	9294	4610
Improved tillage + *in situ* surface mulching	3310	9981	7437

Cover crops and inter crops: Mostly red soils occur on a rolling plateau and a significant portion of rainfall is lost as surface runoff along with considerable soil and nutrient losses. Under such situations, intercropping plays a significant role in enhancing productivity and profitability per unit area and time and through more efficient use of land, water and solar energy besides assuring insurance against crop failure due to failure of one or the other crop due to vagaries of weather or disease/pest epidemics in rainfed agriculture. Intercropping of castor + green gram was found suitable for reducing the erosion losses and providing sustainable production under rainfed conditions in red soils. Experience at the Centre (2002-2006) suggested that at 2% slope in red soils highest runoff loss of 61.2% was recorded under cultivated fallow followed by castor (47.1%). The groundnut and castor + green gram resulted in 31.4 and 37.1 % runoff, respectively (Table 4). The highest soil loss (8.63 t ha^{-1}) was recorded in cultivated fallow plot. Among *kharif* annuals the wide spaced crops *viz.,* castor and maize alone has resulted in 5.10 and 4.69 t ha^{-1} soil loss, respectively. Groundnut and castor + green gram allowed a soil loss of only 3.64 and 3.69 t ha^{-1} which was quite low and within permissible limits thus indicated the effectiveness of these crops for the region. Castor + green gram intercropping recorded highest green gram equivalent yield of 1034 kg ha^{-1} (Table 5). This intercropping was also more remunerative as it recorded the maximum net returns of Rs 20748 ha^{-1} indicating its suitability for *kharif* season under rainfed conditions over other crops.

Table 4: Runoff and soil loss as influenced by different treatments

Crop covers	Runoff (%)	Soil loss (t ha^{-1})
Cultivated fallow	61.2	8.63
Maize	41.7	4.69
Castor	47.1	5.10
Sesame	40.8	4.27
Castor + green gram	37.1	3.69
Clusterbean	39.4	3.96
Groundnut	31.4	3.64

Table 5: Yield and economics of various crops and castor + green gram intercropping system

Sole crop/intercropping system	Green gram equivalent yield (kg ha^{-1})	Net returns (₹ ha^{-1})
Maize	234	(-) 4249
Castor	513	5679
Sesame	902	17799
Castor + green gram (1:2)	1034	20748
Clusterbean	401	2343
Groundnut	647	5382

Contour bunds: A field size runoff plot based study was conducted during rainy season (*kharif*) of 2005 to 2007 in red soils to evaluate the efficiency of some of the mechanical and vegetative measures in reducing runoff, soil and nutrient losses and find out the best suited soil and water conservation measure for slopy agricultural lands. Results indicated that maximum rainwater conservation (87%), as well as reduction in soil (97%) and nutrient loss (organic carbon 96%, nitrogen 93%, phosphorus 89% and potassium 80%) besides considerable higher grain yield (531 kg ha^{-1}, 11% higher than control) of sorghum over farmer's practice was obtained under contour bund at 0.7 m vertical interval in red soils.

Rainwater harvesting for supplemental irrigation: More than 70% area in the region remains fallow during *kharif* season because of low water retention capacity of red soils located at higher elevation. Crops experience water stress even in short dry spells during rainy season. Due to short monsoon season, it is difficult to take a long duration *kharif* crop or even a low water requiring *rabi* oilseed crop without provision for supplemental irrigation as the rainy season generally recedes by middle of September in the region. There is an immense scope for water harvesting and recycling in red soils as it often occur on rolling plateau and significant portion of rainfall is lost as surface runoff. Though the rainfall pattern in the region is very erratic but high intensity showers generate sizeable runoff for storage, which may subsequently be used for recycling it at critical stages of crop growth. The surface runoff generated from sloppy lands during rainy season was harvested and stored into a dug out pond and recycled for providing supplemental or life saving irrigation to selective crops (soybean, toria and mustard) at critical stages of growth for stabilizing and augmenting the yields. The results of 4 years (2002-03 to 2005-06) indicated that the yield of soybean increased by 40% by providing one supplemental irrigation given through harvested rain water at pod filling stage (Narayan *et al.*, 2011). By providing one supplemental irrigation at branching stage (30 days after sowing) the yield of toria increased by 180%. The yield of mustard increased by 130 and 400% with pre-sowing and pre-sowing + one supplemental irrigation given at branching stage, respectively. Soybean and toria with one supplemental irrigation at pod filling and branching stage recorded an additional net return of ₹ 4703 and 9652 ha^{-1}, respectively, over their rainfed crop. Mustard with pre-sowing irrigation recorded an additional net return of Rs 6674 ha^{-1} which further increased to Rs 22623 ha^{-1} with an additional irrigation at branching stage over rainfed crop. Sharda and Ojasvi (2005) also reported potentiality of rainwater harvesting and recycling in increasing the yield of rainfed crops in various parts of the country.

CONCLUSIONS

The results of above studies suggested that soil and water conservation practices *viz.*, contour cultivation, live barriers, cover crops, intercrops, green manuring crops, improved tillage and *in-situ* surface mulching, contour bunds and rainwater harvesting for supplemental irrigation were found effective in resource conservation and enhancing agricultural production on sustainable basis on sloppy agricultural lands under rainfed conditions.

REFERENCES

Anonymous 2010. Annual Report 2010-11, Central Soil and Water Conservation Research and Training Institute, 218,Kaulagarh Road, Dehradun,India.1p.

Dabney, S. M., Liu, Z., Lane, M., Douglas, J, Zhu, J. And Flanagan, D. C. 1999. Landscape benching from tillage erosion between grass hedges. *Soil and Tillage Research,* 51: 219-231.

Narayan, D., Biswas, H. and Tiwari, S. P. 2011. Rainwater harvesting and utilization techniques for red soils of semi-arid region. Tech. Brochure, CSWCRTI, Research Centre, Datia, Maa Sheetla Printers, Datia. pp1-8.

Narayan, D. and Tiwari, S. P. 2011. Improved tillage and mulch practices for resource conservation and higher yield of sorghum in red soils. Tech. Brochure, CSWCRTI, Research Centre, Datia, Maa Sheetla Printers, Datia. pp1-8.

Patil, S.L., Sheelvantar, M. N., Lamani, V. K. and Reddy, R. 2004. Nutrient availability and grain yield of winter sorghum (*Sorghum bicolour* (L.) Moench) as influenced by tillage practices and integrated nutrient management in Vertisols of semi-arid tropics of India. *Acta Agronomica Hungarica,* 52 (2): 173-181.

Saha, R. R., Rai, R. K. and Mukherjee, P. K. 2000. Effect of green manure dhaincha (*Sesbania aculeata*) and phosphorus on growth, yield and phosphorus uptake by wheat (*Triticum aestivum*), *Indian Journal of Agronomy,* 45: 707-710.

Sharda, V.N. and Ojasvi, P.R. 2005. Water harvesting through watershed management in different agro-ecological regions in India. *Indian J. Agric. Sci.,* 75(12): 771-780.

Tiwari, S. P. and Narayan, D. 2010. Soil and water conservation measures for Bundelkhand region.*In* Extension strategy for Bundelkhand region. Zonal Project Directorate, Zone IV (ICAR), Kanpur, pp48-57.

□□□

Chapter 43

Evaluation of Sprinkler and Furrow Irrigation Methods for Summer Groundnut

N. SAHOO AND R.C.DASH

INTRODUCTION

Entire harnessable surface and ground water of our country can bring only about 50% of total cultivable area under irrigation. The only way to increase irrigated area is to go for efficient use of irrigation water. Sprinkler irrigation method is reported to be about 20% more efficient than conventional irrigation method (Sivanappan, 1987). In some crops water saving of 30-50 % has been achieved by using sprinkler irrigation system. Sprinkler irrigation is more popular in closely spaced crops like groundnut, millet and sugarcane. The factors like land leveling of undulated topography, labour scarcity, reduction of water loss due to percolation, conveyance loss in the distribution system and increased water use efficiency have convinced farmers to go for sprinkler irrigation method. The specific objective of the study was to compare different parameters of sprinkler irrigation with that of conventional furrow irrigation for summer groundnut crop.

MATERIALS AND METHODS

Experiment was carried out at OUAT Central Research Station, Bhubaneswar in a sandy soil. The soil has 90.6 % sand, 2.0% silt, 7.4% clay with bulk density of 1.59 gm cc^{-1} and pH of 5.52. Two plots, each measuring 60 m x 60 m were taken where sprinkler and conventional furrow irrigation methods were practiced to grow summer groundnut (variety *'TG-22'*). Each plot was again divided into three subplots, in which, different depths of water were applied. Volume of water supplied to each subplot was measured and recorded yields from these plots were recorded.

Odisha University of Agriculture and Technology, Bhubaneswar – 751 003 Odisha

Determination of ET_0

Daily reference crop evapotranspiration (ET_0) was computed by taking climatological data in to consideration by using modified Penman method. The crop coefficient curve of summer groundnut was developed following the procedures outlined in the guidelines for predicting crop water requirements (Doorenbus and Pruitt, 1977) to determine evapotranspiration of the crop for the growing period.

Determination of Crop Co-efficient

The effect of climate was taken in to consideration while computing ET_0. But the effect of crop characteristics was not incorporated; and hence, to study the effect of crop characteristics, the crop co-efficient was determined from a prepared crop coefficient curve by Giri, 1994. Crop stage period with crop co-efficient (K_c) for groundnut crop are given in Table 1.

Table 1: Duration of different growth stages and crop coefficient of groundnut (TG-22)

Crop variety	Duration in days			
	Initial stage	**Crop development stage**	**Mid season stage**	**Late season stage**
Groundnut				
TG-22	Dec 22 to Jan 15	Jan 16 to Feb 14	Feb 15 to Mar 31	April1 to April 19
Kc	0.26	0.26 - 1.0	1.0	0.58

According to Doorenbus and Pruitt (1977) ET_{crop} was calculated by using the following relation;

$$ET_{crop} = K_c \times ET_0 \quad \text{.... (i)}$$

Where

K_c= Crop co-efficient

ET_0 = Reference crop evapotranspiration, mm day^{-1}

Determination of Uniformity Coefficient

The uniformity co-efficient was computed from the results of sprinkler distribution test using Christiansen's method and was found to be 85 %.

Developmentof Water Production Function

Crop production functions were developed using the empirical relationship between crop yield and the various input parameters. The actual moisture used by the crop during the irrigation interval was gravimetrically determined by measuring the soil moisture up to a depth of 80 cm in the root zone for both the methods of irrigation. The data on depth of water supplied, amount of nitrogen supplied, crop yield and water use efficiency obtained from the experiment were presented in Table 2 for sprinkler and conventional furrow irrigation, respectively. Multivariate production functions were developed using these data.

Table 2: Input-output data for groundnut grown with different methods of irrigation

Types of Irrigation	Treatment	Input (water supplied) (mm)	Nitrogen applied (q ha^{-1})	Output (yield) (q ha^{-1})	WUE (kg ha-mm^{-1})
Sprinkler	S_1	502	20.0	27.30	5.43
	S_2	537	40.0	28.23	5.25
	S_3	572	60.0	28.67	5.02
Conventional	F_1	606	20.0	22.33	3.68
	F_2	653	40.0	24.00	3.66
	F_3	700	60.0	24.00	3.42

The general production function was expressed in the form of :

$$y_c = a_0 + a_1 W_c + a_2 N + a_3 W_c^n + a_4 N^n + a_5 W_c^n \quad \text{... (ii)}$$

Where y_c = Pod yield, q ha^{-1}

W_c = Irrigation water applied, mm

N = Nitrogen level, kg ha^{-1}

a_0 = Combine effect of all the fixed inputs

a_1, a_2, a_3, a_4 and a_5 are coefficients and n is the exponent

$$\text{Again, } Yc = f\,(W_c \times N/P) \quad \text{... (iii)}$$

Where P = Quantity of other fixed inputs, which denotes that only W_c and N are variable inputs.

Equation (ii) was fitted to data for Yc, Wc and N with exponent values of n =1, 2, 1.5 and 0.5 to obtain linear, quadratic, three-halves and square root production function. The data were also fitted to the Cobb-Douglas function or Power Function of the form:

$$Yc = aW_c bN^d \quad \text{... (iv)}$$

Where a, b and d are coefficients

Of all the combinations tried, a linear production function was found to be satisfactory and given in the following format.

$$Y_{gs} = a_c + a_1 W_{gs} \quad \text{... (v)}$$

Where Y_{gs} = Pod yield of groundnut in sprinkler method, q ha^{-1}

W_{gs} = Water applied to groundnut through sprinkler method, mm

$$Y_{gf} = a_c¢ + a_1¢W_{gf} \quad \text{... (vi)}$$

Where Y_{gf} = Pod yield of groundnut in furrow irrigation, q ha^{-1}

W_{gf} = Water applied to groundnut through furrow irrigation method, mm

It is evident from the findings of Sivarao (1987) that application of nitrogenous fertilizer did not influence the pod yield of the groundnut. Hence, the effect of nitrogen on the pod yield of groundnut was neglected and only the effect of irrigation water on the pod yield of groundnut was taken into consideration. Thus, the linear relationship with only water as a variable appears to be justified in a situation, in which, the crop has grown. In conventional furrow irrigation method, water was applied to the plot according to the local practices. Soil samples were taken after 24 hours of irrigation from the depths of 20, 40 and 60 cm to determine the soil moisture percentage by gravimetric method. Yield potential of groundnut was recorded by taking 10 samples from 1 m^2 area from both the plots. It is one of the parameters, which was used for computation of water use efficiency. Water use efficiency is the yield of marketable crop per unit depth of water used in evapotranspiration.

$$WUE = \frac{Y}{ET} \quad \text{... (vii)}$$

Where,

Y = Marketable yield of crop, kg ha^{-1}

ET = Evapotranspiration of the crop, mm day^{-1}

WUE = Water use efficiency, kg ha-mm^{-1}.

Field water use efficiency (E_u) is the ratio of the crop yield to the amount of water used in the field (WR), which includes (G + ET +D).

$$F_u = \frac{Y}{G + ET + D} = \frac{Y}{WR}$$

Where,

G = Water used for metabolic purpose of the crop, mm (negligible)

WR = Water used in the field, mm

D = Deep percolation, mm

In the study, the depth of water needed for proper growth of the crop was determined. The irrigation factor was taken into consideration keeping all other input parameters constant. Yields with respect to water requirement in both the methods were determined.

RESULTS AND DISCUSSION

Crop Evapotranspiration

The weekly crop evapotranspiration for the growing period of summer groundnut has been presented in Figure 1.

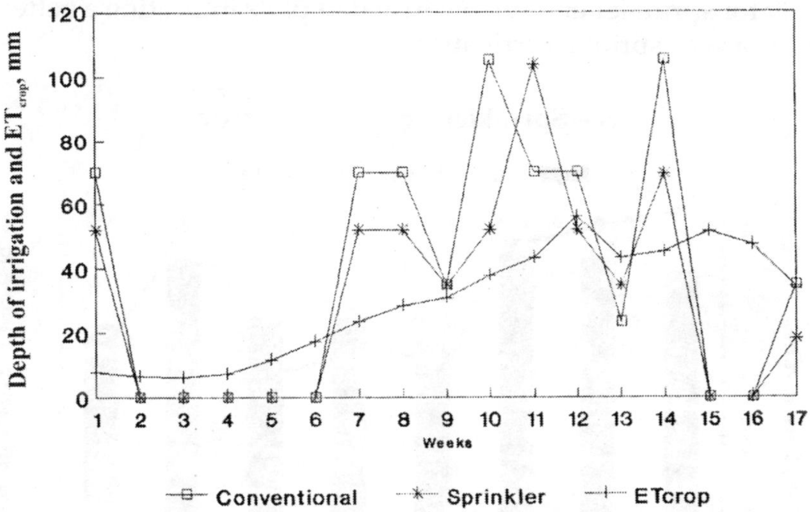

Fig.1: Weekly ET_{crop} and depth of irrigation in sprinkler and furrow method of irrigation

It was observed from the figure that the crop evapotranspiration was maximum during the 12th week of the growing season *i.e.* towards the end of the mid season stage. At this stage, the depth of water applied by the sprinkler method is almost equal to the crop evapotranspiration. From the graph, it is evident that during 12th week, depth of water through sprinkler is 52 mm, whereas, the crop evapotranspiration is 56 mm, which is approximately equal. After 12th week, weekly crop evapotranspiration was decreased gradually.

It is observed from the Figure 1 that during the first five weeks, water supplied through furrow and sprinkler method was much higher than the ET_{crop}. This was due to the fact that soil needed more water for germination of the seeds. Again during this period, the soil temperature was very low because of mild winter, which elongated the germination period. During the entire cropping season, 653 mm of water was applied through furrow method and

537 mm through sprinkler method, whereas, the total crop water requirement during the cropping season was 503 mm. Thus, water supplied through sprinkler method was optimum. In furrow method of irrigation, 22 % excess water was supplied as compared to that of sprinkler method.

Yield Potential

In sprinkler method, the average yield was found to be 27.9 q ha^{-1} where as in furrow method it was only 23.30 q ha $^{-1}$ (Figure 2 and 3). An increase of 20 % yield in sprinkler method was observed over furrow irrigation method. Test weight of pods (1000 pods) was found to be 1620 gm and 1220 gm, respectively for sprinkler and furrow irrigated plots indicating a better quality of kernel in case of sprinkler irrigation.

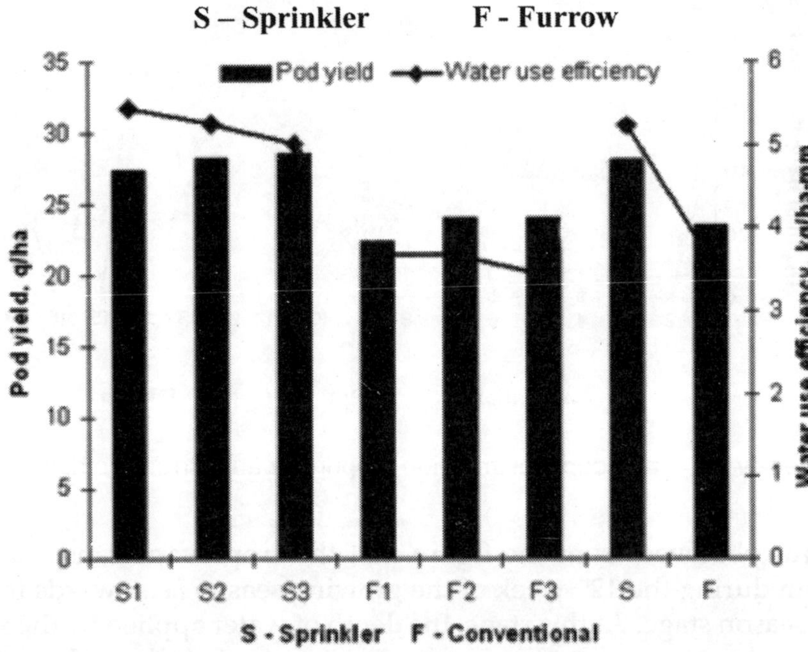

Fig. 2: Pod yield and water use efficiency in sprinkler and furrow method of irrigation

Water Use Efficiency (WUE)

Consumptive water use efficiency obtained from the field study was 5.58 kg ha-mm^{-1} and 4.65 kg ha-mm^{-1} in sprinkler and conventional furrow method of irrigation, respectively (Figure 2). It was noticed that WUE in sprinkler method was 20 % higher than that of conventional furrow method of irrigation.

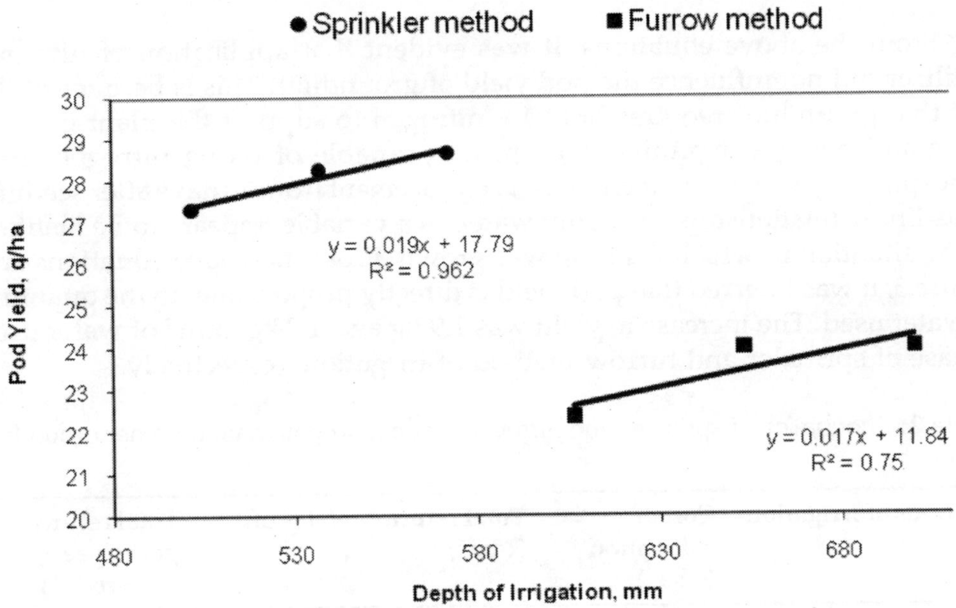

Fig. 3: Water production functions of summer groundnut in sprinkler and furrow method of irrigation

Field Water Use Efficiency

Average field water use efficiency was found to be 5.23 kg ha-mm^{-1} and 3.56 kg ha-mm^{-1} in sprinkler and furrow method of irrigation, respectively. The increase in application of water from 502 mm to 572 mm in sprinkler did not show any appreciable rise in yield as the water use efficiency declined from 5.43 to 5.02 kg ha-mm^{-1} (Table 2). In sprinkler system, consumptive water use efficiency and field water use efficiency was almost same for application of 502 mm depth of water. Hence, 502 mm of water application in sprinkler method was found to be optimum. Water requirement per quintal of produce through sprinkler method was 19.11 ha-mm, which indicated a saving of 32% of water over conventional furrow method of irrigation.

Water Production Function

The relation developed between the yield of groundnut and depth of application of water is given by;

For sprinkler method of irrigation,

$$Y_{gs} = 17.80 + 0.019\, W_{gs} \quad \text{... (viii)}$$

For conventional furrow method of irrigation

$$Y_{gf} = 11.84 + 0.017\, W_{gf} \quad \text{... (ix)}$$

From the above equations, it was evident that application of nitrogen fertilizer did not influence the pod yield of groundnut. This is because of the fact that groundnut requires only 4% nitrogen to support the plant in early stage and being a leguminous crop, it is capable of fixing nitrogen from atmosphere by symbiosis and meets its requirement from 20 days after seeding. Thus linear relationship with only water as a variable appears to be justified in the situation in which the crop was grown. From the above equations and Figure 3, it was inferred that pod yield is directly proportional to the quantum of water used. The increase in yield was 1.9 kg and 1.7 kg mm^{-1} of water used in case of sprinkler and furrow method of irrigation, respectively.

Table 3: Economics of sprinkler and furrow method of irrigation for summer groundnut crop

Methods of Irrigation	Cost of cultivation (₹ ha^{-1})	Total return (₹ ha^{-1})	Net profit (₹)	Total return per rupee invested (₹)
Sprinkler	9,168	59,590	50,422	5.49
Furrow	10,133	48,930	38,597	3.73

While comparing the labour requirement, a saving to the tune of 22 % was observed in sprinkler method over furrow method of irrigation. Water and labour saving could have been enhanced if a large field was taken up for commercial purpose. The return per rupee of investment as shown in Table 3 was ₹ 3.73 in furrow method of irrigation as compared to ₹ 5.49 in sprinkler method of irrigation. The return was 47.18 % higher in sprinkler method in comparison to conventional furrow method of irrigation.

CONCLUSIONS

The following conclusions were drawn from the study.

i. Sprinkler irrigation method has significantly higher water use efficiency as compared to conventional furrow method of irrigation.

ii. There is significant saving of labour in sprinkler method over conventional furrow method of irrigation. Saving of labour is higher where land leveling is involved.

iii. Sprinkler irrigation method is advantageous for groundnut production or any other close growing crops where water is a limiting factor.

iv. Initial investment is higher in sprinkler irrigation method but it will be compensated in the long run.

v. There is significant saving of water and increase in pod yield of groundnut. The yield potential will be enhanced further when a large field is taken in operation.

vi. In most part of the growing season, the water applied through sprinkler method was a little higher than that of ET_{crop} and was found to be optimum.

vii. From the water production functions, it was concluded that the increase in yield per unit depth of water was higher in case of sprinkler method than that of conventional furrow method of irrigation.

REFERENCES

Giri, A. 1994. Conjunctive use of surface and ground water in the Puri canal commands - A case study. Unpublished M.Tech Thesis, Dept. of SWCE., CAET, OUAT, Bhubaneswar.

Doorenbus, J. and Pruitt, W. O. 1977. Guidelines for predicting crop water requirements. Irrigation and Drainage paper No.24, FAO, Rome.

Sivanappan, R. K.1987. Basic data needed /sprinkler design factors. *In*. Sprinkler Irrigation.Oxford and IBH Publishing Pvt. Ltd., New Delhi, pp44-57.

Sivarao, K. S. V. V.1987. Integrated crop management through production function. *Journal of Agricultural Engineering,*ISAE, 24(3): 317-322.

□□□

Section - IV

Chapter 44

Resource Conservation through Participatory Integrated Watershed Management for Enhancing Productivity and Livelihood

ALOK K. SIKKA

INTRODUCTION

Natural resources provide the resource base for production and livelihood support system. Degradation of natural resources coupled with declining per capita land and water availability together with deteriorating soil health and biodiversity are posing serious threat to food, livelihood and environmental securities in the country. Land and water are the prime natural resources. Out of India's net sown area of around 142 M ha, about 60 % is rainfed spanning over several agro-ecologies. About 120 M ha of the country's land area suffers from land degradation problems (Anonymus,2008). The yield fatigue in major food crops like rice and wheat is already visible in the predominantly irrigated areas. Moreover, groundwater now accounts for over 65% of the total irrigated area in the country, and a looming groundwater crisis is evident in most of these areas. The estimates of the Technical Committee on Watershed Development (2006), indicate that even in the best possible scenario of irrigation development, about 40% of the additional supply of food grains will have to come from the rainfed agriculture. Rainfed agriculture has a very important role to play in inclusive growth and sustainable development.

The Eastern region and North–Eastern hilly states together has a geographical area of 92.35 M ha (28 % of the total geographical area of the country) and support a population of 265.78 M. About 45 % of total tribal population of the country lives in this region. Soil erosion and land degradation is a major problem besides large extent of rainfed agriculture in this region. In NEH region alone, about 54.20 % area suffers from degradation due to

Technical Expert (WD), National Rainfed Area Authority (NRAA), Planning Commission Govt of India, NASC Complex, New Delhi - 110 012 deforestation and *jhum* cultivation followed by water erosion (18.08%). It is also observed that around 62.53% of the total geographical area in eastern region is degraded exclusively by water induced soil erosion which in conjunction with salt-affected and acid soils works out to 73.88% (NAAS,2010). Among the eastern states, Jharkhand has maximum area (71.65%) under water erosion followed by Bihar (59.81 %), Odisha (58.46 %), and West Bengal (54.53 %). Thus water induced soil erosion poses a serious threat to sustainable production in the region. Whereas, the region is bestowed with high rainfall, but its potential to support good production and livelihood has not been fully exploited and effective policies, investment decisions and efficient utilization of natural resources in this region can give India a second green revolution.

Climatic variability and climate change are adding new concerns for sustainable development of agriculture, especially in rainfed and degraded areas, which are otherwise risk prone. Conservation and management of natural resources for sustainable development of rainfed agriculture, degraded areas and wastelands is one of the foremost priorities of the Government to unlock their potential for meeting the emerging challenge of food security. The potential of Integrated Watershed Management has been widely recognized for conserving natural resources and enhancing agriculture production and productivity besides empowering rural communities socially and economically. The eastern India has vast untapped potential to meet the challenges of food and livelihood security and increased food demand envisaged in the proposed Food Security Bill.

Rainfed Agriculture

Rainfed agriculture is now emerging as a major opportunity in raising overall agricultural growth. Rainfed agriculture constitutes about 60 % of the net cultivated area (142 M ha). It contributes 40 % to the total food production and supports 40 % of human and 60 % of cattle population (CRIDA, 2007). Rainfed agro ecologies in India cover about 87 % area each of coarse cereals and pulses, 77 % oilseeds, 65 % cotton and 50%cereals besides 80% of mangoes and apples and about 100% of forests.

The distribution of rainfed areas can be broadly classified into arid, semi-arid, dry-sub humid and humid regions. According to estimates of the CRIDA, Hyderabad 15 M ha of rainfed cropped area lies in the arid region and receives less than 500 mm annual rainfall, another 15 M ha is in 500-750 mm annual rainfall zone, and 42 M ha is in the 750-1100 mm rainfall zone. The remaining area lies in annual rainfall zone of greater than 1100 mm. Most of the rainfed agriculture in eastern India falls in the last two categories having a good opportunity with vast potential for development.

Rain-fed areas have also been classified into five major production systems – rice-based, cotton based, coarse cereals based, oilseeds based or pulses based. The rainfed rice production system is mostly prevalent in eastern and north eastern parts of India. About 12 M ha remain fallow after rice and potentials of raising a second crop are tremendous. A sizeable Rice fallow area is in the eastern region. The area has good opportunity for rainwater harvesting to enhance rice productivity and ensure second crop of oilseed, pulses, vegetables, *etc.*

Agricultural activities are more diverse with cultivation of over 34 crops in a year in rainfed areas as compared to 4 or 5 crops in irrigated system (NRAA, 2009). Farming systems of livestock, horticulture, agroforestry, seed spices, special attribute niche crops like guar gum, moth bean *etc*, medicinal & aromatic plants, fishery (in high rainfall areas), bee keeping *etc.*, assume even more importance under rainfed conditions. Contrary to the irrigated production systems that focus on individual crops, rainfed areas provide a good opportunity for diversification to mixed cropping and/or farming systems.

Status and Recent Growth

India ranks first among the rainfed countries in the world in terms of area, but ranks among the lowest in rainfed productivity of one t ha^{-1} as compared to 2.5 tonnes in predominantly irrigated areas. Spanning several agro-ecological regions, the rainfed areas represent the geography with the largest concentration of poverty and backwardness. The key thrust in agricultural policy till now has been to extend technologies to these areas that have evolved to enhance productivity in well endowed areas. This has not led to achieve significant gains. The public investment rate of ₹ 12,000-15,000 ha^{-1} under watershed development schemes is 13 to 16 times less than ₹ 2.0 to 2.5 lakh ha^{-1} under canal command development schemes. Rainfed areas are highly under invested. At the same time, inadequate support for rainfed agriculture in terms of support price, availability of appropriate inputs, credit, market access and agricultural research and extension has caused widespread desperation among farmers.

In spite of the policy neglect, fragmented implementation and low investments, the contribution of rainfed agriculture to the national economy is by no means small. The rainfed crops have shown more impressive growth rates in recent years as compared to irrigated crops like rice and wheat (Table 1). The compound growth rates for the period 1998-99 to 2008-09 reveal that production of coarse cereals increased at a rate of 2.73%; production of pulses continued to be an area of concern. Growth rate in yield of coarse cereals, pulses and oilseeds taken together (2.18%) was much higher than that of rice (1.40%) and wheat (0.46%). Priority should be accorded to bridging of the yield gap in addressing the backwardness of rainfed areas. Priority should be accorded to bridging of the yield gap and exploring other livelihood options in addressing the backwardness of rainfed areas.

Table 1: Growth rate in area, production and yield of major crops in India, 1998-99 to 2008-09

Crop/ Crop group	Compound Annual Growth Rate (%)		
	Area	Production	Yield
Sorghum	-2.82	-1.19	1.77
Pearl millet	0.07	4.11	4.04
Maize	3.00	5.60	2.52
Coarse Cereals	-0.52	2.73	3.26
Chickpea	1.69	2.24	0.54
Pigeonpea	0.34	0.16	-0.17
Pulses	0.61	1.08	0.47
Coarse cereals+ Pulses	-0.03	2.24	2.27
Groundnut	-1.32	0.93	2.29
Castor*	0.22	1.44	1.21
Sunflower	4.94	7.32	2.25
Soyabean	4.39	5.59	1.15
Rapeseed and mustard	1.97	4.24	2.22
Oil seeds	1.60	3.65	2.02
Coarse cereals + Pulses + Oil seeds	0.51	2.70	2.18
Cotton	0.54	10.20	9.60
Rice	-0.12	1.28	1.40
Wheat	0.39	0.85	0.46

*Data used for castor refer to 1997-98 to 2007-08
Source: CRIDA (Adapted from ASI)

Current Efforts and Schemes

A number of government initiatives, policies and programmes directly and/or indirectly address management of natural resources and development of rainfed areas. A variety of programmes/schemes have been initiated by different Ministries/Departments ranging from single component/commodity based sectoral scheme to integrated approach of watershed development. Flagship programme of Mahatma Gandhi National Rural Employment Guarantee Act (MGNREGA), Rashtriya Krishi Vikas Yojana (RKVY), National Horticulture Mission (NHM), Artificial Ground Water Recharge, Artificial

Recharge to Ground water through Dug wells, Repair, Renovation and Restoration of Water Bodies Directly linked to Agriculture, untied Backward Region Grant Fund (BRGF), Integrated Action Plan (IAP) and Corporate Social Responsibilities (CSR) in tribal regions (mostly rainfed) are amongst other schemes for benefit of rainfed areas.

Recently, Rainfed Area Development Programme (RADP) has been launched by Department of Agriculture & Cooperation (DAC) on pilot basis as a sub scheme of RKVY during the year 2011-12 focused on promoting integrated farming system in rainfed areas, besides a scheme on Extending Green Revolution to Eastern India (under RKVY). National Mission on Micro Irrigation, National Afforestation Programme of MoEF, soil health related and oil seed and pulse promotion schemes of DAC are some of the other initiatives/schemes worth mentioning for management of natural resources and rainfed farming. National Mission on Sustainable Agriculture, under the National Action Plan on Climate Change (NAPCC), a new initiative, will address water conservation and enhancing efficiency and productivity of water besides sustainability of natural resources and rainfed agriculture. The National Water Mission and National Mission for a Green India under NAPCC have direct relevance to management of natural resources and rainfed areas. Special packages like that of Bundelkhand, New Land Use Policy of Mizoram, Vidarbha package *etc.*, are some new initiatives.

Watershed Management - A Vehicle for NRM and Rainfed Area Development

Initially started as soil and water conservation programme, watershed management has undergone many paradigm shifts over the years. This process of agricultural and rural development for enhancing productivity and livelihood has travelled a long way from initial protection and conservation oriented piece-meal/sectoral approach of soil and water conservation to restoration of degraded areas for biomass production, eco-restoration, farming system and income generating activities. Earlier generation of watershed programmes were mostly confined to biophysical aspects *viz.*, contour bunding, check dams, afforestation *etc.*, but subsequently, participatory empowerment of communities, transparency, the social and innovative institutional issues came into prominence. Of late, the focus is shifting towards convergence, livelihood security and income generation through watershed development.

Multiple ministries/ departments and agencies have been involved in the growth and development of watersheds with an array of watershed schemes. Three schemes of Drought Prone Area Programme (DPAP), Desert Development Programme (DDP) and Integrated Wasteland Development

Programme (IWDP) of Ministry of Rural Development have now been merged into Integrated Watershed Management Programme (IWMP). National Watershed Development Programme for Rainfed Agriculture (NWDPRA), River Valley Projects (RVPs), Catchment Area Programme in Flood Prone Rivers, Shifting cultivation scheme of Ministry of Agriculture; Watershed Development Fund (WDF) administered by NABARD; Hill Area Development Programme (HADP) and Western Ghats Development Programme (WGDP) of Planning Commission; various afforestation programmes and other national and externally aided projects are amongst the various watershed related schemes. Besides, national flagship program like MGNREGA is spending 65% of its funds on rainwater and Natural Resource Management (NRM) related activities for creating durable assets to build production and livelihood system for sustained income.

In coordination with the Planning Commission, National Rainfed Area Authority (NRAA) in consultation with concerned ministries/departments has come up with the "Common Guidelines for Watershed Development Projects-2008" to have a unified approach and perspective in planning and implementation of all the Government schemes with common approach *w.e.f.*, 1st April, 2008. The new guidelines include innovativeness in approach, decentralization, dedicated institutions at various levels, social, gender and economic equity, involvement of PRIs and competent NGOs, transparency and accountability. Productivity enhancement and livelihood has been given priority along with conservation measures, together with livestock and fisheries management as a central intervention and encourage dairying and marketing of dairy products. These innovative guidelines were adopted by all the states within a year.

Under watershed management programmes (WMPs), since inception up to Tenth Five Year Plan, an area of 50.96 M ha has been developed with an expenditure of Rs.19,036.47 crore. During the first three years of XI Plan, an area of 19.57 lakh ha has been treated under various WMPs of Ministry of Agriculture with expenditure of Rs. 1685.07 crore. Whereas, DoLR of MoRD has sanctioned treatment of 15.13 M ha in 23 states under IWMP during the last 3 years.

Resource Conservation Programmes in NEH and Eastern Regions

The prominent national programmes implemented in both NEH and eastern region are National Watershed Development Programme for Rainfed Areas (NWDPRA), Integrated Watershed Development Programme (IWDP), River Valley Project and Flood Prone Rivers (RVP/FPRs), Watershed Development Project for Shifting Cultivation Areas (WDPSCA) and National Afforestation Programme (NAP). Drought Prone Area Programme (DPAP) was implemented only in Eastern region states. Besides these national programmes/schemes, many state Government funded programmes are also implemented in the states. The maximum area has been treated under IWDP (43.94 lakhs ha) followed by DPAP (25.9 lakh ha), NWDPRA (13.7 lakh ha) and other programmes (Rs.13.55 lakhs ha). A total area of 107.29 lakh ha has been treated by incurring an investment of Rs.3051.83 crores in both the regions. A recent programme has also been initiated by the Government of India for bringing Green Revolution in Eastern Region.

Way Forward for a Paradigm Shift

Mixed rainfed livelihood: Meeting the future demand for food grains (estimated at 280 M tonnes by 2020) would require a step up in the rate of growth of food production where rain-fed agriculture has to play major role. Therefore, a breakthrough in research and development of rain-fed agriculture is a must for poverty alleviation, livelihood promotion and food security in India. A paradigm shift is also required while addressing the rainfed livelihoods because livelihood strategies in rain-fed areas are dependent on a mix of agriculture, livestock, fisheries, horticulture, agro-forestry, alternate income generating activities and wage labour. It will require a holistic and integrated approach that combines natural resource conservation goals with productivity, carrying capacity and livelihood augmentation.

NRM and production system integration: Resource conservation and its sustainable use and management for enhanced productivity and livelihood is to be seen in unison. Programmes and policies for NRM and rainfed farming should therefore address the issues related to convergence and integration, multiple livelihood options, equity and equality (from gender perspective), and vulnerability to climate change besides more intensive research efforts in rainfed crops and livelihoods.

Farming system approach: There is a shift away from a *'single commodity approach'* to *'NRM based farming systems approach'* focusing on producing diverse products that mutually reinforce and can be supported by the available natural resource base without undermining its sustainability.

Efficient rainwater use management: Efforts at enhancing the 'Rain Water Use Efficiency' and water productivity should be central to the paradigmatic shift. Water management in rainfed areas, be it *in-situ* soil moisture or surface water or recharged groundwater, assumes highest priority to increase rain water productivity. Multiple use of harvested water assumes higher priority in eastern India bestowed with medium to high rainfall rainfed areas.

Climate resilient agriculture: Efforts to build resilience of the system along with capacities for adaptation to climate variability and change should be prioritized as the rainfed areas are ecologically fragile and highly vulnerable to vagaries of climate. Provision for adequate safety nets in the form of risk management is to be supported.

Livestock-crop combination: Focus on strengthening and improving productivity of extensive livestock and crop systems through improved service delivery and knowledge management and harness the potential of agronomic and management innovations including conservation agriculture should be prioritized.

Profitability and MSP: Profitability to the farmer besides productivity enhancement has become crucial as the market has become very competitive. This would require careful examination and promotion of interventions that have high efficiency and reduce cost of inputs specially the external ones and should focus more on *in-situ* methods of moisture conservation, use of bio resources, bio pesticides so that the farmers reliance on external inputs is reduced. Conservation agriculture including zero tillage, residue management is some of the options for reducing the cost and increasing the profitability. MSP or prices stabilization for rainfed crops need to be declared and assured to the farmers at the sowing time itself.

Incentivization for conservation: There is need to have a provision for incentivizing farmers for adopting conservation measures on their fields. Declining factor productivity, as a result of physical and agro-chemical soil degradation is a serious concern. Incentivization through labour and/or material subsidies for conservation agriculture, in-situ moisture management, on-farm biomass generation and incorporation, cover cropping with leguminous crops, tree based farming systems, crop-livestock integration and bio-mass recycling should form part of policy and programmes in rainfed areas.

Convergence: There are several programmes and schemes of the ministries of Agriculture, Rural Development, Water Resources, Environment and Forests, Small Scale Industry and Food Processing which are either focused on or have a significant impact on rainfed areas. The programmes and investments of each of these ministries are compartmentalized into silos, with very limited exchange between the silos. All concerned ministries and departments have their own dedicated teams and each believes that it has a unique and exclusive mandate. MoRD has developed guidelines for convergence of MGNREGA with other schemes/programmes. However, despite recognition of the importance of convergence, the current mode of implementation has no effective mechanism through which convergence could be achieved. Convergence across these programmes and departments is a matter of priority for holistic development of rainfed and degraded areas. It is imperative is to create a unified mechanism of convergence amongst different agencies targeting rainfed areas.

Increased investments: Average public expenditure on rain-fed areas has been considerably lower than that in irrigated areas. Over the last 15 years since 1995 (when the watershed programme got a real boost), the average expenditure works out to Rs. 2,130 crore per annum and about Rs. 4,500 per hectare of treated area(Draft Report of Sub-Group on NRAA for 12th Plan). Such low investments are highly inadequate to meet the challenges of bringing about a breakthrough in the production, productivity and livelihoods of rainfed areas. An increased public investment in rain-fed areas will not only revitalise agriculture but also create incentive for private investments.

Rainfed area prioritization for investment priorities: Prioritization of rainfed areas based on its potential for higher marginal returns on investments and niche crops may provide better opportunities for allocating resources in development of rainfed areas. The Eastern region holds considerable potential for enhancing food grain production in the country and a focused priority for agricultural development in this region is called for including extensive rice fallow areas in the country.

Responsive support system: Strengthening responsive support structures and infrastructure such as diverse seed systems, improving the fodder base, technical services, and common facilities for aggregation *etc.*, that involve higher transaction costs is needed.

Promote mechanization: Mechanization in rainfed/dryland areas has become more relevant due to narrow time window of farm operations. Energy consumption is directly related to productivity. However, consumption of total energy and operational energy in irrigated agriculture is 2.2 and 1.9 times of rainfed agriculture, respectively. Promoting mechanization through innovative models like custom hiring of farm machinery, micro enterprises, agri-business *etc.*, need priority focus.

Strong institutions and human capacity: Institutions and capacity building are very weak in rainfed areas. Local level people institutions should be created and their capacity built and strengthened to carry out their roles and responsibilities. Involvement of PRIs for sustainability of the project is to be prioritized. Optimal combination of formal (GO), NGOs and alternative institutions (involving local groups) especially at the micro level should be prioritized in rainfed area schemes to improve delivery efficiency, gap filling and up scaling of experiences.

REFERENCES

Anonymous. 2008. Harmonization of Wastelands/Degraded lands Datasets of India.

CRIDA.2007. Perspective Plan Vision 2025. Central Research Institute for Dryland Agriculture, Hyderabad.

NRAA. 2009. Harnessing Opportunities in Rainfed Areas: A pathway to prosperity – Vision 2025. National Rainfed Area Authority, NASC, DPS Marg, New Delhi – 110012, India

NAAS.2010. Degraded and Wastelands of India; Status and Spatial Distribution, *Directorate of Information and Publications of Agriculture*, ICAR, Pusa, New Delhi. Oxford University Press.

□□□

Chapter 45

Alternate Land Use Systems for Resource Conservation and Livelihood Security

O.P. CHATURVEDI, R. KAUSHAL AND J.M.S. TOMAR

INTRODUCTION

In India, the per capita cropped land and forest land availability is less than 0.2 ha and 0.1 ha, respectively due to which, there is acute shortage of timber, fuelwood and fodder. Further, expanding human and livestock population and fast growing industrial and urbanization are putting ever-increasing pressures on limited natural resources which are leading to land degradation and restricting sustainable production. With no additional land for horizontal expansion of agriculture, there is urgent need to find viable technologies for enhancing productivity and resource conservation on sustainable basis. In this context, Alternate Land Use Systems (ALUS), offers vast potential not only to diversify and increase overall land productivity to meet the demands of fuel, fodder, timber, medicine and other non-woody forest products, but also to enhance green cover, soil and water conservation and reduce the pressure on forest.

Any farm enterprise other than crop production is referred to ALUS. Fodder production for raising livestock is part of crop production system and therefore it does not come under ALUS. Tree farming wood lots, pasture and grassland systems, ley farming and agroforestry systems are different versions of ALUS. (Singh, 1997; Singh and Osman, 1995). ALUS aims at generation of assured income with minimum risk through efficient utilization of available resources and are appropriate for rehabilitating degraded or marginal or wastelands and saving arable lands from further degradation. These systems can withstand vagaries of climate and can efficiently utilize the resources.

Central Soil and Water Conservation Research and Training Institute, 218 Kaulagarh Road, Dehradun Uttarakhand

Agroforestry -The Principal Alternate Land Use : Agroforestry is collective name of land-use systems and technologies where woody perennials (trees, shrubs, palms, bamboos, *etc*) are deliberately used on the same land management units as annual agricultural crops and / or animals, in some form of spatial arrangement or temporal sequence. In agroforestry systems there must be both ecological and economical interactions between the different components (Lundgren and Raintree, 1982). Agroforestry is capable of yielding both wood and food while at the same time conserving and rehabilitating ecosystems. Agroforestry generates high income and minimizes risks through efficient utilization of available resources in cropping enterprises and is thus considered a potential technology for commercial farming, improving degraded and polluted sites, an opportunity for stabilizing fragile ecosystems. The main objective of agroforestry is to take the advantage of complimentary relationships between tree, crops and livestock in such a way so that sustainability, productivity, stability and profitability of system are maintained. Approximately 1.2 billion people (20% of the world's population) depend directly on agroforestry products and services in developing countries (Leakey and Sanchez, 1997). The statistics of the area under agroforestry is not very clear. India is estimated to have between 14,224 M (Ravindranath and Hall, 1995) and 24,602 M (Prasad *et al.*, 2000) trees outside forests, spread over an equivalent area of 17 M ha (GOI, 1999) supplying 49% of the 201 M tons of fuel wood and 48% of the 64 M m^3 of timber consumed annually by the country (Rai and Chakrabarti, 2001). Dhyani *et al.* (2006) reported 7.45 M ha under agroforestry. As per one of the estimate about 27.52 M ha is under various types of tree plantations, which include agroforestry, social forestry, farm forestry and others (Table 1). The task force on Greening India (Anon, 2003) has identified a potential of 10 M ha in irrigated lands and another 128 M ha in rainfed areas that could be developed through agroforestry on watershed basis.

Table 1: Area under various types of tree plantations in India

S. no	Plantation type	Area (M ha)
1	Agroforestry and social forestry	23.0
2.	Externally aided social forestry	
	a. Farm forestry/agroforestry	1.24
	b. Village woodlots	0.50
	c. Strip plantations	0.07
	d. Rehabilitation of degraded lands	0.24
	Sub total	2.06
3	By NGOs/Vas	0.04

4	Tree growers cooperatives	0.04
5	Conservation forestry	1.19
6	Afforestation (windbreaks/shelterbelts), under irrigation region development program of Indira Gandhi Canal project	0.40
	Total	**25.72**

Source: Dhyani (2009)

Traditional agroforestry systems/practices in India : Traditional agroforestry systems are broadly based on indigenous knowledge and the species are selected as a part of the cultural patterns of the community. The farmers integrate a variety of woody perennials in the crop and livestock production systems depending upon the agro-climatic conditions and local requirements (Dhyani, 2011).Traditional agroforestry systems in India ranges from apparently simple forms of shifting cultivation to complex home-gardens, from systems involving sparse stands of trees on farm lands to high-density complex multi storied homesteads of humid lowlands, from systems in which trees play a predominantly 'service' role to those in which they provide main commercial product. Most of these are anecdotal but in some enough research efforts have been carried out in recent times (Anon, 2003; Pathak *et al.*, 2006a). The traditional systems/practices having diverse characteristics of agroforestry are listed in Table 2.

Table 2: Traditional agroforestry systems/ practices prevalent in India

Systems Practice	Agro-ecological adaptation	Features
Shifting Cultivation	In Tropical forest areas	Cyclic slash and burn, growing of arable crops, fallow and back; Public and private lands; Livelihoods
Taungya	All regions	Tree planting, intercropping upto 3 years, moving to next place; Govt. land and trees, intercrops for cultivators; Sharing of land resources between state and landless
Homegardens	Mainly humid tropics	Transitory stage between tropical forest ecosystem and arable cropping; Plantation, fruit, spices, vegetable, fuel and fodder and MPT in backyard; Domestic needs and biodiversity

Large Cardamom and Alder Plantation	Mainly humid tropics	Cardamom supported by alder and other species; Tree harvesting at 15cm basal dia, Lopping after cardamom maturation and pollarding at 2-2.5 m. height for regeneration of younger seedlings and open canopy to regulate light; Soil fertility, shade, fodder, employment for labour and better income.
Plantation Crops and Fruit Orchard	All regions	Intercropped with cereals and other field crops; Short term fruits; Tubers and fodder crops; Enhanced resource use efficiency, Employment, livelihood support and income
Scattered Trees on Farm Lands	In all regions specially semi-arid and arid	Naturally occurring or planted timber, fruit, MPT and main field crops; Income, Wood supply and some fodder in critical period.
Shelterbelts and Wind breaks	In wind prone arid, alpine areas and., coastal areas	Plantations against sea tides, Hot winds etc.; Income; Environmental protection and domestic needs.
Boundary Plantations and Live Hedges	All regions	Trees for various uses on field crops and orchard boundaries; Income, Employment, Supply for domestic needs and industry (e.g. plywood), crop protection and wind break.
Woodlots for Soil Conservation	In hilly areas, along sea coast and ravine	Resource conservation, ground water recharge, microclimatic amelioration
Trees on Rangelands	All regions	Shrubs and bushes in community grazing lands; Resilience against climatic aberrations, Green biomass during lean months and higher production in organized silvi-pastures.
Plantation crops with pastures and animals	Mostly humid and sub-humid regions	Green biomass during lean months and higher production in organized silvi-pastures.

Industrial Plantations with Crops	Intensively cropped areas in north India	For fuel, pole and pulpwood; Income and domestic needs and Industry
Seasonal Forest Grazing	Semi-arid and mountainous ecosystem	Migratory grazing, reduce fire hazards, increase in soil fertility, livelihood support and income.
Aquaforestry	Low lying lands	Fish pond dikes and surface planted with suitable species (e.g. Coconut, Mangroves); Source of Nutrition to aquatic life, Breeding ground, Nutrient cycling; Shade for crustaceans, Employment and higher Income
Apiculture with Trees	All regions	Trees as a medium for apiculture; Employment and Income for landless and resource poor; Better pollination of other crops.

Present agroforestry practices in India : The systematic research in agroforestry geared up after establishment of International Council for Research in Agroforestry (ICRAF) in 1977 (which was renamed in 1991 as the International Centre for Research in Agroforestry). In India, organised research in agroforestry was initiated during 1983 with the establishment of All India Coordinated Research Project on Agroforestry by ICAR at 20 centres and later establishment of National Research Centre for Agroforestry at Jhansi in 1988. At present 39 centres of agroforestry (33 regular, 5 voluntary and I PC unit) are working in the country.

Through ICAR centres, a Diagnosis and Design exercise for existing agroforestry practices in India has been conducted. The prominent agroforestry practices and major emphasis on agroforestry in different regions are presented in Tables 3 and 4 respectively (Pathak *et al.*, 2006a).

Table 3: Prominent agroforestry systems in different agro-climatic regions of India

S.No	Agro-climatic zones	Agroforestry Systems*					
		Agri-Silvi-Culture	Agri-horti Silvi	Agri-horti Culture	Horti-Pastoral	Silvi Pastoral	Specialized
1.	Western Himalayan Region	4	4	5	4	4	-
2.	Eastern Himalayan Region	5	3	5	4	3	Jhum

3.	Lower Gangetic Plains Region	5	-	3	2	2	Energy
4.	Middle Gangetic Plains Region	4	-	2	-	-	Home-stead
5.	Upper Gangetic Plains Region	5	-	5	-	-	Agrihorti silviculture
6.	Trans-Gengetic planis Region	5	-	5	-	-	Agrihorti-silviculture
7.	Eastern Plateau and Hills Region	5	-	4	3	4	Block Plantation
8.	Central Plateau and Hill Region	4	-	5	4	4	Block Plantation
9.	Western Plateau and Hills Region	4	-	5	2	4	-
10.	Southern Plateau and Hills Region	5	-	3	2	4	-
11.	East Coast plains and Hills region	4	5	4	2	2	Home garden/ shelter belt
12.	West Coast Plains and Ghats Region	4	5	4	2	2	Home garden
13.	Gujarat Plains and Hills Regions	4	2	4	3	5	Block plantation
14.	Western Dry Region	5	3	4	-	2	Shelterbelt
15.	The Islands Regions	4	5	4	1	2	Home garden

*No. 1-5 indicate the priority practiced mode1= Least and 5= Highest

Table 4: Main characteristics and Agroforestry emphasis in the major agroecological regions of India

Characteristics	Himalyan region	Indo-Gangetic alluvial plains	Arid and semi-arid region	Humid and sub-humid region	Tropical coastal and Island region
Climate	Varies from temperate cold alpine to sub-tropical; rainfall 8 to 350 cm in west, 200-400 cm in east	Very hot summer and cold winter; rainfall 30 to 200 cm: ground frost common in winter	Dry-humid to very hot, sub-humid; rainfall in arid 40-65 cm; in semi-arid 70-125 cm	Humid to sub-humid; hot: rainfall 100 to 400 cm	Tropical humid, lowlands; rainfall 60-310 cm

Geographical spread	Uttaranchal, Himachal Pradesh, Jammu and Kashmir; elevation 1,000-3,000 m	Below foot hills of Himalaya covering about 47 M ha of Punjab, Haryana, Delhi, Jharkhand Uttar Pradesh, Bihar, Parts of West Bengal; elevation 150 to 600 m	Spreads over Rajasthan, Gujarat, Punjab, Haryana; parts of Uttar Pradesh, Madhya Andhra Pradesh Pradesh, Maharashtra, Karnataka and Tamil Nadu	Spreads over parts of Assam, Meghalaya, Mizoram, Tripura; West Bengal. Orissa, Madhya Pradesh, Chhattisgarh, and the Southern states; elevation 150 to 1,500 m	Coastal regions of West Bengal, Orissa, Andhra Pradesh, Tamil Nadu, Kerala. Karnataka, Maharashtra and the Islands
Soils and vegetation	Light to heavy textured, immature, mostly Inceptisols. Alpine scrub sub-alpine coniferous, tropical evergreen, semi-evergreen forests, alpine grasslands	Mostly alluvial with patches of saline and alkali soils. Plant woodlots of *Eucalyptus*, *Populus*, *Acacia* ,etc xerophytic trees and shrubs occur naturally in degraded soils	Sandy to clay loam; xerophytic thorny scrub forests and woodlands and psamophytic scrub vegetation, grazing lands	Alluvial to clay loam; Oxisols, Alfiosl Evergreen, and deciduous forests	Entisols, Oxisols, Inceptisols, Alfisols. Evergreen, semi-evergreen , moist deciduous, littoral and mangrove forest
Main land-use systems	Grazing lands, forestry, horticultural and forestry species with agricultural crops, shifting cultivation in NE states	Agricultural crops (Wheat, pulses, sugarcane, oil seeds), commercial wood lots, MPTS with crops and along boundaries. degraded grazing lands	Trees and shrubs in crop fields and along Boundaries permanent grazing lands having trees and shrubs with stunt growth	Cash crops (tea and large cardamom), agricultural crops, shifting cultivation forests	Plantation crops and icropping, field crops, fish culture, forestry.
Main land-use and ecological problems	Excessive deforestation, degraded lands. over-grazing, land sliding and shortage of fuel and fodder	Soil salinity, water logging, decline in soil fertility due to over-cultivation, fuel and fodder shortage	Drought, overgrazing, salinity. extension of sand dunes, low water table, soil erosion. degraded lands	Deforestation, soil erosion. Soil acidity and consequent problems, decline in soil fertility, shortage of fuel and fodder. Shortening of fallows	Deforestation & degradation of environment. Coastal erosion, acid-soils and related soil problems, soil erosion along slopes. Shortage of fodder

Major agroforestry emphasis	Planting woodlots on hills. contour farming blended with MPTs. management of pastures. raising orchards	Soil reclamation using MPTs, NFTs, suitable grasses; growing NFTs/ shrubs with crops along field boundary, raising woodlots on degraded lands	Raising shelter belts. wind breaks; growing NFTs in field, raising woodlots (NFTs) on degraded lands	Intercropping with nitrogen fixing trees, alley cropping, improved fallows	Afforestation of coastal regions with littoral and mangrove species, aqua culture with mangroves, multistoreied cropping system; alley cropping

Commercial Agroforestry Practices

The farmers of Punjab, Haryana, Uttar Pardesh and *tarai* belt of Uttarakhand have adopted Poplar based agroforesty in a large way covering an area of 11.48 M ha (Chaturvedi and Khan, 2009). Nearly 15 M poplars plants are raised under agro-forestry plantations in these states every year covering roughly 30000 ha. Similar area is harvested yielding about 3.6 lakh cubic meters wood of poplars yearly supporting the safety match industry and a large number of veneer and plywood units (Lal, 2004). All *rabi* and *kharif* crops can be grown during the first three years except paddy. Poplar can give yield of 150-200 $m^3 ha^{-1}$ in block plantation and 12-20 $m^3 ha^{-1}$ in boundary plantation at 8 years rotation under good care. Poplar planted in agroforestry system is estimated to provide gross returns of 3.27 lakhs $acre^{-1}$ or net discounted return of Rs. 88,749 $acre^{-1}$ in seven years at 12 % rate of interest on investment (B:C ratio = 1:1.192). Without intercropping, the net discounted rate return from poplar is only 53,476 acre $^{-1}$ (B:C ratio = 1:2.13). The results are substantially higher than raising sole crop.

Eucalyptus is grown in Andhra Pradesh, Haryana, Punjab, Uttar Pradesh and Uttarakhand, and many parts of the country to cater the needs of pulpwood, plywood and timber industries Eucalyptus are grown at spacing of 3 m x2 m or 4 m x2 m. Cotton, chillies, blackgram, greengram, rice and groundnut are taken up as intercrops during the first year of plantation cycle. Intercrops are also taken up during the second and third year of plantation, but the yield reduction up to 70 % is observed. Eucalyptus trees are harvested at the end of fourth year for pulpwood and after eight year for timber/ plywood. A minimum of three coppices are taken up. Intercrops are taken up during the second and third coppice also. Average productivity of 20 to 25 cubic meters ha^{-1} $year^{-1}$ has been achieved under un-irrigated conditions. However, productivity standards have been redefined by many farmers by achieving unprecedented record yield of 50 cubic meters $ha^{-1}year^{-1}$ compared to yield of 6 to 10 cubic meters ha^{-1} year^{-1} from normal seedlings. Because of better soils, adequate irrigation facilities and more progressive farmers in Punjab and Haryana productivity of clonal eucalyptus plantations is likely to increase. Farmers can expect net returns of ₹ 4 to 4.5 acre^{-1} at 10 years rotation from irrigated clonal eucalyptus plantations assuming yield of 200 tones and current price of Rs.2500 tonnes^{-1} for eucalyptus logs (Lal, 2004).

Subabul based agroforestry is practiced in Prakassam, Krishna, Guntur, Khammam, Rangareddy, Medak and Nizamabad districts of Andhra Pradesh to cater the needs of pulpwood requirement of ITC, J.K. Sirpur, Orient, West coast and A.P. paper mills. For pulpwood spacing of 1 m x1 m is adopted which can give net returns of ₹ 50,000 ha^{-1} after three years. With intercrops of blackgram, cotton and tabacoo in first year and limited fodder use, a spacing of 2.5 m x 0.75 m is adopted which can give net returns upto ₹ 72000 ha^{-1} after three years. For fodder production in animal husbandry, spacing of 1.25 mx1.25m is followed which can give net returns of ₹ 1,68,632 ha^{-1} after three years. After taking 3 crops (9 yrs), 20-25 t ha $^{-1}$ of root biomass is harvested which is sold in tobacco barns for curing @ ₹ 800 tonne^{-1} (Pathak *et al.*, 2006b).

In regions of high fertility and rainfall like Kerala, small farmers usually maximize returns from land through multistoried cropping where perennial crops such as coconut, arecanut, rubber and pepper are intercropped with seasonal and annual crops like tapiaco, banans, pulses and vegetables.

In North Easter region, agri-horticultural systems involving cultivation of ginger with fruit plants, such as mandarin and guava on mild slopes have been found profitable, besides protecting and conserving the hill soils. Pineapple can be easily associated with multi-purpose trees in paired rows on hill slopes, where it checks soil erosion and also gives considerable cash returns within a short period (Dhyani *et al.*, 1995).

Suitable agri-horticultural systems are being adopted for rehabilitation of degraded lands in different agro-ecological regions. In this system, agricultural crops are grown with *Citrus* species, such as kinnow, lemon, sweet orange, mango, litchi and peach in the initial years, after which, the horticultural crops provide continuous returns. Intercropping of guar, cowpea or pearl millet with peach, turmeric with papaya, *Chrysopogon fulvus* or Napier grass with anola or ber has been identified as an economically viable system for rehabilitation of these lands (Anon, 2000). Intercropping of turmeric with kinnow or peach, and cowpea, blackgram and toria with mango and litchi have also been found to be promising (Dhyani, 2011).

Silvipastoral systems : The importance of silvipastoral system has now been accepted through out the world for increasing the productivity of degraded/problematic lands and rehabilitation of degraded lands. It helps in checking soil erosion, conserving more in-situ moisture, maintaining higher survival per cent, yielding more fodder and fuel wood biomass.

Silvipastoral systems can be raised on marginal lands with suitable fodder grasses like *Dicanthium annulatum, Leptochloa fusca, Brachiaria mutica, Panicum antidotale, Sehima nervosum, Chloris gayana, Cenchrus ciliaris.* A number of MPTs *viz., Grewia optiva, Morus serrata, Celtis australis, Robinia Pseudoacacia, Ulmus wallichiana, Quercus* spp., *Bauhinia variegate, Melia azedarach, Parkinsonia aculeata. Eucalyptus tereticornis,* Cassia *siamea, Acacia auriculiformis, Dalbergia sissoo, Albizia procera, Acacia nilotica, Thespesia populnea, Ailanthus excelsa, Derris indica* etc. have been found suitable for establishment of silvipastoral systems for rehabilitation of degraded land. To improve grass yields and fodder quality of pastures, sowing/planting of legumes like *Stylosanthes hamata, Medicago* and *Trifolium alexandrium* can be done (Singh and Chaturvedi, 2011).

In the Shiwalik foothills, silvipasture systems involving *Eulaliopsis binata, Saccharum munja* or *Vetivaria zizanoides* with *Acacia nilotica* were found to promising. Further, *Eucalyptus tereticornis + Eulaliopsis binata* has been identified as an economically - viable and eco-friendly system for rehabilitation of these lands. The study conducted for 14 years with four multi-purpose tree species *viz., Albizia lebbek, Grewia optiva, Bauhinia purpurea* and *Leucaena leucocephala* and two grass species *viz., Chrysopogon fulvus* and *Eulaliopsis binata.* At CSWCRTI, Dehardun revealed that gravelly and bouldery lands could be effectively utilized by putting them under perennial vegetation for improving soil fertility and biomass production (Vishwanatham *et al.*, 1998; Samra *et al.*, 1999).

A list of the tree/shrubs/grasses suitable for the given area which can be used for fodder is presented in Table 5 (Singh and Chaturvedi, 2011). These species can be grown on a wide variety of sites and lopping can be carried out when the trees have attained sufficient growth.

Table 5: Promising trees, grass and legumes for silvipastoral systems in few regions of India

Agroclimatic zones	Fodder trees	Grass and legumes	Authors
Temperate, (J&K, Himachal Pradesh and NW Himalaya)	**Wet region** *Robinia pseudoacacia, Celtus austratis, Populus alba, Quercus semecarpifolia, Q. dilatala, Acer campbelli.* **Dry region** *Salix fragilis, P. ciliata, P. euphratica, Belula utilis, Acer candatum, Q. ilex*	Lucerne *(Medicago satira)*, Red clover *(Trifolium pratense)*, white clover *(T. repens)*, Cocksfoot *(Dactylis glomerala)*, Timothy *(Pheleum pratense)*	Singh (1986)
Kangra Valley, HP	*Erythrina suberosa Dalbergia sissoo, A. catechu, Albizia stipulata*	*Eulaliopsis binata, Chrysopogonfulvus, Bauhinia purpurea, Albizia lebbek, Napier bajra, Setaria anceps*	Vishwantham *et al.* (1998)
North-eastern Hill region	*i)Albizia lebbek, Bauhinia purpurea, Leucaena leucocephala. ii)Salix* spp., *Populus ciliata, Castanopsis, Celtis australis, Sachima \vallichii, Ficus* spp., *Bauhinia variegata*	*Panicum maximu, Setaria sphacealata, Pennicetum polystachyon, Stylosanthes guyanensis,* Bajra x Hybrid Napier, *Panicum maximum, Doctylis glorneruta, Lolium perenne*	Singh *et al.* (1996)
High rainfall, Doon Valley	*Leucaena leucocephala, Grewia optiva, Acacia sp., Bauhinia retusa, Salix tetrasperma.*	*Eulaliopsis binata, Chrysopogon fulvus*	Dhyani *et al.* (1988)
Saline-alkali soils	*Prosopis juliflora, Acacia nilotica, Terminalia qrjuna, Pongamia pinnata*	*Leptochloa fusca, Chloris gayane, Brachiana mutica, Sporobolus sp.*	Singh *et al.* (1994)

Energy plantations : An energy plantation is one that is grown purely for plant material for their fuel Energy plantations in India has gained emphasis since the inception of the Clean Development Mechanism (CDM) due to various environmental and social benefits.

a) *Woodlots*: In rural areas, majority of the energy requirements are met through biomass from trees and shrubs. Species such as *Casuarina equisetifolia, Eucalyptus,Prosopis juliflora,* and *Leucaena leucocephala* offers high potential for providing wood energy at shorter rotation. In Andhra Pradesh, Tamil Nadu and Karnataka, *Casuarina equistifolia* grown, on degraded lands for energy at 1m ×1m spacing can provide 120 tonnes of firewood at 7 to 10 years rotation. (Anon, 2003). *Eucalyptus* can provide 11.9 t ha^{-1} (3 years old plantation) to 196.5 t ha^{-1} (9 year old plantations) biomass when grown extensively as wood lots at spacing of 2m × 2m. In Ramnathpuram, Distt. of Tamil Nadu, *Prosopis juliflora* is used in cottage industry for making, charcol and high value products.

b) *Plantation for biodiesel production*: The tree borne oilseeds (TBOs) species like Ratanjot, Karanj, Neem, Mahua, Paradise tree and Jojoba offers enormous potential on degraded lands in arid and semi-arid tracts. Research on Jatropha *curcas* and *Pongamia* revealed that these species can be intercropped with annual crops such as cowpea, sesame, sunflower, French bean, black gram, green gram and groundnut *etc.* In suitable plantation *Jatropha* gives about 2 kgs of seed tree^{-1}. If planted in hedges, the reported productivity of *Jatropha* is from 0.8 kg- 1.0 kg. of seed m^{-1} of live fence which is equivalent to seed production of between 2.25 t ha^{-1} and 5 t ha $^{-1}$, depending upon whether the soils are poor or average for plantations and between 2.5 t ha^{-1} annum^{-1} and 3.5 t ha^{-1} annum^{-1}for hedges (Anon, 2004). However, the economics and viability of the *Jatropha* plantation and bio-fuel production are still at initial stage and require long term studies.

Potential of Agroforestry for Sustainable Development

Agroforestry can complement forestry sector efforts in sustainable forest management by providing a set of tree-based conservation and production practices for agriculture lands. Some of the sustainable issues are discussed as under:

a) *Productivity enhancement* : Agroforestry systems primarily involve multiple land use systems for maximizing the productivity of per unit piece of land and time based on land capability classification which in turn is governed by soil characteristics like soil texture, soil depth , slope and extent of erosion. Choices of agroforestry models suitable for different areas vary with land capability classification.

In the last three decades, agroforestry has been promoted in the tropics as a natural resource management strategy that attempts to address the goals of agricultural productivity besides conserving soils, water, local and regional climate and more recently biodiversity (Schroth *et al.*, 2004). Poplar based agroforestry in India are highly productive and can yield 150-200 m^3 ha^{-1} in block plantation and 12-20 m^3 ha^{-1} in boundary plantation at 8 years rotation. Similarly high productivity has also been reported in Eucalyptus and subabul based agroforestry system from

North and South India under unirrigated conditions. Lal (2004) reported, average productivity of 20 to 25 cubic meters ha^{-1} $year^{-1}$ in Eucalyptus under un-irrigated conditions in Punjab and Haryana. With the development of new clones, it is expected that the productivity standard will improve further in the future. Biomass production of some important fast growing agroforestry tree species is given in Table 6.

Table 6: Biomass production of different tree species

Tree species	Productivity (t ha^{-1})
Casuarina equisetifolia	171
Leucaena leucocephala	159
Eucalyptus hybrid	152
Dalbergia sissoo	65

Silvipasture systems are also very helpful in enhancing productivity of degraded lands. Productivity of some MPTs in silvipastoral systems on highly degraded lands in semiarid region of Central India was worked out. *Albizia amara* based silvipastoral system produced 7.95 t $ha^{-1}yr^{-1}$ fodder followed by *Acacia tortilis* (7.6 t $ha^{-1}yr^{-1}$). The production of *Hardvickia binata* based silvipastoral system (800 trees ha^{-1}) didn't significantly affect the growth and yield of black gram in *kharif* and mustard in *rabi* even after six years (Solanki *et al.*, 1999). On the basis of research and potentials of silvipastoral systems under different agro-climatic regions of India, potential species and their production potential are given in Table 7 (Singh and Chaturvedi, 2011).

Table 7 : Potentials of silvipasture system in various regions of India.

Region	Components	Results
Alpine Region	Only pastures with suitable grasses and legumes	*Agrostis* spp., *Chrysopogon gryllus, Dactylis glomerata, Poa pratensis Trifolium repens, T. pratense, Medicago falcata*
Temperate Region	Intensive pasture production, temperate horticultural plant combined with pasture legumes as ground floor, fodder forest legumes	5.7 tha^{-1}, on an av. 8 to 10 $tree^{-1}$ dry matter
North-eastern Region	Silvipasture, *i)Alnus nepalensis+ Stylosanthesiguyensis + Panicum maximum* *ii)Ficus hookerii + Thysanolaena maxima*	13.5 tha^{-1} (DM) fodder yield 12.3t ha^{-1} fodder yield

Outer Himalayas Region	Silvipasture systems, *Dalbergia sissoo* +*Chrysopogon fulvus* *Acacia catechu* + *Eulaliopsis binata* *Leucaena leucocephala* (high density) + Hybrid Napier (NB 5) *Bauhinia purpurea, Albizia lebbek* + *Eulaliopsis binata*	Suitable system for utilization of degraded lands for fuel, fodder and fibre 64-71 t ha^{-1} fuelwood 5.2-5.5 t ha^{-1} yr^{-1} fodder grass for short rotation of 4 -5 years Most suitable for gravelly riverbed lands of Doon Valley.
Shiwalik foothills	Silvipasture, Fodder production from wastelands and watershed management	Fodder production increased by 267%
Indo-Gangetic plains	Tree-grass combination	Annual production of 8-10 t ha^{-1} (4.5-5 t ha^{-1} fuelwood) 7-10t ha^{-1}yr^{-1} biomass production
	Forage production on dry lands	*Leucaena* + sorghum-safflower- Forage yield of 12.3 t ha^{-1}
Humid and sub-humid	Use of quick growing leguminous fodder plants in crop rotations, plantation crops with pastures, Farm forestry. Deenanath (*Pennisetum pedicellautm*) + *Stylosanthes* and Deenanath + *Leucaena*	Best for pasture development in the eroded marginal and sub-marginal hilly area of Southern Bihar
	i)*Leucaena* (Hawaiian, Hybrid-28)+fodder grass. *Glyricidia/Sesbania/*	Suitable combination for plain area in West Bengal
	ii)*Ficus hookerii* *Litsea*+*Thysalaena* /*Andropogon nardus*	Suitable for hilly areas
Coastal area (Kerala *etc.*)	Fodder trees (Subabul/ Agathis/ Desmenthes) + Congo signal + *Stylosanthes or Centrosema*	Total fodder yield of 22-29 t ha^{-1} could be obtained in coconut based fodder production systems.
Arid and semi-arid region	Grazing land and silvipasture and Hortipasture in arid and integration of agri-silvipasture, Forage-cum-coppicing farming on the marginal and sub-marginal lands with intercropping of dryland cereals and legumes, *Acacia tortilis, Albizia lebbek, Prosopis cineraria* with *Chrysopogon fulvus, Cenchrus ciliaris, C. Setigeres*	Very good potential for silvipastoral enterprise

b) *Protection and conservation of natural resources* :Agroforestry interventions reduce erosivity of rainfall/runoff and erodibility of soil through dissipation of rainfall energy by canopy at low height, surface litter, obstructing overland flow, root binding and improving soil physico-chemical conditions (Dhyani, 2011). Soil and water conservation measures coupled with planting of suitable pasture and tree species not only conserve natural resources but also improve productivity by 13 times and income generation by 3 times (Singh and Hazra, 1994). Potential of agroforestry in arresting soil losses have been well documented. A summary of agroforestry practices with potential for the control of soil erosion is given as Table 8 (Young, 1997).

Table 8: Agroforestry practices with potential for control of soil erosion

Agroforestry practices	Suitable climates	Important points
Multistorey including home garden	Mainly in Humid and moist sub-humid climates	Possess an inherent capacity to control erosion through comibation of herbaceous cover with abundant litter
Plantation crop combinations	Humid to moist sub-humid climates	Densely planted combinations of agricultural plantation crops with MPTs appear to control erosion effectively on at least moderate slopes
Hedgerow intercropping, barrier hedge	Humid, sub-humid, possibly in semi-arid climates	A considerable apparent potential to combine erosion control with arable use on gentle to moderate slopes, more speculative potential on steep slopes.
Trees and erosion control structure	All climatic conditions	Supplementary use of trees stabilizes earth structures and gives production from land they occupy.
Windbreaks and shelterbelts	Semi arid zone	Proven potential to reduce wind erosion.
Silvipastoral practices	Semi arid, humid and sub-humid	Opportunities for inclusion of trees and shrubs as part of overall programmes of pastoral improvement
Reclamation forestry	All climatic conditions	Potential for planned design and development.
Combinations of the above in integrated watershed management	All climatic conditions	Substantial opportunities to include agroforestry with other major kinds of land use in integrated planning and management.

Narain *et al.* (1994) compared run off and soil loss at 4% slope under different land use systems in Dehradun and found that maximum run off and soil loss (17.3 tonnes ha^{-1}) incurred under pure maize while the combinations of Subabul and Eucalyptus arrested the soil loss and run off within soil loss tolerance limits (Table 9). Parsad *et al.* (1986) advocated 99.7 % reduction in runoff when bamboo was grown with grass compared to shifting cultivation.

Table 9: Soil loss and run off under different land uses at Dehradun

Land use	Run off (%)	Soil loss (t)	Percent reduction in	
			Run off	Soil loss
Maize alone	18.3	17.7	-	-
Maize + Subabul	8.9	5	51.9	71.8
Maize + Eucalyptus	3.6	0.91	80.3	94.9
Chrysopogon fulvus	1.6	0.33	91.3	98.1
Grass + Subabul	0.6	0.13	96.7	99.3
Subabul	0.4	0.04	97.8	99.8
Grass + Eucalyptus	0.1	0.02	99.5	99.9
Eucalyptus	0.1	0.01	99.5	99.9

Integerated watershed development programme are also very helpful in conserving the natural resources besides improving the productivity. A review on the status of watershed research in India (Samra, 1997) reveals that integrated watershed development can reduce runoff, moderate flooding of downstream area and improves *in-situ* moisture conservation. Further, trees in watershed can improve ground water recharge and raise water table and can improve biomass production. In Tejpura watershed (775.7 ha in Bundelkhand) about 231 ha m of rainwater was stored in 5 check dams which resulted in rise of ground water by 3-7 m (Hazra, 1998) and thus increased underground water table leading to the digging up of wells from five to present number of 315 which increased the irrigation to crop from initial 3.8% to present 100% and also increased cropping intensity (82-209%). The runoff and soil loss was reduced by 66% and 99%, respectively.

c) *Maintenance of soil health* : Trees in agroforestry helps in maintaining the soil fertility by increasing inputs (organic matter, nitrogen fixation, nutrient uptake), reducing losses (organic matter, nutrients) by promoting recycling and checking soil erosion, improving soil physical properties condition primarily due to increased SOM and root activity and secondarily due to increased soil biological activity by soil macrofauna.

Improvements are reflected through lower bulk density (Belsky *et al.*, 1993), lower surface resistance to penetration (Campbell *et al.*, 1994), increased porosity and greater aggregate stability. The changes/ improvement however, may be species-specific and also dependent on size and age of the trees, and site conditions. Several studies have reported improved soil fertility in terms of SOM, extractable P, and exchangeable cations under trees, compared with treeless open areas.

In Doon Valley of Uttarakhand, a decrease in soil pH by 0.4 units, bulk density by 0.1 mg m^{-3}, improvement in organic matter (OM) content by 0.16% and water holding capacity by 15% was observed (Dadhwal and Tomar, 1999). In another study comprising four multi-purpose tree species *viz., Albizia lebbek, Grewia optiva, Bauhinia purpurea* and *Leucaena leucocephala* grown in association with two grass species *viz., Chrysopogon fulvus* and *Eulaliopsis binata,* marked decrease in soil pH and increase in organic carbon, available P and available K in the surface layer was observed (Vishwanatham *et al.*, 1999; Samra *et al.*, 1999).

A study from Bundelkhand region revealed that cropping under trees almost double the OC of soils in three years. *Leucaena* based cropping increased the OC from 0.4 to 0.87% and sesbania based to 0.75%. The tree cropping also improves N and P status of soil but marginally depress K (Hazra, 1994). Positive impact on soil fertility by perennial vegetation was also recorded in semiarid vertisols (Bellary region) where OC increased from 0.03% to 2.16% in ten years (Srivastava, 1994) with *Acacia nilotica.* Singh *et al.* (1997) evaluated agriculture, forestry and agroforestry practices in moderately alkali soils in northwestern India over a period of six year at CSSRI, Karnal. The soil conditions were much improved in terms of soil build up of soil organic matter, N, P and K when trees were associated with trees (Table 10).

Singh *et al.* (1994) reported accumulation of 2.91% organic carbon under arecanut + jackfruit + black pepper + tejpatra followed by 1.85% under arecanut + betelvine + miscellaneous trees as against 0.78% only in a degraded land within 10-15 years of this practice in north east region. A sharp increase in exchangeable Ca, Mg, K and Na was noticed in all the agroforestry interventions over adjoining degraded lands. The exchangeable Al, potential cause of infertility of these lands disappeared completely within 10-15 years of agroforestry practice.

Table 10: Role of agroforestry for changes in some soil properties

Land use system	pH	EC (dsm^{-1})	OC (%)	Available nutrients ($kg\ ha^{-1}$)		
				N	P	K
Initial Soil properties						
Sole crop	9.26	0.56	0.22	65	14.3	238
Eucalyptus based systems	9.32	0.61	0.25	65	13.8	224
Acacia based systems	9.34	0.67	0.28	72	13.6	212
Popular based systems	9.27	0.60	0.26	68	14.4	237
After five years						
Sole crop	8.81	0.45	0.29	75	13.6	208
Eucalyptus based systems	8.65	0.32	0.37	86	18.5	259
Acacia based systems	8.71	0.41	0.48	103	31.9	286
Popular based systems	8.47	0.35	0.43	93	21.7	321

Ameliorative effect of agroforestry has also been reported for problematic soils suffering from salinity, alkalinity and waterlogging problems. Afforestation of *Prosopis chilensis* (Mesquite) was found to improve the physical and chemical properties of degraded sodic soils in Haryana. It reduced pH, electrical conductivity and exchangeable Na levels and increasing infiltration capacity, OC, total N, available P and exchangeable Ca, Mg and K levels (Bhojvaid *et al.*, 1996; Singh, 1999). In studies reported by Singh, 1995, the initial soil pH of 10.3 with organic carbon of 0.12% was restored back to 8.0 and 0.58%, respectively under Prosopis *juliflora*. Some of potential trees for ameliorating alkali soils, saline soils and bio drainage are presented in Table 11.

Table 11: Effective trees for ameliorating soil health

Alkali soils	Saline soils	Bio-drainage
Prosopis juliflora	*Casuarina equisetifolia*	*Populus deltoides*
Acacia nilotica	*Leucaena leucocephala*	*Terminalia arjuna, Eucalyptus tereticornis, Acacia auriculiformis*
Tamarix articulata	*Prosopis juliflora*	*Syzigium cuminii*
Eucalyptus tereticornis	*Acacia nilotica*	*Albizia lebbeck*
Pongamia pinnata	*Tamarix troubii*	*Pongamia pinnata*
Terminalia arjuna	*Tamarix articulata*	*Dalbergia sissoo*
Dalbergia sissoo	*Acacia farnesiana*	
	E. Camaldulensis	

d) *Amelioration of Microclimate* : Trees in agroforestry bring about microclimate changes under their canopies by reducing soil and air temperature, irradiance and wind speed. These changes directly influence the soil water evaporation and humidity, which in turn significantly affect crop growth and productivity. Dela Cruz and Luna (1992) found decrease in light intensity under *Acacia auriculiformis* plantation with relative higher humidity which favoured soil moisture conservation and better soil organic matter content. Further, they reported 2-4-5.9^{0}C lower temperature under tree canopies compared to grassland during day time. In the semi arid tropics savanna of Kenya, *Adansonia digitata* and *Acacia tortilis* trees reduced soil temperature under their crowns by 6^{0}C at 5 cm and 10 cm depth, compared with open areas (Belsky *et al.*, 1993). The reduced radiation load and wind speed under canopies reduce water demand by crops and soil evaporation, which are particularly important during drought periods.

e) *Carbon sequestration* :The practice of eco-restoration of degraded lands/ forests and agroforestry has great potential for mitigating the accumulation of C as CO_2 in the atmosphere by maintaining and increasing the land based carbon sinks. Agroforestry can store carbon through enhancing build up of soil organic matter and by biomass production that can pool more CO_2 from the air. In India, where majority of population is dependent on agriculture; agroforestry provides a long-term strategy to increase the amount of carbon while still allowing for growing of food crops. Average sequestration potential of agroforestry has been estimated to be 25 t C ha^{-1} over 96 M ha of land in India (Sathaye and Ravindranath, 1998). Soil carbon storage and mitigation potential in different land use systems is presented in Table 12, which indicates that agroforestry has vast scope towards mitigating the climate change.

Table 12: Soil organic carbon (SOC) content in some agroforestry systems

Region	Agroforestry system and components	Total storage (t C ha^{-1})	References
Semi-arid	Silvipastoral (age 5 yrs)		Rai *et al*, (2001)
	A. nilotica + natural pasture	9.5-17.0	
	A. nilotica + established pasture	19.7	
	D. sissoo + natural pasture	12.4	
	D. sissoo + established pasture	17.2	
	H. binata + natural pasture	16.2	
	H. binata + established pasture	17.0	
North-western India	Silvipastoral (age 6 yrs)		Kaur *et al.*
	Acacia/ Dalbergia/ Prosopis + *Desmostacya*	6.8-18.6	(2002)
	Acacia/ Dalbergia/ Prosopis + *Sporobolus*	1.5-12.3	

Central India	Block plantations (age 6 yrs) *Gmelina arborea*	24.1-31.1	Swamy *et al*, 2003)
Arid region (Rajasthan)	Agri-silvicultural (age 8 yrs) *E. officinalis* + *V. radiata*	12.7-13.0	Singh (2005)
Semi-arid region	Agri-silvicultural (age 11 yrs) *D. sissoo* + crop	26.0	Anon. (2005)
Semi-arid region	Agri-silvicultural (age 5 yrs) *A. procera* + black-gram-mustard *A. procera* + green-gram-wheat	 34.8 35.1	Ram Newaj *et al.* (2006)
North- western Himalaya	Silvipastoral system Agri-hortipastoral Hortipastoral	2.2 1.2 1.1	Anon. (2006)

f) *Employment generation and livelihood security* : Agroforestry can enhance the income of farmers by generating more employment opportunities thereby providing livelihood security. Numerous profitable combination of tree-crop is available in different Agroclimatic zone. The B:C ratio of some of the important agroforestry system is given in Table 13. In addition to income and profitability, numerous paper and plywood industries like ITC Bhadrachalam WIMCO Ltd., Green Ply Ltd, Shirdi industries Ltd, Oriental paper mill, Ballarpur paper mill, West Coast Paper Mills Ltd. and Hindustan Paper Mills are promoting tree cultivation and are provideing huge employment opportunities. For example in Tamil Nadu, charcoal making from *P. juliflora* is providing employment to the tune of 6.34 M man days and 7.03 M women days per annum. Chaturvedi and Khan, 2009 reported that silvipastoral system on an average cycle of 10 years can generate 120 mandays ha^{-1} employment. Similarly, in Bihar, litchi based agri-horti system has been found to generate employment opportunities of 130-140 man days ha^{-1}.

Table 13: Agroforestry systems with respective B:C ratio in different agroclimatic zones

Zone	Tree	Crops	Yrs	B:C ratio
E. Himalaya	*Anthocephalus kadampa*	Paddy	8	1.6
W. Himalaya	*Prunus persica*	Maize, Soybean	15	1.21
Lower Gangetic	*Eucalyptus hybrid*	Paddy, wheat	10	2.33
Middle Gangetic	*Dalbergia sissoo*	Sesamum	20	1.15
Upper Gangetic	*Populus deltoides*	Paddy, wheat	7	3.31
Trans Gangetic	*Acacia nilotica*	Pearl Millet	20	2.61
Eastern plateau	*Harwickia binata*	Pennisetum	15	2.13
Central plateau	*Emblica officinalis*	Ground nut	8	1.01
Western plateau	*Albbizia amara*	Cenchrus	16	1.56

Southern plateau	*Tamarindus indica*	Chilli	10	2.03
West Coastand Ghat	*Casuarina*	Paddy	7	1.39
Gujarat plains	*Dalbergia sissoo*	caster	20	2.14
Western dry	*Prosopis cineraria*	Pearl millet	20	1.65
Island	*Cocos nucifera*	Paddy	6	1.17

Opportunities and Challenges

- Land use diversification through agroforestry and land management approaches for sustainable production.
- Basic research aimed at understanding process to support balances WUE, NUE, IPM and below and above ground biodiversity for resource management.
- Integrating fisheries – tree crop production system and soil fauna.
- Studies on carbon sequestration, carbon balance and effort to minimize global warming.
- Improved planting material to the farmers.
- Ecological and economical modeling to value the social, cultural and employment (livelihood) systems for their sustained growth.
- Value addition through processing and marketing of tree product.
- Technology transfer.

Strategies to Meet Challenges

(i) Political will

Strong political will and commitment at the level of Central and State Government for:

- Prevention of further degradation of well stocked forests.
- Restoration of forest growing stock and productivity to optimum levels.
- Adequate financial investments.

(ii) Improvement in Productivity

Substantial improvement in productivity of forest resources on sustainable basis thorugh:

- Conserving the forest resources for meeting the national needs.
- Agroforestry practices to achieve productivity levels of 20-30 m^3/ha^{-1}.
- Reforestation of degraded forest lands and marginal agricultural lands.

(iii) Corporate Sector Participation

Linking agroforestry to Corporate Sectors through PPP for:

- Production of quality planting material for plantations.
- Supply and marketing of raw materials to industry for value addition.

(iv) Local Community Co-Operation

Co-operation of the local communities for sustainable development of forest resources absolutely essential. Therefore local communities should be involved in development of tree resources to:

- Conserve forests.
- Supply firewood and fodder on sustainable basis.
- Reduce illicit felling of trees.

(v) Liberalizations of Ceiling Limits/Restrictions

- Statutory ceiling limits on agricultural land holdings for wastelands should be liberalized.
- Tree planting and farm forestry plantations should be treated on par with commercial crops like rubber, tea and coffee etc.
- Restriction on felling of farm growing trees and movement of wood from farm plantations to markets/factories should completely be abolished in all states.
- Support price for wood raw materials of different species needs to be amended at Government level in order to encourage tree farming and provide uniform returns.

REFERENCES

Anon. 2000. Annual Report. 1999-2000. Central Soil water Conservation research and Training Institute. Dehra Dun, India.

Anon. 2003. Report of The Task Force on Greening India for Livelihood Security and Sustainable Development. Planning Commission, New Delhi, Government of India.

Anon. 2004. Report of the committee on development of Biofuels. Planning Commission, New Delhi, Government of India.

Anon. 2005. Annual report, National Research Centre for Agroforestry , Jhansi (U.P.), India

Anon. 2006. All India Coordinated Research Project on Agroforestry. NRCAF, Jhansi.

Bargali, S.S., Singh, S.P. and Singh, R.P. 1992. Structure and function of an age series of eucalyptus plantations in central Himalaya. I. Dry matter dynamics. *Annals of Botany*, 69: 405-411.

Belsky, A.J, Mwonga, S.M. and Duxbury, J.M.1993. Effects of widely spaced trees and livestock grazing on understory environments in tropical savannas. *Agroforestry Systems,* 24: 1–20.

Bhojvaid, P.P., Timmer, V.R. and Singh, Gurbachan.1996. Reclaiming sodic soils for wheat production by *Prosopis juliflora* afforestation in India. *Agroforestry Systems,* 34: 139-150.

Campbell, B.M, Frost, P., King, J.A., Mawanza, M. and Mhlanga, L.1994. The influence of trees on soil fertility on two contrasting semi-arid soil types at Matopos, Zimbabwe. *Agroforestry Systems,* 28: 159–172

Chaturvedi and Khan, I.A. 2009. Agroforestry research Perspectives and Potential. *In:* Agroforestry: natural Resource Sustainability, Livelihood and Climate moderation (Eds. Chaturvedi, O.P., Venkatesh, A., Dwivedi, R.P., Singh, R and Dhyani, S.K. Satish Serial Publishing House, Delhi, India. pp27-42.

Chaturvedi, O.P. and Krishnan, P.R. 2009. Ecosystem development through agroforestry measures. In: Agroforestry: natural Resource Sustainability, Livelihood and Climate moderation (Eds. Chaturvedi, O.P., Venkatesh, A., Dwivedi, R.P., Singh, R and Dhyani, S.K. Satish serial Publishing House, Delhi, India. pp55-70

Dadhwal, K.S. and Tomar, V.P.S. 1999. Enhancing production by managing tree roots in an agroforestry landuse system SAIC Newsletter, Dhaka, Bangladesh. 9: 5&7.

Dela Cruz, L.U. and Luna, A.c. 1992. Effect of *Acacia auriculiformis* and *Gmelina arborea* on soil and microclimate of degraded grassland in Nueva Ecuja, Phillippines. Paper presented at an International symposium on Rehabilitation of Degraded Grasslands held during Sept 15-20, 1992 in Tsukubu, Japan.

Dhyani, S.K. 2009. Challenges and scope of agroforestry research in India. In: Agroforestry: natural Resource Sustainability, Livelihood and Climate moderation (Eds. Chaturvedi, O.P., Venkatesh, A., Dwivedi, R.P., Singh, R and Dhyani, S.K. Satish serial Publishing House, Delhi, India. pp 13-25

Dhyani, S.K. 2011. Alternate Land use and Agroforestry Systems for Resource Conservation and Enhanced Productivity in Hills. Proceedings of Workshop on Mountain Agricultutre in Himalayan Region: Status, Constraints and Potentials. April 2-3, 2011.pp107-132

Dhyani, S.K., Chauhan, D.S., Dinesh Kumar, Kushwaha, R.V. and Lepcha, S.T. 1996. Sericulture based agroforestry systems for hilly areas of north-east India, *Agroforestry Systems,* 34:1-12.

Dhyani, S.K., Chauhan, D.S., Singh, B.P. and Prasad, R.N. 1995. Tree effect on yield and quality of pineapple (kew) in a silvi-horti systems of agroforestry in NEH region of India. *Range Management & Agroforestry,* 16(2): 89-93.

Dhyani, S.K., K.S. Dadhwal and Katiyar, V.S. (1988). Biological measures for rehabilitation of abandoned mine area. *Indian Farming,* 38(7):15-18.16

Dhyani, S.K., Kareemulla, K. and Handa, A.K. 2006. Agroforestry potential and returns across agroclimatic zones. NRCAF, Jhansi.

GOI, 1999. National Forestry Action Programme, Vols. 1/2. Ministry of Environment and Forests, Government of India, New Delhi (Summary).

Hazra, C.R. 1994. Forages from rainshed agroforestry in Bundelkhand region of India and its influence on soil fertility. *In:* Agroforestry systems for degraded lands, (P.Singh. P.S. Pathak and M.M. Roy Eds.), Vol.1, Oxford & I.B.H. Pub. Co. Pvt. Ltd., New Delhi, India, pp140-147.

Hazra, C.R. 1998. Management of rain water resources on watershed basis for sustained agricultural production- an experience of Tejpura Watershed (Jhansi). Agricultural Situation in India. pp55: 120-127

Kaur, B., Gupta, S.R.and Singh, G. 2002. Carbon storage and nitrogen cycling in silvipastoral system on a sodic soil northwestern India. *Agroforestry Systems*, 54: 21-29.

Lal, P. 2004. Integerated development of farm forestry plantation and wood based industries. *Indian Forester*, 1: 71-78

Leakey, R.R.B., Sanchez, P.A. 1997. How many people use agroforestry products? *Agroforestry Today*, 9 (3) : 4–5.

Lundgren and Raintree, 1982. Sustained Agroforestry. In: Agricultural Research and Development-Potential and Challenges in Asia, bnestel (Ed) , ISNAR, The Hague. pp37-49.

Narain, P. Chaudhary, R.S. and Singh, R.K. 1994. Efficacy of conservation measures in northern hilly regions. *Indian Journal of Soil Conservation*, 22 (1-2): 42-62.

Pathak, P.S., Slanki, K.R., Rai, P., Handa, A.K., and Pateria, H.M. 2006[a]. Agroforestry. In: Hanbook of Agriculture. Indian Council of Agriculture Research, N Delhi

Pathak, P.S., Suresh, G., Khan,T.A., and Gupta, S.K. 2006[b]. Subabul in the rural economy of Andhra Pradesh-A case study. Indian Grassland and Fodder Research Institute. Jhansi, U.P. India. pp18-33.

Prasad, R., Pandey, D.N., Kotwal, P.C. 2000. Trees Outside Forests in India: A National Assessment. Indian Institute of Forest Management, Bhopal, India.

Prasad, R.N., Singh, A and Verma, A. 1986. Problems of hill lands and their management in North_East India. *Indian Journal of soil Conservation*, 14(3) : 66-72.

Puri, S., Swamy, S.L. and Jaijwal, A.K. 2002. Evaluation of *Populus deltoides* clones under nursery, field agrisilviculture system in sub-humid tropics of central India. *New Forests*, 23: 45-61.

Rai, P., Yadav, R.S.Solanki, K.R., Rao, and Rajendra Singh. 2001. Growth and pruned biomass production of multipurpose tree species in silvipastoral system on degared lands in semiarid region of Uttar Pradesh, India. *Forest Tree and Livelihood*, 11: 347-364.

Rai, S.N., Chakrabarti, S.K. 2001. Demand and supply of fuelwood and timber in India. *Indian Forester*, 127: 263–279.

Ram Newaj, Shanker, A.K.and Yadav, R.S. 2006. Carbon and Nitrogen dynamics in *Albizia procera* based agroforestry system. Annual report of ICAR ad-hoc Scheme, NRCAF, Jhansi.

Ravindranath, N.H., Hall, D.O., 1995. Biomass, Energy and Environment:ADeveloping Country Perspective from India. Oxford University Press, New York, 376p.

Samra, J.S. 1997. Status of research on watershed management. Presented at 173rd meeting of Governing body of ICAR, New Delhi.

Samra, J.S., Vishwantham, M.K. and Sharma, A.R. 1999. Biomass production of trees and grasses in a silvipasture systems on marginal lands of Doon Valley of north-west India. 2. Performance of grass species. *Agroforestry Systems*. 46:197-212.

Sathaye, J.A., Ravindranath, N.H. 1998. Climate change mitigation in the energy and forestry sectors of developing countries. *Annu. Rev. Energy Environ.*, 23: 387–437.

Schroth G. , Fonseca GAB , Harvey CA, Gascon C., Vascon cellos HL and Izac AMN. 2004. Agroforestry and biodiversity conservation in tropical landscapes. Island Press, Washington. 523p.

Singh, B.P., Dhyani, S.K. and Prasad, R.N. 1994. Traditional agroforestry systems and their soil productivity on degraded alfisols/ultisols in hilly terrain. In: Agroforestry systems for Degraded Lands (Eds. P. Singh, P.S. Pathak & M.M. Roy). Oxford & IBH Pub. Co. Pvt. Ltd., New Delhi. pp205-214.

Singh, C and Chaturvedi, O.P. 2011. Rehabilitation of degraded lands through silvipastoral systems – a balanced approach for alternative land use in the north-western Himalayas. In: Agroforestry Systems for Resources Conservation and Livelihood Security in Lower Himalayas (eds. Panwar, P; Tiwari, A.K. and Dadhwal, K.S.) New India Publishing Agency. New Delhi. pp273-296.

Singh, D.P. and Hazra, C.R. 1994. Rehabilitation of degraded wastelands through soil and water conservation and silvipasture at Gaharawa Watershed. In: Agroforestry systems for degraded lands (P.Singh, P.S. Pathak and M.M. Roy Eds.), Vol.1. Oxford & I.B.H. Pub. Co. Pvt. Ltd., New Delhi, India. pp54-62.

Singh, G. 1995. An agroforestry practice for the development of salt affected soils using *Prosopis juliflora* and *Laptochloa fusca*. *Agroforestry Systems*, 27: 61-75.

Singh, G. Singh T.N., Dager, J.C., Singh, H. and Sharma, V.P. 1997. An evaluation of agriculture, forestry and agroforestry practices in a moderately alkali soil in northwestern India. *Agroforestry Systems*, 37: 219-295

Singh, Gurbachan.1995. An agroforestry practice for the development of salt affected lands using *Prosopis juliflora* and *Leptochloa fusca*. *Agroforestry Systems*, 27 : 61-75.

Singh, Gurbachan.1999. Management of alkaline and saline soils through agroforestry. *In:* Sustainable Rehabilitation of Degraded Lands Through Agroforestry (Eds. K.R. Solanki, A.K. Bisaria and A.K. Handa). National Research Centre for Agroforestry, Jhansi. pp52-61.

Singh, K.A., B.P.S.Yadav and S.N.Goswani.1996. Forage resource management for sustained livestock productivity in Himalayan Agro-ecosystems. *Research Bulletin No.39*, ICAR Research Complex for NEH Region, Barapani. 94p.

Singh, P. 1986. Forage production in India, RMSI, IGFRI, Jhansi, 213p.

Singh, R.P. 1997. Alternate land use systems for sustainable development In: Silvipastoral systems in arid and semi arid ecsystems (Eds Yadav, M.S., Singh, M., Sharma, S.K., Tewari, J.C., and Burman, U) CAZRI, Jodhpur, India. pp. 249-266.

Singh, R.P. and Osman, M. 1995. Alternate land use systems for drylands. In Sustainable Development of dry land Agriculture in India (Ed., Singh, R.P) Scientific Publishers, Jodhpur. pp375-398.

Singh, T.P., Varalakshmi, V. and Ahluwalia, S.K., 2000. Carbon sequestration through farm forestry: case from India. *Indian Forester*, 126, 1257–1264.

Solanki, K.R., Bisaria, A.K. and Handa, A.K. 1999. A decade of research (1988-1998) National Research Centre for Agroforestry, Jhansi. 44p.

Solanki, K.R., Bisaria, A.K. and Handa, A.K. 1999. Multipurpose tree species: Research Retrospect and Prospect. Agribios.

Srivastava, A.K. 1994. Alternative land uses for sustained production in semi-arid vertisols. In: Agroforestry systems for degraded lands (P. Singh, P.S. Pathak and M.M. Roy Eds.) Vol.1, Oxford & I.B.H. Pub. Ltd., New Delhi, India. pp. 128-133.

Swamy, S.L., Puri, S, S. and Singh, A.K. 2003. Growth biomass, carbon storage and nutrient distribution in *Gmelina arborea* Roxb. Stands on red lateritic soils in central India. *Bioresource technology*, 90: 109-126.

Vishwanatham, M.K., Samra, J.S. and Sharma, A.R. 1998. Biomass production of silvi-pasture systems on gravelly lands of Doon Valley. *Bulletin No.T-39/D-26*, CSWCRTI, Dehra Dun,126p.

Vishwanatham, M.K., Samra, J.S. and Sharma, A.R. 1999. Biomass production of trees and grasses in a silvipasture system on marginal lands of Doon Valley of north-west India. 1. Performance of tree species. *Agroforestry Systems*, 46:181-196.

Young, A. 1997. Agroforestry for Soil Management. 2nd Ed. C.A.B. International. Wallingford, UK. 320p.

□□□

Chapter 46

Agricultural Intensification through Crop Diversity and Land Configuration for Maximizing Productivity, Profitability - Needs and Approaches

R. SAHA, R.S. CHAUDHARY, K.M. HATI AND J. SOMASUNDARAM

INTRODUCTION

Stagnating yield and declining input use efficiency in irrigated wheat of the Indo-Gangetic Plain (IGP) coupled with diminishing availability of water for agriculture is a major concern of food security in South Asia. Ensuring food security through sustainable agriculture involves achieving a proper balance between integrated nutrient and water management. Enhancing the crop productivity is a prime objective to feed the ever-growing population from shrinking natural resources. Assuming that the Indian population will stabilize around 1640 M by 2050, the gross per capita water availability will decline from 1820 m^3yr^{-1} in 2001 to as low as 1140 m^3yr^{-1} in 2050 (Gupta and Deshpande, 2004). Compared to the present water availability, it is estimated that by 2025, one third of developing countries including 50% of poor residing in India will be facing physical scarcity of water resources (Molden *et al.*, 2007). Similarly, the demographic projections indicate that per capita land availability has declined from 0.34 ha in 1950 to 0.14 ha now and will be further reduced to 0.10 ha in 2025. This implies that the increase in food productivity has to come by vertical expansion of productivity per unit area. Kanwar and Katyal (1997) estimated that India may need 301 M tons food grains by 2025, if present trend of population growth continues. This scenario shows that food grains production must be increased at the rate of 5-6 M tons yr^{-1} to keep pace with population growth.

Sustainability of the cropping system depends mainly on the soil fertility management which involves adequate and balanced nutrient supply. Therefore, the soil quality is determined by the efficient use of plant nutrients

Indian Institute of Soil Science, Berasia Road, Nabibagh, Bhopal - 462 038
Madhya Pradesh, India

through judicially balanced and integrated use of possible organic resources in conjunction with minimal inorganic chemical fertilizers. In 1950-51, only 0.5 kg nutrients ha^{-1} yr^{-1} was used to produce 51 M tons of food grains, which has arisen to 117 kg ha^{-1} yr^{-1} to produce 230 M tons in 2008-09. Similarly, emphasis of major nutrients has triggered widespread deficiencies of secondary and micronutrients like sulphur (41%), zinc (49%), boron (33%) with other micronutrients, *i.e.,* iron, copper, manganese, molybdenum deficiencies are on the rise (Singh, 2009). The need of accelerating fertilizer response through site-specific and cropping system-specific balanced and adequate nutrient application is, therefore an essential pre-requisite.

Poor soil physical environment require careful management to maintain conditions favorable for plant growth. The key point is that traditional soil physical properties like soil texture, structure, bulk density *etc.,* affect the management practices required to maintain water potential, oxygen diffusion rate, temperature and mechanical resistance in a range suitable for good crop production. The suitable management practices such as tillage, residue management, soil amendments, drainage and irrigation are, therefore, recommended to maintain soil physical health so as to achieve the higher crop production. Intensive tillage in conventional systems leads to gradual decline in soil organic matter through accelerated oxidation, with a consequent reduction in the capacity of the soil to regulate water and nutrient supplies to plants.

In the face of shrinking natural resources and ever increasing demand for larger food and agricultural production arising due to high population and income growths, agricultural intensification is the main course of future growth of agriculture. Because of changing rainfall pattern over years, ground water depletion, hike in labour wages, the existing cropping pattern may be economically viable. It is the time to critically re-design alternative cropping pattern based on agro climatic zone and this must be demonstrated in the farmers' holdings in order to effectively utilize the natural resources and also to stabilize the production and profitability.

Crop diversification -

- From *Low value to High value crops;*
- From *Water Loving crop to Water Saving crop;*
- From *Single crop to Multiple / Mixed crop*
- From C*rop alone to crop with* Crop-livestock-fish-apiculture
- From *Agriculture Production to Production with Processing and Value Addition.*

Crop Diversification & Land Configuration: Indian Perspective : With the advent of modern agricultural technology, especially during the period of the Green Revolution in the late sixties and early seventies, there is a continuous surge for diversified agriculture in terms of crops, primarily on economic considerations. The crop pattern changes, however, are the outcome of the interactive effect of many factors which can be broadly categorized into the following five groups:

i. Resource related factors covering irrigation, rainfall and soil fertility.

ii. Technology related factors covering not only seed, fertilizer, and water technologies but also those related to marketing, storage and processing.

iii. Household related factors covering food and fodder self-sufficiency requirement as well as investment capacity.

iv. Price related factors covering output and input prices as well as trade policies and other economic policies that affect these prices either directly or indirectly.

v. Institutional and infrastructure related factors covering farm size and tenancy arrangements, research, extension and marketing systems and government regulatory policies.

Although the factors that influence the area allocation decision of farmers are all important, they obviously differ in terms of the relative importance both across farm groups and resource regions. While factors such as food and fodder self-sufficiency, farm size, and investment constraints are important in influencing the area allocation pattern among smaller farms, larger farmers with an ability to circumvent resources constraints usually go more by economic considerations based on relative crop prices than by other non-economic considerations. Similarly, economic factors play a relatively stronger role in influencing the crop pattern in areas with a better irrigation and infrastructure potential. In such areas, commercialization and market networks co-evolve to make the farmers more dynamic and highly responsive to economic impulses.

What is most notable is the change in the relative importance of these factors over time. From a much generalized perspective, Indian agriculture is increasingly getting influenced more and more by economic factors. This need not be surprising because irrigation expansion, infrastructure development, penetration of rural markets, development and spread of short duration and drought resistant crop technologies have all contributed to minimizing the role of non-economic factors in crop choice of even small farmers. What is more, the reform initiatives undertaken in the context of the ongoing agricultural liberalization and globalization policies are also going to further strengthen the role of price related economic incentives in determining crop composition both at the micro and macro levels. Obviously, such a changing economic environment will also ensure that government price and trade policies will become still more powerful instruments for directing area allocation decisions of farmers, aligning thereby the crop pattern changes in line with the changing demand-supply conditions. In a condition where agricultural growth results more from productivity improvement than from area expansion, the increasing role that price related economic incentives play in crop choice can also pave the way for the next stage of agricultural evolution where growth originates more and more from value-added production.

Sensible crop diversification should be adopted with ultimate goal to employ economically viable, diversified crop combinations, to help moderate possible weeds, disease and pest problems, to enhance soil bio-diversity, to reduce labour peak and to provide farmers with new risk management opportunities.

Proper land configuration is must for utilizing the residual soil moisture effectively for winter *rabi* crop cultivation. Raised and sunken bed (RSB) system also called 'sorjan' is popular in Java and tidal wetlands in Indonesia, where high value upland crops are grown on raised beds and rice in sunken beds. The land configuration may be permanent or temporary. The temporary raised beds are made after harvesting of lowland rice and used mainly for cultivation of vegetables. Under such land configurations the sunken areas is left unutilized, as a results about 20-40 % land is wasted. Whereas, permanent raised and sunken beds configuration are used for cent per cent land utilization. Here vegetables and other arable crops are grown on raised beds and rice, fish *etc.* are grown on sunken beds. This would increase the cropping intensity as well as the farmers' income. Land levelling is a prerequisite for enhancing the benefits of other resource conservation technologies. In a well levelled field, less amount of water and nutrients are required than an uneven field. The benefits of applying same amount of inputs will be much higher in a levelled field due to their even distribution. In plains of North India, laser land leveller is used that level land to a perfection of ±2 cm from the average elevation. Only, by levelling land, the yield can be increased by 10-25%, saves water to the tune of 40%, increase the nutrient use efficiency by 15-25% and increases land area by 2-6% due to reduction in area required for bunds and channels (Jat *et al.*, 2006)

Crop diversification in rainfed regions aims to make rainfed agriculture achieving nutritional security, more employment and income generating, eco-friendly, poverty alleviation and comparative advantage in new trade regime

(Vittal *et al.*, 2007). Some of the issues and functions provided by diversification in rainfed regions are given below in Table 1.

Table 1: Issues and Functions provided by diversification in rainfed regions

Issue	Functions provided by diversification
Productivity and stability	
High risk and high cost	Risk and cost minimization through yield and income from annual and perennial mixtures
Unabated land degradation	Minimization of kinds, effect and extent of land degradation by appropriate land care through alternate land use systems
Inadequate employment	Staggered employment round the year
Low profitability	High income generation from various Components
Poor energy management	Energy efficient implements.

The crop diversification may results in the following anticipated benefits.

- Alternative crops may enhance profitability
- Diversified rotations can reduce pests
- Labour may be spread out more uniformly
- Different planting and harvesting times can reduce risks from weather
- New crops can be renewable resources of high value products

Effect of Crop Diversification : Diversified crop rotations can increase yield potential by influencing plant diseases, weeds, root distribution, moisture utilization, and nutrient availability. Ideally, crop diversification involves the inclusion of crop species with different resource requirements {*e.g.*, a high N-using crop such as corn combined with low N-using or N-fixing crops such as soybean [*Glycine max* (L.) Merr.] or lentil (*Lens culinaris* Medikus)}. Badaruddin (1994) determined that wheat yields following legumes were higher than yields following wheat and suggested that the benefit was due to a combination of N and non-N benefits. Rotational benefits also were seen in spring wheat grown after fababean (*Vicia faba* L. var. *faba*), lentil, and pea (*Pisum sativum* L.) over a range of N rates in studies conducted in Manitoba.

When water is limiting, rotating crops with different water use patterns and requirements can increase total crop yield and total nutrient removal. Larney (1995) reported that available water in the fall for winter wheat establishment was least after canola (45 mm), followed by continuous winter wheat (59 mm), lentil-flax (74 mm), and fallow (137 mm). Thus, there is potential for higher crop yields following lentil or flax than following canola or continuous winter wheat if the major limiting factor is available water. To optimize crop yield, the nutrient supply rates, whether from chemical fertilizers or organic amendments, must be matched to the yield potential supported by the available water.

Diversified crop rotations will not only influence the demand for nutrients, but may also influence the supply of nutrients to the growing crop. Crops differ substantially in the amount of N returned in the crop residue for use by subsequent crops because N supplied will depend on the amount of crop residue, primarily, and on the concentration of N in the residue. Nitrogen concentration in the residue will determine the net balance between immobilization and mineralization. Over the long term, as decomposition proceeds, all residues will eventually release the minerals they hold. The time required for this to occur will increase as the initial N concentration in the residue decreases and the C/N ratio widens. Straw from a well-fertilized wheat crop could decompose at a similar rate and produce similar amounts of N as a legume residue. Janzen and Kucey (1988) reported that N concentration of lentil, rape (*Brassica napus* L. var. *napus*), and wheat residue had a dominating influence on rate of residue decomposition and nutrient release Increasing the N concentration of the residue by fertilization increased rate of residue decomposition. Some residues also may have more of the nutrient present in a readily soluble inorganic or readily mineralizable organic form and so would release nutrients more readily than those that hold most of the nutrients in a more recalcitrant form. Therefore, species and nutrient management of the preceding crop will influence its nutrient content and the amount of nutrients it will release to the subsequent crop. Placement of residues and method of termination of the crop will also influence N release. Soil incorporation of residues reduces N loss by volatilization, enhances mineralization, and increases the short-term supply of plant-available N.

The amount and type of crop residue and factors influencing its rate of decomposition (*e.g.*, moisture) may also influence the availability of fertilizer N. If fertilizer N is applied with high C/N ratio residues, N immobilization may be substantial because the microbial population will utilize the fertilizer N to build biomass as the residue decomposes. Further, residues can enhance the loss of N by volatilization from broadcast urea [$(NH_2)_2CO$] because the urease enzyme present in the residues can increase the rate of NH_3 release (McInnes *et al.*, 1986). Therefore, where high amounts of crop residues are present, particularly if they contain a wide C/N ratio, separating the fertilizer from the residues by placing N fertilizers below the soil surface can increase fertilizer use efficiency.

Phosphorus dynamics can be affected by cropping intensity and diversification. Intensified cropping in the absence of P inputs from fertilizer or organic amendments will result in a depletion of soil P. McKenzie *et al.* (1992a) (1992b) evaluated the effect of cropping system and fertilizer management on P in two long-term rotation studies in Alberta. They found that without fertilizer addition, continuous cropping resulted in the greatest reduction of almost all soil organic and inorganic P pools. By increasing the frequency of fallowing, the drain on the soil P pools was generally reduced due to less crop uptake of P. Similarly, in a 24 yr study in the Brown soil zone at Swift Current, SK, annual P removal in the grain in fertilized systems decreased with an increase in fallow frequency, increasing soil P from 4.8 to 5.3 kg P ha^{-1} for continuous wheat, 3.5 to 3.8 kg P ha^{-1} for fallow–wheat-wheat, and 3.1 kg P ha^{-1} for fallow–wheat (Selles *et al.*, 1995). However, if continuous cropping was coupled with the addition of N and P fertilizers, there was a positive effect of cropping on P availability, as a balance of P calculation indicated that all P-fertilized systems received more P than was exported in the grain. Bowman and Halvorson (1997) also reported increases in P availability under a continuous cropping system compared with wheat-fallow systems even through P inputs were generally greater in the latter system. The authors attributed the increased P availability to redistribution of soil P from lower depths through biocycling in residue and litter production.

Among the various land surface management practices like raised and sunken bed, ridges and furrow etc. developed for Vertisols, broad-bed and furrow (BBF) system is very promising in controlling surface runoff, reducing the soil loss through erosion and increasing infiltration. The BBF land management system essentially reduces the velocity of runoff water and consequently increases opportunity time for water to infiltrate and reduces sediment losses. Further, during the period of heavy rainfall the furrows allow excess water to drain safely from the plots and thus avoid water congestion to the crop.

Field experiment was conducted on broad bed and furrow (BBF) and flat on grade (FOG) land treatments with five different cropping systems *viz.*, Soybean- chickpea, maize- chickpea, soybean/ maize intercropping- chickpea, soybean/ pigeon pea intercropping and maize/ pigeon pea intercropping and two irrigation levels on a micro-watershed at the experimental farm of Indian Institute of Soil Science, Bhopal. The BBF landform was prepared with the help of a tractor drawn BBF former along the key lines drawn based on a topographic survey. The width of the broad bed was 1.0 m with 0.5 m wide furrows on either side of the bed. In the first year, pigeonpea was taken as mono-crop in lieu of maize/pigeon pea intercropping. In rainy season crops were grown rainfed while in winter season chickpea was grown with two irrigation levels, (i) one pre-sowing (PS) irrigation to chickpea (I_1) and (ii) one PS + one irrigation to chickpea at flowering stage (I_2). The irrigation was provided with the water harvested in a dugout pond at the watershed.

Results showed that irrespective of irrigation to chickpea and cropping systems, results revealed that total system productivity (TSP) as soybean equivalent yield was greater in BBF than FOG; and TSP was higher in I_2 (pre-sowing plus 1 post sowing irrigation) than I_1 (pre-sowing irrigation). Among the 5 cropping systems, there was significant difference in the total productivity of systems (Table 2). Soybean-chickpea system was found to be the least productive except in the first year (2003-04). After 2003-04, system productivity was not favourable for the soybean-chickpea system, because of constantly lower yield of soybean over years, while the maize yield was considerably higher. Consequently, the systems involving maize crop, either as sole or intercrop (as in maize-chickpea, soybean/ maize intercropping-chickpea and maize/ pigeonpea intercropping systems) gave higher productivity than other systems under both BBF and FOG land treatments. Even the TSP was higher in maize/ pigeonpea intercropping systems where there was no subsequent chickpea crop. In the event of non-availability of irrigation water to chickpea, maize/ pigeonpea intercropping is better system than sole soybean. Thus, these three cropping systems *viz.*, maize-chickpea, soybean/ maize intercropping-chickpea and maize/ pigeonpea intercropping *i.e.*, diversification from the sole soybean, hold the promise for increasing productivity in this region.

Table 2: Effect of land configuration and crop diversification on total system productivity expressed as soybean equivalent yield (SEY) in Vertisols

Cropping system	Total system productivity as SEY (kg ha^{-1})							
	Broad-bed and furrow				Flat-on-grade			
	2003-04	2004-05	2005-06	2006-07	2003-04	2004-05	2005-06	2006-07
Irrigation to chickpea								
I_1	2818b	2747b	3857b	3551b	2257b	2425b	3370b	3165b
I_2	2929a	2903a	4196a	3900a	2422a	2542a	3591a	3576a
Cropping systems								
Soybean-chickpea	3530a	2109d	3019c	3044c	2698a	1894c	2613c	2691c
Maize-chickpea	2931b	3457a	4507a	4354a	2295b	3036a	3832a	3959a
Soybean/maize - chickpea	3385a	2807b	4564a	4145ab	2798a	2485b	3933a	3765b
Soybean/pigeonpea	2615c	2369c	3532b	3134c	2262b	2027c	2912b	2778c
Maize/pigeonpea*	1907d	3385a	4513a	3951b	1646c	2975a	4112a	3659b

*There was sole crop of pigeonpea in the year 2003-04

Runoff and soil losses from the field area under broad-bed and furrow (BBF) and flat on grade (FOG) landform treatments were monitored during the rainy seasons. In all the years of our study, total seasonal runoff from the BBF plot was found to be less than that from the FOG (Table 3). This was owing to the reduced speed of runoff from BBF plot due to its uniform slope, which has resulted in higher opportunity time for water to infiltrate more in BBF than FOG treatment. The runoff was 15.4-33.2% and 20.3-57.7% of seasonal rainfall from BBF and FOG landform treatments, respectively during the years of experimentation. The runoff under both BBF and FOG was much higher during the rainy season of 2006 because of unusually high rainfall. The soil losses through runoff from BBF and FOG were higher in high rainfall years; the extent of soil loss was to the tune of 1956 and 2837 kg ha^{-1} from BBF and FOG, respectively in 2003 and 3503 and 6365 kg ha^{-1} in the corresponding treatments in 2006. However, the soil losses were relatively less, 657 and 1466 kg ha^{-1} from BBF and FOG, respectively in 2004 when the seasonal rainfall was quite low.

Table 3: Seasonal rainfall, runoff, and soil loss from broad-bed and furrow (BBF) and flat on grade (FOG) land configurations

Year	Rainfall	Runoff (mm)		Soil loss (kg ha^{-1})	
	(mm)	BBF	FOG	BBF	FOG
2003	1058.0	163.0 (15.4%)	214.9 (20.3%)	1956.0	2836.9
2004	798.2	124.0 (15.5%)	183.3 (23.0%)	657.0	1466.0
2005	946.0	177 (18.7%)	246 (26.1%)	1402.0	3123.0
2006	1513.0	502 (33.2%)	873 (57.7%)	3503.0	6365.0

Values within parentheses indicate the percent of seasonal rainfall

The runoff and soil losses from broad-bed and furrow (BBF) are less than the flat land. Besides this, BBF also helps in safe drainage of excess rainfall and reduces chance of water congestion to the rainy season crops while it retains higher moisture during the later phase of crop growth after withdrawal of monsoon and produced higher crop yield than the traditional flat land sowing system. Farmers may adopt BBF land configuration for growing of crops like soybean, maize, pigeonpea and chickpea. The BBF can also be prepared by bullock-drawn BBF maker, which will be useful to the subsistence farmers who have no tractor to prepare the BBF. The study also provided an option for crop diversification from the present predominantly soybean based cropping systems to maize-chickpea, soybean/ maize intercropping-chickpea and maize/ pigeonpea intercropping in the watershed. The study suggests for the cropping system where maize is a component, either as a sole or intercrops for the region. Water lost as surface run-off should be conserved in watershed ponds and used as supplemental or life-saving irrigation. This will increase system productivity as well as resource-use efficiency in vertisols and maintain soil health by reducing the loss of valuable top soil and nutrients.

Crop Diversification and Land Configuration: Scenario of North-East India

The North eastern region is generally rich in organic matter and nutrient contents as they receive a lot of runoff from surrounding hillocks, which bring a considerable amount of nutrients to the field. The local farmers grow rice without any external nutrient application or in some rare cases very low quantity of organic manure *i.e.* FYM at 5 t ha^{-1} is used once every 3–4 years. However, the land-locked narrow valley lands despite being bestowed with a bountiful of water resources and receiving the highest rainfall, experiences acute shortage of water during post monsoon period. Farmers are not in position to take any crop during *rabi* season because of severe moisture stress. A simple land configuration through raised and sunken bed (RSB) system, in this context, is a useful technology for proper land and water management, inter-plot water harvesting to increase crop intensity (Sharma, 2003). The raised-sunken bed system increased the depth and duration of submergence in sunken bed plots and it remained flooded with water up to 90 - 100 days after transplanting. It was also observed that the narrow raised - sunken bed ratio (1:1 and 1:1.5) was not effective for enhancing the rice and french bean production even though the long duration of water submergence in sunken bed and high moisture status in raised bed (Mishra and Saha, 2007).

Field experiments with rice-vegetable crop sequence showed that rice-tomato gave the highest rice equivalent yield (214.40 q ha^{-1}), followed by rice-carrot (206.4 q ha^{-1}). Highest net return (₹ 66,635 ha^{-1}) was recorded in rice-carrot, closely followed by rice-tomato (₹ 66,139 ha^{-1}). Results showed that improved soil fertility with higher available N (277 kg ha^{-1}), P (8.76 kg ha^{-1}), and K (353 kg ha^{-1}) was noticed in the rice–french bean cropping sequence followed by the rice–potato and rice–tomato sequences (Table 4). The soil biological properties such as population of Rhizobium, bacteria, phosphorus-solubilizing microorganisms, and earthworm activity were all found to be remarkably higher in the experimental field (Das Anup *et al.*, 2008).

Table 4 : Soil fertility status as influenced by in-situ residue management in different cropping sequences (pooled data)

Cropping sequence	Available N (kg ha^{-1})	Available P (kg ha^{-1})	Available K (kg ha^{-1})
Rice-potato	263.7	7.93	349
Rice-tomato	259.5	7.88	346
Rice-frenchbean	277.0	8.76	353
Rice-carrot	266.4	7.86	337
Rice-cabbage	252.2	7.47	328
Rice-fallow	268.0	7.67	348
CD (P = 0.05)	18.24	0.70	11
Initial	264.0	6.97	321

Saha and Ghosh (2010) evaluated the water economy and performance of different crop combinations under standard raised and sunken bed (RSB) system of cultivation under this agro-climatic condition and reported that the profile soil moisture content was significantly higher in rice-tomato/garden pea cropping system (29.3%), lowest being in rice-fallow system (22.5%). The water extraction pattern in the soil profile under different crop combinations reflected that the moisture extraction was maximum (32.2 to 44.5%) from 0-15 cm soil depth irrespective of crop combinations and it also decreased with soil depth. Among the various crop combinations, rice-tomato/garden pea gave the highest rice equivalent yield and production efficiency (18138 kg ha^{-1} and 77.18 kg ha^{-1} day^{-1}, respectively) followed by rice-potato/garden pea system (16982 kg ha^{-1} and 76.50 kg ha^{-1} day^{-1}, respectively).

Photo 1: Land configuration in the form of RSB for increasing cropping intensity, nutrient and water use efficiency

CONCLUSIONS

Diversification and intensification of cropping systems will have major effects on nutrient dynamics. Removal of nutrients will increase because of higher annualized crop yields. This increased removal must be balanced by increased nutrient inputs to ensure optimal crop yield and quality while avoiding soil depletion. Increased return of crop residue from greater annualized crop production can lead to higher soil organic matter levels and a greater potential for nutrient cycling, an effect that will increase with time. Cropping systems that include legumes have a greater potential for contributing N to following crops and may be used to moderate soil NO_3 levels to avoid potential for NO_3 leaching. Phosphorus availability may be influenced by depletion or accumulation of P, based on past cropping and fertilizer management. In addition, preceding crop may influence P availability through residue effects and impacts on VAM activity. Synchrony of nutrient supply with crop demand is essential to ensure optimum crop yield and quality while avoiding negative environmental impacts.

REFERENCES

Badaruddin, M. 1994. Grain legume effects on soil nitrogen, grain yield, and nitrogen nutrition of wheat. *Crop Science*, 34:1304–1309

Bowman, R.A.and Halvorson, A.D.1997. Crop rotation and tillage effects on phosphorus distribution in the central Great Plains. *Soil Science Society of America Journal*, 61:1418–1422

Das, Anup., Patel, D.P., Munda, G.C., Hazarika, U.K., Bordoloi and Jurisandhya. 2008.Nutrient recycling potential in rice-vegetable cropping sequences under in-situ residue management at mid altitude subtropical Meghalaya. *Nutrient Cycling in Agroecosystems*, 82: 251-258.

Gupta, S.K. and Deshpande, R.D. 2004. Water for India in 2050: first order assessment of available options. *Current Science*, 86: 1216-1224.

Janzen, H.H. and Kucey, R.M.N. 1988. Carbon, nitrogen, and sulfur mineralization of crop residues as influenced by crop species and nutrient regime. *Plant and Soil*, 106:35–41

Jat, M.L., Chandna Parvesh, Gupta, Raj, Sharma, S.K. and Gill, M.A. 2006. Laser Land Leveling: A Precursor Technology for Resource Conservation. Rice-Wheat Consortium Technical Bulletin Series 7. New Delhi, India: Rice-Wheat Consortium for the Indo-Gangetic Plains. 48p.

Kanwar, J.C. and Katyal, J.C. 1997. *Proceedings of the Symposium on Plant Nutrient Needs, Supply, Efficiency and policy issues, 2000-2025*. National Academy of Agricultural Sciences, New Delhi, India.

Larney, F.J. 1995. Rotation and tillage effects on available soil water for winter wheat in a semi-arid environment. *Soil and Tillage Research*, 36:111–127.

McInnes, K.J., Ferguson, R.B., Kissel, D.E. and Kanemasu, E.T. 1986. Ammonia loss from applications of urea-ammonium nitrate solution to straw residue. *Soil Science Society of America Journal*, 50:969–974.

McKenzie, R.H, Stewart, J.W.B., Dormaar, J.F. and Schaalje, G.B. 1992a. Long-term crop rotation and fertilizer effects on phosphorus transformations: I. In a Chernozemic soil. *Canadian Journal of Soil Science*, 72:569–579.

McKenzie, R.H, Stewart, J.W.B., Dormaar, J.F. and Schaalje, G.B. 1992b. Long-term crop rotation and fertilizer effects on phosphorus transformations: I. In a Luvisolic soil. *Canadian Journal of Soil Science*,72:581-589.

Mishra, V.K. and Saha, R. 2007. Effect of raised-sunken bed system on inter-plot water harvesting and productivity of rice (*Oryza sativa*) and french bean (*Phaseolous vulgaris*) in Meghalaya. *Indian Journal of Agricultural Sciences*, 77 (2): 73-78.

Molden, D. (Ed.) 2007. *Water for Food, Water for Life – A Comprehensive Assessment of Water management in Agriculture*. London, UK. Comolbo, Sri Lanka: IWMI.

Saha, R. and Ghosh, P.K. 2010. Effect of land configuration on water economy, crop yield and profitability under rice (*Oryza sativa*) -based cropping systems in north-east India. *Indian Journal of Agricultural Sciences*, 80(1): 16-20.

Selles, F., Campbell, C. A. and Zentner, R.P. 1995. Effect of cropping and fertilization on plant and soil phosphorus. *Soil Science Society of America Journal*, 59:140–144.

Sharma P K. 2003. Raised-sunken bed system for increasing productivity of rice-based cropping system in high rainfall areas of Himachal Pradesh. *Journal of the Indian Society of Soil Science*, 51(1): 10-16.

Singh, A. 2009. Development, conservation and utilization of soil resources-random thoughts. *Newsletter, Indian Society of Soil Science*, March No.26.

□□□

Chapter 47

Growth and Economics of Cowpea as Influenced with Fertility Levels in *Eucalyptus Tereticornis* Based Agri-Silviculture System

K.S.AHLAWAT, R.S.PANNU, O.P.LATHWAL AND VINOD KUMAR[1]

INTRODUCTION

India is basically an agricultural country and about 70 % of its people live in villages. Their livelihood is dependent mainly on agriculture and animal husbandry. During the recent years, agroforestry has received the attention of the farmers across the world especially in developing countries, where both forests and agricultural land are under severe stress owing to population pressure and industrialization. Increased food production is the prime concern of planners, extension workers and research scientists engaged in field of agriculture with rapid growing population. The importance of agroforestry for food, fuel, fodder, fruits, fertilizer, timber, etc. and also in conservation of natural resources have been well recognized. However, tree integration in agroforestry system may result in competitive, complementary or supplementary interaction with respect to soil moisture, allelopathy, radiation and plant nutrient balance as well as changes in microclimate affecting the significance of the system and its components. Globally 46 % of agricultural land (over one billion ha) has more than 10% tree cover, while 17 % of farming land has more 30 % tree cover. This land with more than 10 % tree cover is occupied by 31% of the people living in agricultural land (558 M ha) (*Zomer et al*, 2009).

Eucalyptus (*Eucalyptus tereticornis*), is a major species in agroforestry systems in India due to its fast rate of growth, good timber value and coppice ability. Moreover, it has the ability to tolerate a variety of climates ranging

---

Krishi Vigyan Kendra, Kurukshetra-136 118, Haryana, India
[1]Regional Research Station, Bawal

from dry tropical to moist forest zones. For this reason it has been introduced in more than 17 countries including India (Mariani *et al.*, 1981). The initial objective of raising *Eucalyptus tereticornis* was to produce pulpwood; however, now-a-days the wood is also being used for the manufacture of packing cases, poles, crates, furniture and construction of inexpensive dwelling in villages. Keeping in view multipurpose uses and its role in bio-drainage, *Eucalyptus tereticornis* plantation is likely to be adopted on large-scale on government and private lands.

Block plantation of eucalypts has adverse effect on associated crops because being evergreen in nature it continues to compete for light, moisture and nutrient throughout the year and is also allelopathic in nature. However, meager information is available on the magnitude of adverse effect of eucalypts trees on different agriculture crops. The fertilizer requirement of crops intercropped with eucalypts may also vary as compared to sole cropping. Subrahmanyam *et al.* (2001) reported that nitrogen levels increased the yield attributes of cowpea and fodder yields of sorghum increased with increased levels of nitrogen under *Hardwickia binata* plantation. The present study, therefore, was carried out to evaluate the optimum fertilizer requirement of cowpea in eucalypts based agrisilviculture system and influence of fertilizer levels on physiological and economics of cowpea under *Eucalyptus tereticornis* based agrisilviculture system *vis-à-vis* sole crop.

MATERIAL AND METHODS

Present investigation was carried out at the research area of Agroforestry Department, CCS Haryana Agricultural University, Hisar (20°10′ N, 75°46′ E, 215 m above mean sea level), situated in the arid region of north-western India. The climate is subtropical-monsoonic with an average rainfall of 350-400 mm (70-80 % of which occurs during July to September). The soil of the experimental fields was sandy loam in texture and available N, P, K were in low, medium and high range, respectively under both the agrisilviculture and sole cropping system. The total rainfall of 127 mm and 386 mm was received in entire growth period of crop during 2004 and 2005, respectively.

Cowpea was grown in a plot measuring 6m x 4 m during rainy season of 2004 and 2005 in between the rows of *Eucalyptus tereticornis* (planted in July, 1999 at a spacing of 6m x 2m) and as sole crop (without eucalypts) in the field adjacent to eucalypts plantation. The Eucalyptus tree height was 12.38 m and 14.90 m, girth was 40.20 cm and 45.18 cm and canopy width 3.98 m and 4.09 m, respectively during July, 2004 and July, 2005. The experiment consisting of four fertilizer levels *viz.* F_0 (control *i.e.* ,without fertilizer), F_1 (75 % of recommended dose of fertilizer i.e. 18.75 kg N + 46.87 kg $P_2O_5 ha^{-1}$), F_2 (100 % of recommended dose of fertilizer *i.e.*, 25 kg N + 62.5 kg P_2O_5 ha^{-1}) and

F_3 (125 % of recommended dose of fertilizer *i.e.,* 31.25 kg N + 78.12 kg P_2O_5 ha^{-1}), was laid out in randomized block design with four of replications with eucalypts and as sole crop. The recommended dose of fertilizer for cowpea is 25 kg N + 62.5 kg P_2O_5 ha^{-1}. Before the application of pre-sowing irrigation field was ploughed twice. After the application of pre-sowing irrigation two ploughings with disc harrow and one with cultivar followed by planking were given to prepare a good seed bed for sowing of cowpea. The entire amount of nitrogen and phosphorus was drilled at the time of sowing in cowpea as per treatments. The sources of N and P were urea (46% N) and single super phosphate (16 % P_2O_5), respectively. Cowpea (G.C.-3) was sown in north-south direction on 6 July 2004 and 12 July 2005 by hand plough at a row spacing of 30 cm, using seed rate of 50 kg seed/ha in both the systems. The growth parameters of cowpea were recorded at various stages in both the systems. The physiological parameters of cowpea were recorded at various growth stages in both the systems. The crop growth and relative growth rate were measured at 21-40, 41-60, 61-72 and 21-40, 41-60, 61-75 stages during 2004 and 2005, respectively and were calculated by using the following formula :

$$CGR = \frac{w_2 - w_1}{t_2 - t_1}$$

$$RGR = \frac{\log w_2 - w_1}{t_2 - t_1}$$

Where,

w_2 and w_1 = Total dry matter at successive stages

t_2 and t_1 = Time interval

Photosynthesis, transpiration and stomatal conductance were measured at 40 DAS (Days after sowing) and at harvest of cowpea with the help of portable steady photosynthetic apparatus (IRGA, CIRAS-1, PP system, U.K.). Photosynthesis, transpiration and stomatal conductance of fully expended leaves were recorded on clear sky day during 1100-1300 hours of day. Photosynthetic pigments were estimated according to the Hiscox and Isractstman (1979) using dimethyl sulphoxide (DMSO).

Cost of cultivation and gross income of various treatments were calculated on the basis of approved market rates for inputs and outputs. Net returns (₹ ha^{-1}) were worked out by subtracting the total cost of cultivation of each treatment from the gross income of the respective treatment. The cost and return of eucalypts based agrisilviculture system comprising of crop with life cycle of one growing season and eucalyptus with 8.5 years was computed for two experimental years on per hectare basis. The expected establishment and maintenance cost of eucalypts was computed from the first year up to 8.5 years rotation of tree. The expected girth (60 cm) of the eucalyptus after 8.5 years was calculated on the basis of annual girth increment of 5 cm obtained during the course of experiment. The numbers of trees to be harvested (708) were calculated after taking into account 15 % mortality. The expected return from eucalyptus trees ha^{-1} was computed on the basis of market rates per tree (₹ 651) prevailing during December, 2005 and assuming the same rates after 8.5 years. Cost/benefit ratio was computed by, dividing discounted return by discounted cost. The experimental data collected during the course of investigation were subjected to statistical analysis by "Analysis of variance" technique by Fisher (1948) and Cochran and Cox (1959). Critical difference (CD at 5%) was worked out for differentiating the treatment effects from those of chance effects.

RESULT AND DISCUSSION

Crop growth rate (CGR) and relative growth rate (RGR) : CGR of cowpea increased significantly with increase in fertilizer levels upto recommended dose of fertilizer at all crop stages except that at 41-60 days interval it increased significantly upto 125 % of recommended dose of fertilizer during both the years and in both the systems (Figure 1). The maximum RGR in the test crop was recorded at 21-40 days interval and then decreased upto harvest during both the years and in both the systems (Figure 2). At 21-40 days interval and at harvest RGR increased significantly with increase in fertilizer levels upto recommended dose of fertilizer in sole crop during both the years. Whereas, in agroforestry it increased significantly only upto 75 % of recommended dose of fertilizer at harvest. At 41-60 days interval RGR increased significantly with increase in fertilizer levels upto 125 % of recommended dose of fertilizer in both the systems and during both the years. The reduced CGR and RGR in eucalypts sown cowpea as compared to sole cowpea were mainly due to the reduction in incidental solar radiation. The mean percent light available to crop under agroforestry was only 15.7 and 15.1 % at 40 and 60 DAS, respectively as compared to open area.

Photosynthetic pigments, photosynthesis, transpiration and stomatal conductance : Total chlorophyll and carotenoid content increased with the advancement of crop growth upto harvest during both the years and in both the systems. At all crop growth stages except 20 DAS total chlorophyll and carotenoid content increased significantly upto 125 % of recommended dose of fertilizer in agroforestry and sole cowpea during both the years (Figure 3). However, the differences between 100 and 125 % recommended dose of fertilizer were not significant at harvest in both the systems during both the years. Total chlorophyll and carotenoid content was less in agrisilviculture system as compared to sole cowpea. Patidar and Mali (2004) have reported that application of 100 % recommended dose of fertilizer significantly increased chlorophyll content in sorghum as compared to lower levels.

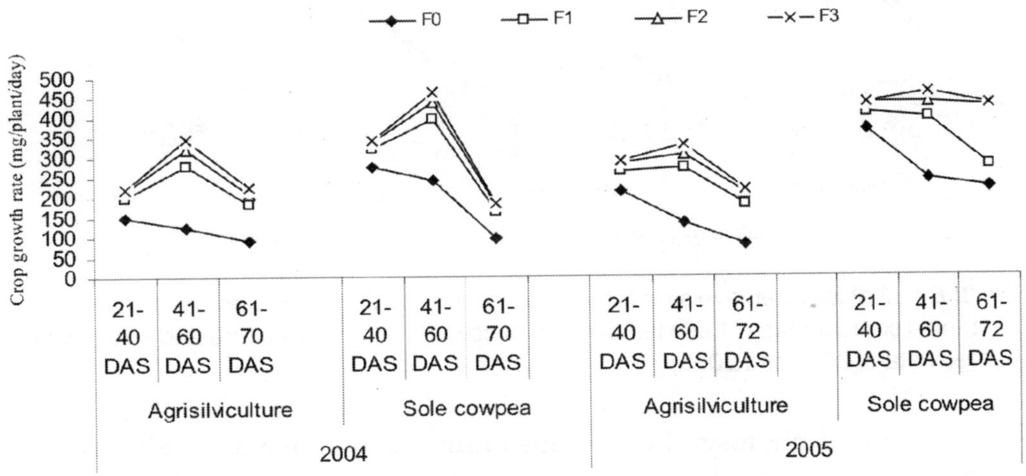

Fig. 1: Effect of fertilizer levels on crop growth rate (mg plant^{-1}day^{-1}) of cowpea in agrisilviculture and sole cowpea (pooled data of 2004 and 2005)

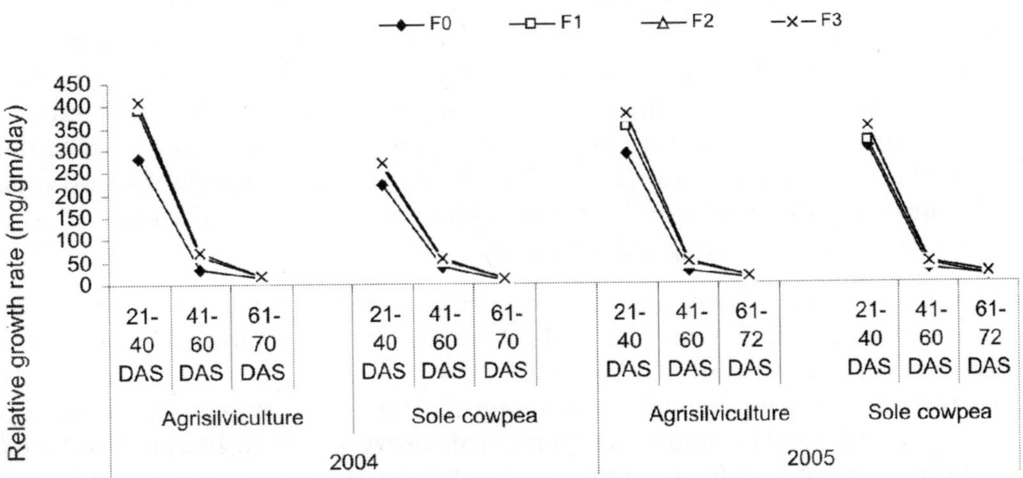

Fig. 2: Effect of fertilizer levels on relative growth rate (mg plant^{-1}day^{-1}) of cowpea in agrisilviculture and sole cowpea (pooled data of 2004 and 2005)

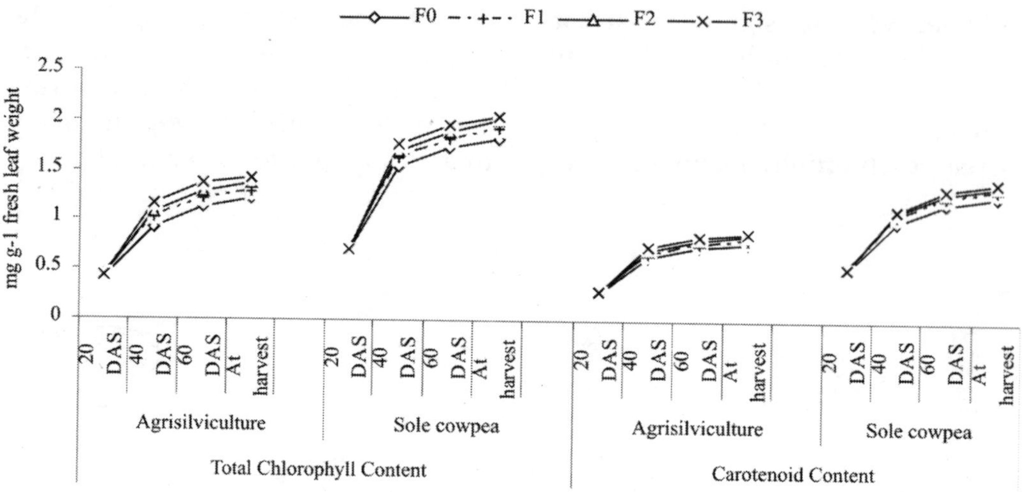

Fig. 3: Effect of fertilizer levels on total chlorophyll content (mg g^{-1} fresh leaf weight) and carotenoid content (mg g^{-1} fresh leaf weight) of cowpea in agrisilviculture and sole cowpea (pooled data of 2004 and 2005)

The rate of photosynthesis, transpiration and stomatal conductance in cowpea, were slightly higher at harvest as compared to 40 DAS of crop in both the systems. Photosynthesis, transpiration and stomatal conductance were more in the sole cowpea as compared to the agroforestry during both the years and at both stages (40 DAS and at harvest). The rate of photosynthesis, transpiration and stomatal conductance increased significantly with increase in fertilizer levels upto recommended dose of fertilizer at 40 DAS of crop and at harvest during both the years and in both the systems (Figure 4). The positive response of these physiological characters to fertilizer particularly nitrogen application was due to its favorable effect on cell enlargement resulting in production of larger leaves which increased the photosynthetic area of plant canopy. Roy (2000) have also reported that the application of nitrogen and its split application influenced photosynthesis, transpiration, stomatal conductance, chlorophyll 'a', chlorophyll 'b' and carotenoids content in flag leaf of barley

On the basis of average of two crop seasons 31.9 and 31.6 %, less photosynthesis, 45.3 and 45.1 % less transpiration, 27.0 and 27.7 % less stomatal conductance at 40 days after sowing and at harvest, respectively was recorded under eucalypts sown cowpea than sole cowpea. The reduction in the physiological parameters under agroforestry was mainly due to reduced sunlight, competition for moisture and nutrients and to some extent effect of allele-chemicals. The higher physiological efficiency of test crop in terms of higher photosynthesis rate, stomatal conductance, transpiration rate and

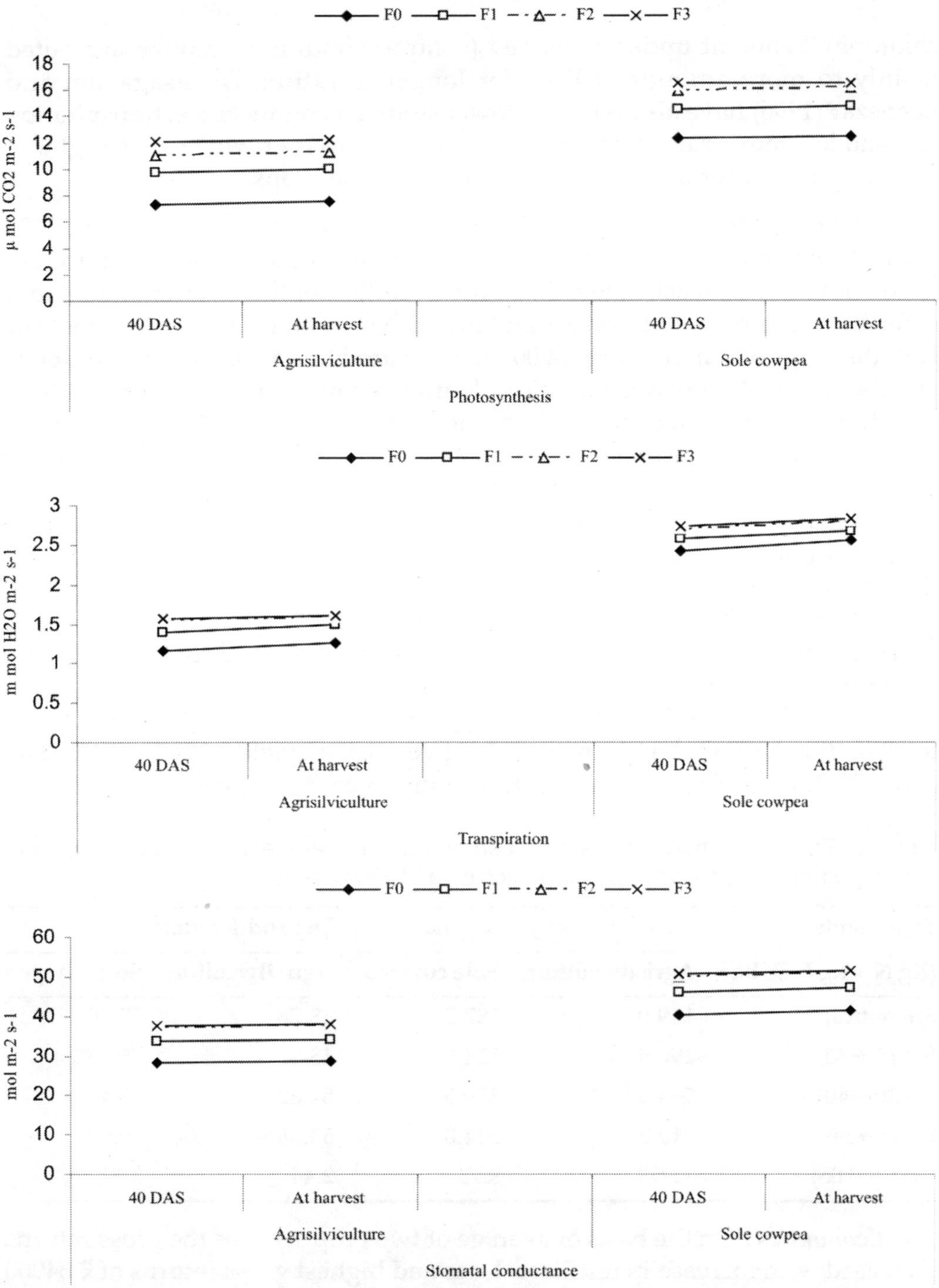

Fig. 4: Effect of fertilizer levels on photosynthesis (μ mol CO_2 m^{-2} s^{-1}), transpiration (m mol H_2O m^{-2} s^{-1}) and stomatal conductance (mol m^{-2} s^{-1}) of cowpea in agrisilviculture and sole cowpea (pooled data of 2004 and 2005

chlorophyll content under unshaded (control) conditions may be attributed mainly to more amount of PAR for longer duration. Devasagayam and Ebenezar (1996) have also reported lower stomatal conductance, transpiration rate and leaf temperature under *Eucalyptus tereticornis* grown crops (sorghum, cowpea, groundnut and blackgram) than the sole crops.

Green and dry fodder yield : Green fodder yield was lower during 2004 than 2005 in both the systems. There was significant increase in green fodder yield with 25 % increase in fertilizer dose over 75 % of the recommended dose of fertilizer during both the years and in both the systems. Dry fodder yield of both the crops was more during 2005 as compared to 2004 in both the systems. Dry fodder yield increased significantly upto recommended dose of fertilizer in both the systems and during both the years (Table 1). Better growth with the application of phosphate was probably due to good root ramification and absorption of plant food material. Besides this phosphorus is also an essential part of many plant constituents *i.e.* phospholipids, ATP *etc.* A favourable supply of phosphorus might have increased the ATP synthesis in plant and thereby might have accelerated the metabolic processes. Nandal and Singh (2001) have reported decreased yield of cowpea and sorghum under different spacings of *Dalbergia sissoo*. Jamadagni *et al.* (2002) reported that fodder yield of cowpea linearly increased with the increase in phosphorus levels. Likewise, Bhilare and Patil (2002) reported that increasing levels of phosphorus from 0 to 30 and 60 kg P_2O_5 ha^{-1} increased the green forage yield of cowpea by 3.45 and 7.72 % and dry matter yield by 4.24 and 8.25 %, respectively.

Table 1: Effect of fertilizer levels on green and dry fodder yield (q ha^{-1}) of cowpea in agrisilviculture and as sole cowpea (pooled data of 2004 and 2005)

Treatments	Green fodder yield (q ha^{-1})		Dry fodder yield (q ha^{-1})	
(Kg N + kg P_2O_5 ha^{-1})	Agrisilviculture	Sole cowpea	Agrisilviculture	Sole cowpea
F_0 (control)	149.0	287.5	38.78	67.60
F_1 (15 + 30)	190.0	324.0	48.83	75.76
F_2 (20 + 40)	204.5	336.5	51.82	78.41
F_3 (25 + 50)	213.5	344.0	53.36	79.98
CD (P=0.05)	11.58	9.73	2.44	2.11

Economics : On the basis of average of two crop seasons the gross returns increased with increase in fertilizer doses and highest gross returns of ₹ 64900 ha^{-1} recorded at 125 % recommended dose of fertilizer in eucalypts based agri-silviculture system (Table 2). In sole cowpea, highest gross returns of ₹ 17225 ha^{-1} were recorded at 125 % recommended fertilizer level, on the basis of two crop seasons. Under both the land use systems maximum net returns were recorded at recommended fertilizer level during both the years. In agrisilviculture system, maximum cost/benefit ratio was recorded in control, however in sole cowpea, maximum BCR was recorded at recommended fertilizer dose during both the years. Kumar *et al.* (2003) have also reported almost double net profit from poplar based agroforestry as compared to sole agricultural crops. Kumar (2006) found maximum net return of ₹ 25146 $ha^{-1}yr^{-1}$ in *Eucalyptus tereticornis* based agroforestry compared to ₹ 16143$ha^{-1}yr^{-1}$ in sole wheat.

Table 2: Economics of cowpea grown with *Eucalyptus tereticornis* and as sole crop influenced by various fertilizer levels (pooled data of 2004 and 2005)

Treatments (kg N + Kg P_2O_5 ha^{-1})	Agrisilviculture				Sole cowpea			
	Gross returns (₹ ha^{-1})	Cost of cultivation (₹ ha^{-1})	Net returns (₹ ha^{-1})	Cost/ benefit ratio	Gross returns (₹ ha^{-1})	Cost of cultivation (₹ ha^{-1})	Net returns (₹ ha^{-1})	Cost/benefit ratio
F_0 (control)	61675	29668	32006	1:2.07	14375	11384	2991	1:1.26
F_1 (15 + 30)	63725	31023	32701	1:2.05	16200	12634	3565	1:1.27
F_2 (20 + 40)	64450	31410	33039	1:2.05	16825	13021	3803	1:1.29
F_3 (25 + 50)	64900	31898	33002	1:2.03	17225	13509	3716	1:1.28
Mean	**63687**	**31000**	**32687**	**1:2.05**	**16156**	**12637**	**3519**	**1:1.27**

CONCLUSIONS

Although all physiological parameters and yield of cowpea in association with eucalypts were low as compared to sole cowpea, however, after accounting for returns from sale of eucalypts economic returns in agrisilviculture system were higher than sole cowpea. It may, therefore be concluded that in eucalypts based agrisilviculture system, there may be loss in terms of crop yield but the loss is well compensated by the sale of eucalypts at the end of rotation.

REFERENCES

Bhilare, R. L. and Patil, V. S. 2002. Response of forage cowpea varieties to phosphorus application. *Forage Res.*, **28** (3):179-180.

Cochran, W.C. and Cox, G.M. 1959. *Experimental Designs.* Asia Publishing House, Bombay.

Devasagayam, M. M. and Ebenezar, E. G. 1996. Physiological changes of arable crops under *Eucalyptus* ecosystem. *Journal of Ecobiology,* **8**(2):149-152.

Fisher, R. A. 1948. Statistical methods for research workers. Ed.10. Oliver and Boyd, Edinburg and London.

Hiscox, J. D. and Isractstman, G. F. 1979. A method for the extraction of chlorophyll from leaf tissue without maceration. *Can. J. Bot.*, 57:1332-1334.

Jamadagni, B. M., Shetye, V. N. and Ingale, B. V. 2002. Response of forage yield of cowpea to different levels of phosphorus. *Forage Res.*, 28 (1): 24-25.

Kumar, A. 2006. *Effect of irrigation and fertilizer levels on wheat (Triticum aestivum L.) production in Eucalyptus tereticornis Sm. based agrisilviculture system*. Ph.D. Thesis, CCS HAU, Hisar, Deptt. of Forestry.

Kumar, S., Gaur, G. S. and Pyare, R. (2003). Economy of poplar based agroforestry system in different agro-ecological regions of U.P. *Progressive Agriculture*. 3: 18-21.

Mariani, E. O., Mariani, C. E. and Lipinsky, S. B. 1981. Tropical eucalypts in: Mc Clure, T. A. and Lipinsky, E. S. (eds.). CRC handbook of biosolar resources Vol. II Resource material CRC Press, Inc Boca Raton, F.L. 373-386.

Nandal, D. P. S. and Singh, R. R. 2001. Productivity of different cropping sequences in *Dalbergia sissoo* Roxb. based agro-silviculture system. *Indian Journal of Forestry*, 24(4):433-436.

Patidar, M. and Mali, A.L. 2004. Effect of farmyard manure, fertility levels and bio-fertilizers on growth, yield and quality of sorghum (*Sorghum bicolor*). *Indian J. Agron.*,49 (2): 117-120.

Roy, D.K. 2000. *Integrated nutrient management in malt Barley*. Ph.D. Thesis, CCS HAU, Hisar, Deptt. of Agronomy.

Subrahmanyam, M. V. R., Sridevi, S., Radhika, K. and Bheemaiah, G. 2001 . Studies on different crop management practices in *Hardwickia binata* based agroforestry system in dry lands. *Indian Journal of Dryland Agricultural Research and Development*, 16(1): 73-77.

Zomer, R.J; Trabucco, A, Coe, R and Place, F. 2009 . Trees on farm. Analysis of global extent and geographical patterns of agroforestry.ICRAF.Working paper no.89.Nairobi Kenya.World Agroforestry Centre.

□□□

Chapter 48

Performance of Inter Cropping of Vegetables in Mango (*Mangifera Indica*) Variety Amrapali

SIDHARTHA KAR

INTRODUCTION

Intercropping is the practice of growing two or more crops in proximity. The most common goal of intercropping is to produce a greater yield on a given piece of land by making use of resources that would otherwise not be utilized by a single crop. Careful planning is required, taking into account the soil, climate, crops, and varieties. It is particularly important not to have crops competing with each other for physical space, nutrients, water, or sunlight. Example of intercropping strategies is planting a tall crop with a shorter crop that requires partial shade.

When crops are carefully selected, other agronomic benefits are also achieved. Lodging prone plants, those that are prone to tip over in wind or heavy rain, may be given structural support by their companion crop. Creepers can also benefit from structural support. Some plants are used to suppress weeds or provide nutrients. Delicate or light sensitive plants may be given shade or protection, or otherwise wasted space can be utilized. So keeping the above view present experiment has been carried out with four treatments & five replications.

MATERIALS AND METHODS

The experiment was laid out in a randomized block design (R.B.D.) having four treatments with five replications under Krishi Vigyan Kendra, Dhenkanal during rabi 2010-11. The soil of the experimental site was red laterite having pH 6.5. Prior to the experiment only mango crop was grown without any space utilization between the rows. Total four numbers of treatments *i.e.*,

Krishi Vigyan Kendra, Dhenkanal – 759 001, Odisha,
Orissa University of Agriculture & Technology, Bhubaneswar, Odisha, India

Mango + Potato, Mango + Tomato, Mango + Onion and only Mango as local check were tested to sea the benefit cost of inter cropping system, net profit and yield of main crop as well as inter crop. Each plot size is about 7 m X 7 m and spacing of inter crops were 45 cm X 20 cm, 15 cm X 10 cm and 60 cm X 45 cm, for potato, onion and tomato, respectively. The plants were grown in red laterite soil & irrigated condition with uniform cultural practices to raise the crop successfully. Yield data has been taken at the time of harvesting and average cost of cultivation, Net profit and benefit cost ratio has been analyzed after harvesting. Data were statistically analyzed through analysis of variance and t-test.

RESULTS AND DISCUSSION

Economics : Analysis of data revealed significant differences among different intercrops with mango. Mango (var. *'amrapali'*) as single crop has lowest cost of cultivation i.e. ₹ 9.67 (000 ha^{-1}) where as mango + Potato (var. *'kufri jyoti'*) has ₹ 72 (000 ha^{-1}.), Mango+Onion (var. ALR) has ₹ 86.3 (000 ha^{-1}) & mango + Tomato (var. *'BT -10'*) has ₹ 86.98 (000 ha^{-1}) In respect to Average net profit Mango+Onion has highest net profit of Rs. 202.37 (000 ha^{-1}) followed with mango + Tomato, mango + Potato & only mango *i.e.,* Rs. 167.22, Rs. 127.40, & Rs. 20.23 (000 ha^{-1}) (Table 1).

Table 1: Intercropping in bearing and non bearing orchards

S no	Treatment	Average Cost of cultivation in Rs. (000 per ha.)	Average Gross Profit in Rs. (000 per ha.)	Average Net Profit in Rs. (000 per ha.)	Yield (q ha^{-1}) Effect of inter-cropping on Yield of mango	Intercrop vegetable	Yield of inter crop	Benefit cost ratio
1	T1 (Mango + Potato)	72.0	199.40	127.40	18.50	Potato	245	2.7
2	T2 (Mango + Tomato)	86.98	254.20	167.22	20.0	Tomato	285	2.92
3	T3 (Mango + Onion)	86.3	288.40	202.37	19.0	Onion	255	3.34
4	T4 (Only Mango) control plot	9.67	29.90	20.23	20.20	-	-	3.09
	SEM ±	2.90	67.10	46.60	-35.35	-	-	-
	CD (0.05)	6.32	146.21	101.54	-77.03 NS	-	-	-
	CD (0.01)				11.76	-	-	-

Yield : Whereas inter cropping has not much significant difference on yield of mango. Yield of mango is highest no intercropping i.e., 20.20 q ha^{-1}. followed with Mango + Tomato, Mango + onion & Mango + potato *i.e.,* 20.0, 19.0 & 18.50 q ha^{-1}. And yield of vegetables like onion as intercrop has highest *i.e.,* 285 q. ha^{-1}). followed with onion & potato *i.e.,* 255 & 245 q ha^{-1} (Table 1).

CONCLUSIONS

Inter cropping of onion in mango orchard has highest return value in mid central table land zone of Odisha. It is also fact that yield of tomato is highest by inter cropping with mango among potato, onion & tomato and mango as single crop has highest yield though availability of nutrient in sufficient quantity. So far inter cropping of onion with mango will give highest return as well as inter cropping of vegetables can utilized the wasted space in orchard. So research need to be geared up for strengthening of crop patterns, crop husbandry, crop irrigation, crop habitat management and crop.

REFERENCES

Antil, R. S., Kumar, V., Narwal, R. P. and Kuhad, M. S. 2001. Nutrient removal and balance in soils of Haryana. Bulletin. Department of Soil Science, CCS Haryana Agricultural.

Brown, J.E, Splittstoesser, W.E. and Gerber, J.M. 1985. Production and economic returns of three vegetables double cropping system, *Journal of the American Society for Horticultural Science*, 110(3): 414-417

Hutabarat, B. Waluyo, Harjadi, S.S., Tjitrosomo, S., Harjadi, W., Widodo, W.D. and Sudarsono 1992. Cropping system of vegetables and its significance on labor use and revenues in Kabupten Bandung, West Java, Center for Agro Socio Economic Research (CASER).

Kumar, N. 1997. Introduction to Horticulture. Rajalakshmi Publications, 28/5 - 693, Vepamoodu Junction, Nagercoil. pp1547- 1550.

Principal Investigator Precision Farming Development Centre (PFDC) Central Institute for Subtropical Horticulture, Rehamankhera, P.O. Kakori, Lucknow - 227 107, Uttar Pradesh.

□□□

Chapter 49

Developing an Integrated Farming Systems using Multiple Criteria Decision Making Tools for the Dryland Alfisols in the Deccan Plateau, India

S.L. PATIL, S.K.N. MATH, R.N. ADHIKARI, PRADEEP DOGRA
K. CHANNABASAPPA AND K.K. REDDY

INTRODUCTION

Green revolution was the beginning in improving crop productivity both in developing and developed countries with adoption of improved cultivars of wheat, maize and rice that were responsive to fertilizers and irrigation. Success of green revolution was only observed in favorable soil and water environments, *i.e.*, fertile soil and irrigated areas, however this success was not observed in poorer agricultural areas (resource poor farmers) and it has led to the evolution of farming system research (FSR) approach with close association between technical and social scientists (Norman, 1993).

In India, with increase in human and bovine population we need to produce more with decreased land area to meet the food demands. It is expected that human population in India may reach around 1370 and 1660 M and we are expected to produce 289 and 349 M tones by 2030 and 2050, respectively. Net cultivable area may reduce from present 142.8 M ha to 137 M ha by 2050 with cultivable area being used for building and gets converted to industrial purposes. This calls for increased crop productivity by nearly 50% to 75% by 2030 and 2050, respectively. This could be possible by putting the available rainfed and irrigated area under intensive cultivation with conservation of natural resources and efficient utilization and recycling of farm waste for greater soil fertility and productivity on sustainable basis (Solaimalai *et al.*, 2005). In India, fortunately most of the cultivable area has an abundant solar energy thus making cropping feasible throughout the year. Income from arable lands is not sufficient for small and marginal farmers and hence, for their sustenance it is essential to adopt/practice dairy, poultry,

Central Soil and Water Conservation Research and Training Institute, Research Centre, Bellary–583 104, Karnataka, India sheep and goat rearing, fish culture, sericulture, biogas production, vermicompost, mushroom cultivation, agro-forestry, agro-horticulture for supplementing farm income. These allied activities fits well within the farm level infrastructure and assures full utilization of by-products. The ultimate solution can be achieved through integrated farming systems (IFS) that improve soil fertility, increases crop productivity, income and nutrition of small and marginal farmers with limited resources availability (Annadurai *et al.*, 1994a and b; Korikantimath and Manjunath, 2008; Panda, 2006). If cropping system alone is practiced, the amount of by-products can be as high as the marketable produce that may go as waste or fetch less price if not utilized in the animal enterprise. Hence, to maintain soil productivity on sustainable basis in fragile rainfed environment especially in red soils (Alfisols) of low rainfall areas of central Karnataka, India, it is essential to adopt suitable IFS that produces enough food to meet the family demands, be profitable to the farmer, conserve natural resources, efficiently utilize the available resources at optimum level including family labour and maintain soil and crop productivity on sustainable basis (Nadagouda, 2000 and Patil *et al.*, 2012).

STUDY AREA

Integrated farming systems study is conducted in red soils of Ramasagara watershed in Chitradurga district of Karnataka State in South India. Ramasagara watershed is located 14°49′31″ N to 14°50′42″ N latitude and 76°47′32″ E to 76°49′16″ E longitude and receives mean annual rainfall around 420 mm on an average. Due to lower rainfall (< 500 mm), this area is classified as Arid even though this region is generally classified as Semi-Arid. In watershed, low total rainfall and its ill distribution from June to October results in 3 severe drought years, 3 drought years and 4 normal rainfall years in a decade. In addition to the low rainfall, the red sandy loams soils in watershed have low water holding capacity and it results in higher drought situations compared to black soils with greater water holding capacity. Age-old field boundary bunds have been totally eroded with reduced size and frequent breaches. Rainwater conservation practices, *i.e.*, *In-situ* rarely adopted to conserve rainwater, top fertile soil and recharge profile prior to sowing. This resulted in accelerated erosion both at field level and at terrace level in this region. Generally in rainfed areas of watershed, mono-cropping of groundnut with traditional cultivars is most prevalent cropping system observed over the years. Lack of adoption of rainwater conservation practices with traditional crop cultivation resulted in lower crop productivity and income of individual farm families in addition to deterioration in soil properties in watershed area. Adoption of rainwater conservation practices reduces soil erosion, recharges soil profile and improves soil properties increases crop and vegetation productivity on sustainable basis, including improving the economic status and livelihood of beneficiaries in the watershed. In Ramasagara watershed, nearly 74% of total area is with cultivation of crops with 63% of area is under rainfed and 11% is under irrigation from bore wells. Cultivated crops under irrigated conditions are supplemented with protective irrigation in border strips from bore wells. Traditional farming systems that is generally adopted in watershed area in both rainfed and protective irrigated conditions results in lower crop productivity and returns is no more sustainable for achieving goals of more food production and conserving natural resources. In view of the present existing situations in watershed, there is a need to develop suitable IFS for farmers practicing subsistence farming from which they need to produce a continuous, reliable and balanced supply of foods, as well as cash. Improved IFS have the advantage (over monocropping systems) of utilizing the synergies of its components that lead to more output from less input use with resource conservation. In IFS studies, three farmers with improved IFS and three control farmers with similar farming situations were selected during 2010-2011 in Ramasagara watershed.

RESULTS AND DISCUSSION

Farmers objectives and weightage : Three farmers were supplemented with inputs to adopt integrated farming systems along with three more farmers of similar farming system as control farmers in Ramasagara watershed, Molakalmur taluk of Chitradurga district in Karnataka State, India, during 2010-2011. Multiple objectives ranked as per preference/importance given by farmers are presented in Table 1. Maximization of farm income was preferred higher by both IFS and control farmers compared to other objectives. Next immediate objective was to increase farm employment, followed by minimization of soil loss and minimization of labour hiring. All farmers have indicated lower value for minimization of labour hiring because majority of field works were done by family members especially by rainfed small farmers, *i.e.,* Mr. Ramesh and control farmer Mr. Karadi Gangappa. Irrigated farmers hired labour to maintain timely field operations in addition to their own family members that was not sufficient to complete field operations/works especially during peak hours.

Table 1: Farmers' objectives and weightage* assigned to each components

Farmer's Particulars		Maximization of farm income	Maximization of on-farm employment	Minimization of soil loss	Minimization of labour hiring	Total
Ramesh S/o Jatingappa	IFS	6	2	2	---	10
Karadi Gangappa	Control	6	2	2	---	10
E. Marenna S/o Jatingappa	IFS	6	3	---	1	10
Kukkala Ganganna	Control	7	2	---	1	10
Guddada Ganganna S/o Bhimappa	IFS	6	2	1	1	10
Sanna Gouda	Control	6	2	1	1	10

*The total sum of the weights given by farmer for all his objectives must be equal to 10.

Family details : Average family size of IFS farmers is 9 and control farmers is 7 with overall average of family size of 8 indicating a joint family culture that exists in this region (Table 2). Female-male ratio is almost 1:1. Family members that contribute to agricultural activities are 29 adults as against 18 children in both IFS and control farmers together.

Table 2: Family details of selected IFS farmers

Farmers		<14 years Children		>14 years Adults		Total
Name	Type of farmers	Male	Female	Male	Female	
Ramesh S/o Jatingappa	IFS	1	3	1	2	7
Karadi Gangappa	Control	1	1	3	2	7
E. Marenna S/o Jatingappa	IFS	2	2	2	3	9
Kukkala Ganganna	Control	1	---	1	2	4
Guddada Ganganna S/o Bhimappa	IFS	4	---	4	3	11
Sanna Gouda	Control	3	---	2	4	9
Total	12	6	13	16	47	

Inputs distribution : Improved cultivars of groundnut (*'TMV-2 and K-6'*), redgram (*'BRG-2'*), maize (*'Super 900M Gold'*) and seed treatment culture, *i.e., Rhizobium* and *Trichoderma* were identified as critical inputs and were distributed to all three IFS farmers (Table 3).

Table 3: Inputs distribution (kg) in 2010-11

Farmer	Groundnut		Redgram	Maize (Super 900M Gold)	Seed Treatment	
	TMV-2	*K-6*	*BRG-2*		*Rhizobium*	*Trichoderma*
Ramesh	90	60	3	-	0.5	0.5
E. Marenna	-	-		10	-	-
Guddada Ganganna	90	60	7	05	0.5	0.5
Total	180	120	10	15	1.0	1.0

Crop component

Cropping pattern: Two IFS farmers, *i.e.*, Ramesh and Guddada Ganganna followed intercropping of groundnut and redgram during *kharif* 2010-11 under rainfed conditions with local and improved groundnut (*'K-6 and TMV-2'*) and redgram (*'BRG-2'*) varieties. Mr. Ramesh cultivated local variety of groundnut and redgram in 0.27 ha, whereas Mr. Guddada Ganganna cultivated local groundnut and improved redgram variety in nearly 5.16 ha under rainfed conditions during *kharif* with rest of the area *i.e.*, 1.01 ha of Ramesh and 1.04 ha of Mr. Guddada Ganganna are being cultivated with improved varieties of groundnut and redgram (Table 4). Small and large control farmers, *i.e.*, Mr. Karadi Gangappa and Mr. Sannagouda sown local varieties of groundnut and redgram as intercrops under 1.42 ha and 2.41 ha, respectively during *kharif* 2010-11.

Table 4: Cropping pattern of selected IFS farmers during 2010-11 (ha)

Crop	Ramesh	Karadi Gangappa	E Marenna	Kukkala Ganganna	Guddada Ganganna	Sanna Gouda
Rainfed						
Groundnut + Redgram	1.28	1.42	--	--	6.20	2.41
Irrigated - *Kharif*						
Bt. Cotton	--	--	1.42	0.81	1.60	1.21
Hybrid Sorghum	--	--	0.33	0.40	--	--
Paddy	--	--	--	--	0.40	0.40
Onion	--	--	--	--	0.30	--
Chillies	--	--	--	--	0.10	--
Hybrid Maize	--	--	--	--	--	0.41
Irrigated - *Rabi*						
Paddy	--	--	0.33	--	--	--
Hybrid Maize	--	--	1.42	0.40	1.60	0.40
Total area (ha)	1.28	1.42	1.75	1.21	8.60	4.43
RF cultivated area (ha)	1.28	1.42	--	--	6.20	2.41
Irrigated area (ha)	0.00	0.00	1.75	1.21	2.40	2.02
Gross cropped area (ha)	1.28	1.42	3.50	1.61	10.20	4.83
Cropping Intensity (%)	100.0	100.0	200.0	133.1	118.6	109.0

Irrigated small farmer, *i.e.,* Mr. E. Marenna cultivated hybrid sorghum and Bt cotton in *kharif* and paddy and hybrid maize during *rabi*/early summer in 0.33 ha and 1.42 ha, respectively during 2010-11. *Rabi*/early summer crops were cultivated after harvest of *kharif* crops. Other control small irrigated farmer, Mr. Kukkala Ganganna cultivated Bt cotton and hybrid sorghum in 0.81 ha and 0.40 ha, respectively during *kharif* 2010. After harvest of *kharif* hybrid sorghum Mr. Kukkala Ganganna cultivated hybrid maize in 0.40 ha during *rabi*. Under irrigated conditions, Mr. Guddada Ganganna cultivated Bt cotton, paddy, onion and chilles in 1.60 ha, 0.40 ha, 0.30 ha and 0.10 ha, respectively during *kharif*, whereas during *rabi*/early summer after harvest of 50% Bt cotton area, 0.80 ha and paddy, onion and chilles cultivated area sown hybrid maize *i.e.,* Super 900M Gold in 1.60 ha during *rabi*/early summer. Control large farmer, Mr. Sannagouda cultivated Bt cotton, paddy and hybrid maize in 1.21 ha, 0.40 ha and 0.41 ha, respectively during *kharif* 2010 under irrigated conditions. During *rabi* 2010-11, cultivated hybrid maize in 0.40 ha. Cultivation of rainfed crops alone during *kharif* resulted in 100% cropping intensity by both rainfed small farmers *i.e.,* Mr. Ramesh and Mr. Karadi Gangappa (Table 4). Among both small irrigated farmers, Mr. Kukkala Ganganna (Control farmer) cultivated hybrid maize in 0.40 ha during *rabi* in addition to 1.21 ha under crops during *kharif* and this resulted in increased cropping intensity to 133%, whereas, Mr. Marenna cultivated crops in 1.75 ha both during *kharif* and *rabi*/early summer resulting in 3.50 ha of gross cultivated area with 200% cropping intensity (Table 4). Cultivation of hybrid maize during *rabi*/early summer in 1.60 ha and 0.40 ha by Mr. Guddada Ganganna and Mr. Sannagouda increased the cropping intensity from 100% to 119% and 109%, respectively. Average cropping intensity of six farmers is 122%.

Livestock component : All six farmers own bullocks. Mr G. Ganganna and Mr Sannagouda maintain buffaloes that yield milk and both Mr G. Ganganna and Mr E. Marenna maintain cows for milk purpose. All three IFS farmers maintain poultry birds and only Mr G. Ganganna has 40 sheep and 20 goats (Table 5).

Rainfall during 2010 : Rainfall that fell during 2010 was above normal (808.4 mm) with a uniform distribution over nine months except during February, March and December when there was no rainfall. Rainfall that received during 2010 was above normal except in September 2010 rainfall was lower (29.6 mm) than normal (112.1 mm). Rainfall of 253.2 mm and 80.8 mm that fell in May and June resulted in complete wetting of soil profile and recharging bore wells. Low rainfall that fell during first forth night of July and only after receipt of 17.4 mm rainfall on 17th July, rainfed crops were sown from 18th-20th July (optimum time of sowing). Because of normal to above normal rainfall and its uniform distribution from sowing to harvest produced normal to slightly higher yields of groundnut, the predominant rainfed crop cultivated during *kharif* in this region. Excess rainfall during maturity produced wilting of groundnut resulting in slightly lower yields than expected higher yields. Normal rainfall that fell during 2010 produced higher crop yields and income from both rainfed and irrigated areas.

Table 5: Livestock particulars of IFS farmers during 2010-11

Farmer		Livestock (nos.)						
		Bullocks	Cows	Buffaloes	Young stock	Goats	Sheep	Poultry birds
Ramesh S/o Jatingappa	IFS	2	1	0	1	0	0	10
Karadi Gangappa	Control	2	0	0	0	0	0	0
E. Marenna S/o Jatingappa	IFS	2	15	0	11	0	0	20
Kukkala Ganganna	Control	2	1	0	1	0	0	0
Guddada Ganganna S/o Bhimappa	IFS	2	2	5	4	40	20	10
Sanna Gouda	Control	2	1	4	3	0	0	0

Crop yields and returns : Grain and straw yields, gross returns, cost of cultivation and net returns were worked out from different crops cultivated by all six farmers in rainfed and irrigated conditions and are presented in Table 6. Greater yields realized from improved cultivars of groundnut (*'K-6 and TMV-2'*) and redgram (*'BRG-2'*), especially K-6 produced greater yields and net returns over local cultivars cultivated by both IFS and control small farmers. Net returns from crops was ₹ 11360/- and ₹ 6120/- from 1.28 ha of Mr. Ramesh and 1.42 ha of Mr Karadi Gangappa, respectively (Table 6). Higher grain and straw yields with 200% cropping intensity resulted in gross and net returns of ₹ 238393/- and ₹ 131959/- by IFS farmer Mr E. Marenna from 1.75 ha irrigated area compared to lower yields and low net returns Rs. 40830/ - realized by Mr Kukkala Ganganna from 1.21 ha.

Net income of individual farmers from crops, livestock and off-farm activities was calculated and presented in Table 8. Net returns from improved varieties of groundnut (*'K-6 and TMV-2'*) and redgram (*'BRG-2'*) were higher (₹ 12696 ha^{-1} and ₹ 8830 ha^{-1}) in rainfed conditions compared to local cultivars (₹ 4400 ha^{-1}) cultivated during *kharif* by Ramesh (Table 8). Higher income for Ramesh was the off farm income (56%) and it was attributed to engagement of Ramesh and his family members in agricultural field works, and next best income was from cultivation of crops that amounted for nearly 35% of his total income and income from animal component was very meager, *i.e.*, 9.5% (Table 8). In case of Mr E. Marenna and Mr G. Ganganna, major income was from cultivation of crops *i.e.* 66.57% and 74.5%, respectively. This clearly indicates that during above normal rainfall situations yields from rainfed crops and irrigated crops are better and income is higher in both irrigated and rainfed situations due to higher yields from crops with uniform rainfall distribution during cropping period in rainfed situations and it was also attributed to recharge of bore wells with greater water availability for irrigated crops. For both irrigated farmers *i.e.,* Mr E. Marenna and Mr G. Ganganna, major income was from cultivation of Bt cotton and maize in irrigated situations (Table 6). Mr E. Marenna maintains more milking animals and hence, income from animal component was higher (29%) compared to Mr G. Ganganna (18%). Income from off-farm activities is lower for both Mr G. Ganganna (6.8%) and Mr E. Marenna (4.5%). Income from crops was also higher with control small irrigated farmer, *i.e.,* K. Ganganna (72.3%) and rainfed and irrigated large farmer, *i.e.,* Mr Sannagouda (73.5%). Income from animal components and off-farm varied from 15.1 to 18.8% and 8.9 to 11.4% for above farmers. It was earlier observed by Shekinah, *et al.* (2005) that farming system with combination of crops, agro-forestry, animal and farm pond was more profitable and sustainable than conventional cropping systems alone at Coimbatore, India. Further, based on farmer participatory research, it is concluded that IFS approach is better over traditional system in its contribution to productivity, profitability, economics and employment generation for small and marginal farmers of Tamil Nadu, India (Jayanthi, *et al.*, 2008).

Table 6: Cropping pattern, yield and returns selected IFS farmers (2010-11)

Farmer	RF/ Irri.	Crop	*kharif/ rabi*	Area (ha)	Total yield (kg) Grain	Straw	Gross returns (₹)	Cost of Cultivation (₹)	Net Returns (₹)
Ramesh	RF	Groundnut + Redgram (K-6 + BRG-2)	*Kharif*	0.61	495+45	1035+175	18465	11243	7222
	RF	Groundnut + Redgram (TMV-2 + BRG-2)	*Kharif*	0.40	276+30	630+105	10583	7391	3191
	RF	Groundnut + Redgram (Local + Local)	*Kharif*	0.27	150+14	325+50	5622	4676	947
		Total		**1.28**			**34670**	**23310**	**11360**
Karadi Gangappa	RF	Groundnut + Redgram (Local + Local)	*Kharif*	**1.42**	780+60	1390+180	**28155**	**22035**	**6120**
E. Marenna	Irri.	Hybrid Sorghum	*Kharif*	0.33	1155	1650	15510	7842	7668
	Irri.	Bt Cotton Hy. Maize Super 900M	*Kharif*	1.42	2600	-	117000	45965	71035

	Irri.	Gold	*Rabi*	1.42	9145	11660	91633	41147	50486
	Irri.	Paddy	*Rabi*	0.33	1250	1750	14250	11480	2770
		Total		**3.50**			**238393**	**106434**	**131959**
Kukkala Ganganna	Irri.	Hybrid Sorghum	*Kharif*	0.40	900	1500	12300	7620	4680
	Irri.	Bt Cotton	*Kharif*	0.81	1245	-	56025	27750	28275
	Irri.	Hybrid Maize (CP-828)	*Rabi*	0.40	1750	2450	17710	9835	7875
		Total		**1.61**			**86035**	**45205**	**40830**
G. Ganganna	RF	Groundnut + Redgram (K-6 + BRG-2)	*Kharif*	**6.20**	2650+270	5700+600	**99750**	**81415**	**18335**
	Irri.	Bt Cotton	*Kharif*	1.60	3800	-	171000	69835	101165
	Irri.	Paddy	*Kharif*	0.40	1500	2300	17300	13055	4245
	Irri.	Onion	*Kharif*	0.30	2475	-	19800	16535	3265
	Irri.	Chillies	*Kharif*	0.10	75	-	7500	5157	2343
	Irri.	Hybrid Maize (CP-828)	*Rabi*	1.60	7500	10500	75900	41020	34880
		Total		**2.40**			**391250**	**227017**	**164233**
Sanna Gouda	RF	Groundnut + Redgram	*Kharif*	**2.41**	1250+100	2500+300	**45700**	**32638**	**13062**
	Irri.	Bt Cotton	*Kharif*	1.21	2350	-	105750	45555	60195
	Irri.	Paddy	*Kharif*	0.40	1250	1800	14300	11230	3070
	Irri.	Hybrid Maize	*Kharif*	0.41	1550	2150	15670	9275	6395
	Irri.	Hybrid Maize	*Rabi*	0.40	1650	2200	16610	9230	7380
		Total					**198030**	**107928**	**90102**

Crop productivity and net returns of IFS Vs Control Farmers **:** Ramesh compared to K. Gangappa realized higher grain yields of groundnut and redgram by 31 and 66% with cultivation of improved cultivars. Higher grain yields resulted in higher net returns by 106% by Ramesh compared to control farmer, *i.e.,* Mr K. Gangappa (Table 7).

Table 7: Crop productivity, gross and net returns per ha in IFS and control farmers

Crop	Productivity Grain (kg ha^{-1})	Straw (t ha^{-1})	Gross Returns (₹)	Net returns (₹)
Mr. Ramesh (IFS)				
Groundnut + Redgram	720+70	1.55+0.26 (31% +66%)	27118	8908 (106%)
Mr. K. Ganagappa (Control)				
Groundnut + Redgram	549+42	0.98+0.13	19827	4309
Mr. Marenna (IFS)				
Hybrid Sorghum	3500 (56%)	5.00	47000	23236 (99%)
Cotton (Bt)	1831 (19%)	-	82395	50025 (43%)
Maize (Super 900M Gold)	6440 (47%)	8.21	64529	35553 (81%)
Paddy	3788	5.30	43183	8395

Mr. K. Ganganna (Control)				
Hybrid Sorghum	2250	3.75	30750	11700
Cotton (Bt)	1537	-	69165	34906
Maize (CP 828)	4375	6.13	44275	19688
Mr. G. Ganganna (IFS)				
Groundnut + Redgram	425+44 (-18%+-7%)	0.92+0.1	16094	2963 (-45%)
Bt Cotton (K)	2375 (22%)	-	106875	63228 (27%)
Paddy (K)	3750 (20%)	5.75	43250	10612 (38%)
Onion (K)	8250	-	66000	10883
Chillies (Byadagi) (K)	750	-	75000	23430
Hybrid Maize (R)	4687 (14%)	6.56	47433	21795 (18%)
Mr. Sanna Gouda (Control)				
Groundnut + Redgram	518+41	1.04+0.12	18921	5379
Bt Cotton (K)	1942	-	87390	49741
Paddy (K)	3125	4.50	35750	7675
Hybrid Maize (K)	3780	5.24	38214	15027
Hybrid Maize (R)	4125	5.50	41525	18450

In irrigated small farmers, productivity of hybrid sorghum increased from 2250 to 3500 kg ha^{-1}, Bt cotton increased from 1537 to 1831 kg ha^{-1} and maize increased from 4375 to 6440 kg ha^{-1} and per cent increase in yields was by 66, 19 and 47, respectively with IFS farmer (Mr. Marenna) compared to control farmer (K. Ganganna). Higher yields realized by IFS farmer was attributed to use of improved hybrids, higher application of farmyard manure, fertilizers and micronutrients with optimum plant population. Greater productivity with IFS farmer produced higher net returns by 99%, 43% and 81% in hybrid sorghum, cotton and maize, respectively (Table 7). Lower yields in groundnut and redgram in rainfed area by IFS farmer (Mr. G. Ganganna) produced lower net returns by 45% compared to Mr. Sanna Gouda (Control farmer). The lower yields in G. Ganganna field was attributed to lesser soil depth with hard soil formation whereas, rainfed fields of Mr. Sanna Gouda are deep and fragile soils with greater water holding capacity producing higher yields. In irrigated conditions, productivity of Bt. Cotton, paddy and maize (*Rabi*) increased by 22, 20 and 14%, respectively thus producing higher net returns by 27, 38 and 18%, respectively by G. Ganganna (IFS) compared to control farmer, *i.e.,* Mr. Sannagouda (Table 7).

Table 8: Income (₹) details of selected IFS farmers

Component	Returns	Rainfed		Irrigated		Rainfed & Irrigated	
		Ramesh	K. Gangappa	E. Marenna	K. Gang-anna	G. Gang-anna	S. Gouda
Crops	Gross returns	34670	28155	238393	86035	391250	198030
	Net returns	11360	6120	131959	40830	164233	90102
	% Total Net returns	35.0	14.1	66.5	72.3	74.5	73.5
Live-stock	Gross returns	20500	18100	109950	31800	109800	55625
	Net returns	3070	16305	57525	10600	41250	18575
	% Total Net returns	9.5	37.5	29.0	18.8	18.7	15.1
Others	Net returns	18000	21000	9000	5000	15000	14000
	% Total Net returns	55.5	48.4	4.5	8.9	6.8	11.4
Total	**Gross returns**	**73170**	**67255**	**357343**	**122835**	**516050**	**267655**
	Net returns	**32430**	**43425**	**198484**	**56430**	**220483**	**122677**

Soil properties : The pH of both rainfed and irrigated farmers field did not vary much and it varied from 6.9 to 8.4 (rainfed) and 7.9 to 8.4 (irrigated). The EC in rainfed area was low and varied from 0.18 to 0.31 dS m^{-1}, whereas it was higher and varied from 0.35 to 0.59dS m^{-1} in irrigated fields. Higher EC in irrigated fields was attributed to use of saline water for irrigation (Table 9). In rainfed area, organic carbon was low and it varied from 0.31 to 0.38% and it was higher for irrigated fields (0.35 to 0.59%). Farmyard manure application at 1-2 t ha^{-1} in rainfed area resulted in lower organic carbon under rainfed situations, whereas with greater quantity application of farmyard manure, i.e., 8-10 t ha^{-1} including sheep penning in irrigated fields produced higher organic carbon.

Application of smaller quantity of fertilizers and farmyard manure produced lower available N in rainfed farmers' fields and it ranged from 186 to 240 kg ha^{-1}, whereas application of higher amounts of N through fertilizers, farmyard manure and sheep penning produced in higher N availability and it varied from 204 to 348 kg ha^{-1} in irrigated farmers' fields. Available N in irrigated farmers' fields was medium. Available P_20_5 varied from 22.0 to 28.0 kg ha^{-1} in rainfed fields to 25.4 to 36.3 in irrigated farmers' fields. Available P_20_5 in rainfed farmers fields was lower whereas, in irrigated farmers fields it was medium and it was attributed to application of higher amounts of DAP along with farmyard manure and sheep penning in irrigated farmers' fields. Available K_20 varied from 248 to 307 kg ha^{-1} in rainfed area to 252 to 309 kg ha^{-1} in irrigated area. In IFS farmers' fields, nutrient status was higher over control farmers fields in rainfed and irrigated conditions and was attributed to application of greater quantity of urea, DAP and farmyard manure with sheep penning (Table 9). In farmers' fields of sandy loam soils of western Tamil Nadu, India, recycling farm and livestock manures improved the soil fertility status in addition to enhanced productivity and profitability (Jayanthi *et al.*, 2008).

Soils of both IFS and control farmers have sufficient calcium, magnesium and sulphur and these soils are not deficient in these elements. Soils of IFS farmers compared to control farmers are richer in calcium, magnesium and sulphur (Table 9). All the farmers' soils are low in zinc with higher zinc availability in IFS farmer fields compared to control farmers and it was attributed to zinc application by IFS farmers (Table 9). All farmers' fields have higher copper content, whereas iron and manganese content ranges from medium to high (Table 9).

Multi-Criteria Decision Making (MCDM) : Initially six farmers (three each for IFS and control) were selected to implement IFS plan with the final objective of the project is to identify and analyze economic and resource use efficiencies of existing farming systems with locally available resources and prevailing socio-economic conditions along with field validation and evaluation through Multi-Criteria Decision Making (MCDM). The MCDM technique handles several objectives simultaneously and optimizes them for developing suitable IFS model for the region under different existing situations. This MCDM technique creates live models of IFS for greater natural resource conservation and improving production on sustainable basis. With shrinking resource base and environmental degradation, it is imperative to rationally analyze existing farming systems under different scenarios and develop through MCDM techniques a set of relevant IFS that balance livelihood requirements and environmental concerns. This project was initiated with the objective to create live models of IFS as per existing resource endowments, agro-climatic and socio-economic conditions to demonstrate their potential for improving production and conservation on sustainable basis in Central Dry Zone of Karnataka, India. To implement IFS, large rainfed and irrigated farmer, Mr. Guddada Ganganna was selected in Ramasagara watershed.

As per the optimum plan, area of crops with lower economic returns were reduced and area with crops with higher economics were increased with due consideration to family needs, soil fertility, fodder requirement and soil and rainwater conservation (Table 10). Further in optimum plan as per MCDM, expected net returns increased by 5%, labour utilization reduced by 2% and soil loss was also reduced by 1%.

Table 9: Soil properties of rainfed, irrigated and rainfed and irrigated farmers fields at sowing and harvest of crops.

Soil Parameter	Unit	Rainfed farmer				Irrigated farmer				Rainfed and irrigated farmer						
										Rainfed area			Irrigated area			
		Ramesh		K. Gangappa		E. Marenna		K. Ganganna		G. Ganganna		S Gouda	G. Ganganna		Sanna Gouda	
		Before Sowing	After Harvest	Before Sowing	After Harvest	Before Sowing	After Harvest	Before Sowing	After Harvest	Before Sowing	After Harvest	After Harvest	Before Sowing	After Harvest	Before Sowing	After Harvest
pH		6.9	7.2	7.3	6.9	8.2	7.9	7.9	8.0	8.2	8.0	8.4	8.2	8.1	8.4	8.2
E.C	(dS m^{-1})	0.18	0.20	0.22	0.20	0.30	0.54	0.39	0.34	0.25	0.31	0.24	0.20	0.19	0.24	0.41
Organic Carbon	(%)	0.37	0.38	0.33	0.32	0.54	0.59	0.41	0.39	0.37	0.34	0.31	0.53	0.46	0.48	0.35
Av. Nitrogen	(kg ha^{-1})	240	230	222	215	289	304	245	204	204	186	189	348	329	269	225
Av. Phosphorus	(kg ha^{-1})	28.0	23.5	26.0	23.0	36.3	32.3	25.5	25.4	26.0	23.0	22.0	35.8	31.5	28.4	29.9
Av. Potassium	(kg ha^{-1})	258	275	254	259	278	298	252	261	263	307	248	293	309	285	268
Calcium	(m.eq./ 100g)	5.4	6.6	5.0	6.9	8.7	9.7	6.2	6.8	5.9	6.6	4.7	7.4	8.8	6.8	6.9
Magnesium	(m.eq./ 100g)	1.9	2.2	1.8	1.9	3.0	3.8	2.2	3.6	2.0	2.5	2.0	3.0	3.0	1.7	1.9
Sulphur	(ppm)	8.9	145	9.0	12.0	12.2	13.2	11.0	12.0	13.8	14.8	12.5	14.2	11.2	12.8	10.5
Copper	(ppm)	1.61	1.62	1.55	1.58	1.69	1.70	1.53	1.55	1.50	1.60	1.40	1.60	1.70	1.50	1.50
Iron	(ppm)	5.50	8.40	5.70	8.00	7.30	5.85	5.14	6.20	5.00	6.70	4.50	5.00	6.70	4.00	5.80
Manganese	(ppm)	8.20	8.00	6.00	6.50	12.71	12.95	10.48	7.98	5.60	8.70	5.00	8.61	6.89	7.71	6.90
Zinc	(ppm)	0.91	1.16	0.91	0.64	0.85	0.72	0.76	0.70	0.82	0.72	0.61	0.82	0.72	0.76	0.60

Table 10: Existing and optimum IFS plan of G. Ganganna (Rainfed and irrigated farmer)

Crop	Season	Irrigated/ Rainfed	Existing Plan	Optimum Plan
Cotton (Long duration)	*Kharif & Rabi*	Irrigated	0.8	1.0
Cotton	*Kharif*	Irrigated	0.8	0.8
Paddy	*Kharif*	Irrigated	0.4	0.3
Onion	*Kharif*	Irrigated	0.3	0.2
chillies	*Kharif*	Irrigated	0.1	0.1
Groundnut + Redgram	*Kharif*	Rainfed	6.2	6.2
Maize	Rabi	Irrigated	1.6	1.4
Buffalo-milking	NA	NA	2.0	2.0
Buffalo-Dry	NA	NA	6.0	6.0
Bullocks	NA	NA	2.0	2.0
All 50 Goats & Sheep	NA	NA	1.0	1.0
Total net returns from existing plan				₹ 168587
Expected total net returns from optimal plan				₹ 176299
Expected % increase in net returns				5%
Labour utilization as per existing plan				2402 mandays
Labour utilization as per optimal plan				2357 mandays
% change in labour utilization				-2%
Soil loss in existing plan				28.80 tonnes
Soil loss in optimal plan				28.60 tonnes
% change in soil loss				-1%

CONCLUSIONS

Introduction of IFS in backward and low rainfall areas of red soils in Chitradurga district, Karnataka, India, improved available resource use, *i.e.*, natural and labour resources, reduced degradation of natural resources, improved knowledge of farmers for crops cultivation, crop and animal productivity and overall income of farmers wherein symbiotic learning and improvement in knowledge of farmer and Scientist with increased overall economic returns to the farmers and availability of research data for publication to Scientists is gain in IFS.

REFERENCES

Annadurai, K., Palniappan, S.P., Chinnusamy, C. and Jayanthi, C. 1994a. Integrated farming system, *Kisan World*, **21**(8): 51.

Annadurai, K., Palniappan, S.P., Chinnusamy, C. and Jayanthi, C. 1994b. Integrated farming system: Need of the hour, *Kisan World*, **21** (9): 39.

Jayanthi, C., Vennila, C. and Nalini, K. 2008. Farmer's participatory research on integrated farming system. "Global Issues. Paddock Action." (Ed. M. Unkovich). Proceedings of 14th Agronomy Conference, 21-25 September, Adelaide, South Australia.

Korikantimath, V.S. and Majunath, B.L. 2008. Integrated farming systems for sustainability in Agricultural Production. Proceedings of National Symposium on "New Paradigms in Agronomic Research". *Indian Society of Agronomy.* Navasari Agricultural University, Gujrat. pp279-281.

Nadagouda, V.B. 2000. What next in Agronomy. Prospects of integrated farming systems. Proceedings of the national seminar, *Agronomy club, UAS Dharwad, Karnataka, India,* pp33-34.

Norman. 1993. FAO. Farm systems management series. *FAO of United Nations, Rome.* 10: 227.

Panda, S.C. 2006. Cropping and Farming Systems. Published by Agrobios(India), Jodhpur, India, 431p.

Patil, S.L., Adhikari, R.N. and Prabhavathi, M. 2012. Multiple Criteria Decision for identifying suitable Integrated Farming Systems (IFS) in different Agro-Ecological Regions for optimizing resource conservation and productivity, Annual Report 2011-12, Central Soil and Water Conservation Research and Training Institute, Research Centre, Bellary, Karnataka, India, pp53-60.

Shekinah, E.D., Jayanthi, C. and Sankaran, N. 2005. Physical indicators of sustainability-A farming systems approach for the small farmer in rainfed Vertisols of the Western zone of Tamil Nadu. *J. Sustainable Agriculture,* 25(3): 43-65.

Solaimalai, A., Ravishankar, N. and Chanderasekar, B. 2005. Farming systems: Theory and practice. 698p. Published by International Book Distributors, Dehra Dun, India.

□□□

Chapter 50

Alternate Agroforestry Systems for Resource Conservation in Shifting Cultivation Areas of North-Eastern Hill Region of India

J.M.S. TOMAR, R.K. AVASHTE, ANUP DAS[1] AND O.P. CHATURVEDI

INTRODUCTION

The Eastern Himalayan ranges cover north-eastern hill (NEH) states of Arunachal Pradesh, Manipur, Meghalaya, Mizoram, Nagaland, Sikkim, Tripura, and parts of Assam and West Bengal. The total geographical area of the region is 26.2 M ha which is about 8% of the country's total area. The region still has more than 64% of the total geographical area under forest cover and continues to be a forest surplus region. The climatic conditions and topography in different North East states varied from plain to high altitude thereby providing congenial altitudinal variations for the production of forest, cereal crops, and vegetables, tropical, sub-tropical and temperate fruits. The north-eastern hill regions are characterized by fragility, marginality, inaccessibility, cultural heterogeneity, ethnicity and rich biodiversity. The economy of the states mainly depends on forest wealth, horticultural crops, agriculture and animal wealth (Borthakur, 1992; Bhatt *et al.*, 2001).

Hilly topography, high rainfall, primitive type of agriculture, excessive exploitation of forest wealth including deforestation, urbanization, population growth at faster rate and its pressure on land etc., are mainly responsible for land degradation in the region. Shifting cultivation, also known as *Jhum* cultivation is most traditional and dominant land use system in this region which is currently responsible for large scale deforestation and degradation of natural resources. Land degradation in the region is 36.64 % of the total geographical area, which is almost double than the national average of 20.17 %. Of various degradation types, water erosion, reduced infiltration,

Central Soil and water Conservation Research and Training Institute , Dehradun, Uttarakhand, India

[1]ICAR Research Complex for NEH Region, Umiam, Meghalaya, india

acidification, nutrient leaching, burning of vegetation, decline in vegetative cover and biodiversity are important in context to the NE region. The annual loss of top soil is much higher (46 t ha^{-1}) than all India permissible limit of 16 t ha^{-1}. Similarly, due to lack of proper water harvesting measures, only 0.88 M ha m water is of use for crop cultivation out of 42.5 M ha m total water resources available in this region (Verma *et al.*, 2001).

Various scientific studies and approaches suggest that mixed land use systems are better in the hilly areas, from the conservation as well as production point of view. In North East region, there is immediate need to develop and disseminate viable and acceptable land use system that can replace shifting cultivation and can check land degradation and enhance productivity and socioeconomic conditions of the small and marginal farmers. Agriculture including, horticulture, fishery, forestry, animal husbandry, *etc.*, thus need to develop on priority basis. However, the present agricultural activities in this region are still not exposed by adequate scientific base in circumventing the land degradation process and scientific exploitation of water resources. Further, the system should be so designed to meet the various needs through effective land and water management techniques *i.e.*, watershed management programme integrating soil conservation measures, land development, agriculture, plantation crops, horticulture, animal husbandry, fishery and forestry. Some of the agroforestry and alternatives land use systems are discussed under the following headings.

Agroforestry systems for north east region : Agroforestry interventions based on judicious combination of forestry, horticulture, livestock and agriculture with erosion control structures, such as half moon terraces, contour bunds, grassed waterways, gully plugging and bench terracing are helpful from conservation and production point of view. Agroforestry reduces risks and compensates the failures by giving income from products other than crops; thereby, making the entire system reasonably stable. These practices help in moderation of micro-climate in the vicinity of trees which benefits field crops. Plantation of tree on field boundaries provide shelter from wind, sun or snow which in turn improve yield of crops and provides wood products. Agroforestry systems offer considerable scope of harnessing advantages of growing trees on marginal lands or with crops on agricultural lands. Agroforestry can restore ecological balance while increasing the area both under agriculture as well as forest cover (Tomar, 2006).

Traditional agroforestry system of north eastern region : Traditional agroforestry systems are broadly based on indigenous knowledge and the species are selected as a part of the cultural patterns of the community. The farmers integrate a variety of woody perennials in the crop and livestock production systems depending upon the agro-climatic conditions and local requirements. Some of the examples include the multitier tree-crops combinations in the home gardens of humid lowlands which generate cash income and meet household requirements. For example, combination of alder (*Alnus nepalensis*) and large cardamom (*Amomum subulatum*) in the humid sub-temperate Himalayas is an excellent example of the commercial but traditional agroforestry systems (Avasthe *et al.*, 2011). Shifting cultivation in is another example of traditional agroforestry which is in vogue since ages. According to one estimate, about 4.8 lakh farming families practice *jhum* cultivation over 1.63 M ha of land (Bujarbaruah, 2004). Prevailing land tenure system also restricts the adoption of farming system approach of land use to a great extent. Shifting cultivation was thought to be good for the time when it emerged but with the reduction of *jhum* cycle to 3 to 5 years in most of the north eastern states, there is sharp decline in crop growth and yield. Besides this, it is responsible for mass scale destruction of forests resources, degradation of lands, losses of valuable *flora* and *fauna* and animal resources besides creation of ecological environment non-conducive for crop production. Rise in pH, temperature and bases of the soil, increased the microbial activity, accelerated mineralization of organic N to inorganic forms are some of the effect of shifting cultivation (Tomar *et al.*, 2012).

Organized agroforestry is not found in the region due to prevailing land tenure system except that of homestead farming, However, there are as many as 23 different tree-crop combinations with highest (10) in mid tropical hills and plain zone, and lowest (4) for temperate and sub-alpine zone. *Pinus kesiya* (Khasi pine) has widest ecological amplitude hence a large number of crops, *e.g.*, paddy, ginger, sweet potato, maize, chillies, beans, turmeric, perilla, dioscorea, seasonal vegetables, *etc.*, are cultivated in the under-story of this tree-crop. Some of the traditional agroforestry systems of the region are given in Table 1. Among various tree-crop combinations, homestead based system like Arecanut + black pepper + pineapple, Arecanut + betel vine + pineapple, and coconut + Arecanut + jackfruit + banana has been found most profitable in terms of economy (Table 2).

Agroforestry as an alternative to shifting cultivation : Three-tier model of land use seems most suitable on watershed basis, where restoration forestry/pure silvi-culture is maintained on hilltop, followed by silvi-pastoral, agri-horticulture and pure agriculture down to the slope, respectively. But this system becomes impractical, when the farmers based on hilltop, used to cultivate the area for his/her subsistence in and around the hilltop itself. Due to this fact, the protective trees cover of hill tops have been removed almost in

Table 1: Traditional agroforestry systems/ practices in NEH region

Agro-climatic zones/ distributions	Trees	Crops	Farmers preference
Temperate sub-alpine zone (1500-2000 m asl)	*Pinus kesiya* + *Schima wallichii, Bambusa pallid, Psidium guajova, Citrus* sp., *Pyrus cummunis, Prunus* sp.	Paddy, Ginger, Colocasia, Chilli, Dioscoria, Pumpkin, sweet potato, Maize, bottle Gourd, Betel vine, Pineapple etc.	*Pinus* + ginger, cabbage, cauliflower, Sweet potato, *Prunus* + cabbage, maize and mustard leaf
Sub-tropical hill and plain zone (400-1500 m asl)	*Citrus reticulata* *Michelia oblonga* *Pinus kesiya* *Schima wallichii* Banana *Erythrina indica* *Bambusa pellida*	Paddy, Ginger, Maize, Turmeric, Colocasia, Chilli, Dioscoria, Pumpkin, Sweet potato, Broom grass, *Setaria* Lady's finger, Sweet potato etc.	Citrus + maize, bean, cauliflower; *Schima* + pineapple, maize and ginger
Mid tropical hill and plain zone (200-800 m asl)	*Artocarpus heterophyllus* *Litchi chinensis* *Psidium guojova* *Banana and Areca*	Paddy, Maize, Maize, Cabbage, Cauliflower, Mustard leaf Potato,	Artocarpus/ banana+ Pine apple; Arecanut + beetle vine, pineapple, coffee and black pepper

Table 2: Estimated annual returns of different components of certain important agroforestry systems from Meghalaya (after Chauhan and Dhyani, 1989)

Tree components and density (ha^{-1})		Crop components and density (ha^{-1})		Yield (t ha^{-1})		Net income (tree + crop) components (₹ ha^{-1})
				Tree	Crop	
Arecanut	1000	Pine apple	40000	1.8	20	24000
Mandarin	400	Pine apple	4000	4.8	20	29000
Arecanut	1000	Black pepper	1000	1.8	4.8	36500
Jack fruit	250	Pine apple	40000	12.8	20	20000
Arecanut	1000	beetle vine	1000	1.8	0.8	18000
Mandarin	400	Ginger/turmeric	40000	4.8	4	25500
Arecanut	1000	Beetle vine + pineapple	1000+20000	1.8	0.8+10	24250
Arecanut	1000	Black pepper + pineapple	1000+20000	1.8	1+10	42750

all the states of NEH region. Under AICRP on Agroforestry, various agroforestry models have been developed as alternative to shifting cultivation. Silvipastoral system comprising *Alnus nepalensis*, pineapple and forage crops like *Panicum maximum* or *Setaria sphacelata* coupled with *Stylosanthes guyanensis* in 1:1 ratio was found to be a sustainable agroforestry practice in soils having 30-60% slope. Forage yield of 13.5 t ha^{-1} was obtained from the combination of Stylo and Setaria. In addition to this, 2.3 t ha^{-1} litter from *Alnus nepalensis* and 4000 fruits ha^{-1} from pineapple were obtained. This system also restored the fertility of these soils (Chauhan *et al.*, 1993). Agri-horticultural systems involving cultivation of ginger with fruit plants, such as mandarin and guava on mild slopes have been found profitable, besides protecting and conserving the hill soils in the NEH region. Pineapple can be easily associated with multi-purpose trees in paired rows on hill slopes, where it checks soil erosion and also gives considerable cash returns within a short period. Under silvi-horti-system with *Acacia auriculiformis* and *Leucaena leucocephala*, pineapple produced an yield of 11.7 t ha^{-1} and 13.4 t ha^{-1}, respectively (Dhyani, *et al.*, 1995). Some of promising agroforestry systems developed under AICRP Agroforestry are given in Table 4. Based on research work concluded in three improved agroforestry systems (Khasi mandarin, alder, Khasi pine) at ICAR research Complex for NEH region, Tomar (2006) reported maximum monetary benefit in ginger plus Khasi mandarin based system (Table 5).

Table 4: Potential agroforestry systems and practices proposed by AICRP on Agroforestry Centres for different agro-ecological regions of India

Centre	System	Agroforestry interventions proposed
ICAR Research Complex for NEH Region, Imphal	Agrisilvipasture Agrihortipasture Silvipasture	♦ Napier grass, maize with *Bauhinia varigata, Pinus roxburghii, Schima wallichii*, and *Artocarpus integrifolia*. ♦ *Citrus rotundifolia*, pineapple with napier. ♦ Lemon with guinea grass, ricebean, groundnut and soybean. ♦ *Eucalyptus sp. L. leucocephola* and *P. pedicellatum*.
ICAR Research Complex for NEH Region, Lembucherra	Agrisilvi-horticulture	♦ Assam lemon is planted in the inter spaces of *Acacia sp.* and *Tectona grandis.* Turmeric is also grown as inter crop. ♦ Pineapple is planted in the inter-spaces of trees such as *A. auriculiformis, M. alba, Gliricidia sepium* etc. ♦ Sesame is a compatible oil seed crop under *A. auriculiformis*.
ICAR Research Compex for	Silviculture (250-400) plants ha^{-1}	♦ *Pinus kesiya, Schima wallichii, Cryptomeria japonica.*

NEH Region, Umiam	Silvipastoral (400 plants ha^{-1}) Agrisilvicultural (400-500 plants ha^{-1}) Sericulture based (i)Agri-silviculture (ii)Silvi-horti-pasture Agri-horticulture Paddy-cum-sericulture Homesteads and Aquaculture	♦ *Alnus nepalensis, Schima wallichii, Michelia oblonga* with *Setaria /Guinea* and native grasses. ♦ *Alnus nepalensis, Paraserianthes falcataria, Michelia champaca, Erythrina indica, Parkia roxburghii, Prunus cerasoides* and Soybean-linseed, Ginger/turmeric followed by groundnut in alternate year. ♦ *Morus alba* (1.80mx0.9m) and groundnut-mustard. ♦ *Morus alba* (0.9m x0.9m) *Syzygium cumini+P.guajava* (6x4m)/Assam lemon (4mx3m)/pear(6mx3m) with pineapple in paired rows with fruit trees and native grasses on bunds. ♦ Guava (5mx2.5m; 5mx5m; 5mx7.5m) with Ginger/ turmeric or groundnut/soybean or chillies+frenchbean; pineapple in guava rows + groundnut. ♦ Khasi mandarin (5mx2.5m, 5mx5m), Assam lemon (4mx3m) with Ginger/turmeric or groundnut/soybean or chillies + frenchbean. ♦ *Morus alba* on bunds of rice field. ♦ Guava and Banana on bunds with groundnut/beans and chillies with Guava on bunds, rice in lowlands and vegetables and fish in farm ponds.
ICAR Research Compex for NEH Region, Gangtok	Silvi-pasture Agri-silviculture	♦ *Ficus hookerii* + *Thysanoleana maxima*. ♦ *Ficus hookerii* + napier ♦ Among 17 fodder trees, *Ficus lacor* gave highest fresh fodder (204 kg tree^{-1}) ♦ Ginger, turmeric, large cardamom and Dinanath grass can be grown successfully (up to 11-15m distance) and rice, soybean, finger millet, Nandi setaria and fine stylo (beyond 15m) from bamboo shade.

Table 5: Estimated monetary benefits of different crops in association with three developed agroforestry systems

Tree component	Density (ha^{-1})	Crop components	Density m^{-2}	Yield (t ha^{-1})	Net income (₹ ha^{-1})
Alder	416	Maize	6	3.50	13500
		Rice	10	1.86	7300
		Ginger	4	0.68	73600
Khasi mandrin	800	Maize	6	5.60	26100
		Rice	10	3.40	19700
		Ginger	4	8.00	88000
Khasi pine	416	Maize	6	17.80	3000
		Rice	10	0.50	(-) 3500
		Ginger	4	5.40	56800

Bamboos for eco-restoration and augmentation of food **:** NEH is known for rich diversity of bamboo resource. Bamboo species cover an area of around 10.03 M ha, which contribute 12.8% of the total forest cover of the country. Out of 126 plant species available in India, nearly half of the variability is available in NEH region. Tribal communities of the region use this potential resource for food, shelter, furniture, handicraft, medicines and various ethno-religious purposes. Tender bamboo shoots of some of the species are edible and sold in the local markets by primary and secondary vendors. There are 12 major edible bamboo species in the region, which could be included in mass afforestation programmes (Bhatt *et al.*, 2003 and Tomar *et al.*, 2009).

Integrated farming system (IFS) **:** Integrated farming system is based on the concept that "there is no waste", and "waste is only a misplaced resource which can become a valuable material for another product" (Edwards *et al.*, 1986). It is generally considered particularly relevant to benefit the rural poor. All the components of agriculture like crop-livestock-fish-forestry-horticulture are integrated in a complimentary way. Soil and water conservation is also given due importance in this type of farming system. Thus, the integrated farming system may provide the environmentally sustainable and economically viable technology that encompasses rational utilization of available resources of the region. The emphasis in such system is on optimizing resource utilization rather than maximization of individual element in the system. The systems provides provide alternative to traditional land use practices including shifting cultivation and helps in maximizing return per unit area per unit time in terms of produce and money (Tomar *et al.*, 2012).

Various location specific integrated farming system models have been developed by various research organizations including ICAR. Rice-fish-cattle; rice-fish-goat; rice-fish-poultry; rice-fish-duck; rice-fish-pig and rice-fish are some of the potential farming systems practiced by the farmers. These integrated farming systems have been found economically viable as evident from the data shown in Table 5.

Table 6: Economic status of different integrated farming systems

S. No.	Farming system	Cost- benefit	Source	Remarks
1.	Agrihortisilvi- pisciculture	1:2.8	Tanjang *et al.*, 2006	Practiced by *Kalita group* in Arunachal Pradesh.
2.	Agrihortisilvi- cultural system	1:1.4	Arunachalam *et al.*, 2002	Practiced by *Nyishi* tribes of Arunachal Pradesh.
3.	Agripisciculture	1:3.4	Tanjang *et al.*, 2006	Practiced by *Apatani* tribes of Arunachal Pradesh
4.	Rice-fish-cattle	1:1.69	Bhatt & Bujarbaruah, 2005	Developed over marshy lands in Meghalaya
5.	Rice-fish-goat	1:1.44	Bhatt & Bujarbaruah, 2005	Developed over marshy lands in Meghalaya
6.	Rice-fish-poultry	1:1.41	Bhatt & Bujarbaruah, 2005	Developed over marshy lands in Meghalaya
7.	Rice-fish-duck	1:1.31	Bhatt & Bujarbaruah, 2005	Developed over marshy lands in Meghalaya
8.	Rice-pig-fish	1:1.31	Bhatt & Bujarbaruah, 2005	Developed over marshy lands in Meghalaya
9.	Agrihorticulture system	1:1.57	Bhatt and Mishra, 2003	Rice was intercropped with fruit trees like Assam lemon, guava and peach over sloppy lands in Meghalaya
10.	Agropastoral system	1:1.45	Bujarbaruah, 2004	The study was conducted in one of the watersheds under farming system research project at Meghalaya

11. Horti agricultural systems	1: 2.79	Tomar and Bhatt, 2004	Improved rice varities was intercropped with fruit tree species viz. Guava, Assam lemon and Peach over sloppy lands in Meghalaya.
12. Agro-aquaculture AFS	1: 3.38	Tomar and Bhatt, 2005	The aquaculture based agroforestry system studied in mid hill situation of the region with all judicious combination of perennial cash crops, production of fruits, crops and livestock.

Horticultural and plantation crops land use system : The agro-climatic condition of North East India is ideal for cultivation of plantation crops. Tea (*Camellia* spp.), as a plantation crop was introduced in the region very early while other plantation crops such as coffee (*Caffea* spp.), rubber (*Hevea brasiliensis*), arecanut (*Areca catechu*), black pepper (*Piper nigrum*), *etc.*, were introduced initially with the idea of providing alternative method of livelihood for the farmers doing shifting cultivation. Various bodies such as National Committees for Agriculture and other commodity committees had recommended that in order to provide a better method of agriculture it would be useful to introduce plantation crops in the region as one of the important alternatives.

Slope of land for horticultural and plantation crops should not be preferably more than 100%. Soil depth must be minimum 1.0 m contour bunds at 2 m vertical interval, half moon or crescent shape circle at the location of planting, a few bench terraces at the lower slope towards foothills for growing vegetables and pineapple are essential conservation measures. Field bunds/ borders planted with economically important NTFPs bearing plants also arrests soil erosion. The studies with pineapple cultivation, even when planted across the slope, resulted in soil loss to the tune of 24.0-62.6 t $ha^{-1}year^{-1}$ during second year (Singh *et al.*, 1996). Some horticultural crops grown on hill slope gave extremely encouraging economic return. Yield potential of newly planted Assam lemon (*Citrus* spp.) orchard was found to be 11300, 12800 and 37200 fruits ha^{-1} during 3rd, 4th and 5th year after planting, respectively (Verma *et al.*, 2001).

Agri-horti-silvipastoral land use system : This system comprises land use at the foothills with agricultural crops, horticulture in the mid portion of the hill slope and silvipastrol land use towards the top of the hill. Land up to 100 % slope having soil depth greater than 1.0 m can be used for agri-horti-silvipastoral system. Contour bunds, bench terraces, half-moon terraces, grassed waterways and stilling basins are the conservation measures required for the treatment of land. Choice of tree species in the system which produces MTFPs such as edible parts, gums, resins, tannins, dyes, medicine, fibers, fodders, etc. would depend on local environment.

An experiment was conducted by dividing watershed into three tier system *viz.,* upper 1/3rd area under pasture and silviculture for rearing livestock (goats and pigs), middle 1/3 area under horticulture including orange (*Citrus* spp.), guava (*Psidium guajava*) and pineapple (*Ananus comosus*) and remaining 1/3 lower area under agriculture for cultivation of cereals, pulses, vegetables, spices, fodder, *etc.* The middle portion of micro watershed of which 50% area was put under orange, 25% under guava and 25% area was put under Assam lemon. The economic evaluation of the agri-horti-silvipastrol system with guava as horticulture component has revealed that the system is economically viable (Rao, 1991). Agri-horti-silvi-pastoral system with livestock, dairy farming and Agro-pastoral system recorded a input output ratio of 1: 2.14, 1: 2.08 and 1: 2.05, respectively and was recommended as viable alternative to shifting cultivation in north east (Bhatt *et al.,* 2001).

Multi-storey cropping land use system : This is highly productive, fully sustainable and very practicable system. To increase the cropping intensity, multi-storey crop combination consisting of crop of varying canopy orientation and rooting have also been developed which entails differential harvesting of solar energy and utilization of soil fertility of variable depth based on the principle of canopy dimension and rooting pattern. One crop combination is coconut (*Cocos nucifera*) + black pepper + pineapple. In such high intensity cropping programme a higher efficiency in utilization of solar energy in incident on a given area is obtained as it is intercepted at vertical intervals by the plant canopies (Singh and Singh, 1997).

Livestock-Based Land Use System : For livestock-based land use system, the land up to 100% slope with minimum 0.5 m soil depth can be utilized for livestock farming. Contour bunds, trenches and grassed waterways are minimum requirement of land treatment. Crops and cropping pattern of such land uses differ depending on the type of enterprise. The fodder production system has to ensure stability of fertility status of soil, availing the opportunity of moisture supply towards maximum fodder production for larger period during the year and conservation of fodder for lean season. Important grasses for the purpose are *Chrysopogon fulvus, Chloris gyana, Dichanthium annulatum, Panicum antidolate, sataria anceps, Sehima nervosum*; legumes- *Atylosia scarabaeoides, Macroptilium atropurpureum, Slylosanthes gracilis, Clycine javanica* and important tree and shrub species for humid tropics- *Albizia chinensis, Albizia lebbek, Alizia procera, Artocarpus heterophyllus, Sesbania grandiflora, Sesbania sesban* trees and shrubs species for temperate and sub temperate areas- *Betula alboides, Celtis australis, Morus serrata, Robinia pseudoacacia* etc., and grasses and legumes for temperate and sub temperate areas- *Bromus inermis, Dactylis glomerata, Poa pratensis, lespedeza ceraces, Lupinus augustifolis, Trifolium incernatum, Trifolium prantensis, Trifolium incernatum, Trifolium response* are suitable for development of livestock based land use systems (Singh and Srivastava, 1990). Carrying capacity of such high land use has been estimated to be 4 to 5 livestock unit^{-1} ha^{-1} with setaria and stylo (1:1) mixture of fodder production. This system has potential for substantial income (1:1.78) from the farmyard manure and self sufficiency in fuel production through biogas plant. In such systems, 90 % of annual rainfall could be retained in the watershed and soil loss was restricted to 2 t ha^{-1} year^{-1} (Verma *et al.*, 2001).

Sericulture-based systems: Sericulture-based systems can help in income generation besides conserving natural resources. Sericulture-based silvi-horti-pastoral system with mulberry, guava, pineapple (in paired rows) and grasses on bunds was found ideal on the hill slopes (30-45%) with 0.6-1.0 m soil depth. The mulberry plantation got fully established in the first year itself and reached its maximum yielding capacity from the second year onwards. The rice-cum-sericulture system in the lowlands provided frequent cash-returns by harvesting at least 3 cocoon crops at an internal of 3-4 months, besides rice crops every year (Dhyani *et al.*, 1996). For the foothill and valley land situations, agro-aquaculture system comprising composite fish culture with azolla, rice, vegetables, groundnut and fruit crops was found profitable and provided employment to 2 persons throughout the year, besides ensuring food supply to the family.

Water-harvesting for irrigation and fish production unit : The water resource potential in the northeastern region is perhaps one of the largest in the entire country. Due to high rainfall there is tremendous potential for agro-aquaculture in the region so that requirement of fish and meat could be fulfilled. The region holds wide varieties of fish fauna ranging from warm water to cold water and hill stream fishes, which are well adapted to the varying weather conditions at different regimes. Of the approximately 806 species inhabiting freshwaters of India, the NEH Region represent 267 fish species (33.13% of the total Indian freshwater fishes) belonging to 112 genera under 38 families and 10 orders. There is maximum scope for exploitation of ground water in Assam, Tripura and Arunachal Pradesh as these three states have very good scope for developing irrigation potential through ground water. In the other states of the north east India, ground water may be developed in the small limited intermountain valleys or foothill areas, where the sediments are predominantly clayey with thin bands of fine sands. Fish production system has very high potential and can be adopted as subsidiary source of income. There is great potential for integrated fish farming with livestock, poultry, piggery and duckery. Marshy lands are particularly suitable to create water harvesting structures and integration of various components like fishery cum piggery, fishery cum duckery, fishery cum poultry and fishery cum livestock, *etc.* (Bhatt and Bujarbaruah, 2005). Most of the water harvesting structures can be created by involvement of local manpower. Dug-out-cum embankment type of ponds can be created as seasonal or perennial water bodies for fish production. At individual and community levels, this programme has tremendous potential not only in boosting the economy but also as immense value in developing water resources of the region.

Future Thrust Areas

- Identification of agroforestry practices and ALUS for resource conservation and its optimization to increase economic return sustaining the traditional practices.
- Role of Agroforestry and ALUS in addressing the emerging issues like global warming (through studies on carbon sequestration, controlled production of methane etc. from the animal component), quality control (SPS), post harvest technology *etc.*
- Development of economically feasible technologies for enhancement of hydro-physical, chemical and microbiological condition of soil to suit the land use for site-specific crop production at different elevations of NEH hills.
- Integrated farming system approaches on watershed basis for restoration of highly degraded land by agro-forestry, horticulture, pasture plantation crops.
- Promotion of the "option enhancement" process like components of crop-tree-animal-fish through which individual household can modify and diversify their production options for nutritional and food security based on their biophysical and socioeconomic condition.

CONCLUSIONS

In recent years there has been considerable emphasis on agroforestry all over the world since it aims at simultaneous production of food and wood and at the same time minimize the risk of crop failure during drought years. An appropriate agroforestry system improves soil physical properties, maintains soil organic matter content and promotes nutrient cycling. However, suitable agroforestry models for different agroclimatic zones of India are yet to be evolved and tested. In the north-eastern hill region of Indian forest ecosystems have been undergoing fast degradation during the last few decades owing to deforestation caused by shifting cultivation and other human disturbances. Therefore, there is a need for developing sustainable food production systems in the hills of northeastern region. Indigenous tribes of this region have been growing crop and trees together in different combinations since time immemorial. In such traditional systems, a great deal of variability in crop diversity, crop composition and crop rotation is observed depending on the area and needs of the farmers. Such diversity is primarily because of differences in topography, climate and soil conditions. The needs of the farmers also influence the composition of mixed cropping. A number of multipurpose tree species are grown with cereals, vegetables and other cash crops.

REFERENCES

Arunachalam, A., Khan, M.L. and Arunachalam, K. 2002. Balancing traditional *juhm* cultivation with modern agroforestry in Eastern Himalaya biodiversity hot spot. *Current Science*, 83: 117-118.

Avasthe, R.K., Singh, K.K. and Tomar J.M.S. Tomar. 2011. Large cardamom (*Amomum subulatum Roxb.*) based agroforestry systemfor production, resources conservation and livelihood security in the Sikkim Himalayas. *Indian J. of Soil conservation*, 39(2): 155-160.

Bhatt, B.P., Singh, R., Mishra, L.K, Tomar, J.M.S., Singh, Matver, Chauhan, D.S., Dhyani, S.K, Singh, K.A., Dhiman R.K. and Datta M. 2001. Agroforestry Practices and Research: An Overview. In: Steps towards modernization of agriculture in NEH Region (Eds. Verma N.D. and Bhatt, B.P.). ICAR Research Complex for NEH Region, Umiam, Meghalaya, pp 365-392.

Bhatt, B.P. and Bujarbaruah, K.M. 2005. *Intensive Integrated Farming System: A Sustainable Approach of Land use in Eastern Himalayas*. Technical Bulletin No. 46, ICAR Research Complex for NEH Region, Umiam, Meghalaya, 43p.

Bhatt, B.P. and Mishra, L.K 2003. Production potential and cost-benefit analysis of agrihorticulture agroforestry system in North-East India. *J. Sustainable Agriculture*, 22: 99-108.

Borthakur, D. N. 1992. Agriculture of the North Eastern Region with special Reference to Hill Agriculture. Beecee Prakashan, Guwahati.

Bujarbaruah, K.M. 2004. Organic farming: Opportunities and challenges. In: Souvenir International Conference on *Organic Food* (Eds. Sarma, B.K.; Bhatt, B.P. and Sahay, C.S.). ICAR Research Complex for NEH Region, Umiam, Meghalaya, 14-17 February, pp13-23.

Chauhan, D.S., Dhyani S.K. and Desai A.R. 1993. Productivity potential of grasses in association with *Alnus nepalensis* and Pineapple under silvopastoral systems on agroforestry systems in Meghalaya. *Indian J. of Dryland Agric. Res.& Dev.*, 8(1): 60-64

Chauhan, D.S. and Dhyani, S.K.1989. Traditional Agro-forestry Practices in North-East Himalayan Region of India. Indian *J. Dry Land Agri. Res. Develop.*, 4(2): 73-81.

Dhyani, S.K., Chauhan, D.S., Dinesh Kumar, Kushwaha, R.V. and Lepcha, S.T. 1996. Sericulture based agroforestry systems for hilly areas of north-east India, *Agroforestry Systems*, 34:1-12.

Dhyani, S.K., Chauhan, D.S., Singh, B.P. and Prasad, R.N. 1995. Tree effect on yield and quality of pineapple (kew) in a silvi-horti systems of agroforestry in NEH region of India. *Range Management & Agroforestry*, 16(2): 89-93.

Edwards, P., Kaewpaitoon, K., McCoy, E.W. and Chantachaeng, C. 1986. Pilot small scale crop/livestock/fish integrated farm. AIT Research report, 184, Bangkok, Thailand, 131 p.

Rao, N.V. 1991. Ecopnomic aspects. Annual report. ICAR Research Complex for NEH Region, Umiam, Meghalaya.

Singh, A.K. and Singh, S.K. 1997. Orchard management of fruit crops. *Farmer and Parliament*, 35 (9): 9-24.

Singh, N. P., Singh, O. P. and Jamir, N. S.1996. *Sustainable agriculture development strategy for North Eastern Hill Region of India.* (*Eds.* S. P. Shukla, and N. Sharma) Pub. Mittal Publication, New Delhi, pp346-351.

Singh, Punjab and Srivastava, A.K. 1990. *Forage production technology*. Pub. Director IGFRI, Jhansi, Uttar Pradesh. pp24-30.

Tanjang, S., Deb, S., Brahma, S., Arunachalam, S., Melkania, U., Arunachalam, K. and Srivastava, K. 2006. Traditional agroforestry systems of *Nyishis, Apatanis* and *Kalitas*. *In*: Bhatt, B.P. and Bujarbaruah, K.M. (eds.) *Agroforestry in North East India: Opportunities and Challenges*, ICAR Research Complex for NEH Region, Umiam, Meghalaya, pp313-321.

Tomar J.M.S. 2006. Ecological interaction between trees and crop components in selected agro-forestry systems of Meghalaya. Ph D thesis. North Eastern Hill University, Shillong.

Tomar J.M.S., Upadhaya, K., Tripathi, O.P. and Pandey, H.N. 2007. Agroforestry Systems and Practices Prevailing in Meghalaya. In: Puri and Panwar (eds.) "Agroforestry Systems and Practices" pp. 357-366. New India Publishing Agency Pitam Pura, New Delhi.

Tomar, J.M.S. and Bhatt, B.P. 2007. Growth performance and biomass production of some agroforestry tree species in nursery in northeastern Himalayan region. *Indian J. of Agroforestry*, 9(1): 38-41.

Tomar, J.M.S. and Bhatt, B.P. 2004. Studies on horti-agriculture systems at mid hill altitude of Meghalaya. *Indian J. of Agroforestry*, 6(2): 35-39.

Tomar, J.M.S. and Bhatt, B.P. 2005. Studies on aquaculture agroforestry system in North eastern Himalayas, India. *Indian J. of Hill farming*, 18 (1&2): 21-27.

Tomar, J.M.S., Anup Das, Lokho Puni O.P. Chaturvedi and G.C.Munda 2012. Shifting Cultivation in North Eastern Region of India – Status and Strategies for Sustainable Development in *The Indian Forester*, 138(1): 52-62.

Tomar, J.M.S., Hore, D.K. and Annadurai, A. 2009. Bamboos and their conservation in North-East India. *The Indian Forester*, June: 817-824.

Verma, N.D., K.K. Satapathy, R.K. Singh, J.L. Singh and K.K. Dutta, 2001. Shifting cultivation and alternative farming systems. In: *Steps towards modernization of agriculture in NEH Region.* (Eds. N.D. Verma and B.P.Bhatt) ICAR Research complex for NEH Region, Umiam, Meghalaya. pp345-364.

□□□

Chapter 51

Sustainable Management of Natural Resource through Different Alternate Agro-Forestry Land Use Systems in Peninsular India

SHARANAPPA JANGANDI, J. NARAYANA, MUDALAGIRIYAPPA AND ANILKUMAR D. TALAGERI

INTRODUCTION

India is endowed with a vast and rich diversity of natural resources particularly soil, climate and agro-biodiversity. To realize the optimum potential of the agricultural production system on a sustained basis, efficient management of natural resources is of paramount importance. According to National Bureau of Soil Survey and Land Use Planning 187.7 M ha land area (57 % of the total geographical area) is degraded. Due to water erosion, about 5334 Mt of top fertile soil is being eroded a annually resulting in the loss of total food production of around 30 million tons per year. Soil resource base is also shrinking at an alarming pace of 0.25 M ha annum^{-1} due to rapid industrialization and urbanization. (Sarkar, 2005).

India's share in land resource of the world is only 2% on which about 17 % of world's human population and 16 % of worlds livestock, survive. However, with its diverse agro-climate topography and soil types. India is capable of producing a wide range of crops and vegetation. The land surface is represented by red and lateritic soils (alfisols, oxisols, ultisols and inceptisols *etc.*, 172.2 M ha) and black soils. Vertisols and their associations (73.5 M ha) alluvial soils (entisols and inceptisols 58.4 M ha) desert soils. (Velautham *and* Bhattacharya, 2000). The key properties of different soil type land use management and agro ecological factors, soil productivity is related to land use and management systems and assessed in relation to some indicators of soil physical, chemical, hydrological and biological properties, including top soil depth, structure, texture, infiltration rate, pH, nutrients status, soil organic

Department of P.G. Studies and Research in Environmental Science, Kuvempu University, Shankaragatta, Shimoga, Karnataka, India matter, soil bio-diversity etc. a scientific approach has been made in the present study under central dry zone of Hiriyur, Karnataka, India.

MATERIALS AND METHODS

In view of paucity of information on degradation of land resources a study was designed in study of different alternative agro forestry land use systems through run-off soil and nutrients loss at Zonal Agricultural Research Station, Babbur Farm, Hiriyur, Karnataka, India during 2009-2010.The predominant land use systems namely natural vegetation, silvi-horti, pastoral, silvi-horti, agri-horti, agricultural systems were identified for the study. The five land use treatments were laid out in an area of one hectare each, the experimental plots were systematically bunded and set the run-off water diverted to one out let so that the run-off soil loss could be quantified with gauging structures like H-flume and Coshocton wheel were installed.

Location and Geology of the Experimental Site : The experimental area was situated 13 57′ 32″ north latitúde and 76 37′ 38″ E East longitude at an elevation of 606.1 metres above MSL. The soil properties consist of Dharwad schists and gnesis and dominated with schistose rocks (dominant chlorite schists) type of parent materials. The soil belongs to chromic Haplusterts order vertisol. The texture of the soil ranges from clay dominated with 2:1 expanding type of montmorillonite clay, minerals. Structure was from subangular blocky to angular blocky with the maximum water holding capacity ranging from 54 to 60 %. The soils were alkaline in reaction (pH 8.5 to 8.7) with organic carbon content ranging from 0.4 to 0.6 % and cation exchange capacity from 60 to 75 Cmol (p+) kg^{-1}. These soils are low in available nitrogen (200-250 kg ha^{-1}) and available phosphorus (5-10 kg ha^{-1}) high in available potassium (150-300 kg ha^{-1}) (Table 1).

In general, the land slope was 2.6 to 2.9 % having soil depth 38 to 40 cm in all the land use systems. The cultivation practices like two time cultivator, one time passing the blade harrow were adopted on contour under agriculture, agri-horti system and one time cultivation in silvi-horti system. The silvi-horti pasture and natural system was maintained as such without any disturbance, sunflower crop was grown during monsoon in agriculture, agri-horti system. The redgram seeds were drilled using seed drill and periodical intercultural operations were followed during cropping season by maintaining furrow in between crop rows to conserve rain water.

Measurement of hydrological parameters : Daily rainfall was recorded by self recording siphon type of rain gauge and run off, soil and nutrients losses were estimated systematically by H-flume, self recording automatic water level recorder and Coshocton wheel were installed at the lower out let point under each system.

Soil Erodibility Indices

Dispersion ratio

Dispersion ratio was calculated by using the expression (Singh, 1980).

$$\text{Dispersion ratio} = \frac{\text{Amount of silt + clay obtained after water dispersion of soil}}{\text{Total amount of silt + clay obtained after complete dispersion of the same soil.}} \times 100$$

Erosion ratio

Erosion ratio was calculated by using the expression (Singh, 1980).

$$\text{Erosion ratio} = \frac{\text{Dispersion ratio}}{\text{\% silt + clay / moisture equivalent}}$$

Erosion index

Erosion index was calculated by using the expression (Sahi *et al.*, 1977).

$$\text{Erosion index} = \frac{\text{Dispersion ratio}}{\text{\% clay/half of the water holding capacity}}$$

RESULTS AND DISCUSSION

Run off in different alternative land use system varied significantly when least total run off of 27.10 mm is associated in natural system (Table 4) whereas the same is highest in silvi-horti system 82.10 mm. This is attributed to the more open space available where splash erosion will occur due to non-hindrance to rain drops. Lesser run off in natural system is due to the fact that the rain drops cannot reach the soil directly which are intern obstructed by canopy of tree species. Also roots which are closely distributed which not allow the flow of water instead to downward percolation. Similar findings were also reported by Kemper and Derpsch, 1981 and Khera *et al.* (2005).

Causes of land degradation : Land degradation depends mainly on the erosivity of external factors and erodibility of soil factors, of which soil erosion ratio, soil erosion index and dispension ratio of soil are important ones. Higher erosion ratio (9.05 and 9.41), erosion index (12.07 and 12.07) and dispersion ratio (16.94 and 18.45) at surface and sub surface soil respectively found in Agri-Horti system of land use which is normally low in natural system of land use. (Table 3). This higher rate of erosion index ratio and dispersion ratio is attributed to direct rain drops, non-hindrance of rain drops by tree canopy, poor root network which are weak because of annual fibrous root system. In case of forest soil, strong root network, tap root system binding of soil mass by humus content do not allow the water to flow but to percolate. Similar findings have been reported by Yadav *et al.* (2006).

Soil organic matter and nutrients loss : Significant variation with respect to loss of soil organic matter and nutrients is observed under different land use system (T-2). Highest and least soil loss was registered in silvi-horti system (2.3 t ha^{-1}) and natural system (0.664 t ha^{-1}), respectively Kanitkar *et al.* (1960). Organic matter loss from the soil was highest in silvi-horti system (45.20 kg ha^{-1}) and the same was least (13.0 kg ha^{-1}) in natural land use system according to Beke *et al.*, 1990. Nitrogen loss from the soil was highest in agricultural land use system (13.00 kg ha^{-1}) and the same was least (4.88 kg ha^{-1}) in natural land use system. Phosphorus content loss was also high in agricultural land use system (0.25 kg ha^{-1}) and least in natural land use system (0.12 kg ha). Potassium loss was highest in silvi-horti system (17.30 kg ha^{-1}) and least (5.0 kg ha^{-1}) in natural system of land use. However highest major nutrients loss (30.30 kg ha^{-1}) was in Silvi-horti land use system, and the same was least in natural system (10.0 kg ha^{-1}) and perceptibly, highest secondary nutrients loss was (250 kg ha^{-1}) in silvi-horti system and least was in natural system (85.80 kg ha^{-1}). Similar trend was recorded with respect to micronutrients loss (Table 2). Whereas major loss of nutrients in silvi-horti system is mainly attributed to lower grass, high soil loss and higher uptake by horticultural crops, similar results were also reported by Sharma and Panwar (1977), However, higher total natural resource soil loss was registered in silvi-horticultural system (2.30 t ha^{-1}) and lowest in natural (0.65 t ha^{-1}) land use system.

In recycling of nutrients, creation of favorable microclimate, condition and higher food production, maintaining balance in oxygen, carbon dioxide in atmospheric importance and humidity. Mainly different Agro-forestry land use system studies can provide sufficient free canopy cover which retards impacts of rain drops which in turn reduces splash erosion. The present study helps the farmers to minimize the fertilizer cost and land degradation by substituting leaf litters they will improve the soil organic matter which results in enhancing maximum moisture holding capacity of soil, run-off and soil loss will reduce significantly and minimizes the nutrients loss.

Table 1: General Characteristics of Soil in selected site.

Soil Type	:	Black soil
Soil texture	:	Clay loamy to clayey
Soil structure	:	Sub angular blacky to angular
Sand	:	29 %
Silt	:	27%
Clay	:	44 %
Maximum water holding capacity	:	54-60 mm
Bulk density	:	1.45 mgm^{-3}
% porosity	:	46.14
Infiltration rate	:	4.30 cm hr^{-1}
pH	:	8.50
Organic carbon (%)	:	0.4-0.6
CEC (C mol p+ kg^{-1})	:	60-75
Available nitrogen (kg ha^{-1})	:	200
Available phosphorus (kg ha $^{-1}$)	:	5-10
Available potassium (kg ha $^{-1}$)	:	150
Dehydrogenase activity	:	5.02 ×g TPF g^{-1} soil

Table 2: Total soil organic matter and nutrient losses in different alternative land use systems.

Particular	Natural Vegetation	Silvi-Horti Pasture system	Silvi-Horti system	Agri-Horti system	Agriculture system
	2009-10	2009-10	2009-10	2009-10	2009-10
Soil loss (t ha $^{-1}$)	0.65	1.20	2.30	2.25	2.22
Organic matter loss (kg ha^{-1})	13.00	44.00	45.20	25.00	33.00
TOTAL	**13.65**	**45.20**	**47.50**	**27.25**	**35.22**
Major nutrient loss (kg ha^{-1})					
Nitrogen	4.88	12.00	12.80	11.50	13.00
Phosphorus	0.12	0.20	0.20	0.16	0.25
Potassium	5.00	16.64	17.30	8.12	13.5
TOTAL	**10.00**	**28.84**	**30.30**	**19.78**	**26.75**
Secondary nutrient loss (kg ha^{-1})					
Sulphur	4.00	10.20	10.90	8.15	11.00
Calcium	57.60	76.80	179.00	95.00	145.00
Magnesium	24.20	65.70	60.10	25.70	62.00
TOTAL	**85.80**	**152.70**	**250.00**	**128.85**	**218.50**
Micronutrient loss (kg ha^{-1})					
Copper	0.10	0.25	0.26	0.18	0.24
Zinc	0.30	0.55	0.70	0.25	0.70
Manganese	2.15	6.35	7.12	2.40	5.40
Iron	0.55	1.30	1.30	0.75	1.30
TOTAL	**3.10**	**8.45**	**9.38**	**3.58**	**7.64**

Table 3: Soil erosion index and dispersion ratio are influenced by different alternative land use system.

Treatment	Erosion Ratio		Erosion Index		Dispersion ratio	
	0-15 cm 2009-10	15-30 cm 2009-10	0-15 cm 2009-10	15-30 cm 2009-10	0-15 cm 2009-10	15-30 cm 2009-10
Natural vegetation	5.90	6.91	8.61	9.75	12.42	14.06
Silvi - Horti. Pasture system	5.94	7.33	8.72	10.31	12.43	14.69
Silvi - Horti. system	7.02	8.26	9.96	11.07	14.86	16.51
Agri - Horti system	9.05	9.41	12.07	12.07	16.94	18.45
Agriculture system	6.99	9.85	9.67	9.67	13.75	18.47
S.Em ±	0.62	0.87	1.04	1.04	1.42	1.37
C.D.	NS	NS	NS	NS	NS	NS

Table 4: Runoff as influenced by rainfall intensity in different alternative land use system.

Rainfall intensity (mm hr $^{-1}$)	Runoff causing through rainfall Amount (mm)	Different land use system			
		Natural system	Silvi-Horti system	Agri-Horti. System	Agricultural system
	2009-10	2009-10	2009-10	2009-10	2009-10
0-10	-	-	-	-	-
10-20	66.20	-	10.06	6.97	7.26
20-30	62.30	5.50	17.04	11.39	12.23
30-40	71.30	12.24	28.00	19.61	27.21
40-50	61.20	9.36	27.00	17.43	20.79
50-60	-	-	-	-	-
Total	261.00	27.10	82.10	55.40	67.49

CONCLUSIONS

In respect of soil and nutrients losses through run-off in natural land use system was found to be least compared to silvi-horticultural land use system. Available nitrogen was maximum in Agricultural land use system and phosphorous and potassium was highest in natural land use system. The results indicated that natural land use system have potential for keeping the soil, water and nutrient loss much below the permissible limits.

REFERENCES

Beke, G.S., Foround, N., Channappa, T.C. and Entz, T. 1990. Runoff and Soil loss from experimental plots in southern Albertes using simulated rainfall. *Can. Agric. Engg.*, 33: 205.

Kanitkar, N.V., Sirur, S.S. and Gokhale, D.M. 1960. *Dry farming in India.* ICAR, New Delhi.

Kemper, B. and Derpsch, H.L. 1981. Effect of cropping pattern on soil erosion, Trans. *Am. Sol. Agri. Engg.*, 67: 235, pp8.

Khera, K. L. and Kahlon, M.S. 2005. impact of land use pattern on soil erosion in sub mount an Punjab. *Indira J. soil cons.*, 33 (3) : 204-206.

Sahi, B.P., Singh, S.N., Sinha, A.C. and Acharya, B. 1977. Erosion index – A new index of soil erodibility. *J. Indian Soc. Soil Sci.*, **25** : 7.

Sarkar, A.K. 2005. Managing natural Resources for increasing Agricultural production in Eastern India. *Journal of the Indian society of soil science*, 53(4): 435-447.

Sharma, S.G. and Panwar, K.S. 1977. Effectiveness of crop cover for reducing splash erosion. *Soil conserv, Digest.*, 5(1): 1-7.

Singh, R.A. 1980. A text book of soil physical analysis. Kalyani Publishers, New Delhi.

Velautham, M. and Bhattacharya, T. 2000. Soil resource management for Agricultural production in India *ICMNR*, Feb. 14-18 New Delhi.

Yadav, R.P., Samra, J.S., Aggarwal, R.K. and Agnihotri, Y. 2006. Effect of soil and water conservation measures on soil quality of degraded lands in Shivaliks. *International journal of Tropical Agriculture*, 24 (1-2).

□□□

Chapter 52

Moisture Pattern and Nutrient Uptake by Greengram (*Vigna radiata*) as Influenced by Nitrogen and Phosphorus Levels in Eucalyptus Based Agri-Silviculture System

K.S.AHLAWAT, O.P.LATHWAL, R.S.PANNU AND JOGENDER SINGH

INTRODUCTION

Agroforestry is dynamic, ecologically based natural resource management practice that, through the integration of trees on farms and in the agriculture landscape, diversifies and sustains production for increased social, economic and environmental benefits for land users at all levels. It will help in maintenance of soil productivity by reducing soil erosion and improving salt affected soils by lowering down the water table. Tree integration in agroforestry system may result in competitive, complementary or supplementary interaction with respect to soil moisture, allelopathy, radiation and plant nutrient balance as well as changes in microclimate affecting the significance of the system and its components. Block plantation of eucalypts has adverse effect on associated crops because being evergreen in nature it continues to compete for light, moisture and nutrient throughout the year. The average yield of pulses including greengram is low because of its cultivation in soils of low in fertility, at prevailing intense heat and chronic moisture stress conditions. The strategy of fertilizer management is one of the important agronomic techniques for enhancing productivity and nitrogen and phosphorus play a pivotal role in the nourishment of greengram. In agroforestry systems trees modify nutrient cycling by increasing the quantity of nutrients in the root zone, thereby increasing nutrient availability to crops.

Eucalypts has more than 600 species, among which *Eucalyptus tereticornis* are cultivated more among the farming communities in India. In north India there is large scale afforrestation of eucalyptus along the canals, roadsides and farmers field on field bunds as well as in blocks. Block plantation of eucalyptus has adverse effect on associated crops because being evergreen in

Krishi Vigyan Kendra, Kurukshetra-136 118, Haryana, India

nature, continues to compete for light, moisture and nutrient throughout the year and is also allelopathic in nature. *Vigna radiata* (L.) Wilczek commonly known as greengram is the most widely distributed of the six cultivated Asiatic *Vigna* species. The average yield of pulses including greengram is low because of its cultivation in soils of low in fertility, at prevailing intense heat and chronic moisture stress conditions. The strategy of irrigation and fertilizer management is one of the important agronomic techniques for enhancing productivity and nitrogen and phosphorus play a pivotal role in the nourishment of greengram. The present study, therefore, was carried out to evaluate the moisture pattern under eucalyptus and nutrient uptake by greengram as influenced by nitrogen and phosphorus levels in eucalyptus based agrisilviculture system *vis-à-vis* sole crop.

MATERIAL AND METHODS

Present investigation was carried out at the research area of Agroforestry department, CCS Haryana Agricultural University, Hisar (20°10′ N, 75°46′E, 215 m above mean sea level), situated in the arid region of north-western India. The climate is subtropical-monsoonic with an average rainfall of 350-400 mm (70-80 % of which occurs during July to September). The soils of the experimental area are derived from Indo-Gangetic alluvium, very deep and sandy loam in texture having some nodules of calcium carbonate in the profile. Before sowing of crops composite soil samples were taken from both the experimental fields at 0-15, 15-30 and 30-45 cm depth during both years. After processing, soil samples were analysed for their physico-chemical properties. The data in Tables 1 and 2 indicate that physical and chemical properties of both the fields were almost similar except the organic carbon content in the surface layer, which was higher in eucalypts field as compared to control field. Available N, P, K were in low, medium and high range, respectively in both the fields. Soils of both the fields were slightly alkaline in nature.

Table 1: Physical properties of soil of experimental fields

Property	Depth of the soil surface (cm)					
	Agroforestry			Sole Crop		
	0-15	15-30	30-45	0-15	15-30	30-45
Sand (%)	64.5	61.3	59.7	64.6	61.5	59.6
Silt (%)	18.3	19.6	20.4	18.2	19.5	20.5
Clay (%)	17.2	19.1	19.9	17.2	19.0	19.9
Textural class	Sandy loam	Sandy loam	Sandy loam	Sandy loam	Sandy loam	Sandy loam

International Pipette Method, Piper, 1966

Table 2: Chemical properties of soil of experimental fields

Particulars	Year	Agroforestry Soil depth (cm)			Sole Crop Soil depth (cm)			Method used
		0-15	15-30	30-45	0-15	15-30	30-45	
Organic carbon	2004	0.48	0.27	0.18	0.36	0.31	0.26	Walkley and Black's wet digestion method (Jackson, 1973)
	2005	0.51	0.29	0.21	0.38	0.32	0.27	
Soil pH	2004	7.72	7.81	7.86	8.35	7.62	7.48	pH meter with glass electrode in 1:2 soil water suspension (Jackson, 1973)
	2005	7.79	7.88	7.92	8.36	7.89	7.62	
EC (dSm^{1} at 25°C)	2004	0.22	0.25	0.27	0.38	0.37	0.35	Conductivity bridge method (Richards, 1954)
	2005	0.25	0.26	0.28	0.35	0.35	0.32	
Available N ($kg\ ha^{-1}$)	2004	93.89	76.39	64.35	92.35	74.38	60.58	Alkaline permanganate method (Subbiah and Asija, 1956)
	2005	96.62	79.58	67.59	94.90	76.59	63.93	
Available P ($kg\ ha^{-1}$)	2004	13.70	11.65	8.70	16.90	11.60	9.00	Olsen's method (Olsen *et al.*, 1954)
	2005	13.90	11.70	8.90	17.00	11.80	9.10	
Available K ($kg\ ha^{-1}$)	2004	439.5	394.6	378.3	450.8	400.4	385.3	Flame photometric method (USDA Handbook No.60., 1954)

Greengram was grown in a plot measuring 6m x 4 m during rainy season of 2004 and 2005 in between the rows of *Eucalyptus tereticornis* (planted in July, 1999 at a spacing of 6m x 2m) and as a sole crop (without eucalyptus) in the field adjacent to eucalypts plantation. The experiment consisting of four fertilizer levels *viz.*, F_0 (control i.e. without fertilizer), F_1 (75 % of recommended dose of fertilizer *i.e.*, 15 kg N + 30 kg P_2O_5 ha^{-1}), F_2 (100 % of recommended dose of fertilizer *i.e.*,20 kg N + 40 kg P_2O_5 ha^{-1}) and F_3 (125 % of recommended dose of fertilizer *i.e.*, 25 kg N + 50 kg P_2O_5 ha^{-1}), was laid out in randomized block design with four of replications with eucalyptus and as sole crop. The recommended dose of fertilizer for greengram is 20 kg N + 40 Kg $P_2O_5 ha^{-1}$. The entire amount of nitrogen and phosphorus was drilled at the time of sowing through urea (46 %) and single super phosphate (16%), respectively. Green gram (cv. '*Asha*') was sown in north-south direction on 6 July 2004 and 12 July 2005 by hand plough at a row spacing of 30 cm, using seed rate of 20 kg seed ha^{-1} in both the system. The uptake of N, P and K was worked out at harvest, separately for grain and stover and was summed up for finding out total uptake of these elements in different treatments. Uptake of nitrogen, phosphorus and potassium was computed as:

$$\text{Uptake of N,P,K (kg ha}^{-1}) = \frac{\text{N, P, K per cent in grain or stover x yield of grain or stover (kg ha}^{-1})}{100}$$

Soil moisture content (per cent on dry weight basis) was determined by gravimetric method by drawing soil samples at sowing and before each irrigation. Soil samples were taken from 1.5 and 3 meter distance from tree row from different soil depth *viz.*, 0-15, 15-30, 30-45, 45-60, 60-75 and 75-90 cm with the help of locally fabricated post hole augar. Soil moisture per cent was determined by drying the soil samples in oven at 105°C till a constant weight attained. The loss in soil moisture was expressed as percent on dry weight basis. The moisture per cent was worked out as follows:

$$\text{Moisture percent} = \frac{\text{Weight of moist soil} - \text{Weight of oven dry soil}}{\text{Weight of oven dry soil}}$$

RESULT AND DISCUSSIONS

Soil Moisture Content : Moisture content was higher in the control plot as compared to agroforestry at all the soil depths and at all the data recorded stages during both the years. The differences in moisture content in agroforestry compared to control were clearly visible before first and second irrigation during both the years. Zonation of soil moisture was determined upto 90 cm depth with the interval of 15 cm. The moisture content increased with increasing depth in both the systems (Tables 3 and 4). Moisture content under eucalyptus in the upper soil depths (0-45 cm) was about 50 % compared to control both before first and second irrigation during both the years. This might be attributed to additional soil water extraction by eucalyptus in addition to test crop in agroforestry as compared to sole test crop. At lower depths also moisture content in agrofrestry was less than control field. Percent decrease in moisture content under eucalyptus over control was higher at 1.5 m distance than 3.0 m distance from the tree line at all soil depths and at all data recorded stages during both the years. Suresh and Rao (1999) have also reported that the moisture in the 0-90 cm soil profile was higher under the sole crop than under trees. The data clearly indicated that eucalyptus trees had severe competition for moisture with test crop. Similar results have also been obtained by Kumar (2003).

Higher soil moisture depletion in *Eucalyptus tereticornis* based agroforestry as compared to sole crops also resulted in reduced value of growth parameters of crops under *Eucalyptus tereticornis* sown crop as compared to sole crop. The reduced value of yield parameters and consequently grain and stover yield under eucalyptus was mainly due to less availability of incidental radiation and competition for moisture as clearly reflected in moisture studies. Reduction in yield of agricultural crops due to moisture competition has also been reported Shukla *et al.* (2000) and Pandey and Sharma (2001) under different agroforestry systems.

Table 3: Moisture content (%) under agroforestry system and control before first irrigation

Soil depth (cm)	Moisture content (%) Agroforestry Distance from tree line								Control	
	1.5 m		% decrease over control		3 m		% decrease over control			
	2004	2005	2004	2005	2004	2005	2004	2005	2004	2005
0-15	3.4	4.0	53.4	56.5	3.8	4.4	47.9	52.1	7.3	9.2
15-30	4.3	5.1	50.0	53.6	4.9	5.5	43.0	50.0	8.6	11.0
30-45	5.4	6.1	48.0	50.4	6.1	6.7	41.3	45.5	10.4	12.3
45-60	12.0	12.8	27.2	28.8	12.8	13.5	22.4	25.0	16.5	18.0
60-75	16.1	17.9	16.5	16.7	17.3	18.9	10.3	12.0	19.3	21.5
75-90	20.6	23.4	7.2	8.9	21.4	23.8	3.6	7.3	22.2	25.7
Mean	10.3	11.5	33.7	35.8	11.0	12.1	28.0	31.9	14.0	16.2

Table 4: Moisture content (%) under agroforestry system and control before second irrigation

Soil depth (cm)	Moisture content (%) Agroforestry Distance from tree line								Control	
	1.5 m		% decrease over control		3 m		% decrease over control			
	2004	2005	2004	2005	2004	2005	2004	2005	2004	2005
0-15	2.6	3.1	59.3	63.0	3.2	3.6	50.0	57.1	6.4	8.4
15-30	3.7	4.2	50.6	57.5	4.0	4.7	46.6	52.5	7.5	9.9
30-45	4.8	5.3	50.0	52.2	5.4	5.9	43.7	46.8	9.6	11.1
45-60	12.6	13.8	24.0	24.5	13.3	14.5	19.8	20.7	16.6	18.3
60-75	16.4	18.4	15.4	15.9	17.6	19.6	9.2	10.5	19.4	21.9
75-90	20.8	24.3	7.5	6.5	21.9	25.2	2.6	3.0	22.5	26.0
Mean	10.1	11.5	34.4	36.6	10.9	12.2	28.6	31.7	13.6	15.9

Effect of Fertilizers on Nutrient Uptake

Nitrogen, Phosphorus and Potassium uptake:N uptake by grain increased significantly with increase in fertilizer level only upto 75 % of recommended dose of fertilizer in both the systems and during both the years. N uptake by stover and total uptake increased significantly with increase in fertilizer levels upto recommended dose of fertilizer in agroforestry during both the years (Table 5). However, in sole greengram N uptake by stover and total uptake increased significantly only upto 75% of recommended dose of fertilizer during both the years. Suresh and Rao (2000) also reported that the nitrogen uptake in different plant parts of cowpea increased with increase in nitrogen levels.

Table 5: Effect of fertilizer levels on nitrogen uptake (kg ha^{-1}) by greengram

Treatment Fertilizer levels (Per cent of recommended dose)	N uptake by grain				N uptake by stover				Total N uptake			
	Agroforestry		Sole Greengram		Agroforestry		Sole Greengram		Agroforestry		Sole Greengram	
	2004	2005	2004	2005	2004	2005	2004	2005	2004	2005	2004	2005
0	14.79	15.76	34.69	36.20	22.54	24.52	50.04	55.41	37.33	40.30	84.74	91.61
75	22.33	23.82	43.42	45.13	34.39	37.38	65.48	71.11	56.76	61.20	108.90	116.25
100	24.89	26.56	46.17	47.90	38.18	41.75	71.26	78.22	63.07	68.31	117.43	126.12
125	26.33	28.17	47.02	48.80	40.49	44.43	73.64	81.70	66.83	72.60	120.66	130.50
CD at 5%	2.75	2.92	3.22	3.64	3.12	4.10	6.58	7.95	5.43	6.48	10.95	11.74

P uptake by greengram grain increased significantly only upto 75% of recommended dose of fertilizer in both the systems during both the years. In agroforestry, P uptake by stover and total phosphorus uptake increased significantly with increase in fertilizer levels upto 100% of recommended dose of fertilizer during both the years (Table 6). However, in sole greengram, P uptake by stover as well as total uptake increased significantly only upto 75% of recommended dose of fertilizer. Under sole cropping, Yakadri *et al.* (2004) reported that N at 20 kg ha^{-1} + P at 60 kg ha^{-1} resulted in the highest N and P uptake in greengram. Application of 75% of recommended dose of fertilizer increased potassium uptake of greengram grain over control during both the years and in both the systems (Table 7). Potassium uptake by stover and total uptake increased significantly upto recommended fertilizer dose in both the systems and during both the years. Wanjari *et al.* (2003) found there was a liner increase in the uptake of N, P and K as the fertilizer level increased in sorghum.

Table 6: Effect of fertilizer levels on phosphorus uptake (kg ha^{-1}) by greengram

Treatment Fertilizer levels (Per cent of recommended dose)	P uptake by grain				P uptake by stover				Total P uptake			
	Agroforestry		Sole Greengram		Agroforestry		Sole Greengram		Agroforestry		Sole Greengram	
	2004	2005	2004	2005	2004	2005	2004	2005	2004	2005	2004	2005
0	1.72	1.86	4.22	4.41	2.87	3.13	6.31	6.92	4.60	4.99	10.53	11.33
75	2.55	2.71	5.21	5.39	4.44	4.85	8.14	8.85	6.99	7.56	13.35	14.24
100	2.83	3.02	5.49	5.70	5.00	5.59	8.94	9.66	7.83	8.61	14.43	15.36
125	2.94	3.13	5.57	5.79	5.40	6.10	9.51	10.28	8.34	9.23	15.08	16.07
CD at 5%	0.34	0.36	0.32	0.39	0.50	0.68	1.00	0.94	0.75	0.95	1.27	1.34

Table 7: Effect of fertilizer levels on potassium uptake (kg ha^{-1}) by greengram

Treatment Fertilizer levels (Per cent of recommended dose)	K uptake by grain				K uptake by stover				Total K uptake			
	Agroforestry		Sole Greengram		Agroforestry		Sole Greengram		Agroforestry		Sole Greengram	
	2004	2005	2004	2005	2004	2005	2004	2005	2004	2005	2004	2005
0	11.73	12.48	28.83	30.09	25.24	26.23	55.32	58.15	36.97	38.71	84.15	88.25
75	17.40	18.12	35.73	37.05	37.08	39.11	68.80	71.57	54.48	57.23	104.53	108.62
100	18.88	19.80	37.41	39.13	41.73	44.29	75.09	78.42	60.61	64.09	112.50	117.55
125	19.66	20.67	37.88	39.91	45.25	48.32	79.31	82.47	64.91	69.00	117.19	122.38
CD at 5%	2.10	2.42	3.00	2.55	3.95	4.34	6.12	6.45	5.15	6.48	7.68	8.74

Increasing levels of fertilizer application increased the N, P, K content both in agroforestry and sole test crops, as the application of fertilizers is known to increase root cation exchange capacity which enhances absorption of nutrients (Elgabaly, 1962). On account of higher grain/stover yield of test crop, NPK uptake was higher under mono crop than agroforestry during both the years. The higher uptake of nitrogen, phosphorus and potassium at higher fertilizer levels was due to increased availability of nutrients in the soil which resulted in rapid growth and extensive root system thereby increasing nutrient absorption from the soil. Kumar (2006) also reported that N, P, K uptake in wheat grain and straw increased significantly with increase in fertilizer level both in agroforestry and sole wheat.

CONCLUSIONS

The moisture in the 0-90 cm soil profile was higher under the sole crop than than under eucalyptus. Moisture content under eucalyptus in the upper soil depths (0-45 cm) was about 50 % compared to control. Eucalyptus trees had severe competition for moisture with test crop. Additional soil water extraction by eucalyptus in addition to test crop in agroforestry as compared to sole test crop. Percent decrease in moisture content under eucalyptus over control was higher at 1.5 m distance than 3.0 m distance from the tree line. So, there is need to assured irrigation supply to crop sown under eucalyptus. N, P and K uptake in greengram grain increased significantly upto 75 % of recommended dose of fertilizer in both the systems. Whereas N, P and K uptake in greengram stover and total uptake increased significantly upto recommended dose of fertilizer in agroforestry, however, in sole greengram N and P uptake in stover and total uptake increased significantly upto 75% of recommended dose of fertilizer. The higher uptake of nitrogen, phosphorus and potassium at higher fertilizer levels was due to increased availability of nutrients in the soil which resulted in rapid growth and extensive root system thereby increasing nutrient absorption from the soil.

REFERENCES

Elgabaly, M. M. 1962. On the mechanism of anion uptake by plant roots - II. Effect of the cation exchange capacity of plant roots on Cl^{-1} uptake. *Soil Science*, 93: 350-361.

Jackson, M. L. 1973. *Soil Chemical Analysis*. Prentice Hall of India Pvt. Ltd., New Delhi.

Kumar, A. 2003. Studies on the production potential of winter season crops with eucalypts (*Eucalyptus tereticornis*) based agri-silviculture system. M.Sc. Thesis, CCS HAU, Hisar, Deptt. of Forestry

Kumar, A.2006. Effect of irrigation and fertilizer levels on wheat (*Triticum aestivum* L.) production in Eucalyptus tereticornis Sm. based agrisilviculture system. Ph.D. Thesis, CCS HAU, Hisar, Deptt. of Forestry.

Olsen, S. R., Cole, C. V., Watenable, F. S. and Dean, L. A.1954. Estimation of available phosphorus by extraction with sodium bicarbonate. U.S. Dept. Agric. Circ. 939, Washington, D.C.

Pandey, C. B. and Sharma, D. K. 2001. *Acacia nilotica* tree/wheat interaction: effect of irrigation and fertilizer in a traditional agroforestry system in central India. *Ind. J. Agroforestry*, 3: 140-146.

Piper, G. S. 1966. *Soil and Plant Analysis*. Hans Publications, Bombay.

Richards, L.A.1954. *Diagnosis and Improvement of Saline and Alkaline Soils* – USDA Hand Book No.60, Washington, D.C.

Shukla, N.P., Shiv Dhar and Dhar, S.2000. Effect of *Sesbania* alley width and moisture regimes on productivity of fodder sorghum. *Range Management and Agroforestry*, 21: 53-61.

Subbiah, B. V. and Asija, G.L.1956. A rapid procedure for the estimation of available nitrogen in soils. *Current Sci.*,25: 258-260.

Suresh, G. and Rao, J.V.1999. Intercropping sorghum with nitrogen fixing trees in semiarid India. *Agroforestry Systems*, 42: 181-194.

Suresh, G. and Rao, J.V. 2000. The influence of nitrogen-fixing trees and fertilizer nitrogen levels on the growth, yield and nitrogen uptake of cowpea on a rainfed alfisol. *Expt. Agric.*, 36:41-50.

U.S.D.A. 1954. *Diagnosis and Improvement of Saline and Alkaline Soils*. USDA Hand Book No. 60. United States Department of Agriculture, Washington, D.C.

Wanjari, S. S., Deshmukh, J. P., Wankhade, S. G. and Atale, S. B.2003. Yield and nutrient uptake of sorghum under rainfed condition. *Agricultural Science Digest*, 23 (2): 146-148.

Yakadri, M., Tahatikunta, R. and Latchanna, A. 2004. Dry matter production and nutrient uptake of greengram (*Vigna radiata* L. Wilczek) as influenced by nitrogen and phosphorus application during wet season. *Legume Res.*, 27 (1):58-61.

□□□

Chapter 53

Effect of Irrigation, Mulch and Frost on Aonla (*Emblica officinalis*) Plantation under Submontane Conditions of Punjab

SATVINDER SINGH, SUKHJINDER PAUL, AMIT SALARIA, S.S. BAWA S.C. SHARMA, SHER SINGH, RAKESH K SHARMA, D.S. RANA, SUKHVINDER SINGH AND BHARAT BHUSHAN[1]

INTRODUCTION

Regional Research Station for Agro-climatic Zone-I of Punjab Agricultural University (PAU), Ludhiana is situated at Ballowal Saunkhri in the north-east region of Punjab. It lies in the region of 10-20 km wide strip lying immediately next to Shiwalik hills locally known as '*Kandi*', meaning area between plains and hills. This region covers nearly 5.38 lakh ha area which is about 10 % of the total area of the state (Figure 1). This is situated at 31°5′5" N latitude and 76°23′20" E longitude with height above mean sea level ranging between 300-500 m. The climate of the region varies between semi-arid to sub-humid. The average monthly maximum and minimum temperature is 38.4 °C in the month of May and 5.2 °C in the month of January, respectively. On the extreme, maximum temperature recorded at Ballowal Saunkhri was 46.8°C during June 1995 and minimum of -3.0 °C during January 2003. The Indian goose berry (*Emblica officinalis* syn. *Phyllanthus emblica*), popularly known as Amla or Aonla an important fruit crop of commercial significance is quite hardy, prolific bearer and highly remunerative with comparatively less care or even without much care. It belongs to the family Euphorbiaceae and is known as *amla, amlaki, amali, ambala, amalakamu* and *nelli* in different parts of India. The fruit is highly nutritive and is the richest source of vitamin C among fruits except Barbados Cherry. It is valued as an antiscorbutic, diuretic, laxative and antibiotic. Amla is said to be indigenous to tropical South-Eastern Asia particularly in central and southern India. It thrives well throughout tropical India and is met with wild or cultivated form in the region extending from the base of Himalaya to Ceylon and from Malacca to South China (Bose and Mitra, 1990).

Regional Research Station (PAU), Ballowal Saunkhri, Dist S B S Nagar, Punjab, India – 144 521
[1]Dept. Soils, Punjab Agricultural University, Ludhiana, Punjab, India – 141 004

It is a potential crop for commercial growing in marginal soils and degraded lands such as salt affected soils, ravines and dry regions. It goes into dormancy from February to mid March under north Indian conditions. Trees of most of the grafted varieties start bearing after four years of planting. Flowering takes place from end of March to April. After fertilization, there is no increase in fruit size and it remains dormant upto end of July. Fruit size starts increasing from end of July and attains full size upto the end of November. Therefore, this fruit plant passes out the hottest season of April to June in semi-dormant stage under rainfed conditions of *Kandi* area, which is hot and dry period. After June, due to arrival of monsoon, the moisture requirement of the tree and the developing fruits is taken care of by rains upto September and thereafter, by the stored rain water in the soil. Thus,

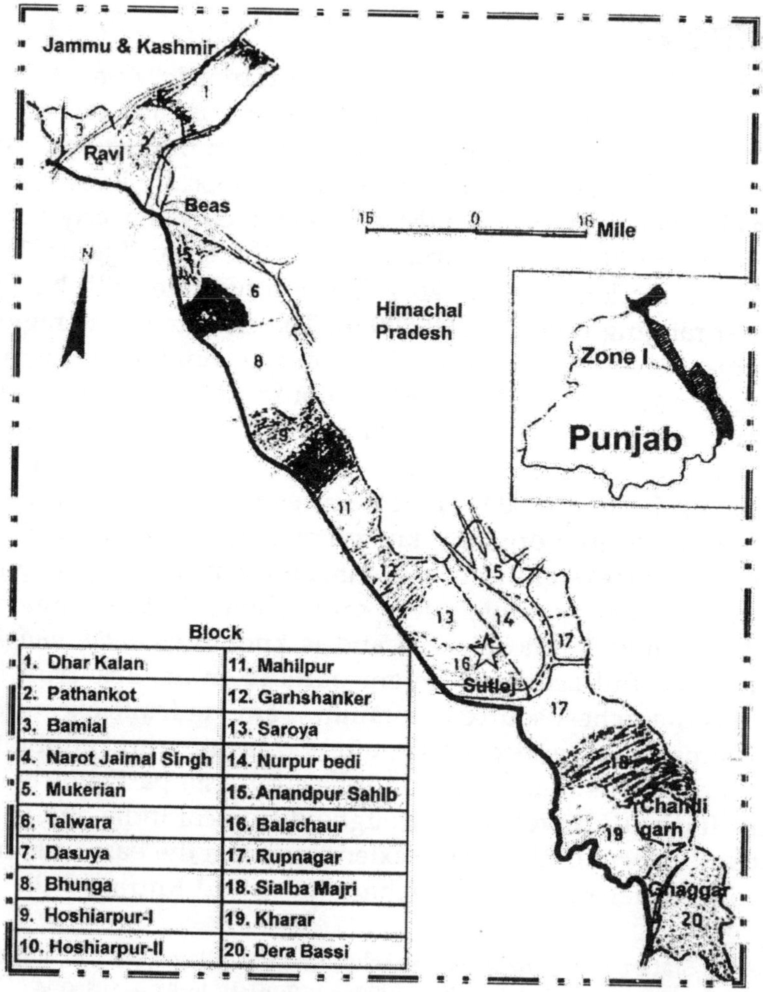

Fig. 1: Location of the study area (Anonymous 1986)

there is a coincidence of growth and fruit characteristics of amla with that of naturally available moisture from monsoon rains. Due to these reasons, amla is one of the very suitable fruit plants which can be grown gainfully (Dhaliwal *et al.*, 2003) in the *Kandi* region of the state. This paper presents the irrigation water requirement with and without polythene mulching coupled with *in-situ* rain water harvesting and conservation technology in Amla.

MATERIALS AND METHODS

Soil of experimental field was sandy loam, alkaline, low in organic carbon & available nitrogen and medium in available phosphorus. Two mulch treatments *viz.* bunding without polythene (M_0) and bunding with polythene (M_1) (black colour 250 micron polythene sheet)covering with 2-3 cm soil layer by providing a ditch on the outer side of the polythene were applied to Amla plants (Chakaiya) sown at a spacing of 7.5 m x 7.5 m. The experiment was designed in randomized block design. The irrigations of 5 cm depth based on the irrigation water to pan evaporation ratios (IW/CPE) of 0.3 (I_1), 0.5 (I_2) and 0.7 (I_3) were applied to new-planted plants.

RESULTS AND DISCUSSION

Approximately 30 % of *kharif* rainfall (903.2 mm) in the region goes as runoff. To conserve the in-situ rain water, the trapezoidal trenches were dug out around the plants to conserve the rain water (Figure 2). The dug-out trenches have the capacity to store approximately 350-400 litres of water.

Fig. 2: Dug-out trenches around amla plants

The growth of the plants was observed in term of increase in stem diameter and plant height (Table 1). Higher % increase in stem diameter and plant height was recorded in the mulched treatment where the irrigations were applied at the IW/CPE ratio of 0.7, followed by the 0.5 and lowest under 0.3 (Figure 3). However, in case of unmulched treatments, the higher increase was observed in treatment with IW/CPE ratio of 0.3 followed by 0.7 and 0.5 (Table 2).

Table 1: Effect of different irrigation levels and mulching on the growth of Amla

Treatment	Stem diameter (cm)			Height (cm)		
	At planting time	After 9 months	After 16 months	At planting time	After 9 months	After 16 months
I_1M_0	0.57	1.62	2.50	36.49	72.63	78.80
I_1M_1	0.67	2.10	3.30	48.75	92.33	110.73
I_2M_0	0.83	1.91	3.00	47.50	82.66	95.20
I_2M_1	0.56	1.72	3.20	33.50	82.37	112.93
I_3M_0	0.70	1.54	2.50	48.67	74.50	100.17
I_3M_1	0.58	1.95	3.60	35.00	119.33	132.10
L.S.D. (P= 0.05)	-	NS	NS		NS	9.84

Table 2: Percent increase in stem diameter and plant height after 16 months of planting

Treatment	Stem diameter (mm)	Height (cm)
I_1M_0	338.6	116.0
I_1M_1	392.5	127.1
I_2M_0	261.4	100.4
I_2M_1	471.4	237.1
I_3M_0	257.1	105.4
I_3M_1	520.7	277.4

Fig. 3: Performance of amla plants

To safeguard the plants during winter (December and January), thatching of young plants (Figure 4), cutting and pruning of older twigs and branches was done. However, during 2006-07, 58.3 % mortality of amla plants was observed due to frost damage. Gap filling was done but treatment wise irrigation was not applied in order to save the plants from further frost damage and proper establishment of new gap filled plants. Again in 2007-08, plants were damaged (88.9 % mortality) by frost (Table 3), as there was prolonged occurrence of frost (20 days) from mid-December 2007 to mid-February 2008 (Figure 5). The plants also experienced sub-zero temperature upto -3.0°C which was 6 °C below normal. Rainfall during winter was approximately 1/3rd of the normal value which further aggravated the chill effect.

Fig. 4: Thatching of amla plants using sesamum stalks

Table 3: Effect of frost damage to amla plants

Treatment		Number of dead plants		Mortality rate (%)	
		2006-07	2007-08	2006-07	2007-08
IW/PAN E = 0.3	M_0	7	11	58.3	91.7
	M_1	10	12	83.3	100.0
IW/PAN E = 0.5	M_0	9	9	75.0	75.0
	M_1	7	11	58.3	91.7
IW/PAN E = 0.7	M_0	4	9	33.3	75.0
	M_1	5	12	41.7	100.0
		Total		**58.3**	**88.9**

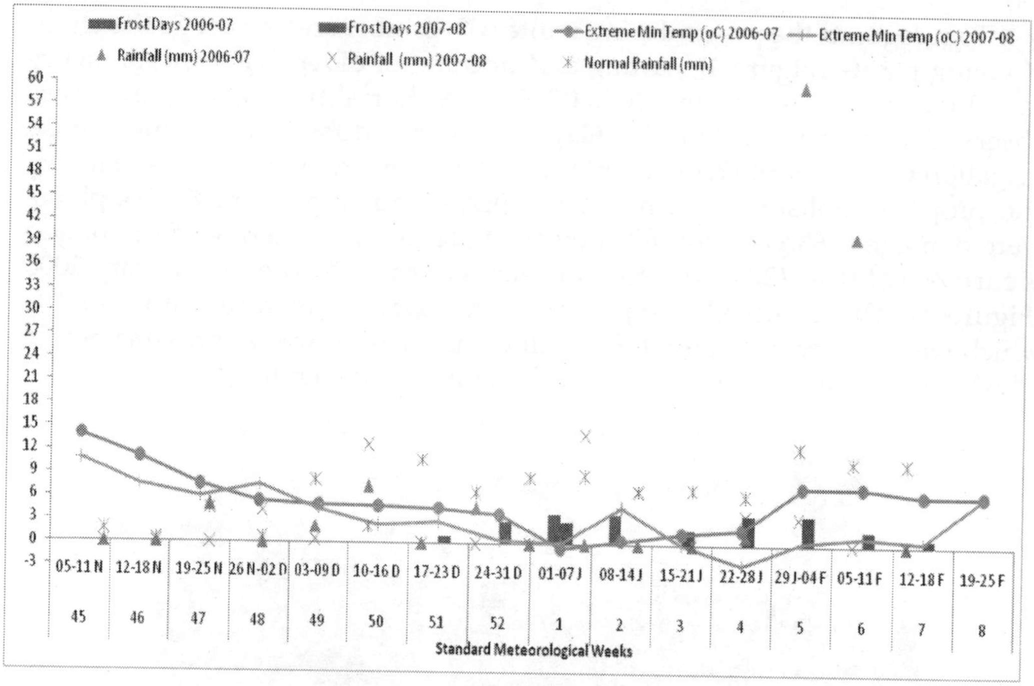

Fig. 5: Week wise occurrence of frost, extreme minimum temperature and rainfall N – November; D – December; J – January; F - February

Least frost damage (37.5 % in 2006-07 and 83.3 % during 2007-08) was observed in IW/CPE = 0.7 and 0.5 irrigation treatment, respectively, signifying the importance of light and frequent irrigations in keeping the soil moisturized. Water acts as buffer in safeguarding soil from low temperatures. This is due to two inherent properties of water having second highest molar specific heat capacity (75.327 J mol^{-1} °K) of any known substance, after ammonia, as well as a high heat of vaporization (40.65 kJ mol^{-1}) (Anonymous, 2012). These two unusual properties are due to extensive hydrogen bonding between its molecules which allows water to buffer large fluctuations in temperature, thereby moderating frost effect. Lesser frost damage was also observed along the boundary of field where there was natural shelter belt. Frost has immediate effect on plants with signs of frost injury on leaves, bark and fruits (Figure 6). Near zero or sub-zero temperatures accompanied with low moisture regime in the soil further aggravates its effect on the vegetation. The mechanism of frost damage is still not completely clear, but there is a fairly general consensus that extra-cellular ice formation results in rapid removal of water from cells and severe dehydration results. The consequent disruption of membrane function and enzyme systems causes widespread necrosis of the affected tissue and results in the typical 'frost burn' symptom (Allard *et al.*, 1998).

Fig. 6: Frost damage to amla plant (left) and fruit (right)

The *Kandi* area due to its undulating and sloppy topography and annual rainfall of approximately 1094.3 mm has huge potential for harvesting of rainwater. About 80 % of the rainfall is received in three months (monsoon period from July to September) and approximately 30-40 % of this rain goes as runoff (Sur *et al.,* 2002). This harvested water can be used during frost susceptible period of December-January for irrigation. Also micro-sprinklers can be used for moisturizing soil and air. Also the rootstock is not affected by frost. In-situ budding of sprouted local rootstock (Figure 7) can be done instead of replantation.

Fig. 7: Sprouting of local rootstock of Amla plants after frost damage

CONCLUSIONS

Occurrence of frost is quite un-predictable and quantification of its severity is not easy. Adequate safeguards against frost protection needs to be taken during the months of December and January. Young amla plantation needs to be protected from frost injuries. The study re-iterates importance of rainwater harvesting. It should be carried on priority basis to make water available during December and January, when there is no or very less rainfall and frost occurrence is more. The study also identifies future research issues like quantification of role of micro-sprinklers in moisturizing soil and air to mitigate frost effect in plantations, determination of threshold soil moisture levels with suitable irrigation intervals which can mitigate frost effect to a minimal level etc. Likewise effect of other practices (smoking fields in night, mulch *etc.*) to minimize frost effect also needs to be measured. Potential use of cost effective synthetic materials for thatching around individual small plants needs to be quantified. The study concluded that frost damages young amla plantation and fruit and prior adequate measures (as stated above) should be adopted in conjunction to minimize its effect.

REFERENCES

Anonymous. 1986. Status Report – Regional Research Station for Kandi Area Ballowal-Saunkhri (Hoshiarpur). Directorate of Research. Punjab Agricultural University, Ludhiana.

Anonymous. 2012. http://en.wikipedia.org/wiki/water Chemical and Physical Properties

Allard, F., Houde, M., Krol. M., Ivanov, A., Huner, N. and Sarhan, F. 1998. Betaine improves freezing tolerance in wheat. *Plant and Cell Physiology*, 39: 1194-1202.

Bose, T. K. and Mitra, S. K. 1990. Fruits: Tropical and Subtropical. Naya Prokash, Calcutta. pp 757-758

Dhaliwal, H. S., Grewal, S. S. and Kanwar, J. S. 2003. Rainfed Horticulture for *Kandi* Area. Research Compilation, Zonal Research Station for Kandi Area (Punjab Agricultural University), Ballowal Saunkhri, Nawanshahr - 144-521

Sur, H. S., Sharma, S. C., Bhardwaj, A., Grewal, S. S. and Bawa, S. S. 2002. Salient Research Findings On-Farm Applied Research under Integrated Watershed Development Project, Zonal Research Station for Kandi Area, Ballowal Saunkhri, Nawanshahr, Punjab, 3p.

□□□

Chapter 54

Influence of Host Species, Organic and Inorganic Nutrition on Growth and Establishment of Sandalwood (*Santalum album*, L.) for Improving Cultivation and Productivity under Forestry and Agroforestry Plantations

KALA. S, K.K. SURESH[1] AND K. KUMARAN[1]

INTRODUCTION

Sandalwood is the fragrant heartwood species of genus Santalum (Family: Santalaceae). In India, the genus is represented by *Santalum album* Linn. Its wood is known commercially as *"Vegetable Gold" or "Queen of Essential Oils" or "East Indian Sandalwood"*. Sandal is one of the five scheduled trees in Tamil Nadu. It is called the *"Royal Tree"*, because of its oil having high value in perfumery and in medicine. Sandal is a semi-root parasite and was first reported by Scott in 1871. As a semi-root parasite, it produces highly valued aromatic heartwood containing the major aromatic constituents, a and b santalol, which are widely used in the perfumery and cosmetic industries.

Due to over exploitation, illicit felling and failure of artificial regeneration methods, the species is considered threatened in its habitats and it has been resulted in decline of potential population and genetic erosion from natural stands Meera *et al.*, 2000). There is no known substitute for sandal oil and the species distribution is also limited. Because of these reasons, Sandal is considered precious and is accorded scheduled tree status. According to ICFRE, Dedradun sandal publication (1997), the sandal family is distributed between 30°N and 40°S from Indonesia in the west to Juan Fernandez Island in the north, to New Zealand in the south. Sandal is indigenous to peninsular India. It is naturally distributed over an area of 9600 km^2 of which over 90 % of the

Central Soil and Water Conservation Research and Training Institute, Research Centre, Chhalesar, Agra-282006, Uttar Pradesh, India.
[1]Forest College and Research Institute (TNAU), Mettupalayam – 641301, Tamil Nadu, India

area is found in Karnataka and Tamil Nadu. In Tamil Nadu, it is distributed over an area of 3,040 sq.kms. Dense population of sandal is found particularly in Javadis hills of Tirupattur forest division and Chitteri hills of Harur forest divisions, are considered as the *'home land of sandal'*. Sandal is also found in Kerala, Andra Pradesh, Orissa, Maharastra, Himachal Pradesh and Rajasthan in the natural forest and private lands.

Sandalwood population in India is voicing concern for the decline of stock mostly due to illicit and distress felling and poor artificial regeneration techniques in establishing sandal at field. Sandal performs differently with different host species, Kushalappa (1995), Rai (1990), Debabrata Swain (1999), Nasi and Ehrhart (1997) reported that sandal response well to lime application. Application of biofertilizers to trees will greatly improve sandal growth (Harley and Smith,1983).

There is no specific research on nourishing sandal in the main field with organic, inorganic, certain chemicals and bio fertilizers. In the artificial establishment of sandal nutrition plays a vital role not only in the nursery but also in the main field. Hence, standardization of nutrition treatments for sandal artificial regeneration is need] of the hour. This paper presents the results of the investigation carried out at Forest College and Research Institute, Mettupalayam during 2004-2005 to standardize the nutrition for field establishment of sandal.

MATERIALS AND METHODS

The field experiment was established at Forest College and Research Institute, Mettupalayam during 2004-2005. The experimental field lies between 11° 19'N latitude and 77° 56' E longitude and at an altitude of 300m above mean sea level. The mean annual rainfall is 895mm, distributed over 49 rainy days with north- east monsoon contributing 60% and summer showers 20% of the total rainfall. The mean maximum and minimum temperatures are 35°C and 20.5°C, respectively.

Six months old sandal seedlings with the height ranging from 47 cm to 51 cm and basal diameter ranging from 4.5 to 5.0 mm were chosen from the nursery and plantation in the main field under randomized block design with seventeen treatments. Each treatment had twenty seedlings and there were three replications. The plant spacing adopted was 2.0 m x 1.5 m. At the time of planting the seedling of secondary host species were planted along with sandal seedling in the same pit. For this study four host different host species which was the best as reported by different scientists were selected as secondary host plants for the main field. These were *Cynodon dactylon* (Kushalappa, 1995); *Albizia Saman* (Surendran *et al.*,1999); *Casuarina equisetifolia* (Rai, 1990;

Taide *et al.*, 1994 & Srivastava, 2003); *Pongamia pinnata* (Nagaveni & Vijay lakshmi, 2003).

The biofertilizers were applied at 15cm depth at the plant root zone at the time of planting. The 15 g of inorganic fertilizers like H_2PO_4, citric acid, orthophospharic acid, fructose and sucrose were diluted with 250 ml of water and it was applied to each plant in the form of soil application 5 days after planting. Spot application of DAP and Micro nutrient mixture were applied at the time of planting in the main field. The foliar spray of 0.5% of micronutrient solution was sprayed over the leaf of sandal seedlings, five days after planting. The following treatments were applied in the main field *viz.*, T_1 - Sandal + *Cyanodon dactylon*, T_2 - Sandal + *Albizia saman*, T_3 - Sandal + *Casuarina equisetifolia*, T_4 - Sandal + *Pongamia pinnata*, T_5- Sandal + Azophos (5 g) + VAM (20 g), T_6 - Sandal + Vermicompost (20 g), T_7- Sandal + Potassium dihydrogen orthophosphate (15 g), T_8 - Sandal + Citric acid (15 g), T_9 - Sandal + Orthophosphoric acid (15 ml), T_{10}- Sandal + Fructose (15 g), T_{11}-Sandal + Sucrose (15 g), T_{12}- Sandal + DAP (25 g), T_{13} - Sandal + Micronutrient mixture (15 g) - soil application, T_{14} - Sandal + Micronutrient mixture (0.5%) – foliar application, T_{15} - Sandal + Azophos (5 g) + VAM (20 g) + DAP (25 g) + Micronutrient Mixture (15 g) – soil application, T_{16} - Sandal + Azophos (5 g) + VAM (20 g) + DAP (25 g) + Micronutrient mixture (0.5%) – foliar application and T_{17} - Control.

The plant management practices like irrigation, weeding, pruning of hosts were properly carried out whenever needed. The biometric observations like shoot length and collar diameter were taken at two month intervals for randomly selected seedlings and their mean was recorded and presented. Fresh weight of shoot and root, dry weight of shoot and root, total dry matter production, shoot / root ratio, root volume were recorded from randomly selected five seedlings from each treatment in a replication. Then mean values were calculated and presented. The chlorophyll content of the leaves was also done. Root length, root volume, dry matter production, shoot / root ratio and chlorophyll content of leaves were recorded at end of 10 month after planting. The details on these are as follows.

The length of shoot was measured from collar region to tip of the shoot and the mean was recorded and expressed in cm. The seedling was removed from the poly bags without damaging the root and the length was measured from the collar region to the tip of the root and expressed in cm. The collar diameter was measured by using vernier calliper at shoot collar region and it was expressed in cm. The seedling after recording all the biometric observations was separated into shoot and root. The fresh weights was taken individually for shoot and root and mean values of fresh shoot and root weight were arrived and expressed in grams (g). The nursery samples collected to record shoot and root fresh weight were then dried at 65-75°C in an electric oven for 17 ± 1 hour, cooled in desicators for 30 minutes and were weighed separately for shoot and root and mean dry weight of shoot and root were calculated and expressed in gram. The total dry matter production (TDMP) per seedling was arrived by summing up shoot and root oven dry weight and the mean of five plants were expressed in gram.

The seedlings after recording all the biometric observations were separated into shoot and root. The fresh root was taken individually and the root volume was calculated by cylinder method and expressed in milli litre (Chadurvedi and Khanna,1990). The volume of roots can be calculated with the help a xylometer which consists of a graduated vessel using the principle of water displacement. Water is poured in the vessel and the initial reading of water level is taken. The washed soil free roots are than immersed and the water level is recorded. The difference between two readings gives the volume of the root submerged.

The shoot / root ratio was calculated by using the formula given by Saravanan (1991).

$$\text{Shoot / Root Ratio} = \frac{\text{Dry weight of shoot (g)}}{\text{Dry weight of root (g)}}$$

To study the effect of treatments on chlorophyll content of the leaves, fresh leaf samples from randomly selected seedlings were analyzed. The chlorophyll content was estimated from fully matured unfurled third leaf from the top of the main branch adopting the method of Yoshida *et al.* (1971) and expressed as mg per gram of fresh weight. Leaf tissue of 250 mg were taken from the fresh leaf samples and washed in distilled water. Then the leaf tissues were ground with 10 ml of 80% acetone using pestle and mortar. The homogenate solution was centrifuged at 5000 rpm for 10 minutes. The supernatant was collected and the volume was made upto 25 ml using 80% acetone. The optical density of the content was measured at 663, 652 and 645nm. Then chlorophyll a, chlorophyll b and total chlorophyll content were calculated using the following formula:

$$\text{Chlorophyll 'a' content (mg g}^{-1}\text{)} = \frac{12.7 \times \text{OD at } 663 - 2.69 \times \text{OD at } 645}{1000 \times w} \times V$$

$$\text{Chlorophyll 'b' content (mg g}^{-1}\text{)} = \frac{22.7 \times \text{OD at } 645 - 4.68 \times \text{OD } 663}{1000 \times w} \times V$$

$$\text{Total Chlorophyll (mg g}^{-1}) = \frac{\text{OD at 652 x 1000 x V}}{\text{34.50 x 1000 x w}}$$

where,

V = volume made (25 ml); w = weight of leaf tissues taken (250 mg)

The recorded values of biometric observations, biomass estimation, chlorophyll estimation and soil analysis in the field experiments were subjected to statiscal analysis, following Panse and Sukhatme (1967).

RESULTS AND DISCUSSION

The growth parameters of sandal significantly varied among the treatments. The beneficial effects of these treatments were observed at 2, 4, 6, 8 and 10 months after planting of seedlings in the field. Among the seventeen treatments evaluated for the rate of shoot length increment and collar diameter increment and number of branches (Table 1), total dry matter production,

Table 1: Effect of nutrition treatments on rate of shoot length increment of sandal in the main field.

Months After planting / Treatment	Rate of Shoot Length Increment (cm)		Rate of Collar Diameter Increment (cm)		Number of branches/ plant
	2 MAP	10 MAP	2 MAP	10 MAP	10 MAP
T_1	5.58	7.78	1.17	1.97	5.0
T_2	4.59	6.30	0.83	1.93	7.0
T_3	8.25	9.80	2.40	3.50	11.3
T_4	5.69	7.11	0.80	1.67	4.0
T_5	7.79	9.56	2.30	3.27	9.7
T_6	4.87	6.15	0.87	1.80	6.0
T_7	4.33	5.77	0.87	2.03	4.3
T_8	4.17	5.83	1.07	2.03	4.0
T_9	4.37	5.86	1.20	2.07	3.7
T_{10}	4.93	6.35	1.40	2.33	6.0
T_{11}	7.82	9.73	2.33	3.33	9.0
T_{12}	4.97	6.52	1.17	2.17	4.0
T_{13}	4.44	6.12	0.93	2.23	5.0
T_{14}	5.11	7.24	0.80	1.97	6.3
T_{15}	7.73	9.49	2.27	3.23	8.7
T_{16}	7.63	9.44	2.23	3.17	7.7
T_{17}	3.73	5.55	0.63	1.60	3.3
SEd	0.54	0.49	0.15	0.15	0.85
CD (p= 0.05)	1.10	1.00	0.30	0.32	1.74

root length and shoot / root ratio and chlorophyll content of the leaves the results are given in Table 2 (Figure 1 & 2). The following four treatments showed significant growth performance over other treatments *viz.*, Sandal along Casuarina (T_3), sandal with sucrose application (T_{11}), Azophos + VAM (T_5), Azophos + VAM + DAP + Micronutrient mixture - soil application (T_{15}) and Azophos + VAM + DAP + Micronutrient mixture-foliar application (T_{16}) without the host plant. The superior performance of sandal along with Casuarina (T_3) may be due to the nitrogen availability to sandal when it was grown along with this N_2 fixing non-leguminous host species. The present findings were in consonance with findings of Ananthapadmanabha *et al.* (1984); Rangaswamy *et al.* (1986); Radomiljac *et al.*(1998) and Nagaveni and Vijaylakshmi (2003).

Table 2: Effect of nutrition treatments on rate of shoot length increment of sandal in the main field

Months After planting / Treatment	Growth and Quality Indices of Sandal Seedlings				
	Root length (cm)	Total Dry Matter Production (g seedling^{-1})	Root volume (ml)	Total Chlorophyll	Shoot/ Root Ratio
T_1	50.0	48.92	22.40	1.52	1.25
T_2	50.6	45.99	19.60	1.60	1.16
T_3	59.5	72.93	33.73	1.82	1.54
T_4	51.9	46.62	18.43	1.57	1.33
T_5	52.9	57.78	31.57	1.74	1.44
T_6	47.1	37.26	18.47	1.47	1.23
T_7	45.6	37.69	17.87	1.44	1.22
T_8	45.9	37.23	16.20	1.46	1.12
T_9	51.6	37.83	18.60	1.50	1.20
T_{10}	52.3	39.83	22.10	1.50	1.08
T_{11}	55.3	59.05	28.33	1.78	1.47
T_{12}	49.2	40.39	18.03	1.58	1.10
T_{13}	48.1	38.76	19.70	1.48	1.07
T_{14}	45.7	38.53	18.67	1.49	1.09
T_{15}	52.4	56.8	27.33	1.71	1.52
T_{16}	50.9	56.08	26.93	1.70	1.50
T_{17}	41.5	33.57	15.43	1.40	1.05
SEd	1.31	1.86	1.18	0.02	0.09
CD (p= 0.05)	2.66	3.79	2.41	0.04	0.19

Fig. 1: Influence of host species / nutrition treatments on growth performance of Sandal at 10 months after planting.

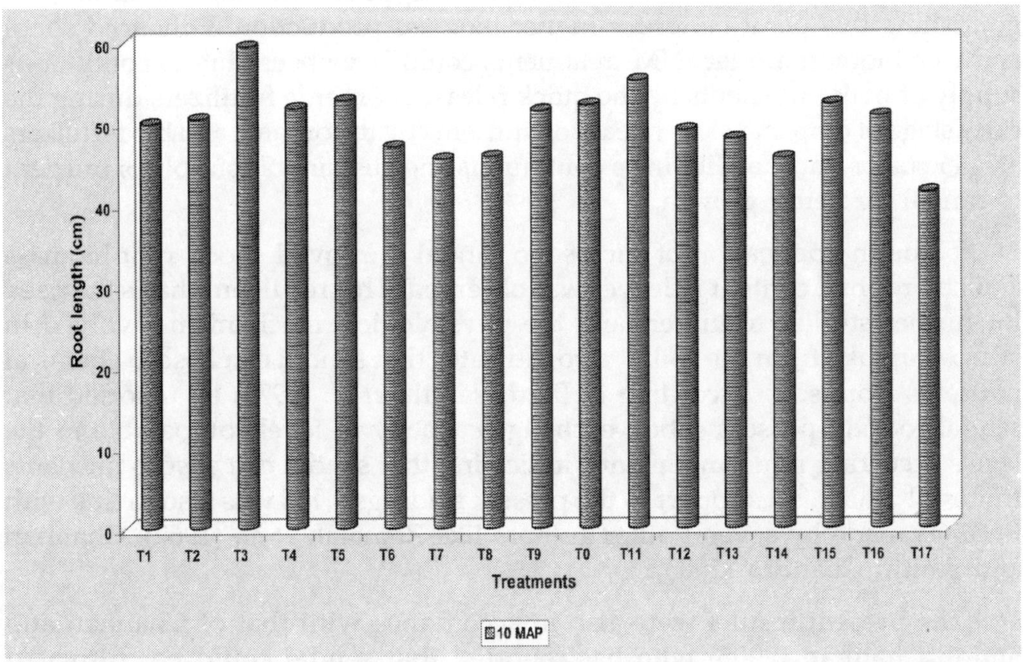

Fig. 2: Effect of nutrition treatments on root length of sandal at 10 months after planting.

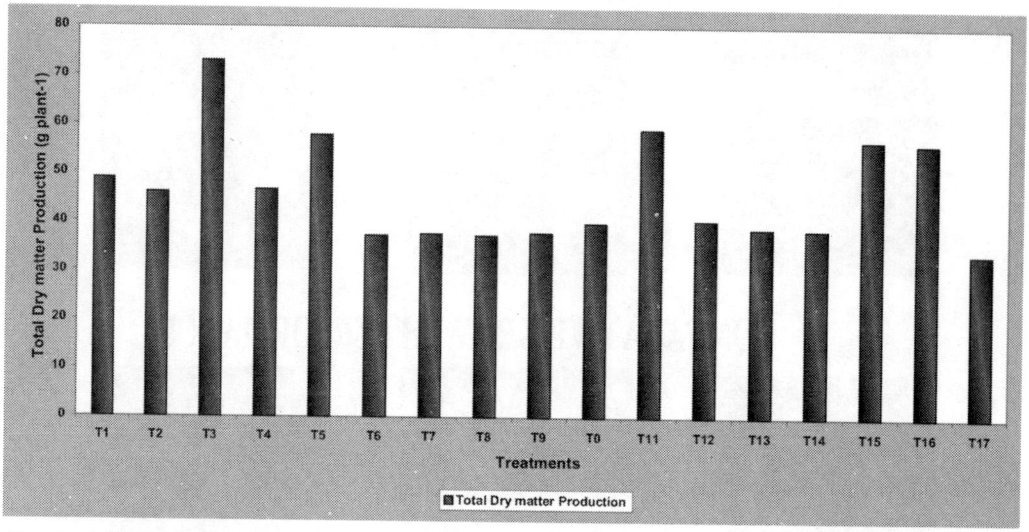

Fig. 3: Effect of nutrition treatments on total dry matter production of the sandal at 10 months after planting.

In general, biofertilizers had tremendous nutrients mobilizing power to the root zone especially by promoting insoluble form of phosphate to soluble form by producing organic acids and by fixing of nitrogen for sandal growth (Somani, 1987). In the absence of host, the combined application of Azophos + VAM + DAP + micronutrients, sucrose application could also help in nourishing this plant better for higher biomass production. Enhanced shoot and root biomass under INM treatments could have been due to continuous supply of major nutrients by the quick release inorganic fertilizers during the early stage of crop and slow release of nutrients by the organic and bio fertilizers at later stages, thus facilitating a continuous and sustainable supply of nutrient to sandal for better growth.

Through application of sucrose to sandal improved shoot, root biomass and chlorophyll content in leaves was observed. This result emphasis the need for further studies to understand the physiological mechanism involved in sucrose uptake from the soil. It also indicates that sandal can absorb chemical nutrients from soil. According to Parthasarathi *et al.* (1974) he reported that sandal root do possess cation exchange capacity of level comparable to the levels occurring in normal plants, indicating that sandal can absorb nutrients from soil which is a support to the present findings. This was also in line with findings made by several other authors like Ramaiah *et al.* (1962); Chauhan and Smithi Chandra (2003).

The present results were also in accordance with that of Chauhan and Smithi Chauhan (2003) who has reported that sandal could be cultivated hydroponically without associations of a host for more than 18 months. These results clearly indicates that sandal can absorb nutrients directly form nutrient medium, if they are available to the roots in a readily absorbable form.

CONCLUSIONS

The present finding clearly concludes that sandal along with host (Casuarina), sandal with sucrose application, sandal with biofertilizers (Azophos + VAM) and INM treatments have equally performed well under the main field. This study clearly states that sandal could respond well to sucrose nutrition, biofertilizers and integrated nutrients application in the main field. The above finding has brought out that sandal could absorb nutrients in the main field, if they are in a readily available form. The result finally states that, sandal can be grown in the main field up to one year of establishment without host through proper nutrition by the combination of organic manure, biofertilizers and inorganic fertilizers. This breakthrough will go a long way in artificial establishment of sandal plants for improving large scale cultivation and productivity under forestry and agroforestry plantations.

REFERENCES

Ananthapadmanabha, A.S., Rangaswamy, C.R., Sarma, Nagaveni, S.H. Jain and Krishnappa, H.P. 1984. Host requirement of sandal. *Indian For.*, 110 (3) : 264-268.

Chauhan, S.V.S. and Smiriti Chandra. 2003. Cultivation of sandal (*Santalum album* L.). Hydroponically without host. *Annals of Forestry*, 11(1): 85-87.

Chadurvedi, S. and L.S. Khanna. 1990. Forest Mensuration. IBD Publicatttions, Dehradun. pp 241.

Debabrata Swain. 1999. Eucalyptus - Bane (or) Boon. *Indian For.*, 123(5),440p.

Harley, J.L. and Smith, S.E. 1983. Mycorrhizal symbiosis. Academic Press. London, 75p.

Kushalappa, K.A. 1995. Sandal in social forestry - recent advances in research and management of sandal in India. pp206-209.

Nagaveni,H.C. and Vijayalakshmi, G.2003. Growth performance of *Santalum album* with different leguminous and non- leguminous host species. *Sandalwood Research News letter*. Issue 18 :1-4.

Nasi, R. Y. Ehrhart. 1997. Sandalwood a perfume of property - Part II plantations. *Bois et Forests des Tropiques* (1996) No. 248: 5-16.

Panse, V.G. and Sukhatme, P.V. 1967. Statistical methods for agricultural workers (2nd eddition) ,Indian council for agricultural research, New Delhi, 368p .

Parthasarathi, K., Gupta, S.K. and Rao, P.S. 1974. Studies on sandal spike cation exchange capacity of sandal in health and disease. *Current Science*, 40: 640-641.

Radomiljac, A.M., McComb, J.A. and McGrath, J.F. 1998. Intermediate host influence on the root hemi-parasite *Santalum album* L. biomass partitioning. *Sandal and Its Products.* pp85-89.

Rai, S.N. 1990. Status and cultivation of sandalwood in India. In: Hamilton, L., Conrad, C.E. eds. Proceedings of the Symposium on sandalwood in the Pacific. April. 9-11, Gen. Tech. Rep. PSW – 122. Pacific Southwest Research Station, Forest Service, US Department of Agriculture, Honolulu. 66-71.

Saravanan, P.P. 1991. Studies on the nutrient amendment on *Acacia* sp. M.Sc (Ag.) Thesis. Tamil Nadu Agriculture University, Coimbatore,205p.

Srivastava, R.K. 2003. Mortality rate and regeneration status in different age groups of *Santalum album* under some species hosts. *Indian For.*,129 (5): 999-1008.

Ramaiah, P.K., Parthasarathi, K. and Rao, P.S. 1962. Studies on sandal spike. A histochemical study of sandal root haustoria in relation to mineral nutrition. *Proc. Indian Acad. Sci.*, 56 B: 111-113.

Rangaswamy, C.R., Jain, S.H. and Parthasarathi, K. 1986. Soil properties of some sandal bearing areas. *Van Vigyan*, 24: 3-4.

Somani, L.L. 1987. Biofertilizers in Indian agriculture (an annotated bibliography, 1906 - 84). Concept Publishing Company ,402p

Surendran, C., Parthiban, K.T., Bhuvaneswaran, C. and Murugesh, M. 1999. Silvicultural strategies for augmentation of sandal regeneration. In: Radomiljac, A.M., H.S. Ananthapadmanabha, R.M. Melbourn, K. Stayanarayan (eds.). *Sandal and its products.* ACIAR Proceedings, Vol. 84, Arawang Communications, Canberra. pp 69-73.

Taide, Y.B., Babu, L.C and Abraham, C.C. 1994. Influence of host species on the initial growth and development of sandal (Santalum album Lin.). *Indian Journal of Forestry,* 174L: 288-292.

Yoshida, S., Forno, D.A. and Cock, J.H. 1971. Laboratory manual for physiological studies of rice. IRRI publication, Philippines. pp36-37.

□□□

Chapter 55

Rehabilitation of Degraded Lands in a Tribal Dominated Watershed for Environmental and Livelihood Security

S.P. TIWARI AND OM PRAKASH

INTRODUCTION

Integrated participatory watershed management has been considered as key for sustainable livelihood and environmental security (Annonymous, 2008). However, management of extensive wastelands particularly with poor vegetative cover, rock out crops and dominated by tribal/asset less community is a great challenge for scientists and farming community (Sastry *et al.*, 2004). Bundelkhand region is one among the most degraded ecosystem characterized by undulated and rugged topography, highly eroded and dissected land, poor soil fertility, scarce ground water resource, erratic distribution of rainfall, heavy biotic pressure, inadequate vegetation cover and frequent crop failure, resulting in scarcity of food and fuel (Palsania *et al.*, 2011). The livelihood of the people in the region is primarily dependent on marginal agriculture and livestock. There is lack of alternate livelihood opportunities in the area resulting exploitation of the natural resources especially community forest/wastelands for fuel, fodder and timber *etc.* According to the report of forest survey of India it is one of the most backward and disadvantaged region of the country spread over in 7.16 M ha area supporting 15.62 M human and 8.36 M animal population with a forest cover of 1.24 M ha (Annonymus, 2005). Land degradation is a complex phenomenon strongly driven by interactions among socio-economic and biophysical factors (Chandra *et al.*, 2008) thus peoples participation in watershed development programme has been recognized as an integral component of the watershed management programme of the country (Sharda *et al.*, 2005). This paper focuses on the achivements of an IWDP undertaken in a tribal dominated watershed occupied largely by wastelands for livelihood

Central Soil and Water Conservation Research & Training Institute,
Research Centre, Datia, Madhya Pradesh, India

and environmental security in Bundelkhand region through people's participation.

MATERIALS AND METHODS

Study site: The watershed area falls between 78° 27' E longitude 25° 43' N latitude above 284 m msl in village Bajni of district Datia, a part of Bundelkhand region. The total area of the watershed is 532 ha, which is dominated by tribals and backward community. More than 65% area is occupied by degraded waste lands. Scarcity of water for drinking as well as irrigation is one of the major problems in the watershed (Om Prakash *et al.*, 2010) Community forest land has very poor canopy cover and is used as grazing grounds resulting in high runoff and soil loss.

Soils and topographic features: The area has both red and black soils which have developed from Vindhyan ranges of rocks. The general slope ranges from 0.3 to 15%. While red soils occurs at higher elevation black soils are restricted to lower reaches, which truly represents the typical red (Rakar and Parwa) and Black (Kabar and Mar) soils of the region and exhibits distinct heterogeneity with regards to texture, depth, fertility status and leads to operational as well as management problems as described by Tiwari *et al.* (2009).

Climate and land use pattern in watershed: The climate in the area has been characterized as short duration rainy season with very erratic rainfall, extreme weather during summer (47°C) and winter (0°C). Average annual rainfall is 860 mm. with a high evaporation rates (140 to 170 cm) leaves large water deficit during most part of the year. A significant portion of the watershed is under rainfed agriculture. Out of the 194 ha cultivable land 64 ha was not cultivated due to want of irrigation water. Further, out of the cultivated (130ha) area nearly 43% was under rainfed. Major cultivated crops are sorghum, black gram, green gram and sesame during kharif and mustard, gram and wheat in *rabi* season.

Socio-economic and demographic set up: The watershed covers two hamlets namely Bajni and Prakash Nagar and dominated by backwards and tribals (*saharia* and *mongia*) population. There are 176 families with average 5 to 6 members in each family. More than 52 % population is engaged in Agriculture and cattle rearing with average land holdings of size 0.17 ha to 2.45 ha in two hamlets. A large population (47.7%) in the watershed is land less and mainly depends on labour for their livelihood.

Works implemented in the watershed: Several need based developmental interventions were executed in the watershed, which include arable land renovation, water harvesting, crop improvement and drainage line treatments.

Mechanical measures undertaken in the watershed include, construction of 13 stone check dams, Creation of 96 ha-m additional water storage capacity through deepening the existing village pond, Construction of percolation tank (6.5 ha m capacity) having 2 ha spread and 40 ha catchment in convergence with state soil conservation department, 17 water outlet structures (waste weirs) on neck points of gullies covering 123 ha field area, minor leveling on 62.7 ha agricultural fields. For the management of wasteland, a model of wasteland management was developed for demonstraions and replication in other areas of the region. Under this a patch of 33 ha community wasteland was put under closure against biotic interference through CPT /stone wall. A rocky patch of 28ha within this protected area was further treated with silvi-pasture techniques combined with mechanical measures and protection *viz.*, contour trenches, gully plugs. *Stylocentus hamata* was sown on the berms of trenches. The existing stock of *Anogesis pendula* (*Kardhai*) and *Butea monosperma* (*Dhak*) was managed and an existing root stock of ber were developed for quality fruits through in-situ budding techniques. Demonstraions of improved agronomic practices under major crops during *kharif* (paddy, maize, jowar, black gram, cowpea sesame soybean) and in *rabi* (wheat and gram) were laid down on farmer's fields. Plantation of 250 trees of 4 species (Shisam, eucalyptus, bamboo, neem) on field boundary / in agro forestry system was also done in the area.

Monitoring of biophysical and socio-economic indicators: Monitoring of biophysical and socio-economic indicators was done in the watershed through out the project period (1998-2002). Three micro-watersheds under different land uses were gauged for runoff and soil loss. Biophysical and socio-economic indicators such as Land leveling index (LLI), Crop diversification index (CDI) Cultivation land utilization index (CLUI), Induced watershed Eco-Index (IWEI), Peoples participation Index (PPI) were employed to assess the efficacy of various activities and using standard formulae as summarized by Sharda *et al.* (2005).

RESULTS AND DISCUSSION

Biophysical Impacts

Impact on runoff and soil loss: Three micro-watersheds were gauged in the watershed, two of which were treated with mechanical and vegetative measures and one was kept as control (untreated). Sharma *et al.* (2004) summarized that due to various soil and water conservation measures undertaken in the watershed, average runoff in whole of the watershed decreased from 25.35% to 16.3% and average soil loss decreased from 12.08 t $ha^{-1}yr^{-1}$ to 8.27 t $ha^{-1}yr^{-1}$ due to imposition of various soil and water conservation treatments imposed in the watershed.

Water Resource Development

Surface water storage: Under common property resources works like renovation of village pond, nallah stabilization and wasteland management were carried out in the WS. The existing village pond (Dharam talaiya) was deepened for increasing its capacity for recharging the wells in vicinity and for miscellaneous community use. Water storage capacity was increased three fold from 48 ha-m to144 ha-m in phased manner during 1998-2002. The structure wise details of water storage capacity created and water stored during the period under implementation is presented in Table 1. In addition to this, a percolation tank of 6.5 ha-m capacity having 2 ha spread and 40 ha catchment area was also constructed by the state soil conservation in the watershed.

Table 1: Ground water storage capacity created and its utilization in the Bajni watershed

S.No	Year	Rainfall (mm)	Storage created (ha-m)		Water harvested during the year (ha-m)
			Yearly	Cumulative	
1	1998	820.8	48.0	48.0	48.0
2	1999	1159.4	42.8	90.8	90.8
3	2000	545.4	5.6	96.4	96.4
4	2001	1125.5	-	96.4	96.4
5	2002	623.0	42.6	144.0	144.0

Irrigation potential and ground water availability: Ground water study revealed that duration of water availability in the WS increased by 50-100% in 29 wells and 12 bore wells which were influenced by conservation structures. The wells which had water for 4 months, increased water availability for 8 months, similarly those that had for 8 month increased water availability for whole year. Irrigation potential of wells in the watershed has increased nearly six fold (Table 2) due to various watershed interventions undertaken in the watershed.

Table 2: Irrigation potential of wells after treatment of Bajni watershed

Irrigation water Potential			
Before		After	
Area irrigated (ha)	Volume withdrawn (ha-m)	Area irrigated (ha)	Volume withdrawn (ha-m)
8.2	1.97	48.0	11.50

Impact on Production System

Impact of land leveling: Minor leveling and bunding was one of the foremost activities in the WS, in which farmers taken keen interest and actively participated by contributing 10-40 % of the cost of leveling. General land slope of the fields undertaken for leveling varied from 2-10% and land leveling indices (LLI) varied from 0.05 to 0.25. Leveling was done on 62.7 ha area covering a total of 71 farmers, resulted an additional increase of wheat yield ranged from 4.69 to 6.25 q ha^{-1} (Table 3) apart from increased and efficient use of rain water.

Table 3: Impact of land leveling on grain yield of wheat (q ha^{-1}) on farmer's fields

Grain yield of Wheat (q ha^{-1})		
Before	**After**	**Additional Grain yield**
9.37 to 14.68 (11.87)*	15.00 to 20.00 (17.31)*	4.69 to6.25 (5.44)*

*Figures in parenthesis are average across farmer's fields

Increase in cultivable land: Cultivated land utilization Index (CLUI) increased by 0.05 and 0.07 during *kharif* and *rabi* seasons, respectively which indicates the impact of watershed interventions on increase in cultivable land area and duration of crop cultivation. A drastic increase in the area of different crops in the WS was also recorded to the tune of 150 % in *kharif* and 34.5% in *rabi* season during post project period with a change in crop diversification Index (CDI) from 0.242 to 0.549 in *rabi* and 0.262 to 0.351 in *kharif* season due to increased availability of water in the WS.

Increase in crop productivity and fertilizer use: Crop development programme undertaken on farmer's field with HYV and improved management practices resulted in higher crop yields in the watershed areas and ranged from 10.1 to 83.9% in *kharif* and 23.4 to 149% in case of existing *rabi* crops with introduction of new crops like soybean in *kharif* and barley and Taramira in *rabi* season (Sharma *et al.*,2004) . These activities could help in increasing the overall crop productivity in the watershed @ 31.2% in five years, *i.e.*, from 10.2q ha^{-1} to 13.4q ha^{-1} (Table-4).

The higher crop fertilization index (0.75) recorded in 2002 against initial CFI value of 0.43 in 1997 this clearly witnessed the increased motivation of WS farmers towards use of balanced fertilization and creating awareness on other production technologies.

Table 4: Impact of watershed treatments on crop productivity in terms of wheat equivalent yield (WEY).

Crop	Before			After		
	Area (ha)	Yield (q ha^{-1})	WEY (q ha^{-1})	Area (ha)	Yield (q ha^{-1})	WEY (q ha^{-1})
Groundnut	17.90	4.23	7.76	26.70	7.56	13.86
Urd	0.20	2.40	4.40	3.50	2.80	5.13
Moong	0.60	1.88	5.01	8.30	2.07	5.52
Jowar+Pigeonpea	3.60	4.22+0.8	4.95	7.02	4.76+0.94	5.68
Soybean	-	-	-	4.80	1.12	1.87
Wheat	85.30	11.66	11.66	95.60	16.40	16.40
Wheat+Mustard	-	-	-	3.00	12.8+1.95	17.68
Gram	10.20	225	5.25	16.50	5.60	13.07
Mustard	8.70	381	9.53	14.80	4.70	11.75
Barley	-	-	-	8.00	7.80	6.50
Sonha Taramira)	-	-	-	2.40	2.60	4.77
Total	126.50		10.21	190.80		13.40

Development of community wasteland for increasing production: Establishment of vegetation and pastures felt difficult in the WS area due to "*Annapratha*" in which farmer's prefer to go for open grazing. Therefore, WS farmers were motivated towards stall feeding by growing fodder crops on their fields particularly during *kharif* season where sizeable area is kept fallow, as well as to maintain closure/protection of 33 ha community wastelands against biotic interferences selected for silvi-pasture development. About 28ha area of this rocky patch was treated with silvi-pasture techniques combined with mechanical measures and protection *viz.*, contour trenches, gully plugs, CPT/stone wall. *Stylocenthus hamata* was sown on the berms of trenches. The existing stock of *Anogesis pendula* (*Kardhai*) and *Butea monosperma* (*Dhak*) was also protected and managed. These combined efforts could help in increasing the forage yield from 0.09 t ha^{-1} to 1.97 t ha^{-1} from the community land. Grasses were sold to cut by villagers @ Re 1.00 per head load. A number of forest and horticulture trees were also planted in this area for increasing the productivity. Existing stock of deshi ber were also improved through budding technique. The study on vegetation status showed that the average cover percentage was increased from 24.7% in 1999 to 47.9% in 2001 due to protection in the WS. Watershed Eco-Index (WEI) worked out to be 0.10 for the watershed in 2002 suggested that an additional 10% of watershed area has rehabilitated through green biomass cover.

Livestock composition and milk production: Farmer's motivation towards cattle's stall feeding and increased on-farm availability of quality fodder further encouraged the farmers to replace their less productive livestock's with high yielding buffalo breed "*Murrah*" as witnessed by number of cows declined from 124 to 71 and buffalo increased from 132 to 195 in the watershed. Hence, due to this shift in milk production in the WS and became almost double in five years with increase in per capita milk availability from 0.385 to 0.947 liter day^{-1}.

Participation of community: Active participation of WS community is an important component of watershed management for its sustainability. People participation index worked out for different stages namely planning, implementation and monitoring, evaluation and maintenance before with drawl of the project suggested a over all peoples participatory index (PPI) value of 58% indicating that the stake holders overall participation was high. The contribution of community towards WS activities/works is considered as a measure of participation. Moreover contribution in terms of cash and kind enhance the responsibility in commitment of the stake holders to maintain the works and activities created under the project. Contribution received from farmers towards various works varied between 10 to 50 % depending upon the nature of work. Also it is important to note that most of the contribution was received in cash rather than kind. This suggests positive impact of the project in terms of community participation. A sum of ₹ 3,45,959 has been deposited in the WDF account that has been handed over to the watershed functionaries for post project maintenance of various works.

Community Organization and Institutional Arrangements

Beneficiaries were organized in to various groups to managing their affairs themselves. Watershed Association (WA), Watershed Committee (WC), in the watershed was constituted. Various User groups (UGs) and Self-help groups (SHGs) were formed by considering common interest/profession, social affinity and compatibility so as to benefit them through best use of technologies as well as to generate employment and impart training. The bottom up approach to maintain transparency of expenditure made in the project was adopted by opening a watershed bank accounts operated with the joint signature of President of watershed association, Secretary of Watershed committee and representative of WDT *i.e.*, Nodal officer of the project from CSWCRTI, RC, Datia. This has helped in maintaining transparency in handling project funds. Watershed development fund account was opened to deposit the money received as a contribution towards various activities and this fund can be used for maintenance of infrastructure created in the watershed.

A sum of ₹ 3, 45,939 was maintained in the WDF at the time of handing over to village panchayat as per withdrawal policy and it increased to ₹ 4,39,265 (till June, 2011) after earning interest from the fixed deposits. As per the guidelines, only interest earned on this WDF amount shall be used by the panchayat for maintenance of assets developed in the watershed, thus this basic amount will remain maintained as corpus fund for sustainable maintenance of watershed assets.

Socio-economic Impacts

Employment and income generation: Additional employment of 12874 man days has been generated through various WS works/activities. Maximum employment was provided to the villagers in terms of watch and ward and through various works such as bund compaction, CPT /stone wall contour trenching digging of pits for plantation, construction of waters harvesting structures *etc.* The maximum increase in income was observed from agriculture in the case of large farmers followed by small and medium farmers. A considerable increase in the income from labour was observed in case of land less. It could be attributed to more employment in the watershed partly due to watershed intervention and partly due to development process. Considering tangible benefits alone the overall B: C ratio worked out was 1.65:1. It indicates that the project seems to be economically feasible. The intangible benefits also include overall improvement of cropping system, employment generation and socio-economic conditions of the people.

CONCLUSIONS

The study highlights that ameliorating benefits of soil and water conservation measures employed in a tribal dominated watershed through IWDP were quite evident. As a follow up action, the experiences from this watershed may be helpful to various stake holders engaged in the field of soil and water conservation and watershed management to focus on further refinement of the technologies, to enhance their suitability and acceptability by end users.

REFERENCES

Annonymous.2005. State of forest report 2005. Forest survey of India, Dehradun 171p

Annonymous.2008. Common guidelines for watershed development projects. National Rainfed Area authority (NRAA) , MoA Govt of India. 52p.

Chandra, R Soni P, and Yadav V.2008 Fuel wood,fodder and livestock status in a Himalayan watershed in Mussoorie hills (Uttrakhand, India). *Indian Forest*,134: 894-906

Om Prakash, Dev Narayan and Biswas, H. 2010. Participatory watershed management for food security and resource conservation in Bundelkhand region-A case study. *In:* Proc. National Seminar on management of natural resources and environment in India held at GAD Instt. of development studies, Amritsar during 23-24 Oct., 44p.

Palsania, D. R., Yadav, R.S., Ramesh Singh, Tiwari, R.K., Chaturvedi, O.P. and Dhyani, S.K. 2011.Resource appraisal of Garh kundar Dabar watershed in Bundelkhand region of central India. 39(3): 263-270

Sastry, G., Reddy, Y.V.R. and Prakash, O.2004. Soil and water conservation interventions in watershed development programmes - economic viability, acceptability and emerging issues, *Indian J. Soil Cons.*, 32 (2) : 81-90.

Sharma, A.K. , Tiwari, A.K., Katiyar, V.S.,Om Prakash, Nitant, H.C., Tiwari, R.K., Bhatt, V.K., Brijlal, R.C., Kaushal, Sant Raj and Sharma, N.K. 2004. Participatory watershed management for sustainable development in Bajni watershed Datia (MP), CSWCRTI, Datia, pp1-69.

Sharda, V.N, Samra, J.S. and Pradeep Dogra. 2005 Participatory watershed management programme for sustainable development: Experiences from IWDP. *Indian J. soil cons.*, 33(2):93-103.

Tiwari, S.P., Dev Narayan and Biswas, H. 2009. Edaphic constraints vis-a-vis conservation practices for increasing productivity of rainfed lands in Bundelkhand Region. *In*: Proc. Conf on Food ad environmental security through Resource conservation in Central India: Challenges and opportunities (FESCO): pp 213-222.

□□□

Chapter 56

Multitier Cropping System for Resource Conservation and Livelihood Augmentation of Small Holders in Southern Odisha

PRAVEEN JAKHAR, B.S. NAIK, HOMBE GOWDA, H.C. AND D. BARMAN[1]

INTRODUCTION

Southern Odisha extends over ten districts of the state (6.6 M ha area) falling under Eastern Ghats region of India and feeds 12.8 M population (Anonymous, 2011). This highland zone is located on Eastern pleatue with red and lateritic soil in sub-humid moist region with mean potential evapotranspiration 1400-1600 mm and mean temperature of 25-28^{0}C. A considerable portion of the upland remains fallow despite their potential for supporting integrated production system under rainfed farming. Eighty percent of the annual rainfall (1500 mm) is received during monsoon season and rest is sporadically distributed during the remaining part of the year. The highland zone of Odisha is bestowed with bounty rainfall and the farmers of the region are fast catching up to innovative idea for growing of vegetables, flowers, and fruits integrated cultivation under multi-tier Agroforestry system to get year around income. Multi-tier or multistory cropping is a system of growing crops together of different height at the same time on the same piece of land and thus using land, water and space most efficiently and economically (Anonymous, 2009). Keeping in view of above points the present study was undertaken to find out suitable Multi-tier cropping system for Eastern Ghats high land zone of Odisha.

MATERIALS AND METHODS

The field experiment was conducted at the Research Farm of Central Soil and Water Conservation Research and Training Institute, Research Centre,

Central Soil and Water Conservation Research and Training Institute, Research Centre, Sunabeda, Koraput, Odisha 763 002

[1]Central Research Institute for Jute and Allied Fibers, Barrackpore, Kolkata-743 101

Sunabeda (latitude 19⁰ 45′ 30½ N, longitude 82⁰ 56′ E and altitude 950 m above mean sea level) in Koraput District of Odisha, during *Kharif* seasons of 2007-10. Soil of the study site was red lateritic *Ustocherpets*, sandy clay loam in texture and acidic in nature. Soil of experimental site was low in organic C (0.39%), available phosphorus (16.7 kg ha^{-1}) and high in available potassium (253.2 kg ha^{-1}). The treatments comprised of boundary plantation of one forest tree *Gliricidia sepium*, one fruit tree Papaya (*Carica papaya*) and four crop combinations (Ginger; *Zingiber offcinale* + Pigeon Pea; *Cajanus cajan* (8:2), Ragi Elusina indica + Pigeon Pea (6:2) and pure ragi, broadcast sowing as farmers' practice) on 2 % sloping land The treatments were set in a randomized complete block design (RBD) replicated thrice. The statistical analysis was carried out using the ANOVA procedures of RBD. Each experimental plot was 12 m wide and 18 m long on a terraced land having 2% slope. The recommended doses of the NPK were applied through Urea, DAP and MOP respectively. The soil samples (0-15 cm) were collected before sowing and after harvesting of the crop from each plot every year and were analysed for pH, EC, N, P and K by following standard procedures. For gauging the soil loss and runoff multi-slot divisors (5:7) were installed. Rainfall intensity and quantity was recorded in research center's 'B' class meteorological laboratory.

RESULTS AND DISCUSSION

Runoff and soil loss :During the experimental period (2007-10), the numbers of yearly rainfall events were 54, 44, 37 and 59 occurred with the corresponding rainfall depth of 835.6, 790.8, 756.8 and 1162 mm causing runoff in all the treatment plots. Runoff caused significant soil loss along the years of study in different multitier cropping systems in comparison to control treatments (Table 1). Across the slope cultivation of ginger and pigeonpea yielded lowest runoff (45.2 mm) and subsequent lowest soil loss (3.6 t ha^{-1}) than ragi cultivation practices. In comparison to farmers' practice of broadcasting; line sowing of ragi decreased runoff and erosion by 23 and 15 %, respectively. Square plantation of papaya in sole ragi and intercropped ragi: pigeonpea (6:2) plots with boundary plantation of *Gliricidia* on bunds reduced net soil loss by 36 and 45 %, respectively in comparison to ragi line sowing. This can be attributed to rainfall interception by introduction of papaya which created physical barrier with its anchorage of soil by roots against smooth flow of runoff on slopes and *Gliricidia* hedgerow planted along the boundary of the plots. Pandey and Chaudhary (2010) in their study in Andaman Island reported that *Gliricidia* hedgerows in vegetable cultivation and as cover crop in coconut plantation reduced soil loss by 61%.

Table 1: Effect of different treatments on resource conservation

Treatment	Runoff (mm)					Erosion (tha^{-1})				
	2007	2008	2009	2010	Average	2007	2008	2009	2010	Average
Ginger	95.5	36.6	67.1	52.3	62.9	5.10	4.20	3.03	2.37	3.68
Pigeonpea	102.8	73.5	62.6	89.5	82.1	6.67	7.13	5.73	2.47	5.50
Ragi Broadcasting (Farmers' practice)	141.5	125.5	122.9	177.4	141.8	15.53	13.40	11.40	7.43	11.94
Ragi line sowing	163.5	136.5	70.1	66.6	109.2	10.37	9.83	9.77	10.63	10.15
Papaya + Pigeonpea + *Gliricidia*	106.4	87.3	37.3	36.0	66.8	5.10	5.87	6.13	3.74	5.14
Papaya + Ginger + *Gliricidia*	65.5	27.9	45.2	63.9	50.6	4.27	3.43	4.27	3.30	3.82
Papaya + Ginger: Pigeonpea (8:2)+ *Gliricidia*	49.3	39.0	44.1	48.4	45.2	4.63	4.43	3.33	2.13	3.63
Papaya+ Ragi + *Gliricidia*	75.2	56.7	34.1	79.8	61.4	7.87	5.17	4.47	8.32	6.46
Papaya + Ragi: Pigeonpea (6:2) + *Gliricidia*	37.6	43.2	30.8	51.1	40.7	5.73	6.07	5.67	4.77	5.53
Cultivated fallow	232.9	141.8	162.5	361.4	224.6	21.17	18.60	19.53	18.37	19.42
SE m±	11.9	10.2	10.6	17.5	12.5	0.73	1.10	0.96	1.05	0.96
CD at 5%	35.8	30.5	31.8	52.4	37.6	2.19	3.29	2.86	3.14	2.87

Nutrient loss and soil properties: Fallow land reported maximum nutrient loss to the tune of 109.9, 0.31 and 10.70 kg ha^{-1} for OC, P and K, respectively (Table 2). Sole ginger and ginger intercropped with pigeonpea in papaya + *Gliricidia* system recorded 58 and 41 % less OC loss than the farmers' practice of ragi broadcasting. The higher magnitude of soil loss was responsible for a greater loss of OC. Although the reduction in nutrient loss is not much high but the long-term effect may result better conservation. The consecutive application of *Gliricidia* green leaf manuring to ginger beds added significant amount of nutrient thus improving the nutrient uptake of different crops under multitier cropping system (Wani *et al,* 2003). Inclusion of legume (Pigeonpea) as intercrop which fixed atmospheric nitrogen in different treatments as well as green leaf manuring by lopping of *Gliricidia* significantly enriched soil OC over the pre-sowing status. Available P in ragi + pigeonpea intercropping showed an improvement of 14.1 % over sole ragi cultivation. Similarly sole ginger as well as ginger intercropping with pigeonpea under papaya multitier cropping recorded 14.7 and 12% increase, respectively. Gangwar *et al.* (2004) found that green manuring of *Leucaena* increased available P and K by 38.40 and 13.90%, respectively over initial value. However, in spite of consequent addition of *Gliricidia* green matter, there was no significant change in soil pH and electrical conductivity (EC).

Yield and Economics : Ginger yield markedly improved under multitier cropping (Table 3). In comparison to sole cultivation under papaya intercropping ginger yield improved by 15q and in ginger + pigeonpea intercropping system by 21 qha^{-1}. This positive change in yield would be attributed better nutrient availability and the shading effect provided by *Gliricidia* hedgerow, as ginger is a heavy feeder and partial shade loving plant. An impressive increase in ragi yield by 80% in line sowing, 70% under multitier cropping and 66% in 6:2 intercropping, respectively *vis-a-vis* broadcasting method shows the scope of improvement in the age-old practice of broadcast sowing prevalent in this region. Dass and Sudhishri (2010) in their study on ragi: pigeonpea intercropping in the region recommended 6:2 ratio with highest LER (land equivalent ratio) and net returns. Among all the cropping systems, papaya + Gliricidia based multitier system along with ginger-pigeonpea intercropping (8:2) gave maximum net returns of ₹ 111927 from one hectare of land with B: C ratio of 1.78 followed by sole ginger with net returns of ₹ 78000 ha^{-1}. Cultivation of pigeonpea along with papaya fruit tree increased net returns by ₹ 16,277. Sole Ragi cultivation on sloping lands was not economical because it gave poor returns as well as it permitted soil erosion. Hence integration of remunerative multitier components is required to make it a profitable enterprise.

Table 2: Effect of different treatments on nutrient loss studies

Treatment	Nutrient loss (kg ha^{-1})			Nutrient availability (kgha^{-1})			pH	EC
	Organic C	Available P	Available K	O C (%)	Available P	Available K		(dsm^{-1})
Ginger	34.8	0.09	1.53	0.52	13.90	251.7	5.35	0.083
Pigeonpea	48.2	0.10	2.03	0.54	17.53	323.6	5.14	0.075
Ragi Broadcasting (Farmers' practice)	76.0	0.23	5.80	0.27	14.50	256.3	5.30	0.076
Ragi line sowing	62.3	0.16	6.51	0.30	17.93	322.3	5.12	0.063
Papaya + Pigeonpea + *Gliricidia*	42.9	0.12	3.87	0.50	15.40	257.0	5.34	0.042
Papaya + Ginger + *Gliricidia*	31.6	0.09	[illegible]	0.47	16.30	253.2	5.22	0.056
Papaya + Ginger: Pigeonpea (8:2)+ *Gliricidia*	44.7	0.12	2.27	0.54	15.77	233.5	5.44	0.037
Papaya+ Ragi + *Gliricidia*	50.9	0.15	3.13	0.36	14.70	260.7	5.60	0.042
Papaya + Ragi: Pigeonpea (6:2) + *Gliricidia*	57.5	0.12	2.57	[illegible]	21.87	204.7	5.40	0.032
Cultivated fallow	109.9	0.31	10.70	0.24	15.57	250.4	5.01	0.061
SE m$\pm$	6.5	0.04	0.71	0.05	1.28	-	-	-
CD at 5%	19.3	0.11	2.12	0.16	3.86	NS	NS	NS
Presowing status	--	--	--	0.39	16.67	253.2	4.62	0.065

Table 3: Yield and economic returns of different treatments

Treatment	Main crop Yield (q ha^{-1})	Cost of Cultivation	Net Returns (₹ ha^{-1})	B:C ratio
Ginger	96	1,14,000	174000	1.68
Pigeonpea	15	17,150	26850	2.80
Ragi Broadcasting (Farmers' practice)	3	8,872	-6172	0.30
Ragi line sowing	15	13,070	430	1.03
Papaya + Pigeonpea + *Gliricidia*	13	37323	16277	1.44
Papaya + Ginger + *Gliricidia*	111	141173	92827	1.66
Papaya + Ginger: Pigeonpea(8:2)+ *Gliricidia*	117 (3)	141673	111927	1.78
Papaya+ Ragi + *Gliricidia*	10	32323	11323	0.65
Papaya + Ragi: Pigeonpea (6:2) + *Gliricidia*	9 (8)	34823	10877	1.31

CONCLUSIONS

Results from the present study revealed that multitier cropping system is more remunerative than the prevailing farmers' practice of ragi monocropping. The region receives good rainfall and this kind of technological intervention can address the issue of sustainable management of sloppy lands of eastern India with good profit to farmers.

REFERENCES

Anonymous. 2009. Agronomic Terminology. Indian Society of Agronomy. Division of Agronomy. Indian Agricultural Research Institute. New Delhi. 127p.

Anonymous. 2011. Census of India.

Dass, A. and Sudhishri, S. 2010. Intercropping in finger millet (*Elusine coracana*) with pulses for enhanced productivity, resource conservation and soil fertility in uplands of southern Odisha. *Indian Journal of Agronomy*, 55 (2): 89-94.

Gangwar, K.S., Sharma S. K. and Tomar O. K. 2004. Alley cropping of Subabul (*Leucaena leucocephala*) for sustaining higher crop productivity and soil fertility of rice (*Oryza sativa*) - wheat (*Triticum aestivum*) system in semi arid conditions. *Indian Journal of Agronomy*, 49 (2): 84-88.

Pandey, C. B. and Chaudhary, S. K. 2010. Soil and nutrient losses studies from different land uses and vegetative methods for their control on hilly terrain on southern Andaman. *Indian Journal of Agricultural Sciences*, 80 (5): 399-404.

Wani, S. P., Singh H. P., Sreedevi T. K., Pathak P., Rego T. J., Shiferraw B. and Iyer S. R. 2003. Farmers participatory integrated watershed management: Adarsha watershed, Kothapally India- an innovative and upscapable approach. Case 7: 123-147.

□□□

Chapter 57

Cropping System Approach for Natural Resource Management

N.K. SHARMA, K. S. DADHWAL AND B.N. GHOSH

INTRODUCTION

Natural resources like soil and water must be carefully conserved and judiciously utilised to sustain the ever increasing requirements of human and animal population. Land suffers from soil and water losses, as it is cultivated very frequently for raising various crops and very often remains exposed to rain and wind resulting in accelerated erosion. In India, it has been estimated that about 5334 M tonnes of soil is getting eroded annually which works out to be 16.4 t ha^{-1} yr^{-1}. Soil erosion reduces the soil moisture storage, depletes soil nutrients, thus, restricting the choice of crops to be grown and ultimately causing reduction in production in the long run. This can be considerably prevented through cropping system approach in combination with adoption of conservation agronomy practices. Introduction of erosion resisting crops either as a pure or inter crop in the prevailing cropping systems with suitable soil conservation agronomic practices help to intercept raindrops and reduce the splash effects. It helps in better intake of water rate in soil by improvement in soil organic matter content and soil structure. It also helps in retarding and reducing overland flow of runoff through the crop geometry, dense growing crops, intercropping/mixed cropping, strip cropping etc. Cropping systems having dense canopy crops prevents the impact of beating rain drops. Study conducted on canopy produced by crops and splash erosion have shown significant correlation between these two parameters. Higher the canopy, lesser is the splash. Thus, with good ground cover, erosion is minimized. Inclusion of legumes either as a pure or inter crop in the cropping system is the cheapest soil conservation measure to combat soil and fertility losses.

Central Soil & Water Conservation Research & Training Institute, 218 Kaulagarh Road Dehradun - 248 195,Uttarakhand, India

Runoff and Soil Loss from Agricultural Land

Cultivation: During farm operations, row direction along contour create small furrows and ridges which reduce the runoff velocity, increases opportunity time and manage rill formation. Contour cultivation is a good conservation measure on lands up to 1.5 - 2.0 % slope. In Doon Valley, contour cultivation of maize reduced runoff and soil loss by 24 and 32%, compared to up down cultivation which recorded 54% runoff and 28t $ha^{-1}yr^{-1}$ soil loss on 8% slope.

Crop choice: Probability of occurrence of rainfall, soil type and corresponding length of growing season are considered as a main factor for choice of crops. Crops with good germination and early establishment are necessary to provide vegetative cover in farm lands. The depth of soil and water holding capacity further influence the plant available water status. Thus, identification of efficient crops and cropping systems for different locations are important keeping in view the agro-climatic condition of the area.

Cover crops: Erosion from cultivated fields can be reduced if the land has enough crop canopy during peak rainy season because good cover canopy acts like umbrella and protects the soil. Cover crop is a close-growing crop raised mainly for protection of the soil. Effectiveness of the cover crop depends upon close spacing and early development of good canopy so that it can intercept the rain drop and minimum soil surface is open for the erosion. Impact of falling rain drops is the primary cause of soil erosion as they disturb the soil particles as soon as they strike on soil surface. The cover crops and cropping systems vary from region to region, depending upon the soil and climatic conditions.

At Dehradun cowpea provided the maximum canopy followed by soyabean, *moong*, *urd* and *dhaincha*. At Vasad (Gujarat) also cowpea provided best vegetative cover for the soil. Sunnhemp was the next best and moong & groundnut were found effective to some extent.

At Kota, velvet bean formed the canopy effective to the maximum extent followed by cowpea. At Kanpur, *moong* served as the most effective canopy followed by *urd* and guar. At Rehmankhera, cowpea formed the best canopy cover followed by *moong*, groundnut, *urd* and soyabean. All the legumes were equally effective in reducing the soil and water losses. Thus, among the legumes, cowpea, soyabean and natural grass cover proved to be important crops for providing good cover for the land during the rainy season (Table 1).

Table 1: Soil loss and runoff under different crops/cropping systems at various places

Place	Cropping system	Soil loss (t ha^{-1})	Runoff (% of rainfall)
Kota (1% slope)	Soyabea + blackgram	6.0	20
	Sorghum + pigeonpea	3.3	14
Dehradun (2% slope)	Maize (45x40 cm^2)	3.4	10
	Maize + *Urd*	3.7	11
	Maize + *Til*	4.0	12
	Maize + *Mandua*	3.6	12
	Maize + Sunnhemp	4.7	13
	Maize alone	6.9	15

'C' factor values : As a component of Universal Soil Loss Equation (USLE), the' C' factor is the ratio of soil loss from the land cropped under specified condition to that tilled continuously and kept fallow under similar conditions. Since the' C' factor takes into account the protection against beating action of rain drops and over land flow to some extent, there is a need to synchronize the cover development with rainfall causing erosion. Study was conducted to find out the 'C' factor values for various crops at different location (Table 2). Data shows that a wide row crop has maximum values, whereas inter cropping with leguminous crops or legumes as pure crops have minimum values.

Table 2: 'C' factor values for various crops at different location

Crop	C Factor (Range)
Maize	0.40-0.50
Bajra	0.41-0.61
Jowar	0.38-0.64
Castor	0.79
Til	0.51
Cowpea	0.17-0.39
Green gram	0.28-0.39
Black gram	0.22-0.54
Red gram	0.38-0.49
Soybean	0.42
Guar	0.42-0.59
Inter cropping Crop	
Maize+ cowpea	0.37
Maize+soybean	0.35
Bajra-cowpea	0.38
Maize+*til*	0.35
Jowar+*arhar*	0.32-0.40
Jowar-green gram	0.44
Til-gram	0.39
Guar-*arhar*	0.35

Crop rotation: In a good crop rotation, the effects of soil improving crops offset the effects of soil depleting crops. Inclusion of legumes in the cropping systems along with use of fertilizers take more effective care of the fertility aspects of the soil under different rotations. In all the years under trial, yield of maize was quite satisfactory and there was a good margin of profit where maize was one of the crops in rotation (Table 3). Among the various rotations tried maize-peas and maize-wheat rotations were found to be quite economical under rainfed conditions (Tejwani *et al.* 1975).

Table 3: Yield of crops (kg ha^{-1}) and monetary return under different crop rotations at Dehradun)

Rotation	*Kharif* (kg ha^{-1})		*Rabi* (kg ha^{-1})		Profit (+) or loss (-) (Rs ha^{-1})		
	Grain	Straw	Grain	Straw	*Kharif*	*Rabi*	Per rotation
Maize-wheat	3,735	12,428	964	2,446	+ 969	+110	+1,080
Cowpea-wheat	-	12,150	938	2,192	- 488	+95	- 393
Maize-linseed	3,754	12,900	484	1,334	+1,007	- 328	+ 679
Maize-peas	4,061	13,262	1,312	3,275	+1,236	+386	+1,623

Source : Tejwani *et al.* (1975)

Singh and Sumer Chand (1979) at Dehradun while testing different crop sequences for their production obtained 68-73 q ha^{-1} (approx.) grain production in case of maize-wheat, maize-barley, paddy-wheat and paddy-barley rotations. In case of sequences involving soybean, yield were low because of low residual soil moisture left for *rabi* crops for obtaining satisfactory germination of *rabi* crops followed soybean.

Crop Geometry and Plant Population

Crop geometry: In a study at Dehradun, the effect of crop geometry on soil and water losses at 4% slope was determined. It is evident from the average of 3 years data that there was a definite reduction in soil loss (from 22.6 to 13.9 t ha^{-1}) and water losses (47.8 to 33% of rainfall) by increasing inter-row spacing from 45 to 90 cm and decreasing intra-row spacing from 40 to 20 cm without effecting the crop yield (Table 4) (Bhardwaj *et al.,* 1980).

Plant population: Increasing plant population to 1.5 lakh ha^{-1} and using 50% plants as mulch after one month of sowing considerably controlled soil and water losses in corn. But considering increase in yield and reduction in runoff and soil loss, 1.0 lakh population should be preferred over 55000 plants ha^{-1} (Narain and Singh, 2000).

Table 4: Effect of crop geometry on soil and water loss and yield of maize (3 years average)

	Inter and intra-row spacing			
	45cm x 40cm	60cmx 30cm	75cm x 24 cm	90cm x20 cm
Runoff as % of rainfall (1282.5 mm)	47.8	39.7	33.9	33.0
Soil loss (t ha^{-1})	22.6	20.6	18.5	13.9
Grain yield (kg ha^{-1})	1547	1538	1577	1446
Straw yield (kg ha^{-1})	4811	4816	4774	4471

Source : Bhardwaj *et al.* (1980)

Tyagi *et al.* (1998) while comparing maize plant populations of 40, 55, 70 and 85 thousand plants per hectare as a sole crops and in combination with cowpea and blackgram as intercrops concluded that maize (70,000 plants) + cowpea gave highest maize equivalent total production over six years of 42.70 q ha^{-1}. Runoff and soil and nutrient loss followed the trend of canopy cover and plant population i.e. decreased with increase in plant population of maize or intercropping with legumes. However, Khola *et al.* (1999b) concluded that maize grain yield was higher by 144 kg ha^{-1} under 60 x 20 cm (83,000 plants ha^{-1}) spacing than 60 x 30 cm (55,000 plants ha^{-1}) but succeeding wheat grain yield was slightly lower by 58 kg ha^{-1} under the former.

Sowing time : Sowing time play a crucial role in success or failure of crops particularly under rainfed conditions. Feasibility of intercropping maize with cowpea, black gram and soybean under early (10 June), normal (25 June) and late (10 July) sowing revealed that normal sowing gave highest maize equivalent yield of 30.10 q ha^{-1} and developed canopy faster than early or late sowing (Khola *et al.*, 1999a).

Strip and Intercropping

Strip cropping :Singh and Bhushan (1976) concluded that maximum production was obtained with 2 rows of maize (60 cm apart) and 8 rows of soybean (30 cm apart). Since there is a major production of soybean in this production (3708 kg ha^{-1}) maximum economic returns were also obtained under this treatment (Table 5).

Table 5: Yield of maize and Soybean under different crop mixtures (Av. 4 years)

Crops	No. of maize rows			No. of soybean rows			
	10:0	8:2	6:4	5:5	4:6	2:8	0:10
Maize	3690	3254	2722	2502	2557	2054	-
Soybean	-	285	358	632	900	1654	2071
Total grain production	3690	3539	3079	3215	3457	3708	2071

Source : Singh and Bhushan (1976)

Inter cropping: Earlier studies on maize revealed that there was no reduction in yield of maize by increasing row to row spacing up to 90 cm without reducing the plant population. The inter row space was, therefore, utilized to grow a thick canopy developing and erosion resisting crop like cowpea. Four years average values (Table 6) have shown that after 50 days of planting, canopy cover under pure crop of maize was 68% while that under double row intercrop of cowpea was 94. The splash erosion was also reduced from 64/57.6 m^2/77 days under pure crop of maize to 22 g/57.6 m^2/77 days under intercrop. The results conclusively suggest that intercropping of cowpea (double row) for vegetable purpose is more advantageous (net return : Rs. 3525 ha^{-1} than intercropping of cowpea (double row) with maize for grain (net return : Rs. 1135 ha^{-1}) both in terms of monetary benefit and soil conservation (Bonde and Mohan, 1984) and Sharma *et al.* (2011).

Table 6: Grain yield (kg ha^{-1}) of maize and cowpea, net income (Rs ha^{-1}) under different intercropping treatments (Av. of 4 years).

Treatment	Grain yield		Net income	Canopy (%) on 50th day	Splash (gm)
	Maize	Cowpea			
Pure maize	2402	-	1657	68	64
Maize + cowpea single row for grain	2104	105	1499	86	40
Maize + cowpea double row for grain	2000	107	1135	87	32
Maize + cowpea single row for vegetable	2007	1302	2930	88	34
Maize + cowpea double row for vegetable	2009	1607	3525	94	22

Source : Bonde and Mohan (1984)

But while testing the suitable legume intercrops, Bonde and Mohan (1985) concluded that intercropping of maize with soybean was superior over cowpea and black gram producing an yield of 2809 kg ha^{-1} of maize and 606 kg ha^{-1} of soybean (Table 7). Intercropping of legumes and sesame in maize at 2% slope in Doon valley showed advantage in reducing soil loss by half

(2.3 t ha^{-1}) (Tomar, 1992). Residual effect of inclusion of legume (cowpea) resulted in higher yield of succeeding wheat crops by 61 %. Khola *et al.* (1999b) also reported that intercropping of cowpea or soybean with maize either for fodder or *in -situ* mulching reduced maize grain yield by 101-451 kg ha^{-1} compared to sole counterpart (2741 kg ha^{-1}). However, succeeding wheat grain yield was higher under mulching of cowpea (2744 kg ha^{-1}) or sorghum (2683 kg ha^{-1}) in maize and lower under sole stands of maize (2291-2349 kg ha^{-1}).

Table 7: Grain yield (kg ha^{-1}) of maize and legumes intercrops under various planting systems

Treatment	Yield (kg ha^{-1}) (4 years Av.)	
	Maize	Legume
Maize sole crop (90cm row spacing)	2606	-
Maize + cowpea (uniform row 30-30-30 cms)	2604	408
Maize + cowpea (paired row 34-22-34 cm)	2605	408
Maize + soybean (uniform row 30-30-30 cms)	2803	409
Maize + soybean (paired row 34-22-34 cm)	2809	606
Maize + blackgram (uniform row 30-30-30 cms)	2508	307
Maize + blackgram (paired row 34-22-34 cm)	2801	308

Source : Bonde and Mohan (1985)

Beneficial effect of mulching of intercropped legumes with maize for mulching/seed production/green pods on canopy cover, soil and water losses and productivity has also been reported by Mohan (1992), Narain and Singh (1996), Khola *et al.* (1999a), Singh *et al.* (2003) and Tomar *et al.* (2003). Sharma *et al.* (2010) found Live Mulching with fast-growing annual green manure legumes like sunnhemp (*Crotalaria juncea*) or prunings of *Leucaena leucocephala* grown as hedge rows can be done for moisture conservation as well as nutrient cycling in the maize-wheat cropping system.

Crop management : Harvesting methods of maize can significantly effect the succeeding rainfed wheat production in maize-wheat system. Sharma and Khola (1999) while comparing the different maize harvesting methods for in-situ moisture conservation concluded that when ploughing was done immediately after maize harvesting followed by spread of stover in the field, conserved maximum moisture in the profile for succeeding wheat crop.

CONCLUSIONS

Soil loss due erosion caused reduction in the soil moisture storage and soil nutrients, thus, restricting the choice of crops to be grown and ultimately causing reduction in production. This can be considerably prevented through cropping system approach in combination with adoption of conservation agronomy practices. Introduction of erosion resisting crops in the prevailing cropping system with suitable soil conservation agronomic practices help to intercept raindrops, reduce splash effects, improve soil organic matter and reduce overland flow. For this, legumes considered to be the best option which not only produce good canopy but also has economic value, pulse crops are generally suitable for such purposes. Inclusion of such type of close growing vegetation in the cropping system not only reduces soil loss and runoff but also add sufficient quantity of organic matter and nutrients in the soil resulting in increase of crop productivity. Thus, cropping system approach in conjunction with suitable agronomic conservation measures is a best strategy to manage the natural resources for getting higher productivity on sustainable basis.

REFERENCES

Bonde, W.C. and Mohan, S.C. 1984. Intercropping in maize for effective soil conservation, moisture utilization and maximizing production in Doon valley. *Annual Report*, CSWCRTI, Dehradun, pp72-73.

Bonde, W.C. and Mohan, S.C. 1985. Intercropping in maize with legumes under rainfed conditions. *Annual Report*, CSWCRTI, Dehradun, 49p.

Bhardwaj, S.P. 1984. Effect of bunding and strip cropping on soil and water losses in large size runoff plots at 4% slope. *Annual Report*, CSWCRTI, Dehradun, pp53-54.

Bhardwaj, S.P., Khybri, M.L., Sewa Ram and Prasad, S.N. 1980. Effect of crop geometry on soil and water losses under maize in large size runoff plots on 4% slope. *Annual Report*, CSWCRTI, Dehradun,pp125-126.

Khola, O.P.S., Dube, R.K. and Sharma, N.K. 1999a. Conservation and productivity ability of maize (*Zea mays* L.)- Legume intercropping systems under varying date of sowing. *Indian Journal of Agromony*, 44 (1): 40-46.

Khola, O.P.S., Sharma, N.K. and Khullar, A.K.1999b. Evaluation of different maize growing systems for resource conservation and biomass production under varying fertility conditions. *Annual Report*, CSWCRTI, 1999-2000, Dehradun, pp32-33.

Mohan, S.C. 1992. Effect of green manuring with intercropping on soil fertility and moisture conservation under rainfed conditions, *Annual Report*, 1992-93, CSWCRTI, Dehradun, 44p.

Narain, P. and Singh, R. 1996. Erosion control and productivity restoration through sunhemp mulching and green manuring. *Annual Report*, CSWCRTI, Dehradun, pp40-41.

Narain, P. and Singh, R.K. 2000. Quantifiction of ground cover effect on runoff and soil loss through artificial and natural rainfall. *Annual Report*, CSWCRTI, Dehradun, pp23-24.

Sharma, A. R., Singh Ratan, Dhyani, S. K., Dube, R. K. 2010. Moisture conservation and nitrogen recycling through legume mulching in rainfed maize (*Zea mays*)-wheat (*Triticum aestivum*) cropping system. *Nutr. Cycl. Agroecosyst*, 87:187–197.

Sharma, N.K. and Khola, O.P.S. 1999. Evaluation and improvement of traditional maize harvesting methods for moisture conservation and wheat production. *Annual Report*, CSWCRTI, Dehradun, pp28-29.

Sharma, N.K., Mandal, D. and Tiwari, A. 2011. Impact of maize based intercropping on resource conservation and productivity. *Annual Report*, CSWCRTI, Dehradun, pp22-23.

Singh, R., Dhyani, S.K. and Dube, R.K. 2003. Conservation tillage and green manure mulching for optimizing productivity in maize-wheat cropping system in the sub-mountainous Himalayan region. *Annual Report*, CSWCRTI, Dehradun, pp21-22.

Singh, Gurmel and Bhushan, L.S. 1976. Crop mixture strip planting of maize and soybean. Annual Report, CSWCRTI, Dehradun,pp163-164.

Singh Gurmel and Sumer Chand. 1979. Crop sequence. Annual Report, CSWCRTI, Dehradun, pp 172-173.

Tejwani, K.G, Bhaskaran, A.R. Thomus, P.K. Verma, B., Dyal, R. Khybri, M.L. and Bains, K.S. 1975. Reseach in agronomy for soil conservation. *In*: Tejwani, K.G., Gupta, S.K. and Mathur, H.N. (Eds). *Soil and water conservation research*, ICAR, New Delhi, pp106-145.

Tomar, D.S.1992. Effect of intercropping of mzie on soil erosion and yield. *Annual Report*, CSWCRTI, Dehradu, 38p.

Tomar, D.S., Singh, R. Joshi, B.P. and Khullar, A.K. 2003. Intercropping studies in rainfed maize-wheat cropping system on sloping lands in Doon valley. *Annual Report*, CSWCRTI, Dehradun, 27p.

Tyagi, P.C., Mohan, S.C. and Joshi, B.P.1998. Crop cover management in maize legume cropping system for sustainable production. *Annual Report*, CSWCRTI, Dehradun, pp26-27.

Chapter 58

Evaluation of *Kharif* Onion Cultivars in Odisha Condition

P.TRIPATHY, B.B. SAHOO, A.PRIYADARSHINI, S.K. DAS AND D.K. DASH

INTRODUCTION

Onion (*Allium cepa* L.) is one of the commercial vegetable and spice crops of India. Onion is the integral components of culinary preparations. Besides being used as condiments, the medicinal properties of the commodities add value to their importance. India produces 139.00 lakh tons of onion from 8.47 lakh ha area (Annon. 2011). India ranks 1st in area, 2nd in production & 3rd in export in the world. In India, onion is predominantly cultivated during *rabi* about 60% followed by 20% each in *kharif* & late *kharif* season. It is mainly grown in the states of Maharashtra, Odisha, Karnataka, Uttar Pradesh, Gujarat, Tamil Nadu & Madhya Pradesh (Mahanthesh *et al.*, 2009). Odisha is one of the important potential states for cultivation of onion, which produces 298.80 ('000) tons of onion from 32.10 ('000 ha) (NHB, 2009-10) with very low productivity of around 91.20 q ha^{-1} as against India's average productivity of 141.2 q ha^{-1}. The higher productivity could be determined by selection of suitable varieties, balanced nutrition, optimum water management as well as need based plant protection measures. Among these factors, selection of suitable variety(s) plays an important role in enhancing the yield as well as productivity. That means, the onion production depends on the cultivation of varieties that are suitable for different conditions exposed by specific environment. Production of onion in *kharif* & late *kharif* season is more important to have continuous supply of onion round the year. Therefore, evaluation of *kharif* onion cultivars is a basic step towards the popularization for its successful cultivation.

Keeping this in view, the present study was initiated to evaluate the cultivars suitable for *kharif* as well as late *kharif* seasons under Odisha condition

All India Network Research Project on Onion & Garlic, College of Horticulture, Orissa University of Agriculture and Technology, Sambalpur, Odisha, India.

not only to increase the production, productivity of onion but also to supply continuously to the market for higher net profit.

MATERIALS AND METHODS

The present field study on," Evaluation of *kharif* onion cultivars in Odisha condition " was carried out at College of Horticulture, Orissa University of Agriculture & Technology (OUAT), Chiplima, Sambalpur, Odisha, India under All India Network Research Project on Onion & Garlic, during *kharif* season of 2009. Sixteen cultivars of onion including five national check varieties supplied by Directorate of Onion & Garlic Research, Rajgurunagar, Pune were evaluated by adopting RBD with three replications. The details of the source of cultivars are presented in Table 1.

Table 1: Details of source of cultivars evaluated during *kharif* season

S. No.	Entries	Source
1.	NRCRO-1(1133)	DOGR, Pune
2.	NRCWO-1 (W-448)	DOGR, Pune
3.	NRCRO-2 (1156)	DOGR, Pune
4.	NRCWO-2 (W El Composite)	DOGR, Pune
5.	NRCRO-3 (RGO-53)	DOGR, Pune
6.	NRCWO-3 (W-302)	DOGR, Pune
7.	NOL. 115	Nirmal Seeds
8.	NOL.103	Nirmal Seeds
9.	Sel. 157	IARI,New Delhi
10.	VG-19	IIHR, Bangalore
11.	Col. 652	NHRDF, Nashik
12.	L-28 (C)	NHRDF, Nashik
13.	Bhima Raj (C)	DOGR, Pune
14.	N-2-4-1(C)	MPKV, Rahuri
15.	Bhima Red (C)	DOGR, Pune
16.	Arka Niketan (C)	IIHR, Bangalore

All the sixteen cultivars were sown in the nursery on 20.07.2009. The seedlings of all the cultivars were transplanted on 04.09.2009 in a plot size of 3m X 2m with a spacing of 15 cm X 10 cm. All the recommended package of practices was adapted to all the cultivars uniformly. The data was recorded on ten randomly selected plants from each treatment on vegetative parameters like plant height, number of leaves plant^{-1}, collar thickness; bulb parameters like average bulb weight, bulb colour & days to bulb maturity; as well as yield attributes like marketable & total bulb yield. The data collected on various aspects were subjected to statistical analysis as per the procedure suggested by Sukhatme & Amble (1995).

RESULTS AND DISCUSSION

The results on performance of *kharif* onion cultivars on vegetative, bulb and bulb yield attributes are presented in Table 2. There were significant differences among the different cultivars tested for various aspects under study. Among the varieties, significantly highest plant height was recorded in Arka Niketan (47.37 cm) & was *at par* with cultivars like *'NRCRO-1', 'NRCRO-3', 'NRCWO-1', 'NRCWO-2', 'NOL-115', 'N-2-4-1'* and *'Bhima Red'* (42.83 cm to 45.73 cm). The variety *'Arka Niketan'* had produced significantly highest number of leaves plant (9.77), closely followed by *'NRCWO-1', 'NRCRO-3', 'NOL-115', 'Sel-157', 'L-28', 'Bhima Red'* and *'N-2-4-1'* (8.80 to 9.25), whereas lowest number of leaves was produced by variety, COLL-652 (7.00). On the other hand, the variety, *'N-2-4-1'* had highest collar thickness (14.73 mm) than other tested lines, except *'NRCRO-1', 'NRCWO-1', 'NRCRO-2', 'NRCWO-3', 'Sel-157'* and *'Arka Niketan'*, which were *at par* statistically. The variety, *'Arka Niketan'* (128 days) took significantly minimum time for harvesting than rest of the cultivars, except *'N-2-4-1'* (130 days). However, the cultivar, Sel-157 took maximum time (136.33 days) for harvesting. The variation might be due to the genetic and environmental effect.

Among the tested cultivars, *'L-28'* had produced significantly the heaviest bulb of 95.74g than rest of the cultivars except *'N-2-4-1'* & *'Sel-157'* (93.74g to 95.63g). A perusal to Table 2 also revealed significant variations among the cultivars for both marketable as well as total bulb yield. From commercial point of view, the marketable bulb yield in onion is more important than that of the total bulb yield. Hence, due consideration should be taken care while selection of suitable variety for commercial cultivation, particularly during *kharif* and late *kharif* seasons, which is treated as off season cultivation of onion. During *kharif* season, the marketable bulb yield varies as low as 78.56 q ha^{-1} (L-28) to as high as 208.36 q ha^{-1} (*Arka Niketan*). The variety, *'Arka Niketan'* had produced significantly highest marketable bulb yield of 208.36 q ha^{-1}than rest of the cultivars tested. However, *statistical parity* was observed in cultivars like *'NRCRO-1', 'NRCWO-1', 'NRCRO-2', 'NRCWO-2', 'NRCWO-3', 'Sel.-157'* & *'N-2-4-1'* (149.73 q $ha^{-1}$to 206.04 q ha^{-1}) with highest yield, *'Arka Niketan'*. With regards to production of total bulb yield under *kharif* season by different cultivars, the range varies from 96.49 q ha^{-1} (*'L-28'*) to 241.44 q ha^{-1} (*'N-2-4-1'*). The variety, *'N-2-4-1'* had produced significantly highest total bulb yield of 241.44 q ha^{-1}than the cultivars like *'VG-19', 'COLL-652', 'L-28', 'Bhima Raj'* & *'Bhima Red'*. However, *statistical parity* was observed with rest of the cultivars under study. Similar results of

Table 1: Performance of *kharif* Onion under Odisha conditions

S. No.	Lines/ Variety (1)	Plant height (cm) (2)	# leaves plant^{-1} (3)	Collar Thickness (mm) (4)	Average Bulb Weight (g) (5)	Marketable yield (q ha^{-1}) (6)	Total yield (q ha^{-1}) (7)	Colour of Bulbs (8)	DTH (days) (9)
1	NRCRO-1	43.77	8.07	14.33	65.39	178.24	209.17	LR	136.00
2	NRCWO-1	43.07	8.80	13.29	80.00	200.86	229.72	White	135.33
3	NRCRO-2	39.80	8.43	12.74	67.82	201.62	239.09	LR	134.00
4	NRCWO -2	45.73	7.13	12.90	63.20	167.91	216.21	White	135.00
5	NRCRO-3	43.07	8.80	9.82	60.96	149.53	185.94	LR	134.33
6	NRCWO-3	34.23	7.47	13.41	70.62	149.73	178.34	White	135.67
7	NOL-115	44.57	9.13	11.92	64.79	148.66	197.28	LR	132.67
8	NOL-103	38.90	7.38	10.75	60.86	149.46	208.9	DR	132.67
9	SEL-157	40.03	9.23	14.21	95.63	163.85	205.73	DR	136.33
10	VG-19	36.20	7.60	9.42	59.03	101.99	126.49	LR	135.67
11	COLL-652	41.00	7.00	11.25	65.93	124.89	164.5	LR	134.67
12	L-28 (C)	41.97	8.23	11.37	95.74	78.56	96.49	DR	134.00
13	Bhima Raj (C)	40.90	7.83	12.04	80.93	83.89	123.79	LR	135.67
14	N-2-4-1(C)	45.23	8.27	14.73	93.87	206.04	241.44	LR	134.00
15	Bhima Red (C)	42.83	8.83	12.91	65.67	128.83	177.11	LR	130.00
16	Arka Niketan (C)	47.37	9.77	11.78	57.35	208.36	238.5	LR	128.00
	Mean	**41.79**	**8.28**	**12.3**	**71.74**	**152.65**	**189.92**		**134.00**
	CD (5%)	5.38	1.25	2.19	14.39	58.81	64.33		3.53
	CV (5%)	7.72	9.06	10.7	12.03	23.11	20.32		1.58

higher bulb yield by *'Arka Niketan'* was also reported under Pune, Maharashtra ; *'NRCWO-1'*, *'NRCWO-2'* & *'N-2-4-1'* under Bihar condition; *'NRCWO-1'*, *'NRCWO-3'* & *'NRCRO-2'* at Hissar ; *'NRCWO-2'* & *'NRCWO-3'* at West Bengal condition; *'NRCWO-2'* at Karnataka; *'NRCRO-2'* & *'NRCWO-2'* at Kalayanpur; *'NRCRO-2'* at Coimbatore; *'NRCRO-1 & NRCWO-1'* at Junagarh as well as *'N-2-4-1'* at IARI condition (Annon., 2011).

The results also indicated that during *Kharif* season more than q ha^{-1} marketable bulb yield was recorded in the varieties like *'NRCWO-1'* (200.86 q ha^{-1}), *'NRCRO-1'* (201.62 q ha^{-1}), *'N-2-4-1'* (206.04 q ha^{-1}) & *'Arka Niketan'* (208.36 q ha^{-1}). On the other hand, total bulb yield of more than 200 q ha^{-1}was recorded in variety *'Sel-157'* (205.73 q ha^{-1}), *'NOL-103'* (208.90 q ha^{-1}), *'NRCRO-1'* (209.17 q/ha), *'NRCRO-2'* (216.21 q/ha), *'NRCWO-1'* (229.72 q ha^{-1}), *'Arka Niketan'* (238.50 q ha^{-1}), *'NRCRO-2'* (239.09 q ha^{-1}) and *'N-2-4-1'* (241.44 q ha^{-1}). The maximum bulb yield of these varieties can be mainly attributed to the growth parameters which are impotant components of growth & had positive as well as significant correlation with yield. The maximum bulb yield from these varieties during *kharif* season may also be du t the accumulation of maximum photosynthates in the bulbs. The variations in the bulb yields of different varieties of onion have also been reported from several places (Patil, *et al.*, 1991; Bhonde, *et al.*, 1992 ; Khan, 1997; Mohanty & Prusty, 2002).

CONCLUSIONS

Thus, it can be concluded that under Odisha condition, the cultivars like *'Arka Niketan'*, *'N-2-4-1'*, *'NRCRO-2'* & *'NRCWO-1'* may be recommended for *Kharif* season cultivation for optimum bulb yield and profit due to their high potential. These cultivars can be further used for crop improvement programme for development of new lines suitable for *kharif* & late *kharif* season cultivation.

ACKNOWLEDGEMENTS

We gratefully acknowledge the Orissa University of Agriculture & Technology, Odisha, India for the research facilities provided and to the Director, Directorate of Onion & Garlic Research (ICAR), Rajgurunagar, Pune, India for providing the financial & other facilities to carry out this study under AINRP on Onion and Garlic.

REFERENCES

Anonymous. 2011.*Annual Report 2010-11 of All India Network Research Project on Onion & Garlic*, Directorate of Onion & Garlic Research, Rajgurunagar, Pune, pp01 & 55 to 79.

Bhonde, S.R.; Srivastava, K.J. and Singh, K.N. 1992. Evaluation of varieties for late *kharif* crop of onion in Nasik area. *News Letter,* 12 (1): 1-2.

Khan, I.A. 1997. Performance of onion (*Allium cepa* L.) cultivars in mid hill conditions of Uttar Pradesh. *Prog.Hort.,* 29 (1-2): 97-99.

Mahanthesh,B; Harshavardhan, M.; Ravi Prasad Sajjan, M; Janardhan, G. and Vishnuvardhana 2009. Evaluation of onion varieties / hybrids for dry matter production and yield I *Kharif* season under irrigated condition in central dry zone of Karnataka. *The Asian Journal of Horticulture (June to November, 2009),* 4 (1): 10-12.

Mohanty, B.K. and Prusty, A.M.2002.Varietal performance of Onion in Rainy season. *Indian J.Agric.Res.,*36 (3): 222-224.

Patil, D.R.; Hebbara,M. and Naganagoud, A. 1991. Performance of onion varieties on black soils. *Karnatake J. Agric. Sci.,* 4 (1-2): 54-56.

Sukhatme P V and Amble V N 1995. *Statistical Methods for Agricultural Workers.* pp 145-156.

□□□

Chapter 59

Nathapodytes Nimmoniana : A Profitable Medicinal Tree for Degraded Lands to Enhance Livelihood Security in the Eastern Ghats Region

H.C. HOMBE GOWDA, R. VASUDEVA[1], PRAVEEN JAKHAR & M. MADHU

INTRODUCTION

Nothapodytes nimmoniana (Grah.) Mabber. (Syn: *N. foetida* Blume; *Mappia foetida* Miers, Fam: Icacinaceae) is a lower canopy tree of evergreen to moist deciduous forest types. It grows up to 12 m height in undisturbed evergreen forests but appears shrubby in secondary / disturbed forests. It is also reported from shola forests of South India (Srivastava, 2000; Ravikumar and Ved, 2000). It is distributed throughout tropical America and Asia and in parts of China and Sri Lanka. In India, it is naturally grown in the Western Ghats and the North Eastern hills. The occurrence at Eastern Ghats is also reported recently. It is commonly known as "*Durvasane mara*" in Kannada (meaning 'foul smelling tree') and as "*Teku lakda*" or as "*kalgur*" in Hindi. In Kerala, it is called "*Pinari*" and the edible fruits are marketed as an NTFP at ₹ 10 kg^{-1}. The seed kernels contain about 44 % non-edible oil (Nair, 2000).

The species has recently gained international importance due to its newly identified pharmacological and curative properties. The wood-extracts of this tree are used in cancer-fighting drugs. The active principle of the wood, Camptothecin (CPT), is a known potent drug in treating mammalian cancer and HIV virus (Wall *et al.*, 1966; Priel *et al.*, 1991). Recently several pharmaceutical companies have registered patents on the pharmacological properties of this species. Hence CPT is the third most important alkaloid required by pharmaceutical companies around the world. In the international market, one kg of CPT fetches around 15,000 US dollars. *Nothapodytes* is the most convenient source for large-scale production of this alkaloid. Strangely,

Central Soil and Water Conservation, Research and Training Institute, Research Centre, Sunabeda, Dist:Koraput, Odisha-763 002,India

[1]Dept. of Forest Biology, College of Forestry, Sirsi, Karnataka

use of *N. nimmoniana,* as a medicine is not reported from any traditional health systems of India. However, a few tribes in Uttara Kannada district use the bark of this tree to treat skin diseases. The leaves of this species are also used as alternative to 'tendu leaves' to wrap tobacco (Hombe Gowda, 2002).

Roots of *N. nimmoniana* contain highest CPT concentration of 0.1%, while the bark and wood contain about 0.08% and 0.06%, respectively. The lowest concentration of 0.01% is recorded in the leaves (Govindachari and Viswanathan, 1972). Hence, in order to extract the alkaloid, essentially the entire tree is chopped. This leads to the death of trees upon harvesting it. In the unorganized sector, one kg of dried wood chips is believed to cost around Rs.100 to 140. In recent years this has led to the large-scale exploitation of the species from its wild habitats, despite the ban on green felling. In one of the reports it is said that about 54 tons of wood chips were shipped from Mumbai port alone during 1995. This has resulted in a sudden loss of populations, which is estimated to be around 20 % (Anon, 1997). Consequently, *N. nimmoniana* has been assigned the threat status of *'Vulnerable/Regionally'* (Vu/R) by the Foundation for Revitalization of Local Health Traditions (Anon., 1997). One way of reducing the pressure on natural forests for this resource is to domesticate it and incorporate into existing agroforestry models. However, this requires a critical assessment of various traits of *N. nimmoniana* and genetic tailoring of the species. Unfortunately, studies on its biology, reproduction and silviculture are very recently attempted.

Reproductive biology

A complete knowledge of reproductive biology and breeding system and propagation techniques of species is a prerequisite for developing an appropriate package of practices before introduction to agroforestry systems.

Breeding system: Recently Hombe Gowda *et al.*, (2002a) reported, based on observations of two large populations (n=60) in Joida and Ulavi forests in Dandeli Forest Division, northern Karnataka, that the species is polygamous in nature. It shows a wide array of breeding types with male, female, hermoprodite, monoecious, andromonoecious, gynomonoecious, trimonoceious at individual level; androdioecious and gynodioecious types at population level. Such baseline data are also important for the genetic improvement and domestication of a species. In this article we report on several aspects of pollination biology, seed biology and standardized method for vegetative propagation.

Reproductive phenology: Twenty mature trees of *N. nimmoniana* around Sirsi were marked and were observed every fortnight to work out the phenological events. Flowering in *N. nimmoniana* begins soon after leaf sprouting, usually coinciding with monsoon season *i.e.*, during July. It is completed in the first week of September. The synchronized blooming among individuals of different sex facilitates gamete exchange. The development of fruits takes about 65 to 70 days after fertilization. Fruits start maturing from first week of October and ends during last week of November (Hombe Gowda *et al.*, 2002b). Generally small birds are attracted because of high pulp content of the fruit, which is sweet in taste. Birds like Red-whiskered Bulbul, Red-vented Bulbul and Pigeon were the major dispersal agents. Ganesh and Davidar. (2001) also reported that these fruits are dispersed by small birds in Kalakad-Mundanthurai Tiger Reserve.

Flower and Pollination in N. nimmoniana: About 250 to 350 flowers are borne on terminal corymbose cymes. Anthesis occurs during the early morning hours (06:00 – 08:00 h) and maximum pollen production at noon (12:00 – 14:00 h). Generally male flowers last for two days, while that of the female and bisexual flowers may extend up to 8 days based on whether or not they are pollinated. On any given day during peak flowering period, an average inflorescence bears about 15-40 opened flowers.

Male flowers are larger than female and bisexual flowers and are dull yellowish in color with rudimentary pistil. Female flowers are small with rudimentary stamens that are devoid of pollen grain production. Bisexual flowers are intermediate to male and female flowers in size. Both stamens and pistils are medium in size and are 3.0 mm and 5.0 mm in length, respectively. In bisexual flowers only 1-2 anthers produce pollen grains and stigma is active. None of the bisexual flowers in the bagged inflorescences (n=20) set into fruits, suggesting that *N. nimmoniana* requires out cross pollen. The flowers show perfect fly-pollination syndrome. About 12 species of dipteran flies were attracted to the flowers mainly because of their characteristic foetid odor. Male flowers produced copious amount of pollen on which all the dipeteran flies actively fed although a small amount of nectar was present in all the flower types. About 7 species of Lepidopteron, 6 species of wasps, 3 species of Hymenopterans, and 2 species of bugs were recorded in the present study. It appears that *N. nimmoniana* is a generalist in terms of insect vector.

Assisted pollination: Assisted pollination experiments showed that fruit set among open pollinated flowers was limited by the lack of pollen transfer. Open pollinated flowers showed only 4.9 % fruit set while that of assisted pollinated flowers was 79.5 %. Since blooming of this species coincides with the rainy season, fruit set is limited by low pollinator activity as well by pollen wash by rains. Such pollination limitation has been observed among several tree species. *eg., Tamarindus indica*. These results suggest that fruit production can be enhanced by artificial pollination or by encouraging pollinators.

In vitro pollen germination: A simple agar based *in-vitro* germination medium was standardized to test the pollen grain viability in *N. nimmoniana,* which contained 15% sucrose and 0.5% agar. This medium can be very reliably and effectively used to monitor the pollen grain viability across time and can also be employed to periodically check the viability in the event of cryopreservation of pollen grains, which is very useful in breeding programs.

Propagation

Seed dormancy and pre-sowing seed treatment: Ripen fruits collected from two populations of *N. nimmoniana* in Dandeli forest division, Karnataka, during the peak fruiting season, were depulped and seeds thus extracted were shade-dried for two days. The bulk seeds were immediately used in the germination test as well as stored in airtight containers for viability studies. To standardize a pre-sowing seed treatment, six different treatments were imposed. Seeds were sown on an acid-washed sand medium in a green house, adopting completely randomized design (CRD) with four replications (n=100 each) and were watered daily with distilled water. Seed germination (radical emergence) was recorded on a daily basis for 80 days to compute per cent germination and germination rate. Fresh seeds of *N. nimmoniana,* without any pre-treatment, initiated germination in about 25 days of sowing and slowly achieved about 89 per cent germination in about 80 days suggesting a mild dormancy. Srivastava (2000) has also reported that, *N. nimmoniana* seeds collected from the shola forests of Nilgiris, south India recorded 80 % germination after 60 days of sowing. However, in Jammu region, *N. nimmoniana* showed a strong seed dormancy, which precluded germination till one year without pre-sowing treatment (Sharma *et al.,* 2000). Although lower temperature and humidity during months of August through March are reported to be the key factors influencing seed germination in Jammu, genetically based provenance variation cannot be precluded. Since natural geographical range of *N. nimmoniana* is wide, covering contrasting climatic regions, adaptive provenance variation in germination may be important. It would be interesting to study such variation among seed sources across wide bio-climatic zones.

Seed longevity: To test the seed viability, bulk seeds stored in airtight containers under normal laboratory conditions were sown in four replications of hundred seeds each (after removing seed coat and soaking in cold water for 12 h) at every fifteen days interval. In this experiment total germination count was done after 80 days of sowing in the respective treatment.

As clearly shown in Figure 1, seeds remained viable up to 150 days under Sirsi conditions. The viability was excellent (near 100%) up to 45 days. In the third month, the viability reduced to 75 % and afterwards, it reduced

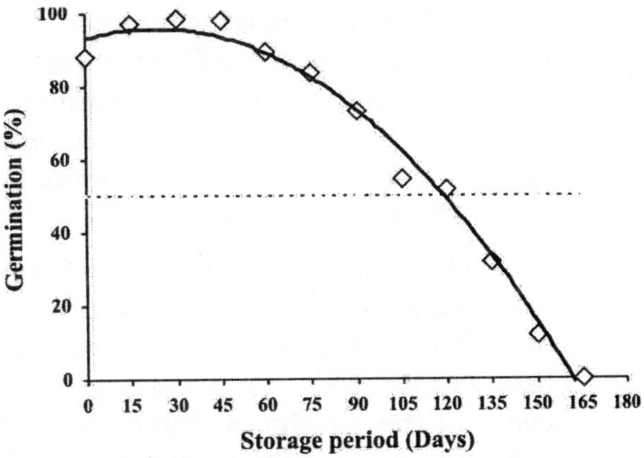

Fig. 1: Seed viability across period in *N.* storage *nimmoniana.*

drastically. While seeds, which were kept for 120 days, had 52 % and the end of five month the viability had crashed to 12 %. At 165th day, the viability was zero (Figure 1) suggesting that *N. nimmoniana* behave like sub-orthodox seed type. This is perhaps because of higher amount of fatty acids present in the seed (about 48 % oil in the kernels as reported by Nadkarni *et al.*, 1944). Such association of lesser longevity and higher oil content has been reported in *Azadiracta indica, Madhuca latifolia* and *Shorea robusta* (Bonner, 1990). Efforts are, however, required standardize methods to increase the seed longevity beyond five months by altering storage conditions.

Vegetative propagation: Stem cuttings of 15-20 cm length, obtained from natural populations during the month of February, were used for standardization of vegetative propagation technique. These cuttings were treated with different plant growth regulators at different concentration and at different combination levels. Growth regulators such as IAA, IBA and Coumarin in solution form were used. This experiment was conducted in a small mist chamber adopting Completely Randomized Design with eight replications. A set of ten cuttings formed one replication. Sand and coir pith was used as a rooting media at 1:1 ratio. Slant cut was given to each cutting before hormonal treatment and cut ends were treated with 0.2% Bavistin for half an hour to avoid fungal attack. Quick dip method was followed to treat cuttings with hormones.

The stem cuttings treated with Coumarin 2000 ppm showed better response, with maximum per cent rooting, per cent sprouting, number of roots and length of roots. Maximum number of sprouts, roots, leaves per cutting and shoot length were observed in IBA 1000 ppm, IAA 3000ppm and IAA+IBA 2000+2000 ppm, respectively. However, these variations were not statistically significant.

Traits of *N. Nimmoniana* Suitable for Agroforestry

The following traits make the species a likely candidate for inclusion as a perennial component of agroforestry systems.

- Firstly, *N. nimmoniana* is a shade tolerant species and comes up very well in regions where soil moisture levels are moderately high. It can be included as a component of existing multistoried systems. Hence, it may be suitable to hill and transition zones of Eastern Ghats region. However, there is an urgent need to evaluate its performance on wider range of agro climatic zones.
- Secondly, the species is a good coppicer, and withstand pruning/ pollarding as evidenced by the fast coppice production and recovery of the felled trees in natural forests. Hence it can be adopted as a part of hedge or live fence to protect main crop from biotic interference. A five-year-old tree produces about 3-4 coppices following pollarding, which can grow to a height of about 1.5 to 2 meters within a year, producing about 0.5-3 kg biomass on dry weight basis. Since the bark to stem ratio would be higher in these coppices, a higher CPT yield may result from this.
- Thirdly, it is not preferred as firewood since the wood emits a strong foetid odour when burnt and the species is non-browsable. Hence, there is no need to give extra care and protection while planted in open field conditions.
- Fourthly, this plant is associated with several other species in natural habitats hence it may not have any allelopathic effects, making it fit for agroforestry.
- Fifthly, propagation of *N. nimmoniana* through seeds as well as vegetative modes have been standardized hence planting material can be easily raised at low cost.
- Finally, due to fairly good market price and possibility of harvesting wood within four to five years, the species ensures early returns. Presently, good market links exist ensures sales of commodity.

By introducing N. nimmoniana into agroforestry systems at Eastern Ghats humid region, camptothecin can be harvested on a sustainable basis since biomass can be collected through pollarding periodically, making the plantation a renewable source of CPT. With careful selection of high-alkaloid yielding-genotypes, the yield of CPT could be even higher. It is also highly desirable to determine if leaves could be used as a repeatedly harvestable source of CPT.

N. nimmoniana as a hedge crop: It can be grown as hedge crop and can be harvested at the age of six to seven years after planting through pollarding. Since bark and wood are the chief renewable sources for alkaloid extraction, coppice-management becomes most important. The first harvest can yield 0.5 kg dry biomass per plant. Subsequent harvesting can be done at every one or two years interval depending on the growth. Two hundred twenty five plants can be planted as a hedge plant, at every two-meter interval in one-hectare square land. If we consider the yield of about 0.5 kg dry biomass per tree after every pollarding, a farmer can potentially earn a minimum of about ₹ 12,000 to 20,000 in the first harvest and ₹ 8,000 to 16,000 per ha per subsequent harvest (considering the range of market rate of ₹ 100 and 140 per kg of wood chips). This additional return would be earned at a negligible initial and annual cost of production. Because of its perennial nature, harvesting can be done for a longer time on sustainable basis.

As a crop for waste and degraded land: Though *N. nimmoniana* tolerates shade, it can come up very well even under open conditions provided some soil moisture is available. Hence it can be grown as a main crop on degraded and wastelands adopting 2m x 2m spacing. At the end of seven years, one may estimate a dry biomass of 1.5 tonns ha^{-1} producing a gross return of ₹ 1,50,000 to 3,00,000 per ha. The subsequent harvesting can generate a gross income of ₹ 1-2 lakh ha^{-1}.

The current annual global demand for CPT is around 15-20 kg and CIPLA Company is the most important producer of this alkaloid. However, the demand for this resource might increase substantially in the years to come. If the links for the marketing of the wood chips are well established and a genuine buy-back system is introduced, more and more farmers would be definitely interested in growing this species.

CONCLUSIONS

By growing *Nothapodytes nimmoniana* as an alternative crop for barren sloppy lands and associated crop in the existing agroforestry system of Eastern Ghat, farmers would be financially benefitted in addition to the natural resource conservation. By growing as a physical barrier in the steep slope farmers can arrest the soil at regular interval in slopping lands of the Eastern Ghats hilly area. Because of these unique features suitable as a component in the agroforestry system it may be included as a perennial component in the already existing multistoried and sloppy areas of the Eastern Ghats to improve the livelihood security of the rural community and efficient utilization of degraded waste lands. Due to fairly good market price and demand, the species may ensure early returns to the farmers with addition to natural resource conservation.

REFERENCES

Anonymous. 1997. Guidelines for National policy and multidimensional conservation programme *Amruth*, 3: 1-17.

Bonner, F.T. 1990. Storage of seeds: potential and limitations for germplasm conservation. *Forest Ecology and Management*, 35: 35-43.

Ganesh, T. and Davidar, P. 2001. Dispersal modes of tree species in the wet forests of southern Western Ghats. *Current Science*, 80(3): 394-399.

Govindachari, T.R. and Viswanathan, N. 1972. Alkaloids of *Mappia foetida*. *Phytochemistry*, 11: 3529-3531.

Hombe Gowda, H. C., Vasudeva, R.,Georgi, P. M, Boraiah, K. T. and Patil, S. K. 2002b. Sexual system and seed biology of *Nothapodytes nimmoniana* (grah.): an important anti-cancer drug yielding medicinal tree. Paper presented at "*National Conference on Recent Trends in Plant Science Research*" on November 14-15, 2002.

Hombe Gowda, H.C. 2002. Reproductive biology of *Nothapodytes nimmoniana* (Grah.) Mabber.: an important anti-cancer drug yielding medicinal tree. M.Sc. (For) thesis submitted to the University of Agricultural Sciences, Dharwad. 78p.

Hombe Gowda, H.C., Vasudeva, R., Georgi P.M., Uma Shanker, R. and Ganeshaiah, K.N. 2002a. Breeding types in *Nothapodytes nimmoniana*: an important medicinal tree. *Current Science*, 83:101-102.

Nadkarni, R.A.; Airan, J.W. and Shah., S.V. 1944. Fatty oils from seeds of *Mappia foetida* (Olacaceae). *Current Science*, 13: 233.

Nair, K.K.N., 2000. Manual of non-wood forest produce plants of Kerala. Kerala Forest Research Institue, Peechi, pp244-246.

Priel, E., Showalter, S.D., Roberts, M., Oroszlan, S. and Blair, D. 1991. Anti-HIV property of Camptothecin. *Journal of Virology*, 65: 4137-4141 .

Ravi, Kumar and Ved, D.K. 2000. *100 Red-listed Medicinal Plants of Conservation Concern in Southern India*. Foundation of Revitalization of Local Health Traditions, Bangalore, pp261-263.

Sharma, S.N., Puri, S.C., Srivastava, T.N., Handa, G. and Kaul, B.L. 2000. Enhancement of seed germination in *Nothapodytes foetida*. *Journal of Medicinal and Aromatic Plant Sciences*, 22: 206-210.

Srivastava, R.K. 2000. Seed germination tests of pioneer shola species in Nilgiris. *Indian Forester*, 126(1): 89-90.

Wall, M.E., Wani, M.C., Cook, C.E., Palmer, K.H., Mcphail, A.T. and Sim, G.A. 1966. Camptothecin and taxol; from discovery to clinic. *Journal of American Chemistry Society*, 88: 3888-3890.

□□□

Chapter 60

Livelihood Improvement through Cashew Processing Success Story of a Tribal Village

P. MURALI MOHAN, JITASATRU MOHANTY[1] & P. SASI BHUSHANA RAO[2]

INTRODUCTION

Gajapati District has 1619 villages. Only 10 % of these villages are implementing Odisha Forestry Sector Development Project (OFSDP) *i.e.,* about 164 nos of VSS/EDC under 316 Self Help Groups (SHG). As a Field Level Expert (FLE) the first author (PMM) has been visiting all the VSSs thoroughly, interacting with departmental field staff and assigned NGOs. After careful study it is found that typical Income Generating Activity (IGA) is not feasible. Therefore with the kind permission of DFO, Gajapati and under his able guidance Centre for Innovation and Voluntary Action (CIVA) selected a more vulnerable/debt burden/poor literacy/inaccessible village like Chadheipada to provide a helping hand to the villagers by initiating an inclusive model IGA through innovative approach which is on the verge of becoming a role model to other parts of the state.

Chadheipada, a tribal village under R.Udayagiri Forest Range, is situated in Parimal Gram Panchayat under Nuaguda Block of Gajapati District. It is surrounded by hills and forest. The people mostly depend upon cultivation and forest produce. Crops like maize, ragi and paddy are grown here. Some of the villagers collect tamarind, cashew and karanja to sustain their livelihood. Apart from that, they are also daily wage earners when ever work is available. Sometimes, they are debt trapped by local traders of forest produce and succumb to distress sale of such product to them in order to get rid of the loan burden. Considering the natural forest resource and the willingness of the tribal people, Odisha Forestry Sector Development Project has implemented

F.L.E, Paralakhemundi, Gajapathi District, Odisha

[1]DFO-cum-DMU Chief, Paralakhemundi

[2]Centre for Innovation and Voluntary Action (CIVA), Rayagada, Odisha, India

this livelihood enhancement project in this village through constitution of a VSS (Vana Samrakshyan Samiti) with an objective to improve the economic condition of the people along with the condition of the Forests. The DFO-cum-DMU Chief and F.L.E of Paralakhemundi Forest Division while trying to enhance their socio-economic conditions a NGO named **"CIVA" (Centre for Innovation in Volunteer Action)** came forwarded to extend its support in this noble endeavour.

Motivation : The FLE and CIVA conducted several interactive meetings with the villagers of Chadheipada to study the ground level situation before any activity was initiated. These meetings in course of time yielded success as the villagers and the Maa Manikeswari SHG of the village showed keen interest to know more and get involved in a new developmental process. They were able to decide what was needed for them and what the activity they could take up on their own under the supervision of FLE and CIVA. This has been an inclusive approach with internalized mechanism of functioning. The activities were aimed at protection of forests, conservation of Biodiversity and Income Generation Activities. The expertise was arranged availing services of field resource persons. During the process CIVA introduced an innovative approach of Fine Art Education Programme (FAEP) an alternative to non formal education/ poster education. The FAEP was specific for the village. It had been very well received by the villagers and the officials. At the end of the day CIVA could see the success. CIVA, however, continued with non-formal education as well as skill development programmes and imparting training for the proposed IGA.

Income generation activities (IGA) : The villagers under the supervision of CIVA introduced many ecofriendly IGA and also cashew processing as a principal IGA with great success. The following activities are going on in the village for the development of the villagers, for the villagers and by the villagers.

Primary activities with direct benefits

- Raising of Vegetable Seedlings
- Compost Pits
- Cashew Business: Collection, Procurement, Processing and Marketing

Secondary activities with indirect benefits

- FAEP, Non Formal Education Programme (NFEP), Training and Awareness Programes
- Use of Smokeless Chullas
- Village pond works
- Inculcation of newer practices

The first three of the above mentioned activities are classified as Primary activities which offer immediate benefits and profits to the villagers. The last four activities are known as secondary and have a bearing on the overall development in the long run bringing in attitudinal change and inculcation of newer practices among the villagers. *FLE of OFSDP and CIVA formulated the integrated developmental strategy which is inclusive with internalized operational mechanism. This is akin to the Mahatma Gandhi's autonomous village concept.*

Raising of vegetable seedlings : The female members of the village are trained in raising of seedlings. The Maa Manikeshwari SHG participated in the day to day programmes. Initially, they have raised Brinjal, and Tomato seedlings in their field. First year, they sold nearly seedlings worth of Rs. 8000/- in the surrounding villages. They never saw such amount in their group. The practice continues. Besides seedlings they also grew their own kitchen garden including Lady Fingure and shared the produce.They are also growing Beans, raw Banana and Papaya fruits.

Raising of Vegetable Seedlings

Humus Compost Pit : CIVA has proposed to start the facilities of humus compost for vegetable seedlings and their own fields. The group members are actively involved in the humus compost as per the laid down procedure. They used the humus compost for themselves in their sample plots.

Humus Compost Pit

Cashew Business

Education and awareness: The Chadheipada villagers have been living in the forest since ages with plenty of natural resources around them. They have dense cashew plantation along the slopes and banks of river flowing nearby. Cashew plantations are cash crops and form the livelihood of the tribes particularly for Souras in Gajapati District as both fruits and seeds yield money. We need to appreciate that cashew plants are primeval and not exotic and hence play significant role in economy and ecology of the village. It is a cash crop of our communities. CIVA educated the SHG of Chadheipada the importance of cashew in their ecology and economy and benefits and profits of collection, procurement processing and trading of cashew.

Collected Raw materials

Earlier the cashew nuts were collected traditionally by the village community and sold as such after drying either in local market or to the vested traders. In order to provide good price through value addition it was decided to install a cashew processing unit so that the local people would not only earn more income but also ultimately become owners of some valuable assets. It now became a perennial source of their income since the demand of cashew in the market as a consumable item is on ascending scale.

Skill Development Programme on Cashew Processing Unit

SDP on Cashew Processing

CIVA is a NGO based at Rayagada working in the area since last few years. They too have good experience and expertise as well as outside contact on cashew processing. In the mean time a proposal submitted by CIVA for capacity Building of SHG members on cashew processing for Chadheipada VSS to NABARD got approved and the field level expert requested the organization to link Maa Manikeswari SHG for their training. Other than Maa Manikeswri SHG of Chadheiapada, Mother Teresa SHG of Nuapalli, Nandini SHG of Hatiguda, Sania SHG of Konisahi & Jyoti SHG of Dhimiri street also underwent comprehensive skill development training for one month.

Training /exposure visit : The preliminary training was found to be inadequate to handle such a unit on sustained basis. As desired by the villagers CIVA arranged to provide a long term capacity building training to some young and energetic members of the SHG at Palasa of Andhra Pradesh where different types of cashew units are functional. Accordingly the FLE arranged training for 12 members in two phases at Divakar & Ranga Reddy Cashew units at Palasa, AP for a period of one month. The SHGs and VSS members learnt cashew value addition process with assistance of CIVA staff.

Procuring Raw Material : The trained community members gave publicity in neighboring villages for procuring cashew nuts. They procured 2689 kg cashew from local and neighboring villages for Rupees One lakh sixty one thousand, three hundred forty only, during March 2010 and December 2010. FLE along with local forest officials conducted awareness programme at Chadheipada and convinced villagers to install a cashew processing unit by explaining the cost benefit analysis of the project and prepared a business development plan as under.

The Maa Manikeswari SHG consisting of 12 tribal ladies was motivated to take up this project. After detailed discussion, the said SHG being convinced with the business development plan, decided to establish a cashew processing unit.

a) Business Development Plan

S. No.	Description	Amount (₹)
1. Fixed Capital		
a)	Heat Chamber	1,34,029/-
b)	Boiler Machine	
c)	Cutter	
d)	Accessories	35,600/-
	Total	**1,69,629/-**
2. Working Capital		
a)	Raw material	2,00,000/-
b)	Labour	2,000/-
c)	Transportation	8,000/-
d)	Electricity	8,500/-
e)	Misc.	33,000/-
	Total	**2,51,500/-**
Grand Total		**4,21,129/-**

b) Profit analysis

S. No.	Description	Amount (₹)
Working capital		
1	Raw materials procured 2689 Kgs. @ 60/-	1,61,340/-
2	Labour	2,000/-
3	Transportation	8,300/-
4	Electricity	8,500/-
5	Misc.	33,000/-
	Total	**2,13,140/-**
6	Finished products marketed 672 Kgs. as per gradation	3,21,076/-
Profit = (Finished products – Working capital)		**1,07,936/-**

Sources of funding : As already indicated above, an amount of 4.21 lakh was necessary for the unit where as the SHG intending to do it is having **no capital rather a debt burden of Rs. 30,000/-** However, the F.L.E contacted various financial institutions and converged different Govt. schemes to arrange the aforesaid capital as under.

SHG received cheque from Panjab National Bank, R.Udayagiri

The Branch Manager, Panjab National Bank, R.Udayagiri came forward along with APD, DRDA, Gajapati for this project.

S. No.	Institution	Amount (₹)
1	Loan from Panjab National Bank	1,10,000/-
2	Subsidy from DRDA (SGSY)	1,20,000/-
3	Loan from VSS (R.F. of OFSDP)	90,000/-
4	CIVA's own contribution	63,000/-
5	SHG own contribution	34,000/-
	Total	**4,17,000/-**

Arrangement of building for the unit : The VSS Office-cum-Meeting hall constructed under OFSDP is initially used for the purpose of storing the raw materials since the poor tribals have no godown for large quantity of materials. For Machine and other accessories, a defunct building of Soil Conservation Department

Raw materials kept in VSS Building

was taken through discussion with the concerned officials by the FLE and with minor repair by the villagers from their own contribution. The said building became habitable and ready for installation of the unit.

Procurement of machinery : The machinery along with its accessories was procured from Rotex, Pune in collaboration with DRDA, Gajapati. The technician from the supplying company installed the machine in the village and conducted preliminary training for its operation to the villagers.

Installation of Cashew machine by Rotex

Skill Development Programme for establishment of cashew processing unit

Value addition of cashew nut : Before initiation of cashew processing unit, the community received special training from experts from Palasa, Andhra Pradesh. Later on, the SHG members worked on their own along with other trainees for cashew processing unit. Lastly, they have gained confidence for processing of available material and grading.

Gradation of finished Cashew nuts

Credit linkages : The Maa Manikeswari SHG members submitted application form along with relevant documents for loan at PNB of R.Udayagiri. The Bank Manager sanctioned loan amount of Rupees One lakh and twenty nine only for procurement of the cashew processing machine and Rupees One lakh and ten thousand for raw material procurement.

Market linkage : The FLE arranged discussion with different traders of Odisha and Andhra Pradesh and persuaded them to visit the unit and assess the quality of product. After a visit by the traders for assessing the quality, the rate was fixed through negotiation at a competitive bidding price. Accordingly the product was sold at the highest value offer. The VSS members particularly the SHG were quite jubilant to experience such huge monitory transaction in their life for the first time.

Manikeswari Group is selling Cashew nuts in value added packets

a) Actual profit accrued

S. No.	Description	Amount (₹)
I	Sale proceed	3,21,077/-
II	**Working capital**	
	a) Repaid loans to Panjab National Bank, R.Udayagiri including interest	1,14,141/-
	b) Repaid loans to VSS including interest	99,000/-
	Total	**2,13,141/-**
	Profit (Sale proceed – Working capital)	**1,07,936/-**

b) Utilization of the profit by the Group

S. No	Description	Amount (₹)
i.	Investment (From Profit) as fixed deposit in Panjab National Bank, R.Udayagiri	50,000/-
ii.	Distribution of profits among 12 members @2000/-each	24,000/-
iii.	Donated 02 Nos. of Ceiling Fans to VSS Building	3,000/-
iv	Invest for Brick making (36000 Nos.)	30,936/-
	Total	**1,07,936/-**

Socio-economic benefits of Groups

S. No.	Description	Amount (₹)
1	Availed Soil Conservation Building for Conducting Training-cum-Cashew boiling	8,00,000/-
2	Amount of grant received from DRDA (SGSY) for construction of Building for Cashew Processing Unit	3,00,000/-
3	The cost of the machine after the clearance of loan	1,46,000/-
	Total Maa Manikeswari SHG's permanent Assets	**12,46,000/-**

Thus at the end of the season for cashew business, the villagers not only cleared the bank loan and loans of the VSS but also had fixed deposits in the Bank besides creating assets worth a few lakhs of rupees. At the end of the day it has been a very satisfying endeavour for the CIVA and FLE of OFSDP under the invaluable guidance of the DFO and advices from several concerned authorities.

Secondary Activities

Role of fine art education programme (FAEP) and conservation facility : Forest Ecosystem Restoration can't be executed unless the tribal communities are aware of its advantages. For protection of forests imparting Environmental Education (EE) is imperative (Patnaik and Sasibhushana Rao, 2000) and it should permeate into all age groups and sections of the society (Sasi Bhushana Rao, 1999) if we consider holistic development. The villagers should be imparted EE which is always different from literacy and schooling programmes. An illiterate can also be imparted EE.

i. Forests are shrinking. Hence, the fauna of that eco-system are dislocated into other eco-systems disturbing the balance of the eco-system and impacting the climate.

ii. Despite under taking various measures and awareness programs, it hasn't been possible to bring change in the mind set up of the tribal communities particularly with regard to the shifting cultivation.

To protect the forests and bio-diversity, a "new process", has to be developed.

i. Tribes continue to depend on forests for their livelihood.

ii. The educated individuals are leaving the villages in search of better earning opportunities and the uneducated remain in the village.

iii. Communication gap between the educated community and the tribes impede exchange of ideas. Thus the interactions are not translated into actions.

iv. Hence, there is a need for Fresh research work and innovation for conducting further successful awareness programs.

v. General banners / posters have not yielded desired results. In most of these cases people need to be literate.

vi. In view of lower percentage of success from the earlier processes, we have chosen Fine Art as an important tool to impart EE. Fine Art is an innovative tool conceived and introduced by CIVA to accomplish the task (Murali Mohan and Sasi Bhushana Rao, 2011) in Chadheipada village. "Fine Art Education Programme (FAEP)" Or Wall Art Education Programme (WAEP) has been found to be very useful , for awareness, imparting EE and enabling them to appreciate their problems and empower them to plan and protect their own natural resources and bio-diversity. Thus the programme is inclusive in nature (Murali Mohan and Sasi Bhushana Rao,2011). The walls of VSS are used for FAEP which could impart EE successfully to all sections and age groups of people in Chadheipada and nearby villages. EE is the basis for Eco-restoration and development. Our FAEP has come for appreciation by all officials and visitors to the place.

Fine Art

Education for Community Members

The Non Formal Education is neither traditional nor obligatory and structured. Here people learn outside the context of a formal school/ Institution and in a leisurely and friendly environment.

This education center is an eye opener for other NFE centers. Here, the SHG members and their children actively participate and learn using floor black boards. FAEP of CIVA is taking the front seat in NFE of Chadheipada. If funds are provided CIVA is prepared to take up FAEP in other villages in the district.

SHG member using Floor Black board

Floor Black board

Smokeless chula : Most of the families cook their food traditionally in open fire with very little concern for the loss of large quantities of firewood. The village woman works for a long time throughout the day in the farm and also collects firewood. After such laborious work, she has to cook food on a traditional Chula, which gives out lot of heat and smoke which would create health problem for the woman.

Community utilizing Smokeless Chulla

CIVA promoted use of Smokeless Chula which is innovative, consumes less firewood and also is eco- friendly and less hazardous to health. More so the fuel efficiency Chulla can be made out of locally available materials and by small scale industries.

The Smokeless Chulla is superior to the traditional Chulla for following reasons:

- Due to higher fuel efficiency fuel wood can be saved by 50%.
- The environment (Carbon gases inside the huts) is almost smoke free.
- The heat in the kitchen is minimized.
- The drudgery can be reduced.
- Vessels will have less soot.
- Fire accidents are minimized.

Renovation of pond : The SHG group is actively involved in cashew business assisted by VSS, but they are unable to pay for the ccashew trading. CIVA organisation has liaison with National Rural Employment Generation Scheme (NREGS) and mobilized fund for renovation of village pond. Thus the community availed employment opportunities and simultaneously earned their wages which were availed by the group members for cashew trading.

Renovation of Pond

Inculcation of Newer Practices

FAEP, motivation meetings, raising seedlings, kitchen garden, cashew business, pond renovation and other group works yielded the following positive results which benefitted the village in general.

- Coming together and spending free time in VSS building practicing clean and green surroundings.
- Sense of ownership and belongingness.
- Communicate with each other and deliberate on issues of concern on their own.
- Development of communication skills with the out siders of the village.
- Trade and market skills.
- Concern and act for protection of the forest and conservation of biodiversity.
- Attitudinal change in villagers and partnership in the development process as they are educated over several aspects even while many of them remained illiterate.

End Result of the Project

- RaFencing of VSS compound by the community
- Rising of vegetable seedlings and selling,
- Planting of 51 Nos. of Teak seedlings alongside the boundary of VSS
- Papaya cultivation around VSS
- Installation of 40 Nos. of smokeless Chulas.

Brick making by SHG members

- Installation of Humus Compost.
- Brick making & marketing of 36000 marketing.
- De-seeding of Tamarind & Trading.
- Maize cultivation & clearance of Old Loan.
- Enhancement of saving contribution.
- Capability of all the members specially Animator, President & all SHG enhanced.

Convergence

- Installation Vermi compost pit and Drip irrigation for vegetable cultivation by Krushi Vikash
- Yam & Ground nut cultivation & Trading by Krushi Bigyan Kendra
- Renovation of village pond by MGNREGS

Inspirational impact

Two SHGs namely Maa Manikeswari and the Merry Mission groups are functioning in Chadheipada village. These two SHGs are BPL categories and Schedule Tribes in the initial period. Now, these two groups are self-sufficient and sustained and also established two manufacturing units in village.

- The Maa Manikeswari group had established a Cashew processing unit.
- The Maa Teressa Mission Sakti group established a Rice Huller in the village.

These two units stand as examples in whole district and bear testimony to the efforts of FLE and an NGO CIVA.

Chadheipada has not only become a role model but also is able to inspire neighboring villages like Parvati Mahila Mission, Libiriguda to emulate our strategy.

Other SHGs operating around our village have too started trading their own available resources after getting training from our group.

CONCLUSIONS

The project execution and results would suggest that during the entire process the activities were never imposed on villagers by our thoughts rather they were allowed to deliberate to come forward with their decisions. Thus, it has been an inclusive approach having all the operational mechanisms internalised. The development is neither at the cost of the ecology nor culture and indigenous knowledge. It is similar to a true Gandhi's trusteeship. Gandhiji's concept of Gram Swaraj / Self Sufficient Villages is the only solution to Sustainable Development and Conservation of Natural Resources.

With dedicated and concerted efforts and perseverance by CIVA , the simple, innocent, poor tribals labored diligently supported by financial and institutional augmentation through convergence of various developmental schemes the debt burdened tribal SHG women became rich worth a few lakhs within a year. CIVA a NGO played a catalytic role in accomplishing this task much to the appreciation by the authorities and the ever thankful villagers of Chadheipada.

Acknowledgement

The first author (PMM) is grateful to Sri J.S.Mohanty, I.F.S., DFO, Paralakhemundi, for his invaluable guidance, Sri A.K.Panda, LDM, Andhra Bank, Gajapati district, Sri P. Chandra Sekhar, MIS Expert, Rayagada, Sri Nursingh Nimolin, Field Coordinator, Rayagada, Sri Umakanta Panda, Educationist, Paralakhemundi, Sri.G.K. Dangua, AGM NABARD, Gajapati district, Sri P.C.Mishra, Branch Manager, Punjab National bank, R.Udayagiri, Sri Abani Kumar Rath, DEO, Devagiri FMU, Paralakhemundi, Sri T.Karji, IPO, DIC, Gajapati district and Sri P. Uma Maheswar, APD (ORMAS), DRDA, Paralakhemundi for their kind support in several ways.

REFERENCES

Murali Mohan,P and Sasi Bhushana Rao, P. 2011. Role of fine Art in biodiversity management. In: Proceedings of the National Seminar on Climate Change - Impact on Environment and Human Health. Ed. P.Sasi Bhushana Rao. Publ. Gayatri Vidya Parishad Degree College (Autonomous). ISBN : 978-81-921799-0-2 pp 86-94.

Naba Kumar Patnaik and P. Sasi Bhushana Rao, 2000.Environmental Education for the 21st Century.

Sasi Bhushana Rao, P and Rao, P.M. 2000. Environment Management and Auditing. Deep and Deep publication, New Delhi. ISBN 81-7629-248-6, pp 201-210.

Sasi Bhushana Rao, P. 1999. Environmental Education – A long way to go. In: *Corporate reporting and accounting practices* Ed. P.M. Rao, Publ: Deep and Deep publication. New Delhi. pp 226-241.

Sasi Bhushana Rao, P. 2011. Inventory of Fauna and Strategies for conservation of Biodiversity of Mahendragiri Range of Hills of Andhra and Orissa Border. Major Research Project Report.101p.

□□□

Chapter 61

Crop Diversification under Irrigation Medium Land Situation of Assam - A Possibility

A.BAISHYA, J.P.HAZARIKA, K.MAHANTA, J.HAZARIKA AND R.HAZARIKA

INTRODUCTION

Indian agriculture presently expose to second generation problem like raising or lowering of water table, nutrient imbalance, soil degradation, salinity, resurgence of pests and diseases, environmental pollution and decline in farm profit. Crop diversification shows lot of promise in alleviating these problems through fulfilling the basic needs and regulating farm income, withstanding weather aberrations, controlling price fluctuation, ensuring balanced food supply, conserving natural resources, reducing the chemical fertilizer and pesticide loads, environmental safety and creating employment opportunity (Gill and Ahlawat,2006). Crop diversification has been recognised as an effective strategy for achieving the objectives of food security, nutrition security, income growth, poverty alleviation, employment generation, judicious use of land and water resources, sustainable agriculture development and environmental improvement (Hedge *et.al.*,2003).

The necessity for crop diversification arise from (i) reducing risk associated with yield, market and prices (ii) arresting the degradation of natural resources and environment and (iii) attaining national goals like employment generation, self-reliance in critical crop products and earning foreign exchange (Reddy and Suresh, 2009). Diversification is the process to advantage of emerging opportunities created by technology, new markets, changes in policy *etc.* to meet certain goals, challenges and threats and to reduce risk (Chand and Chauhan, 2002). In Assam, the total cropped area is 39.94 lakh ha out of which 27.51 lakh ha is net sown area which comprises 5.25 lakh ha irrigated area (Baishya *et al.*,2011). With suitable crop diversification along with incorporation of high valued as well as less water requiring crops Water Productivity can be enhanced.

Assam Agricultural University, Jorhat – 785 013, Assam

MATERIALS AND METHODS

The trial was conducted under All Co-ordinated Research Project on Integrated Farming System at Assam Agricultural University, Jorhat, Assam during the period from 2008 – 09 to 2010 – 11.The soil of the experimental site was Sandy loam in texture (Coarse Sand: 6.68%, Fine Sand: 61.43%, Silt: 15.23% and Clay: 15.50%) the bulk density of the soil was 1.21mg m^{-3}, Particle density of : 2.29mg m^{-3},Maximum Water Holding Capacity of 47.15%, Field Capacity was29.36%, Permanent Wilting Point was 7.60%, Water Stable Aggregate of 81.21% and Mean Weight Diameter was 1.81%. The experiment was laid out in Randomized Block Design with eight treatments and four replications in a plot size of 6m X 5m. Out of the eight different cropping sequences, T_1 and T_2 treatments consist of first and second pre-dominant cropping system of Assam. The eight cropping sequences along with their varieties are presented in Table 1, date of sowing/planting and date of Harvesting in Table 2 and fertilizer application, spacing, duration in Table 3.

Table 1: Treatment details of the experimental study

Treatment	*Kharif*	*Rabi*	Summer
T_1	Winter rice (*Rajendra Suwasini*)	F	Autumn rice (*'Disang'*)
T_2	-do-	Toria (*'TS – 38'*)	F
T_3	-do-	Cabbage (*'Golden Acre'*)	Green gram (*'Pratap'*)
T_4	-do-	Cauliflower (*'Pusa Snowball'*)	Black gram (*'PU – 31'*)
T_5	-do-	Knolkhol (*'White Vienna'*) – French bean (*'Anupam'*)	Cowpea (Fodder) (*'EC – 4216'*)
T_6	-do-	Capsicum (*'California Wonder'*)	Cowpea (Fodder) (*'EC – 4216'*)
T_7	-do-	Chilli (*'Namdhari* $NSF_1$1701')	Black gram (*'PU – 31'*)
T_8	-do-	Radish (*'Pusa Chetki*)' – French bean (*'Anupam'*)	Green gram (*'Pratap'*)

F = Fallow

Table 2: Sowing, planting and harvesting dates.

Tr. no	Date of Sowing in Seed bed			Date of planting/sowing in Main field			Date of Harvesting		
	Kharif	*Rabi*	Summer	*Kharif*	*Rabi*	Summer	*Kharif*	*Rabi*	Summer
T_1	02-06-10	-	02-03-11	30-06-10	-	01.04.11	07.10.10	-	20.06.11
T_2	-do-	-	-	-do-	5-11-10	-	-do-	08.02.11	-
T_3	-do-	30-09-10	-	-do-	5-11-10	15.3.11	-do-	13.01.11	01.06.11
T_4	-do-	30-09-10	-	-do-	5-11-10	15.3.11	-do-	20.12.10	20.05.11
T_5	-do-	30-09-10 -	-	-do-	5-11-10 7-01-10	20.3.11	-do-	17.12.10 20.02.11	15.05.11
T_6	-do-	17.10.10	-	-do-	15.11.10	25.3.11	-do-	27.02.11	15.05.11
T_7	-do-	30-09-10	-	-do-	5-11-10	15.3.11	-do-	17.02.11	19.05.11
T_8	-do-	- -	-	-do-	5-11-10 7-01-11	15.3.11	-do-	19.12.10 17.02.11	01.06.11

Table 3: Fertilizer application, spacing and duration of different crops.

Tr. no.	Fertilizer (N-P -K kg ha^{-1})			Spacing (cm x cm)			Duration (days)		
	Kharif	*Rabi*	Summer	*Kharif*	*Rabi*	Summer	*Kharif*	*Rabi*	Summer
T_1	80-40-40	-	80-40-40	20x15	-	20x15	128	-	110
T_2	-do-	60-40-40	-	-do-	Broad-cast	-	-do-	95	-
T_3	-do-	120-60-60	15-35-10	-do-	60x30	30x10	-do-	74	80
T_4	-do-	80-60-60	15-35-10	-do-	45x45	30x10	-do-	81	69
T_5	-do-	80-60-60 30-40-20	20-40-20	-do-	30x15 45x30	35x10	-do-	78 44	51
T_6	-do-	120-60-60	20-40-20	-do-	30x30	35x10	-do-	133	51
T_7	-do-	120-60-60	15-35-10	-do-	30x30	30x10	-do-	140	68
T_8	-do-	50-100-100 30-40-20	15-35-10	-do-	20x10 45x30	30x10	-do-	44 41	80

Meteorological influences:Meteorological information during the year 2010-11 at Assam Agricultural University, Jorhat revealed maximum temperature up to 32.6^0 C (during August 2010) and minimum temperature up to 10.1^0 C (during January, 2011); highest relative humidity (Morning) up to 97% (during January) and lowest (evening) 54.8% (during March); maximum evaporation (3.1 mm day^{-1}) was during the month of June and total rainfall received during this year was 2173 mm and total number of rainy days were 162 days, bright sunshine hours varies between 5.1 hr day^{-1} to 6.1 hr day^{-1} from November 2010 (5.1 hr day^{-1} during January,2011) to

May 11 (6.1 hrday^{-1} during March 2011) but in between June, 2010 to October, 2010 it varies between 3.1 hrday^{-1} to 4.7 hrday^{-1}.

Irrigation scheduling: Winter rice did not require any irrigation throughout the growing period.Highest number of irrigation were provided to Autumn rice was 6 cm depth one irrigation was provided at the time of puddling and subsequently 4 numbers of irrigations with 4 cm depth was provided during different growth stages of Autumn rice. Our irrigations of 4 cm depth of irrigation were provided to Capsicum and Chilli and three irrigations of 4 cm depth were applied to cabbage. Two irrigations of 4 cm depth were given to Cauliflower, Knolkhol and French bean. For toria crop 4 cm depth of one irrigation was applied at the time of flowering. One irrigation of 2 cm depth was provided to summer Green gram and Black gram before sowing and subsequently another 2 cm depth of irrigation was given at 15 days after germination due to dry weather conditions. Cowpea (fodder) was irrigated only 2 cm depth of irrigation at 20 days after germination of crop.

During the study period observation *viz.*, grain and straw/stover/ head/ fodder yield of different crops were recorded and converted it to rice equivalent yield of different cropping sequences and economic viability of the system were recorded and calculated. To identify the efficient cropping systems and system productivity was calculated. System productivity is yield per unit area per unit time. The equivalent yield was calculated in terms of *kharif* crop of rice in this rice based cropping system using the following formula:

$$REY = \Sigma\, Yi \times Pi\, / P\,(p)$$

Where REY denotes Rice Equivalent Yield; Yi = Yield of different crops; Pi = Price of respective crops and P(p): Price of Paddy.

Land use efficiency was calculatingusing the following method and it refers to the extent of land area used in a year.

$$LUE\,(\%) = TND\,(i) \times 100/\,365$$

Where TND (i) denotes the total number of days field remained occupied under different crops (i = 1- - - - - n).

Apparent nutrient use productivity (ANUP) was calculated on dividing the equivalent yield of the system by the total quantity of nutrients used. However, this measure is an indicator of the nutrient use productivity but not the true measure. The system which utilizes the nutrient in different crops more efficiently per unit of nutrient is considered to be more efficient system. This measure reflects the efficiency of the nutrient use. The profitability of the system was also calculated by accounting cost of production of individual crops. Benefit cost ratio was also worked out by dividing the total returns from the cropping system to the total cost of production. The ratio of more than a unit is considered to be more profitability. Crop profitability (₹ ha^{-1} day^{-1}) was worked out by dividing net return of particular cropping systems to number of days field occupied by the system. Whereas the system profitability (₹ $ha^{-1}day^{-1}$) is the ratio of total net returns to 365 days.

RESULTS AND DISCUSSION

Major weeds:Cyprus and water hyacinth were mostly observed in winter rice - autumn rice sequence.In case of winter rice - Toria sequence, water hyacinth, boreria and commelina were the predominant weeds.Winter vegetable – Summer pulses/ Cowpea (Fodder) sequence brymaria, commelina and centella were mostly found weeds.

Insects and Diseases:The insect's whorl maggot, plant hopper, leaf roller and rice bugs were observed with rice but no diseases were reported from the rice crop. In some vegetables and summer pulses, caterpillars were observed. Bacterial wilt disease was found in capsicum and chilli.

Water quality: Three water samples of irrigation water were analysed for its quality and found that the average pH is 6.7, EC 0.4 ds - m, HCO_3 is 0.36 ppm,chloride 59 ppm and Iron 8.0 ppm.

Yield of different crops and rice equivalent yield (REY):The average yield of different crops and average REY from 2008-09 to 2010-11 are presented in Table 4and revealed that highest REY (218.06 q ha^{-1}) with winter rice - Cabbage - Green gram sequence as compared to other cropping sequences. This might be because of higher yield (188.50 q ha^{-1}) obtained with cabbage as well as with the relatively better price of the produce (₹ 10.00 kg^{-1}). The Winter rice variety '*Rajendra* suwasini' even though produce around 30 q ha^{-1} but it gave a good return because the sale price of the rice grain was ₹ 1100.00 q^{-1} (Table - 4).

Efficient cropping system: The average (for the year 2008-09 to 2010-11) REY, BCR, system productivity (SP)(kg $ha^{-1}day^{-1}$), nutrient use productivity (kg ha^{-1} kg^{-1} nutrient applied), employment generation (man days ha^{-1}) and land use efficiency (%) under eight different cropping sequences are presented in Table 5. The highest REY (218.06q ha^{-1}), BCR (4.42), SP (60kg $ha^{-1}day^{-1}$) and NUP (47.93kg $ha^{-1}kg^{-1}$nutrient applied) were recorded with Winter rice – Cabbage – Green gram sequence followed by Winter rice – Capsicum – Cowpea (Fodder) sequence with 167.61q ha^{-1}, 3.86, 46 kg ha^{-1} day^{-1} and 34.92 kg ha^{-1} kg^{-1} nutrient applied, respectively. This was due to high yield as well as better price obtained for vegetable crops. The finding is in conformity with Baishya *et al.* (2009) and Singh *et al.* (2010).

Table 4: Productivity/yield of crops (Av. of 2008 – 09 to 2010 – 11) and REY of different cropping systems

Cropping sequence	*Kharif (q ha^{-1})*		*Rabi(q ha^{-1})*	Summer(*q ha^{-1}*)		REY (*q ha^{-1}*)
	Grain	Straw	Grain/ Head/Fruit	Grain	Straw	
Winter rice – Fallow – Autumn rice	31.20	40.80	F	24.30	31.10	50.73
Winter rice – Toria – Fallow	29.60	38.80	9.40	-	-	51.80
Winter rice – Cabbage – Green gram	30.00	38.90	188.50	3.90	8.40	218.06
Winter rice – Cauliflower – Black gram	28.70	37.30	67.50	5.60	5.00	127.77
Winter rice – Knolkhol – French bean – Cowpea (Fodder)	30.90	39.60	55.00 43.60	-	160.60	138.33
Winter rice – Capsicum – Cowpea (Fodder)	31.40	41.00	39.80	-	153.50	167.61
Winter rice – Chilli – Black gram	28.00	36.40	31.20	4.60	3.10	123.61
Winter rice – Radish – French bean – Green gram	32.40	42.00	97.30 51.40	3.90	6.30	156.24

But employment generation was highest (309 man days ha^{-1}) with Winter rice – Capsicum – Cowpea (Fodder) sequence followed by Winter rice – Knolkhol – French bean – Cow pea (Fodder) sequence with 307 man days ha^{-1}. This was because of more manpower requirement forintercultural operation with capsicum crop in first case and knolkhol and french bean in second case during the *rabi* season. Highest land use efficiency (81%) was observed with winter rice – radish – french bean – green gram sequence because out of the eight sequences this particular sequence occupy the field for longer number of days (293 days) followed by winter rice – cauliflower – black gram (80%) with the occupation of field by 282 days.

Soil fertility status: Soil fertility status after completion of the third year of different sequences *i.e.,* after harvesting of summer crops of 2011 are presented in Table 6. Soil p^{H} increasedin all the treatments except T_7 and increased OC (%) only in two treatment *i.e.,*T_3 (0.85%) and T_8 (0.82%) compared to the initial value. Available N increased in all the treatments except T_2 (216.75 kg ha^{-1}) and T_4 (170.73 kg ha^{-1}) than that of initial value (217.25 kg ha^{-1}). But available P and K increased to a greater extent in all the treatments (from 26.59 to 38.93 kg ha^{-1} available P and from 68.18 to 92.57 kg ha^{-1}for available K) than that of initial value (16.41 kg ha^{-1} available P and

56.28 kg ha^{-1} available K). Highest quantity of available N (232.97 kg ha^{-1}), available P (38.93 kg ha^{-1}) and available K (92.57 kg ha^{-1}) were found after completion of winter rice – radish – french bean – green gram sequence.Palaniappan and Sivaraman (1994) and Saha and Moharana (2007) who emphasize that rice – pulse system was effective for higher amount of available N, P and K. Obviously pulses, with characteristics promotion of free living microorganisms (*Rhizobium spp.*) solubilisation of insoluble Al – P and Fe – P fractions through exudation of organic substances through their nodules and release of K by mineralization are known to enrich the soil.

Table 5: Average REY ,BCR, system productivity,nutrient use productivity,employment generation and land use efficiency of different sequences.

Cropping sequence	REY (q ha^{-1})	BCR	System Productivity (kgha^{-1}day^{-1})	Nutrient use Productivity (kgha^{-1}kg^{-1} nutrient applied)	Employment Generation (man-days ha^{-1})	LUE (%)
Winter rice – Fallow – Autumn rice	50.73	1.81	14	21.14	217	58
Winter rice – Toria – Fallow	51.08	2.55	14	17.03	152	57
Winter rice – Cabbage – Green gram	**218.06**	**4.42**	60	**47.93**	302	77
Winter rice – Cauliflower – Black gram	127.77	3.06	35	30.79	300	80
Winter rice – Knolkhol – French bean – Cowpea (Fodder)	138.33	2.63	38	25.62	307	79
Winter rice – Capsicum – Cowpea (Fodder)	167.61	3.86	46	34.92	309	77
Winter rice – Chilli – Black gram	123.61	3.55	34	27.17	301	78
Winter rice – Radish – French bean – Green gram	156.24	3.12	43	33.60	263	81

Prices considered:

CropPrice (₹/Q)		CropPrice (₹/ha)	
Winter rice (Grain)	1100/-	Autumn rice(Grain)	800/-
Winter rice (Straw)	20/-	Toria	2400/-
Green gram	4500/-	Cabbage	1000/-
Cauliflower	1000/-	Knolkhol	1000/-
Black gram	3000/-	Frenchbean	1000/-
Capsicum	3500/-	Chilli	3000/-
Radish	700/-		

N, P and K uptake by different sequences: The N, P and K uptake by different cropping sequences are presented in Table 7. Highest uptake of N (237.67 kg ha^{-1}) with winter rice – knolkhol – french bean – cowpea (Fodder) sequence whereaswinter rice – radish – french bean – black gram remove highest amount of available P (272.51 kg ha^{-1}) and available K (223.10 kg ha^{-1}). The uptake of all the three N, P and K nutrients found to be highest during the *rabi* season. The maximum removal of N, P and K may be due to greater biomass production which was also reported by (Sharma *et al.*, 2008). Because highest uptake of N as well as P and K was observed with both the four crop sequences out of eight different sequences and the rest six sequences two are double cropped and four are triple cropped.

Table 6: Final soil fertility status under different cropping systems

Tr.No.	Cropping sequence	pH	O.C. (%)	Available nutrients (kgha^{-1}) N	P	K
T_1	Winter rice – Fallow – Autumn rice	5.81	0.58	218.54	29.45	78.65
T_2	Winter rice – Toria – Fallow	5.76	0.57	216.75	30.67	68.18
T_3	Winter rice – Cabbage – Green gram	5.76	0.85	223.07	34.77	76.28
T_4	Winter rice – Cauliflower – Black gram	5.80	0.61	170.73	31.77	76.72
T_5	Winter rice – Knolkhol – French bean – Cowpea (Fodder)	5.84	0.62	228.52	26.59	89.00
T_6	Winter rice – Capsicum – Cowpea (Fodder)	5.73	0.42	224.50	33.17	76.50
T_7	Winter rice – Chilli – Black gram	5.67	0.50	220.60	28.08	77.74
T_8	Winter rice – Radish – French bean – Green gram	5.76	0.82	**232.97**	**38.93**	**92.57**
Initial value		5.69	0.62	217.25	16.41	56.28

CONCLUSIONS

The most promising cropping systems prevailed in the state of Assam are winter rice – autumn rice and winter rice – toria which are out yielded by all the other cropping systems under the study. Therefore crop intensification and diversification under medium land with irrigation facilities have greater potential and promising in the region. Out of the eight cropping sequences, winter rice - cabbage - green gram sequence is found to be most effective from the point view of productivity, economic viability, sustainability, employment soil sustainability.

ACKNOWLEDGEMENT

We are very much thankful to Dr K. M. Bujarbarua, Hon'ble Vice Chancellor, Assam Agricultural University, Jorhat for the guidance and inspiration received from him to carry out the research work as well as dissemination of research findings through presentation of scientific paper. Our sincere thanks and gratefulness also goes to Dr B. Gangwar, Director, Project Directorate of Farming System Research, Modipuram, Meerut, U. P. for financial support and guidance received to carry out the research programme.

REFERENCES

Baishya,A.,Hazarika,J.P,Medhi,B.K,Gogoi,J.K.,Kalita,M.C.and Ahmed,S.2009. Constraints and opportunities of crop diversification in Assam. *Indian Journal of Agronomy,* 54(2):200-205.

Baishya,A.,Hazarika,J.P., Medhi,B.K.and Kalita,M.C. 2011.Zone 2.Eastern Himalayan Region. Efficient Alternative Cropping Systems edited by B.Gangwar and Anil Kumar Singh.Publish by Project Directorate for Farming Systems Research (ICAR).Modipuram,Meerut,U.P:32-40.Chand, Ramesh and Chauhan, Sonia (2002). Socio-economic factors in agricultural diversification in India. Agricultural situation in India. February, 2002:523 – 529.

Chand,Ramesh and Chauhan,Sonia .2002.Socio-economic factors in agricultural diversification in India. *Agricultural situation in India,* February,:523-529.

Gill,M.S.and Ahlawat,I.P.S.2006.Crop diversification – its role towards sustainability and profitability. *Indian Journal of Fertilizers,*2(9):125-138,150.

Hedge,D.M.,Prakash Tiwari,S.and Rai,M.2003.Crop diversification in Indian Agriculture. *Agricultural Situation in India,* August :255-272.

Palaniappan, S. P. and Sivaraman, K. 1994. Cropping sequence in the tropics. Principles and Management. 211 p. New Age International Limited, Publishers, New Delhi, India.

Reddy,B.N.and Suresh,G.2009.Crop diversification with oilseed crops for maximizing productivity, profitability and resource conservation. *Indian Journal of Agronomy,*54 (2):206-214.

Saha, S. and Moharana, M. 2007. Production potential and economics of different rice based relay cropping systems under rainfed shallow lowlands of coastal Orissa. *Oryza,* 44 (2): 134 – 36.

Sharma, R. P., Pathak, S. K., Haque, M. and Lal Manser.2008. Productivity, Profitability and nutrient balance as influenced by diversification of Rice (*Oryza sativa*) - Wheat (*Triticum aestivum*) cropping system. *Indian Journal of Agronomy,* 53 (2): 97 - 101.

Singh, S. K.,Kumar, D. and Lal, S. S. 2010. Integrated use of crop residues and fertilizers for sustainability of potato (*Solanum tuberosum*) based cropping system in Bihar. *Indian Journal of Agronomy,* 55 (3): 203 - 208.

□□□

Section - V

Chapter 62

Watershed Management Programmes for Food and Environmental Security : An Experience from the Odisha State in Eastern Region

G.B. REDDY

INTRODUCTION

Odisha possesses a varied physiography due to its rather peculiar geographical location and wide range of physical features. The extensive ranges of hilly forests, several lofty peaks, rolling uplands, long stretch of coastline, extensive riverine system, brackish waters, coastal mangroves, and coastal plains together have endowed the state with a wide range of ecological habitats for a diverse and broad spectrum of flora and fauna. The distinctive aspect, which clearly differentiates the agro-ecological conditions of the four zones, is the topographical feature of land. It is significant to note that for the state as a whole, highland constitutes as much as 45.8 % of the cultivated area. This is significantly higher in the Eastern Ghats region (64.2 %). In the Northern Plateau and Central Table Land, highland constitutes half of the cultivated area. On the other hand, in the Coastal Plains, highland constitutes 29.2 % and lowland constitutes 36 % of the cultivated area. The intrinsic fertility of the residual soil of the highlands is low, since these are predominantly lateritic, highly leached, acidic, and low in nitrogen and organic matter. They are also shallow, sandy, and susceptible to erosion. Because of the porous nature of the highland soil and its topography, its moisture retention capacity is rather low. Along with poor intrinsic fertility, this makes the highlands suitable only for low water-intensive crops. The transported alluvial soil of the lowlands is moderately fertile, less acidic, more productive than the residual soil, and has a higher moisture retention capacity. It is, thus, suitable for paddy cultivation. Highland Odisha is drained by a number of major rivers, which have given rise to fertile river valleys suitable for paddy cultivation. The state receives about 1500 mm (60 inches) of rainfall normally,

Director, Odisha Watershed Development Mission (WDM), Bhubaneswar, Odisha, India with a variability of 25–30 %. About 77 % of the rainfall comes from the south-west monsoon (June to September). However, the south-western districts of Kalahandi, Balangir, and Koraput fall in the rain shadow zone of the south-west monsoon and hence receive highly erratic rainfall. Barring the river basins, which comprise of fertile parts of the region, population density in much of the highland Odisha is quite low compared to the coastal plains and this also broadly corresponds to the level of agricultural productivity. The ecological conditions in Odisha have been conducive to large tracts of natural vegetation. Bamboo and sal forests abound in this region. Forests yield a wide range of 'minor' products that provide livelihood support to the original inhabitants of this region almost throughout the year. Beside forest resources, most of the major mineral deposits are also found in this region

The trend in *effective* forest cover (*i.e.*, closed forest area having crown density of more than 40 % of total geographical area) as against *nominal* forest cover (*i.e.*, total forest land as per cent of total geographical area) can be seen from the satellite data on forest cover, available from the early 1970s. In 1972–75, the *effective* forest cover was only 24 %. In the course of the next 24 years or so, the nominal forest cover remained constant at around 30 %, but the *effective* forest cover dwindled to less than 17 %. This is due to the unchecked degradation of forest cover. This has resulted in a decrease in the closed forest area (as per cent of total forest area) from 77 % during 1972–75 to 30 % by 1996. At the level of districts, it is only in the districts of Mayurbhanj and Kandhamal that the *effective* forest cover has remained at around 30 %. Increased degradation of forests has resulted in increased vulnerability to droughts and floods. The land is more prone to droughts due to the increased moisture stress as a result of more rapid run-off of rainwater. On the other hand, due to increased sediment load, the delta channels are unable to discharge the floodwaters to the sea, thus resulting in floods. Moreover, drainage congestion in the coastal deltas renders large areas of land waterlogged throughout the year.

Watershed Context

Watershed is defined as a geohydrological unit which drains its runoff through a common outlet. All the natural resource based interventions in watersheds are governed by the hydrodynamics of water. A manageable watershed is taken as a unit of approximately 500 ha. It constitutes land mass, water resources and vegetation. Therefore, successful watershed development lies with effective management of soil, water and vegetation.

Human resource development also critically affects the management of watersheds. Natural resource development and management is the major component in any watershed programme. It essentially relates to soil and water conservation in the watershed which means proper land use, protecting land against all forms of deterioration, building and maintaining soil fertility, conserving water for farm use, proper management of local water for drainage, flood protection and sediment reduction and increasing productivity from all land uses.

Objectives of Watershed Development

The main objectives of such projects are

1. To promote the economic development of the village community which is directly or indirectly dependent on the watershed through optimum utilization of the watershed's natural resources like land, water, livestock, vegetation, *etc.*, that will mitigate the adverse effects of drought and prevent further ecological degradation.
2. To encourage restoration of ecological balance in the village through:-
   - Sustained community action for the operation and maintenance of assets created and further development of the potential of the natural resources in the watershed.
   - Simple, easy and affordable technological solutions and institutional arrangements that make use of, and build upon, local technical knowledge and available materials.
3. To improve the economic and social condition of the resource-poor and the disadvantaged sections of the Watershed Community such as the asset less and the women through:-
   - Distribution of the benefits of land and water resources development and the consequent biomass production in an equitable and gender sensitive manner.
   - Strengthening the relationship between communities and watershed resources and formalizing their entitlements over common property resources
4. To enhancing the livelihoods options for poor and diversifying the same through
   - Establishing their institutions for development, use and sustainable management of watersheds
   - Developing necessary capacities (knowledge, skills and attitudes) and financial capital of poor members of different watershed based institutions.

- Facilitating convergence of several programs, opportunities at watershed level

5. To reduce regional disparity between irrigated and rainfed areas by
   - Enhancement of agricultural productivity and production in a sustainable manner
   - Enhancement of food security and incomes from sustainable farming practices and use of natural resources
   - Greater access to income generating opportunities and focus on their human resources development.
   - Employment generation and promotion of savings & credit

Reversing Land and Water Degradation in Watershed for Ecological Security

An important part of the watershed is the drainage system and its associated gullies, which drain the land area from the higher elevations into the stream or river outlet. Together with land stabilization treatments, attention to the drainage gully system will also be required both to stabilize the drainage lines within the watershed and to improve downstream water quality by protection from sediment deposition. In many of the areas, drainage lines are deeply gullied and are extending into both the arable and non-arable areas. To tackle the problem of gullies and development of water resources different interventions are taken up in all the three reaches of the drainage lines like Obstruct the flow of water through live check dams, brush wood dams, loose boulder structures etc in upper ridges. Management interventions for medium reaches are, construction of sunken structure on or along the drainage line, stabilize banks primarily with vegetative measures.Technogical interventions for the lower reaches are, stabilize banks with vegetative and structural measures wherever needed, construct bigger dugouts in natural depressions preferably along the drainage line, Series of runoff management structures to store water for recharge of ground water and Construct a series of recharge wells near dugouts to serve as artificial cavities to store water under ground.

Negative trends in resource degradation were a big challenge that needed to be tackled to meet the desired result of watershed project. Generally it was found that poor farmers tend to be associated with marginal lands and low yields. Land degradation and water scarcity were key factors limiting food production and wealth generation for poor people. Efforts to slow or reverse land and water degradation were the biggest challenges. About 80 % of the agricultural land was rainfed. Uncertainty of rainfall and poor socioeconomic conditions of the farmers living in these regions prevent them from making heavy investments in rainfed agriculture, resulting in poor return. In-situ and ex-situ conservation of rainwater through a range of innovative techniques and enhancing rainwater-use efficiency through supplementary irrigation has lead to the creation of a number of bright spots for development.

Environmental Implications of these Interventions

The watershed development interventions result in increased soil moisture in the arable land. Thus the crop loss is minimized due to moisture stress. The increased moisture availability also allows the farmers to take a pre *Rabi* crop like pulses in the residual moisture. The in-situ soil moisture management also increases the soil fertility and thus positively impact on the agricultural productivity.

- Due to soil and water conservation measures the surface water spread area is increased. This reduces the volume of runoff at the down stream and minimizes down stream hazards in terms of sand casting, flooding and crop loss. The stored water is also used for supplemental irrigation in water scarce condition.
- Water resource is used not only for crops but also for drinking by human beings as well as livestock. Hence, under watershed programme adequate attention is given to meet drinking water needs on a priority basis. Specific water harvesting structures for drinking water by animals may be considered in ridge areas, waste land *etc.*, where animals go for grazing
- The water bodies so generated are also put into optimal water use by pisciculture. The productivity of water is also increased by adopting low duty crops.
- Ground water table in the watershed is increased due soil and water conservation measures in the watershed. The ground water is exploited by construction of wells and tube wells. This water is used for irrigation of crops. Thus the cropping intensity and productivity of the land is increased.
- Due to increase in soil moisture regime and better water availability the crop canopy and the vegetation status in watershed is increased. This helps in maintaining the natural resource base in the watershed. Thus the ecosystem is positively impacted.
- Because of improved natural resource base, the natural resource based livelihood opportunities increases for the watershed dwellers. Thus, livelihoods of the community are increased.

Ensuring Food Security in Watershed Programme

The elimination of hunger and food insecurity has been the focus of global and national debates, especially since the 1990s. At the World Food Summit held at Rome in 1996, food security was defined as "access by all people at all times to enough quantities of nutritionally adequate and safe food for an active and healthy life". In the context of Orissa, a combination of economic, social, ecological, and institutional factors contributes to food insecurity. A high level of income poverty, a large tribal population living in remote areas with poor connectivity, and periodic recurrence of drought and floods (sometimes simultaneously in different parts of the state) give rise to a situation of chronic and endemic food insecurity. Taking into account food availability, food access, and food absorption, represented by 19 indicators, Orissa has been put in the category of 'severely food insecure' regions. It has been pointed out that severe food insecurity in Orissa is primarily due to the presence of a vulnerable rural population with poor livelihood access or livelihood susceptible to natural disasters.

Food Security is a state of assuring the physical availability and economic accessibility to enough food (in an environmentally and socially sustainable manner) in terms of quantity (safe, nutritious, balanced), quality (amount, distribution, calories) and cultural acceptability for all people at all times for a healthy and active life. The various measures implemented through the integrated watershed management program, particularly improving the sustainability of groundwater source, have improved the food, fodder and fuel security over a period of time. The participatory watershed management and convergence approach enhanced the agricultural productivity, food security and incomes, decreased poverty of rural poor, reduced labour migration and improved environmental quality.

Interventions for Ensuring Food Security

Reduction in the level of food in security is a manifestation of the multiple factors viz. enhanced coping capacities, agricultural production, diversion of livelihood activities and access to consumption credit (Impact Assessment Report of WORLP). One of the key mandates of watershed programme is targeting the poorest of the poor in the project and improving their livelihoods. Food security can be achieved through interventions like formation of grain banks, growing cash crops, increasing productivity agricultural crop, promoting livestock, managing common property resources, promoting back yard horticulture in the watershed areas.

Food Security has been possible for the poorest of the poor only because of use of appropriate technologies for both progressive and resource poor farmers with availability of inputs at their door steps at affordable prices. Irrigation coverage through water harvesting structures and check dams, financial support through revolving fund and grant funds also helped in boosting the agriculture sector in the project. Further, all this could be possible through involvement of all the stake holders and providing appropriate capacity building inputs.

Grain banks for food security in the watershed villages : Community led grain banks are symbol of community preparedness and long term measures to tackle food insecurity. Grain Banks help address local food insecurity, particularly the problem of transient hunger during lean periods and against starvation during natural calamities. The members borrow grains from the banks and return it with interest. More often the community managed grain banks are comparatively faster than to respond to local food shortages than the centralized food security programmes. During the month of September maximum stock is available and later the women borrow it for local festivities.

Grain banks have developed over time, moving from a simple food security tool to a more comprehensive self-help mechanism for the economic and social development of the rural community. Strategic planning and effective management, involving the local community and especially SHGs has paved path for its sustainability and success. The table below reflects the details of grain banks operational in the WORLP districts:

Details of grain banks in WORLP districts

Name of the District	No.of Grain banks	No.of Grain Banks managed by WSHGs	Types of Grain	Total Qty (In Qts.)
Kalahandi	167	162	Paddy, Pulses,Rice	1360
Bolangir	73	18	Paddy, Rice	123
Bargarh	19	03	Paddy, Rice	165
Nuapada	174	174	Paddy, Rice	419
Total	**433**	**357**		**2067**

Cash Cropping for Food Security

One particular manifestation of commercialization in watershed area is cash cropping. While commercialization in watershed programme as a result of increase in water availability can include cash cropping involves crops like Maize, Sunflower and vegetables produced for cash that have a higher value than those consumed for food within the household. Concerns have long been expressed that the expansion of cash cropping amongst smallholders could undermine their food security. It has been witnessed that cash crops that increased cash incomes resulted in better nutrition. cash crops into farming systems was generally have a positive impact on household-level food security when cash crops are farmed in combination with marketable food and it generates regular rather than periodic cash flows with early rather than delayed income flows.

Increase in productivity for food security : The most effective strategy for ensuring that the poorest people have food supplies in hand is to aim for sustainable increases of agricultural productivity on their land holdings. Key components of such a strategy is capacity building of farmers on new technologies of crop production; ensure supply of quality inputs, promotion of organic farming. All those interventions has resulted in increase in productivity of important food crops in watershed areas which directly benefitted the small farmers in meeting their food requirement.

Food security through livestock promotion : Livestock is the backbone of people's economy and many cultural practices revolve around it. In rural areas, households rely on the sale of livestock to meet their emergency expenses like medical care, veterinary services and purchase of other household consumables. The nutritional benefits for the community also of great importance. Livestock forms a big proportion of people's assets and helps in meeting their long term financial needs. To improve the viability of livestock sector *i.e.*, to decrease vulnerabilities various initiatives are taken in watershed programme like provision of improved buck for breed improvement, regular preventive vaccination to reduce mortality.

Availability of Food through Backyard Horticulture : Backyard fruit trees are a prime source of nutrition to the family. As Entry Point Activity backyard fruit trees plantation was promoted by providing papaya, drumstick, lemon and banana trees to each family. This activity apart from meeting their family's nutritional requirement also provided additional income through sale of surplus production.

Common Property Resources Management – An Important Asset for Food Security

CPRs represent a form of natural and social capital that individuals and communities can draw on in pursuit of their livelihood strategies including addressing food security. They form a part of rural peoples' strategies for adjusting to the harsh and stressful environment. The role of CPRs is particularly important in lean times, acting as a buffer in times of drought or other crises. During droughts, the collection of tubers and wild foods from the forest is an important coping mechanism. CPRs also provide independent sources of income to women, through, for instance, goat rearing, broom making, leaf plate making and collection of tendu leaves and tamarind.

CONCLUSIONS

Watershed management in the country has expanded extensively. Over-exploitation of the water resource in most parts of the country has led to a rapid decline in the groundwater table. This has threatened not only the food security and environment, but also the sustainable development. Further depletion of groundwater resource has been affecting the small and marginal farmers the most, threatening their livelihood in many cases. The sustainability of groundwater use is one of the core areas, which requires major attention for meeting the water requirement and ensuring food security. An important way of addressing the issue is by augmenting the groundwater supplies in the shallow aquifers on micro watershed basis through groundwater recharging and rainwater harvesting system. Our experience from community watersheds showed that recharging can be made much more effective by the use of scientific inputs and analysis than otherwise.

Sustainable groundwater development and management in the overexploited regions needs to be taken up by incorporating artificial recharge to groundwater from in-situ and ex-situ rainwater harvesting through integrated watershed interventions, management of salinity ingress in coastal aquifers, conjunctive use of surface- and groundwater, management of poor/ marginal quality groundwater, water conservation by increasing water-use efficiency, regulation of groundwater development and extraction, etc. Several studies conducted in the community watersheds through integrated watershed management approach have concluded that these technologies have been successful in the sustainable development; and the management of groundwater resource would be the key to achieve breakthrough in agricultural production and food security.

REFERENCES

Alderman, H., Hoddinott, J. and Kinsey, B.2003. Long-term consequences of early childhood malnutrition. IFPRI Discussion Paper 168. International Food Policy Research Institute: Washington DC, USA.

Arya, V. 1998. Collaboration for Innovation in Participatory Natural Resources Management: The PAHAL Project in Dungarpur District. Srijan, New Delhi.

Ashley, C. and Maxwell, S. 2001. Rethinking rural development. *Development Policy Review*, 19 (4): 395–425.

Azad, N.1996. Engendered Mobilization – the Key to Livelihood Security: IFAD's experience in South Asia. International Fund for Agricultural Development: Rome, Italy.

Behrman, J.R., Alderman, H. and Hoddinot, J. 2004. Hunger and malnutrition. Paper for the Copenhagen Consensus – Challenges & Opportunities, 19 February 2004.

Binswanger, H.P. 1986. Agricultural mechanisation: a comparative historical perspective. *The World Bank Research Observer*, 1: 27–56.

Honore, G.1997. Indo-German bilateral project 'watershed management'. Strategy and approach. GTZ, New Delhi.

Human Development Report, Orissa 2004

Impact Assessment Report WORLP (Western Orissa Rural Livelihood Project)-2009-10

Suhas P. Wani, Raghavendra Sudi and Pathak, P. 2005. Sustainable Groundwater Development through Integrated Watershed Management for Food Security, International Crops Research institute for the Semi Arid Tropics (ICRISAT).

❑❑❑

Chapter 63

Resource Conservation Measures for Sustainable Agriculture in NEH Region

K. K. SATAPATHY

INTRODUCTION

Soil and water are amongst the most important natural resources within the ecosystem. Our existence depends largely upon thin layer of soil from which we produce most of our food, fibres and timber. However, it has been estimated that millions of hectares of cultivated land are lost each year because of land degradation. In a general way, land degradation can be defined as a loss of land productivity quantitatively through various processes such as erosion by water or wind, salinization, water logging, depletion of nutrients, deterioration of soil structure, pollution and the like. Historians often relate the vanishing of civilization to loss of productivity of the lands that once supported them. The costs associated with loss of soil nutrients and productivity, and off-farm site costs consisting of problems that are exported off the farmer's land like gully erosion, sedimentation of water ways and reservoirs, contamination of drinking water and so on. Population growth has put, tremendous pressure on the land, denying it the natural inputs of time and space to regenerate.

The North Eastern Hills region of India is prone to high rainfall, low solar radiation, low temperature with undulating hilly terrain having varying altitudes and slope, Shifting cultivation, excessive exploitation of forest wealth, expansion of development activities have affected land and, water resources. The devastation caused to the forest cover leading loss of valuable biodiversity of the region, faulty land use systems leading soil erosion in the hills and flood in the plains, open mining and quarries coupled with rapid unplanned urbanization are some of the management factors responsible for depletion

Director, Natonal Institute of Research on Jute & Allied Fibre Technology, Kolkata – 700 040, West Bengal of flora and fauna and also creating vast areas into unproductive wastelands. Over the past few decades, the North Eastern Hill regions have experienced rapid population growth, which has resulted in expanding cultivation on to marginal lands. Given the remoteness, lack of infrastructure, and difficult topography, questions have been raised as to whether intensive cropping systems are sustainable given the naturally high rates of erosion and the relatively low availability of inputs. It's being increasingly felt that the human use of natural resources exceeds the natural carrying or regeneration capacity and the questions of its adverse impact is being raised. The farmers in the area have adapted their farming systems to these difficult conditions for generations but questions about long term sustainability of different production practices often arises. Resource degradation depleting forest cover and loss of biodiversity are the issues of greatest concern for sustainable development of agriculture in the region. Since the situation is practically irreversible, it would be prudent to adopt a strategy and approach for optimal and sustainable use of natural resources in the region.

Land Use and Soil Erosion

High rainfall and steep hill topography is always associated with problem of severe soil erosion, particularly when the land use systems have biotic interference. Quantitive facts on soil erosion hazards fro soil erosion studies in the region indicate that except forest land use, none of the land uses are safe and lead to land degradation process (Table 1). Micro plot studies on bare fallow slope of 15, 30, 45, 60, 75 and 100 % revealed the soil loss and runoff varied from 13, 15, 10, 21.54 t ha^{-1} yr^{-1} and 18.49 to 26.82% of rainfall, respectively (Singh, 1985). This study also indicates the erosion and runoff increase up to 45% slope, thereafter it decreased linearly with further increase in the slope. Awasthi (1984) reported 17.80 to 205.6 t $ha^{-1}yr^{-1}$ soil loss from 65% slope (are fallow). In hills, development of new sites for building construction, development of roads and area under kitchen gardening are major sources of soil erosion. In a case study, Singh *et al.*, (1981) reported 16.8 t ha^{-1} yr^{-1} soil loss from 6.9 ha hilly watershed under homestead land use. The hydrographic surveys of the Umiam reservoir (Meghalaya) conducted during January 1990, revealed a sediment deposition of 11703 ac. ft during the last 25 years, giving an annual rate of sedimentation of 26.05 ha. m per 100 km^2 of catchment area (Anon., 1990). Biotic interferences in jungle cutting, burning, clearing and dibbling of seeds in shifting cultivation, accounts for a considerable amount of loss of soil material. The soil erosion from hill slopes (60-70%) under first, second year and abandoned jhum was estimated to be 147,170 and 30 t $ha^{-1}yr^{-1}$ (Singh, 1981). Its estimated that 181 M tones of soil is lost every year from this region due to this defective system of farming (Prasad *et al.*, 1981).

Table 1: Soil erosion hazards associated with various land use practices in NEH Region

Land use system/practices	Experimental plot size	Soil loss (t^{-1} ha^{-1} yr^{-1})	Average annual rainfall (mm)
Shifting cultivation	Small	30.2-170.2	1600
Shifting cultivation	Field	5.1-83.3	1600
Tuber crops on raised bed (bun)	Medium	40.0-50.0	1800
Pineapple cultivation along the slope (first 2 years)	Small	24.0-62.6	1600
Homestead areas	Field	16.8	1600
Mixed crop of maize and rice	Small	19.7-21.0	1600
Rice crop on slope	Small	32.9-45.0	1600
Bare fallow	Small	83.8	1600
Cropping systems	Medium	51.0-83.8	1600
Grass cover (planted)	Medium	10.83	1600
Natural bamboo forest	Field	0.04-0.52	1600

Note: Area of small, medium and field size plots were in ranges of 2-5, 16.40 and 69,000 m^2, respectively.
(Source: Prasad, 1986)

Soil Erodibility and Crop Management Factors

In North Eastern Himalayas, according to land capability classification, there is no class-I and II land, little class III and some class IV land. As a consequence large areas of class VI and class VII lands are being used for cultivation. Since only irrigated land is properly bench terraced, rest of the land is grossly mismanaged. No comprehensive and qualitative measurement of soil erodibility (erosion per annum per unit of the erosivity factor R) is done in the region. Awasthi (1984) however, determined soil erodibility factor at Byrnihat (Meghalaya) from uncultivated bare fallow plot at 65% slope. The average soil erodibility factor was worked out to b 5 kg ha^{-1} per unit of rainfall. He also calculated crop management factor C-combined influence of vegetative cover and the various cultural practices on soil erosion used in the universal soil loss equation. The grass cover offered maximum protection with a very low 'C' value (0.133). For mixed, sole cropping of rice and topo-sequential cropping the 'C' value range from 0.568 to 0.644 for widely spaced crop of maize the 'C' value range ranged from 0.857 to 1.0 under different management practices.

Land Capability Classification and Selection of Land Use Systems

Land capability classification is widely employed as the basis for land use planning; land is graded into a number of capability classes usually I-VIII on the basis of its inherent limitations of erosion, wetness, soil and climate, and so on. Class I-IV are arable-suitable for rainfed arable use; class V is applied to special situations such as wet valley floors whilst class VI to VII are non arable and considered suitable for grazing, forestry and the like. According to these classifications, in North Eastern Himalayas there is no class I and II land, little class III land and some class IV land. As a consequence large areas of class VI and class VII lands (to be categorized as non arable classes) are used for cultivation in the region and therefore, the concept comes into conflict with the current land use. Areas of sloping land are already being used to grow subsistence food crops, and families and sometimes whole communities are dependent upon the produce. It would be socially undesirable to attempt to change this situation. Owing to the typical soil and climatic conditions in the region, a modified land capability classification as mentioned below have been tried out (Table 2) in the North Eastern Region (Borthakur, 1992).

Table 2: Modified land capability classification

Class A-1	slopes, 0-5%	Suitable for cultivation without special soil conservation measures.
Class A-2	slopes, 6-50%	Suitable for cultivation with special conservation measures
Class B-1	slopes, 51-100%	With shallow soil depth (less than 1.75 mm), suitable for pasture and fodder.
Class B-2	slopes, 51-100%	Having soil depth of more than 1.75 mm, suitable for orchards, cash crops, plantation crops *etc.*
Class C	slopes, over 100%	Suitable for forest

(Source: Borthakur, 1992)

The major factors deciding a land use in hills are land slope, soil depth, extent of land degradation, agro-climatic conditions and people's participation. The development of land use systems should be based on watershed approach for resource conservation point of view. Among the factors responsible for determining the land use systems, people's participation is a persuasive factor and agro-climatic conditions is only to decide the nature of agricultural crops, trees or grasses.

Soil Conservation Measures

Agronomic measures: Maintaining a dense canopy cover is the most effective biological means for minimizing soil and water losses on hill slopes. Growing of two or more crops in mixed, inter or strip cropping systems provide better canopy cover and obstruct overland flow. Dense foliage of erosion resistant leguminous crops like cowpea and soybean reduces soil loss by preventing the rains from beating the soil surface desirably. Various options relating to the management of crop canopy are yet to be evaluated for NEH region. Some studies carried out on mixed cropping of maize with rice revealed that runoff and soil loss was reduced significantly compared with maize along the slope. Rice and maize based intercropping systems such as rice-groundnut/ soybean in 3:1 row ratio and maize + groundnut / soybean in 1:1 row ratio performed well and resulted in higher net returns in upland mid altitude of Meghalaya and Arunachal Pradesh. If arable cropping is unavoidable on steep lands due to socio-cultural factors, the grass cover should be integrated with arable cropping systems in strips at suitable intervals to minimize runoff and soil erosion.

Contour trenches: These are trenches excavated along the contours to break the slope length for reducing the velocity of surface runoff, the water retained in the trenches help in conserving the moisture. The size of trenches depends upon the soil depth available and its cross section may vary from 100 sq. cm to 2500 sq. cm and are designed according to the rainfall to be retained per unit area. The new contour trench farming system with contour trenching as mechanical conservation measure was initiated under watershed development programmes in the region. The trenches are dug by cutting a trench of 30 cm deep and 45 cm wide along the contour at an interval varying from 20-50 m according to slope percent. The trench interval is longer on the lower slope and smaller on the higher slope. The trench is dug manually and excavated soil from the trench is dumped on the down stream along the trench to form a bund on which grasses, preferably fodder grasses with strong root formation are grown, and to reinforce the trench, banana and sugarcane or pineapple are planted in few rows just below the bund. The uncut area is left to prevent the trench from turning into undesirable gully under onslaught of the run off contour trenching is adopted in *jhum* land practiced on the lower edges of the slope by leaving the higher slope for horticulture and forestry. The function of the contour trench is to break the slope and convert surface water into subsurface water. Topsoil removed due to cultural operations gets accumulated with decayed biomass in the trench, which are scooped out' from the trench and utilized as compost manure to the crops. The new contour trench farming system conforms to the principles of soil and water conservation to conserve rain water in- site and stop it attain erosive velocity leading to severe erosion.

Erosion permitting crops like tapioca, sweet potatoes *etc.*, can safely be cultivated in the new farming system within allowable soil loss (Thansanga, 1997). The digging of trenching involves lesser manpower and expenditure than terracing and can easily be maintained by the members of cultivators' family. Adoption of trenching has increased the net return appreciably from both cases of settled cultivation which will ultimately contribute to replacement of shifting cultivation (Table 3 & 4).

Table 3: Cost of digging and area lost for cultivation due to contour trenching

S.No	Slope group (%)	Length of trench line (m)	Cost of trenching (Rs. ha^{-1})	Area lost per ha (%)
1	0.10	200	2,000	0.90
2	10.30	250	2,500	1.12
3	30.50	334	3,400	1.50
4	>50	500	5,000	2.25

Table 4: Economic analysis of different agriculture practices

Practice	Total revenue (Rs ha^{-1})	Total cost (Rs ha^{-1})	Net revenue (Rs ha^{-1})	Input-output ratio
Shifting cultivation	19,924	16,895	+3,029	1:1.18
Agriculture with trenching	29,176	17,640	+11,536	1:1.65
Agroforestry with trenching	71,100	36,630	+34,470	1:1.94

***Contour bunds*:** These are small embankments or bunds constructed across the slope to decrease the slope length, which reduces soil erosion and diverts the excess runoff to the designed outlet. The eroded soil is retained within the bund interspaces, which get levelled up in the course of 4 to 8 years to form bench terrace. These bunds on steep slopes are created by way of excavated parabolic channels (0.3m top and 0.2m deep) on contours and keeping the dugout soil in the form of a bund at the lower edge of channel. The vertical interval of these bunds may vary from 0.5 to 1.5m depending on the land use and soil depth.

***Bench terraces*:** Bench terraces are series of flat beds constructed across the hill slope separated at regular intervals in a step like formation. Manual labour as well as bulldozer can be engaged to form bench terraces. Bench terraces with inward slopes are adopted in the high rainfall areas. The alignment of bench terraces on slopes should be made to obtain convenient width making deviations wherever necessary for depressions; sharp turn's field boundaries and so forth. However, the loss of surface area due to bench leveling under bunds, risers' outlets is by far the largest loss in terracing. Such measures are normally adopted where soil depth is more than 1.0 m. Terracing of the entire hill slope is not necessary since trees and horticultural crops can be raised without terraces. Only the lower portion of the hills need be terraced for agricultural crops. The terrace risers, which constitute 30-40 % of total area, can be utilised for growing perennial fodder grasses and legumes, which not only help in conservation but also provides enough fodder.

Terracing steep hill slopes: Cultivation on steep slopes in high rainfall areas with out protection measure often causes serious watershed problems. By applying terracing and protected waterways, these steep slopes could be cultivated safely and profitably. Though the most prevalent agricultural practice in the hills is shifting cultivation, there are several pockets which have some kind of irrigation facilities where excellent terrace cultivation is practiced. Under *jhum* control programme, allotment of wetland terraces with assured irrigation has been the most effective means of attracting *jhumias* to switch over to settled agriculture. On the other hand, instances are available of destruction of upland terraces where cultivation under rainfed conditions is practiced. High cost of terracing, loss of area, fall in crop yield during initial years, absence of appropriate production technology for rainfed benches- these appear to be the major constraints in promoting the idea of terrace cultivation among tribal farmers. Studies in ICAR however, have demonstrated that use of the latest technology with bench terrace as conservation base can provide a stable alternative to shifting cultivation.

The small farmers in this region, however, finds difficulty to identify themselves with energy intensive and technology rich methods of terracing such as use of heavy earth cutting equipment curving out bench terraces. In ICAR, a relatively cheaper method for constructing bench terrace has been developed with the use of contour bunds which may be easily acceptable by farmers. Contour bunds on steep slopes are created by way of excavating parabolic channels (3 m top and 2 m deep) on contour and keeping the dug out soil in the form of a bund. The purpose is to divert excess runoff during rains and retain the soil within the bunds. Row crops such as maize *etc.*, can be grown along the slope in the inter space between the bunds. The trench created at the upper edge of lower terraces gets leveled up during tillage operations. Bunds are also to be re-strengthened in the same process after every crop and the trench should illuminated during the tillage/ cultivation practices of every crop. The areas between the two bunds get leveled up and finally take the shape of a terrace in about 4-8 years. In an experiment in the ICAR, after cultivating 3 crops, slope of terrace surface to average 7.8% from original 28.8%.

Bench leveling of paddy lands results in reduction of surface area for cultivation under bunds, risers and channels. In NEH region, bench terrace cultivation exists in slopes up to 100 % having terrace width as low as 3 m and situations on steep hills normally do not permit of longer length of terrace. Level beds with vertical interval of about 1m along with 15 sq.m cross section shoulder bunds and 1:1 batter slope for terrace risers was found satisfactory for movement on the slope for performing various agricultural operations (Prasad *et al.,* 1987).

Bio terracing : Bench terracing although universally acknowledge as an effective soil conservation measure for settled cultivation on the hill slopes is yet to be popularised in the eastern Himalayan region. High cost of construction, decline in soil fertility due to top soil fertility into low crop yields in the initial years and often improper shoulder bunds causing devastating soil erosion are the major constrains in creating bench terraces by cut and fill method. But the application of contour hedgerow technology was able to eliminate many of these constraints. .

Table 5: Rate of slope reduction on terraces under process of terracing by gradual development at Basar, Arunachal Pradesh

Terrace	Original slope	Final slope (%)	Slope reduction crops (%)	No. of slope	Period of operation upto Feb,91 (months)
T-1	27.9	8.5	69.5	4	18
T-2	32.9	6.0	81.7	3	18
T-3	29.8	5.7	80.8	4	17
T-4	26.0	9.6	63.0	3	16
T-5	27.5	9.3	66.2	3	15
Average	28.8	7.8	72.2	3	16.8

(*Source*: Dutta ***et al.***, 1991)

A typical design of two systems of hedgerow entails the construction of hill side ditches and vegetative barrier by planting of a leguminous hedge *(Erythrina indica)* in one system (Bio terracing -I), and in another system a strong, tugged broom grass *(Thysanolaena maxima)* (Bio terracing-II), on the bund (rise) of the ditches which followed contour line of the hill side to create natural terraces through soil redeposition process. The contour hedge rows of both the plant species can function effectively as soil barriers from the second year of planting and their efficacy in controlling soil loss beyond the contour bunded area increases as the hedge plant grew with time. The hedge row provides mechanical and biological barrier that helped to minimize soil erosion by reducing runoff.

A case study on the performance of bio terracing *vis-a-vis* cut & filled method has been present in Table 6. During the five years of study, the original slope reduced annually at the rate of 4 % under bio terracing and contour bunds without hedge. Thus, a hill slope of 28 to 38 % can be terraced within 5-7 years through the application of hedge row technology. The bio terracing is advantageous over cut and fill method of terracing in terms of cost effectiveness. Bio terracing required 75 to 80 mandays as against 497 mandays under cut and fill method. Contour hedge row technology provides an option for farming on the hill slopes on a sustainable basis. (Singh *et al.*, 1981a).

Table 6: Changes in slope, mandays, requirements, cost, crop yield and biomass productivity under different systems of terracing

System of terracing	Original slope (%)	Slope in fifth year	Mandays required ha^{-1} (cost in ₹ ha^{-1})	Crop Productivity q ha^{-1})			Total biomass (q ha^{-1})
				Maize	Ginger	Turmeric	
Cut and fill method of terracing	38.2	-	497	26.55	81.17	98.32	448.7
Bio terracing-I	27.2	7.3	75 (3,750)	26.74 (12.29-34.40)	89.11	76.29	176.5
Bio terracing-II	28.2	8.03	85 (4,250)	22.50 (11.16-24.60)	38.69	26.95	258.5
Contour bund without hedge	28.6	8.41	42 (2,100)	16.10 (10.78-20.20)	60.58	54.48	137.8

Half-moon terrace: Where complete terracing is not desired or feasible productive, multiple use of steep slopes is possible by planting tree samplings/ plants on half-moon terraces. These are level circular beds having 1 to 1.5 m diameters, cut into half moon shape on the hill slope. These beds are used mainly for planting and maintaining. These beds are used mainly for fruit trees like guava, citrus *etc*., in horticultural land uses.

Vegetative bunds: Barrier hedges substantially reduce runoff and increases infiltration. Some of runoff may cross the barrier, whilst the entrained soil will be partly filtered out and deposited. Pineapple plantation has been successfully used as vegetative bunds in hills.

Grassed waterway: These are trapezoidal or parabolic channels planted with suitable close growing grasses constructed along the slope preferably on natural drainage line to act as an outlet for the terrace system. These channels along with silting basins serve the purpose of energy dissipation of flowing runoff water.

Revegetation of badly eroded lands using geojute: Landslides in road risers, agricultural land and sides of meandering streams in hills areas are a common and destructive phenomenon especially during rainy season. Steep slopes, dwindling vegetation, unstable soil mass over a rock with heavy rainfall make the hill slopes prone to landslides. Bringing such degraded areas under permanent vegetation cover is the only possible solution to this problem. But development of permanent vegetative is not possible because of high runoff on such slopes do not allow the plant to stand for the required period of time to stabilize. Mechanical measures that also serve as first line defense against flow of sediments and runoff from the catchments area are invariably needed for protection of the slopes. *Geojute* matting, is a natural product which may be used for erosion control and which decomposes within the ecological cycle. With normal thickness of about 5 mm, jute mesh provides an open area of about 65% through which planting can easily be done. It's bio-degradable and rots in about two years. The main purpose of using *Geojute* matting vegetation over eroded slopes is to provide stability to the upper most layer of soft soil and the seeds sown over the slope till new vegetation are established over the area. The species such as *Arundinaria hirsute Munro (Sejnaka)*, *Ipomea carnea (Besarma)*. *Thysanolaena maxima* (Broom grass) can be also being planted for greening the area. Geojute material prevents surface erosion providing support temporarily for the new vegetation to stabilize the eroded area, up to a period of six months (one rainy season), provided, the material found suitable for use is low and medium slope group of areas where erosion is caused mainly by the surface runoff. The eroded areas can be further guarded with series of Gabion structures constructed with wire mesh and stone materials at the lower reaches of the slopes and across the waterways. The cost of *Geojute* laying would be around Rs 13.32 m^2 at 1994-95 prices.

Protecting stream banks: All rivers and streams while flowing through hilly regions have a tendency to erode away their banks. The erosion destroys productive crop fields, human habitation situated on the river margin. A cheaper way of dealing with the problem of riverbed and bank maintenance is through plantation. The stream banks can be protected by planting water loving vegetation particularly of out sprouting type. The effectiveness of engineering measures may be enhanced by several vegetative measures. ICAR research complex for ER has identified several plants/ species of out sprouting and water loving type which grow profusely near stream beds and are well suited to river bank protected schemes. Some of these species are *Salix species, Vitex negundo, Populous species Arundo donax, Ipomea carnea, Tatrpha curas, Pennisetum purpureum* (napier, *etc.*). Stream bank slopes if not subject to strong current can be well protected from erosion by a vegetal cover. Depending on the severity of the problem, vegetative and other measures may be followed as follows.

- Sloping bank the steep and vertical banks to a grade of about 1.5:1 and side by side planting of live hedges and erecting spurs at their toe.
- Live hedges shoots of *Ipomoea carnea, Arundo donax,* Hybrid napier, *Tabopha carcas* may be planted in trenches to act as retard. Tree plantation are made on the sandy river bed immediately after flood season and which grows to sufficient height next season to be able to retard the flow and aggrade contiguous area.
- Jacks (wooden posts of 6 cm dia and 2-3 m in length) placed in series along the eroding bank tied together.
- Jetted posts (green posts) of 2 to 2.5 m in length and 6 to 12 cm diameter erected along the activity eroding banks in two rows, 1.0 m apart.
- The freshly deposited silt near these spurs and behind the hedges should immediately be stabilized by planting cuttings of *Pennesetum purpurum, Ipomea* or other suitable vegetation in rows laid normally to the general direction of flow as soon as the flow subsides.
- The areas behind the banks should be afforested with suitable tree species like *Dalbergia sisco, Acacia catchu, Albizza stipulalata, etc.* The selection of species however may be finalized after consulting forest department (Sastry, 1997).
- Public awareness however needs to be created to protect the banks against cattle grazing as well as damage by human agencies.

Table 7: In-situ retention of rainfall under watershed based land use systems

Land use	Mean slope of the watershed (%)	Soil and water conservation measures	Per cent of annual rainfall retained	Soil loss (%) (t ha^{-1} yr^{-1})
Umiam, 1000 m amsl, clay loam, 2554 mm annual rainfall**				
Natural fallow	52	Trace-98.46	0.00-00.05	
Agriculture (F)	32	Contour trench	0.00-98.68	0.00-0.16
Agriculture (C)	32	Contour bunds + Bench terrace	98.53-99.77	Trace-0.33
Agri (C)-hort-silvi pastoral	32	Contour bunds	98.27-99.99	Trace-1.22
Horticulture	42	Do	96.18-99.75	Trace-4.37
Agro-forestry	33	Do	96.83-99.52	Trace-0.38
Forestry	38	-	92.79-98.27	Trace-7.80

F = Fodder based, C= Food crop

Hydrological evaluation of land use system

Studies in watershed based farming systems have shown the scope of using steep slopes for crop production. Soil and losses from different farming systems were studied in two different locations, which are summarized in Table 7. Some of the potential farming systems such as agriculture on bench terraces, horticulture, agri-horti-silvi-pastoral systems *etc.*, have been evaluated at the experimental watersheds for their long term runoff, soil losses and so on. Land use practices in micro watersheds with soil and water-conservation measures were found very effective in retention of rainfall. Rainfall retained in-situ varied between 80-100%. The surface runoff from shifting cultivation watershed remained higher in the range of 19.89 mm to 54.99 mm in a year compared to other watersheds. Also as expected the watershed treated with *jhum* (shifting) cultivation yield the highest peak runoff (86.1 mm hr^{-1}) while the one left undisturbed with natural vegetation gave the minimum peak runoff (4.49 mm hr^{-1}). However, provision of trenches in fodder based agriculture most effectively conserved moisture and produced peak runoff 7.81 mm hr^{-1}. Maximum peak flow rate did not differ in mixed block forest, silvi-pastoral system and agri-horti-silvi-pastoral system showing the effectiveness of suitable soil conservation and appropriate land use system on hill slopes. The contributions to stream flow in the watersheds having substantial area under natural forest is primarily by subsurface flow (base flow). The watersheds having continuous stream flow characteristics generated base flow to the extent of 70-90% of its total water yield. The highest base flow was obtained in the pine afforested undisturbed watershed, which constituted on an average 23.5% of annual rainfall.

CONCLUSIONS

Land resources are limited and finite. Sustainability of the land resource base has become great concern in recent years. The resource is degrading due to improper management practices and soil erosion sapping its productive capacity. To counter this downward spiral of mutual destruction of man and nature, development goals have to be tailored to meet indigenous needs. Because restoring top soil is very difficult, land degradation should be prevented before it arises; instead of attempting to cure it afterwards for sustaining the productivity of the soil resource is critically important. The key to effective soil conservation is good land management (land husbandry). The way land is managed is the result of a complex interaction between needs constraints and potential of the farmers within the environments (bio-physical and socio-economics in which they operate). Environmentally sound traditional practice attuned to low population densities have been unable to adjust rapidly enough to the decreasing land per resident ratios, resulting in practices that are environmentally damaging. Therefore soil conservation must be an integral part of agriculture development and should start with improved farming and improved production, and not be considered as a separate activity. Good land husbandry is the key to effective soil conservation. Soil conservation and agricultural production no longer should be regarded as separate activities. Land has been studied too much in soil conservation programs, where the land user (*i.e.*, farmer) has been studied too little. Therefore soil conservation activities must be bottom up rather than 'top down' in orientation and planned and implemented from the outset, with the full knowledge, cooperation and involvement of the farmers and local communities. Long term programmes are more effective than short term programs. It's important to try and link measures that will provide short term production/welfare benefits for farmers with those required for conservation purposes which are likely to have a payoff only in the long run. The acceptance of the new soil conservation approach will depend on the greater integration of soil conservation staff with farmers, researchers, extension workers and planers.

REFERENCES

Anonymous.1976. *Report on the National Commission on Agriculture.* Part XIII, Ministry of Agriculture, Govt. of India, New Delhi.

Anonymous.1987.*Basic Statistics of North Eastern Region.* North Eastern Council Secretariat, Shillong.

Anonymous.1990. Report on hydrographic surveys and sedimentation studies, Umiam Reservoir Project, Meghalaya.

Awasthi, R. P. 1984. Studies on the production potential, runoff soil and nutrient loss under different systems of cropping and management practices on steep slopy *jhum* land. Unpub. Ph.D. Thesis, Bidhan Chandra Krishi Viswa Vidyalaya, Kalyain (WB).

Awasthi, R. P., Prasad, R. N. and Chatterjee, B.N. 1990. Effect of cropping systems on runoff and soil losses from steep slopes from North East India. *Indian J.: Soil Cons.,* 10(1): 23-25.

Beasby, R.S. 1976. Contribution of subsurface flow from the upper slopes of forested watersheds to channel flow. *Soil Sci. Soc. Am. Proc.,* 40: 955-957.

Blackburn, W.H., Knight, R.W., Wood, J.C and Pearson, H.A. 1990. Stream flow and sediment loss from intensely managed forest watersheds in east Texas. *Water Resour. Bull.,* 26(3): 465-477.

Borthakur, D. N. 1992. *Agriculture of the North Eastern Region, with Special Reference to Hill Agriculture.* BEE CEE Prakashan, Guwahati.

Bundela, D.5., Satapathy, K.K,Mazumdar, B. and Jena, S.K 1999. Development of portable hydrological monitoring system to assess land degradation at micro level. National conference on watershed development-A strategy for food security, rural employment and environmental production, Bhubaneswar, Jan. 20-22, Abstracts, 6.

Crow, F.R., Ghermazien, T. and Pathak, C.S. 1983. The effect of land use parameters on runoff simulation by the VSDAHL hydrological model. *Trans. ASAE.,* 26(1): 148-152.

Das, M.M. and Dutta, L. 1986. Regional variation in landuse and agriculture in North East India. *North Eastern Geographer,* 18(1&2): 28-48.

Dutta, K.K, Devivedi, R.N. and Satapathy, K.K. 1991. Development of bench terraces using contour bunds as mechanical barrier. *Annual Report* (1990~ 91), ICAR Research Complex for NEH Region, Shillong.

Goswami, D.C. 1985. Brahmaputra river, Assam, India: Physiography, basin denudation, and cannal aggradation. *Water Resour. Res.,* 2(7): 959-978.

Jones, O.R., Cck, H.V., Smith, R.J., Coleman, G.A. and Hansor, V.L. 1985. Runoff, soil and nutrient losses from rangeland and dry farmed crop land in the southern high plains. Indian J. *Soil and Water Conser.,* 40: 161-164.

Logue, K.M. and Freeze, A. 1985. A comparison of rainfall- runoff modelling techniques on small upland catchments. *Water Resour. Res.,* 21(2): 229 -248.

Mishra, A.K., Gupta, S.K, Jena, S.K. and Agarwal, K.N. 2000. Prospects of micro irrigation in North Eastern region of India. *J. of the North Eastern Council,* 20(3): 3-13.

Mukhopadhyay, S.C 1999. Terrain characteristics of the drainage basins of North East India and their impact on water utilisation. In: *Water and Water Resources Management.* Omsons Publications, New Delhi, pp 8-19.

Neog, A.K. 1997. *Transforming Hill Agriculture in Tribal Areas of North East India. Trends in Agrian Strucutre in the Hills of North East India.* Commonwealth Publishers, New Delhi, pp 99-108.

Pathak, C.R. 1999. Water resources inventory and prospects in India. In: *Water and Water Resources Management.* Omsons Publications, New Delhi, pp1-7.

Prasad, R.D. and Sharma, V.C. 1994. Potential indigenous farming systems of North Eastern Hill Region. *Research Bulletin,* ICAR Research Complex for NEH Region, Barapani.

Prasad, R.N., Singh, A. and Verma, A. 1986. Problems of hill lands and their management in North Eastern India. *Ind. J. Soil Conser.,* 14(3): 66-72.

Prasad, R.N. 1986. Soil fertility under shifting terrace and permanent valley cultivation. In: *Land use Pattern in North East India.* Gayan publishers, Ludhiana.

Prasad, R.N.Singh, A. and Verma. A. 1987. Application of research findings for management of land and water resources in Eastern Himalayan Region. *Research Bulletin,* ICAR Research Complex for NEH Region, Shillong.

Rai, R.N. 1981. Water management is North Eastern Hill Region. *Research Bulletin* No. 14, pp15-18, ICAR Research Complex for NEH Region, Shillong, Meghalaya.

Rai, R.N. and Singh, A. 1986. Effect of hill slope on runoff, soil loss, nutrient loss and rice yield. *J. Soil Conser.,* 14(2): 1-6.

Rai, R.N. and Singh, A. 1987. Studies on topolegumes on steep hill slopes for growing rice and maize crop. *Indian J. Soil Conser.*, 15(2): 91-95.

Ramakrishnan, P.S. 1993. *Shifting Agriculture and Sustainable Development- An Interdisciplinary Study from North-Eastern India.* Oxford University Press, New Delhi.

Sahu, S.D., Singh, A. and Singh, M. D. 1980. Farm power source and their utilisation in NEH Region. *Research Bulletin,* ICAR Research Complex for NEH Region, Shillong.

Sastry, G., Juyal, G.P. and Samra, J.S. 1997. Rehabilitating areas affected by mines, landslides, torrents and ravines. *In: Advances in Wasteland Development.* Soil Conservation Society of India, New Delhi.

Satapathy, K.K., Panda, R.K. and Prasad, R.N. 1995. Hydrologic aspects of water harvesting tank in the North Eastern hilly region of India. *J. of Soil and Water Conser.*, 41(3&4): *201-216.*

Satapathy, K.K. 1996. Hill slope runoff under conservation practices. *Research Bulletin.* ICAR Research Complex for NEH Region, Barapani.

Sethi, R.M., Nair, G.K., Sheshagiri, Rao and Singh, R.R. 1994. Status of land degradation in India. *Jr. Soil and Water Conser.*, 38(3&4): 163-172.

Sharma, K.D. 1983. Indigenous surface drainage system in Meghalaya - A case study. AICRP on Water N!anagement, ICAR Research Complex for NEH Region, Shillong.

Sharma, K.D. 1987. Estimation of soil erosion hazard through rainfall analysis in North East India. *Jr. Agril. Engg.*, 24(1): 47-55.

Sharma, N. 1982. A broad appraisal of the exploitable water resources and the projected water needs of the Brahmaputra basin. In: *India's North East, A Multifaceted View,* Pankaj Thakur Publishing House, Tinsukia, pp169-177.

Sharma, B.R. 1991. Conservation and efficient utilization of water resources for development of Himalayan agriculture. *J. Water Resour. Soc.*, 2(4): 2-5.

Sharma, K.D. 1987. Estimating soil erosion hazard through rainfall in the N.E. India. *J. Agril. Engg.*, 24(1): 55 -64.

Singh, A. 1987. Studies on some aspects of soil and water in relation of resource management in North Eastern Hill Region. Faculty of Agriculture, Bidhan Chandra Krishi Viswa Vidyalaya, West Bengal.

Singh, A and Singh, M.D. 1981. Soil erosion hazard in north eastern hills regions. *Research Bulletin No.* 10. ICAR Research Complex for NEH Region. Shillong.

Singh, A and Singh, M.D. 1981a. Effect of various stages of shifting cultivation on soil erosion from steep hill slopes. *Indian Forester,* 106(2):115–121.

Singh, A. 1989. Bamboo drip irrigation. *Research Bulletin,* ICAR Research Complex for NEH Region, Shillong.

Singh, M.D. and Singh, A 1980. Traditional farm tools of North Eastern Region. *Research Bulletin* No.5, ICAR Research Complex for NEH Region, Shillong.

Singh, M.D. 1985. RaiAfall- runoff and soil loss relationships on varying degree of hill slopes. Unpub. M.E. Thesis, Department of Civil Engg. Assam Engineering College, Guwahati.

Sonawal, O. K., Tripathy, A.K. and Kirekha, V. 1989. Abo - An indigenous ·farming system of Nagaland. *Ind. J. Hill Farming,* 2(1): 1-8.

Taher, M. 1986. Physiographic frame work of North East India. *North Eastern Geographer,* 18(1&2): 1 -19 Thansanga, R. 1997. New contour farming systems. *Ind. Ir. Soil Cons.,* 25(2): 167-169.

Thansanga, R., 1997.New Contour Farming System, Directorate of Agriculture and Minor Irrigation, Aizawl, Government of Mizoram (Booklet).

Thansanga, R. and Saxena, R.S. 2000. Water resource availability and management in the North Eastern.

Tusa Kumoto, Y. and Ohta, T. 1988. Runoff processes on a steep forested slope. *J. Hydrol.,* 2: 165-178.

National seminar on Strategies for Agricultural Research in the North East. Shillong Nov. 10-12, 1999.

Verma, A; Yadav, B.P.S.; Sampat, K.T. and Borthakur, D.N. 1982. *Research Bulletin* No.18, ICAR Research Complex for NEH Region, Shillong.

□□□

Chapter 64

Remote Sensing and GIS Based Land Resource Development Action Plan for Ghodaghaguri Mini-Watershed of Odisha

A. P. SAHU, K. K. P. MOHANTA, B. PANIGRAHI AND A. K. ROUT[1]

INTRODUCTION

Watershed is a geo-hydrological unit draining at a common point by a system of streams. Land and water are the two most important natural resources of a watershed that govern all life supporting systems in that unit and hence needs careful attention for its development and use. The eleventh and twelfth schedules of Constitutional Amendment Acts have empowered the institutions of local governance for planning and implementing the integrated schemes at micro-levels. In this context, planning, assessment, management and utilization of natural resources, especially land and water at local level assume the prime importance owing to the synergetic effect of land and water resources within a watershed. Integrated management of natural resources on watershed basis is the best rational approach for sustainable development in regards to food security, poverty alleviation and welfare of poorer society. Watershed management requires development of a huge data base of the process acting over the watershed and analysis of data and interpretation of results. Remote Sensing (RS) and Geographical Information System (GIS) are the advanced tools used for this purpose.

Many works have been demonstrated use of remote sensing and GIS for development of watershed and its management. Mishra *et al.*, (1997) used Survey of India topo-sheets and IRS-1C-LISS III imagery to study the geomorphology, drainage, wastelands and other terrain condition to produce

Department of Soil and Water Conservation Engineering, College of Agricultural Engineering and Technology, Orissa University of Agriculture and Technology, Bhubaneswar-751 003, Odisha, India.

[1]Directorate of Research, Orissa University of Agriculture and Technology, Bhubaneswar, Odisha, India

the maps at 1:50,000 scale after prioritizing the micro watersheds of Orissa based on remote sensing techniques. Dutt *et al.*, (1986) identified and mapped the watershed at micro level using LANDSAT Thematic Mapper False Colour Composite through visual interpretation in Bangalore district of Karnataka and showed that integration of remotely sensed data with revenue record data yields sufficiently accurate information. Dhinwa (1992) used multi-date remote sensing data to prepare seventeen land use/land cover maps on 1:50,000 scale and ARC/INFO GIS Package to analyse the land use and compare the land use information in Bharatpur district of Rajasthan. Tomar *et al.*, (2002) prepared the land use map of Upper Shipra Watershed of the Ri-Bhai district of Meghalaya using LISS-III and PAN (06/07) data of IRS-IC satellite and revealed that shifting cultivation was practiced in most of the areas. Ahuja *et at.* (1992) prepared soil resource map through visual interpretation of IRS-1A-LISS-II FCC data at 1:50,000 scale of Bhiwani district, Haryana, which was helpful for land use planning of that area. Srivastava and Mishra (1995) mapped the drainage pattern of Raniganj Coal field of Burdwan district, West Bengal on the basis of LANDSAT-TM and IRS-LISS-II FCC imagery and interpreted their characteristics in terms of basin morphology, surface materials and underlying rock types for effective planning of watershed. RS and GIS techniques along with natural resource survey of the watershed was used by Khan *et al.* (2001) to delineate priority watersheds for soil and water conservation measures leading to sustainability of the Guhiya basin of Rajasthan. RS and GIS techniques were applied for the integrated watershed planning of Bijpur watershed in Khurda district of Orissa by Paul *et al.* (2008). IRS-1D, LISS III FCC satellite images and SOI topo-sheets were used as the data sources for land and water management plan preparation of the watershed. This paper presents the action plan for the land resources development of Ghodaghaguri mini-watershed in Karanjia block of Keonjhar district of Odisha through remote sensing and GIS technique for sustainable development of the watershed.

MATERIALS AND METHODS

Ghodaghaguri mini-watershed lies between 21° 46' 8" N to 21° 49' 39" N latitude and 85° 56' 4" E to 86° 00' 4" E longitudes at a distance of 3 kms. from the block head-quarter of Karanjia as well as sub-divisional head-quarter of Pachpir sub-division in the state of Odisha (Figure 1). The mini-watershed has an area of 2013ha spreading over seven villages namely Ghodaghaguri, Sarabhadi, Jhaddumuria, Diajoli, Bagdapha, Pokamunda and Sarasposi. It comes under the North Central Plateau Climatic Zone. The climate of the area is sub-tropical monsoon type having 1437mm annual average rainfall and 27.2°C average annual temperature. The general relief of the terrain is within 120 to 220m and highest being 232m above mean sea level. Denundatial and residual hills of granitic gneiss rocks are present in the watershed along with pediments, shallow weathered buried pediments and valleys. The drainage system of the watershed drains to Deo River leading to Baitarani River and eventually falls at Bay of Bengal. The watershed area predominantly consists of moderately deep to very deep well drained sandy loam soil with p^H value ranging from 5.25 to 5.48.

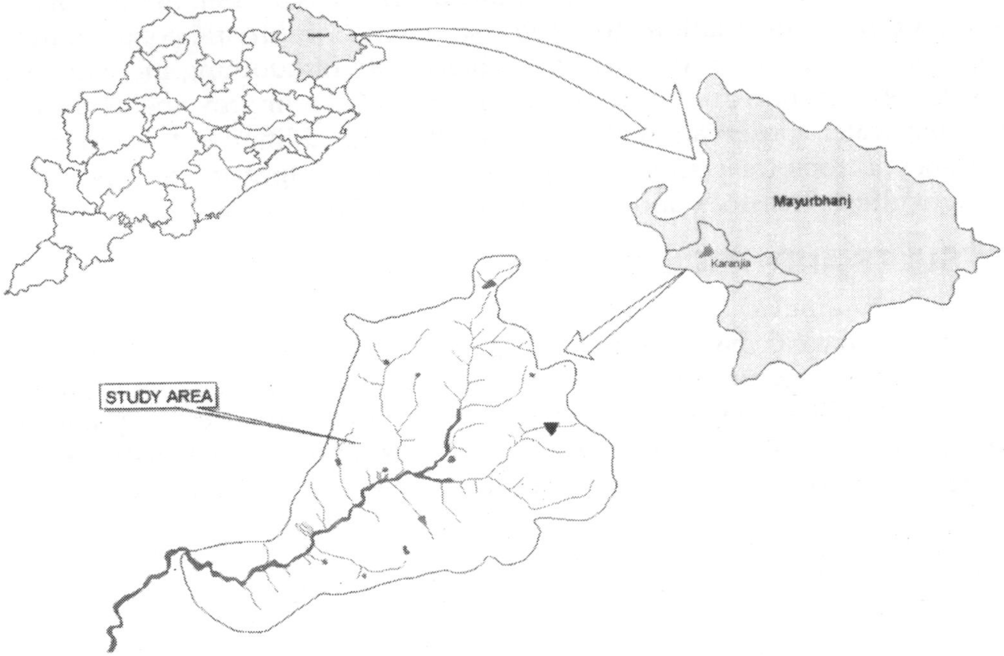

Fig. 1: Location map of Ghodaghaguri mini watershed

IRS-1D-LISS III FCC (1:50,000 scale) of 17th March 2003 and 23rd October 2003 were used for the preparation of thematic maps like land use/land cover map, hydro-geomorphology map, soil resource map and drainage map of the watershed. The Survey of India topographical sheet numbering 73 G/13 on 1:50,000 scales with 20m contour interval was used to prepare the base map and finalization of land use/land cover classes, hydro-geomorphological units and slope map of the watershed.

The Ghodaghaguri mini-watershed was delineated using SOI topo-sheets and latest geo-coded satellite image in 1:50,000 scale. The land use map was prepared using visual interpretation techniques. Initial base map of the study area was prepared from SOI topo-sheets and superimposed on satellite FCC and finalized after ground truthing. The hydro-geomorphological map, drainage map, slope maps, soil resource maps and Digital Elevation Model

(DEM) were also prepared (Mohanta, 2004). ARC/INFO- 7.2 was used for the analysis.

The action plan for the land resources management of the watershed was generated integrating the thematic information like land use/ land cover, hydro-geomorphology, slope, drainage, soil, ground water potential. While generation of action plans optimality of the present land use was considered with respect to sustainable production and quality of ecosystem. While making alternate recommendations for land use practice, futuristic considerations such as exploitation of ground water, possibility of adopting more efficient system of irrigation and water management and site improvements through soil and water conservation were kept in view. In addition to this, improved varieties of crops, trees, shrubs and grasses and advantages of interdependency of agriculture, livestock and other practices have been considered.

RESULTS AND DISCUSSION

The thematic maps like land use/land cover map, hydro-geomorphological map, soil map, slope map and drainage map developed for the Ghodaghaguri mini-watershed have been presented in Figures 2 to 6, respectively. The land use/land cover classification, hydro-geomorphological characteristics, slope attributes and soil resource characteristics of the watershed have been presented in Tables 1 to 4, respectively.

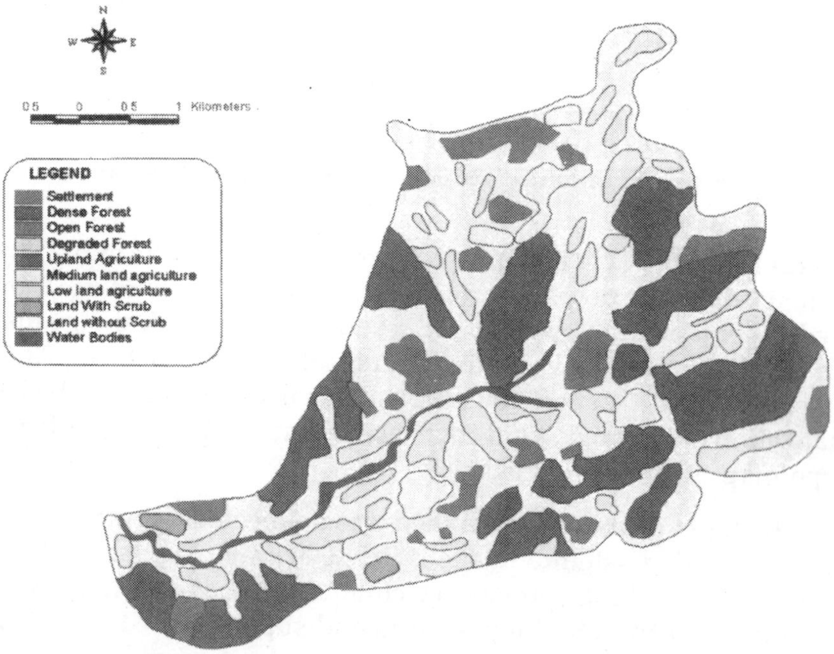

Fig. 2: Land use / land cover map of the watershed

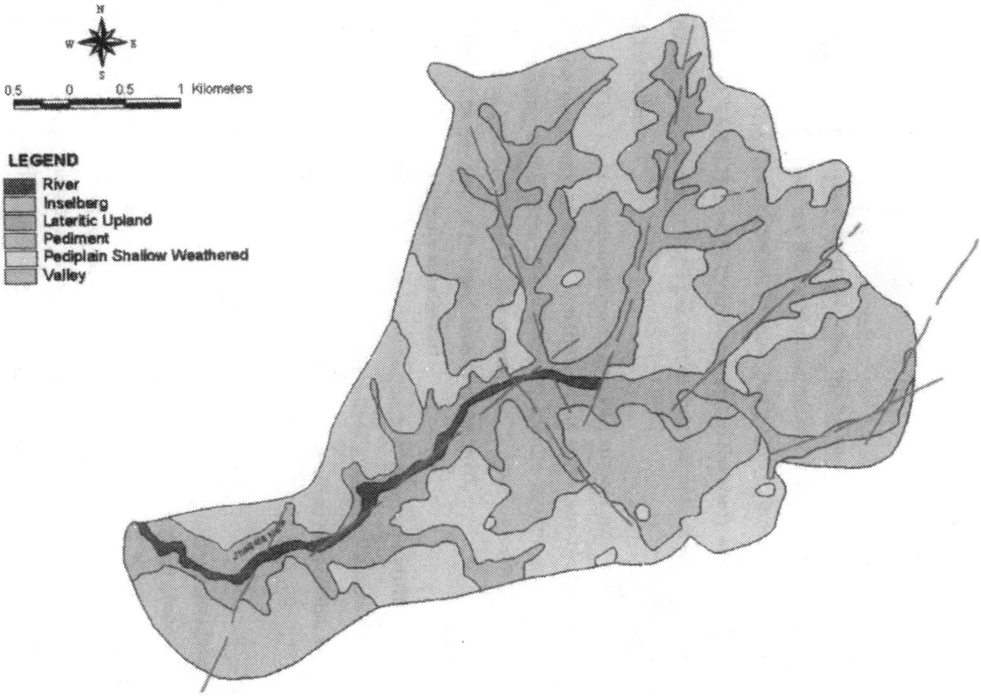

Fig. 3: Hydro-geomorphology map of the watershed

Fig. 4: Soil map of the watershed

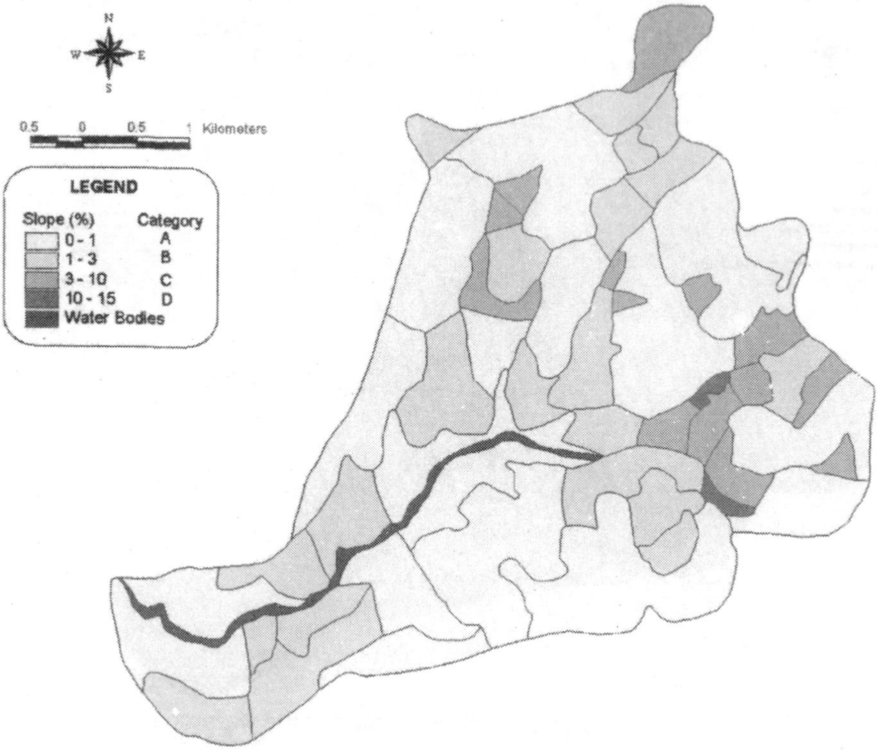

Fig. 5: Slope map of the watershed

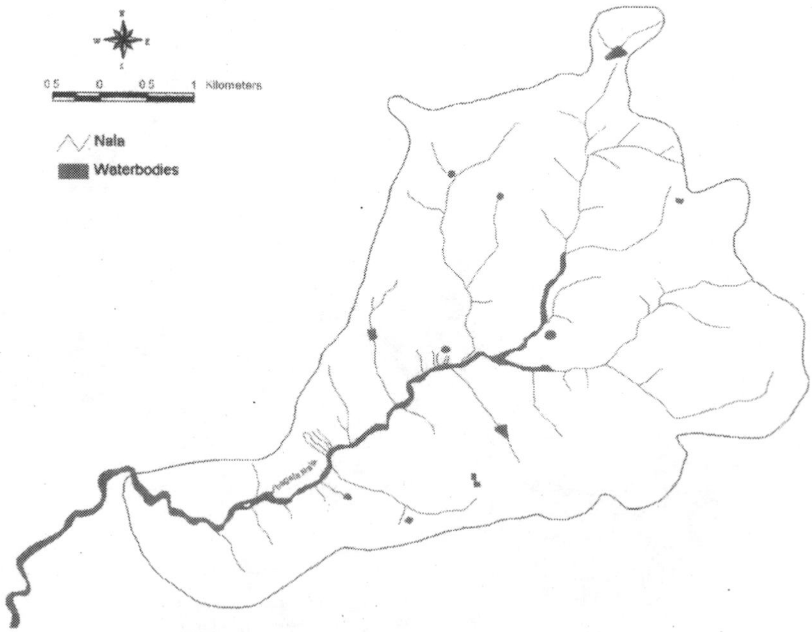

Fig. 6: Drainage map of the watershed

Table 1: Land use/land cover classification of Ghodaghaguri mini-watershed

Map unit	Land use classes	Area (ha)	% of the total geographical area
I	Built up land	145	7.21
II	Forest		
	Dense forest	504	25.07
	Open forest	27	1.34
III	Agricultural land		
	Up land agriculture	40	1.90
	Medium land agriculture	1018	50.64
	Low land agriculture	231	11.49
IV	Wasteland		
	Land with scrub	13	0.65
V	Water bodies		
	Nalas/Reservoirs/Tanks etc.	35	1.59
	Total	**2013**	**100**

Table 2: Hydro-geomorphological characteristics of the watershed

Map unit	Geomorphic unit	Ground water prospects	Area (ha)	% of the total geographical area
I	Inselberg	Negligible	11	0.55
LU	Lateritic up land	Poor	878	44.63
P	Pediment	Moderate to poor	126	5.28
PPS	Pediplain shallow weathered	Moderate	508	25.82
V	Valley	Good	490	23.69
	Total	**2013**	**100**	

Table 3: Slope attributes of the watershed

Map unit	Slope category	Slope (%)	Area (ha)	% of the total geographical area
1.	Nearly level (A)	0-1	1212	60.20
2.	Very gently sloping (B)	1-3	565	28.06
3.	Gently to moderately sloping (C)	3-10	183	9.09
4.	Strongly sloping (D)	10-15	53	2.63
	Total	**2013**	**100**	

Table 4: Soil resource characteristics of the watershed

Map unit	Terrain/Geo-morphic location	Soil characteristics	Area (ha)	% of the total geographical area
1.	Inselberg	Very shallow to shallow excessively drained, loamy sand to sand in surface texture, p^H – 5.21	11	0.55
2.	Lateritic up land	Moderately deep to deep, well drained sandy loam in surface texture, p^H – 5.48	878	44.63
3.	Pediment	Deep to very deep, moderately well drained to well drained, sandy loam to sandy clay loam in surface texture, p^H – 5.05 to 5.40	126	5.28
4.	Pediplain shallow weathered	Very deep moderately well drained, sandy loam in surface texture, p^H – 5.25 to 5.42	508	25.82
5.	Valley	Very deep, imperfectly drained, sandy loam to silty loam in surface texture, p^H – 5.25 to 5.42		
		Total	**2013**	**100**

The recommended action items for land resources development and utilization of the Ghodaghaguri mini-watershed has been tabulated in Table 5. The table reveals that the optimally used land covers 488 ha. The productive potential of some lands under certain land use/land cover practices seem to be good, where dense forests, double cropped areas, agricultural plantations and forest plantations have been included. Plants like bamboo, eucalyptus, teak, karanjia may be adopted in those areas. The lands under scrubs and degraded forests of 140ha area inside the reserve forest should be brought back to vegetation cover through afforestation of species depending upon the local climate, soil, management practice and need of the local people. Gap plantation is suggested in an area of 27ha in the open forest and land area affected by cultivation occurring inside the boundaries of declared forests. Various species of plant like Simaruba, Gambhari, Chakunda, Karanja, Neem, Bamboo and Cashew may be adopted.

Table 5: Recommended action items for land resources development of the Ghodaghaguri mini-watershed

Map unit	Action item	Area (ha)	% of the total geographical area
1.	Optimally used land	488	24.24
2.	Afforestation	140	6.95
3.	Gap plantation	27	1.34
4.	Plantation in fractured zones with peripheral vegetative bunding	16	0.79
5.	Agro-horticulture	993	49.33
6.	Intensive agriculture using ground water	281	13.96
7.	Intensive agriculture through surface water harvesting	13	0.65
8.	Fodder plantation/Dry land horticulture	20	1.00
9.	Pisciculture	35	1.74
	Total	**2013**	**100**

Fractures created due to weathering and erosion in barren rocky areas, wastes and sheet rock areas is seen to have developed soil to certain extent, which can be exploited for plantation. This will increase livestock raising and crop production. Bunding the peripherals of these lands with vegetation is required to check surface runoff. An area of 16ha has been suggested under this action item.

Some pediments in the watershed with 5-10 % slope are, at present, under seasonal and marginal cultivation, which needs improvement in the present practice. Agro-horticulture in these regions will play an important role as production of annual crops is insufficient. The low water requiring crops such as sesamum, green gram, black gram and horse gram may be adopted in this proposed system along with the deep rooted dry land fruit trees to tolerate the monsoon aberrations. Agro-horticulture is recommended in an area of 993 ha covering the maximum area *i.e.*, 49.33 % of the mini watershed.

Single cropped areas on moderately weathered buried pediments within 0-3% slope are generally suitable for intensive agriculture. Since the ground water potential is good and soils on these lands are coarse loamy to fine loamy in texture, very deep and moderately well drained, these lands can be effectively put under agriculture. Presently these lands are mainly used for paddy cultivation with patches of sugarcane, where with appropriate exploitation of ground water, second crop like paddy in *rabi* season can be grown. For this purpose, 13.96 % of the total area *i.e.,* 281 ha is recommended in the action plan.

An area of 13 ha can be allotted under intensive agriculture through surface water harvesting. The single cropped areas located on topographically lower portion like valley fills and fractured valleys have the potential for use under intensive agriculture through efficient surface water harvesting, where crops in two different seasons *i.e., kharif* and *rabi* can be grown.

Shallow weathered buried pediments within 0-5% slope and pediments within 5-10% slope range are presently remaining barren, which can be used for fodder plantation or dry land horticulture. Soils of these two types of lands are coarse loamy in texture with fine gravels within 80 cm of the surface, deep to very deep and well drained. The ground water condition is very poor on the pediments but moderate along the fractures and also on the shallow weathered buried pediments. Lands with lower topography and better soil condition can be used for dry land horticulture in comparison to fodder plantation. Fodder plantation in the rest areas will increase and help in livestock raising and animal husbandry. This action has been suggested in an area of 20 ha in the watershed.

The existing tanks, ponds which retain water for 3-6 months in a year and reservoirs located inside the mini-watershed can be used for pisciculture. New water harvesting structures and water bodies suggested for renovation can be used better by impounding rainwater for fish culture like *rohu, catla* and *mrigal*. The earthen bunds of the tanks and ponds can be used with plantation of horticultural species. A total area of 35 ha in the watershed has been suggested for above developmental and pisciculture.

CONCLUSIONS

The land resources development action plan has been prepared for Ghodaghaguri mini-watershed using Remote Sensing and Geographical Information System with an aim to judiciously develop the region with better soil and water conservation practices. Nine action plan items have been suggested in the watershed covering an area of 2013 ha land. If these recommended action plans are carried out, the land resources of the watershed will improve, which will ultimately enhance the agricultural production and socio-economic status of the people of the watershed.

REFERENCES

Dutt, C.B.S., Nageswara Rao, P. P., Manikiam, B., Gupta, A. K., Behera, G. and Ganesh Raj, K.1986. Identification and mapping of waste lands at micro level using remote sensing technique- An approach. ISRO-NNRMS, *Technical Report*, 58.

Ahuja, R. L., Manchanda, M. L., Sangwan, B. S. and Agarwal, R. P.1992. Utilization of remote sensing data for soil resources mapping and its interpretation for land use planning of Bhiwani district, Haryana. *J. Indian Society of Remote Sensing-Photonirvachak,* 20 (2 & 3): 105-120.

Dhinwa, S. P.1992. Land use change analysis of Bharatpur district using GIS. *J. Indian Society of Remote Sensing-Photonirvachak,* 20 (4): 237-250.

Srivastava, V. K. and Mitra, D.1995. Study of drainage pattern of Raniganj Coal Field (Burdwan district) as observed on LANDSAT-TM-IRS-LISS-II imagery. *J. Indian Society of Remote Sensing-Photonirvachak,* 23 (4): 225-235.

Mishra, P., Mohanty, A. R. and Nayak, B. K.1997. GIS based digital delineation of micro-watersheds of Orissa. *J. Indian Society of Remote Sensing-Photonirvachak,* 22 (3) : 232-239.

Khan, M. A., Gupta, V. P. and Moharana, P. C.2001. Watershed prioritization using remote sensing and GIS- a case study from Guhiya, India. *J. Arid Environment,* 49(3):465-475.

Tomar, J. M. S., Satapathy, K. K. and Dhyani, S. K.2002. Integrated approach of remote sensing and GIS in characterization and evaluation of natural resources for watershed management in Upper Shipra Watershed, Meghalaya. *Indian J. Soil Conservation,* 30(3): 206-213.

Mohanta, K. K. P. 2004. Watershed development plan preparation using remote sensing and GIS technique- A case study. Unpublished M. Tech. thesis, CAET, OUAT, Bhubaneswar.

Paul, J. C., Mishra, J. N., Pradhan, P. L. and Sharma, S. D.2008. Remote sensing and GIS aided land and water management plan preparation of watershed- a case study. *J. Agricultural Engineering,* 45(3): 27-33.

□□□

Chapter 65

Livelihood Security through Farm Mechanization : A Case Study in Rainfed Area of Odisha

S. K. NANDA, B. BEHERA[1] AND P. K. ROUT[2]

INTRODUCTION

Odisha is predominantly an agricultural state where seventy five per cent people depend on agriculture and allied sector for their livelihood. Irrigated area in the state accounts for only thirty five per cent while sixty five per cent is under dry land agriculture. Generally, farmers in rain-fed area of the state complete all their agricultural operations starting from field preparation to seed sowing, in a short period of time depending on the available soil moisture. But, in recent years, the changing weather, unavailability of agricultural labourer and bullock create an unfavourable situation for achieving timeliness in various operations of agriculture. Hence, the focus of the planners now, is to train farmers to get more profit from rain-fed areas through field crops and vegetable cultivation, where small water sources are available for providing life saving irrigation.

Use of bullock and country plough in Odisha is more prevalent particularly in tribal areas where as the expenditure on fodder and cost of maintenance of the bullocks is very high. These factors make agriculture unprofitable. In rural areas especially in tribal areas farmers are unable to invest fund for purchase of costly agricultural machineries like power tiller, tractor, combine harvester *etc.* So, educated and trained youth can boost mechanization of agriculture through establishment of Custom Hiring Centres operated by Kissan Clubs or Self Help Groups (SHG) and supply agricultural machineries to the farmers

Department of farm Machinery and Power, College of Agricultural Engineering and Technology,
[1]College of Agriculture
[2]AICRP on FIM, Orissa University of Agriculture and Technology, Bhubaneswar, Odisha, India

on hiring basis which will give employment to unemployed youth and technical help to the farmers.

The field size in agriculture is gradually becoming smaller making operation of tractor with machinery difficult. In this situation ploughing of field, seed sowing, water pumping and threshing operations etc. can be done in proper time by power tiller with a good benefit cost ratio.

A sub-project entitled "*Sustainable Rural Livelihood and Food Security to Rain fed Farmers in Odisha*" has been operating in three under developed districts of Odisha i.e., Kandhamal, Kalahandi and Dhenkanal under "National Agriculture Innovation Project (NAIP, Component-3)" in PPP mode for the rapid development of agriculture. The project started in 2008 and is expected to continue for five years. The custom hiring centres (CHC) were established to increase farm productivity and profitability with the objectives to increase farm profitability, to promote capacity building through training and demonstration and to improve the socio-economic condition of farming community.

MATERIALS AND METHODS

NAIP-3, OUAT, Bhubaneswar has adopted six village clusters in three rain fed districts of Odisha. This project has taken up fourteen different activities in these clusters of villages and is supported by different scientists of different discipline with the assistance of Senior Research Fellows (SRF) in each cluster and RA at head quarters. Advisory committees such as CAC, CIC and CMU provide guidance and monitor different activities of this project at regular intervals.

On an experimental basis Custom Hiring Centres have been created in all six village clusters to provide support to the farmers with basic agricultural machinery. Agricultural machineries like power tiller, multi fuel water pump, pedal pump, power and knapsack sprayer, paddy thresher, winnower, paddy reaper, chaff cutter and pulse thresher etc. have been made available to the farmers of these villages. These machineries were procured by utilizing the fund of this project after March 2008. These were tested in OUAT and then supplied to cluster level committees in all six clusters of the project. An amount fixed by the cluster committee for each machine has been collected towards sustainability fund from the hiring charges of power tiller, pump set and sprayers and deposited in bank account. As per mandate of the project, this fund shall help in the operation and maintenance of the machineries after withdrawal of the project.

Different training programmes have been conducted by the scientists and technical trainers to mobilise and increase awareness of the villagers about the benefit of machineries and to motivate them to use these machines. The judicious use of machineries was achieved through ten awareness training programmes. Persistent demonstration in the villages were undertaken to select operators and improve their operation skill including minor repairs. Twenty persons have been thoroughly trained in operation, maintenance and minor repair of power tiller, pump sets and power sprayers. Booklets on power tiller and other machineries have been distributed among the educated and trained farmers.

Table 1: Different crops grown in the different village clusters

District	Name of Cluster (Distance from head quartes, Bhubaneswar)	*Kharif*	*Rabi*
Dhenkanal	Odapada (120 km) Sadar (100 km)	Paddy, Groundnut	Groundnut, Onion
Kandhamal	Khajuripada (240 km) G. Udayagiri (220 km)	Paddy, vegetables (Brinjal, beans), Turmeric, Maize	Vegetables (Brinjal, Beans, Peas, Cabbage, Cauliflower, capsicum, tomato, radish)
Kalahandi	Narla (385 km) Golamunda (480 km)	Paddy, cotton	Paddy, field pea, green gram

The villagers of Dhenkanal district get maximum benefit from the cultivation of groundnut and paddy. Similarly the villagers of Kandhamal and Kalahandi get benefit from paddy, vegetable and paddy respectively.

Custom Hiring Centre (CHC)

Power tiller : Power tiller was introduced and used for the first time by the farmers most of the clusters of villages and particularly in Kandhamal district. Villagers are using Power tiller for different purposes starting from tillage using rotary tiller to threshing of paddy.

Pump sets and sprayers : :Multi fuel pump set and Krushak Bandhu (KB) pumps have been supplied to farmers of different clusters as per their demand. These pump sets are mostly used for raising nursery and cultivation of vegetables. However, farmers also hire pump sets in all clusters from other sources as per requirement during peak season. Power sprayer and Knapsack sprayers have been supplied to the village clusters.

Pedal thresher and winnower : The farmers in Kandhamal and Kalahandi do not require straw for thatching of houses as they have traditionally tiled roofs on their houses. Hence, the practice in villages of these districts is to leave the paddy straw in the field after harvesting the panicles of paddy from the top. However, the farmers were motivated to harvest paddy from ground level. Therefore, recovery of whole straw was possible with the introduction of pedal thresher and winnower. The whole straw provided the basic raw material for use in button mushroom cultivation in which womenfolk of the village clusters are engaged in these districts. Mushroom spawn production units have been established in G. Udayagiri, Narla, Golamunda, Odapada, Dhenkanal sadar clusters which have produced mushroom spawn in bottles and sold to the village households mainly through women SHG. However, in village clusters of Dhenkanal district, the whole straw is also used for thatching of houses and for consumption of cattle in addition to mushroom cultivation.

***Paddy reaper and chaff cutter*:** These machineries were introduced among the farmers for the first time in three clusters. The paddy reaper has gained popularity in Narla and Dhenkanal sadar whereas power tiller operated chaff cutter has been used by few farmers.

Farmers are using the whole and cut straw for mushroom cultivation as well as consumption of cattle.

Table 2: Details of operation of farm equipments during 2010-11

Equipments/ Machinery	Households benefited	Running Time	Contribution to sustainable fund (')	Total (')
Power tillers (6 Nos.)	246	366 hr (41.82 ha)	20 / hr	7,320
Multi fuel Pump sets(6Nos)	56	70 hr	10 / hr	700
KB Pumps (12 Nos.)	18	88.6 days	5 / day	443
Sprayers (12 Nos.)	34	69 days (84 ha)	5 / day	345
		8,808		

Table 3: Target and achievement of Power tiller use in all six village clusters

Period	Target (hours)	Achievement (hours)
January'09 to March'09	50	20.9
April'09 to March'10	220	209
April'10 to March'11	350	366
April'11 to September'11	350	334
Total	970	930

From the above Table 3, it can be clearly observed that the annual use of power tiller by the villagers of the different clusters has increased. The reason for this increase is due to creation of awareness amongst farmers about the utility of power tiller through trainings. The confidence of operators and farmers in operation of power tiller has been built up by imparting training on repair and maintenance of this machine at reasonable cost. About twenty operators in six clusters have developed thorough skill in operation of power tiller and above all, the improvement of quality of land preparation with consequent increase in crop yield has increased its use and level of acceptance.

Collection of sustainability fund : An amount of Rs 18, 500 has been collected by the committee member and deposited in bank account towards sustainability fund from the hiring charges of power tiller, pump set and sprayers.

Synergy with other organisations: Training of farmers in repair and maintenance of power tiller and other machineries as well as field demonstration of agricultural machinery for line sowing in some clusters have been taken up by the help of the staff of AICRP on FIM. There is a good synergy between the two projects in motivating farmers for adopting mechanisation in agriculture.

S. No.	Type of machinery	No. of demonstration	Name of cluster	Area
1	Tractor drawn planter	8	Dhenkanal sadar, Narla, Odapada	10 ha
2	Tractor operated post hole digger	2	Dhenkanal sadar	150 pits

In all village clusters there is a synergy with Department of Agriculture, Government of Odisha with the support of Assistant Agriculture Engineers (AAE) and Assistant Agriculture Officers (AAO) for technical input in agriculture.

The NGO partners have come forward to work in collaboration with this project for socio-economic development of villagers through formation of SHGs as both have the same objectives with matching working area. So, the group effort has resulted in productive output.

RESULTS AND DISCUSSION

Increase in crop yield : In the three districts adopted by NAIP-3 as mentioned above both the clusters of Dhenkanal district even though closer to urban area than other two districts the soil in the area under the villages is sandy loam to sandy type with low water holding capacity. The soil type in two clusters of Kandhamal is red sandy loam soil, whereas the soil in Kalahandi is predominantly black soil. The farmers in all the three districts are resource poor and depend on bullock and traditional plough in most of the households, whereas few households hire tractor with tiller for tillage of their land. The source for lifesaving irrigation is also limited and not adequate. Therefore, the average yield level in three districts is low at 3.42 t ha^{-1} as illustrated in Table 4 below. However, when optimum plant population is maintained for paddy cultivation and power tiller is used for land preparation by puddling the high leftover stubbles of paddy (in Kandhamal and Kalahandi), grass and weed etc. get thoroughly incorporated in soil to improve the soil nutrition status and thereby improve the yield of paddy on an average in the three districts by 0.5 t ha^{-1} (14.6 %). Similarly, the yield level increased to 4.4 t ha^{-1} (28.7 %) when HYV seed, adequate dose of fertilizer and land preparation by power tiller was provided.

The groundnut crop is grown only in Dhenkanal district in both *Kharif* and *Rabi*. In other two districts even though this crop is grown in small patches but in the adopted villages this crop is not grown. Therefore the yield data for groundnut is only for Dhenkanal district. The yield increased from 0.72 to 1.1 t ha^{-1} (52.7 %) and increased to 1.23 t ha^{-1} (70.8%) with improve practice when HYV seed, adequate dose of fertilizer and land preparation by power tiller was provided.

Table 4: Impact of farm mechanization on productivity and economics of field crops during 2009-10

Parameters	Traditional practice with bullock & plough		Improved practice with power tiller		Improved practice with PT, HYV seed & fertilizer	
	Rice	Groundnut	Rice	Groundnut	Rice	Groundnut
Productivity (t ha^{-1})	3.42	0.72	3.92	1.10	4.40	1.23
Gross Income (Rs.)	34,200	25,200	39,200	38,500	44,000	43,050
Cost of cultivation with hired labour (Rs.)	16,783	22,641	16,881	23,818	17,520	24,425
Labour employed (man days)	120	163	55	38	62	43
Benefit Cost Ratio	2.0	1.11	2.32	1.61	2.51	1.76

Social impact : The tribal farmers were ploughing their undulated land traditionally with local small sized bullocks to grow paddy, turmeric and some vegetables during Kharif. Earlier these farmers used to be idle during Rabi season. Either they were moving to other states in search of jobs or while away their time in quarrelling and family discord, mostly becoming addicted to country liquor (Mahuli and Salap). Now, all family members in the households are engaged in productive agricultural activities in both Kharif and Rabi season.

Success stories of some farmers of Kandhamal district who have been benefitted through custom hiring centre are given below.

I. G. Udayagiri Cluster

1. Sri Khageswar Pradhan (57)

Village: Katadaganda, Block: G. Udayagiri, Dist: Kandhamal

S. No	Particular	Before	After
1	Area of cultivation	2 acres	5 acres
2	Crops grown	Rice, Turmeric & Beans	Rice, Turmeric, Beans, Peas, Cauliflower, Tomato, capsicum
3	Cost of cultivation	₹ 5,000/- for 5 acre	₹ 1,200/- for 5 acre
4	Income	Able to manage his family only	₹ 75,000/- from 2 acre land by vegetable cultivation
5	Occupation	Agriculture	Farming & operates his rice mill which could not repaired due to poor financial condition
6	Implement owned	Nil	One water pump set

2. Sri Janaka Pradhan (42)

Village: Kilakia, Block: G. Udayagiri, Dist: Kandhamal

S. No	Particular	Before	After
1	Area of cultivation	Nil	4 acres
2	Crops grown	Nil	Rice, Turmeric, Beans, Peas, Cauliflower, Tomato, capsicum
3	Income	Able to manage his family only	₹ 25,000/- from 0.5 acre land by cauliflower cultivation
4	Occupation	Working for Antaranga Club for purchasing & selling of turmeric	Farming as well as working for Antaranga Club

3. Sri Santu Pradhan (26)

Village: Katadaganda, Block: G. Udayagiri, Dist: Kandhamal

S. No	Particular	Before	After
1	Area of cultivation	Nil	1 acres
2	Crops grown	Nil	Beans, Peas, Cauliflower, Tomato, capsicum
3	Income	Able to manage his family only	₹ 14,000/- from 0.25 acre land by tomato cultivation
4	Occupation	Depends upon daily wages	Farmer
5	Implement owned	Nil	1 water pump set

II. Khajuripada cluster

1. Sri Kishore Mallick (31)

Village: Pradhanpada, Block: Khajuripada, Dist: Kandhamal

S. No.	Particulars	Before	After
1	Area of cultivation	2.2 acres	4 acres
2	Crops grown	Rice, turmeric and some vegetable	Rice, Turmeric, Beans, Peas, Cauliflower, Tomato, capsicum
3	Income	Able to manage his family only	₹ 20,000/- from 0.5 acre land by tomato cultivation
4	Occupation	Depends upon daily wages and agriculture	Farming

2. Sri Biswamitra Pradhan (48)

Village: Adasipada, Block: Khajuripada, Dist: Kandhamal

S.No	Particular	Before	After
1	Area of cultivation	Nil	2 acres
2	Crops grown	Nil	Rice, Turmeric, Beans, Peas, Cauliflower, Tomato, capsicum
3	Income	Hardly able to manage his family	₹ 15,000/- from 1 acre land by cauliflower cultivation
4	Occupation	Goes to Surat, Gujarat for searching job (Daily wager)	Farming

3. Smt Gitanjali Pradhan (38)

Village: Adasipada, Block: Khajuripada, Dist: Kandhamal

S. No.	Particulars	Before	After
1	Area of cultivation	1 acre	2 acres
2	Crops grown	Rice and some vegetable	Rice, Turmeric, Beans, Peas, Cauliflower, Tomato, capsicum
3	Income	Hardly able to manage her family	₹ 5,000/- from 0.25 acre land by cauliflower cultivation
4	Occupation	Owned a tea and snacks shop	Farming and management of shop

CONCLUSIONS

It is observed that after intervention of NAIP the villagers of all village clusters have been motivated towards agriculture. Now they have increased their productivity by 28.7 % in paddy, 70.8 % in groundnut, 65 % in vegetables and the area under cultivation has been increased by 60%. Crop diversification has been improved after adoption of crop rotation. Now the cropping intensity is 180% which was 60 % only in earlier case. Village to urban movement of villagers has been minimized.

From the study of above six village clusters of three districts of Odisha in the context of changing climate, severe labour and draught animal scarcity, it can be concluded that introduction of agricultural machinery is very much essential in addition to other essential agricultural inputs for the improvement of agriculture and crop production. It is also observed that farmers in all clusters particularly in the tribal areas, who never knew the use of agricultural machineries, showed great interest and are now getting maximum benefit from these machineries. Hence, increase in awareness and interest of farmers is the most important parameter to judge the progress, whereas the scientific personnel provide motivation and technical guidance to the farmers. The working environment established through the mode of Public, Private and People's Participation (P_4) in this project shall ensure its sustainability.

REFERENCES

Annual report NAIP, Comp-3, OUAT, Bhubaneswar, 2010-11, Livelihood improvement in three disadvantaged districts of Odisha. Page 21-25.

Annual report NAIP, ICAR, 2011-12. Page-64.

□□□

Chapter 66

Site Specific Cropping Plan for Dorika Watershed of Sivasagar District in Assam

ARUNIMA GOGOI, M.C. TALUKDAR, UTPAL BARUAH & ANANTA DUTTA

INTRODUCTION

Sivasagar district of Assam occupies an area of 2, 66,800 ha constituting 3.40 % of the states geographical area (Dutta, 2009). Dorika is one of the major rivers in Sivasagar district. Dorika watershed covers an area of 29,015 ha. Most of the cultivated areas under this watershed remain inundated for most part of the rainy season. Rice is cultivated during the rainy season and thereafter most of the cultivable lands remain fallow. In Assam, single cropped rice culture with a yield of 1.85 t ha^{-1} which is very low compared to 3.1 t ha^{-1} in Tamil Nadu and 3.4 t ha^{-1} in Punjab (Anonymous, 1996). The total food grain production of Assam stands at 36.72 lakh tonnes as against the demand of 45.33 lakh tones (Anonymous, 1997). Thus, there is an urgent need for a continuous and sustained increase of crop production. A critical appraisal of land resources and then allocating them to the various cropping system in both rainy and post rainy seasons based on their potential will improve the cropping intensity and compensate the crop loss suffered during rains and floods (Vadivelu *et al.*, 2003). In such attempts the soil maps were prepared with sufficient ground truth which played a vital role in fitting the site specific cultivation. Therefore, to increase productivity and to encourage double crop, the present investigation was taken up to evaluate the soils of Dorika watershed for growing various crops, particularly introducing cultivation of some important crops during rabi season in the dominantly rice alone cropping system.

MATERIALS AND METHODS

Site description : The Dorika watershed of Sivasagar district is a part of the Upper Brahmaputra Valley Zone of Assam. This watershed is situated

Department of Soil Science, Assam Agricultural University, Jorhat-785 013 Assam, India between 94°27 to 94°52 E longitude and 26°48 to 27°03 N latitude. This watershed has an area of 29,015 ha comprising topography of hill to alluvial plain.

Climate, physiography and geology : The climate of this area is humid subtropical with an average annual rainfall of 2,334 mm. The soil moisture of this study area is Udic and the soil temperature regime is hyperthermic. The watershed area represents the plains of the Brahmaputra valley. The alluvium deposited at the foothill slopes forming the piedmonts. Geology of the area is alluvium of Pleistocene (Wadia, 1966) and recent periods in the plains and tertiary sedimentary rocks in the hill. Some areas in southern side of this watershed have low hills in continuation of the Naga Hills.

Delineation, flooding and drainage : The boundary of Dorika watershed is delineated based on Survey of India toposheet No. 83J/13, J/9; 83I/12 and I/8 (Figure 1) and IRS IC- LISS III (2008) geocoded false colour composite (FCC) products (1:50,000 scale) were visually interpreted for physiographic delineation of Dorika watershed comprising with drainage network. The northern part of this watershed particularly along the river Brahmaputra is often subjected to flooding during the rainy season. The rest of the watershed is often remaining under submerged condition for the most part of the rainy season. The drainage condition is poor during and immediately after flooding.

Soil site suitability evaluation of crops : The suitability of the soils for growing rice, wheat, maize, mustard, potato, banana and tea was evaluated following the method proposed by Sys *et al.* (1991).The soil characteristics were matched with the requirements of the crops at different limitation levels. The suitability classes and subclasses were decided by most limiting soil characteristics. After the evaluation, the soil units comprising two or more soil series were allotted to these crops for which they are highly suitable (S1) and moderately suitable (S2). The crops which were in association with marginally suitable (S3) and presently not suitable (N1) classes are not recommended because their cultivation will not be economical. Thus, the site specific cropping plan was prepared for identified location of Dorika watershed based on soil characteristics and limitation of soil site suitability.

RESULTS AND DISCUSSION

Soils: Eleven soil site locations were identified under this watershed. The texture of the soils varies from sandy loam to clay loam or silty clay to clay in texture (Table 1). The soils of Banmukh and Panbecha which occurs northern side to river Dorika are silty loam in texture. Whereas the soils of Santak are the southeastern part of this watershed lies near the Naga hills and texture varies from sandy loam to loamy sand. Soil pH is ranges from 4.48 to 5.81. High rainfall associated with loss of bases due to leaching and presence of Al^{3+}on exchange complex are largely responsible for low soil pH (Sahoo *et al.* 2010).The organic carbon content is varies from 0.74 to 1.25 %. Organic carbon content is more in the hill soils of Charaideo. This may be attributed to the luxuriant vegetation. The base saturation is less than 50% in the soils of alluvial plain of Santak, Lunpuria, Abhaypuria, Chengmoigaon, Hatimuria, Dorikapar, Chengalibari, Banmukh Panbecha and Katanipar Satra. But in Charaideo, the base saturation is found to be 54.20%.The base saturation increases with increase in pH. Comparatively high base saturation in Charaideo (hill soils) may be due to addition of basses from leaf litter. The cation exchange capacity of the soils under Dorika watershed is ranges between 5.14 to 13.9cmol (p+) $kg^{-1.}$ Relatively low CEC of these soils may be due to dominance of clay minerals with low CEC and presence of hydrous oxide of iron and aluminum in these soils (Sarkar *et al.*, 2002).

Table 1: Physico chemical properties of soils of Dorika watershed of Sivasagar district

Site No	Location	Textural sequence upto 100 cm (%)	Sand (%)	Silt (%)	Clay (%)	pH (1:2.5)	OC (%)	CEC (Cmol p^+ kg^{-1})	BS(%)
1	Santak	Sl, l, sl, sl	72.3	12.5	15.2	5.31	0.84	5.14	37.44
2	Lunpuria	Cl, cl, l	28.4	40.1	31.5	4.48	0.78	10.95	38.80
3	Abhaypuria	Cl, l,cl	30.0	38.8	31.2	4.93	1.20	11.12	48.71
4	Charaideo	Sl, l, l, sl	70.5	11.3	18.2	5.81	1.25	8.61	54.20
5	Chengmoigaon	Sic, sil, c, cl	20.5	38.1	41.6	5.22	1.02	12.32	46.17
6	Hatimuria	Sic, sil, cl, c	12.6	45.4	42.0	5.15	0.91	12.65	42.51
7	Dorikapar	Cl, l, cl	26.3	43.4	30.3	5.4	0.99	13.30	44.50
8	Chengalibari	C, c, cl	30.0	18.6	51.4	5.19	1.14	13.90	45.20
9	Banmukh	Sil, sic, sl	28.9	45.8	25.3	5.21	0.74	8.30	40.70
10	Panbecha	Sil, l, sl	26.5	49.9	23.6	5.16	0.76	10.42	36.60
11	Katanipar satra	Sil, sic, c, cl	36.1	47.2	16.7	5.21	1.12	13.60	46.40

Sandy loam- sl, Clay loam-cl, Silty clay-sic, Clay-c, Silt loam-sil, loam-l,

Field crops : Rice is grown on all eleven cultivated soils during the rainy season between May to November. During this period, 2138.8 mm annual rainfall is received. The drainage condition remains imperfect or poor due to shallow water table and therefore, only rice can be cultivated during the rainy season. All soils under this watershed were found to be moderately suitable for rice (Table 2). But, the soils of Santak and Charaideo were found to be marginally suitable. pH and Soil texture denoted as(s) and Soil pH denoted as (f) are the major limitation. Liming and addition of organic manure or growing green manure crops will improve the suitability of these soils for rice. The climatic condition is suitable for wheat. The soils of Charaideo are moderately suitable (S2) and the other ten identified soil sited area are marginally suitable (S3) due to lower pH. Liming will improve the suitability class for wheat.

Table 2: Suitability class of the soils of Dorika Watershed of Sivasagar District

Site No	Location	Rice	Wheat	Maize	Mustard	Potato	Tea	Banana
1	Santak	S3 sf	S3 fs	S2 sf	S3 f	S2 f	S2 f	N1 sf
2	Lunpuria	S2 f	S3 f	S3 f	S3 fs	S3 sf	S2 l,w	S3 f
3	Abhaypuria	S2 f	S3 f	S3 f	S3 fs	S3 sf	S2 l,w	S3 f
4	Charaideo	S3 s	S2 sf	S2 s	S2 f	S1	S2 f	S3 sf
5	Chengmoigaon	S2 f	S3 sf	S3 sf	S3 fs	N1sf	S3 sf,	N1 sf
6	Hatimuria	S2 f	S3 sf	S3 sf	S3 fs	N1sf	S3 sf,	N1sf
7	Dorikapar	S2 f	S3 f	S2 f	S3 fs	S3 sf	S2 f	S2f
8	Chengalibari	S2 f	S3 fs	S3 sf	N1 fs	N1f	S3	S3sf
9	Banmukh	S2 sf	S3 f	S3 wf	S3 f	S3 sf,	N1	S3sf,w
10	Panbecha	S2 sf	S3 f	S3wf	S3 f	S3 sf	N1	S3sf,w
11	Katanipar satra	S2 sf	S3 f	S3 f	S3 f	S3 sf	S3w	S3sf

Highly suitable-S1; Moderately suitable-S2; Marginally suitable-S3; Presently not suitable-N1

*Limitation, Soil pH-f, soil texture-s, drainage-w, base saturation-l

Maize is acid sensitive and therefore eight soil sited area are marginally suitable (S3) due to strong acidity. Drainage is another constraint in the soils of Banmukh and Panbecha.The remaining area of Santak, Charaideo and Dorikapar are moderately suitable (S2) for maize cultivation. For mustard, the soils of Charaideo is moderately suitable (S2) and except Chengalibari other soil series are marginally suitable (S3).The soils of Chengalibari is presently not suitable(N1) for cultivation of Mustard due to presence of high per cent of clay and low pH.

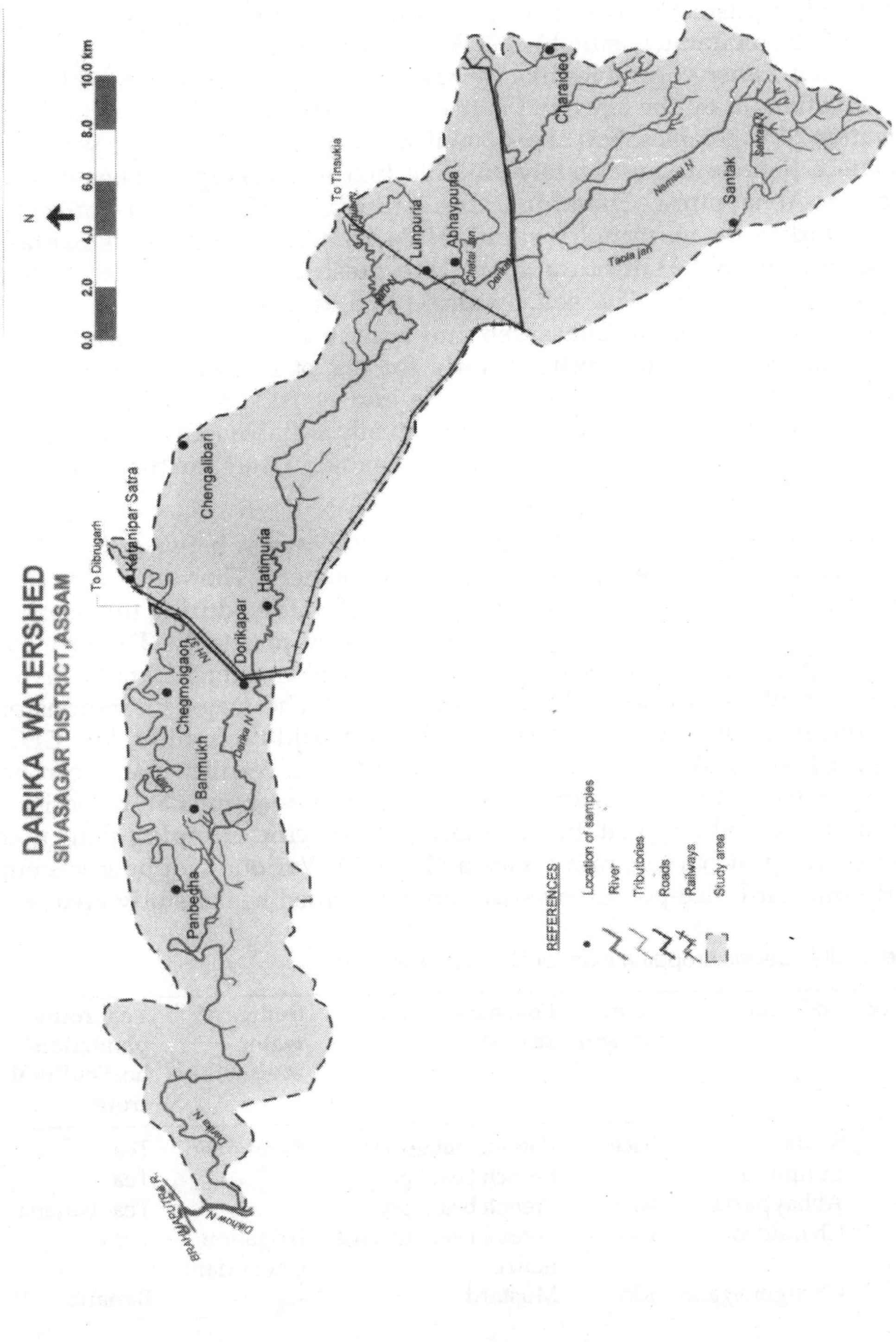

Fig. 1: Delineated of Dorika watershed

Horticultural crops: Potato is highly suitable (S1) in Charaideo but in Santak it is moderately suitable. Other nine soil sites are found to be marginally suitable (S3) to unsuitable with pH and texture is the major constraints. Tomato is found to be moderately suitable (S2) whereas the other sites are marginally suitable (S3). Other vegetables like cabbage, pea are marginally suitable (S3) for cultivation. In all the sites, pH is the major constraints that can limit the cultivation of vegetables. Soil site suitability for horticultural crops was also evaluated. Banana is moderately suitable (S2) in Dorikapar. The soils of Lunpuria, Abhaypuria, Charaideo, Chengalibari, Banmukh and Panbecha were found to be marginally suitable(S3) whereas the soils of Santak, Chengmoigaon and Hatimuria gaon were presently not suitable (N1)for cultivation of Banana. The soil texture, pH and drainage are the major limitations. The soils of Charaideo and Santak are presently under tea cultivation. Soil site suitability criteria for tea were evaluated. Soils of Charaideo, santak,Lunpuria, Abhaypuria and Dorikapar are moderately suitable (S2).The remaining others are marginally suitable (S3) and presently not suitable(N1) with soil limitations of organic matter, texture, pH and Drainage (*w*).

Site specific cropping plan: During the rainy season between May and September, the soils have a poor drainage environment. The water table also fluctuates within one meter of the soil surface .Therefore, during this season, the options are limited and only rice can be grown profitably. The soil site suitability parameters resulted that the soils of Dorika watershed not equally suitable for all crops due to different limitations. The major limitations of these soils lie in the physical properties (texture), acidity, fertility, low CEC, base saturation, and poor drainage. The acidity and fertility status can be increase by use of liming, organic manure and green manure. On the basis of limitation and soil site suitability criteria site specific cropping plan was prepared for post harvest rainy season (Table 3). Various cropping system like rice-mustard, rice-potato, rice-rice can be adopted in the study area.

Table 3: Site specific cropping plan for Dorika watershed

Site No	Location	Rainy season	Post harvest rainy season	Source of water availability	Year round plantation/ horticultural crops
1	Santak	Rice	Potato, maize, pea	Farm pond	Tea
2	Lunpuria	Rice	French bean, pea	-	Tea
3	Abhaypuria	Rice	French bean, pea	-	Tea, banana
4	Charaideo	Rice	Potato, pea, mustard, maize	Irrigation, Check dam	Tea
5	Chengmoigaon	Rice	Mustard	-	Banana

6	Hatimuria	Rice	Rice	Farm pond	Banana
7	Dorikapar	Rice	Maize	-	Tea, Banana
8	Chengalibari	Rice	Rice	-	-
9	Banmukh	Rice	Green gram, pea	Irrigation	Arecanut,
10	Panbecha	Rice	Green gram, pea	-	Arecanut
11	Katanipar satra	Rice	Gram, pea	Farm pond	Arecanut

CONCLUSIONS

The area experiences sufficient rainfall with length of growing period of 290 days and optimum temperatures suitable for a number of crops. But, pH, drainage, physical texture, nutrient constraints become the limiting factors for potential utilization of the land resources. Some agricultural crops like potato, mustard, maize, pea, green gram, French bean can be grown in these soils with liming and fertilizer management. Beside this, the water stored in the soils is in appreciable amount which can be utilized for intensification during Rabi under rainfed condition. The water stored in the ponds can be utilized for life saving irrigation during critical periods particularly for growing wheat and potato. Thus, with judicious rain water and soil fertility management may increase the cropping intensity of this watershed.

REFERENCES

Anonymous. 1996. Agricultural Situation in India. April 1996. Directorate of Economics and Statistics. Ministry of Agriculture. Govt. of India

Anonymous. 1997. Directorate of Economics and Statistics, Government of Assam, Guwahati

Dutta, M. D. 2009. Land Use Planning for Rainfed Agriculture in Tipukjan Watershed of Assam using remote sensing and GIS Techniques. Ph.D Thesis, Assam Agril. University, Jorhat. 3p

Sahoo, A.K.,Sarkar, Dipak., Baruah, U. and Butte, P.S. 2010. Characteristics and classification and Evaluation of Soils of Langol Hill, Manipur for rational land use planning. *Journal of the Indian Society of Soil science,* 58 : 355-362

Sarkar, D., Baruah, U., Gangopadhyay, S.K., Sahoo, A. K.and and Velayutham, M. 2002. Characteristics and classification of soils of Loktak catchment area of Manipur for sustainable land use planning. *Journal of the Indian Society of Soil science,* 50, 196-204

Sys, C., Van Ranst, E and Debaveye, J. 1991. Land Evaluation, Part II, Methods in Land Evaluation. Agricultural Publication No.7, General Administration for Development Cooperation, Belgium.

Vadivelu,S., Baruah, U., Bhaskar, B.P., Thampi, J., Sarkar,D., Walia,C.S. ,Nayak, D.C. and Maji A.K. 2003. A land use plan for Jorhat district of Assam State. *Agropedology,* 13:1-10

Wadia, D. N. 1966. Geology of India. The English Language Book Society and Macmillan Co. Ltd. London, pp76-92.

□□□

Chapter 67

Evaluation and Appraisal of Irrigation Efficiencies Obtained on Farmer's Field for Increased Crop Productivity and Water Conservation in Clay Loam Soils of Jabalpur MP : A Case Study

SATISH SHARMA, M.K. HARDAHA[1], L.D.KOSHTA[2]

INTRODUCTION

One of the most important inputs in the new technology is water. For increase crop production, development and improved utilization of water sources are of great importance. The water resources, which have already been developed, if utilized efficiently, would not only increase in coverage of new project but also increase the crop production per unit area. Therefore knowledge of appropriate irrigation method and handling of stream size is mandatory for farmers. This not only minimizes the loss of water but also maximizes the production without incurring money on expensive leveling work. The border irrigation is a method of irrigation which can be proposed for adoption by cultivator of Bargi command area for getting higher production.

MATERIAL AND METHODS

Study area: The present study has been carried out to evaluate quantitatively the effect of border irrigation on production of wheat crop with moderate land leveling which can be done by farmers with available implements. The project site is 38 km away from Jabalpur on Jabalpur-Gotegaon Road. The present study was carried out in the command of Dabola minor of Rani Awanti Bai Sagar Pariyojana Jabalpur. The results of the study denotes with the computation of irrigation efficiencies on borders prepared by the farmers and comparison of efficiencies and production with control field. Infiltration characteristics of soil of each field were found out by double

College of Agriculture, JNKVV Ganj Basoda Distt.Vidisha -464 221 Madhya Pradesh, India
[1]Jawaharlal Nehru Krishi Vishwavidyalaya, Jabalpur, Madhya Pradesh, India
[2]Deptt. of Forestry, College of Agriculture , JNKVV, Jabalpur, Madhya Pradesh, India

ring infiltrometer. The values of the parameters of infiltration equation are shown in Table 1.

Table 1: Parameters of modified Kostikove equation

Field No.	Values of Co-efficient		
1	0.9136	0.4825	- 0.5210
2	1.6599	0.3887	- 0.4045
3	0.8790	0.4700	0.5924

Bulk density and texture of soils in different fields are shown in Table 2.

Table 2: Mechanical analysis of Soil

Field No.	Average bulk density, γ (g cc^{-1})	Soil textural class
1	1.368	Clay loam
2	1.386	Clay loam
3	1.452	Sandy clay loam

Soil physical parameter *viz.*, field capacity, wilting point and available water were determined using pressure plate apparatus Table 3.

Table 3: Soil physical parameter

Field No.	Field capacity at 1/3 bar (%)	Wilting Point At 15 bar (%)	Depth of available water (in 90cm root zone depth) in cm
1	30	14	19.69
2	30	14	19.95
3'	26	12	18.29

All the three fields were prepared with the bullock drawn indigenous plough and disk harrow followed by leveler and planner. After the field is prepared, line sowing of *'WH-147'* variety of wheat was done with the help of bullock drawn seed drill. Eighteen borders were prepared in field no.1 and nine borders were prepared in field no.2 with the help of bullock drawn bund former and human laborers. The field no.3 was taken as control field, where flood irrigation was followed. In border irrigated fields farmers were guided to prepare the border without much disturbance or minimum manipulation of soil of the field soil. The borders prepared were having the average longitudinal slope ranging from 0.90% to 2.07% and average width of border was 6m and 5m in the field no.1 and 2, respectively. The length of the border was variable, depending on the boundaries of the fields. A source of irrigation water was Dabola minor of LBC of Rani Avanti Bai Sagar Reservoir. Irrigation water was conveyed up to fields by unlined water courses. The stream size measurement was done with 7.5 cm parshall flume. The water was applied to the borders by overflowing of the field channels at upper end. All the borders and control field were given irrigation at the same time.

In case of border irrigation the observation were taken for volume of water applied and length of ponding at tail ends were noted. Soil samples were taken at three different depths before and after irrigation to assess the storage of water in the root zone of the crop. For control field observation were taken for volume of water applied, storage of moisture in the root zone due to irrigation portion of the field which was ponded and volume of water which has escaped from the field as runoff were observed. The application, storage, distribution and crop water use efficiencies were calculated for each field under study.

The longitudinal slopes of the borders are determined by plotting the profiles of the borders and fixing the top surface line by trial and error procedure so that this passes through the average elevation of the profile at centre of gravity of the border. The top surface lines are adjusted such that the earth work involved is minimum and the ratio of cutting and filling is around 1.2 to make provision for the shrinkage of the soil in the filling portion. The stream size for irrigation for individual border are adjusted so that the time of advance of the stream is equal to the opportunity time required for infiltration of the depth of water needed to raise the moisture content of the soil from 50% depletion level or water content at the time of irrigation to field capacity. Advance and recession functions are used to determine the time of advance and recession for different intermediate points of the borders.

Relationship between Efficiencies with Condition of Field

The condition of field is expressed in terms of uniformity index. Higher the uniformity index the profile of the actual border will be more close to the profile of the theoretical borders. The graphs are plotted between uniformity index and differences of efficiencies, difference as a percentage of theoretical efficiencies. Ratio of the actual to theoretical values of time of advance, time of recession , opportunity time, the streame size ,application efficiency, storage efficiency and distribution efficiency. These are shown in Figures. All these values shows decreasing trend with increase of uniformity index. All the values of actual irrigation efficiencies increase with the increase of uniformity index. Thus, the prepared borders which are close to the profile of theoretical borders shows better performance.

RESULTS AND DISCUSSION

The comparative values of efficiencies obtained in all the fields shown in the Table 4.

Table 4: Irrigation efficiencies in different fields of farmers in the command area

Field No.	Method of Irrigation	Average Value (%)			Av. water use efficiency
		Ea	Es	Ed	(q ha-cm^{-1})
1	Border	63.34	77.06	33.79	0.7264
2	Border				
3'	Flood	42.19	71.58	40.45	0.3881

From the Table 4 it could be observed that the increase in application efficiencies in border irrigated field is 21.15% as compared to flood irrigation in field No.3. In flooded field, the water application efficiency is limited to 42.19% because of wastage of water due to deep percolation in ponded portion and runoff from the field. In borders prepared by the farmers without much leveling the water application efficiency has increased to 63.34%, due to better management and distribution of water.

Storage and distribution efficiency cannot be compared due to ponding is both the case. The average value of water use efficiency for border irrigated field is 0.7264 q ha-cm^{-1} and for control field 0.3881 q ha-cm^{-1}. The figure obtained in the study which was carried out at Tawa command of Hoshangabad, MP (1991) are 0.6139 and 0.2852 q ha-cm^{-1} for the same crop obtained in the present study which are comparable with the observed figure. Both of this study indicates increased crop production and water productivity obtained on borders prepared by the farmers in actual field condition. Performance evaluation of the crop productivity undertaken in the selected field are shown as in Table 5.

Table 5: Crop-production or yield of wheat crop

Field No.	Area (ha)	Production (q ha^{-1})
1	0.66	32.50
2	0.41	35.00
3	0.85	25.00

Table 5 gives the average yield of border irrigated field was 33.75 q ha^{-1}, whereas in the field irrigated by flood irrigation the yield was found to be 25 q ha^{-1}. The production in border in comparison to flooded field increased from 25 q ha^{-1}to 33.75 q ha^{-1}. Thus it is evident that border irrigation has increased the crop production by 8.75 q ha^{-1}which is 35% more than the production obtained in the control field. Therefore it is evident that the adoption of the border irrigation produced pronounced effect over the control field in the command area.

CONCLUSIONS

Based on the study following conclusion are drawn:

- Uniformity index have been developed to compare the actual existing profile of the borders with the profile of the average slope. Higher the uniformity index the profile of the actual border will be more close to the profile of the theoretical borders.
- In flooded field the water application efficiency is low because of the wastage of water due to deep percolation in ponded portion and runoff from the field. About 88.33% of the field received 67% of the water needed which has caused the 74% of the production of the border field.
- In Borders prepared by the farmers the water application efficiency has increased to 63.34% and yield has increased by 35% in comparison to the control field.
- The maximum width of the border should be between 5 to 6 m to prevent the water from dispersion and channel formation.
- The average stream size worked out is 0.635 lps per m width of the border. The average irrigation time for the 0.635lps m^{-1} stream size is 98.20 min. This information can be utilized for preparing the schedule for supplying the water to the farmers.
- In borders, the field water use efficiency has increased from 0.3881 q ha-cm^{-1} to 0.7264 q ha-cm^{-1}.
- Thus border irrigation is recommended for the farmers in the Bargi command to obtain better production and higher water use efficiencies which yields in increased water and crop productivity with sustainable crop production in the command area.

REFERENCES

Anonymous. 1961. Evaluation and intervension studies in Tawa Irrigation Project, Hoshangabad (M.P.). Agril. Finance Corporation Ltd., Bombay.

Anonymous. 1991. An action research (Adaptive Research Trial)- a progress report of *Rabi*-1990-91, JNKVV-ZARS, Powarkheda, Hoshangabad (M.P.)

Argal, S.K. 1986. " Development of advance function for design of borders".Unpublished M.tech. Thesis, deptt. of Soil &Water Engg. CAE, JNKVV, Jabalpur MP.

Michael, A.M. 1983. Resource analysis and plan for efficient water management. A case study of Mahi Right Bank Canal Command Area, Chap. 10, pp. 254-256.

Pahalwan, R.N. 1975. Irrigation methods for application of water in the fields, CAE, JNKVV, MP.

Tyagi and Lal. 1970. Predicting water advanced in irrigation furrows & border. *J. Agril. Engg.* 7-4, pp. 16-23.

Wilke,O. and Smerdon,E.T. 1965. A solution of Irrigation advances problem. Jour. Irrigation & Drainage Div. Proc. A.S.C.E., 91, IR 3, pp 23-24.

□□□

Chapter 68

Analytical Approach to Determine the Safe Pumping Limit in Coastal Areas of Odisha

RANU RANI SETHI, M.J.KALEDHONKAR, LAXMIPRIYA MISHRA[1], M. DAS AND A.KUMAR

INTRODUCTION

Indian coastal aquifers constitute the richest groundwater reservoirs second to the Indo-Gangetic alluvial plain, which is one of the world's largest fresh groundwater reservoirs. Groundwater is considered as an important source of fresh water in supplying water for domestic and irrigation use. Groundwater can be regarded as a renewable natural resource, if there exists a balance between the recharge and abstractions of the aquifer. According to World Meteorological Organization only 2.5% of total water volume on earth is fresh water and the remaining is saline water. In coastal areas since ground water system is in contact with saline water, one of the major problems is salt water intrusion. The natural balance between fresh and saline water in coastal aquifers is disturbed by groundwater withdrawals and other human activities. This lowers the groundwater levels and increases saline water flow to fresh water of coastal aquifer.

Few studies have been conducted for coastal aquifer in Nellore district of Andhra Pradesh, in which a finite element based flow and transport simulation model has been implemented. The model was calibrated for two years of time period between July 2000 and July 2002, both in terms of hydraulic heads and salt concentration. The calibrated and partially validated simulation model was used to evaluate the effectiveness of planned pumping strategies to locally control the saltwater intrusion in the study area (Datta *et al.*, 2009). Hany *et al.*, (2011) has developed a cost-effective method to control seawater intrusion in coastal aquifers. Rejani *et al.* (2009) has studied the Balasore coastal

Directorate of Water Management, Bhubaneswar, Odisha, India
[1]College of Agricultural Engineering and Technology, Orissa University of Agriculture and Technology, Bhubaneswar, Odisha, India

groundwater basin in Odisha and found that overdraft has resulted in declining groundwater levels and seawater intrusion in the area a current groundwater withdrawal for irrigation was 71% more than the safe yield. It was concluded that some urgent measures must be taken to ensure sustainable groundwater resources in the basin. In Odisha, coast line has been stretched around 480km. The average groundwater development in Odisha is 26.17% highest in Bhardrak (55.49%), followed by Kendrapara (52.8%), Balasore (48.58%) and Jagatsinghpur (47.37%) (Directorate of Groundwater Survey and Investigation, Orissa). Jagatsinghpur district is located in the East coast of Odisha and comes under East and South-Eastern Coastal Plain Agro-Climatic conditions. The district is congenial and prone to occurrence of various calamities like flood, cyclone and even drought. It is the one of the thickly populated coastal districts of Odisha and main stream of economy comes from agriculture. The district is also having problems of saline soils, water logging area which adversely affects the production and productivity. The available groundwater resources in Jagatsinghpur district for irrigation are estimated to be of the order of 38829 ha-m for the year 1992, based on the norms recommended by the 'Ground Water Estimation Committee' and it has been observed that groundwater level has receded at least 2 meters within a fortnight in Ersama area. As the Coastal area is hydrologically very sensitive to saline intrusion, abstraction of groundwater from coastal aquifers should be restricted with scientifically defined yield limits. The objectives of this paper were to characterize the coastal tract in Erasama block of Odisha and to develop an analytical approach to determine the safe pumping depth within the coastal aquifer.

MATERIALS AND METHODS

Study was carried out for Ersama block in Jagatsinghpur district of Odisha, which forms a part of the Mahandi valley overlain by the Quaternary formation. The river Mahanadi forms the northern boundary and Devi River largely constitutes the western and south- western boundary. Bay of Bengal is the eastern boundary of the district. Various data were collected from different sources like National Atlas Map, Survey Of India, Census Of India, Aerial Photographs, Satellite Imagery (IRS), Central Ground Water Board, Geological Survey of India, India Meteorological Department, District Planning and Development Office, Soil Conservation Office, Field Surveys & Other Govt. agencies and then the thematic maps for Jagatsinghpur has been prepared. Then from the geo-referenced Odisha map, Ersama Block of Jagatsinghpur district was extracted. The area has been delineated and thematic mapping has been prepared by using TNT mips GIS software. Groundwater table depth was monitored from 30 monitoring wells distributed with their distance from the coast line. Contour mapping was prepared by pre and post monsoon water table depth. Analytical approach of Ghyben-Herzberg model was used to determine the freshwater-Saltwater interface, which is expressed as

$$h_S = \frac{y_f}{(y_s - y_f)} h_f = \frac{1}{(1.03 - 1)} h_f$$

Total Head = Pressure Head + Elevation Head

For the seawater and fresh water, separately

$$h_f = \frac{P_f}{Y_f} + h \quad h_s = \frac{P_s}{Y_s} + h$$

Where, P is for pressure; ã is the specific weight; sub-indices f and s stand for fresh and sea, respectively. The elevation head H should be equal to a same value if we discuss at a particular point P. This point is chosen along the salt-fresh water interface. At any point along the salt water-fresh water interface, the hydraulic pressure should be a constant, or the pressure arose from the seawater and the fresh-water side should be equal to each other, so we should have $P_f = P_s$. From the above equation,

Then at the interface,

$$P_f = Y_f(h_f - h) \quad P_s = Y_s(h_s - h)$$

By the geometric relations,

$$h_S = \frac{y_f}{(y_s - y_f)} h_f = \frac{1}{(1.03 - 1)} h_f$$

This equation is Ghyben-Herzberg equation. It shows that the change of seawater thickness in the aquifer depends on the change of fresh-water thickness. However, the increase of saltwater thickness has been amplified by as many as 33 times, depending on the specific weight of the freshwater and the seawater. The distribution of freshwater head along this boundary is a function of saline water depth and the difference in their respective densities. The depth to the freshwater/saltwater interface in a coastal aquifer under steady-state conditions is approximately 33 times the height of the freshwater head (hf) above sea level. In a coastal aquifer freshwater recharge mainly from lateral inflow forms a lens that floats on top of a base of higher density seawater. Changes in recharge and/or induced conditions of pump age and artificial recharge may disturb the hydrostatic equilibrium that results in the movement of the location of the interface and may also create a transition zone. The transition zone is a zone of mixing between salt and fresh water. From this, interface of seawater and freshwater has been determined by slope of land, bed rock position, water table elevation. Fresh water flow was calculated by

$$Q_{Flow} = a \times K \times i$$

Where, a= cross sectional area, *K*= hydraulic conductivity and *i*= hydraulic gradient. In the present study, one unit length along sea coast (*i.e.,* 1 km) was considered and flow to sea was determined. To determine storage contribution, change in water table height and specific yield was used.

$$Q_{Storage} = a \times Specificyield \times \Delta h$$

Then total fresh water flow to sea was calculated by

$$Q_{Total} = Q_{Flow} + Q_{Storage}$$

It was assumed that during summer season, water table depth was at minimum level and hence pumping for irrigation was not allowed. For other months, pumping and water table conditions was considered. The depth of pumping well was considered as half depth of full fluctuation. Pumping volume was taken as half of flow to sea in each month. For the coastal areas, smaller pumps of 1 or 2 hp was considered. Pumping was restricted within 5 hrs per day with break after half time. Daily volume of pumping for each was worked out on the basis of 5 hrs working period. Number of pumps was worked out by dividing total volume by volume of single pump.

RESULTS AND DISCUSSION

Rainfall analysis showed that the district is characterized by hot summer, high humidity and dry winter. The average annual rainfall in the district is 1472mm. The district experiences mild to normal drought 6 to 40 % with rare severe drought 11 to 33 % of the year. Rainfall occurs mainly in the wet period (85%), which normally extends from late July -October, with sparse storm events during the summer months. Nearly 70 % of the total area of the district is utilized as cultivable area and of this 58 % area is being irrigated from various sources like canal and groundwater irrigation. The minimum and maximum temperature ranged between 14°C to 35°C. Annual rainfall and Potential Evapotranspiration (PET) was 1415.63mm and 2052.34mm, respectively. So water demand is always higher than the available water from rainfall except the month of August and September (Figure 1). This indicated that groundwater is being used extensively for growing *rabi* and summer crops in these areas.

The block wise water table depth showed that pre monsoon water table depth in Erasama varied between 2.87m to 4.11m in Erasama and Balikuda blocks. Similarly post monsoon water table depth varied within 1.33m and 2.59m in Jagatsinghpur and Raghunathpur blocks respectively (Figure 2).

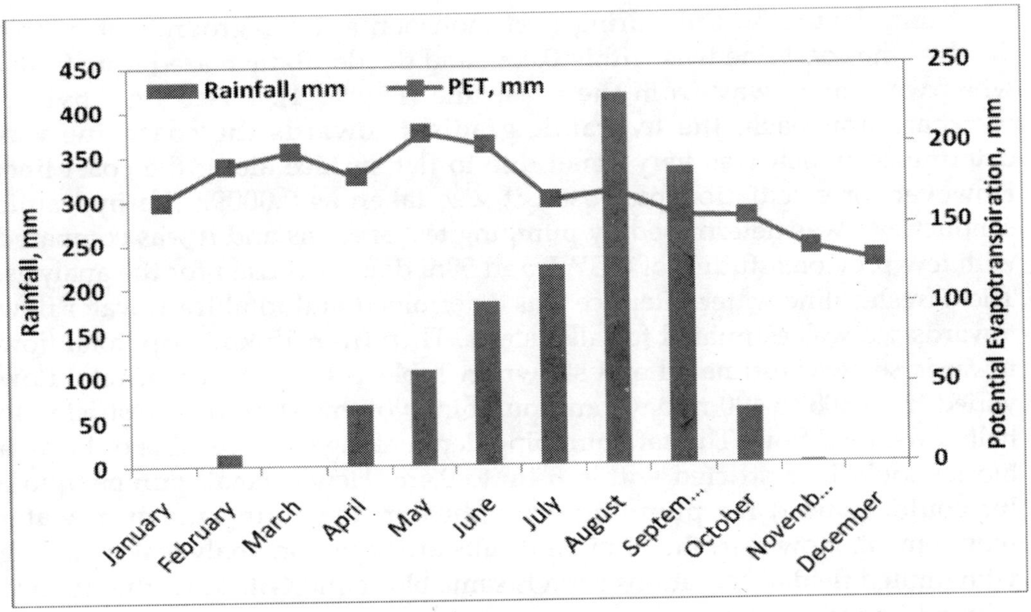

Fig. 1: Water budget in Ersama

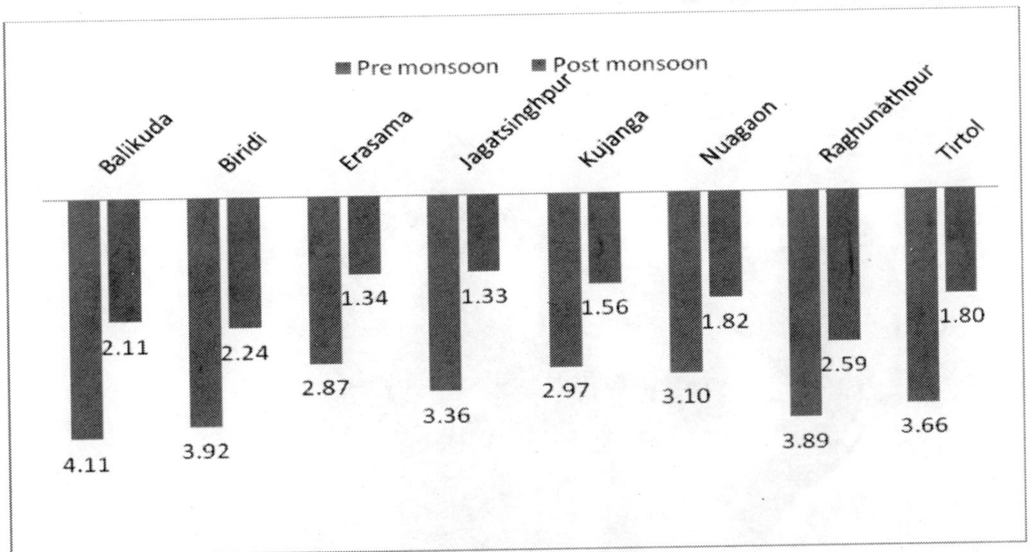

Fig. 2: Average groundwater table depth in Jagatsinghpur

Analytical Approach : Ghyben-Herzberg approach was used for one km width strip which stretched from 1-15 km away from the coast line. Al total 5 strips have been considered to determine the freshwater flow towards sea. The groundwater table depth during pre-monsoon 2010 showed that it varied within 0.13m near to the coast line while it was within 0.71m away from the coast line (Figure 3). But during post monsoon season, groundwater table depth at the coast line was within 0.08m and the depth increased upto 0.40m when we move away from the coast line (Figure 4). Based on Ghyben-Herzberg approach, the hydraulic gradient towards the coast line was determined which was very small due to flat surface along the coast line. However for calculation purposes, it was taken as 0.0009. The hydraulic conductivity was determined by pumping test analysis and it was compared with few previous studies of CGWB and 50m day^{-1} was used for the analysis. Then, fresh-saline water interface was determined and total fresh water flow towards sea was estimated for all 5 strips. Then from 15 km strip, total flow towards sea was estimated and shown in Table 1. It showed that total flow varied from 308 to 400 m^{3}day^{-1} and pumping volume should be restricted to half of the total flow. The safe pumping depth along the coast line of Ersama block should be restricted within 6.4m to 9.4m. Hence, small pumps upto 3 lps could be used for pumping groundwater depending the crop water requirement. However, these initial results are based on analytical modeling with limited field observations from Ersama block and will be further refined with additional field data.

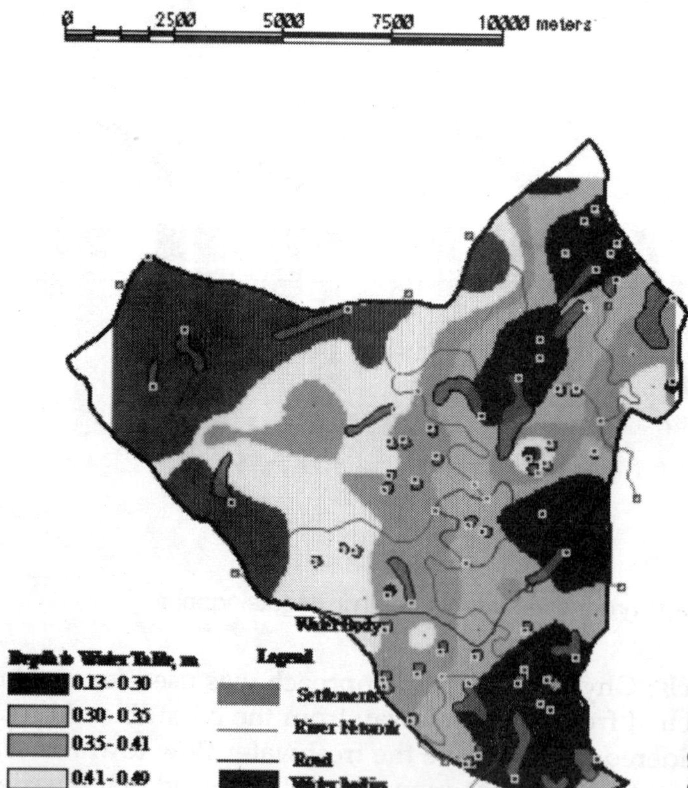

Fig. 3: Spatial variation in groundwater table depth during pre monsoon season in Ersama

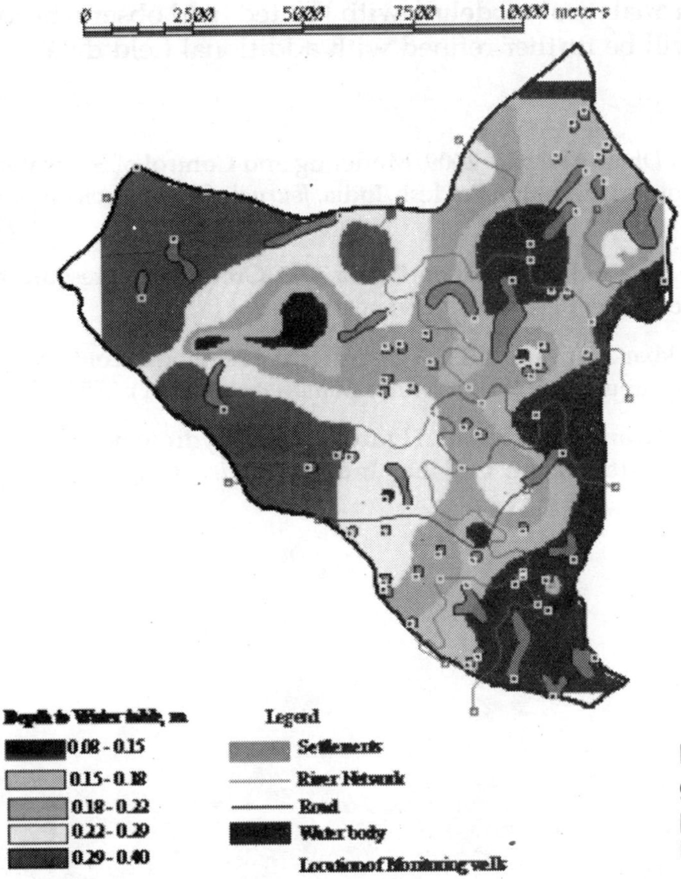

Fig. 4: Spatial variation in groundwater table depth during post monsoon season in Ersama

Table 1: Safe pumping depth along coast line of Ersama by analytical approach

Strip No	Distance from sea (km)	Depth to water table (m)	C/S area for flow (m^2)	Q_{total}, (m^3day^{-1})	Pumping volume (m^3day^{-1})	Pumping depth, (m)
1	15	0.39	6825	307.13	153.56	6.44
2	15	0.42	7315	329.18	164.59	7.35
3	15	0.38	6650	299.25	149.63	6.65
4	15	0.54	8910	400.95	200.48	9.45
5	15	0.42	7350	330.75	165.38	7.35

CONCLUSIONS

In Ersama coastal areas of Jagatsinghpur district, where fresh water is floating over the saline water, safe pumping depth along the coast line should be restricted within 6.4m to 9.4m. Small pumps upto 1-2 lps could be used for pumping groundwater depending the crop water requirement. However, these initial results are based on analytical modeling with limited field observations from Ersama block and will be further refined with additional field data.

REFERENCES

Datta, B., Harikrishna, V. and Dhar, Anirban. 2009. Modeling and Control of Saltwater Intrusion in a Coastal Aquifer of Andhra Pradesh, India. *Journal of Hydro-environment Research*,3(3):148-159.

Directorate of Groundwater Survey and Investigation: Orissa.2011. Groundwater resources of Orissa. Department of Water Resources, Govt of Orissa.

Hany, F. Abd-Elhamid and Akbar A. Javadi. 2011. A cost effective method to control sea water intrusion in coastal aquifers. *Water Resources Management*, 25(11): 2755-2780.

Rejani, R., Jha, M. K., Panda, S.N. and Mull, R.2003. Hydrologic and hydrologic analyses in a coastal Ground water basin, Odisha, India. *Applied Engineering in Agriculture*,19(2):177-186.

□□□

Chapter 69

Enhancing Groundwater Availability in Watersheds of the Dry Zone - A Boon or a Bane ? A Quality Assessment

A. RAIZADA, M. PRABHAVATHI AND R.N. ADHIKARI

INTRODUCTION

Groundwater is the only reliable natural resource for domestic, agriculture and industrial uses. Rapid depletion of groundwater as a consequence of continuously increasing demand and contamination due to various geological and anthropogenic factors threaten the quality of ground water.

Irrigation accounts for over 90% of water consumption in India (Rosegrant *et al.*, 2002). However, investments in irrigated agriculture has begun to decline due to high cost of development of new water resources, inter-sectoral competition and environmental degradation of most assured water sources. Further, productivity levels in the "Green Revolution" areas have begun to stagnate, leading to an increased focus in hitherto less favoured rainfed areas. Integrated rural development in a watershed mode is now being promoted on a massive scale with huge amounts of public investments (Farrington *et al.*, 1999). Numerous reports and papers have indicated that suitable land use practices, in-situ soil and water conservation activities, collection of run off in harvesting structures have led to increases in ground water re-charge (Adhikari *et al.*, 2008; Rao *et al.*, 1996; Sharda *et al.*, 2006). But due to high subsidies provided on irrigation water and energy, ground water exploitation and over use is increasing at alarming rates. This over use is more evident during periods of low rainfall (Table 1).

Besides the problems of availability of sufficient groundwater, water quality issues need significant attention due to their long term impact on the resource base, human health and economic development (Milovanovic, 2007).

Central Soil and Water Conservation Research & Training Institute, Research Centre, Hospet Road, Bellary 583 104, Karnataka 583104, India

For evaluating the suitability of ground water for irrigation or human consumption, it is necessary to be aware about its chemical composition. It is also possible to understand the change in water quality due to rock-water interaction or due to anthropogenic influence (Todd, 1980). Therefore, monitoring the quality of ground water is as important as assessing its quantity. Assessing the risk involved in the use of poor quality water means that the hazard associated with the quality needs to identified (Smith, 2001).

Groundwater contains a wide variety of dissolved inorganic chemical constituents in various concentrations due to interactions between water (a mild acid) and geological material. Chemical composition of ground water varies due to frequency of rains which leach out soils, residence time of water in the root zone, presence of organic matter etc. The water front does not move uniformly, since the strata are heterogeneous. Movement through larger pores is more rapid than finer pores. The overall effect is that the composition of ground water varies spatially and temporally. Various studies have been carried out in the Deccan region of India on the assessment of ground water characteristics (Sujataha and Reddy, 2003; Subramani *et al.*, 2005; Jeevanandam *et al.*, 2006). This paper deals with the assessment of ground water quality obtained through bore wells from two watersheds in the semi-arid parts of Karnataka and their suitability for irrigation.

Table 1: Increase in groundwater use for irrigation with time in K.D.Pally watershed (AP)

Period	Irrigated area	Irrigation quantity required	Water availability*	Over exploitation	Excess exploitation in low RF year over normal year
(Years)	(ha)	(ha m)	(ha m)	(ha m)	(%)
Pre project	37.2	21.42	53.94	0	
Post project					
2005	96.4	61.64	58.41	3.23	
2006	113.5	72.74	58.27	14.47	99.8
2007	113.3	72.55	58.40	14.15	
2008	102.9	65.76	58.52	7.24	
Av. of post - project	106.5	68.17	58.40	9.77	

*Sources of water availability includes rainfall, natural recharge and recharge from WHS both within & outside the watershed. From: Adhikari *et al.*, communicated paper entitled – Augmentation of ground water recharge and water quality improvement by water harvesting structures in the semi-arid Deccan.

MATERIAL AND METHODS

Study area: Two watersheds situated in the Chitradurga district of Karnataka were identified for sampling. One watershed named Netranahalli (14°37′50″ to 14°37′47″ N Lat. & 76°42′46″ to 76°44′51″ E Long.) and the other Ramasagar (14°49′31″ to 14°50′42″ N Lat & 76°44′32″ to76°49′16″) are nearing their implementation phase. Both the watershed lies in the valley of the Tunga and Bhadra rivers which form a part of the Krishna basin. The dominant soils are red sandy loam which has been derived from granite and gneiss. The average annual rainfall of both the watersheds is 417.3 mm and about 80% of the annual precipitation is received in about 31 rainy days between June and October.

The major crops grown are ground nut, sorghum, red gram, cow pea, bajra, ragi, sunflower. Farmers with some irrigation facilities grow crops like maize, paddy, sunflower, chillies, onions and cotton which are normally irrigated by flood irrigation. Due to the risk involved in the highly varied rainfall distribution, the number of bore wells and the command area has increased over time. Bore well water samples weie collected during 2011-12 from 52 locations in Ramasagara and 14 in Netranahalli watersheds. The collected samples were analysed for their physico-chemical constituents such as pH, Electrical conductivity (EC) and important cations such as calcium (Ca^{2+}), magnesium (Mg^{2+}), sodium (Na^{+}), potassium (K^{+}) as well as anions such as chlorides (Cl^{-1}), sulphates (SO_4^{2-}), CO_3^{2-} and HCO_3^{-}. All the physico-chemical parameters mentioned above were assessed using the standard procedures (APHA, 1995). The pH and EC were measured by a Systronics pH meter, while SO_4 was determined by spectrophotometer, Na and K were determined by using flame photometer. Ca, Mg, HCO_3 CO_3 and Cl were analysed by titrimetric method. Sodium adsorption ratio (SAR), soluble sodium percentage (SSP), residual sodium carbonate (RSC), permeability index (PI), magnesium hazard (MH) and Kelly's ratio were analysed by the procedure adapted by Saxena (1990). The major chemical parameters, which decide the suitability of groundwater for irrigation, are EC, SAR, SSP, RSC, PI, MH and Kelly's ratio. The analysis was carried out immediately after the collection of water samples at the centralised soil laboratory.

RESULTS AND DISCUSSION

Evaluation of Water Quality for Irrigation

pH and electrical conductivity (EC): The suitability of groundwater for irrigation depends upon the mineral constituent present in water. Groundwater of study area is neutral to moderately alkaline with pH values from 7.02 to 8.71. The pH of most of the water samples (83%) are within the prescribed Indian standards for irrigation purposes. The electrical conductivity values varied from 0.91 to 7.3 dS m^{-1}. Only 48% of samples were less than permissible limit (2.25 dS m^{-1}) which is normally recommended for agricultural purpose (Table 2).

Table 2: Classification of water based on EC

EC dS m^{-1}	Class	No. of water samples
<0.25	Excellent	Nil
0.25-0.75	Good	Nil
0.75-2.25	Permissible	23 (35)
>2.25	Unsuitable	43 (65)

Figure in parenthesis indicates per cent of water samples

Sodium adsorption ratio (SAR) and soluble sodium percentage (SSP): The sodium adsorption ratio (SAR) is a measure of sodium hazard in the groundwater samples. It is calculated based on Na, Ca and Mg concentrations in the water samples. The SAR value of the water samples varied between 2.6 to 17.1. Saleh *et al* (1999) reported that irrigation water with SAR values greater than 9 will cause permeability problems on shrinking and swelling clayey soils. The SSP content of irrigation water shows that nearly 36% of bore wells are unsuitable for irrigation (Table.3). When sodium concentration is high in irrigation water, sodium ion is absorbed by the clay particles and the soil reduces the permeability and this result in soil with poor internal drainage. Therefore, air and water circulation is restricted during the wet condition and such soils are usually hard when dry (Collins and Jenkins, 1996).

Table 3: Classification of water based on SAR and SSP

SAR	Class	No. of water samples	SSP	Class	No. of water samples
<10	Excellent	52 (79)	<20	Excellent	-
10-18	Good	14 (21)	20-40	Good	8 (12)
18-26	Doubtful	Nil	40-60	Permissible	34 (52)
>26	Unsuitable	Nil	>60	Doubtful	24 (36)

Figure in parenthesis indicates per cent of water samples

Residual sodium carbonate (RSC) and Permeability index (PI): RSC is an important parameter to evaluate the suitability of irrigation water (Siddiqui *et al.*, 2005). RSC of groundwater sample vary from -42.2 to 11meq l^{-1} (Table. 4). A high value of RSC in water leads to an increase in adsorption of sodium on soil. Irrigation water having RSC values greater than 5 meq l^{-1} have been considered harmful to the growth of plants, while irrigation waters with above 2.5 meq l^{-1} are generally considered unsuitable for irrigation. The soil permeability is affected by the long term use of irrigation water as it is influenced by Na, Ca, Mg and HCO_3 content of the soil. PI of groundwater samples vary from 35 to 97%. PI of more than 75% indicates maximum permeability.

Table 4: Classification of water based on RSC and hardness

RSC	Class	No. of water samples	Permeability index	Class	No. of water samples
<1.25	safe	46 (69)	>75	excellent	19 (29)
1.25-2.5	moderate	7 (11)	25-75	good	47 (71)
>2.5	unsuitable	13 (20)	<25	unsuitable	Nil

Figure in parenthesis indicates per cent of water samples

Magnesium hazard (MH) and Kelly's ratio: The relationship between Na, Ca and Mg in the ground water also needs due consideration when evaluating sodicity potential of groundwater for sustained irrigation (Table 5). The MH (Mg/Ca+Mg) and Kelly's ratio (Na/Ca+Mg) were determined which ranged from 49.4 to 90.3 and 0.44 to 5.37, respectively (Table 4). More Mg present in water affects the soil quality converting it to alkaline and decreases crop yield. The ideal Kelly's ratio for irrigation suitability is less than 1. Kelly's ratio describes the proportion of Na with respect to Ca and Mg (Kelly, 1957). Kelly's ratio of more than one indicates an excess level of Na in water, while those with a ratio of more than 3 are unsuitable for irrigation.

Table 5: Classification of water based on Mg hazard and Kelly's ratio

Mg hazard	Class	No.of water samples	Kelly ratio	Class	No. of water samples
<50	Suitable	1 (2)	<1	Suitable	23 (35)
>50	unsuitable	65 (98)	>1	Unsuitable	43 (65)

Figure in parenthesis indicates per cent of water samples

The general dominance of cations is Na>Mg>Ca>K and for anions is HCO_3>SO_4>Cl>CO_3. Total hardness (TH) ranges from 208 to 2252 mg l^{-1}. This is due to weathering of carbonate minerals such as Ca and Mg. Hardness of water limits its use for industrial purposes; it causes scaling of pots and boilers and closure of irrigation pipes.

With the introduction of irrigation in hitherto 'dryland' regions large tracts in Karnataka have either become saline or are increasingly facing problems associated with salinity (Balangevi *et. al.*, 2002). They reported that nearly 80,000 ha is affected by salinity In studies done by them on saline vertisols, they concluded that irrigation with ridge and furrow method at 1.2 crop ET was an optimum irrigation schedule for obtaining higher crop yields.

Surface irrigation methods usually have low water application efficiency leading to water logging and increased problem of salinity and alkalinity. A system of well defined lengths of furrows or borders in relation to the hydraulic gradient, infiltration, discharge and water requirement of crops can enhance the application of efficiency. Open ended furrows are better than closed borders. Frequent irrigation with lesser depth of water per irrigation maintain higher water availability in the upper part of the root zone (Patel & Dastane, 1969). With each irrigation the upper area is leached more than the lower root zone, which reduces osmotic pressure and increases soil wetness.

Since rainfall is a limiting factor in these tracts, it is essential to manage rain water by adopting methods like uniform levelling, *bunding* and tillage operations before the onset of the monsoons. Simple *bunding* and tillage operations can increase yields by 2 to 2.5 times. Dressing with FYM and fertilizers has been observed to be beneficial in increasing yields of wheat and pearl millet grown in salt affected soils (Maliwal and Paliwal, 1972). Application of manure increases nutrient uptake which was reflected in increased biomass and yield of wheat (Punatamkar *et. al.*, 1972). Alternatively, green manuring crops like *Sesbania aculeata* (*Dhaincha*) need to be sown in rotation to reduce the harmful effect of salinity. Application of phosphatic fertilizers to maintain phosphorus availability at levels higher than the normal is also beneficial for enhancing plant growth. Finally, it is the choice of crops which affects yields and productivity, under the given situations, moderately salt tolerating crops like soghum, castor, safflower, cotton, sunhemp, cluster bean, pigeon pea, cowpea, onions can be cultivated.

CONCLUSIONS

Groundwater in the watersheds studied is generally alkaline and hard to very hard in nature. The concentrations of major cations in groundwater are in the order of Na>Mg>Ca>K, while for the anions HCO_3>SO_4>Cl>CO_3. Most of the water samples fall in the unsuitable range for irrigation due to excess of EC, Na and Mg. These irrigated areas need adequate drainage to overcome salinity problems created by continued use of unsuitable water. However, P_I values indicate that almost all the groundwater samples are suitable for irrigation. According to overall assessment of the watershed groundwater quality was found to be suitable for irrigation purposes in only 35% of the wells sampled.

Thus, there is an urgent need to encourage and popularize practices that minimise use of bore well water by selection of suitable crops, crop rotation and educating farmers about the judicious drawal of water to minimize the harmful effect of indiscriminate irrigation. At the policy level there is a need to take up suitable measures for ground water recharge and controlled exploitation simultaneously and development of guidelines for ground water exploitation in the semi-arid regions. Ground water is a precious natural resource and its sustainable management is the need of the hour.

REFERENCES

A.P.H.A. 1995. *Standard methods for the examination of water and waste water* (19th ed.). American Public Health Association, Washington D.C 874.

Adhikari.R.N., Singh.A.K., Math.S.K.N., Mishra.P.K. and Reddy.K.K, 2008. Response of water harvesting structures on ground water recharge in red soil of semi-arid region of Andhra Pradesh, *Jour. of Indian Water Resources Society*, 28 (2): 1-5.

Balangavi, S., Ravishankar, G. and Hebbara, M. 2002. Land and water management strategies for sunflower in saline Vertisols of Tunga Bhadra irrigation command area. *In*: *Resource Conserving Technologies for Social Upliftment* (Ed: V.N.Sharda, K.S.Dadhwal, S.K.Dhyani, G.P.Juyal, P.Dogra, A.Raizada & O.P.S.Khola) pp 320 - 322.

Collins, R. and A.Jenkins. 1996. The impact of agriculture and land use on stream chemistry in the middle hills of the Himalayas. *Nepal J. Hydro.*, 185: 71-86

Farrington J, Turton C and James AJ (eds.). 1999. *Participatory Watershed Development: Challenges for the Twenty-First Century*. New Delhi: Oxford University Press.

Jeevanandam, M.,Kannan, R., Srinivasulu, S., and Rammohan, V. 2006. Hydrochemistry and groundwater quality assessment of lower part of the Ponnaiyar River Basin, Cuddalore, South India. *Environmental Monitoring and Assessment,* 132(1): 263 - 274.

Kelly, W.P.1957. Adsorbed sodium cation exchange capacity and percentage sodium adsorption in alkali soils. *Science,* 84: 473-477

Maliwal, G.L. and Paliwal, K.V. 1972. Effect of fertilizers and manure on the growth and chemical composition of wheat irrigated with saline water. *Agrochemica,* 16: 450- 458.

Milovanovic,M.2007. Water quality assessment and determination of pollution sources along the Axios/Vardar River, Southeastern Europe, *Desalination*, 213: 159 - 173.

Patel, A.S. and Dastane, N.G. 1969. Effect of layout, irrigation frequency and level of irrigation salinity on radish. *Ind. J. Agric. Sci,* 39: 706- 712.

Punatamkar, S.S., Mehta, P.C. and Seth, S.P. 1972. Effect of gypsum and manure on the growth of wheat irrigated with bicarbonate rich water. *J. Indian Soc. Soil Sci., 20*: 281-285.

Rao, M.S.R., Adhikari, R.N., Chittaranjan, S. and Chandrappa, M. 1996. Influence conservation measures on ground water regimes in a semi-arid tract of south India. *Agricultural Water Management*, 30: 301-312.

Rosegrant M. 1997. *Water resources in the 21st Century: Challenges and implications for action.* Food,Agriculture and the Environment Discussion Paper 20. IFPRI, Washington, DC.

Saleh, A., Al-Ruwaih, F., and Shehata,M.1999. Hydro- geochemical processes operating within the main aquifers of Kuwait. J. *Arid Env.*, 42.195-209

Saxena, M.M.1990. *Environmental analysis of water, soil and air*. Agro-botanical Pub. India, 1-16.

Sharda, V.N., Kurothe, R.S., Sena, D.R., Pande. V.C. and Tiwari, S.P. 2006. Estimation of groundwater recharge from water storage structures in semi-arid climate of India. *Journal of Hydrology*, 329: 224-243.

Siddiqui, A., Naseem, S. and Jalil,T. 2005. Groundwater quality assessment in and around Kalu Khuhar Super highway, Sindh. *Pakistan J. Applied Sciences*, 5 (7): 1260-1265.

Smith, K.2001. *Environment Hazards: Assessing risk and Reducing Disaster*, (3rd Edition) Routlege, London.324p.

Subramani, T., Elango, L. and Damodarasamy, S.R. 2005. Groundwater quality and its suitability for drinking and agricultural use in Chithar river Basin, Tamil Nadu, India. *Environmental Geology*, 47: 1099 – 1110.

Sujataha, D. and Reddy, B.R. 2003. Quality characterization of groundwater in the south –eastern part of the Ranga Reddy District, Andhra Pradesh, India. *Environmental Geology*, 44: 579 – 586.

Todd, D.K. 1980. *Ground Water Hydrology*, Wiley, New York. 535p.

□□□

Chapter 70

Small Scale Broiler Farming : An Alternate Livelihood for the Landless in Rural Odisha - A Successful Intervention

N. PANDA, B. PANIGRAHI, S. NAYAK, B. BEHERA AND S.N. DASH

INTRODUCTION

Poultry farming is one of the fastest growing sector in agriculture as compared to other agricultural activities. India ranks the 3rd in egg production and the 4th in broiler production in the world (FAO, 2008). The steady progress of poultry development in the country has been a result of research and development over the years. The various factors for steady progress include nutrition, management and health which directly or indirectly affect the poultry industry.

Livestock in Odisha, like agriculture is the endeavour of the marginal and landless farmers of the state. Around 80% of population of the state in rural areas earn their livelihood from agriculture and allied sectors mostly predominantly by animal husbandry(Kurup, 2003). Backyard poultry has been raised as part of their traditional integrated farming system involving crops and livestock since centuries by the poor, weaker and vulnerable section of the society including tribal. The native colour birds have less genetic potency and are prone to many contagious diseases leading to higher mortality and less production efficiency (Rama Rao *et al.*, 2007). The present commercial poultry germplasm has tremendous growth potential that can be exploited by the small, marginal and landless farmers of the state for livelihood support and employment opportunity. Taking this hypothesis into consideration in NAIP-3, the broiler poultry has been introduced in the six clusters of villages in 3 disadvantaged districts of Odisha namely, Dhenkanal, Kandhamal and Kalahandi.

Orissa University of Agriculture and Technology Bhubaneswar- 751 003

MATERIALS AND METHODS

Twenty farmers were selected in the 6 clusters of NAIP adopted villages in three districts by the cluster committee, farmers from broiler twenty beneficiaries included 9 from Dhenkanal 4 from Kandhamal and 7 from Kalahandi (Table 1). Twenty (20) low cost poultry shed, each of 150 sq ft. area were constructed in the land area owned by beneficiaries with provision of future expansion. The dimension of the shed was 15 foot (4.5 m) length and 10 foot (3.0 m) width. Out of 150 sq. ft area, an area of 50m^2 was partitioned a partition was made for storage of feed, equipment and other accessories and the rest 100 sq ft. was used for rearing of birds. For rearing of chicks the farmers were provided with feeders, waterers and brooders for scientific management.

Table 1: Cluster wise farmers selected for broiler farming

Name of the District	Cluster	No. of farmers	Name of the farmers
Dhenkanal	Dhenkanal Sadar	5	Sribatcha Biswal, Ranjan Nayak, Artabandhu Mallick, Sridhar Bhuyan and Saranga Raj
	Odapada	4	Gadadhar Dehuri, Soubhagini Mohanty, Soudamini Parida and Santosh Nayak
Kandhamal	G.Udayagiri	2	Janaka Pradhan and Pratap Mallick
	Khajuripada	2	Basanta Bindhani and Sanjeeb Sahani
Kalahandi	Narla	6	Prasanna Sa, Trinath Pradhan, Tapi Durga, Barun Majhi, Baladev Deep and Sibaram Sabar
	Golamunda	1	Baishnab Naik

Each farmer was provided with 100 numbers of day old commercial strains of broiler chicks, 3 kg feed (starter and finisher) and medicines and vaccines. The farmers were trained at the farm site about brooding, feeding, medication and vaccination of poultry birds. The birds were fed with starter feed for 2-3 weeks and finisher feed upto marketing. The birds were vaccinated against Ranikhet Disease by Lassota Strain on 7th day, against Gumboro disease on

14th day and lassota vaccination was repeated on 21st day. The birds were purchased from the authorized local supplier and the farmers were linked with the suppliers for sale of birds at marketable age.

RESULTS AND CONCLUSION

The growth of the birds was very good and the average weight of the birds at the age of selling (35- 42 days) ranged from 1.53 kg to 2.01 kg (Table 2). The mortality of the birds upto marketing varied from 2 to 5% depending on the management condition. This mortality is acceptable for small farming. Bhapkar (2007) reported that uniformity of the broiler plays an important role in achieving the market body weight which was seen in this case. The feed conversion ratio (FCR) was 1.54 to 1.82 in different cluster. The farmers sold the birds at a rate of Rs. 60-75 kg^{-1} live weight and net gain of Rs.1100- 1200 $batch^{-1}$ of 100 birds.

S. No	Parameter	District wise achievement			Total/
		Dhenkanal	Kandhamal	Kalahandi	Average
1	No of farmers given (100 birds)	9	4	7	20
2	Age at selling (week)	5.00-6.00	6.00-8.00	5.00-7.00	5.00-6.00
3	Live weight at marketing(kg)	1.58-2.20	1.53-1.90	1.55-1.93	1.68-2.10
4	Mortality upto marketing (%)	3.20-5.00	3.00-9.00	3.00-7.00	2.00-6.00
5	Feed Conversion Ratio (FCR)	1.54-1.75	1.60-2.10	1.58-1.85	1.54-1.82

The farmers are now very well confident to vaccinate the birds at right time and right dose. They are capable of feeding, watering, giving medicines and doing other managemental practises on their own. Most of the beneficiaries are continuing with the existing sheds. One farmer, Prasanna Sa from Ghantamal of Narla cluster reared and sold 7-8 batches per year during 2010-11. By utilising the profit, he constructed one more shed with capacity for rearing of 150 birds. He is rearing 21- 25 days of birds upto marketing and getting regular income. Another farmer from Belapada village, Santosh Nayak was inspired by the net profit. He constructed another shed with capacity for rearing 300 birds and he is rearing both coloured and commercial strains and getting a good profit of Rs. 2100 to 3600 $batch^{-1}$.

CONCLUSIONS

From the above result it is seen that rearing of broilers in small scale at village condition is a profitable livelihood that can be expanded in other villages. The farmers should be given regular training for scientific management, balanced feeding and vaccination of the birds for sustainable production and economic upliftment. A farmer can construct a poultry shed of 1500 sq.ft. area partitioned into 3 compartments of 500 sq.ft. each on rotation basis if he can purchase 500 number of a day old chicks on every 21 days. He can successfully run 15 batches of 500 birds excluding two months of peak summer. He can earn a profit of Rs.75,000 $year^{-1}$ (5,000 $batch^{-1}$) in return he can maintain minimum livelihood of 6 members.

REFERENCES

Bhapkar,P. 2007. Broiler Uniformity- A key economic production. *Poultry Fortune*, 23-28.

Food and Agriculture Organization, 2008

Kurup, M.P.G. 2003. Livestock in Orissa – The socio economic perspective- Manihar Pub. New Delhi.

Rama Rao, S.V., Sharma,R.P., Raju, MVLN and Panda,A.K.,2007. Poultry for poverty alleviation and rural development in India. *Livestock International*, 11:15-23.

□□□

Chapter 71

Adoption of Soil and Water Management Practices in Rainfed Areas by the Beneficiaries of Government Sponsored Programmes

S. K. NATH, B.P.MOHAPATRA[1], R. K. RAJ[2], K.C.BARIK

INTRODUCTION

About 533 M tonns *i.e.*,16.4 t ha^{-1} of soil is eroded in each year in India. From the inception of planning commission, there are a series of programmes implemented in India for soil conservation in co-ordination with rural development, employment and poverty elimination. From 7th five year plan watershed programmes become the major thrust areas of the Ministry of Agriculture and Cooperation. To bring the watershed development in a streamlined manner for a wider vision of soil and water conservation in rain-fed areas in India the Central government launched National Watershed Development Project for Rain fed Areas (NWDPRA) during 1986-87, *i.e.* 8th five year plan. Watershed is a definite geo-hydrological unit or land mass from which water drains through a common outlet into a stream or a river. Watershed management concept refers to rational utilization of land and water for optimum and sustained production with minimum hazard to natural resources. Odisha state lies in the east-coast of India have only 31% irrigation intensity during 2006 in comparison to 44% of the national average. Watershed programme under the NWDPRA project is continuing in all the thirty districts for soil and water conservation since the 8th five year plan. Participation of people and adoption of the various activities is the key indicator of successful implementation of the programme. Keeping this fact in view this study was conducted to find out the adoption of soil and water conservation practices by the beneficiaries of NWDPRA.

Krishi Vigyan Kendra, OUAT, Deogarh, Odisha -768 119

[1]Associate Professor (Extension Education), [2]Joint Director, DEE, OUAT, Bhubaneswar

MATERIALS AND METHODS

Out of the thirty districts, Dhenkanal district was selected randomly which is present in the mid central table land agro-climatic zone. The total cropped area of the district is 248.4 thousand ha of which only 51.74 thousand ha is irrigated. Badajora nallah mini watershed of Dhenkanal sadar block and Benga nallah of Odapada block were purposively selected as NWDPRA programmes were implemented by the soil conservation department of the district in both the watersheds. Both the watersheds were covering two villages each. Jankira village of Dhenkanal sadar block and Kalanga village of Odapada block were taken, respectively for the study purpose. Forty respondents from project area and an equal number of respondents from non-project area belonging to neighboring villages where NWDPRA programmes were not implemented, *i.e.*, Kaimati and Dinabandhupur, were selected randomly to assess the implementation of the programmes. Data was collected through a structured interview schedule. The respondents were categorized into three groups on the basis of knowledge and adoption of the watershed activities, *i.e.* high, medium and low categories. Correlation among the socio-economic variables and knowledge and adoption of NWDPRA programmes were found out. The adoption of practices was measured in a 3-point continuum. Mostly adopted, partially adopted and least adopted of the practices were allotted to 2, 1, 0 scores, respectively. The extent of adoption was ranked according the total score obtained by each practice.

RESULTS AND DISCUSSION

NWDPRA aims at a holistic development of the project areas through soil and water conservation methods. After the introduction of the programme, the adoption level of various watershed management practices by the beneficiaries were studied and presented in table 1.

Table 1: Distribution of respondents according to their adoption of watershed management practices

Level	Project area		Non-project area	
	F	%	F	%
High	14	35.00	04	10.00
Medium	21	52.50	15	37.50
Low	05	12.50	21	52.50

The economic activities taken up by the respondents in project and non-project areas were compared and mentioned in table 2.

Table 2: Distribution of respondents according to their adoption of economic activities

Level	Project area		Non-project area	
	F	%	F	%
High	13	47.50	05	12.50
Medium	19	32.50	6	15.00
Low	8	20.00	29	72.50

It can be observed from the table 1 that 47.50, 32.50, 20 % of the respondents of the project area had high, medium and low level of adoption, respectively. In the non-project area 37.50% respondents were having medium level of adoption of watershed activities while 12.50 and 52.50% of respondents were of high and low category. It is observed from table 2 that the adoption of various economic activities of inhabitants of project area was higher than the non- project area. It may be due to higher socio-economic status of the people of project area. To establish a relationship between some selected socio-economic variables with the knowledge level and adoption of soil and water management practices by the beneficiaries of project area, the spearman's correlation coefficient is calculated and presented below. This study resembles the findings of Bhaskaran and Praveen (1982) and Dhyani *et al.* (1984).

Table 3: Relationship of socio-economic variables with the knowledge and adoption level of respondents in the project area

Variables	R value	
	Knowledge	Adoption
Education	0.427**	0.2235*
Caste	0.014	0.021
Farm size	0.5774**	0.1126
Annual income	0.9435**	0.078
Social participation	0.2123*	0.1269
Extension contact	0.231*	0.623**

*significant at 5% level, **significant at 1% level

From the above table it can be found that education, annual income and farm size has positively and highly significant relationship with knowledge level of respondents where as only extension contact has the same relationship with the adoption of various soil and water conservation methods. Caste of the respondents was found positive but not significantly related with the knowledge and adoption of practices. Educational qualification also had positive and significant relationship with both of the parameters. The above study supports the findings of Rao (1988) and Prabhu and Kandan (1990).

Table 4: Extent of adoption of soil conservation practices in project areas

S. No.	Practices	Adopted	Partially adopted	Not adopted	Total score	Rank
1	Growing vetiver grass on contour hedges	18(45.00)	05(12.50)	17(42.50)	41	V
2	Contour cultivation	24(60.00)	2(5.00)	16(40.00)	50	II
3	Terrace farming in uplands.	20(50.00)	3(7.50)	17(42.50)	43	IV
4	Construction and repair of bonds and hedges	18(45.00)	8(20.00)	14(35.00)	44	III
5	Planting of fast growing trees in barren lands	30(75.00)	5(12.50)	5(12.50)	65	I
6	Planting of trees along the road side	10(25.00)	5(12.50)	25(62.50)	25	VI
7	Planting of wood trees behind house	10(25.00)	5(12.50)	25(62.50)	25	VI

Figures in parentheses represent the percentage

A perusal from Table 4 reveals that the most adopting soil conservation practice in the NWDPRA programme area was the planting of fast growing trees in barren lands (75%). While analyzing rank analysis in the adoption of soil conservation methods it was found that the contour cultivation and construction of bonds and hedges ranked second and third, respectively. Planting of trees along road side and behind the house were the least adopted practices followed.62.50% of people were not adopting both the practices in the project area. It indicates lack of awareness among the beneficiaries of the programme on these important aspects. This may be due to the improper monitoring of the implementation of the programme by the competent authority.

CONCLUSIONS

Watershed development programmes are aimed at a holistic improvement of the project area. From the study it was found that the soil and water conservation practices were not totally adopted among all the beneficiaries of the project area. The implementing agency should try for proper dissemination of knowledge on watershed development and ultimate adoption by the beneficiaries. Proper monitoring system should be developed for effective implementation of the programme and eradication of the loopholes in the system.

REFERENCES

Bhaskaran, K. and Praveen. 1982. Adoption of improved dry land agricultural technology in an integrated dryland agricultural development project in Andhra Pradesh. *Indian Journal of Extension Education.* 18 (1&2): 32-39.

Dhyani, B. L., Ram babu, Agrawal, M. C. and Kumar, N. 1984. Impact of watershed management technology on farm-income. *Indian Journal of Soil Conservation.* 22(3): 22-27.

Prabhu, M. K. and Kandan, K. C. 1990. A study on knowledge level of farmers towards soil conservation practices. *Indian Journal of Extension Education.* 26 (3&4):84-88.

Reddy, H. C. V. 1987. Attitude and adoption behavior of farmers relating to watershed development programme in Bangalore district, M.Sc.(Ag.) Thesis, UAS, Bangalore.

❏❏❏

Chapter 72

Delineation and Prioritization of Suitable Areas for Recharge in State of Tamilnadu

PRADEEP KUMAR ROUT, M. J. KALEDHONKAR[1]

INTRODUCTION

Artificial recharge is also defined as the process by which ground water is augmented at a rate exceeding that under natural condition of replenishment. Artificial recharge aims at augmenting the natural replenishment of ground water storage by some method of construction, spreading of water, or by artificially changing natural conditions. It is useful for reducing overdraft, conserving surface run-off, and increasing available ground water supplies.

Planning of artificial groundwater recharge schemes : CGWB (Central Ground Water Board) (2000) provided guidelines for recharge planning and CGWB (2002) gave master plan for groundwater recharge in India. Proper planning is essential for the successful outcome of any artificial recharge scheme. Planning of artificial recharge schemes involves the formulation of a suitable plan, under a given set of natural conditions, to augment the natural ground water recharge. CGWB (2007) suggested detailed consideration of the following aspects as necessary requirement for evolving a realistic plan for an artificial recharge scheme.

MATERIALS AND METHODS

By learning Arc-GIS software, a map can be created, edited, georeferenced, digitized and contouring can also be done. Weighted sum, Weighted Overlay, Raster Conversion *etc* can also be done by arc-GIS spatial analysis for ground water Recharge.

College of Agricultural Engineering and Technology, Bhubaneswar, Odisha, India
[1]Directorate of Water Management, Bhubaneswar, Odisha, India

Spatial analysis for delineation of suitable areas for groundwater recharge : The CGWB (2007) has identified that climatic, topographic, soil, land-use and hydro-geologic conditions are important factors controlling the suitability of an area for artificial recharge. The climatic conditions broadly determine the spatial and temporal availability of water for recharge, whereas the topography controls the extent of run-off and retention. The prevalent soil and land use conditions determine the extent of infiltration, whereas the hydro-geologic conditions govern the occurrence of potential aquifer systems and their suitability for artificial recharge.

Above discussion suggests that soil properties, which control infiltration process and aquifer properties, which allow storage of recharged water in aquifer are important in identifying suitable areas for recharge. Accordingly, soil infiltration or soil drainage can be considered as important variable influencing recharge process. Soil infiltration is mainly controlled by soil texture, land slope, land use and moisture content of soil. The land slope and land use both can be put together by considering land capability class. Once runoff water enters into soil mass through infiltration process, it will percolate down and reach to aquifer, where it is stored. If aquifer has good storage space in it, the recharged water can be held for longer time. Thus aquifer properties also influence identification of suitable areas for recharge. In view of above discussion, three main properties soil drainage, land capability and geology are considered as important variables to delineate suitable areas for recharge. If both properties infiltration and aquifer storage capacity are given equal importance, then soil drainage and land capability will get 50 percent weightage while geology responsible for aquifer storage will get 50 % weightage. Soil drainage and land capability are treated equally, then each will get 25 % weightage (Table 1).

Table 1 : Weightage to different themes in delineation of suitable areas for recharge

S. No.	Theme	Weightage (%)
1	Geology	50.00
2	Soil drainage	25.00
3	Land capability	25.00

These three variables/themes vary spatially. Spatial variations, within the theme, are considered by dividing the theme into different categories and assigning appropriate scores.

Spatial analysis for prioritization of suitable areas for groundwater recharge **:** In case of delineation of suitable area for recharge, spatial variations in geology, soil drainage and land capability within the state are considered. The depth to water table is added to these themes to decide prioritization of recharge planning. The depth to water table theme is given 50 % weightage to prioritization process. The remaining three themes are given 50 % weightage. The details are provided in Table 2.

Table 2: Weightage to different themes in prioritization process

S. No.	Theme	Weightage (%)
1	Depth to water table	50.00
2	Geology	25.00
3	Soil drainage	12.50
4	Land capability	12.50

CGWB (2007) suggests that prioritization of areas for artificial recharge is normally done by overlaying post monsoon depth to water level maps with maps depicting the long-term trend of ground water levels. From these maps, it is possible to demarcate areas with various combinations of depth to water levels and water level trends. Normally, areas having deeper water levels and declining water level trends are given higher priority identification of area feasible for artificial recharge. Areas having shallow water levels / rising water level trends are not considered.

Availability of water for recharge : A realistic assessment and quantification of the source water is very much essential in recharge planning. Once, the area for recharge planning is prioritized, information about source of water and amount of water is required. In our monsoon type climate, runoff from rain events is important source for recharging water. Amount of runoff water varies with annual rainfall amount, soil type and land use. An appropriate value of runoff coefficient is determined considering land use (or land capability) and soil texture. It helps in estimation amount of runoff available from annual rainfall for recharge purpose. On the basis of different values of annual rainfall, state is divided into 5 categories as given below in Table 3.

Table 3: Different categories on basis annual rainfall

S. No.	Rainfall category	Annual rainfall amount (mm)
1	Very low	<750
2	Low	750-1000
3	Medium	1000-1500
4	High	1500-2500
5	Very High	>2500

The availability of water for recharge is linked with annual rainfall, runoff coefficient and area, where recharge planning is done. In this study, available water for recharge is estimated for Areas with Priority 1 and 2 using following formula.

AvaWR = Area * RC * AR

Where

AvaWR = Available water for recharge

RC = Runoff coefficient

AR = Annual rainfall

By selecting appropriate values of runoff coefficient, the amount of available water for recharge is estimated.

RESULTS AND DISCUSSION

Delineation of suitable areas for groundwater recharge : Three main properties geology, soil drainage and land capability are identified as important variables influencing groundwater recharge. Geo-referenced digitized map of geology, soil drainage and land capability of Tamil Nadu with weightage as defined in Table 1, 2, and Table 3 respectively, and are shown in Figure 1, Figure 2 and Figure 3 respectively

The above shown maps (Figure 1, 2 and 3), with overall weightage as 50, 25 and 25 %, are used in spatial analysis using weighted sum function of ArcGIS and results are shown in Figure 4. In this figure suitable areas are classified into 10 classes on the basis of scores during spatial analysis.

Prioritization of suitable areas for groundwater recharge : Similar to delineation of suitable area for recharge, prioritization of suitable recharge areas is done with geology, soil drainage, land capability and depth to water table theme. The geo-referenced digitized map of depth to water table for state of Tamil Nadu is shown in Figure 5.

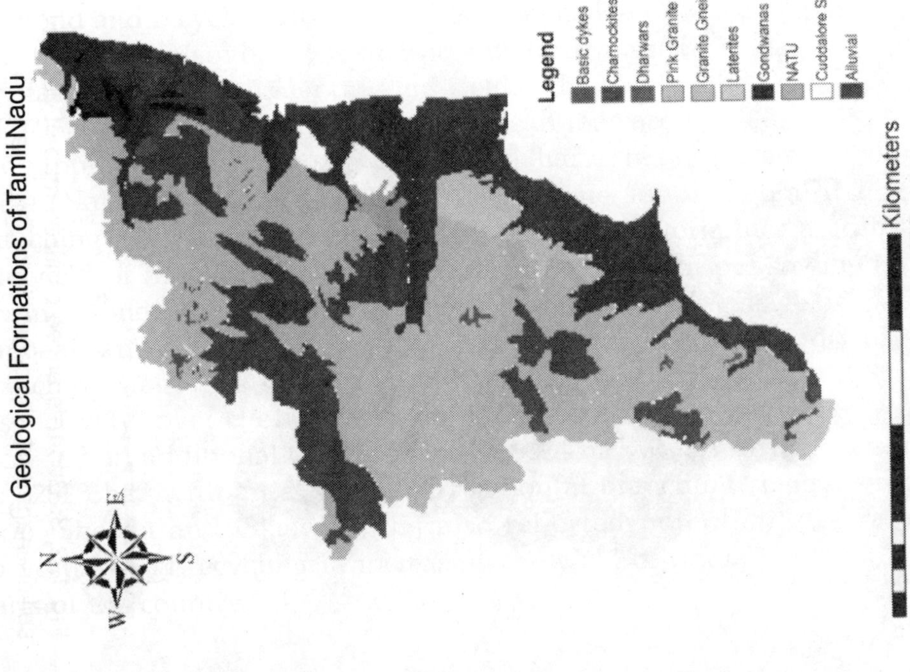

Fig. 1

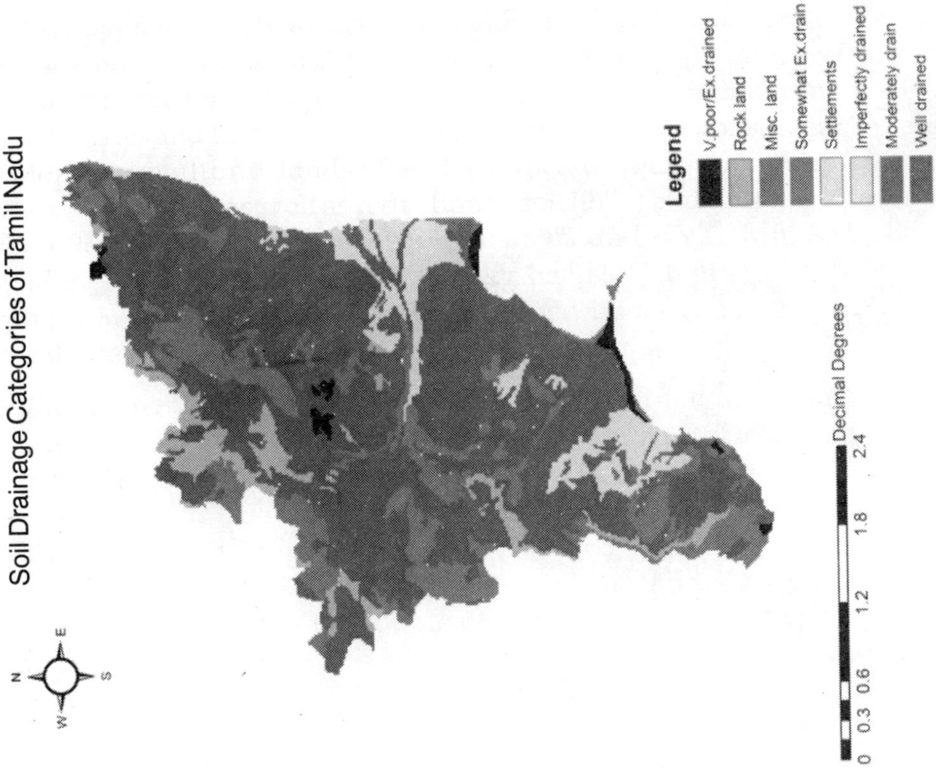

Fig. 2

Fig 1, 2, 3: Geo-referenced digitized map of geology, soil drainage and land capability of Tamil Nadu

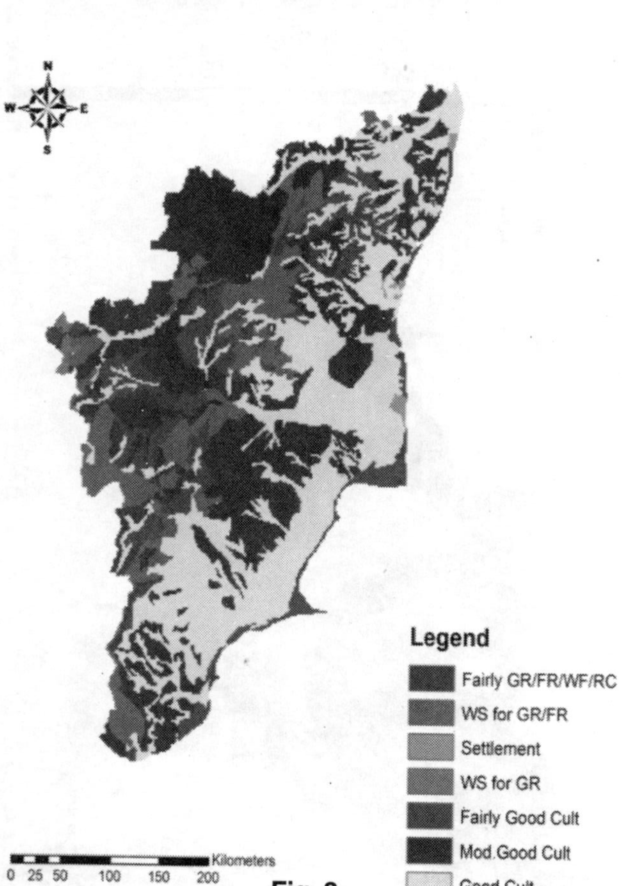

Fig. 3

Fig 1, 2, 3: Geo-referenced digitized map of geology, soil drainage and land capability of Tamil Nadu

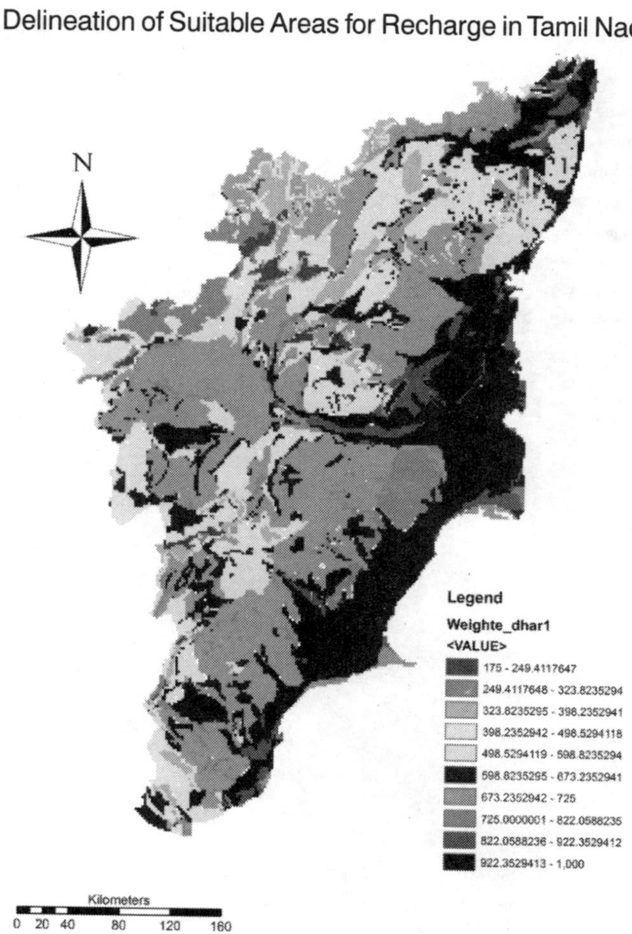

Fig. 4: Delineation of suitable areas for recharge in Tamil Nadu (10 Class)

The digitized maps of different themes are given different weightages as per their importance. The depth to water table (Figure 5) is given 50 % weightage. The remaining three themes are given 50 % weightage; 25 % to geology (Figure 1), 12.5 % to soil drainage (Figure 2) and 12.5 % to land capability (Figure 3). Spatial analysis is performed using weighted sum function of ArcGIS and results of prioritization process are shown in Figure 6.

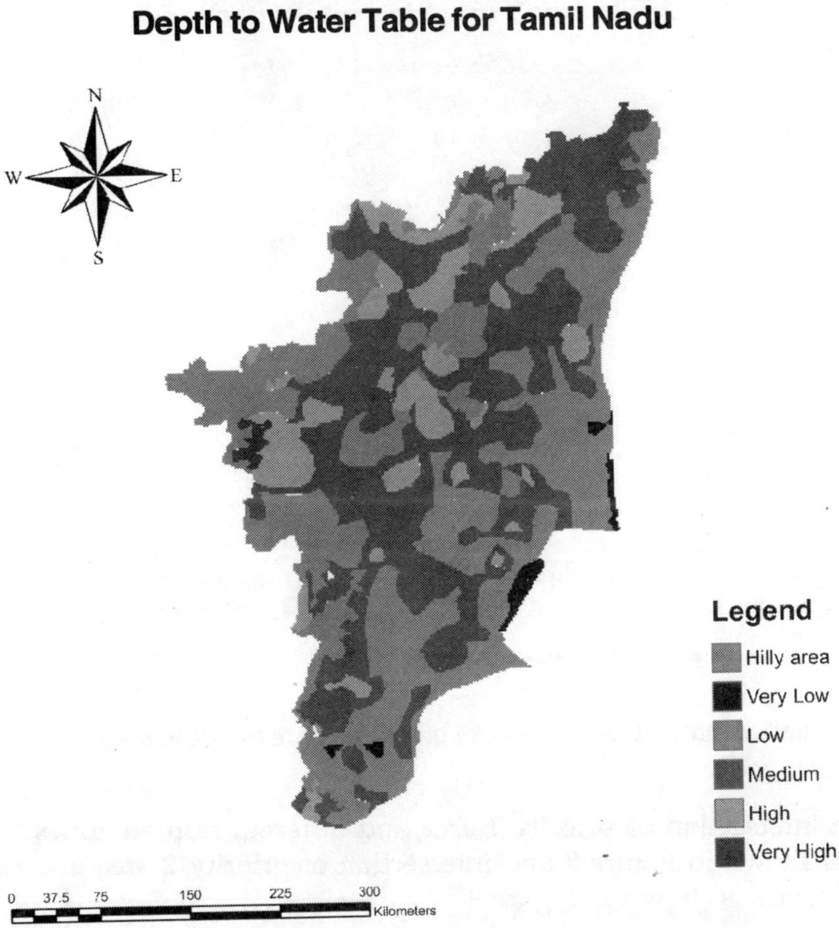

Fig. 5: Digitized map of depth to water table for Tamil Nadu

Availability of water for recharge in prioritized suitable areas **:** Amount of runoff water varies with annual rainfall amount, soil type and land use. A runoff coefficient is influenced by these variables. Therefore, selection of appropriate value of runoff coefficient is necessary for determination of excess amount of runoff for recharge purpose. The geo referenced digitized map of Tamil Nadu is shown in Figure 7.

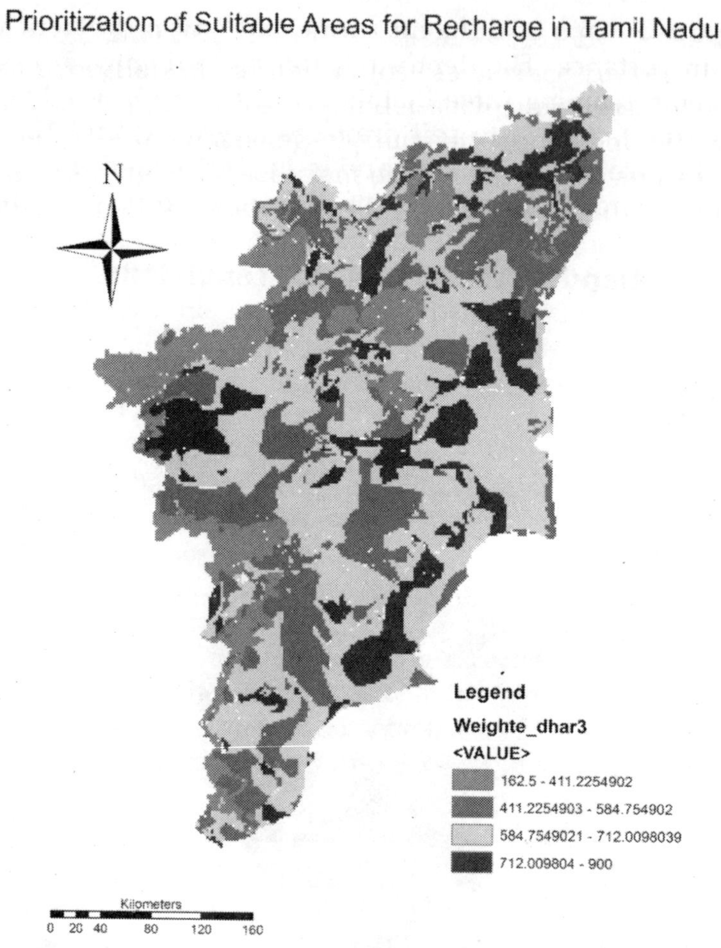

Fig. 6: Prioritization of suitable areas for groundwater recharge in state of Tamil Nadu

An intersection of priority 1 area and different rainfall zones of Tamil Nadu is shown in Figure 8 and intersection of priority 2 area and different rainfall zones is shown in Figure 9.

By using information provided by CGWB runoff coefficients for very low, low and medium categories of rainfall for state of Tamil Nadu are determined as given in Table 4. These three categories are considered because Priority 1 and Priority 2 areas are in these rainfall categories only.

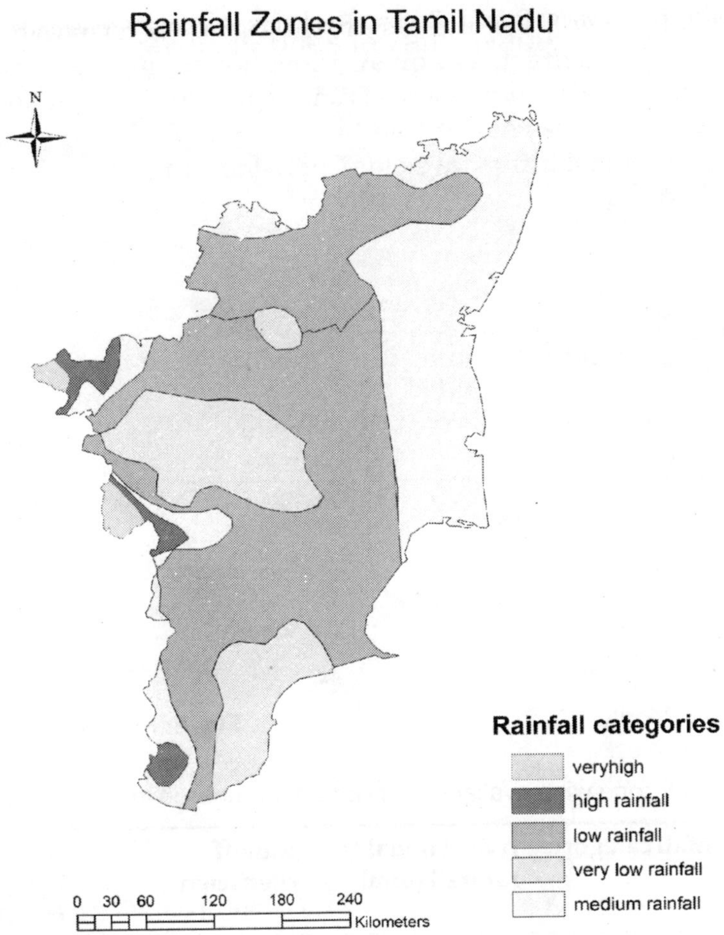

Fig. 7: Digitized map of rainfall zones of Tamil Nadu

Table 4: Runoff coefficient for different categories annual rainfall

S. No.	Rainfall category	Annual rainfall amount (mm)	Runoff coefficient
1	Very low	<750	0.30
2	Low	750-1000	0.35
5	Medium	1000-1500	0.45

Using rainfall zone wise runoff coefficient values given in Table 4, availability of water for recharge on basis of per square km area under different zones is given in Table 5.

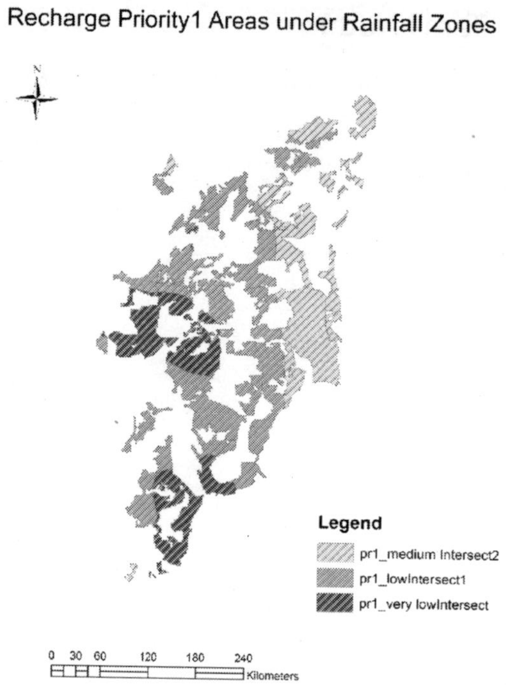

Fig. 8: Priority 1 Areas

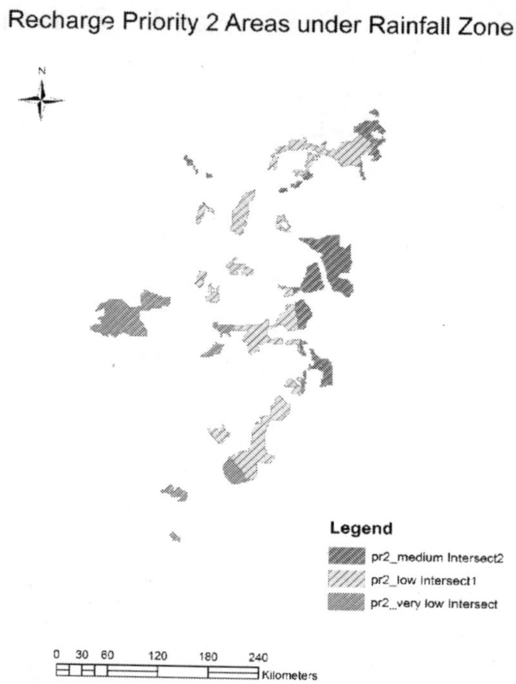

Fig. 9: Priority 2 Areas

Table 3.5: Rainfall zone wise availability of water for recharge in state of Tamil Nadu

S. No.	Rainfall category	Ave. Annual rainfall (mm)	Runoff coefficient	Runoff volume per sq.km (million m^3)
1	Very low	650	0.30	0.195
2	Low	875	0.35	0.31
5	Medium	1250	0.45	0.56

The results in Table 5 suggest that runoff volume for recharge may increase with annual rainfall amount. It is 0.195 million m^3 in very low category, while it is 0.31 and 0.56 million m^3 for low and medium category, respectively. This runoff volume will be useful to decide number of recharge structures in Priority 1 and Priority 2 areas.

CONCLUSIONS

Methodology for delineation of suitable areas for GW Recharge & Prioritization of Recharge planning has been developed & applied for the state of Tamil Nadu. The Estimation of excess water for Recharge in Prioritized areas in different zones was done.

REFERENCES

Ambrish Kumar and Sharma, H.C. 2008. Land and water management for groundwater recharge using RS and GIS. Groundwater Resources Assessment, Recharge and Modeling, Macmillan India Ltd., New Delhi. 179-188.

Central Ground Water Board. 2000. Guide on artificial recharge to groundwater. 93p.

Central Ground Water Board. 2002. Master Plan for artificial recharge to groundwater in India. 116p.

Central Ground Water Board. 2007. Manual on Artificial Recharge to Groundwater, 181p.

Gontia, N.K., Dave, H.N., Gohel, S.J. 2002. Groundwater recharge in hard rock region in Gujarat.

Healy, R.W and Cook, P.G. 2002. "Using groundwater levels to estimate recharge". *Hydrogeology Journa,*10 (1) : 40-51.

Kearns, AK, Hendrickx, JMH. 1998. Temporal variability of diffuse groundwater recharge in New Mexico. NM water Resource Res Inst Tech Completion Report, 309:43.

Khadri, S.F.R. 2008. Geophysical and hydro-geological study of Mandu plateau Dhar district, M.P. India. Groundwater Resources Assessment, Recharge and Modeling, Macmillan India Ltd., New Delhi. 34-44.

Luckey, R.R., Gutentag, E.D., Heimes, F.J. and Weeks, J.B. 1986. Digital simulation of groundwater flow in the High Plains aquifer in parts of Colorado, Kansas, Nebraska, New Mexico, Oklahoma, South Dakota, Texas, and Wyoming. US Geol Surv Prof Paper, 1400-D: 57.

MOWR, 2002. *National Water Policy-2002*, Ministry of Water Resources, Govt. of India, New Delhi, 9p.

Murali, K. 2008. Delineation of suitable sites for artificial recharge system using remote sensing and geographic information system techniques. Groundwater Resources Assessment, Recharge and Modeling, Macmillan India Ltd., New Delhi. 189-204.

□□□

Chapter 73

True Potato Seed (TPS) Technology Brought Thousand Smiles Back : A Case Study of Potato Farming in West Tripura District of Tripura

PIJUSH KANTI BISWAS AND DIPAK NATH

INTRODUCTION

North east India consist of 8 states, namely, Arunachal Pradesh, Assam, Manipur, Meghalaya, Mizoram, Nagaland, Sikkim and Tripura. The NE hill region lies between 21^0-29^0 North latitude and 89^0-98^0 East longitude and is characterized by hot sub humid (moist) to humid agro-climate in the Assam plains and warm per-humid in the hilly areas of the region. Nearly 10%of the total geographical area in India comes under the NEH region and has about 10% of the County's total potato area (Anon.,2005). Compared to national average of 18.2 t ha^{-1}, potato yield in the NE states except Tripura (19.7 t ha^{-1}) has been all time low (4.2- 8.3 t ha^{-1}). The low potato yield in the NEH region could be attributed to many factors. However, per capita availability of potato in the region is higher than the national level. (Singh *et al,* 2003).

Potato is one of the important vegetable crops grown in Tripura. The significance of this crop to the rural economy as well as agriculture of the state could be comprehended from the fact that potato occupies more than 5717 thousand ha of land which accounts for 110 thousand MT production. In contrast the figure with regard to the per unit potato production in the state is very surprising. The potato productivity in Tripura is 18 t ha^{-1} which is higher than the national average of 17.57 t ha^{-1} (FIB,2008) as well as productivity figures of major producing states of the country, *viz.,* Uttar Pradesh (22.63 t ha^{-1}), West Bengal (21.03 t ha^{-1}), and Punjab (18.73 t ha^{-1})

But though figures are satisfactory then also farmers are facing big loss in terms of yield. The main reasons for the low potato yields are adequate and untimely availability of essential crop inputs like healthy seed, fertilizers,

Krishi Vigyan Kendra, Chebri, Khowai, West Tripura District-799 207, Tripura, India pesticides *etc.*, coupled with poor management practices followed by the growers. Prevalence of serious diseases like late blight, brown rot/ bacterial wilt, *etc.*, is also responsible for low productivity in the region. Farmers usually resort to use of highly degenerated seed stocks and bring down the productivity of potatoes in this region. This paper presents a case study of potato farming in West Tripura District of Tripura.

MATERIALS AND METHODS

During the routine survey conducted by KVK, West Tripura, some potato farmers of Khowai sub division were being convinced about the advantages of TPS technology, expressed their keenness to undertake production of seedling tuber by using TPS. Necessary training programmes on production technology of TPS were organized and the selected farmers by the KVK were trained by the experts of the Dept. of Agriculture at Horticulture Research Complex (HRC), Nagicherra, Agartala and scientific staff of KVK, West Tripura. Training on production technology of tuber lets/ seedling tubers from TPS, disease and pest management strategies and post harvest management were imparted. Demonstrations were conducted in individual trained farmer's field consecutively for two years. Thereby they adopted the technology and continuing the TPS production till date. Basic inputs like TPS (50 gms.), fertilizer and plant protection chemicals were supplied by the KVK for conducting demonstration initially for two years. Similarly, women SHGs of West Tripura were also trained for production of seedling tubers from TPS and each group was provided the basic input as mentioned earlier KVK, West Tripura formulated and adopted two methods for conducting the demonstrations for raining of seedling tuber.

Raised bed: TPS were sown in line on finely prepared raised beds of dimension 3m×0.75m×0.15m at a depth of 0.5cm. Fertilizer *viz.*, N:P:K were applied at 50:100:150 kg per ha as a basal dose and another 100kg of N was applied during two earthing up @ 50kg per earthing up with well decomposed farm yard manure mixed at 50:50 ratio with soil on 30th and 45th day after sowing. Haulm cutting was done at 80-85 days and 10 days after harvesting was completed. Harvests were treated with 3% boric acid before storing in cold storage.

Flat bed: TPS were sown on finely prepared main field in rows having different spacing as follows- a) single row with spacing of 20cm×5cm, b) single row with spacing of 25cm×4cm, c) paired row with 10cm×30cm between row to row and two paired row. Seed to seed distance was maintained at 5cm, d) multiple rows with 10cm×5cm between rows and plant to plant with 4-5 rows per bed maintained at distance of 30cm. A total of 100 plants per sq. m were maintained in all cases. Fertilizer dose, earthing up, haulm cutting was same as described for raised bed. Two recommended methods *viz.*, transplantation and seedling tuber were used for production of potato using TPS.

During the crop period, farmers of different area were insisted for *in-situ* field visit and training were also conducted in the field itself on cultivation practices of TPS and management approaches. After successful completion of each crop field day were also organized where farmers of different village, state dept officials, extension functionaries were invited for effective transfer of technology. In the programme farmers growing TPS shared their experience with the experts.

RESULTS AND DISCUSSION

A result of the demonstration of production of seedling tubers is presented in Table 1. Results reviled that, there is a significant increase in production and productivity of seedling tuber in 2nd and 3rd year as compared to 1st year. This might be due to experience gained by the farmers for the new technology on TPS production.

Table 1: Year wise details of production of seedling tubers

Year	No. of farmers	Seed supplied (kg)	Area (ha)	Total production of seedling tuber (tons)	Yield (t ha^{-1})
1st	30	1.50	1.20	20.05	16.71
2nd	58	2.90	2.28	48.59	21.31
3rd	65	3.25	2.71	57.90	29.39

Economics of seed production: The economics of potato seed production is presented in Table 2. The following Table shows the economics of potato cultivation by 3 different SHGs (3×16= 48 nos.) in different location of West Tripura is presented. The average production cost per ha was found to be ₹ 126,833/- and production cost varied between ₹ 1,26,750/- and 1,30,250/- among the SHGs. Simillarly, the minimum gross returns of ₹ 7,65,250/- to a maximum of ₹ 10,52,000/- with an average of ₹ 9,03,000/-. Benefit cost ratio varied between 6.0:1.0 and 8.1:1.0 with the average value of 7.0:1.0. This results clearly shows the benefits of TPS production and thus groups are very well accepted this technology.

Table 2: Group wise cost of production and returns of potato seed production

S. No.	Expenditure (₹)	SHG-I (MilanDaal)	SHG- II (Gitanjali)	SHG- III (Basudev)	Average
A. Cost of Cultivation					
1	Land preparation	7,500	7,500	7,500	7,500
2	Cow dung	7,500	7,500	7,500	7,500
3	Fertilizer application	6,250	7,500	6,500	6,750
4	Seed	20,000	20,000	20,000	20,000
5	Plant protection chemical	16,000	20,000	18,000	18,000
6	Labour cost	25,000	26,500	26,000	25,833
7	Cold storage	41,250	41,250	41,250	41,250
	TOTAL (A)	**1,27,500**	**1,30,250**	**1,26,750**	**126,833**
B. Production of tubers					
1	Production of seedling tuber/ tuber lets from TPS in kg	19,125	26,300	22,300	22,575
C. Gross Returns					
1	Gross income (₹)= (B× Rs. 40.00, i.e., av. market price of tuber lets)	7.65,000	10,52,000	8,92,000	9,03,000
D. Net Returns					
1	Net return (₹) (C- A)	6,50,000	9,21,750	7,65,250	7,79,000
E. Benefit Cost Ratio					
1	Benefit: cost	6.0:1	8.1:1	7.0:1	7.0:1

CONCLUSIONS

Competent techniques for transferring technologies to target users is a key towards and all round development of any sector. TPS could be an alternative technology to increase productivity and reduce the cost of potato production. TPS in place of traditional seed tubers offers a big ray of hope to the potato growers of the state.

REFERENCES

Anonymous. 2005. True potato seed for potato production in NE states of India. Extension folder published by CPRI, Shimla

FIB. 2008. *Farm Guide 2008*. Farm Information Bureau, by Directorate of Horti and Soil Conservation, Dept.ofAgriculture, Government of Tripura.

Singh, P. H., Chandla, V.K., Paulkhurana, S. M., Garg, I. D., Thakur, K. C., Singh, S. and Lal, S. S.2003. Potato cultivation in NE India. Extension bulletin published by CPRI, Shimla.

□□□

Chapter 74

Agricultural Scenario in Indian Himalayan Region Status, Potentials and Strategies

M. MADHU AND V.N. SHARDA[1]

INTRODUCTION

Hills and mountains are spread over the country covering 23 states. However, the largest area is located in the Himalayas and having two distinct geo-political sub-regions *viz.*, North Eastern (NEH) and North-Western Himalayan (NWH) region. The NEH region comprises of eight states *viz.*, Arunachal Pradesh, Assam, Manipur, Meghalaya, Mizoram, Nagaland, Sikkim and Tripura. On the other hand, NWH region comprises of three states namely; Himachal Pradesh, Jammu and Kashmir and Uttarakhand (Samra,*et.al.*,1999).

The region presents a unique geographical area where altitude varies from about 50 meters to 7000 m above mean sea level and the climate, therefore, ranges from subtropical plains to temperate hills. The temperature ranges from less than 0° C in upper ridges of Himalaya to 38-40°C in plains. The two regions - north eastern and north western Himalaya are quite distinct when mean annual rainfall is considered. NE region receives average annual rainfall ranging from 2,800 to 12,000 mm whereas in NW Himalaya, it ranges from 350 to 3,000 mm. Out of the total 21 agro-ecological regions in the country, four regions are covered exclusively and one partly by the Himalayan states. These five zones represent wide varieties of climate ranging from cold arid to warm per-humid. The climate is warm per-humid in NE hills whereas it is warm sub-humid to humid in NW hills which is conducive for growth of a large variety of plants, therefore, almost all the tropical as well as temperate crops are grown. Rice is the staple cereal in the north-east whereas rice and

Central Soil and Water Conservation Research and Training Institute, Research Centre Sunabeda -763 002, Koraput, Odisha, India

[1]Member, Agricultural Scientist Recruitment Board, KAB-II, ICAR, New Delhi, India

wheat both are consumed in north-west. Maize is grown both in NE as well as in NW hills. Almost 94% of the area under cereals in NE hill is occupied by rice and maize whereas rice, wheat and maize occupy 93 % area of the cereals in NW hills. Millets, especially finger millet and barnyard millet are minor cereals of the hills and are confined mainly to Uttaranchal in NW hills. They are also cultivated in some states of NE hills.

Land Use Pattern in the Himalaya region

The major land utilization pattern in the whole Himalayan region and All India is presented in Figure 1 and Table 1. Himalaya has highest percent forest area as compared India (23% and the share of net urea sown in India (46%) is almost four times that of Himalayas (17%). The significant contribution of Himalayan region for the environment betterment of the country at the cost of its food security since percent net area sown to the reported area is less. On the other hand, piteous condition of hill farmers, which is further worsened by small, scattered and fragmented land holdings.

The area under forest accounts about 48 and 57% of the reported area in NEH and NEW, respectively (Figure 2). These is no significant differences in net area sown between the region but NWH having more area under un-cultivable land (may be permanent pasture and grazing lands) in comparison to NEH region which is a hardship on livestock and difficulties in fetching fodder from inaccessible areas.

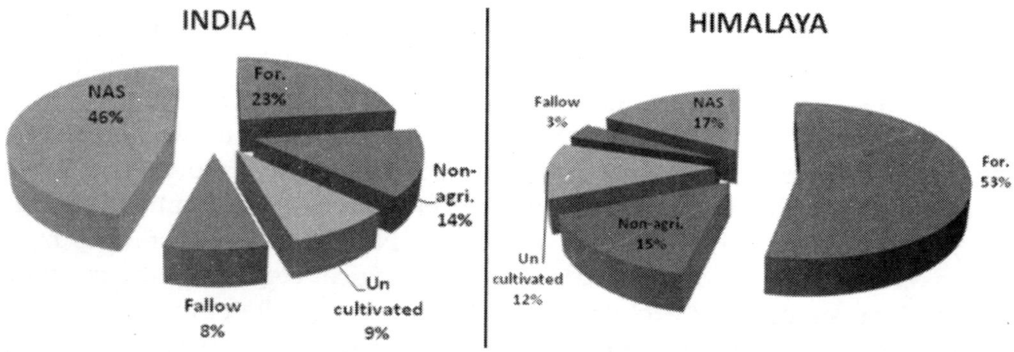

Fig. 1: Land utilisation pattern in India and Himalaya (2000-09)

Among the NEH region, maximum area under forest is in the states of Arunachal Predesh (92%) followed by Manipur (86.8%), Mizoram (77.8%) and Tripura (57.8%) which is more than the region average forest area (57%). All the states of the NEH region have more than 40% area under forest except in the state of Assam. The net sown area is maximum in Assam (35.1%) followed by Tripura (26.7%) and Nagaland (19.6%). The remaining states have less than 15% of their reported area is under net sown area which is below the region average net sown area (18%). The states like Meghalaya (26.8%) and Nagaland (11.4%) poses more area under total uncultivated land and remaining states have less than the region average area of 6%.

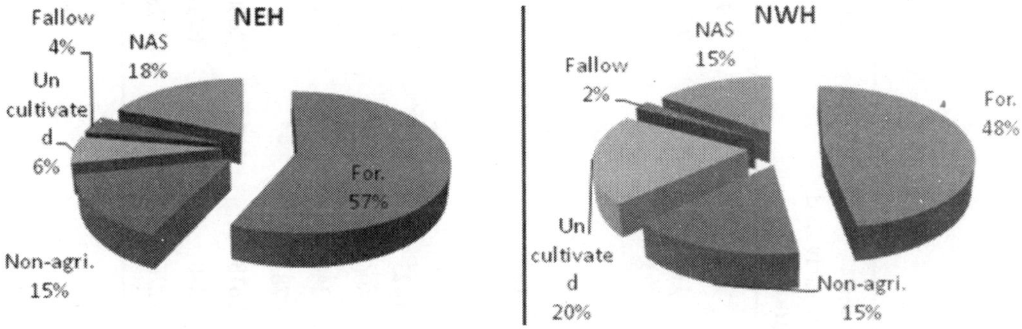

Fig. 2: Land utilisation pattern in NEH and NWH region (2000-09)

In the NWH region, Uttarakhand (61.2%) and J&K (53.5%) have more area under forest which is higher than the Himalayan region but slightly less than the NWH region. Net sown area is maximum in J&K (19.6%) which is higher than the NWH region (14.6%) and Himalayan region (16.8%). The Himachal Pradesh has highest area under uncultivated land (37.3%) which is higher than the NWH (20.1%) and Himalayan region (11.4%).

The table also shows that only Assam in NEH and Himachal Pradesh in NWH have less percent forest area than the recommended 33% forest area by the National Forest policy. It also clear from the table that, the Himalayas are well endowed in respect of forest cover for maintaining the eco-logical balance in the region.

Agricultural Scenarios in the Indian Himalayan Region

The Himalayan region is endowed with a wide range of physiographic, climate, soil and biological diversity, A large number of crops belonging to Cereals, Pulses, Oilseeds , Fruits, Vegetables, Plantation and Spices, Flowers and other crops are being cultivated in the region. The livelihood and food security is the primary objective of the farmers. The crops grown in the Himalayan region is presented in Table 2.

Per capita availability: Per capita availability of gross sown, net sown, gross irrigated and net irrigated area is higher at the country level as compared to the Himalayan region (Table 3). Per capita net sown area is higher by 50 % in NEH compared to NWH region. However per capita net irrigated area is higher in NWH compared to NEH and the Himalayan region.

Table 3.1: State wise land utilization statistics for Himalayan region (Mean of 2000-09) (Area in thousand ha)

S.No	State	Population (million)	TGA	Reported area for land use	Forest	Non-agricultural land use	Total uncultivated land	Fallow	Net area sown	Gross cropped area
1	Arunachal Pradesh	1.1	8374	5594	5154	51	104	96	190	258
2	Assam	26.6	7844	7850	1925	2543	446	178	2757	3951
3	Manipur	2.4	2233	1950	1693	27	8	0	221	221
4	Meghalaya	2.3	2243	2227	947	224	596	228	231	278
5	Mizoram	0.9	2108	2067	1596	134	47	208	92	92
6	Nagaland	2.0	1658	1595	863	74	182	164	312	375
7	Sikkim	0.5	710	725	319	250	11	35	109	120
8	Tripura	3.2	1049	1049	606	134	28	2	280	297
	NEH -TOTAL	**38.0**	**26219**	**23057**	**13103**	**3437**	**1423**	**912**	**4193**	**5592**
9	Himachal Pradesh	6.1	5567	4544	1099	1130	1696	74	545	951
10	Jammu & Kashmir	10.1	22224	3781	2023	583	340	93	741	1108
11	Uttarakhand	8.5	5348	5670	3470	414	782	97	764	1213
	NWH-TOTAL	**24.6**	**33139**	**13995**	**6592**	**2127**	**2818**	**265**	**2050**	**3271**
	HIMALAYA	**62.6**	**59358**	**37052**	**19695**	**5564**	**4240**	**1176**	**6242**	**8863**
	ALL INDIA	**1028.6**	**328**	**305423**	**69560**	**42299**	**27276**	**26161**	**140128**	**189529**

(Source: Indiastat.com and Ministry of Agriculture)

Table 2: Major crops grown in the Himalayan region

S.No	Crops	NEH	NWH
1	Food grains	Rice, Maize, Wheat, Pulses	Wheat, Maize and Rice & Pulses
2	Oil Seeds	Rapeseeds & Mustard, Soybean, Sesamum, Coconut, Linseed, Niger & Groundnut	Rapeseed & Mustard, Soybean, Sesamum and Groundnut
3	Spices and Condiments	Betelnuts, Ginger, Chillies and Black pepper	Ginger
4	Fruits	Banana, Citrus fruits, Mango and Papaya	Mango, Apple, Citrus fruits
5	Vegetables	Potato, Sweet potato, Onions and Tapioca	Potato and other crops

Table 3: Per-capita availability of crop sown and irrigated area in the region

S.No	States/Region	Gross Sown Area	Net Sown Area	Gross Irrigated Area	Net Irrigated Area
1	Arunachal Pradesh	0.23	0.17	0.04	0.04
2	Assam	0.13	0.10	0.01	0.01
3	Manipur	0.10	0.10	0.02	0.02
4	Meghalaya	0.10	0.09	0.03	0.02
5	Mizoram	0.10	0.10	0.02	0.02
6	Nagaland	0.18	0.15	0.05	0.03
7	Sikkim	0.21	0.19	0.02	0.01
8	Tripura	0.09	0.08	0.03	0.02
	NEH	0.14	0.12	0.03	0.02
9	Himachal Pradesh	0.15	0.08	0.03	0.02
10	Jammu & Kashmir	0.10	0.07	0.04	0.03
11	Uttarakhand	0.14	0.08	0.06	0.04
	NWH	0.13	0.08	0.04	0.03
	HIMALAYA	0.14	0.11	0.03	0.02
	India	0.17	0.13	0.07	0.05

Among the states, all the states in NWH and Tripura and Meghalaya in NEH have lowest per capita availability of net sown area (Table 3). However, per capita net irrigated area varied between 0.01 and 0.04 in the Himalayan states which is less than the country per capita availability. This table reveals that, in hilly areas per capita resource availability is much lesser than the plain and has to produce more per unit of land to maintain the self – sufficiency in food requirement.

Arable land available per farmers is low in Himalayan region (1.0 to 1.6 ha) as compared to India (1.5 ha). However, NEH is having higher arable lands per farmers than the NWH region. All the states have higher arable lands per farmer except in state like HP, J&K, Uttrakhand, Tripura, Assam and Manipur.

The Table 4 presents the cropping intensity and other percent selected parameter in Himalayan states. Cropping intensity of the country is 135 % against 128% in the Himalayan region. All the NWH region states (150-175%) have higher cropping intensity than in NEH states (100-143%). The other percent selected parameters indicated that, the Himalayan region is less percent of irrigated as compared to India. The NWH is better position as compared to NEH region with respect to the irrigated to net sown area / gross cropped area. The rain dependant agriculture is being practiced in the Himalayan region as compared to the country.

Area and Productivity of Crops in the Himalayan Region

Average data of 1999-2000 to 2008-09 is analysed and presented under the following headings.

Area under crops

Cereals and millets: Area under food grains occupies the major share (64.7%) out of the total gross cultivated areas in the country in order to maintain the food security in the country. Among the cereals and millets, the majority of the area is under rice (44%) and wheat (26.10%) followed by Jowar (9.10%), Bajra (9.52), Maize (7.36%), Ragi (1.62%) and, other cereals and millets (1.16%) and Barley (1.25%). Among the pulses Red gram (14.9%) and Gram (29.0%) occupies the major share followed by other pulses (56.1%) like cowpea, green gram, black gram etc.

Himalayan region occupies 74.3% of the gross cropped area under food crops (70.9% cereals and 3.4% pulses) which is 9.6 per cent higher than the India (Figure 3). The crops like rice (63.39%), wheat (17.48%) and maize (13.29%) among the cereals and Red gram (5.8%), gram (9%) and other pulses (92.2%) like cowpea, green gram, black gram, horse gram etc. among the pulses are the predominant food crops in the region. The percent share of rice and maize to the total cropped area much higher than India indicating that, these are the major stable food crops in the region. The NWH occupies more area under food crops (83.8%) than the NEH region (68.9%). It is obvious that rice holds the major share (91.25%) among the cereals followed by maize

(4.88%) and wheat (2.31%). On the other hand, wheat holds the major share (39%) among the cereals followed by maize (25.21%) and rice (23.90%) in the NWH region.

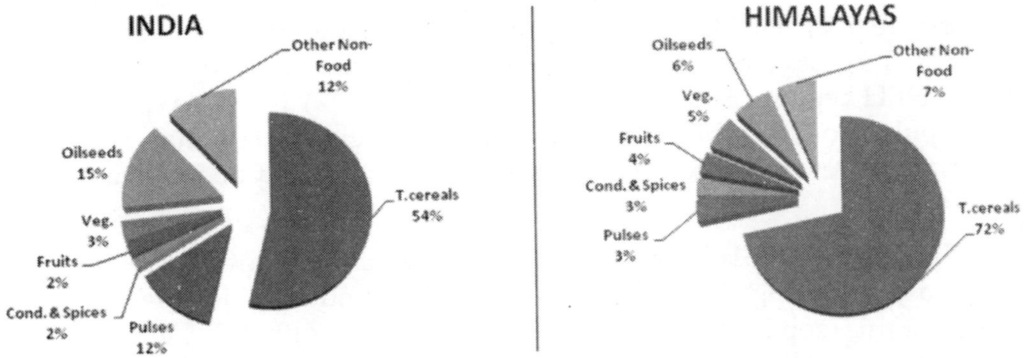

Fig. 3: Crop group wise area in India and Himalaya

All the NEH sates are dominated with rice (65.98%) except the state of Sikkim where maize is the dominant food crop (56.32) followed by rice (22.21%) (Figure 4). Wheat and other cereals are the least crops in all the states of NEH region. Among the NWH states, HP dominated with wheat (46.66%) followed by Uttarakhand (41.13%), where in J&K dominated with maize crop (36.66%) followed by wheat (30%) and rice (29%). All the states in NEW region shares higher percent of wheat (30-46.6%) and maize in HP and J&K as compared to All India share of these crops among the cereals.

Pulses: Pulses are the second to the cereals in importance and supplement protein in our diet in order to maintaining the balanced dietary system and removal of mal nutrition apart from improving soil fertility status. These crops are competing with the cereals and other commercial crops in the region. About 12.2 % of (23.24 m ha) the gross cropped area of the country (189.5 M ha) is under pulses against only 3.4 % of the gross cropped area of Himalayas. The population in the Himalayas has to depend on other protein

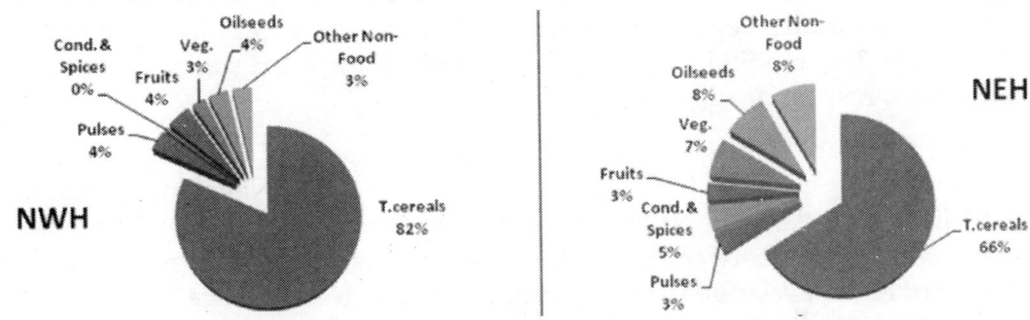

Fig. 4: Crop group wise area in NEH and NWH region

supplementing crops and animal products to meet their protein requirement. There is no significant difference in share of area under pulses in NEH and HWH region they occupies 3.3 and 3.5 % of the gross cropped area, respectively. Among the NEH states, Nagaland (8.8%), Sikkim (6.8%), Mizoram (0.5 %) and Manipur (4.0%) occupies maximum share of area under pulses to the gross cropped area and remaining states have less than the share of the NEH region (3.3%). Uttrakhand state (4.4 %) have maximum area under pulses to the gross cropped area in NWH region (1.21 M ha). The entire region face severe shortage of pulses due to low productivity and less cultivated area.

Oil seeds: In India, next to cereals, oil seed crops share 14.4 % of gross cropped area during the last decade (2000-2009) and considerable increased in area of this crop by 14.9 %. The share of oilseeds is only 6.3 % area of gross cropped area in Himalayas but it was decreased in area during the last decade by 7.5 %. On the other hand, the share of area under oil seeds in the NEH and NWH is 8.0 and 6.3% of the gross cropped area, respectively but decreased in area by 7.5 to 7.8% in both the region during the last decade. Decreased in area under oil seeds in all the status of Himalayas except in the states of Arunachal Pradesh, Mizoram, Nagaland in NEH and Uttarakhand in NWH reigon.

Horticulture: The Himalayas region occupies 5.4% of grass cropped area under vegetables against only 3.3% in India. During the last decade the vegetable gaining importance and increased in area by 32.8 and 26.8 % in the country and Himalayas, respectively. This was mainly attributed to increased irrigated area in the country and in the region. The area under vegetables is two times higher in NEH (7.0%) than in NWH region (2.5%). The NEH region have higher share of area under vegetable (7.0%) to its gross cropped area in comparison to India (3.3%) and Himalayas (5.4%). The share of vegetable area to the gross cropped area is maximum in the states of Meghalaya (21.2%) followed by Manipur (10.2%), Arunachal Pradesh (8.7%), Sikkim (6.1%) and Nagaland (5.5%). All the states in NEH region have higher percent of gross cropped area under vegetables than the NWH region except in the states of Mizoram (1.2%) and Sikkim (2.2%). On overall all, the NEH states performing better with respect of vegetables areas.

The vegetable crops are spreading in all the states of Himalayas except the four states , *viz.*, HP in NWH and Tripura (-6.25%), Nagaland (-63.51%) and Mizoram (- 9.6%) in NEH during the last decade (2000-2009). There is a severe competition between vegetable crops and other food crops for area in the entire Himalayas due to short duration and high economical value of the vegetables particularly from cultivation of off-season vegetables.

Table 4: State wise per cent of selected parameters in the Himalayan region

States/ Region	Cropping Intensity (%)	Net irrigated to agriculture land	Total irrigated area to agriculture land	Net irrigated to net sown area	Total irrigated to total cropped area	Total un-irrigated to total cropped area	Total un-irrigated to net sown area	Net un-irrigated area to Net area sown	Net Un-irrigated to total cropped area
Arunachal Pradesh	123.46	26.88	35.72	40.62	43.63	57.37	69.51	59.38	48.12
Assam	143.28	4.42	5.20	5.19	4.24	95.76	137.23	94.81	66.20
Manipur	100.00	20.87	20.87	21.53	21.53	78.47	78.47	78.47	78.47
Meghalaya	120.21	5.50	6.75	25.41	25.77	74.23	89.24	74.59	62.04
Mizoram	100.00	3.99	4.55	13.92	15.93	84.07	84.07	86.08	86.08
Nagaland	119.81	10.24	14.20	21.65	24.87	75.13	90.09	78.35	65.48
Sikkim	109.80	5.69	7.73	7.89	9.78	90.22	99.04	92.11	83.91
Tripura	106.28	19.02	32.15	21.05	33.39	66.61	70.73	78.95	74.34
NEH	**115.40**	**12.10**	**15.90**	**19.70**	**22.40**	**77.70**	**89.80**	**80.30**	**70.60**
Himachal Pradesh	174.58	13.31	22.66	19.83	19.35	80.65	140.80	80.17	45.92
Jammu & Kashmir	149.55	29.42	43.12	41.62	40.79	59.21	88.50	58.38	39.07
Uttaranchal	158.80	22.60	36.10	44.90	45.20	54.80	175.90	55.10	34.70
NWH	**161.00**	**21.80**	**33.90**	**35.50**	**35.10**	**64.90**	**135.10**	**64.50**	**39.90**
HIMALAYA	**127.80**	**14.70**	**20.80**	**24.00**	**25.90**	**74.20**	**102.10**	**76.00**	**62.20**
INDIA	**135.24**	**32.20**	**44.48**	**42.06**	**42.91**	**57.12**	**77.17**	**57.94**	**42.89**

The Himalayan region shares higher percent of gross cropped area to fruit crops (3.6%) than the National level (2.0%) due to favourable condition for cultivation of different varieties of fruits in the region. The NWH region have higher share of gross cropped area to fruit crops (4.4%) than the NEH region and the states like HP and J&K contribute the lion's share as compared to other states. The maximum percent of gross cropped area under fruits in the states of Meghalaya (11.4%) followed by Mizoram (5.0%), Nagaland (4.8%) and Manipur (3.4%) among the NEH state. Arunachal Pradesh and Sikkim shares negligible area under fruit crops in comparison to other states in Himalayas. In entire Himalayan region, area under fruit crops is increasing except the state like Assam (-1.0%) and Nagaland (-56.0%) where decreasing trend in area was observed during the last decade (2000-09). Like vegetable crops, fruit crops are also competing with other food and non-food crops due to their profitability as compared to food grain crops in the region.

Condiments and spices: In India about 1.6% of the gross cropped area is under condiments and spices which is much lesser than Himalayas (2.9%). The contribution of NEH region (4.3%) is significantly higher compared to NWH region (0.5%) with respect to condiments and spices. All the NEH States (2.2 to 23.2%) shares the maximum gross cropped area to condiments and spices against only 0.2 to 0.7% in NWH region. Growth during the last decade (2000-2009), clearly indicated that at the National level (+ 1.41%) and all the states in the Himalayas showed increasing in area except the two States namely Uttarakhand in NWH and Tripura in NEH region.

Non food crops: About 26.0 % of the gross cropped area is under non-food crops in India against only 12.7% in Himalayan. The NEH (16.1%) shares the maximum gross cropped area under non-food crops compared to NWH region (6.9%). Out of the 11 Himalayas states, only two states namely HP (3.2%) and Manipur (0.9%) shares less area under non-food crops compared to other states in the region (7.3 to 18.5%).

However the growth trend of total non-food crops area in the region is entirely different from other crops. Only four states in the region showed increasing trend in area namely Arunachal Pradesh (+ 16.5%), Manipur (27.3%), Mizoram (6.4%) and Nagaland (25%). Overall, the area under non-food crops is slowly decreasing in the region and these areas slowly diverted to other crops like vegetable, fruits and condiments & species.

Productivity of Crops

Average productivity (2000-09) of all the crops in the Himalayas is significantly lower than the country productivity level except for pulses where 28% higher productivity than the all India average productivity (Table 5). The integrated efforts are needed to address this low productivity of crops in the region.

Table 5: State wise average productivity of crops in Himalayan region

States/Region	Rice	Maize	Wheat	Ragi	Total cereals	Total pulses	Total food grains	Oilseeds	Vegetables
Arunachal Pradesh	1204	1383	1575	-	1218	1071	1212	983	4325
Assam	1482	721	1132	-	1465	556	1427	511	11692
Manipur	2383	2548	-	-	2384	495	2294	1441	7142
Meghalaya	1774	1476	1760	-	1715	836	1686	670	8392
Mizoram	1460	1747	-	-	1503	1125	1510	744	8767
Nagaland	1644	1716	1692	-	1624	1031	1552	744	8367
Sikkim	1461	1510	1342	896	1410	913	1357	1028	4625
Tripura	2360	993	2357	-	2348	643	2294	1614	11633
NEH	**1721**	**1512**	**1644**	**896**	**1708**	**834**	**1666**	**951**	**8130**
Himachal Pradesh	1477	2276	1473	1146	1762	646	1720	495	16350
Jammu & Kashmir	1972	1592	1602		1660	487	1622	640	15500
Uttaranchal	1939	1392	1921	1226	1733	663	1682	817	10875
NWH	**1796**	**1779**	**1665**	**1186**	**1718**	**599**	**1675**	**651**	**14242**
Himalaya	**1741**	**1579**	**1649**	**1021**	**1711**	**770**	**1669**	**868**	**9788**
India	**2051**	**1998**	**2715**	**1393**	**1979**	**600**	**1726**	**934**	**14740**

The average productivity level of rice and maize is much better in NWH region than NEH region (1721 and 1512 kg ha^{-1}). However there is no difference in productivity level of wheat, cereals and total food grains between the regions. The pulses and oilseeds are performing better in NEH region where productivity level higher by 39% and 46% for pulses and oil seeds, respectively. In NEH region, productivity of rice is maximum in the states of Manipur (2383 kg ha^{-1}) followed by Tripura (2360 kg ha^{-1}) and productivity level far below the region productivity level in the remaining states. On the other hand, productivity of rice lower in the state of HP (1477 kg ha^{-1}) compared to the NWH region (1796 kg ha^{-1}).

With respect to the maize in the NEH region, states like Assam (721 kg ha^{-1}), Meghalaya (1476 kg ha^{-1}), Arunachal Pradesh (1383 kg ha^{-1}) and Tripura (993 kg ha^{-1}) have lower productivity than the region average productivity of 1512 kg ha^{-1}. However, maximum productivity of maize was observed in HP (2276 kg ha^{-1}) in NWH region which is higher than the productivity in the region (1779 kg ha^{-1}) and India (1998 kg ha^{-1}). The average productivity of wheat in all the Himalayan states far below the average productivity of national level. The productivity of wheat is ranged between 1132 kg ha^{-1} in Assam and 2357 kg ha^{-1} in Tripura. The average productivity of ragi in the Himalayas is 1021 kg ha^{-1} which is 36% lower than the average productivity of India (1393 kg ha^{-1}). This is one of the crop grown only in NWH region with higher average productivity of 1186 kg ha^{-1} against 1021 kg ha^{-1} in Himalayan region.

The average productivity of total cereals is higher at the National level (1979 kg ha^{-1}) as compared to Himalaya region (1711 kg ha^{-1}). There is no significant difference in yield between NEH and NWH region. All the three states in NWH region have productivity of 1660 kg to 1762 kg ha^{-1}. On the other hand, productivity of total cereals is maximum in Tripura (2348 kg ha^{-1}) and Manipur (2384 kg ha^{-1}) as compared to other states (12180-1503 kg ha^{-1}) in NEH region.

The Himalayan region has upper hand with respect to the average productivity of pulses (770 kg ha^{-1}) against only 600 kg ha^{-1} at National level. Average productivity is better in NEH (834 kg ha^{-1}) compared to NWH region (599 kg ha^{-1}). Among the states, Arunachal Pradesh (1071 kg ha^{-1}), Mizoram (1125 kg ha^{-1}) and Nagaland (1031) have higher productivity level of pulses than other states in the Himalayas. Average productivity of oilseeds in Himalayan region is 868 kg ha^{-1} against 934 kg ha^{-1} in India. The NEH region has higher average productivity level (951 kg ha^{-1}) against only 651 kg ha^{-1} in NWH region. Maximum productivity of oilseeds is observed in the states of Manipur (1441 kg ha^{-1}), Sikkim (1028 kg ha^{-1}) and Tripura (1614 kg ha^{-1}).

Vegetables: The average productivity of vegetable is higher by 34% in India (14740 kg ha^{-1}) compared to Himalayan region (9788 kg ha^{-1}). However, vegetable productivity level is higher in NWH (14242 kg ha^{-1}) than the NEH region (8130 kg ha^{-1}). All the NWH states (10875 - 16350 kg ha^{-1}) and Assam (11692 kg ha^{-1}) and Tripura (11633 kg ha^{-1}) have higher average productivity level than the other states in the region (4325-8767 kg ha^{-1}).

Production and Monetary Loss Due to Water Erosion in Rainfed Areas of the Himalayas

Cereals crops constitute an integral part of Indian agriculture as they are an essential ingredient of Indian diet. Therefore Cereals occupy significantly larger cultivated area compared to oilseeds and pulse crops. Due to water included soil erosion a total of 13.45 million (66%) tones of Cereals, Oilseeds and pulse (COP) production is lost in the country against only 1.12 million tonnes loss of production in Himalayas.

The percent share of loss in production was higher in Himalayas under Cereals (89%) against 66% in the country due to the fact that fragile ecosystem of the Himalayas occupies more area under rainfed cereals crops (Figure 5). But percent loss in production under pulse and oilseeds crops is very low (4 and 7%) in Himalayas compared to 22 and 33% under pulse and oilseeds in India, respectively.

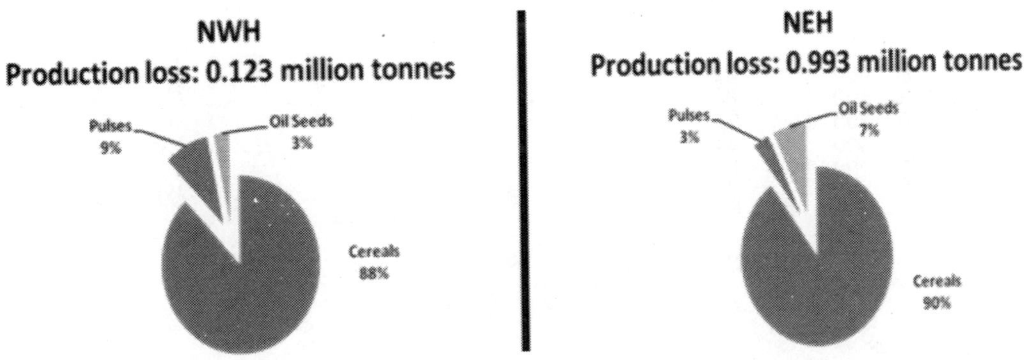

Fig. 5: Production loss due to water erosion in rainfed areas of NEH and NWH region

On the other hand monetary loss in Himalayas accounts only 8.1% of the total monetary loss in the country (₹ 111321 million). The percent monetary losses follow the same trend that of loss in production (Figure 6). The loss in production is seven times higher in NEH (0.993 M tonnes) compared to NWH region (0.123 M tonnes). In both the region (NEH & NWH) production loss in cereals accounts the lion's share of 88 to 90 % against only 9% pulses in NWH and 7% oilseeds in NWH region.

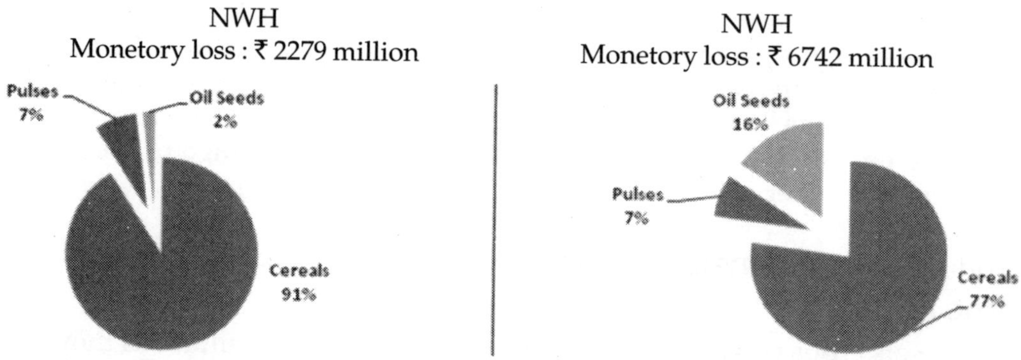

Fig. 6: Monetary loss due to water erosion in rainfed areas of NEH and NWH region

State wise loss in production and monetary loss in the Himalayan region presented in Figure 7. Among the states, maximum loss in production and monetary is in the state of Assam (54 and 46%) followed by Nagaland (12 and 10 %) in the NEH region and Uttarakhand (5.6 and 12 %) in the NWH region. The maximum loss in production in NEH region due to mountainous topography coupled with unscientific land and crop management practices following shifting cultivation leads to high erosion rates resulting in low productivity of crops.

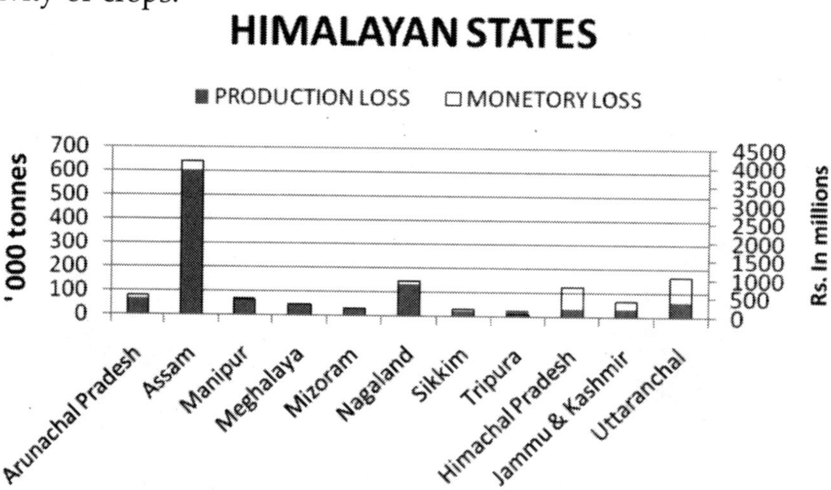

Fig. 7: Production and Monetary loss due to water erosion in rainfed areas Himalayan states

Growth Scenario of Area under Different Crops in the Himalayan Region

In order to study the growth scenario of area under different crops, the comparison was made between average data of first five and last five years during 1999-2000 to 2008-09. The growth trend of area under cereals almost stagnated at the National level but it was negative growth tread in Himalayas (-3.4%). At the National level, positive growth trend was observed in Maize (12.45%) and in Barley (60%) and there was negative growth trend for remaining cereals except for Rice and Bajra where no significant change (Table 6). Area under like Maize (1.4%), Jowar (100%), Bajra (17.6%), ragi (7.3%) are in increasing trend and remaining cereals are having negative growth trend with overall negative growth trend with respect to area under total cereals in the Himalayas.

In NEH region, the over all growth trend of area under cereals was negative (-5.5%) in comparison to NWH where no significant change in the area (Figure 6.2). Only crops like Maize (12.3%) and Ragi (60.3%) are showed positive growth trend under cereals in NEH on the other trend no significant change in the area under all the cereals crops but slight positive trend was observed for rice (1.03%) and wheat (0.52%).

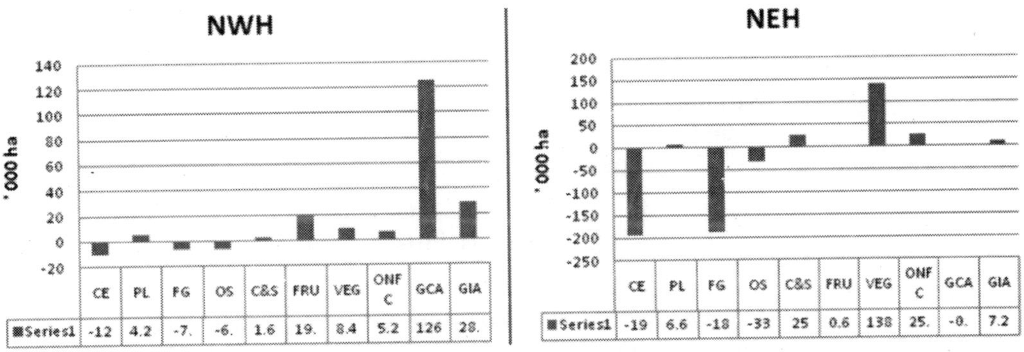

Fig. 8: Growth scinario of crops in NEH and NWH region

The state like, Arunachal Pradesh, Meghalaya, Mizoram and Nagaland in NEH region and J & K state in NWH region have positive growth trend with respect to area under total cereals.

Table 6: State wise and region wise growth scenario of area under different crops

States/UTs	No trend	Increasing	Decreasing
Arunachal Pradesh	Wheat, Pulses	Rice, Maize, Ragi, Total cereals, C&Sp, Veg., Oil Seeds,	Other Cereals, R&M,
Assam		C&Sp, Veg.,	Rice, Maize, Wheat,Ragi, Other cerealsTotal cereals, Pulses, Potato, All oil Seeds,

Manipur	Maize, Oil seeds	Rice, Total cereals, Pulses, C&Sp, Veg.,	
Mizoram		Maize, Total cereals, Pulses, C&Sp, Fruits, Potato,	Rice, Oil seeds
Meghalaya	Rice, other cereals, Oil seeds,	Pulses, C&Sp, Fruits,	Wheat, Veg.
Nagaland		Rice, Maize,Total Cereals, Pulses,C&Sp, potato, R&M, Soybean, Oilseeds,	Ragi, Fruits, Veg.
Sikkim	All cereals crops, Pulses, C&Sp, Oil Seeds,	Pulses, Potato, Veg.,	Oil seeds
Tripura	Maize,C&Sp, Veg.,	Rice, Total cereals	
HP	C&Sp, Mango, Banana, appale, Total Fruits, Veg., Oilseeds,	Pulses,	All Cereals
J&K	C&Sp, Veg.,	Rice,Wheat, Total cereals, Pulses, Total fruits,	Maize,Ragi, R&M, Oil Seeds
UK	Rice, Wheat, Pulses, C&Sp,	Total Fruits, Veg., Oil seeds,	Maize, Ragi, Total Cereals
NWH	C&Sp,	Rice, Wheat, Pulses, Mango, Banana, appale, Total Fruits, Veg.,	Maize, Ragi, Total cereals, All oilseeds
NEH	Mango, Papaya, Total Fruits	Maize, Ragi, Pulses, Ginger, Turmaric, C&Sp, Citrus, Banana, All Veg., Soybean,	Rice,Wheat, other & Total cereals, Cardamum, Betelnut, R&M, Linseed, Coconut, OilSeeds,
HIMALAYA		Maize, Ragi, Pulses, Ginger,Turmaric, C&Sp,Ginger, Turmaric, C&Sp, Mango, Citrus, Banana, Apple, Total fruits, Potato, Veg., Soybean	Rice, Wheat, Total Cereals, Cardamum, Betelnut, All oil seeds except Soybean

India	Rice, Maize, Barley, Pulses, Black pepper, Ginger, Turmaric, Betelnut, C&Sp, All fruits, all vegetables, Sesamum, R&M, Coconut,Soybean, Sunflower, Total Oil Seeds	Jowar, Bajra, Ragi, Wheat, Total cereals, Chillies, Sweet Potato, Groundnut, Castor, Linseed, Niger, Safflower

C&Sp: Condiments & Spices, R&M: Rapeseed & Mustard

The trend analysis of area under pulses at the National level reviled that, there is a significant increased in area by 6 % with registering increased area under gram (15%) and other pulses (2.7%). In Himalayan region, total area under pulses increased by 3.6% with increased area under other pulses like Cowpea, Black gram, Green gram, etc. but significant decreased in the area of gram (-11.11%) and no significant change in the area of redgram.

Both the region (NEH and NWH) have the uniform area under pulses (3.6%). The other pulses like cowpea, green gram, black gram etc are competing with the area of gram (-19%) and redgram (-5.7%) thus registering the negative growth trend in these crops in NEH region. On the other hand, redgram (+19%) and other pulses (3.3%) are competing with the area of cereals and other crops in NEH region. All the states of NEH region having positive growth trend with respect to the area under pulses except in the state of Assam (-5.2%) and Tripura (-13.5%). All the states of NWH region are performing very well with respect to the area under total pulses.

Self Sufficiency of Food

The self sufficiency of cereals, pulses and oil seeds were calculated by taking into account of population during 2001 and average area, production and productivity of crops during 2000-09 (Table 7). The total cereals and millets form the major share of food crops in the Himalayan region and show a surplus in production over 640 thousand tones (Figure 9). Both NEH and NWH regions are also showed surplus in cereals production. However, there is a deficit in pulses over 938 thousand tones and oilseeds by 298 thousands tones in Himalayan region. On the other hand, NEH showed deficit of 385 thousand tones and surplus in oilseeds over 225 thousand tones. Among the NEH states, only three states namely, Assam, Meghalaya and Mizoram showed deficit in cereals by 490, 147 and 45 thousand tones, respectively. Pulses showed deficit in all the states of NEH region and maximum was in Assam followed Tripura. Where in, Assam, Manipur, Meghalaya and Mizoram showed deficit in Oilseeds.

In NWH region, only J&K showed deficit in food grains and oilseeds. Pulses showed deficit in all states and oil seeds was surplus in HP and Uttrakhand. With respect to the food self sufficiency in the Himalayan region, only five states found to be surplus in food grain and these states are Arunachal Pradesh, Nagaland, Sikkim, Tripura in NEH region and HP and Uttrakhand in NWH region.

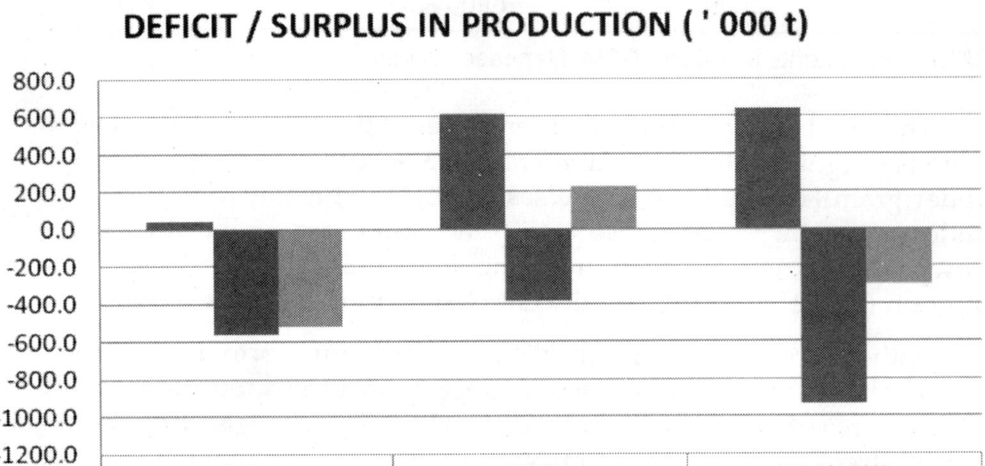

	NEH	NWH	HIMALAYA
Cereals	39.3	609.4	639.5
Pulses	-559.4	-384.8	-937.5
Oilseeds	-520.1	224.6	-298.0

Fig. 9: Deficit / Surplus of production in Himalayan region

Only possible means to achieve the self sufficiency in production from theses crops is possible through horizontal and vertical expansion by increasing area through utilizing fallow and other cultivable waste lands and by increasing their productivity through multi pronged integrated approach. The additional area required to produce these crops at present level of productivity and or additional productivity need to be achieved by considering the present area under theses crops are presented in Figure 10 and 11.

Strategies for Enhancing Productivity and Use Efficiency of Resources

Integrated Nutrient Management: The average consumption of chemical fertilizer is 19.3 kg ha^{-1} in Himalayas against 95 kg ha^{-1} in the country (Anonymous, 2002). In fact, in rainfed areas in the hills, crops are grown without chemical fertilizer and, thus, it is organic by default. However, in the region integrated use of nutrients through manures, fertilisers, green

Table 7: Requirement, actual production, surplus/deficit of food grains and oilseeds in NEH and NWH region (All in thousand tonnes)

State / Region	Cereals & Millets			Pulses			Oilseeds			Food Grain		
	Req.	Pro.	S/D	Req.	Pro.	S/D	Req.	Pro.	S/D	Req.	Pro.	S/D
Arunachal Pradesh	172.4	229.6	57.2	19.9	7.7	-12.2	39.8	26.5	-13.4	192.3	237.3	45.0
Assam	4208.9	3719.1	-489.8	486.2	61.5	-424.6	972.3	143.7	-828.6	4695.0	3780.7	-914.4
Manipur	377.4	397.0	19.6	43.6	4.2	-39.4	87.2	1.7	-85.5	421.0	401.2	-19.8
Meghalaya	366.4	219.7	-146.6	42.3	3.6	-38.7	84.6	6.6	-78.1	408.7	223.3	-185.4
Mizoram	140.8	95.7	-45.1	16.3	4.9	-11.4	32.5	4.3	-28.2	157.0	100.6	-56.4
Nagaland	314.2	379.7	65.5	36.3	33.5	-2.8	72.6	46.3	-26.2	350.5	413.2	62.7
Sikkim	85.4	95.1	9.7	9.9	7.3	-2.5	19.7	10.0	-9.7	95.3	102.4	7.2
Tripura	504.2	581.3	77.1	58.2	5.3	-52.9	116.5	8.0	-108.4	562.4	586.6	24.2
HP	960.2	1375.7	415.5	110.9	20.2	-90.7	221.8	8.0	-213.8	1071.1	1395.9	324.7
J&K	1591.0	1445.2	-145.9	183.8	13.9	-169.9	367.6	42.3	-325.2	1774.8	1459.1	-315.7
UK	1339.8	1677.0	337.3	154.8	31.9	-122.8	309.5	25.7	-283.8	1494.5	1709.0	214.4
NEH	6169.6	6208.9	39.3	712.6	153.3	-559.4	1425.3	382.1	-1043.2	6882.3	6362.2	-520.1
NWH	3891.0	4500.5	609.4	449.4	64.6	-384.8	898.9	74.0	-824.9	4340.5	4565.1	224.6
HIMALAYA	10060.7	10700.2	639.5	1162.1	224.5	-937.5	2324.1	447.4	-1876.7	11222.7	10924.7	-298.0

Req.: Requirement (Cereals & Millets @ 158 kg capita^{-1} annum^{-1}, Pulse @ 18.25 kg capita^{-1} annum^{-1}, oilseeds@36.5 g capita^{-1} annum^{-1})
Pro.: Production, D/S: Deficit / Surplus.

manuring, vermin-compost etc, need to be popularized for harnessing the potential yield of modern high yielding varieties coupled with other resource conservation technologies in the region. There is a greater potential for organic farming in the region and this systems need to be researched upon for improving their productivity and their effect on soil health, soil microbes etc. before their extrapolation to the users. There is a need to develop package of practices for organic farming for all cropping systems, including fruits. Organic cropping system modules need to be developed and validated at multiple locations. A network project on research on organic farming, farming system research, precision farming and resource conservation technologies to be structured under the Directorate of Farming System Research.

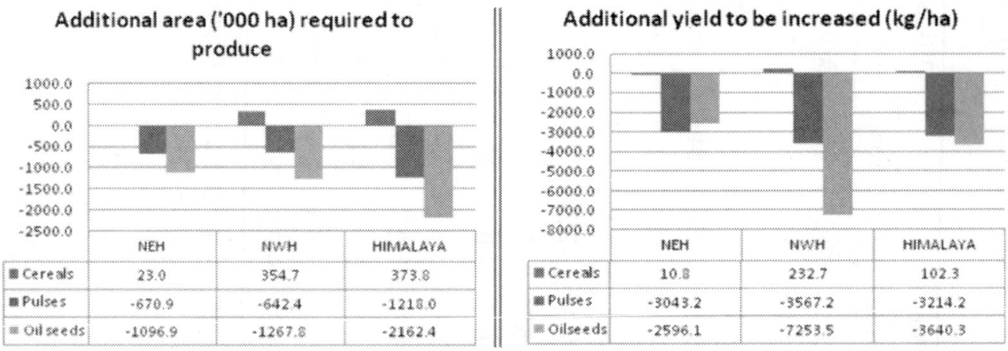

Fig.10: Additional area and yield to be increased to produce in Himalayan region

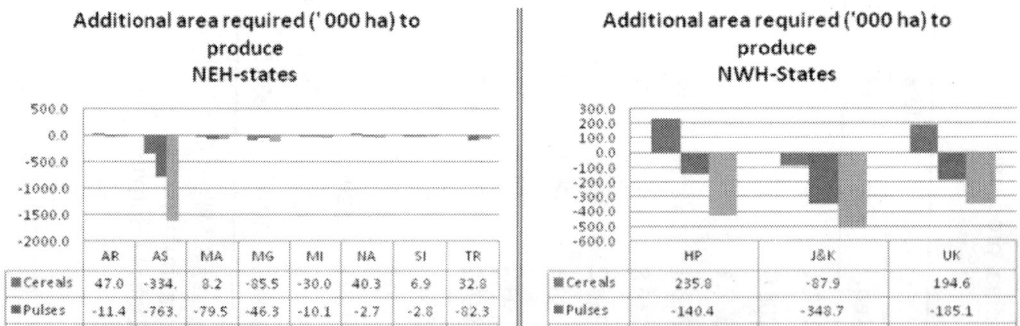

Fig. 11: Additional area required to produce in NEH and NWH states

Cropping systems for Increase cropping intensity: The present cropping intensity ranges from 100-175 %, however, it can be increased to 200% by cultivating the areas that are left fallow. By effective transfer of already developed remunerative cropping systems for rainfed and irrigated areas by incorporating high value, short duration crops like vegetables, flowers and medicinal & aromatic plants expected to increase the cropping intensity to 200 % in the region(Gupta, 2003).

Water harvesting and use of micro irrigation systems: Despite adequate rainfall in both the flanks of Himalayas, there is shortage of water not only for raising crop but for domestic purposes also. This is mainly due to lack of adoption of appropriate soil and water conservation measures in these areas. Presently cost effective water harvesting techniques are available for storing the water for life saving irrigation. There is a greater potential to use sprinklers and drip irrigation techniques for efficient and productive use of harvested rainwater. Research on application part of liquid fertilizers through micro-irrigation systems *etc.*, need to be intensified by the SAUs and ICAR institutes since there is a lot of scope for expansion of micro irrigation systems in the region.

Diversification cropping systems and integrated farming systems: Hills have potential for short duration vegetables and diversified integrated farming systems which are economical and viable to the hill farmers with limited land holdings. These off season vegetables fetch premium price, which enhances the profitability. In addition, the crop sequences encompassing vegetables are capable of increasing productivity per unit area and time. Water mill based integrated farming systems comprising Crops+Fish+ Poultry+Piggery gave an annual net profit of Rs. 26,000-31,000/- as additional benefits over only about Rs.10,000/- from water mill in the North-West Himalayan region(Sharda, *et.al.*, 2009). The region is tremendous potential for medicinal and aromatic plants (MAPs), there is a need for evolving a mechanism for sharing and exchange of planting material with in the Himalayan states which have enough experience in cultivation of MAPs.

Mechanization of hill agriculture for increasing productivity, profitability and reducing drudgery: Mechanisation plays an important role in performing filed operations timely and more efficiently apart reducing drudgery especially in the hilly terrain. For example the cost on weeding can be reduced to just half by selecting the proper spray machine, nozzle, knowing about the prevailing weed species, adding proper surfactant or adjuvant to the herbicide. Timely farm operations add to the profit, as it is a non-monetary item.

Adequate supply of quality seeds of improved crop varieties: In any production system without using the quality seeds we cannot expect maximum productivity even with optimum use of other resources. Therefore timely supply of quality seeds of recommended crop verities plays an important role in enhancing productivity and efficient use of available resources. Cultivation of crops with improved nutritional value will help in alleviating nutritional insecurity too.

A coordinating centre for important crops of this region (rice, rapeseed and mustard) research is needed in the region as there is no centre to coordinate research in the region. For popularization of HYVs of crops like maize and other crops developed by public sector R&D organizations, State Depts. should undertake their large scale multiplication to ensure their availability and timely supply to farmers. Since the region has a large area under rainfed conditions, short duration and drought resistant varieties need to be developed to overcome negative effect of climate change.

Utilisation of waste lands for fodder production: Availability of adequate quantity of good quality fodder will support rearing of livestock. Two of the mountain states *viz.,* Himachal Pradesh has 14 % of the total area under permanent pasture that supports livestock, however, state like Uttaranchal and J&K in NW hill and all the states of NE hill have only negligible area under permanent pasture; consequently they face shortage of fodder (Bisht and Gupta, 2003). This deficit can be met by growing fodder on waste lands, forest floors as well as on the terrace risers. The technology is available and it needs to be put in place.

Soil and water conservation measures and watershed management: In order to mitigate the drought like situation during the cropping season and to conserve precious natural recourses mainly soil, water and vegetation, adoption of appropriate soil and water conservation measures are very much needed in this fragile ecosystem. Soil and water conservation is most important in hills. The average soil loss in Indian Himalayas is 20 t ha/ year. It varies from 21 t ha^{-1} yr^{-1} in maize to 92 t ha^{-1} yr^{-1} in bare fallows. The situation is more serious in NE Himalayas where cultivation is still done on the slopes ranging from 40 to 70 %. According to an estimate the M tonnes. A number of agronomic, mechanical, vegetative and bio-engineering measures have been evaluated for their economical and conservation efficiency in the region by various R&D organisations in the region. The result of Integrated Watershed Development Project in hills has been encouraging and it needs to be replicated (Sharda and Juyal, 2003).

There is a need for convergence of various developmental schemes operating in the states to overcome the shortage of funds for rehabilitation of degraded and wastelands, for which technological packages may be made available to the implementing agencies. The total production loss due to soil erosion is accounts about 1.12 M tonnes in the Himalayan region which is more or less equal to the total deficit in the food grains and oil seeds (1.23 M tonnes) in the region. This suggests that resource conservation measures are long way to go to increasing the productivity of crops in the region.

Post -harvest technologies for value addition: The region is potential for cultivation of diversified crops. Post Harvest Technology (PHT) programmes in the region should be strengthened. Under PHT programmes, in addition to horticultural products, processing of milk and meat products should also be induced. Develop package of practices from harvest to storage to minimize post harvest losses by 70-75 % and value addition for converting by-products into feeds by adding ingredients.

Tackling the problem of wild animals: The wide spread area of forest accompanied with wild animals in the region. Especially in the interface zone of forest and agricultural lands, problem of wild animal is become severe and this forces the farmers to abandon their land from cultivation of agricultural crops. This problem need to be addressed through a team consisting of experts of wild life, animal husbandry, agriculture, horticulture etc. for preparing a roadmap, and all failed/successful measures adopted till date to control monkey menace should need to study.

Fodder production: Livestock are the important component of hill agriculture. For tackling the acute problem of feed and fodder for livestock in hilly areas, there is a need for strengthening of fodder research in the region. Alpine pastures are facing the problem of infestation by weeds. Coordinated effort is needed with full funding and staff.

To overcome the fodder shortage, other suitable plant biomass material available in large quantity may be used for preparing fortified feed blocks. The wasteland area available in the states may be converted into pasture lands. Also demonstration of fodder production technologies should be carried out in the fodder farm of Regional Station of IGFRI at Srinagar and other places. All the KVKs in the region should demonstrate fodder production technology in the field to the progressive farmers.

Transfer of technology: A number of production-cum-conservation technologies are available in the region and this need to be transferred to the field effectively to increasing the productivity level of all the production systems in the region. Integration of research and extension by getting farmers' feedback through KVKs for identifying researchable issues/problems in the region are to be strengthened. Enhancing the knowledge and managerial ability of farmers and the extension workers is important which facilitate in transfer of technology and adoption by the farmers.

For augmenting agricultural production in the region, there is a dire need to bridge the gap between recommendations of research organizations and farmer's practices. For this all out efforts need to be made by all concerned to ensure availability of complete package of practice to the farmers. It should be ensured by the extension agencies of State Govts and SAUs that the complete package of recommended practices is adopted by the farmers.

CONCLUSIONS

Increase in productivity and profitability will remain an important component of hill farming to address the needs of future population pressure. The main reason of slow growth or even stagnant crop productivity in the Himalayan region can be attributed to deteriorating soil physical conditions, declining organic matter and unbalanced fertilization with over reliance on nitrogen fertilizer, declining use of organic sources and widely prevalent moisture stress due to lack of suitable soil and water conservation measures.

Crop intensification or diversification strategies will not only depend on individual circumstances but also on availability of appropriate technologies and inputs. There is a need for close coordination between water management, fertilizer use, pest control and other crop husbandry practices for integrated crop management. Wide range of technologies is required for sound water management under irrigated conditions for highly intensive cropping. In rainfed area, conservation agriculture for better moisture and soil conservation as well as water harvesting is required to diversify into high value crops. The shift to knowledge intensive commercial farming systems, require greatly improved flow of information inputs, skill upgradation and policy support to farmers.

REFERENCES

Anonymous, 2002. http://databank.nedfi.com.

Bisht, J.K, and Gupta,H.S. 2003. Fodder production strategies for Uttranchal. In: Sharma, R., Pandey, R.K., Shah,D., Kumar,A. (Eds.), Sustainable Mountain Agriculture. Pub. Centre for Development Studies, Uttranchal Academy of Administration, Nanital,pp 233-245.

Gupta, H.S. 2003. Agricultural productivity in the hills and mountains of India. In: Sharma, R., Pandey, R.K., Shah,D., Kumar,A. (Eds.), Sustainable Mountain Agriculture. Pub. Centre for Development Studies, Uttranchal Academy of Administration, Nanital,pp 65-68.

Ministry of Agriculture. 2011. http.//dacnet.nic.in

Samra, J.S., Dhyani,B.L. and Sharma, A.R. 1999. Problems and prospects of natural resource manage,ment in Indian Himalayas-A base paper. Hill and Agro-Ecosystem Directorate, Dehradun Under NATP. 146p.

Sharda V.N., Pradeep dogra and Chandra Prakash. 2009. Potential technologies for resource conservation & productivity enhancement. Bulletin No. T-57/D-36, CSWCRTI, Dehradun, India. 150p.

Sharda,V.N. and Juyal, G.P. 2003. Technological interventions for soil and water conservation in the Shivalik areas. In: Sharma, R., Pandey, R.K., Shah,D., Kumar,A. (Eds.), Sustainable Mountain Agriculture. Pub. Centre for Development Studies, Uttranchal Academy of Administration, Nanital,pp 65-68.

□□□

Chapter 75

Management of Rhizome Rot of Ginger through Soil Solarization under Organic Cultivation

N. RAJA KUMAR, K. RAVINDRA KUMAR[1] AND G. SUNIL KUMAR BABU

INTRODUCTION

Ginger is cultivated in several parts of the world, and the most important countries being *viz.*, India, China, Nigeria, Sierra Leone, Indonesia, Bangladesh, Australia, Fiji, Jamaica and Nepal. Among them, India and China are the dominant suppliers to the world market. In terms of quality, Jamaican and Indian ginger are considered of superior quality followed by West African type. Jamaican ginger possesses delicate aroma and flavour and is sometimes through considered as first grade. Indian ginger is being exported to the world market from Cochin and Calicut in Kerala. The refreshing pleasant aroma, biting taste and carminative property of ginger make it an indispensable ingredient of food processing throughout the world. At present, India is the largest producer of ginger in the world accounting for about one-third of the total world output. It's followed by Thailand and Japan. The area under ginger is 70,000 ha with a total production of 230,000 Mt and productivity of 32, 85 kg ha^{-1} (Anon., 2005). In India, a large portion of the ginger produced is consumed domestically as a green ginger or dried ginger. It is cultivated in many states, but major producers are Meghalaya, Kerala, Arunachal Pradesh, Orissa, West Bengal and Mizoram. In Karnataka, the area under ginger is 13,628 ha, with a production of 13,6688 tonnes and productivity of 10 t ha^{-1} (Anon., 2003). The area under ginger went up to 77,610 ha in 1998-99 from 75,570 ha in 1997-98 and the output to 2,63,170 tonnes from 2,52,110 tonnes (Singhal, 2003).

Rhizome rot of ginger caused by *Pythium aphanidermatum* (Edson) Fitz is a major constraint for the production of healthy rhizome, some times causing total failure of crop. Chemical control of this pathogen is not economical

Regional Agricultural research Station, Anakapalle- 531 001, Andhra Pradesh, India.
[1]Horticultural Research Station, Chintapalle, Andhra Pradesh, India

because of high cost of chemicals, breake down of resistance, environmental pollution, and deleterious effect to non-target beneficial soil micro-organism and ultimately the choice of the consumer for a organic product.

MATERIALS AND METHODS

The present research was carried out at Horticultural Research Station, Chintapalle during *kharif,* 2007-08 and 2008-09.The rhizomes were planted in field at 25 cm x 15 cm spacing with three replications in split plot design with a plot size of 2.2m x 2m was prepared after pre-monsoon shower in the first week of April. Two main plots and five sub plots were maintained. The beds in the plot meant for solarization were immediately covered with transparent polythene sheets of thickness 100ìm and the sides were sealed. The non-solarized plots were also maintained without polythene tarping. After solarization the beds were imposed with the five treatments *viz., Trichoderma harzianum* seed treatment + soil application (50g l^{-1}), Bacterial Consortium supplied by Indian Institute of spice research, Calicut, *Trichoderma harzianum* + Bacterial Consortium, Mancozeb seed treatment+ soil drench were applied in treated plots along with control (water soaked/sprayed) in both the solarized and non solarized main plots. Apparently healthy rhizomes of variety Chintapalle local were used. Observations *viz.,* per cent germination, tiller production, per cent disease incidence and yield per bed were recorded. The data obtained in the present investigations for various parameters were subjected to ANOVA for a completely randomized design for *in-vitro* studies and randomized complete block design for *in-vivo* studies by using M-STATC programme.

Soil Solarization :The land to be solarized was thoroughly cultivated and then leveled so as to minimize protrusions as clods, stubble and stones to prevent tearing of the sheet. 40-50 mm irrigation was given before mulching. Clear, transparent polythene sheet of 100ìm thickness was used. The polythene sheet was applied immediately after irrigation and when it was least windy. Solarization was conducted for six weeks during April to May. All free edges were buried and the soil around them compacted so as to prevent escape of heated air or moisture.

Monitoring Soil temperature: Soil temperature in each plot was recorded daily at three different depths *viz.,* 5cm, 15cm 30 cm and 60cm using permanently fixed soil thermometers.

Monitoring microbial populations: Pathogen population in both the plots was analyzed before the commencement of soil solarization and after it for every 15 days. Pythium population was estimated by using serial dilution method on Oat meal agar media.

RESULTS AND DISCUSSION

Observations on management of rhizome rot in ginger showed that highest germination (90.13%), more number of tillers (9.91), low disease incidence (6.34 PDI) and highest yield of 2.68 kg per bed when seed material were solarized in polyethylene bags for 30 min and treated with *Trichoderma harzianum* 50g l^{-1} as seed treatment and soil application. This is followed by solarization of seed material in polyethylene bags for 30 min and Mancozeb (@ 0.25 %) before planting as seed treatment and soil application as drenching where in highest germination (83.18%), more number of tillers (9.24), low disease incidence (8.71 PDI) and highest yield of 2.16 kg per bed were recorded. Soft rot disease was more with low germination (56.08%), less number of tillers (6.56), high disease incidence (43.71 PDI) and low yield of 0.56 kg per bed in non solarized and untreated rhizomes. These results are in agreement

Table 1: Effect of soil solarization along with fungicides, biocontrol and organic amendment on rhizome rot of ginger

	Treatment	Solarized				Non- Solarized			
		Germination (%)	Tiller production (No.)	Disease incidence (PDI)	Yield Bed-1 (Kg)	Germination (%)	Tiller production (No.)	Disease incidence (PDI)	Yield Bed-1 (Kg)
T1	*Trichoderma harzianum* seed treatment + soil application	90.13	9.91	6.34	2.68	83.18	9.24	8.71	2.16
T2	Bacterial Consortium	82.45	8.64	30.25	1.32	75.27	8.78	35.35	1.02
T3	*Trichoderma harzianum* + Bacterial Consortium	84.05	8.42	21.98	1.06	73.35	8.46	26.21	0.96
T4	Mancozeb seed treatment+ soil drench	90.1	10.84	8.14	2.16	78.75	9.84	10.46	2.07
T5	Control	66.66	7.0	39.46	0.65	56.08	6.56	43.71	0.56
				Main plots					
	Sub plots								
	SE.m	0.75	0.39	0.13	0.04	0.61	0.56	0.32	0.046
	CV	4.85	22.40	2.90	13.33	2.49	20.39	4.69	10.06
	CD at 5%	2.97	1.54	0.52	0.15	3.03	2.78	1.68	0.23
	Interaction (MxV)	2.51	NS	1.39	0.19				

with Joseph and Sivaprasad (2000), who isolated native antagonist *T. virde* from healthy ginger plants and evaluated against rhizome rot. Two isolates *viz.*, isolate no. 12 and 7 stimulated plant growth and isolate no. 12 gave higher rhizome yield, better biomass production and disease suppression also. Similarly, pot culture studies on rhizome rot revealed that, seed treatment and soil application of native antagonist *T. harzianum* was effective in reducing disease incidence and increasing germination (Shanmugam *et al.*, 1999). Ram *et al.* (2000) evaluated resident isolates of biocontrol agents *viz.*, *T. harzianum*, *T. aureoviride* and *T. virens* against rhizome rot of ginger. Rhizome treated with these bioagents significantly reduced the pathogens and the disease and also increased the yield.

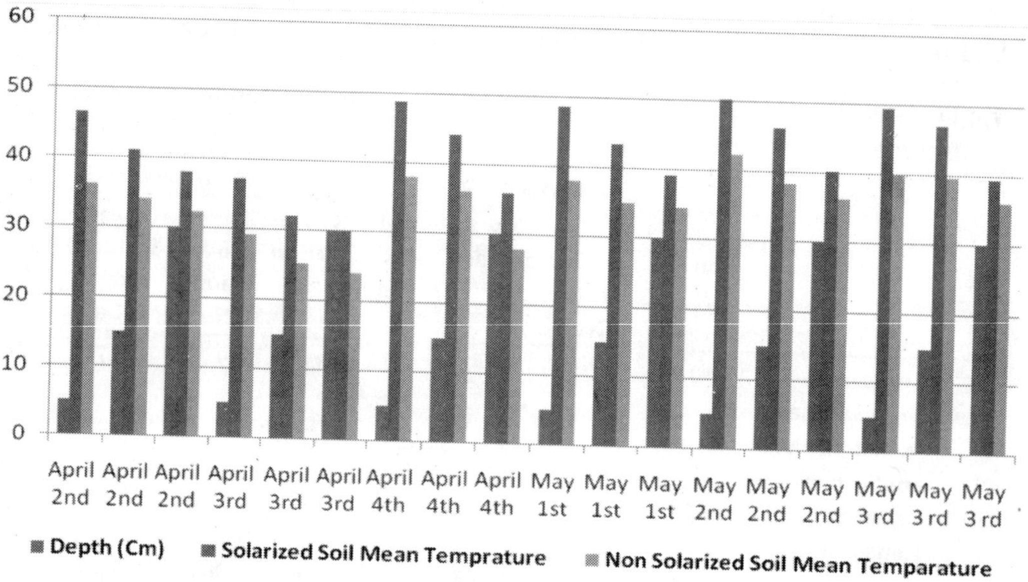

Fig. 1: Comparison between temperature levels in solarized and non- solarized soil

Temperatures build up in solarized plot: Soil temperature, range of daily temperature and difference of average temperature in both the solarized and non solarized plots are given in Figure 1. Increase in soil temperature in solarized plot was observed at all the three depths studied. Soil temperature in solarized plot at 5 cm depth was increased by 10.40°C over non-solarized. In deeper depth too temperature was increased by 4 to 8°C. During April 4th week the highest temperature was recorded of about 48.90°C at 5 cm depth. In the month of May, highest temperature was recorded (50.40°C) during 2nd week of May at 5 cm depth. Rise in soil temperature to the tune of 10°C in solarized field has been reported by Katan (1987). Maximum temperature attained decrease with increasing soil depths. Balakrishnan *et al.* (1996)

reported that, there was increase in temperature in solarized plot by 8°C at 5 cm depth over non-solarized plot. In successful solarization experiments enhancement of soil temperatures to a range of 37-51°C have been obtained in upper soil layers (Katan *et al.*, 1976; Grinstein *et al.*, 1979, Jacobson *et al.*, 1980). This temperature range (37-52°C) is sufficient to kill or inhibit most of the major pathogens provided sufficiently long exposure time for sunlight is allowed (Katan, 1981).

Pathogen population : *Pythium aphanidermatum* population was monitored during solarization period in both the plots. Significant reduction in pathogen population could be noticed in solarized plots (Table 2). *Pythium aphanidermatum* population was almost same before solarization in both the plots. Fifteen days after solarization *P. aphanidermatum* population in solarized plot decreased from 35.42 cfu g^{-1} of air-dried soil to 28.64 cfu g^{-1} whereas in non-solarized plot there was increase in populations of pathogens. Thirty days after solarization, *P. aphanidermatum* population still reduced in solarized plot as compared to non-solarized plots. Significant reduction of *P. aphanidermatum* occurred in solarized plots at the end of solarization period.

Table 2: Effect of soil solarization on population of *Pythium aphanidermatum* (cfu g^{-1} soil x 10^{5})

S.No	Days	Solarized soil	Non-solarized soil
1	0	35.42	36.22
2	15	28.64	38.41
3	30	12.37	39.12
4	45	1.94	41.56
5	60	1.32	42.81

Pythium aphanidermatum population was monitored during solarization period in both the plots. Significant reduction in pathogen population could be noticed in solarized plots. *Pythium aphanidermatum* population was almost same before solarization in both the plots. Significant reduction of *P. aphanidermatum* occurred in solarized plots by the end of solarization period. The reason may be that, during solarization period there was increase in temperature upto 50°C which killed the pathogens. The effectiveness efficacy of soil solarization in reducing the soil-borne plant pathogens is well documented by Katan *et al.* (1976). Usman *et al.* (1996) reported that, *Pythium* population was less in solarized plots compared to non-solarized plots. Balakrishnan *et al.*, (1996) reported that, there was significant reduction in *P. aphanidermatum* in solarized plot when compared to non-solarized plot. Stapleton and De Vay (1984) obtained up to 70 % reduction of *Pythium* species when solarization was conducted for 4-6 weeks and maximum soil temperatures in solarized soils were 7-12°C higher than in non-solarized soil. In solarized plot there was increase in temperature upto 45-52°C and at this temperature most of the soil borne pathogens are susceptible and they were killed. Hence, soil solarization helps in reduction of soil borne pathogens.

CONCLUSIONS

From this study it can be concluded that soil solarization helps the farming community to manage the disease effectively and also enhance the yield levels. This can be recommended to farmers to manage important rhizome rot disease of ginger.

REFERENCES

Anonymous, 2003. Horticultural Crop Statistics of Karnataka State at a Glance, 2002-03. Directorate of Horticulture, Lalbagh, Bangalore, 26p.

Anonymous, 2005. Food and Agricultural Organisation Production Year Book, Published by Economic and Social Department, FAO, Rome.

Balakrishnan, P., Usman, N.M. and Sarma, Y.R. 1996. Management of rhizome rot disease of ginger (Zingiber officinale Rosc.) by soil solarization. *Journal of Plantation Crops*, 24 (Supplement): 192-199.

Grinstein, A., Orion, D., Greenberger, A. and Katan, J. 1979. Solar heating of the soil for the control of Verticillum dahliae and Pratylenchus thornei in potatoes. *In*: Soil Borne Plant Pathogens (Ed. Schhippers B. and Gams, W.). Academic Press. London, pp 431-438.

Jacobson, R., Greenberger, A., Katan, J., Levi, M. and Alon, H. 1980. Control of Egyptian Broomrape (*Orobanche aegyptiaca*) and other weeds by means of solar heating of the soil by polyethylene mulching. *Weed Science*, 28: 312-316.

Joseph, L. and Sivaprasad. 2000. Biological control of rhizome rots of ginger. *In*:Proceedings of Indian Phytopathological Soceity Golden Jubilee and International Conference on Integrated Plant Disease Management for Sustainable Agriculture, Vol. I. at New Delhi, from November, 11-15, 1997, pp339- 340.

Katan, J. 1981, Solar heating of soil for control soil borne pest. *Annual Review of Phytopathology*, 19:211-236

Katan, J. 1987, Soil Solarization, *In* :Innovative approaches to plant disease management, (Ed. I. Chet). John Wiley and Sons Inc., New York. pp77-105

Katan, J., Greenberger, A., Aton, H. and Grinstein, A. 1976.Solar heating by polythene mulching for the control of diseases caused by soil borne pathogens. *Phytopathology*, 66:683-688.

Ram, D., Mathur, K., Lodha, B.C. and Webster, J. 2000. Evaluation of resident bio-control agents as seed treatment against ginger rhizome rot. Indian Phytopathology, 53: 450-454.

Shanmugam, V., Varma, A.S. and Surendran, M. 1999. Management of rhizome rot of ginger by antagonistic microorganisms. *Madras Agriculture Journal,* 86: 339- 341.

Singhal. V. 2003, Indian Agriculture 2003. Indian Economic Data Research Centre, New Delhi, pp593- 598.

Stapleton, J.J. and De Vay.J.E. 1984. Thermal components of soil solarization as related to changes in soil and root microflora and increased plant growth response. *Phytopathology,* 74: 255-259.

Usman, N.M., Balakrishnan, P. and Sarma, Y.R. 1996, Biocontrol of rhizome rot of ginger. *Journal of Plantation Crops,* 24: 184-191.

□□□

Chapter 76

Alley Cropping System for Natural Resource Conservation in Shifting Cultivated Degraded Lands of Eastern Ghats

HOMBE GOWDA, H.C., PRAVEEN JAKHAR, D. BARMAN AND M. MADHU

INTRODUCTION

Shifting cultivation is primitive systems of agriculture, still tribal are practicing in the hilly regions of the Eastern Ghats and it is considered to the most ancient system of agriculture dating back to the lower Neolithic period. It is also known as "Field Forest Rotation"or slash and burn agriculture or *Podu* cultivation. Shifting cultivation is practiced in some form or other in almost all the tribal areas of Odisha. The problem of shifting cultivation is perhaps most acute in Odisha than any other state in the country. The total area under the *Podu* cultivation in Odisha is about 1.6 M ha and 20 lakh tribal families are depending on this cultivation practices for their livelihood (Dash, 2006). This agriculture practice is considered devastative and disadvantageous as it not only causes harm to the eco-system but also exerts negative impact on economy. The major problems from this practice are loss of forest cover, soil erosion, and spring dry up, heavy floods, silting of the tanks and fields and damage to crops. Sahu *et al.* (2005) reported that the due to the practice of shifting cultivation, soil is eroding to a tune of 84-170 t ha^{-1}. Again, due to this practice the productivity of the land becomes very low; this is attributed by the all above said reasons. That's why tribes migrate to new area to cultivate the land. Further, due to the increased population, majority of the shifting cultivation area are under continuous cultivation by causing more damage to soil surface. Even though it is disadvantageous, this system cannot be avoided completely but should be discouraged by providing suitable agro-forestry practices (Dash, 2006).

Agro-forestry practice is also a very old system of farming with ensured productivity, protectively and sustainability. Resource conservation can be

Central Soil and Water Conservation Research and Training Institute, Research Centre, Sunabeda, Dist.: Koraput, Odisha, India obtained by encouraging the practice of Agro-forestry with new dimensions (Jha, 2005). Agro-forestry ensures maximum production, multiple products with soil and water conservation. Alley cropping is one of the most important systems of agro-forestry with ensured productivity, protectively and sustainability. This system ensures maximum production, multiple products with soil and water conservation. Alley intercropping is a farming system in which trees or shrubs are planted in rows, along contour line in sloppy land. Alley cropping helps to minimize soil erosion, decreases surface run-off and increases the nutrient status of soil. Thus, this system helps in increasing the land use efficiency by utilizing tree foliage as green manures to improve the fertility status of the soil. By looking at the present scenario, intensive alley cropping system is undertaken to study the bio-physical performance of Alley crops and to assess the productivity of intercrops like Ragi and upland paddy in Alley cropping.

MATERIALS AND METHODS

The experimented was conducted at Central Soil and Water Conservation Research and Training Institute, Research Centre, Sunabeda, Koraput, which is located in the Central part of the Eastern Ghats. The experiment was executed in one ha area with two tree species *Gliricidia maculata* and *Leucaena leucocephala* on 5 and 10 % slopes with Ragi and upland paddy with two different planting laid in spit plot design. Runoff and soil loss was monitored through multislot devisors connected with tanks. Earthen bunds around each plot were made with drains to channelized runoff water to multislot devisers. Standard package of practices were followed for growing crops. The data on runoff, soil loss and yield was collected for three years and analysed. The average data on all the above parameters was presented and discussed.

RESULTS AND DISCUSSION

Runoff and soil loss: Among all the treatment, runoff (15.64 %) and soil loss (5.02 t ha^{-1}) was minimum under Gliricidia trench planting at 5 % slope with Ragi crop (Table 1). Better rainfall interception by the dense Gliricidia canopy and crop coupled with water retention by trenches attributed to lesser runoff and soil loss under this treatment. On the other hand, maximum runoff (27.13%) and soil loss (20.34 t ha^{-1}) was obtained under control with upland paddy at 10% slope (Table 1). In general, Gliricidia based treatment recorded lesser runoff and soil erosion effectively as compared to Subabul based treatment. Grewal, *et al.* (1992) report that practice of Alley cropping is helps in minimize soil erosion, decreases surface run-off and increases the nutrient status of soil. Apart from soil and water conservation, green biomass produced by *Leucaena leucocephala* and *Gliricidia maculate* helps in soil amelioration and improves the soil fertility (Kang and Gutteridge, 2003). This practice is also effective to control the splash erosion (Kanaujia and Bhatia, 2001) and improves the soil aggregate stability (Mapa and Gunasenah, 2001). Alley cropping system with Nitrogen fixing trees (NFTs) reduces soil erosion to a tune of 6 t ha^{-1} compare to control (69 t ha^{-1}). It also reduces the runoff to a tune of 304% and decreased in the slope gradient by 1.1° per year (Rose, 2001) and found best alternative system to shifting cultivation (*Anonymous*, 2007).

Table 1: Effect of different treatments on runoff and soil loss under Ragi and Up land Paddy during 2010-11.

T. No	Slope (%)	Alley spp.	Planting method	Runoff (%)		Soil loss (t ha^{-1})	
				Ragi	Up land paddy	Ragi	Up land paddy
1	5	Subabul	Normal	19.39	19.90	7.45	8.66
2	5		Trench	19.07	19.43	6.96	7.26
3	10		Normal	21.37	22.21	11.78	13.37
4	10		Trench	20.92	21.36	9.87	10.45
5	5	Gliricidia	Normal	17.53	19.78	5.85	7.38
6	5		Trench	15.64	17.69	5.02	5.96
7	10		Normal	18.97	20.68	7.33	8.26
8	10		Trench	18.07	19.53	6.95	7.50
9	5	Control	nil	22.57	23.63	12.96	14.06
10	10		nil	24.14	27.13	15.39	20.34

Crop Yields

Ragi: Among all the treatments, highest average grain yield of Ragi (18.70 q ha^{-1}) was obtained under Gliricidia normal planting at 5 % slope during the study period. However, intercropping of Ragi with various treatments did not affect the yield significantly (Table 2). In general better crop growth of Ragi was observed in trench planting with irrespective of Alley crops. Better growth of Ragi and upland paddy was observed under Gliricidia was mainly due availability of soil moisture coupled with complimentary interaction of Gliricidia hedge row and crops. Green mannuring of Gliricidia was also influenced to improve the grain yield among two crops. The lowest grain yield was observed under control which was attributed due to high surface runoff resulted in low soil profile contribution to growth of crops.

Upland paddy: Highest yield (18.88 q ha^{-1}) was obtained in Gliricidia trench planting at 5% slope. As we know up land paddy is a more resource requiring crop, which was ensured under Gliricidia trench planting resulting in higher grain yield. Higher yield of upland paddy was also influenced due to the microclimate enhancement by the thick Gliricidia hedge row and green mannuring. Lowest average yield of upland paddy was recorded in control at 10% slope due to poor rain water conservation. Also evident that Gliricidia based treatment effectively controls runoff and soil loss than Subabul based treatment and control.

Table 2: Effect of different treatments on bio-physical characteristics of Ragi and Up land Paddy during 2009-11.

T.No.	Slope (%)	Alley spp.	Planting methods	Grain Yield of Ragi (q ha^{-1})			Avg. Yield	Grain Yield of Upland Paddy (q ha^{-1})			Avg. Yield
				2009	2010	2011		2009	2010	2011	
1	5	Subabul	Normal	10.76	17.84	15.58	14.73	17.01	17.90	13.50	16.14
2	5		Trench	12.90	17.28	16.13	15.44	17.48	19.63	13.37	16.83
3	10		Normal	10.39	15.84	21.60	15.94	16.56	15.30	12.93	14.93
4	10		Trench	10.74	16.27	17.84	14.95	17.11	15.67	13.23	15.34
5	5	Gliricidia	Normal	10.93	21.03	24.15	18.70	17.45	20.93	16.90	18.43
6	5		Trench	12.06	18.38	23.91	18.12	17.55	22.13	16.97	18.88
7	10		Normal	10.80	17.12	22.80	16.91	16.72	20.30	14.50	17.17
8	10		Trench	11.06	16.42	23.80	17.09	17.84	18.87	15.37	17.36
9	5	Control	nil	9.91	14.33	15.04	13.09	16.34	15.33	11.87	14.51
10	10		nil	9.74	13.65	15.10	12.83	16.09	15.03	10.30	13.81

The above results are on par with the *Leucaena leucocephala* based alley cropping model, which reveals that addition of lopping material adds an amount of 80 kg organic nitrogen which intern increase the grain yield of rice and wheat significantly. It also decreases the pH of the soil from 8.74 to 8.43, BD from 1.7 to 1.5 g cc^{-1}, improves the nitrogen content by 24.10%, available P_2O_5 by 13.40%, available K_2O by 18.40% and infiltration rate by 38.20% in sandy-loam soil of Uttar Pradesh (Gangwar *et al.*, 2004). Kaushal Kumar *et al.*, (2003) also reported that higher biomass (39.28%) was achieved in 4 m alley than 6 m alley width. It also showed that this agroforestry practice is more economical in the alluvial soil of Kanpur (Kanaujia and Bhatia, 2001). *Gliricidia maculate* based alley cropping system has increased Maize yield by 54%, soil nitrogen by 22%, and phosphorus by 32% and CEC by 34% (Abunyewa *et al.*, 2004).

CONCLUSIONS

Alley cropping system is one of the best alternate land use system particularly in tribal belt where shifting cultivation is predominant. From this study it could be concluded that, Gliricidia is most promising as an alley crop over subabul to obtain higher biomass production and grain yield of ragi and paddy crop. Trench method of planting was found better for resource conservation at all the slopes.

REFFERENCES

Abunyewa, A., Asiedu, E.K., Nyamekye, A.L. and Cobbina, J. 2004. Alley cropping Gliricidia sepium with Maize: 1. The effect of hedgerow spacing, pruning height and phosphorus application rate on Maize yield. *Journal of Biological Sciences*, 4(2): 81-86.

Anonymous. 2007. Inga species and alley cropping alternative to slash and burn agriculture. *Agroforestry review*, 15:47-53.

Dash Balaram. 2006. Shifting cultivation among the tribes of Orissa, *Orissa Review*, 76-84,

Gangwar, K.S., Sharma, A.K. and Tomar, O.K. 2004. Alley cropping of Subabul (*Leucaena leucocephala*) for sustaining higher crop productivity and soil fertility of Rice (Oryza sativa)- Wheat (*Triticum aestivum*) system in semi- arid conditions. *Indian Journal of Agronomy*, 49 (2): 84-88.

Grewal, S.S., Mittal, S.P., Surjit Dyal and Agnihotri. 1992. Agroforestry systems for soil and water conservation and sustainable production from foot hill areas of North India. *Agroforestry Systems*, 17: 183-191.

Jha, L.K. 2005. *Shifting Cultivation, APH Publishing Corporation*, 113-162.

Kanaujia, V.K. and Bhatia, K.S. 2001. Effect of alley cropping on splash erosion and growth and yield components in eroded alluvial soils. *Journal of Soil and Water Conservation*, 20:109-113.

Kang, B.T. and Gutteridge. 2003. Forage tree legumes in Alley cropping systems. *Journal of Sustainable agriculture*, 26(3): 58-74.

Kaushal Kumar, Kanaujia, V.K., and Gangwar, M.. 2003. *Leucaena* based alley cropping system for Sustainable crop production under rained condition. *Indian Journal Soil Conservation*, 31(3): 265-268.

Mapa, R.B. and Gunasenah, P.M. 2001. Effect of alley cropping on Soil aggregate stability of a tropical alfisol. *Soil science and Plant analysis*, 32: 9-10.

Rose C.W. 2001. Soil erosion models and implementation for conservation of sloping tropical lands, paper presented in *International soil conservation Organization*, Stott, D. E., Mohtar, R. H., and Steinhardt, G. C. (Eds.) : 852-859.

Sahu, S.K., Nayak, R.K. and Sarangi, D. 2005. Sustainable soil and land management under shifting cultivation in Orissa. *Orissa Review*, 49-52.

Section - VI

Chapter 77

Water and Gender Equity

KRISHNA SRINATH

INTRODUCTION

Land, water, climate and biological diversity form the natural base of agriculture is essential to rural development and sustainable livelihoods. The growing demand for food, water, fibre and energy is disrupting agro-ecosystems, eroding biodiversity and depleting land and water. Natural resources must be used in a way that meets today's needs, while conserving them for future generations. That will require action to develop capacities, from global to farm level, for their sustainable management and regulation. Women manage natural resources daily in their roles as farmers and household providers. Typically, they are responsible for growing subsistence crops, and often have unique knowledge of local crop species. To meet family needs, rural women and girls walk long distances to collect fuel wood and water. Despite their reliance on natural resources, women have less access to and control over them than men. Usually it is men who put land, water, plants and animals to commercial use, which is often more valued than women's domestic uses. Women farmers represent more than a quarter of the world's population. Women comprise, on an average, 43% of the agricultural work force in developing countries, ranging from 20 per cent in Latin America to 50% in Eastern Asia and Sub-Saharan Africa. Yet women have less access than men to agriculture related assets, inputs and services. Had they enjoyed the same access to productive resources as men, women could boost yield by 20-30% ; raising the overall agricultural output in developing countries by two and a half to four per cent. This gain in production could lessen the number of hungry people in the world by 12-17%, besides increasing women's income (FAO, 2011). India has a predominantly agrarian economy, 70% of her population is rural; of those households, 60% engage in agriculture as

Directorate of Research on Women in Agriculture, Bhubaneswar their main source of income. It has always been India's most important economic sector. In this important agricultural sector woman plays a vital role, because it is largely a household enterprise. Women in India are major producers of food in terms of value, volume and number of hours worked. Nearly 63 % of all economically active men are engaged in agriculture as compared to 78% of women. Almost 50 % of rural female workers are classified as agricultural labourers and 37% as cultivators. About 70 % of farm work was performed by women. It is observed that women play a significant and crucial role in agricultural development and allied fields including, main crop production, livestock production, horticulture, post-harvesting operations, forestry, fishing *etc.*, it is a fact long taken for granted but ignored since ages.

INTERNATIONAL SCENARIO

Most of the world's 1.2 billion poor people, two thirds of whom are women, live in water scarce countries and do not have access to safe and reliable supplies of water for productive and domestic uses (IFAD, 2001a). The bulk of these rural poor people are dependent on agriculture for their livelihoods and live in sub-Saharan Africa and South Asia, the regions which are also home to most of the world's water poor (Molden, 2007). One third of the world's population is currently experiencing some kind of physical or economic water scarcity. A growing competition for water from different sectors, including industry, agriculture, power generation, domestic use, and the environment, is making it difficult for poor people to access this scarce resource for productive, consumptive and social uses. In water-scarce regions and countries, inequity in access to water resources is increasing because of competition for limited resources and this particularly affects poor rural people, especially women. Women's secure access to water and land is central to achieving the Millennium Development Goals, in particular Goal 1 (reducing by half the proportion of people living in extreme poverty and hunger by 2015) and Goal 3 (promoting gender equality and empowering women). Most of the agricultural water management programmes and projects addressed the need to strengthen water users associations (WUAs), thus achieving one of the fundamental objectives of increasing the participation of beneficiaries in the design and implementation of programmes and projects (IFAD, 2001b). This review examines the impact of water-related projects on women, women's role in managing water resources and the constraints women face in gaining access to water.

Although international policymakers are increasingly recognizing women's roles in agriculture, in general and irrigated agriculture in particular, many women farmers remain poor, vulnerable to food insecurity and marginalized. Already in 1992, the central role of women in water management was recognized in the Dublin Principles (adopted at the International Conference on Water and the Environment, Dublin). Since then, policymakers have made attempts to incorporate gender issues in water development projects, including in the resolution declaring 2005-2015 the International Water for Life Decade. However, these policies have not been adequately translated into practice, and attempts in some projects to involve women in water management initiatives have met with only modest success. The reasons for these disappointing results range from lack of understanding of gender issues by policymakers and project staff, to lack of will and commitment at the project design and implementation phases, to lack of capacity among project staff in skills and the use of relevant tools, to the unavailability of gender disaggregated data, to prevailing cultural norms in the societies. In fact, women generally have limited influence, do not exert political pressure, or are simply not heard or seen: a sign of great insensibility.

Women manage water resources not only for productive uses, but also for domestic purposes. Sanitation and hygiene for good health are their responsibility, and they often play an active role in the construction and preventive maintenance and repair of sanitation facilities. Securing water for both productive and domestic uses is critical in achieving food security and improved rural livelihoods in most parts of the world, but particularly in arid and semi-arid areas. However, despite the role that women play in reducing food insecurity through their knowledge of crop production, local biodiversity, soils and local water resources, they are often excluded from decision-making processes in new agricultural water management approaches and other projects and initiatives on natural resource allocation. Securing access to land among poor farmers, particularly women, can lead to secure water rights. It can then lead to access to other resources such as financial services and investment in farms, offering the potential to improve livelihoods and reduce water wastage. As a result, most of the agriculture and water initiatives that aim at enabling poor and vulnerable farmers to improve their livelihoods and provide access to productive resources fail to take into account women's concerns about the multiple uses of water. Women use water for agriculture, domestic tasks, health and sanitation, while men's water use priorities mainly revolve around agriculture or livestock. Workloads and responsibilities Women all over the world play an active role in agriculture, thus contributing to food security. In many countries, women are involved in rainfed agriculture as well as backyard or irrigated home gardening, while men often are responsible for rainfed commodities and land management aspects of irrigation. Depending on the traditions of the societies they live in, the prevailing norms and the migration patterns of men, women may play different roles in the production cycle. The availability of clean water close to home reduces women's workloads and the time saved in fetching water may be spent on other activities to strengthen livelihood resilience, including productive activities such as crop production. Therefore, women are often interested in using rainfall runoff or irrigation water for purposes other than irrigating field crops. Most of the water supply projects in the past were developed with a single dimension; they either focused on domestic water supply or provided water only for irrigation.

Communities, on the other hand, have diverse uses for water such as for agriculture, fishing, livestock watering, small businesses, kitchen gardening and domestic tasks. In the past, agricultural water management projects have not generally been designed to take into account these multiple uses for water within water management schemes. This trend is changing, and water projects are becoming more multi-purpose, multi-use and multiuser. The involvement of communities, both men and women, in the selection of and planning for such interventions is the key to successful gender mainstreaming. Not addressing the multiple uses of water has been recognized as one of the causes of the lower participation of women in WUAs. In some irrigation systems, the use of irrigation water supplies for domestic purposes is considered illegal. Irrigation has made it easier for women's animals to be watered in convenient places. In particular, providing water for multiple uses reduces drudgery and provides women with more time for other, more productive or livelihoods activities. Access to irrigated land: Understanding the links between land and water governance Natural resources are one of the foundations in the effort to overcome poverty among poor people in rural areas. Thus improvements in the management of these resources are the focus of many development initiatives and projects that seek poverty reduction by empowering poor people to improve their livelihoods. Experience shows that many challenges remain in achieving these goals in an equitable and sustainable manner. One of the main obstacles to improving the livelihoods of poor rural people is the lack of attention given to gender issues and women's access to natural resources, in particular land and water. Although research offers evidence on women's multiple roles in agricultural production, their access to productive resources such as land, water, fertilizer, credit and other inputs remains limited. In most developing countries, access to water for productive use in general and for irrigation in particular is intrinsically linked to access to land. In all parts of the world, relatively few women own land.

NATIONAL SCENARIO

We know that most of the developing countries depend on agriculture for livelihood and food security. Agriculture needs manpower, if the manpower split into gender wise, amazing fact is that the women contribution is greater or equal to men. The rural women are very active in cultivation, dairy, fisheries, crop processing and other allied areas. Nearly 70% of Indian rural women are employed in agriculture and they are responsible for 60-80 % of food production. They play major role in animal husbandry, horticulture and poultry which are their main source of income and it is noticed that they always involved in labour and tolerance intensive works like transplantations and weeding operations. Among the rural women workforce, most of them are agriculture labor and some of cultivators. There are lot of variations in involvement of women in agriculture which is based on their culture, economic status, regions and crop selection. While the men are moving to cities for better occupations the women are taking care of cultivation and sometimes they used to work as farm labor to support their family needs. The weaker section of the women used to market their products such as selling vegetables and other food crops in farmers market or door to door. Most of the women cultivators are involved in food crops rather than cash crops since cash crops need more marketing efforts which are traditionally taken care by men. The women in higher socio economic sections are not involved directly in cultivation or live stock and they used to help in labor administration, supporting activities and accounting. Ironically, we often define the men as farmers, in fact the women also equally involved in agriculture in all phases. If we think beyond the conventional definition of "FARMER", most of the rural women should be considered as farmers but they are ignored. Still Indian society has some gender boundaries, but the rural women involve in multiple productive roles and it is necessary to give equal opportunity in agriculture.

Most of the women farmers are marginal or small farmers, landless tenant farmers and farm labour; they don't have enough land to cultivate, less knowledge, limited access to innovative technologies, low capital and less credit facilities. In order to empower and improve the women farmers' productivity, they need to have proper farm training and capacity building programs to compete various challenges in their rural livelihood. Women land ownership plays a major role in their credit worthiness, if they have right on the agriculture land they will have access to financial resources and savings. New legislations have to be taken place in farm land reforms like allocating the wasteland to women self help groups (SHG). Rural women have been acknowledged as the core food producers and processors, banks should recognize them as entrepreneurs, loans and Kissan credit cards have to be approved for them. Special support and market connectivity has to establish for rural women artisans to sell their agro based handicrafts. The farm labor wages should be the same without gender differences and the equal employment benefits have to be given in all rural schemes like NREG. A sufficient political representation must be required by rural women in local level councils, it directs them as an equal partners in political, economical and legal decision making. We must appreciate Government of India for implementing various rural women welfare schemes and also the NGO's who are working on women empowerment for poverty alleviation

It has become increasingly accepted that women should play an important role in water management and that this role could be enhanced through the strategy of gender mainstreaming.

- In rural Benin, girls ages 6-14 spend an average of one hour a day collecting water compared with 25 minutes for their brothers.
- In Malawi, there are large variations in the amount of time allocated for water collection based on seasonal factors, but women consistently spend four to five times longer than men on this task.
- In Tanzania, a survey found school attendance to be 12% higher for girls in homes located 15 minutes or less from a water source than in homes one hour or more away. Attendance rates for boys appeared to be far less affected by distance from water sources.
- In 12% of households children carry the main responsibility for collecting water, with girls under 15 years of age being twice as likely to carry this responsibility as boys under the age of 15 years.
- Research in sub-Saharan Africa suggests that women and girls in low-income countries spend 40 billion hours a year collecting water—the equivalent of a year's worth of labour by the entire Work force in France.
- In Africa, 90% of the work of gathering water and wood, for the household and for food preparation, is done by women. Providing access to clean water close to the home can dramatically reduce women's workloads, and free up time for other economic activities. For their daughters, this time can be used to attend school
- Around 38.5% families travel at least half a km to fetch drinking water in Odisha villages. The number of such families was 32.5% in 2001, which increased by 6% in 10 years. Director census operations Bishnupada Sethi gave a presentation on the data after it was released here. Around 38.5% families travel at least half a km to fetch drinking water in Odisha villages. The number of such families was 32.5% in 2001, which increased by 6% in 10 years.

Women are more vulnerable to deficit in water resource both at agriculture and household domains. And understanding of the issues in natural resource and water management requires and understanding of roles and relations of men and women as a part of the overall natural resource management systems. Women often have knowledge, expertise and skill in adapting to water scarcity however, the management of scare water results in great amount of drudgery due to long distance to be covered for fetching water and rationing the water for various household, livestock and backyard crop production activities. Climate change has significant impacts on fresh water sources, affecting the availability of water used for domestic and productive tasks. The consequences of the increased frequency in floods and droughts are far reaching, particularly for vulnerable groups, including women who are responsible for water management at the household level. Drought lead to somewhat permanent migration flood causes temporary dislocation of the community.

Potable water resources are scarce in the coastal areas and people depended on scanty public supply for drinking and food preparation. Water scarcity was considered to be one of the atrocities to women as they had to spend a long time, energy and money in some cases to fetch potable water. While mothers went out for fetching water girls managed the household chores which affected their education. Water scarcity led to use of well water for drinking after boiling with Tulsi leaves, ginger and cloves to remove odour and ill effects. Tea and coffee prepared in well water tasted bad. Good quality clothes could not be washed frequently and hence sun dried. Daily bath, an essential routine of Malayalees (people of Kerala State) was restricted in summer months. White cotton clothes, a symbol of Kearalite were also sparingly used due to scarcity of good water for washing.

Strategies for Efficient Water Use

- Targets, such as those in the Millennium Development Goals relating to water, are unlikely to be achieved unless gender perspectives are integrated into planning and implementation activities
- Reduced time, health, and care-giving burdens from improved water services give women more time for productive endeavors, adult education, empowerment activities, leisure
- Convenient access to water and sanitation facilities increase privacy and reduce risk to women and girls of sexual harassment/assault while gathering water
- Higher rates of child survival are a precursor to the demographic transition to lower fertility rates; having fewer children reduces women's household responsibilities and increases their opportunities for personal development.

The importance of involving both women and men in the management of water and sanitation and access-related questions has been recognized at the global level, starting from the 1977 United Nations Water Conference at Mar del Plata, the International Drinking Water and Sanitation Decade (1981-90) and the International Conference on Water and the Environment in Dublin (January 1992), which explicitly recognizes the central role of women in the provision, management and safeguarding of water. Reference is also made to the involvement of women in water management in Agenda 21 (chapter 18) and the Johannesburg Plan of Implementation. Moreover, the resolution establishing the International Decade for Action, 'Water for Life' (2005-2015), calls for women's participation and involvement in water-related development efforts. The differences and inequalities between women and men influence how individuals respond changes in water resources management. Understanding gender roles, relations, and inequalities can help explain the choices people make and their different options. Involving both women and men in integrated water resources initiatives can increase project effectiveness and efficiency. Without specific attention to gender issues and initiatives, projects can reinforce inequalities between women and men and even increase gender disparities.

Technologies for Water Management at Household Level

Household water treatment interventions may play an important role in protecting public health where existing water sources, including those delivered via a piped network or other improved sources, are untreated, are not treated properly or become contaminated during distribution or storage. Household water treatment applications are a range of technologies, devices or methods employed for the purposes of treating water at the household level or at the point of use in other settings, such as schools, health-care facilities and other community locations. Point-of-use water treatment is another term used for household water treatment. While most household expenses tend to increase over time, water expense tends to decrease year after the year though the demands continue to increase. The situation is different in peri-urban and rural areas where expenditure on drinking water increases as bottled water has proliferated into the daily lives of this segment of population.

- Development of simple household technology for efficient water use.
- Development of water carrying devices, *eg.*, Water trolley and water bag developed by AICRP on Home Science.
- Educating the members of the household on water use and management and water budgeting.
- Specific planning for securing available water while developing infrastructure in a locality.

National Water Mission is part of the eight climate change missions of India. Water scarcity from being merely a topic for decision is becoming a harsh reality and if judiciously utilized. Addressing water scarcity requires actions at local, national and river basin levels. It also calls for actions at global and international levels, leading to increased collaboration between nations on shared management of water resources (rivers, lakes and aquifers) and of the benefits thereof. Being intersectoral in its nature, the response to water scarcity requires: collaboration; the sharing of joint visions and policy principles; and joint action in addressing the issue. A major challenge in addressing water scarcity successfully in countries is the institutional fragmentation of responsibilities in the water development sector. A cross boundary partnership to meet the most basic need of human will be fundamental to any strategy in this direction.

CONCLUSIONS

As farmers, agricultural workers and entrepreneurs, women form the backbone of agricultural rural economy in the developing regions, and yet, together with children, they remain one of the most vulnerable groups. With far less access to education and technology, a host of other socio-economic factors have had an adverse impact on the lives of women farmers in recent years. These include the accelerated pace of globalization, the associated policy and institutional changes including economic liberalization and commercialization of agriculture, rapid pace of population growth and urbanization. Rural-urban migration, growing pressure on land, water, agro-biodiversity and firewood and natural disasters associated with climate change have been unfavourable to women in agriculture, and they have often not been able to take advantage of opportunities from new technologies, markets, or contract farming. The constraints and opportunities that women face in agriculture vary across regions and countries, depending on the socio-cultural and agro-ecological contexts. Despite many policy reforms both at the macro and micro level, gender issues have not received the attention they deserve. Hence, the current situation has to urgently change.

REFERENCES

http://www.ose.state.nm.us/water-info/conservation/pdf-manuals/Roof-Reliant-Landscaping/RRL-Chapter-6.pdf.

http://www.indiawaterportal.org/taxonomy/term/12292.

Coping with Water Scarcity Challenges of Twentieth Century, World Water Day 2007, FAO

Srinath, K. 2008. Gender and Coastal Zone Biodiversity, Gender, Technology and Development, Vol 12: 209-227

Human Development Report. 2006. Beyond scarcity: Power, poverty and the global water crisis. United Nations Development Programme (UNDP).

Progress on sanitation and drinking water: 2010 update. WHO/UNICEF Joint Monitoring Programme.

Mainstreaming Gender in Water Management. Resource Guide. 2006. United Nations Development Programme (UNDP), Gender and Water Alliance (GWA).

□□□

Chapter 78

Effect of Conservation Agriculture on Yield, Soil Conservation and Economics in Maize-Wheat Cropping System in North-West Himalayan Region

B.N. GHOSH, PRADEEP DOGRA, N.K. SHARMA AND K.S. DADHWAL

INTRODUCTION

According to FAO (2010), conservation agriculture (CA) is a concept for resource-saving agriculture crop production that strives to achieve acceptable profits together with sustained production levels while concurrently conserving the environment. Conservation agriculture is characterized by three principles which are linked to each other, namely continuous minimum mechanical soil disturbance, permanent organic soil cover and diversification of crop species grown in sequence or associations. It has been reported that a productivity loss of 13.4 m ton of food grain worth rupees Rs.111.3 billion due to soil erosion by water in rainfed area of different states of India (Sharda *et al.*, 2010) and maize crop productivity loss of 8.0 to10.30 kg ha^{-1} mm^{-1} soil loss have been reported in Indian Sub-Himalays (Khybri *et al.* 1982 and Sur *et. al.*, 1998).Vegetative barriers is beneficial to reduce sheet and rill erosion, ephemeral gully erosion, manage water flow, stabilize steep slopes, trap sediment and provide other products like aromatic oil. Minimum tillage is recommended for soils of the Indian Himalayan region due to reduced cost of cultivation, more retention of soil water and physical protection of soil organic carbon (SOC) in the aggregates.

The maize-wheat rotation in N-W Himalayan states is practiced on 2141.8 thousand ha with an average productivity of 14.3 q ha^{-1} in maize and 18.5 q ha^{-1} in wheat under rainfed conditions, having less than the national average productivity of 21.6 q ha^{-1} of maize and 25.8 q ha^{-1} of wheat. Traditionally, resource poor farmers of the NWHR use to applied fertilizers/manures to maize/ *kharif* crops, and the wheat crop is grown on residual fertility. Under

Central Soil and Water Conservation Research and Training Institute, 218 Kaulagarh Road, Dehradun-248 195, Uttarakhand, India this system of cultivation, land degradation due to nutrient mining is aggravated day by day at this productivity level. Considering the present level of productivity and nutrient removal by cereals in the Himalayan states, it is evident that to achieve a reasonable productivity of around 1.5 t ha^{-1}, the nutrient removal will have to be almost one and half times than the present value (Bhattacharyya *et al.*, 2009). Minimum tillage is recommended for the soils of the region due to reduced cost of cultivation, more retention of soil water and physical protection of soil organic carbon in the aggregates (Prakash *et al.*, 2004). Weed cutting and mulching is considered to be one of the effective weed management as well as erosion and moisture conservation techniques (Chandra *et al.*, 2002). But its efficacy in integrated input/nutrient management has not been studied extensively. The efficacy of the grass species *Panicum maximum*, which extensively grows in subtropical mid-hill conditions, in reducing run off and soil loss in slopping land was documented (Ghosh, 2009). Oil yielding aromatic grass specie like *Cymbopogon martini* (palmarosa) is generally recommended for degraded lands of hill and mountain agro ecosystem, but its efficacy as vegetative barrier in reducing run-off and soil loss, and its oil yield and profitability are not tested so far. Therefore, in this context, aromatic grass based conservation agriculture is a challenge to the natural resource manager. Thus, to achieve a sustainable land use with enhanced crop productivity in the N-W Himalayan region, organic input management needs to be intensified considerably. The present study was undertaken with the objectives to evaluate the effect of CA on productivity, resource conservation and economics in maize-wheat cropping sequence in Entisols NWHR.

METERIALS AND METHODS

A fixed plot field study was carried out for four years (2007-08 to 2010-11) at Central Soil and Water Conservation Research and Training Institute, Research Farm, Selakui, Dehradun, Uttarakhand on 2% slope. The soils of the experimental sites were classified as fine mixed hyperthermic family of Typic udorthent and slightly acidic in nature having pH 5.8, organic carbon 6.0 g kg^{-1}, and available N, P and K as 270, 24 and 110 kg ha^{-1}, respectively. The soils were silty clay loam in texture having 12.5% clay and 35.5% silt with bulk density 1.36 g cc^{-1} and basic infiltration rate of 1.42 mm hr^{-1}. The experiment was laid out in a randomized block design with three replication having 100 m x 20 m run off plot comprising four treatments.

T_1: 100:60:40 $N:P_2O_5:K_2O$ + conventional tillage (CT) + chemical weeding + *Panicum* vegetative barrier

T_2 : FYM (5 t ha^{-1}) + minimum tillage (MT) + 1 weed mulch (30 DAS) + Palmarosa vegetative barrier

T_3: FYM (5 t ha^{-1}) + vermi-compost (1.0 t ha^{-1}) + MT + 2 weedmulch (25 and 50 DAS) + Palmarosa vegetative barrier

T_4: FYM (5 t ha^{-1}) + vermi-compost (1.0 t ha^{-1}) + poultry manure (2.5 t ha^{-1}) + MT + 3 weed mulch (20, 40 and 60 DAS) + Palmarosa vegetative barrier

Establishment of vegetative barriers of grass species :A furrow of 10 cm width and 20 cm depth along each contour line at 1 m vertical interval in a paired row at a distance of 75 cm by using a small agricultural implement was opened. Plant root slips containing 2-3 tillers at a spacing of 20 cm in the furrows in paired row were planted. For accommodating a paired row of vegetative barriers, planting was done in a staggered fashion. Planting was done during second week of July (at the start of monsoon) for both the grasses in the first year of experimentation (2007). The soil excavated from contour furrows was heaped on downstream side to form the bund. Panicum was established in T1 treatment whereas Palmarosa was established in T_2, T_3 and T_4 treatment. Grass species were harvested twice in a year (April-May and October-November) for fodder yield of panicum and oil yield of palmarosa.

Tillage, manuring and mulching : In conventional tillage plots, tillage operations were performed four times with tractor drawn cultivator, whereas in minimum tillage plots, two tillage operations were provided with 30% maize and wheat crop residues. Recommended NPK dose (120:60:40 kg ha^{-1}) in T_1 treatment was applied where half of nitrogen and full of phosphorus and potassium were applied at the time of sowing. The remaining nitrogen was top-dressed at tillering, panicle initiation and dough stages. In treatments T_2, T_3 and T_4, the farmyard manure (FYM @ 5 t ha^{-1}), vermi-compost (VC @ 1.0 t ha^{-1}) and poultry manure (PM @ 2.5 t ha^{-1}) were applied at the time of final land preparation before sowing of *kharif* crops. Weed mulching was done at 30 days after sowing (DAS) in T_2 treatment, 25 and 50 DAS in T_3 treatment, whereas in T_4 treatment, weed mulching were done 20, 40 and 60 DAS. The crop rotation of maize–wheat was followed where wheat was grown on residual fertility with one hand weeding in organic treated plots (T_2, T_3 and T_4). Atrazine weedicide for maize and Isoputoron for wheat @ 1.5 *a.i* litre^{-1} were applied to control weeds in T_1 treatment. The maize var. 'kanchan' and wheat var. 'PBW4' were sown as *kharif* and *rabi* crops, respectively.

Measurements and analysis: Run off data was recorded through stage level recorder, whereas soil loss was measured through collection of sediment in runoff water. Data on soil moisture was measured at fortnightly interval for determination of water used by the crop. Water use efficiency (WUE) was calculated as grain yield divided by water use. Bulk density was determined through core sample and infiltration by using double infiltration ring.

Aggregate size distribution was determined by the wet sieving method as described by Yoder (1936). Actual evapotranspiration (ET) was estimated following field water balance method and water use efficiency (WUE) was calculated in terms of yield per unit water loss (ET). A representative portion of each soil sample was air dried, powdered, and passed through 0.2 mm sieve for determination of electrical conductivity (EC) and pH in a 1:2.5 soil: water suspension, organic carbon (OC) by Walkley and Black's method, available soil nitrogen by alkaline-$KMnO_4$ method, Olsen-P, NH_4OAC-K, (using standard procedures as outlined by Jackson, 1973). Zn, Fe, Cu, and Mn were determined using the DTPA (diethylene triamine penta acetic acid) extraction method developed by Lindsay and Norvell (1978) using Atomic Absorption Spectrophotometer (Analytical Jena Model). Microbial biomass carbon (MBC) determinations were made using chloroform fumigation technique as described by Jenkinson and Ladd (1981). Dehydrogenase activity was estimated using TTC (TriphenylTetrazolium Chloride (Casida *et al.*, 1964). One gram samples of dried ground grain, straw, and root of maize and wheat were digested in HNO_3 and $HClO_4$ mixture for measurement of mean N, P and K contents. N, P and K uptake was estimated using mean N, P and K contents and dry matter yields of maize and wheat, respectively. Carbon sequestration potential was calculated from initial soil organic carbon and organic carbon after four years of cultivation using bulk density and soil depth.

Soil quality index (SQI): To determine a soil quality index, four main steps were followed: (i) define the goal, (ii) select a minimum data set (MDS) of indicators that best represent soil function, (iii) score the MDS indicators based on their performance of soil function and (iv) integrate the indicator scores into a comparative index of soil quality. To select a representative minimum data set (MDS) (Andrews *et al.*, 2002a) only those soil properties that showed significant treatment differences were selected. Significant variables were chosen for the next step in MDS formation through principle component analysis (PCA) (Andrews *et al.*, 2002a). We then summed up the weighted MDS variables scores for each observation using the following equation:

$$SQI = \sum_{i=1}^{n} Wi\ Si$$

Where S is the score for the subscripted variable and Wi is the weighing factor derived from the PCA. Here the assumption is that higher index scores meant better soil quality or greater performance of soil function. Further, the percent contribution of each final key indicator was also calculated. The cost of cultivation and net return of the system were worked out at the prevailing market price both for inorganic and organic treatments.

Statistical analysis: Statistical analyses were performed using standard Analysis of Variance (Gomez and Gomez, 1984). Treatment means were compared at the $p < 0.05$ level of significance using least significant difference (LSD).

RESULTS AND DISCUSSION

Effect on yield of maize and wheat : The average yield of maize and wheat was significantly affected with integrated input management (Table 1). On an average, T_4 treatment yielded 16.2% less productivity of maize than T_1 treatment where as wheat yield was recorded 47.8% more in T_4 compared to T_1 treatment. This is ascribed to the fact that slow nutrient supplying capacity of the organic manure in T_4 produced approximately two-third yield of inorganic manure in T_1, but higher residual nutrient and water availability during *rabi* season in T_4 increased wheat yield (Sarkar, 1998). Over the years, maize yield varied significantly with a declining trend depending upon the soil erosion level in all the treatments, whereas wheat yield decreased in T_1 treatment but increased in T_2, T_3 and T_4 treatments. The residual effect of organic manures and interaction affect of organic inputs might be the reasons for higher yield of wheat in the subsequent years for T_2, T_3 and T_4 treatments (Ghosh and Singh, 2003). The wheat equivalent yield (WEY) of T_4 treatment showed 9.0% higher value than T_1 treatment indicating overall residual effect of organic manures and integrated use of organic inputs after four cycles of cropping. But, other organic treatments (T_2 and T_3) showed comparative less equivalent yields at this level of input management compared to T_1 treatment (Table 1).

Table 1: Average (2007-08 to 2010-11) grain yield of maize and wheat (t ha^{-1}) as affected by different treatments

Treatment	Maize	Wheat	Wheat equivalent yield (WEY)	% increase/ decrease in WEY
T_1	2.34	1.15	2.95	-
T_2	1.05	1.20	2.01	-32.0
T_3	1.71	1.45	2.77	-6.0
T_4	1.96	1.70	3.21	9.0
CD (P=0.05)	0.42	0.23	1.22	-

Effect on biomass yield of vegetative barriers : The biomass yield of vegetative barriers of grass species was also significantly affected due to the treatments, which ranged from 4.85 to 9.95 q ha^{-1}. Among the two grass species, panicum showed higher biomass yield than palmarosa (Table2). The variable dry biomass yield is ascribed to the silt deposition behind the grass barriers providing water and nutrient availability. The live weed mulch biomass also significantly varied from 5.23 to 21.77 q ha^{-1} depending upon the number of cutting and mulching, and nutrition level. The nutrient recycling of N, P and K from the weed biomass was 62.8, 134.0 and 190.5 kg ha^{-1} in T_2, T_3 and T_4, respectively (Table 2).

Table 2: Average biomass yield of weeds (for mulching) and vegetative barriers, and annual nutrient cycling through biomass of weed and vegetative barriers (after oil extraction).

Treatment	Vegetative barriers (kg ha^{-1})		Fodder / Oil yield* (q ha^{-1})	Weed (kg ha^{-1})		Annual nutrient cycling (kg ha^{-1})			
	Fresh	Dry		Fresh	Dry	N	P	K	Total
T_1	30.95	9.95	9.95	-	-	-	-	-	
T_2	14.65	4.85	2.40*	16.05	05.23	17.7	2.7	042.4	062.8
T_3	16.20	5.10	2.56*	44.03	14.77	38.1	5.8	090.1	134.0
T_4	18.15	6.10	3.24*	69.87	21.77	54.3	8.3	127.9	190.5
CD (P=0.05)	02.46	1.27	-	06.45	02.12	-	-	-	-

Effect on runoff and soil loss : The integrated use of organic input management showed significant impact on reduction of run off and soil loss (Table 3). The extent of reduction in run off and soil loss in T_4 treatment was 44.8% and 36.8%, respectively compared to inorganic treatments (T_1). The minimum tillage, live mulch and vegetative barriers in T_4 treatment played significant role in reducing runoff and soil loss. The role of vegetative barriers in reducing run off and soil loss has already been reported under similar climatic and land form condition (Ghosh, 2009). The combined effect of minimum tillage and mulch in increasing soil infiltration rate and reducing run-off and soil loss has also been reported by Bhattarcharyya *et al.* (2009).

Table 3: Four years average run of and soil loss

Treatment	Run off (%)	Soil loss (t ha^{-1})
T_1	34.88	5.59
T_2	35.00	5.83
T_3	30.60	4.59
T_4	24.30	3.72
CD (P=0.05)	2.84	1.86

Moisture conservation and water productivity : Addition of organic matter and minimum tillage decreased the percentage of micro-aggregates as is evident from the distribution of different size aggregates (Figure 1) to the tune of 8-9% from its original value, which increased the volume of micro porosity responsible for water retention / conservation. The soil moisture data just before sowing of wheat crops exhibited on an average 58.3 mm more moisture upto 60 cm soil depth in T_4 treatment compared to T_1 treatment (Table 4) indicating the beneficial role of organic input management over inorganic input. The conserved moisture, therefore, helped in higher water use by the crops, which subsequently increased the WUE in the order of T_4>T_3>T_2>T_1 by both the crops (Table 4). Change in micro porosity for conserving soil moisture *vis-a-vis* higher water productivity due to organic matter recycling and minimum tillage have also been reported by Bhattarcharyya *et al.* (2009) and Hati *et al.* (2006).

Table 4: Average run off, water use , water use efficiency of maize and wheat crops and moisture conservation for wheat as affected by different treatment

Treatment	Maize			Wheat		Moisture conservation for wheat (mm)
	Runoff (mm)	WU (mm)	WUE (kg ha-mm^{-1})	WU (mm)	WUE (kg ha-mm^{-1})	
T_1	424	321	5.49	243.7	4.01	-
T_2	452	310	3.81	227.3	4.53	6.13
T_3	378	338	5.25	248.7	5.16	28.3
T_4	305	352	5.96	257.0	5.79	58.3
CD (P=0.05)	12.3	2.34	0.48	11.4	0.64	-

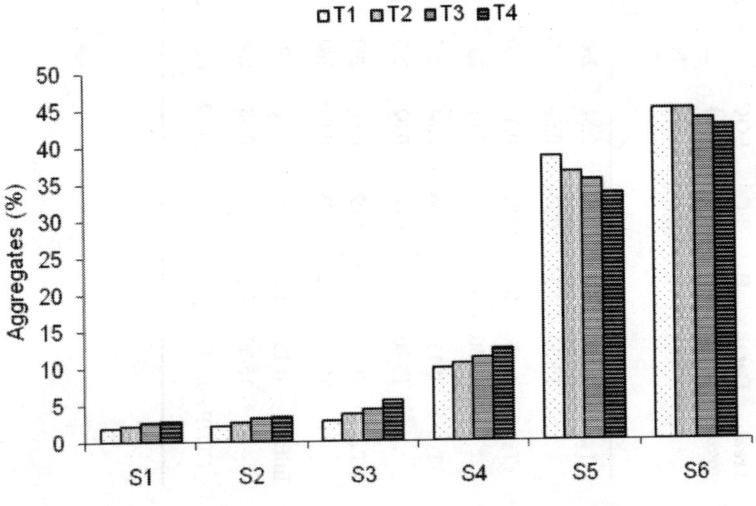

Fig. 1: Effect of different treatments on the percentages of six different aggregate-size fractions at 0-15 cm soil depth.

S1 = >4.75 mm,
S2 = 2-4.75 mm,
S3 = 1-2 mm
S4 = 0.25-1 mm
S5 = 0.1-0.25 mm,
S6 = <0.25 mm

Table 5: Treatment effect on selected soil quality indicators to a depth of 0-15 & 15-30 cm after fourth cycle of cropping

Treat-ment	Depth (cm)	pH	EC (ds m^{-1})	SOC (%)	Available N (kg ha^{-1})	Available P (kg ha^{-1})	Available K (kg ha^{-1})	BD (g cc^{-1})	Basic infiltration rate (mm hr^{-1})	Zn (ppm)	Cu (ppm)	Fe (ppm)	Mn (ppm)	MBC (mg kg^{-1})	Dehydrogenase ug TPF g^{-1} $24h^{-1}$
T_1	0-15	5.8	0.82	0.61	260	26	90	1.38	1.40	0.90	0.56	58.52	25.46	135	33.8
	15-30	5.7	0.78	0.60	250	24	105	1.40		0.46	0.46	48.64	17.34		
T_2	0-15	5.9	0.82	0.62	240	25	98	1.35	1.45	1.09	0.7	58.43	25.08	163	40.8
	15-30	5.9	0.85	0.61	220	28	105	1.37		0.56	0.67	54.76	21.13		
T_3	0-15	6	0.86	0.64	265	27	108	1.34	1.55	1.15	0.62	78.87	26.15	179	44.7
	15-30	6	0.88	0.62	255	31	112	1.36	-	0.86	0.70	50.84	19.99		
T_4	0-15	6.1	1.02	0.66	290	29	118	1.32	1.65	1.18	0.85	93.02	28.11	198	89.1
	15-30	6.2	1.03	0.63	280	35	128	1.34	-	1.44	1.08	51.52	20.7		
Initial	0-15	5.8	0.8	0.60	270	24	110	1.36	1.42	0.95	0.7	105.2	35.42	131	26.1
	15-30	5.6	0.85	0.58	254	25	123	1.38		0.92	0.75	101.4	31.05		
CD (P =0.05)		NS	NS	0.012	1.28	2.43	1.86	0.068	0.014	NS	NS	NS	NS	4.36	2.47

Effect on soil chemical, physical and biological properties : The different organic and inorganic input management showed significant effect on physico-chemical and biological properties of soil. In general, the available nitrogen and potassium decreased in T_1, T_2 and T_3 treatments, whereas it increased in T_4 treatment. However, the available phosphorus increased in all the treatments (Table 5). The nutrient use efficiency of major nutrients (NPK) increased to the tune of 40.9% in T_4 treatment compared to T_1 treatment following the order of $T_4>T_3>T_2>T_1$ owing to less loss of nutrients both by runoff water, sediment and deep percolation. The deep percolation loss of nutrient showed that maximum loss was incurred in T_1 and minimum in T_4 (Table 6) owing to less transmission of water to deeper layers in T_4 treatment compared to T_1 as was also reported by Bhattarcharyya *et al.* (2009). Except Zn in sub-surface (15-30 cm) soils in T_1 and T_2 treatments, which was below the critical level (0.6 ppm), in all other treatments, micronutrient status were sufficient for plant nutrition (above the critical level). Of the physical properties, the bulk density increased in T_1 treatment whereas it decreased in all other treatments increasing the basic infiltration rate (Table 5) indicating the beneficial effects of organic input management (Bhattarcharyya *et al.*, 2009). The biological properties like microbial biomass carbon and dehydrogenase activity were higher in organic input treatments and maximum was recorded in T4 treatment, which might be ascribed to higher substrate level for the microbes in these treatments (Manna *et al.*, 2006).

Table 6: Estimation of percolation loss of NPK as affected by different treatments

Treatment	Nutrient use efficiency by maize (%)				Soil Test balance (kg ha^{-1})	Uptake balance (kg ha^{-1})	Unaccounted/ Percolation loss (kg ha^{-1})
	N	P	K	NPK			
T_1	19.3	15.7	60.1	29.8	-62	41	103
T_2	20.1	15.3	63.2	39.8	-62	10	72
T_3	24.3	17.7	69.1	40.6	15	35	20
T_4	28.6	18.3	75.9	42.0	29	40	11
CD (P= 0.05)	0.83	0.42	1.86	2.46	2.86	4.36	1.85

Effect of carbon sequestration potential (CSP) and soil quality index (SQI) : When these selected soil parameters (physical, chemical and biological) were expressed in terms of SQI, it was found that highest SQI was observed in T_4 treatment and lowest in T_1 treatment indicating the consequences of soil degradation following intensive inorganic agriculture. Further analysis of data revealed that CSP followed the similar trend to that of SQI which commensurate with the WEY (Figure 2.) indicating CSP may be used as indicator of SQI (Rasmussen and Collins, 1991). It was observed that with each unit increase of CSP, the SQI increased by 0.093% and WEY by 4.30 q ha^{-1} of which maize by 1.84 q ha^{-1} and wheat by 2.86 q ha^{-1}. Earlier, Lal *et al.* (2005) have reported maize and wheat yield increased by 1.5-2.0 q ha^{-1} and wheat by 2.0-3.0 q ha^{-1} for each ton of CSP in tropical climatic condition of India.

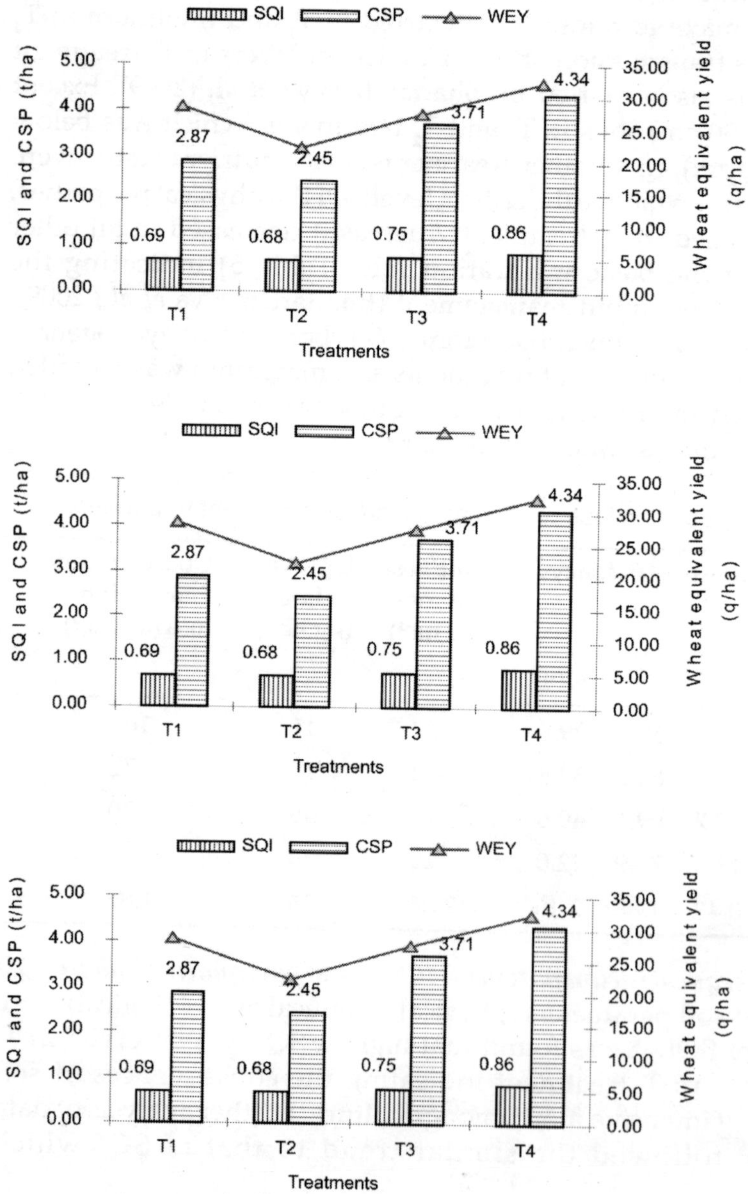

Fig. 2: Relationship between wheat equivalent yields, carbon sequestration potential (CSP) and soil quality index (SQI) as affected by different treatments.

Economics of the system : The total net return from maize-wheat crop rotation, worked out at constant prices, indicated significant difference among the treatments. It also varied significantly with the advancement of years (Table 7). In the initial year, the total net return from the crop rotation under organic input management treatments was negative because of the lower yield of maize. However, during the successive years, the net returns were positive. The average net return after four cycles of cropping indicated that the T_4 treatment generated 13.0% less return than T_1. However, in terms of per ton of soil lost, the T_4 treatment exhibited highest return ₹ 4085 t^{-1} compared to T_1 treatment (₹ 3209 t^{-1}), indicating that organic treatment (T_4) can generate higher return at the cost of soil loss.

Table 7: Net return from maize-wheat (Rs.ha^{-1}) at constant prices in different treatments (2007-08-2010-11)

Treatment	Years				Average (₹ ha^{-1})	Decrease (%)	Net return per tonne of soil lost (₹ t^{-1})
	2007-08	2008-09	2009-10	2010-11			
T_1	5765	15108	22242	21963	16270	-	3209
T_2	-609	9661	11396	18396	9711	-40	1707
T_3	-5042	13565	21397	20648	12642	-22	2886
T_4	-6189	13627	23723	25378	14135	-13	4085
CD (P= 0.05)	-	-	-	-	168	-	-

CONCLUSIONS

Conservation farming with minimum tillage, mulch, vegetative barriers and organic amendments increased crop productivity and soil quality. It is inferred from this study that conservation farming can produce slightly higher (9%) equivalent yield but conserve resources significantly compared to inorganic agriculture. Conservation agriculture has the potential to produce sustained higher yield in long run but gives less net return in short term. It was observed that with each unit increase of CSP, maize yield increased by 1.84 q ha^{-1} and wheat by 2.86 q ha^{-1} which is useful information in carbon credit farming. Adoption of appropriate soil conservation measures can make the conservation farming more remunerative.

REFERENCES

Andrews, S. S., Mitchell, J. P., Mancinelli, R., Larlen, D. L., Hartz, T. K., Horwarth, W. R., Pettygrove, G. S., Scow, K. M. and Munk, D. S. 2002a. On farm assessment of soil quality in California's Central Valley. *Agronomy Journal*, 94: 12–23.

Bhattacharyya, R., VedPrakash, Kundu, S., Srivastava, A.K. and Gupta, H.S. 2009. Soil aggregation and organic matter in a sandy clay loam soil of the Indian Himalayas under different tillage and crop regimes. *Agriculture, Ecosystem & Environment*, 132: 126-134.

Casida, L.E., Klein, D. A. and Santoro, T. 1964. Soil dehydrogenase activity. *Soil Science*, 98:371–376.

Chandra, S., Singh, R. D., Bhatnagar, B. K. and Bisht, J. K. 2002. Effect of mulch and irrigation on tuber size, canopy, temperature, water use and yield of potato. *Indian Journal of Agronomy*, 47 (3): 443-448.

FAO. 2010. What is conservation? Available at: http://www.fao.org/ag/ca.Ia.html accessed 28/10/2010.

Ghosh, B. N. 2009. Vegetative barriers for erosion control in western Himalayan region.Tech. Brochure. CSWCRTI, Dehradun,pp1-8.

Ghosh, B. N., Singh, R. D. and Dadhwal, K. S. 2006. Available organic sources for IPNM in North-western Himalayan states for sustainable crop production. *Indian Farming*, 56(3): 10-12.

Ghosh, B. N. and Singh, R. D. 2003. Effect of conjoint use of farmyard manure and nitrogen on rice (*Oryza sativa*)-wheat (*Triticum aestivum*) system in Uttaranchal mid-hill soils. *Indian Journal of Agricultural Sciences*, 73 (2): 1-4.

Gomez, K. A. and Gomez, A. A. 1984. Statistical procedure for Agricultural Research. 2nd ed. Singapore: John Wiley and Sons.

Hati, K., Swarup, A., Singh, D., Misra, A. K. and Ghosh, P. 2006. Long-term continuous cropping, fertilization, and manuring effects on physical properties and organic carbon content of a sandy loam soil. *Australian Journal of Soil Research*, 44:487-495.

Khybri, M. L., Prasad, S. N. and Sewa Ram. 1982. Effect of top soil removal on the growth and yield of rainfed maize. *Indian Journal of Soil Conservation*, 8(2): 164-169.

Lindsay, W. L., Norvell, W. A. 1978. Development of a DTPA soil test for zinc, iron, manganese and copper. *Soil Science Society America Journal*, 42: 421–428.

Jackson, M. L. 1973. Soil chemical analysis. Indian ed. New Delhi: Prentice Hall of India.

Jenkinson, D. S. and Ladd, J. N. 1981. Microbial biomass in soil: measurement and turnover. In: Paul, E.A., Ladd, J.N. (Eds.), *Soil Biology and Biochemistry* ,5: 415–457.

Lal, R. 2005. Carbon sequestration and climate change with specific reference to India. In: International conference on soil, water and environmental quality-Issues and strategies. Indian Society of Soil Science, New Delhi, pp295-302.

Manna, M. C., Swarup, A., Wanjari, R. H., Singh, Y. V., Ghosh, P. K., Singh, K. N., Tripathi, A. K. and Saha, M. 2006. Soil Organic Matter in a West Bengal Inceptisol after 30 Years of Multiple Cropping and Fertilization. *Soil Science Society America Journal*, 70:121- 129.

Prakash Ved, Bhattacharyya, R. and Srivastava, A. K. 2004. Effect of tillage management on yield and soil properties under soybean based cropping system in mid-hills of north-western Himalayas. *Indian Journal of Agricultural Science*, 74 (11): 573-577.

Rasmussen, P. E. and Collins, H. P. 1991. Long-term impacts of tillage, fertilizer and crop residue on soil organic matter in temperate semi arid regions. *Advances in Agronomy*, 45: 93–134.

Sarkar, A. K. 1998. Intregted nutrient management strategies for sustainable crop production in eastern region. In: *Long-term Soil Fertility Management through Integrated Plant Nutrient Supply. (Swarup, et al. eds.) IISS, Bhopal*, pp112-124.

Sharda,V.N., Dogra, P. and Prakash, C. 2010. Assessment of production losses to water erosion in rainfed area of India. *Journal of Soil and Water Conservation*, 65 (20):79-91.

Sur, H.S., Singh, R. and Malhi, S. S. 1998. Influence of simulated erosion on soil properties and maize yield in north-western India. *Communication in Soil Science and Plant Analysis*, 29 (17 & 18): 2647- 2658.

Yoder, R. E. 1936. A direct method of aggregate analysis of soils and study of the physical nature of soil erosion losses. *American Society of Agronomy Journal*, 28: 337–351.

□□□

Chapter 79

Assessment of Land and Water Resources for Maximizing Annual Net Return from Agricultural Production in Barak Valley of North East India

LAXMI NARAYAN SETHI, PRAVEEN KUMAR AND SUSHIL KUMAR

INTRODUCTION

India is an agricultural based country and its economy is evaluated on the performance of agricultural output. In recent years agricultural productivity has been decreased due to various factors like socioeconomic, environmental and poor management practices. Above all the agricultural production and productivity is determined by two factors such as land and water resources utilization. Thus, a detailed knowledge of these resources is an important prerequisite for agricultural planning.

Barak valley is faced with several fundamental problems *viz.*, an ever-increasing population, a limited available of cultivable land and utilization of available water resources. These problems intensify the importance of developing efficient natural resources use strategies. The future of the region depends on the water stored in the Barak River and rainfall on the land surface for all purposes. To combat the complex problem of availability of land and water resources for optimal crop planning in the Barak valley of Assam it is imperative to go for integrated land and water resources planning and management for agriculture.

One of the recent trends of solution of land and water resources management problems is to aggregate several models into an integrated software that focuses on interaction between the user and the data, models and computers (Andreu *et al.*, 1996). The optimization model has been widely used for drought management (Raman *et al.*, 1992), river basin planning (Jamieson and Fedra, 1996), surface water planning in river basin (Ito *et al.*, 2001), flood prevention (Ford and Killen, 1995), flood warning (Ford, 2001)

Department of Agricultural Engineering, Assam University (A Central University)
Silchar-788 011, Assam, India

and for conjunctive management of surface water and groundwater (Emch and Yeh, 1998).

Several researchers have applied a number of simulation and optimization models to derive planning and operating strategies for irrigation reservoir systems (Vedula and Kumar, 1996; Divakar *et al.*, 2011). In irrigated agriculture, where various crops are competing for a limited quantity of land and water resources, linear programming is one of the best tools for optimal allocation of land and water resources (Sethi *et al.*, 2002; Sethi *et al.*, 2006). The integrated land and water resources planning and management for salinity control (Panda *et al.*, 1996 and Cai *et al.*, 2003) has been studied using various optimization techniques. The present study focuses primarily on the assessment of land and water resources for maximizing the annual return and identifies best alternatives for sustainable crop production in the Barak valley.

MATERIALS AND METHODS

Study Area : The study area (Barak valley) is situated in the southern part of the Indian state of Assam. The geographical position of Barak valley is latitude 24°8' and 25°8' N and longitude 92°15' and 93°15' E and with altitude of 31.40 m from the mean sea level (Figure 1).The zone has a geographical area of 6,922 km² (8.84% of state) with three districts, *viz.*, Cachar, Hailakandi, and Karimganj.Barak valley zone has 38.58 % of total geographical area under forest, 33.08 % as net sown area and 8.48 % as area sown more than once. About 5.47 % of land is used for plantation crops, miscellaneous trees and other. The population of Barak valley is predominantly a rural one and 70.73% of the total work force engaged in agriculture and allied activities. The region has an undulating topography characterized by hills, hillocks, wide plains and low-lying waterlogged areas.

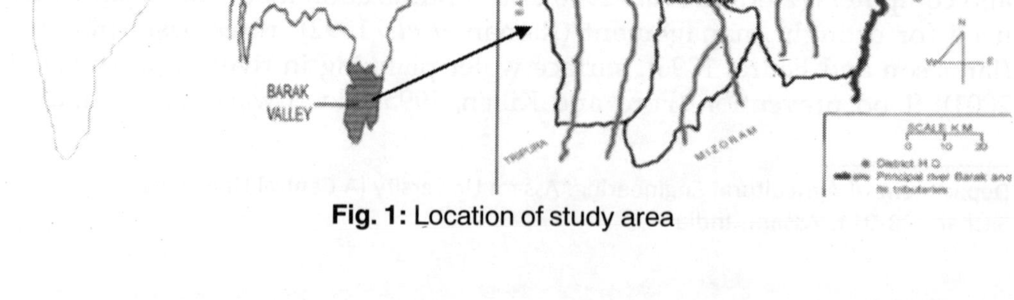

Fig. 1: Location of study area

The type of soil of the zone varying from sandy to clay mostly suitable for field crops including horticultural crops. The soil pH ranges from 4.6 to 5.7. Among all types of soil, the largest area of the zone is covered by non-laterised red soils surrounding the entire alluvial region on all sides. The climate of the Barak valley region is subtropical, warm and humid. The average rainfall of the region is 3180 mm with average rainy days of 146 days per annum. During the summer months the temperature generally varies from 25 to 40°C, while during the winter season the temperature ranges between 10 to 25°C. The relative humidity varies from 58 to 91 % and the sunshine hour varies from 10 to 12 hrs.

Agriculture is the main occupation of the people. Rice is the staple diet of the people and cultivation of rice is the main occupation of those engaged in agriculture. The farmers usually grow crops in two seasons' *viz.*, monsoon (June-September) and winter (January-May) in both rainfed and irrigated areas. Crops grown in the study area during monsoon season are paddy (Oryza *sativa* L.), and pigeon pea (*Cajanus cajan* Milsp.) and in winter season crops are rice, wheat (*Triticum aestivum* L.), mustard (*Brassica juncea* Coss.), black gram (*Phaseolus mungo* Linn.), and garlic (*Allium sativum* L.), oilseed, arhar and vegetables. As a principal crop, the farmers normally grow paddy in most of the cultivable areas (either irrigated or rainfed) in monsoon season and winter season. In summer season, there is no cultivation in the study area due to unavailability of water or flooding of fields.

Rainfall and river water flowing in Barak River and its tributaries are the major sources of water available for irrigation in addition groundwater is also used for irrigation but comparatively vary less proportion. The groundwater is lifted for irrigation by the government and private owned shallow tubewells (depth less than 20 m) and there is no deep tubewells (depth more than 60 m) available for irrigation purpose.

An average annual surface water potential of 585.6 km^3 has been assessed in this basin and out of this 24.0 km^3 is utilizable water. The structures used for irrigating the study area are river lift points, private and government owned shallow tubewells. As canal water supply is not available, the river lifts are considered as the surface water to compute the available irrigation water supply. The surface water (river lifts) and ground water supply (pumping units) were computed using the data of water resources of the study area. It's observed that there is surplus river water of 482.86 Mm^3 (after withdrawing 1.4 Mm^3) and surplus groundwater of 869.28Mm^3(after withdrawing 20.72Mm^3).

The entire Barak valley is represented by unconsolidated, semi-consolidated and consolidated formations. The depth of water level varies from a few meters to 4m below ground level (bgl) in alluvial sediments. Discharge from tube well varies from 5.5 - 8.0 m^3 h^{-1}. The estimated replenishable and utilizable ground water in Barak valley is 852 Mm^3 and 780 Mm^3, respectively.

Optimization Model : The linear-programming model was formulated for optimal land and water resources allocation in different agriculture (rainfed and irrigated), and seasons (monsoon and winter) of three district of Barak valley. The objective of the optimization model is to maximize the net annual return (annual return per hectare from crop yield excluding the cost of cultivation and other equipment). The objective function consists of maximizing the net annual return (Z) subject to the constraints on the availability of land and water resources along with other input parameters.

$$\text{Max}\,Z = \sum_{i=1}^{3}\sum_{j=1}^{2}\sum_{k=1}^{2}\sum_{c=1}^{n}\left[\left(R_{ijkc} \times A_{ijkc}\right) - \left\{\begin{matrix}\left(Crw_{ijkc} \times Qrw_{ijkc}\right) + \\ \left(Cgwp_{ijkc} \times Qgwp_{ijkc}\right) + \\ \left(Cgwg_{ijkc} \times Qgwg_{ijkc}\right)\end{matrix}\right\}\right] \quad \ldots (1)$$

Where i = districts, *i* = 1 for Cachar, *i* = 2 for Hailakandi, *i* = 3 for Karimganj district; *j* = type of agricultural, *j* = 1 for rainfed and *j* = 2 for irrigated agriculture; *k* = crop growing season, *k* = 1 for monsoon and *k* = 2 for winter season; *c* = crop 1, 2, 3,.....*n*; *n* = number of crops grown in a season; R_{ijkc} = net return per unit area, (excluding the cost of irrigation water) for crop c grown in season *k* of jth type of agriculture in district *i*, ($Rsha^{-1}$); A_{ijkc} = area allocated to crop *c* grown in season *k* of jth type of agriculture in district *i*, (*ha*) (decision variable); Crw_{ijkc} = unit cost of river water utilization in season k for j^{th} type of agriculture in district *i* for crop *c*, (Rsm^{-3}); Qrw_{ijkc} = amount of river water allocated in season *k* for j^{th} type of agriculture in district i for crop *c*, (m^3) (decision variable); $Cgwp_{ijkc}$ = unit cost of utilization of groundwater from private tube well in season *k* for j^{th} type of agriculture in district *i* for crop *c*, (Rsm^{-3}); $Qgwp_{ijkc}$ = amount of groundwater utilized from private tube well in season k for jth type of agriculture in district i for crop *c*, (m^3) (decision variable); $Cgwg_{ijkc}$ = unit cost of utilization of ground water from government tube well in season *k* for j^{th} type of agriculture in district *i* for crop *c*, (Rsm^{-3}); and $Qgwg_{ijkc}$ = amount of ground water utilized from government owned tube well in season *k* for j^{th} type of agriculture in district *i* for crop *c*, (m^3) (decision variable).

Constraints : Maximization of objective function is subject to the following constraints.

Water requirements: The calculated net water requirement for the three districts of all crops must be satisfied by the water available from surface water and ground water for all the season.

$$\sum_{i=1}^{3}\sum_{j=1}^{2}\sum_{c=1}^{n} NIR_{ijkc} \times A_{ijkc} - \sum_{i=1}^{2}\sum_{j=1}^{2} \alpha(\beta Qrw_{ijk} + Qpgw_{ijk} + Qggw_{ijk}) \leq 0 \quad \text{For all value of } k \quad \text{... (2)}$$

Where is the net water requirement for the crop *c* in the season *k* of the jth type of agriculture in the ith district *á* and *â* are the field application and conveyance efficiencies respectively. The reference crop evapotranspiration (ETo) was used to calculate the net irrigation requirement for irrigation (Allen *et al.*, 1998). Among the evapotranspiration (*ET*) models, the Hargreaves model was selected for the present study because it requires only two easily accessible parameters, temperature and solar energy.

Land area constraints: The total area allocated to various crops must not be greater than the total cultivable area

$$\sum_{i=1}^{3}\sum_{j=1}^{2}\sum_{c=1}^{n} A_{ijkc} \leq TA_k \quad \text{For all value of } k \quad \text{... (3)}$$

Where denotes the total available land area for cultivation in the season *k*, (*ha*)

Water availability constraints: The water available from both surface and ground water must satisfy the water requirement during the season.

$$\sum_{i=1}^{3}\sum_{j=1}^{2} (Qrw + Qgwg + Qgwp)_{ijk} \leq A(R + G)_k \quad \text{For all value of } k \quad \text{... (4)}$$

Where $(Qrw + Qgwg + Qgwp)_{ijk}$ = total amount of river and ground water utilization in the season *k* for j^{th} type of agriculture in the district *i*, (m^3); and $A(R + G)_k$ = total amount of available water in the both source river and groundwater in season *k*, (m^3)

Allocation of area constraints: The maximum or minimum area allocated to particular crop according to the market requirement

$$A_{ijkc} \leq \mu_{ijkc}^{max} TA_{ijkc} \quad \text{For Maximum Area} \quad \text{... (5)}$$

$$A_{ijkc} \geq \mu_{ijkc}^{min} TA_{ijkc} \quad \text{For Minimum Area} \quad \text{... (6)}$$

Where A_{ijkc} is the area up to which may be allocated μ_{ijkc}^{max} and μ_{ijkc}^{min} are the factor for which allocated land area increased and decreased, TA_{ijkc} is the area allocated as per existing cropping pattern in i^{th} district of jth type of agriculture in the season k for crop *c*.

Non-negativity of constraints

$$A_{ijkc} \geq 0; Qrw_{ijk} \geq 0; Qgwg_{ijk} \geq 0 \text{ and } Qgwp_{gijk} \geq 0 \quad \text{... (7)}$$

For all value of *i, j, k,* and *c.*

Method of Finding Optimal Solution : The data of the study area were collected from different Central and State government agencies such as the Central Water Commission, the State Agriculture, and Statistics & Economics, Assam, India. Based on the land and water resources information received from the different sources were assessed for estimating the input parameters (cropping pattern with its net return, water supply and requirement) of the optimization model.

The computer software package QSB (Quantitative Systems for Business) which is capable of handling 100 constraint 100 numbers of decision variables, is used in computer to solve the linear programming considering the assumptions as followed by Panda *et al.*,(1996), and Sethi *et al.*, (2006) *viz.*, (i)soil of the districts is considered as homogeneous; (ii) farm resources such as land, labour and capital are divisible; (iii) each unit of land receives the same management practices for a particular crop; (iv) time and period of the crop is same in every year; (v) the rainfall is considered as uniformly distributed throughout the area under considered; (vi) all the relationships in the objective function and in the constraints are linear; (vii) uniform horse power electric motors are used for lifting water from specific water resources; (viii) uniform pumping durations are considered for each type of water lifting structures; (ix) the irrigation efficiencies such as conveyance efficiency of river lift system (β), field water application efficiency (α),and Irrigation system efficiencies (αβ), are considered as 75, 65 and 48%, respectively.

Existing Cropping Pattern and Net Return : The existing cropping pattern of the study area in both monsoon and winter seasons and the corresponding net returns (excluding irrigation water cost) considering the corresponding yield, market price, and cost of cultivation has been evaluated a and presented in Table 1. It is observed that the variation of return excluding irrigation water cost depends upon the soil, agriculture, the crops with their corresponding yield, cost of cultivation and market price. The cost of irrigation from government owned river lift and shallow tube wells is ₹ 750 ha^{-1} and from privately owned shallow tube wells is ₹ 1350 per ha^{-1} (based on rates of Government of Assam). Though the paddy crop is grown in maximum area but the highest return is earned from the vegetables cultivation.

Table 1: Existing cropping pattern and net return excluding irrigation cost

Districts (i)	Agriculture (j)	Seasons (K)	Crops (c)	Area (*ha*) A_{ijkc}	Yield (kg ha^{-1})	Cost of cultivation (₹ ha^{-1})	Net return (Excluding cost of irrigation) (₹ ha^{-1})R_{ijkc}
Cachar	Rainfed	Monsoon	Paddy	8035	1794	10400	1745
			Vegetables	4450	2972	8100	11825
		Winter	Paddy	3960	1651	11000	1252
			Arahar	75	500	5290	1629
			Potato	78	2532	5570	1404
			Pea	36	600	6322	1280
	Irrigated	Monsoon	Paddy	4836	1191	12000	1977
			Wheat	394	1146	13000	2617
		Winter	Paddy	2768	1651	12000	1959
			Oilseed	1420	533	5550	10500
Karimganj	Rainfed	Monsoon	Paddy	4367	991	10500	1475
			Vegetables	2225	2972	8100	12800
		Winter	Paddy	198	1551	13500	1075
			Arahar	38	500	5920	1180
			Wheat	100	1046	14000	2190
	Irrigated	Monsoon	Paddy	2418	1794	13000	1663
			Wheat	97	1046	14000	2517
		Winter	Paddy	1384	1551	13300	1865
Hailakandi	Rainfed	Monsoon	Paddy	2791	1651	11050	1700
			Vegetables	1483	2872	7700	10825
			Blackgram	52	500	5000	1320
		Winter	Paddy	1320	1551	12850	1180
			Mustard	40	524	5750	9500
			Vegetables	494	2872	6500	14156
	Irrigated	Monsoon	Paddy	1612	1794	12000	1794
			Potato	26	2972	5770	1504
		Winter	Paddy	922	1651	12500	1688

Existing Water Supply and Requirement : Net irrigation requirement at various crop developmental stages for different crops grown in the study area is computed from the evapotranspiration model based on Food and Agriculture Organization (FAO) recommendations. The FAO recommended crop coefficient (k_c) and number of days for various stages of crop growth and rainfall values were used to predict the evapotranspiration and net irrigation requirement (NIR) at different stages of crops grown in the valley. Existing seasonal water supply through river lift and different pumping units is shown in Table 2. The status of the surface and groundwater resources in Barak valley is presented in Table 3 and 4, respectively.

Table 2: Existing structures used for water supply (draft) from different sources in the Barak valley

Name of the sources	Pumping units	No. of pumping units		Average discharge	Operating hours ($hday^{-1}$)		Total draft (M m^3)	
		Monsoon season	Winter season	(m^3s^{-1})	Monsoon season	Winter season	Monsoon season	Winter season
Surface water	River lift	5	12	0.019	6	9	0.30	1.10
Groundwater	Government	45	78	0.030	8	10	5.80	12.63
	Private	23	39	0.024	8	10	2.38	5.10

Source: District Agriculture Office and Lift Irrigation Department of Assam

Table 3: Status of the surface water availability in Barak valley

Description	Unit	Quantity
Catchment area	Mm^2	41723
Mean flow	Mm^3	483
Utilizable	Mm^3	24000
Water availability per hectare	m^3	11600
Runoff per hectare	m^3	56680
Annual Surface water potential	Mm^3	585600
Annual surface water availability	Mm^3	48000
Annual availability of water per hectare of cultivable area	m^3	43447
Average water potential	Mm^3	484

Source: World Bank, Strategy Report (2007).

Table 4: Status of the groundwater resource in Barak valley

District	GW resources Dynamic (Mm^3)	Utilizable GW resources forirrigation (Mm^3)	Utilizable GW resources for drinking and allied (Mm^3)	Gross Draft (Mm^3)	Balance available (Mm^3)
Cachar	817	694	123	1	693
Hailakandi	98	83	15	3	80
Karimganj	133	113	20	4	109

Source: Central Ground Water Board (2008-09)

RESULTS AND DISCUSSION

Optimal Allocation of Land and Water Resources : Optimization model, developed for obtaining optimum cropping pattern and water resources management, consists of 36 decision variables out of which 27 variables corresponds to crop variables and 9 corresponds to water resources, The optimization model was executed for the existing cropping pattern in three districts of the valley using Quantitative System Business (QSB) package (Chang, 1993). Initially as Case-1, the model was solved without providing the ranges of constraints (no minimum and maximum crop area and water allocation) and optimal allocation of land and water resources. The maximum annual net return obtained is 55.9 Million rupees but the entire cultivable area of three districts has been allocated to vegetables only. The results computed are not realistic in nature as far as food habit and season concerned. Thus it is required to incorporate certain constraints to solve the real study area problems for getting the alternative scenarios for crop planning in the Barak valley. The detailed alternative scenarios and their results are explained in the next section.

Alternative Scenarios for Land and Water Resources Management : In the present study, it is found that the available water resources are much more than the crop water requirements. Hence, the analysis for alternate scenario to solve the real study area problems was carried out varying land allocation to different crops, agricultures and growing seasons considering the available water resources. The system was executed for three different cropping situations based on local requirements *viz.*, **Case 1:** Without Area Constraints *i.e.*, there are no lower and upper bounds on the area cultivated for each crop. The analysis is required to ascertain the model response in the absence of any restriction on maximum/minimum cropped area for any specific crop; **Case 2:** With paddy area constraints *i.e.*, paddy cultivation is restricted to existing paddy area while there are no bounds for all other crops. This alternative was examined since paddy cultivation is essential for the farmers of the area for their food habits even if it is not economically viable; **Case 3:** Existing cropping pattern. Since, no data is available regarding the minimum requirements of crops for the farmers as well as the food targets of the region. So, the model was executed at four different ranges of the cultivable area considering the existing cropping pattern to satisfy the basic food targets of the region. These ranges with fixed deviations from the existing area for each crop for all *i, j, k* and *c* are (1) Allowing 20% deviation i.e. $\mu_{ijkc}^{min} = 0.8$ and $\mu_{ijkc}^{max} = 1.2$; (2) Allowing 30% deviation *i.e.*, $\mu_{ijkc}^{min} = 0.7$ and $\mu_{ijkc}^{max} = 1.3$; (3) Allowing 40% deviation *i.e.*, $\mu_{ijkc}^{min} = 0.6$ and $\mu_{ijkc}^{max} = 1.4$; (4) Allowing 50%

deviation *i.e.*, $\mu_{ijkc}^{min} = 0.5$ and $\mu_{ijkc}^{max} = 1.5$. Fifty percent of the existing cropped area is considered as the minimum required area to be cultivated for each crop to satisfy the food habit and farmer can get more return from other crops with the less water resources. District wise optimal cropping pattern with optimal water resources allocation, and maximal annual return obtained for the said cases and ranges of cropping situation are shown in Table 5. However, seasonal optimal allocation of water resources for with the existing cropping situation (Case 2) is presented in Figure 2.

The optimal water allocation for three different cropping situations attains the maximum level at case 1 and decreases in case 2 and case 3 after which increases for range 1, 2, 3 and 4. The optimal annual returns at for three different cropping situations (Case 1, Case 2 and Case 3) are 55.9, 33.6 and 17.8 Million rupees, respectively. For the case 1, the crops with most economic benefit are preferred for cultivation in the entire area. The cropping pattern with vegetables only was found as optimal in monsoon season in Cachar and Karimganj and in winter season in Hailakandi district. Although the crops

Table 5: Optimum cropping pattern, water resources allocation and net annual return for different cropping situations.

District	Agriculture	Season	Crop	Optimal Area Allocation (ha)						
				Case1	Case2	Case3	Range1	Range2	Range3	Range4
Cachar	Rainfed	Monsoon	Paddy	0	8035	8035	6428	5624.5	4821	4017.5
			Vegetable	26052	0	4450	5340	5785	6230	6675
		Winter	Paddy	0	3960	3960	3168	2772	2376	1980
			Arahar	0	0	75	60	52.5	45	37.5
			Potato	0	0	78	62.4	54.6	46.8	39
			Pea	0	0	36	28.8	25.2	21.6	18
	Irrigated	Monsoon	Paddy	0	4836	4836	5803.2	6286.8	6770.4	7254
			Wheat	0	0	394	472.8	512.2	551.6	591
		Winter	Paddy	0	2768	2768	2984.8	3093.2	3201.6	3310
		Oilseed	0	0	1420	1704	1846	1988	2130	
Karimganj	Rainfed	Monsoon	Paddy	0	4367	4367	3493.6	3056.9	2620.2	2183.5
			Vegetable	10827	0	2225	2670	2892.5	3115	3337.5
		Winter	Paddy	0	198	198	158.4	138.6	118.8	99
			Arahar	0	0	38	30.4	26.6	22.8	19
			Wheat	0	0	100	120	130	140	150
	Irrigated	Monsoon	Paddy	0	2418	2418	2577.4	2657.1	2736.8	2816.5
			Wheat	0	0	97	116.4	126.1	135.8	145.5
		Winter	Paddy	0	1384	1384	1660.8	1799.2	1937.6	2076
Hailakandi	Rainfed	Monsoon	Paddy	0	2791	2791	2529.2	3365.5	2267.4	2136.5
			Vegetable	0	0	1483	1779.6	1927.9	2076.2	2224.5
			Blackgram	0	0	52	41.6	36.4	31.2	26
		Winter	Paddy	0	1320	1320	1056	924	792	660
			Mustard	0	0	40	48	52	56	60
			Vegetable	8740	0	494	592.8	642.2	691.6	741
	Irrigated	Monsoon	Paddy	0	1612	1612	1934.4	1128.4	2256.8	2418
			Potato	0	0	26	20.8	18.2	15.6	13
		Winter	Paddy	0	922	922	737.6	645.4	553.2	461
Water allocation (Mm³)				427.88	163.34	492.32	488.61	487.3	484.86	483.57
Net annual return (Million ₹)				55.9	33.6	17.8	20.0	21.1	22.2	23.3

may yield maximum annual return it may not be feasible due to the local food requirements. It's observed that with the increase in allowable deviation from the existing cropping pattern, more economic benefit is obtained. The farmers depending on their prevalent situation may choose a suitable cropping pattern from the above alternatives.

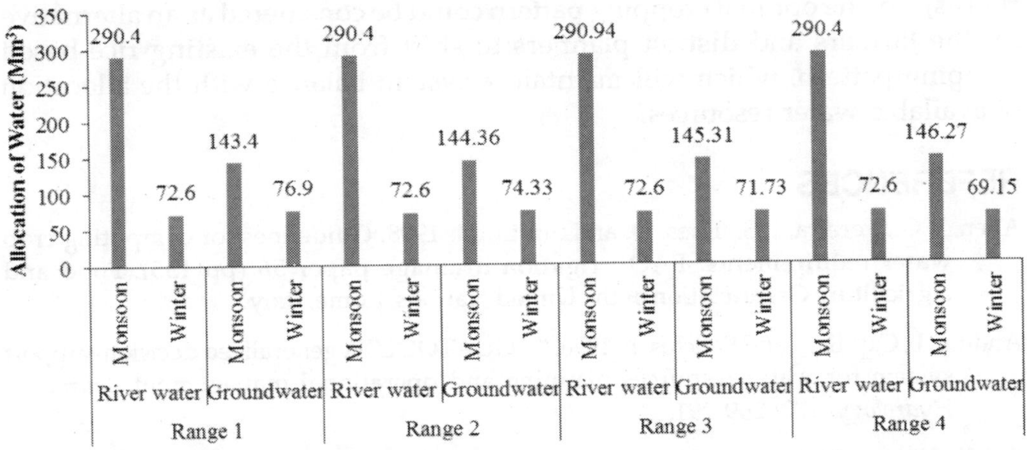

Fig. 2: Optimal seasonal water resources allocation at different ranges of cropping pattern in Barak valley

The possible scenarios obtained from the various ranges of land allocation strategies shows the increase in net annual return with the increase in ranges of deviation due to reduction of acreage for high water requirement crops with the low water requirement crops. Considering the minimum required crop acreage (50% of existing cropping pattern), the 50% deviation from the existing crop acreage for each crop was found as optimal in the region. It is observed that the annual return obtained from the optimal allocation of cropping pattern for the range 4 is 31% higher than the existing return (17.8 Million rupees). So, the optimal cropping scenarios of range 4 allocation could be considered as an alternative for the farmers and district planners to shift from the existing rice-based cropping pattern, which will maintain a system balance with the allocation of available water resources.

CONCLUSIONS

The linear programming model formulated for maximization of annual net return with optimal water and cropping pattern allocation for the monsoon and winter seasons in the Barak valley was found to be an effective tool for land and water resources allocation. The cropping pattern obtained for seven different ranges of deviation from the existing area for the cropping pattern also gives significant output in the form of alternative cropping pattern. The results of alternative cropping patterns and water resources allocations for maximizing net annual return are found significant as per the expectation under changing scenario. Optimal cropping pattern found for the Barak valley is the 50% deviation from the existing high water requirement crop (paddy) with the 31% higher annual return than the existing return (17.8 Million rupees). So, the optimal cropping pattern could be considered as an alternative for the farmers and district planners to shift from the existing rice-based cropping pattern, which will maintain a system balance with the allocation of available water resources.

REFERENCES

Allen, R.G., Pereira, L.S., Raes, D. and Smith, M. 1998. Guidelines for computing crop water requirements. FAO irrigation drainage paper 56 (pp 135). Food and Agriculture Organization of the United Nations, Rome, Italy.

Andreu, J., Capilla, J. and Sanchis, E. 1996. "AGUATOOL" A generalized decision-support system for water resources planning and operational management. *Journal of Hydrology*, 177: 269-291.

Cai, X., Kinney, D.C. and Lasdon, L.S. 2003. Integrated hydrologic-agronomic- economic model for river basin management. *Journal of Water Resources Planning and Management*, 129(1): 4-17.

Divakar, L. 2011.Optimal alocarían of bulk wáter supplies to competing use sectors based on economic criterion-an application to the chao phraya river basin, Thailand, *Journal of Hydrology*, 40 (1-2): 22-35.

Emch, P.G. and Yeh, W.G. 1998. Management model for conjuctive use of coastal surface water and groundwater. *Journal of Water Resources Planning and Management*, 124(3):129-139.

Ford, D.T. 2001. Flood-warming decision support system for sacramento, California. *Journal of Water Resources Planning and Management*, 127(4): 254-260.

Ford, D.T. and Killen, R.J. 1995. PC-based Decision Support System for Trinity River, Texas. *Journal of Water Resources Planning and Management*, 121(5): 375-381.

Ito, K., Xu, Z.X., Kojiri, T. and Kawamura, A. 2001. Decision support system for surface water planning in river basins. *Journal of Water Resources Planning and Management*, 127(4): 272-276.

Jamieson, D.G. and Fedra, K. 1996. The "water ware" decision support system for river planning. *Journal of Hydrology*, 177: 199-211.

Panda, S.N., Khepar, S.D. and Kaushal, S.N. 1996. Inter-seasonal irrigation system planning for water-logged sodic soils. *Journal of Irrigation and Drainage Engineering*, 122(3): 135-144.

Raman, H., Mohan, S. and Rangcharya, N.C.V. 1992. Decision support for crop planning during droughts. *Journal of Irrigation and Drainage Engineering,* 118(2): 229-241.

Sethi, L.N., Kumar, D.N., Panda, S.N. and Mal, B.C. 2002. Optimal cropping and conjuctive use of water resources of a coastal river basin. *Journal of Water Resources Planning and Management,* 16(2):145-164.

Sethi, L. N., Panda, S. N. and Nayak, M. K. 2006. Optimal crop planning and water resources allocation in a coastal groundwater basin Orissa, India. *Agricultural Water Management,* 83:209-220.

Vendula ,S. and Kumar, D.N. 1996. An integrated model for optimal reservoir operation for irrigation of multiple crops. *Water Resources Research,* 32(4): 1101-1108.

□□□

Chapter 80

Tribal Agriculture in Eastern Ghats of Odisha - A Perspective Study

B. S. NAIK, PRAVEEN JAKHAR, M. MADHU, H. GOWDA
B.K. DASH AND G.B. NAIK

INTRODUCTION

The term "Tribal agriculture "is nothing but practice of shifting cultivation and it is considered to be the most ancient system of cultivation dating back to the lower Neolithic period between 13000 and 30000 BC. It is also known as slash and burn agriculture. It is an extravagant and unscientific form of land use. In India, shifting cultivation is prevalent in Eastern and North Eastern Regions, and about 85 % of the total cultivation in northeast India is by shifting cultivation (Singh and Singh, 1992). Earlier the shifting cultivation cycle followed by leaving land fallow was 25-30 years but due to increasing pressure on land for cultivation, this has been reduced drastically now to 2-3 years. Previously, the fallow cycle was of 20-30 years duration, thereby permitting the land to return to natural condition (Patro and Panda, 1994). As the cycle is reduced to 2-3 years, the land is under high risk of degradation leading to large-scale deforestation, soil and nutrient loss, and invasion by weeds and other species. Due to shifting cultivation on steep slopes (44-53%), the soil loss reported was about 50 t ha^{-1} $year^{-1}$ with corresponding nutrient loss 703 kg of organic carbon,144 kg of phosphorus and 7 kg of potash annually (Goswami *et al.*, 2009). The indigenous biodiversity has also been affected to a large extent and resilience of ecosystem has broken down. Though it is well known, the tribal people are still practicing shifting cultivation on steep hill slopes.

Implication of Shifting Cultivation

In this practice, cultivation is done on a patch of cleaned forest area surrounding the human settlement. The existing trees and bushes on the

Central Soil and Water Conservation Research and Training Institute, Research Centre, Sunabeda, Koraput, Odisha, India selected site are cut during November-January, allowed to dry and burnt by fire. Before the onset of monsoon, seeds are sown or dibbled on the cleaned area. Under this practice, millets, cereals, pulses and oilseeds crops normally grown. In one patch, this process continues for 2-3 years till the fertility goes down. Then, the patch is abandoned and a new site is selected for the purpose. The abandoned patch regenerates, become fertile but the original forest flora, fauna and ecology are never restored.

Extent of Shifting Cultivation at a Glance in India

In India, about 5.0 million tribal families are practicing this system of agriculture on 4.37 million hectares of land covering 11 states (Sahu *et al.*, 2005). The details of state wise dependant tribal families and area under shifting cultivation is given below in Table 1.

Table 1: Shifting cultivation scenario in different states of India

State	Dependant Tribal families (M)	Area under shifting cultivation (M ha)	% of total area in India
Andhra Pradesh	0.11	0.15	3.4
Arunachal Pradesh	0.43	0.21	4.8
Assam	0.31	0.31	7.1
Bihar	0.23	0.19	4.3
Madhya Pradesh	0.19	0.38	8.7
Manipur	0.36	0.26	5.9
Meghalaya	0.61	0.47	10.8
Mizoram	0.40	0.19	4.3
Nagaland	0.19	0.12	2.7
Odisha	2.00	1.60	36.6
Tripura	0.19	0.49	11.2
Total (India)	**5.02**	**4.37**	**100.0**

Shifting Cultivation in Odisha

The state Odisha is situated in the eastern part of India having geographical area of 15.57 M ha comprising 4.75 % geographical area of India. It lies in the tropical zone between lat 17° 47' and 22°34' N and long 81° 22' and 87° 29' E and is blessed with varied bio-diversities and rich natural resources. Odisha in the Eastern Region alone accounts for the largest area of 1.6 m ha (36.6%) under shifting cultivation in India. The total population of the state is 36.7 million of which 22.12 % is tribal population (Census, 2001) constituting 62 tribal communities. Shifting cultivation is extensively practiced by tribals in Odisha and about 2 lakh tribal families from more than 20 tribal communities are engaged in this practice The shifting cultivation is locally called as *podu* by the tribals in Odisha.

Characteristics of Shifting Cultivation in Odisha

The characteristics of shifting cultivation in Odisha can be broadly divided according to the local topography (slope, altitude), climate (rainfall, soil type and vegetation cover), population density, land use and cropping pattern, community involved in shifting cultivation, and ownership of land. The topography and climatic conditions largely influence the land uses and cropping pattern. Besides these the social structures of the community and population density also largely govern the method of shifting cultivation. In varied situations shifting cultivation is practiced in Odisha by tribals.

Based on topography, climate and population density of shifting cultivation area, Odisha is divided in to three zones namely North Zone, South Zone and Central Zone (District statistical hand books of respective districts of Odisha, 1995). North zone consists of 7 Blocks under shifting cultivation. It has less diversity in topographic character and its areas lie between 300- 900 meters of altitude with average slope of the land varies from 1-8%. It receives average annual rainfall of 1178-2508 mm. Here the population density ranges between 26-250 per km^2.

The southern zone consists of 6 Blocks and it has two separate areas, one with mountainous character nearing to boarder of Andhra Pradesh and other with plateau character near the boarder of Chhattisgarh. Most of the area in South Zone falls within the plateau topography, where the altitude ranges from 100-300 m. The average land slope found found is gentle and varies between 1-8%. This zone receives average annual rainfall of 1198 -1610 mm. The population density ranges between 11-50 per sq km.

The Central Zone of Odisha consists of 36 Blocks under shifting cultivation is largely mountainous due to the presence of Eastern Ghats. Here the topographic condition varies from place to place and in most cases altitude varies within the range of 300-900 meters. The slope variation is between 2 - 60 % but large areas fall within the range of 30-60 %. In comparison with the northern and southern zone, the slope of this region is steep. The population density is very high (251- 450 per sq km) to very low (11-25 per sq km). Highest rainfall of 2508 and 2311 mm is received at Th. Rampur and Kashipur due to mountainous character of the location.

It is observed from the land use study at community level that the whole area under shifting cultivation in Odisha can be divided in to two regions namely paddy and oilseed zone of the north and millet and pulses zone of the central and southern parts of Odisha. It also seems that distribution of rainfall largely determines the cropping pattern. Beside rainfall, the condition of slope of the land is also responsible for selection of crop. By and large area having more than 1400mm rainfall is suitable for hilly paddy. However on gentle slope area, due to more retention capacity of water, paddy can be cultivated with less rainfall. On the contrary in a steep slope area, having less retention capacity of water, even higher rainfall may not be helpful for paddy cultivation. This seems to be the single most reason for preferring millet crops in the central and southern zones of shifting cultivation area.

Area under Shifting Cultivation in Odisha

Shifting cultivation particularly is being practiced by the tribals in the Eastern Ghats region (Patro and Panda, 1994). Odisha alone accounts for the largest area of 1.6 M ha (36.6%) under shifting cultivation in India and about 2 lakh tribal families are engaged in such practice. It has been estimated that about 5298 km^2 area annually is under this primitive agriculture practice. The areas affected by this practice were about 300 sq. miles approximately prior to 1936. It became nearly 12,000 sq. miles in 1948 after annexation of princely States. In the pre-plan periods, attempts were made to make an estimate of the area under shifting cultivation in the State. Dr. H.F. Mooney had made an estimation about the area under shifting cultivation in 1951 as 12,770 sq. miles or about one-fifth of the total land surface in the state was affected by shifting cultivation and nearly one million tribal people depended upon this method of raising crops for their living (Mohapatra and Devi, 1973).The best source of information now available regarding coverage under shifting cultivation are the vegetation map of India, prepared by institute, Francis, Pondicherry, India. The area under shifting cultivation in Odisha available from different sources area given in Table 2.

Table 2: Area under shifting cultivation in Odisha from various sources

S. No.	Source	Year	Pop.Size	Area affected Hectares
1.	H.F. Mooney	1951	9,35,700	32,68,120
2.	Forest enquiry committee report	1959	9,27,900	
3.	ICAR	1958	10,00,000	8,60,000
4.	Dhebor Comm	1960-61	9,35,700	8,33,335
5	French Inst. Pondichery & ICAR	1967	7,06,412	30,23,358
6	FAO/UNFPA	1980	7,06,400	26,49,000
7	Task Force	1983	-	26,49,000
8	F.R.T.S.-I Imaginary	1984	-	920092.30

Source: Dash, 2006

Based on the task force of Government of India report, more than 30,000 km^2 of land in Odisha is under such cultivation. Lenka (2001) stated 184018.46 ha (1.18%) is under shifting cultivation in Odisha. Shifting cultivation is prevalent in Sambalpur, Keonjhar,Koraput, Phulbani, Ganjam, Kalahandi and other southern and western districts of Odisha, covering 119 blocks. The details of area under shifting cultivation district wise covering number of Blocks in Odisha is given below in table 3. Out of total shifting cultivation area in Odisha, highest area (22.7%) is under Sambalpur district and lowest area (1%) is under Bolangir district.

Table 3: District wise area under shifting cultivation in Odisha

District	No of Blocks under shifting cultivation	Area covered (ha)	Total geographical area (%)	% of total shifting cultivation area (15.57 M ha)
Mayurbhanj	10	7303.40	4.0	0.05
Keonjhar	7	40638.20	22.1	0.26
Sundargarh	6	8190.74	4.5	0.05
Sambalpur	12	41781.00	22.7	0.27
Bolangir	6	1795.40	1.0	0.01
Dhenkanal	11	5121.20	2.8	0.03
Koraput	22	30651.16	16.7	0.20
Phulbani	15	20995.66	11.4	0.13
Kalahandi	12	5861.40	3.2	0.04
Ganjam	18	21680.30	11.8	0.14
Total	**119**	**184018.46**	**100.0**	**1.18**

Source: D. Lenka, 2001

According to the assessment made by NRSA, India, the total area under shifting cultivation in Odisha is 1177.28 km^2. The district wise split of it is given in Table 4. The Districts having highest area under shifting cultivation are Phulbani (610.03 km^2) followed by Gajapati (207.32 km^2) and Kalahandi (145.52 km^2).

Table 4: District wise area under shifting cultivation (abdonded jhum and current Jhum)

District	Shifting cultivation area (km^2)		Total area under shifting cultivation (km^2)
	Abdoned Jhum	Current Jhum	
Debgarh	3.93	0	3.93
Gajapati	27.01	180.31	207.32
Ganjam	19.91	0	19.91
Jharsuguda	0.05	0	0.05

Kalahandi	94.82	50.70	145.52
Keonjhar	18.75	1.59	20.34
Koraput	16.81	26.22	43.03
Malkangiri	26.88	25.62	52.5
Nayagarh	21.33	12.20	33.53
Phulbani	306.03	304	610.03
Rayagada	2.40	35.24	37.64
Sambalpur	3.10	0	3.1
Sundargarh	0	0.38	0.38
Total	**541.02**	**636.26**	**1177.28**

Source: Waste Land Atlas of India (2005)

The Tribal Perception of Shifting Cultivation

Different tribes of Odisha practicing shifting cultivation have different names for this kind of subsistence activity. These are usually expressed by the names given to their swiddens among different categories of land under cultivation. The Juang of Keonjhar call it Toila chasa and their brother tribe, the Paudi Bhiyan identify it by *Toila, Biringa* or *Kaman, Bagada, Saraba* or *Baru* is the term used by the Saora of Ganjam to distinguish their swiddens. The Dongria Kondh of Rayagada district call it Haru and the Desia Kondh of Rairakhol and Bamra sub-division, 'Rama'. The Kondh of Kalahandi district use different term, '*Dongar Chasa*' and livang or Kunda Chasa is the term prevalent among the Bonda of Malkangiri district. Shifting cultivation practiced by the *Koya* of Malkangiri district in the foot hills is called Lanka podesanad.

This practice is followed by more than 20 types of tribal communities in Odisha. The major tribal communities viz. Paudi Bhuiyan, Buniya,Jatapa, Kamar, Juanga, *Kondha, Kutia Kondha, Dongaria Kondha, Lanjia Sauras, Paraja, Godaba, Koya, Didayi, Bonda, Jhang* and *Pauri Bhuyan, Peranga* and *Erenga Kolha* are involved in this practice (Dash, 2006).Many festivals and other such rituals revolve around the *podu* fields, because the tribals view *podu* cultivation not just as a means of their livelihood, but as a way of life. The list of tribes practicing shifting cultivation in different areas of the state is given in Table 5.

Table 5: Tribes practicing shifting cultivation in different areas of Odisha

Tribe	District (undivided)	Area
Bondo	Koraput	Khairput area of Malkangiri Sub-Division
Didayi	-do-	Kudumulgumma area of Malkangiri Sub-division
Koya	do-	Malkangiri sub-Division
Gadaba	-do-	Similiguda & Pottangi area of Koraput Sub-Division
Paroja	-do-	Dasamantapur area of Koraput Sub-division

Lanjia Soura	-do-	Puttasingi area of Gunupur Sub-Division
	Ganjam	Parlakhemundi Sub-division
(a) Kondh	Koraput	Koraput Sub- Division, Rayagada Sub-division
	Phulbani	Kandhamal Subdivision
	Kalahandi	Th. Rampur and Lanjigarh of Kalahandi Sub-division
	Sambalpur	Bamara area of Deogarh Sub-Division, Rairakhol Sub-Division
(b) Kutia Kondh	Koraput	Gudari, Ramanguda Muniguda, Chandrapur area of Gunupur Sub-Division.
	Phulbani	Balliguda Sub-Division
(c) Dongria Kondh	Koraput	Bisamcuttack area of Gunupur Sub-Division.
Juang	Keonjhar	Telkoi, Harichandrapur area of Keonjhar Subdivision
Paudi Bhuyan	Keonjhar	Telkoi, Banspal area of Keonjhar Sub-division
	Sundargarh	Bonai Sub-division
	Dhenkanal	Pallahara Sub- Division
	Sambalpur	Bamara area of Deogarh sub- Division
Erenga Kolha	Sundargarh	Bonai Sub- division
Parenga	Koraput	Pottangi area of Koraput Sub-Division

Source : *Odisha Review* (2006)

Ownership of *Podu* Field

According to the tribal customary rules, the type of ownership of swiddens varies among the tribes from individual ownership to communal ownership. Among the *Juang* and *Pauri Bhuiyan* of Keonjhar district the land under shifting cultivation is the communal property owned by all the villagers in common. Every year in the month of Magha (January-February) the village headman and the priest select the land for cultivation. Generally the patch which has completed its rotation cycle and is sufficiently covered with trees and bushes is considered suitable to be cleared for shifting cultivation. After the patch is selected, the headman demarcates the boundary lines by putting mark on trees by axe and the land is sub-divided into several plots which are allot ted to the households (Mohapatra and Devi, 1973). The ownership of land after allotment is transferred to the head of the household for a period of 2 to 3 years till he cultivates it actively. In Koraput and Ganjam areas the communal ownership of the village over *Podu* land is completely absent. Each household owns a number of sites on hill slopes and uses these in rotation. In course of time Podu land has become private property which can be owned and inherited by customary right. There is no legal sanction behind it. A *Bonda* even sells his plots under shifting cultivation to another and mortgages it whenever he is in need (Mohapatra and Devi, 1973). Among the Koya the unreserved forest land, is treated as private property and is also inherited by the legal heirs.

Detrimental Effects of Shifting Cultivation

Shifting cultivation causes loss of flora and fauna which includes precious species of tree plants, shrubs, medicinal plants and minor forest products. With reduction in its cycle from 20–30 years to 2–3 years, the land under shifting cultivation looses its nutrients and the top soil. Frequent shifting from one land to the other also affects the ecology of the region. The area under natural forest has declined; the fragmentation of habitat, local disappearance of native species and invasion by exotic weeds and other plants are some of the other ecological consequences of shifting cultivation. During the last fifty years, the forest cover of Odisha has been drastically reduced. For instance, the forest area in 1962 was about 65868.9 km^2 which was reduced to 57184 km^2 in 1997 and in the tribal districts of Odisha, 25760 km^2 (1993) of forests were reduced to 25424 km^2 in 1997(Forest Survey of India, 1997). In Odisha due to shifting cultivation, the districts found badly affected are Korput, Keonjhar,Phulbani and Ganjam. Due to this unscientific practice, the soil quality becomes poor and infertile with low water holding capacity.

The problem of soil erosion due to shifting cultivation is very serious in Odisha. The quantum of soil erosion varies from 84 to 170 t ha^{-1} $year^{-1}$ in bare fallow. The intensity of erosion depends on slope, soil characteristics, crop coverage, depth of soil and rainfall pattern etc. It has been estimated from an experiment that, in an average 7-10 t ha^{-1} $year^{-1}$ surface soil is lost along with plant nutrients due to such practice (Dash, 2006).

Strategies to Curb Shifting Cultivation

The important objectives for controlling shifting cultivation should include restoring ecological balance in the hill areas and improving socio-economic conditions of tribal *podu* practicing families by weaning them away from *podu* cultivation. The following initiatives and efforts should be made for controlling shifting cultivation and keeping the tribal people away from such unscientific practice.

1. Creation of employment opportunities for the resource poor tribals for income generation on a regular basis through proper utilization of the land resources.
2. Encouragement for the cooperative efforts for carrying out forest-based activities, i.e. basket making, rope making, cane furniture processing of minor forest produce, honey collection, etc. have to be made

commercially viable by providing proper marketing facilities. This will not only discourage tribals from practicing shifting cultivation but will also help them monetarily.

3. Formation of Village Forest Committees for the protection and development of the degraded forests. These committees by providing suitable incentives to the tribals, after the time of harvest can divert some of the tribals away from the shifting cultivation. Generating employment opportunities during the lean season of forestry operations will also prevent tribals from shifting to other areas. Employing the tribals for collection of *kendu* leaves and *sal* seeds and also involving the tribals in the various rural employment schemes is also the need of the hour.

4. Conducting mass awareness programme and educating tribal men, women and children about the menace of shifting cultivation. The services of various non-Governmental organizations and voluntary agencies, besides the regular Government machinery, are required for undertaking such work on sustainable basis.

5. Implementation of various land development programmes for tribals i.e land/jhola land development, development of irrigation, agriculture development, raising of orchards and horticulture plantations, raising of plantation crops, forest plantation, pasture development, animal husbandry, development of Pisiculture and soil conservation measures.

6. Creation of subsidiary occupation like fishery, sericulture, poultry, piggery etc. to generate additional income and employment.

7. Creation of food surplus will reduce the area under shifting cultivating.

REFERENCES

Census. 2001. http://Odishagov.nic.in.

Dash, B. 2006. Shifting cultivation among the tribes of Odisha. *Odisha Review*, 76-84.

Forest Survey of India. 1997. *State of Forest Report. 1997*, p.29 and p.61.

Goswami, S.N., Mandal, D.K. and others. 2009. Shifting cultivation: Boon or Bane? Green Hope. pp. 35-37.

Lenka, D. 2001. "Agriculture in Odisha", Kalyani Publishers, New Delhi.

Mohapatra, K. and Devi, K.B. 1973. Shifting Cultivation in Odisha, *ADIBASI*, Vol.XIV, No.4, pp.11-28.

Patro, S. N. and Panda, G. K. 1994. *Eastern Ghat in Odisha: Environment, Resources and Development*, Odisha Environment Society.

Sahu, S.K., Nayak, R.K., and Sarangi, D. 2005. Sustainable soil and land management under shifting cultivation in Odisha. *Odisha Review*, 49-52.

Singh, J.S. and Singh, S.P. 1992. *Forests of Himalaya*, Gyanodaya Prakashan, Nainital, pp.294.

Waste land Atlas of India. 2005. Ministry of Rural Development, Department of Land Resources, Government of India and National Remote Sensing Agency (NRSA), Department of Space, Government of India, Hyderabad.

□□□

Chapter 81

Potential of on Farm Employment Generation through Natural Resource Conservation Activities in Watersheds of the Deccan Region of India

R.N.ADHIKARI, S.L.PATIL, A.RAIZADA, M.PRABHAVATHI, D.RAMAJAYAM, N.LOGANANDHAN, B.MONDAL, W.MURALIDHAR AND B.N.SESHADRI

INTRODUCTION

In India, out of 142 M ha of arable lands, 60% (85.2 M ha) is rainfed. Karnataka has the second largest area (5 M ha) under rainfed agriculture after Rajasthan in the country. Crop yields in dryland areas are 1-1.5 t ha^{-1} which is lower by two to five folds of the achievable potential yield (Wani *et al.*, 2012). Potential of dryland agriculture could be unlocked using available scientific technologies for improving rural livelihoods through watershed development programmes and sustainable intensification of rainfed areas. Watershed management activities, along with providing several tangible and intangible benefits, generates lot of casual and regular employment opportunities during the implementation phase itself as well as during post-intervention period. In a study carried out by NABARD in Giridih, Jharkhand, revealed that more than 80% funds allotted for land development work such as soil and water conservation, soil health enhancement programmes, creation of water harvesting structures, check dam, micro-lift, dug well etc., and large scale horticulture development led to huge employment generation (Anonymous, 2008). In another report by Prem Singh *et al.* (2010), it was reported that integrated watershed development led to increased employment opportunities for the community members with better wage earnings in construction work during the implementation phase and engagement in agricultural fields during the post-watershed programmes, but no specific formal mechanisms were developed to enhance the opportunities.

Due to the harsh climatic situations, uncertain rainfall, traditional crop cultivation methods, poor yields and poor marketing avenues, most farmers

Central Soil & Water Conservation Research & Training Institute, Research Centre Bellary, Karnataka, India in the dry zones face acute hardships in obtaining regular employment. Hence, execution of watershed development programmes should be done involving various types of land based and non-land based activities for creation of employment for the rural masses. This paper reports the temporary employment generated in a watershed of the semi-arid region during the initial stages of development phase in various conservation activities.

MATERIALS AND METHODS

Study area : The watershed is located in the north of Chitradurga district (14°37′50″ to 14°38′47″ N Lat. and 76°42′46″ to 76°44′51″ E Long.) with elevation ranging from 550 to 570 m above mean sea level. The total area of the watershed is about 479.17 ha including arable and non-arable land with variable slopes of 3 to 5%. The watershed is to be treated with proven technologies of soil and water conservation, rainwater harvesting and it's recycling, dissemination of improved packages of production to increase production in terms of food, fodder, fuel and fruits and non-land based activities for the land less. Most of the stakeholders in the watershed belong to the marginalized sections of society. Their socio-economic status is poor and annual family income ranges between ₹ 16, 580 to ₹ 36, 756 for different categories of households and 15% of available workforce seasonally migrates in search of work. Since most of the arable land in the watershed is rainfed, total employment for 187 farm families in crop enterprises is only for 17533 man days which is equal to 94 man days employment availability per family per year which is hardly sufficient to engage them and provide livelihood. For improving the production level in dry land semi-arid and generating gainful employment, adoption of soil and water conservation on watershed basis has been initiated in this area. Participatory planning revealed that farmers were keen to be involved in NRM activities either collectively or individually, so that they were not forced to migrate outside in search of employment.

Soil and Water Conservation Activities Undertaken

Arable lands: In order to conserve soil and water in-situ in arable lands, contour and field *bunding* was taken up on a large scale using manual labour, from the ridge to the valley. The activity was taken up during the lean season when the crops were harvested and produce disposed off. An area of 150 ha was planned to be *bunded* over three years. Contour/field *bunding* of 0.81 m^2 cross section taken up in the fields and loose stone waste weirs were constructed in order to dispose off the excess runoff safely through these waste weirs. Necessary loose stone revetment was also provided on either side of the waste weirs where the natural waterway is shallow, in order to protect the bund on either side of waste weirs. Where bunds existed, field boundaries were also renovated with the same cross section. This activity will minimize soil erosion; enhance moisture conservation in the fields, resulting in improved yields. The details of the measures adopted are given in Figure 1.

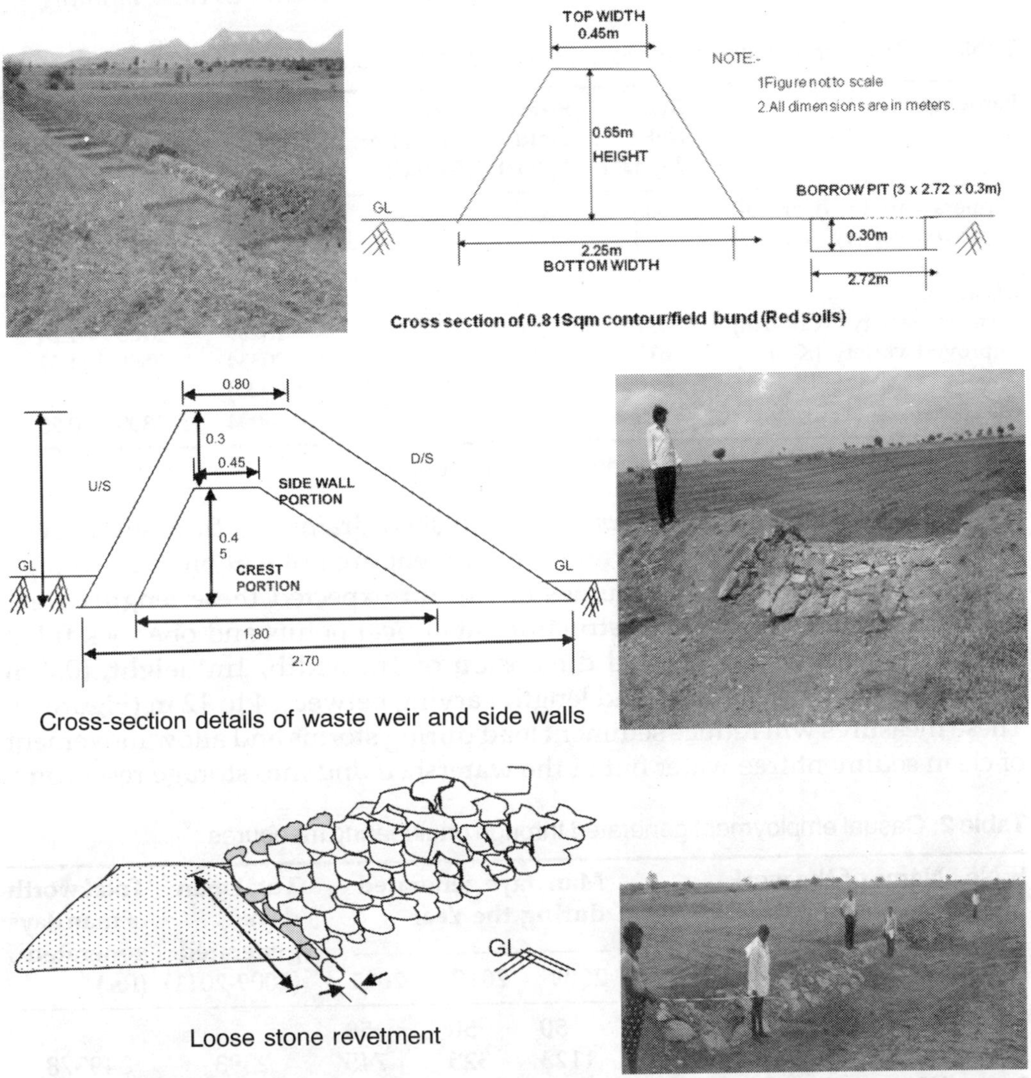

Fig. 1: Field bunding and construction of waste weirs and side walls in the watershed, along with their dimensions.

Impact of bunding on groundnut yields: Introduction of high yielding cultivars of the dominant crop in the watershed along with soil conservation measures led to significant increase in groundnut yields and net returns

(Table 1). Grain yield increased by 30 and 35% in case of cultivar TMV2 and K6 over control, leading to increase in gross returns by 35% and 33%, respectively over control. Accordingly B: C ratio increased by 24 and 27%, respectively. These increases can be attributed to the improved performance of the cultivars and improved soil moisture retention due to field *bunding.*

Table 1: Groundnut yields of improved cultivars in *bunded* areas

Particular	Grain yield (kg ha^{-1})	Straw yield (q ha^{-1})	Cost of cultivation (Rs.ha^{-1})	Gross returns (Rs.ha^{-1})	Net returns (Rs.ha^{-1})	B-C ratio
Farmers variety (Control)	403	8.02	11268	13290	2023	1.17
Improved variety (TMV-2)	545	10.67	12273	17962 (35)	5689	1.45
	(35)	(33)			(181)	(24)
CD at 5%	80.2	1.71	–	2646	2160	0.15
Farmers variety (Control)	471	7.36	12056	15239	3183	1.24
Improved variety (K-6)	613	13.00	12776	20334	7558	1.57
	(30)	(77)		(33)	(137)	(27)
CD at 5%	125	2.48	–	4094	3398	0.25

Values in parenthesis are percent increase over control

Treatment of drainage lines: The watershed drains out through a single drain with a length of 7600 m and a catchment area of 479 ha. Due to high intensity storms and large volumes of run off expected these lengths were treated by a series of gabionic structures at critical points and one rock filled dam. These are with a general dimension of 1m width, 1m height, (0.7 m above GL, 0.3 m below G.L.) and length varying between 4 to 12 m (Figure 2). These measures will reduce sediment load during storms and allow movement of clean sediment free water out of the watershed and into storage reservoirs.

Table 2: Casual employment generated through engineering measures

S. No	Name of the work	Man days generated during the Year			Total man days generated (2009-2011)	Total worth of man days generated (Rs.)
		2009	2010	2011		
1.	Area bunded (ha)	50	50	50		
2.	*Bunding*	1123	525	745	2393	349378
3	Waste weirs	362	306	232	900	131400
4	Revetments	77	23	55	155	22630
5	Construction of drinking water tank (1no.)	85	-	-	85	12410
6	Gabion structures (5 no's)	-	90	-	90	13140
	Total	1647	944	1032	3623	528,958

*@ Rs.146 per day for un-skilled labour

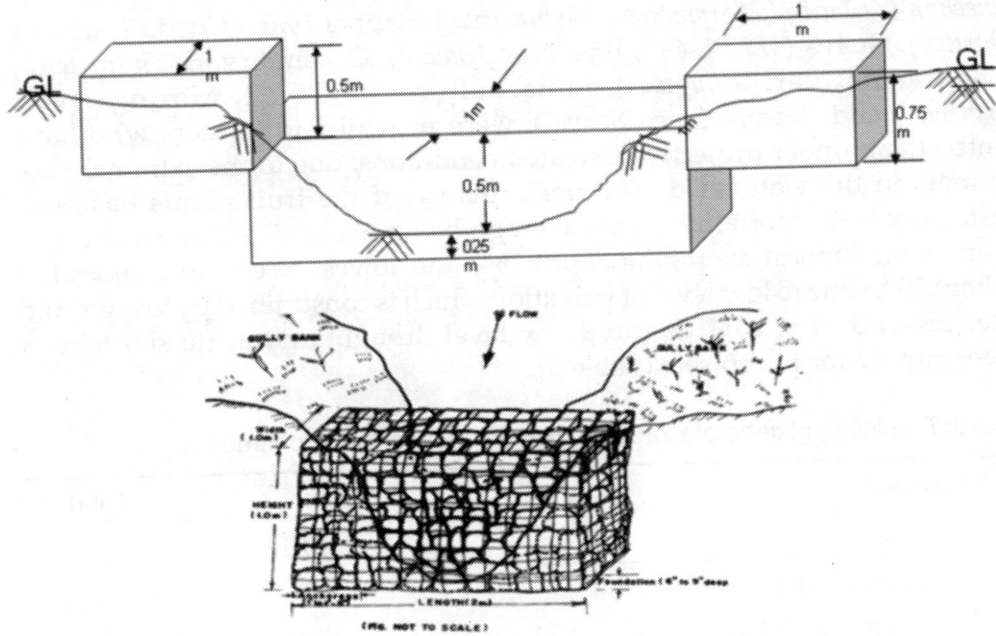

Drawing gabion structure

Fig. 2: Construction of gabionic structures as drainage line treatment measures.

Dryland horticulture developed : Development of horticulture has potential to create both casual and regular employment generation and also provide nutritious food supplement to farmers. Establishment of dry land horticulture was carried out over 45 ha covering more than 50 farmers. Prior to taking up the activities, selection of site and choice of fruit plants was decided. It was also agreed that fruit plants would be provided only when the farmer had excavated pits of required size and spacing in the selected site. Once this was over, one year old healthy plants of Mango (variety

'Baneshan', *'Mallika'*, *'Alphanso'*, *'Aishwarya'*), Sapota (var. *'Cricket ball'* and *'kalipatti'*), Guava (var. *'L-49'*), lime (var. *'Balaji'*), Coconut (var. *'Tiptur tall'*), Pomegranate (var. *'Bhagwa'*), tamarind (var. *'DTS-1'*), Moringa (var. *'Dhanraaj'*) and Papaya (var. *'Surya'*), were provided to farmers who have planted them under protective irrigated conditions, due to the extremely hot summers in the watershed. The performance of the fruit plants has been satisfactory with mortality on an average being about 27%.The survival of sapota is the highest averaging about 75% and lowest in coconut (mortality of about 50%) due to low level of irrigation which is constrained by low ground water levels during 2011 due to a low level drought and acute shortage of power supply for pump sets (Table 3).

Table 3: Employment generated through development of fruit orchards

S. No.	Particular	Year 2010-11	Year 2011-12	Total
1.	Area covered (ha)	10.9	34.4	45.3
2.	No. of pits excavated	1603	5524	7127
3.	Man days utilized (no.s) (@ 3 pits per day of 1 cu m each)	534.3	1841.33	2375
4.	Average man days per ha generated (no.)	49	53.52	52.42
5.	Employment generated per ha. of fruit trees planted (Rs.)	7154	7814	
6.	Total worth of employment*	78008	268834	346842

*@Rs.146 per day

Employment generation in NRM activities : Different types of watershed based treatment activities like soil and moisture conservation measures in agricultural lands, drainage line treatment measures, horticulture plantation, etc. were executed by manual labour only. Construction of mechanical measures in the watershed was carried out with the active participation and willingness of farmers who were able to get casual employment in the watershed. Construction of bunds has been done as per survey number of fields. The results indicated that around 3623 and 2375 man days of temporary employment at the worth of ₹ 5.29 lakhs and ₹ 3.47 lakhs were generated for engineering and horticultural works, respectively during 2009-11 period. Year wise break-up of man days generated under different engineering activities is presented in Table 2. At current day prices (₹ 156 day^{-1} of 2011) this would be equal to ₹ 5.65 lakhs.

The magnitude of employment is not in-significant; further these activities will help in generating recurring employment in the future when yields from fruit trees will be regularly available. Available literature suggests that introduction of watershed technologies increased the cropping intensity, production levels and shifts the farming activities from less labour intensive (low value) to more labour intensive (high value) crops, livestock and other enterprises which in turn increases the labour absorption per ha of cultivated area or per head of livestock or income per household (Math *et al.*, 1997 and Ram Babu *et al.*, 1997). Mishra and Mondal (2008) also observed from an ex-post analysis of three Operational Research Projects based watersheds at Karnataka, that not only temporary employment but regular employment opportunities in agriculture increased significantly during the post-intervention period.

CONCLUSIONS

The analysis clearly indicates that participatory planning and implementation of watershed development programmes can generates significant employment opportunities through various NRM activities. Hence, efforts should be made to develop a plan in future watershed programmes which aims for maximizing the total on-farm and off-farm employment by integration of various land-based and non-land-based measures in carrying out the watershed programmes.

REFERENCES

Anonymous. 2008. *District Agriculture Plan*, Giridih District, NABARD Consultancy Services, Jharkhand Regional Office, Ranchi.

Math, S.K.N., Rao, M. S. R. and Padmaiah, M. 1997. Joladarasi Model Watershed Development Programme in Bellary District of Karnataka: A Diagnostic Evaluation. *Journal of Rural Development*, **16**(2): 313-328.

Mishra, P.K. and Mondal, B. 2008. Employment Generation in Watershed Development Programme - A Micro Analysis. In: *Advances in Soil and Water Conservation in Indian Perspective*, Soil Conservation Society of India and IGKV, Raipur (C.G.). pp256-263.

Prem Singh, Behera, H.C. and Aradhana Singh. 2010. *Impact and effectiveness of Watershed development programmes in India*. Centre for Rural Studies, National Institute of Administrative Research, Lal Bahadur Shastri National Academy of Administration (LBSNAA), Mussoorie.

Ram Babu, Dhyani, B.L., Agarwal, M.C. and Samra, J.S. 1997. *Economic Evaluation of Watershed Management Projects*. Central Soil and Water Conservation Research and Training Institute, Bulletin No. T-33/D-23, Dehradun.

Wani, S.P., Sarvesh, K.V., Krishnappa, K., Dharmarajan, B.K. and Deepaja, S.M. 2012. *Bhoochetana: Mission to Boost Productivity of Rainfed Agriculture through Science-led Interventions in Karnataka*. ICRISAT, Hyderabad, India, 84p.

□□□

Chapter 82

Drought Induced Shocks and Coping Strategies by Small Farmers in the Dry Tropics

BISWAJIT MONDAL, N. LOGANANDHAN AND A. RAIZADA

INTRODUCTION

Agriculture, being a means of livelihood of almost two third of the population in the country represents India's most important economic sector. It provides livelihood to about 70% of the population, employs about 65% of the work-force, contributes about 24% to national Gross Domestic Product (GDP) and nearly 16% to export earnings (DARE, 2003). Nearly 68% area of the country is perennially reeling under recurring drought. The 'chronically drought-prone areas' around 33% receive less than 750 mm of rainfall, while 35%, classified as 'drought-prone' receive a rainfall of 750-1,125 mm. The drought-prone areas of the country are confined to peninsular and western India - primarily arid, semi-arid and sub-humid regions. An analysis of 100 years of rainfall data reveals that the frequency of 'below-normal rainfall' in arid, semi-arid and sub-humid regions is 54-57%, while severe and rare droughts occurred once every eight to nine years in arid and semi-arid zones (Samra, 2004).

The concept of drought varies from one region to another depending upon normal climatic conditions, available water resources, agricultural practices and the various socio-economic activities of the region (Prasad, 1998). According to Gautam and Rao (2007), the risk involved in successful cultivation of crops depends on the nature of drought (chronic and contingent), its probable duration, and periodicity of occurrence within the season. In the arid region where the mean annual rainfall is less than 500 mm, drought is almost an inevitable phenomenon in most of the years. In semi-arid regions (mean annual rainfall 500-750 mm), droughts occur in 40 to 60% of the years due to deficit seasonal rainfall or inadequate soil moisture availability between

Central Soil and Water Conservation Research and Training Institute,
Research Centre, Bellary, Karnataka, India

two successive rainfall events. Even in dry sub-humid regions (annual rainfall 750-1200 mm), contingent drought situations occur due to break in monsoon conditions.

Drought produces a complex set of highly differentiated adverse impacts that ripple through many sectors of the economy. It affects the biophysical, socio-economic and environmental sectors of the region (Pandey *et al.*, 2007) and leads to food insecurity, malnutrition, starvation, poverty, disinvestment in human capital and reduction in the financial resources. Wilhite (1993) and Alston and Kent (2004) argued that the problem of the dryland agriculture has received attention in recent times not because of the increase in the frequency and the severity of drought, but because the vulnerability of the society to drought has increased due to growth in population, changing farming practices, *etc.*

Indian drought management systems presently equipped to view drought as a transient phenomenon to be dealt with on a piecemeal basis, as and when they occur. Existing practices are essentially fire-fighting models centred on relief, when efforts over the years should have been directed at drought-proofing the country by involving community/civil society and *panchayati raj* institutions. This paper accounted the rainfall distribution and quantified the drought during the year 2011, assess its' impact on small farmers and prioritize the coping strategies by the small and marginal farmers of two watersheds in dry tropics of semi-arid region.

MATERIALS AND METHODS

Study area : Karnataka state was selected purposively for the study as major portion of the State comes under hot and semi-arid eco-region. Karnataka stands second after Rajasthan, in terms of total geographical area prone to drought (Wani *et al.*, 2012). During the last decade, the state experienced a severe drought for three consecutive years (2001-02, 2002-03 and 2003-04) and 159 taluks were listed as drought affected. During these periods, the State received 23% less rainfall than the normal (Biradar & Sridhar, 2009). Two watersheds namely; Netrenahalli and Ramasagara located in Molkalmuru taluk of Chitradurga district of Karnataka, which is one of the worst affected districts in this region, were selected for detailed investigation.

Data and analysis : In this study, monthly rainfall data pertaining to Molkalmuru taluk, where the watersheds are located, for the period of 18 years (1994-2011) were used for drought analysis. Primary data were collected by personal interview of the 60 farm households belonging to two watersheds as well as focus group discussions covering socio-economic conditions of farmers, crops and animal details, impact of drought and their coping strategies.

The identification and assessment of drought severity were done using Standardized Precipitation Index (SPI) (McKee *et al.*, 1993). The SPI for an item indicates how far and in what direction that item deviates from its distribution's mean, expressed in units of its distribution's standard deviation. In this study, SPI was obtained from the monthly precipitation record by calculating the z-score (standard normal variate) which is similar to SPI that will have a mean zero and a standard deviation of one. The z-score for precipitation is the standardization of a given time series, X_i as $X_1, X_2, ..., X_n$, which was calculated as:

$$z-\text{score} = \frac{xi - \overline{X}}{Sx}$$

where, $\overline{X}$ is the arithmetic mean and *Sx* is the standard deviation of the rainfall series.

Garett's ranking technique (Garett and Woodworth, 1969) was used to identify and rank the constraints faced by the farmers due to incidences of droughts as well as prioritize the coping strategies to mitigate the drought induced shocks using the formula:

$$G = \frac{100(Rij - 0.50)}{Nj}$$

where,

G = Percentage position

R_{ij} = Rank given for the ith constraint/ coping strategy by the jth respondent

N_j = Number of constraints/ coping strategies ranked by the jth respondent.

The percentage position of each rank was converted into scores using Garett's table for preparation of final rank.

RESULTS AND DISCUSSION

General features of the study area and rainfall distribution : The study area located in the central dry zone of Karnataka bordering arid condition. It receives low to moderate rainfall (long-term average is 486.7 mm) and is one of the drought prone districts. The watersheds cover an area of 479 and 480 ha and small and marginal farms dominate the farmer households (78 and 64%), respectively. Major portion of the cultivated area is rainfed (87 and 85%) and only 13 and 15% get irrigation from bore-wells. The area experiences a tropical wet and dry climate. The wet season (June to September) is much shorter and receives low precipitation from the south-west monsoon. The remaining months of the year are generally dry, because the area does not receive any precipitation from the north-east monsoon. Rainfed agriculture dominates, with only 20.5% of crop land area being irrigated, largely by ground water. Groundnut (26%), maize (18%), ragi (10%), sunflower (6%) and jowar (6%) are the major crops growing in the region.

Monthly rainfall data for 18 years period was considered for calculation of SPI for the selected watersheds. The analysis showed that monthly SPI was negative for 69% of cases and extreme drought was experienced during October 2008 with an SPI value of -1.56 (severe drought) (McKee *et al.*, 1993) (Table 1). June to October was the most critical months experiencing maximum number of droughts which results into inadequate soil moisture during the crop growing season causing extreme crop stress and wilting of crops in the region.

Table 1: Monthly SPI values for 18 years rainfall at the selected watersheds

Year	Jan.	Feb.	Mar.	Apr.	May	Jun.	Jul.	Aug.	Sep.	Oct.	Nov.	Dec.
1994	-0.42	-0.23	-0.35	-0.79	-0.79	-1.00	-0.16	-0.94	-0.96	1.38	-0.40	-0.37
1995	3.56	-0.44	-0.35	-0.83	-0.28	-0.71	0.23	-0.24	-0.69	-0.52	-0.66	-0.37
1996	-0.42	-0.44	-0.31	0.95	-0.21	1.98	-0.72	0.44	1.33	1.30	-0.47	3.22
1997	0.03	-0.44	-0.14	0.49	-0.79	1.10	-0.43	-0.67	-0.56	-0.47	2.38	2.17
1998	-0.42	-0.44	-0.35	-0.48	-0.79	1.09	0.06	-0.07	0.38	0.16	-0.39	-0.04
1999	-0.42	-0.20	-0.35	1.38	1.58	-0.82	0.36	-0.05	1.07	2.04	-0.58	-0.37
2000	-0.42	1.25	-0.35	-0.91	-0.31	-0.12	0.02	1.98	-0.55	0.80	-0.66	-0.37
2001	-0.42	-0.44	-0.35	1.13	-0.85	0.12	-0.78	-0.30	1.49	-0.46	-0.41	-0.37
2002	0.72	1.91	-0.35	-1.00	0.39	-0.58	0.61	-0.68	-0.61	-0.11	-0.22	-0.37
2003	-0.42	-0.44	-0.24	-0.31	-0.91	-0.92	-0.55	-0.24	-1.00	-0.10	-0.66	-0.37
2004	-0.42	-0.44	0.28	2.63	0.77	-0.49	0.79	-1.09	0.23	-0.43	-0.66	-0.37
2005	1.21	-0.44	-0.35	0.02	0.09	-0.34	3.20	0.25	-0.43	0.11	-0.22	-0.37
2006	-0.42	-0.44	0.77	-0.78	-0.58	-0.47	-0.78	-1.21	-0.38	-0.85	2.05	-0.37
2007	-0.42	-0.44	-0.35	0.04	-0.39	2.22	-0.52	-0.98	1.85	1.38	-0.56	-0.18
2008	-0.42	3.02	3.83	-0.78	-0.63	-1.28	-0.97	0.05	-1.09	-1.56	-0.66	-0.37
2009	-0.42	-0.44	-0.35	-1.00	1.06	0.33	-1.03	1.45	1.45	-1.26	1.50	-0.37
2010	-0.01	-0.44	-0.35	-0.03	2.84	0.06	0.98	2.26	-0.53	-0.66	1.04	-0.37
2011	-0.42	-0.44	-0.35	0.26	-0.18	-0.18	-0.29	0.07	-0.99	-0.75	-0.40	-0.37

SPI calculated from annual rainfall showed that 4 years *viz.*, 1995, 2003, 2008 and 2011 experienced moderate to severe drought with SPI values less than -1 (Figure 1). The year 2011 was found to be the driest year with a negative SPI value for all the months except April and August (Figure 2).

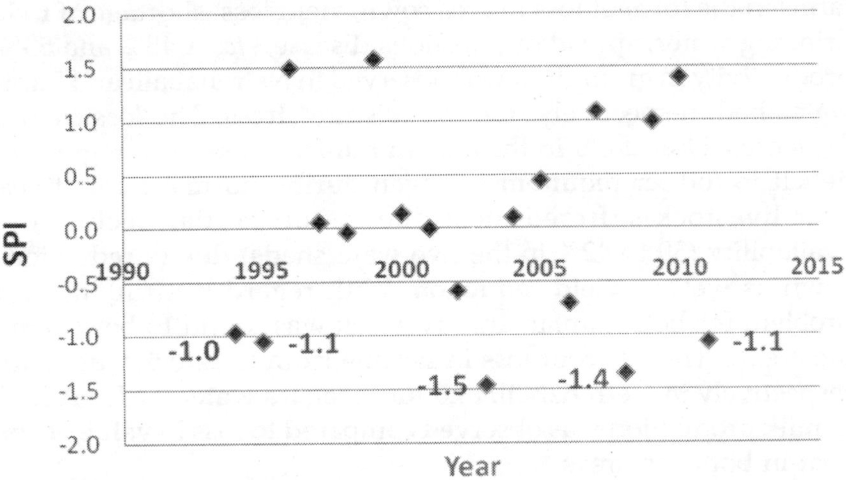

Fig. 1: Annual SPI values for 18 years rainfall at the selected watersheds

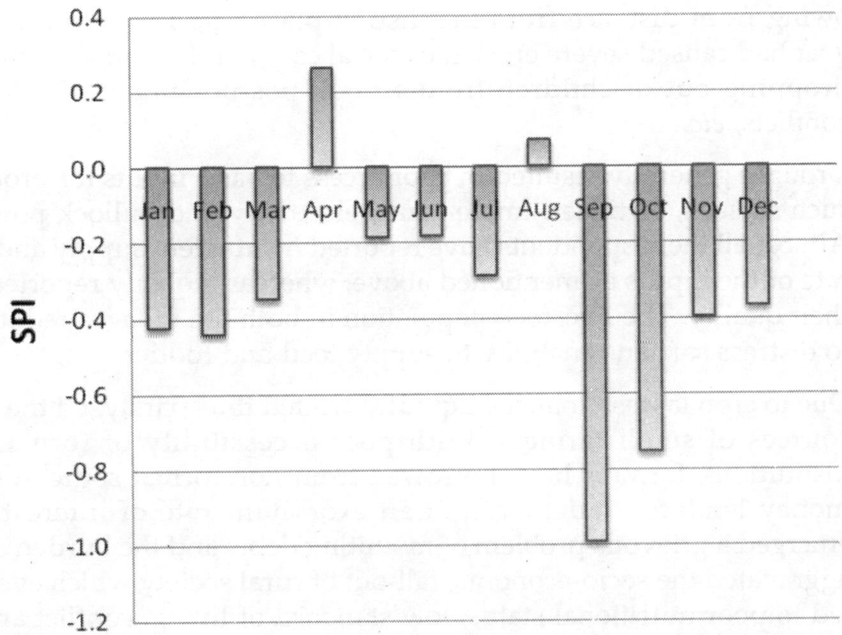

Fig. 2: Monthly SPI values for rainfall during 2011 at the selected watersheds

Impact of drought incidence in the study area : Impact of drought incidence during 2011 on the small farms in the study area was assessed with the help of primary data collected from the two watersheds are described below.

I. Drought during 2011 caused heavy reduction of natural capital of the farms in the form of loss in area sown, yield loss, shortage of fodder and drinking water, spread of infectious disease, *etc.* A 45% and 53% loss in productivity in groundnut was observed in Netrenahalli and Ramasagara watershed, respectively due to moisture stress. The loss in area sown registered 11 and 6% in the area in rainfed crops. The region remained deficit in fodder requirements even during normal rainfall years; this year livestock suffered most due to substantial decline in fodder availability (30 to 42% in the two watersheds) due to reduction in area sown as well as yield reduction. With regard to drinking water, the problem for both human and livestock was found to be severe in both the areas. The per cent loss in income from livestock was 35 and 33%, respectively in Netrenahalli and Ramasagara watersheds. A higher loss in milk production was observed compared to loss in value of sheep and goat in both the areas.

II. As such there is a less social capital build-up at the study area due to frequent droughts; low standard of living due to low productivity and lack of business opportunities; low education, medical and other facilities owing from distance from the district place, *etc.* Drought during last year had caused severe erosion of social capital reflecting in instances of dropping out of children from school, postponement of marriages, conflicts, *etc.*

III. Drought generally resulted in poor access to basic inputs for production such as seed, fertilizer, implements, electricity and bullock power, *etc.* Almost all the respondents have reported insufficient supply and higher rate of the inputs as mentioned above, whereas, nobody reported about their quality. The livestock population in both the villages reduced due to distress sale and inability to supply feed and fodder.

IV. Due to crop losses, financial liquidity eroded thus paralyzed the income sources of small farmers. With poor accessibility of formal credit institutions, farmers have borrowed from non-formal sources, such as money lenders, traders, *etc.* at an exorbitant rate of interest. There emerged a grievous problem of mounting debts and the burden of debts aggravated the socio-economic fall-out of rural society which eventually led to poor nutritional status, low standard of living, conflict and even changes in behaviour of people.

Constraint analysis : Garrett ranking technique was used to identify and rank the various constraints faced by the respondents due to incidences of drought. Table 2 presents the per cent position worked out from the ranks given by the respondents to each of the constraints which further converted to Garett scores from the standard table (Garett and Woodworth, 1969) and final rank has been prepared. The results indicated that moisture stress due to scanty rainfall, low in-situ moisture and drying of bore-wells which led to the decline in productivity of crops was the major constraint faced by the farmers in the region. Fodder shortage aggravated during drought period that led to the low productivity and distress selling of animals. Due to reduction in cultivated area and yield loss, employment opportunities lost which in turn resulted into forced migration. The other constraints faced by them were low consumption levels, increase in indebtedness, scarcity of drinking water, *etc.*

Table 2: Constraint analysis by Garrett ranking technique

S. No.	Constraints	Per cent position	Garett score	Rank
1.	Moisture stress	9.67	75	I
2.	Fodder shortage	11.67	73	II
3.	Depletion of farm assets/ animals due to distress sale	54.33	47	V
4.	Low purchasing power	52.33	48	IV
5.	Increase indebtedness	59.00	45	VI
6.	Unemployment	31.00	59	III
7.	Conflict, social unrest, etc.	75.67	36	IX
8.	Scarcity of drinking water	61.67	44	VII
9.	Low standard of living	80.33	32	X
10.	Forced migration	64.33	42	VIII

Drought coping strategies : The people in the dryland areas have their own strategies to cope with the drought situations. The farmers in the drought prone areas can predict the seasonal rainfall fluctuations based on historical averages and formulate coping strategies to deal with the distress situations (Chen, 1991). The various coping mechanisms adopted by the farmers in the study watersheds as came out during focus group discussion under the present investigation have been listed and farmers have been asked to assign ranks in terms of efficiency to safeguard to drought impact as per their perceptions and practices. The results indicated that small and marginal farmers in the area adopt mainly three types of coping strategies like: changing and diversifying their farming practices, postponement of farming activities and borrowing (mainly from non-formal sources) (Table 3). Other practices they follow are lowering their consumption levels, migration to nearby town/ cities, sale of livestock and other household assets, postponement of social activities like marriages, *etc.*

Table 3: Prioritization of coping strategies by Garrett ranking technique

S. No.	Constraint	Per cent position	Garett score	Rank
1.	Changing/diversifying farming practices	11.33	73	I
2.	Borrowing	30.33	60	III
3.	Maintaining buffer of grains and fodder	73.33	37	IX
4.	Sale of livestock	56.67	46	VI
5.	Lowering consumption	51.67	49	IV
6.	Sale of assets	60.67	44	VII
7.	Postponing cropping and resort to wage labour	15.00	70	II
8.	Migration	54.00	47	V
9.	Postponement of social activities	64.00	42	VIII
10.	Others	83.00	30	X

CONCLUSIONS

The drought of 2011 had a very discernible impact on the farmers in the semi-arid region who are largely dependent on agriculture and its subsidiary occupations for their livelihoods. Productivity of crops as well as animals reduced owing to low income of households, poor employment opportunities in agriculture, lowering the consumption levels, *etc.* were the perceptible impact on the small and marginal farmers. As such, there used to be a positive correlation between the drought prone area and poverty. Drought leads to poverty, which in turns makes the situation conducive for drought to strike. Water and food for people and water and feed for livestock are the major concern for the drought affected region in general and small farmers in particular. Traditionally, various adjustments are done and various strategies are followed in order to overcome the shocks induced by drought situations. In the present study, most of the coping strategies were found to be in terms of adjustments in consumption pattern, social activities and earning more income from non-farm activities as well as borrowing. Prasad (1998) also indicated coping measures for drought induced famine are reduction in consumption spending, migration and borrowing. Thus, for policy framework and management intervention, apart from proper understanding of the local people's awareness, responses and priorities, the following measures need to be adopted to mitigate the impact of drought.

a) Input subsidy, assured power supply and quality extension services for the farmers in drought affected areas.

b) Supply of cereals at subsidized rates and other essentials including safe drinking water to overcome the problem of deficit consumptions.

c) Necessary assistance for fodder, feed and health facilities for livestock which would discourage distress selling of the animals. Besides, raising of fodder trees on the farm boundaries and fruit trees, grasses on the field bunds can be suggested to mitigate the impact of drought on livestock and livelihoods.

d) Contingency crop planning involving choice of suitable crops and/ or crop varieties, alternative crop strategies, mid-season corrections and crop life saving measures.

e) Institutional credit for diversifying activities like dairy development, sheep/ goat rearing, small enterprises, issue of loans for revolving fund for efficient functioning of SHG groups, medium and long term loan at a nominal rate of interest for purchase of tools, instruments and raw materials to the small entrepreneurs/ artisans.

f) Water resources management through augmentation of water supply, rationing of water use, efficient utilization and management of water resources through sprinkler, drips, *etc.*

g) Emphasizing community participation for drought mitigation through government policies and programmes both at micro and macro level.

h) Popularizing crop insurance to reduce the adverse financial impact of drought on the farmers to increasing its' coverage and reach.

ACKNOWLEDGEMENTS

Authors wish to acknowledge the support rendered by Mr. K. Channabasappa, TO (T-5) and Mr. Umesh Haloodar, Project Assistant for field data collection.

REFERENCES

Alston, M. and Kent, J. 2004. *Social Impact of Drought: A Report of NSW Agriculture*. Centre for Rural Social Research. New South Wales. Charels Sturt University.

Biradar, N. and Sridhar, K. 2009. Consequences of 2003 Drought in Karnataka with Particular Reference to Livestock and Fodder. *Journal of Human Ecology*, **26**(2): 123-130.

Chen, M. A. 1991. *Coping with Seasonality and Drought*. Sega Publication. New Delhi.

DARE 2003. Annual Report of Department of Agriculture Research and Education, Ministry of Agriculture, Government of India.

Garett, E. H. and Woodworth, R.S. 1969. *Statistics in Psychology and Education*. Vakils, Feffer and Simons Pvt. Ltd., Bombay, 329p.

Gautam R.C. and Rao, J.V. 2007. Concepts of Rainfed Agriculture. In: *Integrated Water Management*. Pusa, New Delhi.

McKee, T. B., Doesken, N. J. and Kleist, J. 1993. The Relationship of Drought Frequency and Duration to Time Scales. Preprints, Eighth Conference on Applied Climatology, Anaheim, California, USA, pp179–184.

Pandey S., Bhandari, H. and Hardy, B. 2007. *Economic Costs of Drought and Rice Farmers Coping Mechanisms: Cross-Country Comparatives Analysis*. International Rice Research Institute, Phillipines.

Samra, J.S. 2004. Review and Analysis of Drought Monitoring, Declaration and Management in India. Working Paper 84. Drought Series Paper 2. International Water Management Institute, Colombo.

Wani, S.P., Sarvesh, K.V., Krishnappa, K., Dharmarajan, B.K. and Deepaja, S.M. 2012. *Bhoochetana: Mission to Boost Productivity of Rainfed Agriculture through Science-led Interventions in Karnataka*. ICRISAT, Hyderabad, India, 84p.

Wilhite D.A. 1993. *Drought Assessment, Management and Planning: Theory and Case Studies*. Mass. Kluwer Academic Publishers, Boston.

□□□

Chapter 83

Case Study of Economic Assessment of Soil Erosion Damage : Application to Smallholder Farms in Marginal Lands of River Systems

V. C. PANDE, R. S. KUROTHE, H. B. SINGH, S. P. TIWARI[1], GOPAL KUMAR, B. K. RAO, A. K. VISHWAKARMA AND G. L. BAGDI

INTRODUCTION

The problems of land degradation and soil erosion have drawn the attention of the policy makers since long. The annual average loss of nutrients from land due to soil erosion has been estimated as 5.37 - 8.4 M tones (Mt) and the loss of production due to non-development of ravines has been estimated to be 3 Mt annum^{-1} (Fertilizer Statistics, 2007-08). Soil erosion, world over, is becoming a serious problem because of considerable economic damage it causes to society at large. It has intrigued the conservationists due to the difficulties in estimating the economic damages of soil erosion (Herath, 2001). Formulation of erosion control policies and prioritization of resource allocation to address the problem warrants assessment of bio-physical causes and economic effects. While scholars have drawn attention to the soil erosion magnitude (Kurothe *et al.*, 1997), only few attempts have been made to assess the economic damage (Ananda *et al.*, 2001, Herath, 2001, Ananda, 1998). Complexities between soil erosion, crop yield and economic loss have been highlighted by many scholars (Thampapillai and Anderson, 1994). While efforts have been made to examine the impact of soil erosion on crop yields using data from field plot trial (Miranowski, 1984; Van Kooten *et al.*, 1989), scholars have also used subjective elicitation procedure to examine soil erosion damage (Ananda *et al.*, 2001, Herath, 2001, Segarra and Taylor, 1987). Little empirical work, however, has been carried out on the economics of soil conservation on smallholder farms in developing countries (Eaton, 1996).

The present study uses a hybrid approach incorporating the two methods. Using the on-farm trial data from research farm in Mahi ravines, yield-erosion

CSWCRTI, Research Centre, Vasad 388 306, Distt. Anand, Gujarat,
[1]CSWCRTI, Research Centre, Datia, Madhya Pradesh, India

relationship was developed. This was projected for the smallholder farms in Mahi ravines and erosion damage function was developed. The study, then, tried to incorporate on-site data of existing production system into farmers' decision making by employing erosion damage function developed. It is contended that such a hybrid approach, where applicable, can give appropriate relationships between soil erosion, crop yield and economic loss for assessment of soil erosion economic damage. Application of the approach in assessing erosion damage for large degraded lands can help policy makers design suitable policies and allocate resources to address the problem.

MATERIAL AND METHODS

Economic damage due to soil erosion was assessed for a scenario representing the agricultural practice followed on a typical smallholder farm whose land is adjacent to ravine. The study used both primary and secondary data. The primary data pertained to the smallholders' farm in Mahi ravines of Gujarat. The secondary data on soil erosion and crop yields were collected from the Research Project Records of Central Soil & Water Conservation Research & Training Institute, Research Centre, Vasad. The research farm of the center, located in the vicinity of Mahi ravines, represented the marginal lands cultivated by the smallholders' farms in Mahi ravines. It, therefore, justified developing the soil erosion and crop yield relationship and projecting it for the soil erosion scenario being examined for Mahi ravines.

Study area, sample size and data collection : Of the total ravine-affected land in Gujarat, the largest gullied area of 61,888 ha is along the river Mahisagar (Sharma *et al.*, 1981) and, therefore, these ravines were selected for the study. Two districts- Vadodara and Anand- were selected along the left and right banks of the river, where most of the ravine lands are spread. Five villages, two in the Anand district and three in the Vadodara district, were selected based on the ravine area in the districts. A list of farms, comprising marginal, small and medium farm-categories, with lands adjacent to ravines was finalized and data on land-use and crops were collected. However, the smallholder farms, for which complete and consistent information was not available, were dropped from the sample. Thus, a sample of 100 farms was retained for the analysis. A well-structured questionnaire was used to collect information from the fields. Surveys were conducted to elicit primary information on soil conservation problem, crop and cropping system, perception of farmers about erosion and soil losses from their fields, its intensity and measures of soil and water conservation. The secondary data on area, production and farm harvest price were also collected for the district.

Methodology

The data collected on different economic variables were analyzed by computing statistical averages and economics was worked out following the standard discount cash flow technique, *viz.* present value of net benefit. Using secondary data from Research Farm located in Mahi ravines, relationship between erosion and yield was examined. The erosion - yield relationship was then used to estimate erosion damage function following Walker (1982) for the representative smallholder farm using the crop input-output data collected from farmers and local prices of input and output.

Yield erosion relationship : This model captures the relationship between crop yield and loss of top soil providing insights to the magnitude of the loss of physical output to soil erosion. Soil erosion alters the physical, chemical and biological properties of soil which determines crop productivity. This function considers yield of crop as the dependent variable and land degradation measure as the independent variable. The degradation measure, in literature, has been taken either topsoil depth or cumulative soil loss. Soil erosion alters the physical, chemical and biological properties of soil, which affect the crop productivity. Further, interactions among factors such as soil properties, climate and input levels also affect the crop productivity (Pierce and Lal, 1994). Field observations have indicated that input management, on small and marginal farms do not vary drastically over a given span of time. This is particularly so in the face of financial and infrastructure constraints like market, extension services etc with which such smallholder farms operate in the rural areas. Assuming that climate, crop management/ land management are given for this region, the yield damage function would give the relationship between crop yield and soil loss. Various functional forms including linear, curvilinear and exponential functions have been used to express the relationship between soil loss and crop yield (Smith and Shaykewich, 1990). Following Lal (1987) data on crop yields and soil loss was used to develop relationship between crop yield and cumulative soil loss in absence of the data on soil depth.

Damage function analysis on marginal farms : The yield erosion relationship developed from research farm data were used for damage function analysis along with field data from smallholder farms in Mahi ravines. The yield erosion relationships for conventional (erosive) practice and conservation practice were projected for the erosion and non-erosion scenarios on marginal farms. These relationships were used to compute the rate of yield decline on smallholder farms under the two scenarios at different soil loss. The average yield levels of crops for erosion and non-erosion farms were taken as those realized on conservation adopted and non-adopted marginal farms in Mahi ravines. Using the average yield levels and rate of yield declines, the average yield levels to be realized in time frame were projected under the two scenarios. This information along with information on conservation measures taken by adopters in the study area was used to compute erosion damage function.

Erosion damage function : This refers to the private economic benefits forgone when an erosive practice is adopted instead of a conservation practice (Walker, 1982). An erosion damage function thus represents the monetary losses associated with soil erosion. This is vital to internalize the on-site effects of soil erosion to the farm economy. Various approaches have been used to estimate erosion damage function using yield erosion relationship (Burt, 1981; Walker, 1982; Miranowski, 1984; Pagiola, 1995; Segarra and Taylor, 1987; Sinden and Yapp, 1987; Taylor and Young, 1985; Walker and Young, 1986; Smith and Shakewich, 1990; Gunathilaka and Abeygunawardena, 1993). While most of these studies used yield as the dependent variable and top soil depth as the independent variable, the present study used cumulative soil loss as explanatory variable. Walker (1982) defined the damage function as the difference between the present value of net revenue streams of erosive farming practice and a conservation practice,

$$DFV_t = PNB_e - PNB_c \quad \ldots 1$$

Where, DFV_t is the value of the damage function in year t, PNB_c is the private profitability of choosing the conservation practice in the current year and PNB_e is the private profitability of choosing the erosive (conventional) practice currently and postponing the conservation decision to another year.

$$PNB_e = P.\,Y_e\,(t, D_{t-1}) - C_e\,(t, D_{t-1}) + \sum_{i=1}^{T-1} \frac{P.Y_c(t+i, D_t) - C_c(t+i, D_t)}{(1+r)^i} \quad \ldots 2$$

$$PNB_c = P.Y_c\,(t, D_{t-1}) - C_c(t, D_{t-1}) + \sum_{i=1}^{T-1} \frac{P.Y_c(t+i, D_{t-1}) - C_c(t+i, D_{t-1})}{(1+r)^i} \quad \ldots 3$$

Where,

P = Price of crop

Y_e = crop yield with erosive practice as function of cumulative soil loss and time

Y_c = crop yield with conservation practice

D_t = cumulative soil loss

C_c = variable cost of production with conservation practice

C_e = variable cost of production with erosive practice

T = Number of years in time horizon

r = real private rate of discount

The yield equations developed in the previous section were used to estimate value of damage function by incorporating local prices and costs. This was justified in view of the localized nature of yield erosion equation developed for the area. This relationship is highly site specific and would depend, among other things, on initial soil depth. The reason for estimating value of damage function was to determine the optimal year for a farmer to switch over from an erosive practice to a less erosive practice. The damage function was assessed in respect of ridge and furrow system of conservation practice as this was found to be quite effective on research farm at Vasad.

RESULTS AND DISCUSSION

Profile of farms, land use and cropping system : Among the sample farms, the land mostly belonged to marginal, small and medium farm categories, with average holding size of 1.3 ha. About 22.4% landholdings were being cultivated by tenants and they belonged to small farm category. The farm holdings studied were adjacent to ravines. As the holding-size increased, the proportion of ravine land also increased in different classes of holdings. The land-use pattern of farms revealed that the distribution of arable and non-arable land was not markedly different within different landholding classes. Bajra (*Pearl millet*) and bajra-based cropping system (Pearl millet + Pigeon pea) was most prevalent across all the categories of farms. The farm holdings lied adjacent to ravine with varying slopes.

Problem of soil water loss : Most of the farmers reported the incidence of run-off and consequent soil loss during rains. The farmers without conservation history reported a higher incidence than those with conservation history; though the difference was non-significant. The farms in the former group reported a higher effect of run-off on loss of top soil and the overall effect on yield loss though effect on loss of fertilizer and seed from fields was reported to be of similar trend. It was inferred that the loss of top soil on farms with no conservation measures was of relatively high magnitude having effect on yield loss.

Estimation of yield-erosion relationship : For developing this function in respect of farms in Mahi ravines, experimental plot data at Vasad research farm was used for farmers's practice as well as improved conservation practice (Ridge & Furrow Technique) for pearl millet + pigeon pea cropping system taken in the experiment. In absence of data on loss of top soil depth and other parameters, crop yield was fitted to the cumulative soil loss data following Lal (1987). Other researchers examining the economic effects of soil erosion have also adopted the equation derived by Lal (Bishop and Allen, 1989; Bishop,

1990; Ehui *et al.*, 1990). It has, however, been contested that the relationship is site specific (Eaton, 1996). Transferring this relationship to other sites could be fraught with problems if not accounted for soil type and profile of erosion. The present study tried several forms in addition to the exponential function in view of the limited data availability. The following exponential form was used for further analysis to develop two scenarios *viz.*, erosive practice and conservation practice in view of its inherent qualities (Eaton, 1996),

(I) Farmers' practice (Conventional tillage)

$Y = 39.89e^{(-0.00567X))}$

This explains an erosive production system.

(II) Conservation practice: Ridge and furrow tillage practice was considered to develop scenario of conservation practice. An exponential function of the form is used,

$Y=15.74\ e^{(-0.015X)}$

Where, Y is yield ($kgha^{-1}$) and X is cumulative soil loss.

Temporal impact of soil erosion on yield : The yield-damage functions developed above were used to explore the effects of soil erosion on yield over time (Figure 1). The yield loss over time depends on the initial soil depth and

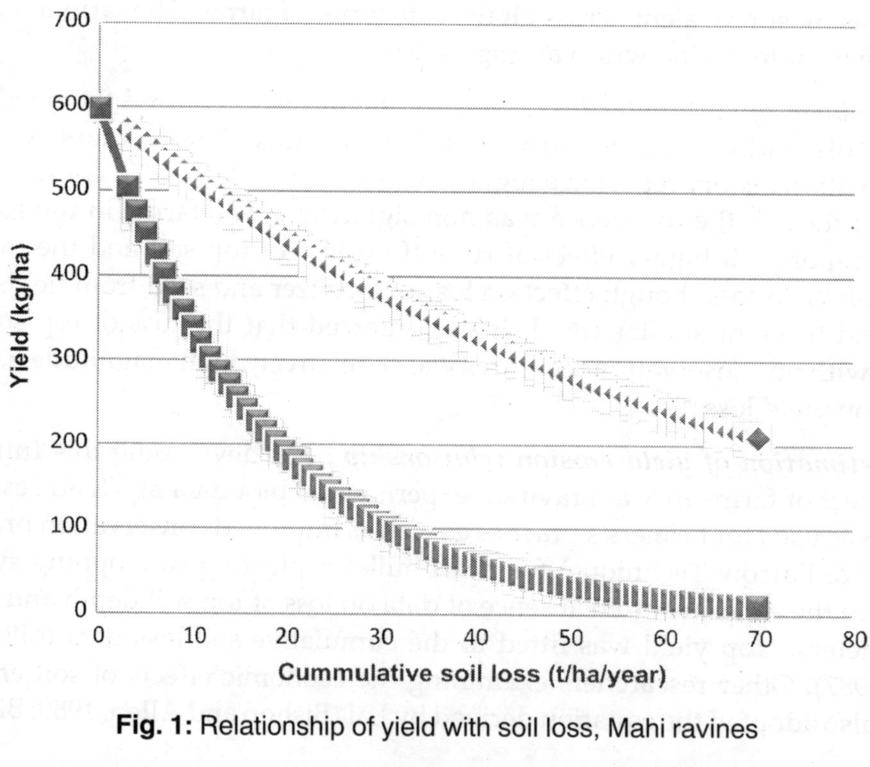

Fig. 1: Relationship of yield with soil loss, Mahi ravines

the rate of soil erosion over time. Data on soil loss and yield, available from experimental farm, were used to simulate yield loss from cumulative soil loss over time. This information was used to know the time period when 50 % and 75 % reduction of crop yield would occur with non-conservation and conservation practices adopted on farm (Table 1).

Under conventional (erosive) production scenario, it would take 13 and 24 years to realize reduction in yield to the tune of 50 % and 75 %, respectively. But if a conservation practice is adopted, this period is extended to 46 and more than 70 years, respectively. In other words, the conservation production practice defers yield reduction in time scale substantially.

Table 1: The long-term productivity effects of soil erosion

	Production practice		Years differed by conservation practice (to realize adverse effect on yield)
	Erosive practice	Conservation practice (RFT)	
Time taken to reach 50 % of yield reduction (Years)	13	46	33
Time taken to reach 75 % of yield reduction (Years)	24	> 70	> 44

RFT – Ridge & Furrow Technique

Estimation of erosion damage : The value of damage function was estimated for a decision period of 30 years considering a 10% discount rates with output price of Pearl millet as Rs. 5 kg^{-1} (Table 2).

Table 2: Present value of damage function on smallholder farms farms, Mahi ravines

Soil loss (t $ha^{-1}year^{-1}$)	Damage function value (Rs. ha^{-1})	Decision year for switching over to conservative practice (years)
3	-36	25
5	-23	15
8	-16	8

At 3% soil loss, farmer switches over to conservative practice in 25th year, the decision year reducing to 15 and 8 years at soil losses of 5 and 8 t $ha^{-1}year^{-1}$, respectively. A change in discount rate lower than 10% was also used to examine the effect of varying discount rates on decision to switch over.

Table 3: Sensitivity of discount rate to decision year for switching over to conservative practice

Soil loss (t $ha^{-1}year^{-1}$)	Discount rate - 7%		Discount rate - 5%	
	Damage value function (Rs ha^{-1})	Decision year	Damage value function (Rs ha^{-1})	Decision year
3	-40	22	-28	15
5	-25	10	-64	6
8	-16	8	-5	8

The results revealed sensitivity of switch over decision to discount rate. Depending upon the discount rate considered by smallholder farm, at 3 t $ha^{-1}year^{-1}$ soil loss, farmer tends to switch over to conservation practice in 22 and 15 years at 7% and 5% discount rates, respectively (Table 3). The optimal time frame similarly changes with different soil loss scenarios against pearl millet output price of Rs 5 kg^{-1}, an increase by Rs. 1 kg^{-1} enhances the decision year drastically. At 5% discount rate, the decision year switches to 29 years. Beyond that the decision year to switch over to conservative practice is more than 30 year, the planning horizon for which the analysis was done (Table 4).

Table 4: Crop price sensitivity analysis, *Pearl millet* price Rs 6 kg^{-1}

Discount rate (%)	Decision year (Years)
5	29
7	> 30
10	> 30

DISCUSSION

Assessing the productivity effects of soil erosion is a challenging task because crop yield is an integrative effect of numerous variables. Further, effect of erosion on crop yield is a function of changes in land quality with respect to the parameters specific to the region under study. A long time data on understanding this relationship is a pre requisite to the estimation of the damage function for a region. Application of a hybrid approach in this study was demonstrated with a limited data set available for Mahi ravine region. The exponential form of yield erosion relationship was considered because of its inherent qualities as the limited data available for developing yield erosion relationship did not give a satisfactory result. The estimation of damage function value using local prevailing price of output revealed the sensitivity of the decision year for a farmer to switch over from existing erosion practice of cultivation to conservative practice to factors like output price and discount rate. This is supported by works from other scholars like Collins and Headley, 1983; Eaton, 1996. Further, the damage function approach used here is specific to a crop or cropping system. The cropping system used in this study is pearl millet + pigeon pea which is largely followed on smallholder farms in Mahi ravines. A different cropping system in other part of the region would yield a different scenario. Yet the results of the study could be used to suggest policy prescriptions. Farmers' choice of discount rate and the relevant planning horizon might affect the returns to conservation activities. This is affected by property rights existing with farmer. Lack of well-defined property rights, as prevalent in India, over land may lead to an increase in the rate at which the farmers discount future returns to conservation activities (Bishop, 1992). In effect, farmers may be less willing to invest in conservation efforts if they are uncertain of reaping the future benefits. Similarly, a favourable output price scenario would greatly alter farmer's decision making approach incentivizing him to invest on conservation (Pande *et al.*, 2011) in general and switch over to conservation practice in particular.

CONCLUSIONS

Economic damage due to soil erosion was assessed for a scenario representing the agricultural practice followed on a typical smallholder farm whose land is adjacent to ravine. Using the on-farm trial data from research farm in Mahi ravines, yield-erosion relationship was developed. This was projected for the smallholder farms in Mahi ravines and erosion damage function was developed with local output prices and cost of production realized by the smallholder farms. Scenarios were developed regarding farmer's decision to switch over to conservation practice from the existing erosive practice at 10% discount rate considering 30 years of production horizon for different soil loss as case study. Pearl millet + Pigeon pea cropping system, the most prominent cropping system on small holder farms in Mahi ravines, was taken for study. Farmers' existing erosive practice was compared with a conservation practice *viz.*, ridge & furrow technique (RFT). Results revealed that decline in yield to the extent of 50% and 75% took much longer in case of conservative practice than the time span realized in case of erosive practice. The decision year to switch over to conservative practice reduced from 25 years to 15 years if soil loss increased from 3 t $ha^{-1}year^{-1}$ to 5 t $ha^{-1}year^{-1}$. Sensitivity analysis performed for different discount rates and output prices found the switch over decision year to be quite sensitive to these parameters, suggesting thereby, that policy prescriptions such as property rights and output price might play role in the decision making process of these farming groups. Favourable rights and price scenario could help in internalization of the user cost of erosion by incentivizing farmers to switch over to conservative practice early in time frame.

ACKNOWLEDGEMENT

The data used for this study were collected in a research project approved by the Institute Research Council. Financial help and encouragement provided by the Director of the institute is acknowledged. The support provided by the head of the center in conducting field surveys and technical support in data collection are duly acknowledged.

REFERENCES

Ananda, J; Herath, G and Chisholm, A. 2001. Determination of yield and erosion damage functions using subjectively elicited data: application to smallholder tea in Sri Lanka. *The Australian Journal of Agricultural and Resource Economics*, 45(2): 275-289.

Ananda, J. 1998. Soil Erosion Damage Function for Smallholder Tea in Sri Lanka: An Empirical Estimation. Paper presented at the 1st World Congress of Environmental and Resource Economists, June 24-27, 1998, Isola di San Giorgio, Venice, Italy.

Bishop, J. and Allen, J. 1989. The On-Site Costs of Soil Erosion in Mali. Environment Department Working Paper No. 21. Washington D.C.: World Bank.

Bishop, J. 1992. Economic Analysis of Soil Degradation. Gatekeeper Series No. LEEC 92-01. London: IIED/ UCL London Environmental Economics Centre.

Bishop, J. 1990. The Cost of Soil Erosion in Malawi. Draft Report for Malawi Country Operations Division, World Bank.

Burt, O.R. 1981. Farm Level Economics of Soil Conservation in the Palouse Area of the Northwest, *American Journal of Agricultural Economics*, 63 (1): 83-92.

Collins, R.A and Headley, J.C. 1983. Optimal Investment to Reduce the Decay Rate of an Income Stream: The Case of Soil Conservation. *Journal of Environmental Economics and Management*, 10:60-71.

Eaton, E. 1996. The economics of soil erosion: a model of farm decision-making. Discussion Paper, DP 96-01, International Institute for Environment and Development.

Ehui, S.K., Kang, B.T. and Spencer, D.S.C. 1990. Economic Analysis of Soil Erosion Effects in Alley Cropping, No-till and Bush Fallow Systems in South Western Nigeria. *Agricultural Systems*, 34:349-68.

Fertiliser Statistics. 2007-08. The Fertilizer Association of India, New Delhi.

Gunatilleke, H.M. and Abeygunawardena, P. 1993. An Economic Analysis of Soil Conservation in Tobacco Lands in the Hanguranketha Area of Sri Lanka. *Indian Journal of Agricultural Economics*, 48 (1): 106-112.

Herath, G. 2001. Estimating the user cost of erosion in tea small holdings in Sri Lanka. *Australasian Journal of Regional Studies*, 7(1): 97-111.

Kurothe, R.S., Butta, R.K. and Sharma, J.P. 1997. Soil erosion map of Gujarat. *Indian Journal of Soil Conservation*, 25(1): 9-13.

Lal, R. 1987. Effects of Erosion on Crop Productivity. *Critical Reviews in Plant Sciences*, 5(4):303-67.

Miranowski, J.A. 1984. Impacts of productivity loss on crop production and management in a dynamic economic model. *American Journal of Agricultural Economics*, 66: 61-71.

Pagiola, S. 1995. Price Policy and Returns to Soil Conservation in Kitui and Machakos, Kenya. Environment Department, The World Bank, Washington, DC.

Pande, V.C., Kurothe, R.S., Singh, H.B. and Tiwari, S.P. 2011 Incentives for Soil and Water Conservation on Farm in Ravines of Gujarat: Policy Implications for Future adoption. *Agricultural Economics Research Review*, 24(1): 109-118.

Pierce, F.J. and R. Lal. 1994. Monitoring the Impact of Soil Erosion on Crop Productivity. In R. Lal (ed), Soil Erosion: Research Methods, (2nd Edition), Soil and Water Conservation Society.

Segarra, E. and Taylor, D.D. 1987. Farm level analysis of soil conservation: An application to the Piedmont area of Virginia. *Southern Journal of Agricultural Economics*, 19: 61-67.

Sharma, A.K., Pradhan, I.P., Nema, J.P. and Tejwani, K.G. 1981. *25 years of research on Soil & Water Conservation in ravine lands of Gujarat*. Monograph No. 2. Central Soil & Water Conservation Research & Training Institute, Research Centre, Vasad 388 306, Gujarat.

Sinden, J.A. and Yapp, T.P. 1987. The Opportunity Cost of Land Degradation in New South Wales: A Case Study, Paper Presented to the Australian Agricultural Economics Society Conference, Adelaide.

Smith, E.G. and Shaykevich, C.F. 1990. The economics of soil erosion and conservation on six soil groupings in Manitoba. *Canadian Journal of Agricultural Economics*, 38: 215-231.

Taylor, D.B. and Young, D.L. 1985. The Influence of Technological Progress on the Long Run Farm Level Economics of Soil Conservation. *Western Journal of Agricultural Economics*, Vol. 1, No. 1, pp: 63-76.

Thampapillai, D.J. and Anderson, J.R. 1994. A review of the socio-economic analysis of soil degradation problems for developed countries. *Review of marketing and Agricultural Economics*, 62: 291-315.

Van Kooten, G.C., Weisensel, W.P. and Jong, E.de. 1989. Estimating the costs of soil erosion in Saskatchwan. *Canadian Journal of Agricultural Economics*, 37: 63-75.

Walker, David J. 1982. A Damage Function to Evaluate Erosion Control Economics. *American Journal of Agricultural Economics*, 64 (4): 690-698.

Walker, D.J. and Young, D.L. 1986. The Effect of Technical Progress on Soil Erosion Damage and Economic Incentives for Soil Conservation. *Land Economics*, 62 (1): 83-93.

□□□

Chapter 84

Impact of Groundwater Development for Agriculture and Constraints in its Utilization in Odisha State of Eastern India

S.K. SRIVASTAVA, RANU RANI SETHI, ASHWANI KUMAR AND A.K. NAYAK

INTRODUCTION

Groundwater plays a vital role in agricultural development by enhancing the productivity of other inputs and by providing assured irrigation to the farmers. Because of the land augmenting character and assured irrigation water, groundwater development has always been the priority area by the policy makers. In agriculture sector, importance of groundwater has been increasing manifold due to factors *inter alia* technological breakthrough in extraction technology, soft loans for installation of groundwater extraction mechanism and remunerative relative price ratio in favour of water intensive, commercial and horticultural crops. Groundwater accounts for more than 55 % of India's net irrigated area and major concern nowadays is the sustainable agricultural growth through equitable development of groundwater.

In Odisha state of Eastern India, groundwater development is poor (26.14%) as compared to other parts of the country due to variety of the reasons. Yield enhancing nature of groundwater irrigation, offers a great scope to improve the agricultural productivity in the state through sustainable groundwater development. The present paper attempts to diagnose the trend and pattern in the groundwater development, examines its impact on agricultural income and analyses the constraints in its utilization to provide a feedback to the policy makers and academicians for improved livelihood security in the state.

DATA AND METHODOLOGY

The study uses secondary data on different aspects of groundwater development in Odisha. District-wise data on groundwater availability, draft

Directorate of Water Management, Bhubaneswar, Odisha, 751 023, India and development has been collected from Directorate of Groundwater Survey and Investigation, Government of Odisha for the year 2009-10. Further, district wise net domestic product for agriculture has been collected from Economic Survey, 2009-10, Directorate of Economics and Statistics (DES), Government of Odisha to study the impact of groundwater on agricultural income. Finally, data to study pattern of groundwater development and constraints analysis was collected from 3rd (2000-01) and 4th (2006-07) minor irrigation census, DES. Tabular analysis has been used to examine the trend and pattern of groundwater development. To study the impact of groundwater development on agriculture, log-linear regression function was fitted in which per hectare net district domestic production was regressed on groundwater development. Log form of the variables directly gives the elasticity of agricultural income with respect to groundwater development.

RESULTS AND DISCUSSION

Status of groundwater resources in Odisha : Gross groundwater recharge has been estimated as 17.77 billion cubic meter (BCM) in Odisha through different sources (Table 1). Rainfall contributes 71% in gross groundwater recharge. With 1.09 BCM of natural groundwater losses, net groundwater availability in Odisha is 16.69 BCM. Out of this, only 4.36 BCM groundwater is drafted with irrigation sector extracting the highest (79.6%). Thus, overall groundwater development in Odisha stands only 26.14%. However, there exists wide variability in its development across different districts ranging from 8.76% in Malkangiri to 55% in Bhadrak due to varying hydro-geological

Table 1: Status of groundwater resources in Odisha in 2009-10

Particular	Value
Gross Groundwater recharge (BCM)	17.77
Natural Groundwater losses (BCM)	1.09
Net Groundwater resources (BCM)	16.69
Groundwater Draft (BCM)	4.36
Irrigation	3.47 (79.6)
Domestic	0.74 (16.99)
Industry	0.15 (3.41)
Groundwater Development (%)	26.14
Maximum	55% (Bhadrak)
Minimum	8.76% (Malkangiri)

Figures within parentheses includes share in total groundwater draft

Source: Directorate of Groundwater Survey and Investigation, Government of Odisha, 2011 and socio-economic conditions. Among the 314 blocks, groundwater development in 23 coastal blocks is more than 50%. At the same time, in 25 hard rock blocks, groundwater development in less than 10 %. Further, 36 and 6 coastal blocks are partially and fully affected by salinity problem, respectively. Temporally, groundwater development in Odisha has increased from 14.79% in 1999 to 26.14% in 2009. For future, after reserving groundwater for industry and domestic uses for the next 25 years, 1.19 BCM and 0.84 BCM groundwater is available for creation of additional irrigation potential at 100% and 70 % level of groundwater development, respectively.

Irrigation potential created (IPC) from groundwater and its utilization : Odisha's agricultural economy is primarily rainfed with only 35% (31.77 M ha) of the gross cropped area irrigated through different sources (Govt. of Odisha, 2010). Groundwater resources constitute only 6.14% (1.95 M ha) of gross irrigated area (4th MI Census). Although the irrigation potential created through groundwater in the state is 5.5 M ha (Table 2), its utilization (IPU) is only 35.22% (1.95 M ha) in 2006-07 raising several efficiency issues in its utilization. The pace at which IPC through groundwater increased between 2000-01 (3rd MI census) and 2006-07 (4th MI census), due to massive Government support and incentives, could not continue for IPU resulting into decline in the per cent utilization from 51.53% to 35.22% during the same period. Among different sources of groundwater irrigation, dugwells predominated contributing 86%, 39.24% and 43.08% of the total groundwater structures, total IPC and total IPU, respectively in 2006-07. However, in utilization of created potential, shallow tubeswells outpaced dugwells with the per cent utilization of 45.62 in 2006-07. The per cent utilization of created

Table 2: Creation of irrigation potential and its utilization through groundwater between 3rd (2000-01) and 4th (2006-07) Minor Irrigation (MI) Census in Odisha

Particulars	Dugwells		Shallow tuebwells		Deep tubewells		Total Groundwater	
	2000-01	2006-07	2000-01	2006-07	2000-01	2006-07	2000-01	2006-07
Number (000)	378 (88.52)	406 (86.02)	44 (10.30)	61 (12.92)	5 (1.17)	6 (1.27)	427 (100)	472 (100)
IPC (000, ha)	146 (48.67)	217 (39.24)	122 (40.67)	204 (36.89)	32 (10.67)	133 (24.05)	300 (100)	553 (100)
IPU (000, ha)	77 (49.68)	84 (43.08)	66 (42.58)	93 (47.69)	11 (7.10)	18 (9.23)	155 (100)	195 (100)
% utilization	52.73	38.90	54.43	45.62	35.03	13.26	51.53	35.22

Figures within parentheses are share of respective GW device in total GW devices
Data Source: 3rd (2000-01) and 4th (2006-07) Minor Irrigation Census, Government of India

potential declined for all the sources with deep tubewells witnessing highest decline during 2000-2006. Poor utilization of the irrigation potential through groundwater might be due to several factors such as unfavourable hydro-geological conditions, frequent failure of wells, poor maintenance, lack of assured electricity supply, saline water intrusion, etc.

Impact of groundwater development on agriculture : Although groundwater constitutes a small share in GIA, it has a significant role in improving agricultural productivity in Odisha. Estimated log-linear regression function revealed that about a quarter (R^2) of the agricultural productivity (₹ ha^{-1}) is dependent on groundwater (Table 3). Estimated coefficient for groundwater development was positive (0.598) and significant at 1% level of significance and represents elasticity of agricultural productivity with respected to groundwater development. Thus, groundwater bears a positive relationship with the agricultural income and opens scope for improving agricultural income in the state through its proper development.

Table 3: Results of regression analysis between agricultural income and groundwater development in Odisha

Particular	Estimates
Constant	8.569* (0.631)
Groundwater development (log)	0.598* (0.199)
R^2	0.244
No. of observations	30

Dependent variable: Per hectare net district domestic product from agriculture (₹ ha^{-1}) in Logarithmic form

*significant at 1% degree of significance

Figures within parentheses are standard errors of estimated coefficients

Constraints in groundwater utilization in Odisha : Under-development and inefficient utilization of groundwater resources in Odisha might be because of variety of hydro-geological, agro-climatic and socio-economic constraints. A perusal of Table 4 reveals that about a quarter of total groundwater structures in Odisha are not in use due to temporary or permanent reasons in 2006-07. 62.68% structures are not in use temporarily and 37.32% permanently. Non-functioning of groundwater structure resulted into loss of 136.37 thousand hectare (24.66% of IPC) irrigation potential annually in Odisha. About 60% (80.09 thousand ha) loss in the IPC is due to non-functioning of deep tubewells alone which irrigate highest command area as compared to other sources. It is to be noted that permanent reasons in general and destruction of deep tubewells in particular (Table 5) was the prime reason for the non-functioning. Dugwells and shallow tubewells contributed 25.10% and 18.18% loss in IPC in 2006-07.

Table 4: Proportion of GW structures not in use and loss in the irrigation potential (2006-07) (%)

Wells type	In-use	Not-in use	Temporary not- in use*	Permanently not- in use*	IPC lost
Dugwells	74.90	25.10	63.81	36.19	25.10 (54.43)
Shallow TW	81.82	18.18	59.86	40.14	18.18 (36.99)
Deep TW	39.62	60.38	39.76	60.24	60.38 (80.09)
Total	75.34	24.66	62.68	37.32	24.66 (136.37)

Figures within parentheses are absolute loss in IPC (000 ha) due to non-functioning of wells

* % of total not-in-use

Data source: 4th Minor Irrigation Census, Government of India

Different sources of groundwater exhibited different reasons of non-functioning (temporary or permanent) in Odisha (Table 5). Less discharge was the prime reason for temporary non-functionality of dugwells as 33.49 % of total temporarily non-functional dugwells were due to it. It is to be noted that dugwells are the most feasible source of groundwater in hard rock regions which comprises of more than 80% of state's geographical area and characterized by less discharge, high draw-down and uncertain water bearing zones. Artificial recharge structures can be installed in these regions to improve the discharge rate of the wells. In case of shallow and deep tubewells, mechanical breakdown was the prime reason for their temporary non-functionality. These wells can be targeted and incentives can be provided to make these wells functional. Among the permanent reasons, destruction of wells was the single most important factor of non-functionality of all type of structures. Repairing and reinstallation of these structures will go a long way to improve the groundwater utilization and bring additional land under groundwater irrigation. Drying-up of the wells was the second important reason for permanent non-functioning which can be removed through artificial recharge structures.

Table 5: Reason-wise distribution of groundwater structure in Odisha in 2006-07 (%)

Reasons	Dugwells	Shallow tubewells	Deep tubewells	Total
Reasons for temporary not-in-use*				
Inadequate Power	0.23	5.15	8.07	0.83
Mechanical Breakdown	3.66	30.11	39.13	6.76
Less discharge	33.49	15.69	12.56	31.46
Others	62.62	49.05	40.23	60.95
Reasons for permanently not-in-use*				
Salinity	0.17	2.61	2.10	0.52
Destroyed	63.73	51.23	71.16	62.83
Dried up	15.39	14.44	9.43	14.99
Seawater intrusion	1.86	1.78	1.28	1.82
Industrial effluents	0.13	0.29	0.09	0.14
Others	18.71	29.65	15.95	19.69

*share of respective reason in total temporary and permanently not-in-use structures

CONCLUSIONS

The groundwater development in Odisha, though improved over the years, is still low (26.14%) with wide inter-regional variability as compared to other parts of the country. Further, inspite of small share in gross irrigated area, groundwater bears a significant and positive role in improving the agricultural income of the state. Low groundwater development in the state is accompanied by poor utilization of already created irrigation potential ranging from 13.26% for deep tubewells to 45.62% in shallow tubewells. In the light of good groundwater reserve (0.84 BCM at 70% level of development) for future, it can be further developed by providing economic incentives (subsidy and easy access to credit for groundwater structures) and policy supports (cheap and reliable electricity supply, better support price, better marketing infrastructure etc for groundwater dependent crops). On the other hand, utilization of the already created irrigation potential can be improved substantially by installing suitable artificial recharge structures especially in hard rock regions, repairing and reinstallation of non-functional groundwater structures. This will bring additional land under irrigation and subsequently contribute positively for the improvement of agricultural income in the state.

REFERENCES

Government of India. 2011. 3rd and 4th minor irrigation census, 2006-07, Ministry of Water Resources, Government of India.

Government of Odisha. 2010. Economic Survey, 2009-10, Directorate of Economics and Statistics.

Government of Odisha. 2010. Orissa Agricultural Statistics, 2008-09, Directorate of Agriculture and Food Production.

Government of Odisha. 2011. Groundwater Resources of Orissa, Directorate of Groundwater Survey and Investigation, Government of Odisha.

□□□

Chapter 85

Selection of Centrifugal Pumps for Various Categories of Farmers in Odisha

JANAKI BALLAV SWAIN*, B. B. SAHOO[1] AND B. C. SAHOO[2]

INTRODUCTION

Odisha is bestowed with plenty of rainfall amounting to 1450 mm per annum. Out of this about 78 % occurs during rainy season (June to September) due to south-west monsoon and remaining 22 % during rest of the year. Apart from the temporal variation of rainfall, spatial variation is also evidenced across the state varying from 1200 mm in southern coastal plain to 1700 mm in northern plateau region (Anonymous, 2009). Above all, the state receives on an average 230.76 billion cubic meters (BCM) of rainfall per annum. In spite of that drought and flood are two natural phenomena responsible for frequent crop failure in the state. Erratic distribution pattern of rainfall is one of the vital reasons for such catastrophe. Rainfed agriculture that spreads over 65 % of the net cultivated area of the state is no more reliable to cater to the food grain demand of growing population as a result of the concept of seasonal irrigated high yielding agriculture has come into being in the recent years.

It is reported that the state Odisha is endowed with 10 % of water resources of the country. However, the groundwater potential of the state has remained largely unexploited so far. The ground water potential of the state as per the assessment made in 2001 is 21.01 BCM out of which 60% is safe and usable. So, enough scope is there to tap groundwater sources to meet the demand of agriculture and other sectors. In Odisha, about 18.80

*N.I.T., Rourkela, Odisha, India

[1] College of Agricultural Engineering and Technology, Orissa University of Agriculture and Technology, Bhubaneswar, Odisha, India

[2]Department of Soil and Water Conservation engineering, College of Agricultural Engineering and Technology, Orissa University of Agriculture and Technology, Bhubaneswar, Odisha, India.

lakh ha out of the net cropped area of 61.8 lakh ha are irrigated. About 76% of the total working force is engaged in agriculture and allied sectors (Anonymous, 2009). The demand pattern of water is changing rapidly with increase in population, urbanization and industrialization. The constraints of water availability and increasing pressure on it from various sectors have urged upon evolving means and ways to allocate the scarce resource properly in the state. Agricultural growth is essential for fostering the food grain demand of the growing population along with the economic development of the state.

Lifting of water from the sources at lower to higher elevation can be accomplished by various human and animal powered devices. With advent of time, reciprocating pumps have become popular in lieu of low and medium water lifting devices for resource poor farmers in remote areas. On the other hand, the small and marginal farmers constitute 82% of the farming community in the state. Apart from monsoon aberrations in rainy season, increase in food grain demand has emerged as a challenge before the scientists and stakeholders. In order to increase agricultural production and productivity, attempt is being made to bring more and more land under irrigation. When it is not possible to make provision for flow irrigation, lift irrigation practices are being adopted through the use of mechanically operated centrifugal pumps. Throughout India, the operation of lift irrigation systems is grossly inefficient, resulting in massive over-consumption of energy and inadequate irrigation, limiting agricultural productivity and resulting in reduced welfare of common farmer (Anonymous, 1983). Most of the farmers do not have adequate knowledge about selection of right size centrifugal pumps that can lift required amount of water to their crop fields. Buying a pump of higher capacity not only increases the financial burden on the farmer but also aids unnecessary consumption of electrical energy and sometimes invites problem for the underground water source. Similarly, buying a lower size pump makes the water supply system handicapped. Installation of diesel operated pump sets for lifting water makes the operation and maintenance cost of these schemes exorbitant. However, for lifting small order of discharge by individual cultivators, portable diesel pump sets are feasible as they provide greater flexibility and mobility for installation at different points of the water source(s). Again, the season-based cropping pattern also has a key role to play in deciding the capacity of the pumps. This paper presents the selection of right size of centrifugal pumps for various categories of farmers in different zones of Odisha.

MATERIAL AND METHODS

The present study on selection of centrifugal pumps has been accomplished through three stages such as i) collection of data on existing cropping pattern, average depth of ground water table and demographic features district-wise; ii) interpretation of the data with respect to available lifts for pumping water, irrigation requirement for the cropping pattern; and iii) selection of a suitable pump for each category of farmer basing upon its efficiency and power requirement.

Study area : The state Odisha lies over 17.49^0 to 22.34^0 N latitude and 81.27^0 to 87.29^0 E longitude spreading over an area of 1, 55, 707 sq.km. Geographically, the state is surrounded by Chhatisgarh in west, Jharkhand and West Bengal in north, Andhra Pradesh in south and Bay of Bengal at its east. The population of the state has gone up to 4.21 crores (2011census) and out of them 47.66 lakh are farm families. The state is comprised of 30 districts out of which 9 districts *viz.*, Balasore, Bhadrak, Kendrapara, Jagatsinghpur, Cuttack, Jajpur, Puri, Ganjam and Khurda come under the eastern deltaic or coastal zone and rest 21 in western interior zone. Rice is the principal crop grown in the state followed by maize in *kharif* season. Cropping during *rabi* season is mainly confined to irrigated areas with residual soil moisture. Other important crops produced in the state are pulses (Arhar, black gram, green gram and kulthi), oilseeds (groundnut, mustard, til and niger), fibres (jute, mesta and cotton), sugarcane, vegetables and spices. In addition, mango, banana, coconut and cashewnut are the main horticultural crops.

METHODOLOGY

The state has been divided into two zones such as zone I comprising of coastal districts and zone II comprising of western interior districts. The parameters such as crop water requirement, depth of groundwater table and land holding size of the farmers are collected from both the zones and used in selection of right size of pump for various categories of farmers.

Cropping Pattern : Rice is found to be the predominant crop in both zone-I and II in *kharif* season. It is grown extensively in all topo-sequences under both rainfed and irrigated conditions. Non-rice crops such as maize among cereals, arhar among pulses and niger among oilseeds are grown in sporadic patches in uplands of zone-II during rainy season. Irrigation need of the crops other than rice in *kharif* season is not taken into account. Recommended depth of water in each irrigation and the irrigation interval for some predominant crops in the state are presented in Table 1. While rice-vegetable and rice-oilseeds are the extensively practiced cropping sequences in zone-I, rice-pulses and rice-oilseeds are the dominant cropping sequences in zone-II. In estimation of the water requirement of the crops in different zones of the state, the depth of water and rotation period presented in Table 1 are used. Duration of operation of pumps in both the zones is assumed to be 6 hours in a day.

Table 1: Depth and interval of irrigation for variety of crops

Crop	Season	Depth of water in each irrigation (cm)	Irrigation interval (Days)
Rice	*kharif*	10.0	7
Maize	*kharif/Rabi*	7.5/5.0	15
Vegetables	*Rabi*	7.5	10
Oilseeds	*Rabi*	5.0	15
Pulses	*Rabi*	5.0	15

The design discharge rate of centrifugal pump is estimated by the following equation:

$$Q = \frac{27.78}{T} \sum_{n=1}^{n} \frac{A_n D_n}{R_n} \quad \text{... (1)}$$

where, Q = Rate of discharge of pump, $l\ s^{-1}$; A = area of land under n^{th} crop, ha; D = depth of water applied for n^{th} crop per irrigation, cm; R = irrigation interval for n^{th} crop, days; and T = duration of pumping, hr day^{-1}. 'n' is an index for the crop type in the operational holding of the farmer and it starts from 1 and goes up to n.

Total Head of the Pump : Water table depth of unconfined aquifers in cropping area is considered to determine the average depth of water table. The depth of groundwater table in pre- and post-monsoon periods are taken together to determine the average depth of water table district-wise. Further, the average water table depths in the districts under each zone are averaged together to estimate the mean depth of water table in the zone. Average depth of water table in both the zones is considered as the static suction lift for the centrifugal pumps. Frictional head loss in the pipes and static suction and delivery head are considered to estimate the total head required to operate the centrifugal pumps (Church,1972). The total head for pumping (H) is expressed as:

$$H = h_s + h_{sf} + \frac{v_s^2}{2g} + h_d + h_{df} + \frac{v_d^2}{2g} \quad \text{... (2)}$$

where, h_s = static suction head, m; h_{sf} = headloss due to friction in suction line including entrance losses, m; $v_s^2/2g$ = velocity head in suction line, m; h_d = static delivery head, m; h_{df} = headloss due to friction in discharge side of the pump, m; and $v^2_d/2g$ = velocity head in discharge side. The velocity of flow in suction line as well as delivery pipe is to be determined basing upon the cross sectional area of the pipes.

The head loss due to friction in the suction side is expressed as:

$$h_{sf} = \frac{4fl_s v_s^2}{2gd} + k_s \frac{v_s^2}{2g} + k_f \frac{v_s^2}{2g} \quad \text{... (3)}$$

where, f = coefficient of friction in suction line, fraction; l_s = length of suction pipe (assumed to be same as the static suction lift, m; g = acceleration due to gravity, m s^{-2}; d = diameter of the suction pipe, m; k_s = coefficient of friction in strainer, fraction; and k_f = coefficient of friction in foot valve, fraction. Average depth of groundwater table, as obtained from the collected data is considered as the length of suction pipe (l_s). Values of k_s and k_f are assumed to be 0.95 and 0.8, respectively. The coefficient of friction in both suction and discharge line is assumed to be same and its value taken in the present study is 0.005.

The head loss due to friction in the supply line is calculated as:

$$h_{df} = \frac{4fl_d v_d^2}{2gd} \quad \text{... (4)}$$

Where, l_d = length of supply pipe from the centre line of the pump to the free discharge line, m. For all the cases of irrigation pumps, the length of supply pipe is assumed to be 1 m, the diameter of the suction pipe is 8 cm and that of the discharge pipe is 7.5 cm for all types of pumps in the present study.

Selection of Centrifugal Pump : The parameters for selection of suitable centrifugal pumps for various categories of farmers in coastal and interior districts of the state include head, capacity, various efficiencies of the pumps, power requirement and economics *etc.* Careful study of head conditions and pump location may be useful in saving power over long period without a major increase in the initial cost of the project. Appropriate size of pipe based on estimated future loads is another example of how careful planning can be made to pay off in terms of operating economies. Selection of centrifugal pump for various categories of farmers has been carried out based on the criteria such as (i) power requirement by centrifugal pump for the designed discharge, (ii) comparison of pump characteristic curves of similar pumps of various manufacturers and (iii) matching of the performance curve with the system head curve to select the most efficient one.

Power requirement of centrifugal pump : Power requirement of a pump depends on the weight of water to be lifted per unit time against the total head and various pump efficiencies. Thus, the input horse power (IHP) of a pump is expressed as:

$$\text{IHP} = \frac{\text{WHP}}{\eta_{\text{pump}} \times \eta_{\text{drive}} \times \eta_{\text{motor}}} \qquad \text{... (5)}$$

where, (6)

where, $\text{WHP} = \frac{\gamma \times Q \times H}{75}$

where, WHP = water horse power, hp; γ = unit weight of water, 1000 kg m^{-3}; Q = volume of water lifted per unit time, $l\ s^{-1}$; H = total head of pumping, m.

Pump characteristic curve : Operating properties of the pumps are expressed either graphically through performance curves or through standard tables. These curves are developed basing upon the relationship of head with discharge, overall efficiency and power requirement under either a constant impeller diameter or at constant impeller speed. In absence of these curves, the performance of the pump can be read from standard table values supplied by the manufacturers. The pump among the lot which requires minimum power to operate at the design total head and capacity efficiently is selected. Apart from this, selection of the right size pump requires the knowledge of the use of the system head curve and pump performance curve. Static suction lift, well drawdown, head loss due to friction and operating head together form the gross total head of the pumping system. The system head curve superimposed on the pump performance curve helps selecting the most efficient pump with maximum efficiency (75-80%) and desired discharge at the estimated total head (Reddy *et al.*, 2001).

RESULTS AND DISCUSSION

Selection of centrifugal pumps has been made for the coastal (zone-I) and interior districts (zone-II) separately basing upon the cropping pattern, design discharge rate, total head available for pumping along with the operating parameters of various pumps obtained from different manufacturers as presented below.

Design discharge rate : Basing upon the depth of irrigation and rotation period of the crops grown during *kharif* and *rabi* seasons in both the zones, the design discharge rate is derived using Eq. 1. Further, the land holding size of each category of farmer is taken into account to derive the design discharge rate for the land holding size of marginal, small, medium and large farmers in respective zones (Table 2).

Table 2: Pump discharge rate required for various categories of farmers

Category of farmer	Land holding (ha)	Pump discharge rate (l s^{-1})	Final pump discharge rate adding 20% loss in conveyance (l s^{-1})
Marginal	1.0	6.61	7.93
Small	2.0	13.23	15.92
Medium	4.0	26.45	31.88
Large	5.0	33.06	39.68

Total Head for Pumping : Average depths of groundwater table based on its fluctuation in pre- and post-monsoon periods are found to be 3.57 and 5.03 m in zone-I and zone-II, respectively. Total head required for pumping has been estimated taking into account the suction lift, the discharge head (assumed as 1.0 m irrespective of the zones), head loss due to friction and velocity head in suction and delivery line (Table 3). The length of suction pipe is assumed to be equal to the average depth of groundwater table in the zone. The velocity of flow in both suction and delivery side are derived by dividing the cross sectional area of the pipe with the design discharge. The total head required for pumping is found to vary with the category of the farmer as shown in Table 3. It is found to be 5.03, 6.43, 11.98 and 16.11 m for the marginal, small, medium and large category farmers, respectively, in zone-I and 6.69, 8.73, 16.90 and 22.84 m, respectively, in zone-II. It is found that the head required for pumping is larger for each category of farmer in zone-II as compared to his counterpart in zone-I. The reason may be the higher suction lift in the zone arising out of the variation in depth of the groundwater table.

Table 3: Total head for pumping in both the zones of Odisha

Farmer category	Zone	Suction lift (m)	Discharge head (m)	Head loss due to friction (m)	Entrance losses (m)	Velocity head (m)	Total head (m)
Marginal	Zone-I	3.57	1.0	0.11	0.22	0.13	5.03
Small				0.46	0.89	0.51	6.43
Medium				1.83	3.53	2.05	11.98
Large				2.83	5.54	3.17	16.11
Marginal	Zone-II	5.03	1.0	0.22	0.28	0.16	6.69
Small				0.89	1.15	0.66	8.73
Medium				3.56	4.65	2.66	16.90
Large				5.51	7.19	4.11	22.84

The power requirement of centrifugal pumps in the form of BHP and input horse power (IHP) of the motor for various categories of farmers in both zones is given in Table 4. It may be seen that both the BHP and IHP are increasing from marginal to large farmer in both the zones. However, the power requirement of the pumps in zone-I is found to be lower than that of the pumps in zone-II for the same operational holdings. The reason may be attributed to the higher suction lift in zone-II than that in zone-I.

Table 4: Power requirement for the pump for various categories of farmers

Zone	Category of farmer	Design discharge ($l\,s^{-1}$)	Total head (m)	BHP requirement (hp)		Input hp to motor (hp)	
				Close coupled	Flexibly coupled	Close coupled	Flexibly coupled
I	Marginal	7.93	5.03	0.65	0.94	0.81	1.18
	Small	15.92	6.43	1.70	2.43	2.13	3.04
	Medium	31.88	11.98	6.35	9.08	7.94	11.35
	Large	39.68	16.11	10.64	15.21	13.30	19.01
II	Marginal	7.93	6.69	0.87	1.25	1.10	1.56
	Small	15.92	8.73	2.31	3.30	2.88	4.12
	Medium	31.88	16.90	8.97	12.82	11.21	16.02
	Large	39.68	22.84	15.09	21.57	18.86	26.96

Selection of centrifugal pump : Selection of right size of centrifugal pump has been made using the standard pump performance tables collected from the local dealers of five national level pump manufacturers. The manufacturers have been named as A, B, C, D and E, instead of their names in order to avoid personifying of the results. With the design discharge rate and total pumping head estimated for various categories of farmers of each zone, the corresponding power requirement and efficiency of the pumps has been computed and discussed below.

Size of pump for marginal farmers : Design discharge rate for the operational holding of a marginal farmer is 7.93 *lps* for both zone I and II. Thus, the input horse power requirement in case of a mono-block centrifugal pump to meet the irrigation need of the crops grown by a marginal farmer in zone-I and zone-II is 0.81 and 1.1 HP against total head of 5.03 and 6.69 m, respectively. Manufacturer 'A' claims that its one *HP* (input HP) size pump has a discharge capacity of 7.81 *lps* when the total head is 6 m. So, a centrifugal pump having input hp of one HP may be recommended for the marginal farmers of Odisha.

Size of pump for small farmers : The head requirement for small farmers in zone-I and zone-II is 6.43 and 8.73 m, respectively, and design discharge rate for the crops grown in their operational holdings is 15.92 *lps* irrespective of the zone. A comparison of performance tables of different manufacturers shows that a two HP pump of manufacturer 'A' gives a discharge rate of 15.97 *lps* against 6 m head whereas, a two HP pump of manufacturer 'E' gives 11.10 *lps* against 8 m head. Further, a three HP pump of manufacturer 'B' gives 16.5 *lps* discharge rate against a total head of 9 m. Thus, the manufacturer 'E' does not meet the water requirement of the farmer. However, the other two pumps show less variation in discharge rate with change in total head. Hence, the two HP centrifugal pump of the manufacturer 'A' and 3 HP centrifugal pump of the manufacturer 'B' are recommended for the small farmers in zone-I and zone-II, respectively, as they are capable of meeting the design pumping rate.

Size of pump for medium farmers : Total head available for the medium category farmers in zone-I and zone-II is 11.98 and 16.90 m, respectively. The design pumping rate for the crops grown in the operational holdings is 31.88 *lps* irrespective of the zones in Odisha. Under these conditions, a 7.5 HP pump of manufacturer 'B' gives 33 *lps* of water, a three hp pump of manufacturer 'D' gives 32 *lps* of water against 12 m of head. Between these two pumps, the three *HP* pump of manufacturer 'D' requires lower power input and its discharge variation is minimal with change in head. Hence, it is recommended for the medium category of farmers in zone-I.

Under total head of 18 m, a 12.5 HP centrifugal pump of manufacturer 'B' shows 33.83 *lps* discharge whereas a five *HP* pump of manufacturer 'C' delivers a discharge of 34 *lps*. Also the variation in discharge with fluctuation in head in case of five hp pump is within acceptable range. Hence, the five hp pump of manufacturer 'C' is recommended for the medium farmers of zone-II of Odisha.

Size of pump for large farmers : The design pumping rate required for the operational holdings of large farmers irrespective of the zones in the state is 39.68 *lps*. Total operating head available for the pumps is 16.11 and 22.84 m in zone-I and zone-II, respectively. The manufacturers other than 'C' and 'E' have not furnished data for the total available head and design discharge of large farmer in their performance tables. The manufacturer 'C' has developed a pump of 7.5 *HP* size which delivers a discharge of 41.7 *lps* against a head of 22 m. On the other hand, the manufacturer 'E' develops a pump of 20 *HP* that discharges at the rate of 42.7 *lps* against a head of 22 m. Thus, the 7.5 *HP* pump of manufacturer 'C' is recommended for the large farmers in both the zones of Odisha.

CONCLUSIONS

A marginal farmer of Odisha should possess a centrifugal pump of size one *HP* in order to lift water from underground unconfined aquifers for growing crops in his available operational holding. While a small farmer in coastal belt of the state should have a pump of size two *HP*, the same category of farmer in interior districts of Odisha should have a three *HP* pump to meet the irrigation requirement of the crop area in his operational holding. Again, while three *HP* pumps are recommended for the medium farmers in coastal belt, a five *HP* pump is required for the same category farmer in the interior districts. For large farmers in the state, a pump of size of at least 7.5 *HP* is recommended.

REFERENCES

Anonymous, 1983. Parliamentary committee report on energy consumption in agricultural pumpsets, Govt. of India, New Delhi.

Anonymous, 2009. Agricultural Statistics. Department of Agriculture, Government of Odisha, Bhubaneswar.

Church, A.H. 1972. Centrifugal Pump and Blower. John Wiley & sons, Inc. Newyork, 465p.

Reddy, K S., Gharde, K D.,Varshney, A C. and Bhandarkar, D M. 2001. Assessment of agricultural pumping systems in Chhattisgarh state. *Agric. Engng.*, 38(1): 64-69.

□□□

Chapter 86

Strategic Planning for Resource Conservation and Sustainable Rural Livelihood Security in Eastern India

BASUDEV BEHERA

INTRODUCTION

Livelihood comprises abilities, assets (stores, resources, claims and access) and activities required for a means of living. Livelihood is sustainable if it copes up with and recovers from stress and shocks, maintains or enhances its capabilities and assets and provides livelihood opportunities to the next generation and contributes net benefits to other livelihoods at the local and global level in the short and long term basis. Livelihood security refers to meeting the expenses of family on food, clothing, shelter, health, education for children, ceremonies, recreations, aesthetic needs of members so as to maintain a decent standard of living in the society. A rural family having six members including two children has to generate net income of more than Rs. 56,940 $annum^{-1}$ for livelihood security and comfortable living. The states in Eastern India viz. Odisha, West Bengal, Bihar, Jharkhand, Eastern UP and Chhattisgarh have population of 41947358, 80221171, 82998509, 26909428, 76500000 and 20795956, respectively. The zone receives 2-3 times more rainfall compared to the Northwest States. Rice based production system is predominant in the zone. This zone has unexploited good quality ground water aquifers and vast resources of social capital for sustainable production of crops. In spite of availability of abundant natural resources, the agricultural productivity in the zone is low due to several biophysical, technological, socio-economic and institutional constraints affecting the agricultural production. The various on farm and off farm livelihood includes agriculture, animal husbandry, handicraft, collection & sale of minor forest products, employment in public & private sector, daily wage earning and business strategies for improving livelihood in agriculture include varietal substitution, production of quality seed, crop diversification, soil water conservation, soil health

Department of Agronomy, College of Agriculture Orissa University of Agriculture and Technology, Bhubaneswar-751 003, Odisha, India management and farm mechanization. Farmers grow low yielding stress susceptible local genotypes of rice which give very low yield. Varietal substitution is essential. There is need to develop and promote biotic- stress (flood, drought and salt) tolerant rice varieties and their management practices. Availability of good quality seed is a serious issue. This can be partly addressed by training of farmers for quality seed production for their own use and also for sale or sharing with other farmers. Seed Village Programme in some of the states could be strengthened. Farmers should adopt remunerative crop(s) or cropping systems in place of less remunerative crop(s) or cropping systems. In upland areas, there is needed to grow deep-rooted non-paddy crops. Soil erosion and run off loss accounts for 25-40 % of the rainfall. Construction of on-farm water harvesting structures amounting to 10% of the farm land should be promoted to harvest run-off water. Soil acidity problems, nutrient deficit and toxicity, inadequate and imbalanced fertilizer application drastically reduce crop yield. Soil amelioration measures viz. use of paper mill sludge (PMS) for reclaiming soil acidity, Farm Yard Manure(FYM), proper method of fertilizer application, biofertilizer application and supplement of deficit nutrients should be adopted for enhancing land productivity. Animal productivity is low due to rearing of indigenous breeds, inadequate and imbalanced feed, disease incidence and mortality, inadequate health care facilities, uncongenial environment and faulty management and decrease in area of pasture land and shortage of green fodder. Strategies for enhancing animal productivity include breeding, feeding, health care and management.

Promotion of some ancillary activities can provide gainful employment. Enough rice straw is available for mushroom production. The climate is also congenial for raising paddy straw mushroom during *kharif* (8 months) and oyster (dhingri) mushroom during *rabi*(4 months). Bee keeping is a low cost, eco-friendly non-land based industry of immense importance for income generation in agriculture sector through increase in production and productivity of more than 80% of crop plants through efficient pollination. Honey bee is the best pollinator among all the bio-agents available in the globe. Forest based enterprises viz. leaf plate and cup-making, bamboo craft and incense –stick (*agarbathi* making) can provide gainful employment to the under-employed in rice based farming system. Farming system involving crop-tree-animal integration has become essential for small and marginal farmers due to shrinkage in area under cropping and need for meeting household requirement from limited holding, small and fragmented land holdings, seasonal nature of income and employment in rainfed belt and out-migration and deterioration of resource base. Promotion of Self help groups can play a major role in alleviation of poverty and improving the livelihood of rural poor. In Eastern India, SHGs have been successfully adopted women-friendly technologies like mushroom cultivation, apiculture, leaf-tray making, vegetable cultivation, bamboo-craft and incense-stick making.

MATERIALS AND METHODS

Integrated livelihood index : Planning Commission, Govt. of India have identified 150 districts as disadvantaged. The districts have been identified by developing an Integrated Livelihood Index (ILI) based on six indices (in turn based on 53 parameters). The six component indices are:

i) *Infrastructure Index*: It includes percentage of inhabited villages having different types of communication facilities, percentage of inhabited village having *pucca* approach roads and percentage villages not linked with roads. Other indices considered are percentage of inhabited villages having post and telegraph offices and telephone connection and per thousand villages having different government development programmes / schemes.

ii) *Agricultural Status Index*: It includes productivity of rice, wheat, pulses, oilseeds, cotton, sugarcane, fruits vegetables per hectare. Other parameters included are productivity of meat, milk and eggs per animal. Cropping intensity, irrigation and fertilizer are also considered

iii) *Nutritional Status Index*: It includes consumption of rice, wheat, cereals, pulses, eggs, fish, milk and milk products per capita.

iv) *Economic Status Index*: It includes per capita income and percentage of population below poverty line.

v) *Health and Sanitation Status Index*: It includes per capita expenditure on health, water supply, sanitation and family welfare. Per thousand households having *katcha*, semi *pucca* dwelling units and at a distance less than 0.5 km from hospitals/ health centers are other parameters including percentage of villages having medical institutions.

vi) *Food Availability Status*: It includes per capita availability of rice, wheat, pulses, oilseeds, sugarcane, fruits, vegetables, meat, milk and eggs.

RESULTS AND DISCUSSION

Classification of livelihood : Various means of livelihood are included under three sectors *viz.* production, service and trade. The livelihoods producing goods *viz.* agriculture, animal husbandry and handicraft come under production sector. Tailoring, black smithy, gold smithies are included under service sector. People engaged in trade sector generate income through marketing of produce/goods. Livelihoods may be on-farm or off - farm, depending on whether it is based on land or not. Livelihood options in rural areas of Eastern India include agriculture, animal husbandry, handicraft, employment in public and private sector, collection and sale of minor forest products, daily wage earning and business.

Agricultural scenario in Eastern India : The states in Eastern India receive 2-3 times more rainfall compared to the Northwest States. Rice based production system is predominant in the zone. This zone has unexploited good quality ground water aquifers and vast resources of social capital for sustainable production of crops. In spite of availability of abundant natural resources, the agricultural productivity in the zone is low due to several biophysical, technological, socio-economic and institutional constraints affecting the agricultural production. Biological constraints like use of long duration low yielding crop varieties, disease and insect pest susceptibility of crops and problem of wild animal's *viz.* elephants, monkeys, and jackals reduce crop productivity. Physical constraints are climatic and edaphic. Climatic constraints viz. high variability in commencement and cessation of SW monsoon, erratic rainfall distribution- spatially & temporally, frequent and prolonged dryspells causing drought and depressions and cyclones leading to high intensity rainfalls and floods make agriculture risk prone and uncertain. Predominant edaphological constraints are soil crusting and poor crop emergence, rapid surface drying and poor crop establishment, low WHC and profile storage, light texture, leaching of nutrients, low organic carbon and cation exchange capacity, nutrient deficit and toxicity, dry and hard soil in uplands causing difficulty in harvest of root crops and intercultural operations, hard pan, sloping topography, high runoff loss (26-40%), soil erosion - 10t ha^{-1} annually and washing away of applied fertilizers, salinity and water logging problems in deltaic areas,. Important technological constraints include improper and soil water conservation practices, shallow and along the slope tillage, improper crop rotation and lack of organic matter recycling, untimely sowing and poor crop management and lack of intercultivation and weed management. Socio- economical constraints viz. fragmented and scattered holdings, low borrowing and investment capacity of farmers, low literacy, low draft power, lack of subsidiary enterprises, over grazing, deforestation and lack of vegetation cover restrict adoption of high-tech agriculture. The institutional constraints viz. inadequate infrastructural facility(credit, input supply / market insurance etc), timely non-adoptability / non-availability of technology, disparity in public investment, sectored approach among different developmental agencies should be tackled for making a breakthrough in agriculture in Eastern India.

The hike in food grains production from 97.32 million tons in 1950-51 to 233.88 M t during 2008-09 in India was mainly from the traditional food bowl areas of Punjab, Haryana and western Uttar Pradesh. The sustainability of livelihood and food security in long run from the zone is under threat due to over exploitation of natural resources like ground water and soil along with excessive consumption of energy. There is urgent need to reduce the over exploitation of the natural resources in North West region and exploit untapped potential of the resource rich Eastern India to increase total agriculture production of India.

Strategies for making agriculture livelihood more remunerative varietal substitution : Farmers grow low yielding stress susceptible local genotypes of rice which give very low yield. Varietal substitution is essential. There is need to develop and promote biotic- stress (flood, drought and salt) tolerant rice varieties and their management practices. Out of 266.2 lakh ha rice area in eastern India (UP, Bihar, Jharkhand, West Bengal, Assam, Orissa and Chhattisgarh), approximately 143.4 lakh ha is rain fed and prone to different abiotic stresses like flood, drought or soil salinity/sodicity. Approximately 50 lakh ha and 87.8 lakh ha rice area in eastern India frequently encounters flash-flood and drought, respectively. Approximately 6 lakh ha rice area is affected by coastal and 7.5 lakh ha by inland salinity/sodicity. These abiotic stresses are the most important yield limiting factors for eastern India. Rice productivity of eastern India is very low (1.1 to 2.5 tons ha^{-1}) and also fragile. Apprehending these stresses, farmers use little inputs which further aggravate the situation. In recent years there has been major advancements in developing high yielding rice varieties which can tolerate the flash flood, drought or high soil salt concentration. Also some nutrient and crop management technologies have been identified which can further enhance the stress tolerance and productivity of stress tolerant rice varieties both under normal and stress conditions. Promotion of these stress tolerant rice varieties along with their management practices will enhance and stabilize the rice productivity. This will also encourage the farmers to use more inputs in stress-prone areas. Different strategies include characterization of stress-prone areas up to village level in different states using GIS & remote sensing, ground information, revenue maps etc, rapid seed multiplication of stress tolerant rice varieties by involving research institutions, seed corporations/companies, state government farms and through schemes like seed village etc, rapid targeted dissemination of flood, drought or salt tolerant rice varieties as seed minikits or other methods like subsidized seed cost, demonstration etc, development of management practices for these varieties as leaflets/information brochure for distribution with minikits and other extension programmes and tracking of the varieties and impact assessment. Promotion of stress tolerant rice varieties along with their management practices can increase rice productivity by 1 t ha^{-1} in eastern India.

Production of quality seed : Availability of good quality seed is a serious issue. This can be partly addressed by training of farmers for quality seed production for their own use and also for sale or sharing with other farmers. Seed Village Programme in some of the states could be strengthened.

Crop diversification : Farmers should adopt remunerative crop(s) or cropping systems in place of less remunerative crop(s) or cropping systems. In upland areas, there is needed to grow deep-rooted non-paddy crops. Non-paddy crops *viz.* cereals (maize and millets), pulses, oilseeds, vegetable crops, tuber crops, spices, fibers, fruit trees, flowers, medicinal and bio-diesel crops are more remunerative under rain fed high land situation as compared to rice. Tuber crops like yam, elephant foot yam, sweet potato, tapioca, colocassia and arrowroot tolerate prolonged drought, besides giving high productivity. Further, a sustained economic growth, rising per capita income and growing urbanization are apparently causing a shift in the consumption patterns towards high-value food commodities like fruits, vegetables, mushroom, dairy, poultry, meat and fish products from staple food such as rice, wheat and coarse cereals. Such a shift in consumption patterns towards high-value food commodities indicates an on-going process of agricultural diversification the eastern states of India. Extent of tree based farming is an indicator of stability in rainfed agriculture. Rain fed land in Eastern India is suitable for growing various fruit crops viz. mango, jackfruit, litchi, sapota, lemon, custard apple, pomegranate, ber, guava, tamarind, anola, blackberry, cashew nut and pine apple. Growing of leguminous crops (green gram, black gram, pigeon pea and groundnut) helps in weed control and soil fertility build-up besides giving yield and monetary return during initial gestation period 5-6 years. Silvi-pastoral systems comprising trees viz. subabul and shisham and fodder crops as *Cenchrus* and *Stylo* provide round the year green fodder for the cattle (Zhang *et al.*, 2008).

Soil water conservation and soil health management : Soil erosion and run off loss, soil acidity problems, nutrient deficit and toxicity, inadequate and imbalanced fertilizer application drastically reduce crop yield. Construction of on-farm water harvesting structures amounting to 10% of the farm land should be promoted to harvest run-off water. Soil amelioration measures *viz.* use of paper mill sludge(PMS) for reclaiming soil acidity, Farm Yard Manure (FYM), proper method of fertilizer application, bio fertilizer application and supplement of deficit nutrients should be adopted for enhancing land productivity.

Farm mechanization : Establishment of machinery hubs in KVKs and Custom Hiring Centres in villages will ensure timeliness of agricultural operations and profitability from farms.

Strategies for making animal husbandry livelihood more remunerative : Animal productivity is low due to rearing of indigenous breeds, inadequate and imbalanced feed, disease incidence and mortality, inadequate health care facilities, uncongenial environment and faulty management and decrease in area of pasture land and shortage of green fodder. Strategies for enhancing animal productivity include breeding, feeding, health care and management (Tesfaye *et al.*, 2005). Broiler farming with proper feeding, health care and management under National Agricultural Innovation Project (Component 3) sub-project in three disadvantaged districts of Odisha, namely, Kandhamal, Dhenkanal and Kalahandi proved remunerative. Twenty beneficiaries including women were given 100 broiler birds each with feed, medicine and vaccines. The average weight of the birds at the age selling (35-42 days) ranged between 1.68 to 2.01 kg. The food conversion ratio (FCR) varied from 1.54 to 1.82. The mortality of the bird's upto marketing varied from 2 to 5%. The farmers got net profit of Rs. 1100 to 1200 per batch of 100 birds. Rearing of 'Black Bengal' breed of goat in Kandhamal and Kalahandi of Odisha under semi-intensive management system was profitable.

Ancillary activities : Rice is the staple food crop in Eastern India. Enough straw is available for mushroom production. The climate is also congential for raising paddy straw mushroom during *kharif* (8 months) and oyster (*dhingri*) mushroom during *rabi* (4 months). A shed of 25′ x 12 ′with tier system can accommodate 120 paddy straw mushroom beds and 225 oyster bags. Production from 120 paddy straw mushroom beds will be 180kg @1.5 kg/ bed/month. The net return will be ₹ 3850 per month. Production from 225 oyster bags will be 450 kg per 2 months and the net return will be ₹ 8200 per two months or ₹ 4100 per month. Bee keeping is a low cost, eco-friendly non-land based industry of immense importance for income generation in agriculture sector through increase in production and productivity of more than 80% of crop plants through efficient pollination. Honey bee is the best pollinator among all the bio-agents available in the globe. Forest based enterprises viz. leaf plate and cup-making, bamboo craft and incense –stick (*agarbathi* making) can provide gainful employment to the under-employed in rice based farming system.

Farming system approach : A farming system is defined as the population of individual farm units that have broadly similar, resource bases, enterprise patterns, household livelihoods and constraints, and for which similar development strategies and interventions would be appropriate. Farming system involving crop-tree-animal integration has become essential for small and marginal farmers due to shrinkage in area under cropping and need for meeting household requirement from limited holding, small and fragmented land holdings, seasonal nature of income and employment in rainfed belt and out-migration and deterioration of resource base.

Promotion of self help groups : Self help groups can play a major role in alleviation of poverty and improving the livelihood of rural poor. These are small economically homogenous affinity groups of rural poor voluntarily coming together to save small amount regularly which are deposited in a common fund to meet members' emergency needs and to provide collateral free loans as decided by the group. They also deposit their savings in banks and acquire easy access to bank loans with low rate of interest to start profitable enterprises. These are informal associations of 10 to 20 persons from the underprivileged sections of the society. The members voluntarily come together based on affinity with a common objective of achieving their socio-economic development. They promote participation, savings and credit to achieve the set objectives, The SHGs are developed in pattern of Grameen Bank of Bangladesh founded by Mohammed Yunus.In Eastern India, SHGs have been successfully adopted I women-friendly technologies like mushroom cultivation, apiculture, leaf-tray making, vegetable cultivation, bamboo-craft and incense-stick making.

REFERENCES

Tesfaye, A., Jemal, I., Ferede, S. and Curran M.M. 2005. Technology Transfer Pathways and Livelihood Impact Indicators in Central Ethiopia. *Tropical Animal Health and Production*, 37: 101-122.

Zhang, L., Zhang, Y., Yan, J. and Wu, Y. 2008. Livelihood diversification and cropland use pattern in agro-pastoral mountainous region of eastern Tibetan Plateau. *Journal of Geographical Sciences*, 18(4): 499-509.

□□□

Chapter 87

Confidence Building Innovative Entry Point Activities for Sustained Participation of Marginalized Societies in Watershed Development Programmes

S.L. PATIL, S.K.N. MATH, R.N. ADHIKARI, A. RAIZADA, K.K. REDDY AND K. CHANNABASAPPA

INTRODUCTION

Natural resource management is a challenging task in the present world of climate change especially in the fragile environment of Semi-Arid Tropics (SAT) of world and in the Vertisols and Alfisols of India with rainfall less than 750 mm annually. Farming in the SAT in south India is complex, diverse, risk prone and cultivation of crops is largely rainfed with many areas experiencing problems associated with variable rainfall, land degradation and poor soils. Land degradation in SAT of India affects over 166 M ha and it accounts for nearly 50% of the total geographical area of the country (Sehgal and Abrol,1992). Erosion by water is the single most important factor affecting an area of 86.9 M ha and run-off water carries fine silt and clay particles rich in plant nutrients and organic carbon leaving behind impoverished soils. Conservation of scarce rainfall and top fertile soil becomes more critical in arable lands for greater productivity on sustainable basis especially during low rainfall and drought years.

Development of environmentally sustainable farming practices that requires to be implemented with resource constrained small farmers that gives them greater profits is a real challenge and arduous task in the climate change/ resilience in the SAT region. Further, greater concern in Asia and other developing countries is the need to increase food production to keep pace with population growth. Large percentage of available land is marginal in nature and much of the land holdings are managed by small farmers who experience severe economic constraints. The efficient use of rainfall is the key to maximizing agricultural production in the SAT. Greater increases in food

---

Central Soil and Water Conservation Research and Training Institute, Research Centre Bellary–583 104, Karnataka, India production are possible by using optimum management practices and productivity on farmers' fields and it could be doubled in Asia and quadrupled in Africa using available technologies especially in the watersheds as reported by Wani *et al.* (2003) and Rockstrom *et al.* (2007).

Research and development works in SAT mainly emphasized on increasing the availability of water for crops through rainwater conservation measures besides reducing soil erosion with implementation of various soil and water conservation measures/structures within the watersheds but without participatory approach through farmers' involvement. This approach of watershed developmental works without farmers participation has neither provided a positive impact on the productivity nor encouraged the farmers to participate in development and management of watersheds and maintain these structures when the implementing agency withdrew the support mainly because only a few resourceful farmers benefited from the programme (Wani *et al.,* 2003a). Recent experiences by both researchers and development functionaries in conservation of natural resources through watershed management indicated that participatory approach of beneficiaries, *i.e.,* people's participation leads to successful of implementation of watershed (Olson, 1971; Wani *et al.,* 2003; Sreedevi *et al.,* 2004).

In view of this, under new guidelines of National Watershed Development Project for Rainfed Areas (NWDPRA) programme subsumed under Macro Management of Agriculture (MMA) Scheme funded by Department of Agriculture & Cooperation under Ministry of Agriculture, Government of India in XI Five Year Plan, a Ramasagara watershed project covering four villages *i.e.,* Ramasagara, Hanumapur, Venktapur and Devasamudra in Molakalmur taluk of Chitradurga district in Karnataka State was implemented in 480 ha from 2008 to March 2012, in a participatory mode with greater peoples participation.

Entry point activities : Under the new guidelines of NWDPRA, nearly 5% of the total watershed budget is earmarked for Entry point activities (EPA) to be spent for the needs of the watershed/village beneficiaries in the first year of preparatory phase and to develop their confidence in implementing all watershed activities planned under watershed development project.

The EPA is envisaged to build a rapport between project implementing agency (PIA) and rural people before initiating watershed programs. Usually, entry point intervention/activity is identified through PRA and Grama Sabha meetings. It involves building rapport with community, strengthening and sustaining it through out the program and thereafter. In reality it involves a lot of time and resources of project staff and could determine the success or failure of program. A carefully laid out EPA could address community's aspiration and capitalize on it by ensuring larger peoples' participation in all watershed activities. Certain goodwill activities such as creating drinking water facility and sanitation to the community, conducting health awareness camps, construction of community halls or bus shelters, promotion of kitchen gardens were carried out as EPAs'. Support to group income-generating activities such as fish farming in village tanks and providing power threshers with community contribution were some other rapport building measures practiced by the implementing agencies (Fernandes, 2000). In Pakistan construction of a *bundh* (embankment), repairing of roads, construction of small culverts, supply of school furniture, chemical control of ipomoea (a weed) in a river and construction of culvert were taken up as EPA in watershed development program. Generally EPAs were restricted to community needs for a short period, and was not fully successful to provide greater incentives for motivating people to participate in long-term conservation activities that provide no tangible benefit in short run as critically analyzed by FAO (www.fao.org). Detail analysis of watershed projects in India revealed that a major reason for low impact of projects was the lack of equity of the benefits shared by beneficiaries (Wani *et al.*, 2003). Till the date, reasons for failure of watershed projects were top-down approach as per project plan without considering needs of beneficiaries that led to low community level participation especially small and marginal farmers and landless who form a major part of stakeholder community and this approach affected project sustainability (Wani *et al.*, 2005).

MATERIALS AND METHODS

Study area : Ramasagara watershed is located at 14⁰49'31" N to 14⁰50'42" N latitude and 76⁰47'32" E to 76⁰49'16" E longitude. Micro watershed of Ramasagara is located in Venkatapura mini watershed and is a sub watershed of Kanapuram watershed and is a unit of Chinnahagri main watershed (Figure 1). Perimeter of the Ramasagara watershed is 10667.7 m. The Ramasagara watershed Code is 4D3C6A2c. Watershed has diversified land uses *i.e.,* both rainfed and irrigated, pasture land (open scrub) and hillock with degraded forest. Total cultivated area in the watershed is 355.48 ha (74% of area) of which 63.2% (303.48 ha) is under rainfed and 10.8% (52 ha) is under irrigation from bore wells. Cultivated crops are supplemented with protective irrigation in border strips. Beneficiaries of watershed belong to Ramasagara (86 beneficiaries), Devasamudra (50 beneficiaries), Hanumapur (19 beneficiaries), and Venkatapur (20 beneficiaries). Nearly 20 landless beneficiaries exist in watershed.

Identification of entry-point-activities (EPA) **:** Generally EPAs' were identified through *Grma Sabha* and watershed stakeholders meet to meet the immediate needs of the community instead of individual needs. Moreover EPAs' are so chosen and done as to develop the rapport between the project implementing agency (PIA) and the rural people before initiating watershed programs. While deciding the activities of EPA following criteria was considered for successful implementation of the Ramasagara watershed project.

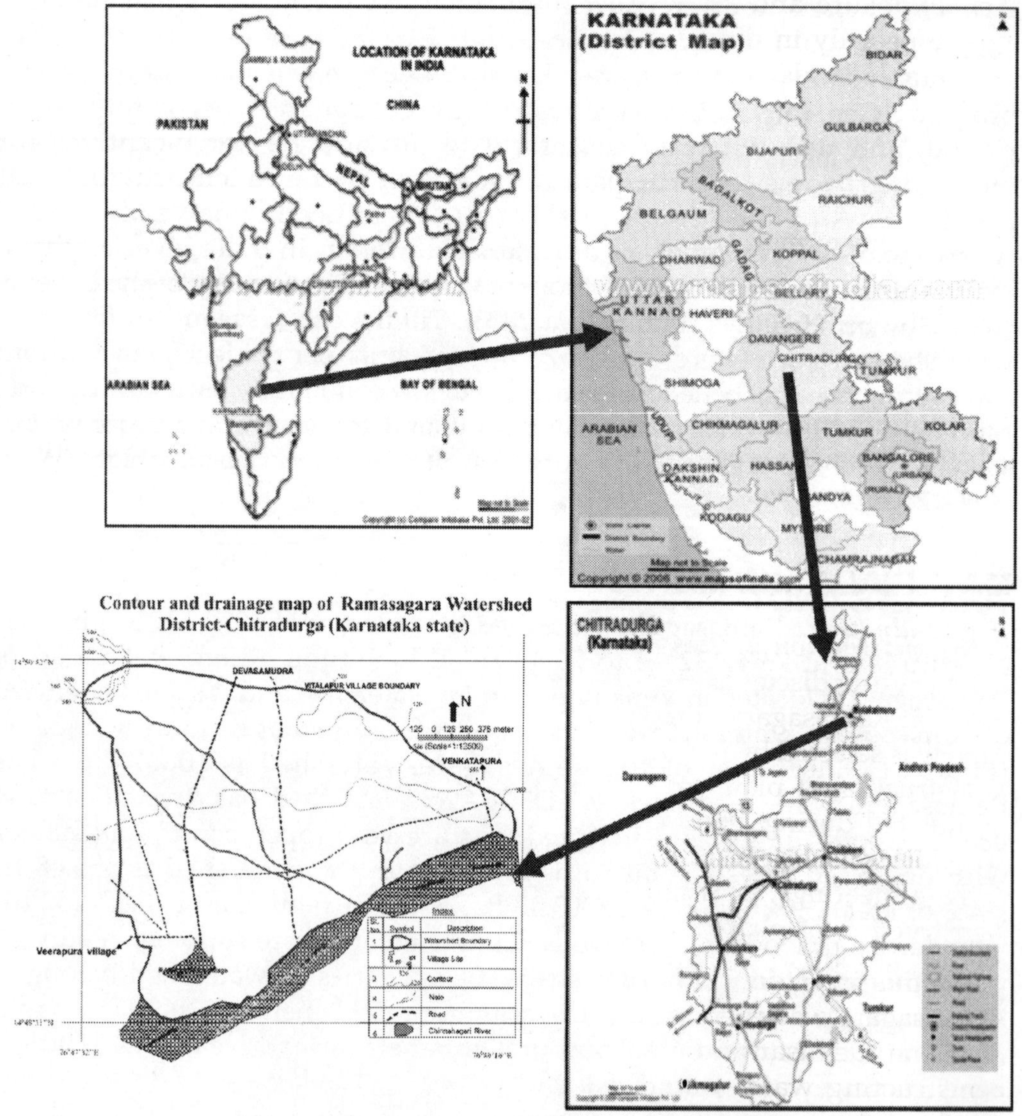

Fig. 1: General View of Ramasagara watershed

- Community need of the village that involve higher stakeholders in watershed.
  - Village platform was constructed in front of community hall.
- Community need of the area where SC/ST colony was located in the watershed village with special emphasis on daily basic needs of animals.
  - Two animal water troughs at Ramasagara village and SC/ST colony of Devasamudra village.
- Exposure and knowledge improvement of watershed beneficiaries especially in the land based activities covering both arable and non-arable lands.
- Exposure visit to Research Farm and training programme on soil and water conservation and watershed management at CSWCRTIs' Research Centre, Bellary, Karnataka State, India.
  - Both human and animal health was considered as priority.
- Animal Health camps at Ramasagara and Devasamudra villages and Human Health camp at Ramasagara.
  - Soil health and irrigation water quality was also given top priority.
- Representative soil samples in upper, middle and lower reaches of watershed were analyzed for soil health and all the bore wells in the watershed were analyzed for their quality and its suitability for irrigation to crops.

RESULTS AND DISCUSSION

Construction of masonry platform for village meetings at Ramasagara village: After discussion with beneficiaries of watershed and meetings of *Grama Sabha* at Ramasagara village, site in front of Ramasagara Village Community Hall was selected by Ramasagara villagers for constructing masonry platform. Construction of platform was initiated in the last week of October and was completed during December 2009 with measurement of 12.6 m length, 6.10 m width and height of 0.60 m. Platform after completion is being used for village meetings, dramas and marriages (Photo 1). Total cost incurred for construction of platform was ₹ 1,14,172/-.

Construction of water trough for animals : As per resolution of Ramasagara watershed society, one site each was selected at Harijan colony (S.C.) at Devasamudra village and other site at Ramasagara village for constructing water trough for cattle. Construction of water troughs was initiated during second week of December 2009 and completed in January 2010. Length, width and height of the water troughs are 4.8 m, 2.1 m and 0.6

m, respectively at both the villages, with stone slab pitching at Ramasagara village only (Photo 2). Water troughs served as drinking water points for cattle and cost incurred for construction water troughs were ₹ 19,366/- at Devasamudra and ₹ 21,115/- at Ramasagara villages. On an average 100 animals in both villages use this facility for drinking water daily.

Photo 1: A view of platform during construction phase and after its construction at Ramasagara

Devasamudra village Ramasagara village

Photo 2: Final view of the water troughs at the Ramasagara watershed

Animal health camp : Animal Health Camp was organized by Research Centre, Bellary on 29th June 2009 at Ramasagara village as an entry point activity in collaboration with Animal Husbandry Department, Molkalmur, Chitradurga district, Karnataka State (Photo 3). In this program, 31 infertile cows and buffaloes were vaccinated for *Haemorrhagia septicimia*. In Ramasagara village more than 2000 sheep and goats were vaccinated with PPR and were treated for parasite de-worming. Nearly 600 cattle and buffaloes were also vaccinated. During Animal Health Camp, Dr Shivarudrappa, Assistant Director, Veterinary, enlightened farmers about common seasonal diseases that affect animal health especially in June with onset of monsoon and their remedial measures for maintaining animal health. On this occasion, Dr S.L.Patil, Project leader of the Ramasagara watershed has emphasized that animal component in watershed management plays a major role on the income of the farmers' in these villages/region.

Photo 3: Animal Health Camp at Ramasagara watershed as entry point activity

Human health camp at Ramasagara village : Human Health Camp was organized in Ramasagara Watershed by CSWCRTI, Research Centre, Bellary in collaboration with Primary Health Centre, Rampur and Ashok Siddapur and Ayurvedic Hospital Devasamudra on 22.12.2009 at Ramasagara village (Photo 4). Dr S.L. Patil, the PI of the watershed explained farming community that the human health is important for greater productivity in the region and human health is also given greater importance in Ramasagara Watershed Programme. Hence, Human Health Camp was included in the watershed activity as per the new NWDPRA guidelines of the Ministry of Agriculture, New Delhi. The Physicians examined 100 farm families and treated them for malaria, acute respiratory tract infection, acute diarrhoea diseases, worm infestation, anti-natal cases, *Pyoderma*, viral fever, scabies and sexually transmitted diseases. Blood smear examination was also conducted for patients, and the medicines were distributed to all the farm families who had undergone medical check up and required treatment. Prevalence of some widespread diseases in patients was due to lack of hygienic and clean drinking water and it is a major cause of poor health among watershed farmers. Dr R.K. Farahana, enlightened watershed beneficiaries on the cleanliness, personnel hygiene, importance of nutritious food to the women especially to the pregnant women and children and in addition she also said that lack of education and poverty has resulted in ignorance in health awareness among the farmers in the Ramasagara watershed.

Photo 4: Doctors treating patients and blood smear tests in Human health camp at Ramasagara

Soil analysis : Soil samples were collected from 35 farmers fields representing both irrigated and rainfed areas in the upper, middle and lower reaches in top soil (0–15 cm) and sub soil (15–30 cm) were collected and analyzed for soil chemical properties *i.e.*, pH, E.C, Organic Carbon, available Nitrogen, available Phosphorus and available Potassium.

Soil chemical properties in 0–15 cm soil depth : The pH of soil in rainfed area varied from 8.0 in the middle reach to 8.4 in upper reach indicating that these soils are normal in pH, whereas in irrigated area, soil pH varied from 8.2 in middle reach to 8.5 in upper reach and 9.0 in lower reach. Continuous irrigation with bore wells has resulted in increased in pH in irrigated area from normal to alkaline (Table 1). Electricity conductivity (E.C) of these soils in rainfed area was normal and varied from 0.08 dS m^{-1} in upper reach to 0.12 dS m^{-1} in lower reach. Continuous irrigation has resulted in accumulation of salts in the soil profile in irrigated area in all reaches of watershed. The E.C has increased from 0.33 dS m^{-1} in upper reach to 0.54 dS m^{-1} in middle reach and to 0.79 dS m^{-1} in lower reaches clearly indicating that the salts have leached from upper reach to lower reach. Organic carbon of soils in rainfed area was low and varied from 0.32% in lower reach to 0.44% in middle reach. Organic carbon was higher in irrigated area and was medium and varied from 0.44% in upper reach to 0.66% in middle reach. Available nitrogen of soil was medium both in rainfed and irrigated areas in upper, middle and lower reaches. In general, medium available N in these soils of watershed was attributed to mono-cropping of groundnut in rainfed area and cultivation of crops that contribute more root biomass with application of farmyard manure resulted in medium available of N in irrigated area even through farmers apply little to medium N to the crops in rainfed and irrigated areas respectively. Available N varied from as low as 297 kg ha^{-1} in rainfed of lower reach to as high as 387 kg ha^{-1} in irrigated area of middle reach. Available Phosphorus varied from low to medium in both irrigated and rainfed area of watershed. Available phosphorus was low in lower reach and varied from

14.7 to 19.5 kg ha^{-1} in lower reach and was higher in middle reach and varied 31.8 kg ha^{-1} (rainfed) to 45.4 kg ha^{-1} in middle reach. Available potassium was low to medium with greater available K observed in middle reach.

Table 1: The average values of soil properties of soil samples collected from representative fields of beneficiaries in Ramasagara watershed (0–15 cm)

Quality	Upper reach		Middle reach		Lower reach	
	Irrigated (Bore wells)	Rainfed	Irrigated (Bore wells)	Rainfed	Irrigated (Bore wells)	Rainfed
pH	8.5	8.4	8.2	8.0	9.0	8.3
E.C. (dS m^{-1})	0.33	0.08	0.54	0.09	0.79	0.12
Organic carbon (%)	0.48	0.35	0.66	0.44	0.65	0.32
Available N (kg ha^{-1})	320	353	387	334	365	297
Available P_2O_5 (kg ha^{-1})	19.2	29.5	45.4	31.8	19.5	14.7
Available K_2O (kg ha^{-1})	229	378	430	405	471	218

Soil chemical properties in 15–30 cm soil depth :In general, values of pH, E.C and nutrient status was lower in sub soils compared to top soil. The pH of soil was normal in both irrigated and rainfed areas in upper, middle and lower reach, except, in irrigated area of lower reach, pH was alkaline (8.9). Continues irrigation with bore wells might have resulted in increased pH in irrigated area from normal to alkaline (Table 2). Electricity conductivity of soil in rainfed area was normal and varied from 0.09 dS m^{-1} in lower reach to 0.11 dS m^{-1} in upper reach. Irrigation from bore wells with higher EC resulted in accumulation of salts in soil profile in irrigated area in all reaches of watershed. The E.C. in soil increased from 0.22 dS m^{-1} in upper reach to 0.41 dS m^{-1} in middle reach and further increased to 0. 59 dS m^{-1} in farmers fields of lower reach. Increase in E.C. of soil from upper to lower reaches indicates that the salts have leached from upper reach to lower reach. Organic carbon of soil in rainfed area was low and varied from 0.21% in lower reach to 0.35% in middle reach. Organic carbon was higher in irrigated area compared to rainfed areas in watershed and it was attributed to application of greater quantity of farmyard manure and incorporation of greater root biomass and leaves fall into the soil. Organic carbon was low and varied from 0.35% in lower reach to 0.43% in middle reach. Available nitrogen of soil was medium in both rainfed and irrigated areas in upper, middle and lower reaches. In general, higher available N, *i.e.*, medium available N was attributed to cultivation of mono-cropping of groundnut in rainfed area and cultivation of crops that contribute more root biomass with application of greater quantity of farmyard manure as a resulted in medium available of N in irrigated area even through farmers apply lower N rates to the crops both in rainfed and irrigated areas. Available N varied from as low as 258 kg ha^{-1} in rainfed area of middle reach to 271 kg ha^{-1} in irrigated area of upper reach. Available phosphorus varied from low to medium in both irrigated and rainfed areas of watershed. Available phosphorus was low in lower reach and varied from 14.6 to 19.3 kg ha^{-1} in lower reach and was higher in middle reach and varied 31.4 kg ha^{-1} (rainfed) to 43.9 kg ha^{-1} (irrigated). Available K was medium with higher available K was observed in middle reach compared to upper and lower reaches of farmers fields in watershed.

Table 2: Average values of soil properties of the soil samples collected from representative fields of beneficiaries in Ramasagara watershed (15–30 cm)

Quality	Upper reach		Middle reach		Lower reach	
	Irrigated (Bore wells)	Rainfed	Irrigated (Bore wells)	Rainfed	Irrigated (Bore wells)	Rainfed
pH	8.0	8.3	8.2	8.0	8.9	8.3
E.C. (dS m^{-1})	0.22	0.11	0.41	0.10	0.59	0.09
Organic carbon(%)	0.37	0.32	0.43	0.35	0.35	0.21
Available N(kg ha^{-1})	267	272	259	258	271	271
Available P_2O_5(kg ha^{-1})	19.0	29.1	43.9	31.4	19.3	14.6
Available K_2O(kg ha^{-1})	376	394	420	383	332	334

Quality of irrigation water in bore wells : Water samples from all 33 bore wells were collected from upper, middle and lower reaches of watershed and analyzed for quality. Water quality parameters that were analyzed include *i.e.*, pH, E.C, calcium, magnesium, potassium, sodium, carbonate, bicarbonate, chlorine, fluorine and nitrate. From these chemical properties of water, Sodium Adsorption Ratio (SAR) and Residual Sodium Carbonate (RSC) were calculated. The pH of irrigation bore well water was less than 8 (7.9) in middle reach indicating that it is good quality water (Table 3). The pH of bore wells water in upper reach is 8.1 and in lower reach is 8.3 and the quality of water is classified as fair as it is more than 8. Lower salt concentration in irrigation bore well water in upper reach has resulted in E.C. of 1.22 indicating that water quality from these bore wells is good and is safe for irrigation to crops having less detoriation in soil quality with less salts accumulation in soil profile after irrigation. Quality of irrigation water from bore wells of middle reach are classified as fair with E.C. of 1.82 that have higher salt concentration than the quality of water from bore wells of upper reach. Water from bore wells of lower reach can only be used as a protective irrigation to crops under unavoidable circumstances as the salt concentration of lower reach is higher *i.e.*, 3.27 and quality of irrigation water is classified as poor. Even if crops are irrigated the quantity of irrigation water to be used as a protective irrigation may be reduced so that the soil quality is not detoriated over years. Chloride content in irrigation water varies from 2.74 m eq l^{-1} (upper reach) to 6.30 m eq l^{-1} (lower reach). It is safe to irrigate when the chloride content of the irrigation water is less than 4.0. Sodium adsorption ratio (SAR) of irrigation water is <0.43 in irrigation water of bore wells located in upper, middle and lower reaches. The SAR of irrigation water in all the bore wells in watershed is excellent. Residual sodium carbonate (RSC) of irrigation bore well water in upper reach is excellent (-0.58), where as RSC of bore well water in middle and lower reaches were 1.76 and 1.86 indicating that quality of irrigation water is fair and trends towards poor quality. Sodium percentage of quality of irrigation water of all bore wells in watershed is excellent as values for percentage for sodium varies from 11.4 in middle reach to 12.9 in lower reach.

Table 3: Average values of water quality in bore wells of beneficiaries in Ramasagara watershed

Quality	Upper reach	Middle reach	Lower reach
pH	8.01	7.09	8.03
E.C. (dS m^{-1})	1.22	1.82	3.27
Cl (m eq l^{-1})	2.74	4.16	6.30
SAR	0.43	0.42	0.45
RSC	-0.58	1.76	1.86
% Sodium	12.07	11.04	12.09

Training programmes : Two training programmes were conducted from 17th to 18th March and 19th to 20th March for 40 beneficiaries with 20 beneficiaries in each training programme. In training program, beneficiaries of watershed were exposed to soil and rainwater conservation and watershed management. Beneficiaries were exposed to different terrace and inter-terrace rainwater conservation measures in arable and non arable lands for control of soil erosion. Interactions with farmers on integrated nutrient management, improved crops and cropping systems in watershed for higher yields were beneficial. Beneficiaries expressed that they were willing to adopt improved crop cultivation practices that results in greater conservation of natural resources and higher crop yields. Farmers were also exposed to the live soil and water conservation measures adopted for black soils at the Research Farm.

CONCLUSIONS

Nearly 5% of earmarked budget for innovative entry point activities (EPA) under new NWDPRA guidelines was effectively utilized to meet the needs of beneficiaries of watershed and to build up a good rapport for smooth implementation of watershed and this endeavour was fruitful. Masonry platform constructed in front of the Ramasagara village community hall is being used for village meetings, dramas, village functions. The two animal water troughs constructed in Ramasagara and Devasamudra village are used for drinking water for nearly 100 animals daily. Representative soil samples in upper, middle and lower reaches of 35 farmers' fields in irrigated and rainfed areas indicated that these soils were low to medium in available N, P and K and neutral in soil pH. Water samples of all 33 bore wells of lower, middle and upper reaches in watershed indicated that quality of water in all the reaches was generally saline while better quality of water in upper reach and lower reaches bore well water can be used as protective irrigation. Through animal and human health camps, awareness of human and animal health was made in addition to the treatment. Through training programmes the beneficiaries of watershed were exposed to soil and water conservation measures in black soils.

REFERENCES

Dixit, S., Wani, S, P., Rego, T. J. and Pardhasardhi, G. 2007. Knowledge-based Entry Point and Innovative Up scaling Strategy for Watershed Development Projects. *Indian J. Dryland Agric. Res. & Dev.*, 22(1): 22-31.

Fernandes, A.P. 2000. Equity in watershed management: The MYRADA experience in social and institutional issues in watershed management in India. New Delhi, India: OIKOS; and Philippines: IIRR.

Olson, M. 1971. The logic of collective action: public goods and theory of groups (revised edition). New York, USA: Schocken Books.

Rockström, J., Hatibu, N. Oweis T. Y. and Wani, S. P. 2007. Managing Water in Rainfed Agriculture In: *Water for Food, Water for Life*. David Molden, (Ed.) Earthscan, London and International Water Management Institute, Colombo.

Sehgal, J. andAbrol, I.P. 1992. Land Degradation Status: India, Desertification Control *Bulletin No. 21*.UNEP, Nairobi.

Sreedevi, T.K., Shiferaw, B. and Wani, S.P. 2004. Adarsha watershed in Kothapally: Understanding the drivers of higher impact. *Global Theme on Agroecosystems Report no. 10*. Patancheru 502 324, Andhra Pradesh, India: International Crops Research Institute for the Semi-Arid Tropics. 24 p.

Wani, S.P., and Ramakrishna, Y. S. 2005. Sustainable Management of Rainwater Through Integrated Watershed Approach for Improved Rural Livelihoods. *In "Watershed ManagementChallenges: Improved Productivity, Resources and Livelihoods"* (eds. Bharat R. Sharma, J.S. Samra, C.A. Scott and Suhas P. Wani). IWMI, Sri Lanka . pp39-60.

Wani, S.P., Pathak, P., Sachan, R. C. and Pande, S. 2003a. Improved management of Vertisols in the semiarid tropics for increased productivity and soil carbon sequestration. *Soil Use and Management*, 19: 217-222.

Wani, S.P., Singh, H. P. Sreedevi, T. K. Pathak, P. Rego, T. J. Shiferaw, B. and Iyer, S. R. 2003. Farmer participatory integrated watershed management: Adarsha watershed, Kothapally India. An innovative and upscalable approach.- AQ case Study. In: R.R. Harwood and A.H. Kassam, Editors, *Research Towards Integrated Natural Resources Management: Examples of Research Problems, Approaches and Partnerships in Action in the CGIAR*, Interim Science Council, Consultative Group on International Agricultural Research, Washington, DC, USA (2003) pp123-148.

□□□

Chapter 88

Profitable Alternate Cropping Systems for Rainfed Sugarcane in North-Coastal Zone of Andra Pradesh

V.GOURI, T.CHITKALADEVI AND R.ANKAIAH

INTRODUCTION

In Andhra Pradesh sugarcane is growing in an area of 0.16 M ha with a production of 11.6 M tons (Directorate of Economics and Statistics, 2010). North Coastal Zone of Anahra Pradesh comprising of Srikakulam, Vizianagaram and Visakhapatnam districts having total net cultivated area of 0.47 M ha out of which 82.6% area is under rainfed condition. The cropping system in the zone is manly dependent on seasonal rain fall. During the last ten years both area and production of sugarcane fluctuate considerably from year to year, due to variations in climatic conditions, the vulnerability of areas cultivated under rainfed conditions, fluctuations in prices and changes in returns from competing crops. Cost of production of sugarcane is increasing and the farmers are unable to harvest the crop due to labour shortage as well as hike in labour charges. The problem is going to further deteriorate unless sugarcane farmers are provided with options of alternate cropping systems particularly under rainfed conditions. Cropping system approach involves sequential cropping intercropping and mixed cropping for efficient use of natural as well as man made resources for increasing the cropping intensity and production (Shinde *et al.*, 2009). Under rainfed conditions in Bidar district of Karnataka redgram + blackgram cropping system found to be most profitable compared to other cropping systems (Rajkumar, 2007). However sugarcane and Bengal gram cropping system gave higher monetory returns in three dry zones of Northern Karnataka (Wadear, 2003). Keeping these facts in view the present investigation was taken up and this paper presents the profitable cropping systems for rainfed sugarcane in North Coastal Zone of Andhra Pradesh.

Regional Agricultural Research Station, Anakapalle, Visakhapatnam, Andhra Pradesh, India

MATERIALS AND METHODS

The present study was conducted at Regional Agricultural Research Station, Anakapalle, Andhra Pradesh during *Kharif*, 2008-09 and 2009-10. The soils of the experimental field are sandy loam in texture, neutral in pH, normal in EC, medium in available nitrogen, medium in available phosphorous and high in available potassium (Table. 1).

Table 1: Physico-chemcial properties of experimental site

pH	7.60
EC (ds m^{-1})	0.37
Organic carbon	0.42
Available Nitrogen (kg ha^{-1})	298
Available Phosphorous (kg ha^{-1})	30
Available Potassium (kg ha^{-1})	290
Texture	Sandy loam

Rainfall : Rainfall received during 2008-09 and 2009-10 is presented in Figure 1. An amount of 450 mm rainfall received in 33 rainy days during the first year of study where as an amount of 718.4 mm received in 38 rainy days during second year of study. Amount and distribution of rainfall is normal during both the years except higher amount of rainfall (260.4 mm) received during October and November months of 2009-10 which coincided with flowering and pod formation stage of redgram cropping systems helped to get higher yield.

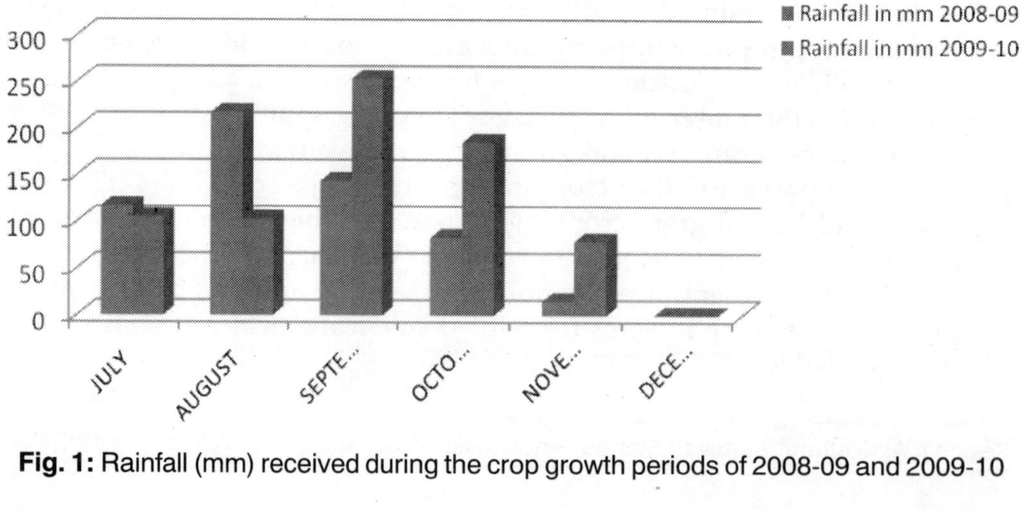

Fig. 1: Rainfall (mm) received during the crop growth periods of 2008-09 and 2009-10

The experiment was designed in Randomized Block Design (RBD) with three replications comprising twelve cropping systems. Different cropping systems are two sole cropping systems (rainfed sugarcane and rainfed rice), four intercropping systems (castor+bajra, redgram+maize, redgram+rice, tapioca+Greengram) and six sequence cropping systems(Maize-greengram, Sesamum-greengram, Soyabean–greengram, Marigold–greengram and Bhendi -greengram). The rainfed crops were sown with the onset of monsoon during *kharif* season in both the years. All the agronomic practices including fertilizer application were carried out as per the recommendations. The crops were harvested as and when they matured. After harvesting of maize, groundnut, sesamum, soyabean, marigold and bhendi, greengram crop was sown as sequence crop. Yield parameters such as number of pods per plant were recorded in groundnut, sesamum, soyabean, castor and greengram. Cob weight was recorded in maize and number of flowers and flower weight were recorded in marigold. After each picking of bhendi weight was recorded and expressed in q ha^{-1}. In sugarcane, yield attributing parameters like number of millable canes, cane length were recorded at the time of harvest. Cane yield was recorded after stripping the leaves and detopping. Other crop yields were converted into sugarcane equivalent yields by using the following formula :

$$\text{Sugarcane equivalent yield} = \frac{\text{Yield of other crop x Price of other crop}}{\text{Price of sugarcane crop}} + \text{Yield of sugarcane crop}$$

After converting the yield of other crops into sugarcane equivalent yield, data is statistically analysed by the analysis of variance method as per procedure out lined by Panse and Sukhatme (1967). Economics of each cropping system in terms of gross monetary returns, net monetary returns and cost benefit ratio are calculated based on value of economic produce of both sugarcane and other crops as per the market price at the time of harvest.

RESULTS AND DISCUSSION

Sugarcane equivalent yields **:** Cane Equivalent Yield (CEY) was calculated on the basis of existing market price of component crops (Singh *et al.*, 2011). Most of the studies on sugarcane based cropping systems indicated that higher cane equivalent yield was recorded in intercropping of sugarcane with different crops compared to sole sugarcane (Hasure,2005; Mahadeva Swamy and Martin,2002;). As the objective of this study is to find out alternate cropping systems for rainfed sugarcane, different cropping systems were compared with sole rainfed sugarcane. Significantly higher sugarcane yield equivalents were recorded in sole rainfed sugarcane during both the years due to favorable weather conditions prevailed during crop growth period. Among different cropping systems bhendi-greengram and marigold-green gram cropping systems during 2008-09 and redgram + rice, redgram+maize during 2009-10 (Table 2) recorded highest sugarcane yield equivalents . Higher yield coupled with increased market price of redgram resulted in higher sugarcane yield equivalents among different cropping systems (Rajkumar, 2007) during the second year of study.

Table 2: Yield of different cropping systems and Sugarcane yield equivalent yield under rainfed situation during 2008-09 and 2009-10.

S. No.	Treatment	Yield of base crop (q ha^{-1})		Yield of intercrop (q ha^{-1})		Sugarcane yield equivalents (q ha^{-1})	
		2008-09	2009-10	2008-09	2009-10	2008-09	2009-10
1	Sugarcane	594	556	-	-	594	556
2	Maize-Greengram	20.8	20	1.5	1.8	181	152.9
3	Groundnut-Greengram	7.2	6.3	1.8	1.3	250.5	84.6
4	Sesame-Greengram	1.9	1.7	1.6	2.3	99.3	107.6
5	Castor + Bajra	3.3	10.2	1.3	1.7	55.9	96
6	Redgram + Maize	11.2	20	7.1	6.1	264.3	383.4
7	Redgram + Rice	10.6	22	7.4	5	264.8	420.6
8	Soybean- Greengram	2.2	1.7	1.9	3	62.9	99.2
9	Marigold- Greengram	19.7	11.8	1.5	1	280	133.5
10	Bhendi - Greengram	46.5	39	1.5	2.8	421.3	311.8
11	Rainfed rice	21.4	18	-	-	178	180
12	Tapioca + Greengram	62.5	75	1.6	1.1	98.5	76.9
	SEm$\pm$					10.545	4.35
	CD (0.05)					30.92	12.75

Economics of cropping systems : Higher gross monetary returns were obtained with sole rainfed sugarcane during both the years followed by bhendi-greengram and groundnut-greengram cropping systems during the first year and redgram+rice, redgram+maize cropping systems during the second year of study. However, in terms of net monetary returns, with the prevailed monsoon conditions bhendi-greengram (₹ 27,950) and marigold-green gram (₹ 21,665) cropping systems gave higher returns during 2008-09 where as redgram+rice (₹ 55,775) and redgram+maize (₹ 51,040) cropping systems recorded higher net monetary returns during 2009-10. Rajkumar (2007) observed redgram+Blackgram cropping system gave higher cost benefit ratio of 1.91 than other cropping systems under rainfed situation of Bidar district of Karnataka.

Though higher sugarcane equivalent yields and gross monetary returns were obtained with the rainfed sugarcane, due to long duration and higher cultivation expenses particularly seed cost, labour cost for planting and harvesting, it was not fond to be profitable in terms of cost benefit ratio. The cropping systems which gave higher cost benefit ratio such as redgram+maize (4.6), redgram+rice (4.55) and bhendi-greengram (2.25) found to be more remunerative compared to the rainfed sugarcane. Sandeep (2002) found that under rainfed situation the net profit was higher in redgram cropping system than sugarcane. Comparatively higher yields are recorded in redgram+maize and redgram+rice cropping systems during second year of study over the first year because of copious of rainfall (Figure 1.) received during later part of the crop growing stage resulted in increased flowering, pod formation and pod development. Higher yield of redgram due to rainfall received during the months of October and November which coincides with flowering and pod formation of redgram crop was also reported by Gouri *et al.* (1997).

Table 3: Economics of different cropping systems and Sugarcane yield equivalent yield under rainfed situation during 2008-09 and 2009-10

S. No.	Treatment	Gross monetory returns (Rs. ha^{-1})		Net monetory returns (Rs. ha^{-1})		BC ratio	
		2008-09	2009-10	2008-09	2009-10	2008-09	2009-10
1	Sugarcane	71280	94520	21460	44720	1.40	1.90
2	Maize-Greengram	21713	26000	8723	13010	1.70	2.00
3	Groundnut-Greengram	36060	14375	16500	815	2.20	1.06
4	Sesame-Greengram	11920	18300	2720	9100	1.30	1.99
5	Castor + Bajra	6705	13260	-1795	4760	0.80	1.56
6	Redgram + Maize	31795	65185	17650	51040	2.20	4.60
7	Redgram + Rice	31780	71500	16055	55775	2.00	4.55
8	Soybean- Greengram	7550	16870	-350	8970	0.96	2.14
9	Marigold- Greengram	33600	22700	21665	10765	2.80	1.90
10	Bhendi - Greengram	50550	53000	27950	30400	2.20	2.35
11	Rainfed rice	21400	30600	9150	18350	1.70	2.50
12	Tapioca + Greengram	11820	13070	1470	2720	1.10	1.26

CONCLUSIONS

Higher cost benefit ratio of redgram+maize (4.6), redgram + rice (4.55) and bhendi-greengram (2.35) indicated that these cropping systems are more profitable than sugarcane under rainfed conditions, though significantly higher sugarcane equivalent yields were recorded in rainfed sugarcane in North Coastal Zone of A.P.

REFERENCES

Directorate of Economics and Statistics, 2010. An outline of Agriculture Situation in Andhra Pradesh, Hyderabad.

Gouri, V. Buchareddy, B. and Bhaskar Reddy, B.1997. Effect of intercropping sunflower with legumes on yield and economics. *J. of Oil Seeds Res.*, 22(4):318-320.

Hasure, R.R. Kadam, U.A. Nigade, R.D and More, S.M.2005. Studies on Planting geometry and intercrops on yield contribution characters, cane yield, quality and economics of seasonal sugar cane (Co 86032). Cooperative Sugar. 36(12):983-988.

Mahadeva Swamy, M. and Martin, G.J. 2002. Production potential of Wide row sugarcane intercropped with aggregatum onion (*Allium cepa*) under different row ratios, fertilizer levels and population densities. *Indian J. Agron.* 47(3):361-366.

Rajkumar, B., 2007, Economics of redgram based cropping systems in Bidar district. *M.Sc (Agri.) Thesis*, Univ. Agric. Sci., Dharwad

Panse, V.G. and Sukhatme, P.V.1967. Statistical methods for agricultural workers (2nd eddition) ,Indian council for agricultural research, New Delhi, 368p .

Sandeep, B. 2002. An economic analysis of cropping systems in Bidar district of Karnataka. M.Sc (Ag) Thesis. Uni. of Agric.Sci., Dharwad, India.

Shinde, N. Patil, B. L. Murthy, C. and Desai, N. R.M.2009. Profitability analysis of Sugarcane based intercropping systems in Belgaum district of Karnataka. *Karnataka J. of Agric. Sci.*, 22(4):820-823.

Singh, A. Menhilal. and Singh, S.N.2011. System productivity and yield advantage of autumn sugarcane (*Saccharum* spp. hybrids) and linseed (*Linum usitatissimum)* intercropping under subtropics. *Indian Journal of Sugarcane Technology,* 26(2): 7-9

Wadear, P.R., 2003, Animal based farming systems for long-term sustainability in northern Karnataka – A socio-economic assessment. *Ph.D Thesis,* Univ. Agric. Sci.,Dharwad.

□□□

Chapter 89

Integrated Farming Systems for Livelihood Security in Rainfed Areas of Odisha

P.J. MISHRA AND S. S. NANDA

INTRODUCTION

The rainfed areas of Odisha are subjected to droughts very frequently. Once the *kharif* crop is lost due to drought, farmers not only face scarcity of food grains but also their social security became at stake. There is regular and heavy migration of medium, small and marginal farmers and landless labours to distant cities in search of employment that largely depends on farming as their livelihood. Their livelihood security can be achieved if the income from all sources is able to meet a reasonable standard of living of a family. The main strategies to improve livelihood are: intensification of existing production patterns; diversification of production and processing to make it market-oriented with value-addition and integration of several components in a farming system to increase income. Unlike irrigated areas where intensive farming could be practiced with crop diversification with year round cropping, in rainfed areas the farming has to be heavily dependent on monsoon. Crops and varieties should be selected having tolerant to short duration dry spells in the initial, med season or terminal phase of cropping. In rainfed areas there is ample scope for inclusion of vegetables as a profitable enterprise in up and medium lands during rainy and post rainy seasons. Poultry is also a profitable short rotation enterprise that could supplement the off season farm income in a big way and generate employment for farm family as well as land less labourers in the locality. Therefore there is urgent need to develop profitable IFS model for rainfed paddy based farming situations for livelihood security of farmers.

MATERIALS AND METHODS

A study was conducted on integrated farming system at on-farm research unit at, Nayagarh, Odisha during 2006-07 and 2007-08 involving paddy,

Odisha University of Agriculture and Technology, Bhubaneswar – 751003 Odisha vegetables and poultry as the components of integrated farming system and paddy-greengram and local vegetables as the traditional farming practice. In 1.2 ha area of integrated farming system module, paddy occupied 0.2 ha, an area of 0.72 ha was allotted for vegetables and rest area was allotted to banana, papaya and poultry shed in which poultry birds (*Vanaraja*) were maintained. The soil of the experiment site was sandy clay loam with pH 5.1. The available N, P and K content of the soil was 185, 22 and 320 kg ha^{-1}, respectively. To sustain the productivity the residues obtained in the system was recycled.

In IFS module high yielding paddy (*Pratikhya*) was grown in 0.2 ha of medium land with N 80 kg ha^{-1}, P_2O_5 40 kg ha^{-1} and K_2O 40 kg ha^{-1} and other recommended package of practices. Hybrid papaya (*Red lady*) 100 plants were grown along the vegetable border line which starts fruiting at eight months. Tissue culture banana both for table purpose (DC) and culinary purpose (*bantala*) which was more in demand throughout the year were grown with proper management. High value vegetables e.g. pointed gourd, arrowroot are grown to give economic return to the farmer from August to December. In the poultry unit 50 no's of one day old chicks of dual purpose breed (*Vanaraja*) were maintained in each batch with proper nutrition and health care for 45 days . Six batches of poultry were raised during the year and the poultry litters were used in the crop field as manure. In another 1.2 ha area conventional cropping system was followed with paddy-greengram system in 0.4 ha and local vegetables in 0.8 ha area. Local paddy was grown with N 40 kg ha^{-1}, P_2O_5 20 kg ha^{-1} and K_2O 20 kg ha^{-1} and green gram, local vegetables grown without any chemical fertilizers. Observations on the productivity and economics of components and the farming system as a whole and employment generation were recorded as per the standard procedure.

RESULTS AND DISCUSSION

Integrated Farming System module produced higher income and profit than traditional paddy-pulse and vegetable systems in this study. In IFS module from 1.2 ha land holding during 2006-07 (Table 1) total income, net return and benefit cost ratio (BCR) were ₹ 163200, ₹ 91800 and 2.29 as against net return of ₹ 70600, ₹ 32600 and BCR 1.86 with conventional farming. Similarly during 2007-08 total income, net return and benefit cost ratio (BCR) from Integrated Farming System module was ₹ 193000, ₹ 110500 and 2.34 as against net return of ₹ 75400, ₹ 35900 and BCR 1.91 with conventional farming. The total income, net profit and BCR are higher during 2007-08 than 2006-07 mainly due to harvest of table purpose banana (*bantala*) during second year of planting. Inclusion of high yielding varieties of paddy, tissue culture banana, hybrid papaya, hybrid varieties of vegetables, high value vegetables *e.g.*, pointed gourd, arrowroot and backyard poultry with dual purpose Vanaraja breed not only increase the overall income but also generate additional employment for rural farm families and minimizes the risk and uncertainty of conventional farming.

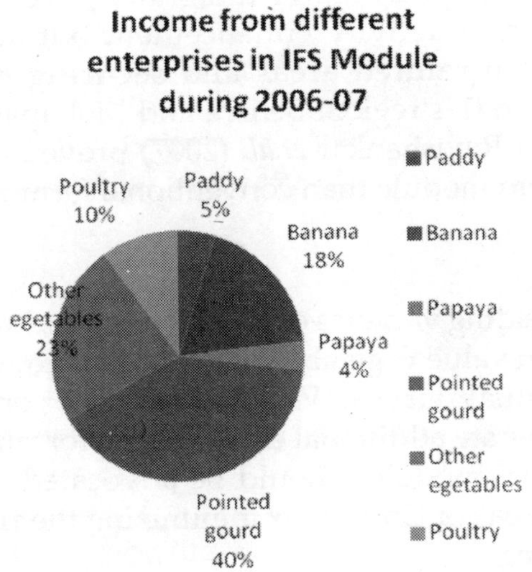

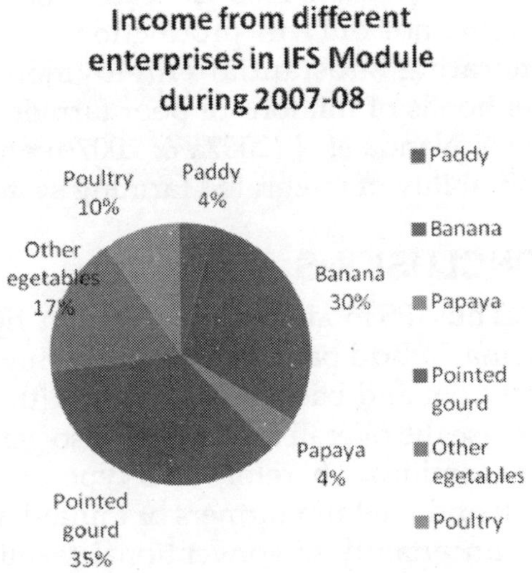

Table 1: Comparison of integrated farming system module with traditional farming.

Crop/Enterprise	Variety	Area/No. of Plants	Income (₹ year^{-1})	
			2006-07	2007-08
Farming system approach				
Paddy	*Pratikhya*	0.20 ha	8000	8000
Banana	*DC*	300 plants	30000	30000
	Bantala	150 plants	-	27000
Papaya	*Red lady*	100 plants	6000	8000
Pointed Gourd	*Balangir Local*	0.40 ha	65000	68000
Other vegetables*		0.32 ha	38200	33500
Poultry	*Vanaraja*	300 Birds	16000	18500
Total Income		1.2 ha	163200	193000
Expenditure			71400	82500
Net Return			91800	110500
BCR			2.29	2.34
Employment generated (man days)			320	350
Traditional Farming				
Rice-greengram	Local	0.4 ha	15600	16400
Vegetables	Local	0.8 ha	55000	59000
Total income		1.2 ha	70600	75400
Expenditure			38000	39500
Net return			32600	35900
BCR			1.86	1.91
Employment generated (man days)			115	130

* Onion (*Nasik red*), Cucumber (*Annapurna)*, Chilli, Radish, Pumpkin, Garlic and Arrowroot

The additional employment generation by the IFS module was to the tune of 205 and 220 man days per year over traditional farming of same land holdings (1.2ha). This indicates the huge potential of integrated farming systems not only in production and productivity enhancement but also generating substantial employment in rainfed areas and securing the livelihoods of millions of poor farmers in this region. Behera and Mohapatra (1999), Nanda *et.al* (2007a & 2007b) and Ravishankar *et.al.* (2007) proved the profitability of integrated farming system module than conventional farming.

CONCLUSIONS

Thus IFS module involving high yielding varieties of paddy, tissue culture banana, hybrid papaya, hybrid and high value vegetables, e.g. pointed gourd, arrowroot and backyard poultry with dual purpose *Vanaraja* breed not only increase the overall income but also generate additional employment for rural farm families. Therefore this type of IFS modules should be advocated for small and medium farmers of rainfed areas of Odisha for minimizing the risk and uncertainty of conventional farming.

REFERENCES

Behera, U.K. and Mahapatra, I.C. 1999. Income and employment of small and marginal farmers through integrated farming systems. *Indian J. of Agron.*, 44(3): 431–439.

Nanda, S.S., Mishra, S.N. and Mohanty, M. 2007a. On-farm farming system for income and employment generation in Kalahandi district of Orissa. *Journal of Farming Systems Research and Development*, 13(1): 56–59.

Nanda, S.S., Mohanty, M., Mishra, P.J., Mishra, S.N. and Jena, B.K. 2007b. Integrated farming systems for small farmers of east and south eastern coastal plain of Orissa. In: Extended Summery, 3rd National Symposium on Integrated Farming System held during 26-28 October at Jaipur, Rajastan. FS.4, pp12–14.

Ravishankar, N., Pramanik, S. C., Rai, R.B., Nawaz., S., R. B., Biswas., T.K. and Bibi., N. 2007. Study on integrated farming system in hilly upland areas of Bay Islands. *Indian J. Agron.*, 52(1): 7-10.

□□□

About the Editors

Dr M. Madhu, presently Principal Scientist and Head, Central Soil and Water Conservation Research & Training Institute (CSWCRTI), Research Centre, Sunabeda -763 002, Koraput, Odisha, India. He completed Ph.D in Agronomy from UAS, GKVK, Bengaluru, Karnataka. He is a recipient of UAS Gold Medal during M.Sc and Ph.D, "Dr H.R. Arakeri Gold Medal" for the best Ph.D Student in Agronomy-1995, "ARIC Award for Contribution to Crop Research-1997", IASWC Young Soil Conservationists Award-2004 and IASWC Gold Medal-2010. He is recipient of prestigious USDA Global Research Alliance (GRA) Fellowship Programme-2011 and visited USA, Nepal and Bangladesh on study and research assignments. Dr Madhu has vast experience of research, training and outreach activity related to NRM. He was leader and associated in more than 15 consultancy projects related to Integrated Watershed Management and resource conservation-cum-livelihood programmes. He has to his credit more than 100 publications including research papers in reputed national and international journals, edited books, book chapters, manuals, technical reports and bulletins, symposium / workshop/ seminar papers, popular articles *etc.*, and member of many professional bodies. He has wide work experiences in different capacities as a Senior Scientist and In-charge Head at CSWCRTI, Research Centre, Udhagamandalam, TN, India, Senior Scientist and Secretary of IASWC at CSWCRTI, Dehradun, UK, India.

Praveen Jakhar is presently Scientist (Agronomy) at CSWCRTI-Research Centre, Sunabeda, Koraput, Odisha. He did his post-graduation from Haryana Agricultural University, Hisar in Agronomy and joined Agricultural Research Service in 2007. He has worked in different institute funded and other collaborative research and consultancy projects as leader and associate. Sh Jakhar has specialization in Conservation Agronomy, Watershed Management, Technology Transfer and Human Resource Development. He has published 12 research papers in reputed national and international journals apart from being author of popular articles, bulletins and book chapters.

Dr Partha Pratim Adhikary is presently Scientist (Soil Physics and SWC) at CSWCRTI, Research Centre, Sunabeda, Koraput (Odisha). He obtained his Master's degree in Soil Science from Chaudhury Charan Singh Haryana Agricultural University, Hisar, Haryana and Ph.D (Agricultural Physics) from Indian Agricultural Research Institute (IARI), New Delhi, India. He is recipient of National Scholarship, Junior and Senior Research Fellowships from ICAR and CSIR. He is also recipient of S.D. Nijhawan Gold Medal from Chaudhury Charan Singh Haryana Agricultural University, Hisar for his brilliant performance during M.Sc programme. He is also recipient of Jawahar Lal Nehru Award by ICAR. His field of research work includes development of pedotransfer functions of soil hydraulic parameters for Indian soils, modeling of groundwater pollution, GIS and Spatial Variability, watershed management and soil erodibility factor. He has more than 20 publications to his credit which include research papers in journals of national and international repute, book chapters, compendium, reports etc.

Hombe Gowda, H. C. is Scientist of Forestry at Central Soil and Water Conservation Research and Training Institute (CSWCRTI), Research Centre, Sunabeda, Odisha, India, since 2007. He is post graduated in Forestry from College of Forestry, Sirsi, Karnataka. Mr Gowda has a long standing interest on Agroforestry ecosystem dynamics, ecosystem services and watershed management. He is the author and co-author of 2 books and more than 25 scientific research articles. He is a Herpetologist by interest and saved many poisonous reptiles from human habitats. Presently he is pursuing his doctoral degree from University of Gottingen, Germany.

Dr V. N. Sharda, presently Member, Agricultural Scientist Recruitment Board has a brilliant career and vast experience of more than 34 years of research, training and extension in the fields of Hydrology, Soil and Water Conservation and Watershed Management. Dr Sharda critically evaluated the impact of planting Eucalyptus in natural grasslands, evolved the concept of conservation bench terrace system and studied hydrological behavior of mountains and ravenous watershed besides estimating the production losses in major rainfed crops of India across different states and prioritization of erosion risk areas. He is recipient of prestigious Fulbright Senior Post-Doctoral Research Fellowship of United States Information Agency in the area of environmental studies. In recognition of his outstanding contributions in the fields of Watershed Management, Water Conservation and Erosion Modeling, he has been bestowed with several awards which include the prestigious Rafi Ahmed Kidwai Award, Dr Rajendra Prasad Puruskar, Jawahar Lal Nehru Award for the best Ph. D. Thesis, Vasant Rao Naik Award of Indian Council of Agricultural Research, Intellectual's Honour-2003 "Uttaranchal Ratan", Bharat Singh Award of Ministry of Water Resources, Eminent Agricultural Engineer Award of IEI(I), Team Award of ICAR, a Citation by USDA-ARS, Fellowship and Recognition Award of National Academy of Agricultural Sciences, New Delhi and Fellowships of Institution of Engineers (India), Indian National Academy of Engineers, Indian Society of Agricultural Engineers and Indian Association of Soil & water Conservationists. He is a member of several national and state level committees and has more than 270 publications to his credit which include research papers in journals of national and international repute, books, book chapters, reports, etc.

Dr P. K. Mishra is presently the Director of Central Soil & Water Conservation Research and Training Institute, Dehradun. He has an exemplary academic career as a student, researcher and a professional in Soil and Water Conservation Engineering and 30 years of field and research experience in different capacities. He was trained in Professional Development Program in UK and in the field of GIS and Hydrologic Modelling in USA and Australia. He studied at University of Karlsruhe, Germany for his Ph.D work under DAAD fellowship from 1986-1988. His major contributions are in the area of drought management, documentation of ITKs relating to Soil and Water Conservation, Common Pool Resources (CPR) management, development of research watersheds, compilation of rainfed technologies and climate change research initiatives. Dr. Mishra is the recipient of a number of awards which include DAAD fellowship from Germany, best undergraduate student award by ICAR, New Delhi, Student Award from Indian Society of Agricultural Engineers (ISAE), New Delhi, Jawaharlal Nehru Award of ICAR, New Delhi for Best Ph.D work in Engineering and Technology, Fellowship Award and Dr. K.G. Tejwani Award for Management of Natural Resources of Indian Association of Soil & Water Conservationists (IASWC), Dehradun, SCSI "Gold Medal" Award by Soil Conservation Society of India (SCSI), New Delhi, Vasant Rao Naik Award 2010 of ICAR and Ground water Augmentation Award (as a team member) from MoWR, GoI. He has about 245 publications to his credit including research papers in National and International journals of repute, books, book chapters, papers in national and international proceedings, popular articles, bulletins and reports and short communications.

Dr Mohammed Khalid Khan presently working as Dean, College of Agricultural Engineering and Technology, Orissa University of Agriculture and Technology, Bhubaneswar. He did his M. Tech (Agricultural Processing and Food Engg.) from PAU, Ludhiana and Ph.D from IIT, Kharagpur. He has got more than 30 years of experience in teaching, research and extension activities in the field of Agricultural Processing and Food Engineering and Post Harvest Technology. He has worked as Research Engineer and Officer-in-Charge of AICRP on Post Harvest Technology, OUAT Bhubaneswar Centre for 14 years. Under his leadership the centre developed many post harvest equipment/machines and many innovative process technologies and received the 'BEST CENTRE AWARD' from ICAR and the 'BEST TEAM AWARD' from ISAE, New Delhi. Dr Khan is also recipient of the 'SAMANTA CHANDRA SEKHAR AWARD for the year 2009 in recognition of his original and creative research work in the field of Engineering and Technology. He has guided 3 Ph.D, 25 M. Tech and 40 B. Tech students and published more than 60 research papers in different National and International Journals and written 15 Technical Bulletins and Books to his credit.